Metals

Metalloids

Nonmetals

10 8B	11 1B	12 2B	13 3A	14 4A	15 5A	16 6A	17 7A	18 8A
								2 **He** 4.00 helium
			5 **B** 10.81 boron	6 **C** 12.01 carbon	7 **N** 14.01 nitrogen	8 **O** 16.00 oxygen	9 **F** 19.00 fluorine	10 **Ne** 20.18 neon
			13 **Al** 26.98 aluminum	14 **Si** 28.09 silicon	15 **P** 30.97 phosphorus	16 **S** 32.07 sulfur	17 **Cl** 35.45 chlorine	18 **Ar** 39.95 argon
28 **Ni** 58.69 nickel	29 **Cu** 63.55 copper	30 **Zn** 65.39 zinc	31 **Ga** 69.72 gallium	32 **Ge** 72.61 germanium	33 **As** 74.92 arsenic	34 **Se** 78.96 selenium	35 **Br** 79.90 bromine	36 **Kr** 83.80 krypton
46 **Pd** 106.42 palladium	47 **Ag** 107.87 silver	48 **Cd** 112.41 cadmium	49 **In** 114.82 indium	50 **Sn** 118.71 tin	51 **Sb** 121.75 antimony	52 **Te** 127.60 tellurium	53 **I** 126.90 iodine	54 **Xe** 131.29 xenon
78 **Pt** 195.08 platinum	79 **Au** 196.97 gold	80 **Hg** 200.59 mercury	81 **Tl** 204.38 thallium	82 **Pb** 207.2 lead	83 **Bi** 208.98 bismuth	84 **Po** (209) polonium	85 **At** (210) astatine	86 **Rn** (222) radon
110 — (269)	111 — (272)	112 — (277)		114 — (285)		116 — (289)		

64 **Gd** 157.25 gadolinium	65 **Tb** 158.93 terbium	66 **Dy** 162.50 dysprosium	67 **Ho** 164.93 holmium	68 **Er** 167.26 erbium	69 **Tm** 168.93 thulium	70 **Yb** 173.04 ytterbium	71 **Lu** 174.97 lutetium
96 **Cm** (247) curium	97 **Bk** (247) berkelium	98 **Cf** (251) californium	99 **Es** (252) einsteinium	100 **Fm** (257) fermium	101 **Md** (258) mendelevium	102 **No** (259) nobelium	103 **Lr** (260) lawrencium

Introductory Chemistry

Introductory Chemistry

SECOND EDITION

Nivaldo J. Tro

WESTMONT COLLEGE

Upper Saddle River, New Jersey 07458

Library of Congress Cataloging-in-Publication Data

Tro, Nivaldo J.
Introductory chemistry / Nivaldo J. Tro.—2nd ed.
p. cm.
Includes index.
ISBN 0-13-147058-2 (alk. paper)
1. Chemistry I. Title.

QD33.2.T76 2006
540--dc22 2004053457

Senior Editor: Kent Porter Hamann
Editor in Chief: John Challice
Development Editor: Dan Schiller
Editor in Chief, Development: Ray Mullaney
Executive Managing Editor: Kathleen Schiaparelli
Assistant Managing Editor: Beth Sweeten
Senior Marketing Manager: Steve Sartori
Managing Editor, Science Media: Nicole Jackson
Assistant Managing Editor, Science Supplements: Becca Richter
Media Editor: Michael J. Richards
Project Manager: Kristen Kaiser
Director of Creative Services: Paul Belfanti
Art Director: Jonathan Boylan
AV Art Editor: Connie Long
Editorial Assistant: Jacquelyn Howard
Cover Image: Quade Paul
Interior & Cover Design: Kenny Beck
Cover and Chapter Opening Illustrations: Quade Paul
Assistant Manufacturing Manager: Michael Bell
Manufacturing Buyer: Alan Fischer
Director, Image Resource Center: Melinda Reo
Manager, Rights and Permissions: Zina Arabia
Manager, Visual Research: Beth Brenzel
Manager, Cover Visual Research & Permissions: Karen Sanatar
Image Permission Coordinator: Nancy Seise
Art Studio: Artworks
Production Manager: Ronda Whitson
Manager, Production Technologies: Matthew Haas
Illustrators: Royce Copenheaver, Jay McElroy, Mark Landis
Art Quality Assurance: Timothy Nguyen, Stacy Smith, Pamela Taylor
Managing Editor of AV Management and Production: Patricia Burns
Production Supervision/Composition: Preparé, Inc.

Pearson Prentice Hall
Pearson Education, Inc.
Upper Saddle River, New Jersey 07458

Printed in the United States of America

10 9 8 7 6 5 4 3 2 1

ISBN 0-13-147058-2

Pearson Education Ltd., *London*
Pearson Education Australia Pty. Limited, *Sydney*
Pearson Education Singapore Pte. Ltd.
Pearson Education North Asia Ltd., *Hong Kong*
Pearson Education Canada, Ltd., *Toronto*
Pearson Educación de Mexico, S.A. de C.V
Pearson Education—Japan, *Tokyo*
Pearson Education Malaysia, Pte. Ltd.

To Annie

ABOUT THE AUTHOR

Niva Tro has been a faculty member at Westmont College in Santa Barbara, California, since 1990. He received his B.A. degree in chemistry from Westmont College in 1985 and earned a Ph.D. from Stanford University in 1989, after which he performed postdoctoral research at the University of California at Berkeley. Honored as Westmont's outstanding teacher of the year in 1994 and again in 2001, he was also named the college's outstanding researcher of the year in 1996.

Professor Tro lives in Santa Barbara with his wife, Ann, and their four children, Michael, Alicia, Kyle, and Kaden. For leisure, he enjoys reading, writing, snowboarding, camping, and other outdoor activities with his family.

BRIEF CONTENTS

CONTENTS

6 Chemical Composition 153

7 Chemical Reactions 189

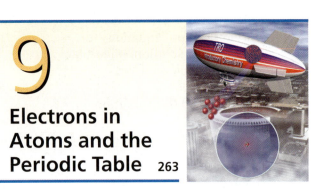

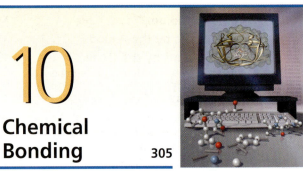

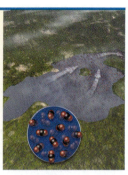

15 Chemical Equilibrium 513

16 Oxidation and Reduction 559

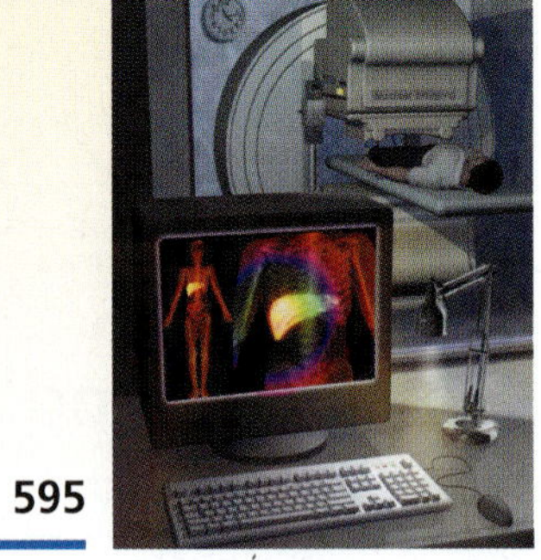

19 Biochemistry 677

Problem-Solving Procedures

TO THE STUDENT

This book is for *you*, and every text feature has you in mind. I have two main goals for you in this course: to see chemistry as you never have before, and to develop the problem-solving skills you need to succeed in chemistry.

I want you to experience chemistry in a new way. Each chapter of this book is written to show you that chemistry is not just something that happens in a laboratory; chemistry surrounds you at every moment. I have worked with several outstanding artists to develop photographs and art that will help you visualize the molecular world. From the opening example to the closing chapter, you will *see* chemistry. I hope that when you finish this course, you think differently about your world because you understand the molecular interactions that underlie everything around you.

I also want you to develop problem-solving skills. No one succeeds in chemistry—or in life, really—without the ability to solve problems. I can't give you a formula for problem solving, but I can give you strategies that will help you develop the *chemical intuition* you need to understand chemical reasoning. Look for several recurring structures throughout this book designed to help you master problem solving. The most important ones are (1) the solution map, a visual aid that helps you navigate your way through a problem; (2) the two-column Examples, in which the left column explains in clear and simple language the purpose of each step of the solution shown in the right column; and (3) the three-column Examples, which describe a problem-solving procedure while demonstrating how it is applied to two different Examples.

Lastly, know that chemistry is not reserved only for those with some special talent or special capability. With the right amount of effort and some clear guidance, anyone can master chemistry, including you.

Sincerely,

Nivaldo J. Tro
tro@westmont.edu

TO THE INSTRUCTOR

I thank all of you who used the first edition of *Introductory Chemistry*—you made this book a success from its first year of publication. The preparation of the second edition has enabled me to further refine and strengthen the text to even better fulfill its foundational mission: teaching chemical skills in the context of relevance.

Introductory Chemistry is designed for a one-semester, college-level, introductory or preparatory chemistry course. Students taking this course need to develop problem-solving skills—but they also must see *why* these skills are important to them and to their world. *Introductory Chemistry* extends chemistry from the laboratory to the student's world. It motivates students to learn chemistry by demonstrating how it plays out in their daily lives.

This is a visual book. Today's students often learn by seeing, so wherever possible, I have used images to help communicate the subject. In developing chemical principles, for example, I worked with several artists to develop multipart images that show the connection between everyday processes visible to the eye and the molecular interactions responsible for those processes. This art has been further refined and improved in the second edition, making the visual impact sharper and more targeted to student learning.

For the same reason, in teaching problem-solving skills, I use a *solution map* to help students see the general logic of working through a multistep problem. Extensive flow charts are also incorporated throughout the book, allowing students to visualize the organization of chemical ideas and concepts. In this edition, I have reworked the color scheme of both the solution maps and the flow charts to increase their pedagogical value, making them more intuitive and more consistent throughout the text. Thus, the solution maps now utilize the colors of the visible spectrum—always in the same order, from violet to red.

At key points throughout this book, I use a *three-column* layout, in which students learn a general procedure for solving problems of a particular type as they see it applied to two worked Examples simultaneously. In this format, the *explanation* of how to solve a problem is placed directly beside the actual steps in the *solution* of the problem. Your positive comments about the pedagogical benefits of this approach in the first edition prompted me to carry the strategy one step further and convert many of the *single-column* worked Examples to a *two-column* format for the second edition. In these Examples, one column describes how the problem is solved and explains the rationale for each step, while the other column shows the actual steps. Many of you said that you use a similar technique in lecture and office hours.

I have also added a new feature entitled *Conceptual Checkpoints* to this edition. These checkpoints are short questions that students can use to test their mastery of key concepts as they read through a chapter. Emphasizing understanding rather than calculation, they are designed to be easy to answer if the student has grasped the essential concept but difficult if he or she has not. The answers to these checkpoints, like the answers to the *Skillbuilder* exercises that follow each worked Example, are given at the end of each chapter.

Some significant improvements have been made to key content areas as well. These include:

- an enhanced discussion of the scientific method in Chapter 1
- a new section on physical separation methods in Chapter 3
- a more intuitive explanation of the mole, focusing on the underlying logic of the concept, in Chapter 6
- a clearer organization of the solubility rules for compounds in Chapter 7

Finally, the real-world relevance of the text has been further strengthened by the addition of half a dozen new interest boxes, dealing with such topics as ion size and nerve impulse transmission, artificial sweeteners, and Kevlar.

The media package that accompanies this book has been greatly enhanced for the second edition. To help both you and your students achieve your goals, we have provided innovative, book-specific, interactive media elements for all of your major instructional needs:

- classroom presentation
- student tutorial
- online assessment and course management

Marginal icons in the text now identify the Student Tutorials, interactive Live Examples, and manipulable 3-dimensional molecular images available to students in OneKey, the Accelerator CD, and the Companion Website that accompany the book. These components (as well as other available supplements) are fully described in the Print and Media Resources section of the Preface that follows.

To help you prepare your lectures more efficiently and teach most effectively, this text is now available in an *Annotated Instructor's Edition*. In addition to the entire student text, the *AIE* includes marginal annotations containing teaching tips, suggestions for dealing with common student misconceptions, ideas for classroom demonstrations, icons that identify the art and tables reproduced as color acetates in the Transparency Pack, and references to relevant instructional articles on a wide range of chemical topics.

I hope this book and the changes in the second edition support you in your mission of teaching students chemistry. Ours is a worthwhile cause, even though it requires constant effort. Please feel free to email me with any questions or comments you might have. I look forward to hearing from you as you use this book in your course.

Sincerely,

Nivaldo J. Tro
tro@westmont.edu

PREFACE

The design and features of this text have been conceived to work together as an integrated whole with a single purpose: to help students understand chemical principles and master problem-solving skills in a context of relevance. Students must not only be able to grasp chemical concepts and solve chemical problems, but also understand how those concepts and problem-solving skills are relevant to their other courses, their eventual career paths, and their daily lives.

Teaching Principles

The development of basic chemical principles—such as those of atomic structure, chemical bonding, chemical reactions, and the gas laws—is one of the main goals of this text. Students must acquire a firm grasp of these principles in order to succeed in the general chemistry sequence or the chemistry courses that support the allied health curriculum. To that end, the book integrates qualitative and quantitative material, and proceeds from concrete concepts to more abstract ones.

Organization of the Text

The main divergence in topic ordering among instructors teaching introductory and preparatory chemistry courses is the placement of electronic structure and chemical bonding. Should these topics come early, at the point where models for the atom are being discussed? Or should they come later, after the student has been exposed to chemical compounds and chemical reactions? Early placement gives students a theoretical framework within which they can understand compounds and reactions. However, it also presents students with abstract models before they understand why they are necessary. I have chosen a later placement for the following reasons:

1. *A later placement seems more flexible.* An instructor who wants to cover atomic theory and bonding earlier can simply cover Chapters 9 and 10 after Chapter 4. However, if atomic theory and bonding were placed earlier, it would be more difficult for the instructor to skip these chapters and come back to them later.
2. *A later placement allows earlier coverage of topics that students can more easily visualize.* Coverage of abstract topics too early in a course can lose some students. Chemical compounds and chemical reactions are more tangible than atomic orbitals, and the relevance of these is easier to demonstrate to the beginning student.
3. *A later placement gives students a reason to learn an abstract theory.* Once students learn about compounds and reactions, they are more easily motivated to learn a theory that explains them in terms of underlying causes.
4. *A later placement follows the scientific method.* In science, we normally make observations, form laws, *and then* build models or theories that explain our observations and laws. A later placement follows this ordering.

Nonetheless, I know that every course is unique and that each instructor chooses to cover topics in his or her own way. Consequently, I have written each chapter for maximum flexibility in topic ordering. In addition, the book is offered in two formats. The full version, *Introductory Chemistry*, contains 19 chapters, including organic chemistry and biochemistry. The shorter version, *Introductory Chemistry Essentials*, contains 17 chapters and omits these topics.

Improving Conceptual Understanding

Students often have difficulty in developing a true understanding of the concepts that underlie chemical behavior. These concepts are important not only as the foundation for problem solving but also as the basis for insight into how the chemical world works. To address this need, we have incorporated a new set of questions at the end of appropriate sections that encourage students to test their grasp of the key principles in the section.

NEW Conceptual Checkpoints

These strategically located conceptual questions

- enhance understanding of chemical principles,
- encourage students to stop and think about the ideas just presented before going on to new material, and
- provide a tool for self-assessment.

Presented in multiple-choice format, the questions are designed so that an attentive student can answer them quickly, with little or no calculation.
Answers to Conceptual Checkpoints are given at the end of each chapter, along with explanations of the reasoning involved.

CONCEPTUAL CHECKPOINT 6.1

Which of these statements is *always* true for samples of atomic elements, regardless of the type of element present in the samples?

(a) If two samples of different elements contain the same number of atoms, they must contain the same number of moles.
(b) If two samples of different elements have the same mass, they must contain the same number of moles.
(c) If two samples of different elements have the same mass, they must contain the same number of atoms.

Answers to Conceptual Checkpoints are given at the end of each chapter, along with explanations of the reasoning involved.

Answers to Conceptual Checkpoints

6.1 (a) The mole is a counting unit; it represents a definite number (Avogadro's number, 6.02×10^{23}). Therefore, a given number of atoms always represents a precise number of moles, regardless of what atom is involved. Atoms of different elements have different masses, so if samples of different elements have the same mass, they *cannot* contain the same number of atoms or moles.

6.2 (a) Avogadro's number was defined so as to make the molar mass of a compound (the mass of one mole, or Avogadro's number of molecules or formula units), *in grams*, numerically equal to the formula mass of the compound in amu. In order for the molar mass *in kilograms* to be numerically equal to the formula mass in amu, we would need 1000 times as many molecules or formula units in a mole. In other words, Avogadro's number would have to be 1000 times larger, or 6.02×10^{26}.

6.3 (b) This compound has the highest ratio of oxygen atoms to chromium atoms, and so must have the greatest mass percent of oxygen.

Developing Problem-Solving Skills

The development of problem-solving skills is the main goal of the text. To this end, *Introductory Chemistry* develops a systematic approach in which problem-solving skills are

- emphasized throughout each chapter,
- developed through many in-chapter Examples,
- reinforced with *Skillbuilder* exercises immediately following each Example,
- reviewed in unique chapter summaries, and
- practiced and synthesized in end-of-chapter exercises.

Starting with the simplest problems involving unit conversions, the text presents students with a basic procedure for solving most chemical problems. This procedure and the Examples that demonstrate its use have a number of special pedagogical features, designed to help students master problem-solving skills through understanding rather than mechanical repetition.

UNIQUE Solution Maps

Many of the Examples employ a unique visual approach in which students are shown how to draw a *solution map* for the problem. They are taught to outline the steps—using conversion factors and equations—needed to get from the information they are given to the information they are trying to find. This technique encourages students to think through the problem and formulate a solution strategy, and helps them see the relationship between the big picture and the individual steps. In response to requests from users of the text, there is a more consistent use of color for the buttons in the solution maps in this edition.

Converting between Grams of a Compound and Grams of a Constituent Element

Now, we have everything we need to solve our sodium problem. Suppose we want to know how many grams of sodium there are in 15 g of NaCl. The chemical formula gives us the relationship between moles of Na and moles of NaCl:

$$1 \text{ mol Na} \equiv 1 \text{ mol NaCl}$$

To use this relationship, we need *mol* NaCl. But, we have *g* NaCl. We can, however, use the *molar mass* of NaCl to convert from g NaCl to mol NaCl. Then we use the conversion factor from the chemical formula to convert to mol Na. Finally, we use the molar mass of Na to convert to g Na. The solution map is:

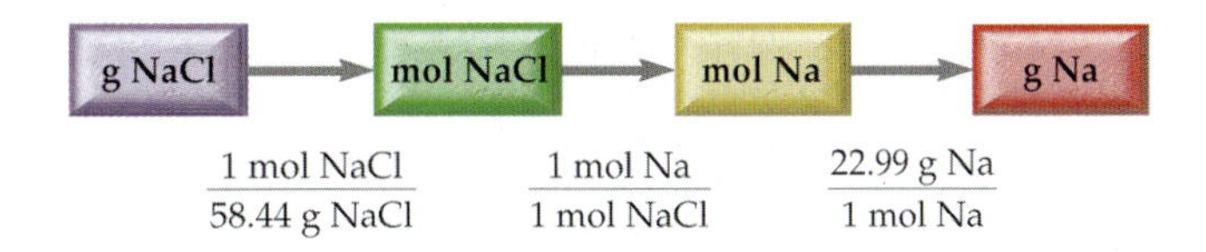

Notice that we must convert from g NaCl to mol NaCl *before* we can use the chemical formula as a conversion factor.

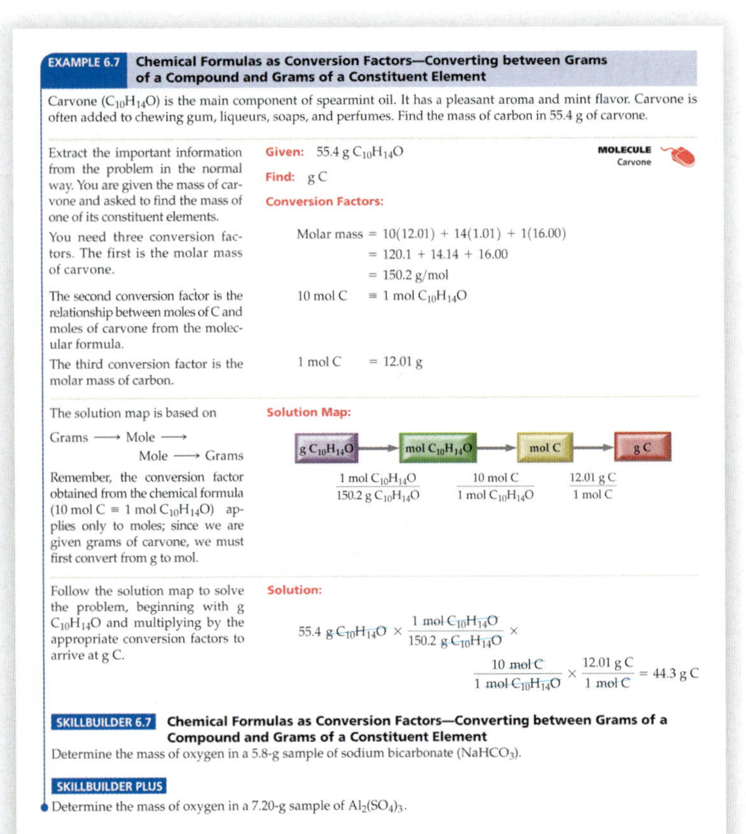

EXAMPLE 6.7 Chemical Formulas as Conversion Factors—Converting between Grams of a Compound and Grams of a Constituent Element

Carvone ($C_{10}H_{14}O$) is the main component of spearmint oil. It has a pleasant aroma and mint flavor. Carvone is often added to chewing gum, liqueurs, soaps, and perfumes. Find the mass of carbon in 55.4 g of carvone.

Extract the important information from the problem in the normal way. You are given the mass of carvone and asked to find the mass of one of its constituent elements.

Given: 55.4 g $C_{10}H_{14}O$

Find: g C

MOLECULE Carvone

Conversion Factors:

You need three conversion factors. The first is the molar mass of carvone.

$$\text{Molar mass} = 10(12.01) + 14(1.01) + 1(16.00) = 120.1 + 14.14 + 16.00 = 150.2\ \text{g/mol}$$

The second conversion factor is the relationship between moles of C and moles of carvone from the molecular formula.

$$10\ \text{mol C} \equiv 1\ \text{mol}\ C_{10}H_{14}O$$

The third conversion factor is the molar mass of carbon.

$$1\ \text{mol C} = 12.01\ \text{g}$$

The solution map is based on Grams ⟶ Mole ⟶ Mole ⟶ Grams

Remember, the conversion factor obtained from the chemical formula (10 mol C ≡ 1 mol $C_{10}H_{14}O$) applies only to moles; since we are given grams of carvone, we must first convert from g to mol.

Solution Map:

g $C_{10}H_{14}O$ ⟶ mol $C_{10}H_{14}O$ ⟶ mol C ⟶ g C

$$\frac{1\ \text{mol}\ C_{10}H_{14}O}{150.2\ \text{g}\ C_{10}H_{14}O} \qquad \frac{10\ \text{mol C}}{1\ \text{mol}\ C_{10}H_{14}O} \qquad \frac{12.01\ \text{g C}}{1\ \text{mol C}}$$

Follow the solution map to solve the problem, beginning with g $C_{10}H_{14}O$ and multiplying by the appropriate conversion factors to arrive at g C.

Solution:

$$55.4\ \text{g}\ C_{10}H_{14}O \times \frac{1\ \text{mol}\ C_{10}H_{14}O}{150.2\ \text{g}\ C_{10}H_{14}O} \times \frac{10\ \text{mol C}}{1\ \text{mol}\ C_{10}H_{14}O} \times \frac{12.01\ \text{g C}}{1\ \text{mol C}} = 44.3\ \text{g C}$$

SKILLBUILDER 6.7 Chemical Formulas as Conversion Factors—Converting between Grams of a Compound and Grams of a Constituent Element

Determine the mass of oxygen in a 5.8-g sample of sodium bicarbonate ($NaHCO_3$).

SKILLBUILDER PLUS

Determine the mass of oxygen in a 7.20-g sample of $Al_2(SO_4)_3$.

NEW Two-Column Examples

In this edition, all but the simplest Examples are presented in a unique two-column format. The left column explains each step, while the right column shows how the step is executed. This format encourages students to think about the reason for each step in the solution and to fit the steps together in the context of an overall plan.

UNIQUE *Skillbuilder* Exercises

Every worked Example in *Introductory Chemistry* is followed by at least one similar but unworked *Skillbuilder* problem, allowing students to make an immediate test of the problem-solving techniques they have just learned. Many important Examples also have an additional *Skillbuilder Plus* problem, affording students further opportunity for practice. For immediate reinforcement, answers to all *Skillbuilder* and *Skillbuilder Plus* problems are given at the end of each chapter.

UNIQUE Three-Column Problem-Solving Procedures

Among the distinctive features of this book are the numerous specific procedures for solving particular types of problems, presented in a unique three-column format. The first column outlines the problem-solving procedure and explains the reasoning that underlies each step. The second and third columns show how the steps are implemented for two typical Examples. Seeing the method applied to solve two related but slightly different problems helps students understand the general procedure in a way that no single Example could convey.

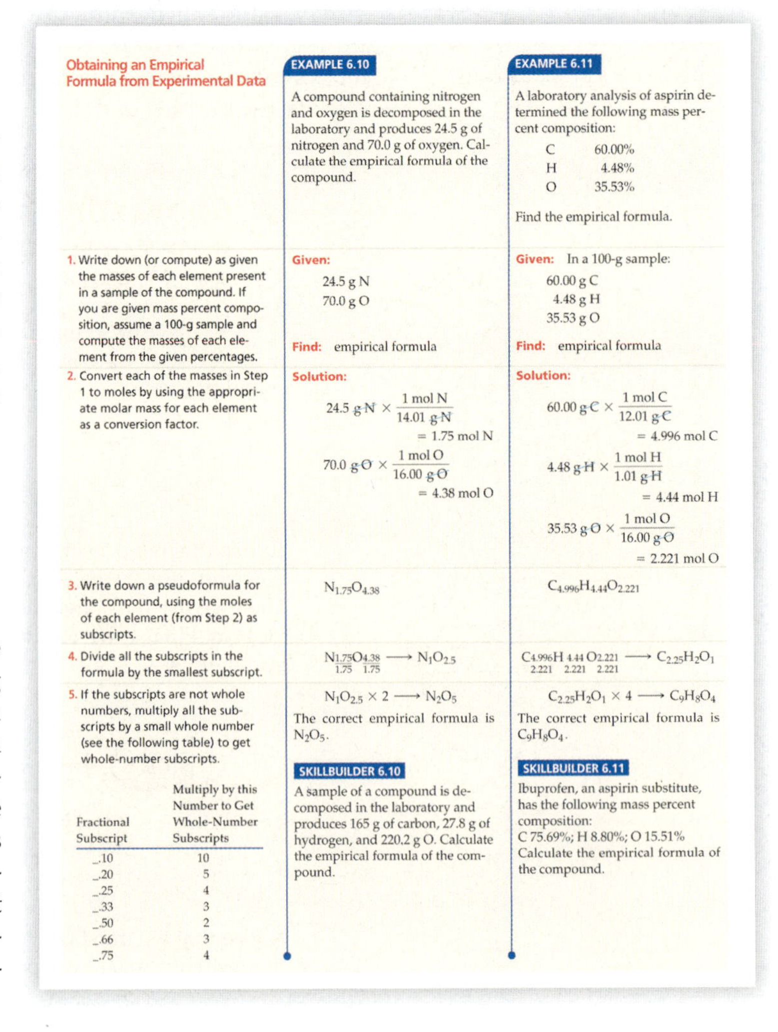

Obtaining an Empirical Formula from Experimental Data	EXAMPLE 6.10	EXAMPLE 6.11
	A compound containing nitrogen and oxygen is decomposed in the laboratory and produces 24.5 g of nitrogen and 70.0 g of oxygen. Calculate the empirical formula of the compound.	A laboratory analysis of aspirin determined the following mass percent composition: C 60.00%; H 4.48%; O 35.53%. Find the empirical formula.
1. Write down (or compute) as given the masses of each element present in a sample of the compound. If you are given mass percent composition, assume a 100-g sample and compute the masses of each element from the given percentages.	**Given:** 24.5 g N; 70.0 g O **Find:** empirical formula	**Given:** In a 100-g sample: 60.00 g C; 4.48 g H; 35.53 g O **Find:** empirical formula
2. Convert each of the masses in Step 1 to moles by using the appropriate molar mass for each element as a conversion factor.	**Solution:** $24.5\ \text{g N} \times \frac{1\ \text{mol N}}{14.01\ \text{g N}} = 1.75\ \text{mol N}$; $70.0\ \text{g O} \times \frac{1\ \text{mol O}}{16.00\ \text{g O}} = 4.38\ \text{mol O}$	**Solution:** $60.00\ \text{g C} \times \frac{1\ \text{mol C}}{12.01\ \text{g C}} = 4.996\ \text{mol C}$; $4.48\ \text{g H} \times \frac{1\ \text{mol H}}{1.01\ \text{g H}} = 4.44\ \text{mol H}$; $35.53\ \text{g O} \times \frac{1\ \text{mol O}}{16.00\ \text{g O}} = 2.221\ \text{mol O}$
3. Write down a pseudoformula for the compound, using the moles of each element (from Step 2) as subscripts.	$N_{1.75}O_{4.38}$	$C_{4.996}H_{4.44}O_{2.221}$
4. Divide all the subscripts in the formula by the smallest subscript.	$N_{\frac{1.75}{1.75}}O_{\frac{4.38}{1.75}} \longrightarrow N_1O_{2.5}$	$C_{\frac{4.996}{2.221}}H_{\frac{4.44}{2.221}}O_{\frac{2.221}{2.221}} \longrightarrow C_{2.25}H_2O_1$
5. If the subscripts are not whole numbers, multiply all the subscripts by a small whole number (see the following table) to get whole-number subscripts.	$N_1O_{2.5} \times 2 \longrightarrow N_2O_5$ The correct empirical formula is N_2O_5.	$C_{2.25}H_2O_1 \times 4 \longrightarrow C_9H_8O_4$ The correct empirical formula is $C_9H_8O_4$.

Fractional Subscript	Multiply by this Number to Get Whole-Number Subscripts
_.10	10
_.20	5
_.25	4
_.33	3
_.50	2
_.66	3
_.75	4

SKILLBUILDER 6.10

A sample of a compound is decomposed in the laboratory and produces 165 g of carbon, 27.8 g of hydrogen, and 220.2 g O. Calculate the empirical formula of the compound.

SKILLBUILDER 6.11

Ibuprofen, an aspirin substitute, has the following mass percent composition: C 75.69%; H 8.80%; O 15.51%. Calculate the empirical formula of the compound.

Visualization: Connecting the Macroscopic and Microscopic Worlds

Today's students often absorb information best when it is presented in graphic form or accompanied by some visual reinforcement. While pictures cannot replace clear explanations, they can supplement them in crucial ways. The art program of this book was expressly conceived to help students visualize connections between molecular processes and the behavior of macroscopic objects.

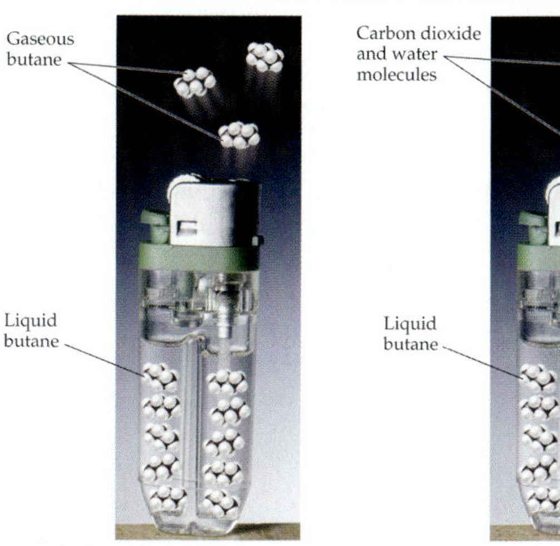

▲ **Figure 3.11 Vaporization: a physical change** If you push the button on a lighter without turning the flint, some of the liquid butane vaporizes to gaseous butane. Since the liquid butane and the gaseous butane are both composed of butane molecules, this is a physical change.

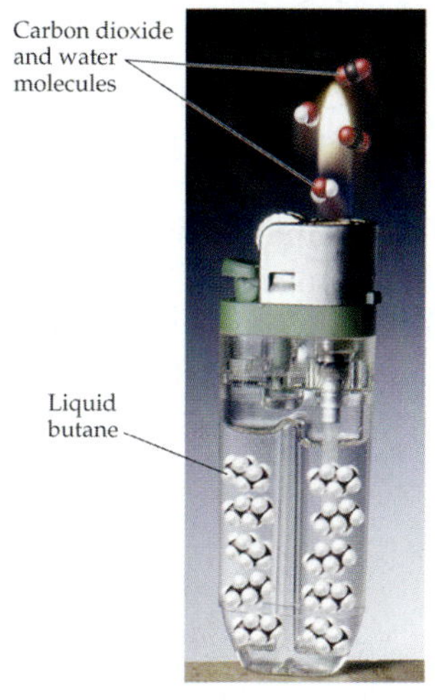

▲ **Figure 3.12 Burning: a chemical change** If you push the button *and* turn the flint to create a spark, you produce a flame. The butane molecules react with oxygen molecules in air to form new molecules, carbon dioxide and water. This is a chemical change.

◄ Many concepts are illustrated using a two-part visual image: a photograph of a real-world object or process, and a depiction of what is taking place on the molecular level, either superimposed or shown as a magnified breakout.

▼ Many molecular formulas in the text are depicted not only with structural formulas, but also with space-filling drawings of the molecules for greater clarity and vividness.

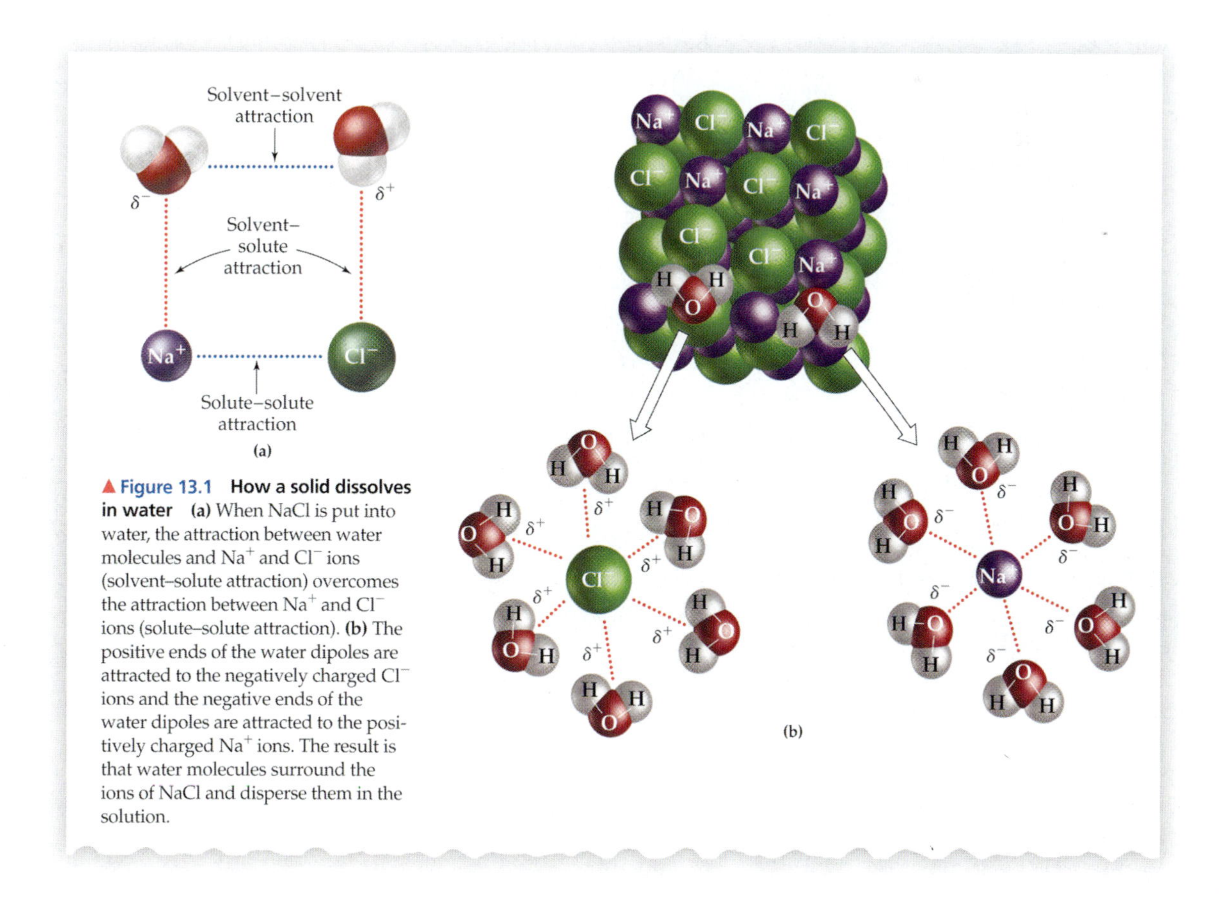

▲ **Figure 13.1 How a solid dissolves in water** **(a)** When NaCl is put into water, the attraction between water molecules and Na^+ and Cl^- ions (solvent–solute attraction) overcomes the attraction between Na^+ and Cl^- ions (solute–solute attraction). **(b)** The positive ends of the water dipoles are attracted to the negatively charged Cl^- ions and the negative ends of the water dipoles are attracted to the positively charged Na^+ ions. The result is that water molecules surround the ions of NaCl and disperse them in the solution.

Creating Interest in Chemistry

Students learn best when they feel that the material is interesting and relevant to their lives. Interest in the field of chemistry and in the topic under study is enhanced by means of two recurring features.

UNIQUE Chapter Openers

Every chapter opens with a description of an everyday situation or practical application that clearly demonstrates the importance of the material covered in that chapter. These openers often involve topics that are particularly relevant to the lives of students, such as health, consumer, environmental, or societal issues. Each chapter introductory narrative is accompanied by a piece of art portraying the content in a striking visual image that combines macroscopic and molecular views. Each chapter introductory narrative is accompanied by a unique and striking visual image, combining macroscopic and molecular views, that brings the content to life for the student.

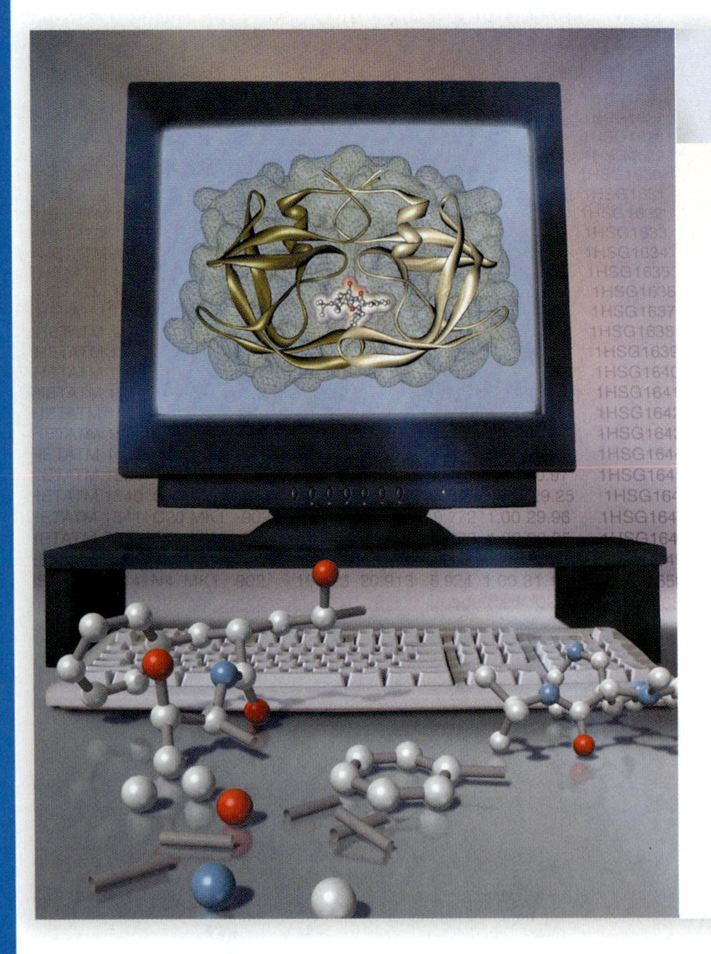

CHAPTER 10

Chemical Bonding

"The fascination of a growing science lies in the work of the pioneers at the very borderland of the unknown, but to reach this frontier one must pass over well traveled roads ..."

Gilbert N. Lewis (1875–1946)

10.1 Bonding Models and AIDS Drugs

In 1989, researchers discovered the structure of a molecule called HIV-protease. HIV-protease is a protein (a class of biological molecules) synthesized by the human immunodeficiency virus (HIV), which causes AIDS. HIV-protease is crucial to the virus's ability to replicate itself. Without HIV-protease, HIV could not spread in the human body because the virus could not copy itself, and AIDS would not develop.

Proteins are discussed in more detail in Chapter 19.

With knowledge of the HIV-protease structure, drug companies set out to design a molecule that would disable protease by sticking to the working part of the molecule (called the *active site*). To design such a molecule, researchers used **bonding theories**—models that predict how atoms bond together to form molecules—to simulate how potential drug molecules would interact with the protease molecule. By the early 1990s, these companies had developed several drug molecules that seemed to work. Since these molecules inhibit the action of HIV-protease, they are called *protease inhibitors*. In human trials, protease inhibitors in combination with other drugs have decreased the viral count in HIV-infected individuals to undetectable levels. Many AIDS patients are still alive today because of the development of these drugs.

Bonding theories are central to chemistry because they predict how atoms bond together to form compounds. They predict what combinations of atoms form compounds and what combinations do not. For example, bonding theories predict why salt is NaCl and not $NaCl_2$ and why water is H_2O and not H_3O. Bonding theories also explain the shapes of molecules, which in turn determine many of their physical and chemical properties. The bonding theory you will learn in this chapter is called **Lewis theory,** named after the American chemist who developed it, G. N. Lewis (1875–1946). It involves representing electrons as dots and drawing what are called *dot structures* or *Lewis structures* to represent molecules. These structures, which are fairly simple to draw, have tremendous predictive power. It takes just a few minutes to

◀ The gold-colored structure on the computer screen is a representation of HIV-protease. The molecule shown in the center is Indinavir, a protease inhibitor.

305

Interest Boxes

Introductory Chemistry has four types of interest boxes: Everyday Chemistry, Chemistry in the Media, Chemistry in the Environment, and Chemistry and Health.

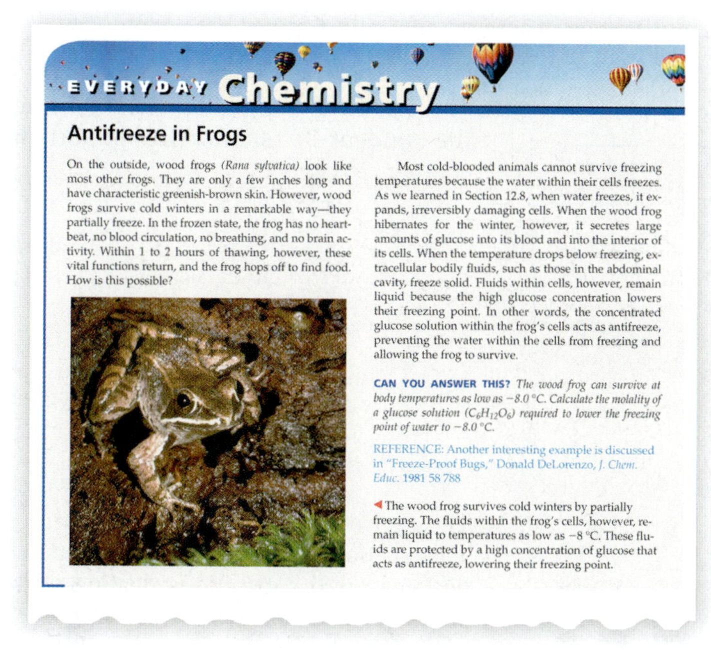

EVERYDAY Chemistry

Antifreeze in Frogs

On the outside, wood frogs *(Rana sylvatica)* look like most other frogs. They are only a few inches long and have characteristic greenish-brown skin. However, wood frogs survive cold winters in a remarkable way—they partially freeze. In the frozen state, the frog has no heartbeat, no blood circulation, no breathing, and no brain activity. Within 1 to 2 hours of thawing, however, these vital functions return, and the frog hops off to find food. How is this possible?

Most cold-blooded animals cannot survive freezing temperatures because the water within their cells freezes. As we learned in Section 12.8, when water freezes, it expands, irreversibly damaging cells. When the wood frog hibernates for the winter, however, it secretes large amounts of glucose into its blood and into the interior of its cells. When the temperature drops below freezing, extracellular bodily fluids, such as those in the abdominal cavity, freeze solid. Fluids within cells, however, remain liquid because the high glucose concentration lowers their freezing point. In other words, the concentrated glucose solution within the frog's cells acts as antifreeze, preventing the water within the cells from freezing and allowing the frog to survive.

CAN YOU ANSWER THIS? *The wood frog can survive at body temperatures as low as −8.0 °C. Calculate the molality of a glucose solution ($C_6H_{12}O_6$) required to lower the freezing point of water to −8.0 °C.*

REFERENCE: Another interesting example is discussed in "Freeze-Proof Bugs," Donald DeLorenzo, *J. Chem. Educ.* 1981 58 788

◀ The wood frog survives cold winters by partially freezing. The fluids within the frog's cells, however, remain liquid to temperatures as low as −8 °C. These fluids are protected by a high concentration of glucose that acts as antifreeze, lowering their freezing point.

- *Everyday Chemistry* boxes describe in chemical terms what is happening on the atomic or molecular level in common, everyday processes, such as the bleaching of hair, and explain the properties of familiar materials, such as Kevlar.
- *Chemistry in the Media* boxes discuss chemical topics that have been in the news, such as the controversy over oxygenated fuels.
- *Chemistry in the Environment* boxes deal with environmental issues that are closely tied to chemistry, such as acid rain and the ozone hole.
- *Chemistry and Health* boxes focus on biomedical topics as well as those related to personal health and fitness, such as drug dosage, ulcers and antacids, and isoosmotic solutions for transfusions.

The interest boxes in *Introductory Chemistry* all contain questions that relate directly to the chapter material, helping students apply what they have just learned.

Review and Assessment

UNIQUE Chapter in Review

Each chapter ends with a review consisting of two sections, the first focusing on chemical principles and the second on chemical skills. Each section is itself divided into two columns.

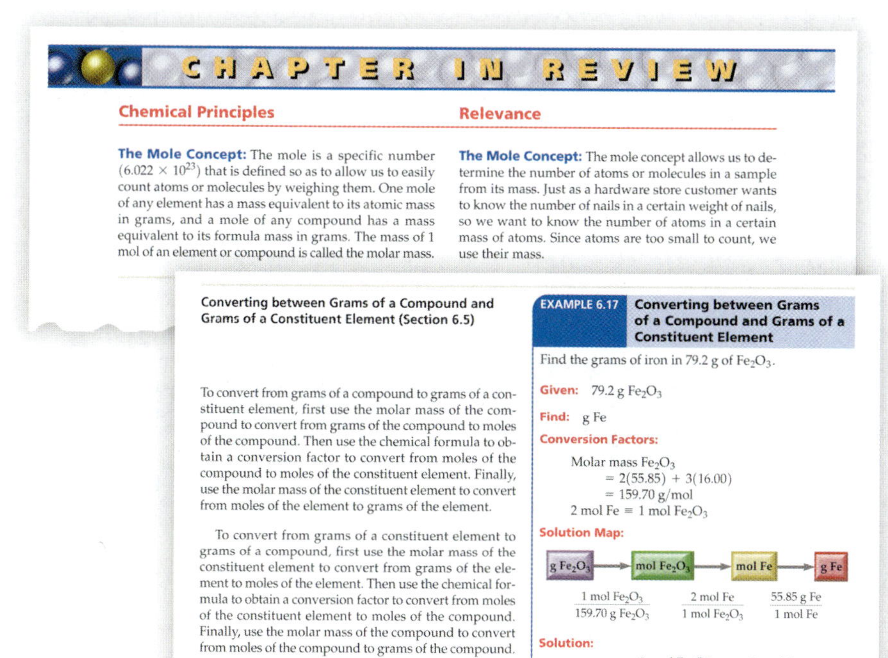

CHAPTER IN REVIEW

Chemical Principles	Relevance
The Mole Concept: The mole is a specific number (6.022×10^{23}) that is defined so as to allow us to easily count atoms or molecules by weighing them. One mole of any element has a mass equivalent to its atomic mass in grams, and a mole of any compound has a mass equivalent to its formula mass in grams. The mass of 1 mol of an element or compound is called the molar mass.	**The Mole Concept:** The mole concept allows us to determine the number of atoms or molecules in a sample from its mass. Just as a hardware store customer wants to know the number of nails in a certain weight of nails, so we want to know the number of atoms in a certain mass of atoms. Since atoms are too small to count, we use their mass.

Converting between Grams of a Compound and Grams of a Constituent Element (Section 6.5)

To convert from grams of a compound to grams of a constituent element, first use the molar mass of the compound to convert from grams of the compound to moles of the compound. Then use the chemical formula to obtain a conversion factor to convert from moles of the compound to moles of the constituent element. Finally, use the molar mass of the constituent element to convert from moles of the element to grams of the element.

To convert from grams of a constituent element to grams of a compound, first use the molar mass of the constituent element to convert from grams of the element to moles of the element. Then use the chemical formula to obtain a conversion factor to convert from moles of the constituent element to moles of the compound. Finally, use the molar mass of the compound to convert from moles of the compound to grams of the compound.

EXAMPLE 6.17 Converting between Grams of a Compound and Grams of a Constituent Element

Find the grams of iron in 79.2 g of Fe_2O_3.

Given: 79.2 g Fe_2O_3

Find: g Fe

Conversion Factors:

Molar mass Fe_2O_3
$= 2(55.85) + 3(16.00)$
$= 159.70$ g/mol

2 mol Fe ≡ 1 mol Fe_2O_3

Solution Map:

g Fe_2O_3 → mol Fe_2O_3 → mol Fe → g Fe

$$\frac{1 \text{ mol } Fe_2O_3}{159.70 \text{ g } Fe_2O_3} \qquad \frac{2 \text{ mol Fe}}{1 \text{ mol } Fe_2O_3} \qquad \frac{55.85 \text{ g Fe}}{1 \text{ mol Fe}}$$

Solution:

$$79.2 \text{ g } Fe_2O_3 \times \frac{1 \text{ mol } Fe_2O_3}{159.70 \text{ g } Fe_2O_3} \times \frac{2 \text{ mol Fe}}{1 \text{ mol } Fe_2O_3} \times \frac{55.85 \text{ g Fe}}{1 \text{ mol Fe}} = 55.4 \text{ g Fe}$$

- In the Chemical Principles section, the left column summarizes the key principles while the right column tells why each is important.
- In the Chemical Skills section, the left column describes the key skills while the right column contains a worked Example illustrating each skill.

Student Exercises

All chapters contain exercises divided into four types: questions, problems, cumulative problems, and highlight problems. Many new exercises have been added in this edition to provide a full range of assessment opportunities.

Questions

1. Why is reaction stoichiometry important? Can you give some examples?
2. What does it mean to say that two quantities in a chemical reaction are equivalent?
3. Write conversion factors showing the relationships between moles of each of the reactants and products in the following reaction.

   $N_2(g) + 3\,H_2(g) \longrightarrow 2\,NH_3(g)$

4. For the reaction in Problem 3, how many molecules of H_2 are required to completely react with two molecules of N_2? How many moles of H_2 are required to completely react with 2 mol of N_2?
5. Write the conversion factor that you would use to convert from moles of Cl_2 to moles of NaCl in the following reaction.

If you have 7 cups of noodles, 27 tomatoes, and 9 cloves of garlic, how many servings of pasta can you make? Which ingredient limits the amount of pasta that is possible?

9. In a chemical reaction, what is the limiting reactant?
10. In a chemical reaction, what is the theoretical yield?
11. In a chemical reaction, what are the actual yield and percent yield?
12. If you are given a chemical equation and specific amounts of each reactant in grams, how would you determine how much product can possibly be made?
13. Consider the following generic chemical reaction.

    $A + 2\,B \longrightarrow C + D$

    Suppose you have 12 g of A and 24 g of B. Which of the following statements are true?

◀ The *Questions* are qualitative, requiring the student to summarize important chapter concepts.

Mass Percent Composition

67. A 2.45-g sample of strontium completely reacts with oxygen to form 2.89 g of strontium oxide. Use this data to calculate the mass percent composition of strontium in strontium oxide.
68. A 4.78-g sample of aluminum completely reacts with oxygen to form 6.67 g of aluminum oxide. Use this data to calculate the mass percent composition of aluminum in aluminum oxide.
69. A 1.912-g sample of calcium chloride is decomposed into its constituent elements and found to contain 0.690 g Ca and 1.222 g Cl. Calculate the mass percent composition of Ca and Cl in calcium chloride.
70. A 0.45-g sample of aspirin is decomposed into its constituent elements and found to contain 0.27 g C, 0.020 g H, and 0.16 g O. Calculate the mass percent composition of C, H, and O in aspirin.
71. Copper(II) fluoride contains 37.42% F by mass. Use this percentage to calculate the mass of fluorine in grams contained in 28.5 g of copper(II) fluoride.
72. Silver chloride, often used in silver plating, contains 75.27% Ag. Calculate the mass of silver chloride in grams required to make 4.8 g of silver plating.

◀ The *Problems* are quantitative in nature. This section, the longest, is divided by headings into subsections dealing with the major topical areas of the chapter.

▶ *Cumulative Problems* require students to synthesize several of the skills they have learned in the chapter and in previous chapters.

Cumulative Problems

97. A pure copper cube has an edge length of 1.42 cm. How many copper atoms does it contain? (volume of a cube = (edge length)^3; density of copper = 8.96 g/cm^3)
98. A pure silver sphere has a radius of 0.886 cm. How many silver atoms does it contain? (Volume of a sphere = $\frac{4}{3}\pi r^3$; density of silver = 10.5 g/cm^3)
99. A drop of water has a volume of approximately 0.05 mL. How many water molecules does it contain? (density of water = 1.0 g/cm^3).
100. Fingernail-polish remover is primarily acetone (C_3H_6O). How many acetone molecules are in a bottle of acetone with a volume of 325 mL? (density of acetone = 0.788 g/cm^3).
101. Complete the following table:

Substance	*Mass*	*Moles*	*Number of Particles (atoms or molecules)*
Ar	___	4.5×10^{-4}	___
NO_2	___	___	1.09×10^{20}
K	22.4 mg	___	___
C_8H_{18}	3.76 kg	___	___

102. Complete the following table:

Substance	*Mass*	*Moles*	*Number of Particles (atoms or molecules)*
$C_6H_{12}O_6$	15.8 g	___	___
Pb	___	___	9.04×10^{21}
CF_4	22.5 kg	___	___
C	___	0.0388	___

▶ *Highlight Problems* are set within a context that will be of particular interest to the student because of its timeliness, familiarity, or relevance to some important issue. Such problems often include photographs or art.

101. A major event affecting global climate is the El Niño/La Niña cycle. In this cycle, equatorial Pacific Ocean waters warm by several degrees Celsius above normal (El Niño) and then cool by several degrees Celsius below normal (La Niña). This cycle affects weather not only in North and South America, but as far away as Africa. Why does a seemingly small change in ocean temperature have such a large impact on weather?

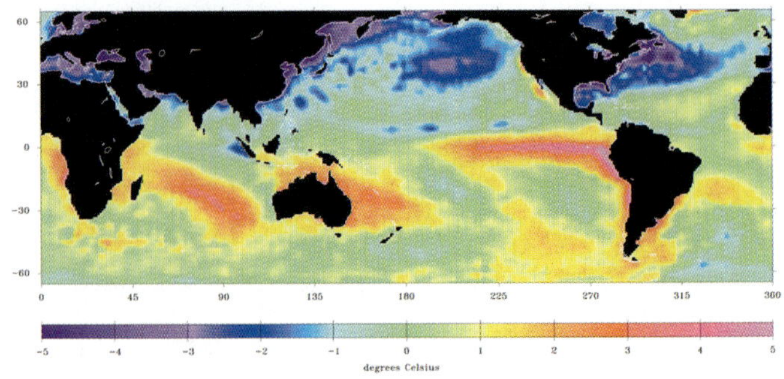

▲ Temperature anomaly plot of the world's oceans for December 23, 1997. The red section off the western coast of South America is the El Niño effect, a warming of the Pacific Ocean along the equator.

All Problems and Cumulative Problems are paired: A problem that is answered in the back of the book (identified by a blue number) is followed by a similar problem without an answer. All answers are available in the instructor's full *Solutions Manual*.

Print and Media Resources

For the Instructor

Annotated Instructor's Edition (0–13–149513–5) with annotations by Carl Hoeger, University of California, San Diego. This special edition contains the entire student text plus marginal annotations to aid instructors in preparing their lectures. Included are suggestions for classroom demonstrations, teaching tips, Transparency Pack icons, and references to articles on the effective presentation of many chemical topics.

Instructor's Resource and Full Solutions Manual (0–13–147062–0) by Mark Ott, Jackson Community College, and Matthew Johll, Illinois Valley Community College. This manual features lecture outlines with presentation suggestions, teaching tips, suggested in-class demonstrations, and topics for classroom discussion. It also contains full solutions to all the end-of-chapter problems from the text.

Test Item File (0–13–147064–7) by Kuruvilla Zachariah, Ohio University. This printed test bank includes more than 1300 questions. A computerized version of the test item file is available in the Instructor's Resource Center.

Transparency Pack (0–13–147069–8) This set contains 175 full-color transparencies from the text. Chosen specifically to put principles into visual perspective, it can save you time while you are preparing your lectures.

Instructor's Resource Center on CD/DVD (0–13–147063–9) This fully searchable and integrated collection of resources is designed to help you make efficient and effective use of your lecture preparation time as well as to enhance your classroom presentations and assessment efforts. This package features the following:

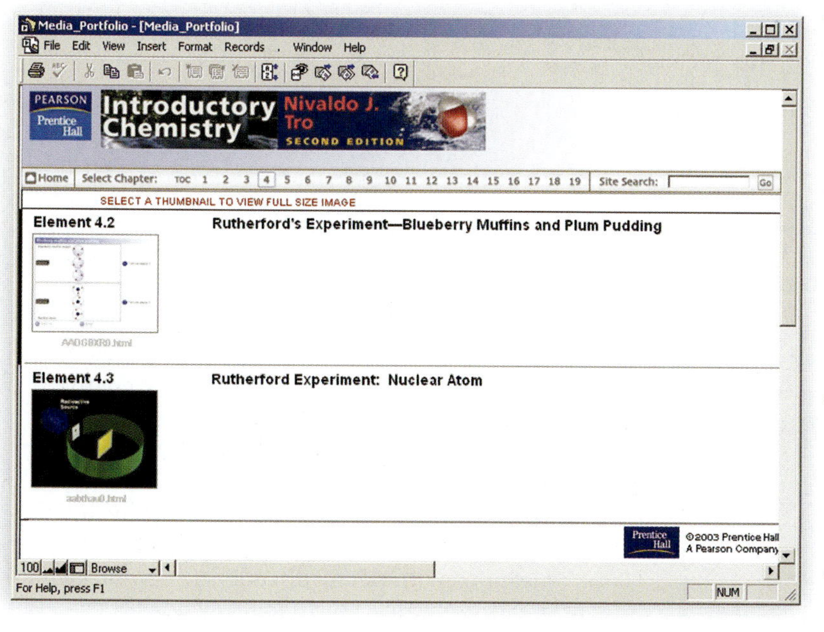

- Nearly all the art from the text, including tables, in JPG and PDF formats; movies; animations; interactive activities; and the *Instructor's Resource Manual* Word™ files.
- Four PowerPoint™ presentations: (1) a lecture outline presentation for each chapter, (2) all the art from the text, (3) the worked Examples from the text, and (4) CRS (Classroom Response System) questions.
- A search engine tool that lets you find relevant resources via a number of different parameters, such as key terms, learning objectives, figure numbers, and resource type.
- The TestGen, a computerized version of the Test Item File that allows you to create and tailor exams to your needs.

ONEKEY (Access at http://chem.prenhall.com/tro)

OneKey offers the best teaching and learning resources all in one place.

- OneKey for *Introductory Chemistry, Second Edition*, is all your *students* need for access to your course materials anytime, anywhere.
- OneKey is all *you* need to plan and administer your course.

Conveniently organized by text chapter, these compiled resources help you save time and help your students reinforce and apply what they have learned in class.

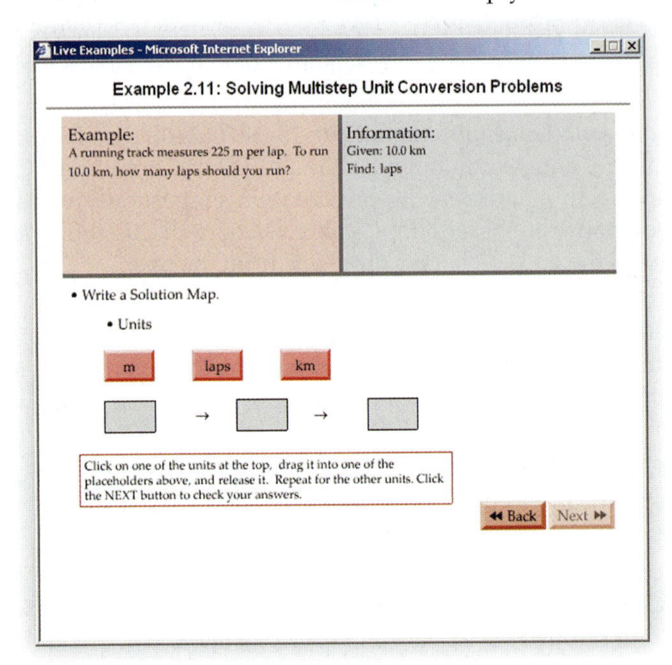

Student Resources

- The new **WebBook**, a full electronic version of the text with embedded Live Examples, animations, movies, and molecules
- **Live Examples**, an interactive version of selected Examples from the text
- A **3-D Molecule Gallery** to help students visualize many important molecules introduced in the course
- **Quizzes** of multiple-choice and true/false questions featuring hints and answer-specific feedback
- More online resources such as **Research Navigator, Link Library,** and the **Math Tutorial**

Instructor Resources

- Resources from the **Instructor's Resource Center on CD/DVD** (described on the previous page) are also included in OneKey.

Assessment Content

- Multiple choice **Quiz Questions** with hints and feedback
- **Student Tutorials** featuring animations, movies, and molecules with multiple-choice follow-up questions
- **Test Item File** questions

PHGradeAssist 2

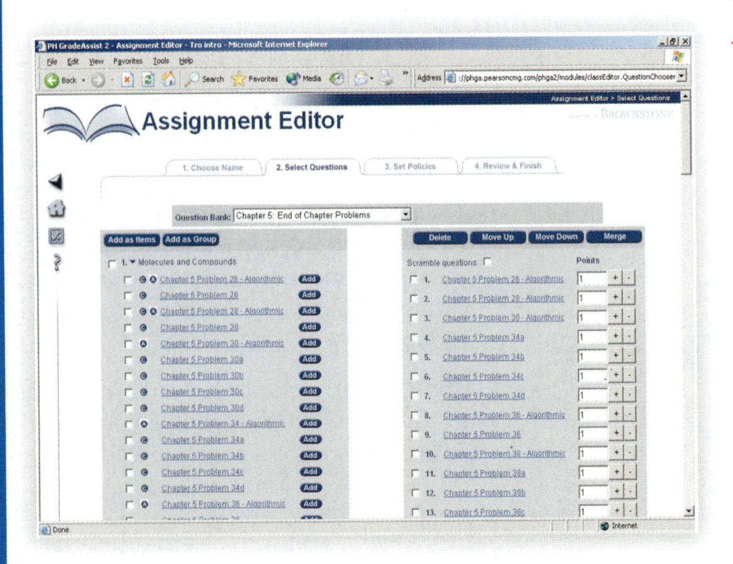

PH GradeAssist Available only through OneKey, PH GradeAssist is a timesaving homework and assessment system that allows students unlimited practice with algorithmically generated problems in an online environment. Instructors can administer quizzes and assignments, control the content and assignment parameters, and receive assignments and view performance statistics with the built-in gradebook. This customizable system has a volume built specifically for this text. Content includes selected end-of-chapter questions with detailed feedback as well as links to appropriate **WebBook** sections. The PH GradeAssist platform is going through continual, seamless upgrades, to offer instructors more options and to make the program easier for students to use.

With contributions by Roy Kennedy, Massachusetts Bay Community College; Mark E. Ott, Jackson Community College; Michael Hauser, St. Louis Community College–Meramec; Jeffrey Lehman, Grossmont College; Cary Willard, Grossmont College; and Bette Kreuz, University of Michigan–Dearborn.

COURSE MANAGEMENT

All of the content in **OneKey** is also available in **WebCT** and **Blackboard**. Prentice Hall offers content cartridges for these text-specific Classroom Management Systems. Visit **http://cms.prenhall.com** or contact your Prentice Hall sales representative for details. More basic courses, loaded with just the Test Item file, are also available.

For the Student

Study Guide (0–13–147071-X) by Donna Friedman, St. Louis Community College–Florissant Valley. This book assists students through the text material with chapter overviews and practice problems for each major concept in the text, followed by two or three self-tests with answers at the end of each chapter.

Selected Solutions Manual (0–13–147085–2) by Matthew Johll, Illinois Valley Community College. This book provides solutions only to those problems that have a short answer in the text's Answers section (problems numbered in blue in the text).

Math Review Toolkit (0–13–147066–3) by Gary L. Long, Virginia Tech. This print resource, free when packaged with the text, reinforces the skills necessary to succeed in chemistry. Keyed specifically to chapters in *Introductory Chemistry, Second Edition*, it includes additional mathematics review, problem-solving tools, and examples.

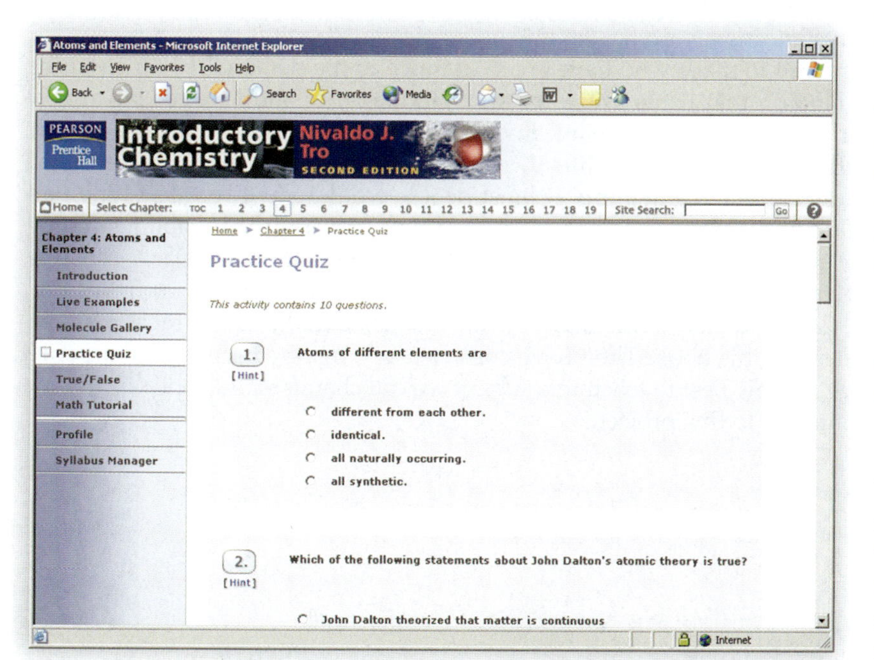

◀ *Companion Website*
(Access at **http://chem.prenhall.com/tro**) For students who wish to utilize a Website for additional practice, the Companion Website contains the **Chemical Principles**, **Live Examples**, **Molecule Gallery**, **Practice Quiz**, **True/False**, and **Math Tutorial** modules included in the Course Management Systems described previously.

Student Accelerator CD-ROM
This student CD—which contains animations, movies, and molecules from the Companion Website—accelerates the performance of the Website when students download high-bandwidth media, so that students are not restricted by slow connections. It can also be used apart from the Companion Website if a student doesn't have a live Internet connection. Many of the media elements available on the CD are referenced in the text by means of marginal icons and titles. The *Introductory Chemistry, Second Edition* Accelerator CD comes packaged with every new student textbook for no additional charge.

Acknowledgments

This book has been a group effort, and there are many people whose help has meant a great deal to me. First and foremost, I would like to thank my editor, Kent Porter Hamann, who took a chance on a little-known author from a small liberal-arts college when she signed me to write the first edition of this book. She has believed in me and encouraged me from day one—I am forever indebted to her. I also owe much to Ray Mullaney, whose wisdom and gentle guidance have steered this project throughout, as well as to John Challice and Paul Corey for their unwavering commitment and support.

New to this edition is Dan Schiller, a development editor with an author's instinct and an incredibly sharp eye for detail. I am grateful to Dan for his ideas, his persistence, his deep and broad knowledge, and especially for the many hours he has poured into this project. This book would be much different without him. I deeply appreciate the expertise and professionalism of my copy editor, Amy Schneider, as well as the skill and diligence of Fran Daniele, Simone Lukashov, and their colleagues at Prepare. Thanks are also due to my media editor, Michael Richards, ably assisted by Ed Dodd III and Bridget K. Page; my project manager, Kristen Kaiser; executive marketing manager Steve Sartori; art director Jon Boylan; Patti Burns, Connie Long, and the staff of Artworks; editorial assistant extraordinaire Jackie Howard; and the rest of the Prentice Hall team—they are part of a first-class operation. This text has benefited immeasurably from their talents and hard work. I owe a special debt of gratitude to Quade Paul, who made my ideas come alive in his art.

I am happy to acknowledge the help of my colleagues Allan Nishimura and David Marten, who have supported me in my department while I worked on this book. I am also deeply obligated to my provost, Shirley Mullen, who gives me the freedom to be who I am and encourages me to develop in ways unique to myself. She is an outstanding faculty leader and an inspiration to me.

I am grateful to those who have given so much to me personally while writing this book. First on that list is my wife, Ann. Her patience and love for me are beyond description. I must also thank my children, Michael, Ali, Kyle, and Kaden, whose smiling faces and love of life always inspire me. I come from a large Cuban family, whose closeness and support most people would envy. Thanks to my parents, Nivaldo and Sara; my siblings, Sarita, Mary, and Jorge; my siblings-in-law, Jeff, Nachy, Karen, and John; my nephews and nieces, Germain, Danny, Lisette, Sara, and Kenny. These are the people with whom I celebrate life.

Lastly, I am indebted to the many reviewers, listed next, whose ideas are scattered throughout this book. They have corrected me, inspired me, and sharpened my thinking on how best to teach this subject we call chemistry. I deeply appreciate their commitment to this project.

Reviewers of the 2nd Edition

David S. Ballantine, Jr., *Northern Illinois University*
Colin Bateman, *Brevard Community College*
Michele Berkey, *San Juan College*
Steven R. Boone, *Central Missouri State University*
Bryan E. Breyfogle, *Southwest Missouri State University*
Morris Bramlett, *University of Arkansas–Monticello*
Frank Carey, *Wharton County Junior College*
Robbey C. Culp, *Fresno City College*
Michelle Driessen, *University of Minnesota–Minneapolis*
Donna G. Friedman, *St. Louis Community College–Florissant Valley*
Crystal Gambino, *Manatee Community College*
Steve Gunther, *Albuquerque Technical Vocational Institute*

Michael Hauser,
St. Louis Community College–Meramec
Newton P. Hillard, Jr.,
Eastern New Mexico University
Carl A. Hoeger,
University of California–San Diego
Donna K. Howell,
Angelo State University
Nichole Jackson,
Odessa College
T. G. Jackson,
University of South Alabama
Donald R. Jones,
Lincoln Land Community College
Roy Kennedy,
Massachusetts Bay Community College
Blake Key,
Northwestern Michigan College
Kirk Kawagoe,
Fresno City College
Rebecca A. Krystyniak,
St. Cloud State University
Laurie LeBlanc,
Cuyamaca College
Ronald C. Marks,
Warner Southern College
Carol A. Martinez,
Albuquerque Technical Vocational Institute
Charles Michael McCallum,
University of the Pacific
Robin McCann,
Shippensburg University
Victor Ryzhov,
Northern Illinois University
Theodore Sakano,
Rockland Community College
Deborah G. Simon,
Santa Fe Community College
Mary Sohn,
Florida Institute of Technology
Peter-John Stanskas,
San Bernardino Valley College
James G. Tarter,
College of Southern Idaho
Ruth M. Topich,
Virginia Commonwealth University
Eric Trump,
Emporia State University
Mary Urban,
College of Lake County
Richard Watt,
University of New Mexico
Lynne Zeman,
Kirkwood Community College

Reviewers of the 1st Edition

Lori Allen
University of Wisconsin—Parkside
Laura Andersson
Big Bend Community College
Danny R. Bedgood
Arizona State University
Christine V. Bilicki
Pasadena City College
Warren Bosch
Elgin Community College
Bryan E. Breyfogle
Southwest Missouri State University
Carl J. Carrano
Southwest Texas State University
Donald C. Davis
College of Lake County
Donna G. Friedman
St. Louis Community College at Florissant Valley
Leslie Wo-Mei Fung
Loyola University of Chicago
Dwayne Gergens
San Diego Mesa College
George Goth
Skyline College
Jan Gryko
Jacksonville State University
Roy Kennedy
Massachusetts Bay Community College
C. Michael McCallum
University of the Pacific
Kathy Mitchell
St. Petersburg Junior College
Bill Nickels
Schoolcraft College
Bob Perkins
Kwantlen University College
Mark Porter
Texas Tech University
Caryn Prudenté
University of Southern Maine
Connie M. Roberts
Henderson State University
Rill Ann Reuter
Winona State University
Jeffery A. Schneider
SUNY–Oswego
Kim D. Summerhays
University of San Francisco
Ronald H. Takata
Honolulu Community College
Calvin D. Tormanen
Central Michigan University
Eric L. Trump
Emporia State University

CHAPTER 1

The Chemical World

"Imagination is more important than knowledge."

Albert Einstein (1879–1955)

1.1 Soda Pop Fizz

Open a can of soda pop and you hear the familiar "chchchch" of pressure release. Take a sip and you feel the carbon dioxide bubbles on your tongue. Shake the can before you open it and you will be sprayed with the liquid. A can of soda pop, like most things, is a chemical mixture. Soda pop consists primarily of sugar, water, and carbon dioxide. It is the unique combination of these chemicals that gives soda pop its properties. Do you want to know why soda pop tastes sweet? You need to know about sugar and solutions of sugar with water. We learn about solutions in Chapter 13. Do you want to know why soda fizzes when you open it? You need to understand gases and their ability to dissolve in liquids and how that ability changes with changing pressure. We learn about gases in Chapter 11. Do you want to know why drinking too much soda pop makes you gain weight? You need to understand energy and the production of energy by chemical reactions. We discuss energy in Chapter 3 and chemical reactions in Chapter 7. You need not go any farther than your own home and your own everyday experiences to discover chemical questions. Chemicals compose virtually everything: the soda; this book; your pencil; indeed, even your own body.

► Virtually everything around you is composed of chemicals.

◄ Soda pop is a mixture of carbon dioxide and water and a few other substances that give flavor and color. When soda pop is poured into a glass, some of the carbon dioxide molecules come out of the mixture, producing the familiar fizz.

Chemists are particularly interested in the connections between the properties of substances and the particles that compose them. For example, why does soda pop fizz? Like all substances, soda pop is composed of tiny particles called atoms. Atoms are so small that a single drop of soda pop contains about one billion trillion of them. In soda pop, as in most substances, these atoms are bound together to form several different types of molecules. The molecules important to fizzing are carbon dioxide and water. Carbon dioxide molecules consist of three atoms—one carbon and two oxygen—held together in a straight line by chemical bonds.

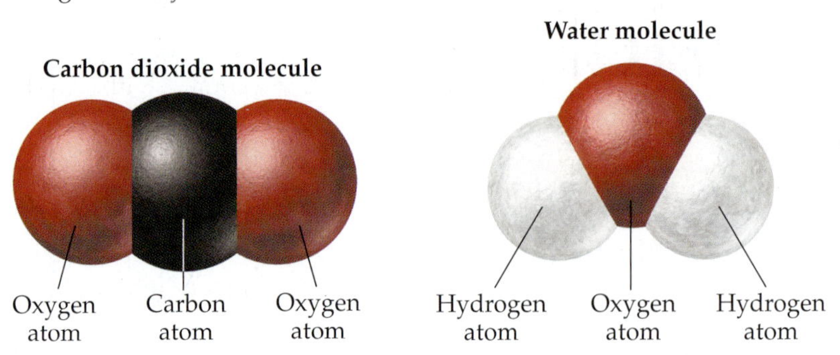

Water molecules also consist of three atoms—one oxygen and two hydrogen—bonded together, but rather than being straight like carbon dioxide, the water molecule is bent.

We will explore the nature of atoms, molecules, and chemical bonds more fully in later chapters. For now, think of atoms and molecules as tiny particles that compose all matter, and chemical bonds as the attachments that hold atoms together.

The details of how atoms bond together to form a molecule—straight, bent, or some other shape—as well as the type of atoms in the molecule, determine *everything* about the substance that the molecule composes. The characteristics of water molecules make water a liquid at room temperature. The characteristics of carbon dioxide molecules make carbon dioxide a gas at room temperature. The characteristics of sugar molecules allow them to interact with our taste buds to produce the sensation of sweetness.

The makers of soda pop use pressure to force gaseous carbon dioxide molecules to mix with liquid water molecules. As long as the can of soda is sealed, the carbon dioxide molecules remain mixed with the water molecules, held there by pressure. When the can is opened, the pressure is released and carbon dioxide molecules escape out of the soda mixture (▼ Figure 1.1). As they do, they create bubbles—the familiar fizz of soda pop.

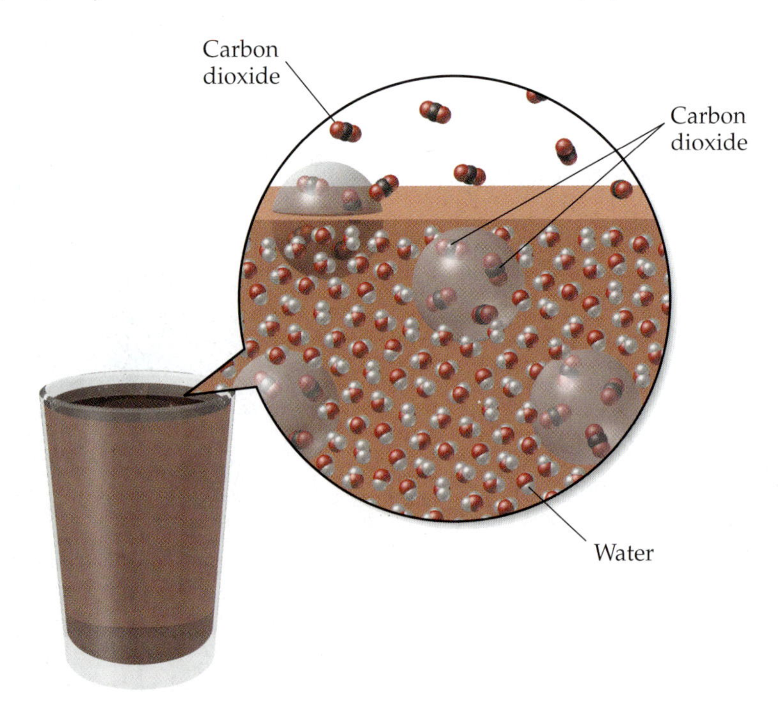

▶ **Figure 1.1 Where the fizz comes from** Bubbles in soda pop are pockets of carbon dioxide gas molecules escaping out of the liquid water.

1.2 Chemicals Compose Ordinary Things

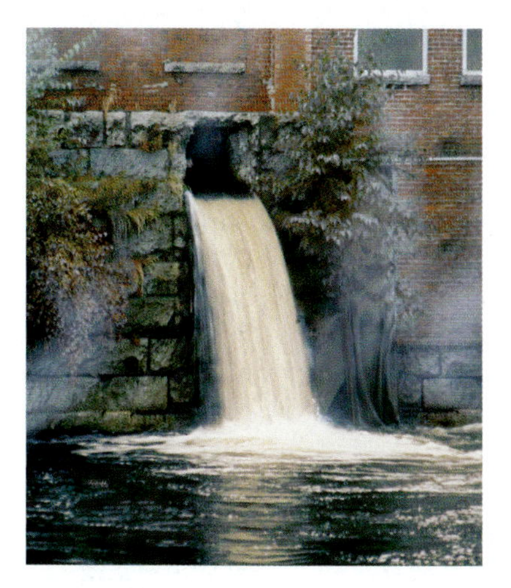

▲ The public often has a very narrow view of chemicals, thinking of them only as dangerous poisons or pollutants.

Is soda pop composed of chemicals? In the broad definition of chemicals, yes. In fact, there is nothing you can hold or touch that is *not* made of chemicals. Unfortunately, when most people think of chemicals, they envision a can of paint thinner in their garage marked with a skull and crossbones, or they recall headlines about rivers polluted by industrial compounds. But chemicals compose more than just these things—they compose ordinary things, too. Chemicals compose the air we breathe and the water we drink. They compose toothpaste, Tylenol, and toilet paper. Chemicals make up virtually everything we come into contact with. This broader notion of chemicals is what chemists use. Chemistry explains the properties and behavior of chemicals, in the broadest sense, by helping us understand the molecules that compose them.

As you experience the world around you, molecules are interacting to create your experience. Imagine watching a sunset. Molecules are involved in every step. Molecules in air interact with light from the sun, scattering away the blue and green light and leaving the red and orange light to create the color. Molecules in your eyes absorb that light, and as a result are altered in a way that sends a signal to your brain. Molecules in your brain then interpret the signal to produce images and emotions. All of this—mediated by molecules—creates the experience of seeing a sunset.

Chemists are interested in why ordinary things are the way they are. Why is water a liquid? Why is salt a solid? Why does soda fizz? Why is a sunset red? Throughout this book you will learn the answers to these questions and many others. You will learn the connections between the behavior of matter and the behavior of the particles that compose it.

◀ Chemists are interested in knowing why ordinary things, such as water, are the way they are. When a chemist sees a pitcher of water, she thinks of the molecules that compose the liquid and how they determine its properties.

TUTORIAL
Chemistry Examples

1.3 All Things Are Made of Atoms and Molecules

Professor Richard Feynman, in a lecture to first-year physics students at the California Institute of Technology, said that the most important idea in all human knowledge is that *all things are made of atoms*. Since atoms are usually bound together to form molecules, however, a chemist might add the concept of *molecules* to Feynman's bold assertion. This simple idea—that all

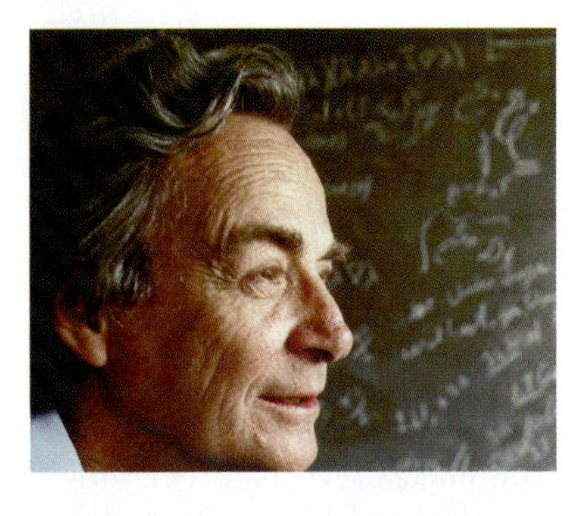

▲ Richard Feynman (1918–1988), Nobel Prize–winning physicist and popular professor at California Institute of Technology.

things are made of atoms and molecules—explains much about our world and our experience of it. Atoms and molecules determine how matter behaves—if they were different, matter would be different. Water molecules, for example, determine how water behaves. Sugar molecules determine how sugar behaves, and the molecules that compose humans determine much about how we behave.

There is a direct connection between the world of atoms and molecules and the world you and I experience every day. Chemists explore this connection. They seek to understand it. A good, simple definition of **chemistry** is the science that tries to understand what matter does by studying what atoms and molecules do.

Chemistry—The science that seeks to understand what matter does by studying what atoms and molecules do.

1.4 The Scientific Method: How Chemists Think

Chemists are scientists and use the **scientific method**—a way of learning that emphasizes observation and experimentation—to understand the world. The scientific method stands in contrast to ancient Greek philosophies that emphasized *reason* as the way to understand the world. Although the scientific method is not a rigid procedure that automatically leads to a definitive answer, it does have key characteristics that distinguish it from other ways of acquiring knowledge. These key characteristics include the observation of nature, the formulation of hypotheses, the testing of hypotheses by experiment, and the formulation of laws and theories.

The first step in acquiring scientific knowledge (▼ Figure 1.2) is often the **observation** or measurement of some aspect of nature. Some observations are simple, requiring nothing more than the naked eye. Other observations emerge from experiments that rely on the use of increasingly sensitive instrumentation. Occasionally, an important observation happens entirely by chance. Alexander Fleming, for example, discovered penicillin when he observed a bacteria-free circle around mold that had accidentally grown on his culture plate. Regardless of how the observation occurs, it usually involves the measurement or description of some aspect of the physical world. For example, Antoine Lavoisier (1743–1794), a French chemist who studied combustion, made careful measurements on the mass of objects before and after burning them in closed containers. He noticed that there was no change in the mass during combustion. Lavoisier made an *observation* about the physical world.

The mass of an object is a measure of the quantity of matter within it.

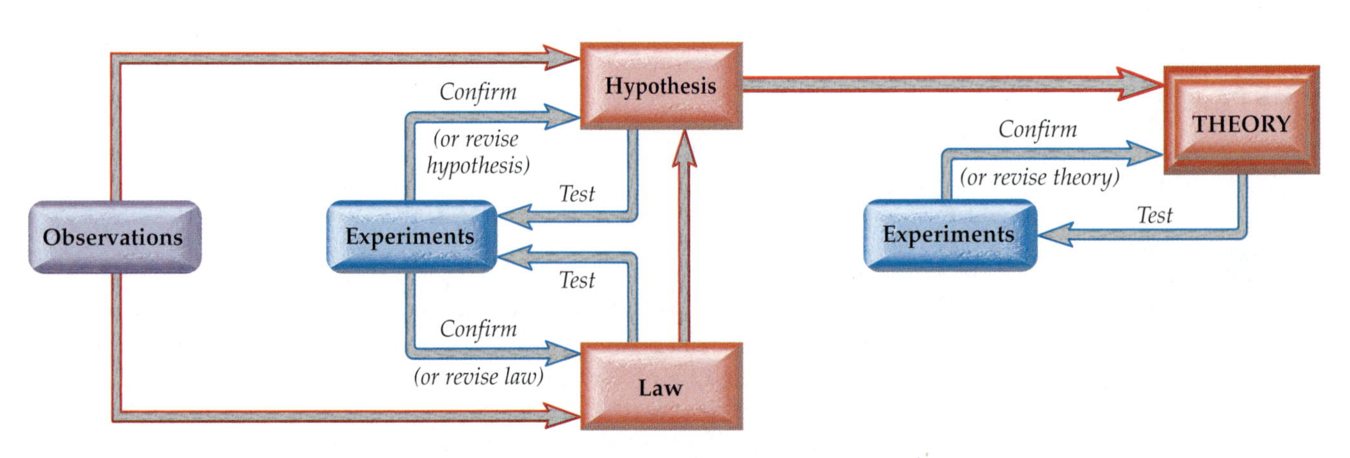

▲ **Figure 1.2 The scientific method**

▶ (Right) Painting of the French chemist Antoine Lavoisier and his wife, Marie, who helped him in his work by illustrating his experiments, recording results, and translating scientific articles from English. (The Metropolitan Museum of Art)
(Far right) John Dalton, the English chemist who formulated the atomic theory.

Observations often lead scientists to formulate a **hypothesis**, a tentative interpretation or explanation of the observations. For example, Lavoisier explained his observations on combustion by hypothesizing that combustion involved the combination of a substance with a component of air. A good hypothesis is *falsifiable*, which means that further testing has the potential to prove it wrong. Hypotheses are tested by **experiments**, highly controlled observations designed to validate or invalidate hypotheses. The results of an experiment may confirm a hypothesis or show it to be mistaken in some way. In the latter case, the hypothesis may have to be modified, or even discarded and replaced by an alternative. Either way, the new or revised hypothesis must then be tested through further experimentation.

Sometimes a number of similar observations can lead to the development of a **scientific law**, a brief statement that synthesizes past observations and predicts future ones. For example, based on his observations of combustion, Lavoisier developed the **law of conservation of mass** that states, "In a chemical reaction matter is neither created nor destroyed." This statement grew out of Lavoisier's observations, but more importantly, it predicted the outcome of similar experiments on any chemical reaction. Laws are also subject to experiments, which can prove them wrong or validate them.

Scientific theories are also called *models*.

One or more well-established hypotheses may form the basis for a scientific **theory**. Theories try to provide a broader and deeper explanation for what we observe in terms of underlying causes. They are models of the way nature is, and often predict behavior that extends well beyond the observations and laws on which they are founded. A good example of a theory is the **atomic theory** of John Dalton (1766–1844). Dalton explained the law of conservation of mass, as well as other laws and observations, by proposing that all matter was composed of small, indestructible particles called atoms. Dalton's theory was a model of the physical world—it went beyond the laws and observations of the time to explain these laws and observations.

Theories are also tested and validated by experiments. Notice that the scientific method begins with observation, forms laws, hypotheses, and theories based on those observations, and then returns to observation to determine their validity. If a law, hypothesis, or theory is inconsistent with the findings of an experiment, it must be revised and new experiments must be conducted to test the revisions. Over time, poor theories are eliminated, and good theories—those consistent with experiments—remain. Established theories with strong experimental support are the most powerful pieces of scientific knowledge. People unfamiliar with science sometimes say, "That is just a theory," as if theories were mere speculations. However, well-tested theories are as close to truth as we get in science. For example, the idea that all matter is made of atoms is "just a theory," but it is a theory with two hundred years of experimental evidence to support it, including the recent imaging of atoms themselves (◀ Figure 1.3). Established theories should not be taken lightly—they are the pinnacle of scientific understanding.

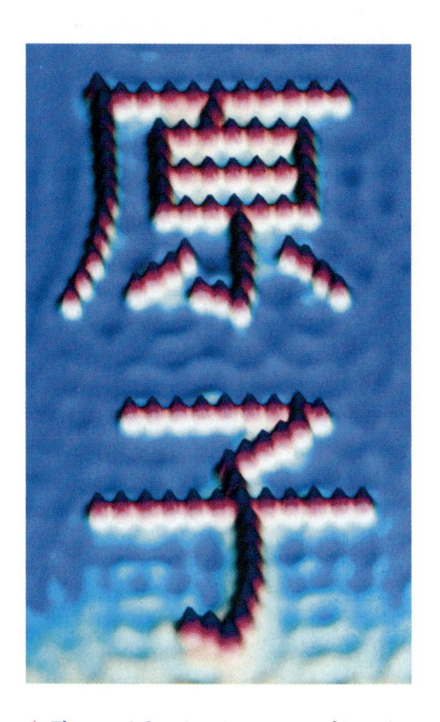

▲ **Figure 1.3 Are atoms real?** The atomic theory has two hundred years of experimental evidence to support it, including recent images, such as this one, of atoms themselves. This image shows the Japanese characters for "atom" written with individual iron atoms on top of a copper surface.

EVERYDAY Chemistry

Combustion and the Scientific Method

Early chemical theories attempted to explain common phenomena such as combustion. Why did things burn? What was happening to a substance when it burned? Could something that was burned be unburned? Early chemists burned different substances and made observations to try to answer these questions. They observed that things would stop burning if placed in a closed container. They found that many metals would burn to form a white powder that they called a *calx* (now we know that these white powders are oxides of the metal), and that the metal could be recovered from the calx, or unburned, by combining it with charcoal and heating it.

Chemists in the first part of the eighteenth century formed a theory about combustion to explain these observations. In this theory, combustion involved a fundamental substance that they called *phlogiston*. Phlogiston was present in anything that burned and was released during combustion. Flammable objects were flammable because they contained phlogiston. When things were burned in a closed container, they didn't burn for very long because the space within the container became saturated with phlogiston. When things burned in the open, they continued to burn until all of the phlogiston within them was gone. This theory also explained how metals that had burned could be unburned. Charcoal was a phlogiston-rich material—they knew this because it burned so well—and when it was combined with a calx, which was a metal that had been emptied of its phlogiston, it transferred some of its phlogiston into the calx, converting it back into the unburned form of the metal. The phlogiston theory was consistent with all of the observations and was widely accepted as valid.

Like any theory, the phlogiston theory had to be tested continually by experiment. One set of experiments, conducted in the mid-eighteenth century by Louis-Bernard Guyton de Morveau (1737–1816), consisted of weighing metals before and after burning them. In every case the metals *gained* weight when they were burned. However, the phlogiston theory predicted that they should *lose* weight because phlogiston was supposed to be lost during combustion. The phlogiston theory needed modification.

The first modification was to suppose that phlogiston was a very light substance so that it actually "buoyed up" the materials that contained it. When phlogiston was released, the material actually became heavier. Such a modification seemed to fit the observations but also seemed far-fetched. Antoine Lavoisier developed a more likely explanation by devising a completely new theory of combustion. According to Lavoisier, when a substance burned, it actually took something *out* of the air, and when it unburned, it released something back into the air. Lavoisier said that burning objects *fixed* (attached or bonded) the air and that the *fixed* air was released when unburning. In a confirming experiment (▼ Figure 1.4), Lavoisier roasted a mixture of calx and charcoal with the aid of sunlight focused by a giant burning lens, and found that a huge volume of "fixed air" was released in the process. The scientific method had worked. The phlogiston theory was proven wrong, and a new theory of combustion took its place—a theory that, with a few refinements, is still valid today.

CAN YOU ANSWER THIS? *What is the difference between a law and a theory? How does the preceding story demonstrate this difference?*

◀ **Figure 1.4 Focusing on combustion** The great burning lens belonging to the Academy of Sciences. Lavoisier used a similar lens to show that a mixture of *calx* (metal oxide) and charcoal released a large volume of *fixed air* (oxygen) when heated.

1.5 A Beginning Chemist: How to Succeed

▲ To succeed as a scientist, you must have the curiosity of a child.

You are a beginning chemist. This may be your first chemistry course, but it is probably not your last. To succeed as a beginning chemist, keep these things in mind. First, chemistry requires curiosity and imagination. If you are content knowing that the sky is blue, but don't care *why* it is blue, then you may have to rediscover your curiosity. I say "rediscover" because even children—or better said, *especially* children—have this kind of curiosity. To succeed as a chemist, you must have the curiosity and imagination of a child—you must want to know the *why* of things.

Second, chemistry requires calculation. Throughout this book, I will ask you to calculate things and to quantify information. *Quantification* involves measurement as part of observation—it is one of the most important tools in science. Quantification allows us to go beyond merely saying that this object is hot and that one is cold or that this one is large and that one is small. It allows us to specify the difference precisely. For example, two samples of water may feel equally hot to our hands, but when we measure their temperatures, we find that one is 40 °C and the other is 44 °C. Even small differences can be important in a calculation or experiment, so assigning numbers to observations and manipulating those numbers becomes very important in chemistry.

Lastly, chemistry requires commitment. To succeed in this course, you must commit yourself to learning chemistry. Roald Hoffman, winner of the 1981 Nobel Prize for chemistry, said,

> *I like the idea that human beings can do anything they want to. They need to be trained sometimes. They need a teacher to awaken the intelligence within them. But to be a chemist requires no special talent, I'm glad to say. Anyone can do it, with hard work.*

Professor Hoffman is right. The key to success in this course is hard work, and that requires commitment. You must do your work regularly and carefully. If you do, you will succeed, and you will be rewarded by seeing a whole new world—the world of molecules and atoms. This world exists beneath the surface of nearly everything you encounter. I welcome you to this world and consider it a privilege, together with your professor, to be your guide.

CHAPTER IN REVIEW

Chemical Principles

Matter and Molecules: Chemists are interested in all matter, even ordinary matter such as water or air. You need not go to a chemical storeroom to find chemical questions because chemicals are all around you. Chemistry is the science that tries to understand what matter does by understanding what molecules do.

Relevance

Matter and Molecules: Chemists want to understand matter for several reasons. First, chemists are simply curious—they want to know why. Why are some substances reactive and others not? Why are some substances gases, some liquids, and others solids? Chemists are also practical; they want to understand matter so that they can control it and produce substances that are useful to society and to humankind.

The Scientific Method: Chemists use the scientific method, which makes use of observations, hypotheses, laws, theories, and experiments. Observations involve measuring or observing some aspect of nature. Hypotheses are tentative interpretations of the observations. Laws summarize the results of a large number of observations, and theories are models that explain and give the underlying causes for observations and laws. Hypotheses, laws, and theories are tested and validated by experiment. If they are not confirmed, they must be revised and tested through further experimentation.

The Scientific Method: The scientific method is important because it works as a way to understand the world. Since its inception, knowledge about the natural world and corresponding technologies using that knowledge have grown rapidly. The application of the scientific method has produced knowledge and technology that has raised living standards throughout the world with advances such as increased food production, rapid transportation, unparalleled access to information, and longer life spans.

Success as a Beginning Chemist: To succeed as a beginning chemist, you must be curious and imaginative, be willing to do calculations, and be committed to learning the material.

Success as a Beginning Chemist: Understanding chemistry will give you a deeper appreciation for the world in which you live, and if you choose science as a career, it will be a foundation upon which you will continue to build.

Key Terms

atomic theory [1.4]
chemistry [1.3]
experiment [1.4]
hypothesis [1.4]
law of conservation of mass [1.4]
observation [1.4]
scientific law [1.4]
scientific method [1.4]
theory [1.4]

Exercises

Questions

Answers to all questions numbered in blue appear in the Answers section at the back of the book.

1. Why does soda fizz?
2. What are chemicals? Give some examples.
3. What do chemists try to do? How do they understand the natural world?
4. What is meant by the statement, "Matter does what molecules do"? Give an example.
5. Define chemistry.
6. How is chemistry connected to everyday life? Is chemistry relevant only in the chemistry laboratory?
7. Explain the scientific method.
8. Give an example from this chapter of the scientific method at work.
9. What is the difference between a law and a theory?
10. What is the difference between a hypothesis and a theory?
11. What is wrong with the statement, "It is just a theory"?
12. What is the law of conservation of mass and who discovered it?
13. What is the atomic theory and who formulated it?
14. What are three things you need to do to succeed in this course?

Problems

Note: The exercises in the Problems section are paired, and the answers to the odd-numbered exercises (numbered in blue) appear in the Answers section at the back of the book.

15. Examine the opening figure of this chapter. Use the information in Section 1.1 to identify the two molecules sitting next to the cola glass and identify each of the atoms within each molecule.
16. Examine Figure 1.1 and, from a molecular point of view, explain why soda pop fizzes. What molecules are inside a bubble within soda pop?

17. Classify each of the following as an observation, a law, or a theory.
    (a) When a metal is burned in a closed container, the mass of the container and its contents does not change.
    (b) Matter is made of atoms.
    (c) Matter is conserved in chemical reactions.
    (d) When wood is burned in a closed container, its mass does not change.

18. Classify each of the following as an observation, a law, or a theory.
    (a) The star closest to Earth is moving away from Earth at high speed.
    (b) A body in motion stays in motion unless acted upon by a force.
    (c) The universe began as a cosmic explosion called the big bang.
    (d) A stone dropped from an altitude of 450 meters falls to the ground in 9.6 seconds.

19. A chemist in an imaginary universe does an experiment that attempts to correlate the size of an atom with its chemical reactivity. The results are as follows.

Size of Atom	*Chemical Reactivity*
small	low
medium	intermediate
large	high

(a) Can you formulate a law from this data?
(b) Can you formulate a theory to explain the law?

20. A chemist decomposes several samples of water into hydrogen and oxygen and weighs (or more correctly "measures the mass of") the hydrogen and the oxygen obtained. The results are as follows:

Sample Number	*Grams of Hydrogen*	*Grams of Oxygen*
1	1.5	12
2	2	16
3	2.5	20

(a) Can you summarize these observations in a short statement?

Next, the chemist decomposes several samples of carbon dioxide into carbon and oxygen. The results are as follows:

Sample Number	*Grams of Carbon*	*Grams of Oxygen*
1	0.5	1.3
2	1.0	2.7
3	1.5	4.0

(b) Can you summarize these observations in a short statement?
(c) Can you formulate a law from the observations in **(a)** and **(b)**?
(d) Can you formulate a theory that might explain your law in **(c)**?

Weights and Measures
SCALE
Weighs 25 pounds
by ounces

CHAPTER 2

Measurement and Problem Solving

"The important thing in science is not so much to obtain new facts as to discover new ways of thinking about them."

Sir William Lawrence Bragg (1890–1971)

2.1 Measuring Global Temperatures

As scientists grow increasingly concerned about changes in Earth's climate, global warming has become a household term. Average global temperatures affect everything from agriculture to weather and ocean levels. The media have reported that global temperatures are increasing. These reports are based on the work of scientists who—after analyzing records from thousands of temperature-measuring stations around the world—concluded that average global temperatures have risen by 0.6 °C in the last century.

Notice how the scientists reported their results. What if they had reported a temperature increase of simply 0.6 without any *units*? The result would be unclear. Units are extremely important in reporting and working with scientific measurements, and they must always be included. Suppose that the scientists had reported additional zeros with their results—for example, 0.60 °C? or 0.600 °C—or that they had reported the number their computer displayed after averaging many measurements, something like 0.58759824 °C? Would all of these convey the same information? Not really. Scientists agree to a standard way of reporting measured quantities in which the number of reported digits reflects the precision in the measurement—more digits, more precision; fewer digits, less precision. Numbers are usually written so that the uncertainty is in the last reported digit. For example, by reporting a temperature increase of 0.6 °C, the scientists mean 0.6 ± 0.1 °C (± means plus or minus). The temperature rise could be as much as 0.7 °C or as little as 0.5 °C, but it is not 1.0 °C. The degree of certainty in this particular measurement is critical, influencing political decisions that directly affect people's lives.

A unit is a standard, agreed-on quantity by which other quantities are measured.

◀ Measurement is part of our daily lives. In this illustration, a pumpkin is weighed for pricing. **Question:** Can you read the weight on the scale?

 2.2 Scientific Notation: Writing Large and Small Numbers

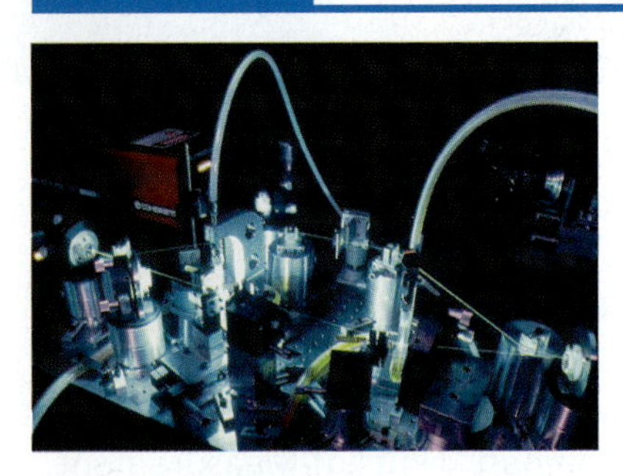

▲ Lasers such as this one can measure time periods as short as 1×10^{-15} s.

Science has constantly pushed the boundaries of the very large and the very small. We can, for example, now measure time periods as short as 0.000000000000001 seconds and distances as great as 14,000,000,000 light-years. The many zeros in these numbers are cumbersome to write, so scientists often use **scientific notation** to write them more compactly. In scientific notation, 0.000000000000001 becomes 1×10^{-15}, and 14,000,000,000 becomes 1.4×10^{10}. A number written in scientific notation consists of a **decimal part**, a number that is usually between 1 and 10, and an **exponential part**, 10 raised to an **exponent**, *n*.

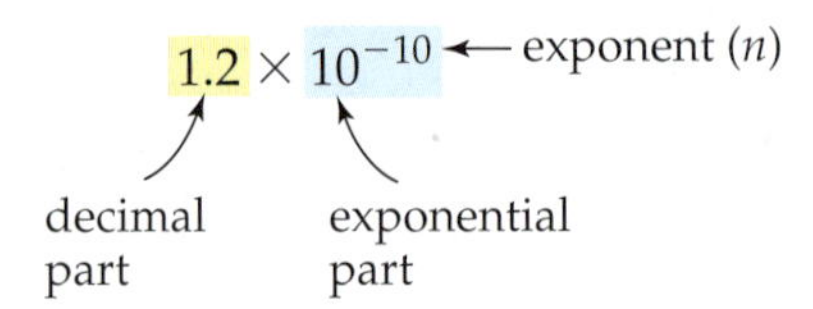

A positive exponent means 1 multiplied by 10 *n* times.

$$10^0 = 1$$
$$10^1 = 1 \times 10$$
$$10^2 = 1 \times 10 \times 10 = 100$$
$$10^3 = 1 \times 10 \times 10 \times 10 = 1{,}000$$

A negative exponent $(-n)$ means 1 divided by 10 *n* times.

$$10^{-1} = \frac{1}{10} = 0.1$$
$$10^{-2} = \frac{1}{10 \times 10} = 0.01$$
$$10^{-3} = \frac{1}{10 \times 10 \times 10} = 0.001$$

To convert a number to scientific notation, we move the decimal point to obtain a number between 1 and 10 and then multiply by 10 raised to the appropriate power. For example, to write 5983 in scientific notation, we move the decimal point to the left three places to get 5.983 (a number between 1 and 10) and then multiply by 1000 to make up for moving the decimal point.

$$5983 = 5.983 \times 1000$$

Since 1000 is 10^3, we write:

$$= 5.983 \times 10^3$$

We can do this in one step by counting how many places we move the decimal point to obtain a number between 1 and 10 and then writing the decimal part multiplied by 10 raised to the number of places we moved the decimal point.

$$5983 = 5.983 \times 10^3$$

3 2 1

If the decimal point is moved to the left, as in the previous example, the exponent is positive. If the decimal is moved to the right, the exponent is negative.

$$0.00034 = 3.4 \times 10^{-4}$$

1 2 3 4

To express a number in scientific notation:

1. Move the decimal point to obtain a number between 1 and 10.
2. Write the result from Step 1 multiplied by 10 raised to the number of places you moved the decimal point.
   - The exponent is positive if you moved the decimal point to the left.
   - The exponent is negative if you moved the decimal point to the right.

EXAMPLE 2.1 Scientific Notation

The U.S. population in 2004 was estimated to be 293,168,000 people. Express this number in scientific notation.

To obtain a number between 1 and 10, move the decimal point to the left 8 decimal places; therefore the exponent is 8. Since the decimal point was moved to the left, the sign of the exponent is positive.

Solution:

$$293{,}168{,}000 \text{ people} = 2.93168 \times 10^{8} \text{ people}$$

SKILLBUILDER 2.1 Scientific Notation

The total federal debt in 2004 was approximately $7,132,000,000,000. Express this number in scientific notation.

Note: The answers to all Skillbuilders appear at the end of the chapter.

EXAMPLE 2.2 Scientific Notation

The radius of a carbon atom is approximately 0.000000000070 m. Express this number in scientific notation.

To obtain a number between 1 and 10, move the decimal point to the right 11 decimal places; therefore the exponent is 11. Since the decimal point was moved to the right, the sign of the exponent is negative.

Solution:

$$0.000000000070 \text{ m} = 7.0 \times 10^{-11} \text{ m}$$

SKILLBUILDER 2.2 Scientific Notation

Express the number 0.000038 in scientific notation.

CONCEPTUAL CHECKPOINT 2.1

The radius of a dust speck is 4.5×10^{-3} mm. In decimal notation, this would be expressed as:

(a) 4500 mm
(b) 0.045 mm
(c) 0.0045 mm
(d) 0.00045 mm

Note: The answers to all Conceptual Checkpoints appear at the end of the chapter.

2.3 Significant Figures: Writing Numbers to Reflect Precision

▲ Since pennies come in whole numbers, 7 pennies means 7.00000 pennies. It is an exact number and therefore never limits significant figures in calculations.

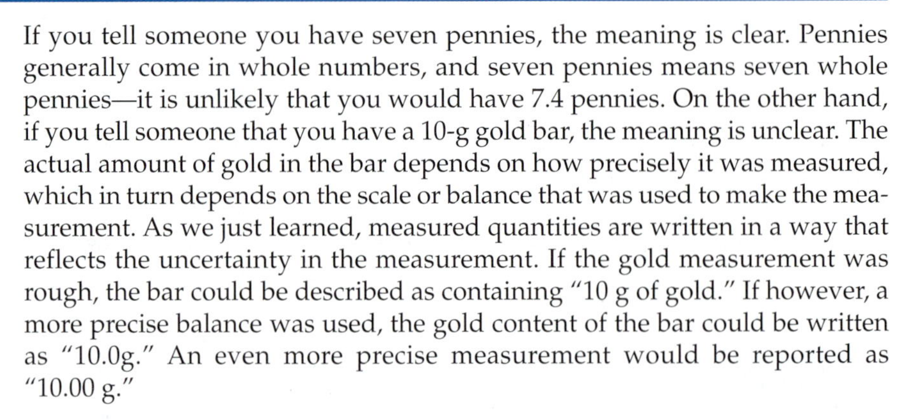

If you tell someone you have seven pennies, the meaning is clear. Pennies generally come in whole numbers, and seven pennies means seven whole pennies—it is unlikely that you would have 7.4 pennies. On the other hand, if you tell someone that you have a 10-g gold bar, the meaning is unclear. The actual amount of gold in the bar depends on how precisely it was measured, which in turn depends on the scale or balance that was used to make the measurement. As we just learned, measured quantities are written in a way that reflects the uncertainty in the measurement. If the gold measurement was rough, the bar could be described as containing "10 g of gold." If however, a more precise balance was used, the gold content of the bar could be written as "10.0g." An even more precise measurement would be reported as "10.00 g."

> Scientific numbers are reported so that every digit is certain except the last, which is estimated.

For example, suppose a reported number is:

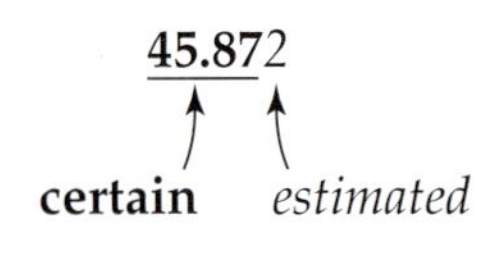

▲ The amount of gold in a 10-g gold bar depends on how precisely it was measured.

The first four digits are certain; the last digit is estimated.

Suppose that we weigh an object on a balance with marks at every 1 g, and suppose that the pointer is between the 1-g mark and the 2-g mark (▼ Figure 2.1) but much closer to the 1-g mark. We mentally divide the space between the 1- and 2-g marks into ten equal spaces and estimate that the pointer is at about 1.2 g. We then write the measurement as 1.2 g, indicating that we are sure of the "1" but have estimated the ".2."

A balance with marks every tenth of a gram requires us to write the result with more digits. For example, suppose that on this more precise balance the pointer is between the 1.2-g mark and the 1.3-g mark (▼ Figure 2.2). We again divide the space between the two marks into ten equal spaces and estimate the third digit. For the figure shown, we report 1.26 g.

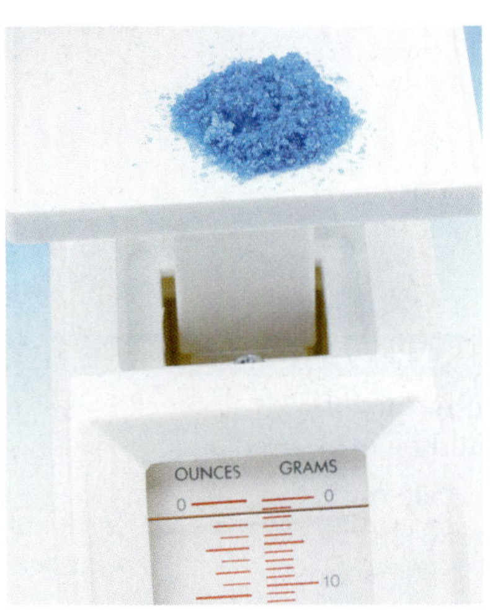

► **Figure 2.1 Estimating tenths of a gram** This balance has markings every 1 g, so we estimate to the tenths place. To estimate between markings, mentally divide the space into ten equal spaces and estimate the last digit. This reading is 1.2 g.

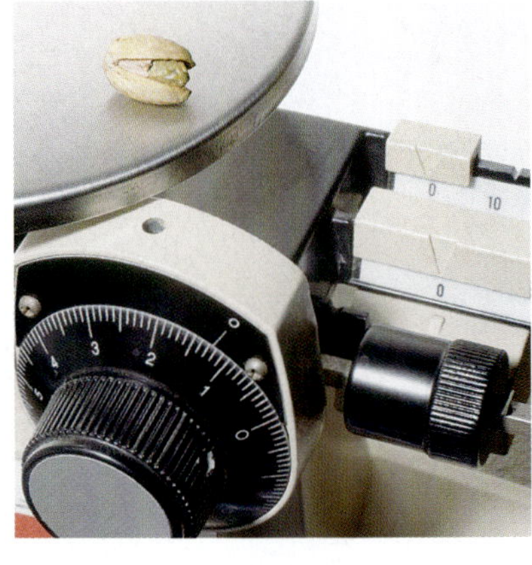

▲ **Figure 2.2 Estimating hundredths of a gram** Since this scale has markings every 0.1 g, we estimate to the hundredths place. The correct reading is 1.26 g.

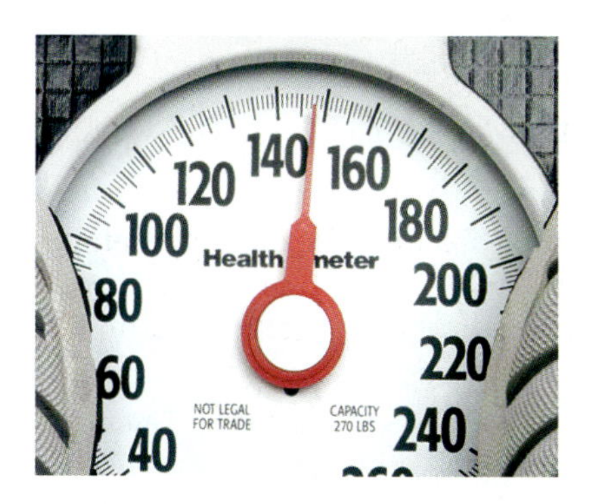

▲ **Figure 2.3** Reading a bathroom scale

▲ **Figure 2.4** Reading a thermometer

EXAMPLE 2.3 Reporting the Right Number of Digits

The bathroom scale in ◀ Figure 2.3 has markings at every 1 lb. Report the reading to the correct number of digits.

Solution:

Since the pointer is between the 147- and 148-lb markings, we mentally divide the space between the markings into ten equal spaces and estimate the next digit. In this case, the result should be reported as:

147.7 lb

What if you estimated a little differently and wrote 147.6 lb? In general, one unit of difference in the last digit is acceptable because the last digit is estimated and different people might estimate it slightly differently. However, if you wrote 147.2 lb, you would clearly be wrong.

SKILLBUILDER 2.3 Reporting the Right Number of Digits

A thermometer is used to measure the temperature of a backyard hot tub, and the reading is shown in ◀ Figure 2.4. Write the temperature reading to the right number of digits.

Counting Significant Figures

The non-place-holding digits in a reported measurement are called **significant figures** (or **significant digits**) and, as we have seen, represent the precision of a measured quantity. The greater the number of significant figures, the greater the precision of the measurement. We can determine the number of significant figures in a written number fairly easily; however, if the number contains zeros, we must distinguish between the zeros that are significant and those that simply mark the decimal place. In the number 0.002, for example, the leading zeros simply mark the decimal place; they *do not* add to the precision of the measurement. In the number 0.00200, the trailing zeros *do* add to the precision of the measurement.

To determine the number of significant figures in a number, follow these rules:

When a number is expressed in scientific notation, all trailing zeros are counted as significant.

1. All nonzero digits are significant.

   1.05   0.0110

2. Interior zeros (zeros between two numbers) are significant.

   4.0208   50.1

3. Trailing zeros (zeros after a decimal point) are significant.

   5.10   3.00

4. Leading zeros (zeros to the left of the first nonzero number) are not significant. They only serve to locate the decimal point.

   Thus, the number 0.0005 has only one significant digit.

5. Zeros at the end of a number but before a decimal point are ambiguous and should be avoided by using scientific notation.

   For example, does 350 have two or three significant figures? Write the number as 3.5×10^2 to indicate two significant figures or as 3.50×10^2 to indicate three.

Exact Numbers

Exact numbers have an unlimited number of significant figures. Exact numbers originate from three sources:

- from the accurate counting of discrete objects. For example, 3 atoms means 3.00000 . . . atoms.
- from *defined quantities*, such as the number of centimeters in 1 m. Because 100 cm is defined as 1 m,

  $$100 \text{ cm} = 1 \text{ m} \qquad \text{means} \qquad 100.00000\ldots \text{cm} = 1.0000000\ldots \text{m}$$

  Note that some conversion factors are defined quantities while others are not.
- from integral numbers that are part of an equation. For example, in the equation, $radius = \frac{diameter}{2}$, the number 2 is exact and therefore has an unlimited number of significant figures.

EXAMPLE 2.4 Determining the Number of Significant Figures in a Number

How many significant figures are in each of the following numbers?

(a) 0.0035
(b) 1.080
(c) 2371
(d) 2.97×10^5
(e) 1 dozen = 12
(f) 100,000

Solution:

The 3 and the 5 are significant. The leading zeros only mark the decimal place and are not significant.	**(a)** 0.0035 two significant figures
The interior zero and the trailing zero are significant, as are the 1 and the 8.	**(b)** 1.080 four significant figures
All digits are significant.	**(c)** 2371 four significant figures
All digits in the decimal part are significant.	**(d)** 2.97×10^5 three significant figures
Defined numbers have an unlimited number of significant figures.	**(e)** 1 dozen = 12 unlimited significant figures
This number is ambiguous. Write as 1×10^5 to indicate one significant figure or as 1.00000×10^5 to indicate six significant figures.	**(f)** 100,000 ambiguous

SKILLBUILDER 2.4 Determining the Number of Significant Figures in a Number

How many significant figures are in each of the following numbers?

(a) 58.31
(b) 0.00250
(c) 2.7×10^3
(d) 1 cm = 0.01 m
(e) 0.500
(f) 2100

CONCEPTUAL CHECKPOINT 2.2

A researcher reports that the Spirit rover on the surface of Mars recently measured the temperature to be −25.49 °F. This means that the actual temperature can be assumed to be:

(a) between −25.490 °F and −25.499 °F
(b) between −25.48 °F and −25.50 °F
(c) between −25.4 °F and −25.5 °F
(d) exactly −25.49 °F

The COBE Satellite and Very Precise Measurements That Illuminate Our Cosmic Past

Since the earliest times, humans have wondered about the origins of our planet. Science has slowly probed this question and has developed theories for how the universe and the Earth began. The most accepted theory today for the origin of the universe is the big bang theory. According to the big bang theory, the universe began in a tremendous expansion about 13.7 billion years ago and has been expanding ever since. A measurable prediction of this theory is the presence of a remnant "background radiation" from the expansion of the universe itself. That remnant is characteristic of the current temperature of the universe. When the big bang occurred, the temperature of the universe was very hot and the associated radiation very bright. Today, 16 billion years later, the temperature of the universe is very cold and the background radiation very faint.

In the early 1960s, Robert H. Dicke, P. J. E. Peebles, and their coworkers at Princeton University began to build a device that might measure this background radiation and thus take a direct look into the cosmological past and provide evidence for the big bang theory. At about the same time, quite by accident, Arno Penzias and Robert Wilson of Bell Telephone Laboratories measured excess radio noise on one of their communications satellites. As it turned out, this noise was the background radiation that the Princeton scientists were looking for. The two groups published papers together in 1965 reporting their findings along with the corresponding temperature of the universe, about 3 degrees above absolute zero, or 3 K. (We will define temperature measurement scales in Chapter 3. For now, know that 3 K is an extremely low temperature, several hundred degrees below zero on the Fahrenheit scale.)

In 1989, the Cosmic Background Explorer (COBE) satellite was developed by NASA's Goddard Space Flight Center to measure the background radiation more precisely. It was launched in late 1989 and determined that the background radiation corresponded to a universe with a temperature of 2.735 K. (Notice the difference in significant figures from the previous measurement.) The satellite went on to measure tiny fluctuations in the background radiation that amount to temperature differences of 1 part in 100,000. These fluctuations, though small, are an important prediction of the big bang theory. Scientists announced that the COBE satellite had produced the strongest evidence yet for the big bang theory of the creation of the universe. This is the way that science works. Measurement, and precision in measurement, are important to understanding the world—so important that we dedicate most of this chapter just to the concept of measurement.

CAN YOU ANSWER THIS? *How many significant figures are there in each of the preceding temperature measurements (3 K, 2.735 K)?*

▲ The COBE Satellite, launched in 1989 to measure background radiation. Background radiation is a remnant of the big bang that is believed to have formed the universe.

2.4 Significant Figures in Calculations

When we use measured quantities in calculations, the results of the calculation must reflect the precision of the measured quantities. We should not lose or gain precision during mathematical operations.

Multiplication and Division

In multiplication or division, the result carries the *same number of significant figures* as the factor with the fewest significant figures.

For example:

$$\underset{(3\text{ sig. figures})}{5.02} \times \underset{(5\text{ sig. figures})}{89.665} \times \underset{(2\text{ sig. figures})}{0.10} = 45.0118 = \underset{(2\text{ sig. figures})}{45}$$

The intermediate result (in blue) is rounded to two significant figures to reflect the least precisely known number (0.10), which has two significant figures.

For division, we follow the same rule.

$$\underset{(4\text{ sig. figures})}{5.892} \div \underset{(3\text{ sig. figures})}{6.10} = 0.96590 = \underset{(3\text{ sig. figures})}{0.966}$$

The intermediate result (in blue) is rounded to three significant figures to reflect the least precisely known number (6.10), which has three significant figures.

Rounding

When rounding to the correct number of significant figures:

Round down if the last (or leftmost) digit dropped is 4 or less;
round up if the last (or leftmost) digit dropped is 5 or more.

For example, consider rounding each of the following to two significant figures.

2.33 rounds to 2.3
2.37 rounds to 2.4
2.34 rounds to 2.3
2.35 rounds to 2.4

Be certain to use only the *last (or leftmost) digit being dropped* to decide in which direction to round—ignore all digits to the right of it. For example, to round 2.349 to two significant figures, only the 4 in the hundredths place (2.349) determines which direction to round—the 9 is irrelevant.

2.349 rounds to 2.3

For calculations involving multiple steps, round only the final answer—do not round off between steps. This prevents small rounding errors from affecting your final answer.

EXAMPLE 2.5 **Significant Figures in Multiplication and Division**

Perform the following calculations to the correct number of significant figures.

(a) $1.01 \times 0.12 \times 53.51 \div 96$
(b) $56.55 \times 0.920 \div 34.2585$

Round the intermediate result (in blue) to two significant figures to reflect the two significant figures in the least precisely known quantities (0.12 and 96).

Solution:

(a) $1.01 \times 0.12 \times 53.51 \div 96 = 0.067556 = 0.068$

Round the intermediate result (in blue) to three significant figures to reflect the three significant figures in the least precisely known quantity (0.920).

(b) $56.55 \times 0.920 \div 34.2585 = 1.51863 = 1.52$

SKILLBUILDER 2.5 **Significant Figures in Multiplication and Division**

Perform the following calculations to the correct number of significant figures.

(a) $1.10 \times 0.512 \times 1.301 \times 0.005 \div 3.4$
(b) $4.562 \times 3.99870 \div 89.5$

Addition and Subtraction

In addition or subtraction, the result carries the *same number of decimal places* as the quantity carrying the fewest decimal places.

For example:

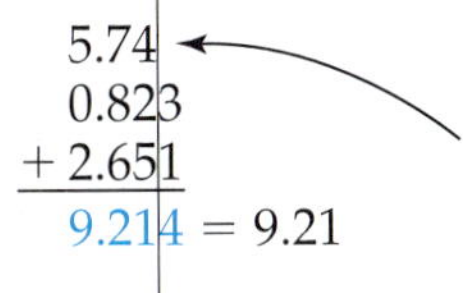

It is sometimes helpful to draw a vertical line directly to the right of the number with the fewest decimal places. The line shows the number of decimal places that should be in the answer.

We round the intermediate answer (in blue) to two decimal places because the quantity with the fewest decimal places (5.74) has two decimal places.

For subtraction, we follow the same rule. For example:

$$\begin{array}{r} 4.8 \\ -\,3.965 \\ \hline 0.835 \end{array} = 0.8$$

We round the intermediate answer (in blue) to one decimal place because the quantity with the fewest decimal places (4.8) has one decimal place. Remember: *For multiplication and division, the quantity with the fewest* significant figures *determines the number of significant figures in the answer. For addition and subtraction, the quantity with the fewest* decimal places *determines the number of decimal places in the answer*. In multiplication and division we focus on significant figures, but in addition and subtraction we focus on decimal places.

EXAMPLE 2.6 Significant Figures in Addition and Subtraction

Perform the following calculations to the correct number of significant figures.

(a)
$$\begin{array}{r} 0.987 \\ +125.1 \\ -1.22 \\ \hline \end{array}$$

(b)
$$\begin{array}{r} 0.765 \\ -3.449 \\ -5.98 \\ \hline \end{array}$$

Solution:

Round the intermediate answer (in blue) to one decimal place to reflect the quantity with the fewest decimal places (125.1). Notice that 125.1 is not the quantity with the fewest significant figures—it has four while the other quantities only have three—but because it has the fewest decimal places, it determines the number of decimal places in the answer.

(a)
$$\begin{array}{r} 0.987 \\ +125.1 \\ -1.22 \\ \hline 124.867 \end{array} = 124.9$$

Round the intermediate answer (in blue) to two decimal places to reflect that the quantity with the fewest decimal places (5.98).

(b)
$$\begin{array}{r} 0.765 \\ -3.449 \\ -5.98 \\ \hline -8.664 \end{array} = -8.66$$

SKILLBUILDER 2.6 Significant Figures in Addition and Subtraction

Perform the following calculations to the correct number of significant figures.

(a)
$$\begin{array}{r} 2.18 \\ +5.621 \\ +1.5870 \\ -1.8 \\ \hline \end{array}$$

(b)
$$\begin{array}{r} 7.876 \\ -0.56 \\ +123.792 \\ \hline \end{array}$$

Calculations Involving Both Multiplication/Division and Addition/Subtraction

In calculations involving both multiplication/division and addition/subtraction, do the steps in parentheses first; determine the number of significant figures in the intermediate answer; then do the remaining steps.

For example:

$3.489 \times (5.67 - 2.3)$

We do the subtraction step first.

$$5.67 - 2.3 = 3.37$$

We use the subtraction rule to determine that the intermediate answer (3.37) has only one significant decimal place. To avoid small errors, it is best not to round at this point; instead, underline the least significant figure as a reminder.

$$= 3.489 \times 3.3\underline{7}$$

We then do the multiplication step.

$$3.489 \times 3.3\underline{7} = 11.758$$
$$= 12$$

We use the multiplication rule to determine that the intermediate answer (11.758) rounds to two significant figures (12) because it is limited by the two significant figures in $3.\underline{3}7$.

EXAMPLE 2.7 Significant Figures in Calculations Involving Both Multiplication/Division and Addition/Subtraction

(a) $6.78 \times 5.903 \times (5.489 - 5.01)$
(b) $19.667 - (5.4 \times 0.916)$

Solution:

Do the step in parentheses first. Use the subtraction rule to mark 0.4790 to two decimal places since 5.01, the number in the parentheses with the least number of decimal places, has two.

(a) $6.78 \times 5.903 \times (5.489 - 5.01)$
$= 6.78 \times 5.903 \times (0.4790)$
$= 6.78 \times 5.903 \times 0.47\underline{9}0$

Then perform the multiplication and round the answer to two significant figures since the number with the least number of significant figures has two.

$6.78 \times 5.903 \times 0.47\underline{9}0 = 19.1707$
$= 19$

Do the step in parentheses first. The number with the least number of significant figures within the parentheses (5.4) has two, so mark the answer to two significant figures.

(b) $19.667 - (5.4 \times 0.916)$
$= 19.667 - (4.9464)$
$= 19.667 - 4.\underline{9}464$

Then perform the subtraction and round the answer to one decimal place since the number with the least number of decimal places has one.

$19.667 - 4.\underline{9}464 = 14.7206$
$= 14.7$

SKILLBUILDER 2.7 Significant Figures in Calculations Involving Both Multiplication/Division and Addition/Subtraction

(a) $3.897 \times (782.3 - 451.88)$
(b) $(4.58 \div 1.239) - 0.578$

CONCEPTUAL CHECKPOINT 2.3

Which of the following calculations would have its result reported to the *greater* number of significant figures?

(a) $3 + (15/12)$
(b) $(3 + 15)/12$

2.5 The Basic Units of Measurement

By themselves, numbers have little meaning. Read this sentence: When my son was 7 he walked 3, and when he was 4 he threw his baseball 8 and said his school was 5 away. The sentence is confusing because we don't know what the numbers mean—the **units** are missing. The meaning becomes clear, however, when we add the missing units to the numbers: When my son was 7 *months old* he walked 3 *steps*, and when he was 4 *years old* he threw his baseball 8 *feet* and said his school was 5 *minutes* away. Units make all the difference. In chemistry, units are critical. Never write a number by itself; always use its associated units—otherwise your work will be as confusing as the preceding sentence.

The two most common unit systems are the **English system**, used in the United States, and the **metric system**, used in most of the rest of the world. The English system uses units such as inches, yards, and pounds, while the metric system uses centimeters, meters, and kilograms. The most convenient system for science measurements is based on the metric system and is called the **International System** of units or **SI units**. SI units are a set of standard units agreed on by scientists throughout the world.

The abbreviation *SI* comes from the French *le Système International*.

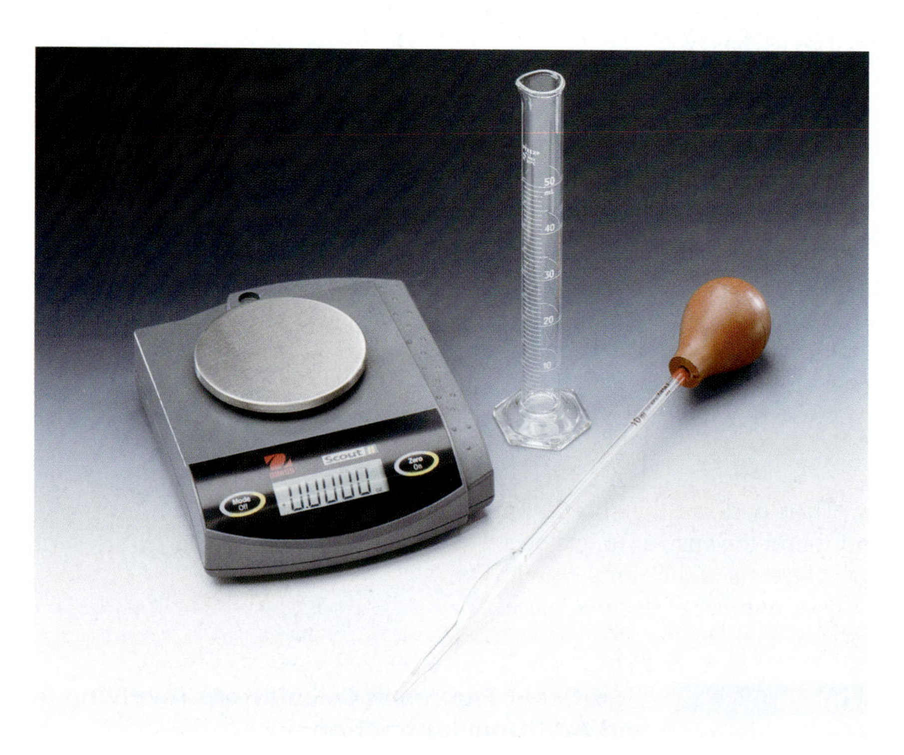

▲ Science uses instruments to make measurements. Every instrument is calibrated in a particular unit without which the measurements would be meaningless.

TABLE 2.1 Important SI Standard Units

Quantity	Unit	Symbol
length	meter	m
mass	kilogram	kg
time	second	s
temperature*	kelvin	K

*Temperature units are discussed in Chapter 3.

The Standard Units

The standard units in the SI system are shown in Table 2.1. They include the **meter (m)** as the standard unit of length; the **kilogram (kg)** as the standard unit of mass; and the **second (s)** as the standard unit of time. Each of these standard units is precisely defined. The meter is defined as the distance light travels in a certain period of time: 1/299,792,458 s. (▼ Figure 2.5). (The speed of light is 3.0×10^8 m/s) The kilogram is defined as the mass of a block of metal kept at the International Bureau of Weights and Measures at Sèvres, France (▼ Figure 2.6). The second is defined using an atomic standard (▼ Figure 2.7).

Most people are familiar with the SI standard unit of time, the second. However, if you live in the United States, you may be less familiar with the meter and the kilogram. The meter is slightly longer than a yard (a yard is 36 inches while a meter is 39.37 inches). Thus, a 100-yd football field measures only 91.4 m.

The kilogram is a measure of mass, which is different from weight. The **mass** of an object is a measure of the quantity of matter within it, while the weight of an object is a measure of the gravitational pull on that matter. Consequently, weight depends on gravity while mass does not. If you were to weigh yourself on Mars, for example, the lower gravity would pull you toward the scale less than Earth's gravity would, resulting in a lower weight. A 150-lb person on Earth would weigh 57 lb on Mars. However, the person's mass, the quantity of matter in his or her body, would remain the same. A kilogram of mass is the equivalent of 2.205 lb of weight on Earth, so if we express mass in kilograms, a 150-lb person on Earth has a mass of approximately 68 kg. A second common unit of mass is the gram (g), defined as follows:

$$1000 \text{ g} = 10^3 \text{ g} = 1 \text{ kg}$$

A nickel (5¢) has a mass of about 5 g.

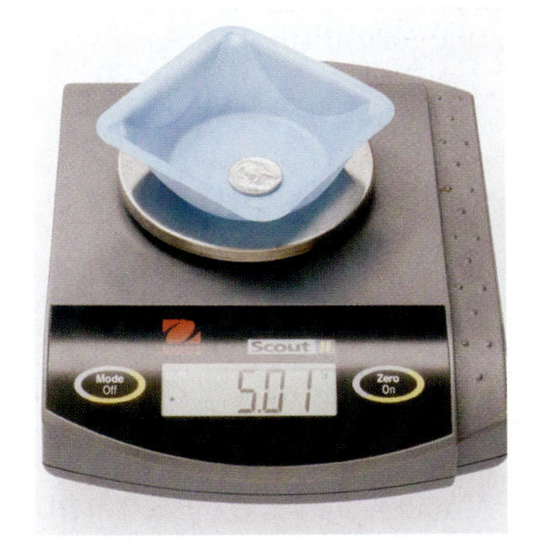

▲ A nickel has a mass of about 5 g.

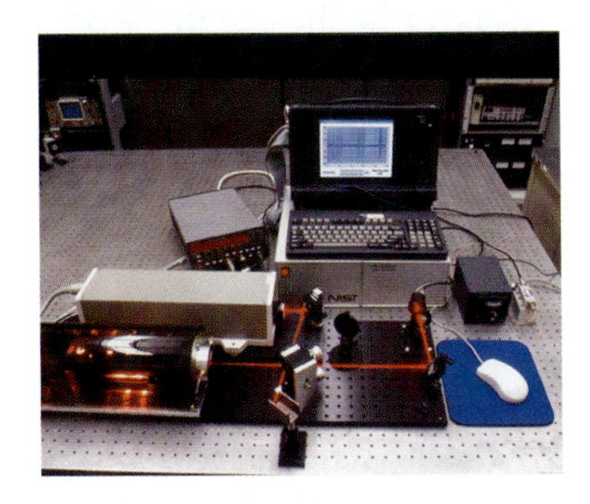

▲ **Figure 2.5 The standard of length** The definition of a meter, established by international agreement in 1983, is the distance that light travels in vacuum in 1/299,792,458 s. **Question:** What reasons can you think of why such a precise standard is necessary?

▲ **Figure 2.6 The standard of mass** A duplicate of the international standard kilogram, called kilogram 20, is kept at the National Institute of Standards and Technology near Washington, D.C.

▲ **Figure 2.7 The standard of time** The second is defined, using an atomic clock, as the duration of 9,192,631,770 periods of the radiation emitted from a certain transition in a cesium-133 atom.

TABLE 2.2 SI Prefix Multipliers

Prefix	Symbol	Multiplier	
tera-	T	1,000,000,000,000	(10^{12})
giga-	G	1,000,000,000	(10^{9})
mega-	M	1,000,000	(10^{6})
kilo-	k	1,000	(10^{3})
deci-	d	0.1	(10^{-1})
centi-	c	0.01	(10^{-2})
milli-	m	0.001	(10^{-3})
micro-	μ	0.000001	(10^{-6})
nano-	n	0.000000001	(10^{-9})
pico-	p	0.000000000001	(10^{-12})
femto-	f	0.000000000000001	(10^{-15})

▲ The diameter of a jelly donut is about 8 cm. **Question:** Why would you *not* use meters to make this measurement?

Prefix Multipliers

The SI system uses **prefix multipliers** (Table 2.2) with the standard units. These multipliers change the value of the unit by powers of 10. For example, the kilometer (km) has the prefix *kilo-*, meaning 1000 or 10^3. Therefore:

$$1 \text{ km} = 1000 \text{ m} = 10^3 \text{ m}$$

Similarly, the millisecond (ms) has the prefix *milli-*, meaning 0.001 or 10^{-3}.

$$1 \text{ ms} = 0.001 \text{ s} = 10^{-3} \text{ s}$$

The prefix multipliers allow us to express a wide range of measurements in units that are similar in size to the quantity we are measuring. Choose the prefix multiplier that is most convenient for a particular measurement. For example, to measure the size of a jelly doughnut, use centimeters because jelly doughnuts have a diameter of about 8 cm. A centimeter is a common metric unit and is about equivalent to the width of your pinky finger (2.54 cm = 1 in.). You could have also chosen the decimeter to express the diameter of a jelly doughnut; then the doughnut would measure 0.8 dm. But you should not choose the kilometer since, in that unit, the diameter is 0.00008 km. Pick a unit similar in size to (or smaller than) the quantity you are measuring. Consider expressing the length of a chemical bond, about 1.2×10^{-10} m. Which prefix multiplier should you use? The most convenient one is probably the picometer (pico $= 10^{-12}$). Chemical bonds measure about 120 pm.

TABLE 2.3 Some Common Units and Their Equivalents

Length

1 kilometer (km) = 0.6214 mile (mi)
1 meter (m) = 39.37 inches (in.)
= 1.094 yards (yd)
1 foot (ft) = 30.48 centimeters (cm)
1 inch (in.) = 2.54 centimeters (cm) (exact)

Mass

1 kilogram (kg) = 2.205 pounds (lb)
1 pound (lb) = 453.59 grams (g)
1 ounce (oz) = 28.35 grams (g)

Volume

1 liter (L) = 1000 milliliters (mL)
= 1000 cubic centimeters (cm^3)
1 liter (L) = 1.057 quarts (qt)
1 U.S. gallon (gal) = 3.785 liters (L)

Derived Units

A derived unit is formed from other units. Common derived units include those for **volume**, which is a measure of space. Any unit of length, when cubed (raised to the third power), becomes a unit of volume. Thus, cubic meters (m^3), cubic centimeters (cm^3), and cubic millimeters (mm^3) are all units of volume. In these units, a three-bedroom house has a volume of about 630 m^3, a can of soda pop has a volume of about 350 cm^3, and a rice grain has a volume of about 3 mm^3. We also use the **liter (L)** and milliliter (mL) to express volume. A gallon is equal to 3.785 L. A milliliter is equivalent to 1 cm^3. Table 2.3 lists some common units and their equivalents.

CONCEPTUAL CHECKPOINT 2.4

To express the dimensions of a polio virus, which is about 2.8×10^{-8} m in diameter, the most convenient unit would probably be:

(a) Mm
(b) mm
(c) μm
(d) nm

2.6 Converting from One Unit to Another

Units are critical in calculations. Knowing how to work with and manipulate units in calculations is one of the most important skills you will learn in this course. In calculations, units help determine correctness. Units should always be included in calculations, and we will think of many calculations as converting from one unit to another. Units are multiplied, divided, and canceled like any other algebraic quantity.

Using units as a guide to solving problems is often called *dimensional analysis.*

Remember:

1. Always write every number with its associated unit. Never ignore units; they are critical.
2. Always include units in your calculations, dividing them and multiplying them as if they were algebraic quantities. Do not let units magically appear or disappear in calculations. Units must flow logically from beginning to end.

Consider converting 17.6 in. to centimeters. We know from Table 2.3 that 1 in. = 2.54 cm. How many centimeters are in 17.6 in.? We perform the conversion as follows:

$$17.6 \text{ in.} \times \frac{2.54 \text{ cm}}{1 \text{ in.}} = 44.7 \text{ cm}$$

The unit *in.* cancels and we are left with *cm* as our final unit. The quantity $\frac{2.54 \text{ cm}}{1 \text{ in.}}$ is a **conversion factor** between *in.* and *cm*—it is a fraction with *cm* on top and *in.* on bottom.

For most conversion problems, we are given a quantity in some units and asked to convert the quantity to another unit. These calculations take the following form:

information given × conversion factor(s) = information sought

Conversion factors are constructed from any two quantities known to be equivalent. In our example, 2.54 cm = 1 in., so we construct the conversion factor by dividing both sides of the equality by 1 in. and cancelling the units.

$$2.54 \text{ cm} = 1 \text{ in.}$$

$$\frac{2.54 \text{ cm}}{1 \text{ in.}} = \frac{1 \text{ in.}}{1 \text{ in.}}$$

$$\frac{2.54 \text{ cm}}{1 \text{ in.}} = 1$$

The quantity $\frac{2.54 \text{ cm}}{1 \text{ in.}}$ is equal to 1 and can be used to convert between inches and centimeters.

Suppose we want to perform the conversion the other way, from centimeters to inches. If we try to use the same conversion factor, the units do not cancel correctly.

$$44.7 \text{ cm} \times \frac{2.54 \text{ cm}}{1 \text{ in.}} = \frac{114 \text{ cm}^2}{\text{in.}}$$

The units in the answer, as well as the value of the answer, are incorrect. The unit $\text{cm}^2/\text{in.}$ is not correct, and, based on our knowledge of centimeters and inches, we know that 44.7 cm cannot be equivalent to 114 in. In solving problems, always look at the final units to see if they are correct and also look at the magnitude of the answer to see if it makes sense. In this case, our mistake was in how we used the conversion factor. It must be inverted.

$$44.7 \cancel{\text{cm}} \times \frac{1 \text{ in.}}{2.54 \cancel{\text{cm}}} = 17.6 \text{ in.}$$

Conversion factors can be inverted because they are equal to 1 and the inverse of 1 is 1.

$$\frac{1}{1} = 1$$

Therefore,

$$\frac{2.54 \text{ cm}}{1 \text{ in.}} = 1 = \frac{1 \text{ in.}}{2.54 \text{ cm}}.$$

In this book, we diagram conversions such as the preceding ones using a **solution map**. A solution map is a visual outline that shows the strategic route required to solve a problem. For unit conversion, the solution map focuses on units and how to convert from one unit to another. The solution map for converting from inches to centimeters is:

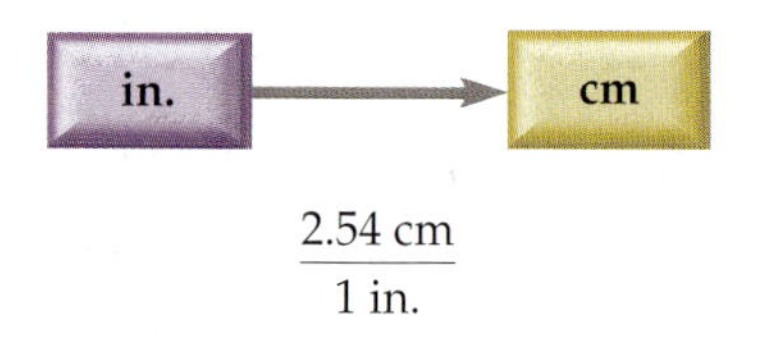

$$\frac{2.54 \text{ cm}}{1 \text{ in.}}$$

The solution map for converting from centimeters to inches is:

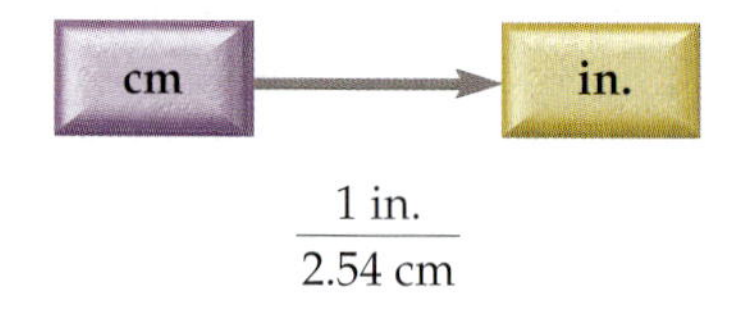

$$\frac{1 \text{ in.}}{2.54 \text{ cm}}$$

Each arrow in a solution map for a unit conversion has an associated conversion factor with the units of the previous step in the denominator and the units of the following step in the numerator. For one-step problems such as these, the solution map is only moderately helpful, but for multistep problems, it becomes a powerful way to develop a problem-solving strategy. The procedure for solving unit conversion problems using the solution map is given in the left column of the following box. The middle and right columns demonstrate examples of how to use the procedure.

Solving Unit Conversion Problems

1. Write down the *given* quantity and its unit(s).
2. Write down the quantity that you are asked to *find* and its unit(s).
3. Write down the appropriate *conversion factor(s)*. Some of these will be given in the problem. Others you find in tables within the text.
4. Write a *solution map* for the problem. Begin with the *given* quantity and draw an arrow symbolizing each conversion step. Below each arrow, write the appropriate conversion factor for that step. Focus on the units. The solution map should end at the *find* quantity.
5. Follow the solution map to solve the problem. Begin with the *given* quantity and its units. Multiply by the appropriate conversion factor(s), canceling units, to arrive at the *find* quantity.
6. Round the answer to the correct number of significant figures. Follow the significant-figure rules in Sections 2.3 and 2.4. Remember that exact conversion factors do not limit the number of significant figures in your answer.
7. Check your answer. Make certain that the units are correct and that the magnitude of the answer makes physical sense.

EXAMPLE 2.8

Unit Conversion

Convert 7.8 km to miles.

Given: 7.8 km

Find: mi

Conversion Factors:
1 km = 0.6214 mi
(This conversion factor is from Table 2.3.)

Solution Map:

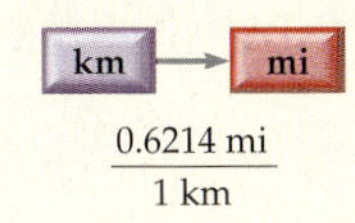

$$\frac{0.6214 \text{ mi}}{1 \text{ km}}$$

The conversion factor is written so that km, the unit we are converting *from*, is on the bottom and mi, the unit we are converting *to*, is on the top.

Solution:

$$7.8 \text{ km} \times \frac{0.6214 \text{ mi}}{1 \text{ km}} = 4.84692 \text{ mi}$$

$$4.84692 \text{ mi} = 4.8 \text{ mi}$$

We round to two significant figures, since the quantity given has two significant figures. (If possible, obtain conversion factors to enough significant figures so that they do not limit the number of significant figures in the answer.)

The units, mi, are correct. The magnitude of the answer is reasonable. A mile is longer than a kilometer, so the value in miles should be smaller than the value in kilometers.

SKILLBUILDER 2.8

Unit Conversion

Convert 56.0 cm to inches.

EXAMPLE 2.9

Unit Conversion

Convert 0.825 m to millimeters.

Given: 0.825 m

Find: mm

Conversion Factors:
1 mm = 10^{-3} m
(This conversion factor is from Table 2.2.)

Solution Map:

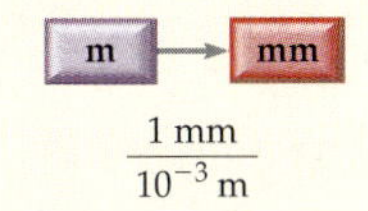

$$\frac{1 \text{ mm}}{10^{-3} \text{ m}}$$

The conversion factor is written so that m, the unit we are converting *from*, is on the bottom and mm, the unit we are converting *to*, is on the top.

Solution:

$$0.825 \text{ m} \times \frac{1 \text{ mm}}{10^{-3} \text{ m}} = 825 \text{ mm}$$

$$825 \text{ mm} = 825 \text{ mm}$$

We leave the answer with three significant figures, since the quantity given has three significant figures and the conversion factor is a definition, which therefore does not limit the number of significant figures in the answer.

The units, mm, are correct and the magnitude is reasonable. A millimeter is shorter than a meter, so the value in millimeters should be larger than the value in meters.

SKILLBUILDER 2.9

Unit Conversion

Convert 5,678 m to kilometers.

CONCEPTUAL CHECKPOINT 2.5

If you needed to convert distance in meters to distance in kilometers, which of these conversion factors would you use?

(a) $1\ \text{m}/10^3\ \text{km}$
(b) $10^3\ \text{m}/1\ \text{km}$
(c) $1\ \text{km}/10^3\ \text{m}$
(d) $10^3\ \text{km}/1\ \text{m}$

2.7 Solving Multistep Conversion Problems

When solving multistep conversion problems, we follow the preceding procedure but simply add more steps to the solution map. Every step in the solution map must have a conversion factor with the units of the previous step in the denominator and the units of the following step in the numerator. For example, suppose we want to convert 194 cm to feet. We set up the problem as we outlined earlier.

Given: 194 cm

Find: ft

Conversion Factors:

$$2.54\ \text{cm} = 1\ \text{in.}$$
$$12\ \text{in.} = 1\ \text{ft}$$

Now we can build our solution map. We start with the quantity given and focus on the units. We use the first conversion factor to convert centimeters to inches and the second to convert inches to feet.

Solution Map:

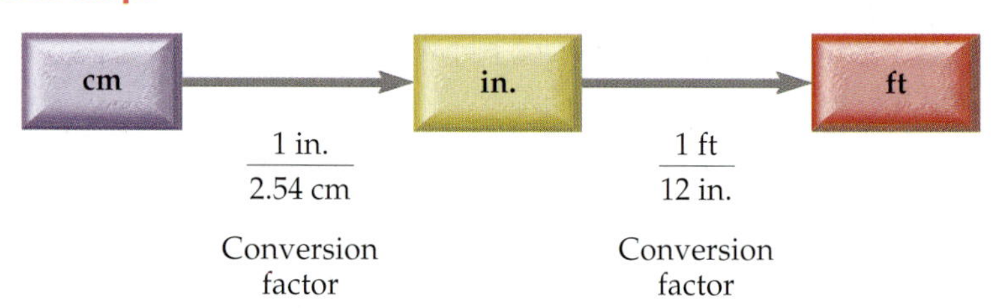

Once the solution map is complete, we follow it to solve the problem.

Solution:

$$194\ \cancel{\text{cm}} \times \frac{1\ \cancel{\text{in.}}}{2.54\ \cancel{\text{cm}}} \times \frac{1\ \text{ft}}{12\ \cancel{\text{in.}}} = 6.3648\ \text{ft}$$

We then round to the correct number of significant figures—in this case, three (from 194 cm, which has three significant figures).

$$6.3648\ \text{ft} = 6.36\ \text{ft}$$

Since one foot is *defined* as 12 in., it does not limit significant figures.

Finally, we check the answer. The units of the answer, ft, are the correct ones and the magnitude seems about right. Since a foot is larger than a centimeter, it is reasonable that the value in feet is smaller than the value in centimeters.

EXAMPLE 2.10 Solving Multistep Unit Conversion Problems

An Italian recipe for making creamy pasta sauce calls for 0.75 L of cream. Your measuring cup measures only in cups. How many cups of cream should you use?

$$4\ \text{cups} = 1\ \text{quart}$$

1. Write down the quantity that is *given* and its unit(s).

Given: 0.75 L

2. Write down the quantity that you are asked to *find* and its unit(s).

Find: cups

3. Write down the appropriate *conversion factor(s)*. The second conversion factor is from Table 2.3.

Note: It is okay if you don't have all the necessary conversion factors at this point. You will discover what you need as you work through the problem.

Conversion Factors:

4 cups = 1 qt

1.057 qt = 1 L

4. Write a *solution map*. Focus on the units, finding the appropriate conversion factors with the units of the previous step in the denominator and the units of the next step in the numerator.

Solution Map:

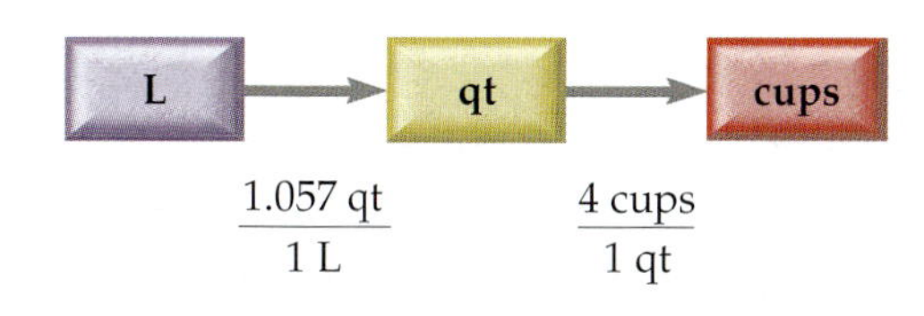

$$\frac{1.057\ \text{qt}}{1\ \text{L}} \qquad \frac{4\ \text{cups}}{1\ \text{qt}}$$

5. Follow the solution map. Starting with the quantity that is given and its units, multiply by the appropriate conversion factors, canceling units, to arrive at the quantity that you are trying to find in the desired units.

Solution:

$$0.75\ \cancel{\text{L}} \times \frac{1.057\ \cancel{\text{qt}}}{1\ \cancel{\text{L}}} \times \frac{4\ \text{cups}}{1\ \cancel{\text{qt}}} = 3.171\ \text{cups}$$

6. Round the final answer. Round to two significant figures, since the quantity given has two significant figures.

3.171 cups = 3.2 cups

7. Check your answer. The answer has the right units (cups) and seems reasonable. We know that a cup is smaller than a liter, so the value in cups should be larger than the value in liters.

SKILLBUILDER 2.10 Solving Multistep Unit Conversion Problems

Suppose a recipe calls for 1.2 cups of oil. How many liters is this?

LIVE EXAMPLE 

EXAMPLE 2.11 Solving Multistep Unit Conversion Problems

A running track measures 255 m per lap. To run 10.0 km, how many laps should you run?

Set up the problem in the standard way. Notice that the quantity 255 m per lap is a conversion factor between m and laps.

Given: 10.0 km

Find: number of laps

Conversion Factors: 1 lap = 255 m

1 km = 10^3 m

Build the solution map beginning with km and ending at laps. Focus on the units.

Solution Map:

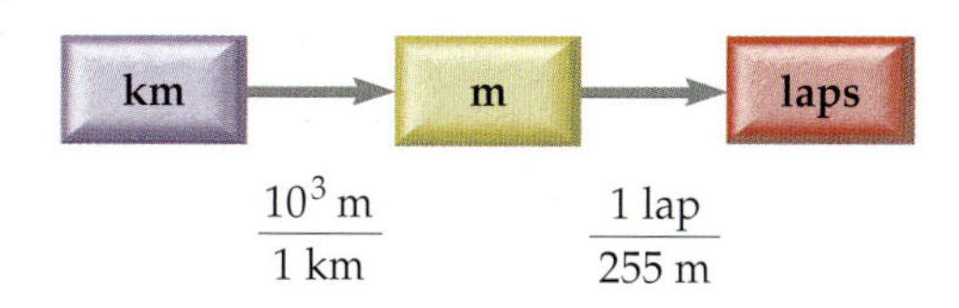

$$\frac{10^3\ \text{m}}{1\ \text{km}} \qquad \frac{1\ \text{lap}}{255\ \text{m}}$$

Follow the solution map to solve the problem. The intermediate answer (in blue) is rounded to three significant figures, because it is limited by the three signficant figures in the given quantity, 10.0 km.

Solution:

$$10.0\ \cancel{\text{km}} \times \frac{10^3\ \cancel{\text{m}}}{1\ \cancel{\text{km}}} \times \frac{1\ \text{lap}}{255\ \cancel{\text{m}}} = 39.216\ \text{laps} = 39.2\ \text{laps}$$

The units of the answer are correct and the value of the answer makes sense—if a lap is 255 m, there are about 4 laps to each km (1000 m), so you would have to run about 40 laps to cover 10 km.

SKILLBUILDER 2.11 Solving Multistep Unit Conversion Problems

A running track measures 1056 ft per lap. To run 15.0 km, how many laps should you run? (1 mi = 5280 ft)

SKILLBUILDER PLUS

An island is 5.72 nautical mi from the coast. How far is the island in meters? (1 nautical mile = 1.151 mi)

Chemistry AND HEALTH

Drug Dosage

The unit of choice in specifying drug dosage is the milligram (mg). Pick up a bottle of aspirin, Tylenol, or any other common drug and the label tells you how many milligrams of the active ingredient are contained in each tablet, as well as the number of tablets to take per dose. The following table shows the mass of the active ingredient in several common pain relievers, all reported in milligrams. The rest of the tablet is composed of inactive ingredients such as cellulose (or fiber) and starch.

The recommended adult dose for many of these pain relievers is one or two tablets every four to eight hours (depending on the specific pain reliever). Notice that the extra-strength version of each pain reliever just contains a higher dose of the same compound found in the regular-strength version. For the pain relievers listed, three regular-strength tablets are the equivalent of two extra-strength tablets (and probably cost less).

The dosages given in the table are fairly standard for each drug, regardless of the brand. For example, when you look on your drugstore shelf, you will find many different brands of regular-strength ibuprofen, some sold under the generic name and others sold under their brand names (such as Advil). However, if you look closely at the labels, you will find that they all contain the same thing: 200 mg of the compound ibuprofen. There is no difference in the compound or in the amount of the compound. Yet these pain relievers will most likely all have different prices. Choose the least expensive. Why pay more for the same thing?

CAN YOU ANSWER THIS? *Convert each of the doses in the table to ounces. Why are drug dosages not listed in ounces?*

Drug Mass per Pill for Common Pain Relievers

Pain Reliever	Mass of Active Ingredient per Pill
aspirin	325 mg
aspirin, extra strength	500 mg
ibuprofen (Advil)	200 mg
ibuprofen, extra strength	300 mg
acetaminophen (Tylenol)	325 mg
acetaminophen, extra strength	500 mg

2.8 Units Raised to a Power

When converting quantities with units raised to a power, such as cubic centimeters (cm^3), the conversion factor must also be raised to that power. For example, suppose we want to convert the size of a motorcycle engine reported as 1255 cm^3 to cubic inches. We know that

The unit cm^3 is often abbreviated as cc.

$$2.54 \text{ cm} = 1 \text{ in.}$$

Most tables of conversion factors do not include conversions between cubic units, but we can derive them from the conversion factors for the basic units. We cube both sides of the preceding equality to obtain the proper conversion factor.

$$(2.54 \text{ cm})^3 = (1 \text{ in.})^3$$
$$(2.54)^3 \text{ cm}^3 = 1^3 \text{ in.}^3$$
$$16.387 \text{ cm}^3 = 1 \text{ in.}^3$$

We can do the same thing in fractional form.

$$\frac{1 \text{ in.}}{2.54 \text{ cm}} = \frac{(1 \text{ in.})^3}{(2.54 \text{ cm})^3} = \frac{1 \text{ in.}^3}{16.387 \text{ cm}^3}$$

2.54 cm = 1 in. is an exact conversion factor. After cubing, we retain five significant figures so that the conversion factor does not limit the four significant figures of our original quantity (1255 cm^3).

We then proceed with the conversion in the normal way.

Solution Map:

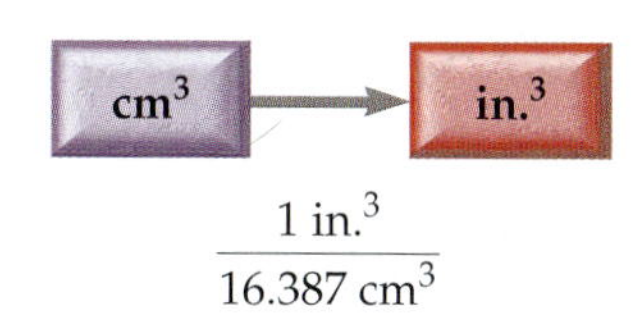

$$\frac{1 \text{ in.}^3}{16.387 \text{ cm}^3}$$

Solution: $1255 \text{ cm}^3 \times \frac{1 \text{ in.}^3}{16.387 \text{ cm}^3} = 76.5851 \text{ in.}^3 = 76.59 \text{ in.}^3$

EXAMPLE 2.12 Converting Quantities Involving Units Raised to a Power

A circle has an area of 2,659 cm^2. What is its area in square meters?

Set up the problem in the standard way.

Given: 2,659 cm^2

Find: m^2

Conversion Factor: 1 cm = 0.01 m

Draw a solution map beginning with cm^2 and ending with m^2. Notice that you must square the conversion factor.

Solution Map:

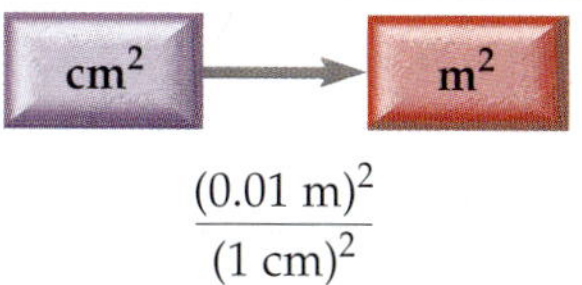

$$\frac{(0.01 \text{ m})^2}{(1 \text{ cm})^2}$$

Follow the solution map to solve the problem. Round the answer to four significant figures to reflect the four significant figures in the given quantity. The conversion factor is exact and therefore does not limit the number of signficant figures.

Solution:

$$2{,}659 \text{ cm}^2 \times \frac{(0.01 \text{ m})^2}{(1 \text{ cm})^2}$$
$$= 2{,}659 \text{ cm}^2 \times \frac{10^{-4} \text{ m}^2}{1 \text{ cm}^2}$$
$$= 0.265900 \text{ m}^2$$
$$= 0.2659 \text{ m}^2$$

The units of the answer are correct, and the magnitude makes physical sense. A square meter is much larger than a square centimeter, so the value in square meters should be much smaller than the value in square centimeters.

SKILLBUILDER 2.12 Converting Quantities Involving Units Raised to a Power

An automobile engine has a displacement (a measure of the size of the engine) of 289.7 $in.^3$. What is its displacement in cubic centimeters?

LIVE EXAMPLE

EXAMPLE 2.13 Solving Multistep Problems Involving Units Raised to a Power

The average annual per person oil consumption in the United States is 15,615 dm^3. What is this value in cubic inches?

Set up the problem in the standard way. The conversion factors come from Tables 2.2 and 2.3.

Given: 15,615 dm^3

Find: volume in $in.^3$

Conversion Factors:

$$1 \text{ dm} = 0.1 \text{ m}$$
$$1 \text{ cm} = 0.01 \text{ m}$$
$$2.54 \text{ cm} = 1 \text{ in.}$$

Write a solution map beginning with dm^3 and ending with in^3. Each of the conversion factors must be cubed, since the quantities involve cubic units.

Solution Map:

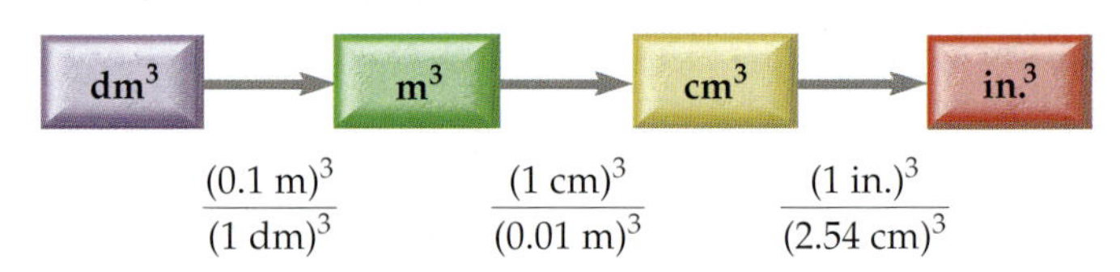

$$\frac{(0.1 \text{ m})^3}{(1 \text{ dm})^3} \qquad \frac{(1 \text{ cm})^3}{(0.01 \text{ m})^3} \qquad \frac{(1 \text{ in.})^3}{(2.54 \text{ cm})^3}$$

Follow the solution map to solve the problem.

Round the answer to five significant figures to reflect the five significant figures in the least precisely known quantity (15,615 dm^3). The conversion factors are all exact and therefore do not limit the number of significant figures.

Solution:

$$15{,}615 \text{ dm}^3 \times \frac{(0.1 \text{ m})^3}{(1 \text{ dm})^3} \times \frac{(1 \text{ cm})^3}{(0.01 \text{ m})^3} \times \frac{(1 \text{ in.})^3}{(2.54 \text{ cm})^3}$$
$$= 9.5289 \times 10^5 \text{ in.}^3$$

The units of the answer are correct and the magnitude makes sense. A cubic inch is smaller than a cubic decimeter, so the value in cubic inches should be larger than the value in cubic decimeters.

SKILLBUILDER 2.13 Solving Multistep Problems Involving Units Raised to a Power

How many cubic inches are there in 3.25 yd^3?

CONCEPTUAL CHECKPOINT 2.6

You know that there are 3 ft in a yard. How many cubic feet are there in a cubic yard?

(a) 3
(b) 6
(c) 9
(d) 27

2.9 Density

Why do some people pay more than $3000 for a bicycle made of titanium? A steel frame would be just as strong for a fraction of the cost. The difference between the two, of course, is mass—for a given volume of metal, titanium has less mass, and is therefore lighter than the steel bike. We describe this

▲ Top-end bicycle frames are made of titanium because of its low density and high relative strength. Titanium has a density of 4.50 g/cm^3, while iron, for example, has a density of 7.86 g/cm^3.

property by saying that titanium is *less dense* than steel. The **density** of a substance is the ratio of its mass to its volume.

$$\text{Density} = \frac{\text{Mass}}{\text{Volume}} \quad \text{or} \quad d = \frac{m}{V}$$

Density is a fundamental property of materials that differs from one substance to another. The units of density are those of mass divided by those of volume, most conveniently expressed in grams per cubic centimeter or grams per milliliter. See Table 2.4 for a list of the densities of same common substances. Aluminum is among the least dense structural metals with a density of 2.70 g/cm^3, while platinum is among the densest with a density of 21.4 g/cm^3. Titanium has a density of 4.50 g/cm^3.

Remember that cubic centimeters and milliliters are equivalent units.

TABLE 2.4 Densities of Some Common Substances

Substance	Density (g/cm^3)
water	1.0
ice	0.92
ethanol	0.789
lead	11.4
copper	8.96
gold	19.3
aluminum	2.7
platinum	21.4
iron	7.86
titanium	4.50
charcoal, oak	0.57
glass	2.6

Calculating Density

The density of a substance is calculated by dividing the mass of a given amount of the substance by its volume. For example, a sample of liquid has a volume of 22.5 mL and a mass of 27.2 g. To find its density, we use the equation highlighted above.

$$d = \frac{m}{V} = \frac{27.2\text{ g}}{22.5\text{ mL}} = 1.21\text{ g/mL}$$

We can use a solution map for solving problems involving equations, but it takes a slightly different form than for pure conversion problems. In a solution map for a problem involving an equation, the solution map shows how the *equation* takes you from the *given* quantities to the *find* quantity. The solution map for this problem is:

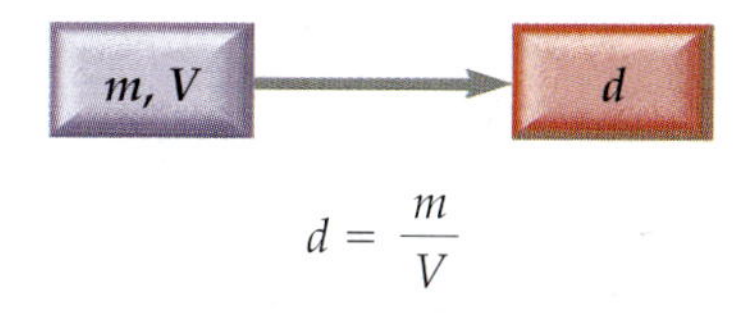

$$d = \frac{m}{V}$$

The solution map shows how the values of m and V, when substituted into the equation $d = \frac{m}{V}$, give the desired result, d.

EXAMPLE 2.14 Calculating Density

A jeweler offers to sell a ring to a woman, telling her that it is made of platinum. Noting that the ring felt a little light, the woman decides to perform a test to determine the ring's density. She places the ring on a balance and finds that it has a mass of 5.84 g. She then finds that the ring *displaces* 0.556 cm^3 of water. Is the ring made of platinum? The density of platinum is 21.4 g/cm^3. (The displacement of water is a common way to measure the volume of irregularly shaped objects. To say that an object *displaces* 0.556 cm^3 of water means that when the object is submerged in a container of water filled to the brim, 0.556 cm^3 overflows. Therefore, the volume of the object is 0.556 cm^3.)

If the ring is platinum, its density should match that of platinum. You are given the mass and volume of the ring. Find the density using the density equation.

Given:

$m = 5.84$ g

$V = 0.556\text{ cm}^3$

Find: density in g/cm^3

Equation: $d = \frac{m}{V}$

Draw a solution map. Begin with mass and volume and end with density.

Solution Map:

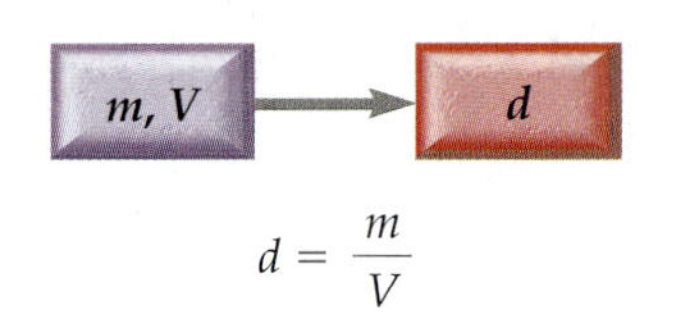

$$d = \frac{m}{V}$$

Substitute the given values into the density equation and compute the density. The density of the ring is much too low to be platinum, and therefore the ring is a fake.

Solution:

$$d = \frac{m}{V} = \frac{5.84\ \text{g}}{0.556\ \text{cm}^3} = 10.5\ \text{g/cm}^3$$

SKILLBUILDER 2.14 Calculating Density

The woman takes the ring back to the jewelry shop, where she is met with endless apologies. They accidentally had made the ring out of silver rather than platinum. They give her a new ring that they promise is platinum. This time when she checks the density, she finds the mass of the ring to be 9.67 g and its volume to be 0.452 cm^3. Is this ring genuine?

Density as a Conversion Factor

Density can also be thought of as a conversion factor between mass and volume. We may know the mass of an object and want to calculate its volume or vice versa. For example, suppose we need 68.4 g of a liquid with a density of 1.32 g/cm^3 and only have a graduated cylinder to measure its volume. How much volume should we measure? We can set up this problem like a standard conversion problem.

Given: 68.4 g

Find: volume in mL

Conversion Factors:

$$1.32\ \text{g/cm}^3$$

$$1\ \text{cm}^3 = 1\ \text{mL}$$

Density is a conversion factor between mass and volume. To convert from g to mL we need to invert the density because we want g, the unit we are converting from, to be on the bottom and cm^3, the unit we are converting to, on the top. Our solution map is:

Solution Map:

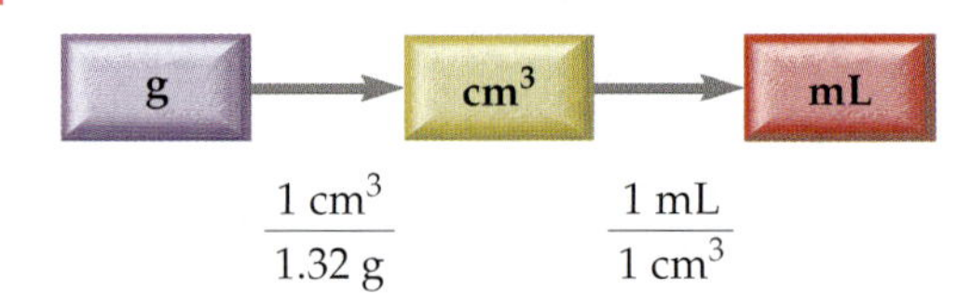

$$\frac{1\ \text{cm}^3}{1.32\ \text{g}} \qquad \frac{1\ \text{mL}}{1\ \text{cm}^3}$$

Solution: $68.4\ \cancel{\text{g}} \times \frac{1\ \cancel{\text{cm}^3}}{1.32\ \cancel{\text{g}}} \times \frac{1\ \text{mL}}{1\ \cancel{\text{cm}^3}} = 51.8\ \text{mL}$

We must measure 51.8 mL to obtain 68.4 g of the liquid.

EXAMPLE 2.15 Density as a Conversion Factor

The gasoline in an automobile gas tank has a mass of 60.0 kg and a density of 0.752 g/cm^3. What is its volume in cm^3?

Set up the problem as usual. You are given the mass in kilograms and asked to find the volume in cubic centimeters. Density is the conversion factor between mass and volume.

Given: 60.0 kg

Find: volume in cm^3

Conversion Factors:

$$0.752\ \text{g/cm}^3$$

$$1000\ \text{g} = 1\ \text{kg}$$

Build the solution map starting with kg and ending with cm^3.

Solution Map:

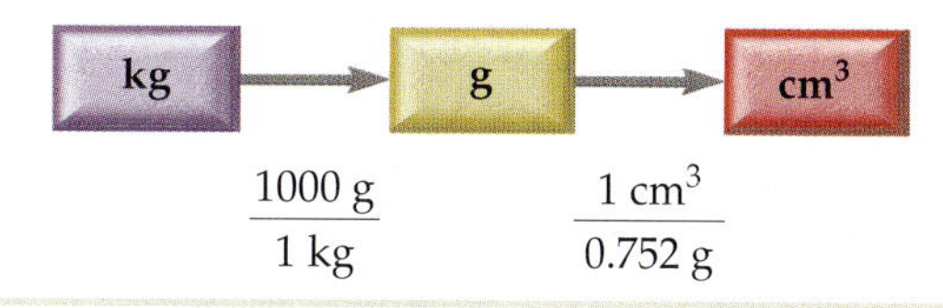

$$\frac{1000\ g}{1\ kg} \qquad \frac{1\ cm^3}{0.752\ g}$$

Follow the solution map to solve the problem.

Solution: $60.0\ \cancel{kg} \times \frac{1000\ \cancel{g}}{1\ \cancel{kg}} \times \frac{1\ cm^3}{0.752\ \cancel{g}} = 7.98 \times 10^4\ cm^3$

SKILLBUILDER 2.15 Density as a Conversion Factor

A drop of acetone (nail polish remover) has a mass of 35 mg and a density of 0.788 g/cm^3. What is its volume in cubic centimeters?

SKILLBUILDER PLUS

A steel cylinder has a volume of 246 cm^3 and a density of 7.93 g/cm^3. What is its mass in kilograms?

Chemistry AND HEALTH

Density, Cholesterol, and Heart Disease

Substances are often separated and classified according to their density. A class of substances classified in this way are lipoproteins, carriers of cholesterol in the blood. Cholesterol is found in animal-derived foods such as beef, eggs, fish, poultry, and milk products and is used by the body for several purposes. However, excessive amounts of cholesterol in the blood—which can be caused by both genetic factors and diet—may result in the deposition of cholesterol in arterial walls, leading to a condition called atherosclerosis, or blocking of the arteries. These blockages are dangerous because they inhibit blood flow to important organs, causing heart attacks and strokes. The risk of stroke and heart attack increases with increasing blood cholesterol levels (Table 2.5).

Cholesterol is carried in the blood by lipoproteins, which are classified by their density. The main carriers of blood cholesterol are low-density lipoproteins (LDLs). LDLs, also called bad cholesterol, have a density of 1.04 g/cm^3. They are bad because they tend to deposit cholesterol on arterial walls and therefore increase the risk of stroke and heart attack. Cholesterol is also carried by high-density lipoproteins (HDLs). HDLs, also called good cholesterol, have a density of 1.13 g/cm^3. HDLs transport cholesterol to the liver for processing and excretion and therefore have a tendency to reduce cholesterol on arterial walls. Too low a level of HDLs (below 35 mg/100 mL) is considered a risk factor for heart disease. Exercise, along with a diet low in saturated fats, is believed to raise HDL levels in the blood while lowering LDL levels.

CAN YOU ANSWER THIS? *What mass of low-density lipoprotein is contained in a cylinder that is 1.25 cm long and 0.50 cm in diameter? (The volume of a cylinder, V, is given by $V = \pi r^2 \ell$, where r is the radius of the cylinder and ℓ is its length.)*

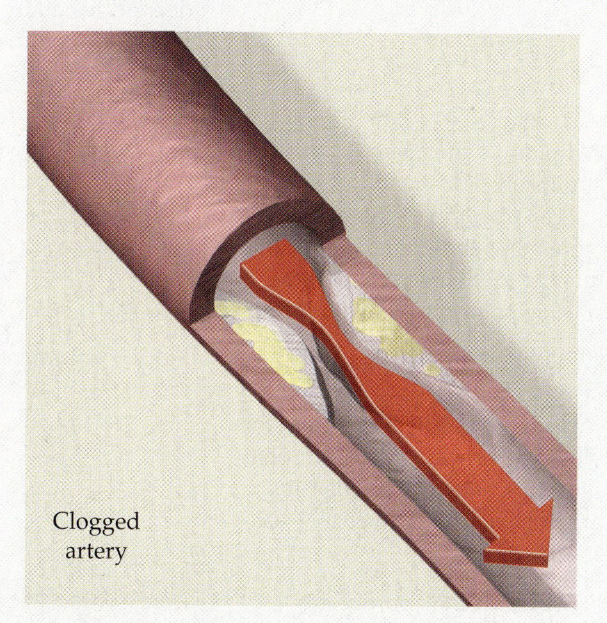

▲ Too many low-density lipoproteins in the blood can lead to blocking of arteries.

TABLE 2.5 Risk of Stroke and Heart Attack vs. Blood Cholesterol Level

Risk Level	Total Blood Cholesterol (mg/100 mL)	LDL (mg/100 mL)
low	<200	<130
borderline	200–239	130–159
high	240+	160+

2.10 Numerical Problem-Solving Strategies and the Solution Map

In this chapter, you have seen a few examples of how to solve numerical problems. The ability to solve a numerical problem is one of the most important things you will learn in this course. When you first look at a problem, what should you do? Some students simply get stuck—they have trouble getting started. To avoid this, develop a strategy to attack and solve problems. What should you do first? What next? In Section 2.6, we developed a strategy or a procedure to solve simple unit conversion problems. We then learned how to modify that procedure to work with multistep unit conversion problems and problems involving an equation. We will now summarize these procedures and generalize our strategy to apply to most of the numerical problems we encounter in this book. As we did in Section 2.6, we provide the general procedure for solving numerical problems in the left column of the following box. The center and right columns show two examples where this procedure is applied.

Solving Numerical Problems	**EXAMPLE 2.16** **Unit Conversion**	LIVE EXAMPLE **EXAMPLE 2.17** **Unit Conversion with Equation**
	A 23.5-kg sample of ethanol is needed for a large-scale reaction. What volume in liters of ethanol should be used? The density of ethanol is 0.789 g/cm^3.	A 55.9-kg person displaces 57.2 L of water when submerged in a water tank. What is the density of the person in grams per cubic centimeter?
1. Write down the *given* quantity and its units. Scan the problem for one or more numbers and their units. This number (or numbers) is (are) the starting point(s) of the calculation.	**Given:** 23.5 kg ethanol	**Given:** $m = 55.9$ kg $V = 57.2$ L
2. Write down the *find* quantity and its units. Scan the problem to determine what you are asked to find. Sometimes the units of this quantity are implied; other times they are specified.	**Find:** volume in L	**Find:** density in g/cm^3
3. Write down the appropriate *conversion factors* and/or *equations*. Sometimes these conversion factors are given in the problem. Other times they are in tables. You may not get them all at this point—that's okay. For problems involving equations, familiarity and practice will help you know what equation to use.	**Conversion Factors:** 0.789 g/cm^3 1000 g = 1 kg 1000 mL = 1 L 1 mL = 1 cm^3	**Conversion Factors and Equation:** 1000 g = 1 kg 1000 mL = 1 L 1 cm^3 = 1 mL $d = \frac{m}{V}$

4. Write a *solution map* for the problem.

- For problems involving only *conversions*, focus on units. The solution map shows how to get from the units in the *given* quantity to the units in the *find* quantity.
- For problems involving *equations*, focus on the equation. The solution map shows how the equation takes you from the *given* quantity (or quantities) to the *find* quantity.

Solution Map:

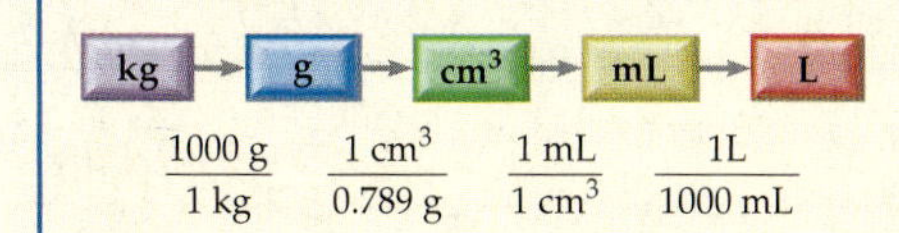

Solution Map:

m, V → d

$$d = \frac{m}{V}$$

5. Follow the solution map to solve the problem.

- For problems involving only *conversions*, begin with the *given* quantity and its units. Multiply by the appropriate conversion factor(s), canceling units, to arrive at the *find* quantity.
- For problems involving *equations*, solve the equation for the *find* quantity. (Use algebra to rearrange the equation so that the *find* quantity is isolated on one side.) Gather each of the quantities that must go into the equation in the correct units. (Convert to the correct units using additional solution maps if necessary.) Finally, substitute the numerical values and their units into the equation and compute the answer.

Solution:

$$23.5\ \text{kg} \times \frac{1000\ \text{g}}{1\ \text{kg}} \times \frac{1\ \text{cm}^3}{0.789\ \text{g}} \times \frac{1\ \text{mL}}{1\ \text{cm}^3} \times \frac{1\ \text{L}}{1000\ \text{mL}} = 29.7845\ \text{L}$$

Solution:

$$d = \frac{m}{V}$$

The equation is already solved for the *find* quantity. Convert mass from kilograms to grams.

$$m = 55.9\ \text{kg} \times \frac{1000\ \text{g}}{1\ \text{kg}} = 5.59 \times 10^4\ \text{g}$$

Convert volume from liters to cubic centimeters.

$$57.2\ \text{L} \times \frac{1000\ \text{mL}}{1\ \text{L}} \times \frac{1\ \text{cm}^3}{1\ \text{mL}} = 57.2 \times 10^3\ \text{cm}^3$$

Compute density.

$$d = \frac{m}{V} \times \frac{55.9 \times 10^3}{57.2 \times 10^3\ \text{cm}^3} = 0.9772727\ \frac{\text{g}}{\text{cm}^3}$$

6. Round the answer to the correct number of significant figures.

Use the significant-figure rules from Sections 2.3 and 2.4.

$$29.7845\ \text{L} = 29.8\ \text{L}$$

$$0.9772727\ \frac{\text{g}}{\text{cm}^3} = 0.977\ \frac{\text{g}}{\text{cm}^3}$$

7. Check both the magnitude and units of the answer for correctness.

Does the magnitude of the answer make sense? Are the units correct?

The units are correct (L) and the magnitude is reasonable. Since the density is less than 1 g/cm^3, the computed volume (29.8 L) should be greater than the mass (23.5 kg).

The units are correct. Since the mass in kilograms and the volume in liters were very close to each other in magnitude, it makes sense that the density is close to 1 g/cm^3.

SKILLBUILDER 2.16

Unit Conversion

A pure gold metal bar displaces 0.82 L of water. What is its mass in kilograms? (The density of gold is 19.3 g/cm^3.)

SKILLBUILDER 2.17

Unit Conversion with Equation

A gold-colored pebble is found in a stream. Its mass is 23.2 mg and its volume is 1.20 mm^3. What is its density in grams per cubic centimeter? Is it gold? (density of gold = 19.3 g/cm^3)

CHAPTER IN REVIEW

Chemical Principles

Uncertainty: Measured quantities are reported so that the number of digits reflects the certainty in the measurement. Write measured quantities so that every digit is certain except the last, which is estimated.

Units: Measured quantities usually have units associated with them. The most convenient unit system for scientific use is the International System. The SI unit for length is the meter; for mass, the kilogram; and for time, the second. Prefix multipliers such as *kilo-* or *milli-* are often used in combination with these basic units. The SI units of volume are units of length raised to the third power, but liters or milliliters are often used as well.

Density: The density of a substance is its mass divided by its volume, $d = m/V$, and is usually reported in units of grams per cubic centimeter or grams per milliliter. Density is a fundamental property of all substances and generally differs from one substance to another.

Relevance

Uncertainty: Measurement is a hallmark of science, and the precision of a measurement must be communicated with the measurement so that others know how reliable the measurement is. When you write or manipulate measured quantities you must show and retain the precision with which the measurement was made.

Units: The units in a measured quantity communicate what the quantity actually is. Without an agreed-on system of units, scientists could not communicate their measurements. Units are also important in calculations, and the tracking of units throughout a calculation is essential.

Density: The density of substances is an important consideration in choosing materials from which to make things. Airplanes, for example, are made of low-density materials while bridges are made of higher-density materials. Density is also important as a conversion factor between mass and volume and vice versa.

Chemical Skills

Scientific Notation (Section 2.2)

To express a number in scientific notation:

- Move the decimal point to obtain a number between 1 and 10.
- Write the decimal part multiplied by 10 raised to the number of places you moved the decimal point.
- The exponent is positive if you moved the decimal point to the left and negative if you moved the decimal point to the right.

Reporting Measured Quantities to the Right Number of Digits (Section 2.3)

Report measured quantities so that every digit is certain except the last, which is estimated.

Examples

EXAMPLE 2.18 Scientific Notation

Express the number 45,000,000 in scientific notation.

45,000,000
7 6 5 4 3 2 1

4.5×10^7

EXAMPLE 2.19 Reporting Measured Quantities to the Right Number of Digits

Record the volume of liquid in the graduated cylinder to the correct number of digits. Laboratory glassware is calibrated—and should therefore be read—from the bottom of the meniscus (see figure).

Since the graduated cylinder has markings every 0.1 mL, the measurement should be recorded to the nearest 0.01 mL. In this case, that is 4.57 mL.

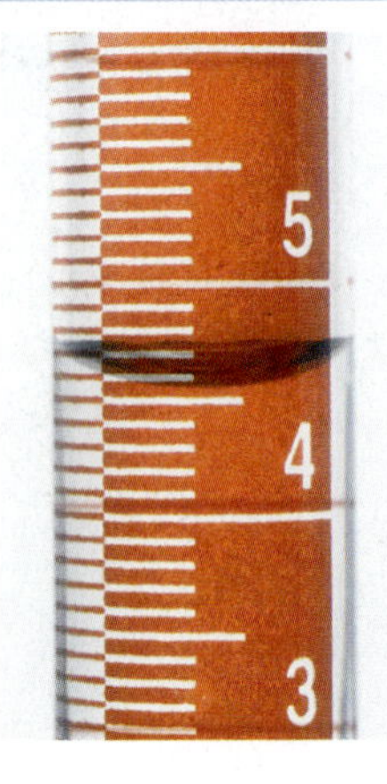

Counting Significant Digits (Section 2.3)

The following digits should always be counted as significant:

- nonzero digits
- interior zeros
- trailing zeros after a decimal point

The following digits should never be counted as significant:

- zeros to the left of the first nonzero number

The following digits are ambiguous and should be avoided by using scientific notation:

- zeros at the end of a number, but before a decimal point

EXAMPLE 2.20 Counting Significant Digits

How many significant figures are in the following numbers?

1.0050	five significant figures
0.00870	three significant figures
5,400	Write as 5.4×10^3, 5.40×10^3, or 5.400×10^3, depending on the number of significant figures intended.

Rounding (Section 2.4)

When rounding numbers to the correct number of significant figures, round down if the last digit dropped is 4 or less; round up if the last digit dropped is 5 or more.

EXAMPLE 2.21 Rounding

Round 6.442 and 6.456 to two significant figures.

6.442 rounds to 6.4
6.456 rounds to 6.5

Significant Figures in Multiplication and Division (Section 2.4)

The result of a multiplication or division should carry the same number of significant figures as the factor with the least number of significant figures.

EXAMPLE 2.22 Significant Figures in Multiplication and Division

Perform the following calculation and report the answer to the correct number of significant figures.

$$8.54 \times 3.589 \div 4.2$$
$$= 7.297$$
$$= 7.3$$

Round the final result to two significant figures to reflect the two significant figures in the factor with the least number of significant figures (4.2).

Significant Figures in Addition and Subtraction (Section 2.4)

The result of an addition or subtraction should carry the same number of decimal places as the quantity carrying the least number of decimal places.

EXAMPLE 2.23 Significant Figures in Addition and Subtraction

Perform the following operation and report the answer to the correct number of significant figures.

$$\begin{array}{r} 3.098 \\ 0.67 \\ -0.9452 \\ \hline 2.8228 \end{array} = 2.82$$

Round the final result to two decimal places to reflect the two decimal places in the quantity with the least number of decimal places (0.67).

Significant Figures in Calculations Involving Both Addition/Subtraction and Multiplication/Division (Section 2.4)

In calculations involving both addition/subtraction and multiplication/division, do the steps in parentheses first, keeping track of how many significant digits are in the answer by underlining the least significant digit, then do the remaining steps. It is best to not round off until the very end.

EXAMPLE 2.24 Significant Figures in Calculations Involving Both Addition/Subtraction and Multiplication/Division

Perform the following operation and report the answer to the correct number of significant figures.

$$8.16 \times (5.4323 - 5.411)$$
$$= 8.16 \times 0.02\underline{1}3$$
$$= 0.1738$$
$$= 0.17$$

Unit Conversion (Sections 2.6, 2.7)

Solve unit conversion problems by following these steps.

1. Write down the given quantity and its units.
2. Write down the quantity to find and/or its units.
3. Write down the appropriate conversion factor(s).
4. Draw a solution map.
5. Follow the solution map. Starting with the given quantity and its units, multiply by the appropriate conversion factor(s), canceling units, to arrive at the quantity to find in the desired units.
6. Round the final answer to the correct number of significant figures.
7. Check the answer.

EXAMPLE 2.25 Unit Conversion

Convert 108 ft to meters.

Given: 108 ft

Find: m

Conversion Factors:

1 m = 39.37 in.
1 ft = 12 in.

Solution Map:

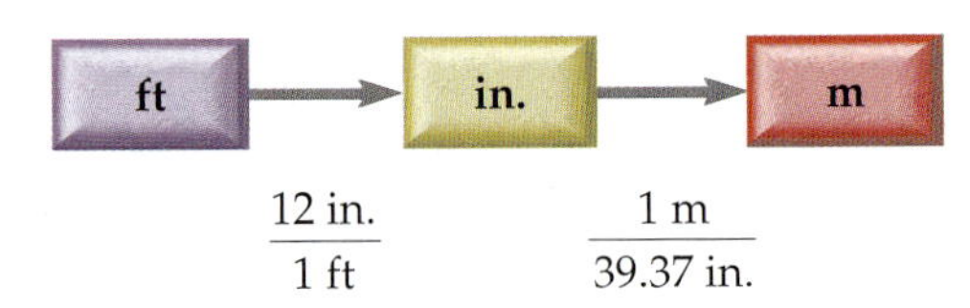

$$\frac{12 \text{ in.}}{1 \text{ ft}} \qquad \frac{1 \text{ m}}{39.37 \text{ in.}}$$

Solution:

$$108 \text{ ft} \times \frac{12 \text{ in.}}{1 \text{ ft}} \times \frac{1 \text{ m}}{39.37 \text{ in.}}$$
$$= 32.918 \text{ m}$$
$$= 32.9 \text{ m}$$

Unit Conversion Involving Units Raised to a Power (Section 2.8)

When working problems involving units raised to a power, raise the conversion factors to the same power.

EXAMPLE 2.26 Unit Conversion Involving Units Raised to a Power

How many square meters are in 1.0 km^2?

Given: 1.0 km^2

Find: m^2

Conversion Factors: 1 km = 1000 m

Solution Map:

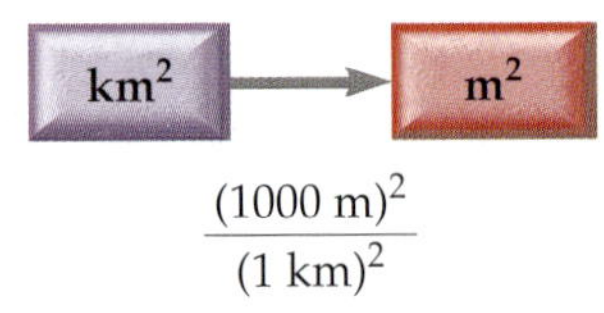

$$\frac{(1000 \text{ m})^2}{(1 \text{ km})^2}$$

Solution:

$$1.0 \text{ km}^2 \times \frac{(1000 \text{ m})^2}{(1 \text{ km})^2}$$
$$= 1.0 \text{ km}^2 \times \frac{1 \times 10^6 \text{ m}^2}{1 \text{ km}^2}$$
$$= 1.0 \times 10^6 \text{ m}^2$$

Calculating Density (Section 2.10)

The density of an object or substance is its mass divided by its volume.

$$d = \frac{m}{V}$$

EXAMPLE 2.27 Calculating Density

An object has a mass of 23.4 g and displaces 5.7 mL of water. What is its density in grams per milliliter?

Given:

$m = 23.4\ \text{g}$
$V = 5.7\ \text{mL}$

Find: density in g/mL

Solution Map:

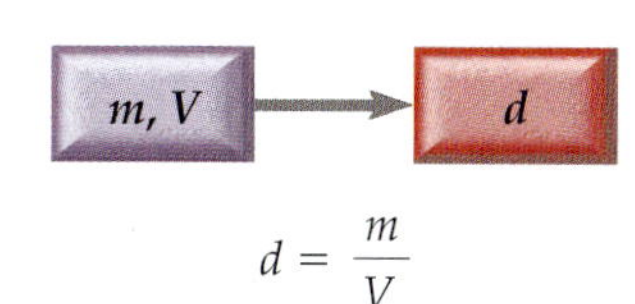

$$d = \frac{m}{V}$$

Solution:

$$d = \frac{m}{V} = \frac{23.4\ \text{g}}{5.7\ \text{mL}} = 4.11\ \text{g/mL} = 4.1\ \text{g/mL}$$

Density as a Conversion Factor (Section 2.10)

Density can be used as a conversion factor from mass to volume or from volume to mass. To convert between volume and mass, use density directly. To convert between mass and volume, you must invert the density.

EXAMPLE 2.28 Density as a Conversion Factor

What is the volume in liters of 321 g of a liquid with a density of 0.84 g/mL?

Given: 321 g

Find: volume in L

Conversion Factors:

0.84 g/mL
1 L = 1000 mL

Solution Map:

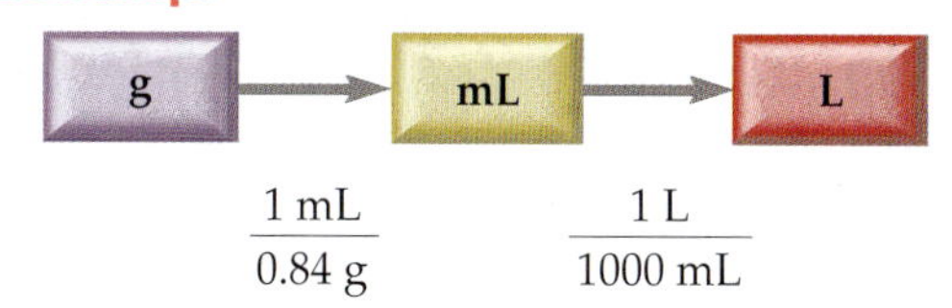

$$\frac{1\ \text{mL}}{0.84\ \text{g}} \qquad \frac{1\ \text{L}}{1000\ \text{mL}}$$

Solution:

$$321\ \text{g} \times \frac{1\ \text{mL}}{0.84\ \text{g}} \times \frac{1\ \text{L}}{1000\ \text{mL}} = 0.382\ \text{L} = 0.38\ \text{L}$$

Key Terms

conversion factor [2.6]
decimal part [2.2]
density [2.9]
English system [2.5]
exponent [2.2]
exponential part [2.2]
International System [2.5]
kilogram (kg) [2.5]
liter (L) [2.5]
mass [2.5]
meter (m) [2.5]
metric system [2.5]
prefix multipliers [2.5]
scientific notation [2.2]
second (s) [2.5]
SI units [2.5]
significant figures (digits) [2.3]
solution map [2.6]
units [2.5]
volume [2.5]

Exercises

Questions

Answers to all questions numbered in blue appear in the Answers section at the back of the book.

1. Why is it important to report units with scientific measurements?
2. Why is it important to report a certain number of digits with scientific measurements?
3. Why is scientific notation necessary?
4. If a measured quantity is written correctly, which digits are certain? Which are uncertain?
5. Explain when zeros count as significant digits and when they do not.
6. How many significant digits are there in exact numbers? What kinds of numbers are exact?
7. What limits the number of significant digits in a calculation involving only multiplication and division?
8. What limits the number of significant digits in a calculation involving only addition and subtraction?
9. What are the rules for rounding numbers? What happens if the last digit being dropped is 5 or more? Less than 5?
10. How do significant figures work in calculations involving both addition/subtraction and multiplication/division?
11. What are the basic SI units of length, mass, and time?
12. What are common units of volume?
13. Suppose you are trying to measure the diameter of a Frisbee. What unit and prefix multiplier would you use?
14. What is the difference between mass and weight?
15. Obtain a metric ruler and measure the following objects to the correct number of significant figures.
    - **(a)** quarter (diameter)
    - **(b)** dime (diameter)
    - **(c)** notebook paper (width)
    - **(d)** this book (width)
16. Obtain a stopwatch and measure each of the following to the correct number of significant figures.
    - **(a)** time between your heartbeats
    - **(b)** time it takes you to do the next problem
    - **(c)** time between your breaths
17. Explain why units are important in calculations.
18. How are units treated in a calculation?
19. What is a conversion factor?
20. Write the conversion factor that converts a measurement in inches to feet.
21. Draw a solution map to convert a measurement in grams to pounds.
22. Draw a solution map to convert a measurement in milliliters to gallons.
23. Draw a solution map to convert a measurement in meters to feet.
24. Draw a solution map to convert a measurement in ounces to grams. (1 lb = 16 oz)
25. What is density? Explain why density can work as a conversion factor. Between what quantities does it convert?
26. Explain how you would calculate the density of a substance. Include a solution map in your explanation.

Problems

Note: The exercises in the Problems section are paired, and the answers to the odd-numbered exercises (numbered in blue) appear in the Answers section at the back of the book.

Scientific Notation

27. Express each of the following numbers in scientific notation.
    - **(a)** 32,667,000 (population of California)
    - **(b)** 1,193,000 (population of Hawaii)
    - **(c)** 18,175,000 (population of New York)
    - **(d)** 481,000 (population of Wyoming)
28. Express each of the following numbers in scientific notation.
    - **(a)** 6,364,262,000 (population of the world)
    - **(b)** 1,236,915,000 (population of China)
    - **(c)** 11,051,000 (population of Cuba)
    - **(d)** 3,619,000 (population of Ireland)
29. Express each of the following numbers in scientific notation.
    - **(a)** 0.00000000007461 m (length of a hydrogen–hydrogen chemical bond)
    - **(b)** 0.0000158 mi (number of miles in an inch)
    - **(c)** 0.000000632 m (wavelength of red light)
    - **(d)** 0.000015 m (diameter of a human hair)
30. Express each of the following numbers in scientific notation.
    - **(a)** 0.000000001 s (time it takes light to travel 1 ft)
    - **(b)** 0.143 s (time it takes light to travel around the world)
    - **(c)** 0.000000000001 s (time it takes a chemical bond to undergo one vibration)
    - **(d)** 0.000001 m (approximate size of a dust particle)

31. Express each of the following in decimal notation.
   (a) 6.022×10^{23} (number of carbon atoms in 12.01 g of carbon)
   (b) 1.6×10^{-19} C (charge of a proton in coulombs)
   (c) 2.99×10^{8} m/s (speed of light)
   (d) 3.44×10^{2} m/s (speed of sound)

32. Express each of the following in decimal notation.
   (a) 450×10^{-9} m (wavelength of blue light)
   (b) 13.7×10^{9} years (approximate age of the universe)
   (c) 5×10^{9} years (approximate age of Earth)
   (d) 4.1×10^{1} years (approximate age of this author)

33. Express each of the following in decimal notation.
   (a) 3.89×10^{9}
   (b) 5.9×10^{-4}
   (c) 8.68×10^{12}
   (d) 7.86×10^{-8}

34. Express each of the following in decimal notation.
   (a) 1.90×10^{8}
   (b) 1.4×10^{-2}
   (c) 3.2×10^{3}
   (d) 4.53×10^{-14}

35. Complete the following table.

Decimal Notation	*Scientific Notation*
2,000,000,000	_____
_____	1.211×10^{9}
0.000874	_____
_____	3.2×10^{11}

36. Complete the following table.

Decimal Notation	*Scientific Notation*
_____	4.2×10^{-3}
315,171,000	_____
_____	1.8×10^{-11}
1,232,000	_____

Significant Figures

37. Read each of the following to the correct number of significant figures. Laboratory glassware should always be read from the bottom of the meniscus.

(a)

(b)

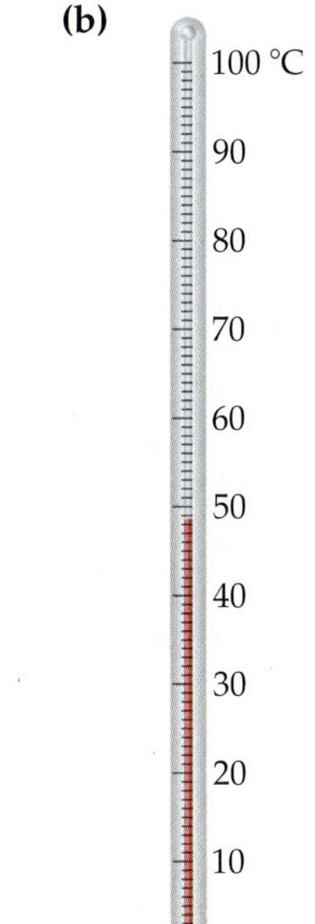

(c)

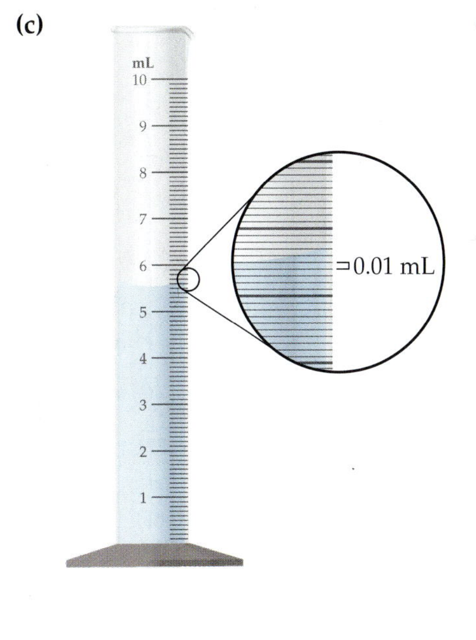

(d)

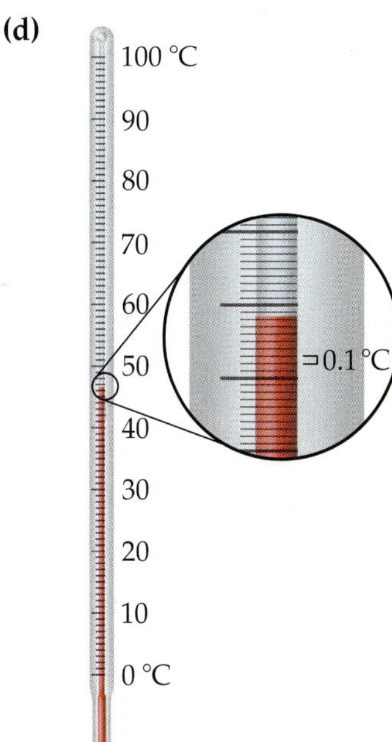

38. Read each of the following to the correct number of significant figures. Laboratory glassware should always be read from the bottom of the meniscus.

(a)

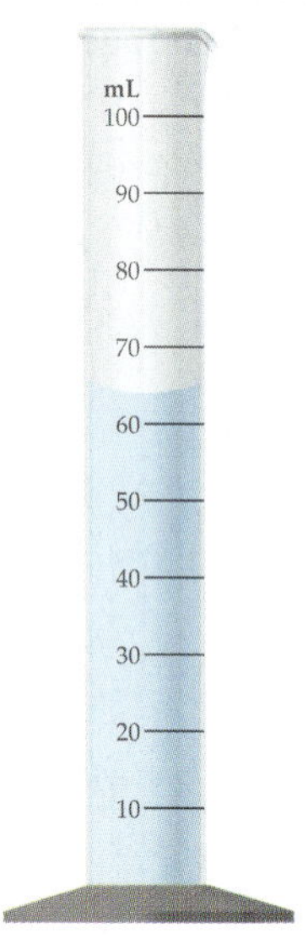

(b)

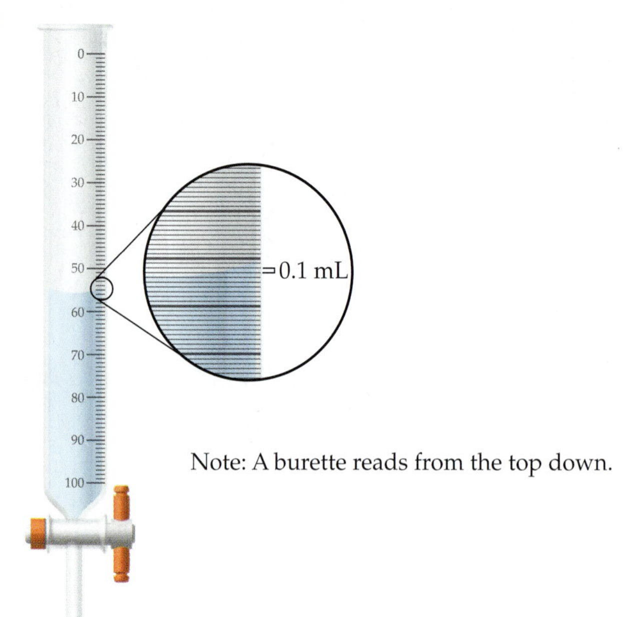

Note: A burette reads from the top down.

(c)

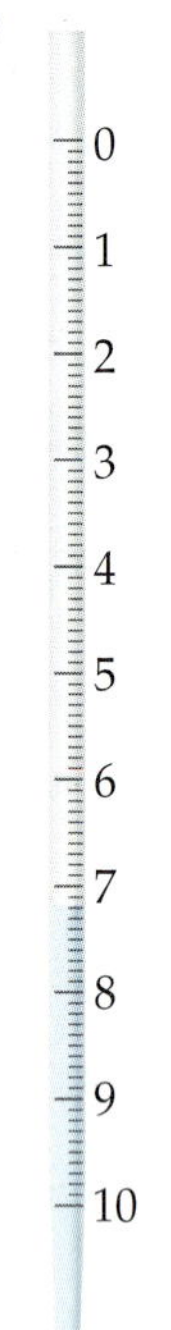

Note: A pipette reads from the top down.

(d)

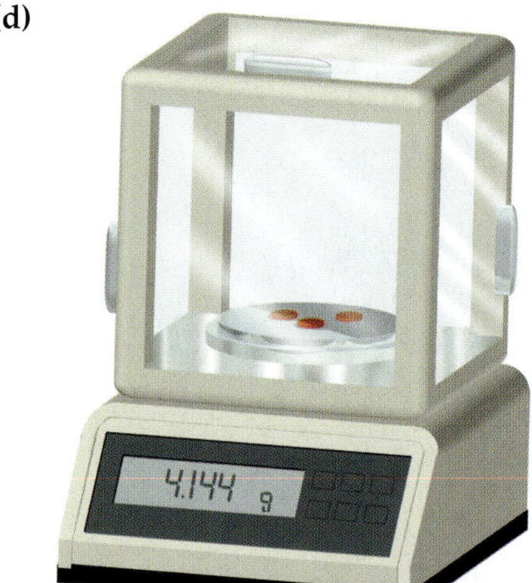

Note: Digital balances normally display mass to the correct number of significant figures for that particular balance.

39. For each of the following measured quantities, underline the zeros that are significant and draw an X through the zeros that are not.
(a) 0.005050 m
(b) 0.0000000000000060 s
(c) 220,103 kg
(d) 0.00108 in.

40. For each of the following measured quantities, underline the zeros that are significant and draw an X through the zeros that are not.
(a) 0.00010320 s
(b) 1,322,600,324 kg
(c) 0.0001240 in.
(d) 0.02061 m

41. How many significant figures are in each of the following measured quantities?
(a) 0.001125 m
(b) 0.1125 m
(c) 1.12500×10^4 m
(d) 11205 m

42. How many significant figures are in each of the following numbers?
(a) 13001 kg
(b) 13111 kg
(c) 1.30×10^4 kg
(d) 0.00013 kg

43. Determine whether each of the entries in the following table is correct. Correct the ones that are wrong.

Quantity	*Significant Figures*
(a) 895675 m	6
(b) 0.000869 kg	6
(c) 0.5672100 s	5
(d) 6.022×10^{23} atoms	4

44. Determine whether each of the entries in the following table is correct. Correct the ones that are wrong.

Quantity	*Significant Figures*
(a) 24 days	2
(b) 5.6×10^{-12} s	3
(c) 3.14 m	3
(d) 0.00383 g	5

Rounding

45. The following numbers are displayed on a calculator. Round each to four significant figures.
(a) 342.985318
(b) 0.009650901
(c) $3.52569241 \times 10^{-8}$
(d) 1,127,436,092

46. The following numbers are displayed on a calculator. Round each to three significant figures.
(a) 9,845.87492
(b) $1.548937483 \times 10^{12}$
(c) 2.3499999995
(d) 0.000045389

47. Round each of the following to two significant figures.
(a) 2.34
(b) 2.35
(c) 2.349
(d) 2.359

48. Round each of the following to three significant figures.
(a) 65.74
(b) 65.749
(c) 65.75
(d) 65.750

49. Each of the following numbers was supposed to be rounded to three significant figures. Find the ones that were incorrectly rounded and correct them.
(a) 42.3492 to 42.4
(b) 56.9971 to 57.0
(c) 231.904 to 232
(d) 0.04555 to 0.046

50. Each of the following numbers was supposed to be rounded to two significant figures. Find the ones that were incorrectly rounded and correct them.
(a) 1.249×10^3 to 1.3×10^3
(b) 3.999×10^2 to 40
(c) 56.21 to 56.2
(d) 0.009964 to 0.010

Significant Figures in Calculations

51. Perform the following calculations to the correct number of significant figures.
(a) $4.5 \times 0.03060 \times 0.391$
(b) $5.55 \div 8.97$
(c) $7.890 \times 10^{12} \div 6.7 \times 10^4$
(d) $67.8 \times 9.8 \div 100.04$

52. Perform the following calculations to the correct number of significant figures.
(a) $89.3 \times 77.0 \times 0.08$
(b) $5.01 \times 10^5 \div 7.8 \times 10^2$
(c) $4.005 \times 74 \times 0.007$
(d) $453 \div 2.031$

53. Determine whether each of the following calculations is performed to the correct number of significant figures. If it is not, correct it.
(a) $34.00 \times 567 \div 4.564 = 4.2239 \times 10^3$
(b) $79.3 \div 0.004 \times 35.4 = 7 \times 10^5$
(c) $89.763 \div 22.4581 = 3.997$
(d) $4.32 \times 10^{12} \div 3.1 \times 10^{-4} = 1.4 \times 10^{16}$

54. Determine whether each of the following calculations is performed to the correct number of significant figures. If it is not, correct it.
(a) $45.3254 \times 89.00205 = 4034.05$
(b) $0.00740 \times 45.0901 = 0.334$
(c) $49857 \div 904875 = 0.05510$
(d) $0.009090 \times 6007.2 = 54.605$

55. Perform the following calculations to the correct number of significant figures.
(a) $87.6 + 9.888 + 2.3 + 100.77$
(b) $43.7 - 2.341$
(c) $89.6 + 98.33 - 4.674$
(d) $6.99 - 5.772$

56. Perform the following calculations to the correct number of significant figures.
(a) $1459.3 + 9.77 + 4.32$
(b) $0.004 + 0.09879$
(c) $432 + 7.3 - 28.523$
(d) $2.4 - 1.777$

57. Determine whether each of the following calculations is performed to the correct number of significant figures. If not, correct it.
(a) $3.8 \times 10^5 - 8.45 \times 10^5 = -4.7 \times 10^5$
(b) $0.00456 + 1.0936 = 1.10$
(c) $8475.45 - 34.899 = 8440.55$
(d) $908.87 - 905.34095 = 3.5291$

58. Determine whether each of the following calculations is performed to the correct number of significant figures. If not, correct it.
(a) $78.9 + 890.43 - 23 = 9.5 \times 10^2$
(b) $9354 - 3489.56 + 34.3 = 5898.74$
(c) $0.00407 + 0.0943 = 0.0984$
(d) $0.00896 - 0.007 = 0.00196$

59. Perform each of the following calculations to the correct number of significant figures.
(a) $(78.4 - 44.889) \div 0.0087$
(b) $(34.6784 \times 5.38) + 445.56$
(c) $(78.7 \times 10^5 \div 88.529) + 356.99$
(d) $(892 \div 986.7) + 5.44$

60. Perform each of the following calculations to the correct number of significant figures.
(a) $(1.7 \times 10^6 \div 2.63 \times 10^5) + 7.33$
(b) $(568.99 - 232.1) \div 5.3$
(c) $(9443 + 45 - 9.9) \times 8.1 \times 10^6$
(d) $(3.14 \times 2.4367) - 2.34$

61. Determine whether each of the following calculations is performed to the correct number of significant figures. If not, correct it.
(a) $(78.56 - 9.44) \times 45.6 = 3152$
(b) $(8.9 \times 10^5 \div 2.348 \times 10^2) + 121 = 3.9 \times 10^3$
(c) $(45.8 \div 3.2) - 12.3 = 2$
(d) $(4.5 \times 10^3 - 1.53 \times 10^3) \div 34.5 = 86$

62. Determine whether each of the following calculations is performed to the correct number of significant figures. If not, correct it.
(a) $(908.4 - 3.4) \div 3.52 \times 10^4 = 0.026$
(b) $(1206.7 - 0.904) \times 89 = 1.07 \times 10^5$
(c) $(876.90 + 98.1) \div 56.998 = 17.11$
(d) $(455 \div 407859) + 1.00098 = 1.00210$

Unit Conversion

63. Perform each of the following conversions within the metric system.
(a) 2.14 kg to grams
(b) 6172 mm to meters
(c) 1316 mg to kilograms
(d) 0.0256 L to milliliters

64. Perform each of the following conversions within the metric system.
(a) 8.95 cm to meters
(b) 1298.4 g to kilograms
(c) 129 cm to millimeters
(d) 2452 mL to liters

65. Perform each of the following conversions within the metric system.
(a) 5.88 dL to liters
(b) 3.41×10^{-5} g to micrograms
(c) 1.01×10^{-8} s to nanoseconds
(d) 2.19 pm to meters

66. Perform each of the following conversions within the metric system.
(a) 1.08 Mm to kilometers
(b) 4.88 fs to picoseconds
(c) 7.39×10^{11} m to gigameters
(d) 1.15×10^{-10} m to picometers

67. Perform the following conversions between the English and metric systems.
(a) 22.5 in. to centimeters
(b) 126 ft to meters
(c) 825 yd to kilometers
(d) 2.4 in. to millimeters

68. Perform the following conversions between the English and metric systems.
(a) 78.3 in. to centimeters
(b) 445 yd to meters
(c) 336 ft to centimeters
(d) 45.3 in. to millimeters

69. Perform the following conversions between the metric and English systems.
(a) 40.0 cm to inches
(b) 27.8 m to feet
(c) 10.0 km to miles
(d) 3845 kg to pounds

70. Perform the following conversions between the metric and English systems.
(a) 254 cm to inches
(b) 89 mm to inches
(c) 7.5 L to quarts
(d) 122 kg to pounds

71. Complete the following table:

m	*km*	*Mm*	*Gm*	*Tm*
5.08×10^8 m	____	508 Mm	____	____
____	____	27,976 Mm	____	____
____	____	____	____	1.77 Tm
____	1.5×10^5 km	____	____	____
____	____	____	423 Gm	____

72. Complete the following table:

s	*ms*	*μs*	*ns*	*ps*
1.31×10^{-4} s	____	131 μs	____	____
____	____	____	12.6 ps	____
____	____	____	155 ns	____
____	1.99×10^{-3} ms	____	____	____
____	____	8.66×10^{-5} μs	____	____

73. Convert 2.255×10^{10} g to each of the following units:
(a) kg
(b) Mg
(c) mg
(d) metric tons (1 metric ton = 1000 kg)

74. Convert 1.88×10^{-6} g to each of the following units.
(a) mg
(b) cg
(c) ng
(d) μg

75. A student loses 2.4 lb in one month. How many grams did she lose?

76. A student gains 1.6 lb in two weeks. How many grams did he gain?

77. A runner wants to run 10.0 km. She knows that her running pace is 7.5 mi/h. How many minutes must she run? Hint: Use 7.5 mi/h as a conversion factor between distance and time.

78. A cyclist rides at an average speed of 24 mi/h. If she wants to bike 195 km, how long (in hours) must she ride?

79. A recipe calls for 5.0 qt of milk. What is this quantity in cubic centimeters?

80. A gas can holds 2.0 gal of gasoline. What is this quantity in cubic centimeters?

Units Raised to a Power

81. Fill in the blanks.
(a) $1.0\ km^2 =$ ____ m^2
(b) $1.0\ cm^3 =$ ____ m^3
(c) $1.0\ mm^3 =$ ____ m^3

82. Fill in the blanks.
(a) $1.0\ ft^2 =$ ____ $in.^2$
(b) $1.0\ yd^2 =$ ____ ft^2
(c) $1.0\ m^2 =$ ____ yd^2

83. The hydrogen atom has a volume of approximately $6.2 \times 10^{-31}\ m^3$. Convert this volume into each of the following units:
(a) cubic picometers
(b) cubic nanometers
(c) cubic angstroms (1 angstrom = 10^{-10} m)

84. Earth has a surface area of 197 million square miles. Convert this area into each of the following units:
(a) square kilometers
(b) square megameters
(c) square decimeters

85. A modest-sized house has an area of 215 m^2. What is its area in the following units?
(a) km^2 **(b)** dm^2 **(c)** cm^2

86. A classroom has a volume of 285 m^3. What is its volume in the following units?
(a) km^3 **(b)** dm^3 **(c)** cm^3

87. Total U.S. farmland occupies 954 million acres. How many square miles is this? (1 acre = 43,560 ft^2; 1 mi = 5280 ft)

88. The average U.S. farm occupies 435 acres. How many square miles is this? (1 acre = 43,560 ft^2; 1 mi = 5280 ft)

Density

89. A sample of an unknown metal has a mass of 35.4 g and a volume of 3.11 cm^3. Calculate its density and identify the metal by comparison to Table 2.4.

90. A new penny has a mass of 2.49 g and a volume of 0.349 cm^3. Is the penny made of pure copper?

91. Glycerol is a syrupy liquid often used in cosmetics and soaps. A 2.50-L sample of pure glycerol has a mass of 3.15×10^3 g. What is the density of glycerol in grams per cubic centimeter?

92. An aluminum engine block has a volume of 4.77 L and a mass of 12.88 kg. What is the density of the aluminum in grams per cubic centimeter?

93. A supposedly gold crown is tested to determine its density. It is found to displace 10.7 mL of water and has a mass of 206 g. Could the crown be made of gold?

94. A vase is said to be solid platinum. It is found to displace 18.65 mL of water and has a mass of 157 g. Could the vase be solid platinum?

95. Ethylene glycol (antifreeze) has a density of 1.11 g/cm^3.
  (a) What is the mass in grams of 417 mL of this liquid?
  (b) What is the volume in liters of 4.1 kg of this liquid?

96. Acetone (fingernail-polish remover) has a density of 0.7857 g/cm^3.
  (a) What is the mass in grams of 28.56 mL of acetone?
  (b) What is the volume in milliliters of 6.54 g of acetone?

Cumulative Problems

97. A thief uses a bag of sand to replace a gold statue that sits on a weight-sensitive, alarmed pedestal. The bag of sand and the statue have exactly the same volume, 1.75 L.
  (a) Calculate the mass of each object.
  (density of gold =19.3 g/cm^3;
  density of sand = 3.00 g/cm^3)
  (b) Did the thief set off the alarm? Explain.

98. One of the particles that compose atoms is called the proton. The proton has a radius of approximately 1.0×10^{-13} cm and a mass of 1.7×10^{-24} g. Determine the density of a proton. (volume of a sphere $= \frac{4}{3}\pi r^3$; $\pi = 3.14$)

99. A block of metal has a volume of 13.4 $in.^3$ and weighs 5.14 lb. What is its density in grams per cubic centimeter?

100. A log is known to be either oak or pine. It displaces 2.7 gal of water and weighs 19.8 lb. Is the log oak or pine? (density of oak = 0.9 g/cm^3; density of pine = 0.4 g/cm^3)

101. The density of aluminum is 2.7 g/cm^3. What is its density in kilograms per cubic meter?

102. The density of platinum is 21.4 g/cm^3. What is its density in pounds per cubic inch?

103. A typical backyard swimming pool holds 150 yd^3 of water. What is the mass in pounds of the water?

104. An iceberg has a volume of 8975 ft^3. What is the mass in kilograms of the iceberg?

105. The mass of fuel in an airplane must be carefully accounted for before takeoff. If a 747 contains 155,211 L of fuel, what is the mass of the fuel in kilograms? Assume the density of the fuel to be 0.768 g/cm^3.

106. A backpacker carries 2.5 L of white gas as fuel for her stove. How many pounds does the fuel add to her load? Assume the density of white gas to be 0.79 g/cm^3.

107. Honda produces a hybrid electric car called the Honda Insight. The Insight has both a gasoline-powered engine and an electric motor and has an EPA gas mileage rating of 70 mi/gal on the highway. What is the Insight's rating in kilometers per liter?

108. You rent a car in Germany with a gas mileage rating of 10.3 km/L. What is its rating in miles per gallon?

Highlight Problems

109. In 1999, NASA lost a $94 million orbiter because one group of engineers used metric units in their calculations while another group used English units. Consequently, the orbiter descended too far into the Martian atmosphere and burned up. Suppose that the orbiter was to have established orbit at 155 km and that one group of engineers specified this distance as 1.55×10^5 m. Suppose further that a second group of engineers programmed the orbiter to go to 1.55×10^5 ft. What was the difference in kilometers between the two altitudes? How low did the probe go?

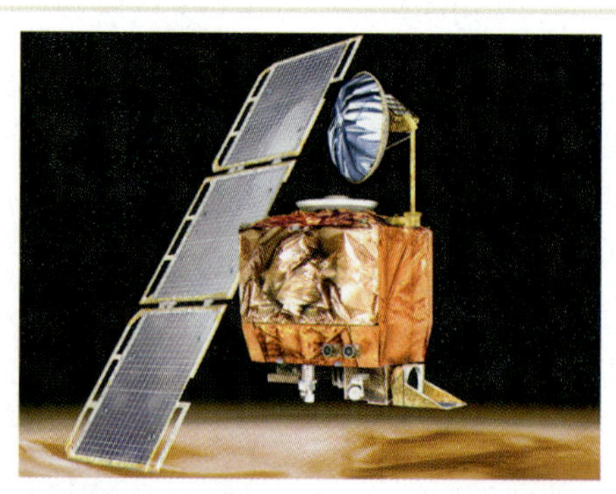

▲ The $94 million Mars Climate Orbiter was lost in the Martian atmosphere in 1999 because two groups of engineers failed to communicate to each other the units that they used in their calculations.

110. A NASA satellite showed that in 2003 the ozone hole was much larger than it was in 2002. The 2003 hole measured 1.04×10^7 mi^2 in diameter, while the diameter of the 2002 hole was 6.9×10^6 mi^2. Calculate the difference in size between the two holes in square meters.

▶ A layer of ozone gas (a form of oxygen) in the upper atmosphere protects Earth from harmful ultraviolet radiation in sunlight. Human-made chemicals react with the ozone and deplete it, especially over the Antarctic at certain times of the year (the so-called ozone hole). The region of low ozone concentration in 2003 (represented here by the dark blue color) was larger than in 2002.

111. In 1999, scientists discovered a new class of black holes with masses 100 to 10,000 times the mass of our sun, but occupying less space than our moon. Suppose that one of these black holes has a mass of 1×10^3 suns and a radius equal to one-half the radius of our moon. What is its density in grams per cubic centimeter? The mass of the sun is 2.0×10^{30} kg, and the radius of the moon is 2.16×10^3 mi. (Volume of a sphere $= \frac{4}{3}\pi r^3$).

112. A titanium bicycle frame contains the same amount of titanium as a titanium cube measuring 6.8 cm on a side. Use the density of titanium to calculate the mass in kilograms of titanium in the frame. What would be the mass of a similar frame composed of iron?

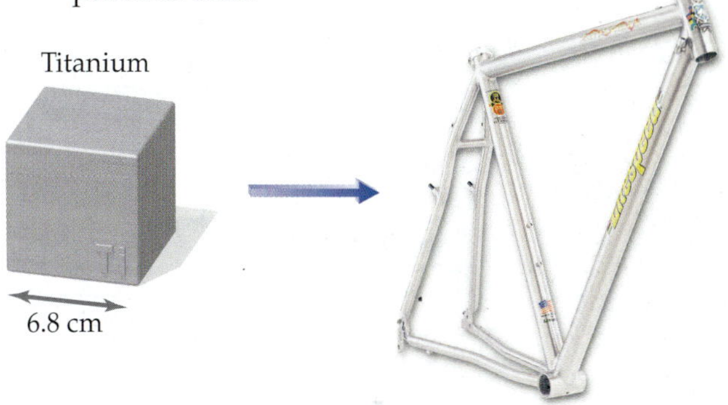

▶ A titanium bicycle frame contains the same amount of titanium as a titanium cube measuring 6.8 cm on a side.

Answers to Skillbuilder Exercises

Skillbuilder 2.1 $\$7.132 \times 10^{12}$

Skillbuilder 2.2 3.8×10^{-5}

Skillbuilder 2.3 103.4 °F

Skillbuilder 2.4 **(a)** four significant figures **(b)** three significant figures **(c)** two significant figures **(d)** unlimited significant figures **(e)** three significant figures **(f)** ambiguous

Skillbuilder 2.5 **(a)** 0.001 or 1×10^{-3} **(b)** 0.204

Skillbuilder 2.6 **(a)** 7.6 **(b)** 131.11

Skillbuilder 2.7 **(a)** 1288 **(b)** 3.12

Skillbuilder 2.8 22.0 in.

Skillbuilder 2.9 5.678 km

Skillbuilder 2.10 0.29 L

Skillbuilder 2.11 46.6 laps

Skillbuilder Plus, p. 30 1.06×10^4 m

Skillbuilder 2.12 4747 cm^3

Skillbuilder 2.13 1.52×10^5 in.^{3}

Skillbuilder 2.14 Yes, the density is 21.4 g/cm^3 and matches that of platinum.

Skillbuilder 2.15 4.4×10^{-2} cm^3

Skillbuilder Plus, p. 35 1.95 kg

Skillbuilder 2.16 16 kg

Skillbuilder 2.17 $d = 19.3$ g/cm^3; yes, it is gold.

Answers to Conceptual Checkpoints

2.1 (c) Multiplying by 10^{-3} is equivalent to moving the decimal point three places to the left.

2.2 (b) The last digit is considered to be uncertain by ± 1.

2.3 (c) The result of the calculation in **(a)** would be reported as 4; the result of the calculation in **(b)** would be reported as 1.5.

2.4 (d) The diameter would be expressed as 28 nm.

2.5 (c) Kilometers must appear in the numerator and meters in the denominator, and the conversion factor in **(d)** is incorrect (10^3 km $\neq$ 1 m).

2.6 (d) $(3 \text{ ft}) \times (3 \text{ ft}) \times (3 \text{ ft}) = 27 \text{ ft}^3$

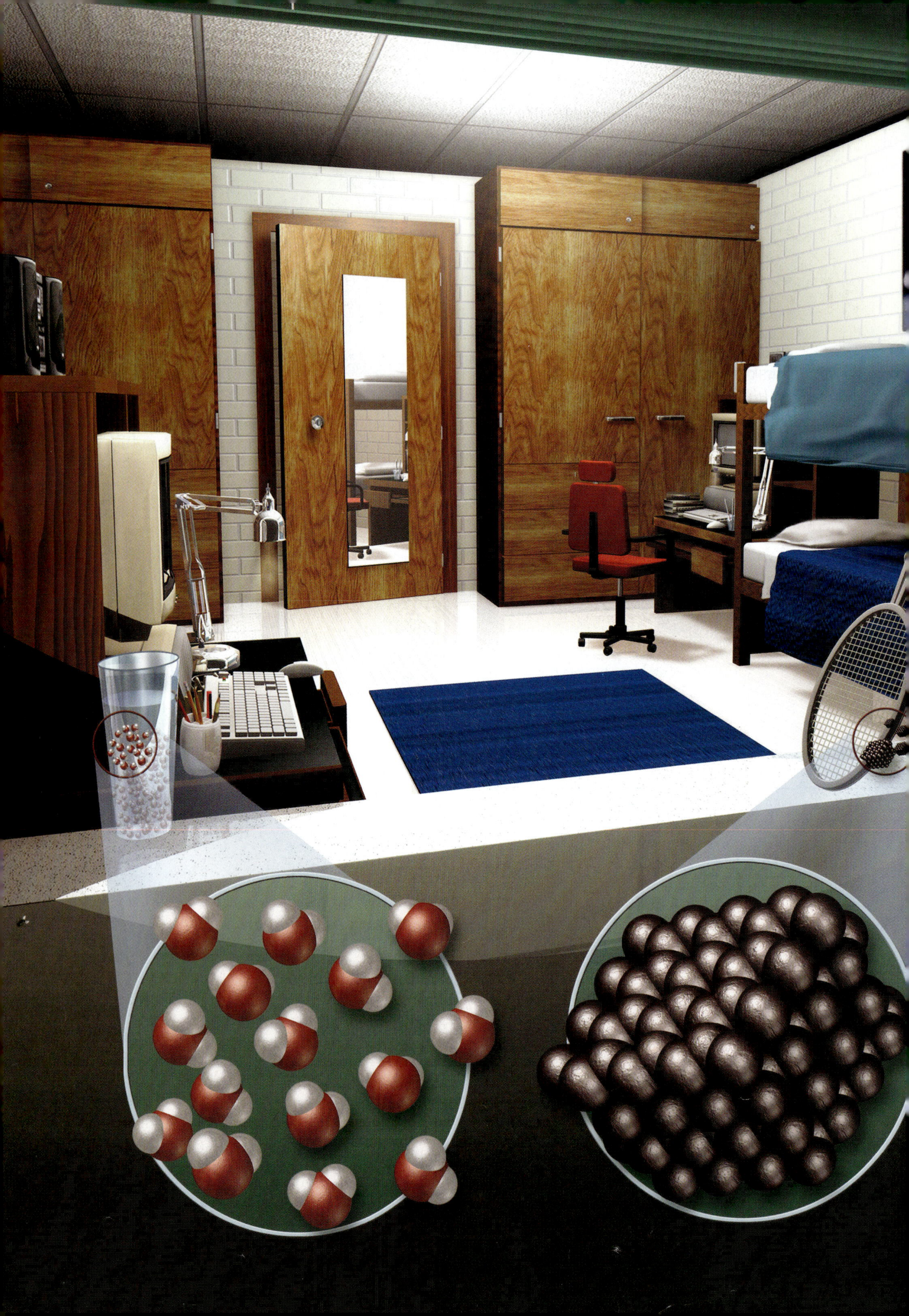

CHAPTER 3

Matter and Energy

"Thus, the task is, not so much to see what no one has yet seen; but to think what nobody has yet thought, about that which everybody sees."

Erwin Schrödinger (1887–1961)

3.1 In Your Room

Look around your room—what do you see? You might see your desk, your bed, or a glass of water. Maybe you have a window and can see trees, grass, or mountains. You can certainly see this book and possibly the table it sits on. What are these things made of? They are made of matter, which we will define more carefully shortly. For now, know that all you see is matter—your desk, your bed, the glass of water, the trees, the mountains, and this book. Some of what you don't see is matter as well. For example, you are constantly breathing air, which is also matter, into and out of your lungs. You feel the matter in air when you feel wind on your skin. Virtually everything is made of matter.

Think about the differences between different kinds of matter. Air is different from water, and water is different from wood. One of our first tasks as we learn about matter is to identify the similarities and differences among different kinds of matter. How are sugar and salt similar? How are air and water different? Why are they different? Why is a mixture of sugar and water similar to a mixture of salt and water but different from a mixture of sand and water? As students of chemistry, we are particularly interested in the similarities and differences between various kinds of matter and how these reflect the similarities and differences between their component atoms and molecules. We want to understand the connection between the macroscopic world and the microscopic one.

◀ Everything that you can see in this room is made of matter. As chemists, we are interested in how the differences between different kinds of matter are related to the differences between the molecules and atoms that compose it. The molecular structures shown here are water molecules on the left and carbon atoms in graphite on the right.

3.2 What Is Matter?

Matter is defined as anything that occupies space and has mass. Some types of matter—such as steel, water, wood, and plastic—are easily visible to our eyes. Other types of matter—such as air or microscopic dust—are impossible to see without magnification. Matter may sometimes appear smooth and continuous, but actually it is not. Matter is ultimately composed of **atoms**, tiny particles too small to see (▼ Figure 3.1). In many cases, these atoms are bonded together to form **molecules**, two or more atoms joined to one another in specific arrangements. Recent advances in microscopy have allowed us to image the atoms (▼ Figure 3.2) and molecules (▼ Figure 3.3) that compose matter, sometimes with stunning clarity.

(a)

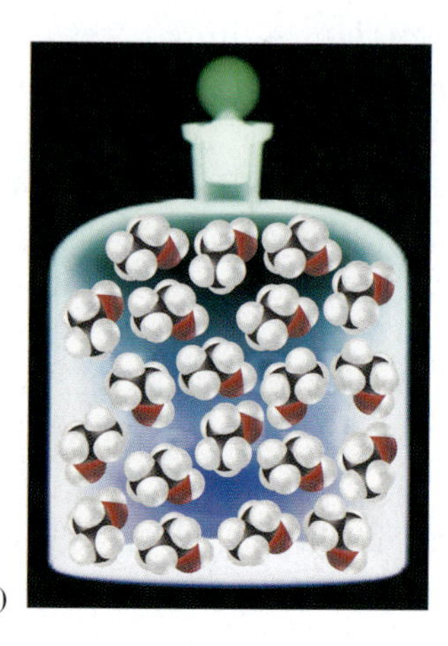
(b)

▲ **Figure 3.1 Atoms and molecules** All matter is ultimately composed of atoms. **(a)** In some substances, such as aluminum, the atoms exist as independent particles. **(b)** In other substances, such as in alcohol, several atoms bond together in well-defined structures called molecules.

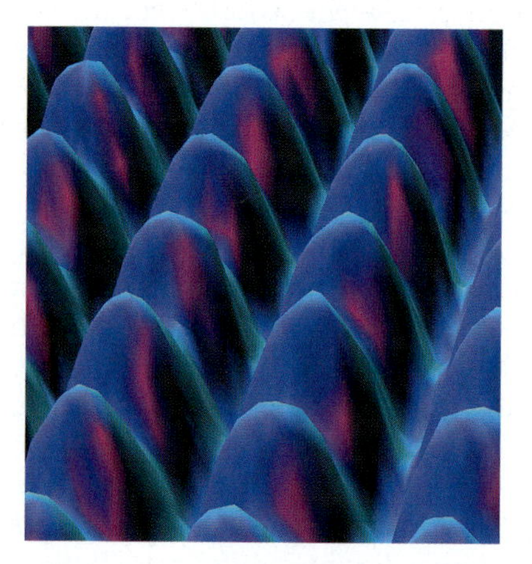

▲ **Figure 3.2 Scanning tunneling microscope image of nickel atoms** A scanning tunneling microscope (STM) creates an image by scanning a surface with a tip of atomic dimensions. It can distinguish individual atoms, as in this photo.

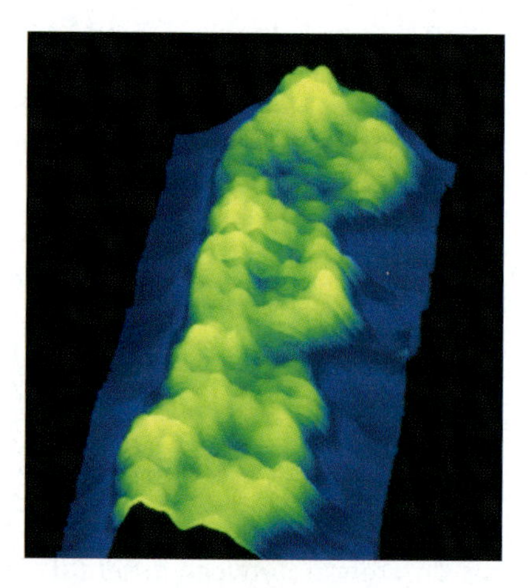

▲ **Figure 3.3 Scanning tunneling microscope image of a DNA molecule** DNA is the hereditary material that encodes the operating instructions for most cells in living organisms.

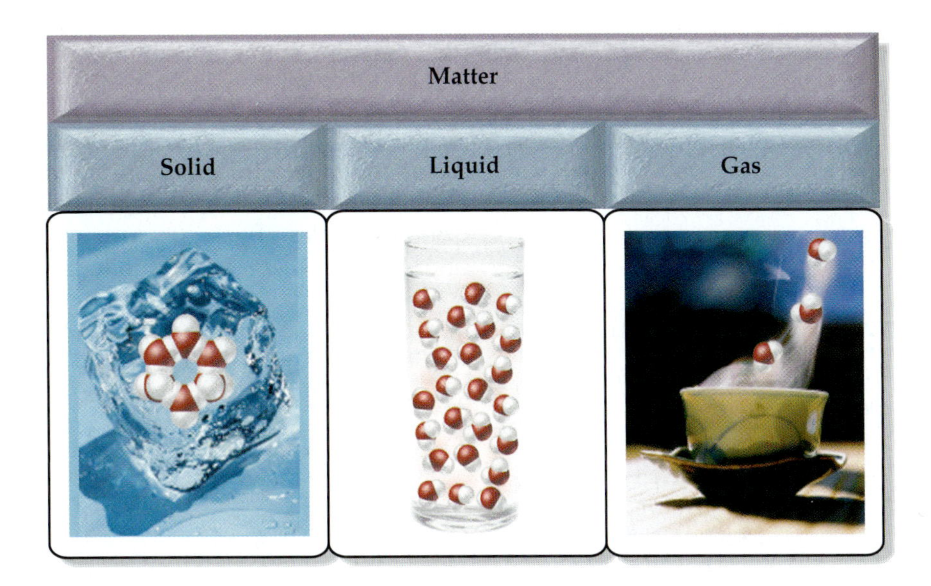

▶ **Figure 3.4 Three states of matter** Water exists as ice (solid), water (liquid), and steam (gas). In ice, the water molecules are closely spaced and do not move relative to one another. In liquid water, the water molecules are closely spaced but are free to move around and past each other. In steam, water molecules are separated by large distances and do not interact significantly with one another.

3.3 Classifying Matter According to Its State: Solid, Liquid, and Gas

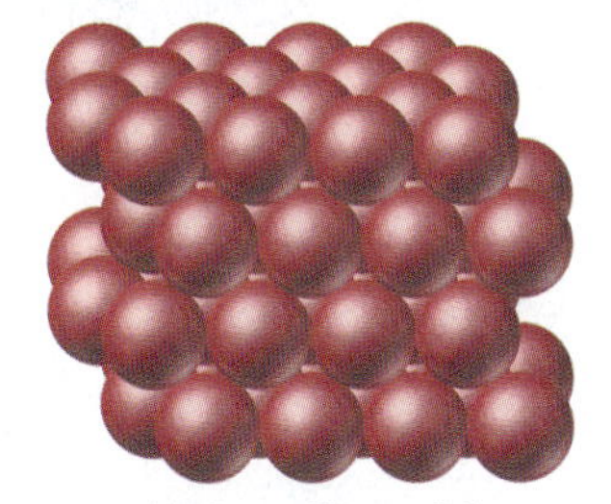

(a) Crystalline solid

(b) Amorphous solid

▲ **Figure 3.5 Types of solid matter** **(a)** In a crystalline solid, atoms or molecules occupy specific positions to create a well-ordered, three-dimensional structure. **(b)** In an amorphous solid, atoms do not have any long-range order.

There are three common **states of matter: solid, liquid**, and **gas** (▲ Figure 3.4). In solid matter, atoms or molecules pack close to each other in fixed locations. Although neighboring atoms or molecules may vibrate or oscillate, they do not move around each other, giving solids their familiar fixed volume and rigid shape. Ice, diamond, quartz, and iron are all good examples of solid matter. Solid matter may be **crystalline**, in which case its atoms or molecules arrange in geometric patterns with long-range, repeating order (◀ Figure 3.5a), or it may be **amorphous**, in which case its atoms or molecules do not have long-range order (Figure 3.5b). Good examples of *crystalline* solids include salt (▼ Figure 3.6) and diamond; the well-ordered, geometric shapes of salt and diamond crystals reflect the well-ordered geometric arrangement of their atoms. Good examples of *amorphous* solids include glass, rubber, and plastic.

In liquid matter, atoms or molecules are close to each other (about as close as in a solid) but are free to move around and by each other. Like solids, liquids have a fixed volume because their atoms or molecules are in close contact. Unlike solids, however, liquids assume the shape of their container because the atoms or molecules are free to move relative to one another. Water, gasoline, alcohol, and mercury are all good examples of liquid matter.

▶ **Figure 3.6 Salt: a crystalline solid** Sodium chloride is an example of a crystalline solid. The well-ordered, cubic shape of salt crystals is due to the well-ordered, cubic arrangement of its atoms.

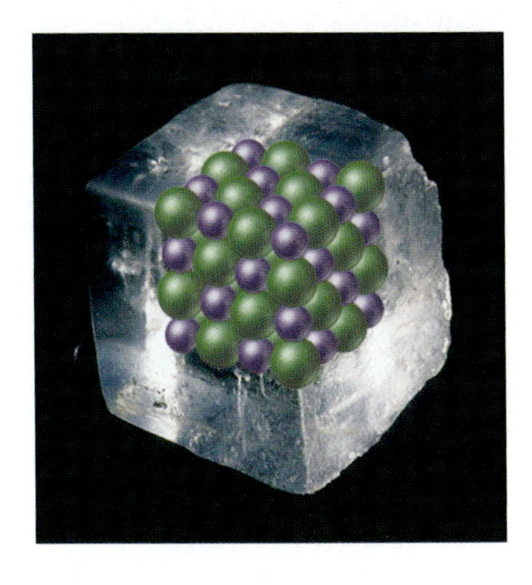

MOLECULE
Sodium Chloride

In gaseous matter, atoms or molecules are separated by large distances and are free to move relative to one another. Since the atoms or molecules that compose gases are not in contact with one another, gases are **compressible** (▼ Figure 3.7). When you inflate a bicycle tire, for example, you push more atoms and molecules into the same space, compressing them and making the tire harder. Gases always assume the shape and volume of their containers. Good examples of gases include oxygen, helium, and carbon dioxide. The properties of solids, liquids, and gases are summarized in Table 3.1.

TABLE 3.1 Properties of Liquids, Solids, and Gases

State	Atomic/Molecular Motion	Atomic/Molecular Spacing	Shape	Volume	Compressibility
solid	oscillation/vibration about fixed point	close together	definite	definite	incompressible
liquid	free to move relative to one another	close together	indefinite	definite	incompressible
gas	free to move relative to one another	far apart	indefinite	indefinite	compressible

▶ **Figure 3.7 Why gases are compressible** Since the atoms or molecules that compose gases are not in contact with one another, gases are compressible.

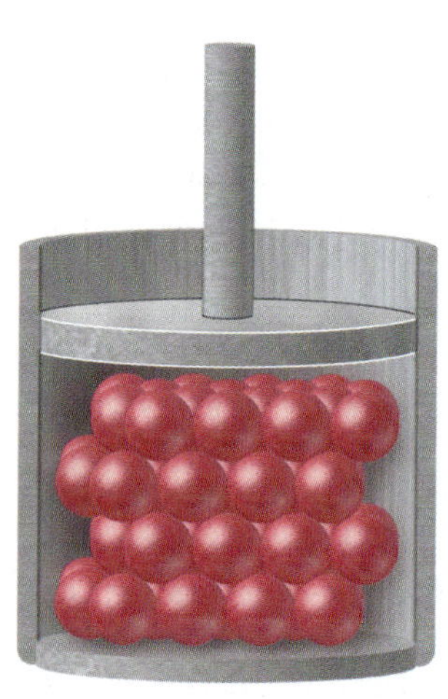

Solid–not compressible

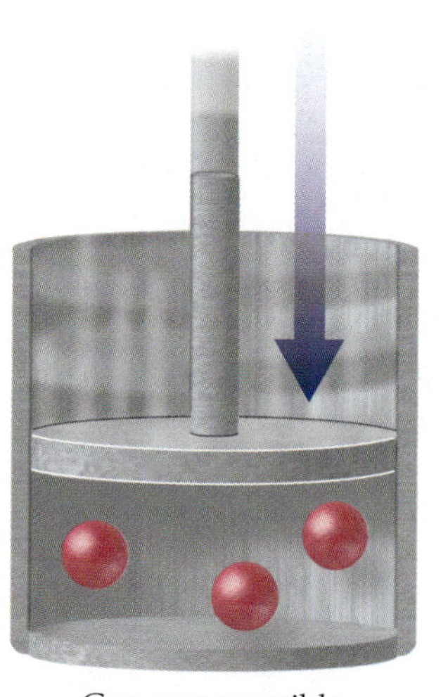
Gas–compressible

3.4 Classifying Matter According to Its Composition: Elements, Compounds, and Mixtures

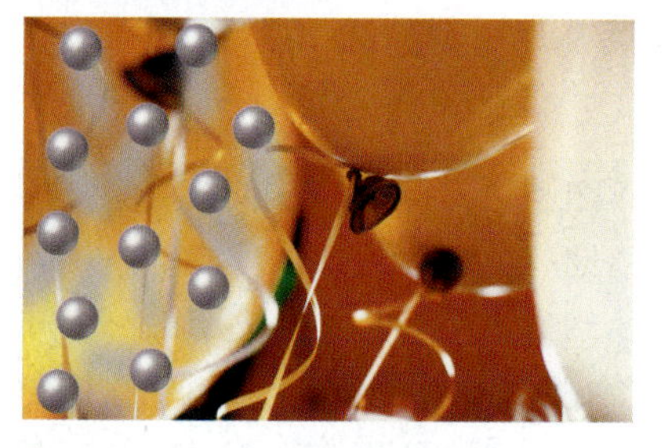

▲ Helium is a pure substance composed only of helium atoms.

In addition to classifying matter according to its state, we can classify it according to its composition (▶ Figure 3.8). Matter may be either a **pure substance**, composed of only one type of atom or molecule, or it may be a **mixture**, composed of two or more different types of atoms or molecules combined in variable proportions.

Pure substances are those composed of only one type of atom or molecule. Helium and water are both good examples of pure substances. The atoms that compose helium are all helium atoms, and the molecules that compose water are all water molecules—no other atoms or molecules are mixed in.

Pure substances can themselves be divided into two types: elements and compounds. Helium is a good example of an **element**, a substance that cannot be broken down into simpler substances. The graphite in pencils is also an element—carbon. No chemical transformation can decompose graphite into simpler substances; it is pure carbon. All known elements are listed in the periodic table in the inside front cover of this book and in alphabetical order on the inside back cover of this book.

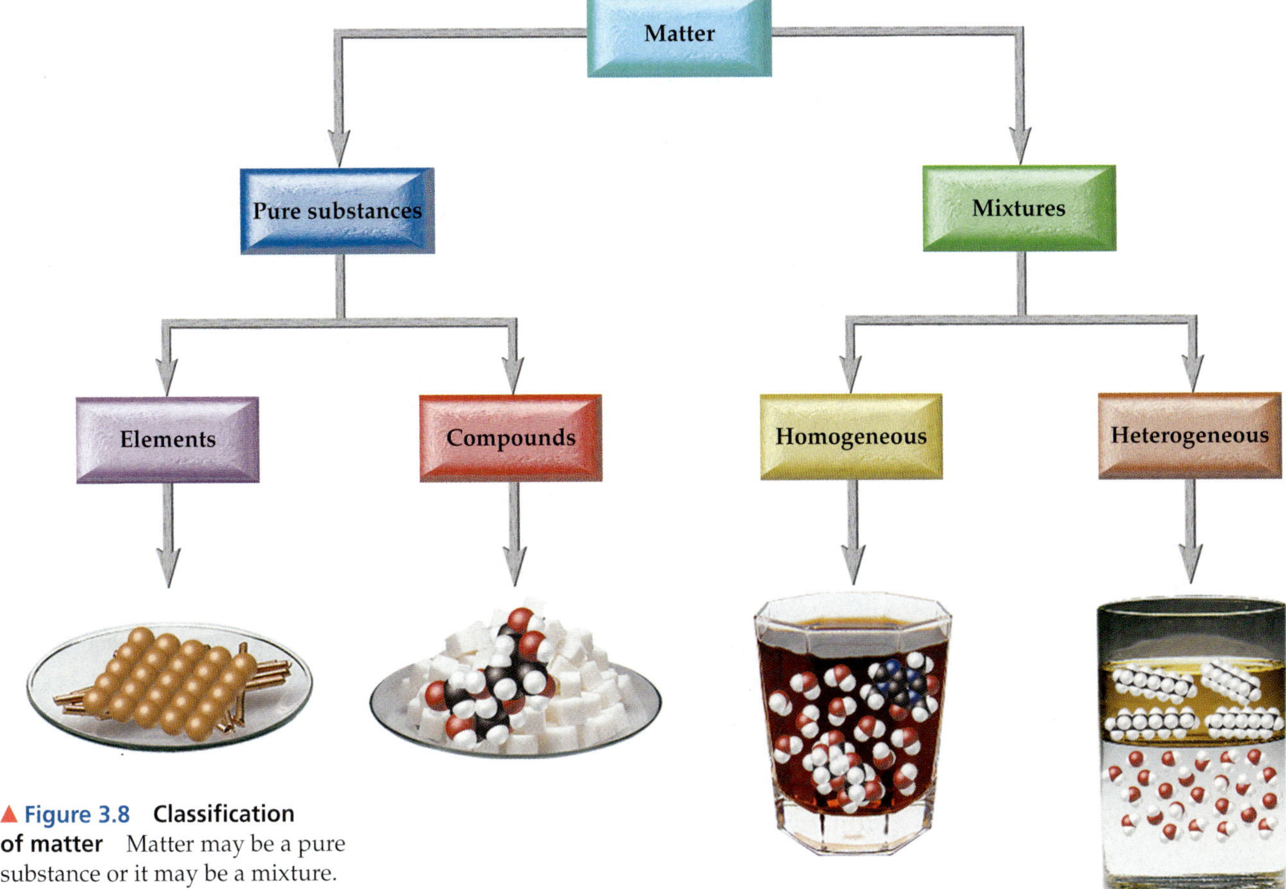

▲ **Figure 3.8 Classification of matter** Matter may be a pure substance or it may be a mixture. A pure substance may be either an element (such as copper) or a compound (such as sugar), and a mixture may be either homogeneous (such as soda pop) or heterogeneous (such as hydrocarbon and water).

A compound is composed of different atoms that are chemically united (bonded). A mixture is composed of different substances that are not chemically united, but simply mixed together.

A pure substance can also be a **compound**, a substance composed of two or more elements in fixed definite proportions. Compounds are more common than pure elements because most elements are chemically reactive and combine with other elements to form many different compounds. Water, table salt, and sugar are good examples of compounds; they can all be decomposed into simpler substances. For example, if you heat sugar on a pan over a flame, you decompose it into carbon (an element) and gaseous water (a new compound). The black substance left on your pan after burning is the carbon; the water escapes into the air as steam.

The majority of matter that we encounter is in the form of mixtures. A cup of apple juice, a flame, salad dressing, and soil are all examples of mixtures; they contain several substances with proportions that vary from one sample to another. Other common mixtures include air, seawater, and brass. Air is a mixture composed primarily of nitrogen and oxygen gas. Seawater is a mixture composed primarily of salt and water, and brass is a mixture composed of copper and zinc. Each of these mixtures can have different proportions of its constituent

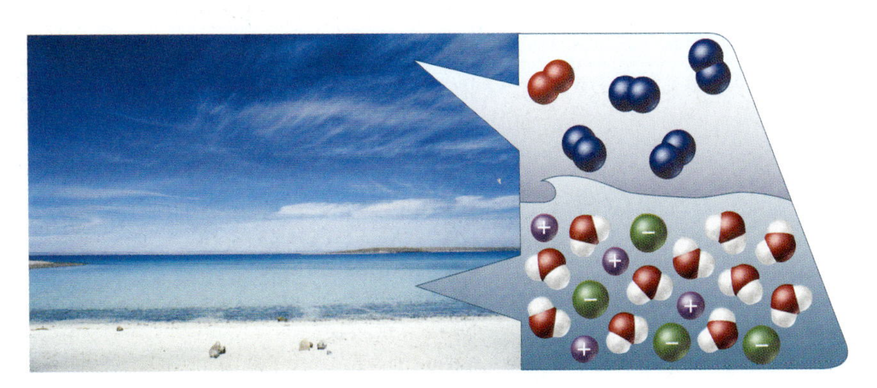

► Air and seawater are good examples of mixtures. Air contains primarily nitrogen and oxygen. Seawater contains primarily salt and water.

components. For example, metallurgists vary the relative amounts of copper and zinc in brass to tailor the metal's properties to its intended use—the higher the zinc content relative to the copper content, the more brittle the brass.

Mixtures themselves can be classified according to how uniformly the substances within them mix. In a **heterogeneous mixture**, such as oil and water, the composition varies from one region to another. In a **homogeneous mixture**, such as salt water, the composition is the same throughout. Homogeneous mixtures have uniform compositions because the atoms or molecules that compose them mix uniformly. Remember that the properties of matter are determined by the atoms or molecules that compose it.

To summarize, as shown in Figure 3.8:

- Matter may be a pure substance or it may be a mixture.
- A pure substance may be either an element or a compound.
- A mixture may be either homogenous or heterogeneous.
- Mixtures may be composed of two or more elements, two or more compounds, or a combination of both.

EXAMPLE 3.1 Classifying Matter

Classify each of the following as a pure substance or a mixture. If it is a pure substance, classify it as an element or a compound; if it is a mixture, classify it as homogeneous or heterogeneous.

(a) lead fishing weight
(b) seawater
(c) distilled water
(d) Italian salad dressing

Solution:

Begin by examining the alphabetical listing of pure elements inside the back cover of this text. If the substance appears in that table, it is a pure substance and an element. If it is not in the table, but is a pure substance, then it is a compound. If the substance is not a pure substance, then it is a mixture. Use your knowledge about the mixture to determine whether it is homogeneous or heterogeneous.

▲ Water is a pure substance composed only of water molecules.

(a) Lead is listed in the table of elements. It is a pure substance and an element.

(b) Seawater is composed of several substances including salt and water; it is a mixture. It has a uniform composition, so it is a homogeneous mixture.

(c) Distilled water is not listed in the table of elements, but it is a pure substance (water); therefore it is a compound.

(d) Italian salad dressing contains a number of substances and is therefore a mixture. It usually separates into at least two distinct regions with different composition and is therefore a heterogeneous mixture.

SKILLBUILDER 3.1 Classifying Matter

Classify each of the following as a pure substance or a mixture. If it is a pure substance, classify it as an element or a compound. If it is a mixture, classify it as homogeneous or heterogeneous.

(a) mercury in a thermometer
(b) exhaled air smoke, breathe
(c) minestrone soup
(d) sugar

Note: The answers to all Skillbuilders appear at the end of the chapter.

3.5 How We Tell Different Kinds of Matter Apart: Physical and Chemical Properties

▲ **Figure 3.9 A physical property** The boiling point of water is a physical property, and boiling is a physical change. When water boils, it turns into a gas, but the water molecules are the same in both the liquid water and the gaseous steam.

In our daily life, we distinguish one substance from another based on the substance's properties. For example, we distinguish water from alcohol based on their different smells, or we distinguish ketchup from mustard based on their different colors and flavors. The characteristics we use to distinguish one substance from another are called **properties**. Different substances have unique properties that characterize them and distinguish them from other substances.

In chemistry, we differentiate between **physical properties**, those that a substance displays without changing its composition, and **chemical properties**, those that a substance displays only through changing its composition. For example, the characteristic odor of gasoline is a physical property—gasoline does not change its composition when it exhibits its odor. On the other hand, the flammability of gasoline is a chemical property—gasoline does change its composition when it burns.

The atoms or molecules that compose a substance do not change when the substance displays its physical properties. For example, the boiling point of water—a physical property—is 100 °C. When water boils, it changes from a liquid to a gas, but the gas is still water (◀ Figure 3.9). On the other hand, the susceptibility of iron to rust is a chemical property—iron must change into iron oxide to display this property (▼ Figure 3.10). Physical properties include odor, taste, color, appearance, melting point, boiling point, and density. Chemical properties include corrosiveness, flammability, acidity, toxicity, and other chemical characteristics.

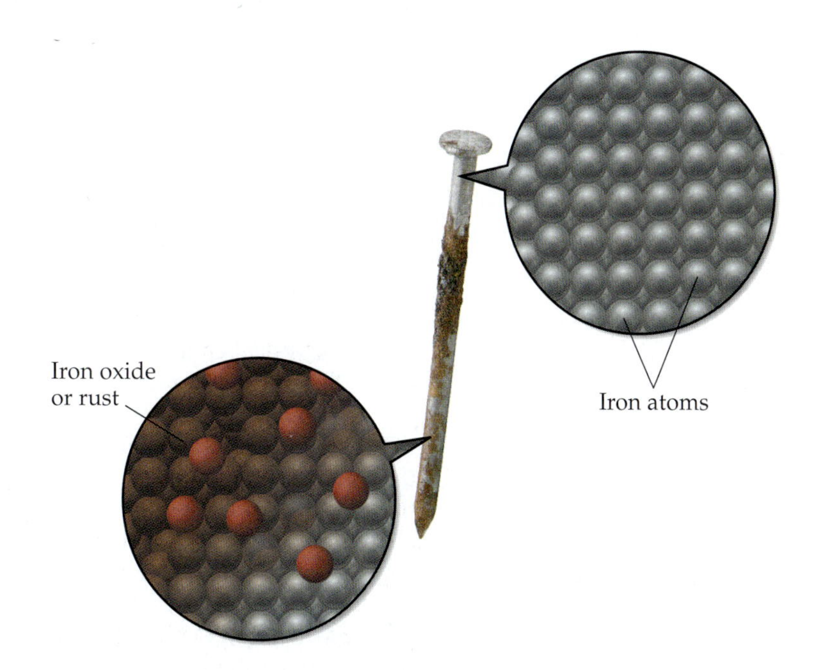

▶ **Figure 3.10 A chemical property** The susceptibility of iron to rusting is a chemical property, and rusting is a chemical change. When iron rusts, it turns from iron to iron oxide.

EXAMPLE 3.2 Physical and Chemical Properties

Determine whether each of the following is a physical or chemical property.

(a) the tendency of copper to turn green when exposed to air
(b) the tendency of automobile paint to dull over time
(c) the tendency of gasoline to evaporate quickly when spilled
(d) the low mass (for a given volume) of aluminum relative to other metals

Solution:

(a) Copper turns green because it reacts with gases in air to form compounds; this is a chemical property.

(b) Automobile paint dulls over time because it can fade (decompose) due to sunlight or it can react with oxygen in air. In either case, this is a chemical property.

(c) Gasoline evaporates quickly because it has a low boiling point; this is a physical property.

(d) Aluminum's low mass (for a given volume) relative to other metals is due to its low density; this is a physical property.

SKILLBUILDER 3.2 Physical and Chemical Properties

Determine whether each of the following is a physical or chemical property.

(a) the explosiveness of hydrogen gas
(b) the bronze color of copper
(c) the shiny appearance of silver
(d) the ability of dry ice to vaporize

3.6 How Matter Changes: Physical and Chemical Changes

Every day, we witness changes in matter: Ice melts, iron rusts, and fruit ripens, for example. What happens to the atoms and molecules that compose these substances during the change? The answer depends on the kind of change. In a **physical change**, matter changes its appearance but not its composition. For example, when ice melts, it looks different—water looks different from ice—but its composition is the same. Solid ice and liquid water are both composed of water molecules, so melting is a physical change. Similarly, when glass shatters, it looks different, but its composition remains the same—it is still glass, so this is again a physical change. On the other hand, in a **chemical change**, matter *does* change its composition. For example, copper turns green upon continued exposure to air because it reacts with gases in air to form new compounds. This is a chemical change.

Phase changes are transformations from one state of matter (such as solid or liquid) to another.

The differences between physical and chemical changes are not always apparent. Only chemical examination of the substances before and after the change can verify whether the change is physical or chemical. For many cases, however, we can identify chemical and physical changes based on what we know about them. Phase changes, such as melting or boiling, or changes that involve merely appearance, such as those produced by cutting or crushing, are always physical changes. Changes involving chemical reactions—often evidenced by heat exchange or color changes—are chemical changes.

The main difference between chemical and physical changes is related to the changes at the molecular and atomic level. In physical changes, the atoms that compose the matter *do not* change their fundamental associations, even though the matter may change its appearance. In chemical changes, atoms do change their fundamental associations, resulting in matter with a new identity. *A physical change results in a different form of the same substance, while a chemical change results in a completely new substance.*

TUTORIAL Mixtures and Compounds

Consider physical and chemical changes in liquid butane, the substance used to fuel butane lighters. In many lighters, you can see the liquid butane through the plastic case of the lighter. If you push the fuel button on the lighter without turning the flint, some of the liquid butane vaporizes to gaseous butane—you can usually hear hissing as it leaks out (▶ Figure 3.11). Since the liquid butane and the gaseous butane are both composed of butane molecules, this is a physical change. On the other hand, if you push the button *and* turn the flint to create a spark, a chemical change occurs. The butane molecules react with oxygen molecules in air to form new molecules, carbon dioxide and water (▶ Figure 3.12). The change is chemical because the molecules that compose the butane have changed into different molecules.

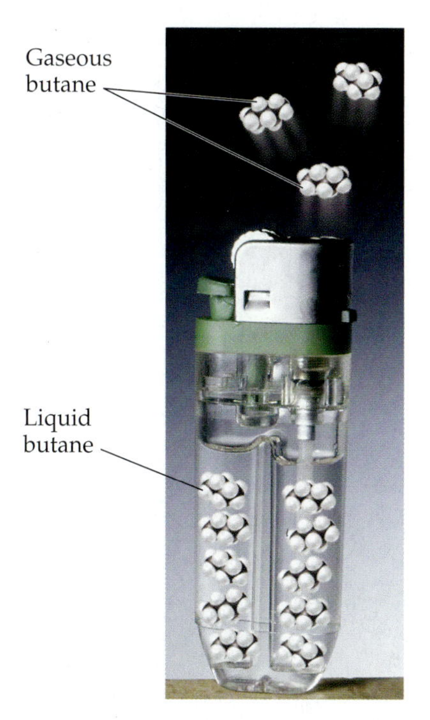

▲ **Figure 3.11 Vaporization: a physical change** If you push the button on a lighter without turning the flint, some of the liquid butane vaporizes to gaseous butane. Since the liquid butane and the gaseous butane are both composed of butane molecules, this is a physical change.

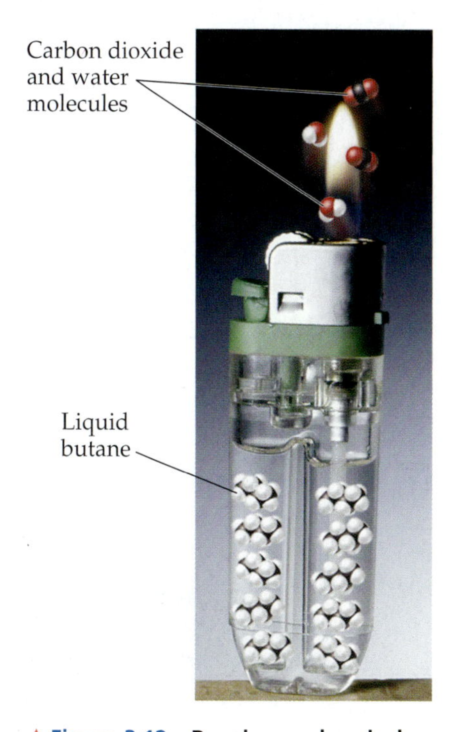

▲ **Figure 3.12 Burning: a chemical change** If you push the button *and* turn the flint to create a spark, you produce a flame. The butane molecules react with oxygen molecules in air to form new molecules, carbon dioxide and water. This is a chemical change.

EXAMPLE 3.3 Physical and Chemical Changes

Determine whether each of the following is a physical or chemical change.

(a) the rusting of iron c
(b) the evaporation of fingernail-polish remover (acetone) from the skin p
(c) the burning of coal c
(d) the fading of a carpet upon repeated exposure to sunlight p

Solution:

(a) Iron rusts because it reacts with oxygen in air to form iron oxide; therefore this is a chemical change.

(b) When fingernail-polish remover (acetone) evaporates, it changes from liquid to gas, but it remains acetone; therefore, this is a physical change.

(c) Coal burns because it reacts with oxygen in air to form carbon dioxide; this is a chemical change.

(d) A carpet fades on repeated exposure to sunlight because the molecules that give the carpet its color are decomposed by sunlight; this is a chemical change.

SKILLBUILDER 3.3 Physical and Chemical Changes

Determine whether each of the following is a physical or chemical change.

(a) copper metal forming a blue solution when it is dropped into colorless nitric acid C
(b) a passing train flattening a penny placed on a railroad track P
(c) ice melting into liquid water P
(d) a match igniting a firework P

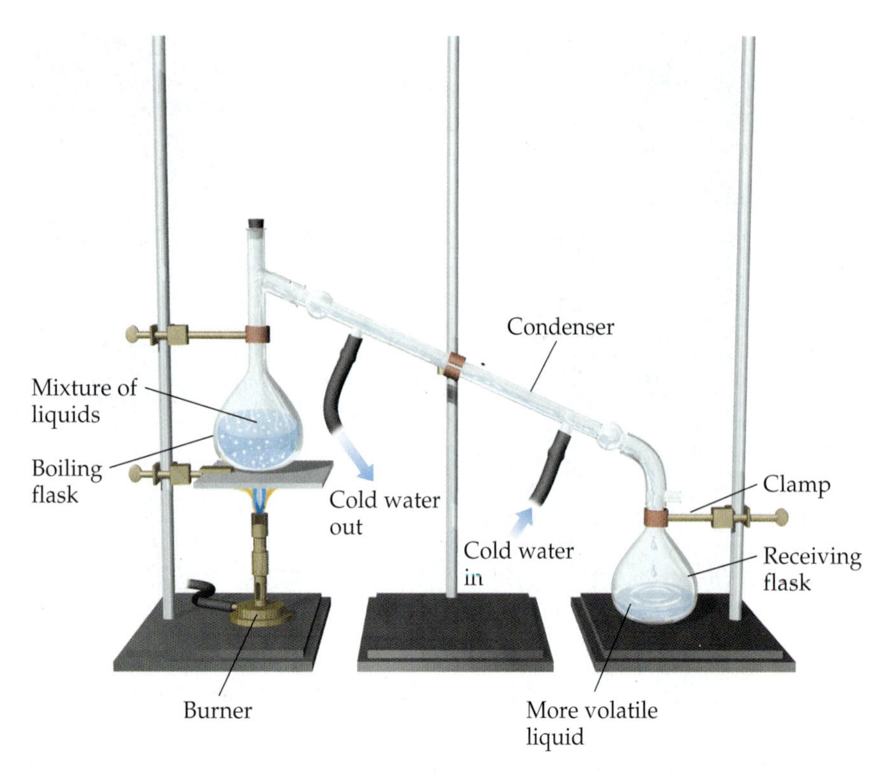

▲ **Figure 3.13 Separating a mixture of two liquids by distillation** The liquid with the lower boiling point vaporizes first. The vapors are collected and cooled until they condense back into liquid form.

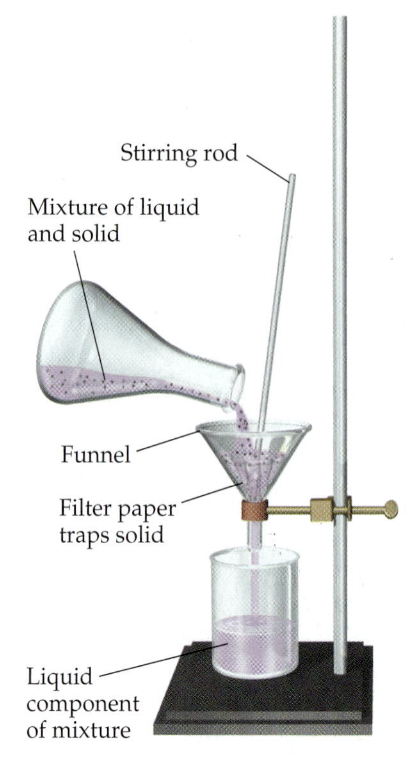

▲ **Figure 3.14 Separating a solid from a liquid by filtration**

Separating Mixtures through Physical Changes

TUTORIAL
Paper Chromatography of Ink

Chemists often want to separate mixtures into their components. Such separations can be easy or difficult, depending on the components in the mixture. In general, mixtures are separable because the different components have different properties. Various techniques that exploit these differences can be used to achieve separation. For example, oil and water are immiscible (do not mix) and have different densities. For this reason, oil floats on top of water and can be separated from water by **decanting**—carefully pouring off—the oil into another container. Mixtures of miscible liquids can usually be separated by **distillation**, a process in which the mixture is heated to boil off the more **volatile** (easily vaporizable) liquid. The volatile liquid is then recondensed in a condenser and collected in a separate flask (▲ Figure 3.13). If a mixture is composed of a solid and a liquid, the two can be separated by **filtration**, in which the mixture is poured through filter paper usually held in a funnel (▲ Figure 3.14).

3.7 Conservation of Mass: There Is No New Matter

As we have seen, our planet, our air, and even our own bodies are composed of matter. Chemical changes do not destroy matter, nor do they create new matter. Recall from Chapter 1 that Antoine Lavoisier, by studying combustion, established the law of conservation of mass, which states:

> Matter is neither created nor destroyed in a chemical reaction.

During chemical changes, the total amount of matter remains constant. How does this happen? When we burn butane in a lighter, the butane slowly disappears. Where does it go? It combines with oxygen to form carbon dioxide and water that go into the surrounding air. The mass of the carbon dioxide and water that form, however, must exactly equal the mass of the butane and oxygen that combined.

We examine the quantitative relationships in chemical reactions in Chapter 8.

For example, 58 g of butane will react with 208 g of oxygen to form 176 g of carbon dioxide and 90 g of water.

$$\underbrace{\begin{array}{c}\text{Butane} + \text{Oxygen} \\ 58\text{ g} + 208\text{ g}\end{array}}_{266\text{ g}} \longrightarrow \underbrace{\begin{array}{c}\text{Carbon dioxide} + \text{Water} \\ 176\text{ g} + 90\text{ g}\end{array}}_{266\text{ g}}$$

The sum of the masses of the butane and oxygen, 266 g, is equal to the sum of the masses of the carbon dioxide and water, which is also 266 g. Matter is conserved.

EXAMPLE 3.4 Conservation of Mass

A chemist forms 16.6 g of potassium iodide by combining 3.9 g of potassium with 12.7 g of iodine. Show that these results are consistent with the law of conservation of mass.

Solution:

The sum of the masses of the potassium and iodine is:

$$3.9\text{ g} + 12.7\text{ g} = 16.6\text{ g}$$

The sum of the masses of potassium and iodine equals the mass of the product, potassium iodide. The results are consistent with the law of conservation of mass.

SKILLBUILDER 3.4 Conservation of Mass

Suppose 12 g of natural gas combines with 48 g of oxygen in a flame. The chemical change produces 33 g of carbon dioxide and how many grams of water?

3.8 Energy

▲ Water behind a dam contains potential energy. As the water flows through the dam, the potential energy is converted to kinetic energy (the energy associated with motion) and then to electrical energy.

Matter is one of the two major components of our universe. The other major component is **energy**, defined as *the capacity to do work*. You may at first think that chemistry is concerned only with matter, but the behavior of matter is driven in large part by energy. As we shall see in later chapters, it is the flow of energy that determines what chemical changes will take place and the direction of chemical reactions. So understanding energy is critical to understanding chemistry. Like matter, energy is conserved. The **law of conservation of energy** states:

Energy can neither be created nor destroyed.

The total amount of energy is constant and cannot change; energy can be changed from one form to another or transferred from one object to another, but not created.

An object possessing energy can do work on another object—it can cause the other object to move. For example, a moving billiard ball contains **kinetic energy**, energy associated with its motion. It can collide with another billiard ball and cause it to move. Water behind a dam contains **potential energy**, energy associated with its position. The water can flow from a higher position to a lower position through the dam, turn a turbine, and produce electrical energy. **Electrical energy** is the energy associated with the flow of electrical charge. Chemical systems contain **chemical energy**, energy associated with potential chemical changes—such as the burning of gasoline—that can move or heat other objects. When we drive a car, we use chemical energy stored in gasoline to move the car forward. When we heat a home, we use chemical energy stored in natural gas to produce heat and warm the air in the house.

Units of Energy

Since energy can be converted from one form to another, all forms of energy can be quantified in the same units. The SI unit of energy is the joule (J), named after the English scientist James Joule (1818–1889), who demonstrated that energy could be converted from one type to another as long as the total energy was conserved. A second unit of energy in common use is the **calorie (cal)**, the amount of energy required to raise the temperature of 1 g of water by 1 °C. A calorie is a larger unit than a joule: 1 cal = 4.184 J. A related energy unit is the nutritional or capital *C* **Calorie (Cal)**, equivalent to 1000 little *c* calories. Electricity bills usually come in yet another energy unit, the **kilowatt-hour (kWh)**. Electricity costs about $0.15 per kilowatt-hour. Table 3.2 shows various energy units and their conversion factors. Table 3.3 shows the amount of energy required for various processes in each of these units.

TABLE 3.2 Energy Conversion Factors

1 calorie (cal)	=	4.184 joules (J)
1 Calorie (Cal)	=	1000 calories (cal)
1 kilowatt-hour (kWh)	=	3.60×10^6 joules (J)

TABLE 3.3 Energy Use in Various Units

Unit	Energy Required to Raise Temperature of 1 g of Water by 1 °C	Energy Required to Light 100-W Bulb for 1 Hour	Energy Used by Average U.S. Citizen in 1 Day
joule (J)	4.18	3.6×10^5	9.0×10^8
calorie (cal)	1	8.6×10^4	2.15×10^8
Calorie (Cal)	0.001	86	215,000
kWh	1.1×10^{-6}	0.10	250

Perpetual Motion

The law of conservation of energy has significant implications for energy use. The best we can do with energy is break even; we can't continually draw energy from a device without putting energy into it. A device that supposedly produces energy without the need for energy input is called a *perpetual motion machine* (▶ Figure 3.15) and, according to the law of conservation of energy, cannot exist. Occasionally, the media report or speculate on the discovery of a system that appears to produce more energy than it consumes. For example, I once heard a radio talk show on the subject of energy and gasoline costs. The reporter suggested that we simply design an electric car that recharges itself while you drive. The battery in the electric car would charge during operation in the same way that the battery in a conventional car recharges, except the electric car would run with energy from the battery. People have dreamed of perpetual motion machines such as this for decades. However, such ideas violate the law of conservation of energy because they produce energy without any energy input. In the case of the perpetually moving electric car, the fault lies in the idea that driving the electric car can recharge the battery—it can't.

The reason that the battery in a conventional car recharges is that energy from gasoline combustion is converted into electrical energy that then charges the battery. The electric car needs energy to move forward and the battery would eventually discharge. Some new hybrid cars (electric and gasoline-powered) can capture energy from braking and use that energy to recharge the battery.

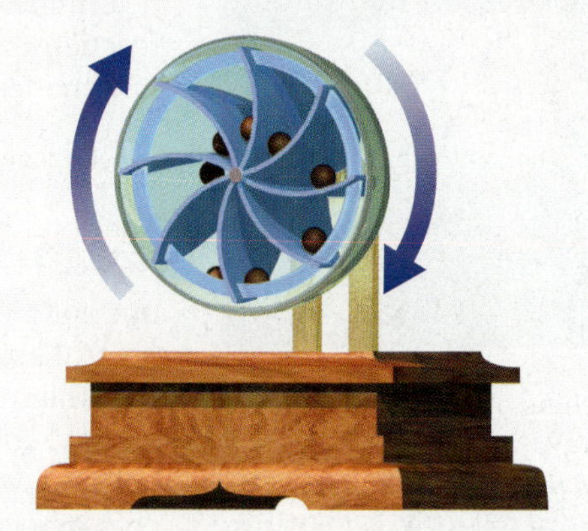

▲ **Figure 3.15 A proposed perpetual motion machine** The rolling balls supposedly keep the wheel perpetually spinning. **Question:** Can you explain why this would not work?

However, they could never run indefinitely without adding fuel. Our society has a continual need for energy, and as our current energy resources dwindle, new energy sources will be required. Unfortunately, those sources must also follow the law of conservation of energy—energy must be conserved.

CAN YOU ANSWER THIS? *A friend asks you to invest in a new flashlight he invented that never needs batteries. What kind of questions should you ask before writing a check?*

EXAMPLE 3.5 Conversion of Energy Units

A candy bar contains 225 Cal of nutritional energy. How many joules does it contain?

Solve this problem using the procedure to solve numerical problems from Chapter 2 (Section 2.10). Here you are given energy in Calories and asked to convert it to energy in joules.

Given: 225 Cal

Find: J

Conversion Factors:

1000 calories = 1 Cal

4.184 J = 1 cal

Draw a solution map. Begin with Cal, convert to cal, and then convert to J.

Solution Map:

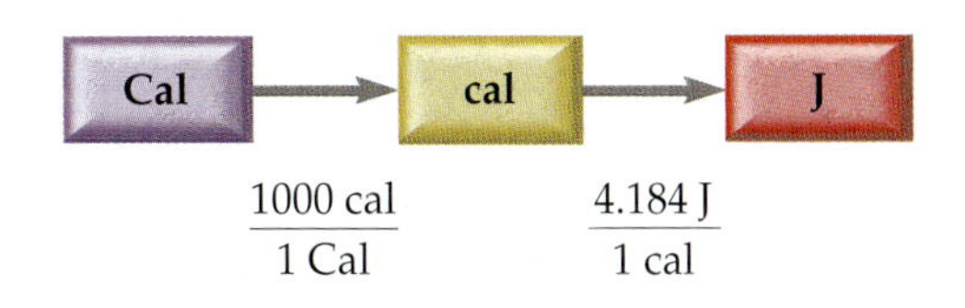

Follow the solution map to solve the problem and round to the correct number of significant digits.

Solution:

$$225\ \text{Cal} \times \frac{1000\ \text{cal}}{1\ \text{Cal}} \times \frac{4.184\ \text{J}}{1\ \text{cal}} = 9.41 \times 10^5\ \text{J}$$

SKILLBUILDER 3.5 Conversion of Energy Units

The complete combustion of a small wooden match produces approximately 512 cal of heat. How many kilojoules are produced?

SKILLBUILDER PLUS

Convert 2.75×10^4 kJ to calories.

CONCEPTUAL CHECKPOINT 3.1

Suppose a salesperson wants to make an appliance seem as efficient as possible. In which of the following units would the yearly energy consumption of the appliance have the lowest numerical value and therefore seem most efficient?

(a) J
(b) cal
(c) Cal
(d) kWh

Note: The answers to all Conceptual Checkpoints appear at the end of the chapter.

3.9 Temperature: Random Molecular and Atomic Motion

The temperature of matter is related to the random motion of the atoms and molecules that compose it. The hotter an object is, the greater the motion and the higher the temperature. We should be careful to not confuse temperature and heat. Heat, which has the units of energy, refers to the *exchange* of thermal energy caused by a temperature difference. Temperature is a *measure* of the thermal energy of matter. There are three different temperature scales in common use. The most familiar in the United States is the **Fahrenheit (°F) scale**. On the Fahrenheit scale, water freezes at 32 °F and boils at 212 °F. Room temperature is approximately 75 °F. The Fahrenheit scale was initially set up by assigning 0 °F to the freezing point of a concentrated saltwater solution and 100 °F to body temperature.

The scale often used by scientists is the **Celsius (°C) scale**. On this scale, water freezes at 0 °C and boils at 100 °C. Room temperature is approximately 25 °C.

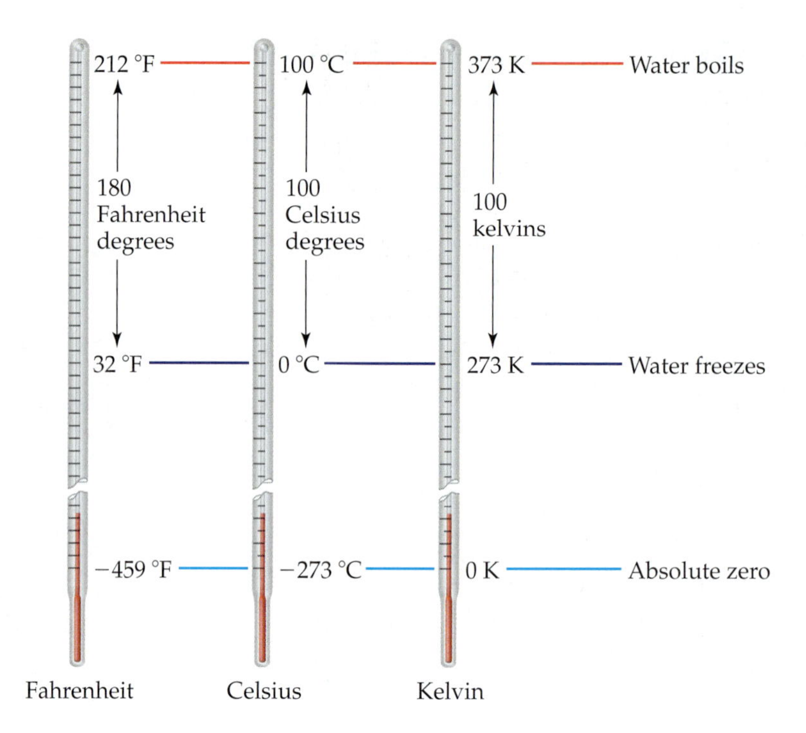

▶ **Figure 3.16 Comparison of the Fahrenheit, Celsius, and Kelvin temperature scales** The Fahrenheit degree is five-ninths the size of a Celsius degree. The Celsius degree and the kelvin degree are the same size.

The Fahrenheit and Celsius scales differ in both the size of their respective degrees and the temperature each calls "zero" (▲ Figure 3.16). Both the Fahrenheit and Celsius scales contain negative temperatures. However, a third temperature scale, called the **Kelvin (K) scale**, avoids negative temperatures by assigning 0 K to the coldest temperature possible, absolute zero. Absolute zero (−273 °C or −459 °F) is the temperature at which molecular motion virtually stops. There is no lower temperature. The kelvin degree, or **kelvin**, is the same size as the Celsius degree—the only difference is the temperature that each scale calls zero.

Conversions between these temperature scales are achieved using the following formulas.

$$K = °C + 273$$

$$°C = \frac{(°F - 32)}{1.8}$$

For example, suppose we want to convert 212 K to Celsius. Following the procedure for solving numerical problems (Section 2.10), we write:

Given: 212 K

Find: °C

Equation: The equation that relates the given quantity (K) to the find quantity (°C) is:

$$K = °C + 273$$

Solution Map: We then build a solution map.

In a solution map involving a formula, the formula establishes the relationship between the variables. However, the formula under the arrow is not necessarily solved for the correct variable until later, as is the case here.

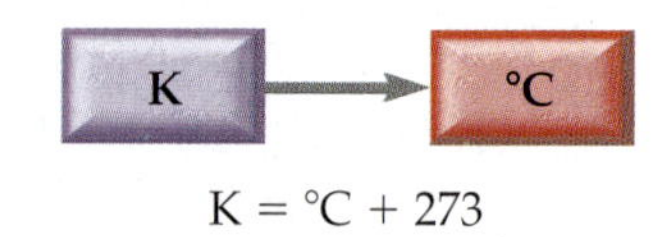

Solution: The equation below the arrow shows the relationship between K and °C, but it is not in the correct form. We must solve the equation for °C.

$$K = °C + 273$$

$$°C = K - 273$$

Finally, we substitute the given value for K and compute the answer to the correct number of significant figures.

$$°C = 212 - 273$$
$$= -61\ °C$$

EXAMPLE 3.6 Converting between Celsius and Kelvin Temperature Scales

Convert −25 °C to kelvin.

Set up the problem in the normal way. You are given a temperature in °C and asked to find K. Use the equation that shows the relationship between the given quantity (°C) and the find quantity (K).

Given: −25 °C

Find: K

Equation:

$$K = °C + 273$$

Build the solution map.

Solution Map:

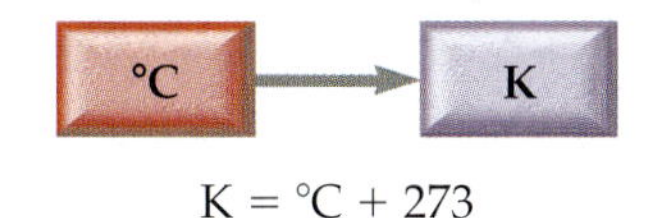

$K = °C + 273$

Follow the solution map to solve the problem by substituting the correct value for °C and computing the answer to the correct number of significant figures.

Solution:

$$K = °C + 273$$
$$K = -25\ °C + 273 = 248\ K$$

SKILLBUILDER 3.6 Converting between Celsius and Kelvin Temperature Scales

Convert 358 K to Celsius.

EXAMPLE 3.7 Converting between Fahrenheit and Celsius Temperature Scales

Convert 55 °F to Celsius.

Set up the problem in the normal way. You are given a temperature in °F and asked to find °C. Use the equation that shows the relationship between the given quantity (°C) and the find quantity (K).

Given: 55 °F

Find: °C

Equation: $°C = \frac{(°F - 32)}{1.8}$

Build the solution map.

Solution Map:

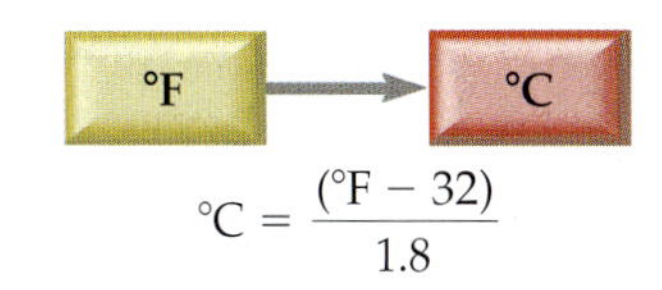

$°C = \frac{(°F - 32)}{1.8}$

Substitute the given value into the equation and compute the answer to the correct number of significant figures.

Solution:

$$°C = \frac{(°F - 32)}{1.8}$$
$$°C = \frac{(55 - 32)}{1.8} = 12.778\ °C = 13\ °C$$

SKILLBUILDER 3.7 Converting between Fahrenheit and Celsius Temperature Scales

Convert 139 °C to Fahrenheit.

EXAMPLE 3.8 Converting between Fahrenheit and Kelvin Temperature Scales

Convert 310 K to Fahrenheit.

Set up the problem in the normal way. You are given a temperature in K and asked to find °F. This problem requires two equations: one relating K and °C and the other relating °C and °F.

Given: 310 K

Find: °F

Equations:

$$K = °C + 273$$

$$°C = \frac{(°F - 32)}{1.8}$$

Build the solution map, which requires two steps: one to convert from K to °C and one to convert from °C to °F.

Solution Map:

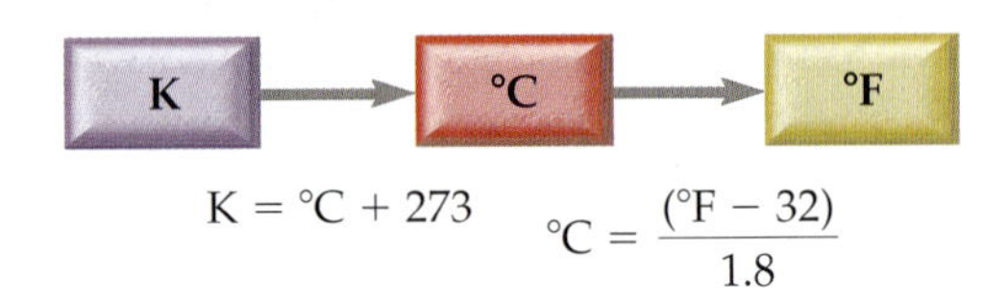

The first equation must be solved for °C. Substitute into this equation to convert from K to °C.

Solution:

$$K = °C + 273$$

$$°C = K - 273$$

$$°C = 310 - 273 = 37\ °C$$

The second equation must be solved for °F. Substitute into this equation to convert from °C to °F and compute the answer to the correct number of significant digits.

$$°C = \frac{(°F - 32)}{1.8}$$

$$1.8\ °C = (°F - 32)$$

$$°F = 1.8\ °C + 32$$

$$°F = 1.8(37) + 32 = 98.6\ °F = 99\ °F$$

SKILLBUILDER 3.8 Converting between Fahrenheit and Kelvin Temperature Scales

Convert −321 °F to kelvin.

CONCEPTUAL CHECKPOINT 3.2

Which of the following temperatures is identical on both the Celsius and the Fahrenheit scale?

(a) 100°
(b) 32°
(c) 0°
(d) −40°

Omitted

3.10 Temperature Changes: Heat Capacity

All substances change temperature when they are heated, but how much they change for a given amount of heat varies significantly from one substance to another. For example, if you put a steel skillet on a flame, its

It is simply by coincidence that lead and gold have the same value of specific heat capacity.

TABLE 3.4 Specific Heat Capacities of Some Common Substances

Substance	Specific Heat Capacity (J/g °C)
lead	0.128
gold	0.128
silver	0.235
copper	0.385
iron	0.449
aluminum	0.903
ethanol	2.42
water	4.184

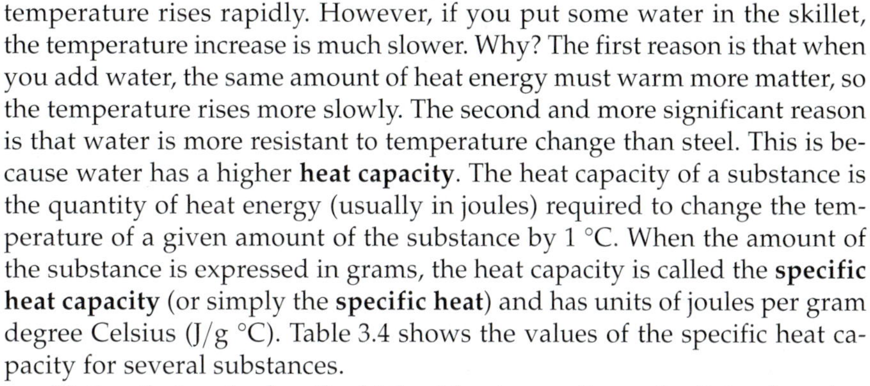

temperature rises rapidly. However, if you put some water in the skillet, the temperature increase is much slower. Why? The first reason is that when you add water, the same amount of heat energy must warm more matter, so the temperature rises more slowly. The second and more significant reason is that water is more resistant to temperature change than steel. This is because water has a higher **heat capacity**. The heat capacity of a substance is the quantity of heat energy (usually in joules) required to change the temperature of a given amount of the substance by 1 °C. When the amount of the substance is expressed in grams, the heat capacity is called the **specific heat capacity** (or simply the **specific heat**) and has units of joules per gram degree Celsius (J/g °C). Table 3.4 shows the values of the specific heat capacity for several substances.

▲ San Francisco enjoys cool weather even in summer months because of the high heat capacity of the surrounding ocean.

Notice that water has the highest heat capacity on the list—changing its temperature requires a lot of heat. If you have traveled from an inland geographical region to a coastal one and have felt the drop in temperature, you have experienced the effects of water's high heat capacity. On a summer day in California, for example, the temperature difference between Sacramento (an inland city) and San Francisco (a coastal city) can be 30 °F; San Francisco enjoys a cool 68 °F, while Sacramento bakes at near 100 °F. Yet the intensity of sunlight falling on these two cities is the same. Why the large temperature difference? The difference between the two locations is the presence of the Pacific Ocean, which practically engulfs San Francisco. Water, with its high heat capacity, absorbs much of the sun's heat without undergoing a large increase in temperature, keeping San Francisco cool. The land surrounding Sacramento, on the other hand, with its low heat capacity, cannot absorb a lot of heat without a large increase in temperature—it has a lower *capacity* to absorb heat without a large temperature increase.

Similarly, only two U.S. states have never recorded a temperature above 100 °F. One of them is obvious: Alaska. It is too far north to get that hot. The other one, however, may come as a surprise. It is Hawaii. The water that surrounds the only island state, having a high heat capacity, moderates the temperature, keeping Hawaii from ever getting too hot.

CONCEPTUAL CHECKPOINT 3.3

If you wanted to heat a metal plate to as high a temperature as possible for a given energy input, you would make the plate out of:

(a) copper
(b) iron
(c) aluminum
(d) it would make no difference

EVERYDAY **Chemistry**

Coolers, Camping, and the Heat Capacity of Water

Have you ever loaded a cooler with ice and then added room-temperature drinks? If you have, you know that the ice quickly melts. In contrast, if you load your cooler with chilled drinks, the ice lasts for hours. Why the difference? The answer is related to the high heat capacity of water within the drinks. As we just learned, water must absorb a lot of heat to raise its temperature, but it must also release a lot of heat to lower its temperature. When the warm drinks are placed into the ice, they release heat, which then melts the ice. The chilled drinks, on the other hand, are already cold, so they do not release much heat. It is always better to load your cooler with chilled drinks—that way, the ice will last the rest of the day.

CAN YOU ANSWER THIS? *Suppose you are cold-weather camping and decide to heat some objects to bring into your sleeping bag for added warmth. You place a large water jug and a rock of equal mass close to the fire. Over time, both the rock and the water jug warm to about 38 °C (100 °F). If you could bring only one into your sleeping bag, which one should you bring to keep you the warmest? Why?*

▲ The ice in a cooler loaded with cold drinks lasts much longer than the ice in a cooler loaded with warm drinks. **Question:** Can you explain why?

Omitted

3.11 Energy and Heat Capacity Calculations

The specific heat capacity of a substance can be used to quantify the relationship between the amount of heat added to a given amount of the substance and the corresponding temperature increase. The equation that relates these quantities is:

$$\text{Heat} = \text{Mass} \times \text{Specific heat capacity} \times \text{Temperature change}$$
$$q = m \times C \times \Delta T$$

where q is the amount of heat in joules, m is the mass of the substance in grams, C is the specific heat capacity in joules per gram per degree Celsius, and ΔT is the temperature change in Celsius. The symbol Δ means *the change in*, so ΔT means *the change in temperature*. For example, suppose you are making a cup of tea and want to know how much heat energy will warm 235 g of water (about 8 oz) from 25 °C to 100.0 °C (boiling). We set up this problem as follows:

ΔT in °C is equal to ΔT in K but is not equal to ΔT in °F.

Given: 235 g water

25 °C initial temperature (T_i)

100.0 °C final temperature (T_f)

Find: amount of heat needed, q

Equation: The equation that relates the given and find quantities is:

$$q = m \cdot C \cdot \Delta T$$

Solution Map:

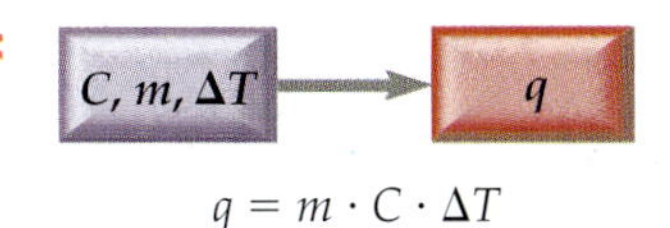

$$q = m \cdot C \cdot \Delta T$$

In addition to m and ΔT, the equation requires C, the heat specific capacity of water. Our next step is to gather all of the required quantities for the equation (C, m, and ΔT) in the correct units. These are:

$$C = 4.18\ \text{J/g °C}$$
$$m = 235\ \text{g}$$

The other required quantity is ΔT. The change in temperature is the difference between the final temperature (T_f) and the initial temperature (T_i).

$$\Delta T = T_f - T_i$$
$$= 100.0\ °\text{C} - 25\ °\text{C} = 75\ °\text{C}$$

Solution: Finally, we substitute the correct values into the equation and compute the answer to the correct number of significant figures.

$$q = m \times C \times \Delta T$$
$$= 235\ \cancel{\text{g}} \times 4.18 \frac{\text{J}}{\cancel{\text{g}}\ \cancel{°\text{C}}} \times 75\ \cancel{°\text{C}}$$
$$= 7.367 \times 10^4\ \text{J} = 7.4 \times 10^4\ \text{J}$$

When discussing energy transfer, we often define the object of our study (such as a flask in which a chemical reaction is occurring) as *the system* and the parts of the universe with which the system exchanges energy as *the surroundings*. Energy is then transferred between the system and its surroundings

It is critical that you substitute each of the correct variables into the equation in the correct units and cancel units as you compute the answer. If, during this process, you learn that one of your variables is not in the correct units, convert it to the correct units using the skills you learned in Chapter 2. Notice that the sign of q is positive (+) if the substance is increasing in temperature (heat going into the substance) and negative (−) if the substance is decreasing in temperature (heat going out of the substance).

EXAMPLE 3.9 Relating Heat Energy to Temperature Changes

Gallium is a solid metal at room temperature but melts at 29.9 °C. If you hold gallium in your hand, it melts from body heat. How much heat must 2.5 g of gallium absorb from your hand to raise its temperature from 25.0 °C to 29.9 °C? The specific heat capacity of gallium is 0.372 J/g °C.

You are given the mass of gallium and its initial and final temperature and asked to find the amount of heat absorbed.

Given: 2.5 g gallium

$$T_i = 25.0\ °\text{C}$$
$$T_f = 29.9\ °\text{C}$$
$$C = 0.372\ \text{J/g °C}$$

Find: q

The equation that relates the given and find quantities is the specific heat capacity equation.

Equation: $q = m \cdot C \cdot \Delta T$

The solution map shows that the specific heat capacity equation relates the given and find quantities.

Solution Map:

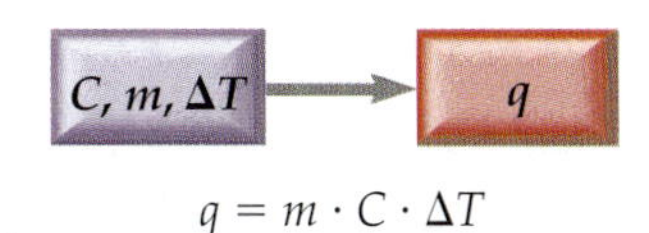

$$q = m \cdot C \cdot \Delta T$$

Before solving the problem, you must gather the necessary quantities—C, m, and ΔT—in the correct units.

Solution:

$C = 0.372\ \text{J/g °C}$

$m = 2.5\ \text{g}$

$\Delta T = 29.9\ \text{°C} - 25.0\ \text{°C} = 4.9\ \text{°C}$

Then substitute the correct variables into the equation, canceling units, and compute the answer to the right number of significant figures.

$$q = m \times C \times \Delta T$$

$$= 2.5\ \cancel{\text{g}} \times 0.372 \frac{\text{J}}{\cancel{\text{g}}\ \cancel{\text{°C}}} \times 4.9\ \cancel{\text{°C}} = 4.557\ \text{J} = 4.6\ \text{J}^*$$

*This is the amount of heat required to raise the temperature to the melting point. Actually melting the gallium requires additional heat.

SKILLBUILDER 3.9 Relating Heat Energy to Temperature Changes

You find a copper penny (pre-1982) in the snow and pick it up. How much heat is absorbed by the penny as it warms from the temperature of the snow, −5.0 °C, to the temperature of your body, 37.0 °C? Assume the penny is pure copper and has a mass of 3.10 g. You can find the heat capacity of copper in Table 3.4 (p. 67).

SKILLBUILDER PLUS

The temperature of a lead fishing weight rises from 26 °C to 38 °C as it absorbs 11.3 J of heat. What is the mass of the fishing weight in grams?

EXAMPLE 3.10 Relating Heat Capacity to Temperature Changes

A chemistry student finds a shiny rock that she suspects is gold. She weighs the rock on a balance and obtains the mass, 14.3 g. She then finds that the temperature of the rock rises from 25 °C to 52 °C upon absorption of 174 J of heat. Find the heat capacity of the rock and determine whether the value is consistent with the heat capacity of gold.

You are given the mass of the "gold" rock, the amount of heat absorbed, and the initial and final temperature. You are asked to find the heat capacity.

Given: 14.3 g

174 J of heat absorbed

$T_i = 25\ \text{°C}$

$T_f = 52\ \text{°C}$

The equation that relates the given and find quantities is the heat capacity equation.

Find: C

Equation: $q = m \cdot C \cdot \Delta T$

The solution map shows how the heat capacity equation relates the given and find quantities.

Solution Map:

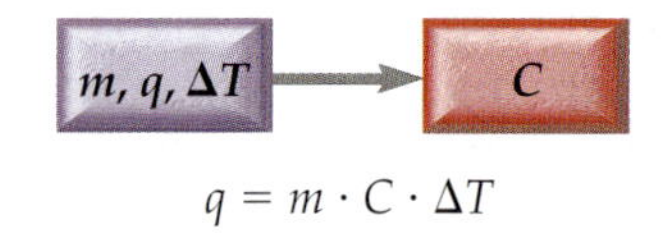

$q = m \cdot C \cdot \Delta T$

Before solving the problem, you must gather the necessary quantities—m, q, and ΔT—in the correct units.

Solution:

$m = 14.3\ \text{g}$

$q = 174\ \text{J}$

$\Delta T = 52\ \text{°C} - 25\ \text{°C} = 27\ \text{°C}$

Solve the equation for C and then substitute the correct variables into the equation, canceling units, and compute the answer to the right number of significant figures.

$$q = m \cdot C \cdot \Delta T$$

$$C = \frac{q}{m\ \Delta T}$$

$$C = \frac{174\ \text{J}}{14.3\ \text{g} \times 27\ \text{°C}} = 0.4506 \frac{\text{J}}{\text{g °C}} = 0.45 \frac{\text{J}}{\text{g °C}}$$

By comparing the computed value of the heat capacity (0.45 J/g °C) with the heat capacity of gold from Table 3.4 (0.128 J/g °C), we conclude that the rock could not be pure gold.

SKILLBUILDER 3.10 Relating Heat Capacity to Temperature Changes

A 328 g sample of water absorbs 5.78×10^3 J of heat. Find the change in temperature for the water. If the water is initially at 25.0 °C, what is its final temperature?

CHAPTER IN REVIEW

Chemical Principles

Matter: Matter is anything that occupies space and has mass. It is composed of atoms, which are often bonded together as molecules. Matter can exist as a solid, a liquid, or a gas. Solid matter can be either amorphous or crystalline.

Classification of Matter: Matter can be classified according to its composition. Pure matter is composed of only one type of substance; that substance may be an element (a substance that cannot be decomposed into simpler substances) or it may be a compound (a substance composed of two or more elements in fixed definite proportions.) Mixtures are composed of two or more different substances whose proportions may vary from one sample to the next. Mixtures can be either homogeneous, having the same composition throughout, or heterogeneous, having a composition that varies from region to region.

Properties and Changes of Matter: The properties of matter can be divided into two types: physical and chemical. The physical properties of matter are those that are displayed without a change in composition. The chemical properties of matter can be displayed only with a change in composition. The changes in matter can themselves be divided into physical and chemical. In a physical change, the appearance of matter may change but its composition does not. In a chemical change, the composition of matter changes.

Conservation of Mass: Whether the changes in matter are chemical or physical, matter is always conserved. In a chemical change the masses of the matter undergoing the chemical change must equal the sum of the masses of matter resulting from the chemical change.

Energy: Besides matter, energy is the other major component of our universe. Like matter, energy is conserved—it can be neither created nor destroyed. Energy comes in various different types, and these can be converted from one to another. Some common units of energy are the joule (J), the calorie (cal), the nutritional Calorie (Cal), and the kilowatt-hour (kWh).

Relevance

Matter: Understanding matter is relevant because everything is made of matter—you, me, the chair you sit on, and the air we breathe. The physical universe basically contains only two things: matter and energy. So we begin our study of chemistry by defining and classifying these two building blocks of the universe.

Classification of Matter: Since ancient times, humans have tried to understand matter and harness it for their purposes. The earliest humans shaped matter into tools and used the transformation of matter—especially fire—to keep warm and to cook food. To manipulate matter, we must understand it. Fundamental to this understanding is the connection between the properties of matter and the molecules and atoms that compose it. If matter is composed of a single type of atom or molecule, it is a pure substance. If it is composed of two or more types of atoms or molecules, it is a mixture.

Properties and Changes of Matter: The physical and chemical properties of matter make the world around us the way it is. For example, a physical property of water is its boiling point at sea level—100 °C. The physical properties of water—and all matter—are determined by the atoms and molecules that compose it. If water molecules were different—even slightly different—water would boil at a different temperature. Imagine a world where water boiled at room temperature.

Conservation of Mass: The conservation of matter is relevant to, for example, pollution. We often think that we create pollution, but actually, we are powerless to create anything. Matter cannot be created. So pollution is simply misplaced matter—matter that has been put into places it does not belong.

Energy: Our society's energy sources will not last forever because as we burn fossil fuels—our primary energy source—we convert chemical energy, stored in molecules, to kinetic and thermal energy. The kinetic and thermal energy is not readily available to be used again. Consequently, our energy resources are dwindling, and the conservation of energy implies that we will not be able simply to create new energy—it must come from somewhere.

Temperature: The temperature of matter is related to the random motions of the molecules and atoms that compose it—the greater the motion, the higher the temperature. Temperature is commonly measured on three scales: Fahrenheit (°F), Celsius (°C), and Kelvin (K).

Temperature: The temperature of matter and its measurement is relevant to many everyday phenomena. Humans are constantly interested in the weather, and air temperature is a fundamental part of weather. We use body temperature as one measure of human health and global temperature as one measure of the planet's health.

Heat Capacity: The temperature change that a sample of matter undergoes upon absorption of a given amount of heat is related to the heat capacity of the substance composing the matter. Water has one of the highest heat capacities, meaning that it is most resistant to rapid temperature changes.

Heat Capacity: The heat capacity of water explains why it is cooler in coastal areas, which are near large bodies of high-heat-capacity water, than in inland areas, which are surrounded by low-heat-capacity land. It also explains why it takes longer to cool a refrigerator filled with liquids than an empty one.

Chemical Skills

Classifying Matter (Sections 3.3, 3.4)

Begin by examining the alphabetical listing of elements in the back of this text. If the substance is listed in that table, it is a pure substance and an element.

If the substance is not listed in that table, use your knowledge about the substance to determine whether it is a pure substance. If it is a pure substance not listed in the table, then it is a compound.

If it is not a pure substance, then it is a mixture. Use your knowledge about the mixture to determine whether it has uniform composition throughout (homogeneous) or nonuniform composition (heterogeneous).

Examples

EXAMPLE 3.11 Classifying Matter

Classify each of the following as a pure substance or a mixture. If it is a pure substance, classify it as an element or compound. If it is a mixture, classify it as homogeneous or heterogeneous.

(a) pure silver
(b) swimming-pool water
(c) dry ice (solid carbon dioxide)
(d) blueberry muffin

Solution:

(a) Pure element; silver appears in the element table.
(b) Homogeneous mixture; pool water contains at least water and chlorine in variable proportions, and it is uniform throughout.
(c) Compound; dry ice is a pure substance (carbon dioxide), but it is not listed in the table.
(d) Heterogeneous mixture; a blueberry muffin is a mixture of several things and has nonuniform composition.

Physical and Chemical Properties (Section 3.5)

To distinguish between physical and chemical properties, ask whether the substance changes composition while displaying the property. If it *does not* change composition, the property is physical; if it *does,* the property is chemical.

EXAMPLE 3.12 Physical and Chemical Properties

Determine whether each of the following is a physical or chemical property.

(a) the tendency for platinum jewelry to scratch easily
(b) the ability of sulfuric acid to burn the skin
(c) the ability of hydrogen peroxide to bleach hair
(d) the density of lead relative to other metals

Solution:

(a) Physical; scratched platinum is still platinum.
(b) Chemical; the acid chemically reacts with the skin to produce the burn.
(c) Chemical; the hydrogen peroxide chemically reacts with hair to bleach it.
(d) Physical; the heaviness can be felt without changing the lead into anything else.

Physical and Chemical Changes (Section 3.6)

To distinguish between physical and chemical changes, ask whether the substance changes composition during the change. If it *does not* change composition, the change is physical; if it *does*, the change is chemical.

EXAMPLE 3.13 Physical and Chemical Changes

Determine whether each of the following is a physical or chemical change.

(a) the explosion of gunpowder in the barrel of a gun
(b) the melting of gold in a furnace
(c) the bubbling that occurs upon mixing baking soda and vinegar
(d) the bubbling that occurs when water boils

Solution:
(a) Chemical; the gunpowder reacts with oxygen during the explosion.
(b) Physical; the liquid gold is still gold.
(c) Chemical; the bubbling is a result of a chemical reaction between the two substances to form new substances, one of which is carbon dioxide released as bubbles.
(d) Physical; the bubbling is due to liquid water turning into gaseous water, but it is still water.

Conservation of Mass (Section 3.7)

The sum of the masses of the substances involved in a chemical change must be the same before and after the change.

EXAMPLE 3.14 Conservation of Mass

An automobile runs for 10 minutes and burns 47 g of gasoline. The gasoline combined with oxygen from air and formed 132 g of carbon dioxide and 34 g of water. How much oxygen was consumed in the process?

Solution:
The total mass after the chemical change is:

$$132\text{ g} + 34\text{ g} = 166\text{ g}$$

The total mass before the change must also be 166 g.

$$47\text{ g} + \text{g oxygen} = 166\text{ g}$$

So, the mass of oxygen consumed is the total mass (166 g) minus the mass of gasoline (47 g).

$$\text{grams of oxygen} = 166\text{ g} - 47\text{ g} = 119\text{ g}$$

Conversion of Energy Units (Section 3.8)

Solve energy unit conversion problems using the problem-solving procedure that we learned in Section 2.10.

- Write down the given quantity and its units.
- Write down the quantity to find and/or its units.
- Write down the appropriate conversion factor(s).
- Write a solution map.

LIVE EXAMPLE 

EXAMPLE 3.15 Conversion of Energy Units

Convert 1.7×10^3 kWh (the amount of energy used by the average U.S. citizen in one week) into calories.

Given: 1.7×10^3 kWh

Find: cal

Conversion Factors:

$$1\text{ kWh} = 3.60 \times 10^6\text{ J}$$
$$1\text{ cal} = 4.18\text{ J}$$

Solution Map:

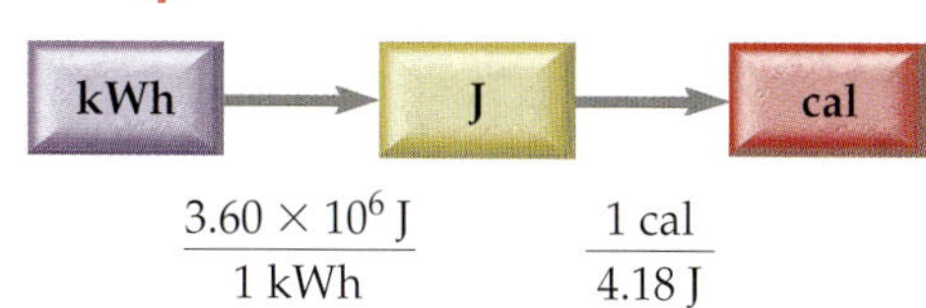

$$\frac{3.60 \times 10^6\text{ J}}{1\text{ kWh}} \qquad \frac{1\text{ cal}}{4.18\text{ J}}$$

- Starting with the given quantity and its units, multiply by the appropriate conversion factor(s), canceling units, to arrive at the quantity to find in the desired units.
- Round the final answer to the correct number of significant figures.
- Check the answer.

Solution:

$$1.7 \times 10^3 \text{ kWh} \times \frac{3.60 \times 10^6 \text{ J}}{1 \text{ kWh}} \times \frac{1 \text{ cal}}{4.18 \text{ J}}$$

$$= 1.464 \times 10^9 \text{ cal}$$

$$1.464 \times 10^9 \text{ cal} = 1.5 \times 10^9 \text{ cal}$$

The unit of the answer, cal, is correct. The magnitude of the answer makes sense since cal is a smaller unit than kWh.

Converting between Celsius and Kelvin Temperature Scales (Section 3.9)

Solve temperature conversion problems using the problem-solving procedure that we learned in Section 2.10. Take the steps appropriate for equations.

- Write down the given and find quantities.
- Write down the appropriate equation that relates K and °C.
- Draw a solution map. Use the appropriate equation that relates the given quantities to the find quantities.
- Solve the equation for the find quantity (°C).
- Substitute the appropriate quantity into the equation and compute the answer to the correct number of significant figures.
- Check the answer.

EXAMPLE 3.16 Converting between Celsius and Kelvin Temperature Scales

Convert 257 K to Celsius.

Given: 257 K

Find: °C

Equation: K = °C + 273

Solution Map:

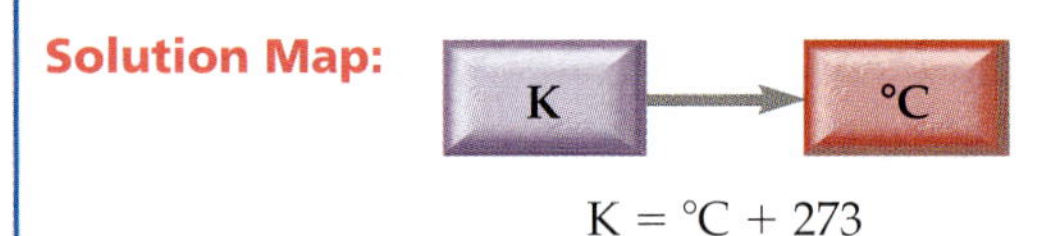

Solution:

$$K = °C + 273$$
$$°C = K - 273$$
$$°C = 257 - 273 = -16\ °C$$

The answer has the correct unit, and its magnitude seems correct (see Figure 3.14).

Converting between Fahrenheit and Celsius Temperature Scales (Section 3.9)

- Write down the given and find quantities.
- Write down the appropriate equation that relates °C and °F.
- Draw a solution map.
- Solve the equation for the find quantity (°F).
- Substitute the appropriate quantity into the equation and compute the answer to the correct number of significant figures.
- Check the answer.

EXAMPLE 3.17 Converting between Fahrenheit and Celsius Temperature Scales

Convert 62.0 °C to Fahrenheit.

Given: 62.0 °C

Find: °F

Equation: $°C = \frac{(°F - 32)}{1.8}$

Solution Map:

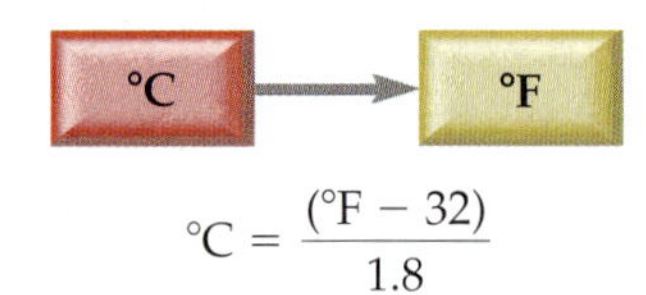

Solution:

$$°C = \frac{(°F - 32)}{1.8}$$
$$1.8\,(°C) = °F - 32$$
$$°F = 1.8\,(°C) + 32$$
$$°F = 1.8\,(62.0) + 32 = 143.60\ °F = 144\ °F$$

The answer has the correct unit, and its magnitude seems correct (see Figure 3.14).

Energy, Temperature Change, and Heat Capacity Calculations (Sections 3.10, 3.11)

LIVE EXAMPLE

EXAMPLE 3.18 Energy, Temperature Change, and Heat Capacity Calculations

What is the temperature change in 355 mL of water upon absorption of 34 kJ of heat?

- Write down the given and find quantities.

Given: 355 mL water; 34 kJ of heat

Find: ΔT

- Write down the equation that relates the given and find quantities.

Equation: $q = m \cdot C \cdot \Delta T$

- Build a solution map, writing the appropriate equation—the one showing the relationship between the given and find quantities—below the arrow.

Solution Map:

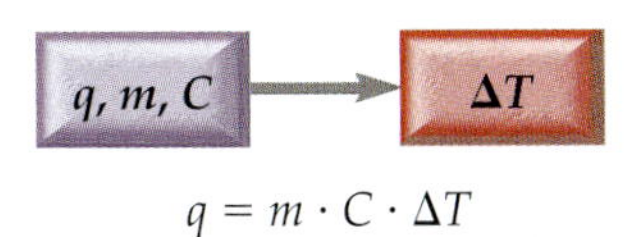

$q = m \cdot C \cdot \Delta T$

- Collect each of the quantities required in the equation in the correct units.

The value for q must be converted from kJ to J:

$$q = 34\ \text{kJ} \times \frac{1000\ \text{J}}{1\ \text{kJ}} = 3.4 \times 10^4\ \text{J}$$

The value for m must be converted from milliliters to grams; use the density of water, 1.0 g/mL, to convert milliliters to grams.

$$m = 355\ \text{mL} \times \frac{1.0\ \text{g}}{1\ \text{mL}} = 355\ \text{g}$$

$$C = 4.18\ \text{J/g °C}$$

- Solve the equation for the find quantity (ΔT).

Solution:

$$q = m \times C \times \Delta T$$

$$\Delta T = \frac{q}{mC}$$

- Substitute the appropriate quantities into the equation and compute the answer to the correct number of significant figures.

$$\Delta T = \frac{3.4 \times 10^4\ \text{J}}{355\ \text{g} \times 4.18\ \text{J/g °C}}$$

$$= 22.91\ \text{°C} = 23\ \text{°C}$$

- Check the answer.

The answer has the correct units and the magnitude seems correct. If the magnitude of the answer were a huge number—such as 3×10^6, for example—we would go back and look for a mistake. If water were to go above 100 °C, it would boil, so such a large answer would be unlikely.

Key Terms

amorphous [3.3]
atoms [3.2]
calorie (c) [3.8]
Calorie (C) [3.8]
Celsius (°C) scale [3.9]
chemical changes [3.6]
chemical energy [3.8]
chemical properties [3.5]
compound [3.4]
compressible [3.3]
crystalline [3.3]
decanting [3.6]
distillation [3.6]
electrical energy [3.8]
element [3.4]
energy [3.8]
Fahrenheit (°F) scale [3.9]
filtration [3.6]
gas [3.3]
heat capacity [3.10]
heterogeneous mixture [3.4]
homogeneous mixture [3.4]
kelvin (K) [3.9]
Kelvin (K) scale [3.9]
kilowatt-hour (kWh) [3.8]
kinetic energy [3.8]
law of conservation of energy [3.8]
liquid [3.3]
matter [3.2]
mixture [3.4]
molecules [3.2]
physical changes [3.6]
physical properties [3.5]
potential energy [3.8]
properties [3.5]
pure substance [3.4]
solid [3.3]
specific heat [3.10]
state of matter [3.3]
volatile [3.6]

Exercises

Questions

Answers to all questions numbered in blue appear in the Answers section at the back of the book.

1. Define matter and give some examples.
2. What is matter composed of?
3. What are the three states of matter?
4. What are the properties of a solid?
5. What is the difference between a crystalline solid and an amorphous solid?
6. What are the properties of a liquid?
7. What are the properties of a gas?
8. Why are gases compressible?
9. What is a mixture?
10. What is the difference between a homogeneous mixture and a heterogeneous mixture?
11. What is a pure substance?
12. What is an element? A compound?
13. What is the difference between a mixture and a compound?
14. What is a physical property? What is a chemical property?
15. What is the difference between a physical change and a chemical change?
16. What is the law of conservation of mass?
17. What is the definition of energy?
18. What is the law of conservation of energy?
19. Name some different kinds of energy.
20. What are three common units for energy?
21. What are three common units for measuring temperature?
22. How do the three temperature scales differ?
23. What is heat capacity?
24. Why are coastal geographic regions normally cooler in the summer than inland geographic regions?
25. The following equation can be used to convert Fahrenheit temperature to Celsius temperature.

$$°C = \frac{(°F - 32)}{1.8}$$

Use algebra to change the equation to convert Celsius temperature to Fahrenheit temperature.

26. The following equation can be used to convert Celsius temperature to kelvin temperature.

$$K = °C + 273$$

Use algebra to change the equation to convert kelvin temperature to Celsius temperature.

Problems

Note: The exercises in the Problems section are paired, and the answers to the odd-numbered exercises (numbered in blue) appear in the Answers section at the back of the book.

Classifying Matter

27. Classify each of the following pure substances as an element or a compound.
    (a) aluminum
    (b) sulfur
    (c) methane
    (d) acetone

28. Classify each of the following pure substances as an element or a compound.
    (a) carbon
    (b) baking soda (sodium bicarbonate)
    (c) nickel
    (d) gold

29. Classify each of the following mixtures as homogeneous or heterogeneous.
    (a) coffee
    (b) chocolate sundae
    (c) apple juice
    (d) gasoline

30. Classify each of the following mixtures as homogeneous or heterogeneous.
    (a) baby oil
    (b) chocolate chip cookie
    (c) water and gasoline
    (d) wine

31. Classify each of the following as a pure substance or a mixture. If it is a pure substance, classify it as an element or a compound. If it is a mixture, classify it as homogeneous or heterogeneous.
    (a) helium gas
    (b) clean air
    (c) rocky road ice cream
    (d) concrete

32. Classify each of the following as a pure substance or a mixture. If it is a pure substance, classify it as an element or a compound. If it is a mixture, classify it as homogeneous or heterogeneous.
    (a) urine
    (b) pure water
    (c) Snickers bar
    (d) soil

Physical and Chemical Properties and Physical and Chemical Changes

33. Classify each of the following properties as physical or chemical.
   - (a) the tendency of silver to tarnish
   - (b) the shine of chrome
   - (c) the color of gold
   - (d) the flammability of propane gas

34. Classify each of the following properties as physical or chemical.
   - (a) the boiling point of ethyl alcohol
   - (b) the temperature at which dry ice evaporates
   - (c) the flammability of ethyl alcohol
   - (d) the smell of perfume

35. The following list contains several properties of ethylene (a ripening agent for bananas). Which are physical properties and which are chemical?
   - colorless
   - odorless
   - flammable
   - gas at room temperature
   - 1 L has a mass of 1.260 g under standard conditions
   - mixes with acetone
   - polymerizes to form polyethylene

36. The following list contains several properties of ozone (a pollutant in the lower atmosphere but part of a protective shield against UV light in the upper atmosphere). Which are physical and which are chemical?
   - bluish color
   - pungent odor
   - very reactive
   - decomposes on exposure to ultraviolet light
   - gas at room temperature

37. Determine whether each of the following changes is physical or chemical.
   - (a) A balloon filled with hydrogen gas explodes upon contact with a spark.
   - (b) The liquid propane in a barbecue evaporates away because the user left the valve open.
   - (c) The liquid propane in a barbecue ignites upon contact with a spark.
   - (d) Copper metal turns green on exposure to air and water.

38. Determine whether each of the following changes is physical or chemical.
   - (a) Sugar dissolves in hot water.
   - (b) Sugar burns in a pot.
   - (c) A metal surface becomes dull because of continued abrasion.
   - (d) A metal surface becomes dull on exposure to air.

39. A block of aluminum is **(a)** ground into aluminum powder and then **(b)** ignited. It then emits flames and smoke. Classify **(a)** and **(b)** as chemical or physical changes.

40. Several pieces of graphite from a mechanical pencil are **(a)** broken into tiny pieces. Then the pile of graphite is **(b)** ignited with a hot flame. Classify **(a)** and **(b)** as chemical or physical changes.

The Conservation of Mass

41. An automobile gasoline tank holds 42 kg of gasoline. When the gasoline burns, 168 kg of oxygen are consumed and carbon dioxide and water are produced. What is the total combined mass of carbon dioxide and water that is produced?

42. In the explosion of a hydrogen-filled balloon, 0.50 g of hydrogen reacted with 4.0 g of oxygen to form how many grams of water vapor? (Water vapor is the only product.)

43. Are the following data sets on chemical changes consistent with the law of conservation of mass?
   - (a) A 7.5-g sample of hydrogen gas completely reacts with 60.0 g of oxygen gas to form 67.5 g of water.
   - (b) A 60.5-g sample of gasoline completely reacts with 243 g of oxygen to form 206 g of carbon dioxide and 88 g of water.

44. Are the following data sets on chemical changes consistent with the law of conservation of mass?
   - (a) A 12.8-g sample of sodium completely reacts with 19.6 g of chlorine to form 32.4 g of sodium chloride.
   - (b) An 8-g sample of natural gas completely reacts with 32 g of oxygen gas to form 17 g of carbon dioxide and 16 g of water.

45. In a butane lighter, 9.7 g of butane combine with 34.7 g of oxygen to form 29.3 g carbon dioxide and how many grams of water?

46. A 56-g sample of iron reacts with 24 g of oxygen to form how many grams of iron oxide?

Conversion of Energy Units

47. Perform each of the following conversions.
    (a) 32.5 J to calories
    (b) 562 cal to joules
    (c) 35.7 kJ to calories
    (d) 287 cal to kilojoules

48. Perform each of the following conversions.
    (a) 456 cal to joules
    (b) 22.3 J to Calories
    (c) 234 kJ to Calories
    (d) 12.4 Cal to joules

49. Perform each of the following conversions.
    (a) 25 kWh to joules
    (b) 249 cal to Calories
    (c) 113 cal to kilowatt-hours
    (d) 44 kJ to calories

50. Perform each of the following conversions.
    (a) 345 Cal to kilowatt-hours
    (b) 23 J to calories
    (c) 5.7×10^3 J to kilojoules
    (d) 326 kJ to joules

51. Complete the following table:

J	*cal*	*Cal*	*kWh*
225 J	____	5.38×10^{-2} Cal	____
____	8.21×10^5 cal	____	____
____	____	____	295 kWh
____	____	155 Cal	____

52. Complete the following table:

J	*cal*	*Cal*	*kWh*
7.88×10^6 J	1.88×10^6 cal	____	____
____	____	1154 Cal	____
____	88.4 cal	____	____
____	____	____	125 kWh

53. An energy bill indicates that the customer used 955 kWh in July. How many joules did the customer use?

54. A television uses 25 kWh of energy per year. How many joules does it use?

55. An adult eats food whose nutritional energy totals approximately 2.2×10^3 Cal per day. The adult burns 2.0×10^3 Cal per day. How much excess nutritional energy, in kilojoules, does the adult consume per day? If 1 lb of fat is stored by the body for each 14.6×10^3 kJ of excess nutritional energy consumed, how long will it take this person to gain 1 lb?

56. How many joules of nutritional energy are in a bag of chips whose label reads 245 Cal? If 1 lb of fat is stored by the body for each 14.6×10^3 kJ of excess nutritional energy consumed, how many bags of chips contain enough nutritional energy to result in 1 lb of body fat?

Converting between Temperature Scales

57. Perform each of the following temperature conversions.
    (a) 212 °F to Celsius (temperature of boiling water)
    (b) 77 K to Fahrenheit (temperature of liquid nitrogen)
    (c) 25 °C to kelvins (room temperature)
    (d) 98.6 °F to kelvins (body temperature)

58. Perform each of the following temperature conversions.
    (a) 102 °F to Celsius
    (b) 0 K to Fahrenheit
    (c) −48 °C to Fahrenheit
    (d) 273 K to Celsius

59. The coldest temperature ever measured in the United States was −80 °F on January 23, 1971, in Prospect Creek, Alaska. Convert that temperature to Celsius and kelvin (assume that −80 °F is accurate to two significant figures).

60. The warmest temperature ever measured in the United States was 134 °F on July 10, 1913, in Death Valley, California. Convert that temperature to Celsius and kelvin.

61. Vodka will not freeze in the freezer because it contains a high percentage of ethanol. The freezing point of pure ethanol is −114 °C. Convert that temperature to Fahrenheit and kelvin.

62. Liquid helium boils at 4.2 K. Convert this temperature to Fahrenheit and Celsius.

63. Complete the following table:

Kelvin	*Fahrenheit*	*Celsius*
0.0 K	____	−273.0 °C
____	82.5 °F	____
____	____	8.5 °C

64. Complete the following table:

Kelvin	*Fahrenheit*	*Celsius*
273.0 K	____	0.0 °C
____	−40.0 °F	____
385 K	____	____

Energy, Heat Capacity, and Temperature Changes

65. Calculate the amount of heat required to raise the temperature of a 97-g sample of water from 42 °C to 67 °C.

66. Calculate the amount of heat required to raise the temperature of a 27-g sample of water from 5 °C to 22 °C.

67. Calculate the amount of heat required to heat a 45-kg sample of ethanol from 11.0 °C to 19.0 °C.

68. Calculate the amount of heat required to heat a 3.5-kg gold bar from 21 °C to 67 °C.

69. If 89 J of heat are added to a pure gold coin with a mass of 12 g, what is its temperature change?

70. If 57 J of heat are added to an aluminum can with a mass of 17.1 g, what is its temperature change?

71. An iron nail with a mass of 12 g absorbs 15 J of heat. If the nail was initially at 28 °C, what is its final temperature?

72. A 45-kg sample of water absorbs 345 kJ of heat. If the water was initially at 22.1 °C, what is its final temperature?

73. Calculate the temperature change that occurs when 248 cal of heat are added to 24 g of water.

74. A lead fishing weight with a mass of 57 g absorbs 146 cal of heat. If its initial temperature is 47 °C, what is its final temperature?

75. An unknown metal with a mass of 28 g absorbs 58 J of heat. Its temperature rises from 31.1 °C to 39.9 °C. Calculate the heat capacity of the metal and identify it using Table 3.4.

76. An unknown metal is suspected to be gold. When 2.8 J of heat are added to 5.6 g of the metal, its temperature rises by 3.9 °C. Are these data consistent with the metal being gold?

77. When 56 J of heat are added to 11 g of a liquid, its temperature rises from 10.4 °C to 12.7 °C. What is the heat capacity of the liquid?

78. When 47.5 J of heat are added to 13.2 g of a liquid, its temperature rises by 1.72 °C. What is the heat capacity of the liquid?

79. Two identical coolers are packed for a picnic. Each cooler is packed with eighteen 12-oz soft drinks and 3 lb of ice. However, the drinks that went into cooler A were refrigerated for several hours before they were packed in the cooler, while the drinks that went into cooler B were packed at room temperature. When the two coolers are opened three hours later, most of the ice in cooler A is still ice, while nearly all of the ice in cooler B has melted. Explain this difference.

80. A 100-g block of iron metal and 100 g of water are each warmed to 75 °C and placed into two identical insulated containers. Two hours later, the two containers are opened and the temperature of each substance is measured. The iron metal has cooled to 38 °C while the water has cooled only to 69 °C. Explain this difference.

Cumulative Problems

81. Calculate the final temperature of 245 mL of water initially at 32 °C upon absorption of 17 kJ of heat.

82. Calculate the final temperature of 32 mL of ethanol initially at 11 °C upon absorption of 562 J of heat. (density of ethanol $= 0.789$ g/mL)

83. A pure gold ring with a volume of 1.57 cm^3 is initially at 11.4 °C. When it is put on, it warms to 29.5 °C. How much heat did the ring absorb? (density of gold $= 19.3$ g/cm^3)

84. A block of aluminum with a volume of 98.5 cm^3 absorbs 67.4 J of heat. If its initial temperature was 32.5 °C, what is its final temperature? (density of aluminum $= 2.70$ g/cm^3)

85. How much heat in kilojoules is required to heat 56 L of water from 85 °F to 212 °F?

86. How much heat in joules is required to heat a 43-g sample of aluminum from 72 °F to 145 °F?

87. What is the temperature change in Celsius when 29.5 L of water absorbs 2.3 kWh of heat?

88. If 1.45 L of water is initially at 25.0 °C, what will its temperature be after absorption of 9.4×10^{-2} kWh of heat?

89. A water heater contains 55 gal of water. How many kilowatt-hours of energy are necessary to heat the water in the water heater by 25 °C?

90. A room contains 48 kg of air. How many kilowatt-hours of energy are necessary to heat the air in the house from 7 °C to 28 °C? The heat capacity of air is 1.03 J/g °C.

91. A backpacker wants to carry enough fuel to heat 2.5 kg of water from 25 °C to 100.0 °C. If the fuel he carries produces 36 kJ of heat per gram, how much fuel should he carry? (For the sake of simplicity, assume that the transfer of heat is 100% efficient.)

92. A cook wants to heat 1.35 kg of water from 32.0 °C to 100.0 °C. If he uses natural gas to heat the water, how much natural gas will he need to burn? Natural gas produces 49.3 kJ of heat per gram. (For the sake of simplicity, assume that the transfer of heat is 100% efficient.)

93. Evaporating sweat cools the body because evaporation absorbs 2.44 kJ per gram of water evaporated. Estimate the mass of water that must evaporate from the skin to cool a body by 0.50 °C, if the mass of the body is 95 kg and its heat capacity is 4.0 J/g °C. (Assume that the heat transfer is 100% efficient.)

94. When ice melts, it absorbs 0.33 kJ per gram. How much ice is required to cool a 12.0-oz drink from 75 °F to 35 °F, if the heat capacity of the drink is 4.18 J/g °C? (Assume that the heat transfer is 100% efficient.)

95. A 15.7-g aluminum block is warmed to 53.2 °C and plunged into an insulated beaker containing 32.5 g of water initially at 24.5 °C. The aluminum and the water are allowed to come to thermal equilibrium. Assuming that no heat is lost, what is the final temperature of the water and aluminum?

96. 25.0 mL of ethanol (density = 0.789 g/mL) initially at 7.0 °C is mixed with 35.0 mL of water (density = 1.0 g/mL) initially at 25.3 °C in an insulated beaker. Assuming that no heat is lost, what is the final temperature of the mixture?

Highlight Problems

97. Classify each of the following molecular pictures as a pure substance or a mixture.

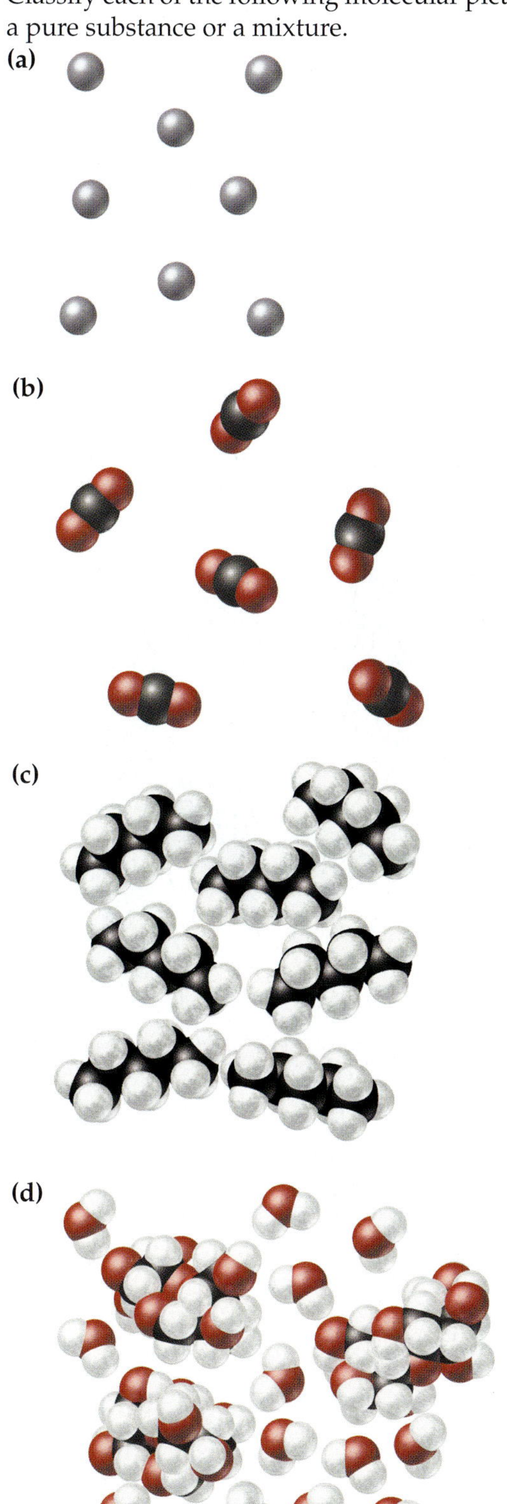

98. Classify each of the following molecular pictures as a pure substance or a mixture. If it is a pure substance, classify it as an element or a compound. If it is a mixture, classify it as homogeneous or heterogeneous.

(a)

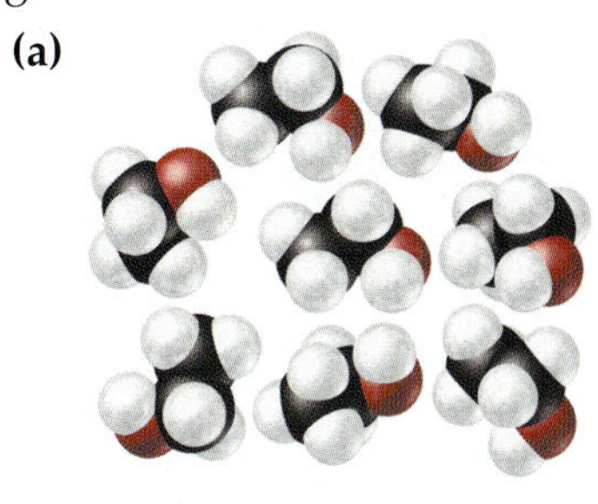

(b)

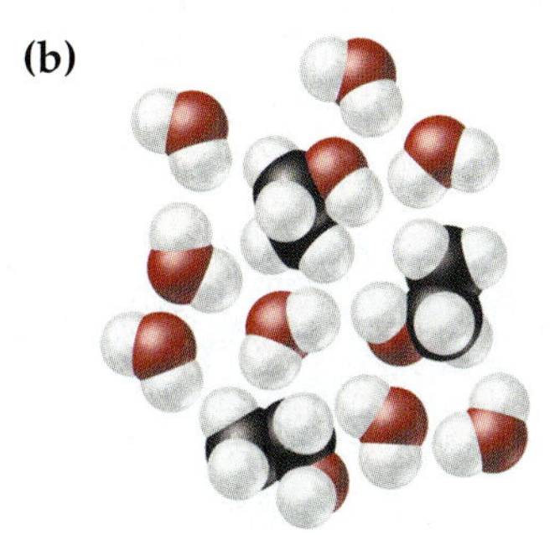

(c)

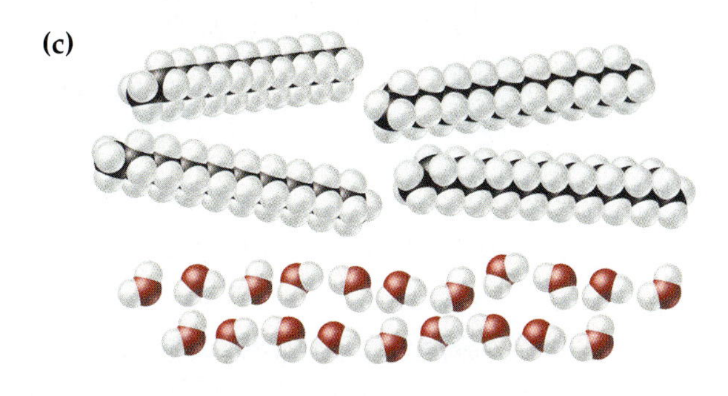

(d)

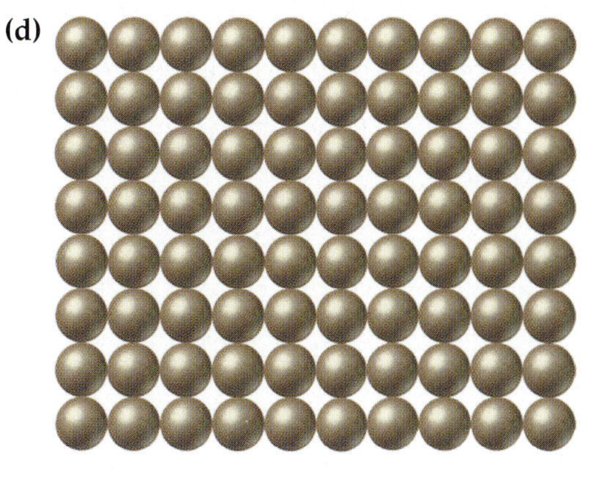

99. The following molecular drawing shows images of acetone molecules before and after a change. Was the change chemical or physical?

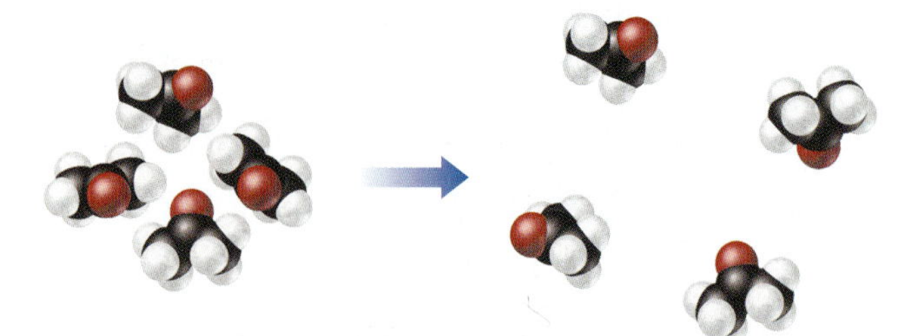

100. The following molecular drawing shows images of methane molecules and oxygen molecules before and after a change. Was the change chemical or physical?

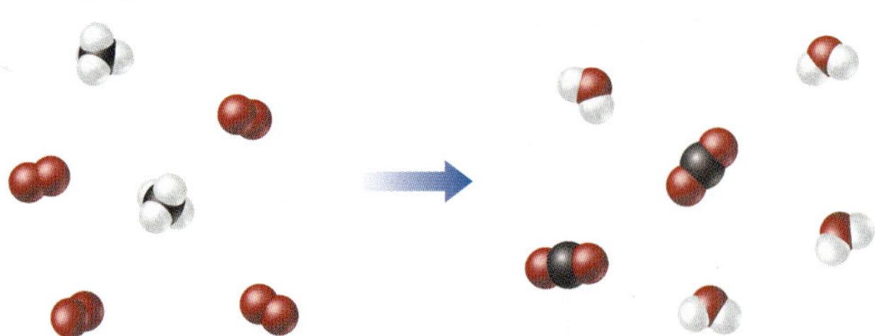

101. A major event affecting global climate is the El Niño/La Niña cycle. In this cycle, equatorial Pacific Ocean waters warm by several degrees Celsius above normal (El Niño) and then cool by several degrees Celsius below normal (La Niña). This cycle affects weather not only in North and South America, but as far away as Africa. Why does a seemingly small change in ocean temperature have such a large impact on weather?

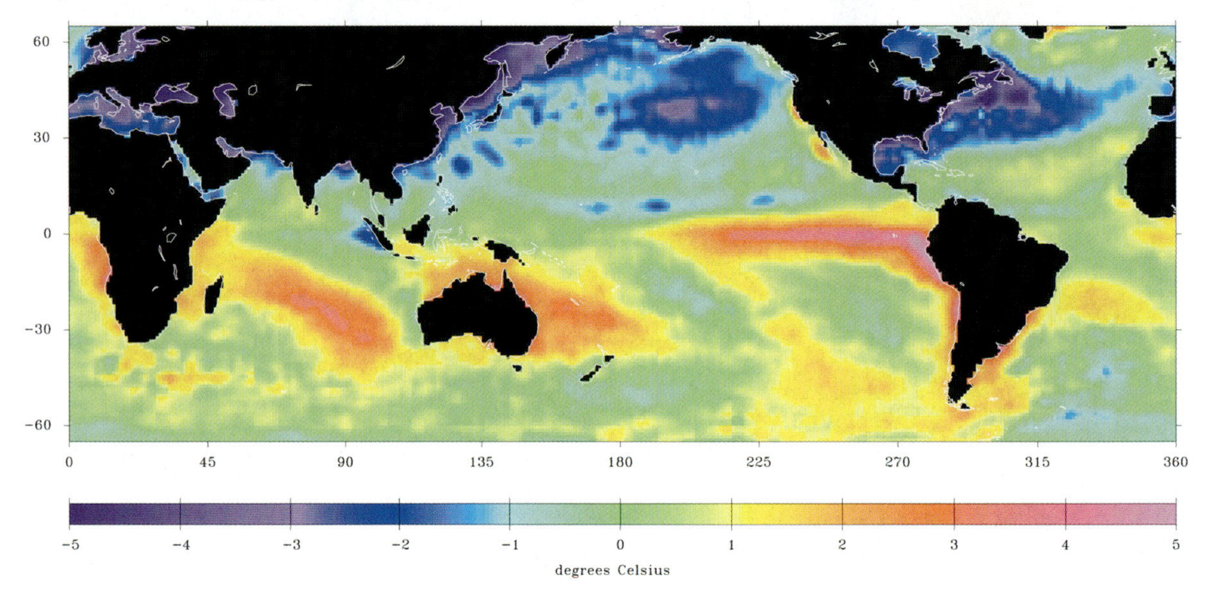

▲ Temperature anomaly plot of the world's oceans for December 23, 1997. The red section off the western coast of South America is the El Niño effect, a warming of the Pacific Ocean along the equator.

102. Global warming refers to the rise in average global temperature due to the increased concentration of certain gases, called greenhouse gases, in our atmosphere. The earth's oceans, because of their high heat capacity, can absorb heat and therefore act to slow down global warming. How much heat would be required to warm the earth's oceans by 1.0 °C? Assume that the volume of earth's oceans is 137×10^7 km^3 and that the density of seawater is 1.03 g/cm^3. Also assume that the heat capacity of seawater is the same as that of water.

▲ The earth's oceans moderate temperatures by absorbing heat during warm periods.

103. Examine the following data for the maximum and minimum average temperatures of San Francisco and Sacramento in the summer and in the winter.

San Francisco (Coastal City)

January		*August*	
High	*Low*	*High*	*Low*
57.4 °F	43.8 °F	64.4 °F	54.5 °F

Sacramento (Inland City)

January		*August*	
High	*Low*	*High*	*Low*
53.2 °F	37.7 °F	91.5 °F	57.7 °F

(a) Notice the difference between the August high in San Francisco and Sacramento. Why is it much hotter in the summer in Sacramento?

(b) Notice the difference between the January low in San Francisco and Sacramento. How might the heat capacity of the ocean contribute to this difference?

Answers to Skillbuilder Exercises

Skillbuilder 3.1 (a) pure substance, element (b) mixture, homogeneous (c) mixture, heterogeneous (d) pure substance, compound

Skillbuilder 3.2 (a) chemical (b) physical (c) physical (d) physical

Skillbuilder 3.3 (a) chemical (b) physical (c) physical (d) chemical

Skillbuilder 3.4 27 g

Skillbuilder 3.5 2.14 kJ

Skillbuilder Plus, p. 63 6.57×10^6 cal

Skillbuilder 3.6 85 °C

Skillbuilder 3.7 282 °F

Skillbuilder 3.8 77 K

Skillbuilder 3.9 50.1 J

Skillbuilder Plus, p. 70 7.4 g

Skillbuilder 3.10 $\Delta T = 4.21$ °C; $T_f = 29.2$ °C

Answers to Conceptual Checkpoints

3.1 (d) kWh is the largest of the four units listed, so the numerical value of the yearly energy consumption would be lowest if expressed in kWh.

3.2 (d) You can confirm this by substituting each of the Fahrenheit temperatures into the equation in Section 3.9 and solving for the Celsius temperature.

3.3 (a) Because copper has the lowest specific heat capacity of the three metals, it will experience the greatest temperature change for a given energy input.

CHAPTER 4

Atoms and Elements

"Nothing exists except atoms and empty space; everything else is opinion."

Democritus (460–370 B.C.)

4.1 Experiencing Atoms at Tiburon

My wife and I recently enjoyed a visit to the northern California seaside town of Tiburon. Tiburon sits next to San Francisco Bay with views of the water, the city of San Francisco, and the surrounding mountains. As we walked along the waterside path, I felt the breeze as it blew over the bay. I could hear the water lapping on the shore, and I could smell the sea air. What was the cause of these sensations? The answer is simple—atoms.

As we learned in Chapter 3, many atoms exist not as free particles but as groups of atoms bound together to form molecules. Nevertheless, all matter is ultimately made of atoms.

Since all matter is made of atoms, it is atoms that we experience in our sensations. The atom is the fundamental building block of everything we hear, feel, see, and experience. When we feel the wind on our skin, we are feeling atoms. When we hear the lapping water, we are hearing atoms. When we touch a shoreside rock, we are touching atoms, and when we smell the sea air, we are smelling atoms. We eat atoms, we breathe atoms, and we excrete atoms. Atoms are the building blocks of matter; they are the basic units from which nature builds. They are all around us and compose everything, including our own bodies.

The exact number of naturally occurring elements is controversial because some elements previously considered only synthetic may actually occur in nature in very small quantities.

Atoms are incredibly small. A single pebble from the shoreline contains more atoms than we could ever count. The number of atoms in a single pebble far exceeds the number of pebbles on the bottom of San Francisco Bay. To get an idea of how small atoms are, imagine that every atom within a small pebble were the size of the pebble itself; then the pebble would be larger than Mt. Everest (▶ Figure 4.1) Atoms are small—yet they compose everything.

◀ All matter is composed of atoms. Seaside rocks are often composed of silicates, compounds of silicon and oxygen atoms. Seaside air, like all air, contains nitrogen and oxygen molecules, but it may also contain substances called amines. The amine shown here is triethylamine, which is emitted by decaying fish. Triethylamine is one of the compounds responsible for the fishy smell of the seaside.

The key to connecting the microscopic world with the macroscopic world is the atom. Atoms compose matter; their properties determine matter's properties. An *atom* is the smallest identifiable unit of an element. There are about ninety-one different elements in nature, and consequently about ninety-one different kinds of atoms. In addition, scientists have succeeded in making about twenty synthetic elements (not found in nature). In this chapter, we learn about atoms: what they are made of, how they differ from one another, and how they are structured. We also learn about the elements that atoms compose and some of the properties of those elements.

▲ **Figure 4.1 The size of the atom** If every atom within a pebble were the size of the pebble itself, then the pebble would be larger than Mt. Everest.

4.2 Indivisible: The Atomic Theory

▲ Diogenes and Democritus, as imagined by a medieval artist. Democritus is the first person on record to have postulated that matter was composed of atoms.

If we simply examine matter, even under a microscope, it is not obvious that matter is composed of tiny particles. In fact, it appears to be just the opposite. If we divide a sample of matter into smaller and smaller pieces, it seems that we could divide it forever. From our perspective, matter seems continuous. The first person recorded as thinking otherwise was Democritus (460–370 B.C.), a Greek philosopher who theorized that matter was ultimately composed of small, indivisible particles he called *atomos,* or "atoms," meaning "indivisible." Democritus suggested that if you divided matter into smaller and smaller pieces, you would eventually end up with tiny, indestructible particles—atoms.

Democritus' ideas were not widely accepted, and it was not until the early nineteenth century that John Dalton formalized a theory of atoms that gained broad acceptance. Dalton's atomic theory has three parts:

1. Each element is composed of tiny indestructible particles called atoms.
2. All atoms of a given element have the same mass and other properties that distinguish them from the atoms of other elements.
3. Atoms combine in simple, whole-number ratios to form compounds.

Today, the evidence for the atomic theory is overwhelming. Recent advances in microscopy have allowed scientists not only to image individual atoms but also to pick them up and move them (▼ Figure 4.2). Matter is indeed composed of atoms.

▶ **Figure 4.2 Writing with atoms** Scientists at IBM used a special microscope, called a scanning tunneling microscope (STM), to move xenon atoms to form the letters *IBM*. The cone shape of these atoms is due to the peculiarities of the instrumentation. Atoms are, in general, spherical in shape.

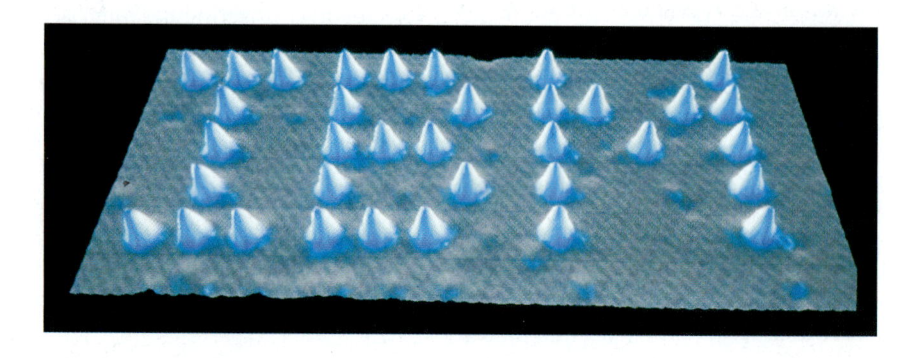

Atoms and Humans

All matter is composed of atoms. What does that mean? What does it imply? It means that everything before you is composed of tiny particles too small to see. It means that even you and I are composed of these same particles. We acquired those particles from the food we have eaten over the years. The average carbon atom in our own bodies has been used by twenty other living organisms before we got it and will be used by other organisms after we die.

The idea that all matter is composed of atoms has far-reaching implications. It implies that our bodies, our hearts, and even our brains are composed of atoms acting according to the laws of chemistry and physics. Some have viewed this as a devaluation of human life. We have always wanted to distinguish ourselves from everything else, and the idea that we are made of the same basic particles as all other matter takes something away from that distinction . . . or does it?

CAN YOU ANSWER THIS? *Do you find the idea that you are made of atoms disturbing? Why or why not?*

4.3 The Nuclear Atom

By the end of the nineteenth century, scientists were convinced that matter was composed of atoms, the permanent, indestructible building blocks from which all substances are constructed. However, an English physicist named J. J. Thomson (1856–1940) complicated the picture by discovering an even smaller and more fundamental particle called the **electron**. Thomson discovered that electrons are negatively charged, that they are much smaller and lighter than atoms, and that they are uniformly present in many different kinds of substances. The indestructible building block called the atom could apparently be "chipped."

Electric charge is more fully defined in Section 4.4. For now, think of it as an inherent property of electrons that causes them to interact with other charged particles.

The discovery of negatively charged particles within atoms raised the question of a balancing positive charge. Atoms were known to be charge-neutral, so they must contain positive charge that balanced the negative charge of electrons. But how did the positive and negative charges within the atom fit together? Were atoms just a jumble of even more fundamental particles? Were they solid spheres, or did they have some internal structure? Thomson proposed that the negatively charged electrons were small particles held within a positively charged sphere. This model, the most popular of the time, became known as the plum pudding model (plum pudding is an English dessert) (◀ Figure 4.3). The picture suggested by Thomson was—to those of us not familiar with plum pudding—like a blueberry muffin, where the blueberries are the electrons and the muffin is the positively charged sphere.

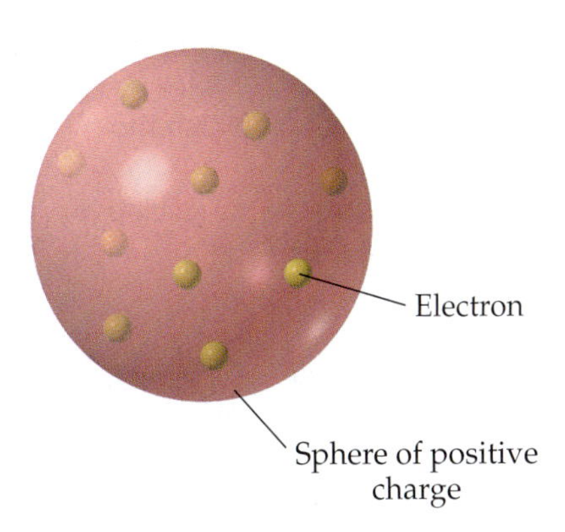

▲ **Figure 4.3 Plum pudding model of the atom** In the model suggested by J. J. Thomson, negatively charged electrons (yellow) were held in a sphere of positive charge (red).

In 1909, Ernest Rutherford (1871–1937), who had worked under Thomson and adhered to his plum pudding model, performed an experiment in an attempt to confirm it. His experiment proved it wrong instead. In his experiment, Rutherford directed tiny, positively charged particles—called alpha-particles—at an ultrathin sheet of gold foil (▶ Figure 4.4). These particles were to act as probes of the gold atoms' structure. If the gold atoms were indeed like blueberry muffins or plum pudding—with their mass and charge spread throughout the entire volume of the atom—these speeding probes should pass right through the gold foil with minimum deflection. Rutherford performed the experiment, but the results were not as he expected. A majority of the particles did pass directly through the foil, but some particles were deflected, and some (1 in 20,000) even bounced back. The results puzzled Rutherford, who found them "about as credible as if you had fired a 15-inch shell at a piece of tissue paper and it came back and hit you." What must the structure of the atom be in order to explain this odd behavior?

TUTORIAL
Rutherford's Experiment: Blueberry Muffins and Plum Pudding

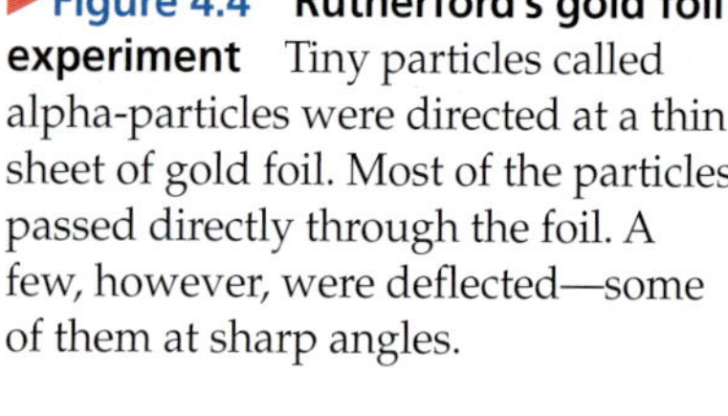

▶ **Figure 4.4 Rutherford's gold foil experiment** Tiny particles called alpha-particles were directed at a thin sheet of gold foil. Most of the particles passed directly through the foil. A few, however, were deflected—some of them at sharp angles.

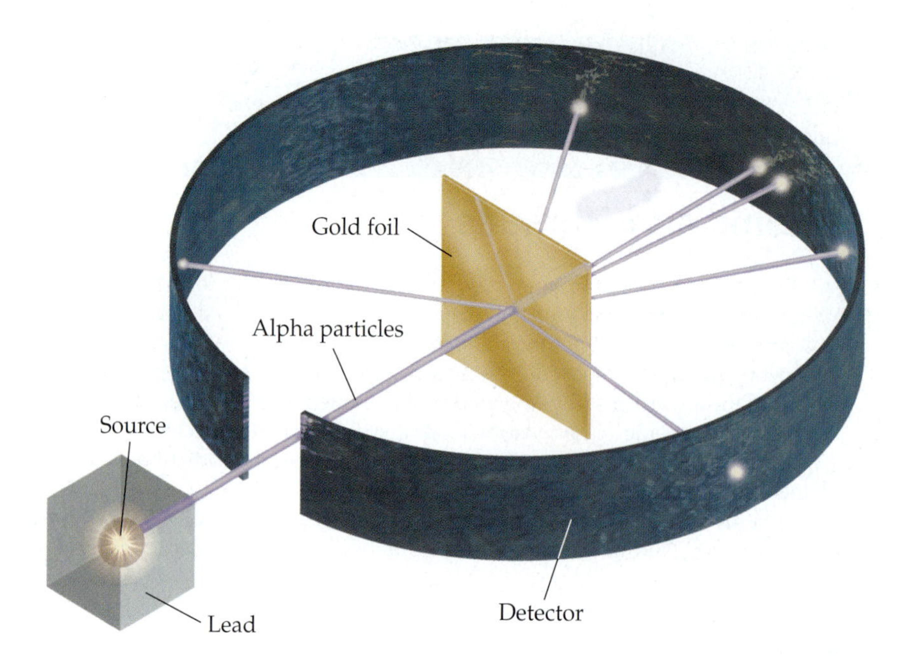

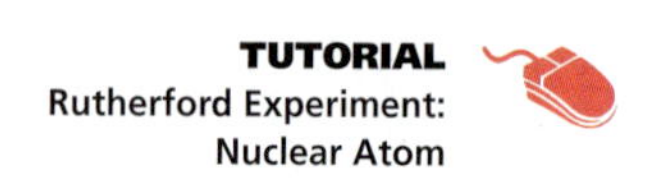

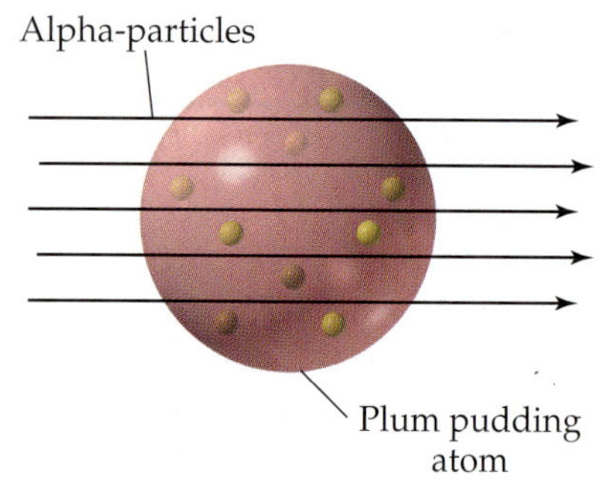

(a) Rutherford's expected result

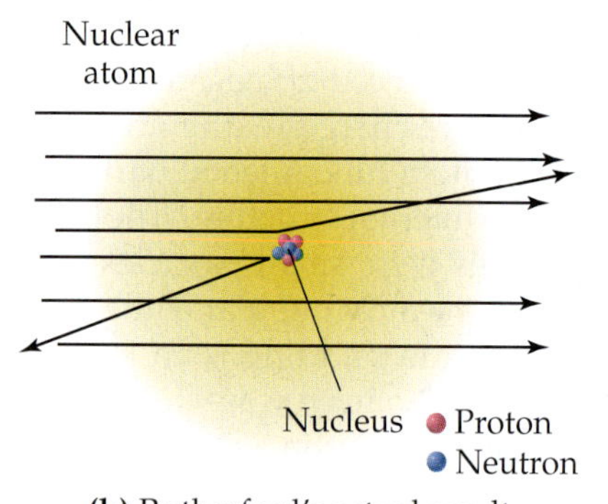

(b) Rutherford's actual result

▲ **Figure 4.5 Discovery of the atomic nucleus** **(a)** Expected result of Rutherford's gold foil experiment. If the plum pudding model were correct, the alpha-particles would pass right through the gold foil with minimal deflection. **(b)** Actual result of Rutherford's gold foil experiment. A small number of alpha-particles were deflected or bounced back. The only way to explain the deflections was to suggest that most of the mass and all of the positive charge of an atom must be concentrated in a space much smaller than the size of the atom itself.

Rutherford created a new model to explain his results (◀ Figure 4.5). He concluded that matter must not be as uniform as it appears. It must contain large regions of empty space dotted with small regions of very dense matter. In order to explain the deflections he observed, the mass and positive charge of an atom must all be concentrated in a space much smaller than the size of the atom itself. Using this idea, he proposed the **nuclear theory of the atom**, which has three basic parts:

1. Most of the atom's mass and all of its positive charge are contained in a small core called the *nucleus*.
2. Most of the volume of the atom is empty space through which the tiny, negatively charged electrons are dispersed.
3. There are as many negatively charged electrons outside the nucleus as there are positively charged particles (*protons*) inside the nucleus, so that the atom is electrically neutral.

Later work by Rutherford and others demonstrated that the atom's **nucleus** contains both positively charged **protons** and neutral particles called **neutrons**. The dense nucleus makes up more than 99.9% of the mass of the atom, but occupies only a small fraction of its volume. The electrons are distributed through a much larger region, but don't have much mass (▶ Figure 4.6). For now, you can think of these electrons like the water droplets that make up a cloud—they are dispersed throughout a large volume, but weigh almost nothing.

Rutherford's nuclear theory was a success and is still valid today. The revolutionary part of this theory is the idea that matter—at its core—is much less uniform than it appears. If the nucleus of the atom were the size of this dot ·, the average electron would be about 10 m away. Yet the dot would contain almost the entire mass of the atom. Imagine what matter would be like if atomic structure broke down. What if matter were composed of atomic nuclei piled on top of each other like marbles? Such matter would be incredibly dense; a single grain of sand composed of solid atomic nuclei would have a mass of 5 million kg (or a weight of about 10 million lb). Astronomers believe there are some places in the universe where such matter exists: neutron stars and black holes.

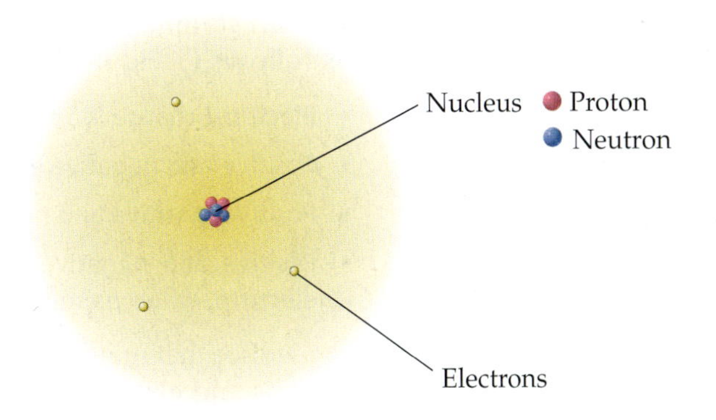

▶ **Figure 4.6 The nuclear atom** In this model, 99.9% of the atom's mass is concentrated in a small, dense nucleus that contains protons and neutrons. The rest of the volume of the atom is mostly empty space occupied by negatively charged electrons. The number of electrons outside the nucleus is equal to the number of protons inside the nucleus. In this image, the nucleus is greatly enlarged and the electrons are portrayed as particles.

4.4 The Properties of Protons, Neutrons, and Electrons

▲ If a proton had the mass of a baseball, an electron would have the mass of a rice grain. The proton is nearly 2000 times as massive as an electron.

Protons and neutrons have very similar masses. In SI units, the mass of the proton is 1.67262×10^{-27} kg, and the mass of the neutron is a close 1.67493×10^{-27} kg. A more common unit to express these masses, however, is the **atomic mass unit (amu)**, defined as one-twelfth of the mass of a carbon atom containing six protons and six neutrons. In this unit, a proton has a mass of 1.0073 amu and a neutron has a mass of 1.0087 amu. Electrons, by contrast, have an almost negligible mass of 0.00091×10^{-27} kg, or approximately 0.00055 amu.

The proton and the electron both have electrical **charge**. The proton's charge is +1 and the electron's charge is −1. The charge of the proton and the electron are equal in magnitude but opposite in sign, so that when the two particles are paired the charges exactly cancel. The neutron has no charge.

What is electrical charge? Electrical charge is a fundamental property of protons and electrons, just as mass is a fundamental property of matter. Most matter is charge-neutral because protons and electrons occur together and their charges cancel. However, you have probably experienced excess electrical charge when brushing your hair on a dry day. The brushing action results in the accumulation of electrical charge on the hair strands, which then repel each other, causing your hair to stand on end.

EVERYDAY Chemistry

Solid Matter?

If matter really is mostly empty space as Rutherford suggested, then why does it appear so solid? Why can I tap my knuckles on the table and feel a solid thump? Matter appears solid because the variation in the density is on such a small scale that our eyes can't see it. Imagine a jungle gym one hundred stories high and the size of a football field. It is mostly empty space. Yet if you viewed it from an airplane, it would appear as a solid mass. Matter is similar. When you tap your knuckle on the table, it is much like one giant jungle gym (your finger) crashing into another (the table). Even though they are both primarily empty space, one does not fall into the other.

CAN YOU ANSWER THIS? *Use the jungle gym analogy to explain why most of Rutherford's alpha-particles went right through the gold foil and why a few bounced back. Remember that his gold foil was extremely thin.*

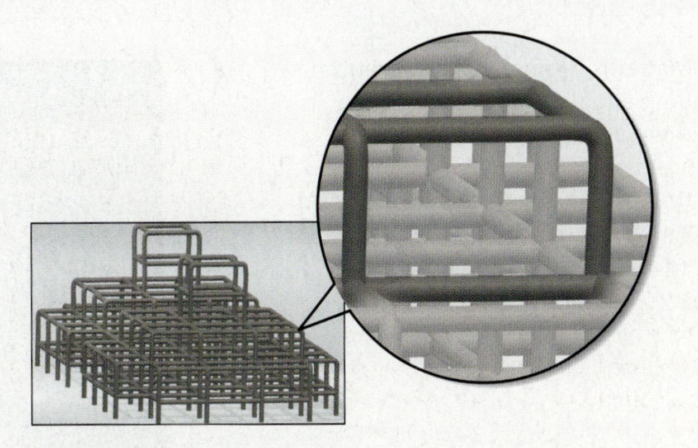

▲ Matter appears solid and uniform because the variation in density is on a scale too small for our eyes to see. Just as this scaffolding appears solid at a distance, so matter appears solid to us.

Positive (red) and negative (yellow) charges attract.

Positive–positive and negative–negative charges repel.

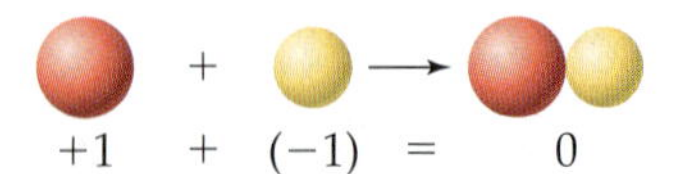

+1 + (−1) = 0

Positive and negative charges cancel.

▲ **Figure 4.7 The properties of electrical charge**

We can summarize the nature of electrical charge as follows (◀ Figure 4.7):

- Electrical charge is a fundamental property of protons and electrons.
- Positive and negative electrical charges attract each other.
- Positive–positive and negative–negative charges repel each other.
- Positive and negative charges cancel each other so that a proton and an electron, when paired, are charge-neutral.

Notice that matter is usually charge-neutral due to the canceling effect of protons and electrons. When matter does acquire charge imbalances, these imbalances usually equalize quickly, often in dramatic ways. For example, the shock you receive when touching a doorknob during dry weather is the equalization of a charge imbalance that developed as you walked across the carpet. Lightning is an equalization of charge imbalances that develop during electrical storms.

If you had a sample of matter—even a tiny sample, such as a sand grain—that was composed of only protons or only electrons, the forces around that matter would be extraordinary, and the matter would be unstable. Luckily, that is not the way matter is—protons and electrons exist together, canceling each other's charge and making matter charge-neutral. The properties of protons, neutrons, and electrons are summarized in Table 4.1.

TABLE 4.1 Subatomic Particles

	Mass (kg)	Mass (amu)	Charge
proton	1.67262×10^{-27}	1.0073	+1
neutron	1.67493×10^{-27}	1.0087	0
electron	0.00091×10^{-27}	0.00055	−1

CONCEPTUAL CHECKPOINT 4.1

An atom with which of these compositions would have a mass of approximately 12 amu and be charge-neutral?

(a) 6 protons and 6 electrons

(b) 3 protons, 3 neutrons, and 6 electrons

(c) 6 protons, 6 neutrons, and 6 electrons

(d) 12 neutrons and 12 electrons

▶ Matter is normally charge-neutral, having equal numbers of positive and negative charges that exactly cancel. When the charge balance of matter is disturbed, as in an electrical storm, it quickly rebalances, often in dramatic ways such as lightning.

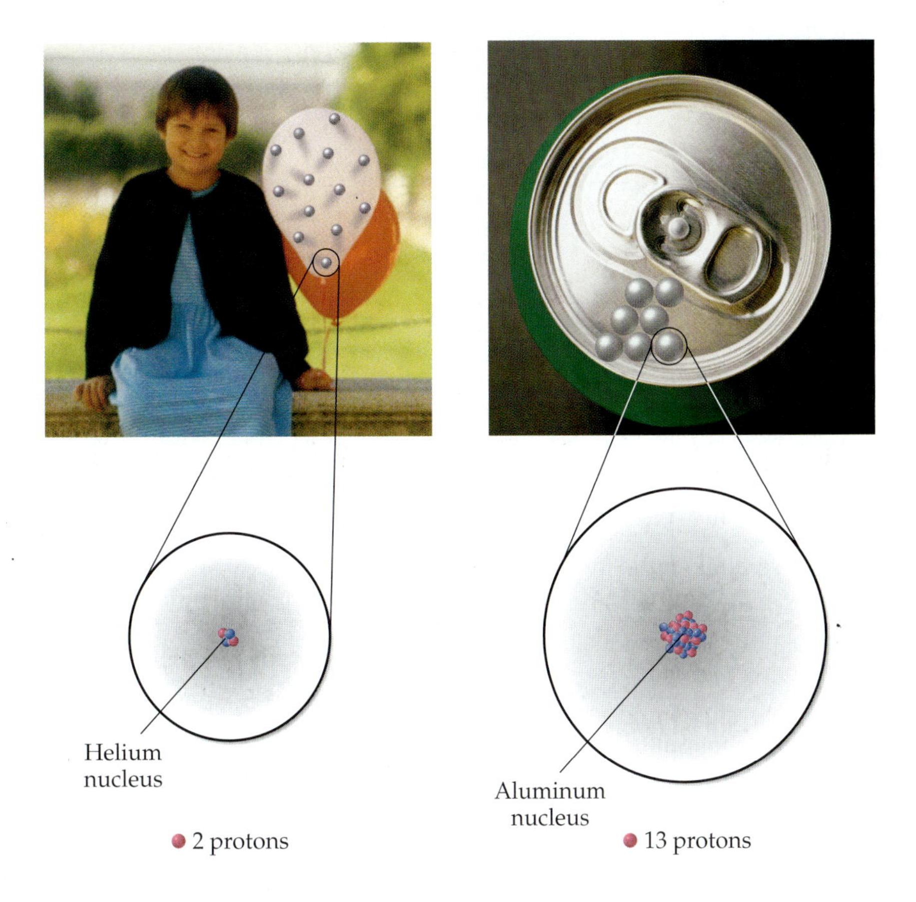

▶ **Figure 4.8 The number of protons in the nucleus defines the element**

4.5 Elements: Defined by Their Numbers of Protons

We have seen that atoms are composed of protons, neutrons, and electrons. However, it is the number of protons in the nucleus of an atom that identifies it as a particular element. For example, atoms with 2 protons in their nucleus are helium atoms, atoms with 13 protons in their nucleus are aluminum atoms, and atoms with 92 protons in their nucleus are uranium atoms. The number of protons in an atom's nucleus defines the element (▲ Figure 4.8). Every aluminum atom has 13 protons in its nucleus—if it had a different number of protons, it would be a different element. The number of protons in the nucleus of an atom is called the **atomic number** and is given the symbol **Z**.

The periodic table of the elements (▶ Figure 4.9) lists all known elements according to their atomic numbers. Each element is represented by a unique **chemical symbol**, a one- or two-letter abbreviation for the element that appears directly below its atomic number on the periodic table. The chemical symbol for helium is He; for aluminum, Al; and for uranium, U. The chemical symbol and the atomic number always go together. If the atomic number is 13, the chemical symbol must be Al. If the atomic number is 92, the chemical symbol must be U. This is just another way of saying that the number of protons defines the element.

Most chemical symbols are based on the English name of the element. For example, the symbol for carbon is C; for silicon, Si; and for bromine, Br. Some elements, however, have symbols based on their Latin names. For example, the symbol for potassium is K, from the Latin *kalium*, and the symbol for sodium is Na, from the Latin *natrium*. Other elements with symbols based on their Greek or Latin names include the following:

lead	Pb	*plumbum*
mercury	Hg	*hydrargyrum*
iron	Fe	*ferrum*
silver	Ag	*argentum*
tin	Sn	*stannum*
copper	Cu	*cuprum*

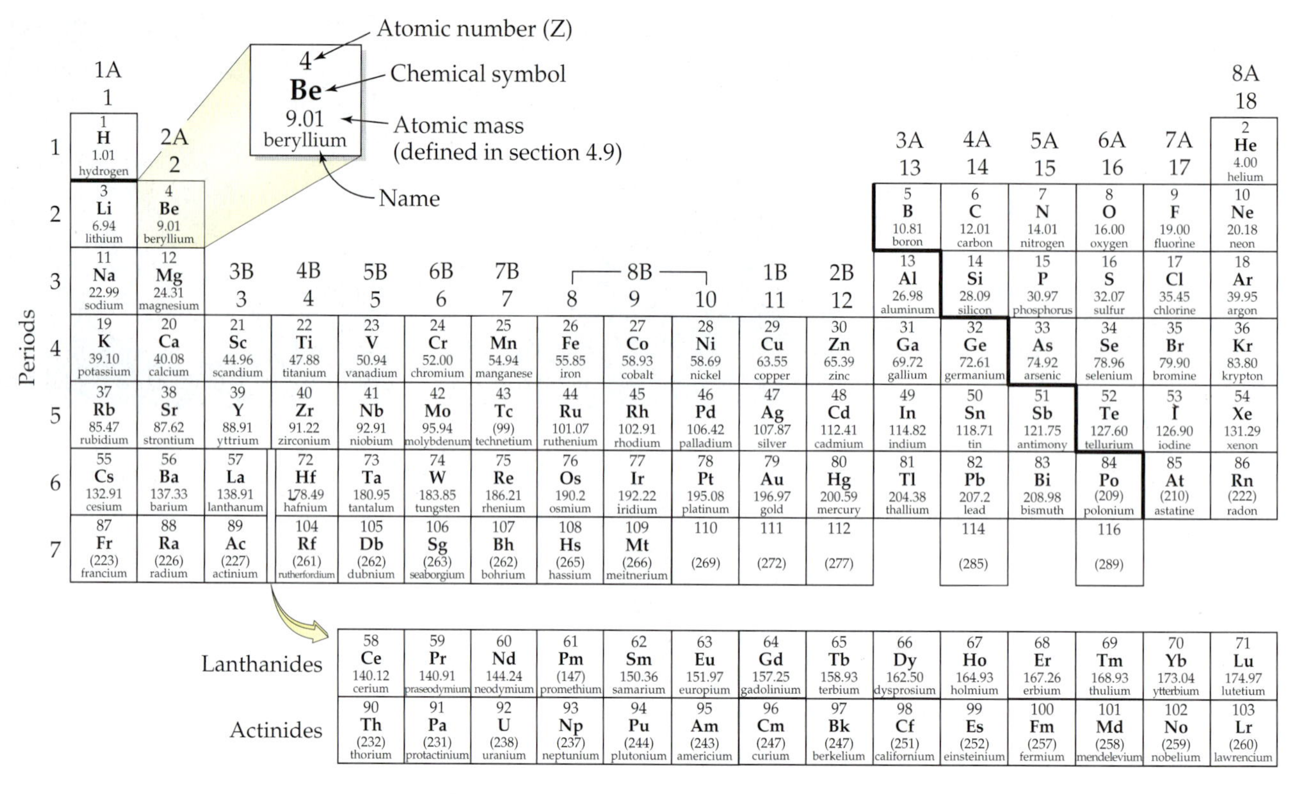

Period	1A 1	2A 2	3B 3	4B 4	5B 5	6B 6	7B 7	8	9	10	1B 11	2B 12	3A 13	4A 14	5A 15	6A 16	7A 17	8A 18
1	1 H 1.01 hydrogen																	2 He 4.00 helium
2	3 Li 6.94 lithium	4 Be 9.01 beryllium											5 B 10.81 boron	6 C 12.01 carbon	7 N 14.01 nitrogen	8 O 16.00 oxygen	9 F 19.00 fluorine	10 Ne 20.18 neon
3	11 Na 22.99 sodium	12 Mg 24.31 magnesium											13 Al 26.98 aluminum	14 Si 28.09 silicon	15 P 30.97 phosphorus	16 S 32.07 sulfur	17 Cl 35.45 chlorine	18 Ar 39.95 argon
4	19 K 39.10 potassium	20 Ca 40.08 calcium	21 Sc 44.96 scandium	22 Ti 47.88 titanium	23 V 50.94 vanadium	24 Cr 52.00 chromium	25 Mn 54.94 manganese	26 Fe 55.85 iron	27 Co 58.93 cobalt	28 Ni 58.69 nickel	29 Cu 63.55 copper	30 Zn 65.39 zinc	31 Ga 69.72 gallium	32 Ge 72.61 germanium	33 As 74.92 arsenic	34 Se 78.96 selenium	35 Br 79.90 bromine	36 Kr 83.80 krypton
5	37 Rb 85.47 rubidium	38 Sr 87.62 strontium	39 Y 88.91 yttrium	40 Zr 91.22 zirconium	41 Nb 92.91 niobium	42 Mo 95.94 molybdenum	43 Tc (99) technetium	44 Ru 101.07 ruthenium	45 Rh 102.91 rhodium	46 Pd 106.42 palladium	47 Ag 107.87 silver	48 Cd 112.41 cadmium	49 In 114.82 indium	50 Sn 118.71 tin	51 Sb 121.75 antimony	52 Te 127.60 tellurium	53 I 126.90 iodine	54 Xe 131.29 xenon
6	55 Cs 132.91 cesium	56 Ba 137.33 barium	57 La 138.91 lanthanum	72 Hf 178.49 hafnium	73 Ta 180.95 tantalum	74 W 183.85 tungsten	75 Re 186.21 rhenium	76 Os 190.2 osmium	77 Ir 192.22 iridium	78 Pt 195.08 platinum	79 Au 196.97 gold	80 Hg 200.59 mercury	81 Tl 204.38 thallium	82 Pb 207.2 lead	83 Bi 208.98 bismuth	84 Po (209) polonium	85 At (210) astatine	86 Rn (222) radon
7	87 Fr (223) francium	88 Ra (226) radium	89 Ac (227) actinium	104 Rf (261) rutherfordium	105 Db (262) dubnium	106 Sg (263) seaborgium	107 Bh (262) bohrium	108 Hs (265) hassium	109 Mt (266) meitnerium	110 (269)	111 (272)	112 (277)		114 (285)		116 (289)		

Lanthanides	58 Ce 140.12 cerium	59 Pr 140.91 praseodymium	60 Nd 144.24 neodymium	61 Pm (147) promethium	62 Sm 150.36 samarium	63 Eu 151.97 europium	64 Gd 157.25 gadolinium	65 Tb 158.93 terbium	66 Dy 162.50 dysprosium	67 Ho 164.93 holmium	68 Er 167.26 erbium	69 Tm 168.93 thulium	70 Yb 173.04 ytterbium	71 Lu 174.97 lutetium
Actinides	90 Th (232) thorium	91 Pa (231) protactinium	92 U (238) uranium	93 Np (237) neptunium	94 Pu (244) plutonium	95 Am (243) americium	96 Cm (247) curium	97 Bk (247) berkelium	98 Cf (251) californium	99 Es (252) einsteinium	100 Fm (257) fermium	101 Md (258) mendelevium	102 No (259) nobelium	103 Lr (260) lawrencium

▲ **Figure 4.9 The periodic table of the elements**

The names of elements were often given to describe their properties. For example, *argon* originates from the Greek word *argos,* meaning "inactive," referring to argon's chemical inertness (it does not react with other elements). *Bromine* originates from the Greek word *bromos,* meaning "stench," referring to bromine's strong odor. Other elements were named after countries. For example, polonium was named after Poland, francium was named after France, and americium was named after the United States of America. Still other elements were named after scientists. Curium was named after Marie Curie, and mendelevium was named after Dmitri Mendeleev. Every element's name, symbol, and atomic number are listed in the periodic table (inside front cover) and in an alphabetical listing (inside back cover) in this book.

Curium
96
Cm
(247)

▶ The name *bromine* originates from the Greek word *bromos,* meaning "stench." Bromine vapor, seen as the red-brown gas in this photograph, has a strong odor.

◀ Curium is named after Marie Curie, a chemist who helped discover radioactivity and also discovered two new elements. Curie won two Nobel Prizes for her work.

EXAMPLE 4.1 Atomic Number, Atomic Symbol, and Element Name

Find the atomic symbol and atomic number for each of the following elements.

(a) silicon
(b) potassium
(c) gold
(d) antimony

Solution:

As you become familiar with the periodic table, you will be able to quickly locate elements on it. For now, it might be easier to find them in the alphabetical listing on the inside back cover of this book, but you should also find their position in the periodic table.

Element	*Symbol*	*Atomic Number*
silicon	Si	14
potassium	K	19
gold	Au	79
antimony	Sb	51

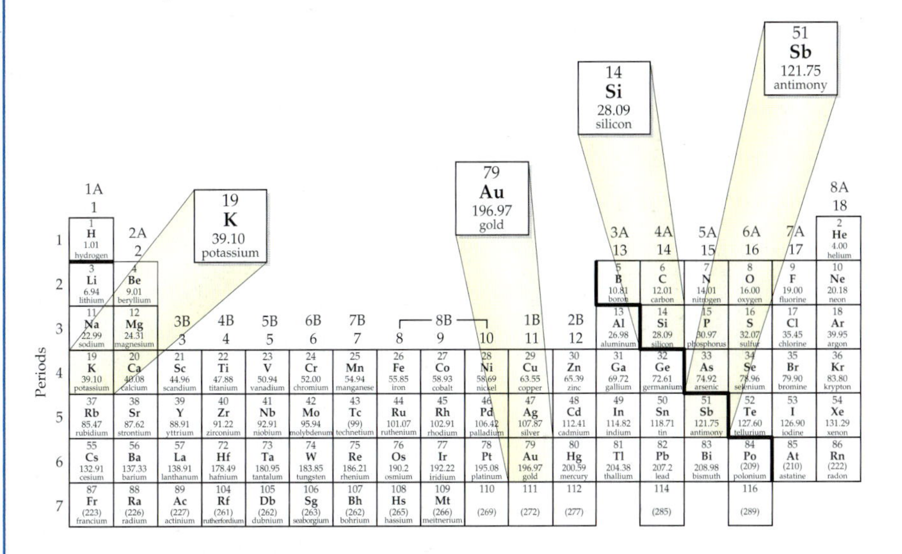

SKILLBUILDER 4.1 Atomic Number, Atomic Symbol, and Element Name

Find the name and atomic number for each of the following elements.

(a) Na
(b) Ni
(c) P
(d) Ta

4.6 Looking for Patterns: The Periodic Law and the Periodic Table

The organization of the periodic table has its origins in the work of Dmitri Mendeleev (1834–1907), a nineteenth-century Russian chemistry professor. In his time, about sixty-five different elements had been discovered. Through the work of a number of chemists, much was known about each of these elements, including their relative masses, chemical activity, and some of their physical properties. However, there was no systematic way of organizing them.

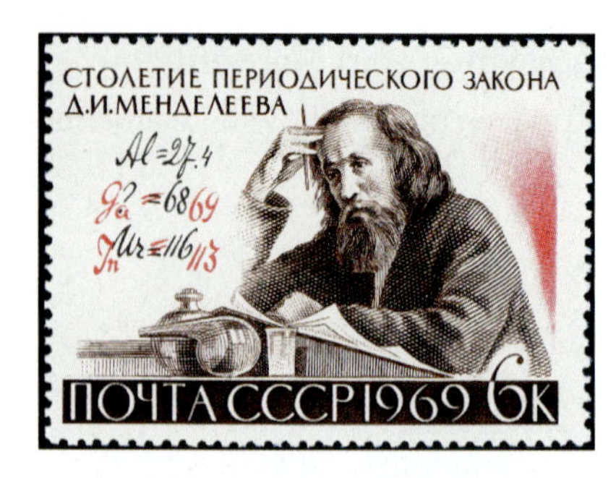

▲ Dmitri Mendeleev, a Russian chemistry professor who proposed the periodic law and arranged early versions of the periodic table, shown on a Russian postage stamp.

1	2	3	4	5	6	7	8	9	10	11	12	13	14	15	16	17	18	19	20
H	He	Li	Be	B	C	N	O	F	Ne	Na	Mg	Al	Si	P	S	Cl	Ar	K	Ca

▲ **Figure 4.10 Recurring properties** The elements shown are listed in order of increasing atomic number (Mendeleev used relative mass, which is similar). The color of each element represents its properties. Notice that the properties (colors) of these elements form a repeating pattern.

In 1869, Mendeleev noticed that certain groups of elements had similar properties. Mendeleev found that if he listed the elements in order of increasing relative mass, those similar properties recurred in a regular pattern (▲ Figure 4.10). Mendeleev summarized these observations in the **periodic law**:

> When the elements are arranged in order of increasing relative mass, certain sets of properties recur periodically.

Periodic means "recurring regularly." The properties of the elements, when listed in order of increasing relative mass, formed a *repeating pattern.*

Mendeleev then organized all the known elements in a table in which relative mass increased from left to right and elements with similar properties were aligned in the same vertical columns (◄ Figure 4.11). Since many elements had not yet been discovered, Mendeleev's table contained some gaps, which allowed him to predict the existence of yet-undiscovered elements. For example, Mendeleev predicted the existence of an element he called *eka-silicon*, which fell below silicon on the table and between gallium and arsenic. In 1886, eka-silicon was discovered by German chemist Clemens Winkler (1838–1904), and found to have almost exactly the properties that Mendeleev had anticipated. Winkler named the element germanium, after his home country.

1 H							2 He
3 Li	4 Be	5 B	6 C	7 N	8 O	9 F	10 Ne
11 Na	12 Mg	13 Al	14 Si	15 P	16 S	17 Cl	18 Ar
19 K	20 Ca						

▲ **Figure 4.11 Making a periodic table** If we place the elements from Figure 4.10 in a table, we can arrange them in rows so that similar properties align in the same vertical columns. This is similar to Mendeleev's first periodic table.

Mendeleev's original listing has evolved into the modern **periodic table**. In the modern table, elements are listed in order of increasing atomic number rather than increasing relative mass. The modern periodic table also contains more elements than Mendeleev's original table because many more have been discovered since his time.

Mendeleev's periodic law was based on observation. Like all scientific laws, the periodic law summarized many observations but did not give the underlying reason for the observation—only theories do that. For now, we accept the periodic law as it is, but in Chapter 9 we examine a powerful theory that explains the law and gives the underlying reasons for it.

The elements in the periodic table can be broadly classified as metals, nonmetals, and metalloids (► Figure 4.12). **Metals** occupy the left side of the periodic table and have similar properties: They are good conductors of heat and electricity; they can be pounded into flat sheets (malleability); they can be drawn into wires (ductility); they are often shiny; and they tend to lose electrons when they undergo chemical changes. Good examples of metals include iron, magnesium, chromium, and sodium.

Nonmetals occupy the upper right side of the periodic table. The dividing line between metals and nonmetals is the zigzag diagonal line running from boron to astatine. Nonmetals have more varied properties—some are solids at room temperature, others are gases—but as a whole they tend to be poor conductors of heat and electricity and they all tend to gain electrons when they undergo chemical changes. Good examples of nonmetals include oxygen, nitrogen, chlorine, and iodine.

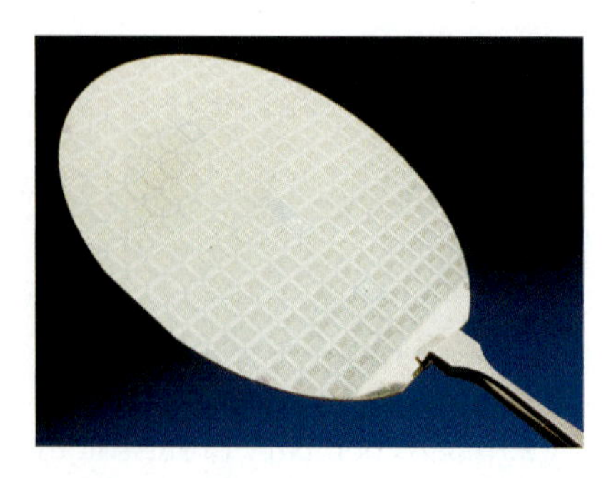

▲ Silicon, a metalloid used extensively in the computer and electronics industries. This silicon wafer is being etched to make computer chips.

Most of the elements that lie along the zigzag diagonal line dividing metals and nonmetals are called **metalloids**, or *semimetals*, and show mixed properties. Metalloids are also called **semiconductors** because of their intermediate electrical conductivity, which can be changed and controlled. This property makes semiconductors useful in the manufacture of the electronic devices central to computers, cell phones, and many other modern gadgets. Good examples of metalloids include silicon, arsenic, and germanium.

Metals

Nonmetals

Metalloids

	1A 1	2A 2	3B 3	4B 4	5B 5	6B 6	7B 7	8B 8	8B 9	8B 10	1B 11	2B 12	3A 13	4A 14	5A 15	6A 16	7A 17	8A 18
1	1 H																	2 He
2	3 Li	4 Be											5 B	6 C	7 N	8 O	9 F	10 Ne
3	11 Na	12 Mg											13 Al	14 Si	15 P	16 S	17 Cl	18 Ar
4	19 K	20 Ca	21 Sc	22 Ti	23 V	24 Cr	25 Mn	26 Fe	27 Co	28 Ni	29 Cu	30 Zn	31 Ga	32 Ge	33 As	34 Se	35 Br	36 Kr
5	37 Rb	38 Sr	39 Y	40 Zr	41 Nb	42 Mo	43 Tc	44 Ru	45 Rh	46 Pd	47 Ag	48 Cd	49 In	50 Sn	51 Sb	52 Te	53 I	54 Xe
6	55 Cs	56 Ba	57 La	72 Hf	73 Ta	74 W	75 Re	76 Os	77 Ir	78 Pt	79 Au	80 Hg	81 Tl	82 Pb	83 Bi	84 Po	85 At	86 Rn
7	87 Fr	88 Ra	89 Ac	104 Rf	105 Db	106 Sg	107 Bh	108 Hs	109 Mt	110	111	112		114		116		

Lanthanides	58 Ce	59 Pr	60 Nd	61 Pm	62 Sm	63 Eu	64 Gd	65 Tb	66 Dy	67 Ho	68 Er	69 Tm	70 Yb	71 Lu
Actinides	90 Th	91 Pa	92 U	93 Np	94 Pu	95 Am	96 Cm	97 Bk	98 Cf	99 Es	100 Fm	101 Md	102 No	103 Lr

▲ **Figure 4.12 Metals, nonmetals, and metalloids** The elements in the periodic table can be broadly classified as metals, nonmetals, and metalloids.

EXAMPLE 4.2 Classifying Elements as Metals, Nonmetals, or Metalloids

Classify each of the following elements as a metal, nonmetal, or metalloid.

(a) Ba
(b) I
(c) O
(d) Te

Solution:

(a) Barium is on the left side of the periodic table; it is a metal.

(b) Iodine is on the right side of the periodic table; it is a nonmetal.

(c) Oxygen is on the right side of the periodic table; it is a nonmetal.

(d) Tellurium is in the middle-right section of the periodic table, along the line that divides the metals from the nonmetals; it is a metalloid.

SKILLBUILDER 4.2 Classifying Elements as Metals, Nonmetals, or Metalloids

Classify each of the following elements as a metal, nonmetal, or metalloid.

(a) S
(b) Cl
(c) Ti
(d) Sb

The periodic table can also be broadly divided into **main-group elements**, whose properties tend to be more predictable based on their position in the periodic table, and **transition elements** or **transition metals**, whose properties

▶ **Figure 4.13 Main-group and transition elements** The periodic table can be broadly divided into main-group elements, whose properties can generally be predicted based on their position, and transition elements, whose properties tend to be less predictable based on their position.

Main group elements | Transition elements | Main group elements (Group number; Periods 1–7)

Period	1A	2A	3B	4B	5B	6B	7B	8B	8B	8B	1B	2B	3A	4A	5A	6A	7A	8A
1	1 H																	2 He
2	3 Li	4 Be											5 B	6 C	7 N	8 O	9 F	10 Ne
3	11 Na	12 Mg											13 Al	14 Si	15 P	16 S	17 Cl	18 Ar
4	19 K	20 Ca	21 Sc	22 Ti	23 V	24 Cr	25 Mn	26 Fe	27 Co	28 Ni	29 Cu	30 Zn	31 Ga	32 Ge	33 As	34 Se	35 Br	36 Kr
5	37 Rb	38 Sr	39 Y	40 Zr	41 Nb	42 Mo	43 Tc	44 Ru	45 Rh	46 Pd	47 Ag	48 Cd	49 In	50 Sn	51 Sb	52 Te	53 I	54 Xe
6	55 Cs	56 Ba	57 La	72 Hf	73 Ta	74 W	75 Re	76 Os	77 Ir	78 Pt	79 Au	80 Hg	81 Tl	82 Pb	83 Bi	84 Po	85 At	86 Rn
7	87 Fr	88 Ra	89 Ac	104 Rf	105 Db	106 Sg	107 Bh	108 Hs	109 Mt	110	111	112		114		116		

Main-group elements are in columns labeled with a number and the letter A. Transition elements are in columns labeled with a number and the letter B.

A competing numbering system does not use letters, but only the numbers 1–18. Both numbering systems are shown in Figure 4.12.

The noble gases are inert (or unreactive) compared to other elements. However, some noble gases, especially the heavier ones, will form a limited number of compounds with other elements under special conditions.

are less easily predictable based simply on their position in the periodic table (▲ Figure 4.13). Each column within the main-group elements in the periodic table is called a **family** or **group** of elements and is designated with a number and a letter printed directly above the column.

The elements within a group usually have similar properties. For example, the Group 8A elements, called the **noble gases**, are chemically inert gases. The most familiar noble gas is probably helium, used to fill buoyant balloons. Helium, like the other noble gases, is chemically stable—it won't combine with other elements to form compounds—and is therefore safe to put into balloons. Other noble gases include neon, often used in neon signs; argon, which makes up a small percentage of our atmosphere; krypton; and xenon. The Group 1A elements, called the **alkali metals**, are all very reactive metals. A marble-sized piece of sodium explodes violently when dropped into water. Other alkali metals include lithium, potassium, and rubidium. The Group 2A elements, called the **alkaline earth metals**, are also fairly reactive, although not quite as reactive as the alkali metals. Calcium, for example, reacts fairly vigorously when dropped into water but will not explode as easily as sodium. Other alkaline earth metals include

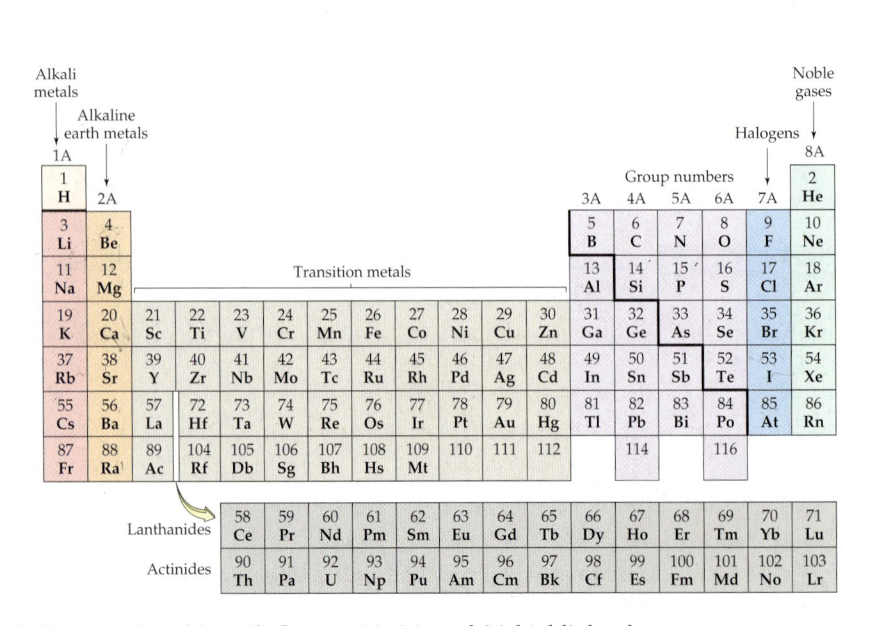

▲ The periodic table with Groups 1A, 2A, and 8A highlighted.

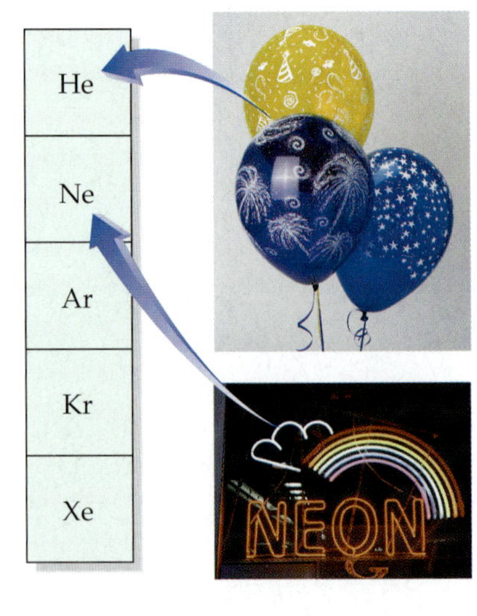

▲ The noble gases include helium (used in balloons), neon (found in neon signs), argon, krypton, and xenon.

▲ The alkali metals include lithium (shown in the first photo), sodium (shown in the second photo reacting with water), potassium, rubidium, and cesium.

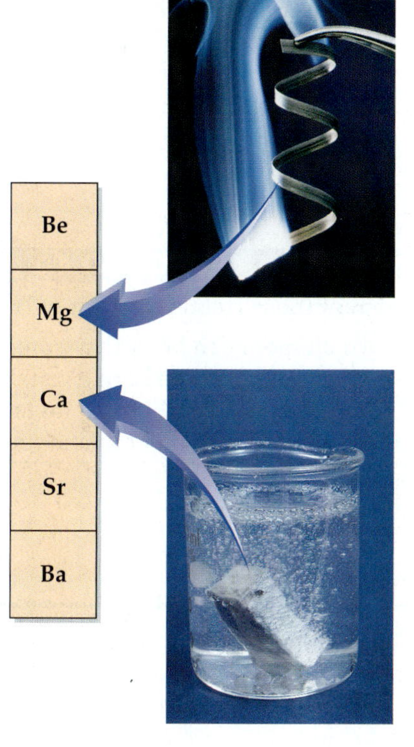

▲ The alkaline earth metals include beryllium, magnesium (shown burning in the first photo), calcium (shown reacting with water in the second photo), strontium, and barium.

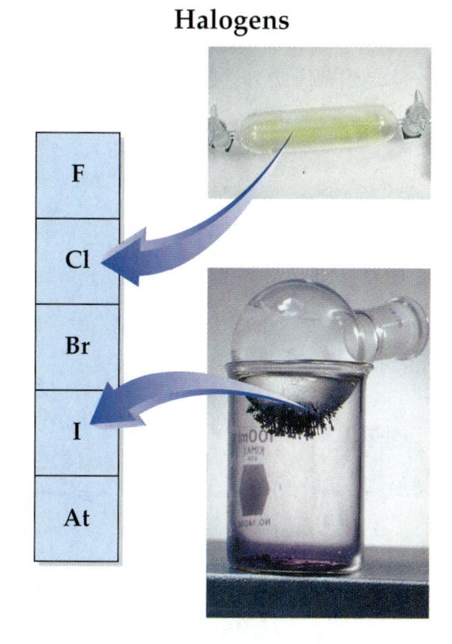

▲ The halogens include fluorine, chlorine (shown in the first photo), bromine, iodine (shown in the second photo), and astatine.

TUTORIAL
Interactive Periodic Table

magnesium, a common low-density structural metal; strontium; and barium. The Group 7A elements, called the **halogens**, are very reactive nonmetals. The most familiar halogen is probably chlorine, a greenish-yellow gas with a pungent odor. Because of its reactivity, chlorine is often used as a sterilizing and disinfecting agent. Other halogens include bromine, a red-brown liquid that easily evaporates into a gas; iodine, a purple solid; and fluorine, a pale yellow gas.

EXAMPLE 4.3 Groups and Families of Elements

To which group or family of elements does each of the following elements belong?

(a) Mg
(b) N
(c) K
(d) Br

Solution:

(a) Mg is in Group 2A; it is an alkaline earth metal.

(b) N is in Group 5A.

(c) K is in Group 1A; it is an alkali metal.

(d) Br is in Group 7A; it is a halogen.

SKILLBUILDER 4.3 **Groups and Families of Elements**

To which group or family of elements does each of the following elements belong?

(a) Li
(b) B
(c) I
(d) Ar

CONCEPTUAL CHECKPOINT 4.2

Which of these statements can NEVER be true?

(a) An element can be both a transition element and a metal.
(b) An element can be both a transition element and a metalloid.
(c) An element can be both a metalloid and a halogen.
(d) An element can be both a main-group element and a halogen.

4.7 Ions: Losing and Gaining Electrons

In chemical reactions, atoms often lose or gain electrons to form charged particles called **ions**. For example, neutral lithium (Li) atoms contain 3 protons and 3 electrons; however, in reactions, lithium atoms lose one electron (e^-) to form Li^+ ions.

$$Li \longrightarrow Li^+ + e^-$$

The charge of an ion is shown in the upper right corner of the symbol.

The Li^+ *ion* contains 3 protons but only 2 electrons, resulting in a net charge of 1+. Ion charges are usually written with the magnitude of the charge first followed by the sign of the charge. For example, a positive two charge is written as 2+ and a negative two charge is written as 2−. The charge of an ion depends on how many electrons were gained or lost and is given by the following formula:

$$\text{Ion charge} = \text{number of protons} - \text{number of electrons}$$
$$= \#p - \#e^-$$

where "p" stands for *proton* and "e^-" stands for *electron*.

For the Li^+ ion:

$$\text{Ion charge} = 3 - 2 = 1+$$

Neutral fluorine (F) atoms contain 9 protons and 9 electrons; however, in chemical reactions fluorine atoms gain one electron to form F^- ions:

$$F + e^- \longrightarrow F^-$$

The F^- *ion* contains 9 protons and 10 electrons, resulting in a −1 charge.

$$\text{Ion charge} = 9 - 10$$
$$= 1-$$

Positively charged ions, such as Li^+, are called **cations**, and negatively charged ions, such as F^-, are called **anions**. Ions behave very differently than the atoms from which they are formed. Neutral sodium atoms, for example, are extremely reactive, reacting violently with most things they contact. Sodium cations (Na^+), on the other hand, are relatively inert—we eat them all the time in sodium chloride (table salt). In nature, cations and anions always occur together so that, again, matter is charge-neutral.

EXAMPLE 4.4 Determining Ion Charge from Numbers of Protons and Electrons

Determine the charge of each of the following ions.

(a) a magnesium ion with 10 electrons
(b) a sulfur ion with 18 electrons
(c) an iron ion with 23 electrons

Solution:

To determine the charge of each ion, we use the ion charge equation.

$$\text{Ion charge} = \#p - \#e^-$$

The number of electrons is given in the problem. The number of protons is obtained from the element's atomic number in the periodic table.

(a) magnesium with atomic number 12

$$\text{Ion charge} = 12 - 10 = 2+ \ (Mg^{2+})$$

(b) sulfur with atomic number 16

$$\text{Ion charge} = 16 - 18 = 2- \ (S^{2-})$$

(c) iron with atomic number 26

$$\text{Ion charge} = 26 - 23 = 3+ \ (Fe^{3+})$$

SKILLBUILDER 4.4 Determining Ion Charge from Numbers of Protons and Electrons

Determine the charge of each of the following ions.

(a) a nickel ion with 26 electrons
(b) a bromine ion with 36 electrons
(c) a phosphorus ion with 18 electrons

LIVE EXAMPLE 

EXAMPLE 4.5 Determining the Number of Protons and Electrons in an Ion

Find the number of protons and electrons in the Ca^{2+} ion.

From the periodic table, find that the atomic number for calcium is 20, so calcium has 20 protons. The number of electrons can be found using the ion charge equation.

Solution:

$$\text{Ion charge} = \#p - \#e^-$$
$$2+ = 20 - \#e^-$$
$$\#e^- = 20 - 2 = 18$$

Therefore the number of electrons is 18. The Ca^{2+} ion has 20 protons and 18 electrons.

SKILLBUILDER 4.5 Determining the Number of Protons and Electrons in an Ion

Find the number of protons and electrons in the S^{2-} ion.

Ions and the Periodic Table

We have learned that in chemical reactions metals have a tendency to lose electrons and nonmetals have a tendency to gain them. For many main-group elements, we can use the periodic table to predict how many electrons tend to be lost or gained. The number above each *main group* column in the periodic table—1 through 8–gives the number of *valence electrons* for the elements

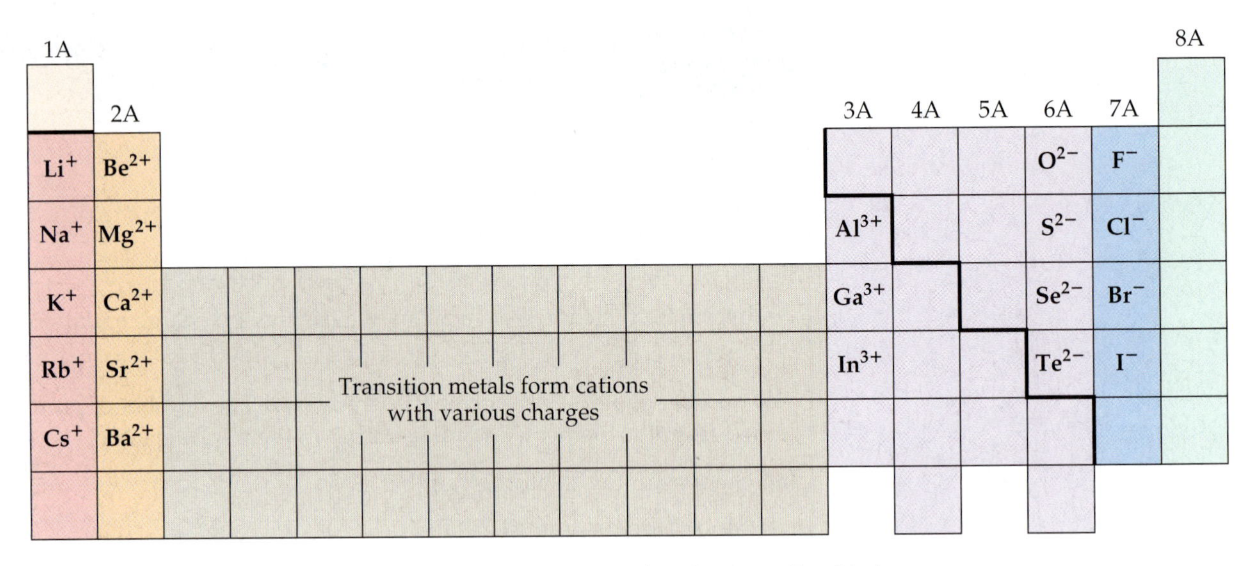

▲ **Figure 4.14 Elements that form predictable ions**

An important exception to this rule is helium—it is in column 8A, but has only 2 valence electrons.

in that column. We discuss the concept of valence electrons more fully in Chapter 9, but for now, think of valence electrons as the outermost electrons in an atom. For example, since oxygen is in column 6A, we know it has 6 valence electrons; since magnesium is in column 2A, we know it has 2 valence electrons, and so on. Valence electrons are particularly important because, as we shall see in Chapter 10, it is these electrons that take part in chemical bonding.

The key to predicting the charge acquired by a particular metal or nonmetal when it ionizes is as follows:

> Main-group elements tend to form ions that have the same number of valence electrons as the nearest noble gas.

For example, the closest noble gas to oxygen is neon. When oxygen ionizes, it acquires two additional electrons for a total of 8 valence electrons—the same number as neon. The closest noble gas to magnesium is also neon. Therefore magnesium loses its 2 valence electrons to attain the same number of valence electrons as neon.

In accordance with this principle, the alkali metals (Group 1A) tend to lose 1 electron and therefore form 1+ ions, while the alkaline earth metals (Group 2A) tend to lose 2 electrons and therefore form 2+ ions. The halogens (Group 7A) tend to gain 1 electron and therefore form 1− ions. The groups in the periodic table that form predictable ions are shown in ▲ Figure 4.14 . Be familiar with these groups and the ions they form. In Chapter 9, we examine a theory that more fully explains why these groups form ions as they do.

EXAMPLE 4.6 Charge of Ions from Position in Periodic Table

Based on their position in the periodic table, what ions will barium and iodine tend to form?

Solution:

Since barium is in Group 2A, it will tend to form a cation with a 2+ charge (Ba^{2+}). Since iodine is in Group 7A, it will tend to form an anion with a 1− charge (I^{-}).

SKILLBUILDER 4.6 Charge of Ions from Position in Periodic Table

Based on their position in the periodic table, what ions will potassium and sulfur tend to form?

CONCEPTUAL CHECKPOINT 4.3

Which of these pairs of ions would have the same total number of electrons?

(a) Na^{+} and Mg^{2+}
(b) F^{-} and Cl^{-}
(c) O^{-} and O^{2-}
(d) Ga^{3+} and Fe^{3+}

4.8 Isotopes: When the Number of Neutrons Varies

All atoms of a given element have the same number of protons; however, they do not necessarily have the same number of neutrons. Since neutrons have nearly the same mass as protons (1 amu), this means that—contrary to what John Dalton originally proposed in his atomic theory—all atoms of a given element *do not* have the same mass. For example, all neon atoms in nature contain 10 protons, but they may have 10, 11, or 12 neutrons (▼ Figure 4.15). All three types of neon atoms exist, and each has a slightly different mass. Atoms with the same number of protons but different numbers of neutrons are called **isotopes**. Some elements, such as beryllium (Be) and aluminum (Al), have only one naturally occurring isotope, while other elements, such as neon (Ne) and chlorine (Cl), have two or more.

Fortunately, the relative amount of each different isotope in a naturally occurring sample of a given element is always the same. For example, in any natural sample of neon atoms, 90.48% of them are the isotope with 10 neutrons, 0.27% are the isotope with 11 neutrons, and 9.25% are the isotope with 12 neutrons. This means that out of 10,000 neon atoms, for example, 9048 have 10 neutrons, 27 have 11 neutrons, and 925 have 12 neutrons. These percentages are called the **percent natural abundance** of the isotopes. The preceding numbers are for neon, but all elements have their own unique percent natural abundance of isotopes.

Percent means "per hundred." 90.48% means that 90.48 atoms out of 100 are the isotope with 10 neutrons.

The sum of the number of neutrons and protons in an atom is called the **mass number** and is given the symbol **A**.

$$\mathbf{A = Number\ of\ protons + Number\ of\ neutrons}$$

For neon, with 10 protons, the mass numbers of the three different naturally occurring isotopes are 20, 21, and 22, corresponding to 10, 11, and 12 neutrons, respectively.

▼ **Figure 4.15 Isotopes of neon** Naturally occurring neon contains three different isotopes, Ne-20 (with 10 neutrons), Ne-21 (with 11 neutrons), and Ne-22 (with 12 neutrons).

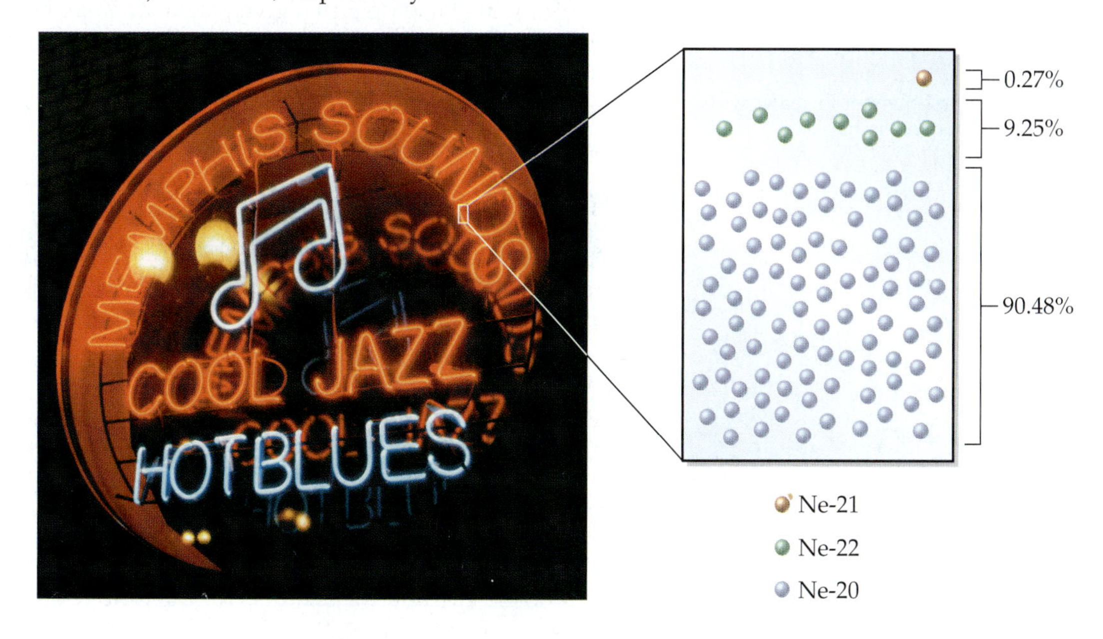

TABLE 4.2 Neon Isotopes

Symbol	Number of Protons	Number of Neutrons	A (Mass Number)	Percent Natural Abundance
Ne-20 or $^{20}_{10}Ne$	10	10	20	90.48%
Ne-21 or $^{21}_{10}Ne$	10	11	21	0.27%
Ne-22 or $^{22}_{10}Ne$	10	12	22	9.25%

Isotopes are often symbolized in the following way:

TUTORIAL
Atomic Number Activity

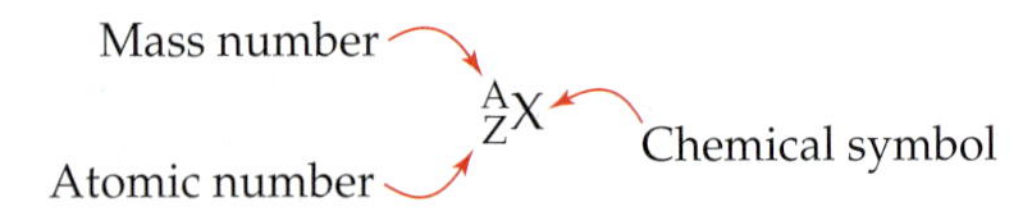

where X is the chemical symbol, A is the mass number, and Z is the atomic number.

For example, the symbols for the neon isotopes are:

$^{20}_{10}Ne$ $^{21}_{10}Ne$ $^{22}_{10}Ne$

Notice that the chemical symbol, Ne, and the atomic number, 10, are redundant: If the atomic number is 10, the symbol must be Ne, and vice versa. The mass numbers, however, are different, reflecting the different number of neutrons in each isotope.

A second common notation for isotopes is the chemical symbol (or chemical name) followed by a hyphen and the mass number of the isotope.

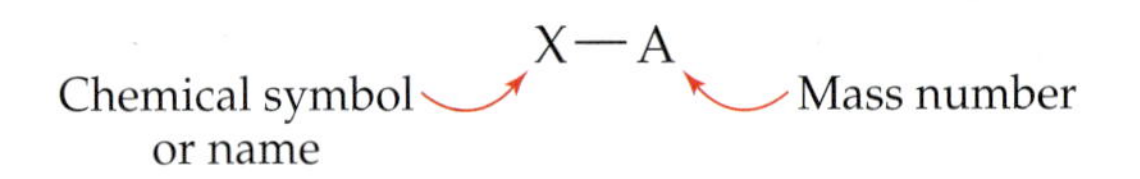

In this notation, the neon isotopes are:

Ne-20 neon-20

Ne-21 neon-21

Ne-22 neon-22

We summarize what we have learned about the neon isotopes in Table 4.2.

Notice that all isotopes of a given element have the same number of protons (otherwise they would be a different element). Notice also that the mass number is the *sum* of the number of protons and the number of neutrons. The number of neutrons in an isotope is the difference between the mass number and the atomic number.

In general, mass number increases with increasing atomic number.

EXAMPLE 4.7 Atomic Numbers, Mass Numbers, and Isotope Symbols

What are the atomic number (Z), mass number (A), and symbols of the carbon isotope with 7 neutrons?

Solution:

From the periodic table, we find that the atomic number (Z) of carbon is 6, so carbon atoms have 6 protons. The mass number (A) for the isotope with 7 neutrons is the sum of the number of protons and the number of neutrons.

$$A = 6 + 7 = 13$$

So, Z = 6, A = 13, and the symbols for the isotope are C-13 and $^{13}_{6}C$.

SKILLBUILDER 4.7 Atomic Numbers, Mass Numbers, and Isotope Symbols

What are the atomic number, mass number, and symbols for the chlorine isotope with 18 neutrons?

EXAMPLE 4.8 Numbers of Protons and Neutrons from Isotope Symbols

How many protons and neutrons are in the following chromium isotope?

$^{52}_{24}\text{Cr}$

Solution:

The number of protons is equal to Z (lower left number).

$$\#p = Z = 24$$

The number of neutrons is equal to A (upper left number) − Z (lower left number).

$$\begin{aligned}\#n &= A - Z \\ &= 52 - 24 \\ &= 28\end{aligned}$$

SKILLBUILDER 4.8 Numbers of Protons and Neutrons from Isotope Symbols

How many protons and neutrons are in the following potassium isotope?

$^{39}_{19}\text{K}$

CONCEPTUAL CHECKPOINT 4.4

If an atom with a mass number of 27 has 14 neutrons, it must be an isotope of which element?

(a) silicon
(b) aluminum
(c) cobalt
(d) niobium

4.9 Atomic Mass

Some books call this *average atomic mass* or *atomic weight* instead of simply *atomic mass*.

An important part of Dalton's atomic theory was that all atoms of a given element have the same mass. However, we just learned that because of isotopes, the atoms of a given element may have different masses, so Dalton was not completely correct. We can, however, calculate an average mass—called the **atomic mass**—for each element. The atomic mass of each element is listed in the periodic table directly beneath the element's symbol; it represents the average mass of the atoms that compose that element. For example, the periodic table lists the atomic mass of chlorine as 35.45 amu. Naturally occurring chlorine consists of 75.77% chlorine-35 (mass 34.97 amu) and 24.23% chlorine-37 (mass 36.97 amu). Its atomic mass is:

$$\begin{aligned}\text{Atomic mass} &= (0.7577 \times 34.97\ \text{amu}) + (0.2423 \times 36.97\ \text{amu}) \\ &= 35.45\ \text{amu}\end{aligned}$$

Notice that the atomic mass of chlorine is closer to 35 than 37, because naturally occurring chlorine contains more chlorine-35 atoms than chlorine-37 atoms. Notice also that when percentages are used in calculations, they must always be converted to their decimal value. To convert a percentage to its decimal value, simply divide by 100. For example:

$$75.77\% = 75.77/100 = 0.7577$$
$$24.23\% = 24.23/100 = 0.2423$$

Chemistry IN THE ENVIRONMENT

Radioactive Isotopes at Hanford, Washington

Nuclei of the isotopes of a given element are not all equally stable. For example, naturally occurring lead is composed primarily of Pb-206, Pb-207, and Pb-208. Other isotopes of lead also exist, but their nuclei are unstable. Scientists can make some of these other isotopes, such as Pb-185, in the laboratory. However, within seconds of their synthesis, Pb-185 atoms emit a few energetic subatomic particles from their nuclei and change into different isotopes of different elements (which are themselves unstable). These emitted subatomic particles are called **nuclear radiation**, and the isotopes that emit them are termed **radioactive.** Nuclear radiation, always associated with unstable nuclei, can be harmful to humans and other living organisms because the energetic particles interact with and damage biological molecules. Some isotopes, such as Pb-185, emit significant amounts of radiation only for a very short time. Others, however, remain radioactive for a long time—in some cases millions or even billions of years.

The nuclear power and nuclear weapons industries both produce by-products containing unstable isotopes of several different elements. Many of these isotopes emit nuclear radiation for a long time, and their disposal is an environmental problem. For example, in Hanford, Washington, which for fifty years produced fuel for nuclear weapons, 177 underground storage tanks contain 55 million gallons of high-level nuclear waste. Certain radioactive isotopes within that waste will produce nuclear radiation for the foreseeable future. Unfortunately, some of the underground storage tanks are aging and leaks have allowed some of the waste to seep into the environment. While the danger from short-term external exposure to this waste is minimal, ingestion of the waste through contamination of drinking water or food supplies would pose significant health risks. Consequently, Hanford is now the site of the largest environmental cleanup project in U.S. history. The U.S. government expects the project to take more than twenty years and cost about $10 billion.

Radioactive isotopes are not always harmful, however, and many have beneficial uses. For example, technetium-99 (Tc-99) is often given to patients to diagnose disease. The radiation emitted by Tc-99 helps doctors image internal organs or detect infection.

CAN YOU ANSWER THIS? *Give the number of neutrons in each of the following isotopes: Pb-206, Pb-207, Pb-208, Pb-185, Tc-99.*

▲ Storage tanks at Hanford, Washington, contain 55 million gallons of high-level nuclear waste. Each tank pictured here holds 1 million gallons.

In general, the atomic mass is calculated according to the following equation:

$$\text{Atomic mass} = (\text{Fraction of isotope 1} \times \text{Mass of isotope 1}) + (\text{Fraction of isotope 2} \times \text{Mass of isotope 2}) + (\text{Fraction of isotope 3} \times \text{Mass of isotope 3}) + \cdots$$

where the fractions of each isotope are the percent natural abundances converted to their decimal values. Atomic mass is useful because it allows us to assign a characteristic mass to each element, and as we will see in Chapter 6, it allows us to quantify the number of atoms in a sample of that element.

EXAMPLE 4.9 Calculating Atomic Mass

Gallium has two naturally occurring isotopes: Ga-69 with mass 68.9256 amu and a natural abundance of 60.11%, and Ga-71 with mass 70.9247 amu and a natural abundance of 39.89%. Calculate the atomic mass of gallium.

Solution:

Convert the percent natural abundances into decimal form by dividing by 100.

$$\text{Fraction Ga-69} = \frac{60.11}{100} = 0.6011$$

$$\text{Fraction Ga-71} = \frac{39.89}{100} = 0.3989$$

Use the fractional abundances and the atomic masses of the isotopes to compute the atomic mass according to the atomic mass definition given earlier.

$$\begin{aligned}\text{Atomic mass} &= (0.6011 \times 68.9256\ \text{amu}) + (0.3989 \times 70.9247\ \text{amu}) \\ &= 41.4\underline{3}12\ \text{amu} + 28.2\underline{9}19\ \text{amu} \\ &= 69.7231 = 69.72\ \text{amu}\end{aligned}$$

SKILLBUILDER 4.9 Calculating Atomic Mass

Magnesium has three naturally occurring isotopes with masses of 23.99, 24.99, 25.98 amu and natural abundances of 78.99%, 10.00%, and 11.01%. Calculate the atomic mass of magnesium.

CHAPTER IN REVIEW

Chemical Principles

The Atomic Theory: Democritus, an ancient Greek philosopher, is the first person on record to assert that matter is ultimately composed of small, indestructible particles. It was not until 2000 years later, however, that John Dalton introduced a formal atomic theory stating that matter is composed of atoms; atoms of a given element have unique properties that distinguish them from atoms of other elements; and atoms combine in simple, whole-number ratios to form compounds.

Relevance

The Atomic Theory: The concept of atoms is important because it explains the physical world. You and everything you see are made of atoms. To understand the physical world, we must begin by understanding atoms. Atoms are the key concept—they determine the properties of matter.

Discovery of the Atom's Nucleus: Rutherford's gold foil experiment probed atomic structure and his results led to the nuclear model of the atom, which, with minor modifications to accommodate neutrons, is still valid today. In this model, the atom is composed of protons and neutrons—which compose most of the atom's mass and are grouped together in a dense nucleus—and electrons, which compose most of the atom's volume. Protons and neutrons have similar masses (1 amu) while electrons have a much smaller mass (≈ 0 amu).

Discovery of the Atom's Nucleus: We can see why this is relevant by asking, what if it were otherwise? What if matter were *not* mostly empty space? While we cannot know for certain, it seems probable that such matter would not form the diversity of substances required for life—and then, of course, we would not be around to ask the question.

Charge: Protons and electrons both have electrical charge; the charge of the proton is $+1$ and the charge of the electron is -1. The neutron has no charge. When protons and electrons combine in atoms, their charges cancel.

Charge: Electrical charge is relevant to much of our modern world. Many of the machines and computers we depend on are powered by electricity, which is simply the movement of electrical charge.

The Periodic Table: The periodic table tabulates all known elements in order of increasing atomic number. The periodic table is arranged so that similar elements are grouped in columns. Elements on the left side of the periodic table are metals and tend to lose electrons in their chemical changes. Elements on the upper right side of the periodic table are nonmetals and tend to gain electrons in their chemical changes. Elements between the two are called metalloids. Columns of elements in the periodic table have similar properties and are called groups or families.

The Periodic Table: The periodic table helps us organize the elements in ways that allow us to predict their properties. Helium, for example, is not toxic in small amounts because it is an inert gas—it does not react with anything. The gases in the column below it on the periodic table are also inert gases and form a family or group of elements called the noble gases. By tabulating the elements and grouping similar ones together, we begin to understand their properties.

Atomic Number: The most fundamental characteristic of an atom is the number of protons in its nucleus; this number is called the atomic number (Z) and defines the element.

Ions: When an atom gains or loses electrons it becomes an ion. Positively charged ions are called cations and negatively charged ions are called anions. Cations and anions occur together so that matter is ordinarily charge-neutral.

Isotopes: While all atoms of a given element have the same number of protons, they do not necessarily have the same number of neutrons. Atoms of the same element with different numbers of neutrons are called isotopes. Isotopes are characterized by their mass number (A), the sum of the number of protons and the number of neutrons in their nucleus.

Each naturally occurring sample of an element has the same percent natural abundance of each isotope. These percentages, together with the mass of each isotope, are used to compute the atomic mass of the element, a weighted average of the masses of the individual isotopes.

Isotopes: Isotopes are relevant because they influence tabulated atomic masses. To understand these masses, we must understand the presence and abundance of isotopes. In nuclear processes—processes in which the nuclei of atoms actually change—the presence of different isotopes becomes even more important. Some isotopes are not stable—they lose subatomic particles and are transformed into other elements. The emission of subatomic particles by unstable nuclei is called radioactive decay. In many situations, such as in diagnosing and treating certain diseases, nuclear radiation is extremely useful. In other situations, such as in the disposal of radioactive waste, it can pose environmental problems.

Chemical Skills

Examples

Determining Ion Charge from Numbers of Protons and Electrons (Section 4.7)

- From the periodic table or from the alphabetical list of elements, find the atomic number of the element; this number is equal to the number of protons.
- Use the ion charge equation to compute charge.

$$\text{Ion charge} = \#p - \#e^-$$

EXAMPLE 4.10 Determining Ion Charge from Numbers of Protons and Electrons

Determine the charge of a selenium ion with 36 electrons.

Solution:
Selenium is atomic number 34; therefore it has 34 protons.

$$\text{Ion charge} = 34 - 36 = -2$$

Determining the Number of Protons and Electrons in an Ion (Section 4.7)

- From the periodic table or from the alphabetical list of elements, find the atomic number of the element; this number is equal to the number of protons.
- Use the ion charge equation and substitute in the known values.

$$\text{Ion charge} = \#p - \#e^-$$

- Solve the equation for the number of electrons.

EXAMPLE 4.11 Determining the Number of Protons and Electrons in an Ion

Find the number of protons and electrons in the O^{2-} ion.

Solution:
The atomic number of O is 8; therefore it has 8 protons.

$$\text{Ion charge} = \#p - \#e^-$$
$$-2 = 8 - \#e^-$$
$$\#e^- = 8 + 2 = 10$$

The ion has 8 protons and 10 electrons.

Determining Atomic Numbers, Mass Numbers, and Isotope Symbols for an Isotope (Section 4.8)

- From the periodic table or from the alphabetical list of elements, find the atomic number of the element.
- The mass number (A) is equal to the atomic number plus the number of neutrons.
- Write the symbol for the isotope by writing the symbol for the element with the mass number in the upper left corner and the atomic number in the lower left corner.
- The other symbol for the isotope is simply the chemical symbol followed by a hyphen and the mass number.

EXAMPLE 4.12 Determining Atomic Numbers, Mass Numbers, and Isotope Symbols for an Isotope

What are the atomic number (Z), mass number (A), and symbols for the iron isotope with 30 neutrons?

Solution:
The atomic number of iron is 26.

$$A = 26 + 30 = 56$$

The mass number is 56.

$$^{56}_{26}\text{Fe}$$

Fe-56

Number of Protons and Neutrons from Isotope Symbols (Section 4.8)

- The number of protons is equal to Z (lower left number).
- The number of neutrons is equal to A (upper left number) − Z (lower left number).

EXAMPLE 4.13 Number of Protons and Neutrons from Isotope Symbols

How many protons and neutrons are in $^{62}_{28}\text{Ni}$?

Solution:
28 protons

$$\#n = 62 - 28 = 34 \text{ neutrons}$$

Calculating Atomic Mass from Percent Natural Abundances and Isotopic Masses (Section 4.9)

LIVE EXAMPLE

EXAMPLE 4.14 Calculating Atomic Mass from Percent Natural Abundances and Isotopic Masses

Copper has two naturally occurring isotopes: Cu-63 with mass 62.9395 amu and a natural abundance of 69.17%, and Cu-65 with mass 64.9278 amu and a natural abundance of 30.83%. Calculate the atomic mass of copper.

Solution:

- Convert the natural abundances from percent to decimal values by dividing by 100.

$$\text{Fraction Cu-63} = \frac{69.17}{100} = 0.6917$$

$$\text{Fraction Cu-65} = \frac{30.83}{100} = 0.3083$$

- Find the atomic mass by multiplying the fractions of each isotope by their respective masses and adding.

$$\begin{aligned}\text{Atomic mass} &= (0.6917 \times 62.9395 \text{ amu}) \\ &\quad + (0.3083 \times 64.9278 \text{ amu}) \\ &= 43.5\underline{3}53 \text{ amu} + 20.0\underline{1}07 \text{ amu} \\ &= 63.5\underline{4}60 \text{ amu} \\ &= 63.55 \text{ amu}\end{aligned}$$

- Round to the correct number of significant figures.
- Check your work.

Key Terms

alkali metals [4.6]
alkaline earth metals [4.6]
anions [4.7]
atomic mass [4.9]
atomic mass unit (amu) [4.4]
atomic number (Z) [4.5]
cations [4.7]
charge [4.4]
chemical symbol [4.5]
electron [4.3]
family (of elements) [4.6]
group (of elements) [4.6]
halogens [4.6]
ions [4.7]
isotopes [4.8]
main-group elements [4.6]
mass number (A) [4.8]
metalloids [4.6]
metals [4.6]
neutrons [4.3]
noble gases [4.6]
nonmetals [4.6]
nuclear radiation [4.9]
nuclear theory of the atom [4.3]
nucleus [4.3]
percent natural abundance [4.8]
periodic law [4.6]
periodic table [4.6]
protons [4.3]
radioactive [4.8]
semiconductor [4.6]
transition elements [4.6]
transition metals [4.6]

Exercises

Questions

1. Why is it important to understand atoms?
2. What is an atom?
3. What did Democritus contribute to our modern understanding of matter?
4. What are three main ideas in Dalton's atomic theory?
5. Describe Rutherford's gold foil experiment and the results of that experiment. How did these results contradict the plum pudding model of the atom?
6. What are the main ideas in the nuclear theory of the atom?
7. List the three subatomic particles and their properties.
8. What is electrical charge?
9. Is matter usually charge-neutral? How would matter be different if it were not charge-neutral?
10. What does the atomic number of an element specify?
11. What is a chemical symbol?
12. Give some examples of how elements got their names.
13. What was Dmitri Mendeleev's main contribution to our modern understanding of chemistry?
14. What is the main idea in the periodic law?
15. How is the periodic table organized?
16. What are the properties of metals? Where are metals found on the periodic table?
17. What are the properties of nonmetals? Where are nonmetals found on the periodic table?
18. Where on the periodic table do you find metalloids?
19. What is a family or group of elements?

20. Locate each of the following on the periodic table and give their group number.
    (a) alkali metals
    (b) alkaline earth metals
    (c) halogens
    (d) noble gases
21. What is an ion?
22. What is an anion? What is a cation?
23. Locate each of the following groups on the periodic table and list the charge of the ions they tend to form.
    (a) Group 1A
    (b) Group 2A
    (c) Group 3A
    (d) Group 6A
    (e) Group 7A
24. What are isotopes?
25. What is the percent natural abundance of isotopes?
26. What is the mass number of an isotope?
27. What notations are commonly used to specify isotopes? What do each of the numbers in these symbols mean?
28. What is the atomic mass of an element?

Problems

Atomic and Nuclear Theory

29. Which of the following statements are *inconsistent* with Dalton's atomic theory as it was originally stated? Why?
    (a) All carbon atoms are identical.
    (b) Helium atoms can be split into two hydrogen atoms.
    (c) An oxygen atom combines with 1.5 hydrogen atoms to form water molecules.
    (d) Two oxygen atoms combine with a carbon atom to form carbon dioxide molecules.
30. Which of the following statements are *consistent* with Dalton's atomic theory as it was originally stated? Why?
    (a) Calcium and titanium atoms have the same mass.
    (b) Neon and argon atoms are the same.
    (c) All cobalt atoms are identical.
    (d) Sodium and chlorine atoms combine in a 1:1 ratio to form sodium chloride.
31. Which of the following statements are *inconsistent* with Rutherford's nuclear theory as it was originally stated? Why?
    (a) Helium atoms have two protons in the nucleus and two electrons outside the nucleus.
    (b) Most of the volume of hydrogen atoms is due to the nucleus.
    (c) Aluminum atoms have 13 protons in the nucleus and 22 electrons outside the nucleus.
    (d) The majority of the mass of nitrogen atoms is due to their 7 electrons.
32. Which of the following statements are *consistent* with Rutherford's nuclear theory as it was originally stated? Why?
    (a) Atomic nuclei are small compared to the size of atoms.
    (b) The volume of an atom is mostly empty space.
    (c) Neutral potassium atoms contain more protons than electrons.
    (d) Neutral potassium atoms contain more neutrons than protons.

Protons, Neutrons, and Electrons

33. Which of the following statements about electrons are true?
    (a) Electrons repel each other.
    (b) Electrons are attracted to protons.
    (c) Some electrons have a charge of −1 and some have no charge.
    (d) Electrons are much lighter than neutrons.
34. Which of the following statements about electrons are false?
    (a) Most atoms have more electrons than protons.
    (b) Electrons have a charge of −1.
    (c) If an atom has an equal number of protons and electrons, it will be charge-neutral.
    (d) Electrons experience an attraction to protons.
35. Which of the following statements about protons are true?
    (a) Protons have twice the mass of neutrons.
    (b) Protons have the same magnitude of charge as electrons, but are opposite in sign.
    (c) Most atoms have more protons than electrons.
    (d) Protons have a charge of +1.
36. Which of the following statements about protons are false?
    (a) Protons have about the same mass as neutrons.
    (b) Protons have about the same mass as electrons.
    (c) Some atoms don't have any protons.
    (d) Protons have the magnitude of charge as neutrons, but opposite in sign.
37. How many electrons would it take to equal the mass of a proton?
38. A helium nucleus has two protons and two neutrons. How many electrons would it take to equal the mass of a helium nucleus?

39. What mass of electrons would be required to just neutralize the charge of 1.0 g of protons?

40. What mass of protons would be required to just neutralize the charge of 1.0 g of electrons?

Elements, Symbols, and Names

41. Find the atomic number (Z) for each of the following elements.
(a) Co
(b) Ir
(c) U
(d) Si
(e) Be

42. Find the atomic number (Z) for each of the following elements.
(a) Al
(b) Sn
(c) Ca
(d) Fr
(e) Xe

43. How many protons are in the nucleus of each of the following atoms?
(a) Mn
(b) Ag
(c) Au
(d) Pb
(e) S

44. How many protons are in the nucleus of each of the following atoms?
(a) Y
(b) N
(c) Ne
(d) K
(e) Mo

45. Give the symbol and atomic number corresponding to each of the following elements.
(a) carbon
(b) nitrogen
(c) sodium
(d) potassium
(e) copper

46. Give the symbol and atomic number corresponding to each of the following elements.
(a) boron
(b) neon
(c) silver
(d) mercury
(e) curium

47. Give the name and the atomic number corresponding to the symbol for each of the following elements.
(a) Au
(b) Si
(c) Ni
(d) Zn
(e) W

48. Give the name and the atomic number corresponding to the symbol for each of the following elements.
(a) Ta
(b) H
(c) S
(d) As
(e) Br

49. Fill in the blanks to complete the following table.

Element Name	*Element Symbol*	*Atomic Number*
_____	Au	79
Tin	_____	_____
_____	As	_____
Copper	_____	29
_____	Fe	_____
_____	_____	80

50. Fill in the blanks to complete the following table.

Element Name	*Element Symbol*	*Atomic Number*
_____	Al	13
Iodine	_____	_____
_____	Sb	_____
Sodium	_____	_____
_____	Rn	86
_____	_____	82

The Periodic Table

51. Classify each of the following elements as a metal, nonmetal, or metalloid.
(a) Sr
(b) Mg
(c) F
(d) N
(e) As

52. Classify each of the following elements as a metal, nonmetal, or metalloid.
(a) Na
(b) Ge
(c) Si
(d) Br
(e) Ag

53. Which of the following elements would you expect to lose electrons in chemical changes?
(a) potassium
(b) sulfur
(c) fluorine
(d) barium
(e) copper

54. Which of the following elements would you expect to gain electrons in chemical changes?
(a) nitrogen
(b) iodine
(c) tungsten
(d) strontium
(e) gold

55. Which of the following elements are main-group elements?
(a) Te
(b) K
(c) V
(d) Re
(e) Ag

56. Which of the following elements are *not* main-group elements?
(a) Al
(b) Br
(c) Mo
(d) Cs
(e) Pb

57. Which of the following elements are alkaline earth metals?
(a) sodium
(b) aluminum
(c) calcium
(d) barium
(e) lithium

58. Which of the following elements are alkaline earth metals?
(a) rubidium
(b) tungsten
(c) magnesium
(d) cesium
(e) beryllium

59. Which of the following elements are alkali metals?
(a) sodium
(b) potassium
(c) silver
(d) magnesium
(e) cesium

60. Which of the following elements are alkali metals?
(a) barium
(b) niobium
(c) lithium
(d) rubidium
(e) calcium

61. Classify each of the following as a halogen, a noble gas, or neither.
(a) Cl
(b) Kr
(c) F
(d) Ga
(e) He

62. Classify each of the following as a halogen, a noble gas, or neither.
(a) Ne
(b) Br
(c) S
(d) Xe
(e) I

63. To what group number does each of the following elements belong?
(a) oxygen
(b) aluminum
(c) silicon
(d) tin
(e) phosphorus

64. To what group number does each of the following elements belong?
(a) germanium
(b) nitrogen
(c) sulfur
(d) carbon
(e) boron

65. Which of the following elements do you expect to be most like sulfur? Why?
(a) nitrogen
(b) oxygen
(c) fluorine
(d) lithium
(e) potassium

66. Which of the following elements do you expect to be most like magnesium? Why?
(a) potassium
(b) silver
(c) bromine
(d) calcium
(e) lead

67. Which of the following pairs of elements do you expect to be most similar? Why?
   (a) Si and P
   (b) Cl and F
   (c) Na and Mg
   (d) Mo and Sn
   (e) N and Ni

68. Which of the following pairs of elements do you expect to be most similar? Why?
   (a) Ti and Ga
   (b) N and O
   (c) Li and Na
   (d) Ar and Br
   (e) Ge and Ga

69. Fill in the blanks to complete the following table.

Chemical Symbol	*Group Number*	*Group Name*	*Metal or Nonmetal*
K	____	____	metal
Br	____	halogens	____
Sr	____	____	____
He	8A	____	____
Ar	____	____	____

70. Fill in the blanks to complete the following table.

Chemical Symbol	*Group Number*	*Group Name*	*Metal or Nonmetal*
Cl	7A	____	____
Ca	____	____	metal
Xe	____	____	nonmetal
Na	____	alkali metal	____
F	____	____	____

Ions

71. Complete each of the following.
   (a) $Na \longrightarrow Na^+ +$ ____
   (b) $O + 2e^- \longrightarrow$ ____
   (c) $Ca \longrightarrow Ca^{2+} +$ ____
   (d) $Cl + e^- \longrightarrow$ ____

72. Complete each of the following.
   (a) $Mg \longrightarrow$ ____ $+ 2e^-$
   (b) $Ba \longrightarrow Ba^{2+} +$ ____
   (c) $I + e^- \longrightarrow$ ____
   (d) $Al \longrightarrow$ ____ $+ 3e^-$

73. Determine the charge of each of the following ions.
   (a) oxygen ion with 10 electrons
   (b) aluminum ion with 10 electrons
   (c) titanium ion with 18 electrons
   (d) iodine ion with 54 electrons

74. Determine the charge of each of the following ions.
   (a) tungsten ion with 68 electrons
   (b) tellurium ion with 54 electrons
   (c) nitrogen ion with 10 electrons
   (d) barium ion with 54 electrons

75. Determine the number of protons and electrons in each of the following ions.
   (a) K^+
   (b) S^{2-}
   (c) Sr^{2+}
   (d) Cr^{3+}

76. Determine the number of protons and electrons in each of the following.
   (a) Ga^{3+}
   (b) Se^{2-}
   (c) Al^{3+}
   (d) Br^-

77. Determine whether each of the following is true or false. If false, correct it.
   (a) The Ti^{2+} ion contains 22 protons and 24 electrons.
   (b) The I^- ion contains 53 protons and 54 electrons.
   (c) The Mg^{2+} ion contains 14 protons and 12 electrons.
   (d) The O^{2-} ion contains 8 protons and 10 electrons.

78. Determine whether each of the following is true or false. If false, correct it.
   (a) The Fe^+ ion contains 29 protons and 26 electrons.
   (b) The Cs^+ ion contains 55 protons and 56 electrons.
   (c) The Se^{2-} ion contains 32 protons and 34 electrons.
   (d) The Li^+ ion contains 3 protons and 2 electrons.

79. Predict the ion formed by each of the following:
   (a) Rb
   (b) K
   (c) Al
   (d) O

80. Predict the ion formed by each of the following:
   (a) F
   (b) N
   (c) Mg
   (d) Na

81. Predict how many electrons will most likely be gained or lost by each of the following:
   (a) Ga
   (b) Li
   (c) Br
   (d) S

82. Predict how many electrons will most likely be gained or lost by each of the following:
   (a) I
   (b) Ba
   (c) Cs
   (d) Se

83. Fill in the blanks to complete the following table.

Symbol	Ion Commonly Formed	Number of Electrons in Ion	Number of Protons in Ion
Te	_____	54	_____
In	_____	_____	49
Sr	Sr^{2+}	_____	_____
_____	Mg^{2+}	_____	12
Cl	_____	_____	_____

84. Fill in the blanks to complete the following table.

Symbol	Ion Commonly Formed	Number of Electrons in Ion	Number of Protons in Ion
F	_____	_____	9
_____	Be^{2+}	2	_____
Br	_____	36	_____
Al	_____	_____	13
O	_____	_____	_____

Isotopes

85. What are the atomic number and mass number for each of the following isotopes?
   (a) the hydrogen isotope with 1 neutron
   (b) the chromium isotope with 27 neutrons
   (c) the calcium isotope with 20 neutrons
   (d) the tantalum isotope with 108 neutrons

86. How many neutrons are in an atom with the following atomic numbers and mass numbers?
   (a) $Z = 28, A = 58$
   (b) $Z = 92, A = 238$
   (c) $Z = 21, A = 45$
   (d) $Z = 18, A = 40$

87. Write isotopic symbols of the form $^{A}_{Z}X$ for each of the following isotopes.
   (a) the oxygen isotope with 8 neutrons
   (b) the fluorine isotope with 10 neutrons
   (c) the sodium isotope with 12 neutrons
   (d) the aluminum isotope with 14 neutrons

88. Write isotopic symbols of the form X-A (for example, C-13) for each of the following isotopes.
   (a) the iodine isotope with 74 neutrons
   (b) the phosphorus isotope with 16 neutrons
   (c) the uranium isotope with 234 neutrons
   (d) the argon isotope with 22 neutrons

89. Write the symbol of each of the following in the form $^{A}_{Z}X$.
   (a) cobalt-60
   (b) neon-22
   (c) iodine-131
   (d) plutonium-244

90. Write the symbol of each of the following in the form $^{A}_{Z}X$.
   (a) U-235
   (b) V-52
   (c) P-32
   (d) Xe-144

91. Determine the number of protons and neutrons in each of the following:
   (a) $^{23}_{11}Na$ **(b)** $^{266}_{88}Ra$
   (c) $^{208}_{82}Pb$ **(d)** $^{14}_{7}N$

92. Determine the number of protons and neutrons in each of the following:
   (a) $^{33}_{15}P$ **(b)** $^{40}_{19}K$
   (c) $^{222}_{86}Rn$ **(d)** $^{99}_{43}Tc$

93. Carbon-14, present within living organisms and substances derived from living organisms, is often used to establish the age of fossils and artifacts. Determine the number of protons and neutrons in a carbon-14 isotope and write its symbol in the form $^{A}_{Z}X$.

94. Plutonium-239 is used in nuclear bombs. Determine the number of protons and neutrons in plutonium-239 and write its symbol in the form $^{A}_{Z}X$.

Atomic Mass

95. Rubidium has two naturally occurring isotopes: Rb-85 with mass 84.9118 amu and a natural abundance of 72.17%, and Rb-87 with mass 86.9092 amu and a natural abundance of 27.83%. Calculate the atomic mass of rubidium.

96. Silicon has three naturally occurring isotopes: Si-28 with mass 27.9769 amu and a natural abundance of 92.21%, Si-29 with mass 28.9765 amu and a natural abundance of 4.69%, and Si-30 with mass 29.9737 amu and a natural abundance of 3.10%. Calculate the atomic mass of silicon.

97. Bromine has two naturally occurring isotopes (Br-79 and Br-81) and an atomic mass of 79.904 amu.
   (a) If the natural abundance of Br-79 is 50.69%, what is the natural abundance of Br-81?
   (b) If the mass of Br-81 is 80.9163 amu, what is the mass of Br-79?

98. Silver has two naturally occurring isotopes (Ag-107 and Ag-109).
   (a) Use the periodic table to find the atomic mass of silver.
   (b) If the natural abundance of Ag-107 is 51.84%, what is the natural abundance of Ag-109?
   (c) If the mass of Ag-107 is 106.905, what is the mass of Ag-109?

99. An element has two naturally occurring isotopes. Isotope 1 has a mass of 120.9038 amu and a relative abundance of 57.4%, and isotope 2 has a mass of 122.9042 amu and a relative abundance of 42.6%. Find the atomic mass of this element and, by comparison to the periodic table, identify it.

100. Copper has two naturally occurring isotopes. Cu-63 has a mass of 62.939 amu and relative abundance of 69.17%. Use the atomic weight of copper to determine the mass of the other copper isotope.

Cumulative Problems

101. Electrical charge is sometimes reported in coulombs (C). On this scale, 1 electron has a charge of -1.6×10^{-19} C. Suppose your body acquires −125 mC (millicoulombs) of charge on a dry day. How many excess electrons has it acquired? (Hint: Use the charge of an electron in coulombs as a conversion factor between charge and electrons.)

102. How many excess protons are in a positively charged object with a charge of +398 mC (millicoulombs)? The charge of 1 proton is $+1.6 \times 10^{-19}$ C.

103. The hydrogen atom contains 1 proton and 1 electron. The radius of the proton is approximately 1.0 fm (femtometers) and the radius of the hydrogen atom is approximately 53 pm (picometers). Calculate the volume of the nucleus and the volume of the atom for hydrogen. What percentage of the hydrogen atom's volume is occupied by the nucleus?

104. Carbon-12 contains 6 protons and 6 neutrons. The radius of the nucleus is approximately 2.7 fm and the radius of the atom is approximately 70 pm. Calculate the volume of the nucleus and the volume of the atom. What percentage of the carbon atom's volume is occupied by the nucleus?

105. Prepare a table such as Table 4.2 for the four different isotopes of Sr that have the following natural abundances and masses.

Sr-84	0.56%	83.9134 amu
Sr-86	9.86%	85.9093 amu
Sr-87	7.00%	86.9089 amu
Sr-88	82.58%	87.9056 amu

Use your table and the preceding atomic masses to calculate the atomic mass of strontium.

106. Determine the number of protons and neutrons in each of the following isotopes of chromium and use the following natural abundances and masses to calculate its atomic mass.

Cr-50	4.345%	49.9460 amu
Cr-52	83.89%	51.9405 amu
Cr-53	9.50%	52.9407 amu
Cr-54	2.365%	53.9389 amu

107. Fill in the blanks to complete the following table.

Symbol	*Z*	*A*	*Number of Protons*	*Number of Electrons*	*Number of Neutrons*	*Charge*
Zn^{+}	___	___	___	___	34	1+
___	25	55	___	22	___	___
___	___	___	15	15	16	___
O^{2-}	___	16	___	___	___	2−
___	___	___	16	18	18	___

108. Fill in the blanks to complete the following table.

Symbol	*Z*	*A*	*Number of Protons*	*Number of Electrons*	*Number of Neutrons*	*Charge*
Mg^{2+}	___	25	___	___	13	2+
___	22	48	___	18	___	___
___	16	___	___	___	16	2−
Ga^{3+}	___	71	___	___	___	___
___	___	___	82	80	125	___

109. Europium has two naturally occurring isotopes: Eu-151 with a mass of 150.9198 and a natural abundance of 47.8%, and Eu-153. Use the atomic mass of europium to find the mass and natural abundance of Eu-153.

110. Rhenium has two naturally occurring isotopes: Re-185 with a natural abundance of 37.40%, and Re-187 with a natural abundance of 62.60%. The sum of the masses of the two isotopes is 371.9087 amu. Find the masses of the individual isotopes.

111. In Chapter 1, we learned the difference between observations, laws, and theories. Give two examples of theories from this chapter and explain why they are theories.

112. In Chapter 1, we learned the difference between observations, laws, and theories. Give one example of a law from this chapter and explain why it is a law.

Highlight Problems

113. The figure is a representation of fifty atoms of a fictitious element with the symbol Nt and atomic number 120. Nt has three isotopes represented by the following colors: Nt-304 (red), Nt-305 (blue), and Nt-306 (green).

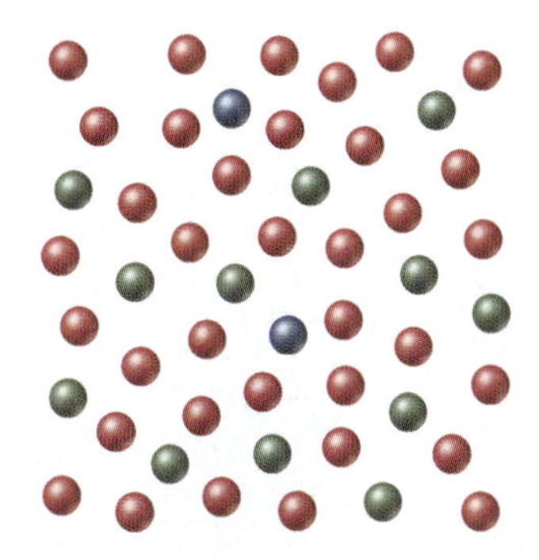

(a) Assuming that the figure is statistically representative of naturally occurring Nt, what is the percent natural abundance of each Nt isotope?

(b) Use the following masses of each isotope to calculate the atomic mass of Nt. Then draw a box for the element similar to the boxes for each element shown in the periodic table in the inside front cover of this book. Make sure your box includes the atomic number, symbol, and atomic mass. (Assume that the percentages from part (a) are good to four significant figures)

Nt-304	303.956 amu
Nt-305	304.962 amu
Nt-306	305.978 amu

114. Neutron stars are believed to be composed of solid nuclear matter, primarily neutrons.

(a) If the radius of a neutron is 1.0×10^{-13} cm, calculate the density of a neutron in g/cm^3. (volume of a sphere $= \frac{4}{3}\pi r^3$)

(b) Assuming that a neutron star has the same density as a neutron, calculate the mass in kilograms of a small piece of a neutron star the size of a spherical pebble with a radius of 0.10 mm.

Answers to Skillbuilder Exercises

Skillbuilder 4.1 (a) sodium, 11 (b) nickel, 28 (c) phosphorus, 15 (d) tantalum, 73

Skillbuilder 4.2 (a) nonmetal (b) nonmetal (c) metal (d) metalloid

Skillbuilder 4.3 (a) alkali metal (b) Group 3A (c) halogen (d) noble gas

Skillbuilder 4.4 (a) +2 (b) −1 (c) −3

Skillbuilder 4.5 16 protons, 18 electrons

Skillbuilder 4.6 K^+ and S^{2-}

Skillbuilder 4.7 Z = 17, A = 35, Cl-35 and $^{35}_{17}Cl$

Skillbuilder 4.8 19 protons, 20 neutrons

Skillbuilder 4.9 24.31 amu

Answers to Conceptual Checkpoints

4.1 (c) The mass in amu is approximately equal to the number of protons plus the number of neutrons. In order to be charge-neutral, the number of protons must equal the number of electrons.

4.2 (b) All of the metalloids are main-group elements (see Figures 4.12 and 4.13).

4.3 (a) Both these ions have 10 electrons.

4.4 (b) This atom must have (27 − 14) = 13 protons; the element with an atomic number of 13 is Al.

Across
1. A type of sugar
16. Element N
26. Thirsty?
42. Part of water
51. Not real sugar
66. ______composition

CHAPTER 5

Molecules and Compounds

"Almost all aspects of life are engineered at the molecular level, and without understanding molecules, we can only have a very sketchy understanding of life itself."

Francis Harry Compton Crick (1916–2004)

5.1 Sugar and Salt

Sodium, a shiny metal (▼ Figure 5.1) that dulls almost instantly upon exposure to air, is extremely reactive and poisonous. If you were to consume any appreciable amount of elemental sodium, you would need immediate medical help. Chlorine, a pale yellow gas (▼ Figure 5.2), is equally reactive and poisonous. Yet the compound formed from these two elements, sodium chloride, is the

◀ Ordinary table sugar is a compound called sucrose. A sucrose molecule, such as the one shown here, contains carbon, hydrogen, and oxygen atoms. However, the properties of sucrose are very different from those of carbon, hydrogen, and oxygen. The properties of a compound are, in general, different from the properties of the elements that compose it.

▲ **Figure 5.1 Elemental sodium** Sodium is an exremely reactive metal that dulls almost instantly upon exposure to air.

▲ **Figure 5.2 Elemental chlorine** Chlorine is a yellow gas with a pungent odor. It is highly reactive and poisonous.

▶ **Figure 5.3** **Sodium chloride** The compound formed by sodium and chlorine is table salt.

relatively harmless flavor enhancer that we call table salt (▲ Figure 5.3). When elements combine to form compounds, their properties change completely.

Consider ordinary sugar. Sugar is a compound composed of carbon, hydrogen, and oxygen. Each of these elements has its own unique properties. Carbon is most familiar to us as the graphite found in pencils or as the diamonds in jewelry. Hydrogen is an extremely flammable gas used as a fuel for the space shuttle, and oxygen is one of the gases that compose air. When these three elements combine to form sugar, however, a sweet, white, crystalline solid results.

In Chapter 4, we learned how protons, neutrons, and electrons compose different elements, each with its own properties and its own chemistry, each different from the other. In this chapter, we learn how these elements combine with each other to form different compounds, each with its own properties and its own chemistry, each different from all the others and different from the elements that compose it. Here is the great wonder of nature, how from such simplicity—protons, neutrons, and electrons—we get such great complexity. It is exactly this complexity, however, that makes life possible. Life could not exist with just ninety-one different elements if they did not combine to form compounds. It takes compounds in all of their diversity to make living organisms.

5.2 Compounds Display Constant Composition

Although some of the substances we encounter in everyday life are elements, most are not—they are compounds. Free atoms are rare in nature. As we learned in Chapter 3, a compound is different from a mixture of elements. In a compound, the elements combine in fixed, definite proportions, while in a mixture, they can have any proportions whatsoever. For example, consider the difference between a mixture of hydrogen and oxygen gas (▶ Figure 5.4), and the compound water (▶ Figure 5.5). A mixture of hydrogen and oxygen gas can have any proportions of hydrogen and oxygen. Water, on the other hand, is composed of water molecules that consist of two hydrogen atoms bonded to one oxygen atom. Consequently, water has a definite proportion of hydrogen to oxygen.

The first chemist to formally state the idea that elements combine in fixed proportions to form compounds was Joseph Proust (1754–1826) in the **law of constant composition**, which states:

> All samples of a given compound have the same proportions of their constituent elements.

◀ **Figure 5.4 A mixture** This balloon is filled with a mixture of hydrogen and oxygen gas. The relative amounts of hydrogen and oxygen are variable. We could easily add either more hydrogen or more oxygen to the balloon.

▶ **Figure 5.5 A chemical compound** This balloon is filled with water, composed of molecules that have a fixed ratio of hydrogen to oxygen. (*Source:* JoLynn E. Funk.)

For example, if we decompose water, we find 16.0 g of oxygen to every 2.0 g of hydrogen, or an oxygen-to-hydrogen mass ratio of:

$$\text{Mass ratio} = \frac{16.0\ \text{g O}}{2.0\ \text{g H}} = 8.0$$

This is true of any sample of pure water, no matter what its origin (▼ Figure 5.6). The law of constant composition applies not only to water, but to every compound. Consider ammonia, a compound composed of nitrogen and hydrogen. Ammonia contains 14.0 g of nitrogen to every 3.0 g of hydrogen, or a nitrogen-to-hydrogen mass ratio of:

$$\text{Mass ratio} = \frac{14.0\ \text{g N}}{3.0\ \text{g H}} = 4.7$$

Even though atoms combine in whole-number ratios, their *mass* ratios are not necessarily whole numbers.

Again, this ratio is the same for every sample of ammonia—the composition of each compound is constant.

▶ **Figure 5.6 Compounds have fixed composition** Water will always have a constant ratio of hydrogen to oxygen, no matter what its source.

2 H atoms (○) to every 1 O atom (●)

2 H atoms (○) to every 1 O atom (●)

EXAMPLE 5.1 Constant Composition of Compounds

Two samples of carbon dioxide, obtained from different sources, were decomposed into their constituent elements. One sample produced 4.8 g of oxygen and 1.8 g of carbon, and the other sample produced 17.1 g of oxygen and 6.4 g of carbon. Show that these results are consistent with the law of constant composition.

To show this, compute the mass ratio of one element to the other by dividing the larger mass by the smaller one. For the first sample:

Solution:

$$\frac{\text{Mass oxygen}}{\text{Mass carbon}} = \frac{4.8 \text{ g}}{1.8 \text{ g}} = 2.7$$

For the second sample:

$$\frac{\text{Mass oxygen}}{\text{Mass carbon}} = \frac{17.1 \text{ g}}{6.4 \text{ g}} = 2.7$$

Since the ratios are the same for the two samples, these results are consistent with the law of constant composition.

SKILLBUILDER 5.1 Constant Composition of Compounds

Two samples of carbon monoxide, obtained from different sources, were decomposed into their constituent elements. One sample produced 4.3 g of oxygen and 3.2 g of carbon, and the other sample produced 7.5 g of oxygen and 5.6 g of carbon. Are these results consistent with the law of constant composition?

 5.3 Chemical Formulas: How to Represent Compounds

The reason that compounds have constant composition with respect to mass (as we learned in the previous section) is because they are composed of atoms in fixed ratios.

We represent a compound with a **chemical formula**, which indicates the elements present in the compound and the relative number of atoms of each. For example, H_2O is the chemical formula for water; it indicates that water consists of hydrogen and oxygen atoms in a 2:1 ratio. The formula contains the symbol for each element, accompanied by a subscript indicating the number of atoms of that element. When the subscript is 1, it is omitted.

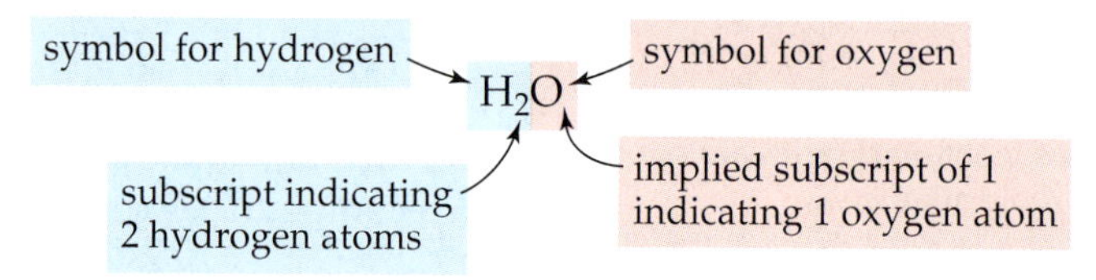

MOLECULE
Carbon Monoxide

▲ The subscripts in a chemical formula are part of what define the compound—if you change a subscript, you change the compound.

Some other common chemical formulas include NaCl for table salt, indicating sodium and chlorine atoms in a 1:1 ratio; CO_2 for carbon dioxide, indicating carbon and oxygen atoms in a 1:2 ratio; and $C_2H_{22}O_{11}$ for table sugar (sucrose), indicating carbon, hydrogen, and oxygen atoms in a 12:22:11 ratio. The subscripts in a chemical formula are part of the compound's definition—if they change, the formula no longer specifies the same compound. For example, CO is the chemical formula for carbon monoxide, an air pollutant with adverse health effects on humans. When inhaled, carbon monoxide interferes with the blood's ability to carry oxygen and can be fatal. CO is the primary substance responsible for deaths of people who inhale too much automobile exhaust. If you change the subscript of the O in CO from 1 to 2, however, you get the formula for a totally different compound. CO_2 is the chemical formula for carbon dioxide, the relatively harmless product of combustion and human respiration. We breathe small amounts of CO_2 all the time with no harmful effects. So, remember:

> The subscripts in a chemical formula represent the relative numbers of each type of atom in a chemical compound; they never change for a given compound.

Chemical formulas normally list the most metallic elements first. Therefore, the formula for table salt is NaCl, not ClNa. For compounds that do not include a metal, the more metal-like element is listed first. Recall from Chapter 4 that metals are found on the left side of the periodic table and nonmetals on the upper right side. Therefore, among nonmetals, those to the left in the periodic table are more metal-like than those to the right and are normally listed first. Therefore we write CO_2 and NO, not O_2C and ON. Within a single column in the periodic table, elements toward the bottom are more metal-like than elements toward the top. Therefore we write SO_2, not O_2S. The specific order for listing nonmetal elements in a chemical formula is shown in Table 5.1.

There are a few historical exceptions to this rule, such as the hydroxide ion, which is written as OH^-.

TABLE 5.1 Order of Listing Nonmetal Elements in a Chemical Formula

C	P	N	H	S	I	Br	Cl	O	F

Elements on the left are generally listed before elements on the right.

EXAMPLE 5.2 Writing Chemical Formulas

Write a chemical formula for each of the following:

(a) the compound containing two aluminum atoms to every three oxygen atoms
(b) the compound containing three oxygen atoms to every sulfur atom
(c) the compound containing four chlorine atoms to every carbon atom

Solution:

Since aluminum is the metal, it is listed first.

(a) Al_2O_3

Since sulfur is below oxygen on the periodic table and since it occurs before oxygen in Table 5.1, it is listed first.

(b) SO_3

Since carbon is to the left of chlorine on the periodic table and since it occurs before chlorine in Table 5.1, it is listed first.

(c) CCl_4

SKILLBUILDER 5.2 Writing Chemical Formulas

Write a chemical formula for each of the following:

(a) the compound containing two silver atoms to every sulfur atom
(b) the compound containing two nitrogen atoms to every oxygen atom
(c) the compound containing two oxygen atoms to every titanium atom

Many times these groups have a charge associated with them and are called *polyatomic ions*. For example, the NO_3 group has a negative charge, NO_3^-. Polyatomic ions are described in more detail in Section 5.7.

Some chemical formulas contain groups of atoms that act as a unit. When several groups of the same kind are present, their formula is set off in parentheses, with a subscript to indicate how many there are. For example, $Mg(NO_3)_2$ indicates a compound containing one magnesium atom and two NO_3 groups.

symbol for NO_3 group
symbol for magnesium → $Mg(NO_3)_2$ ← subscript indicating 2 NO_3 groups
implied subscript indicating 1 magnesium atom
subscript indicating 3 oxygen atoms per NO_3 group
implied subscript indicating 1 nitrogen atom per NO_3 group

To determine the total number of each type of atom within the parentheses, multiply the subscript outside the parentheses by the subscript for each atom inside the parentheses. Therefore, the preceding formula has the following numbers of each type of atom.

Mg: 1 Mg

N: $1 \times 2 = 2$ N (implied 1 inside parentheses times 2 outside parentheses)

O: $3 \times 2 = 6$ O (3 inside parentheses times 2 outside parentheses)

EXAMPLE 5.3 Total Number of Each Type of Atom in a Chemical Formula

Determine the number of each type of atom in $Mg_3(PO_4)_2$.

Solution:

Mg: There are three Mg atoms, as indicated by the subscript 3.

P: There are two P atoms, as we see by multiplying the subscript outside the parentheses (2) by the subscript for P inside the parentheses, which is 1 (implied).

O: There are eight O atoms, as we see by multiplying the subscript outside the parentheses (2) by the subscript for O inside the parentheses (4).

SKILLBUILDER 5.3 Total Number of Each Type of Atom in a Chemical Formula

Determine the number of each type of atom in K_2SO_4.

SKILLBUILDER PLUS

Determine the number of each type of atom in $Al_2(SO_4)_3$.

CONCEPTUAL CHECKPOINT 5.1

Which of the following formulas represents the greatest number of atoms?

(a) $Al(C_2H_3O_2)_3$
(b) $Al_2(Cr_2O_7)_3$
(c) $Pb(HSO_4)_4$
(d) $Pb_3(PO_4)_4$
(e) $(NH_4)_3PO_4$

5.4 A Molecular View of Elements and Compounds

In Chapter 3, we learned that pure substances could be divided into elements or compounds. We can further subdivide elements and compounds according to the basic units that compose them (▶ Figure 5.7). Pure substances may be elements, or they may be compounds. Elements may be either atomic or molecular. Compounds may be either molecular or ionic.

▶ **Figure 5.7 A molecular view of elements and compounds**

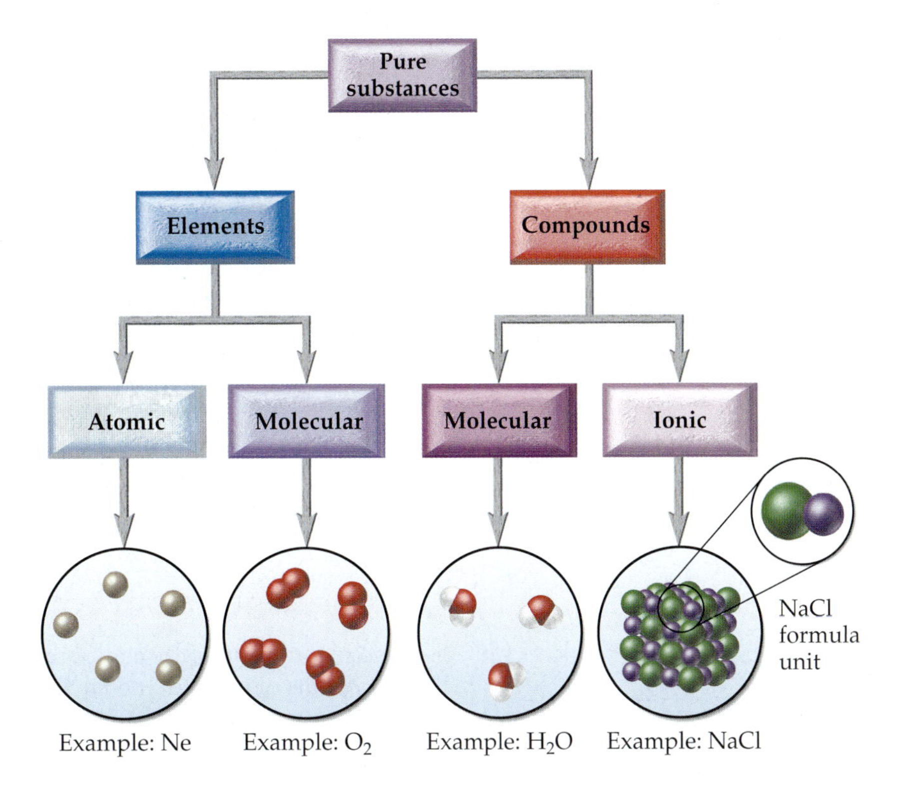

Atomic Elements

Atomic elements are those that exist in nature with single atoms as their basic units. Most elements fall into this category. For example, helium is composed of helium atoms, copper is composed of copper atoms, and mercury is composed of mercury atoms (▼ Figure 5.8).

Molecular Elements

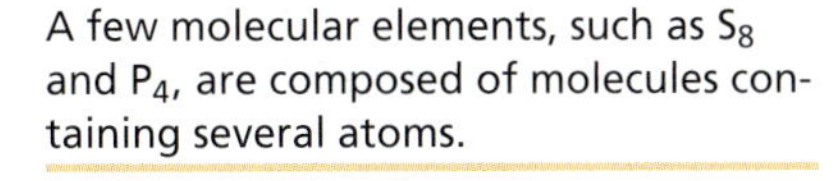

A few molecular elements, such as S_8 and P_4, are composed of molecules containing several atoms.

Molecular elements do not normally exist in nature with single atoms as their basic units. Instead, these elements exist as *diatomic molecules*—two atoms of that element bonded together—as their basic units. For example, hydrogen is composed of H_2 molecules, oxygen is composed of O_2 molecules, and chlorine is composed of Cl_2 molecules (▼ Figure 5.9). Elements that exist as diatomic molecules are shown in Table 5.2 and in ▶ Figure 5.10.

◀ **Figure 5.8 An atomic element** The basic units that compose mercury, an atomic element and a metal, are single mercury atoms.

▶ **Figure 5.9 A molecular element** The basic units that compose chlorine, a molecular element, are diatomic chlorine molecules, each composed of two chlorine atoms.

Period	1A 1	2A 2	3B 3	4B 4	5B 5	6B 6	7B 7	8B 8	8B 9	8B 10	1B 11	2B 12	3A 13	4A 14	5A 15	6A 16	7A 17	8A 18
1	1 H																	2 He
2	3 Li	4 Be											5 B	6 C	7 N	8 O	9 F	10 Ne
3	11 Na	12 Mg											13 Al	14 Si	15 P	16 S	17 Cl	18 Ar
4	19 K	20 Ca	21 Sc	22 Ti	23 V	24 Cr	25 Mn	26 Fe	27 Co	28 Ni	29 Cu	30 Zn	31 Ga	32 Ge	33 As	34 Se	35 Br	36 Kr
5	37 Rb	38 Sr	39 Y	40 Zr	41 Nb	42 Mo	43 Tc	44 Ru	45 Rh	46 Pd	47 Ag	48 Cd	49 In	50 Sn	51 Sb	52 Te	53 I	54 Xe
6	55 Cs	56 Ba	57 La	72 Hf	73 Ta	74 W	75 Re	76 Os	77 Ir	78 Pt	79 Au	80 Hg	81 Tl	82 Pb	83 Bi	84 Po	85 At	86 Rn
7	87 Fr	88 Ra	89 Ac	104 Rf	105 Db	106 Sg	107 Bh	108 Hs	109 Mt	110	111	112		114		116		

Lanthanides	58 Ce	59 Pr	60 Nd	61 Pm	62 Sm	63 Eu	64 Gd	65 Tb	66 Dy	67 Ho	68 Er	69 Tm	70 Yb	71 Lu
Actinides	90 Th	91 Pa	92 U	93 Np	94 Pu	95 Am	96 Cm	97 Bk	98 Cf	99 Es	100 Fm	101 Md	102 No	103 Lr

Main groups; Transition metals; Periods; Elements that exist as diatomic molecules (H, N, O, F, Cl, Br, I highlighted)

▶ **Figure 5.10 Elements that form diatomic molecules** Elements that normally exist as diatomic molecules are highlighted in yellow on this periodic table. Note that six of the seven are nonmetals, including four of the halogens.

TABLE 5.2 Elements That Occur as Diatomic Molecules

Name of Element	Formula of Basic Unit
hydrogen	H_2
nitrogen	N_2
oxygen	O_2
fluorine	F_2
chlorine	Cl_2
bromine	Br_2
iodine	I_2

Molecular Compounds

Molecular compounds are compounds formed from two or more nonmetals. The basic units of molecular compounds are molecules composed of the constituent atoms. For example, water is composed of H_2O molecules, dry ice is composed of CO_2 molecules (▼ Figure 5.11), and acetone (finger nail-polish remover) is composed of C_3H_6O molecules.

Ionic Compounds

Ionic compounds are compounds formed from a metal and one or more nonmetals. When a metal, which has a tendency to lose electrons, combines with a nonmetal, which has a tendency to gain electrons, one or more electrons transfer from the metal to the nonmetal, creating positive and negative ions that are then attracted to each other. The basic unit of ionic compounds is the **formula unit**, the smallest electrically neutral collection of ions. Formula units are different from molecules in that they do not exist as discrete entities, but rather as part of a larger lattice. For example, salt (NaCl) is composed of Na^+ and Cl^- ions in a 1:1 ratio. In table salt, Na^+ and Cl^- ions exist in an alternating three-dimensional array (▼ Figure 5.12). However, any one Na^+ ion does

Remember that nonmetals occupy the upper right side of the periodic table.

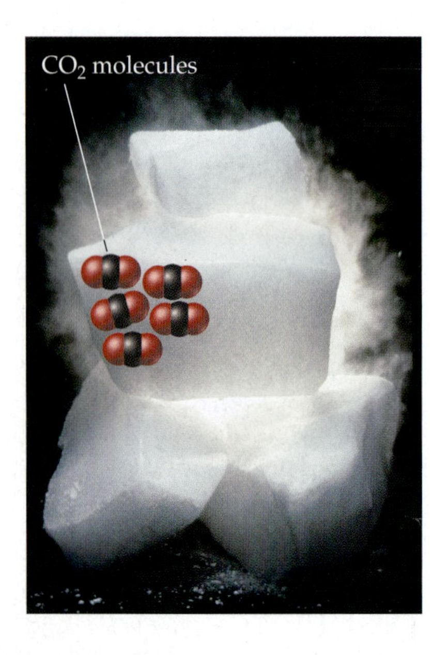

◀ **Figure 5.11 A molecular compound** The basic units that compose dry ice, a molecular compound, are CO_2 molecules.

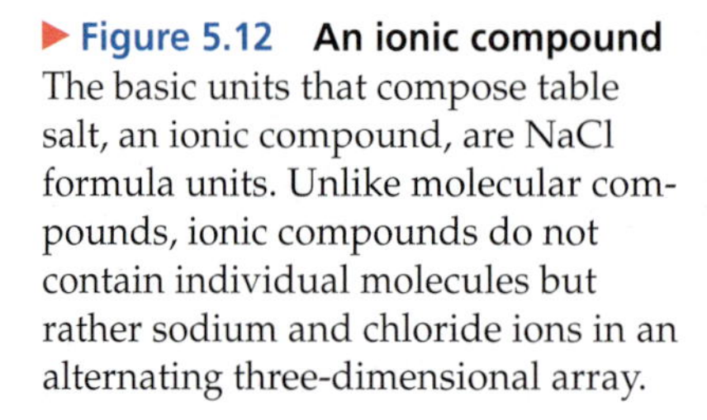
▶ **Figure 5.12 An ionic compound** The basic units that compose table salt, an ionic compound, are NaCl formula units. Unlike molecular compounds, ionic compounds do not contain individual molecules but rather sodium and chloride ions in an alternating three-dimensional array.

not pair with one specific Cl^- ion. Sometimes chemists refer to formula units as molecules, but this is not strictly correct since ionic compounds do not contain distinct molecules.

EXAMPLE 5.4 Classifying Substances as Atomic Elements, Molecular Elements, Molecular Compounds, or Ionic Compounds

Classify each of the following substances as an atomic element, molecular element, molecular compound, or ionic compound.

(a) krypton
(b) $CoCl_2$
(c) nitrogen
(d) SO_2
(e) KNO_3

Solution:

(a) Krypton is an element that is not listed as diatomic in Table 5.2; therefore, it is an atomic element.

(b) $CoCl_2$ is a compound composed of a metal (left side of periodic table) and nonmetal (right side of the periodic table); therefore, it is an ionic compound.

(c) Nitrogen is an element that is listed as diatomic in Table 5.2; therefore, it is a molecular element.

(d) SO_2 is a compound composed of two nonmetals; therefore, it is a molecular compound.

(e) KNO_3 is a compound composed of a metal and two nonmetals; therefore, it is an ionic compound.

SKILLBUILDER 5.4 Classifying Substances as Atomic Elements, Molecular Elements, Molecular Compounds, or Ionic Compounds

Classify each of the following substances as an atomic element, molecular element, molecular compound, or ionic compound.

(a) chlorine
(b) NO
(c) Au
(d) Na_2O
(e) $CrCl_3$

5.5 Writing Formulas for Ionic Compounds

Review Section 4.7 and Figure 4.14 to learn the elements that form ions with a predictable charge.

Since ionic compounds must be charge-neutral and since many elements form only one type of ion with a predictable charge, the formulas for many ionic compounds can be determined based on their constituent elements. For example, the formula for the ionic compound composed of sodium and chlorine must be NaCl and not anything else because in compounds Na always forms 1+ cations and Cl always forms 1− anions. In order for the compound to be charge-neutral, it must contain one Na^+ cation to every Cl^- anion. The formula for the ionic compound composed of magnesium and chlorine, however, must be $MgCl_2$ because Mg always forms 2+ cations and Cl always forms 1− anions. In order for the compound to be charge-neutral, it must contain one Mg^{2+} cation to every two Cl^- anions. In general:

TUTORIAL
Building Blocks for Naming Ionic Compounds

- Ionic compounds always contain positive and negative ions.
- In the chemical formula, the sum of the charges of the positive ions (cations) must always equal the sum of the charges of the negative ions (anions).

To write the formula for an ionic compound, follow the procedure in the left column below. Two examples of how to apply the procedure are provided in the center and right columns.

Writing Formulas for Ionic Compounds	EXAMPLE 5.5	EXAMPLE 5.6
	Write a formula for the ionic compound that forms from aluminum and oxygen.	Write a formula for the ionic compound that forms from magnesium and oxygen.
1. Write the symbol for the metal and its charge followed by the symbol of the nonmetal and its charge. For many elements, these charges can be determined from their group number in the periodic table (refer to Figure 4.14).	**Solution:** Al^{3+} O^{2-}	**Solution:** Mg^{2+} O^{2-}
2. Make the magnitude of the charge on each ion (without the sign) become the subscript for the other ion.	Al^{3+} O^{2-} ↓ Al_2O_3	Mg^{2+} O^{2-} ↓ Mg_2O_2
3. Reduce the subscripts to give a ratio with the smallest whole numbers.	In this case, the numbers cannot be reduced any further; the correct formula is Al_2O_3.	To reduce the subscripts, divide both subscripts by 2. $Mg_2O_2 \div 2 = MgO$
4. Check that the sum of the charges of the cations exactly cancels the sum of the charges of the anions.	Cations: $2(3+) = 6+$ Anions: $3(2-) = 6-$ The charges cancel.	Cations: $2+$ Anions: $2-$ The charges cancel.
	SKILLBUILDER 5.5 Write a formula for the compound formed from cesium and oxygen.	**SKILLBUILDER 5.6** Write a formula for the compound formed from aluminum and nitrogen.

EXAMPLE 5.7 Writing Formulas for Ionic Compounds

Write a formula for the compound that forms from potassium and oxygen.

Solution:

We first write the symbol for each ion along with its appropriate charge from its group number in the periodic table.

K^+ O^{2-}

We then make the magnitude of each ion's charge become the subscript for the other ion.

K^+ O^{2-} becomes K_2O

No reducing of subscripts is necessary in this case. Finally, we check to see that the sum of the charges of the cations $[2(1+) = 2+]$ exactly cancels the sum of the charges of the anion $(2-)$. The correct formula is K_2O.

SKILLBUILDER 5.7 Writing Formulas for Ionic Compounds
Write a formula for the compound that forms from calcium and bromine.

5.6 Nomenclature: Naming Compounds

TUTORIAL
Chemistry Nomenclature Tools

Since there are so many different compounds, chemists have developed systematic ways to name them. If you learn these, you can simply examine a compound's formula to determine its name or vice versa. Many compounds, however, also have a common name. For example, H_2O has the common name *water* and the systematic name *dihydrogen monoxide*. A common name is like a nickname for a compound, used by those who are familiar with it. Since water is such a familiar compound, everyone uses its common name and not its systematic name. In the sections that follow, we learn how to systematically name simple ionic and molecular compounds. Keep in mind, however, that some compounds also have common names that are often used instead of the systematic name. Common names can be learned only through familiarity.

5.7 Naming Ionic Compounds

The first step in naming an ionic compound is identifying it as one. Remember, *ionic compounds are formed from metals and nonmetals;* any time you see a metal and one or more nonmetals together in a chemical formula, it is an ionic compound. Ionic compounds can be divided into two types (▼ Figure 5.13) depending on the metal in the compound. If the metal, in all of its different compounds, always forms a cation with the same charge, then the charge is implied and the compound is a **Type I compound**. Sodium, for instance, has a 1+ charge in all of its compounds and therefore forms Type I compounds. Some examples of Type I metals are listed in Table 5.3. For most of these metals, their charge can be inferred from their group number in the periodic table (see Figure 4.14).

If, on the other hand, the metal forms compounds in which its charge is *not* always the same, then the charge is not implied and the compound is a **Type II compound**. Iron, for instance, has a 2+ charge in some of its compounds and a 3+ charge in others. Such metals are usually found in the section of the periodic table known as the **transition metals** (Figure ▶ 5.14). (Some examples of metals that form Type II ionic compounds are shown in Table 5.5.)

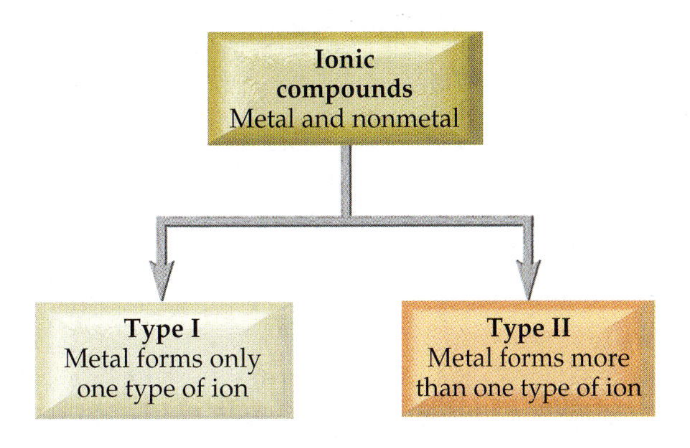

▶ **Figure 5.13 Classification of ionic compounds** Ionic compounds can be divided into two types, depending on the metal in the compound. If the metal, in all of its compounds, always forms an ion with the same charge, it is a Type I ionic compound. If the metal, in all of its compounds, forms ions with different charges, it is a Type II ionic compound.

TABLE 5.3 Metals That Form Type I Ionic Compounds

Metal	Ion	Name	Group Number
Li	Li^{+}	lithium	1A
Na	Na^{+}	sodium	1A
K	K^{+}	potassium	1A
Rb	Rb^{+}	rubidium	1A
Cs	Cs^{+}	cesium	1A
Be	Be^{2+}	beryllium	2A
Mg	Mg^{2+}	magnesium	2A
Ca	Ca^{2+}	calcium	2A
Sr	Sr^{2+}	strontium	2A
Ba	Ba^{2+}	barium	2A
Al	Al^{3+}	aluminium	3A
Zn	Zn^{2+}	zinc	*
Ag	Ag^{+}	silver	*

*The charge of these metals cannot be inferred from their group number.

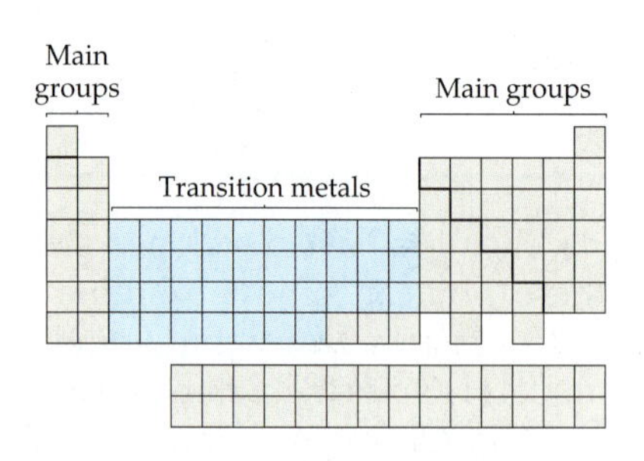

▲ **Figure 5.14 The transition metals** The elements that form Type II ionic compounds are usually transition metals.

Naming Type I Binary Ionic Compounds

Binary compounds are those containing only two different elements. The names for binary Type I ionic compounds have the following form:

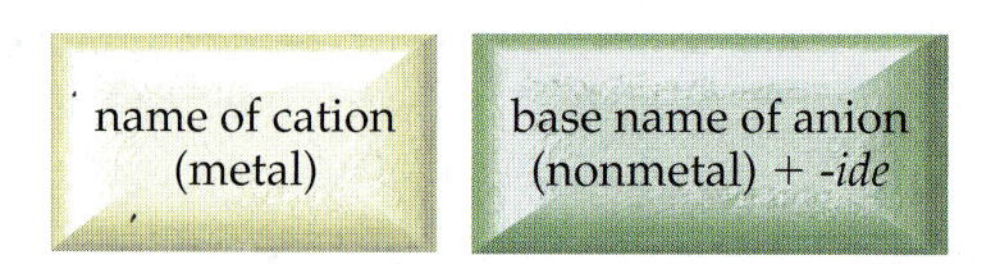

For example, the name for NaCl consists of the name of the cation, *sodium*, followed by the base name of the anion, *chlor*, with the ending *-ide*. The full name is *sodium chloride*.

NaCl sodium chloride

The name of the cation in ionic compounds is the same as the name of the metal.

The name for MgO consists of the name of the cation, *magnesium*, followed by the base name of the anion, *ox*, with the ending *-ide*. The full name is *magnesium oxide*.

MgO magnesium oxide

The base names for various nonmetals and their most common charges in ionic compounds are shown in Table 5.4.

TABLE 5.4 Some Common Anions

Nonmetal	Symbol for Ion	Base Name	Anion Name
fluorine	F^{-}	fluor-	fluoride
chlorine	Cl^{-}	chlor-	chloride
bromine	Br^{-}	brom-	bromide
iodine	I^{-}	iod-	iodide
oxygen	O^{2-}	ox-	oxide
sulfur	S^{2-}	sulf-	sulfide
nitrogen	N^{3-}	nitr-	nitride

EXAMPLE 5.8 Naming Type I Ionic Compounds

Give the name for the compound MgF_2.

Solution:

The cation is magnesium. The anion is fluorine, which becomes *fluoride*. The correct name is *magnesium fluoride*.

SKILLBUILDER 5.8 Naming Type I Ionic Compounds

Give the name for the compound KBr.

SKILLBUILDER PLUS

Give the name for the compound Zn_3N_2.

Naming Type II Binary Ionic Compounds

The names for binary (two-element) Type II ionic compounds have the following form:

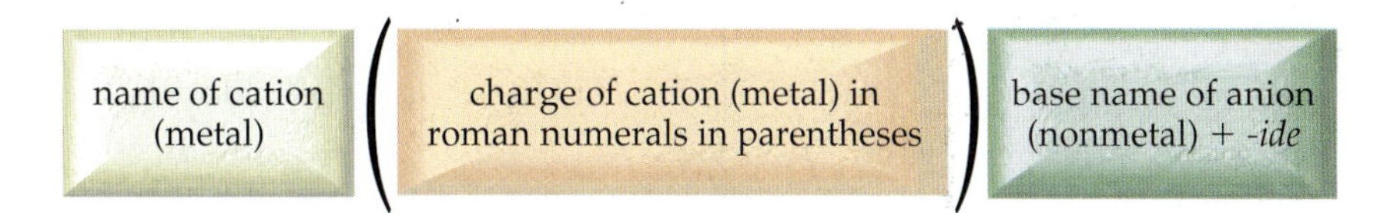

See Table 5.5 for some examples of metals that form Type II ionic compounds. The charge of the metal cation is obtained by inference from the sum of the charges of the nonmetal anions—remember that the sum of all the charges must be zero. For example, the name for $FeCl_3$ consists of the name of the cation, *iron*, followed by the charge of the cation in parentheses *(III)*, followed by the base name of the anion, *chlor*, with the ending *-ide*. The full name is *iron(III) chloride*.

$FeCl_3$ iron(III) chloride

TABLE 5.5 Some Metals That Form Type II Ionic Compounds and Their Common Charges

Metal	Symbol Ion	Name	Older Name*
chromium	Cr^{2+}	chromium(II)	chromous
	Cr^{3+}	chromium(III)	chromic
iron	Fe^{2+}	iron(II)	ferrous
	Fe^{3+}	iron(III)	ferric
cobalt	Co^{2+}	cobalt(II)	cobaltous
	Co^{3+}	cobalt(III)	cobaltic
copper	Cu^{+}	copper(I)	cuprous
	Cu^{2+}	copper(II)	cupric
tin	Sn^{2+}	tin(II)	stannous
	Sn^{4+}	tin(IV)	stannic
mercury	Hg_2^{2+}	mercury(I)	mercurous
	Hg^{2+}	mercury(II)	mercuric
lead	Pb^{2+}	lead(II)	plumbous
	Pb^{4+}	lead(IV)	plumbic

*An older naming system substitutes the names found in this column for the name of the metal and its charge. Under this system, chromium(II) oxide is named chromous oxide. We will *not* use this older system in this text.

The charge of iron must be 3+ in order for the compound to be charge neutral with three Cl^- anions. Likewise, the name for CrO consists of the name of the cation, *chromium*, followed by the charge of the cation in parentheses *(II)*, followed by the base name of the anion, *ox-*, with the ending *-ide*. The full name is *chromium(II) oxide*.

CrO chromium(II) oxide

The charge of chromium must be 2+ in order for the compound to be charge-neutral with one O^{2-} anion.

EXAMPLE 5.9 Naming Type II Ionic Compounds

Give the name for the compound $PbCl_4$.

Solution:

The name for $PbCl_4$ consists of the name of the cation, *lead*, followed by the charge of the cation in parenthesis *(IV)*, followed by the base name of the anion, *chlor-*, with the ending *-ide*. The full name is *lead(IV) chloride*. We know the charge on Pb is 4+ because the charge on Cl is 1−. Since there are 4 Cl^- anions, the Pb cation must be Pb^{4+}.

$PbCl_4$ lead(IV) chloride

SKILLBUILDER 5.9 Naming Type II Ionic Compounds

Give the name for the compound PbO.

Naming Ionic Compounds Containing a Polyatomic Ion

Some ionic compounds contain ions that are themselves composed of a group of atoms with an overall charge. These ions are called **polyatomic ions** and are shown in Table 5.6. Ionic compounds containing polyatomic ions are named using the same procedure we apply to other ionic compounds, except that the name of the polyatomic ion is used whenever it occurs. For example, KNO_3

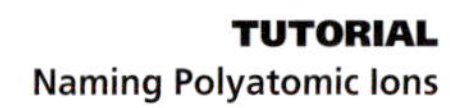

TABLE 5.6 Some Common Polyatomic Ions

Name	Formula	Name	Formula
acetate	$C_2H_3O_2^-$	hypochlorite	ClO^-
carbonate	CO_3^{2-}	chlorite	ClO_2^-
hydrogen carbonate (or bicarbonate)	HCO_3^-	chlorate	ClO_3^-
hydroxide	OH^-	perchlorate	ClO_4^-
nitrate	NO_3^-	permanganate	MnO_4^-
nitrite	NO_2^-	sulfate	SO_4^{2-}
chromate	CrO_4^{2-}	sulfite	SO_3^{2-}
dichromate	$Cr_2O_7^{2-}$	hydrogen sulfite (or bisulfite)	HSO_3^-
phosphate	PO_4^{3-}	hydrogen sulfate (or bisulfate)	HSO_4^-
hydrogen phosphate	HPO_4^{2-}	peroxide	O_2^{2-}
ammonium	NH_4^+	cyanide	CN^-

is named according to its cation, K^+, *potassium*, and its polyatomic anion, NO_3^-, *nitrate*. The full name is *potassium nitrate*.

KNO_3 potassium nitrate

$Fe(OH)_2$ is named according to its cation, *iron*, its charge *(II)*, and its polyatomic ion, *hydroxide*. The full name is *iron(II) hydroxide*.

$Fe(OH)_2$ iron(II) hydroxide

If the compound contains both a polyatomic cation and a polyatomic anion, simply use the names of both polyatomic ions. For example, NH_4NO_3 is named *ammonium nitrate*.

NH_4NO_3 ammonium nitrate

MOLECULES
Sulfite Ion
Sulfate Ion

You must be able to recognize polyatomic ions in a chemical formula, so be familiar with Table 5.6. Most polyatomic ions are **oxyanions**, anions containing oxygen. Notice that when a series of oxyanions contain different numbers of oxygen atoms, they are named systematically according to the number of oxygen atoms in the ion. If there are two ions in the series, the one with more oxygen atoms is given the ending *-ate* and the one with fewer is given the ending *-ite*. For example, NO_3^- is called *nitrate* and NO_2^- is called *nitrite*.

NO_3^- nitrate
NO_2^- nitrite

MOLECULES
Perchlorate Ion
Potassium Permanganate

If there are more than two ions in the series, then the prefixes *hypo-*, meaning "less than," and *per-*, meaning "more than," are used. So ClO^- is called *hypochlorite*, meaning "less oxygen than chlorite," and ClO_4^- is called *perchlorate*, meaning "more oxygen than chlorate."

ClO^- hypochlorite
ClO_2^- chlorite
ClO_3^- chlorate
ClO_4^- perchlorate

EXAMPLE 5.10 Naming Ionic Compounds That Contain a Polyatomic Ion

Give the name for the compound K_2CrO_4.

The name for K_2CrO_4 consists of the name of the cation, *potassium*, followed by the name of the polyatomic ion, *chromate*.

Solution:

K_2CrO_4 potassium chromate

SKILLBUILDER 5.10 Naming Ionic Compounds That Contain a Polyatomic Ion

Give the name for the compound $Mn(NO_3)_2$.

EVERYDAY Chemistry

Polyatomic Ions

A glance at the labels of household products reveals the importance of polyatomic ions in everyday compounds. For example, the active ingredient in household bleach is sodium hypochlorite, which acts to destroy color-causing molecules in clothes (bleaching action) and to kill bacteria (disinfection). A box of baking soda contains sodium bicarbonate (sodium hydrogen carbonate), which acts as an antacid when consumed in small amounts and also as a source of carbon dioxide gas in baking. The pockets of carbon dioxide gas make baked goods fluffy rather than flat.

▲ Compounds containing polyatomic ions are present in many consumer products.

Calcium carbonate is the active ingredient in many antacids such as Tums and Alka-Mints. It neutralizes stomach acids, relieving the symptoms of indigestion and heartburn. Too much calcium carbonate, however, can cause constipation, so Tums should not be overused. Sodium nitrite is a common food additive used to preserve packaged meats such as ham, hot dogs, and bologna. Sodium nitrite inhibits the growth of bacteria, especially those that cause botulism, an often fatal type of food poisoning.

CAN YOU ANSWER THIS? *Write a formula for each of these compounds, which contain polyatomic ions: sodium hypochlorite, sodium bicarbonate, calcium carbonate, sodium nitrite.*

▲ The active ingredient in bleach is sodium hypochlorite.

5.8 Naming Molecular Compounds

The first step in naming a molecular compound is identifying it as one. Remember, molecular compounds form from two or more nonmetals. In this section, we learn how to name binary (two-element) molecular compounds. Their names have the following form:

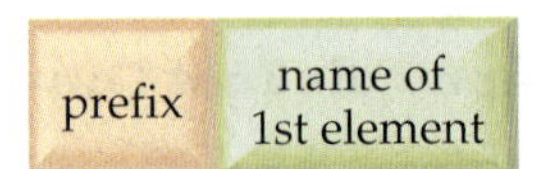

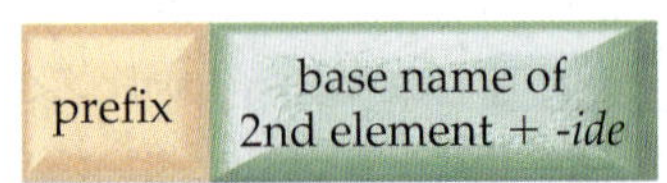

When writing the name of a molecular compound, as when writing the formula, the first element is the more metal-like one (see Table 5.1). The prefixes given to each element indicate the number of atoms present.

mono- 1	*penta-* 5
di- 2	*hexa-* 6
tri- 3	*hepta-* 7
tetra- 4	*octa-* 8

If there is only one atom of the *first element* in the formula, the prefix *mono-* is normally omitted. For example, CO_2 is named according to the first element, *carbon*, with no prefix because *mono-* is omitted for the first element, followed by the prefix *di-*, to indicate two oxygen atoms, followed by the base name of the second element, *ox*, with the ending *-ide.*

carbon di- ox -ide

The full name is *carbon dioxide.*

CO_2 carbon dioxide

When the prefix ends with a vowel and the base name starts with a vowel, the first vowel is normally dropped.

The compound N_2O, also called laughing gas, is named according to the first element, *nitrogen*, with the prefix *di-*, to indicate that there are two of them, followed by the base name of the second element, *ox*, prefixed by *mono-*, to indicate one, and the suffix *-ide*. Since *mono-* ends with a vowel and *oxide* begins with one, an *o* is dropped and the two are combined as *monoxide*. The entire name is *dinitrogen monoxide.*

N_2O dinitrogen monoxide

EXAMPLE 5.11 Naming Molecular Compounds

Name each of the following: CCl_4 BCl_3 SF_6

Solution:

CCl_4

The name of the compound is the name of the first element, *carbon*, followed by the base name of the second element, *chlor*, prefixed by *tetra-* to indicate four, and the suffix *-ide*. The entire name is *carbon tetrachloride.*

BCl_3

The name of the compound is the name of the first element, *boron*, followed by the base name of the second element, *chlor*, prefixed by *tri-* to indicate three, and the suffix *-ide*. The entire name is *boron trichloride.*

SF_6

The name of the compound is the name of the first element, *sulfur*, followed by the base name of the second element, *fluor*, prefixed by *hexa-* to indicate six, and the suffix *-ide*. The entire name is *sulfur hexafluoride.*

SKILLBUILDER 5.11 Naming Molecular Compounds

Name the compound N_2O_4.

5.9 Naming Acids

HCl(g) refers to HCl molecules in the gas phase.

Acids are molecular compounds that dissolve in water to form H^+ ions. They are composed of hydrogen, usually written first in their formula, and one or more nonmetals, written second. Acids are characterized by their sour taste and their ability to dissolve some metals. For example, HCl is a molecular compound that, when dissolved in water, forms H^+ ions. It is an acid. To emphasize that acids occur in water solutions, HCl in water is sometimes represented as HCl(aq), where (aq) means "aqueous" or "dissolved in water." HCl has a characteristically sour taste. Since HCl is present in stomach fluids, its sour taste becomes painfully obvious during vomiting. HCl also dissolves some metals. For example, if you drop a strip of zinc into a beaker of HCl, it will slowly disappear as the HCl converts the zinc metal into Zn^{2+} cations.

Acids are present in many foods, such as lemons and limes, and they are used in some household products such as toilet bowl cleaner and Lime-A-Way. In this section, we simply learn how to name them, but in Chapter 14 we will learn more about their properties. Acids can be divided into two categories: **binary acids**, those containing only hydrogen and a nonmetal, and **oxyacids**, those containing hydrogen, a nonmetal, and oxygen (▼ Figure 5.15).

 Naming Binary Acids

Binary acids are composed of hydrogen and a nonmetal. The names for binary acids have the following form:

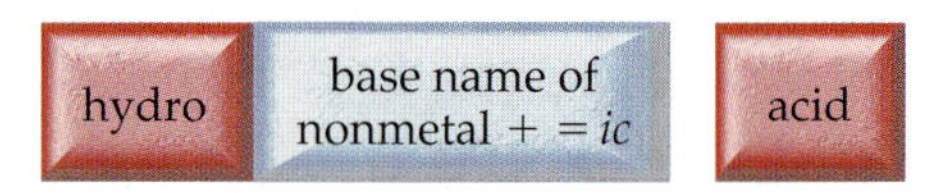

For example, HCl is named hydro*chlor*ic acid and HBr is named hydro*brom*ic acid.

HCl hydrochloric acid HBr hydrobromic acid

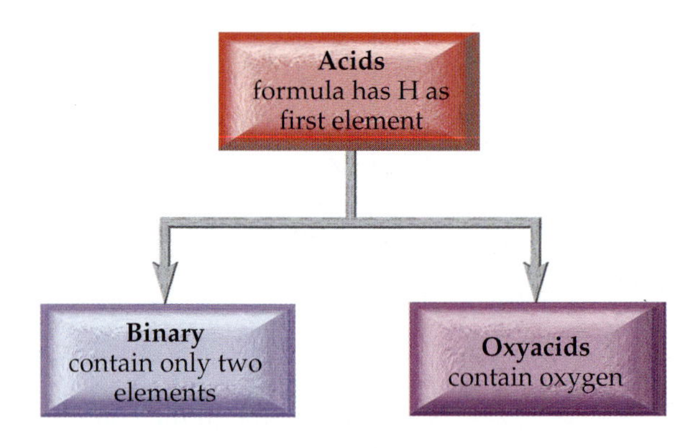

▶ **Figure 5.15 Classification of acids** Acids can be divided into two types, depending on the number of elements in the acid. If the acid contains only two elements, it is a binary acid. If it contains oxygen, it is an oxyacid.

EXAMPLE 5.12 Naming Binary Acids

Give the name of H_2S.

The base name of S is *sulfur*, so the name is *hydrosulfuric acid*.

Solution:

H_2S hydrosulfuric acid

SKILLBUILDER 5.12 Naming Binary Acids

Give the name of HI.

Naming Oxyacids

Oxyacids are derived from the oxyanions found in the table of polyatomic ions (Table 5.6). For example, HNO_3 is derived from the nitrate (NO_3^-) ion, H_2SO_3 is derived from the sulfite (SO_3^{2-}) ion, and H_2SO_4 is derived from the sulfate (SO_4^{2-}) ion. Notice that these acids are simply a combination of one or more H^+ ions with an oxyanion. The number of H^+ ions depends on the charge of the oxyanion, so that the formula is always charge-neutral. The names of oxyacids depend on the ending of the oxyanion (▼ Figure 5.16) and have the following forms:

Oxyanions ending with *-ate*

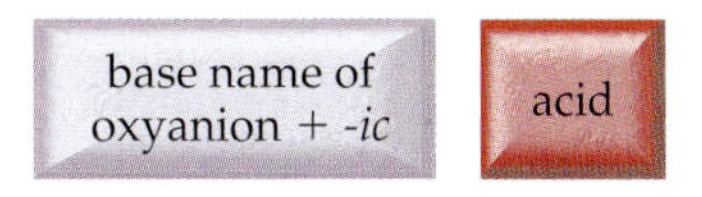

Oxyanions ending with *-ite*

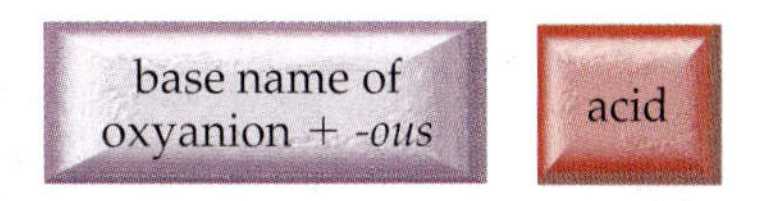

So HNO_3 is named *nitric acid* (oxyanion is nitrate), and H_2SO_3 is named *sulfurous acid* (oxyanion is sulfite).

HNO_3 nitric acid H_2SO_3 sulfurous acid

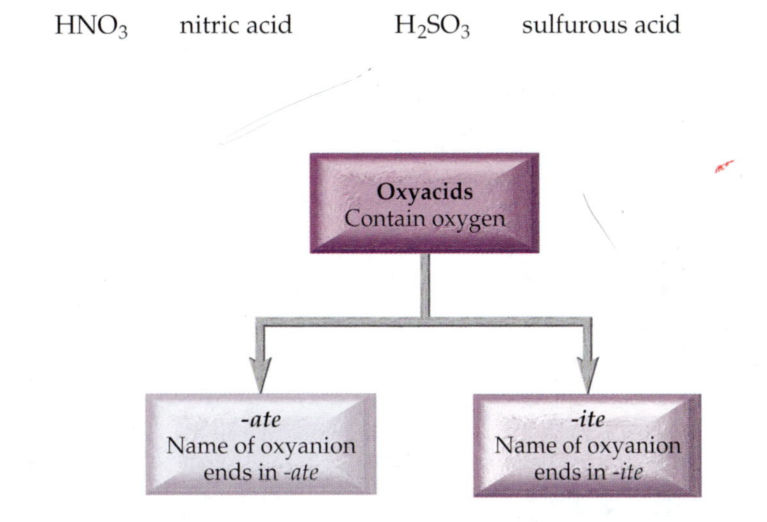

▲ **Figure 5.16 Classification of oxyacids** Oxyacids can be divided into two types, depending on the ending of the oxyanion from which they are derived.

EXAMPLE 5.13 Naming Oxyacids

Give the name of $HC_2H_3O_2$.

Solution:

The oxyanion is acetate, which ends in *-ate*; therefore, the name of the acid is *acetic acid*.

$HC_2H_3O_2$ acetic acid

SKILLBUILDER 5.13 Naming Oxyacids

Give the name of HNO_2.

5.10 Nomenclature Summary

Acids are technically a subclass of molecular compounds; that is, they are molecular compounds that form H^+ ions when dissolved in water.

Naming compounds requires several steps. The flow chart in ▼ Figure 5.17 summarizes the different categories of compounds that we have learned and how to identify and name them. The first step is to decide whether the compound is ionic, molecular, or an acid. You can recognize ionic compounds by the presence of a metal and a nonmetal, molecular compounds by two or more nonmetals, and acids by the presence of hydrogen (written first) and one or more nonmetals.

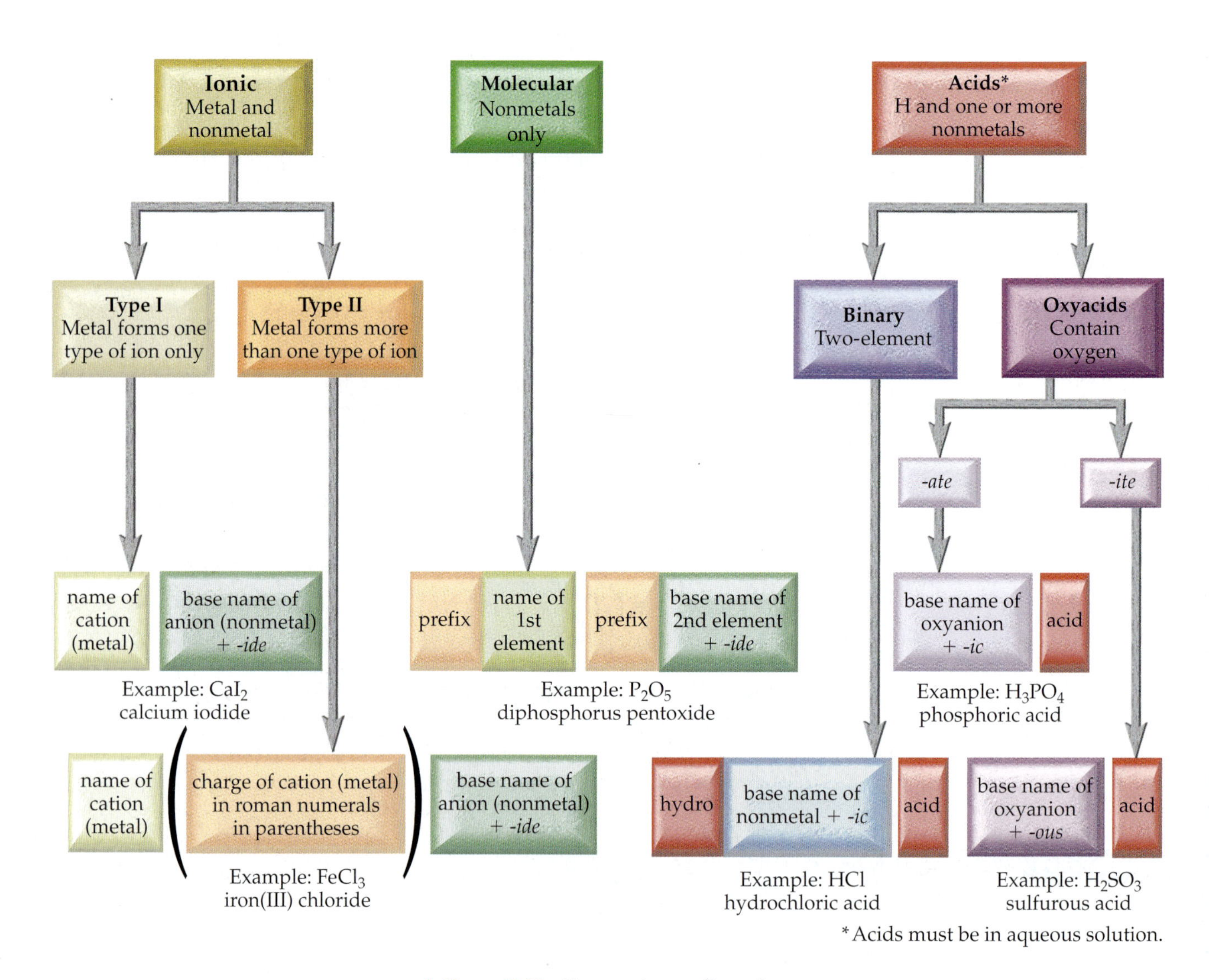

▲ Figure 5.17 Nomenclature flow chart

Ionic Compounds

For an ionic compound, you must next decide whether it is a Type I or Type II ionic compound. Group 1A (alkali) metals, Group 2A (alkaline earth) metals, and aluminum will always form Type I ionic compounds (Figure 4.14). Most of the transition metals will form Type II ionic compounds. Once you have identified the type of ionic compound, name it according to the scheme in the chart. If the ionic compound contains a polyatomic ion—something you must recognize by familiarity—insert the name of the polyatomic ion in place of the metal (positive polyatomic ion) or the nonmetal (negative polyatomic ion).

Molecular Compounds

We have learned how to name only one type of molecular compound, the binary (two-element) compound. If you identify a compound as molecular, name it according to the scheme in Figure 5.17.

Acids

To name an acid, you must first decide whether it is a binary (two-element) acid or an oxyacid (an acid containing oxygen). Binary acids are named according to the scheme in Figure 5.17. Oxyacids must be further subdivided based on the name of their corresponding oxyanion. If the oxyanion ends in *-ate*, use one scheme; if it ends with *-ite*, use the other.

EXAMPLE 5.14 Nomenclature Using Figure 5.17

Name each of the following: CO, CaF_2, HF, $Fe(NO_3)_3$, $HClO_4$, H_2SO_3

Solution:

For each compound, the following table shows how to use Figure 5.17 to arrive at a name for the compound.

Formula	Flow Chart Path	Name
CO	Molecular	carbon monoxide
CaF_2	Ionic ⟶ Type I ⟶	calcium fluoride
HF	Acid ⟶ Binary ⟶	hydrofluoric acid
$Fe(NO_3)_3$	Ionic ⟶ Type II ⟶	iron(III) nitrate
$HClO_4$	Acid ⟶ Oxyacid ⟶ *-ate* ⟶	perchloric acid
H_2SO_3	Acid ⟶ Oxyacid ⟶ *-ite* ⟶	sulfurous acid

Chemistry IN THE ENVIRONMENT

Acid Rain

Acid rain occurs when rainwater mixes with air pollutants—such as NO, NO_2 and SO_2—that form acids. NO and NO_2, primarily from vehicular emissions, combine with water to form HNO_3. SO_2, primarily from coal-powered electricity generation, combines with water to form H_2SO_4. HNO_3 and H_2SO_4 both cause rainwater to become acidic. The problem is greatest in the northeastern United States, where pollutants from midwestern electrical power plants combine with rainwater to produce rain with acid levels that are up to ten times as high as normal.

When acid rain falls or flows into lakes and streams, it makes them more acidic. Some species of aquatic animals—such as trout, bass, snails, salamanders, and clams—cannot tolerate the increased acidity and die. This then disturbs the ecosystem of the lake, resulting in imbalances that may lead to the death of other aquatic species. Acid rain also weakens trees by dissolving nutrients in the soil and by damaging their leaves. Appalachian red spruce trees have been the hardest hit, with many forests showing significant acid rain damage.

▲ A forest damaged by acid rain.

Acid rain also damages building materials. Acids dissolve $CaCO_3$ (limestone), a main component of marble and concrete, and iron, the main component of steel. Consequently, many statues, buildings, and bridges in the northeastern United States show significant deterioration, and some historical gravestones made of limestone are barely legible due to acid rain damage.

▲ Acid rain harms many materials, including the limestone often used for tombstones, buildings, and statues. The problem affects many countries around the world—these damaged gargoyles are part of Notre Dame cathedral in Paris.

Although acid rain has been a problem for many years, recent legislation has offered hope for change. In 1990, Congress passed several amendments to the Clean Air Act that included provisions requiring electrical utilities to reduce SO_2 emissions. Since then, SO_2 emissions have decreased, and rain in the northeastern United States has become somewhat less acidic. With time, and continued enforcement of the acid rain program, lakes, streams, and forests damaged by acid rain should recover.

CAN YOU ANSWER THIS? *Provide the names for each of the following compounds, given here as formulas:*

NO, NO_2, SO_2, HNO_3, $CaCO_3$.

5.11 Formula Mass: The Mass of a Molecule or Formula Unit

Also in common use are the terms *molecular mass* and *molecular weight*, which have the same meaning as formula mass.

In Chapter 4, we learned about atoms and elements, and we designated the average mass of the atoms that compose an element as the atomic mass for that element. In this chapter, we learned about molecules and compounds, and so we designate the average mass of the molecules (or formula units) that compose a compound as the **formula mass**.

For any compound, the formula mass is simply the sum of the atomic masses of all the atoms in its chemical formula:

Formula mass = (# atoms of 1st element in chemical formula × Atomic mass of 1st element) + (# atoms of 2nd element in chemical formula × Atomic mass of 2nd element) + . . .

Like atomic mass for atoms, formula mass characterizes the average mass of a molecule or formula unit. For example, the formula mass of water, H_2O, is:

$$\begin{aligned}\text{Formula mass} &= 2(1.01\ \text{amu}) + 16.00\ \text{amu} \\ &= 18.02\ \text{amu}\end{aligned}$$

and that of sodium chloride, NaCl, is:

$$\begin{aligned}\text{Formula mass} &= 22.99\ \text{amu} + 35.45\ \text{amu} \\ &= 58.44\ \text{amu}\end{aligned}$$

In addition to giving a characteristic mass to the molecules or formula units of a compound, formula mass—as we will learn in Chapter 6—allows us to quantify the number of molecules or formula units in a sample of a given mass.

EXAMPLE 5.15 Calculating Formula Mass

Calculate the formula mass of carbon tetrachloride, CCl_4.

Solution:

To find the formula mass, we sum the atomic masses of each atom in the chemical formula.

$$\begin{aligned}\text{Formula mass} &= 1 \times (\text{Formula mass C}) + 4 \times (\text{Formula mass Cl}) \\ &= 12.01\ \text{amu} + 4(35.45\ \text{amu}) \\ &= 153.81\ \text{amu}\end{aligned}$$

SKILLBUILDER 5.15 Calculating Formula Masses

Calculate the formula mass of dinitrogen monoxide, N_2O, also called laughing gas.

CONCEPTUAL CHECKPOINT 5.2

Which of the following has the greatest formula mass?

(a) O_2
(b) O_3
(c) H_2O
(d) H_2O_2

CHAPTER IN REVIEW

Chemical Principles

Compounds: Matter is ultimately composed of atoms, but those atoms are often combined in compounds. The most important characteristic of a compound is its constant composition. The elements that compose a particular compound are in fixed, definite proportions in all samples of the compound.

Relevance

Compounds: Most of the matter we encounter is in the form of compounds. Water, salt, and carbon dioxide are all good examples of common simple compounds. More complex compounds include caffeine, aspirin, acetone, and testosterone.

Chemical Formulas: Compounds are represented by chemical formulas, which indicate the elements present in the compound and the relative number of atoms of each. These formulas represent the basic units that compose a compound.

Pure substances can be divided according to the basic units that compose them. Elements can be composed of atoms or molecules. Compounds can be molecular, in which case their basic units are molecules, or ionic, in which case their basic units are ions. The formulas for many ionic compounds can be written simply by knowing the elements in the compound.

Chemical Formulas: To understand compounds we must understand their composition, which is represented by a chemical formula. The connection between the microscopic world and the macroscopic world hinges on the particles that compose matter. Since most matter is in the form of compounds, the properties of most matter depend on the molecules or ions that compose it. Molecular matter does what its molecules do; ionic matter does what its ions do. The world we see and experience is governed by what these particles are doing.

Chemical Nomenclature: The names of simple ionic compounds, molecular compounds, and acids can all be written by examining their chemical formula. The nomenclature flow chart (Figure 5.17) shows the basic procedure for determining these names.

Chemical Nomenclature: Since there are so many compounds, we need a systematic way to name them. By learning these few simple rules, you will be able to name thousands of different compounds. The next time you look at the label on a consumer product, try to identify as many of the compounds as you can.

Formula Mass: The formula mass of a compound is the sum of the atomic masses of all the atoms in the chemical formula for the compound. Like atomic mass for elements, formula mass characterizes the average mass of a molecule or formula unit.

Formula Mass: Besides being the characteristic mass of a molecule or formula unit, formula mass is important in many calculations involving the composition of compounds and quantities in chemical reactions.

Chemical Skills

Constant Composition of Compounds (Section 5.2)

The law of constant composition states that all samples of a given compound should have the same ratio of their constituent elements.

To determine whether experimental data is consistent with the law of constant composition, compute the ratios of the masses of each element in all samples. When computing these ratios, it is most convenient to put the larger number in the numerator (top) and the smaller one in the denominator (bottom); that way, the ratio is greater than 1. If the ratios are the same, then the data is consistent with the law of constant composition.

Examples

EXAMPLE 5.16 Constant Composition of Compounds

Two samples said to be carbon disulfide (CS_2) are decomposed into their constituent elements. One sample produced 8.08 g S and 1.51 g C, while the other produced 31.3 g S and 3.85 g C. Are these results consistent with the law of constant composition?

Solution: Sample 1

$$\frac{\text{Mass S}}{\text{Mass C}} = \frac{8.08\text{ g}}{1.51\text{ g}} = 5.35$$

Sample 2

$$\frac{\text{Mass S}}{\text{Mass C}} = \frac{31.3\text{ g}}{3.85\text{ g}} = 8.13$$

These results are not consistent with the law of constant composition and the data given must therefore be in error.

Writing Chemical Formulas (Section 5.3)

Chemical formulas indicate the elements present in a compound and the relative number of atoms of each. When writing formulas, put the more metallic element first.

EXAMPLE 5.17 Writing Chemical Formulas

Write a chemical formula for the compound containing one nitrogen atom for every two oxygen atoms.

Solution:

NO_2

Total Number of Each Type of Atom in a Chemical Formula (Section 5.3)

The numbers of atoms not enclosed in parentheses are given directly by their subscript.

The numbers of atoms within parentheses are found by multiplying their subscript within the parentheses by their subscript outside the parentheses.

EXAMPLE 5.18 Total Number of Each Type of Atom in a Chemical Formula

Determine the number of each type of atom in $Pb(ClO_3)_2$.

Solution:

One Pb atom

Two Cl atoms
Six O atoms

Classifying Elements as Atomic or Molecular (Section 5.4)

Most elements exist as atomic elements, their basic units in nature being individual atoms. However, several elements (H_2, N_2, O_2, F_2, Cl_2, Br_2, and I_2) exist as molecular elements, their basic units in nature being diatomic molecules.

EXAMPLE 5.19 Classifying Elements as Atomic or Molecular

Classify each of the following elements as atomic or molecular: Na, I, N.

Solution:

Na: atomic
I: molecular (I_2)
N: molecular (N_2)

Classifying Compounds as Ionic or Molecular (Section 5.4)

Compounds containing a metal and a nonmetal are ionic. Metals that form more than one ion, typical of transition metals, are Type II.

Compounds containing a metal and one or more nonmetals are ionic. Metals that form one type of ion—typical of Group I, Group II, and aluminum—are Type I.

Compounds composed of nonmetals are molecular.

EXAMPLE 5.20 Classifying Compounds as Ionic or Molecular

Classify each of the following compounds as ionic or molecular. If they are ionic, classify them as Type I or Type II ionic compounds:

$FeCl_3$, K_2SO_4, CCl_4

Solution:

$FeCl_3$: ionic, Type II

K_2SO_4: ionic, Type I

CCl_4: molecular

Writing Formulas for Ionic Compounds (Section 5.5)

1. Write the symbol for the metal ion followed by the symbol for the nonmetal ion (or polyatomic ion) and their charges. These charges can be deduced from the group numbers in the periodic table. (In the case of polyatomic ions, the charges come from Table 5.6.)
2. Make the magnitude of the charge on each ion become the subscript for the other ion.
3. Check to see if the subscripts can be reduced to simpler whole numbers. Subscripts of 1 can be dropped, since they are normally implied.
4. Check that the sum of the charges of the cations exactly cancels the sum of the charges of the anions.

EXAMPLE 5.21 Writing Formulas for Ionic Compounds

Write a formula for the compound that forms from lithium and sulfate ions.

Solution:

Li^+ SO_4^{2-}

$Li_2(SO_4)$

In this case, the subscripts cannot be further reduced.

Li_2SO_4

Cations: $2(1+) = 2+$

Anions: $2-$

Naming Type I Binary Ionic Compounds (Section 5.7)

The name of the metal is unchanged. The name of the nonmetal is its base name with the ending *-ide*.

EXAMPLE 5.22 Naming Type I Binary Ionic Compounds

Give the name for the compound Al_2O_3.

Solution:

aluminum oxide

Naming Type II Binary Ionic Compounds (Section 5.7)

Since the name of Type II compounds includes the charge of the metal ion, we must first find that charge. To do this, compute the total charge of the nonmetal ions.

The total charge of the metal ions must equal the total charge of the nonmetal ions, but have the opposite sign.

The name of the compound is the name of the metal ion, followed by the charge of the metal ion, followed by the base name of the nonmetal + *-ide*.

EXAMPLE 5.23 Naming Type II Binary Ionic Compounds

Give the name for the compound Fe_2S_3.

Solution:

3 sulfide ions × (2−) = 6−

2 iron ions × (?) = 6+

? = 3+

Charge of each iron ion = 3+

iron(III) sulfide

Naming Compounds Containing a Polyatomic Ion (Section 5.7)

Name Type I and Type II ionic compounds containing a polyatomic ion in the normal way, except substitute the name of the polyatomic ion (from Table 5.6) in place of the nonmetal. (This example is Type II.)

The charge on the metal ion must be equal in magnitude to the sum of the charges of the polyatomic ions but opposite in sign.

The name of the compound is the name of the metal ion, followed by the charge of the metal ion, followed by the name of the polyatomic ion.

EXAMPLE 5.24 Naming Compounds Containing a Polyatomic Ion

Give the name for the compound $Co(ClO_4)_2$.

Solution:

2 perchlorate ions × (1−) = 2−

Charge of cobalt ion = 2+

cobalt(II) perchlorate

Naming Molecular Compounds (Section 5.8)

The name consists of a prefix indicating the number of atoms of the first element, followed by the name of the first element, and a prefix for the number of atoms of the second element followed by the base name of the second element plus the suffix *-ide*. When *mono-* occurs on the first element, it is normally dropped.

EXAMPLE 5.25 Naming Molecular Compounds

Name the compound NO_2.

Solution:

nitrogen dioxide

Naming Binary Acids (Section 5.9)

The name begins with *hydro-*, followed by the base name of the nonmetal, plus the suffix *-ic* and then the word *acid*.

EXAMPLE 5.26 Naming Binary Acids

Name the acid HI.

Solution:

hydroiodic acid

Naming Oxyacids with an Oxyanion Ending in *-ate* (Section 5.9)

The name is the base name of the oxyanion + *-ic*, followed by the word *acid* (sulfate violates the rule somewhat, since in strict terms, the base name would be *sulf*).

EXAMPLE 5.27 Naming Oxyacids with an Oxyanion Ending in *-ate*

Name the acid H_2SO_4.

Solution:

The oxyanion is sulfate. The name of the acid is *sulfuric acid*.

Naming Oxyacids with an Oxyanion Ending in *-ite* (Section 5.9)

The name is the base name of the oxyanion + *-ous*, followed by the word *acid*.

EXAMPLE 5.28 Naming Oxyacids with an Oxyanion Ending in *-ite*

Name the acid $HClO_2$.

Solution:

The oxyanion is chlorite. The name of the acid is *chlorous acid*.

Calculating Formula Mass (Section 5.11)

The formula mass is the sum of the atomic masses of all the atoms in the chemical formula. In determining the number of each type of atom, don't forget to multiply subscripts inside parentheses by subscripts outside parentheses.

EXAMPLE 5.29 Calculating Formula Mass

Calculate the formula mass of $Mg(NO_3)_2$

Solution:

$$\text{Formula mass} = 24.31 + 2(14.01) + 6(16.00) = 148.33 \text{ amu}$$

Key Terms

acid [5.9]
atomic element [5.4]
binary acid [5.9]
binary compound [5.7]
chemical formula [5.3]
formula mass [5.11]
formula unit [5.4]
ionic compound [5.4]
law of constant composition [5.2]
molecular compound [5.4]
molecular element [5.4]
oxyacid [5.9]
oxyanion [5.7]
polyatomic ion [5.7]
transition metals [5.7]
Type I compound [5.7]
Type II compound [5.7]

Exercises

Questions

1. Do the properties of an element change when it combines with another element to form a compound? Explain.
2. How would the world be different if elements did not combine to form compounds?
3. What is the law of constant composition? Who discovered it?
4. What is a chemical formula? Give some examples.
5. In a chemical formula, which element is listed first?
6. In a chemical formula, how do you calculate the number of atoms of an element within parentheses? Give an example.
7. What is the difference between a molecular element and an atomic element? List the elements that occur as diatomic molecules.
8. What is the difference between an ionic compound and a molecular compound?
9. What is the difference between a common name for a compound and a systematic name?
10. List the metals that form Type I ionic compounds. What are the group numbers of these metals?
11. Find the block in the periodic table of elements that tend to form Type II ionic compounds. What is the name of this block?
12. What is the basic form for the names of Type I ionic compounds?
13. What is the basic form for the names of Type II ionic compounds?
14. Why are numbers needed in the names of Type II ionic compounds?
15. How are compounds containing a polyatomic ion named?
16. What polyatomic ions have a 2− charge? What polyatomic ions have a 3− charge?
17. What is the basic form for the names of molecular compounds?
18. How many atoms does each of the following prefixes specify? *mono-*, *di-*, *tri-*, *tetra-*, *penta-*, *hexa-*.
19. What is the basic form for the names of binary acids?
20. What is the basic form for the name of oxyacids whose oxyanions end with *-ate*?
21. What is the basic form for the name of oxyacids whose oxyanions end with *-ite*?
22. What is the formula mass of a compound?

Problems

Constant Composition of Compounds

23. Two samples of sodium chloride were decomposed into their constituent elements. One sample produced 4.65 g of sodium and 7.16 g of chlorine, and the other sample produced 7.45 g of sodium and 11.5 g of chlorine. Are these results consistent with the law of constant composition? Show why or why not.

24. Two samples of carbon tetrachloride were decomposed into their constituent elements. One sample produced 32.4 g of carbon and 373 g of chlorine, and the other sample produced 12.3 g of carbon and 112 g of chlorine. Are these results consistent with the law of constant composition? Show why or why not.

25. Upon decomposition, one sample of magnesium fluoride produced 1.65 kg of magnesium and 2.57 kg of fluorine. A second sample produced 1.32 kg of magnesium. How much fluorine (in grams) did the second sample produce?

26. The mass ratio of sodium to fluorine in sodium fluoride is 1.21:1. A sample of sodium fluoride produced 34.5 g of sodium upon decomposition. How much fluorine (in grams) was formed?

27. Use the law of constant composition to complete the following table summarizing the amounts of nitrogen and oxygen produced upon the decomposition of several samples of dinitrogen monoxide.

	Mass N_2O	*Mass N*	*Mass O*
Sample A	2.85 g	1.82 g	1.03 g
Sample B	4.55 g	_____	_____
Sample C	_____	_____	1.35 g
Sample D	_____	1.11 g	_____

28. Use the law of constant composition to complete the following table summarizing the amounts of iron and chlorine formed produced upon the decomposition of several samples of iron(III) chloride.

	Mass $FeCl_3$	*Mass Fe*	*Mass Cl*
Sample A	3.785 g	1.302 g	2.483 g
Sample B	2.175 g	_____	_____
Sample C	_____	2.012 g	_____
Sample D	_____	_____	2.329 g

Chemical Formulas

29. Write a chemical formula for the compound containing one nitrogen atom for every three bromine atoms.

30. Write a chemical formula for the compound containing one carbon atom for every four chlorine atoms.

31. Write chemical formulas for compounds containing each of the following:
    - **(a)** two iron atoms for every three oxygen atoms
    - **(b)** one phosphorus atom for every three chlorine atoms
    - **(c)** one phosphorus atom for every five chlorine atoms
    - **(d)** two silver atoms for every oxygen atom

32. Write chemical formulas for compounds containing each of the following:
    - **(a)** one calcium atom for every two chlorine atoms
    - **(b)** two nitrogen atoms for every four oxygen atoms
    - **(c)** one silicon atom for every two oxygen atoms
    - **(d)** one zinc atom for every two chlorine atoms

33. How many oxygen atoms are in each of the following chemical formulas?
    - **(a)** H_3PO_4
    - **(b)** Na_2HPO_4
    - **(c)** $Ca(HCO_3)_2$
    - **(d)** $Ba(C_2H_3O_2)_2$

34. How many hydrogen atoms are in each of the formulas in Problem 33?

35. Determine the number of each type of atom in each of the following formulas.
    - **(a)** $MgCl_2$
    - **(b)** $NaNO_3$
    - **(c)** $Ca(NO_2)_2$
    - **(d)** $Sr(OH)_2$

36. Determine the number of each type of atom in each of the following formulas.
    - **(a)** NH_4Cl
    - **(b)** $Mg_3(PO_4)_2$
    - **(c)** $NaCN$
    - **(d)** $Ba(HCO_3)_2$

Molecular View of Elements and Compounds

37. Classify each of the following elements as atomic or molecular.
    - **(a)** helium
    - **(b)** chlorine
    - **(c)** oxygen
    - **(d)** sodium

38. Which of the following elements have molecules as their basic units?
    - **(a)** hydrogen
    - **(b)** argon
    - **(c)** bromine
    - **(d)** mercury

39. Classify each of the following compounds as ionic or molecular.
    - **(a)** CS_2
    - **(b)** CuO
    - **(c)** KI
    - **(d)** PCl_3

40. Classify each of the following compounds as ionic or molecular.
    - **(a)** PtO_2
    - **(b)** CF_2Cl_2
    - **(c)** CO
    - **(d)** SO_3

41. Match the substances on the left with the basic units that compose them on the right.

helium	molecules
CCl_4	formula units
K_2SO_4	diatomic molecules
bromine	single atoms

42. Match the substances on the left with the basic units that compose them on the right.

NI_3	molecules
copper metal	single atoms
$SrCl_2$	diatomic molecules
nitrogen	formula units

43. What are the basic units—single atoms, molecules, or formula units—that compose each of the following substances?
(a) $BaBr_2$
(b) Ne
(c) I_2
(d) CO

44. What are the basic units—single atoms, molecules, or formula units—that compose each of the following substances?
(a) Rb_2O
(b) N_2
(c) $Fe(NO_3)_2$
(d) N_2F_4

45. Classify each of the following compounds as ionic or molecular. If it is ionic, classify it as a Type I or Type II ionic compound.
(a) KCl
(b) CBr_4
(c) NO_2
(d) $Sn(SO_4)_2$

46. Classify each of the following compounds as ionic or molecular. If it is ionic, classify it as a Type I or Type II ionic compound.
(a) $CoCl_2$
(b) CF_4
(c) $BaSO_4$
(d) NO

Writing Formulas for Ionic Compounds

47. Write a formula for the ionic compound that forms from each of the following pairs of elements.
(a) strontium and oxygen
(b) sodium and oxygen
(c) aluminum and sulfur
(d) magnesium and bromine

48. Write a formula for the ionic compound that forms from each of the following pairs of elements.
(a) aluminum and oxygen
(b) beryllium and chlorine
(c) calcium and sulfur
(d) strontium and iodine

49. Write a formula for the compound that forms from potassium and
(a) acetate
(b) chromate
(c) phosphate
(d) cyanide

50. Write a formula for the compound that forms from calcium and
(a) hydroxide
(b) carbonate
(c) phosphate
(d) hydrogen phosphate

51. Write formulas for the compounds that form from the element on the left and each element on the right.
(a) K N, O, F
(b) Ba N, O, F
(c) Al N, O, F

52. Write formulas for the compounds that form from the element on the left and each polyatomic ion on the right.
(a) Rb NO_3^-, SO_4^{2-}, PO_4^{3-}
(b) Sr NO_3^-, SO_4^{2-}, PO_4^{3-}
(c) In NO_3^-, SO_4^{2-}, PO_4^{3-}
(Assume In charge is 3+).

Naming Ionic Compounds

53. Name each of the following Type I ionic compounds.
(a) CsCl
(b) $SrBr_2$
(c) K_2O
(d) LiF

54. Name each of the following Type I ionic compounds.
(a) LiI
(b) MgS
(c) BaF_2
(d) NaF

55. Name each of the following Type II ionic compounds.
(a) $CrCl_2$
(b) $CrCl_3$
(c) SnO_2
(d) PbI_2

56. Name each of the following Type II ionic compounds.
(a) $HgBr_2$
(b) Fe_2O_3
(c) CuI_2
(d) $SnCl_4$

57. Determine whether each of the following ionic compounds is Type I or Type II and give each an appropriate name.
(a) Cr_2O_3
(b) NaI
(c) $CaBr_2$
(d) SnO

58. Determine whether each of the following ionic compounds is Type I or Type II and give each an appropriate name.
(a) FeI_3
(b) $PbCl_4$
(c) SrI_2
(d) BaO

59. Name each of the following ionic compounds containing a polyatomic ion.
(a) $Ba(NO_3)_2$
(b) $Pb(C_2H_3O_2)_2$
(c) NH_4I
(d) $KClO_3$
(e) $CoSO_4$
(f) $NaClO_4$

60. Name each of the following ionic compounds containing a polyatomic ion.
(a) $Ba(OH)_2$
(b) $Fe(OH)_3$
(c) $CuNO_2$
(d) $PbSO_4$
(e) KClO
(f) $Mg(C_2H_3O_2)_2$

61. Provide a name for each of the following polyatomic ions.
(a) BrO^-
(b) BrO_2^-
(c) BrO_3^-
(d) BrO_4^-

62. Provide a name for each of the following polyatomic ions.
(a) IO^-
(b) IO_2^-
(c) IO_3^-
(d) IO_4^-

63. Write a formula for each of the following ionic compounds.
(a) copper(II) bromide
(b) silver nitrate
(c) potassium hydroxide
(d) sodium sulfate
(e) potassium hydrogen sulfate
(f) sodium hydrogen carbonate

64. Write a formula for each of the following ionic compounds.
(a) copper(I) chlorate
(b) potassium permanganate
(c) lead(II) chromate
(d) calcium fluoride
(e) iron(II) phosphate
(f) lithium hydrogen sulfite

Naming Molecular Compounds

65. Name each of the following molecular compounds.
(a) SO_2
(b) NI_3
(c) BrF_5
(d) NO
(e) N_4Se_4

66. Name each of the following molecular compounds.
(a) XeF_4
(b) PI_3
(c) SO_3
(d) $SiCl_4$
(e) I_2O_5

67. Write a formula for each of the following molecular compounds.
(a) carbon monoxide
(b) disulfur tetrafluoride
(c) dichlorine monoxide
(d) phosphorus pentafluoride
(e) boron tribromide
(f) diphosphorus pentasulfide

68. Write a formula for each of the following molecular compounds.
(a) chlorine monoxide
(b) xenon tetroxide
(c) xenon hexafluoride
(d) carbon tetrabromide
(e) diboron tetrachloride
(f) tetraphosphorus triselenide

69. Determine whether the name shown for each of the following molecular compounds is correct. If not, give the compound the correct name.
 (a) PBr_5 phosphorus(V) pentabromide
 (b) P_2O_3 phosphorus trioxide
 (c) SF_4 monosulfur hexafluoride
 (d) NF_3 nitrogen trifluoride

70. Determine whether the name shown for each of the following molecular compounds is correct. If not, give the compound the correct name.
 (a) NCl_3 nitrogen chloride
 (b) CI_4 carbon(IV) iodide
 (c) CO carbon oxide
 (d) SCl_4 sulfur tetrachloride

Naming Acids

71. Name each of the following acids.
 (a) $HClO_2$
 (b) HI
 (c) H_2SO_4
 (d) HNO_3

72. Name each of the following acids.
 (a) H_2CO_3
 (b) $HC_2H_3O_2$
 (c) HNO_2
 (d) HCl

73. Write a formula for each of the following acids.
 (a) phosphoric acid
 (b) hydrobromic acid
 (c) sulfurous acid

74. Write a formula for each of the following acids.
 (a) hydrofluoric acid
 (b) hydrocyanic acid
 (c) chlorous acid

Formula Mass

75. Calculate the formula mass for each of the following compounds.
 (a) HNO_2
 (b) $CaCl_2$
 (c) CCl_4
 (d) $Mg(NO_3)_2$

76. Calculate the formula mass for each of the following compounds.
 (a) CO_2
 (b) $C_6H_{12}O_6$
 (c) $Fe(NO_3)_3$
 (d) C_8H_{18}

77. Arrange the following compounds in order of decreasing formula mass.

Ag_2O, PtO_2, $Al(NO_3)_3$, PBr_3

78. Arrange the following compounds in order of decreasing formula mass.

WO_2, Rb_2SO_4, $Pb(C_2H_3O_2)_2$, RbI

Cumulative Problems

79. Write a chemical formula for each of the following molecular models. (White = hydrogen; red = oxygen; black = carbon; blue = nitrogen; yellow = sulfur)

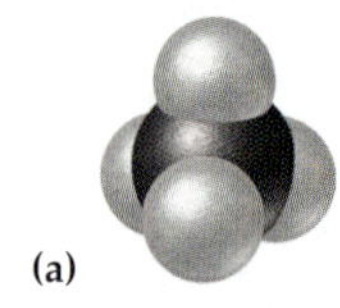
(a)

(b)

(c)

80. Write a chemical formula for each of the following molecular models. (White = hydrogen; red = oxygen; black = carbon; blue = nitrogen; yellow = sulfur)

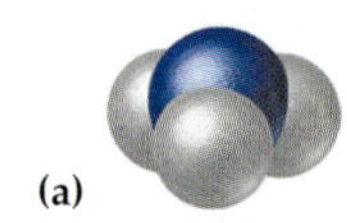
(a)

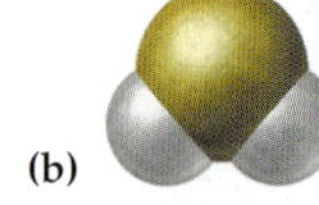
(b)

(c)

81. How many chlorine atoms are there in each of the following?
   (a) three carbon tetrachloride molecules
   (b) two calcium chloride formula units
   (c) four phosphorous trichloride molecules
   (d) seven sodium chloride formula units

82. How many oxygen atoms are there in each of the following?
   (a) four dinitrogen monoxide molecules
   (b) two calcium carbonate formula units
   (c) three sulfur dioxide molecules
   (d) five perchlorate ions

83. Specify the number of hydrogen atoms (white) represented in each of the following:

(a)

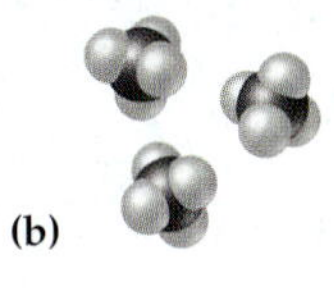
(b)

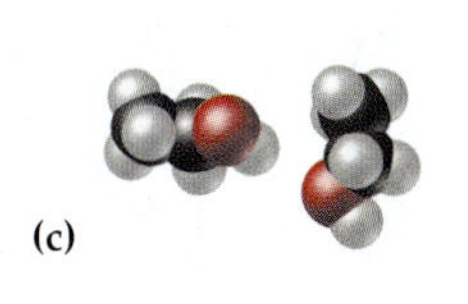
(c)

84. Specify the number of oxygen atoms (red) represented in each of the following:

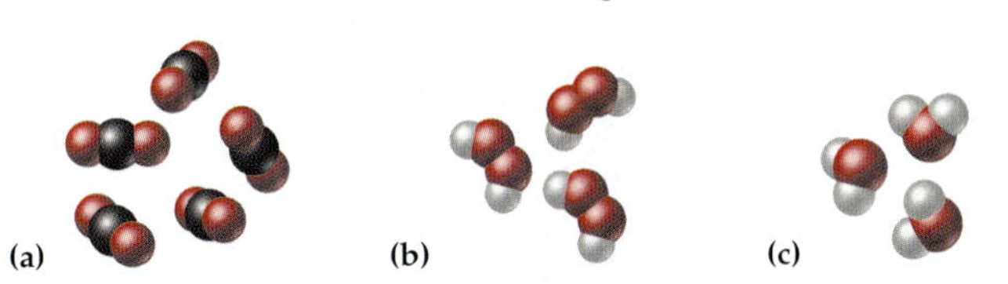
(a) (b) (c)

85. Complete the following table:

Formula	*Type of Compound (Ionic, Molecular, Acid)*	*Name*
N_2H_4	molecular	_____
_____	_____	potassium chloride
H_2CrO_4	_____	_____
_____	_____	cobalt(III) cyanide

86. Complete the following table:

Formula	*Type of Compound (Ionic, Molecular, Acid)*	*Name*
$K_2Cr_2O_7$	ionic	_____
HBr	_____	hydrobromic acid
_____	_____	dinitrogen pentoxide
PbO_2	_____	_____

87. Determine whether each of the following names is correct for the given formula. If not, give the correct name.
   (a) $Ca(NO_2)_2$ calcium nitrate
   (b) K_2O dipotassium monoxide
   (c) PCl_3 phosphorus chloride
   (d) $PbCO_3$ lead(II) carbonate
   (e) KIO_2 potassium hypoiodite

88. Determine whether each of the following names is correct for the given formula. If not, give the correct name.
   (a) HNO_3 hydrogen nitrate
   (b) NaClO sodium hypochlorite
   (c) CaI_2 calcium diiodide
   (d) $SnCrO_4$ tin chromate
   (e) $NaBrO_3$ sodium bromite

89. For each of the following compounds, give the correct formula and calculate the formula mass.
   (a) tin(IV) sulfate
   (b) nitrous acid
   (c) sodium bicarbonate
   (d) phosphorus pentafluoride

90. For each of the following compounds, give the correct formula and calculate the formula mass.
   (a) barium bromide
   (b) dinitrogen trioxide
   (c) copper(I) sulfate
   (d) hydrobromic acid

91. Name each of the following compounds and calculate its formula mass.
   (a) PtO_2
   (b) N_2O_5
   (c) $Al(ClO_3)_3$
   (d) PBr_5

92. Name each of the following compounds and calculate its formula mass.
   (a) $Al_2(SO_4)_3$
   (b) P_2O_3
   (c) HClO
   (d) $Cr(C_2H_3O_2)_3$

Highlight Problems

93. Examine each of the following substances and their molecular views and classify each as an atomic element, a molecular element, a molecular compound, or an ionic compound.

(a)

(b)

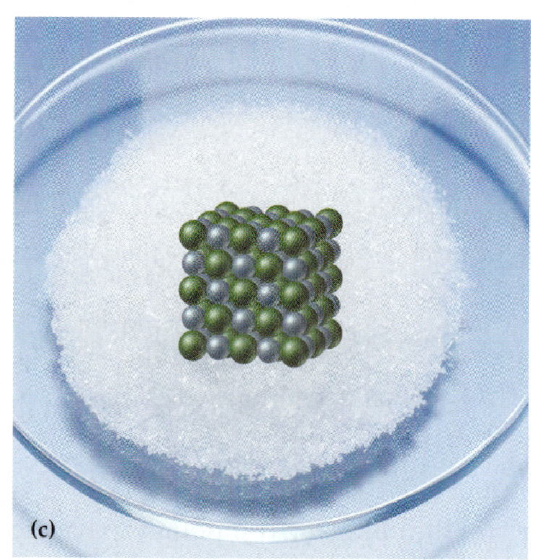
(c)

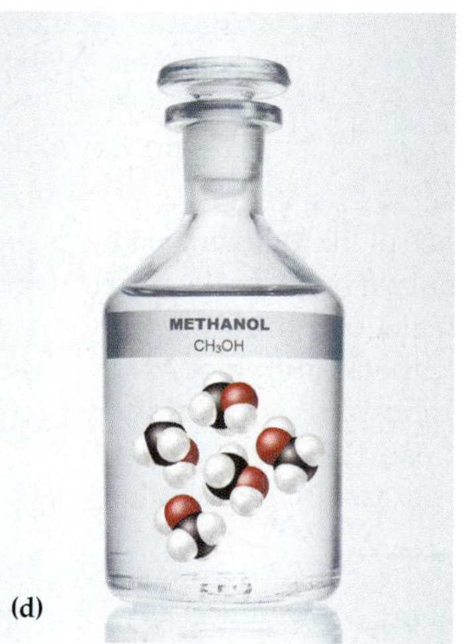

(d)

94. Molecules can be as small as a few atoms or as large as thousands of atoms. In 1962, Max F. Perutz and John C. Kendrew were awarded the Nobel Prize for their discovery of the structure of hemoglobin, a very large molecule that transports oxygen from the lungs to cells through the bloodstream. The chemical formula of hemoglobin is $C_{2952}H_{4664}O_{832}N_{812}S_8Fe_4$. Calculate the formula mass of hemoglobin.

▲ Max F. Perutz and John C. Kendrew won a Nobel Prize in 1962 for determining the structure of hemoglobin by X-ray diffraction.

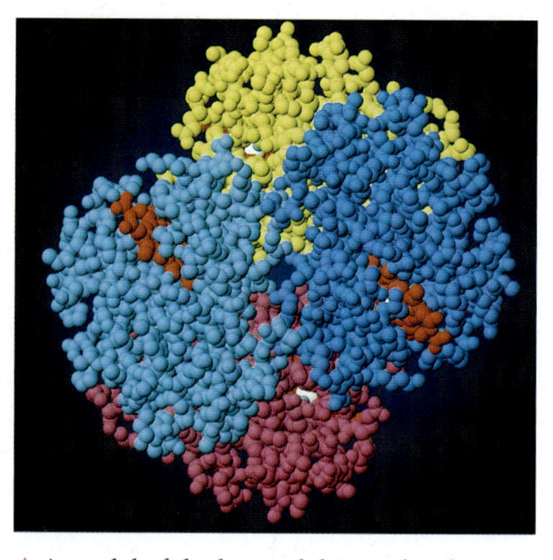
▲ A model of the hemoglobin molecule.

95. Examine each of the following consumer product labels. Write chemical formulas for as many of the compounds as possible based on what we have learned in this chapter.

(a)

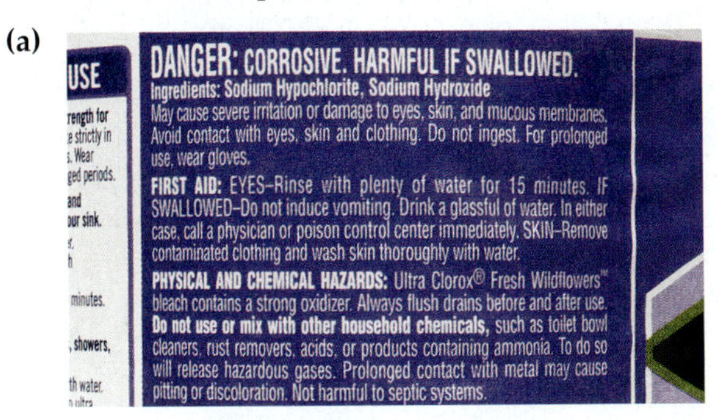
DANGER: CORROSIVE. HARMFUL IF SWALLOWED.
Ingredients: Sodium Hypochlorite, Sodium Hydroxide
May cause severe irritation or damage to eyes, skin, and mucous membranes. Avoid contact with eyes, skin and clothing. Do not ingest. For prolonged use, wear gloves.
FIRST AID: EYES–Rinse with plenty of water for 15 minutes. IF SWALLOWED–Do not induce vomiting. Drink a glassful of water. In either case, call a physician or poison control center immediately. SKIN–Remove contaminated clothing and wash skin thoroughly with water.
PHYSICAL AND CHEMICAL HAZARDS: Ultra Clorox® Fresh Wildflowers™ bleach contains a strong oxidizer. Always flush drains before and after use. Do not use or mix with other household chemicals, such as toilet bowl cleaners, rust removers, acids, or products containing ammonia. To do so will release hazardous gases. Prolonged contact with metal may cause pitting or discoloration. Not harmful to septic systems.

(b)

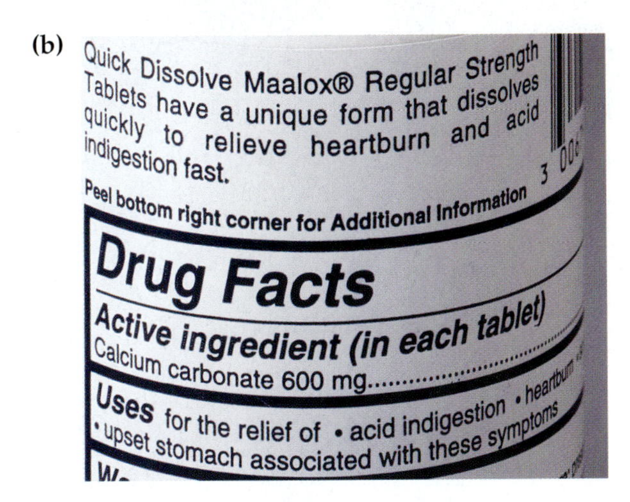
Quick Dissolve Maalox® Regular Strength Tablets have a unique form that dissolves quickly to relieve heartburn and acid indigestion fast.
Peel bottom right corner for Additional Information
Drug Facts
Active ingredient (in each tablet)
Calcium carbonate 600 mg
Uses for the relief of • acid indigestion • heartburn
• upset stomach associated with these symptoms

(c)

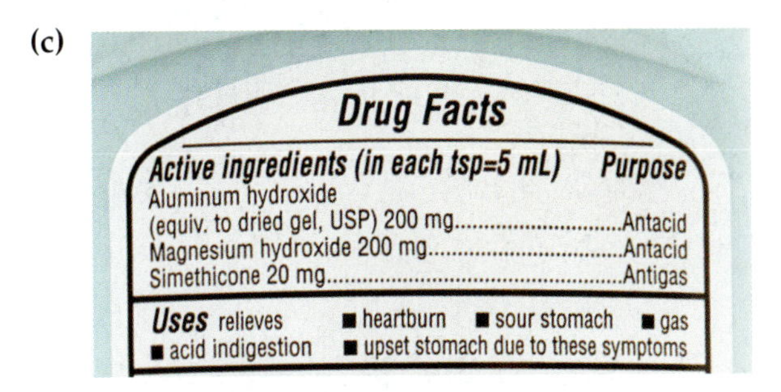
Drug Facts
Active ingredients (in each tsp=5 mL) Purpose
Aluminum hydroxide
(equiv. to dried gel, USP) 200 mg..........Antacid
Magnesium hydroxide 200 mg..........Antacid
Simethicone 20 mg..........Antigas
Uses relieves ■ heartburn ■ sour stomach ■ gas
■ acid indigestion ■ upset stomach due to these symptoms

(d)

Nutrition Facts
Serving Size 1/4 tsp (1g)
Servings Per Container about 340
Amount Per Serving
Calories 0
% Daily Value*
Total Fat 0g 0%
Sodium 95mg 4%
Total Carbohydrate 0g 0%
Protein 0g
Calcium 2%
* Percent Daily Values are based on a 2,000 calorie diet.
INGREDIENTS: CORNSTARCH, SODIUM BICARBONATE, CALCIUM PHOSPHATE, SODIUM ALUMINUM SULFATE.

Answers to Skillbuilder Exercises

Skillbuilder 5.1 Yes, because in both cases

$$\frac{\text{Mass O}}{\text{Mass C}} = 1.3$$

Skillbuilder 5.2 **(a)** Ag_2S **(b)** N_2O **(c)** TiO_2

Skillbuilder 5.3 two K atoms, one S atom, four O atoms

Skillbuilder Plus, p. 122 two Al atoms, three S atoms, twelve O atoms

Skillbuilder 5.4 **(a)** molecular element **(b)** molecular compound **(c)** atomic element **(d)** ionic compound **(e)** ionic compound

Skillbuilder 5.5 Cs_2O

Skillbuilder 5.6 AlN

Skillbuilder 5.7 $CaBr_2$

Skillbuilder 5.8 potassium bromide

Skillbuilder Plus, p. 129 zinc nitride

Skillbuilder 5.9 lead(II) oxide

Skillbuilder 5.10 manganese(II) nitrate

Skillbuilder 5.11 dinitrogen tetroxide

Skillbuilder 5.12 hydroiodic acid

Skillbuilder 5.13 nitrous acid

Skillbuilder 5.15 44.02 amu

Answers to Conceptual Checkpoints

5.1 (b) This formula represents 2 Al atoms + 3(2 Cr atoms + 7 O atoms) = 29 atoms.

5.2 (b)

CHAPTER 6

Chemical Composition

"By definition, a chemist believes in the existence of a new substance only when he has seen the substance, touched it, weighed and examined it, confronted it with acids, bottled it, and when he has determined its 'atomic weight.'"

Eve Curie (1904–)

6.1 How Much Sodium?

Sodium is an important dietary mineral consumed primarily as sodium chloride (table salt) and involved in bodily fluid regulation. Overconsumption of dietary sodium causes elevated body fluid levels that can lead to high blood pressure. High blood pressure, in turn, increases the risk of stroke and heart attack. Consequently, people with high blood pressure should limit their sodium intake. The FDA recommends that a person consume less than 2.4 g (2400 mg) of sodium per day. However, since sodium is usually consumed as sodium chloride, it is not always easy to know how much sodium is consumed. How many grams of sodium chloride can you consume and still stay below the FDA recommendation?

To answer this question, we need to know the *chemical composition* of sodium chloride. From Chapter 5, we know its formula, NaCl, so we know that there is one sodium ion to every chloride ion. However, since the masses of sodium and chlorine are different, the relationship between the mass of sodium and the mass of sodium chloride is not clear from just the chemical formula. In this chapter, we learn how to use the information in a chemical formula, together with atomic and formula masses, to calculate the amount of a constituent element in a given amount of a compound (or vice versa).

◀ Ordinary table salt is a compound called sodium chloride. The sodium within sodium chloride is linked to high blood pressure. In this chapter, we learn how to determine how much sodium is in a given amount of sodium chloride.

▲ The mining of iron requires knowing how much iron is in a given amount of iron ore.

▲ Estimating the threat of ozone depletion from chlorofluorocarbons requires knowing how much chlorine is in a given amount of a particular chlorofluorocarbon.

Chemical composition is important not just for assessing dietary sodium, but for addressing many other issues as well. A company that mines iron, for example, wants to know how much iron it can recover from a given amount of iron ore; a company interested in developing hydrogen as a potential fuel wants to know how much hydrogen it can extract from a given amount of water. Many environmental issues also require knowledge of chemical composition. For example, an estimate of the threat of ozone depletion requires knowing how much chlorine is in a given amount of a particular chlorofluorocarbon such as freon-12 (CF_2Cl_2). Answering these questions requires understanding the relationships inherent in a chemical formula and the relationship between numbers of atoms or molecules and their masses. In this chapter, we learn these relationships.

6.2 Counting Nails by the Pound

Some hardware stores sell nails by the pound. This is easier than selling them individually because customers often need hundreds of nails and counting them takes too long. However, a customer may still want to know the number of nails contained in a given weight of nails. This problem is similar to asking how many atoms are in a given mass of an element. With atoms, however, we *must* use their mass as a way to count them because atoms are too small and too numerous to count individually. Even if you could see atoms and counted them twenty-four hours a day as long as you lived, you would barely begin to count the number of atoms in something as small as a sand grain. However, just as the hardware store customer wants to know how many nails are in a given weight, we want to know the number of atoms in a given mass. How do we do that?

Suppose the hardware store customer buys 2.60 lb of medium-sized nails. Suppose further that a dozen nails weigh 0.150 lb. How many nails did the customer buy? This calculation requires two conversions: one between pounds and dozens and another between dozens and number of nails. The conversion factor for the first part is the weight per dozen nails.

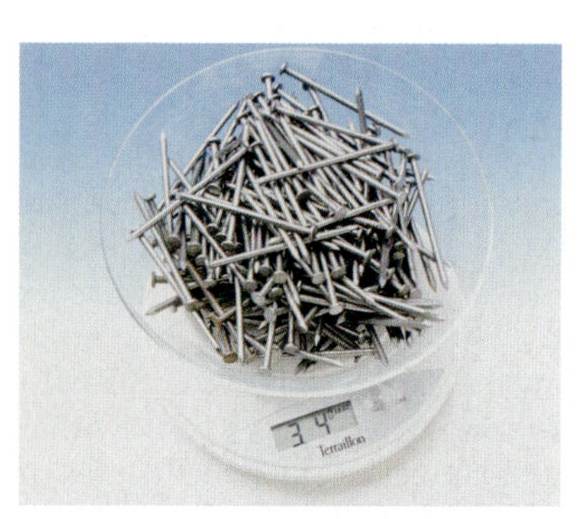

3.4 lbs nails

$$0.150 \text{ lb nails} = 1 \text{ doz nails}$$

The conversion factor for the second part is the number of nails in one dozen.

$$1 \text{ doz nails} = 12 \text{ nails}$$

The solution map for the problem is:

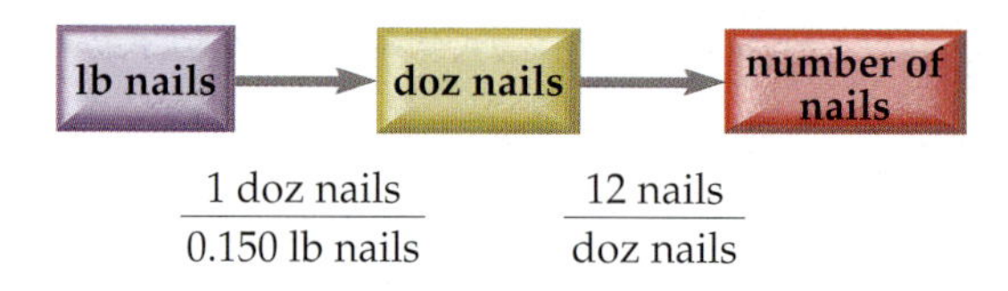

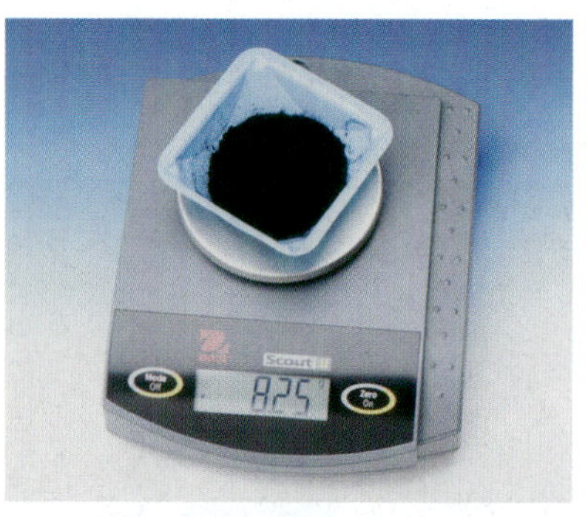

8.25 grams carbon

▲ Asking how many nails are in a given weight of nails is similar to asking how many atoms are in a given mass of an element. In both cases, we count the objects by weighing them.

Beginning with 2.60 lb and using the solution map as a guide, we convert from lb to number of nails.

$$2.60 \text{ lb nails} \times \frac{1 \text{ doz nails}}{0.150 \text{ lb nails}} \times \frac{12 \text{ nails}}{1 \text{ doz nails}} = 208 \text{ nails}$$

The customer who bought 2.60 lb of nails got 208 nails. He counted the nails by weighing them. If the customer purchased a different size of nail, the first conversion factor—relating pounds to dozens—would change, but the second conversion factor would not. One dozen corresponds to twelve nails, regardless of their size.

6.3 Counting Atoms by the Gram

▲ Twenty-two copper pennies contain approximately 1 mol of copper atoms.

Beginning in 1982, pennies became almost all zinc with only a copper plating. Prior to 1982, pennies were mostly copper.

The size of the mole is actually an empirically measured quantity.

Determining the number of atoms in a sample with a certain mass is similar to determining the number of nails in a sample with a certain weight, but the numbers are different. With nails, we used a dozen as a convenient number in our conversions, but a dozen is too small to use with atoms. We need a larger number because atoms are so small. The chemist's "dozen" is called the **mole (mol)** and has a value of 6.022×10^{23}.

$$1 \text{ mol} = 6.022 \times 10^{23}$$

This number is also called **Avogadro's number**, named after Amadeo Avogadro (1776–1856).

The first thing to understand about the mole is that it can specify Avogadro's number of anything. For example, one mole of marbles corresponds to 6.022×10^{23} marbles, and one mole of sand grains corresponds to 6.022×10^{23} sand grains. *One mole of anything is* 6.022×10^{23} *units of that thing*. One mole of atoms, ions, or molecules generally makes up objects of reasonable size. For example, twenty-two copper pennies contain approximately 1 mol of copper (Cu) atoms, and a couple of large helium balloons contain approximately 1 mol of helium (He) atoms.

The second thing to understand about the mole is how it gets its specific value. *The numerical value of the mole is defined as being equal to the number of atoms in exactly 12 g of pure carbon-12.*

This definition of the mole gives us a relationship between mass (grams of carbon) and number of atoms (Avogadro's number). This relationship, as we will see shortly, allows us to count atoms by weighing them.

Converting between Moles and Number of Atoms

▲ Two large helium balloons contain approximately 1 mol of helium atoms.

Converting between moles and number of atoms is similar to converting between dozens and number of nails. To convert between moles of atoms and number of atoms we simply use the conversion factors:

$$\frac{1 \text{ mol atoms}}{6.022 \times 10^{23} \text{ atoms}} \quad \text{or} \quad \frac{6.022 \times 10^{23} \text{ atoms}}{1 \text{ mol atoms}}$$

For example, suppose we want to convert 3.5 mol of helium to the number of helium atoms. We set up the problem in the standard way.

Given: 3.5 mol He

Find: He atoms

Conversion Factor: 1 mol He = 6.022×10^{23} He atoms

Solution Map: We then draw a solution map showing the conversion from moles of He to He atoms.

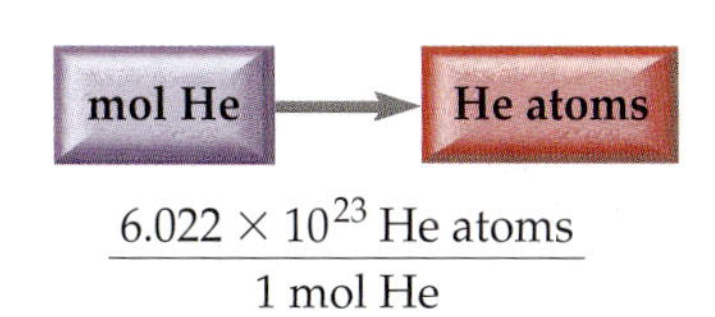

Solution: Beginning with 3.5 mol He, we use the conversion factor to get to He atoms.

$$3.5 \text{ mol He} \times \frac{6.022 \times 10^{23} \text{ He atoms}}{1 \text{ mol He}} = 2.1 \times 10^{24} \text{ He atoms}$$

EXAMPLE 6.1 Converting between Moles and Number of Atoms

A silver ring contains 1.1×10^{22} silver atoms. How many moles of silver are in the ring?

Set up the problem in the normal way. You are given the number of silver atoms and asked to find the number of moles. The conversion factor is Avogadro's number.

Given: 1.1×10^{22} Ag atoms

Find: mol Ag

Conversion Factor: 1 mol Ag = 6.022×10^{23} Ag atoms

Build the solution map, beginning with silver atoms and ending at moles.

Solution Map:

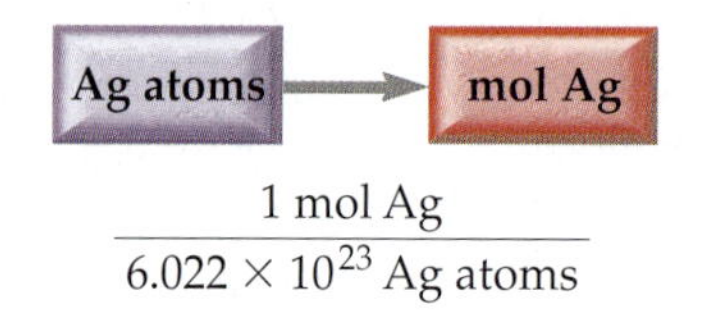

$$\frac{1 \text{ mol Ag}}{6.022 \times 10^{23} \text{ Ag atoms}}$$

Follow the solution map to solve the problem. Beginning with 1.1×10^{22} Ag atoms, use the conversion factor to get to moles of Ag.

Solution:

$$1.1 \times 10^{22} \text{ Ag atoms} \times \frac{1 \text{ mol Ag}}{6.022 \times 10^{23} \text{ Ag atoms}} = 1.8 \times 10^{-2} \text{ mol Ag}$$

SKILLBUILDER 6.1 Converting between Moles and Number of Atoms

How many gold atoms are in a pure gold ring containing 8.83×10^{-2} mol Au?

Converting between Grams and Moles of an Element

We just learned how to convert between moles and number of atoms. As in our nail analogy, we need one more conversion factor to convert from the mass of a sample to the number of atoms in the sample. For nails, we used the weight of one dozen nails; for atoms, we need the mass of one mole of atoms.

> The mass of 1 mol of atoms of an element is called the ***molar mass***. The value of an element's molar mass in grams per mole is numerically equal to the element's atomic mass in atomic mass units.

Since Avogadro's number is defined as the number of atoms in exactly 12 g of carbon-12, and since the amu is defined as 1/12 of the mass of carbon-12, it follows that the molar mass—the mass of 1 mol of atoms in grams—is numerically equal to the atomic mass in amu. For example, copper has an atomic mass of 63.55 amu; therefore, 1 mol of copper atoms has a mass of 63.55 g, and the molar mass of copper is 63.55 g/mol. Just as the weight of 1 doz nails changes for different nails, so the mass of 1 mol of atoms changes for different elements: 1 mol of sulfur atoms (sulfur atoms are lighter than copper atoms) has a mass of 32.07 g; 1 mol of carbon atoms (lighter than sulfur) has a mass of 12.01 g; and 1 mol of lithium atoms (lighter yet) has a mass of 6.94 g.

$$32.07 \text{ g sulfur} = 1 \text{ mol sulfur} = 6.022 \times 10^{23} \text{ S atoms}$$
$$12.01 \text{ g carbon} = 1 \text{ mol carbon} = 6.022 \times 10^{23} \text{ C atoms}$$
$$6.94 \text{ g lithium} = 1 \text{ mol lithium} = 6.022 \times 10^{23} \text{ Li atoms}$$

▶ **Figure 6.1 The mass of 1 mol**
(a) Each of these pictures shows the same number of nails: twelve. As you can see, twelve large nails have more weight and occupy more space than twelve small nails. The same is true for atoms. **(b)** Each of these samples has the same number of atoms: 6.022×10^{23}. Since sulfur atoms are more massive and larger than carbon atoms, 1 mol of S atoms is heavier and occupies more space than 1 mol of C atoms.

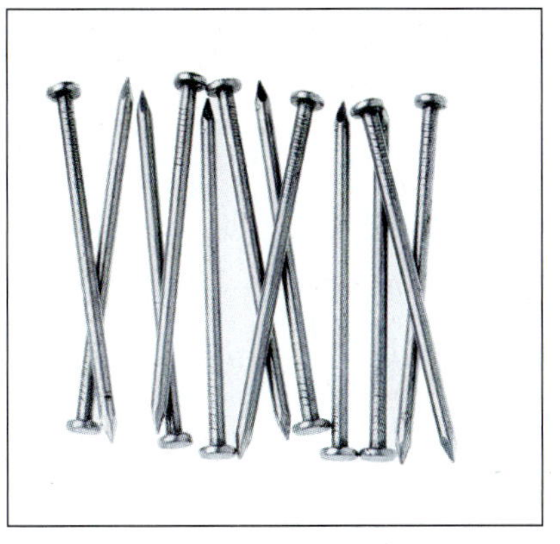

1 dozen large nails

1 dozen small nails

(a)

1 mole S (32.07 g)

1 mole C (12.01 g)

(b)

The lighter the atom, the less mass in a mole of that atom (▲ Figure 6.1).

Therefore, the molar mass of any element becomes a conversion factor between grams of that element and moles of that element. For carbon:

$$12.01 \text{ g C} = 1 \text{ mol C} \quad \text{or} \quad \frac{12.01 \text{ g C}}{1 \text{ mol C}} \quad \text{or} \quad \frac{1 \text{ mol C}}{12.01 \text{ g C}}$$

A 0.58-g diamond would be about a three-carat diamond.

Suppose we want to calculate the number of moles of carbon in a 0.58-g diamond (pure carbon). We set up the problem in the standard way.

Given: 0.58 g C

Find: mol C

Conversion Factor:

$$12.01 \text{ g C} = 1 \text{ mol C}$$

Solution Map: We then draw a solution map showing the conversion from grams of C to moles of C. The conversion factor is the molar mass of carbon.

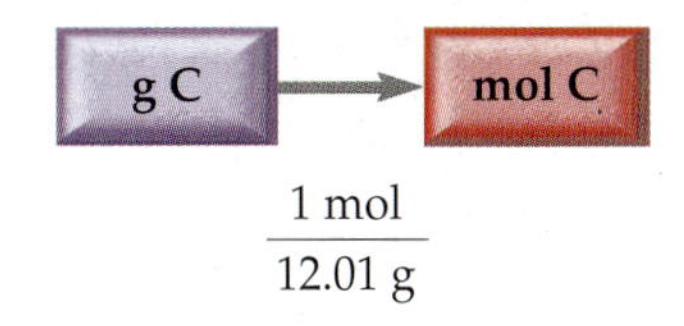

Solution: Beginning with 0.58 g C, we use the conversion factor to get to mol C.

$$0.58 \cancel{\text{g C}} \times \frac{1 \text{ mol C}}{12.01 \cancel{\text{g C}}} = 4.8 \times 10^{-2} \text{ mol C}$$

EXAMPLE 6.2 The Mole Concept—Converting between Grams and Moles

Calculate the number of moles of sulfur in 57.8 g of sulfur.

Set up the problem in the normal way. You are given the mass of sulfur and asked to find the number of moles. The conversion factor is the molar mass of sulfur.

Given: 57.8 g S

Find: mol S

Conversion Factor: 32.07 g S = 1 mol S

Draw a solution map showing the conversion from g S to mol S. The conversion factor is the molar mass of sulfur.

Solution Map:

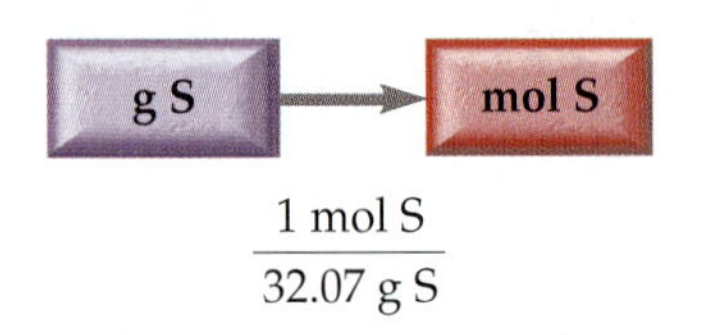

$$\frac{1 \text{ mol S}}{32.07 \text{ g S}}$$

Follow the solution map to solve the problem. Begin with 57.8 g S and use the conversion factor to get to mol S.

Solution:

$$57.8 \cancel{\text{g S}} \times \frac{1 \text{ mol S}}{32.07 \cancel{\text{g S}}} = 1.80 \text{ mol S}$$

SKILLBUILDER 6.2 The Mole Concept—Converting between Grams and Moles

Calculate the number of grams of sulfur in 2.78 mol of sulfur.

Converting between Grams of an Element and Number of Atoms

Now, suppose we want to know the number of carbon *atoms* in the 0.58-g diamond. We first convert from grams to moles and then from moles to number of atoms. The solution map is:

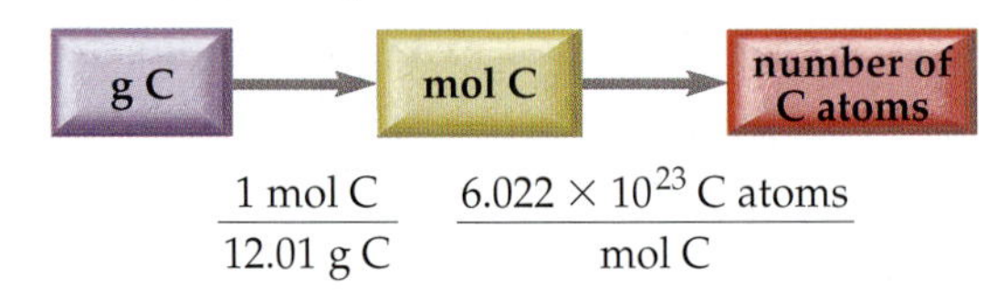

$$\frac{1 \text{ mol C}}{12.01 \text{ g C}} \qquad \frac{6.022 \times 10^{23} \text{ C atoms}}{\text{mol C}}$$

Notice the similarity between this solution map and the one we used for nails:

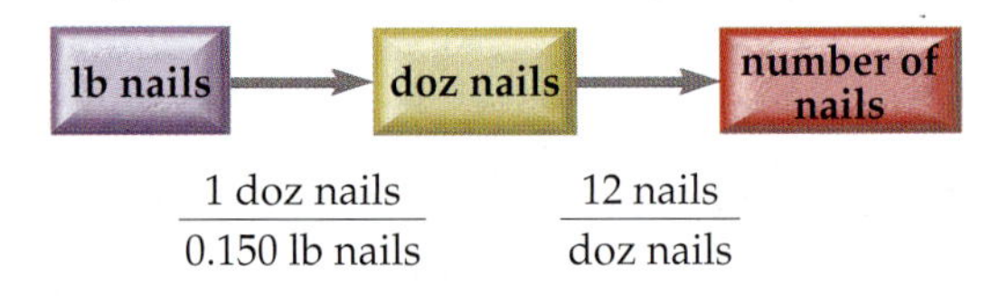

$$\frac{1 \text{ doz nails}}{0.150 \text{ lb nails}} \qquad \frac{12 \text{ nails}}{\text{doz nails}}$$

Beginning with 0.58 g carbon and using the solution map as a guide, we convert to the number of carbon atoms.

$$0.58 \cancel{\text{g C}} \times \frac{1 \cancel{\text{mol C}}}{12.01 \cancel{\text{g C}}} \times \frac{6.022 \times 10^{23} \text{ C atoms}}{\cancel{\text{mol C}}} = 2.9 \times 10^{22} \text{ C atoms}$$

EXAMPLE 6.3 The Mole Concept—Converting between Grams and Number of Atoms

How many aluminum atoms are in an aluminum can with a mass of 16.2 g?

Set up the problem in the normal way. You are given the mass of aluminum and asked to find the number of aluminum atoms. The required conversion factors are the molar mass of aluminum and the number of atoms in a mole.

Given: 16.2 g Al

Find: Al atoms

Conversion Factors:

26.98 g Al = 1 mol Al (molar mass of Al)

$6.022 \times 10^{23} = 1$ mol

The solution map has two steps. In the first step, convert from g Al to mol Al. In the second step, convert from mol Al to the number of Al atoms.

Solution Map:

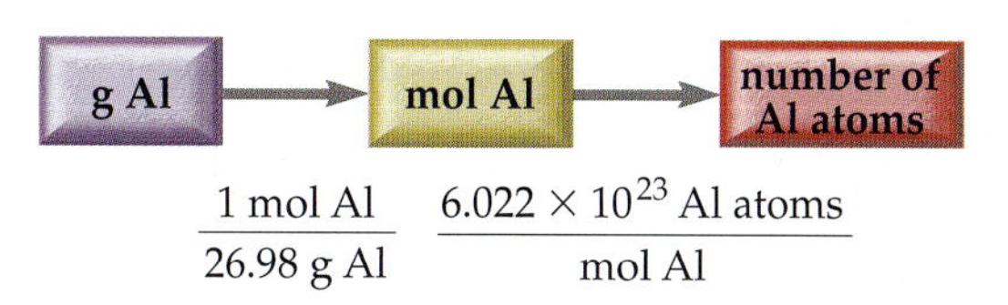

$$\frac{1 \text{ mol Al}}{26.98 \text{ g Al}} \qquad \frac{6.022 \times 10^{23} \text{ Al atoms}}{\text{mol Al}}$$

Follow the solution map to solve the problem, beginning with 16.2 g Al and multiplying by the appropriate conversion factors to arrive at Al atoms.

Solution:

$$16.2 \text{ g Al} \times \frac{1 \text{ mol Al}}{26.98 \text{ g Al}} \times \frac{6.022 \times 10^{23} \text{ Al atoms}}{1 \text{ mol Al}} = 3.62 \times 10^{23} \text{ Al atoms}$$

SKILLBUILDER 6.3 The Mole Concept—Converting between Grams and Number of Atoms

Calculate the mass of 1.23×10^{24} helium atoms.

Before we move on, notice that numbers with large exponents, such as 6.022×10^{23}, are deceptively large. Twenty-two copper pennies contain 6.022×10^{23} or 1 mol of copper atoms, but 6.022×10^{23} pennies would cover Earth's entire surface to a depth of 300 m. Even objects small by everyday standards occupy a huge space when we have a mole of them. For example, one crystal of granulated sugar has a mass of less than 1 mg and a diameter of less than 0.1 mm, yet 1 mol of sugar crystals would cover the state of Texas to a depth of several feet. For every increase of 1 in the exponent of a number, the number increases by 10. So a number with an exponent of 23 is incredibly large. A mole has to be a large number, however, because atoms are so small.

CONCEPTUAL CHECKPOINT 6.1

Which of these statements is *always* true for samples of atomic elements, regardless of the type of element present in the samples?

(a) If two samples of different elements contain the same number of atoms, they must contain the same number of moles.

(b) If two samples of different elements have the same mass, they must contain the same number of moles.

(c) If two samples of different elements have the same mass, they must contain the same number of atoms.

6.4 Counting Molecules by the Gram

The calculations we just performed for atoms can also be applied to molecules for covalent compounds or formula units for ionic compounds. We first look at converting between the mass of a compound and moles of the compound, then we turn to calculating the number of molecules (or formula units) from mass.

Remember, ionic compounds do not contain individual molecules. In loose language, the smallest electrically neutral collection of ions is sometimes called a molecule, but is more correctly called a formula unit.

Converting between Grams and Moles of a Compound

For elements, the molar mass is the mass of 1 mol of atoms of that element. For compounds, the molar mass is the mass of 1 mol of molecules or formula units of that compound. The molar mass of a compound in grams per mole is numerically equal to the formula mass of the compound in atomic mass units. For example, the formula mass of CO_2 is:

Remember, the formula mass for a compound is simply the sum of the atomic masses of all of the atoms in a chemical formula.

$$\begin{aligned}\text{Formula mass} &= 1(\text{Atomic mass of C}) + 2(\text{Atomic mass of O}) \\ &= 1(12.01 \text{ amu}) + 2(16.00 \text{ amu}) \\ &= 44.01 \text{ amu}\end{aligned}$$

The molar mass is therefore:

$$\text{Molar mass} = 44.01 \text{ g/mol}$$

Just as the molar mass of an element serves as a conversion factor between grams and moles of that element, so the molar mass of a compound serves as a conversion factor between grams and moles of that compound. For example, suppose we want to find the number of moles in a 22.5-g sample of dry ice (solid CO_2). We set up the problem in the normal way.

Given: 22.5 g CO_2

Find: mol CO_2

Conversion Factor: 44.01 g CO_2 = 1 mol CO_2 (molar mass of CO_2)

Solution Map: The solution map shows how the molar mass converts grams of the compound to moles of the compound.

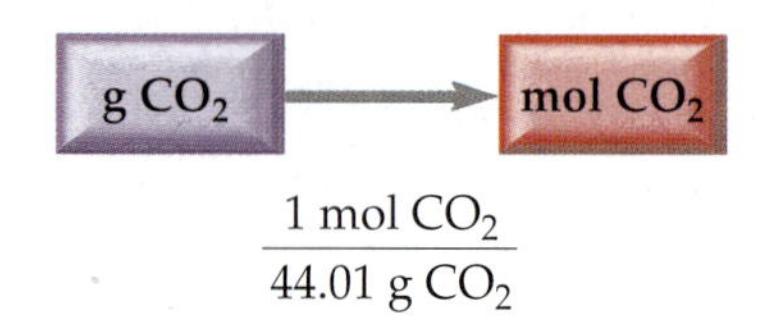

Solution:

$$22.5 \text{ g} \times \frac{1 \text{ mol } CO_2}{44.01 \text{ g}} = 0.511 \text{ mol } CO_2$$

EXAMPLE 6.4 The Mole Concept—Converting between Grams and Moles for Compounds

Calculate the mass (in grams) of 1.75 mol of water.

Set up the problem in the normal way. You are given moles of water and asked to find the mass. The conversion factor is the molar mass of water.

Given: 1.75 mol H_2O

Find: g H_2O

Conversion Factor:

$$\begin{aligned} H_2O \text{ molar mass} &= 2(\text{Atomic mass H}) + 1(\text{Atomic mass O}) \\ &= 2(1.01) + 1(16.00) \\ &= 18.02 \text{ g/mol} \end{aligned}$$

Draw a solution map showing the conversion from mol H_2O to g H_2O.

Solution Map:

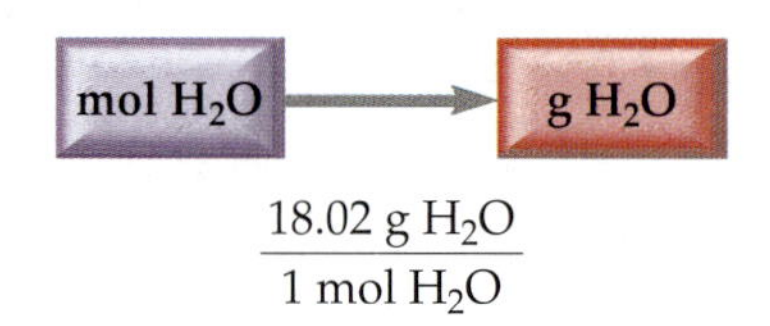

Follow the solution map to solve the problem. Begin with 1.75 mol of water and use the molar mass to convert to grams of water.

Solution:

$$1.75 \text{ mol } H_2O \times \frac{18.02 \text{ g } H_2O}{\text{mol } H_2O} = 31.5 \text{ g } H_2O$$

SKILLBUILDER 6.4 The Mole Concept—Converting between Grams and Moles

Calculate the number of moles of NO_2 in 1.18 g of NO_2.

Converting between Grams of a Compound and Number of Molecules

Suppose that we want to find the *number of CO_2 molecules* in a sample of dry ice (solid CO_2) with a mass of 22.5 g.

The solution map for the problem is:

g CO_2 → mol CO_2 → CO_2 molecules

$$\frac{1 \text{ mol } CO_2}{44.01 \text{ g } CO_2} \qquad \frac{6.022 \times 10^{23} \text{ } CO_2 \text{ molecules}}{1 \text{ mol } CO_2}$$

Notice that the first part of the solution map is identical to computing the number of moles of CO_2 in 22.5 g of dry ice. The second part of the solution map shows the conversion from moles to number of molecules. Following the solution map, we get:

$$22.5 \text{ g } CO_2 \times \frac{1 \text{ mol } CO_2}{44.01 \text{ g } CO_2} \times \frac{6.022 \times 10^{23} \text{ } CO_2 \text{ molecules}}{1 \text{ mol } CO_2} = 3.08 \times 10^{23} \text{ } CO_2 \text{ molecules}$$

EXAMPLE 6.5 The Mole Concept—Converting between Mass of a Compound and Number of Molecules

Find the mass of 4.78×10^{24} NO_2 molecules.

Set up the problem in the normal way. You are given the number of NO_2 molecules and asked to find the mass. The required conversion factors are the molar mass of NO_2 and the number of molecules in a mole.

Given: 4.78×10^{24} NO_2 molecules

Find: g NO_2

Conversion Factor: $6.022 \times 10^{23} = 1$ mol

$$NO_2 \text{ molar mass} = 1(\text{Atomic mass N}) + 2(\text{Atomic mass O}) = 14.01 + 2(16.00) = 46.01 \text{ g/mol}$$

The solution map has two steps. In the first step, convert from molecules of NO_2 to moles of NO_2. In the second step, convert from moles of NO_2 to mass of NO_2.

Solution Map:

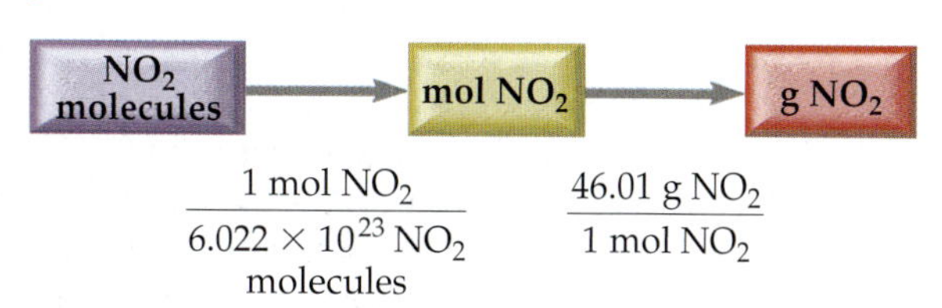

Using the solution map as a guide, begin with molecules of NO_2 and multiply by the appropriate conversion factors to arrive at g NO_2.

Solution:

$$4.78 \times 10^{24} \text{ } NO_2 \text{ molecules} \times \frac{1 \text{ mol } NO_2}{6.022 \times 10^{23} \text{ } NO_2 \text{ molecules}} \times \frac{46.01 \text{ g } NO_2}{1 \text{ mol } NO_2} = 365 \text{ g } NO_2$$

SKILLBUILDER 6.5 The Mole Concept—Converting between Mass and Number of Molecules

How many H_2O molecules are in a sample of water with a mass of 3.64 g?

CONCEPTUAL CHECKPOINT 6.2

If we wanted the formula mass of a compound in amu to equal the molar mass of the compound in kilograms rather than in grams,

(a) the size of the mole would have to be larger than Avogadro's number.
(b) the size of the mole would have to be smaller than Avogadro's number.
(c) the size of the mole would not be affected.

6.5 Chemical Formulas as Conversion Factors

▲ From our knowledge of clovers, we know that each clover has three leaves. We can express that as an equivalence: 3 leaves ≡ 1 clover.

We are almost ready to address the sodium problem in our opening example. To determine how much of a particular element (such as sodium) is in a given amount of a particular compound (such as sodium chloride), we must understand the numerical relationships inherent in a chemical formula. We can understand these relationships with a simple analogy: Asking how much sodium is in a given amount of sodium chloride is much like asking how many leaves are on a given number of clovers. For example, suppose we want to know how many leaves are on 14 clovers. We need a conversion factor between leaves and clovers. For clovers, the conversion factor comes from our knowledge about them—we know that each clover has 3 leaves. We write:

$$3 \text{ leaves} \equiv 1 \text{ clover}$$

Like other conversion factors, this equivalence gives the relationship between leaves and clovers. We use the equivalence sign (≡) because while 3 leaves do not equal 1 clover, 3 leaves are equivalent to one clover, meaning that each clover must have 3 leaves to be complete. With this conversion factor, we can easily find the number of leaves in 14 clovers. The solution map is:

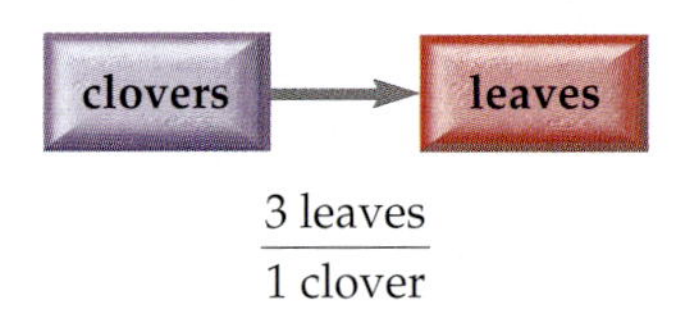

$$\frac{3 \text{ leaves}}{1 \text{ clover}}$$

We solve the problem by beginning with clovers and converting to leaves.

$$14 \text{ clovers} \times \frac{3 \text{ leaves}}{1 \text{ clover}} = 42 \text{ leaves}$$

Similarly, a chemical formula gives us equivalences between elements for a particular compound. For example, the formula for carbon dioxide (CO_2) means there are two O atoms per CO_2 molecule. We write this as:

$$2 \text{ O atoms} \equiv 1 \text{ CO}_2 \text{ molecule}$$

Just as 3 leaves ≡ 1 clover can also be written as 3 dozen leaves ≡ 1 dozen clovers, for molecules we can write:

$$2 \text{ doz O atoms} \equiv 1 \text{ doz CO}_2 \text{ molecules}$$

However, for atoms and molecules, we normally work in moles.

$$2 \text{ mol O} \equiv 1 \text{ mol CO}_2$$

With conversion factors such as these—which come directly from the chemical formula—we can determine the amounts of the constituent elements present in a given amount of a compound.

Chemical formulas are covered in Chapter 5.

▶ Each of these shows an equivalence.

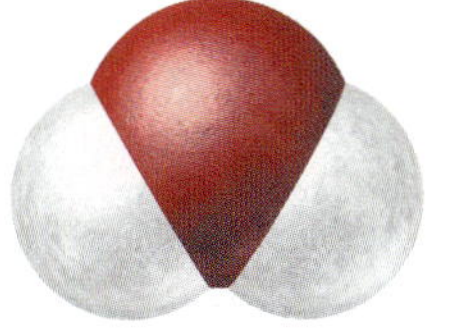

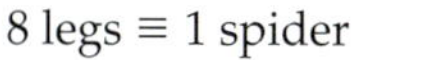

8 legs ≡ 1 spider   4 legs ≡ 1 chair   2 H atoms ≡ 1 H_2O molecule

Converting between Moles of a Compound and Moles of a Constituent Element

Suppose we want to know the number of moles of O in 18 mol of CO_2. Our solution map is:

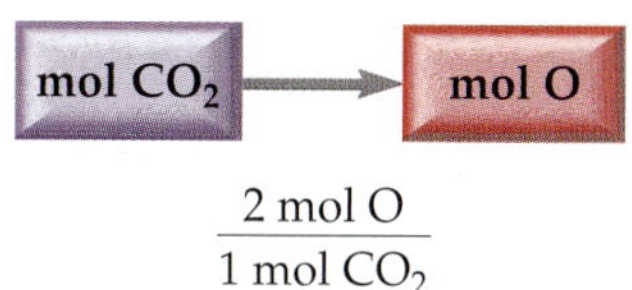

$$\frac{2 \text{ mol O}}{1 \text{ mol } CO_2}$$

We can then calculate the moles of O.

$$18 \cancel{\text{mol } CO_2} \times \frac{2 \text{ mol O}}{1 \cancel{\text{mol } CO_2}} = 36 \text{ mol O}$$

EXAMPLE 6.6 Chemical Formulas as Conversion Factors—Converting between Moles of a Compound and Moles of a Constituent Element

Determine the number of moles of O in 1.7 mol of $CaCO_3$.

Set up the problem in the normal way. You are given the number of moles of $CaCO_3$ and asked to find the number of moles of O. The conversion factor is obtained from the chemical formula, which indicates three O atoms for every $CaCO_3$ unit.

Given: 1.7 mol $CaCO_3$

Find: mol O

Conversion Factor: 3 mol O ≡ 1 mol $CaCO_3$

The solution map shows how the conversion factor from the chemical formula converts moles of $CaCO_3$ into moles of O.

Solution Map:

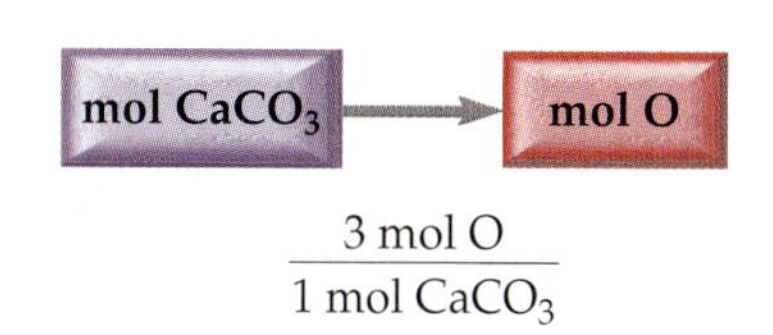

$$\frac{3 \text{ mol O}}{1 \text{ mol } CaCO_3}$$

Follow the solution map to solve the problem. The subscripts in a chemical formula are exact, so they never limit significant figures.

Solution:

$$1.7 \cancel{\text{mol } CaCO_3} \times \frac{3 \text{ mol O}}{1 \cancel{\text{mol } CaCO_3}} = 5.1 \text{ mol O}$$

SKILLBUILDER 6.6 Chemical Formulas as Conversion Factors—Converting between Moles of a Compound and Moles of a Constituent Element

Determine the number of moles of O in 1.4 mol of H_2SO_4.

Converting between Grams of a Compound and Grams of a Constituent Element

Now, we have everything we need to solve our sodium problem. Suppose we want to know how many grams of sodium there are in 15 g of NaCl. The chemical formula gives us the relationship between moles of Na and moles of NaCl:

$$1 \text{ mol Na} \equiv 1 \text{ mol NaCl}$$

To use this relationship, we need *mol* NaCl. But, we have *g* NaCl. We can, however, use the *molar mass* of NaCl to convert from g NaCl to mol NaCl. Then we use the conversion factor from the chemical formula to convert to mol Na. Finally, we use the molar mass of Na to convert to g Na. The solution map is:

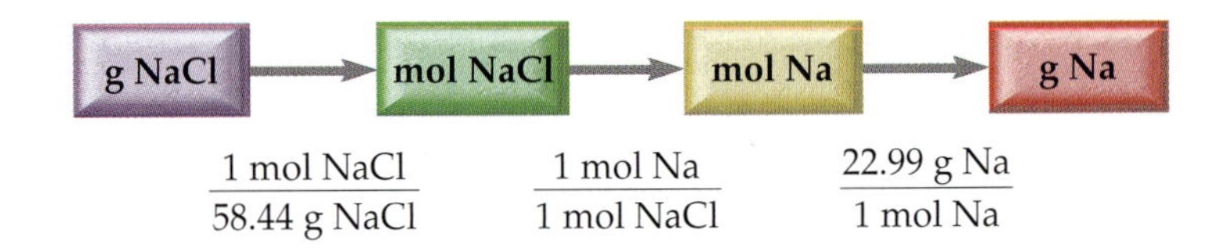

$$\frac{1 \text{ mol NaCl}}{58.44 \text{ g NaCl}} \qquad \frac{1 \text{ mol Na}}{1 \text{ mol NaCl}} \qquad \frac{22.99 \text{ g Na}}{1 \text{ mol Na}}$$

Notice that we must convert from g NaCl to mol NaCl *before* we can use the chemical formula as a conversion factor.

The chemical formula gives us a relationship between moles of substances, not between grams.

We follow the solution map to solve the problem.

$$15 \cancel{\text{g NaCl}} \times \frac{1 \cancel{\text{mol NaCl}}}{58.44 \cancel{\text{g NaCl}}} \times \frac{1 \cancel{\text{mol Na}}}{1 \cancel{\text{mol NaCl}}} \times \frac{22.99 \text{ g Na}}{1 \cancel{\text{mol Na}}} = 5.9 \text{ g Na}$$

The general form for solving problems where you are asked to find the mass of an element present in a given mass of a compound is:

Mass compound ⟶ **Moles** compound ⟶ **Moles** element ⟶ **Mass** element

The conversions between mass and moles are accomplished using the atomic or molar mass, and the conversion between moles and moles is accomplished using the relationships inherent in the chemical formula (▼ Figure 6.2).

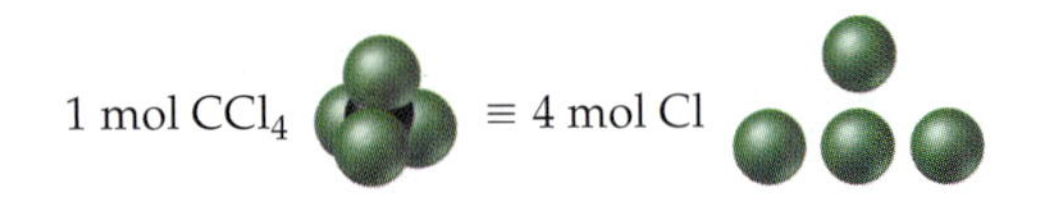

▲ **Figure 6.2** **Mole relationships from a chemical formula** The relationships inherent in a chemical formula allow us to convert between moles of the compound and moles of a constituent element (or vice versa).

EXAMPLE 6.7 Chemical Formulas as Conversion Factors—Converting between Grams of a Compound and Grams of a Constituent Element

Carvone ($C_{10}H_{14}O$) is the main component of spearmint oil. It has a pleasant aroma and mint flavor. Carvone is often added to chewing gum, liqueurs, soaps, and perfumes. Find the mass of carbon in 55.4 g of carvone.

Extract the important information from the problem in the normal way. You are given the mass of carvone and asked to find the mass of one of its constituent elements.

Given: 55.4 g $C_{10}H_{14}O$

Find: g C

MOLECULE Carvone

Conversion Factors:

You need three conversion factors. The first is the molar mass of carvone.

$$\text{Molar mass} = 10(12.01) + 14(1.01) + 1(16.00) = 120.1 + 14.14 + 16.00 = 150.2 \text{ g/mol}$$

The second conversion factor is the relationship between moles of C and moles of carvone from the molecular formula.

$$10 \text{ mol C} \equiv 1 \text{ mol } C_{10}H_{14}O$$

The third conversion factor is the molar mass of carbon.

$$1 \text{ mol C} = 12.01 \text{ g}$$

The solution map is based on

Grams ⟶ Mole ⟶ Mole ⟶ Grams

Remember, the conversion factor obtained from the chemical formula ($10 \text{ mol C} \equiv 1 \text{ mol } C_{10}H_{14}O$) applies only to moles; since we are given grams of carvone, we must first convert from g to mol.

Solution Map:

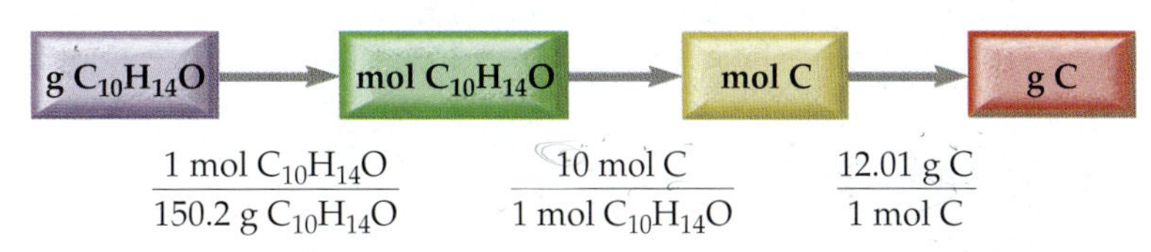

$$\frac{1 \text{ mol } C_{10}H_{14}O}{150.2 \text{ g } C_{10}H_{14}O} \qquad \frac{10 \text{ mol C}}{1 \text{ mol } C_{10}H_{14}O} \qquad \frac{12.01 \text{ g C}}{1 \text{ mol C}}$$

Follow the solution map to solve the problem, beginning with g $C_{10}H_{14}O$ and multiplying by the appropriate conversion factors to arrive at g C.

Solution:

$$55.4 \text{ g } C_{10}H_{14}O \times \frac{1 \text{ mol } C_{10}H_{14}O}{150.2 \text{ g } C_{10}H_{14}O} \times \frac{10 \text{ mol C}}{1 \text{ mol } C_{10}H_{14}O} \times \frac{12.01 \text{ g C}}{1 \text{ mol C}} = 44.3 \text{ g C}$$

SKILLBUILDER 6.7 Chemical Formulas as Conversion Factors—Converting between Grams of a Compound and Grams of a Constituent Element

Determine the mass of oxygen in a 5.8-g sample of sodium bicarbonate ($NaHCO_3$).

SKILLBUILDER PLUS

Determine the mass of oxygen in a 7.20-g sample of $Al_2(SO_4)_3$.

6.6 Mass Percent Composition of Compounds

Another way to express how much of an element is in a given compound is to use the element's mass percent composition for that compound. The **mass percent composition** or simply **mass percent** of an element is the element's percentage of the total mass of the compound. For example, the mass percent composition of sodium in sodium chloride is 39%; a 100-g sample of sodium

Chemistry IN THE ENVIRONMENT

Chlorine in Chlorofluorocarbons

About twenty-five years ago, scientists began to suspect that synthetic compounds known as chlorofluorocarbons (or CFCs) were destroying a vital compound called ozone (O_3) in Earth's upper atmosphere. Upper atmospheric ozone is important because it acts as a shield to protect life on Earth from harmful ultraviolet light (▼ Figure 6.3). CFCs are chemically inert molecules—they do not readily react with other substances—used primarily as refrigerants and industrial solvents. Their inertness, however, has allowed them to leak into the atmosphere and stay there for many years. In the upper atmosphere sunlight breaks bonds within CFCs, resulting in the release of chlorine atoms. The chlorine atoms then react with ozone and destroy it, converting it into O_2.

In 1985, scientists discovered a large hole in the ozone layer over Antarctica that has since been attributed to CFCs. The amount of ozone depletion over Antarctica was a startling 50 percent. The ozone hole is transient, existing only in the Antarctic spring, from late August to November. Examination of data from previous years showed that this gradually expanding ozone hole has formed each spring since 1977 (▼ Figure 6.4), and it continues to form today. (In 2002 the ozone hole was a bit smaller than the year before, but in 2003 it again grew larger.)

A similar hole has been observed during some years over the North Pole, and a smaller, but still significant, drop in ozone has been observed over more populated areas such as the northern United States and Canada. The thinning of ozone over these areas is dangerous because ultraviolet light can harm living things and induce skin cancer in humans. Based on this evidence, most developed nations banned the production of CFCs on January 1, 1996. However, CFCs still lurk in most older refrigerators and air conditioning units and can leak into the atmosphere and destroy ozone.

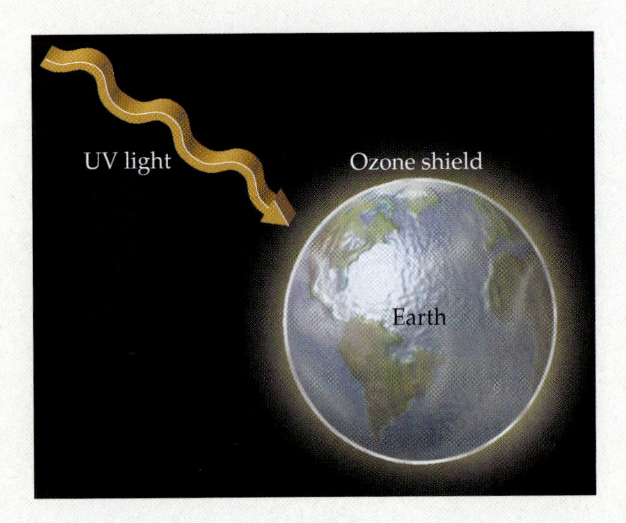

▲ **Figure 6.3 The ozone shield** Atmospheric ozone shields life on Earth from harmful ultraviolet light.

CAN YOU ANSWER THIS? *Suppose a car air conditioner contains 2.5 kg of freon-12 (CCl_2F_2), a CFC. How many kilograms of Cl are contained within the freon?*

▲ **Figure 6.4 Growth of the ozone hole** Antarctic ozone levels in three Septembers from 1979 to 2000. The darkest blue colors indicate the lowest ozone levels.

chloride contains 39 g of sodium. The mass percent composition can be determined from experimental data using the following formula.

$$\text{Mass percent of element } X = \frac{\text{Mass of } X \text{ in a sample of the compound}}{\text{Mass of the sample of the compound}} \times 100\%$$

For example, suppose a 0.358-g sample of chromium reacts with oxygen to form 0.523 g of the metal oxide. Then the mass percent of chromium is:

$$\text{Mass percent Cr} = \frac{\text{Mass Cr}}{\text{Mass metal oxide}} \times 100\%$$

$$= \frac{0.358 \text{ g}}{0.523 \text{ g}} \times 100\% = 68.5\%$$

Mass percent composition can be used as a conversion factor between grams of a constituent element and grams of the compound. For example, we saw that the mass percent composition of sodium in sodium chloride is 39%. This can be written as:

$$39 \text{ g sodium} \equiv 100 \text{ g sodium chloride}$$

or in fractional form:

$$\frac{39 \text{ g Na}}{100 \text{ g NaCl}} \quad \text{or} \quad \frac{100 \text{ g NaCl}}{39 \text{ g Na}}$$

These fractions are conversion factors between g Na and g NaCl.

EXAMPLE 6.8 Using Mass Percent Composition as a Conversion Factor

The FDA recommends that you consume less than 2.4 g of sodium per day. How many grams of sodium chloride can you consume and still be within the FDA guidelines? Sodium chloride is 39% sodium by mass.

Set up the problem in the normal way. The conversion factor is the mass percent of sodium in sodium chloride.

Given: 2.4 g Na

Find: g NaCl

Conversion Factor: 39 g Na ≡ 100 g NaCl

The solution map starts with g Na and uses the mass percent as a conversion factor to get to g NaCl.

Solution Map:

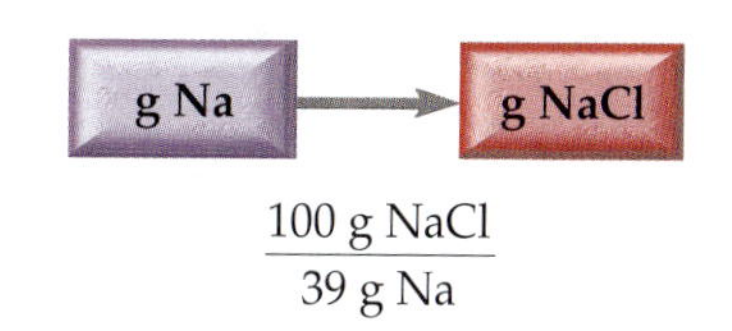

Follow the solution map to solve the problem, beginning with grams Na and ending with grams of NaCl. You can consume 6.2 g NaCl and still be within the FDA guideline.

Solution:

$$2.4 \cancel{\text{g Na}} \times \frac{100 \text{ g NaCl}}{39 \cancel{\text{g Na}}} = 6.2 \text{ g NaCl}$$

▶ Twelve and a half salt packets contain 6.2 g NaCl.

SKILLBUILDER 6.8 Using Mass Percent Composition as a Conversion Factor

If someone consumes 22 g of sodium chloride, how much sodium does that person consume? Sodium chloride is 39% sodium by mass.

6.7 Mass Percent Composition from a Chemical Formula

In the previous section, we saw how to compute mass percent composition from experimental data and how to use mass percent composition as a conversion factor. We can also compute the mass percent of any element in a compound from the chemical formula for the compound. Based on the chemical formula, the mass percent of element X in a compound is:

$$\text{Mass percent of element } X = \frac{\text{Mass of element } X \text{ in 1 mol of compound}}{\text{Mass of 1 mol of compound}} \times 100\%$$

Suppose, for example, that we want to calculate the mass percent composition of Cl in the chlorofluorocarbon CCl_2F_2. The mass percent of Cl is given by:

CCl_2F_2

$$\text{Mass percent Cl} = \frac{2 \times \text{Molar mass Cl}}{\text{Molar mass } CCl_2F_2} \times 100\%$$

The molar mass of Cl must be multiplied by 2 because the chemical formula has a subscript of 2 for Cl, meaning that 1 mol of CCl_2F_2 contains 2 mol of Cl atoms. The molar mass of CCl_2F_2 is computed as follows:

$$\begin{aligned} \text{Molar mass} &= 1(12.01) + 2(35.45) + 2(19.00) \\ &= 120.91 \text{ g/mol} \end{aligned}$$

So the mass percent of Cl in CCl_2F_2 is

$$\begin{aligned} \text{Mass percent Cl} &= \frac{2 \times \text{Molar mass Cl}}{\text{Molar mass } CCl_2F_2} \times 100\% \\ &= \frac{2 \times 35.45 \text{ g}}{120.91 \text{ g}} \times 100\% \\ &= 58.64\% \end{aligned}$$

EXAMPLE 6.9 Mass Percent Composition

Calculate the mass percent of Cl in freon-114 ($C_2Cl_4F_2$).

You are asked to find the mass percent of Cl given the molecular formula of freon-114. The required equation, given earlier, shows how to compute the mass percent of an element in a compound based on the formula for the compound.

Given: $C_2Cl_4F_2$

Find: Mass % Cl

Equation:

$$\text{Mass percent of element } X = \frac{\text{Mass of element } X \text{ in 1 mol of compound}}{\text{Mass of 1 mol of compound}}$$

The solution map simply shows how the mass of Cl in 1 mol of $C_2Cl_4F_2$ and the molar mass of $C_2Cl_4F_2$, when substituted into the mass percent equation, yield the mass percent of Cl.

Solution Map:

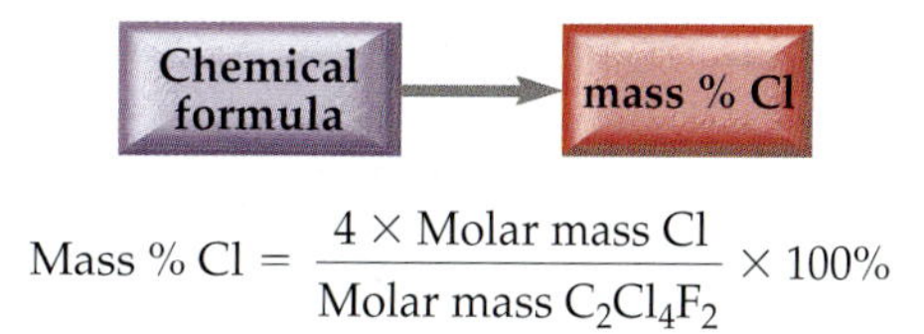

$$\text{Mass \% Cl} = \frac{4 \times \text{Molar mass Cl}}{\text{Molar mass } C_2Cl_4F_2} \times 100\%$$

Calculate the necessary parts of the equation and substitute the values into the equation to find mass percent Cl.

Solution:

$$4 \times \text{Molar mass Cl} = 4(35.45\text{ g}) = 141.8\text{ g}$$

$$\text{Molar mass } C_2Cl_4F_2 = 2(12.01) + 4(35.45) + 2(19.00)$$

$$= 24.02 + 141.8 + 38.00$$

$$= \frac{203.8\text{ g}}{\text{mol}}$$

$$\text{Mass \% Cl} = \frac{4 \times \text{Molar mass Cl}}{\text{Molar mass } C_2Cl_4F_2} \times 100\%$$

$$= \frac{141.8\text{ g}}{203.8\text{ g}} \times 100\%$$

$$= 69.58\%$$

SKILLBUILDER 6.9 Mass Percent Composition

Acetic acid ($HC_2H_3O_2$) is the active ingredient in vinegar. Calculate the mass percent composition of O in acetic acid.

CONCEPTUAL CHECKPOINT 6.3

Which of the following compounds has the highest mass percent of O? (You should not have to perform any calculations to answer this question.)

(a) CrO
(b) CrO_2
(c) Cr_2O_3

Chemistry AND HEALTH

Fluoridation of Drinking Water

In the early 1900s, scientists discovered that people whose drinking water naturally contained fluoride (F^-) ions had fewer cavities than people whose water did not. At the proper levels, fluoride strengthens tooth enamel, which prevents tooth decay. In an effort to improve public health, fluoride has been artificially added to drinking water supplies since 1945. In the United States today, about 62% of the population drinks artificially fluoridated drinking water. The American Dental Association and public health agencies estimate that water fluoridation reduces tooth decay by 40% to 65%.

The fluoridation of public drinking water, however, is often controversial. Some opponents argue that fluoride is available from other sources—such as toothpaste, mouthwash, drops, and pills—and therefore should not be added to drinking water. Anyone who wants fluoride can get it from these optional sources, they argue, and the government should not impose fluoride on the general population. Other opponents argue that the risks associated with fluoridation are too great. Indeed, too much fluoride can cause teeth to become brown and spotted, a condition known as dental fluorosis. Extremely high levels can lead to skeletal fluorosis, a condition in which the bones become brittle and arthritic.

The scientific consensus is that, like many minerals, fluoride shows some health benefits at certain levels—about 1–4 mg/day for adults—but can have detrimental effects at higher levels. Consequently, most major cities fluoridate their drinking water at a level of about 1 mg/L. Since adults drink between 1 and 2 L of water per day, they should receive the beneficial amounts of fluoride from the water.

CAN YOU ANSWER THIS? *Fluoride is often added to water as sodium fluoride (NaF). What is the mass percent composition of F^- in NaF? How many grams of NaF should be added to 1500 L of water to fluoridate it at a level of 1.0 mg F^-/L?*

6.8 Calculating Empirical Formulas for Compounds

In the previous section we learned how to calculate mass percent composition from a chemical formula. But can we go the other way? Can we calculate a chemical formula from mass percent composition? This is important because laboratory analyses of compounds do not often give chemical formulas directly; rather, they give the relative masses of each element present in a compound. For example, if we decompose water into hydrogen and oxygen in the laboratory, we could measure the masses of hydrogen and oxygen produced. Can we get a chemical formula from this kind of data?

▶ We just learned how to go from the chemical formula of a compound to its mass percent composition. Can we also go the other way?

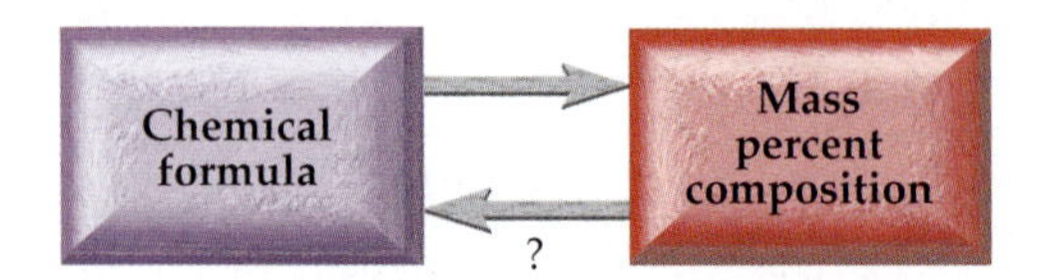

The answer is a qualified yes. We can get a chemical formula, but it is an **empirical formula**. In contrast to a **molecular formula**, which gives the specific number of each type of atom in a molecule, an empirical formula only gives the smallest whole-number ratio of each type of atom. An empirical formula *does not* necessarily represent a molecule. For example, hydrogen peroxide is a compound that contains hydrogen and oxygen with the molecular formula H_2O_2, meaning that hydrogen peroxide molecules consist of two hydrogen atoms and two oxygen atoms:

Hydrogen peroxide
Molecular formula H_2O_2
Empirical formula HO

However, the empirical formula of hydrogen peroxide is simply HO, the smallest whole-number ratio of hydrogen to oxygen atoms. Some other examples of empirical and molecular formulas, as well as the molecules they represent, are listed in the following table.

Empirical Formula	*Molecular Formula*	*Common Name*	*molecule*
CH	C_6H_6	Benzene	
CH	C_2H_2	Acetylene	
CH_2O	$C_6H_{12}O_6$	Glucose	
CO_2	CO_2	Carbon dioxide	

Notice that in some cases, different compounds with different molecular formulas can have the same empirical formula. Notice also that in some cases the empirical and molecular formula for a given compound are identical.

The molecular formula will always be a whole-number multiple of the empirical formula.

$$\text{Molecular formula} = \text{Empirical formula} \times n, \text{ where } n = 1, 2, 3 \ldots$$

For example, H_2O_2 is HO × 2 and C_6H_6 is CH × 6.

$$HO \times 2 \longrightarrow H_2O_2 \qquad CH \times 6 \longrightarrow C_6H_6$$

Calculating an Empirical Formula from Experimental Data

A chemical formula represents a ratio of atoms or moles of atoms, not a ratio of masses.

Suppose we decompose a sample of water in the laboratory and find that it produces 3.0 g of hydrogen and 24 g of oxygen. How do we get an empirical formula from these data?

▶ Water can be decomposed by an electric current into hydrogen and oxygen. How can we find the empirical formula for water from the masses of its component elements?

We know that an empirical formula represents a ratio of atoms or a ratio of moles of atoms, but it *does not* represent a ratio of masses. So the first thing we must do is convert our data from grams to moles. How many moles of each element do we have? To convert to moles, simply divide each mass by the molar mass of that element.

$$\text{Moles H} = 3.0 \, \cancel{\text{g H}} \times \frac{1 \text{ mol H}}{1.01 \, \cancel{\text{g H}}} = 3.0 \text{ mol H}$$

$$\text{Moles O} = 24 \, \cancel{\text{g O}} \times \frac{1 \text{ mol O}}{16.00 \, \cancel{\text{g O}}} = 1.5 \text{ mol O}$$

From these data, we know there are 3 mol of H for every 1.5 mol of O. We can now write a pseudoformula for water:

$$H_3O_{1.5}$$

TUTORIAL Empirical Formula

To get whole-number subscripts in our formula, we simply divide all the subscripts by the smallest one, in this case 1.5.

$$H_{\frac{3}{1.5}}O_{\frac{1.5}{1.5}} = H_2O$$

Our empirical formula for water, which in this case also happens to be the molecular formula, is H_2O. The procedure below can be used to obtain the empirical formula of any compound from experimental data. The left column outlines the procedure, and the center and right columns show two examples of how to apply the procedure.

Obtaining an Empirical Formula from Experimental Data

EXAMPLE 6.10

A compound containing nitrogen and oxygen is decomposed in the laboratory and produces 24.5 g of nitrogen and 70.0 g of oxygen. Calculate the empirical formula of the compound.

EXAMPLE 6.11

A laboratory analysis of aspirin determined the following mass percent composition:

C	60.00%
H	4.48%
O	35.53%

Find the empirical formula.

1. Write down (or compute) as given the masses of each element present in a sample of the compound. If you are given mass percent composition, assume a 100-g sample and compute the masses of each element from the given percentages.

Example 6.10 — Given: 24.5 g N; 70.0 g O

Find: empirical formula

Example 6.11 — Given: In a 100-g sample: 60.00 g C; 4.48 g H; 35.53 g O

Find: empirical formula

2. Convert each of the masses in Step 1 to moles by using the appropriate molar mass for each element as a conversion factor.

Example 6.10 — Solution:

$$24.5 \text{ g N} \times \frac{1 \text{ mol N}}{14.01 \text{ g N}} = 1.75 \text{ mol N}$$

$$70.0 \text{ g O} \times \frac{1 \text{ mol O}}{16.00 \text{ g O}} = 4.38 \text{ mol O}$$

Example 6.11 — Solution:

$$60.00 \text{ g C} \times \frac{1 \text{ mol C}}{12.01 \text{ g C}} = 4.996 \text{ mol C}$$

$$4.48 \text{ g H} \times \frac{1 \text{ mol H}}{1.01 \text{ g H}} = 4.44 \text{ mol H}$$

$$35.53 \text{ g O} \times \frac{1 \text{ mol O}}{16.00 \text{ g O}} = 2.221 \text{ mol O}$$

3. Write down a pseudoformula for the compound, using the moles of each element (from Step 2) as subscripts.

Example 6.10: $N_{1.75}O_{4.38}$

Example 6.11: $C_{4.996}H_{4.44}O_{2.221}$

4. Divide all the subscripts in the formula by the smallest subscript.

Example 6.10: $N_{\frac{1.75}{1.75}}O_{\frac{4.38}{1.75}} \longrightarrow N_1O_{2.5}$

Example 6.11: $C_{\frac{4.996}{2.221}}H_{\frac{4.44}{2.221}}O_{\frac{2.221}{2.221}} \longrightarrow C_{2.25}H_2O_1$

5. If the subscripts are not whole numbers, multiply all the subscripts by a small whole number (see the following table) to get whole-number subscripts.

Fractional Subscript	Multiply by this Number to Get Whole-Number Subscripts
_.10	10
_.20	5
_.25	4
_.33	3
_.50	2
_.66	3
_.75	4

Example 6.10: $N_1O_{2.5} \times 2 \longrightarrow N_2O_5$

The correct empirical formula is N_2O_5.

Example 6.11: $C_{2.25}H_2O_1 \times 4 \longrightarrow C_9H_8O_4$

The correct empirical formula is $C_9H_8O_4$.

SKILLBUILDER 6.10

A sample of a compound is decomposed in the laboratory and produces 165 g of carbon, 27.8 g of hydrogen, and 220.2 g O. Calculate the empirical formula of the compound.

SKILLBUILDER 6.11

Ibuprofen, an aspirin substitute, has the following mass percent composition:
C 75.69%; H 8.80%; O 15.51%
Calculate the empirical formula of the compound.

EXAMPLE 6.12 Calculating an Empirical Formula from Reaction Data

A 3.24-g sample of titanium reacts with oxygen to form 5.40 g of the metal oxide. What is the formula of the oxide?

Begin by setting up the problem in the normal way. You must recognize this problem as one requiring a special procedure.

Given:

3.24 g Ti
5.40 g oxide

Find: empirical formula

1. Write down (or compute) the masses of each element present in a sample of the compound.

   In this case, we are given the mass of the initial Ti sample and the mass of its oxide after the sample reacts with oxygen. The mass of oxygen is the difference between the mass of the oxide and the mass of titanium.

Solution:

3.24 g Ti

$$\text{Mass O} = \text{Mass oxide} - \text{Mass titanium}$$
$$= 5.40 \text{ g} - 3.24 \text{ g}$$
$$= 2.16 \text{ g O}$$

2. Convert each of the masses in Step 1 to moles by using the appropriate molar mass for each element as a conversion factor.

$$3.24 \cancel{\text{g Ti}} \times \frac{1 \text{ mol Ti}}{47.88 \cancel{\text{g Ti}}} = 0.0677 \text{ mol Ti}$$

$$2.16 \cancel{\text{g O}} \times \frac{1 \text{ mol O}}{16.00 \cancel{\text{g O}}} = 0.135 \text{ mol O}$$

3. Write down a pseudoformula for the compound, using the moles of each element obtained in Step 2 as subscripts.

$$Ti_{0.0677}O_{0.135}$$

4. Divide all the subscripts in the formula by the smallest subscript.

$$Ti_{\frac{0.0677}{0.0677}}O_{\frac{0.135}{0.0677}} \longrightarrow TiO_2$$

5. If the subscripts are not whole numbers, multiply all the subscripts by a small whole number to get whole-number subscripts.

Since the subscripts are already whole numbers, this last step is unnecessary. The correct empirical formula is TiO_2.

SKILLBUILDER 6.12 Calculating an Empirical Formula from Reaction Data

A 1.56-g sample of copper reacts with oxygen to form 1.95 g of the metal oxide. What is the formula of the oxide?

6.9 Calculating Molecular Formulas for Compounds

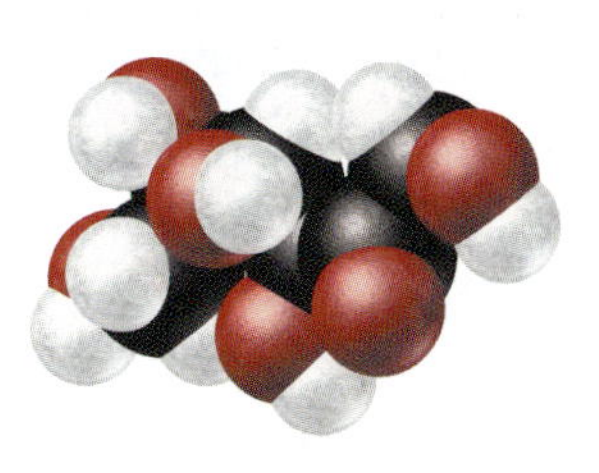

▲ Fructose, a sugar found in fruit.

You can find the molecular formula of a compound from the empirical formula if you also know the molar mass of the compound. Recall from Section 6.8 that the molecular formula is always a whole-number multiple of the empirical formula.

$$\text{Molecular formula} = \text{Empirical formula} \times n, \text{ where } n = 1, 2, 3 \ldots$$

Suppose we want to find the molecular formula for fructose (a sugar found in fruit) from its empirical formula, CH_2O, and its molar mass, 180.2 g/mol. We know that the molecular formula is a whole-number multiple of CH_2O.

$$\text{Molecular formula} = CH_2O \times n$$

We also know that the molar mass is a whole-number multiple of the **empirical formula molar mass**, the sum of the masses of all the atoms in the empirical formula.

$$\text{Molar mass} = \text{Empirical formula molar mass} \times n$$

For a particular compound, the value of n in both cases is the same. Therefore, we can find n by computing the ratio of the molar mass to the empirical formula molar mass.

$$n = \frac{\text{Molar mass}}{\text{Empirical formula mass}}$$

For fructose, the empirical formula molar mass is:

$$\text{Empirical formula molar mass} = 1(12.01) + 2(1.01) + 16.00 = 30.03 \text{ g/mol}$$

Therefore, n is:

$$n = \frac{180.2 \text{ g/mol}}{30.03 \text{ g/mol}} = 6$$

We can then use this value of n to find the molecular formula.

$$\text{Molecular formula} = CH_2O \times 6 = C_6H_{12}O_6$$

EXAMPLE 6.13 Calculating Molecular Formula from Empirical Formula and Molar Mass

Naphthalene is a compound containing carbon and hydrogen that is often used in mothballs. Its empirical formula is C_5H_4 and its molar mass is 128.16 g/mol. Find its molecular formula.

Set up the problem in the normal way. You are given the empirical formula and the molar mass of a compound and asked to find its molecular formula.

Given:

empirical formula = C_5H_4

molar mass = 128.16 g/mol

Find: molecular formula

The molecular formula is n times the empirical formula. To find n, divide the molar mass by the empirical formula molar mass.

Solution:

$$\text{Empirical formula molar mass} = 5(12.01) + 4(1.01) = 64.09 \text{ g/mol}$$

$$n = \frac{\text{Molar mass}}{\text{Empirical formula mass}} = \frac{128.16 \text{ g/mol}}{64.09 \text{ g/mol}} = 2$$

Therefore, the molecular formula is 2 times the empirical formula.

$$\text{Molecular formula} = C_5H_4 \times 2 = C_{10}H_8$$

SKILLBUILDER 6.13 Calculating Molecular Formula from Empirical Formula and Molar Mass

Butane is a compound containing carbon and hydrogen that is used as a fuel in butane lighters. Its empirical formula is C_2H_5 and its molar mass is 58.12 g/mol. Find its molecular formula.

SKILLBUILDER PLUS

A compound with the following mass percent composition has a molar mass of 60.10 g/mol. Find its molecular formula.

C	39.97%
H	13.41%
N	46.62%

CHAPTER IN REVIEW

Chemical Principles

The Mole Concept: The mole is a specific number (6.022×10^{23}) that is defined so as to allow us to easily count atoms or molecules by weighing them. One mole of any element has a mass equivalent to its atomic mass in grams, and a mole of any compound has a mass equivalent to its formula mass in grams. The mass of 1 mol of an element or compound is called the molar mass.

Chemical Formulas and Chemical Composition: Chemical formulas give us the relative number of each kind of element in a compound. These numbers are based on atoms or moles. By using molar masses, we can use the information in a chemical formula to determine the relative masses of each kind of element in a compound. We can then relate the mass of a sample of a compound to the masses of the elements contained in the compound.

Empirical and Molecular Formulas from Laboratory Data: The relative masses of the elements within a compound can be used to determine the empirical formula of the compound. If the chemist also knows the molar mass of the compound, he or she can also determine its molecular formula.

Relevance

The Mole Concept: The mole concept allows us to determine the number of atoms or molecules in a sample from its mass. Just as a hardware store customer wants to know the number of nails in a certain weight of nails, so we want to know the number of atoms in a certain mass of atoms. Since atoms are too small to count, we use their mass.

Chemical Formulas and Chemical Composition: The chemical composition of compounds is important because it lets us determine how much of a particular element is contained within a particular compound. For example, an assessment of the threat to the ozone from chlorofluorocarbons (CFCs) requires knowing how much chlorine is in a particular CFC.

Empirical and Molecular Formulas from Laboratory Data: The first thing a chemist wants to know about an unknown compound is its chemical formula, because the formula reveals the compound's composition. The way chemists often get formulas is by analyzing the compounds in the laboratory—either by decomposing them or by synthesizing them—to determine the relative masses of the elements they contain.

Chemical Skills

Converting between Moles and Number of Atoms (Section 6.3)

To convert between moles and number of atoms, use Avogadro's number, 6.022×10^{23} atoms = 1 mol, as a conversion factor.

Examples

EXAMPLE 6.13 Converting between Moles and Number of Atoms

Calculate the number of atoms in 4.8 mol of copper.

Given: 4.8 mol Cu

Find: Cu atoms

Conversion Factor: 1 mol Cu = 6.022×10^{23} Cu atoms

Solution Map:

mol Cu → Cu atoms

$$\frac{6.022 \times 10^{23}\ \text{Cu atoms}}{1\ \text{mol Cu}}$$

Solution:

$$4.8\ \text{mol Cu} \times \frac{6.022 \times 10^{23}\ \text{Cu atoms}}{1\ \text{mol Cu}} = 2.9 \times 10^{24}\ \text{Cu atoms}$$

Converting between Grams and Moles (Section 6.3)

To convert between grams of an element and moles of an element, use that element's molar mass. The molar mass of any element in atomic mass units is its atomic mass with the units of grams per mole. Therefore the molar mass is a conversion factor between grams and moles. To convert between grams of a *compound* and moles of a *compound*, use the compound's molar mass.

EXAMPLE 6.14 Converting between Grams and Moles

Calculate the mass of aluminum (in grams) of 6.73 moles of aluminum.

Given: 6.73 mol Al

Find: g Al

Conversion Factor: 26.98 g Al = 1 mol Al

Solution Map:

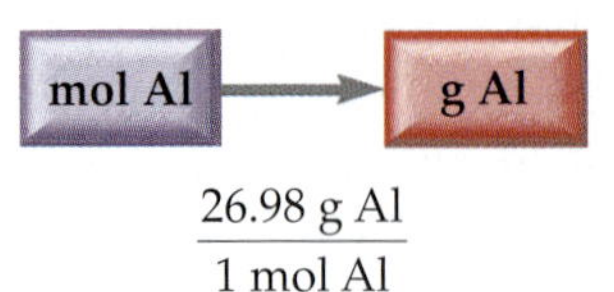

$$\frac{26.98 \text{ g Al}}{1 \text{ mol Al}}$$

Solution:

$$6.73 \cancel{\text{mol Al}} \times \frac{26.98 \text{ g Al}}{1 \cancel{\text{mol Al}}} = 182 \text{ g Al}$$

Converting between Grams and Number of Atoms or Molecules (Section 6.4)

To convert from grams of an element to the number of atoms, first use the molar mass of the element to convert from grams to moles, and then use Avogadro's number to convert moles to number of atoms.

To convert from the number of atoms of an element to grams of the element, first use Avogadro's number to convert from number of atoms to moles, and then use the molar mass to convert from moles to grams.

For compounds, the molar mass is simply the formula mass with the units of grams per mole.

LIVE EXAMPLE

EXAMPLE 6.15 Converting between Grams and Number of Atoms or Molecules

Determine the number of atoms in a 48.3-g sample of zinc.

Given: 48.3 g Zn

Find: Zn atoms

Conversion Factors:

65.38 g Zn = 1 mol Zn

1 mol = 6.022×10^{23} atoms

Solution Map:

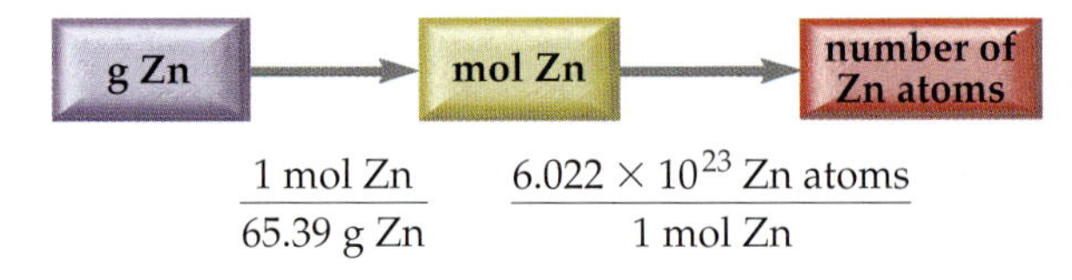

$$\frac{1 \text{ mol Zn}}{65.39 \text{ g Zn}} \qquad \frac{6.022 \times 10^{23} \text{ Zn atoms}}{1 \text{ mol Zn}}$$

Solution:

$$48.3 \cancel{\text{g Zn}} \times \frac{1 \cancel{\text{mol Zn}}}{65.39 \cancel{\text{g Zn}}} \times \frac{6.022 \times 10^{23} \text{ Zn atoms}}{1 \cancel{\text{mol Zn}}} =$$

$$4.45 \times 10^{23} \text{ Zn atoms}$$

Converting between Moles of a Compound and Moles of a Constituent Element (Section 6.5)

To convert between moles of a compound and moles of a constituent element, use the chemical formula of the compound as a conversion factor. The subscripts of each element in the formula specify the number of moles of that element that is equivalent to 1 mol of the compound.

EXAMPLE 6.16 Converting between Moles of a Compound and Moles of a Constituent Element

Determine the number of moles of oxygen in 7.20 mol of H_2SO_4.

Given: 7.20 mol H_2SO_4

Find: mol O

Conversion Factor: 4 mol O ≡ 1 mol H_2SO_4

Solution Map:

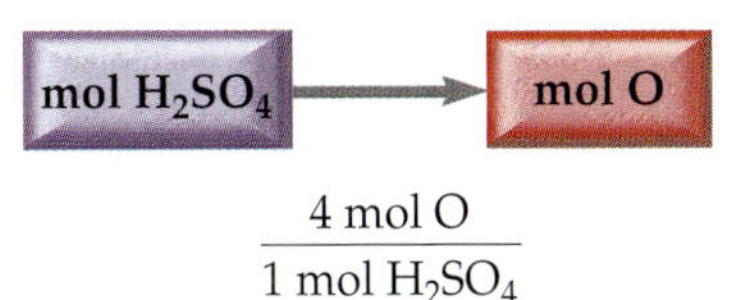

$$\frac{4\text{ mol O}}{1\text{ mol H}_2\text{SO}_4}$$

Solution:

$$7.20\ \text{mol H}_2\text{SO}_4 \times \frac{4\text{ mol O}}{1\text{ mol H}_2\text{SO}_4} = 28.8\text{ mol O}$$

Converting between Grams of a Compound and Grams of a Constituent Element (Section 6.5)

To convert from grams of a compound to grams of a constituent element, first use the molar mass of the compound to convert from grams of the compound to moles of the compound. Then use the chemical formula to obtain a conversion factor to convert from moles of the compound to moles of the constituent element. Finally, use the molar mass of the constituent element to convert from moles of the element to grams of the element.

To convert from grams of a constituent element to grams of a compound, first use the molar mass of the constituent element to convert from grams of the element to moles of the element. Then use the chemical formula to obtain a conversion factor to convert from moles of the constituent element to moles of the compound. Finally, use the molar mass of the compound to convert from moles of the compound to grams of the compound.

EXAMPLE 6.17 Converting between Grams of a Compound and Grams of a Constituent Element

Find the grams of iron in 79.2 g of Fe_2O_3.

Given: 79.2 g Fe_2O_3

Find: g Fe

Conversion Factors:

Molar mass Fe_2O_3
$= 2(55.85) + 3(16.00)$
$= 159.70$ g/mol

2 mol Fe $\equiv$ 1 mol Fe_2O_3

Solution Map:

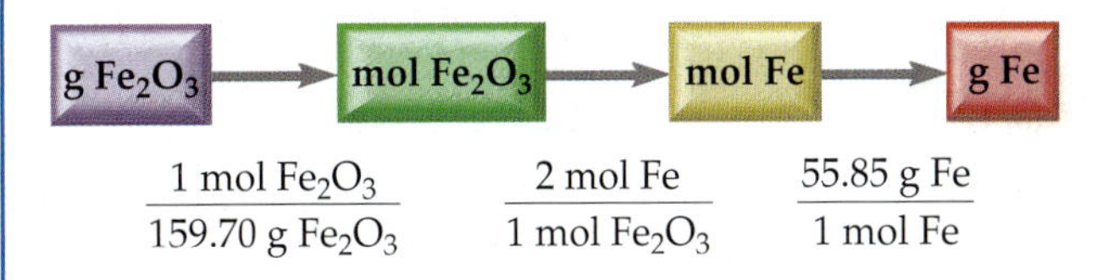

$$\frac{1\text{ mol Fe}_2\text{O}_3}{159.70\text{ g Fe}_2\text{O}_3} \qquad \frac{2\text{ mol Fe}}{1\text{ mol Fe}_2\text{O}_3} \qquad \frac{55.85\text{ g Fe}}{1\text{ mol Fe}}$$

Solution:

$$79.2\text{ g Fe}_2\text{O}_3 \times \frac{1\text{ mol Fe}_2\text{O}_3}{159.70\text{ g Fe}_2\text{O}_3} \times \frac{2\text{ mol Fe}}{1\text{ mol Fe}_2\text{O}_3} \times \frac{55.85\text{ g Fe}}{1\text{ mol Fe}} = 55.4\text{ g Fe}$$

Calculating Mass Percent Composition from Experimental Data (Section 6.6)

The mass percent composition of an element in a compound can be determined by dividing the mass of the element within the compound by the mass of the compound and then multiplying by 100%.

LIVE EXAMPLE

EXAMPLE 6.18 Calculating Mass Percent Composition from Experimental Data

A 3.52-g sample of chromium reacts with fluorine to produce 7.38 g of the metal fluoride. What is the mass percent composition of chromium in the fluoride?

Given:

3.52 g Cr
7.38 g metal fluoride

Find: Cr mass percent

Solution:

$$\%\text{ Cr} = \frac{\text{Mass Cr}}{\text{Mass metal fluoride}} \times 100\%$$

$$= \frac{3.52\text{ g}}{7.38\text{ g}} \times 100\% = 47.7\%$$

Using Mass Percent Composition as a Conversion Factor (Section 6.6)

The mass percent composition of an element in a compound can be used as a conversion factor between grams of the element and grams of the compound. Since percent means per hundred, the mass percent composition of an element in a compound is equal to the number of grams of that element per 100 g of the compound.

LIVE EXAMPLE

EXAMPLE 6.19 Using Mass Percent Composition as a Conversion Factor

Determine the mass of titanium in 57.2 g of titanium(IV) oxide. The mass percent of titanium in titanium(IV) oxide is 59.9%.

Given: 57.2 g TiO_2

Find: g Ti

Conversion Factor: 59.9 g Ti ≡ 100 g TiO_2

Solution Map:

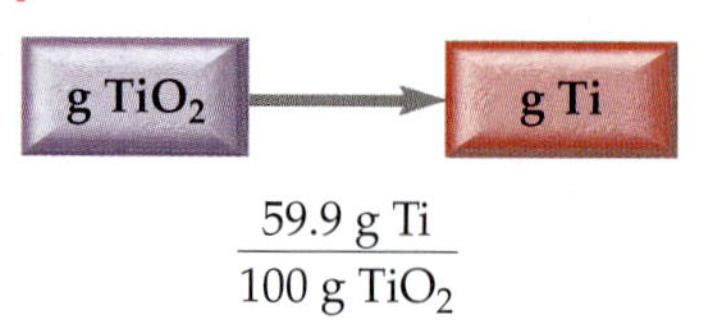

$$\frac{59.9 \text{ g Ti}}{100 \text{ g TiO}_2}$$

Solution:

$$57.2 \text{ g TiO}_2 \times \frac{59.9 \text{ g Ti}}{100 \text{ g TiO}_2} = 34.3 \text{ g Ti}$$

Determining Mass Percent Composition from a Chemical Formula (Section 6.7)

You are asked to find the mass percent of K given the formula of K_2O. The required equation shows how to compute the mass percent of an element in a compound based on the formula for the compound.

The solution map simply shows how the mass of K in 1 mol of K_2O and the molar mass of K_2O, when substituted into the mass percent equation, yield the mass percent of K.

Calculate the necessary parts of the equation and substitute the values into the equation to find the mass percent of K.

EXAMPLE 6.20 Determining Mass Percent Composition from a Chemical Formula

Calculate the mass percent composition of potassium in potassium oxide (K_2O).

Given: K_2O

Find: mass % K

Equation:

$$\text{Mass percent of element } X = \frac{\text{Mass of element } X \text{ in 1 mol of compound}}{\text{Mass of 1 mol of compound}} \times 100\%$$

Solution Map:

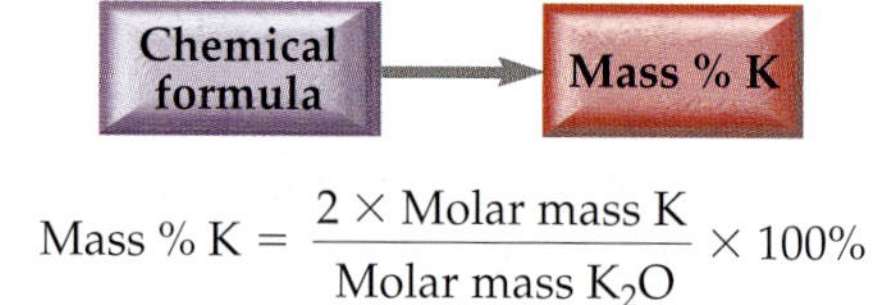

$$\text{Mass \% K} = \frac{2 \times \text{Molar mass K}}{\text{Molar mass K}_2\text{O}} \times 100\%$$

Solution:

$$2 \times \text{Molar mass K} = 2(39.10 \text{ g}) = 78.20 \text{ g}$$

$$\text{Molar mass K}_2\text{O} = 2(39.10) + 16.00 = 94.20 \text{ g/mol}$$

$$\text{Mass \% K} = \frac{78.20 \text{ g K}}{94.20 \text{ g K}_2\text{O}} \times 100\% = 83.01\%$$

Determining an Empirical Formula from Experimental Data (Section 6.8)

To determine an empirical formula from experimental data, follow these steps.

1. Write down (or compute) as given the masses of each element present in a sample of the compound. If you are given mass percent composition, assume a 100-g sample and compute the masses of each element from the given percentages.
2. Convert each of the masses in Step 1 to moles by using the appropriate molar mass for each element as a conversion factor.
3. Write down a pseudoformula for the compound using the moles of each element (from Step 2) as subscripts.
4. Divide all the subscripts in the formula by the smallest subscript.
5. If the subscripts are not whole numbers, multiply all the subscripts by a small whole number to get whole-number subscripts.

LIVE EXAMPLE

EXAMPLE 6.21 Determining an Empirical Formula from Experimental Data

A laboratory analysis of vanillin, the flavoring agent in vanilla, determined the following mass percent composition: C, 63.15%; H, 5.30%; O, 31.55%. Determine the empirical formula of vanillin.

Given: In a 100-g sample, we have 63.15 g C, 5.30 g H, and 31.55 g O.

Find: empirical formula

Solution:

$$63.15 \cancel{\text{g C}} \times \frac{1 \text{ mol C}}{12.01 \cancel{\text{g C}}} = 5.258 \text{ mol C}$$

$$5.30 \cancel{\text{g H}} \times \frac{1 \text{ mol H}}{1.01 \cancel{\text{g H}}} = 5.25 \text{ mol H}$$

$$31.55 \text{ g O} \times \frac{1 \text{ mol O}}{16.00 \text{ g O}} = 1.972 \text{ mol O}$$

$$C_{5.258}H_{5.25}O_{1.972}$$

$$C_{\frac{5.258}{1.972}}H_{\frac{5.25}{1.972}}O_{\frac{1.972}{1.972}} \longrightarrow C_{2.67}H_{2.66}O_1$$

$$C_{2.67}H_{2.66}O_1 \times 3 \longrightarrow C_8H_8O_3$$

The correct empirical formula is $C_8H_8O_3$.

Calculating a Molecular Formula from an Empirical Formula and Molar Mass (Section 6.9)

To calculate the molecular formula from the empirical formula and the molar mass, first compute the empirical formula molar mass, which is the sum of the masses of the all the atoms in the empirical formula.

Next, find n, the ratio of the molar mass to empirical mass.

Finally, multiply the empirical formula by n to get the molecular formula.

EXAMPLE 6.22 Calculating a Molecular Formula from an Empirical Formula and Molar Mass

Acetylene, a gas often used in welding torches, has the empirical formula CH and a molar mass of 26.04 g/mol. Find its molecular formula.

Given:

empirical formula = CH
molar mass = 26.04 g/mol

Find: molecular formula

Solution:

Empirical formula molar mass

$$= 12.01 + 1.01$$

$$= 13.02 \text{ g/mol}$$

$$n = \frac{\text{Molar mass}}{\text{Empirical formula molar mass}} = \frac{26.04 \text{ g/mol}}{13.02 \text{ g/mol}} = 2$$

$$\text{Molecular formula} = CH \times 2 \longrightarrow C_2H_2$$

Key Terms

Avogadro's number [6.3]
empirical formula [6.8]
empirical formula molar mass [6.9]
mass percent (composition) [6.6]
molar mass [6.3]
mole (mol) [6.3]
molecular formula [6.8]

Exercises

Questions

1. Why is chemical composition important?
2. How can you determine the number of atoms in a sample of an element? Why is counting them not an option?
3. How many atoms are in 1 mol of atoms?
4. How many molecules are in 1 mol of molecules?
5. What is the mass of 1 mol of atoms for an element?
6. What is the mass of 1 mol of molecules for a compound?
7. What is the mass of 1 mol of atoms of each of the following elements?
   (a) P **(b)** Pt
   (c) C **(d)** Cr
8. What is the mass of 1 mol of molecules of each of the following compounds?
   (a) CO_2 **(b)** CH_2Cl_2
   (c) $C_{12}H_{22}O_{11}$ **(d)** SO_2
9. The subscripts in a chemical formula give relationships between moles of the constituent elements and moles of the compound. Explain why these subscripts *do not* give relationships between grams of the constituent elements and grams of the compound.
10. Write the conversion factors between moles of each constituent element and moles of the compound for $C_{12}H_{22}O_{11}$.
11. Mass percent composition can be used as a conversion factor between grams of a constituent element and grams of the compound. Write the conversion factor (including units) inherent in each of the following mass percent compositions.
    (a) Water is 11.19% hydrogen by mass.
    (b) Fructose, also known as fruit sugar, is 53.29% oxygen by mass.
    (c) Octane, a component of gasoline, is 84.12% carbon by mass.
    (d) Ethanol, the alcohol in alcoholic beverages, is 52.14% carbon by mass.
12. What is the mathematical formula for computing mass percent composition from a chemical formula?
13. How are the empirical formula and the molecular formula of a compound related?
14. Why is it important to be able to calculate an empirical formula from experimental data?
15. What is the empirical formula mass of a compound?
16. How are the molar mass and empirical formula mass for a compound related?

Problems

The Mole Concept

17. How many mercury atoms are in 5.9 mol of mercury?
18. How many moles of gold atoms do 5.8×10^{24} gold atoms constitute?
19. How many atoms are in each of the following?
    (a) 3.4 mol Cu
    (b) 9.7×10^{-3} mol C
    (c) 22.9 mol Hg
    (d) 0.215 mol Na
20. How many moles of atoms are in each of the following?
    (a) 4.6×10^{24} Pb atoms
    (b) 2.87×10^{22} He atoms
    (c) 7.91×10^{23} K atoms
    (d) 4.41×10^{21} Ca atoms
21. Complete the following table:

Element	*Moles*	*Number of Atoms*
Ne	0.552	____
Ar	____	3.25×10^{24}
Xe	1.78	____
He	____	1.08×10^{20}

22. Complete the following table:

Element	*Moles*	*Number of Atoms*
Cr	____	9.61×10^{23}
Fe	1.52×10^{-5}	____
Ti	0.0365	____
Hg	____	1.09×10^{23}

23. Consider the following definitions.

1 doz = 12

1 gross = 144

1 ream = 500

1 mol = 6.022×10^{23}

Suppose you have 872 sheets of paper. How many _____ of paper do you have?
(a) dozens
(b) gross
(c) reams
(d) moles

24. A pure copper penny contains approximately 3.0×10^{22} copper atoms. Use the definitions in Problem 23 to determine how many _____ of copper atoms are in a penny.
(a) dozens
(b) gross
(c) reams
(d) moles

25. How many moles of tin atoms are in a pure tin cup with a mass of 42.3 g?

26. A lead fishing weight contains 0.12 mol of lead atoms. What is its mass?

27. A pure gold coin contains 0.145 mol of gold. What is the mass of the coin?

28. A helium balloon contains 0.55 g of helium. How many moles of helium does it contain?

29. How many moles of atoms are in each of the following?
(a) 1.54 g Zn
(b) 22.8 g Ar
(c) 86.2 g Ta
(d) 0.034 g Li

30. What is the mass in grams of each of the following?
(a) 7.8 mol W
(b) 0.943 mol Ba
(c) 43.9 mol Xe
(d) 1.8 mol S

31. Complete the following table:

Element	*Moles*	*Mass*
Ne	____	22.5 g
Ar	0.117	____
Xe	____	1.00 kg
He	1.44×10^{-4}	____

32. Complete the following table:

Element	*Moles*	*Mass*
Cr	0.00442	____
Fe	____	73.5 mg
Ti	1.009×10^{-3}	____
Hg	____	1.78 kg

33. A pure silver ring contains 0.0134 mmol (millimol) Ag. How many silver atoms does it contain?

34. A pure gold ring contains 0.0102 mmol (millimol) Au. How many gold atoms does it contain?

35. How many aluminum atoms are in 3.78 g of aluminum?

36. What is the mass of 4.91×10^{21} platinum atoms?

37. How many atoms are in each of the following?
(a) 12.8 g Sr
(b) 45.2 g Fe
(c) 9.87 g Bi
(d) 36.1 g P

38. Calculate the mass in grams of each of the following:
(a) 7.9×10^{21} uranium atoms
(b) 3.82×10^{22} zinc atoms
(c) 5.8×10^{23} lead atoms
(d) 2.1×10^{24} silicon atoms

39. How many carbon atoms are in a diamond (pure carbon) with a mass of 52 mg?

40. How many helium atoms are in a helium blimp containing 536 kg of helium?

41. How many titanium atoms are in a pure titanium bicycle frame with a mass of 1.28 kg?

42. How many copper atoms are in a pure copper statue with a mass of 133 kg?

43. Complete the following table:

Element	*Mass*	*Moles*	*Number of Atoms*
Na	38.5 mg	____	____
C	____	1.12	____
V	____	____	214
Hg	1.44 kg	____	____

44. Complete the following table:

Element	*Mass*	*Moles*	*Number of Atoms*
Pt	____	0.0449	____
Fe	____	____	1.14×10^{25}
Ti	23.8 mg	____	____
Hg	____	2.05	____

45. Determine the number of moles of molecules (or formula units) in each of the following:
(a) 55.2 g sodium chloride
(b) 27.1 g nitrogen monoxide
(c) 4.81 kg carbon dioxide
(d) 1.33 mg carbon tetrachloride

46. Determine the mass of each of the following:
(a) 1.55 mol carbon tetrafluoride
(b) 0.724 mol magnesium fluoride
(c) 1.59 mmol carbon disulfide
(d) 3.87 kmol sulfur trioxide

47. Complete the following table:

Compound	*Mass*	*Moles*
H_2O	112 kg	____
N_2O	6.33 g	____
SO_2	____	2.44
CH_2Cl_2	____	0.0643

48. Complete the following table:

Compound	*Mass*	*Moles*
CO_2	____	0.0153
CO	____	0.0150
BrI	23.8 mg	____
CF_2Cl_2	1.02 kg	____

49. A mothball, composed of naphthalene ($C_{10}H_8$), has a mass of 1.32 g. How many naphthalene molecules does it contain?

50. Calculate the mass in grams of a single water molecule.

51. How many molecules are in each of the following?
(a) 3.5 g H_2O
(b) 56.1 g N_2
(c) 89 g CCl_4
(d) 19 g $C_6H_{12}O_6$

52. Calculate the mass in grams of each of the following:
(a) 5.94×10^{20} H_2O_2 molecules
(b) 2.8×10^{22} SO_2 molecules
(c) 4.5×10^{25} O_3 molecules
(d) 9.85×10^{19} CH_4 molecules

53. A sugar crystal contains approximately 1.8×10^{17} sucrose ($C_{12}H_{22}O_{11}$) molecules. What is its mass in milligrams?

54. A salt crystal has a mass of 0.12 mg. How many NaCl formula units does it contain?

Chemical Formulas as Conversion Factors

55. Determine the number of moles of Cl in 4.7 mol $CaCl_2$.

56. How many moles of O are in 14.8 mol $Fe(NO_3)_3$?

57. Which of the following contains the greatest number of moles of O?
(a) 2.3 mol H_2O
(b) 1.2 mol H_2O_2
(c) 0.9 mol $NaNO_3$
(d) 0.5 mol $Ca(NO_3)_2$

58. Which of the following contains the greatest number of moles of Cl?
(a) 3.8 mol HCl
(b) 1.7 mol CH_2Cl_2
(c) 4.2 mol $NaClO_3$
(d) 2.2 mol $Mg(ClO_4)_2$

59. Determine the number of moles of C in each of the following:
(a) 3.8 mol CH_4
(b) 0.273 mol C_2H_6
(c) 4.89 mol C_4H_{10}
(d) 22.9 mol C_8H_{18}

60. Determine the number of moles of H in each of the following:
(a) 3.15 mol H_2O
(b) 9.88 mol NH_3
(c) 0.0737 mol N_2H_4
(d) 54.1 mol $C_{10}H_{22}$

61. For each of the following, write a relationship between moles of hydrogen and moles of molecules. Then determine the total number of hydrogen atoms present. (H—white; O—red; C—black; N—blue)

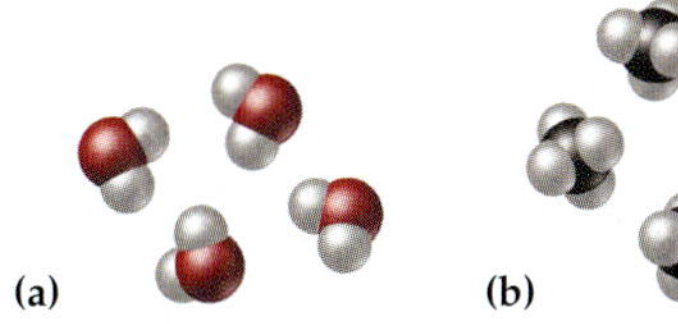
(a)

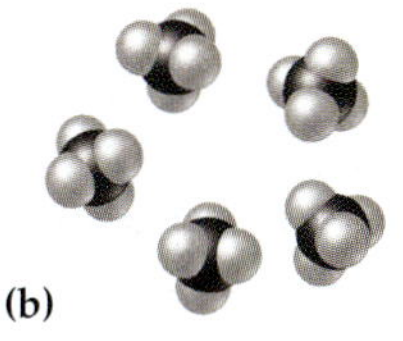
(b)

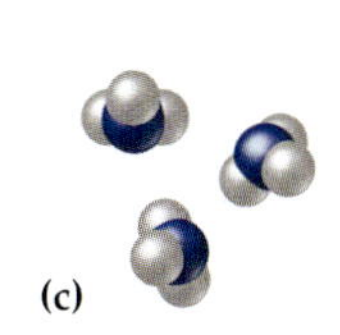
(c)

62. For each of the following, write a relationship between moles of oxygen and moles of molecules. Then determine the total number of oxygen atoms present. (H—white; O—red; C—black; S—yellow)

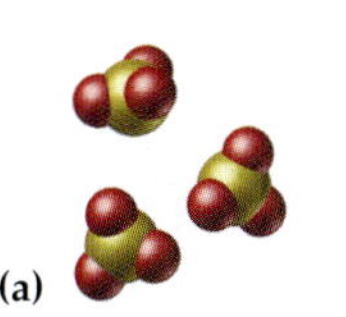
(a)

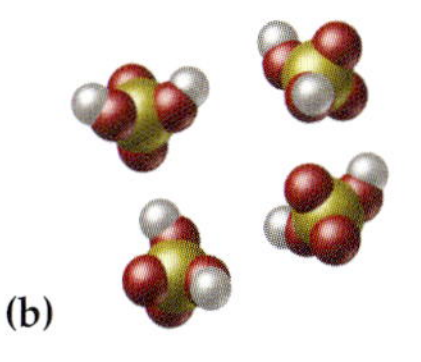
(b)

(c)

63. How many grams of Cl are in 55 g of each of the following chlorofluorocarbons (CFCs)?
(a) CF_2Cl_2
(b) $CFCl_3$
(c) $C_2F_3Cl_3$
(d) CF_3Cl

64. Calculate the number of grams of sodium in 5.0 g of each of the following sodium-containing food additives.
(a) NaCl (table salt)
(b) Na_3PO_4 (sodium phosphate)
(c) $NaC_7H_5O_2$ (sodium benzoate)
(d) $Na_2C_6H_6O_7$ (sodium hydrogen citrate)

65. Iron is found in Earth's crust as several iron compounds often called iron ores. Calculate the mass in kilograms of the amount of each of the following iron compounds that contains 1.0×10^3 kg of iron.
(a) Fe_2O_3 (hematite)
(b) Fe_3O_4 (magnetite)
(c) $FeCO_3$ (siderite)

66. Lead is often found in Earth's crust as several lead compounds called lead ores. Calculate the mass in kilograms of the amount of each of the following lead compounds that contains 1.0×10^3 kg of lead.
(a) PbS (galena)
(b) $PbCO_3$ (cerussite)
(c) $PbSO_4$ (anglesite)

Mass Percent Composition

67. A 2.45-g sample of strontium completely reacts with oxygen to form 2.89 g of strontium oxide. Use this data to calculate the mass percent composition of strontium in strontium oxide.

68. A 4.78-g sample of aluminum completely reacts with oxygen to form 6.67 g of aluminum oxide. Use this data to calculate the mass percent composition of aluminum in aluminum oxide.

69. A 1.912-g sample of calcium chloride is decomposed into its constituent elements and found to contain 0.690 g Ca and 1.222 g Cl. Calculate the mass percent composition of Ca and Cl in calcium chloride.

70. A 0.45-g sample of aspirin is decomposed into its constituent elements and found to contain 0.27 g C, 0.020 g H, and 0.16 g O. Calculate the mass percent composition of C, H, and O in aspirin.

71. Copper(II) fluoride contains 37.42% F by mass. Use this percentage to calculate the mass of fluorine in grams contained in 28.5 g of copper(II) fluoride.

72. Silver chloride, often used in silver plating, contains 75.27% Ag. Calculate the mass of silver chloride in grams required to make 4.8 g of silver plating.

73. In small amounts, the fluoride ion (often consumed as NaF) prevents tooth decay. According to the American Dental Association, an adult female should consume 3.0 mg of fluorine per day. Calculate the amount of sodium fluoride (45.24% F) that should be consumed to get the recommended amount of fluorine.

74. The iodide ion, usually consumed as potassium iodide, is a dietary mineral essential to good nutrition. In countries where potassium iodide is added to salt, iodine deficiency or goiter has been almost completely eliminated. The recommended daily allowance (RDA) for iodine is 150 μg/day. How much potassium iodide (76.45% I) should be consumed to meet the RDA?

Mass Percent Composition from Chemical Formula

75. Calculate the mass percent composition of nitrogen in each of the following nitrogen compounds.
(a) N_2O
(b) NO
(c) NO_2
(d) N_2O_5

76. Calculate the mass percent composition of carbon in each the following carbon compounds.
(a) C_2H_2
(b) C_3H_6
(c) C_2H_6
(d) C_2H_6O

77. Calculate the mass percent composition of each element in each of the following compounds.
(a) $C_2H_4O_2$
(b) CH_2O_2
(c) C_3H_9N
(d) $C_4H_{12}N_2$

78. Calculate the mass percent composition of each element in each of the following compounds.
(a) $FeCl_3$
(b) TiO_2
(c) H_3PO_4
(d) HNO_3

79. Iron ores have different amounts of iron per kilogram of ore. Calculate the mass percent composition of iron for the following iron ores: Fe_2O_3 (hematite), Fe_3O_4 (magnetite), $FeCO_3$ (siderite). Which ore has the highest iron content?

80. Plants need nitrogen to grow, so many fertilizers consist of nitrogen-containing compounds. Calculate the mass percent composition of nitrogen in each of the following nitrogen-containing compounds used as fertilizers: NH_3, $CO(NH_2)_2$, NH_4NO_3, $(NH_4)_2SO_4$. Which fertilizer has the highest nitrogen content?

Calculating Empirical Formulas

81. A compound containing nitrogen and oxygen is decomposed in the laboratory and produces 1.78 g of nitrogen and 4.05 g of oxygen. Calculate the empirical formula of the compound.

82. A compound containing selenium and fluorine is decomposed in the laboratory and produces 2.231 g of selenium and 3.221 g of fluorine. Calculate the empirical formula of the compound.

83. Samples of several compounds are decomposed and the following are the masses of their constituent elements. Calculate the empirical formula for each compound.
(a) 1.245 g Ni, 5.381 g I
(b) 1.443 g Se, 5.841 g Br
(c) 2.128 g Be, 7.557 g S, 15.107 g O

84. Samples of several compounds are decomposed and the following are the masses of their constituent elements. Calculate the empirical formula for each compound.
(a) 2.677 g Ba, 3.115 g Br
(b) 1.651 g Ag, 0.1224 g O
(c) 0.672 g Co, 0.569 g As, 0.486 g O

85. The rotten smell of a decaying animal carcass is partially due to a nitrogen-containing compound called putrescine. Elemental analysis of putrescine showed that it consisted of 54.50% C, 13.73% H, and 31.77% N. Calculate the empirical formula of putrescine.

86. Citric acid, the compound responsible for the sour taste of lemons, has the following elemental composition: C, 37.51%; H, 4.20%; O, 58.29%. Calculate the empirical formula of citric acid.

87. Calculate the empirical formula for each of the following compounds often found in many natural flavors and smells.
(a) ethyl butyrate (pineapple oil): C, 62.04%; H, 10.41%; O, 27.55%
(b) methyl butyrate (apple flavor): C, 58.80%; H, 9.87%; O, 31.33%
(c) benzyl acetate (oil of jasmine): C, 71.98%; H, 6.71%; O, 21.31%

88. Calculate the empirical formula for each of the following over-the-counter pain relievers.
(a) acetaminophen (Tylenol): C, 63.56%; H, 6.00%; N, 9.27%; O, 21.17%
(b) naproxen (Aleve): C, 73.03%; H, 6.13%; O, 20.84%

89. A 1.45-g sample of phosphorus burns in air and forms 2.57 g of a phosphorus oxide. Calculate the empirical formula of the oxide.

90. A 2.241-g sample of nickel reacts with oxygen to form 2.852 g of the metal oxide. Calculate the empirical formula of the oxide.

91. A 0.77-mg sample of nitrogen reacts with chlorine to form 6.61 mg of the chloride. What is the empirical formula of the nitrogen chloride?

92. A 45.2-mg sample of phosphorus reacts with selenium to form 131.6 mg of the selenide. What is the empirical formula of the phosphorus selenide?

Calculating Molecular Formulas

93. A compound containing carbon and hydrogen has a molar mass of 56.11 g/mol and an empirical formula of CH_2. Find its molecular formula.

94. A compound containing phosphorus and oxygen has a molar mass of 219.9 g/mol and an empirical formula of P_2O_3. Find its molecular formula.

95. The following are the molar masses and empirical formulas of several compounds containing carbon and chlorine. Find the molecular formula of each compound.
(a) 284.77 g/mol, CCl
(b) 131.39 g/mol, C_2HCl_3
(c) 181.44 g/mol, C_2HCl

96. The following are the molar masses and empirical formulas of several compounds containing carbon and nitrogen. Find the molecular formula of each compound.
(a) 163.26 g/mol, $C_{11}H_{17}N$
(b) 186.24 g/mol, C_6H_7N
(c) 312.29 g/mol, C_3H_2N

Cumulative Problems

97. A pure copper cube has an edge length of 1.42 cm. How many copper atoms does it contain? (volume of a cube = (edge length)^3; density of copper = 8.96 g/cm^3)

98. A pure silver sphere has a radius of 0.886 cm. How many silver atoms does it contain? (Volume of a sphere = $\frac{4}{3}\pi r^3$; density of silver = 10.5 g/cm^3)

99. A drop of water has a volume of approximately 0.05 mL. How many water molecules does it contain? (density of water = 1.0 g/cm^3).

100. Fingernail-polish remover is primarily acetone (C_3H_6O). How many acetone molecules are in a bottle of acetone with a volume of 325 mL? (density of acetone = 0.788 g/cm^3).

101. Complete the following table:

Substance	*Mass*	*Moles*	*Number of Particles (atoms or molecules)*
Ar	____	4.5×10^{-4}	____
NO_2	____	____	1.09×10^{20}
K	22.4 mg	____	____
C_8H_{18}	3.76 kg	____	____

102. Complete the following table:

Substance	*Mass*	*Moles*	*Number of Particles (atoms or molecules)*
$C_6H_{12}O_6$	15.8 g	____	____
Pb	____	____	9.04×10^{21}
CF_4	22.5 kg	____	____
C	____	0.0388	____

103. Determine the chemical formula of each of the following compounds and then use it to calculate the mass percent composition of each constituent element.
(a) copper(II) iodide **(b)** sodium nitrate
(c) lead(II) sulfate **(d)** calcium fluoride

104. Determine the chemical formula of each of the following compounds and then use it to calculate the mass percent composition of each constituent element.
(a) nitrogen triiodide **(b)** xenon tetrafluoride
(c) phosphorus trichloride **(d)** carbon monoxide

105. The rock in a particular iron ore deposit contains 78% Fe_2O_3 by mass. How many kilograms of the rock must be processed to obtain 1.0×10^3 kg of iron?

106. The rock in a lead ore deposit contains 84% PbS by mass. How many kilograms of the rock must be processed to obtain 1.0 kg of Pb?

107. A leak in the air conditioning system of a corporate building releases 12 kg of CHF_2Cl per month. If the leak were allowed to continue, how many kilograms of Cl would be emitted into the atmosphere each year?

108. A leak in the air conditioning system of an older car releases 55 g of CF_2Cl_2 per month. How much Cl is emitted into the atmosphere each year by this car?

109. Hydrogen is a possible future fuel. However, elemental hydrogen is rare, so it must be obtained from a hydrogen-containing compound. One candidate is water. If hydrogen were obtained from water, how much hydrogen in grams could be obtained from 1.0 L of water? (density of water = 1.0 g/cm^3).

110. Hydrogen, a possible future fuel mentioned in Problem 109, can also be obtained from other compounds such as ethanol. Ethanol can be made from the fermentation of crops such as corn. How much hydrogen in grams can be obtained from 1.0 kg of ethanol (C_2H_5OH)?

111. Complete the following table consisting of compounds that contain only carbon and hydrogen.

Formula	*Molar Mass*	*% C (by mass)*	*% H (by mass)*
C_2H_4	____	____	____
____	58.12	82.66%	____
C_4H_8	____	____	____
____	44.09	____	18.29%

112. Complete the following table consisting of compounds that contain only chromium and oxygen

Formula	*Name*	*Molar Mass*	*% Cr (by mass)*	*% O (by mass)*
____	Chromium(III) oxide	____	____	____
____	____	84.00	61.90 %	____
____	____	100.00	____	48.00%

113. Butanedione, a component of butter and body odor, has a cheesy smell. Elemental analysis of butanedione gave the following mass percent composition: C, 55.80%; H, 7.03%; O, 37.17%. The molar mass of butanedione is 86.09 g/mol. Find the molecular formula of butanedione.

114. Caffeine, a stimulant found in coffee and soda pop, has the following mass percent composition: C, 49.48%; H, 5.19%; N, 28.85%; O, 16.48%. The molar mass of caffeine is 194.19 g/mol. Find the molecular formula of caffeine.

115. Nicotine, a stimulant found in tobacco, has the following mass percent composition: C, 74.03%; H, 8.70%; N, 17.27%. The molar mass of nicotine is 162.23 g/mol. Find the molecular formula of nicotine.

116. Estradiol is a female sexual hormone that causes maturation and maintenance of the female reproductive system. Elemental analysis of estradiol gave the following mass percent composition: C, 79.37%; H, 8.88%; O, 11.75%. The molar mass of estradiol is 272.37 g/mol. Find the molecular formula of estradiol.

Highlight Problems

117. You can use the concepts in this chapter to obtain an estimate of the number of atoms in the universe. The following steps will guide you through this calculation.
   (a) Begin by calculating the number of atoms in the sun. Assume that the sun is pure hydrogen with a density of 1.4 g/cm^3. The radius of the sun is 7×10^8 m and the volume of a sphere is given by $V = \frac{4}{3}\pi r^3$.
   (b) Since the sun is an average-sized star, and since stars are believed to compose most of the mass of the visible universe (planets are so small they can be ignored), we can estimate the number of atoms in a galaxy by assuming that every star in the galaxy has the same number of atoms as our sun. The Milky Way galaxy is believed to contain 1×10^{11} stars. Use your answer from part (a) to calculate the number of atoms in the Milky Way galaxy.
   (c) The universe is estimated to contain approximately 1×10^{11} galaxies. If each of these galaxies contains the same number of atoms as the Milky Way galaxy, what is the total number of atoms in the universe?

▲ Our sun is one of the 100 billion stars in the Milky Way galaxy. The universe is estimated to contain about 100 billion galaxies.

118. Because of increasing evidence of damage to the ozone layer, chlorofluorocarbon (CFC) production was banned in 1996. However, there are about 100 million auto air conditioners that still use CFC-12 (CF_2Cl_2). These air conditioners are recharged from stockpiled supplies of CFC-12. If each of the 100 million automobiles contains 1.1 kg of CFC-12 and leaks 25% of its CFC-12 into the atmosphere per year, how much Cl in kilograms is added to the atmosphere each year due to auto air conditioners? (Assume two significant figures in your calculations.)

EP/TOMS total ozone for Oct. 1, 2001

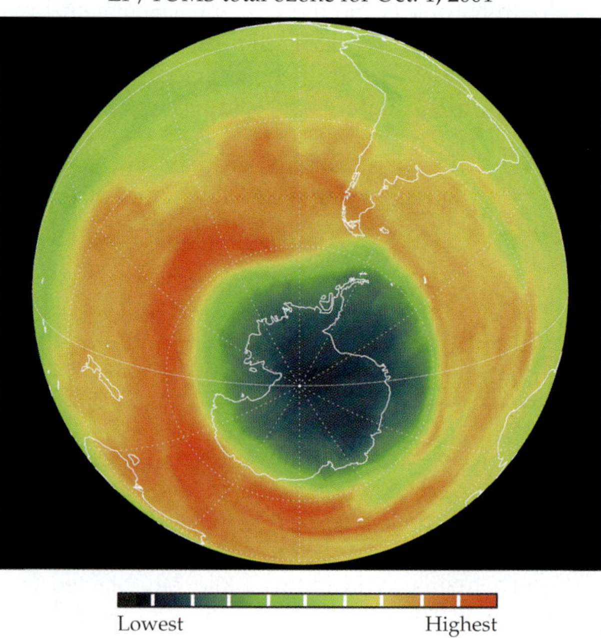

▲ The ozone hole over Antarctica, October 1, 2001. The red areas indicate the highest concentrations of ozone, while the dark green regions near the South Pole have the lowest.

119. In 1996, the media reported that possible evidence of life on Mars was found on a meteorite called Allan Hills 84001 (AH 84001). The meteorite was discovered in Antarctica in 1984 and is believed to have originated on Mars. Elemental analysis of substances within its crevices revealed carbon-containing compounds that normally derive only from living organisms. Suppose that one of those compounds had a molar mass of 202.23 g/mol and the following mass percent composition: C, 95.02%; H, 4.98%. What is the molecular formula for the carbon-containing compound?

▶ The Allan Hills 84001 meteorite. Elemental analysis of the substances within the crevices of this meteorite revealed carbon-containing compounds that normally originate from living organisms.

Answers to Skillbuilder Exercises

Skillbuilder 6.1	5.32×10^{22} Au atoms
Skillbuilder 6.2	89.2 g S
Skillbuilder 6.3	8.17 g He
Skillbuilder 6.4	2.56×10^{-2} mol NO_2
Skillbuilder 6.5	1.22×10^{23} H_2O molecules
Skillbuilder 6.6	5.6 mol O
Skillbuilder 6.7	3.3 g O
Skillbuilder Plus, p. 165	4.04 g O
Skillbuilder 6.8	8.6 g Na
Skillbuilder 6.9	53.28% O
Skillbuilder 6.10	CH_2O
Skillbuilder 6.11	$C_{13}H_{18}O_2$
Skillbuilder 6.12	CuO
Skillbuilder 6.13	C_4H_{10}
Skillbuilder Plus, p. 174	$C_2H_8N_2$

Answers to Conceptual Checkpoints

6.1 (a) The mole is a counting unit; it represents a definite number (Avogadro's number, 6.02×10^{23}). Therefore, a given number of atoms always represents a precise number of moles, regardless of what atom is involved. Atoms of different elements have different masses, so if samples of different elements have the same mass, they *cannot* contain the same number of atoms or moles.

6.2 (a) Avogadro's number was defined so as to make the molar mass of a compound (the mass of one mole, or Avogadro's number of molecules or formula units), *in grams*, numerically equal to the formula mass of the compound in amu. In order for the molar mass *in kilograms* to be numerically equal to the formula mass in amu, we would need 1000 times as many molecules or formula units in a mole. In other words, Avogadro's number would have to be 1000 times larger, or 6.02×10^{26}.

6.3 (b) This compound has the highest ratio of oxygen atoms to chromium atoms, and so must have the greatest mass percent of oxygen.

CHAPTER 7

Chemical Reactions

"Chemistry . . . is one of the broadest branches of science if for no other reason that, when we think about it, everything is chemistry."

Luciano Caglioti (1933–)

7.1 Kindergarten Volcanoes, Automobiles, and Laundry Detergents

Did you ever make a clay volcano in kindergarten that—when filled with vinegar, baking soda, and red food coloring for effect—erupted? Have you pushed the gas pedal of a car and felt the acceleration as the car moved forward? Have you wondered why laundry detergents work better than normal soap to clean your clothes? Each of these processes depends on a **chemical reaction**—the change of one or more substances into different substances.

In the classic kindergarten volcano, the baking soda (which is sodium bicarbonate) reacts with acetic acid in the vinegar to form carbon dioxide gas, water, and sodium acetate. The newly formed carbon dioxide bubbles out of the mixture, causing the eruption. Reactions that occur in liquids and form a gas are called *gas evolution reactions*. A similar reaction causes the fizzing of antacids such as Alka-Seltzer.

Hydrocarbons are covered in detail in Chapter 18.

When you drive your car, hydrocarbons such as octane (in gasoline) react with oxygen from the air to form carbon dioxide gas and water (▶ Figure 7.1). This reaction produces heat, which is used to expand the gases in the car's cylinders and accelerate the car forward. Reactions such as this one—in which a substance reacts with oxygen, emitting heat and forming one or more oxygen-containing compounds—are called *combustion reactions*. Combustion reactions are a subcategory of *oxidation–reduction reactions*, in which electrons are transferred from one substance to another. The formation of rust and the dulling of automobile paint are other examples of oxidation–reduction reactions.

One reason why laundry detergents work better than soap to wash clothes is that they contain substances that soften hard water. Hard water contains dissolved calcium (Ca^{2+}) and magnesium (Mg^{2+}) ions. These ions interfere with

◀ In the space shuttle's main engines, hydrogen molecules, H_2 (white) and oxygen molecules, O_2 (red), which are stored in the central fuel tank, react violently to form water molecules, H_2O. The reaction emits the energy that helps propel the shuttle into space.

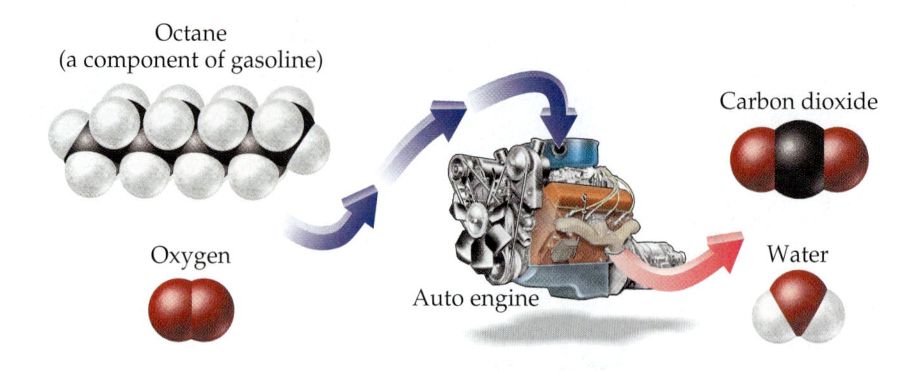

▶ **Figure 7.1 A combustion reaction** In an automobile engine, hydrocarbons such as octane (C_8H_{18}) from gasoline combine with oxygen from the air and react to form carbon dioxide and water.

▲ **Figure 7.2 Soap and water** Soap forms suds with pure water (left), but reacts with the ions in hard water (right) to form a gray residue that adheres to clothes.

the action of soap by reacting with it to form a gray, slimy substance called *curd* or *soap scum* (◀ Figure 7.2). If you have ever washed your clothes in ordinary soap, you may have noticed soap scum as a gray residue on your clothes.

Laundry detergents inhibit curd formation because they contain substances such as sodium carbonate (Na_2CO_3) that remove calcium and magnesium ions from the water. When sodium carbonate dissolves in water, it *dissociates*, or separates into sodium ions (Na^+) and carbonate ions (CO_3^{2-}). The dissolved carbonate ions then react with calcium and magnesium ions in the hard water to form solid calcium carbonate ($CaCO_3$) and solid magnesium carbonate ($MgCO_3$). These solids simply settle to the bottom of the laundry mixture, resulting in the removal of the ions from the water. In other words, laundry detergents contain substances that react with the ions in hard water to immobilize them. Reactions such as these—that form solid substances in water—are called *precipitation reactions*. Precipitation reactions are also used to remove dissolved toxic metals in industrial wastes.

Chemical reactions take place all around us and even inside us. They are involved in many of the products we use daily and in many of our experiences. Chemical reactions can be relatively simple, like the combination of hydrogen and oxygen to form water, or they can be complex, like the synthesis of a protein molecule from thousands of simpler molecules. In some cases, such as the neutralization reaction that occurs in a swimming pool when acid is added to adjust the water's acidity level, chemical reactions are not noticeable to the naked eye. In other cases, such as the combustion reaction that produces a pillar of smoke and fire under the space shuttle during liftoff, chemical reactions are very obvious. In all cases, however, chemical reactions produce changes in the arrangements of the molecules and atoms that compose matter. Many times, these molecular changes cause macroscopic changes that we experience.

7.2 Evidence of a Chemical Reaction

If we could see the atoms and molecules that compose matter, we could easily identify a chemical reaction. Are atoms combining with other atoms to form compounds? Are new molecules forming? Are the original molecules decomposing? Are atoms in one molecule changing places with atoms in another? If one or more of these are happening, a chemical reaction is occurring. Of course, we don't normally see atoms and molecules, so we need other ways to identify a chemical reaction.

Fortunately, many chemical reactions produce easily detectable changes when they occur. For example, when the color-causing molecules in a brightly colored shirt decompose with repeated exposure to sunlight, the color of the shirt fades. Similarly, when the molecules imbedded in the plastic of a child's temperature-sensitive spoon transform upon warming, the color of the spoon changes. These *color changes* are evidence that a chemical reaction has occurred.

▲ **Figure 7.3 A precipitation reaction** The formation of a solid in a previously clear solution is evidence of a chemical reaction.

▲ **Figure 7.4 A gas-evolution reaction** The formation of a gas is evidence of a chemical reaction.

Other changes that identify chemical reactions include the *formation of a solid* in a previously clear solution (▲ Figure 7.3) or the *formation of a gas* (▲ Figure 7.4). Dropping Alka-Seltzer tablets into water or combining baking soda and vinegar (as in our opening example of the kindergarten volcano) are both good examples of chemical reactions that produce a gas—the gas is visible as bubbles in the liquid.

Heat absorption or *emission*, as well as *light emission*, is also evidence for reactions. For example, a natural gas flame produces heat and light. A chemical cold pack becomes cold when the plastic barrier separating two substances is broken. Both of these changes suggest that a chemical reaction is occurring.

▲ A child's temperature-sensitive spoon changes color upon warming due to a reaction induced by the higher temperature.

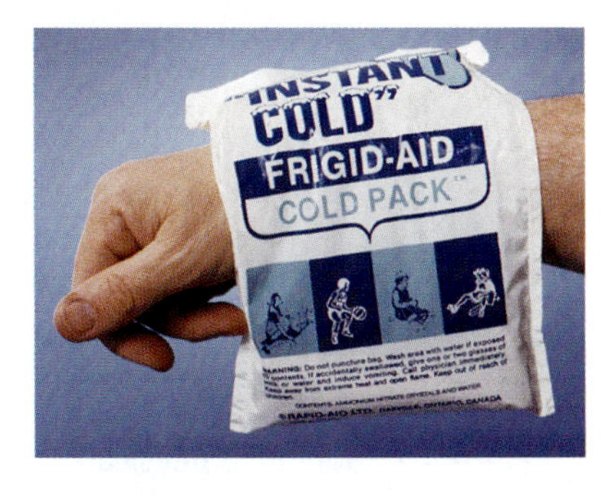

▲ A change in temperature due to absorption or emission of heat is evidence of a chemical reaction. This chemical cold pack becomes cold when the barrier separating two substances is broken.

In **summary**, each of the following provides *evidence of a chemical reaction.*

- a *color change*

- the *formation of a solid* in a previously clear solution

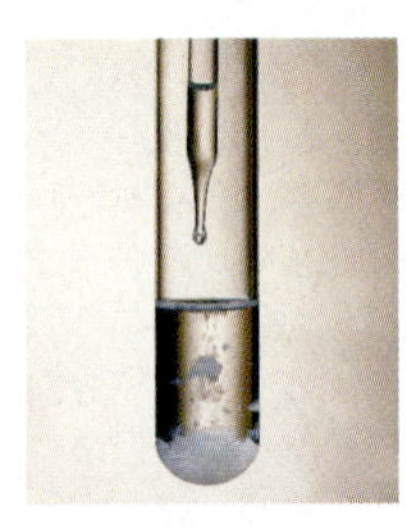

- the *formation of a gas* when you add a substance to a solution

- the *emission of light*

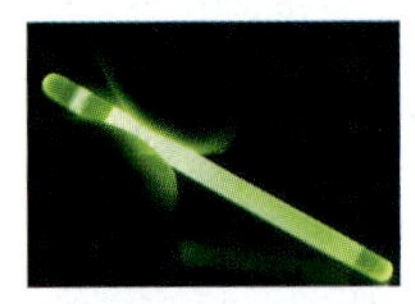

- the *emission* or *absorption of heat*

▲ **Figure 7.5 Boiling: a physical change** When water boils, bubbles are formed and a gas is evolved. However, no chemical change has occurred because the gas, like the liquid water, is also composed of water molecules.

While these changes provide evidence of a chemical reaction, they are not *definitive* evidence. Only chemical analysis showing that the initial substances have changed into other substances conclusively proves that a chemical reaction has occurred. We can be fooled. For example, when water boils, bubbles form, but no chemical reaction has occurred. Boiling water forms gaseous steam, but both water and steam are composed of water molecules—no chemical change has occurred (◀ Figure 7.5). On the other hand, chemical reactions may occur without any obvious signs, yet chemical analysis may show that a reaction has indeed occurred. It is the changes occurring at the atomic and molecular level that determine whether a chemical reaction has occurred. Fortunately, much of the time these molecular changes also result in changes that we can perceive.

EXAMPLE 7.1 Evidence of a Chemical Reaction

Which of the following is a chemical reaction? Why?

(a) ice melting upon warming
(b) an electric current is passed through water, resulting in the formation of hydrogen and oxygen gas which appear as bubbles rising in the water
(c) iron rusting
(d) bubbles forming when a soda can is opened

Solution:

(a) Not a chemical reaction; melting ice forms water, but both the ice and water are composed of water molecules.

(b) Chemical reaction; water decomposes into hydrogen and oxygen, as evidenced by the bubbling.

(c) Chemical reaction; iron changes into iron oxide, changing color in the process.

(d) Not a chemical reaction; even though there is bubbling, it is just carbon dioxide coming out of the liquid.

SKILLBUILDER 7.1 Evidence of a Chemical Reaction

Which of the following is a chemical reaction? Why?

(a) butane burning in a butane lighter
(b) butane evaporating out of a butane lighter
(c) wood burning
(d) dry ice evaporating

7.3 The Chemical Equation

Chemical reactions are represented by **chemical equations**. For example, the reaction occurring in a natural-gas flame, such as the flame on your kitchen stove, is methane (CH_4) reacting with oxygen (O_2) to form carbon dioxide (CO_2) and water (H_2O). This reaction is represented by the following equation.

$$\underbrace{CH_4 + O_2}_{\text{reactants}} \longrightarrow \underbrace{CO_2 + H_2O}_{\text{products}}$$

The substances on the left side of the equation are called the **reactants** and the substances on the right side are called the **products**. We often specify the state of each reactant or product in parentheses next to the formula. If we add these to our equation, it becomes:

$$CH_4(g) + O_2(g) \longrightarrow CO_2(g) + H_2O(g)$$

The (g) indicates that these substances are gases in the reaction. The common states of reactants and products and their symbols used in chemical reactions are summarized in Table 7.1.

TABLE 7.1 Abbreviations Indicating the States of Reactants and Products in Chemical Equations

Abbreviation	State
(*g*)	gas
(*l*)	liquid
(*s*)	solid
(*aq*)	aqueous (water solution)*

*The (*aq*) designation stands for *aqueous*, which means a substance dissolved in water. When a substance dissolves in water, the mixture is called a *solution* (see Section 7.5).

Let's look more closely at our equation for the burning of natural gas. How many oxygen atoms are on each side of the equation?

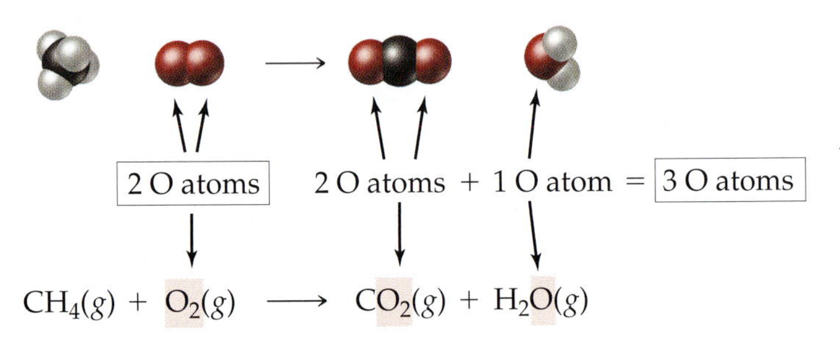

In chemical equations, atoms cannot change from one type to another—hydrogen atoms cannot change into oxygen atoms, for example. Nor can atoms disappear (recall the law of conservation of mass from Section 3.7).

There are two oxygen atoms on the left side and three on the right. Where did this additional oxygen atom on the right side come from? Since chemical equations represent real chemical reactions, atoms cannot simply appear or disappear because, as we know, atoms don't simply appear or disappear in nature. Notice also that there are four hydrogen atoms on the left and only two on the right.

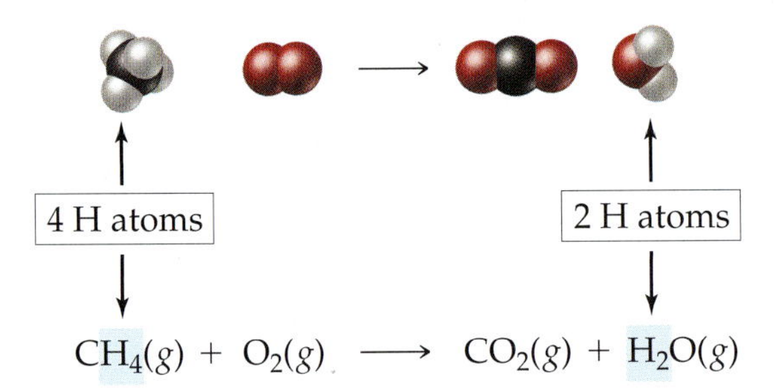

To correct these problems, we must create a **balanced equation**; that is, we must add coefficients—not subscripts—to ensure that the number of each type of atom on both sides of the equation is equal. New atoms do not form during a reaction, nor do atoms vanish—matter must be conserved.

We balance chemical equations by adding coefficients as needed in front of the formulas of the reactants and products. This changes the number of molecules in the equation, but it does not change the *kinds* of molecules. To balance the preceding equation, for example, we put the coefficient 2 before O_2 in the reactants, and the coefficient 2 before H_2O in the products.

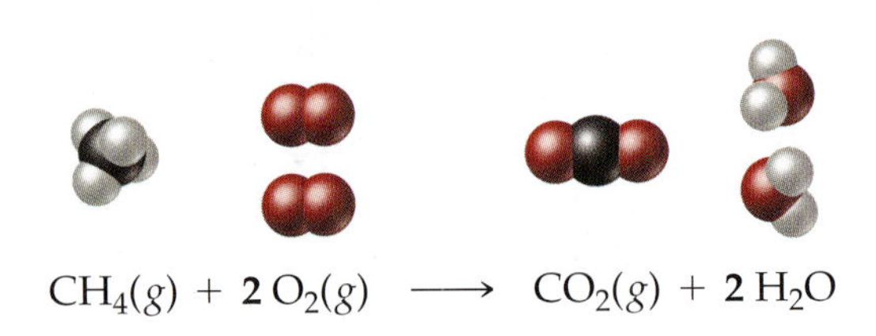

TUTORIAL
Balancing Chemical Equations 

The equation is now balanced because the numbers of each type of atom on either side of the equation are equal. We can verify this by summing the number of each type of atom.

> The number of a particular type of atom within a chemical formula embedded in an equation is obtained by multiplying the subscript for the atom by the coefficient for the chemical formula.

If there is no coefficient or subscript, a 1 is implied. So, for the combustion of natural gas we have:

$$CH_4(g) + 2\,O_2(g) \longrightarrow CO_2(g) + 2\,H_2O(g)$$

Reactants	Products
1 C atom ($1 \times \underline{C}H_4$)	1 C atom ($1 \times \underline{C}O_2$)
4 H atoms ($1 \times C\underline{H_4}$)	4 H atoms ($2 \times \underline{H_2}O$)
4 O atoms ($2 \times \underline{O_2}$)	4 O atoms ($1 \times C\underline{O_2} + 2 \times H_2\underline{O}$)

The numbers of each type of atom on both sides of the equation are equal—the equation is balanced.

▶ A balanced chemical equation represents a chemical reaction. In this image, methane molecules combine with oxygen to form carbon dioxide and water.

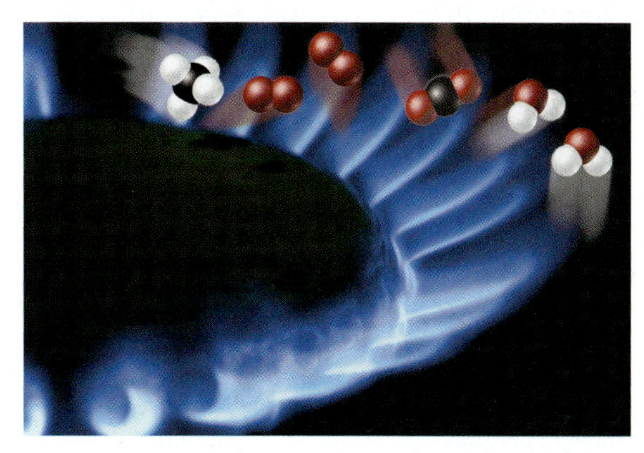

$$CH_4(g) + 2\,O_2(g) \longrightarrow CO_2(g) + 2\,H_2O(g)$$

CONCEPTUAL CHECKPOINT 7.1

In photosynthesis, plants make the sugar glucose, $C_6H_{12}O_6$, from carbon dioxide and water. The equation for the reaction is

$$6\,CO_2 + 6\,H_2O \longrightarrow C_6H_{12}O_6 + x\,O_2$$

In order for this equation to be balanced, the coefficient x must be

(a) 3
(b) 6
(c) 9
(d) 12

7.4 How to Write Balanced Chemical Equations

The following procedure box details the steps for writing balanced chemical equations. As in other procedures, we show the steps in the left column and examples of applying each step in the center and right columns. Remember, change only the *coefficients* to balance a chemical equation, *never the subscripts*.

Writing Balanced Chemical Equations	EXAMPLE 7.2	EXAMPLE 7.3
	Write a balanced equation for the reaction between solid silicon dioxide and solid carbon to produce solid silicon carbide and carbon monoxide gas.	Write a balanced equation for the combustion of liquid octane (C_8H_{18}), a component of gasoline, in which it combines with gaseous oxygen to form gaseous carbon dioxide and gaseous water.
1. Write a skeletal equation by writing chemical formulas for each of the reactants and products. Review Chapter 5 for nomenclature rules. (If a skeletal equation is provided, skip this step and go to Step 2.)	**Solution:** $SiO_2(s) + C(s) \longrightarrow SiC(s) + CO(g)$	**Solution:** $C_8H_{18}(l) + O_2(g) \longrightarrow CO_2(g) + H_2O(g)$
2. If an element occurs in only one compound on both sides of the equation, balance it first. If there is more than one such element, balance metals before nonmetals.	**Begin with Si:** $SiO_2(s) + C(s) \longrightarrow SiC(s) + CO(g)$ **1 Si atom $\longrightarrow$ 1 Si atom** Si is already balanced. **Balance O next:** $SiO_2(s) + C(s) \longrightarrow SiC(s) + CO(g)$ **2 O atoms $\longrightarrow$ 1 O atom** To balance O, put a 2 before $CO(g)$. $SiO_2(s) + C(s) \longrightarrow SiC(s) + 2\,CO(g)$ **2 O atoms $\longrightarrow$ 2 O atoms**	**Begin with C:** $C_8H_{18}(l) + O_2(g) \longrightarrow CO_2(g) + H_2O(g)$ **8 C atoms $\longrightarrow$ 1 C atom** To balance C, put an 8 before $CO_2(g)$. $C_8H_{18}(l) + O_2(g) \longrightarrow 8\,CO_2(g) + H_2O(g)$ **8 C atoms $\longrightarrow$ 8 C atoms** **Balance H next:** $C_8H_{18}(l) + O_2(g) \longrightarrow 8\,CO_2(g) + H_2O(g)$ **18 H atoms $\longrightarrow$ 2 H atoms** To balance H, put a 9 before $H_2O(g)$. $C_8H_{18}(l) + O_2(g) \longrightarrow 8\,CO_2(g) + \mathbf{9}\,H_2O(g)$ **18 H atoms $\longrightarrow$ 18 H atoms**
3. If an element occurs as a free element on either side of the chemical equation, balance it last. Always balance free elements by adjusting the coefficient *on the free element*.	**Balance C:** $SiO_2(s) + C(s) \longrightarrow SiC(s) + 2\,CO(g)$ **1 C atom** $\longrightarrow$ 1 C + 2 C **= 3 C atoms** To balance C, put a 3 before $C(s)$. $SiO_2(s) + \mathbf{3}\,C(s) \longrightarrow SiC(s) + 2\,CO(g)$ **3 C atoms** $\longrightarrow$ 1 C + 2 C **= 3 C atoms**	**Balance O:** $C_8H_{18}(l) + O_2(g) \longrightarrow 8\,CO_2(g) + 9\,H_2O(g)$ **2 O atoms** $\longrightarrow$ 16 O + 9 O **= 25 O atoms** To balance O, put a $\frac{25}{2}$ before $O_2(g)$. $C_8H_{18}(l) + \mathbf{\frac{25}{2}}\,O_2(g) \longrightarrow 8\,CO_2(g) + 9\,H_2O(g)$ **25 O atoms** $\longrightarrow$ 16 O + 9 O **= 25 O atoms**
4. If the balanced equation contains coefficient fractions, clear these by multiplying the entire equation by the appropriate factor.	This step is not necessary in this example. Proceed to Step 5.	$[C_8H_{18}(l) + \frac{25}{2}O_2(g) \longrightarrow 8\,CO_2(g) + 9\,H_2O(g)] \times 2$

$$2\,C_8H_{18}(l) + 25\,O_2(g) \longrightarrow 16\,CO_2(g) + 18\,H_2O(g)$$

5. Check to make certain the equation is balanced by summing the total number of each type of atom on both sides of the equation.

$$SiO_2(s) + 3\,C(s) \longrightarrow SiC(s) + 2\,CO(g)$$

Reactants		Products
1 Si atom	⟶	1 Si atom
2 O atoms	⟶	2 O atoms
3 C atoms	⟶	3 C atoms

The equation is balanced.

SKILLBUILDER 7.2

Write a balanced equation for the reaction between solid chromium(III) oxide and solid carbon to produce solid chromium and carbon dioxide gas.

$$2\,C_8H_{18}(l) + 25\,O_2(g) \longrightarrow 16\,CO_2(g) + 18\,H_2O(g)$$

Reactants		Products
16 C atoms	⟶	16 C atoms
36 H atoms	⟶	36 H atoms
50 O atoms	⟶	50 O atoms

The equation is balanced.

SKILLBUILDER 7.3

Write a balanced equation for the combustion of C_4H_{10} in which it combines with gaseous oxygen to form gaseous carbon dioxide and gaseous water.

EXAMPLE 7.4 Balancing Chemical Equations

Write a balanced equation for the reaction of solid aluminum with aqueous sulfuric acid to form aqueous aluminum sulfate and hydrogen gas.

Use your knowledge of chemical nomenclature from Chapter 5 to write a skeletal equation containing formulas for each of the reactants and products. The formulas for each compound MUST BE CORRECT before you begin to balance the equation.

Solution:

$$Al(s) + H_2SO_4(aq) \longrightarrow Al_2(SO_4)_3(aq) + H_2(g)$$

Since both aluminum and hydrogen occur as pure elements, balance those last. Sulfur and oxygen occur in only one compound on each side of the equation, so balance these first. Sulfur and oxygen are also part of a polyatomic ion that stays intact on both sides of the equation. *Balance polyatomic ions such as these as a unit*. There are 3 SO_4^{2-} ions on the right side of the equation, so put a 3 in front of H_2SO_4.

$$Al(s) + \mathbf{3}\,H_2SO_4(aq) \longrightarrow Al_2(SO_4)_3(aq) + H_2(g)$$

Balance Al next. Since there are 2 Al atoms on the right side of the equation, place a 2 in front of Al on the left side of the equation.

$$\mathbf{2}\,Al(s) + 3\,H_2SO_4(aq) \longrightarrow Al_2(SO_4)_3(aq) + H_2(g)$$

Balance H next. Since there are 6 H atoms on the left side, place a 3 in front of $H_2(g)$ on the right side.

$$2\,Al(s) + 3\,H_2SO_4(aq) \longrightarrow Al_2(SO_4)_3(aq) + \mathbf{3}\,H_2(g)$$

Finally, sum the number of atoms on each side to make sure that the equation is balanced.

The equation is balanced.

$$2\,Al(s) + 3\,H_2SO_4(aq) \longrightarrow Al_2(SO_4)_3(aq) + 3\,H_2(g)$$

Reactants		Products
2 Al atoms	⟶	2 Al atoms
6 H atoms	⟶	6 H atoms
3 S atoms	⟶	3 S atoms
12 O atoms	⟶	12 O atoms

SKILLBUILDER 7.4 Balancing Chemical Equations

Write a balanced equation for the reaction of aqueous lead(II) acetate with aqueous potassium iodide to form solid lead(II) iodide and aqueous potassium acetate.

EXAMPLE 7.5 Balancing Chemical Equations

Balance the following chemical equation.

$$Fe(s) + HCl(aq) \longrightarrow FeCl_3(aq) + H_2(g)$$

Solution:

Since Cl occurs in only one compound on each side of the equation, balance it first. There is 1 Cl atom on the left side of the equation and 3 Cl atoms on the right side. Balance Cl by placing a 3 in front of HCl.

$$Fe(s) + \mathbf{3}\,HCl(aq) \longrightarrow FeCl_3(aq) + H_2(g)$$

Since H and Fe occur as free elements, balance them last. There is 1 Fe atom on the left side of the equation and 1 Fe atom on the right, so Fe is balanced. There are 3 H atoms on the left and 2 H atoms on the right. Balance H by placing a $\frac{3}{2}$ in front of H_2. (That way you don't alter other elements that are already balanced.)

$$Fe(s) + 3\,HCl(aq) \longrightarrow FeCl_3(aq) + \mathbf{\frac{3}{2}}H_2(g)$$

Since the equation contains a coefficient fraction, clear it by multiplying the entire equation by 2.

$$[Fe(s) + 3\,HCl(aq) \longrightarrow FeCl_3(aq) + \frac{3}{2}H_2(g)] \times 2$$

$$2\,Fe(s) + 6\,HCl(aq) \longrightarrow 2\,FeCl_3(aq) + 3\,H_2(g)$$

Finally, sum the number of atoms on each side to check that the equation is balanced.

The equation is balanced.

$$2\,Fe(s) + 6\,HCl(aq) \longrightarrow 2\,FeCl_3(aq) + 3\,H_2(g)$$

Reactants		Products
2 Fe atoms	⟶	2 Fe atoms
6 Cl atoms	⟶	6 Cl atoms
6 H atoms	⟶	6 H atoms

SKILLBUILDER 7.5 Balancing Chemical Equations

Balance the following chemical equation.

$$HCl(g) + O_2(g) \longrightarrow H_2O(l) + Cl_2(g)$$

7.5 Aqueous Solutions and Solubility: Compounds Dissolved in Water

Reactions occurring in aqueous solution are among the most common and important. An **aqueous solution** is a homogeneous mixture of a substance with water. For example, sodium chloride (NaCl) solutions, also called saline solutions, are composed of sodium chloride dissolved in water. They are common both in the oceans and in living cells. You may have formed a NaCl solution yourself by adding table salt to water. As you stir the NaCl into the water, it seems to disappear. However, you know the NaCl is still there because you can taste its saltiness in the water. How does NaCl dissolve in water?

When ionic compounds such as NaCl dissolve in water they usually dissociate into their component ions. A NaCl solution, represented as NaCl(*aq*), does not contain any NaCl units, but rather dissolved Na^+ ions and Cl^- ions.

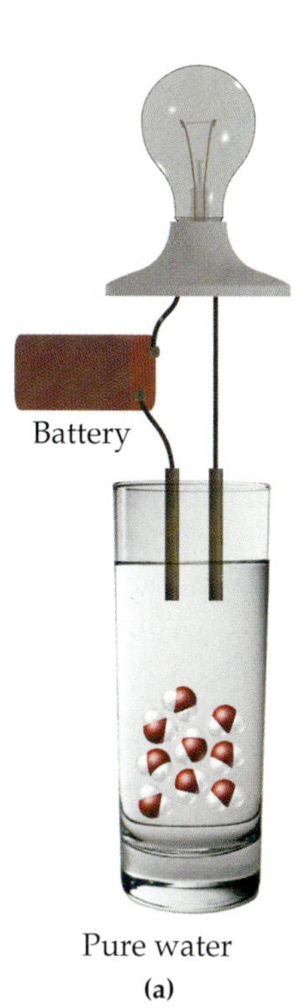

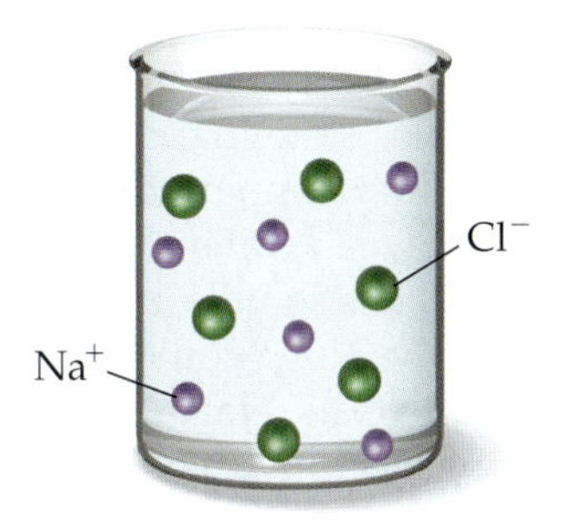

A sodium chloride solution contains independent $\mathbf{Na^+}$ and $\mathbf{Cl^-}$ ions.

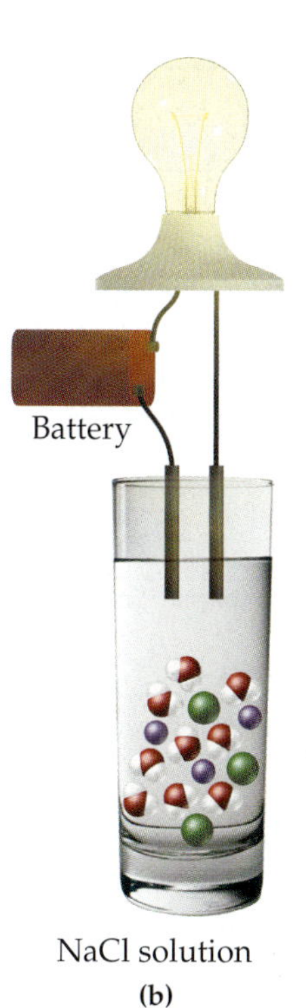

▲ **Figure 7.6 Ions as conductors**
(a) Pure water will not conduct electricity. **(b)** Ions in a sodium chloride solution conduct electricity, causing the bulb to light. Solutions such as these are called strong electrolyte solutions.

We know that NaCl is present as independent sodium and chloride ions in solution because sodium chloride solutions conduct electricity, which requires the presence of freely moving charged particles. Substances (such as NaCl) that completely dissociate into ions in solution are called *strong electrolytes* and the resultant solutions are called **strong electrolyte solutions** (◀ Figure 7.6). Similarly, a $AgNO_3$ solution, represented as $AgNO_3$(*aq*), does not contain any $AgNO_3$ units, but rather dissolved Ag^+ ions and NO_3^- ions. It too is a strong electrolyte solution. When compounds containing polyatomic ions such as NO_3^- dissolve, the polyatomic ions usually dissolve as intact units.

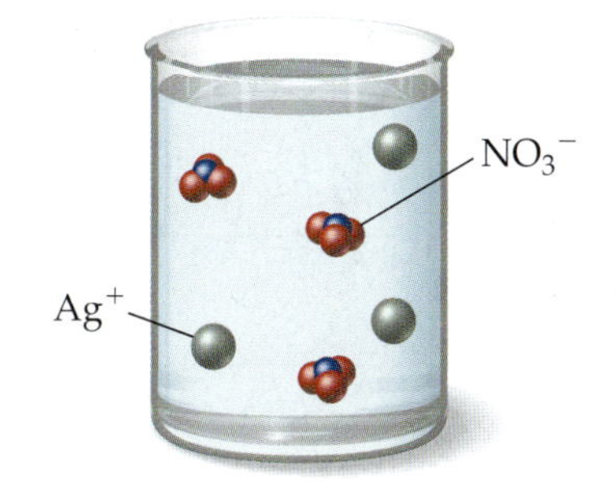

A silver nitrate solution contains independent $\mathbf{Ag^+}$ and $\mathbf{NO_3^-}$ ions.

Not all ionic compounds, however, dissolve in water. AgCl, for example, does not dissolve in water. If we add AgCl to water, it remains as solid AgCl and appears as a white solid at the bottom of the water.

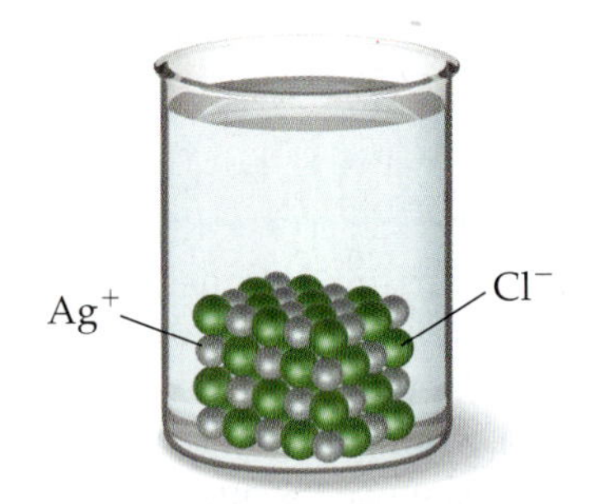

When silver chloride is added to water, it remains as solid AgCl—it does not dissolve into independent ions.

 Solubility

A compound is **soluble** in a particular liquid if it dissolves in that liquid. A compound is **insoluble** if it does not dissolve in the liquid. NaCl, for example, is soluble in water. If we mix solid sodium chloride into water, it dissolves and forms a strong electrolyte solution. AgCl, on the other hand, is insoluble in water. If we mix solid silver chloride into water, it remains as a solid within the liquid water.

There is no easy way to tell whether a particular compound will be soluble or insoluble in water. For ionic compounds, however, there are empirical rules that have been deduced from observations of many compounds. These are called **solubility rules** and are summarized in Table 7.2. For example, the solubility rules state that compounds containing the lithium ion are *soluble*. That means that compounds such as LiBr, $LiNO_3$, Li_2SO_4, LiOH, and Li_2CO_3 will all dissolve in water to form strong electrolyte solutions. If a compound contains Li^+, it is soluble. Similarly, the solubility rules state that compounds containing the NO_3^- ion are soluble. That means that compounds such as $AgNO_3$, $Pb(NO_3)_2$, $NaNO_3$, $Ca(NO_3)_2$ and $Sr(NO_3)_2$ all dissolve in water to form strong electrolyte solutions.

The solubility rules apply only to the solubility of the compounds in water

TABLE 7.2 Solubility Rules

Compounds Containing the Following Ions Are Mostly Soluble	Exceptions
Li^+, Na^+, K^+, NH_4^+	None
NO_3^-, $C_2H_3O_2^-$	None
Cl^-, Br^-, I^-	When any of these ions pairs with Ag^+, Hg_2^{2+}, or Pb^{2+}, the compound is insoluble
SO_4^{2-}	When SO_4^{2-} pairs with Sr^{2+}, Ba^{2+}, Pb^{2+}, or Ca^{2+}, the compound is insoluble
Compounds Containing the Following Ions Are Mostly Insoluble	**Exceptions**
OH^-, S^{2-}	When either of these ions pairs with Li^+, Na^+, K^+, or NH_4^+, the compound is soluble
	When S^{2-} pairs with Ca^{2+}, Sr^{2+}, or Ba^{2+}, the compound is soluble
	When OH^- pairs with Ca^{2+}, Sr^{2+}, or Ba^{2+}, the compound is slightly soluble.*
CO_3^{2-}, PO_4^{3-}	When either of these ions pairs with Li^+, Na^+, K^+, or NH_4^+, the compound is soluble

*For many purposes these can be considered insoluble.

The solubility rules also state that, with some exceptions, compounds containing the CO_3^{2-} ion are *insoluble*. Therefore, compounds such as $CuCO_3$, $CaCO_3$, $SrCO_3$, and $FeCO_3$ do not dissolve in water. Note that the solubility rules contain many exceptions. For example, compounds containing CO_3^{2-} are *soluble when paired with* Li^+, Na^+, K^+, or NH_4^+. Thus Li_2CO_3, Na_2CO_3, K_2CO_3, and $(NH_4)_2CO_3$ are all soluble.

EXAMPLE 7.6 Determining Whether a Compound Is Soluble

Determine whether each of the following compounds is soluble or insoluble.

(a) $AgBr$
(b) $CaCl_2$
(c) $Pb(NO_3)_2$
(d) $PbSO_4$

Solution:

(a) Insoluble; compounds containing Br^- are normally soluble, but Ag^+ is an exception.

(b) Soluble; compounds containing Cl^- are normally soluble, and Ca^{2+} is not an exception.

(c) Soluble; compounds containing NO_3^- are always soluble.

(d) Insoluble; compounds containing SO_4^{2-} are normally soluble, but Pb^{2+} is an exception.

SKILLBUILDER 7.6 Determining Whether a Compound Is Soluble

Determine whether each of the following compounds is soluble or insoluble.

(a) CuS
(b) $FeSO_4$
(c) $PbCO_3$
(d) NH_4Cl

7.6 Precipitation Reactions: Reactions in Aqueous Solution That Form a Solid

In Section 7.1, we learned how sodium carbonate is added to laundry detergent to react with dissolved Mg^{2+} and Ca^{2+} ions to form solids that precipitate (or come out of) solution. These reactions are examples of **precipitation reactions**, reactions that form a solid or **precipitate** upon mixing two aqueous solutions.

Precipitation reactions are common in chemistry. Potassium iodide and lead nitrate, for example, both form colorless, strong electrolyte solutions when dissolved in water (see the solubility rules). When the two solutions are combined, however, a brilliant yellow precipitate forms (▶ Figure 7.7). This precipitation reaction can be described with the following chemical equation.

$$2\,KI(aq) + Pb(NO_3)_2(aq) \longrightarrow PbI_2(s) + 2\,KNO_3(aq)$$

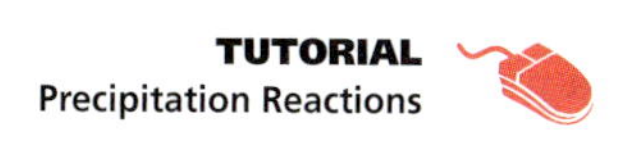

Precipitation reactions do not always occur when mixing two aqueous solutions. For example, if solutions of $KI(aq)$ and $NaCl(aq)$ are combined, nothing happens (▶ Figure 7.8).

$$KI(aq) + NaCl(aq) \longrightarrow \text{NO REACTION}$$

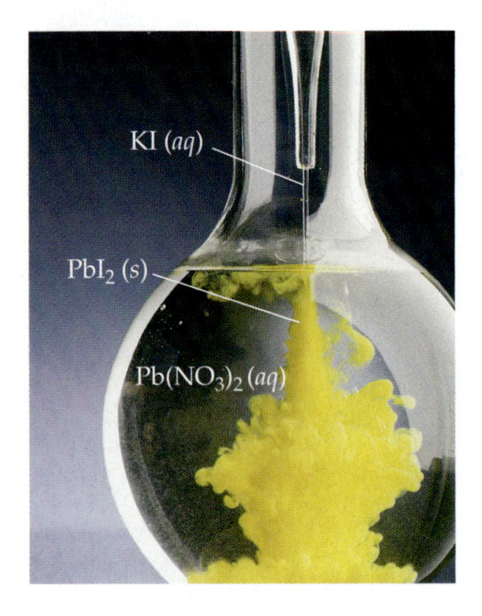

▲ **Figure 7.7 Precipitation** When a potassium iodide solution is mixed with a lead(II) nitrate solution, a brilliant yellow precipitate of $PbI_2(s)$ forms.

▲ **Figure 7.8 No reaction** When a potassium iodide solution is mixed with a sodium chloride solution, no reaction occurs.

Predicting Precipitation Reactions

The key to predicting precipitation reactions is to understand that *only insoluble compounds form precipitates*. In a precipitation reaction, two solutions containing soluble compounds combine and an insoluble compound precipitates. For example, consider the precipitation reaction from Figure 7.7.

$$\underset{\text{soluble}}{2\,KI(aq)} + \underset{\text{soluble}}{Pb(NO_3)_2(aq)} \longrightarrow \underset{\text{insoluble}}{PbI_2(s)} + \underset{\text{soluble}}{2\,KNO_3(aq)}$$

KI and $Pb(NO_3)_2$ are both soluble, but the precipitate, PbI_2, is *insoluble*. Before mixing, $KI(aq)$ and $Pb(NO_3)_2(aq)$ are both dissociated in their respective solutions.

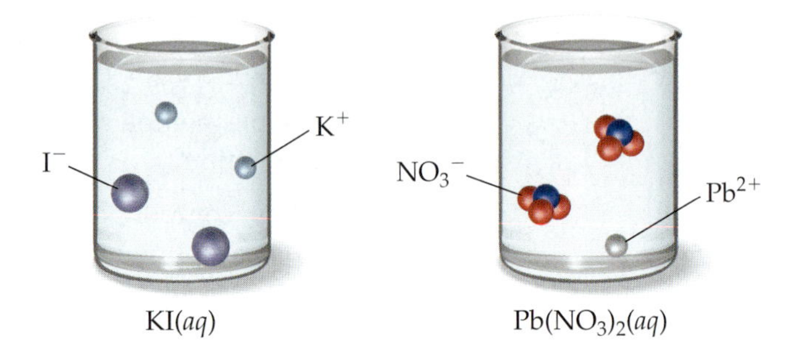

The instant that the solutions are mixed, all four ions are present.

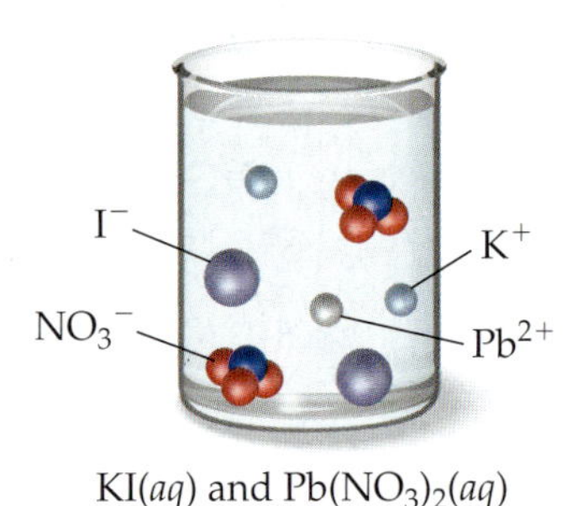

However, new compounds—potentially insoluble ones—are now possible. Specifically, the cation from one compound can now pair with the anion from the other compound to form new (and potentially insoluble) products.

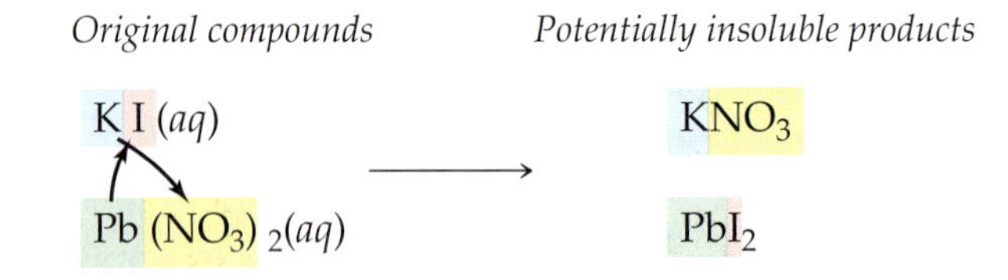

If the *potentially insoluble* products are both *soluble*, then no reaction occurs. If, on the other hand, one or both of the potentially insoluble products are *indeed insoluble*, a precipitation reaction occurs. In this case, KNO_3 is soluble, but PbI_2 is insoluble. Consequently, PbI_2 precipitates.

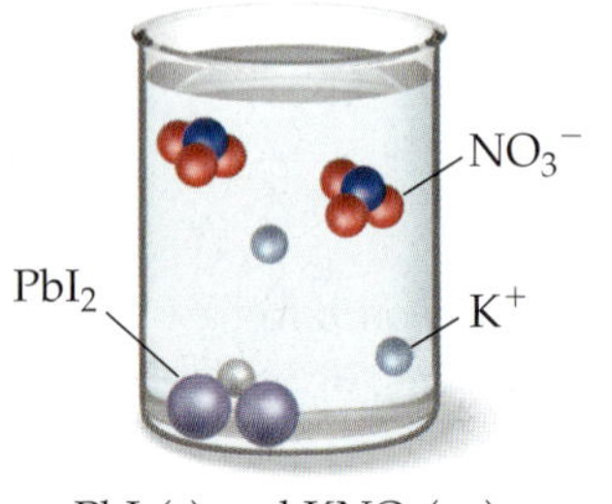

To predict whether a precipitation reaction will occur when two solutions are mixed and to write an equation for the reaction, follow the steps in the procedure box. As usual, the steps are shown in the left column, and two examples of applying the procedure are shown in the center and right column.

Writing Equations for Precipitation Reactions

EXAMPLE 7.7

Write an equation for the precipitation reaction that occurs (if any) when solutions of sodium carbonate and copper(II) chloride are mixed.

EXAMPLE 7.8

Write an equation for the precipitation reaction that occurs (if any) when solutions of lithium nitrate and sodium sulfate are mixed.

1. Write the formulas of the two compounds being mixed as reactants in a chemical equation.

Solution (7.7):

$$Na_2CO_3(aq) + CuCl_2(aq) \longrightarrow$$

Solution (7.8):

$$LiNO_3(aq) + Na_2SO_4(aq) \longrightarrow$$

2. Below the equation, write the formulas of the potentially insoluble products that could form from the reactants. Obtain these by combining the cation from one reactant with the anion from the other. Make sure to write correct formulas for these ionic compounds as described in Section 5.5.

Example 7.7:

$$Na_2CO_3(aq) + CuCl_2(aq) \longrightarrow$$

Potentially Insoluble Products:

NaCl $\quad CuCO_3$

Example 7.8:

$$LiNO_3(aq) + Na_2SO_4(aq) \longrightarrow$$

Potentially Insoluble Products:

$NaNO_3 \quad Li_2SO_4$

3. Use the solubility rules to determine whether any of the new products are indeed insoluble.

Example 7.7: NaCl is *soluble* (compounds containing Cl^- are usually soluble and Na^+ is not an exception).
$CuCO_3$ is *insoluble* (compounds containing CO_3^{2-} are usually insoluble and Cu^{2+} is not an exception).

Example 7.8: $NaNO_3$ is *soluble* (compounds containing NO_3^- are soluble and Na^+ is not an exception).
Li_2SO_4 is *soluble* (compounds containing SO_4^{2-} are soluble and Li^+ is not an exception).

4. If all of the potentially insoluble products are soluble, there will be no precipitate. Write *NO REACTION* next to the arrow.

Example 7.7: Since this example has an insoluble product, we proceed to the next step.

Example 7.8:

$$LiNO_3(aq) + Na_2SO_4(aq) \longrightarrow \text{NO REACTION}$$

5. If one or both of the potentially insoluble products are indeed insoluble, write their formula(s) as the product(s) of the reaction, using (*s*) to indicate solid. Write any soluble products with (*aq*) to indicate aqueous.

Example 7.7:

$$Na_2CO_3(aq) + CuCl_2(aq) \longrightarrow CuCO_3(s) + NaCl(aq)$$

6. Balance the equation. Remember to adjust only coefficients here, not subscripts.

Example 7.7:

$$Na_2CO_3(aq) + CuCl_2(aq) \longrightarrow CuCO_3(s) + \mathbf{2}\,NaCl(aq)$$

SKILLBUILDER 7.7

Write an equation for the precipitation reaction that occurs (if any) when solutions of potassium hydroxide and nickel(II) bromide are mixed.

SKILLBUILDER 7.8

Write an equation for the precipitation reaction that occurs (if any) when solutions of ammonium chloride and iron(III) nitrate are mixed.

EXAMPLE 7.9 Predicting and Writing Equations for Precipitation Reactions

Write an equation for the precipitation reaction (if any) that occurs when solutions of lead(II) acetate and sodium sulfate are mixed. If no reaction occurs, write *NO REACTION.*

1. Write the formulas of the two compounds being mixed as reactants in a chemical equation.

Solution:

$$Pb(C_2H_3O_2)_2(aq) + Na_2SO_4(aq) \longrightarrow$$

2. Below the equation, write the formulas of the potentially insoluble products that could form from the reactants. These are obtained by combining the cation from one reactant with the anion from the other. Make sure to adjust the subscripts so that all formulas are charge-neutral.

$$Pb(C_2H_3O_2)_2(aq) + Na_2SO_4(aq) \longrightarrow$$

Potentially insoluble products

$NaC_2H_3O_2$ $PbSO_4$

3. Use the solubility rules to determine whether any of the potentially insoluble products are indeed insoluble.

$NaC_2H_3O_2$ is *soluble* (compounds containing Na^+ are always soluble).
$PbSO_4$ is *insoluble* (compounds containing SO_4^{2-} are normally soluble, but Pb^{2+} is an exception).

4. If all of the potentially insoluble products are soluble, there will be no precipitate. Write *NO REACTION* next to the arrow.

Since we have an insoluble product, we proceed to the next step.

5. If one or both of the potentially insoluble products are indeed insoluble, write their formula(s) as the product(s) of the reaction, using (*s*) to indicate solid. Write any soluble products with (*aq*) to indicate aqueous.

$$Pb(C_2H_3O_2)_2(aq) + Na_2SO_4(aq) \longrightarrow PbSO_4(s) + NaC_2H_3O_2(aq)$$

6. Balance the equation.

$$Pb(C_2H_3O_2)_2(aq) + Na_2SO_4(aq) \longrightarrow PbSO_4(s) + \mathbf{2}\, NaC_2H_3O_2(aq)$$

SKILLBUILDER 7.9 Predicting and Writing Equations for Precipitation Reactions

Write an equation for the precipitation reaction (if any) that occurs when solutions of potassium sulfate and strontium nitrate are mixed. If no reaction occurs, write *NO REACTION.*

CONCEPTUAL CHECKPOINT 7.2

Which of these reactions would result in the formation of a precipitate?

(a) $NaNO_3 + CaS$
(b) $MgSO_4 + CaS$
(c) $NaNO_3 + MgSO_4$

7.7 Writing Chemical Equations for Reactions in Solution: Molecular, Complete Ionic, and Net Ionic Equations

Consider the following equation for a precipitation reaction.

$$AgNO_3(aq) + NaCl(aq) \longrightarrow AgCl(s) + NaNO_3(aq)$$

This equation is written as a **molecular equation**, an equation showing the complete neutral formulas for every compound in the reaction. However, equations for reactions occurring in aqueous solution may be written to show that aqueous ionic compounds normally dissociate in solution. For example, the previous equation can be written as follows:

$$Ag^+(aq) + NO_3^-(aq) + Na^+(aq) + Cl^-(aq) \longrightarrow AgCl(s) + Na^+(aq) + NO_3^-(aq)$$

When writing complete ionic equations, separate only *aqueous* ionic compounds into their constituent ions. Do NOT separate solid, liquid, or gaseous compounds.

Equations such as this one, showing the reactants and products as they are actually present in solution, are called **complete ionic equations**.

Notice that in the complete ionic equation, some of the ions in solution appear unchanged on both sides of the equation. These ions are called **spectator ions** because they do not participate in the reaction.

$$Ag^+(aq) + NO_3^-(aq) + Na^+(aq) + Cl^-(aq) \longrightarrow AgCl(s) + Na^+(aq) + NO_3^-(aq)$$

Spectator ions

To simplify the equation, and to more clearly show what is happening, spectator ions can be omitted.

$$Ag^+(aq) + Cl^-(aq) \longrightarrow AgCl(s)$$

Species refers to a kind or sort of thing. In this case, the species are all the different molecules and ions that are present during the reaction.

Equations such as this one, which show only the species that actually participate in the reaction, are called **net ionic equations**.

As another example, consider the following reaction between $HCl(aq)$ and $NaOH(aq)$.

$$HCl(aq) + NaOH(aq) \longrightarrow H_2O(l) + NaCl(aq)$$

HCl, NaOH, and NaCl exist in solution as independent ions. The *complete ionic equation* for this reaction is:

$$H^+(aq) + Cl^-(aq) + Na^+(aq) + OH^-(aq) \longrightarrow H_2O(l) + Na^+(aq) + Cl^-(aq)$$

To write the *net ionic equation*, we remove the spectator ions, those that are unchanged on both sides of the equation.

$$H^+(aq) + \cancel{Cl^-}(aq) + \cancel{Na^+}(aq) + OH^-(aq) \longrightarrow H_2O(l) + \cancel{Na^+}(aq) + \cancel{Cl^-}(aq)$$

Spectator ions

The net ionic equation is $H^+(aq) + OH^-(aq) \longrightarrow H_2O(l)$

To summarize:

- A *molecular equation* is a chemical equation showing the complete, neutral formulas for every compound in a reaction.
- A *complete ionic equation* is a chemical equation showing all of the species as they are actually present in solution.
- A *net ionic equation* is an equation showing only the species that actually participate in the reaction.

EXAMPLE 7.10 Writing Complete Ionic and Net Ionic Equations

Consider the following precipitation reaction occurring in aqueous solution.

$$Pb(NO_3)_2(aq) + 2\,LiCl(aq) \longrightarrow PbCl_2(s) + 2\,LiNO_3(aq)$$

Write a complete ionic equation and a net ionic equation for this reaction.

Write the complete ionic equation by separating aqueous ionic compounds into their constituent ions. The $PbCl_2(s)$ remains as one unit.

Solution:

Complete ionic equation:

$$Pb^{2+}(aq) + 2\,NO_3^-(aq) + 2\,Li^+(aq) + 2\,Cl^-(aq) \longrightarrow PbCl_2(s) + 2\,Li^+(aq) + 2\,NO_3^-(aq)$$

Write the net ionic equation by eliminating the spectator ions, those that are not changing during the reaction.

$$Pb^{2+}(aq) + \cancel{2\,NO_3^-(aq)} + \cancel{2\,Li^+(aq)} + 2\,Cl^-(aq) \longrightarrow PbCl_2(s) + \cancel{2\,Li^+(aq)} + \cancel{2\,NO_3^-(aq)}$$

Net ionic equation:

$$Pb^{2+}(aq) + 2\,Cl^-(aq) \longrightarrow PbCl_2(s)$$

SKILLBUILDER 7.10 Writing Complete Ionic and Net Ionic Equations

Consider the following reaction occurring in aqueous solution.

$$2\,HBr(aq) + Ca(OH)_2(aq) \longrightarrow 2\,H_2O(l) + CaBr_2(aq)$$

Write a complete ionic equation and net ionic equation for this reaction.

7.8 Acid–Base and Gas Evolution Reactions

Two other kinds of reactions that occur in solution are **acid–base reactions**—reactions that form water upon mixing of an acid and a base—and **gas evolution reactions**—reactions that evolve a gas. Like precipitation reactions, these reactions occur when the cation of one reactant combines with the anion of another. As we will see in the next section, many gas evolution reactions also happen to be acid–base reactions.

▶ Foods such as lemons, limes, and vinegar contain acids.

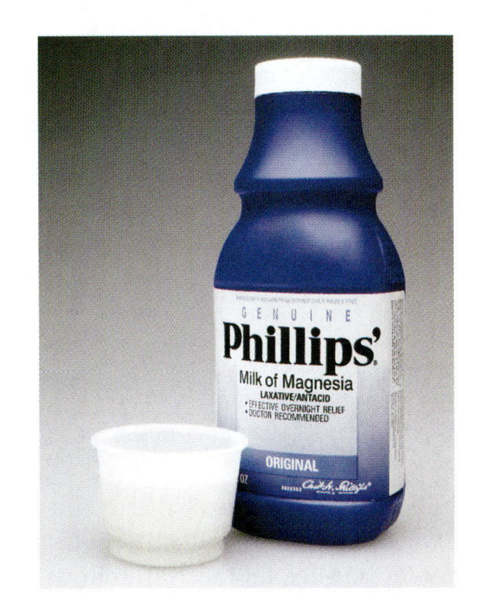

▶ Milk of magnesia is basic and tastes bitter.

Acid–Base (Neutralization) Reactions

We learned in Chapter 5 that an acid is a compound characterized by its sour taste, its ability to dissolve some metals, and its tendency to form H^+ ions in solution. A base is a compound characterized by its bitter taste, its slippery feel, and its tendency to form OH^- ions in solution. Some common acids and bases are listed in Table 7.3. Acids and bases are also found in many everyday substances. Foods such as lemons, limes, and vinegar contain acids. Soap, coffee, and milk of magnesia all contain bases.

Even though coffee itself is acidic overall, it contains some naturally occurring bases (such as caffeine) that give it a bitter taste.

When an acid and base are mixed, the $H^+(aq)$ from the acid combines with the $OH^-(aq)$ from the base to form $H_2O(l)$. For example, consider the reaction between hydrochloric acid and sodium hydroxide mentioned earlier.

$$\underset{\text{Acid}}{HCl(aq)} + \underset{\text{Base}}{NaOH(aq)} \longrightarrow \underset{\text{Water}}{H_2O(l)} + \underset{\text{Salt}}{NaCl(aq)}$$

Acid–base reactions (also called **neutralization reactions**) generally form water and an ionic compound—called a **salt**—that usually remains dissolved in the solution. The net ionic equation for many acid–base reactions is:

$$H^+(aq) + OH^-(aq) \longrightarrow H_2O(l)$$

Another example of an acid–base reaction is that between sulfuric acid and potassium hydroxide.

$$\underset{\text{acid}}{H_2SO_4(aq)} + \underset{\text{base}}{2\,KOH} \longrightarrow \underset{\text{water}}{2\,H_2O(l)} + \underset{\text{salt}}{K_2SO_4(aq)}$$

TABLE 7.3 Some Common Acids and Bases

Acid	Formula	Base	Formula
hydrochloric acid	HCl	sodium hydroxide	$NaOH$
hydrobromic acid	HBr	lithium hydroxide	$LiOH$
nitric acid	HNO_3	potassium hydroxide	KOH
sulfuric acid	H_2SO_4	calcium hydroxide	$Ca(OH)_2$
perchloric acid	$HClO_4$	barium hydroxide	$Ba(OH)_2$
acetic acid	$HC_2H_3O_2$		

Again, notice the pattern of acid and base reacting to form water and a salt.

$$\text{Acid} + \text{Base} \longrightarrow \text{Water} + \text{Salt} \qquad \text{(acid–base reactions)}$$

When writing equations for acid–base reactions, write the formula of the salt using the procedure for writing formulas of ionic compounds given in Section 5.5.

EXAMPLE 7.11 Writing Equations for Acid–Base Reactions

Write a molecular and net ionic equation for the reaction between aqueous HNO_3 and aqueous $Ca(OH)_2$.

Solution:

You must recognize these substances as an acid and a base. First write the skeletal reaction following the general pattern of acid plus base goes to water plus salt.

$$\underset{\text{acid}}{HNO_3(aq)} + \underset{\text{base}}{Ca(OH)_2(aq)} \longrightarrow \underset{\text{water}}{H_2O(l)} + \underset{\text{salt}}{Ca(NO_3)_2(aq)}$$

Next, balance the equation.

$$\mathbf{2}\,HNO_3(aq) + Ca(OH)_2(aq) \longrightarrow \mathbf{2}\,H_2O(l) + Ca(NO_3)_2(aq)$$

Write the net ionic equation by eliminating those ions that remain the same on both sides of the equation.

$$2\,H^+(aq) + 2\,OH^-(aq) \longrightarrow 2\,H_2O(l)$$

or simply

$$H^+(aq) + OH^-(aq) \longrightarrow H_2O(l)$$

SKILLBUILDER 7.11 Writing Equations for Acid–Base Reactions

Write a molecular and net ionic equation for the reaction that occurs between aqueous H_2SO_4 and aqueous KOH.

Gas Evolution Reactions

Some aqueous reactions form a gas as a product. These reactions, as we learned in the opening section of this chapter, are called gas evolution reactions. Some gas evolution reactions form a gaseous product directly when the cation of one reactant reacts with the anion of the other. For example, when sulfuric acid reacts with lithium sulfide, dihydrogen sulfide gas is formed.

$$H_2SO_4(aq) + Li_2S(aq) \longrightarrow \underset{\text{Gas}}{H_2S(g)} + Li_2SO_4(aq)$$

Many gas evolution reactions such as this one are also acid–base reactions. In Chapter 14 we learn how ions such as CO_3^{2-} act as bases in aqueous solution.

Other gas evolution reactions form an intermediate product that then decomposes into a gas. For example, when aqueous hydrochloric acid is mixed with aqueous sodium bicarbonate, the following reaction occurs.

$$HCl(aq) + NaHCO_3(aq) \longrightarrow H_2CO_3(aq) + NaCl(aq) \longrightarrow H_2O(l) + \underset{\text{Gas}}{CO_2(g)} + NaCl(aq)$$

The intermediate product, H_2CO_3, is not stable and decomposes to form H_2O and gaseous CO_2. This reaction is almost identical to the reaction in the

▲ A gas evolution reaction: vinegar (a dilute solution of acetic acid) and baking soda (sodium bicarbonate) produce carbon dioxide.

kindergarten volcano of Section 7.1, which involves the mixing of acetic acid and sodium bicarbonate.

$$HC_2H_3O_2(aq) + NaHCO_3(aq) \longrightarrow H_2CO_3(aq) + NaC_2H_3O_2(aq) \longrightarrow H_2O(l) + CO_2(g) + NaC_2H_3O_2(aq)$$

The bubbling is caused by the newly formed carbon dioxide gas. Other important gas evolution reactions form either H_2SO_3 or NH_4OH as intermediate products.

$$HCl(aq) + NaHSO_3(aq) \longrightarrow H_2SO_3(aq) + NaCl(aq) \longrightarrow H_2O(l) + SO_2(g) + NaCl(aq)$$

$$NH_4Cl(aq) + NaOH(aq) \longrightarrow NH_4OH(aq) + NaCl(aq) \longrightarrow H_2O(l) + NH_3(g) + NaCl(aq)$$

MOLECULES
Bicarbonate Ion, Carbonate Ion, Carbonic Acid

The main types of compounds that form gases in aqueous reactions, as well as the gases that they form, are listed in Table 7.4.

TABLE 7.4 Types of Compounds That Undergo Gas Evolution Reactions

Reactant Type	Intermediate Product	Gas Evolved	Example
sulfides	none	H_2S	$2\,HCl(aq) + K_2S(aq) \longrightarrow H_2S(g) + 2\,KCl(aq)$
carbonates and bicarbonates	H_2CO_3	CO_2	$2\,HCl(aq) + K_2CO_3(aq) \longrightarrow H_2O(l) + CO_2(g) + 2\,KCl(aq)$
sulfites and bisulfites	H_2SO_3	SO_2	$2\,HCl(aq) + K_2SO_3(aq) \longrightarrow H_2O(l) + SO_2(g) + 2\,KCl(aq)$
ammonium	NH_4OH	NH_3	$NH_4Cl(aq) + KOH(aq) \longrightarrow H_2O(l) + NH_3(g) + KCl(aq)$

EXAMPLE 7.12 Writing Equations for Gas Evolution Reactions

Write a molecular equation for the gas evolution reaction that occurs when you mix aqueous nitric acid and aqueous sodium carbonate.

Solution:

Begin by writing a skeletal equation that includes the reactants and products that form when the cation of each reactant combines with the anion of the other.

$$HNO_3(aq) + Na_2CO_3(aq) \longrightarrow H_2CO_3(aq) + NaNO_3(aq)$$

You must recognize that $H_2CO_3(aq)$ decomposes into $H_2O(l)$ and $CO_2(g)$ and write the corresponding equation.

$$HNO_3(aq) + Na_2CO_3(aq) \longrightarrow H_2O(l) + CO_2(g) + NaNO_3(aq)$$

Finally, balance the equation.

$$\mathbf{2}\,HNO_3(aq) + Na_2CO_3(aq) \longrightarrow H_2O(l) + CO_2(g) + \mathbf{2}\,NaNO_3(aq)$$

SKILLBUILDER 7.12 Writing Equations for Gas Evolution Reactions

Write a molecular equation for the gas evolution reaction that occurs when you mix aqueous hydrobromic acid and aqueous potassium sulfite.

SKILLBUILDER PLUS

Write a net ionic equation for the previous reaction.

Chemistry AND HEALTH

Neutralizing Excess Stomach Acid

Your stomach normally contains acids that are involved in food digestion. Certain foods and stress, however, can increase the acidity of your stomach to uncomfortable levels, causing acid stomach or heartburn. Antacids are over-the-counter medicines that work by reacting with and neutralizing stomach acid. Antacids employ different bases as neutralizing agents. Tums, for example, contains $CaCO_3$; milk of magnesia contains $Mg(OH)_2$; and Mylanta contains $Al(OH)_3$. They all, however, have the same effect of neutralizing stomach acid and relieving heartburn.

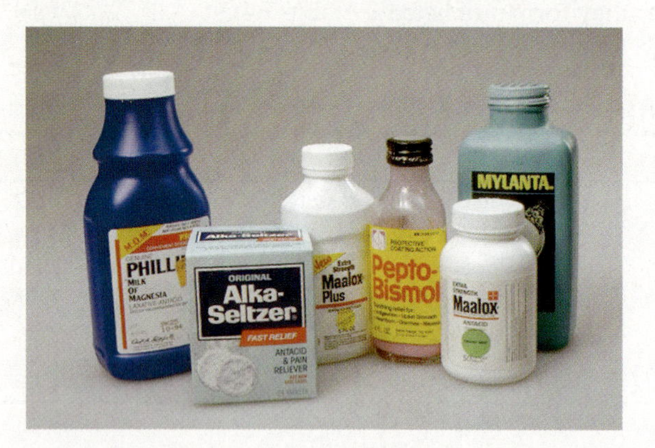

▲ Antacids contain bases such $Mg(OH)_2$, $Al(OH)_3$, and $NaHCO_3$.

CAN YOU ANSWER THIS? *Assume that stomach acid is HCl and write equations showing how each of these antacids neutralizes stomach acid.*

▲ The base in an antacid neutralizes excess stomach acid, relieving heartburn and acid stomach.

7.9 Oxidation–Reduction Reactions

We cover oxidation–reduction reactions more thoroughly in Chapter 16.

Reactions involving the transfer of electrons are called **oxidation–reduction reactions** or **redox reactions**. Redox reactions are responsible for the rusting of iron, the bleaching of hair, and the production of electricity in batteries. Many redox reactions involve the reaction of a substance with oxygen.

$$2\,H_2(g) + O_2(g) \longrightarrow 2\,H_2O(g)$$
(reaction that powers the space shuttle)

$$4\,Fe(s) + 3\,O_2(g) \longrightarrow 2\,Fe_2O_3(s)$$
(rusting of iron)

$$CH_4(g) + 2\,O_2(g) \longrightarrow CO_2(g) + 2\,H_2O(g)$$
(combustion of natural gas)

However, redox reactions need not involve oxygen. Consider, for example, the reaction between sodium and chlorine to form table salt (NaCl).

$$2\,Na(s) + Cl_2(g) \longrightarrow 2\,NaCl(s)$$

This reaction is similar to the reaction between sodium and oxygen to form sodium oxide.

$$4\,Na(s) + O_2(g) \longrightarrow 2\,Na_2O(s)$$

What do these two reactions have in common? In both cases, sodium (a metal with a tendency to lose electrons) reacts with a nonmetal (that has a tendency to gain electrons). In both cases, sodium atoms lose electrons to nonmetal atoms. A fundamental definition of oxidation is *the loss of electrons*, and a fundamental definition of reduction is *the gain of electrons*.

Helpful mnemonics: OIL RIG—**O**xidation **I**s **L**oss; **R**eduction **I**s **G**ain. LEO GER—**L**ose **E**lectrons **O**xidation; **G**ain **E**lectrons **R**eduction

Notice that oxidation and reduction must occur together. If one substance loses electrons (oxidation), then another substance must gain electrons (reduction). For now, be able to identify redox reactions.

A reaction can be classified as a redox reaction if it meets *any one* of these requirements.

Redox reactions are those in which:

- A substance reacts with elemental oxygen.
- A metal reacts with a nonmetal.
- More generally, one substance transfers electrons to another substance.

EXAMPLE 7.13 Identifying Redox Reactions

Which of the following is a redox reaction?

(a) $2\,Mg(s) + O_2(g) \longrightarrow 2\,MgO(s)$
(b) $2\,HBr(aq) + Ca(OH)_2(aq) \longrightarrow 2\,H_2O(l) + CaBr_2(aq)$
(c) $Ca(s) + Cl_2(g) \longrightarrow CaCl_2(s)$
(d) $Zn(s) + Fe^{2+}(aq) \longrightarrow Zn^{2+}(aq) + Fe(s)$

Solution:

(a) Redox reaction; Mg reacts with elemental oxygen.

(b) Not a redox reaction; it is an acid–base reaction.

(c) Redox reaction; a metal reacts with a nonmetal.

(d) Redox reaction; Zn transfers two electrons to Fe^{2+}.

SKILLBUILDER 7.13 Identifying Redox Reactions

Which of the following is a redox reaction?

(a) $2\,Li(s) + Cl_2(g) \longrightarrow 2\,LiCl(s)$
(b) $2\,Al(s) + 3\,Sn^{2+}(aq) \longrightarrow 2\,Al^{3+}(aq) + 3\,Sn(s)$
(c) $Pb(NO_3)_2(aq) + 2\,LiCl(aq) \longrightarrow PbCl_2(s) + 2\,LiNO_3(aq)$
(d) $C(s) + O_2(g) \longrightarrow CO_2(g)$

Combustion Reactions

Combustion reactions are a type of redox reaction. They are important because most of our society's energy is derived from combustion reactions. Combustion reactions are characterized by the reaction of a substance with O_2 to form one or more oxygen-containing compounds, often including water. Combustion reactions also emit heat. For example, as we saw in Section 7.3, natural gas (CH_4) reacts with oxygen to form carbon dioxide and water.

The water formed in combustion reactions may be gaseous (*g*) or liquid (*l*) depending on the reaction conditions.

$$CH_4(g) + 2\,O_2(g) \longrightarrow CO_2(g) + 2\,H_2O(g)$$

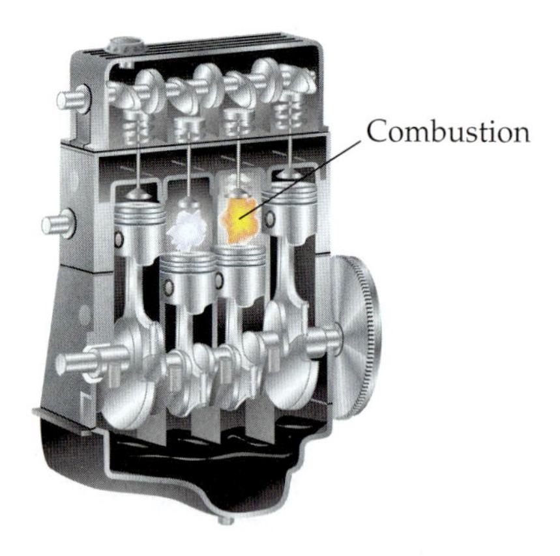

▲ Combustion of octane occurs in the cylinders of an automobile engine.

As we learned in the opening section of this chapter, combustion reactions power automobiles. For example, octane, a component of gasoline, reacts with oxygen to form carbon dioxide and water.

$$2\,C_8H_{18}(l) + 25\,O_2(g) \longrightarrow 16\,CO_2(g) + 18\,H_2O(g)$$

Ethanol, the alcohol in alcoholic beverages, also reacts with oxygen in a combustion reaction to form carbon dioxide and water.

$$C_2H_5OH(l) + 3\,O_2(g) \longrightarrow 2\,CO_2(g) + 3\,H_2O(g)$$

Compounds containing carbon and hydrogen—or carbon, hydrogen, and oxygen—always form carbon dioxide and water upon combustion. Other combustion reactions include the reaction of carbon with oxygen to form carbon dioxide:

$$C(s) + O_2(g) \longrightarrow CO_2(g)$$

and the reaction of hydrogen with oxygen to form water:

$$2\,H_2(g) + O_2(g) \longrightarrow 2\,H_2O(g)$$

EXAMPLE 7.14 Writing Combustion Reactions

Write a balanced equation for the combustion of liquid methyl alcohol (CH_3OH).

Begin by writing a skeletal equation showing the reaction of CH_3OH with O_2 to form CO_2 and H_2O.

Solution:

$$CH_3OH(l) + O_2(g) \longrightarrow CO_2(g) + H_2O(g)$$

Balance the skeletal equation using the rules in Section 7.4.

$$\mathbf{2}\,CH_3OH(l) + \mathbf{3}\,O_2(g) \longrightarrow \mathbf{2}\,CO_2(g) + \mathbf{4}\,H_2O(g)$$

SKILLBUILDER 7.14 Writing Combustion Reactions

Write a balanced equation for the combustion of liquid pentane (C_5H_{12}), a component of gasoline.

SKILLBUILDER PLUS

Write a balanced equation for the combustion of liquid propanol (C_3H_7OH).

7.10 Classifying Chemical Reactions

Throughout this chapter, we have examined different types of chemical reactions. We have seen examples of precipitation reactions, acid–base reactions, gas evolution reactions, oxidation–reduction reactions, and combustion reactions. We can organize these different types of reactions with the following flow chart.

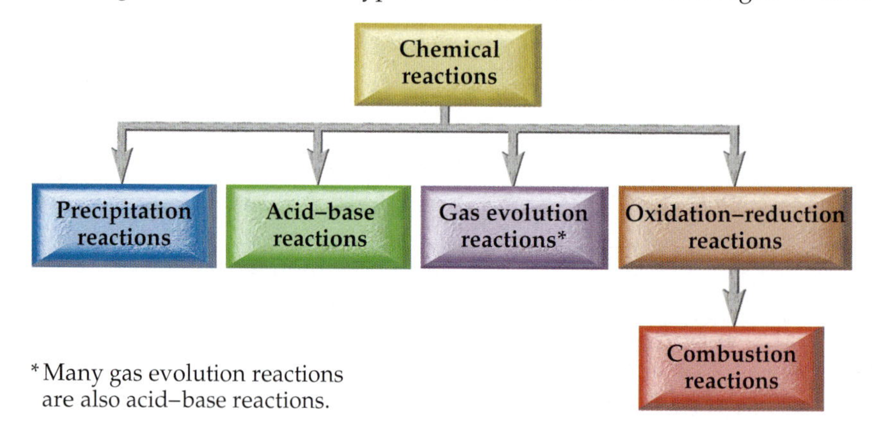

This classification scheme focuses on the type of chemistry or phenomenon that is occurring during the reaction (such as the formation of a precipitate or the transfer of electrons). However, another way to classify chemical reactions is by what atoms or groups of atoms do during the reaction.

Classifying Chemical Reactions by What Atoms Do

Many chemical reactions can be classified into one of the following four categories. In this classification scheme, the letters (A, B, C, D) represent atoms or groups of atoms.

Type of Reaction	Generic Equation
synthesis or combination	$A + B \longrightarrow AB$
decomposition	$AB \longrightarrow A + B$
displacement	$A + BC \longrightarrow AC + B$
double-displacement	$AB + CD \longrightarrow AD + CB$

SYNTHESIS OR COMBINATION REACTIONS. In a **synthesis** or **combination reaction**, simpler substances combine to form more complex substances. The simpler substances may be elements, such as sodium and chlorine combining to form sodium chloride.

$$2\,Na(s) + Cl_2(g) \longrightarrow 2\,NaCl(s)$$

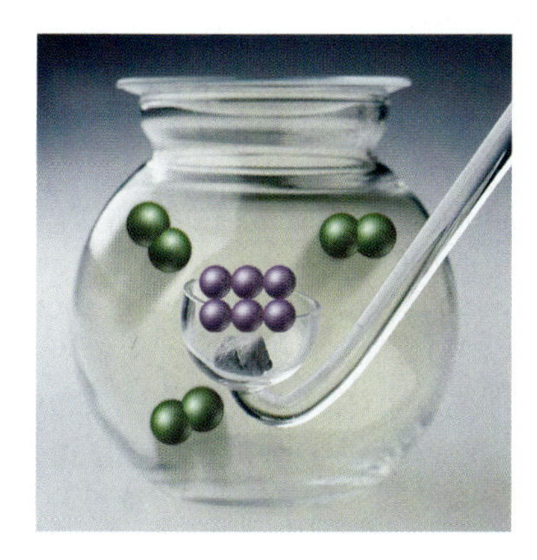

$$2\,Na(s) + Cl_2(g) \longrightarrow 2\,NaCl(s)$$

▲ In a synthesis reaction, two simpler substances combine to make a more complex substance. In this series of photographs we see sodium metal and chlorine gas. When they combine, a chemical reaction occurs that forms sodium chloride.

The simpler substances may also be compounds, such as calcium oxide and carbon dioxide combining to form calcium carbonate.

$$CaO(s) + CO_2(g) \longrightarrow CaCO_3(s)$$

In either case, a synthesis reaction follows the general equation:

$$A + B \longrightarrow AB$$

Other examples of synthesis reactions include:

$$2\,H_2(g) + O_2(g) \longrightarrow 2\,H_2O(l)$$
$$2\,Mg(s) + O_2(g) \longrightarrow 2\,MgO(s)$$
$$SO_3(g) + H_2O(l) \longrightarrow H_2SO_4(aq)$$

Note that the first two of these reactions are also redox reactions.

$$2\,H_2O(l) \longrightarrow 2\,H_2(g) + O_2(g)$$

▲ When electrical current is passed through water, the water undergoes a decomposition reaction to form hydrogen gas and oxygen gas.

DECOMPOSITION REACTIONS. In a **decomposition reaction**, a complex substance decomposes to form simpler substances. The simpler substances may be elements, such as the hydrogen and oxygen gases that form upon the decomposition of water when electrical current passes through it.

$$2\,H_2O(l) \xrightarrow[\text{electrical current}]{} 2\,H_2(g) + O_2(g)$$

The simpler substances may also be compounds, such as the calcium oxide and carbon dioxide that form upon heating calcium carbonate.

$$CaCO_3(s) \xrightarrow[\text{heat}]{} CaO(s) + CO_2(g)$$

In either case, a decomposition reaction follows the general equation:

$$AB \longrightarrow A + B$$

Other examples of decomposition reactions include:

$$2\,HgO(s) \xrightarrow[\text{heat}]{} 2\,Hg(l) + O_2(g)$$

$$2\,KClO_3(s) \xrightarrow[\text{heat}]{} 2\,KCl(s) + 3\,O_2(g)$$

$$CH_3I(g) \xrightarrow[\text{light}]{} CH_3(g) + I(g)$$

TUTORIAL Chemical Reactions in Automotive Airbags 

Notice that most decomposition reactions require energy in the form of heat, electrical current, or light to make them happen. This is because compounds are normally stable and energy must be used to decompose them. *Ultraviolet* or *UV light* is light in the ultraviolet region of the spectrum. UV light carries more energy than visible light and can therefore initiate the decomposition of many compounds. (Light is discussed in more detail in Chapter 9.)

DISPLACEMENT REACTIONS. In a **displacement** or **single-displacement reaction**, one element displaces another in a compound. For example, when metallic zinc is added to a solution of copper(II) chloride, the zinc replaces the copper.

$$Zn(s) + CuCl_2(aq) \longrightarrow ZnCl_2(aq) + Cu(s)$$

A displacement reaction follows the general equation:

$$A + BC \longrightarrow AC + B$$

Other examples of displacement reactions include:

$$Mg(s) + 2\,HCl(aq) \longrightarrow MgCl_2(aq) + H_2(g)$$
$$2\,Na(s) + 2\,H_2O(l) \longrightarrow 2\,NaOH(aq) + H_2(g)$$

TUTORIAL Sodium and Potassium in Water

The last reaction can be seen more easily if we write water as $HOH(l)$.

$$2\,Na(s) + 2\,HOH(l) \longrightarrow 2\,NaOH(aq) + H_2(g)$$

▶ In a single-displacement reaction, one element displaces another in a compound. When zinc metal is immersed in a copper(II) chloride solution, the zinc atoms displace the copper atoms in solution.

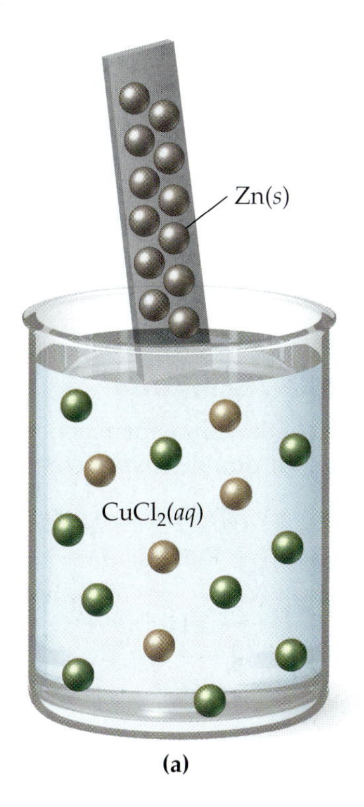

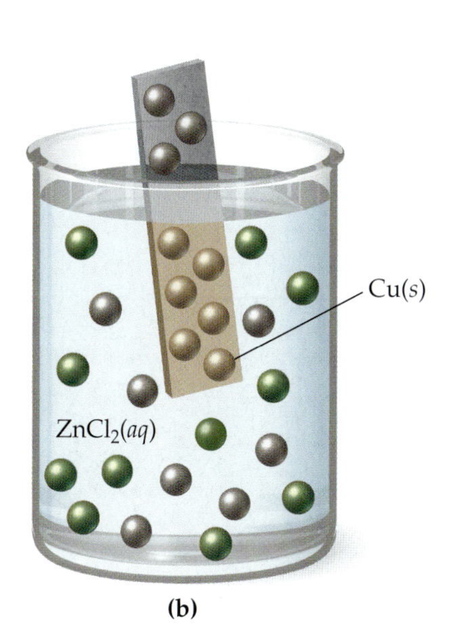

DOUBLE-DISPLACEMENT REACTIONS. In a **double-displacement reaction**, two elements or groups of elements in two different compounds exchange places to form two new compounds. For example, in aqueous solution, the silver in silver nitrate changes places with the sodium in sodium chloride to form solid silver chloride and aqueous sodium nitrate.

This double-displacement reaction is also a precipitation reaction.

$$AgNO_3(aq) + NaCl(aq) \longrightarrow AgCl(s) + NaNO_3(aq)$$

A double-displacement reaction follows the general form:

$$AB + CD \longrightarrow AD + CB$$

Other examples of double-displacement reactions include:

These double-displacement reactions are also acid–base reactions.

$$HCl(aq) + NaOH(aq) \longrightarrow H_2O(l) + NaCl(aq)$$
$$2\,HCl(aq) + Na_2CO_3(aq) \longrightarrow H_2CO_3(aq) + 2\,NaCl(aq)$$

As we learned in Section 7.7, $H_2CO_3(aq)$ is not stable and decomposes to form $H_2O(l) + CO_2(g)$, so the overall equation is:

This double-displacement reaction is also a gas evolution reaction and an acid–base reaction.

$$2\,HCl(aq) + Na_2CO_3(aq) \longrightarrow H_2O(l) + CO_2(g) + 2\,NaCl(aq)$$

Classification Flow Chart

A flow chart for this classification scheme of chemical reactions is as follows:

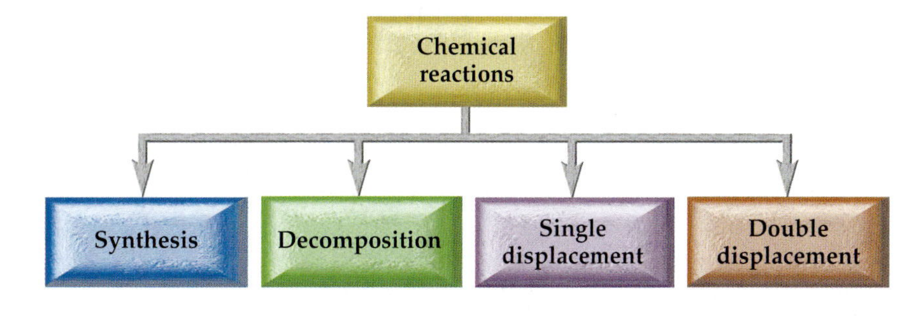

Of course no single classification scheme is perfect because all chemical reactions are unique in some sense. However, both classification schemes—one that focuses on the type of chemistry occurring and the other that focuses on what atoms or groups of atoms are doing—are helpful because they help us see differences and similarities among chemical reactions.

EXAMPLE 7.15 Classifying Chemical Reactions According to What Atoms Do

Classify each of the following reactions as a synthesis, decomposition, single-displacement, or double-displacement reaction.

(a) $Na_2O(s) + H_2O(l) \longrightarrow 2\,NaOH(aq)$
(b) $Ba(NO_3)_2(aq) + K_2SO_4(aq) \longrightarrow BaSO_4(s) + 2\,KNO_3(aq)$
(c) $2\,Al(s) + Fe_2O_3(s) \longrightarrow Al_2O_3(s) + 2\,Fe(l)$
(d) $2\,H_2O_2(aq) \longrightarrow 2\,H_2O(l) + O_2(g)$
(e) $Ca(s) + Cl_2(g) \longrightarrow CaCl_2(s)$

Solution:

(a) Synthesis; a more complex substance forms from two simpler ones.

(b) Double-displacment; Ba and K switch places to form two new compounds.

(c) Single-displacement; Al displaces Fe in Al_2O_3.

(d) Decomposition; a complex substance decomposes into simpler ones.

(e) Synthesis; a more complex substance forms from two simpler ones.

SKILLBUILDER 7.15 Classifying Chemical Reactions According to What Atoms Do

Classify each of the following reactions as a synthesis, decomposition, single-displacement, or double-displacement reaction.

(a) $2\,Al(s) + 2\,H_3PO_4(aq) \longrightarrow 2\,AlPO_4(aq) + 3\,H_2(g)$
(b) $CuSO_4(aq) + 2\,KOH(aq) \longrightarrow Cu(OH)_2(s) + K_2SO_4(aq)$
(c) $2\,K(s) + Br_2(l) \longrightarrow 2\,KBr(s)$
(d) $CuCl_2(aq) \xrightarrow[\text{current}]{\text{electrical}} Cu(s) + Cl_2(g)$

CONCEPTUAL CHECKPOINT 7.3

Precipitation reactions and acid–base reactions can both also be classified as:

(a) synthesis reactions
(b) decomposition reactions
(c) single-displacement reactions
(d) double-displacement reactions

Chemistry IN THE ENVIRONMENT

The Reactions Involved in Ozone Depletion

In Chapter 6, *Chemistry in the Environment: Chlorine in Chlorofluorocarbons*, we learned that chlorine atoms from chlorofluorocarbons deplete the ozone layer, which normally protects life on Earth from harmful ultraviolet light. Through research, chemists have discovered the reactions by which this depletion occurs.

Ozone normally forms in the upper atmosphere according to the following reaction.

(a) $O_2(g) + O(g) \longrightarrow O_3(g)$

When chlorofluorocarbons drift to the upper atmosphere, they are exposed to ultraviolet light and undergo the following reaction.

(b) $CF_2Cl_2(g) \xrightarrow{\text{UV light}} CF_2Cl(g) + Cl(g)$

UV light

Atomic chlorine then reacts with and depletes ozone according to the following cycle of reactions.

(c) $Cl(g) + O_3(g) \longrightarrow ClO(g) + O_2(g)$

(d) $O_3(g) \xrightarrow{\text{UV light}} O_2(g) + O(g)$

UV light

(e) $O(g) + ClO(g) \longrightarrow O_2(g) + Cl(g)$

Notice that in the final reaction, atomic chlorine is regenerated and can go through the cycle again to deplete more ozone. Through this cycle of reactions, a single chlorofluorocarbon molecule can deplete thousands of ozone molecules.

CAN YOU ANSWER THIS? *Classify each of these reactions (a–e) as a synthesis, decomposition, single-displacement, or double-displacement reaction.*

CHAPTER IN REVIEW

Chemical Principles	Relevance
Chemical Reactions: In a chemical reaction, one or more substances—either elements or compounds—change into different substances.	**Chemical Reactions:** Chemical reactions are central to many processes including transportation, energy generation, manufacture of household products, vision, and even life.
Evidence of a Chemical Reaction: The only absolute evidence for a chemical reaction is chemical analysis showing that one or more substances have changed into other substances. However, one or more of the following often accompany a chemical reaction: a color change; the formation of a solid or precipitate; the formation of a gas; the emission of light; and the emission or absorption of heat.	**Evidence of a Chemical Reaction:** We can often perceive the changes that accompany chemical reactions. In fact, we often use chemical reactions for the changes they produce. For example, we use the heat emitted by the combustion of fossil fuels to heat our homes, drive our cars, and generate electricity.
Chemical Equations: Chemical equations represent chemical reactions. They include formulas for the reactants (the substances present before the reaction) and for the products (the new substances formed by the reaction). Chemical equations must be balanced to reflect the conservation of matter in nature; atoms do not spontaneously appear or disappear.	**Chemical Equations:** Chemical equations allow us to represent and understand chemical reactions. For example, the equations for the combustion reactions of fossil fuels let us see that carbon dioxide, a gas that contributes to global warming, is one of the products of these reactions.

Aqueous Solutions and Solubility: Aqueous solutions are mixtures of a substance dissolved in water. If a substance dissolves in water it is soluble. Otherwise, it is insoluble.

Aqueous Solutions and Solubility: Aqueous solutions are common. For example, oceans, lakes, and most of the fluids in our bodies are aqueous solutions.

Some Specific Types of Reactions:

Precipitation reaction: A solid or precipitate forms upon mixing two aqueous solutions.
Acid–base reaction: Water forms upon mixing an acid and a base.
Gas evolution reaction: A gas forms upon mixing two aqueous solutions.
Redox reaction: Electrons are transferred from one substance to another.
Combustion reaction: A substance reacts with oxygen, emitting heat, and forming an oxygen-containing compound and, many times, water.

Some Specific Types of Reactions: Many of the specific types of reactions discussed here occur in aqueous solutions and are therefore important to living organisms. Acid–base reactions, for example, constantly occur in the blood of living organisms to maintain constant blood acidity levels. In humans, a small change in blood acidity levels would result in death, so the body carries out chemical reactions to prevent this. Combustion reactions are important because they are the main energy source for our society.

Classifying Chemical Reactions: Many chemical reactions can be classified into one of the following four categories according to what atoms or groups of atoms do:

synthesis ($A + B \longrightarrow AB$),
decomposition ($AB \longrightarrow A + B$),
single-displacement ($A + BC \longrightarrow AC + B$), and
double-displacement ($AB + CD \longrightarrow AD + CB$).

Classifying Chemical Reactions: We classify chemical reactions to better understand them and to better see similarities and differences among reactions.

Chemical Skills

Identifying a Chemical Reaction (Section 7.2)

To identify a chemical reaction, determine whether one or more of the initial substances changed into different substances. If so, a chemical reaction occurred. One or more of the following often accompanies a chemical reaction: a color change; the formation of a solid or precipitate; the formation of a gas; the emission of light; and the emission or absorption of heat.

Examples

EXAMPLE 7.16 Identifying a Chemical Reaction

Which of the following is a chemical reaction?

(a) Copper turns green on exposure to air.
(b) When sodium bicarbonate is combined with hydrochloric acid, bubbling is observed.
(c) Liquid water freezes to form solid ice.
(d) A pure copper penny forms bubbles of a dark brown gas when dropped into nitric acid. The nitric acid solution turns blue.

Solution:

(a) Chemical reaction, as evidenced by the color change.
(b) Chemical reaction, as evidenced by the evolution of a gas.
(c) Not a chemical reaction; solid ice is still water.
(d) Chemical reaction, as evidenced by the evolution of a gas and by a color change.

Writing Balanced Chemical Equations (Sections 7.3, 7.4)

To write balanced chemical equations, follow these steps.

1. Write a skeletal equation by writing chemical formulas for each of the reactants and products. (If a skeletal equation is provided, proceed to Step 2.)
2. If an element occurs in only one compound on both sides of the equation, balance that element first. If there is more than one such element, and the equation contains both metals and nonmetals, balance metals before nonmetals.
3. If an element occurs as a free element on either side of the chemical equation, balance that element last.
4. If the balanced equation contains coefficient fractions, clear these by multiplying the entire equation by the appropriate factor.
5. Check to make certain the equation is balanced by summing the total number of each type of atom on both sides of the equation.

Reminders

- Change only the *coefficients* to balance a chemical equation, *never the subscripts*. Changing the subscripts would change the compound itself.
- If the equation contains polyatomic ions that stay intact on both sides of the equation, balance the polyatomic ions as a group.

LIVE EXAMPLE 

EXAMPLE 7.17 Writing Balanced Chemical Equations

Write a balanced chemical equation for the reaction of solid vanadium(V) oxide with hydrogen gas to form solid vanadium(III) oxide and liquid water.

$$V_2O_5(s) + H_2(g) \longrightarrow V_2O_3(s) + H_2O(l)$$

Solution:

Vanadium occurs in only one compound on both sides of the equation. However, it is balanced, so we proceed to balance oxygen by placing a 2 in front of H_2O on the right side.

$$V_2O_5(s) + H_2(g) \longrightarrow V_2O_3(s) + \mathbf{2}\,H_2O(l)$$

Hydrogen occurs as a free element, so we balance it last by placing a 2 in front of H_2 on the left side.

$$V_2O_5(s) + \mathbf{2}\,H_2(g) \longrightarrow V_2O_3(s) + 2\,H_2O(l)$$

Finally, we check the equation.

$$V_2O_5(s) + 2\,H_2(g) \longrightarrow V_2O_3(s) + 2\,H_2O(l)$$

Reactants	Products
2 V atoms	2 V atoms
5 O atoms	5 O atoms
4 H atoms	4 H atoms

Determining Whether a Compound Is Soluble (Section 7.5)

To determine whether a compound is soluble, use the solubility rules in Table 7.2. It is easiest to begin by looking for those ions that always form soluble compounds (Li^+, Na^+, K^+, NH_4^+, NO_3^-, and $C_2H_3O_2^-$.) If a compound contains one of those, it is soluble. If it does not, look at the anion and determine whether it is mostly soluble (Cl^-, Br^-, I^-, or SO_4^{2-}) or mostly insoluble (OH^-, S^{2-}, CO_3^{2-}, or PO_4^{3-}). Look also at the cation to determine whether it is one of the exceptions.

EXAMPLE 7.18 Determining Whether a Compound Is Soluble

Determine whether each of the following compounds is soluble.

(a) $CuCO_3$
(b) $BaSO_4$
(c) $Fe(NO_3)_3$

Solution:

(a) Insoluble; compounds containing CO_3^{2-} are insoluble, and Cu^{2+} is not an exception.
(b) Insoluble; compounds containing SO_4^{2-} are usually soluble, but Ba^{2+} is an exception.
(c) Soluble; all compounds containing NO_3^- are soluble.

Predicting Precipitation Reactions (Section 7.6)

To predict whether a precipitation reaction occurs when two solutions are mixed and to write an equation for the reaction, follow these steps.

1. Write the formulas of the two compounds being mixed as reactants in a chemical equation.
2. Below the equation, write the formulas of the potentially insoluble products that could form from the reactants. These are obtained by combining the cation from one reactant with the anion from the other. Make sure to adjust the subscripts so that all formulas are charge-neutral.
3. Use the solubility rules to determine whether any of the potentially insoluble products are indeed insoluble.
4. If all of the potentially insoluble products are soluble, there will be no precipitate. Write *NO REACTION* next to the arrow.
5. If one or both of the potentially insoluble products are indeed insoluble, write their formula(s) as the product(s) of the reaction using (*s*) to indicate *solid*. Write any soluble products with (*aq*) to indicate *aqueous*.
6. Balance the equation.

LIVE EXAMPLE

EXAMPLE 7.19 Predicting Precipitation Reactions

Write an equation for the precipitation reaction that occurs, if any, when solutions of sodium phosphate and cobalt(II) chloride are mixed.

Solution:

$$Na_3PO_4(aq) + CoCl_2(aq) \longrightarrow$$

Potentially Insoluble Products:

$NaCl$ $\quad Co_3(PO_4)_2$

$NaCl$ is soluble.
$Co_3(PO_4)_2$ is insoluble.

$$Na_3PO_4(aq) + CoCl_2(aq) \longrightarrow Co_3(PO_4)_2(s) + NaCl(aq)$$

$$\mathbf{2}\,Na_3PO_4(aq) + \mathbf{3}\,CoCl_2(aq) \longrightarrow Co_3(PO_4)_2(s) + \mathbf{6}\,NaCl(aq)$$

Writing Complete Ionic and Net Ionic Equations (Section 7.7)

To write a complete ionic equation from a molecular equation, separate all aqueous ionic compounds into independent ions. Do not separate solid, liquid, or gaseous compounds.

To write a net ionic equation from a complete ionic equation, eliminate all species that do not change (spectator ions) in the course of the reaction.

EXAMPLE 7.20 Writing Complete Ionic and Net Ionic Equations

Write a complete ionic and net ionic equation for the following reaction.

$$2\,NH_4Cl(aq) + Hg_2(NO_3)_2(aq) \longrightarrow Hg_2Cl_2(s) + 2\,NH_4NO_3(aq)$$

Solution:

Complete ionic equation:

$$2\,NH_4^+(aq) + 2\,Cl^-(aq) + Hg_2^{2+}(aq) + 2\,NO_3^-(aq) \longrightarrow Hg_2Cl_2(s) + 2\,NH_4^-(aq) + 2\,NO_3^-(aq)$$

Net ionic equation:

$$2\,Cl^-(aq) + Hg_2^{2+}(aq) \longrightarrow Hg_2Cl_2(s)$$

Writing Equations for Acid–Base Reactions (Section 7.8)

When you see an acid and a base (see Table 7.3) as reactants in an equation, write a reaction in which the acid and the base react to form water and a salt.

EXAMPLE 7.21 Writing Equations for Acid–Base Reactions

Write an equation for the reaction that occurs when aqueous hydroiodic acid is mixed with aqueous barium hydroxide.

Solution:

$$\underset{\text{acid}}{2\,HI(aq)} + \underset{\text{base}}{Ba(OH)_2(aq)} \longrightarrow \underset{\text{water}}{2\,H_2O(l)} + \underset{\text{salt}}{BaI_2(aq)}$$

Writing Equations for Gas Evolution Reactions (Section 7.8)

See Table 7.4 to identify gas evolution reactions.

EXAMPLE 7.22 Writing Equations for Gas Evolution Reactions

Write an equation for the reaction that occurs when aqueous hydrobromic acid is mixed with aqueous potassium bisulfite.

Solution:

$HBr(aq) + KHSO_3(aq) \longrightarrow$
$H_2SO_3(aq) + KBr(aq) \longrightarrow H_2O(l) + SO_2(g) + KBr(aq)$

Identifying Redox Reactions (Section 7.9)

Redox reactions are those in which any of the following occurs:

- a substance reacts with elemental oxygen
- a metal reacts with a nonmetal
- one substance transfers electrons to another substance

EXAMPLE 7.23 Identifying Redox Reactions

Which of the following is a redox reaction?

(a) $4\,Fe(s) + 3\,O_2(g) \longrightarrow 2\,Fe_2O_3(s)$
(b) $CaO(s) + CO_2(g) \longrightarrow CaCO_3(s)$
(c) $AgNO_3(aq) + NaCl(aq) \longrightarrow AgCl(s) + NaNO_3(aq)$

Solution:

Only **(a)** is a redox reaction.

Writing Equations for Combustion Reactions (Section 7.9)

In a combustion reaction, a substance reacts with O_2 to form one or more oxygen-containing compounds and, in many cases, water.

EXAMPLE 7.24 Writing Equations for Combustion Reactions

Write a balanced equation for the combustion of gaseous ethane (C_2H_6), a minority component of natural gas.

Solution:

The skeletal equation is:

$$C_2H_6(g) + O_2(g) \longrightarrow CO_2(g) + H_2O(g)$$

The balanced equation is:

$$2\,C_2H_6(g) + 7\,O_2(g) \longrightarrow 4\,CO_2(g) + 6\,H_2O(g)$$

Classifying Chemical Reactions (Section 7.10)

Chemical reactions can be classified by inspection. The four major categories are:

Synthesis or combination

$$A + B \longrightarrow AB$$

Decomposition

$$AB \longrightarrow A + B$$

Single-displacement

$$A + BC \longrightarrow AC + B$$

Double-displacement

$$AB + CD \longrightarrow AD + CB$$

EXAMPLE 7.25 Classifying Chemical Reactions

Classify each of the following chemical reactions as a synthesis, decomposition, single-displacement, or double-displacement reaction.

(a) $2\,K(s) + Br_2(g) \longrightarrow 2\,KBr(s)$
(b) $Fe(s) + 2\,AgNO_3(aq) \longrightarrow Fe(NO_3)_2(aq) + 2\,Ag(s)$
(c) $CaSO_3(s) \xrightarrow[\text{heat}]{} CaO(s) + SO_2(g)$
(d) $CaCl_2(aq) + Li_2SO_4(aq) \longrightarrow CaSO_4(s) + 2\,LiCl(aq)$

Solution:

(a) Synthesis; KBr, a more complex substance, is formed from simpler substances.
(b) Single-displacement; Fe displaces Ag in $AgNO_3$.
(c) Decomposition; $CaSO_3$ decomposes into simpler substances.
(d) Double-displacement; Ca and Li switch places to form new compounds.

Key Terms

acid–base reaction [7.8]
aqueous solution [7.5]
balanced equation [7.3]
chemical equation [7.3]
chemical reaction [7.1]
combination reaction [7.10]
combustion reaction [7.9]
complete ionic equation [7.7]
decomposition reaction [7.10]
displacement reaction [7.10]
double-displacement reaction [7.10]
gas evolution reaction [7.8]
insoluble [7.5]
molecular equation [7.7]
net ionic equation [7.7]
neutralization reaction [7.8]
oxidation–reduction (redox) reaction [7.9]
precipitate [7.6]
precipitation reaction [7.6]
products [7.3]
reactants [7.3]
salt [7.8]
single-displacement reaction [7.10]
solubility rules [7.5]
soluble [7.5]
spectator ions [7.7]
strong electrolyte solution [7.5]
synthesis reaction [7.10]

Exercises

Questions

1. What is a chemical reaction? Give some examples.
2. If you could see atoms and molecules, what would you look for as conclusive evidence of a chemical reaction?
3. What are the main evidences of a chemical reaction?
4. What is a chemical equation? Give an example and identify the reactants and products.
5. What do each of the following abbreviations, often used in chemical equations, represent?
   (a) (g)
   (b) (l)
   (c) (s)
   (d) (aq)
6. To balance a chemical equation, adjust the _____ as necessary to make the numbers of each type of atom on both sides of the equation equal. Never adjust the _____ to balance a chemical equation.
7. List the number of each type of atom on both sides of each of the following equations. Are the equations balanced?
   (a) $2\,Ag_2O(s) + C(s) \longrightarrow CO_2(g) + 4\,Ag(s)$
   (b) $Pb(NO_3)_2(aq) + 2\,NaCl(aq) \longrightarrow PbCl_2(s) + 2\,NaNO_3(aq)$
   (c) $C_3H_8(g) + O_2(g) \longrightarrow 3\,CO_2(g) + 4\,H_2O(g)$
8. What is an aqueous solution? Give two examples.
9. What does it mean for a compound to be soluble? Insoluble?
10. How do ionic compounds dissolve in water?
11. How do polyatomic ions within ionic compounds dissolve in water?
12. What is a strong electrolyte solution?
13. What are the solubility rules and how are they useful?
14. What is a precipitation reaction? Give an example and identify the precipitate.
15. Will the precipitate in a precipitation reaction always be a compound that is soluble or insoluble? Explain.
16. Describe the differences between a molecular equation and a complete ionic equation. Give an example to illustrate the difference.
17. Describe the differences between a complete ionic and net ionic equation. Give an example to illustrate the difference.
18. What is an acid–base reaction? Give an example and identify the acid and the base.
19. What are the properties of acids and bases?
20. What is a gas evolution reaction? Give an example.
21. What is a redox reaction? Give an example.
22. What is a combustion reaction? Give an example.
23. What are two different ways to classify chemical reactions presented in Section 7.10? Explain the differences between them.
24. Describe each of the following general types of chemical reactions and write the generic equation for each.
    (a) synthesis
    (b) decomposition
    (c) single-displacement
    (d) double-displacement
25. Give two examples of a synthesis reaction.
26. Give two examples of a decomposition reaction.
27. Give two examples of a single-displacement reaction.
28. Give two examples of a double-displacement reaction.

Problems

Evidence of Chemical Reactions

29. Which of the following is a chemical reaction? Why?
    - (a) Solid copper deposits on a piece of aluminum foil when the foil is placed in a blue copper nitrate solution. The blue color of the solution fades.
    - (b) Liquid ethyl alcohol turns into a solid when placed in a low-temperature freezer.
    - (c) A white precipitate forms when solutions of barium nitrate and sodium sulfate are mixed.
    - (d) A mixture of sugar and water bubbles when yeasts are added. After several days, the sugar is gone and ethyl alcohol is found in the water.

30. Which of the following is a chemical reaction? Why?
    - (a) Propane forms a flame and emits heat as it burns.
    - (b) Acetone feels cold as it evaporates from the skin.
    - (c) Bubbling is observed when potassium carbonate and hydrochloric acid solutions are mixed.
    - (d) Heat is felt when a warm object is placed in your hand.

31. Vinegar forms bubbles when it is poured onto the calcium deposits on a faucet, and some of the calcium dissolves. Has a chemical reaction occurred? Why or why not?

32. When a chemical drain opener is added to a clogged sink, bubbles form and the water in the sink gets warmer. Has a chemical reaction occurred? Why or why not?

33. When a commercial hair bleaching mixture is applied to brown hair, it turns blond. Has a chemical reaction occurred? Why or why not?

34. When water is boiled in a pot, it bubbles. Has a chemical reaction occurred? Why or why not?

Writing and Balancing Chemical Equations

35. Write a balanced chemical equation for each of the following:
    - (a) Solid copper reacts with solid sulfur to form solid copper(I) sulfide.
    - (b) Sulfur dioxide gas reacts with oxygen gas to form sulfur trioxide gas.
    - (c) Aqueous hydrochloric acid reacts with solid manganese(IV) oxide to form aqueous manganese(II) chloride, liquid water, and chlorine gas.
    - (d) Liquid benzene (C_6H_6) reacts with gaseous oxygen to form carbon dioxide and liquid water.

36. Write a balanced chemical equation for each of the following:
    - (a) Solid lead(II) sulfide reacts with aqueous hydrochloric acid to form solid lead(II) chloride and dihydrogen sulfide gas.
    - (b) Gaseous carbon monoxide reacts with hydrogen gas to form gaseous methane (CH_4) and liquid water.
    - (c) Solid iron(III) oxide reacts with hydrogen gas to form solid iron and liquid water.
    - (d) Gaseous ammonia (NH_3) reacts with gaseous oxygen to form gaseous nitrogen monoxide and gaseous water.

37. Write a balanced chemical equation for each of the following:
    - (a) Solid magnesium reacts with aqueous copper(I) nitrate to form aqueous magnesium nitrate and solid copper.
    - (b) Gaseous dinitrogen pentoxide decomposes to form nitrogen dioxide and oxygen gas.
    - (c) Solid calcium reacts with aqueous nitric acid to form aqueous calcium nitrate and hydrogen gas.
    - (d) Liquid methanol (CH_3OH) reacts with oxygen gas to form gaseous carbon dioxide and gaseous water.

38. Write a balanced chemical equation for each of the following:
    - (a) Gaseous acetylene (C_2H_2) reacts with oxygen gas to form gaseous carbon dioxide and gaseous water.
    - (b) Chlorine gas reacts with aqueous potassium iodide to form solid iodine and aqueous potassium chloride.
    - (c) Solid lithium oxide reacts with liquid water to form aqueous lithium hydroxide.
    - (d) Gaseous carbon monoxide reacts with oxygen gas to form carbon dioxide gas.

39. Hydrogen has been widely suggested as a potential fuel to replace fossil fuels. Some scientists are trying to foresee any potential problems that might be associated with a hydrogen-based economy. One group of scientists have calculated that the amount of atmospheric hydrogen could increase by a factor of four due to leaks in hydrogen transport and storage. Upper atmospheric hydrogen gas reacts with oxygen gas to form liquid water. An increase in upper atmospheric water would enhance processes that release atmospheric chlorine atoms. The gaseous chlorine atoms would then react with gaseous ozone (O_3) to form gaseous chlorine monoxide and gaseous oxygen, resulting in the depletion of ozone. Write balanced chemical equations for the two reactions described here.

40. Waste water from certain industrial chemical processes contains aqueous Hg_2^{2+} ions. Since the mercury ion is toxic, it is removed from the waste water by reaction with aqueous sodium sulfide. The products of the reaction are solid mercury(I) sulfide and aqueous sodium ions. Write a balanced equation for this reaction.

41. When solid sodium is added to liquid water, it reacts with the water to produce hydrogen gas and aqueous sodium hydroxide. Write a balanced chemical equation for this reaction.

42. When iron rusts, solid iron reacts with gaseous oxygen to form solid iron(III) oxide. Write a balanced chemical equation for this reaction.

43. Sulfuric acid in acid rain forms when gaseous sulfur dioxide pollutant reacts with gaseous oxygen and liquid water to form aqueous sulfuric acid. Write a balanced chemical equation for this reaction.

44. Nitric acid in acid rain forms when gaseous nitrogen dioxide pollutant reacts with gaseous oxygen and liquid water to form aqueous nitric acid. Write a balanced chemical equation for this reaction.

45. Write a balanced chemical equation for the reaction of solid vanadium(V) oxide with hydrogen gas to form solid vanadium(III) oxide and liquid water.

46. Write a balanced chemical equation for the reaction of gaseous nitrogen dioxide with hydrogen gas to form gaseous ammonia and liquid water.

47. Write a balanced chemical equation for the fermentation of sugar ($C_{12}H_{22}O_{11}$) by yeasts in which the aqueous sugar reacts with water to form aqueous ethyl alcohol (C_2H_5OH) and carbon dioxide gas.

48. Write a balanced chemical equation for the photosynthesis reaction in which gaseous carbon dioxide and liquid water react in the presence of chlorophyll to produce aqueous glucose ($C_6H_{12}O_6$) and oxygen gas.

49. Balance each of the following chemical equations.

(a) $N_2H_4(l) \longrightarrow NH_3(g) + N_2(g)$

(b) $H_2(g) + N_2(g) \longrightarrow NH_3(g)$

(c) $Cu_2O(s) + C(s) \longrightarrow Cu(s) + CO(g)$

(d) $H_2(g) + Cl_2(g) \longrightarrow HCl(g)$

50. Balance each of the following chemical equations.

(a) $Na_2S(aq) + Cu(NO_3)_2(aq) \longrightarrow NaNO_3(aq) + CuS(s)$

(b) $HCl(aq) + O_2(g) \longrightarrow H_2O(l) + Cl_2(g)$

(c) $H_2(g) + O_2(g) \longrightarrow H_2O(l)$

(d) $FeS(s) + HCl(aq) \longrightarrow FeCl_2(aq) + H_2S(g)$

51. Balance each of the following chemical equations.

(a) $BaO_2(s) + H_2SO_4(aq) \longrightarrow BaSO_4(s) + H_2O_2(aq)$

(b) $Co(NO_3)_3(aq) + (NH_4)_2S(aq) \longrightarrow Co_2S_3(s) + NH_4NO_3(aq)$

(c) $Li_2O(s) + H_2O(l) \longrightarrow LiOH(aq)$

(d) $Hg_2(C_2H_3O_2)_2(aq) + KCl(aq) \longrightarrow Hg_2Cl_2(s) + KC_2H_3O_2(aq)$

52. Balance each of the following chemical equations.

(a) $MnO_2(s) + HCl(aq) \longrightarrow Cl_2(g) + MnCl_2(aq) + H_2O(l)$

(b) $CO_2(g) + CaSiO_3(s) + H_2O(l) \longrightarrow SiO_2(s) + Ca(HCO_3)_2(aq)$

(c) $Fe(s) + S(l) \longrightarrow Fe_2S_3(s)$

(d) $NO_2(g) + H_2O(l) \longrightarrow HNO_3(aq) + NO(g)$

53. Determine whether each of the following chemical equations is correctly balanced. If not, correct it.
(a) $Rb(s) + H_2O(l) \longrightarrow RbOH(aq) + H_2(g)$
(b) $2\,N_2H_4(g) + N_2O_4(g) \longrightarrow 3\,N_2(g) + 4\,H_2O(g)$
(c) $NiS(s) + O_2(g) \longrightarrow NiO(s) + SO_2(g)$
(d) $PbO(s) + 2\,NH_3(g) \longrightarrow Pb(s) + N_2(g) + H_2O(l)$

54. Determine whether each of the following chemical equations is correctly balanced. If not, correct it.
(a) $SiO_2(s) + 4\,HF(aq) \longrightarrow SiF_4(g) + 2\,H_2O(l)$
(b) $2\,Cr(s) + 3\,O_2(g) \longrightarrow Cr_2O_3(s)$
(c) $Al_2S_3(s) + H_2O(l) \longrightarrow 2\,Al(OH)_3(s) + 3\,H_2S(g)$
(d) $Fe_2O_3(s) + CO(g) \longrightarrow 2\,Fe(s) + CO_2(g)$

55. Human cells obtain energy from a reaction called respiration. Balance the skeletal equation for respiration.

$$C_6H_{12}O_6(aq) + O_2(g) \longrightarrow CO_2(g) + H_2O(l)$$

56. Propane camping stoves produce heat by the combustion of gaseous propane (C_3H_8). Balance the skeletal equation for the combustion of propane.

$$C_3H_8(g) + O_2(g) \longrightarrow CO_2(g) + H_2O(g)$$

57. Catalytic converters work to remove nitrogen oxides and carbon monoxide from exhaust. Balance the skeletal equation for one of the reactions that occurs in a catalytic converter.

$$NO(g) + CO(g) \longrightarrow N_2(g) + CO_2(g)$$

58. Billions of pounds of urea are produced annually for use as a fertilizer. Balance the skeletal equation for the synthesis of urea.

$$NH_3(g) + CO_2(g) \longrightarrow CO(NH_2)_2(s) + H_2O(l)$$

Solubility

59. Determine whether each of the following compounds is soluble or insoluble. For the soluble compounds, write the ions present in solution.
(a) $NaNO_3$
(b) $Pb(C_2H_3O_2)_2$
(c) $CuCO_3$
(d) $(NH_4)_2S$

60. Determine whether each of the following compounds is soluble or insoluble. For the soluble compounds, write the ions present in solution.
(a) AgI
(b) $AgNO_3$
(c) $CoPO_4$
(d) Na_3PO_4

61. Pair each cation on the left with an anion on the right that will form an *insoluble* compound and write a formula for the insoluble compound. Use each anion only once.

Ag^+	SO_4^{2-}
Ba^{2+}	Cl^-
Cu^{2+}	CO_3^{2-}
Fe^{3+}	S^{2-}

62. Pair each cation on the left with an anion on the right that will form a *soluble* compound and write a formula for the soluble compound. Use each anion only once.

Na^+	NO_3^-
Sr^{2+}	SO_4^{2-}
Co^{2+}	S^{2-}
Pb^{2+}	CO_3^{2-}

63. Determine whether each of the following compounds is in the correct column. If not, move it to the correct column.

Soluble	*Insoluble*
K_2S	K_2SO_4
$PbSO_4$	Hg_2I_2
BaS	$Cu_3(PO_4)_2$
$PbCl_2$	MgS
Hg_2Cl_2	$CaSO_4$
NH_4Cl	SrS
Na_2CO_3	Li_2S

64. Determine whether each of the following compounds is in the correct column. If not, move it to the correct column.

Soluble	*Insoluble*
LiOH	$CaCl_2$
Na_2CO_3	$Cu(OH)_2$
AgCl	$Ca(C_2H_3O_2)_2$
K_3PO_4	$SrSO_4$
CuI_2	Hg_2Br_2
$Pb(NO_3)_2$	$PbBr_2$
$CoCO_3$	PbI_2

Precipitation Reactions

65. Complete and balance each of the following equations. If no reaction occurs, write *NO REACTION*.
   (a) $NH_4Cl(aq) + AgNO_3(aq) \longrightarrow$
   (b) $NaCl(aq) + CaS(aq) \longrightarrow$
   (c) $CrCl_2(aq) + Li_2CO_3(aq) \longrightarrow$
   (d) $KOH(aq) + FeCl_3(aq) \longrightarrow$

66. Complete and balance each of the following equations. If no reaction occurs, write *NO REACTION*.
   (a) $KNO_3(aq) + NaCl(aq) \longrightarrow$
   (b) $KCl(aq) + Hg_2(C_2H_3O_2)_2(aq) \longrightarrow$
   (c) $(NH_4)_2SO_4(aq) + BaCl_2(aq) \longrightarrow$
   (d) $LiI(aq) + SrS(aq) \longrightarrow$

67. Write a molecular equation for the precipitation reaction (if any) that occurs when the following solutions are mixed. If no reaction occurs, write *NO REACTION*.
   (a) sodium carbonate and lead(II) nitrate
   (b) potassium sulfate and lead(II) acetate
   (c) copper(II) nitrate and barium sulfide
   (d) calcium nitrate and sodium iodide

68. Write a molecular equation for the precipitation reaction (if any) that occurs when the following solutions are mixed. If no reaction occurs, write *NO REACTION*.
   (a) potassium chloride and lead(II) acetate
   (b) lithium sulfate and strontium chloride
   (c) potassium bromide and calcium sulfide
   (d) chromium(III) nitrate and potassium phosphate

69. Determine whether each of the following equations for precipitation reactions is correct. If not, write the correct equation. If no reaction occurs, write *NO REACTION*.
   (a) $Ba(NO_3)_2(aq) + (NH_4)_2SO_4(aq) \longrightarrow BaSO_4(s) + 2\,NH_4NO_3(aq)$
   (b) $BaS(aq) + 2\,KCl(aq) \longrightarrow BaCl_2(s) + K_2S(aq)$
   (c) $2\,KI(aq) + Pb(NO_3)_2(aq) \longrightarrow PbI_2(s) + 2\,KNO_3(aq)$
   (d) $Pb(NO_3)_2(aq) + 2\,LiCl(aq) \longrightarrow 2\,LiNO_3(s) + PbCl_2(aq)$

70. Determine whether each of the following equations for precipitation reactions is correct. If not, write the correct equation. If no reaction occurs, write *NO REACTION*.
   (a) $AgNO_3(aq) + NaCl(aq) \longrightarrow NaCl(s) + AgNO_3(aq)$
   (b) $K_2SO_4(aq) + Co(NO_3)_2(aq) \longrightarrow CoSO_4(s) + 2\,KNO_3(aq)$
   (c) $Cu(NO_3)_2(aq) + (NH_4)_2S(aq) \longrightarrow CuS(s) + 2\,NH_4NO_3(aq)$
   (d) $Hg_2(NO_3)_2(aq) + 2\,LiCl(aq) \longrightarrow Hg_2Cl_2(s) + 2\,LiNO_3(aq)$

Ionic and Net Ionic Equations

71. Identify the spectator ions in the following complete ionic equation.

$$Pb^{2+}(aq) + 2\,C_2H_3O_2^-(aq) + 2\,K^+(aq) + 2\,Br^-(aq) \longrightarrow PbBr_2(s) + 2\,C_2H_3O_2^-(aq) + 2\,K^+(aq)$$

72. Identify the spectator ions in the following complete ionic equation.

$$2\,Li^+(aq) + S^{2-}(aq) + Mg^{2+}(aq) + 2\,Cl^-(aq) \longrightarrow MgS(s) + 2\,Li^+(aq) + 2\,Cl^-(aq)$$

73. Write balanced complete ionic and net ionic equations for each of the following reactions.
   (a) $AgNO_3(aq) + KCl(aq) \longrightarrow AgCl(s) + KNO_3(aq)$
   (b) $CaS(aq) + CuCl_2(aq) \longrightarrow CuS(s) + CaCl_2(aq)$
   (c) $NaOH(aq) + HNO_3(aq) \longrightarrow H_2O(l) + NaNO_3(aq)$
   (d) $2\,K_3PO_4(aq) + 3\,NiCl_2(aq) \longrightarrow Ni_3(PO_4)_2(s) + 6\,KCl(aq)$

74. Write balanced complete ionic and net ionic equations for each of the following reactions.
   (a) $HI(aq) + KOH(aq) \longrightarrow H_2O(l) + KI(aq)$
   (b) $Na_2SO_4(aq) + CaI_2(aq) \longrightarrow CaSO_4(s) + 2\,NaI(aq)$
   (c) $2\,HC_2H_3O_2(aq) + Na_2CO_3(aq) \longrightarrow H_2O(l) + CO_2(g) + 2\,NaC_2H_3O_2(aq)$
   (d) $NH_4Cl(aq) + NaOH(aq) \longrightarrow H_2O(l) + NH_3(g) + NaCl(aq)$

75. Mercury (I) ions (Hg_2^{2+}) can be removed from solution by precipitation with Cl^-. Suppose a solution contains aqueous $Hg_2(NO_3)_2$. Write complete ionic and net ionic equations to show the reaction of aqueous $Hg_2(NO_3)_2$ with aqueous sodium chloride to form solid Hg_2Cl_2 and aqueous sodium nitrate.

76. Lead ions can be removed from solution by precipitation with sulfate ions. Suppose a solution contains lead(II) nitrate. Write a complete ionic and net ionic equation to show the reaction of aqueous lead(II) nitrate with aqueous potassium sulfate to form solid lead(II) sulfate and aqueous potassium nitrate.

77. Write complete ionic and net ionic equations for each of the reactions in Problem 67.

78. Write complete ionic and net ionic equations for each of the reactions in Problem 68.

Acid–Base and Gas Evolution Reactions

79. When a hydrochloric acid solution is combined with a potassium hydroxide solution, an acid–base reaction occurs. Write a balanced molecular equation and a net ionic equation for this reaction.

80. A beaker of nitric acid is neutralized with calcium hydroxide. Write a balanced molecular equation and a net ionic equation for this reaction.

81. Complete and balance each of the following equations for acid–base reactions.
   (a) $HCl(aq) + Ba(OH)_2(aq) \longrightarrow$
   (b) $H_2SO_4(aq) + KOH(aq) \longrightarrow$
   (c) $HClO_4(aq) + NaOH(aq) \longrightarrow$

82. Complete and balance each of the following equations for acid–base reactions.
   (a) $HC_2H_3O_2(aq) + Ca(OH)_2(aq) \longrightarrow$
   (b) $HBr(aq) + LiOH(aq) \longrightarrow$
   (c) $H_2SO_4(aq) + Ba(OH)_2(aq) \longrightarrow$

83. Complete and balance each of the following equations for gas evolution reactions.
   (a) $HBr(aq) + NaHCO_3(aq) \longrightarrow$
   (b) $NH_4I(aq) + KOH(aq) \longrightarrow$
   (c) $HNO_3(aq) + K_2SO_3(aq) \longrightarrow$
   (d) $HI(aq) + Li_2S(aq) \longrightarrow$

84. Complete and balance each of the following equations for gas evolution reactions.
   (a) $HClO_4(aq) + K_2CO_3(aq) \longrightarrow$
   (b) $HC_2H_3O_2(aq) + LiHSO_3(aq) \longrightarrow$
   (c) $(NH_4)_2SO_4(aq) + Ca(OH)_2(aq) \longrightarrow$
   (d) $HCl(aq) + ZnS(s) \longrightarrow$

Oxidation–Reduction and Combustion

85. Which of the following reactions are redox reactions?
   (a) $Ba(NO_3)_2(aq) + K_2SO_4(aq) \longrightarrow BaSO_4(s) + 2\,KNO_3(aq)$
   (b) $Ca(s) + Cl_2(g) \longrightarrow CaCl_2(s)$
   (c) $HCl(aq) + NaOH(aq) \longrightarrow H_2O(l) + NaCl(aq)$
   (d) $Zn(s) + Fe^{2+}(aq) \longrightarrow Zn^{2+}(aq) + Fe(s)$

86. Which of the following reactions are redox reactions?
   (a) $Al(s) + 3\,Ag^+(aq) \longrightarrow Al^{3+}(aq) + 3\,Ag(s)$
   (b) $4\,K(s) + O_2(g) \longrightarrow 2\,K_2O(s)$
   (c) $SO_3(g) + H_2O(l) \longrightarrow H_2SO_4(aq)$
   (d) $Mg(s) + Br_2(l) \longrightarrow MgBr_2(s)$

87. Complete and balance each of the following equations for combustion reactions.
   (a) $C_2H_6(g) + O_2(g) \longrightarrow$
   (b) $Ca(s) + O_2(g) \longrightarrow$
   (c) $C_3H_8O(l) + O_2(g) \longrightarrow$
   (d) $C_4H_{10}S(l) + O_2(g) \longrightarrow$

88. Complete and balance each of the following equations for combustion reactions.
   (a) $S(s) + O_2(g) \longrightarrow$
   (b) $C_7H_{16}(l) + O_2(g) \longrightarrow$
   (c) $C_4H_{10}O(l) + O_2(g) \longrightarrow$
   (d) $CS_2(l) + O_2(g) \longrightarrow$

89. Write a balanced chemical equation for the synthesis reaction of $Cl_2(g)$ with each of the following:
   (a) K(s)
   (b) Ca(s)
   (c) Al(s)
   (d) Zn(s)

90. Write a balanced chemical equation for the synthesis reaction of $F_2(g)$ with each of the following:
   (a) Sr(s)
   (b) Ga(s)
   (c) Li(s)
   (d) Ag(s)

Classifying Chemical Reactions by What Atoms Do

91. Classify each of the following chemical reactions as a synthesis, decomposition, single-displacement, or double-displacement reaction.
   (a) $K_2S(aq) + Co(NO_3)_2(aq) \longrightarrow 2\,KNO_3(aq) + CoS(s)$
   (b) $3\,H_2(g) + N_2(g) \longrightarrow 2\,NH_3(g)$
   (c) $Zn(s) + CoCl_2(aq) \longrightarrow ZnCl_2(aq) + Co(s)$
   (d) $CH_3Br(g) \xrightarrow[\text{UV light}]{} CH_3(g) + Br(g)$

92. Classify each of the following chemical reactions as a synthesis, decomposition, single-displacement, or double-displacement reaction.
   (a) $CaSO_4(s) \xrightarrow[\text{heat}]{} CaO(s) + SO_3(g)$
   (b) $2\,Na(s) + O_2(g) \longrightarrow Na_2O_2(s)$
   (c) $Pb(s) + 2\,AgNO_3(aq) \longrightarrow Pb(NO_3)_2(aq) + 2\,Ag(s)$
   (d) $HI(aq) + NaOH(aq) \longrightarrow H_2O(l) + NaI(aq)$

93. NO is a pollutant emitted by motor vehicles. It is formed by the following reaction.
(a) $N_2(g) + O_2(g) \longrightarrow 2\,NO(g)$
Once in the atmosphere, NO (through a series of reactions) adds one oxygen atom to form NO_2. NO_2 then interacts with UV light according to the following reaction.
(b) $NO_2(g) \xrightarrow[\text{UV light}]{} NO(g) + O(g)$
These freshly formed oxygen atoms then react with O_2 in the air to form ozone (O_3), a main component of photochemical smog.
(c) $O(g) + O_2(g) \longrightarrow O_3(g)$
Classify each of the preceding reactions as a synthesis, decomposition, single-displacement, or double-displacement reaction.

94. A main source of sulfur oxide pollutants are smelters where sulfide ores are converted into metals. The first step in this process is the reaction of the sulfide ore with oxygen in reactions such as:
(a) $2\,PbS(s) + 3\,O_2(g) \longrightarrow 2\,PbO(s) + 2\,SO_2(g)$
Sulfur dioxide can then react with oxygen in air to form sulfur trioxide.
(b) $2\,SO_2(g) + O_2(g) \longrightarrow 2\,SO_3(g)$
Sulfur trioxide can then react with water from rain to form sulfuric acid that falls as acid rain.
(c) $SO_3(g) + H_2O(l) \longrightarrow H_2SO_4(aq)$
Classify each of the preceding reactions as a synthesis, decomposition, single-displacement, or double-displacement reaction.

Cumulative Problems

95. Predict the products of each of these reactions and write balanced complete ionic and net ionic equations for each. If no reaction occurs, write *NO REACTION.*
(a) $NaI(aq) + Hg_2(NO_3)_2(aq) \longrightarrow$
(b) $HClO_4(aq) + Ba(OH)_2(aq) \longrightarrow$
(c) $Li_2CO_3(aq) + NaCl(aq) \longrightarrow$
(d) $HCl(aq) + Li_2CO_3(aq) \longrightarrow$

96. Predict the products of each of these reactions and write balanced complete ionic and net ionic equations for each. If no reaction occurs, write *NO REACTION.*
(a) $LiCl(aq) + AgNO_3(aq) \longrightarrow$
(b) $H_2SO_4(aq) + Li_2SO_3(aq) \longrightarrow$
(c) $HC_2H_3O_2(aq) + Ca(OH)_2(aq) \longrightarrow$
(d) $HCl(aq) + KBr(aq) \longrightarrow$

97. Predict the products of each of these reactions and write balanced complete ionic and net ionic equations for each. If no reaction occurs, write *NO REACTION.*
(a) $BaS(aq) + NH_4Cl(aq) \longrightarrow$
(b) $NaC_2H_3O_2(aq) + KCl(aq) \longrightarrow$
(c) $KHSO_3(aq) + HNO_3(aq) \longrightarrow$
(d) $MnCl_3(aq) + K_3PO_4(aq) \longrightarrow$

98. Predict the products of each of these reactions and write balanced complete ionic and net ionic equations for each. If no reaction occurs, write *NO REACTION.*
(a) $H_2SO_4(aq) + HNO_3(aq) \longrightarrow$
(b) $NaOH(aq) + LiOH(aq) \longrightarrow$
(c) $Cr(NO_3)_3(aq) + LiOH(aq) \longrightarrow$
(d) $HCl(aq) + Hg_2(NO_3)_2(aq) \longrightarrow$

99. Predict the type of reaction (if any) that occurs between each of the following substances. Write balanced molecular equations for each. If no reaction occurs, write *NO REACTION.*
(a) aqueous potassium hydroxide and aqueous acetic acid
(b) aqueous hydrobromic acid and aqueous potassium carbonate
(c) gaseous hydrogen and gaseous oxygen
(d) aqueous ammonium chloride and aqueous lead(II) nitrate

100. Predict the type of reaction (if any) that occurs between each of the following substances. Write balanced molecular equations for each. If no reaction occurs, write *NO REACTION.*
(a) aqueous hydrochloric acid and aqueous copper(II) nitrate
(b) liquid pentanol ($C_5H_{12}O$) and gaseous oxygen
(c) aqueous ammonium chloride and aqueous calcium hydroxide
(d) aqueous strontium sulfide and aqueous copper(II) sulfate

101. Classify each of the following reactions in as many ways as possible.
(a) $2\ Al(s) + 3\ Cu(NO_3)_2(aq) \longrightarrow 2\ Al(NO_3)_3(aq) + 3\ Cu(s)$
(b) $HBr(aq) + KHSO_3(aq) \longrightarrow H_2O(l) + SO_2(g) + NaBr(aq)$
(c) $2\ HI(aq) + Na_2S(aq) \longrightarrow H_2S(g) + 2\ NaI(aq)$
(d) $K_2CO_3(aq) + FeBr_2(aq) \longrightarrow FeCO_3(s) + 2\ KBr(aq)$

102. Classify each of the following reactions in as many ways as possible.
(a) $NaCl(aq) + AgNO_3(aq) \longrightarrow AgCl(s) + NaNO_3(aq)$
(b) $2\ Rb(s) + Br_2(g) \longrightarrow 2\ RbBr(s)$
(c) $Zn(s) + NiBr_2(aq) \longrightarrow Ni(s) + ZnBr_2(aq)$
(d) $Ca(s) + 2\ H_2O(l) \longrightarrow Ca(OH)_2(aq) + H_2(g)$

103. Hard water often contains dissolved Ca^{2+} and Mg^{2+} ions. One way to soften water is to add phosphates. The phosphate ion forms insoluble precipitates with calcium and magnesium ions, removing them from solution. Suppose that a solution contains aqueous calcium chloride and aqueous magnesium nitrate. Write molecular, complete ionic, and net ionic equations showing how the addition of sodium phosphate precipitates the calcium and magnesium ions.

104. Lakes that have been acidified by acid rain (HNO_3 and H_2SO_4) can be neutralized by a process called *liming*, in which limestone ($CaCO_3$) is added to the acidified water. Write ionic and net ionic equations to show how limestone reacts with HNO_3 and H_2SO_4 to neutralize them. How would you be able to tell if the neutralization process was working?

105. What solution can you add to each of the following cation mixtures to precipitate one cation while keeping the other cation in solution? Write a net ionic equation for the precipitation reaction that occurs.
(a) $Fe^{2+}(aq)$ and $Pb^{2+}(aq)$
(b) $K^{+}(aq)$ and $Ca^{2+}(aq)$
(c) $Ag^{+}(aq)$ and $Ba^{2+}(aq)$
(d) $Cu^{2+}(aq)$ and $Hg_2{}^{2+}(aq)$

106. What solution can you add to each of the following cation mixtures to precipitate one cation while keeping the other cation in solution? Write a net ionic equation for the precipitation reaction that occurs.
(a) $Sr^{2+}(aq)$ and $Hg_2{}^{2+}(aq)$
(b) $NH_4{}^{+}(aq)$ and $Ca^{2+}(aq)$
(c) $Ba^{2+}(aq)$ and $Mg^{2+}(aq)$
(d) $Ag^{+}(aq)$ and $Zn^{2+}(aq)$

107. A solution contains one or more of the following ions: Ag^{+}, Ca^{2+}, and Cu^{2+}. When sodium chloride is added to the solution, no precipitate occurs. When sodium sulfate is added to the solution, a white precipitate occurs. The precipitate is filtered off and sodium carbonate is added to the remaining solution, producing a precipitate. Which ions were present in the original solution? Write net ionic equations for the formation of each of the precipitates observed.

108. A solution contains one or more of the following ions: $Hg_2{}^{2+}$, Ba^{2+}, and Fe^{2+}. When potassium chloride is added to the solution, a precipitate forms. The precipitate is filtered off and potassium sulfate is added to the remaining solution, producing no precipitate. When potassium carbonate is added to the remaining solution, a precipitate occurs. Which ions were present in the original solution? Write net ionic equations for the formation of each of the precipitates observed.

Highlight Problems

109. The following are molecular views of two different possible mechanisms by which an automobile air bag might function. One of these mechanisms involves a chemical reaction and the other does not. By looking at the molecular views, can you tell which mechanism operates via a chemical reaction?

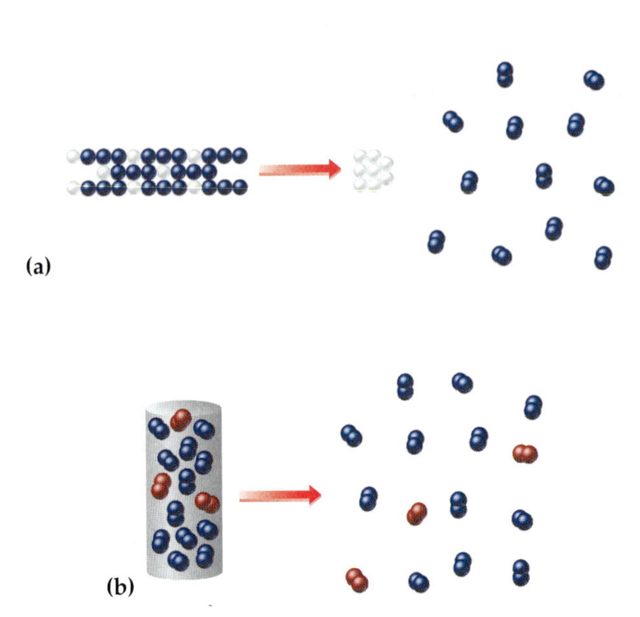

▲ When an airbag is detonated, the bag inflates. These figures show two possible ways in which the inflation may happen.

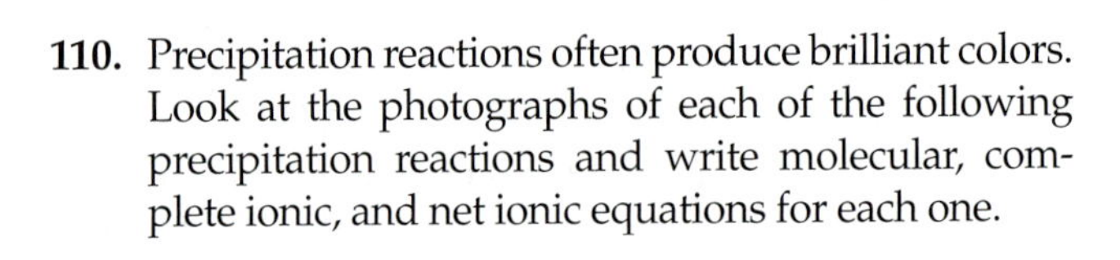

110. Precipitation reactions often produce brilliant colors. Look at the photographs of each of the following precipitation reactions and write molecular, complete ionic, and net ionic equations for each one.

▲ **(a)** The precipitation reaction that occurs when aqueous iron(III) nitrate is added to aqueous sodium hydroxide.

▲ **(b)** The precipitation reaction that occurs when aqueous cobalt(II) chloride is added to aqueous potassium hydroxide.

▲ **(c)** The precipitation reaction that occurs when aqueous $AgNO_3$ is added to aqueous sodium iodide.

Answers to Skillbuilder Exercises

Skillbuilder 7.1 **(a)** Chemical reaction; heat and light are emitted. **(b)** Not a chemical reaction; gaseous and liquid butane are both butane. **(c)** Chemical reaction; heat and light are emitted. **(d)** Not a chemical reaction; solid dry ice is made of carbon dioxide, which sublimes (evaporates) as carbon dioxide gas.

Skillbuilder 7.2
$2\,Cr_2O_3(s) + 3\,C(s) \longrightarrow 4\,Cr(s) + 3\,CO_2(g)$

Skillbuilder 7.3
$2\,C_4H_{10}(g) + 13\,O_2(g) \longrightarrow 8\,CO_2(g) + 10\,H_2O(g)$

Skillbuilder 7.4
$Pb(C_2H_3O_2)_2(aq) + 2\,KI(aq) \longrightarrow PbI_2(s) + 2\,KC_2H_3O_2(aq)$

Skillbuilder 7.5
$4\,HCl(g) + O_2(g) \longrightarrow 2\,H_2O(l) + 2\,Cl_2(g)$

Skillbuilder 7.6 **(a)** insoluble **(b)** soluble **(c)** insoluble **(d)** soluble

Skillbuilder 7.7
$2\,KOH(aq) + NiBr_2(aq) \longrightarrow Ni(OH)_2(s) + 2\,KBr(aq)$

Skillbuilder 7.8
$NH_4Cl(aq) + Fe(NO_3)_3(aq) \longrightarrow$ NO REACTION

Skillbuilder 7.9
$K_2SO_4(aq) + Sr(NO_3)_2(aq) \longrightarrow SrSO_4(s) + 2\,KNO_3(aq)$

Skillbuilder 7.10
Complete ionic equation:
$2\,H^+(aq) + 2\,Br^-(aq) + Ca^{2+}(aq) + 2\,OH^-(aq) \longrightarrow 2\,H_2O(l) + Ca^{2+}(aq) + 2\,Br^-(aq)$
Net ionic equation:
$2\,H^+(aq) + 2\,OH^-(aq) \longrightarrow 2\,H_2O(l)$, or simply
$H^+(aq) + OH^-(aq) \longrightarrow H_2O(l)$

Skillbuilder 7.11
Molecular equation:
$H_2SO_4(aq) + 2\,KOH(aq) \longrightarrow 2\,H_2O(l) + K_2SO_4(aq)$
Net ionic equation:
$2\,H^+(aq) + 2\,OH^-(aq) \longrightarrow 2\,H_2O(l)$

Skillbuilder 7.12
$2\,HBr(aq) + K_2SO_3(aq) \longrightarrow H_2O(l) + SO_2(g) + 2\,KBr(aq)$

Skillbuilder Plus, p. 209
$2\,H^+(aq) + SO_3^{2-}(aq) \longrightarrow H_2O(l) + SO_2(g)$

Skillbuilder 7.13 **(a)**, **(b)**, and **(d)** are all redox reactions; **(c)** is a precipitation reaction.

Skillbuilder 7.14
$C_5H_{12}(l) + 8\,O_2(g) \longrightarrow 5\,CO_2(g) + 6\,H_2O(g)$

Skillbuilder Plus, p. 212
$2\,C_3H_7OH(l) + 9\,O_2(g) \longrightarrow 6\,CO_2(g) + 8\,H_2O(g)$

Skillbuilder 7.15 **(a)** single-displacement **(b)** double-displacement **(c)** synthesis **(d)** decomposition

Answers to Conceptual Checkpoints

7.1 (b) There are 18 oxygen atoms on the left side of the equation, so the same number is needed on the right: 6 + 6(2) = 18.

7.2 (b) Both of the possible products, MgS and $CaSO_4$, are insoluble. The possible products of the other reactions—Na_2S, $Ca(NO_3)_2$, Na_2SO_4, and $Mg(NO_3)_2$—are all soluble.

7.3 (d) In a precipitation reaction, cations and anions "change partners" to produce at least one insoluble product. In an acid–base reaction, H^+ and OH^- combine to form water and their partners pair off to form a salt.

Octane
750

CHAPTER 8

Quantities in Chemical Reactions

"Man masters nature not by force but by understanding. That is why science has succeeded where magic failed: because it has looked for no spell to cast."

Jacob Bronowski (1908–1974)

8.1 Global Warming: Too Much Carbon Dioxide

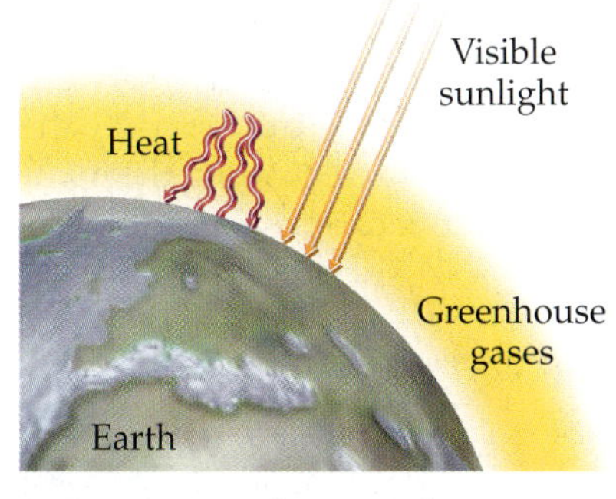

▲ **Figure 8.1 The greenhouse effect** Greenhouse gases act like glass in a greenhouse, allowing visible-light energy to enter the atmosphere but preventing heat energy from escaping.

Average global temperatures depend on the balance between incoming sunlight, which warms Earth, and outgoing heat lost to space, which cools it. Certain gases in Earth's atmosphere, called **greenhouse gases**, affect that balance because they trap heat. These gases act like glass in a greenhouse; they allow sunlight into the atmosphere to warm Earth, but prevent heat from escaping (◀ Figure 8.1). Without greenhouse gases, more heat would escape, and Earth's average temperature would be about 60 °F colder than it is now. Caribbean tourists would freeze at an icy 21 °F, instead of baking at a tropical 81 °F. On the other hand, if the concentration of greenhouse gases in the atmosphere were to increase, Earth's average temperature would rise.

In recent years scientists have become concerned because the atmospheric concentration of carbon dioxide (CO_2)—Earth's most significant greenhouse gas in terms of its contribution to climate—is rising. This rise in CO_2 concentration enhances the atmosphere's ability to hold heat and may therefore lead to **global warming**, an increase in Earth's average temperature. Since 1860, atmospheric CO_2 levels have risen by 25%, and Earth's average temperature has increased by 0.6 °C (about 1.1 °F) (▶ Figure 8.2).

The primary cause of rising CO_2 concentration is the burning of fossil fuels. Fossil fuels—natural gas, petroleum, and coal—provide 90% of our society's energy. However, their combustion produces CO_2. As an example, consider the combustion of octane (C_8H_{18}), a component of gasoline.

$$2\,C_8H_{18}(l) + 25\,O_2(g) \longrightarrow 16\,CO_2(g) + 18\,H_2O(g)$$

◀ The combustion of fossil fuels such as octane (shown here) produces water and carbon dioxide as products. Carbon dioxide is a greenhouse gas that is believed to be causing global warming.

The balanced chemical equation shows that 16 mol of CO_2 are produced for every 2 mol of octane burned. Since we know the world's annual fossil fuel consumption, we can estimate the world's annual CO_2 production. A simple calculation shows that the world's annual CO_2 production—from fossil fuel combustion—matches the measured annual atmospheric CO_2 increase, implying that fossil fuel combustion is indeed responsible for increased atmospheric CO_2 levels.

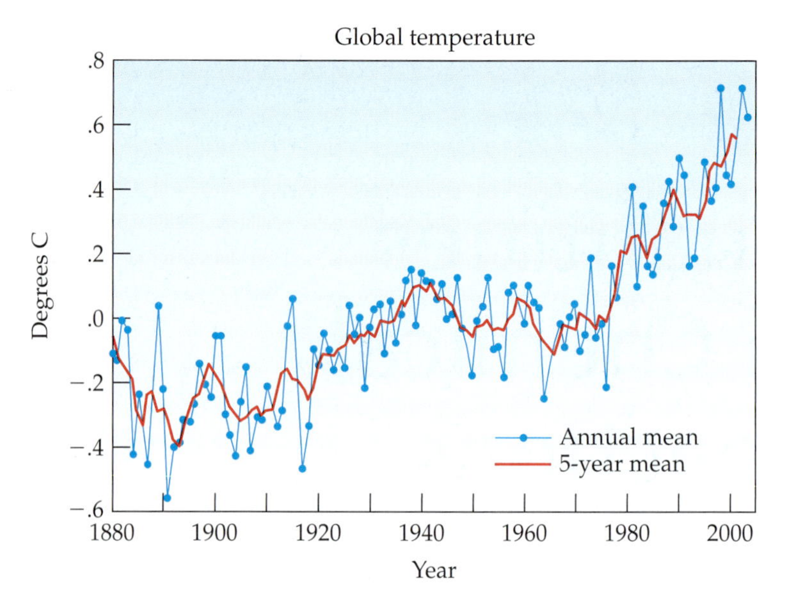

▶ **Figure 8.2 Global warming** Yearly temperature differences from the 120-year average temperature. Earth's average temperature has increased by about 0.6 °C since 1880.

The numerical relationship between chemical quantities in a balanced chemical equation is called reaction **stoichiometry**. Stoichiometry allows us to predict the amounts of products that form in a chemical reaction based on the amounts of reactants. Stoichiometry also allows us to predict how much of the reactants is necessary to form a given amount of product. These calculations are central to chemistry, allowing chemists to plan and carry out chemical reactions to obtain products in the desired quantities.

8.2 Making Pancakes: Relationships between Ingredients

The concepts of stoichiometry are similar to the concepts we use in following a cooking recipe. Calculating the amount of carbon dioxide produced by the combustion of a given amount of a fossil fuel is similar to calculating the number of pancakes that can be made from a given number of eggs. For example, suppose you use the following pancake recipe.

$$\text{1 cup flour} + \text{2 eggs} + \tfrac{1}{2}\text{ tsp baking powder} \longrightarrow \text{5 pancakes}$$

▲ A recipe gives numerical relationships between the ingredients and the number of pancakes.

The recipe shows the numerical relationships between the pancake ingredients. It says that if you have 2 eggs—and enough of everything else—you can make 5 pancakes. We can write this relationship as an equivalence.

$$\text{2 eggs} \equiv \text{5 pancakes}$$

The ≡ sign, which we first encountered in Chapter 6, means "is **equivalent** to." Therefore, 2 eggs are equivalent to 5 pancakes, meaning that for this recipe they must occur in that ratio for the pancakes to turn out right. What if you have 8 eggs? Assuming that you have enough of everything else, how many pancakes can you make? Using the preceding equivalence as a conversion factor, 8 eggs are sufficient to make 20 pancakes.

$$8 \text{ eggs} \times \frac{5 \text{ pancakes}}{2 \text{ eggs}} = 20 \text{ pancakes}$$

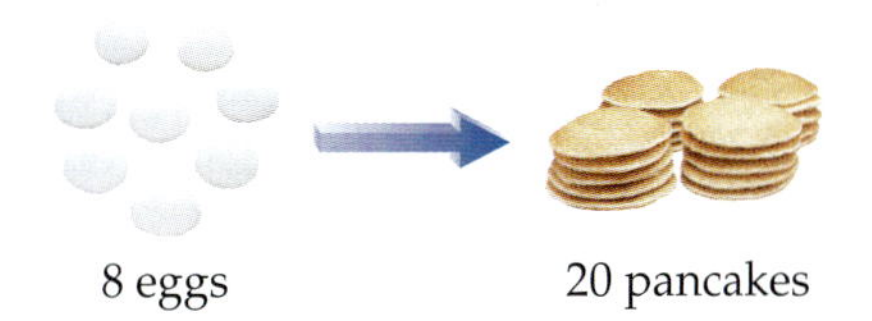

8 eggs 20 pancakes

The pancake recipe contains numerical conversion factors between the pancake ingredients and the number of pancakes. Other conversion factors from this recipe include:

$$1 \text{ cup flour} \equiv 5 \text{ pancakes}$$
$$\tfrac{1}{2} \text{ tsp baking powder} \equiv 5 \text{ pancakes}$$

The recipe also gives us relationships among the ingredients themselves. For example, how much baking powder is required to go with 3 cups of flour? From the recipe:

$$1 \text{ cup flour} \equiv \tfrac{1}{2} \text{ tsp baking powder}$$

With this conversion factor, we can find the appropriate amount of baking powder.

$$3 \text{ cups flour} \times \frac{\frac{1}{2} \text{ tsp baking powder}}{1 \text{ cup flour}} = \tfrac{3}{2} \text{ tsp baking powder}$$

8.3 Making Molecules: Mole-to-Mole Conversions

In a balanced chemical equation, we have a "recipe" for how reactants combine to form products. For example, the following equation shows how hydrogen and nitrogen combine to form ammonia (NH_3).

$$3\, H_2(g) + N_2(g) \longrightarrow 2\, NH_3(g)$$

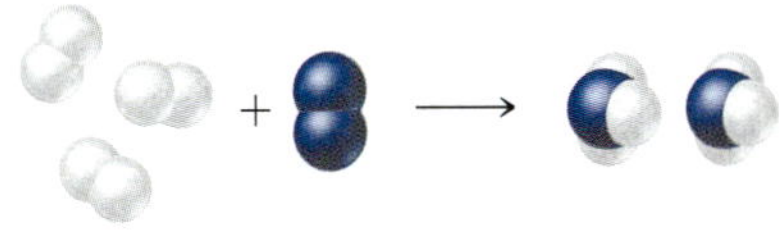

The balanced equation shows that 3 H_2 molecules react with 1 N_2 molecule to form 2 NH_3 molecules. We can express this as the following equivalence.

$$3\, H_2 \text{ molecules} \equiv 1\, N_2 \text{ molecule} \equiv 2\, NH_3 \text{ molecules}$$

Since we do not ordinarily deal with individual molecules, we can express the same equivalence in moles.

$$3 \text{ mol } H_2 \equiv 1 \text{ mol } N_2 \equiv 2 \text{ mol } NH_3$$

If we have 3 mol of N_2, and more than enough H_2, how much NH_3 can we make? We set up this problem in the standard format:

Given: 3 mol N_2

Find: mol NH_3

Conversion Factor: 1 mol N_2 ≡ 2 mol NH_3

We use this conversion factor from the balanced chemical equation in the same way that we used the conversion factors from the pancake recipe.

Solution Map: The solution map begins with mol N_2 and ends with mol NH_3. The conversion factor comes from the balanced chemical equation.

Solution:

mol N_2 → mol NH_3

$$\frac{2 \text{ mol } NH_3}{1 \text{ mol } N_2}$$

We can then do the conversion.

$$3 \cancel{\text{mol } N_2} \times \frac{2 \text{ mol } NH_3}{1 \cancel{\text{mol } N_2}} = 6 \text{ mol } NH_3$$

We have enough N_2 for 6 mol of NH_3.

EXAMPLE 8.1 Mole-to-Mole Conversions

Sodium chloride, NaCl, forms by the following reaction between sodium and chlorine.

$$2\,Na(s) + Cl_2(g) \longrightarrow 2\,NaCl(s)$$

How many moles of NaCl result from the complete reaction of 3.4 mol of Cl_2? Assume that there is more than enough Na.

You are given the number of moles of a reactant (Cl_2) and asked to find how many moles of product (NaCl) will result if the reactant completely reacts. The conversion factor comes from the balanced chemical equation.

Given: 3.4 mol Cl_2

Find: mol NaCl

Conversion Factor: 1 mol Cl_2 ≡ 2 mol NaCl

The solution map begins with moles of chlorine and uses the stoichiometric conversion factor to obtain moles of sodium chloride.

Solution Map:

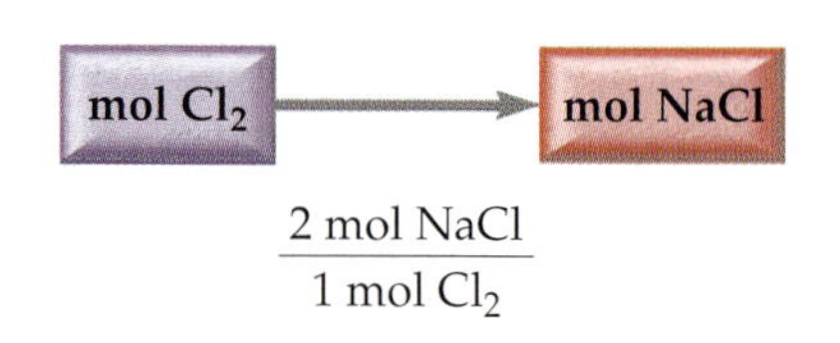

Follow the solution map to solve the problem.

There is enough Cl_2 to produce 6.8 mol of NaCl.

Solution:

$$3.4 \cancel{\text{mol } Cl_2} \times \frac{2 \text{ mol NaCl}}{1 \cancel{\text{mol } Cl_2}} = 6.8 \text{ mol NaCl}$$

SKILLBUILDER 8.1 Mole-to-Mole Conversions

Water is formed when hydrogen gas reacts explosively with oxygen gas according to the following balanced equation.

$$O_2(g) + 2\,H_2(g) \longrightarrow 2\,H_2O(g)$$

How many moles of H_2O result from the complete reaction of 24.6 mol of O_2? Assume that there is more than enough H_2.

Chemistry IN THE MEDIA

The Controversy over Oxygenated Fuels

We have seen that the balanced chemical equation for the combustion of octane, a component of gasoline, is:

$$2\,C_8H_{18}(l) + 25\,O_2(g) \longrightarrow 16\,CO_2(g) + 18\,H_2O(g)$$

We have also learned how balanced chemical equations give numerical relationships between reactants. The preceding equation shows that 25 mol of O_2 are required to completely react with 2 mol of C_8H_{18}. What if there were not enough O_2 in the cylinders of an automobile engine to fully react with the amount of octane flowing into them? For many reactions, a shortage of one reactant simply means that less product forms. We will learn about that later in this chapter. However, for some reactions, a shortage of one reactant causes other reactions—called *side reactions*—to occur along with the desired reaction. In the case of octane and the other major components of gasoline, those side reactions result in pollutants such as carbon monoxide (CO) and ozone (O_3).

In 1990, the U.S. Congress, in efforts to lower air pollution, passed amendments to the Clean Air Act requiring oil companies to add substances to gasoline that prevent these side reactions. Since these additives have the effect of increasing the amount of oxygen during combustion, the resulting gasoline is called oxygenated fuel. The additive of choice among oil companies was a compound called MTBE (methyl tertiary butyl ether). The immediate results were positive. Carbon monoxide and ozone levels in many major cities decreased significantly.

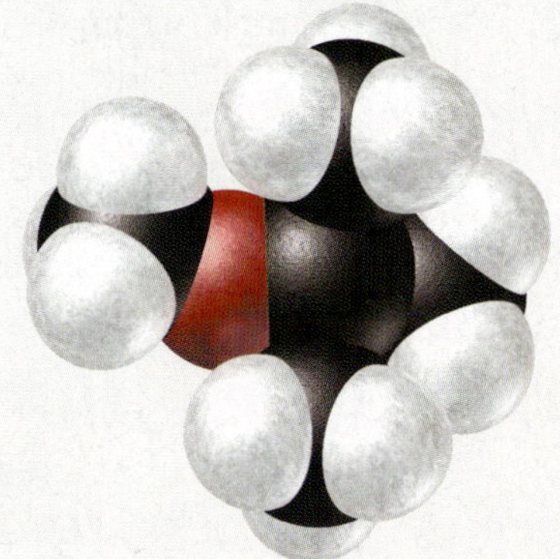

▲ MTBE was the additive of choice.

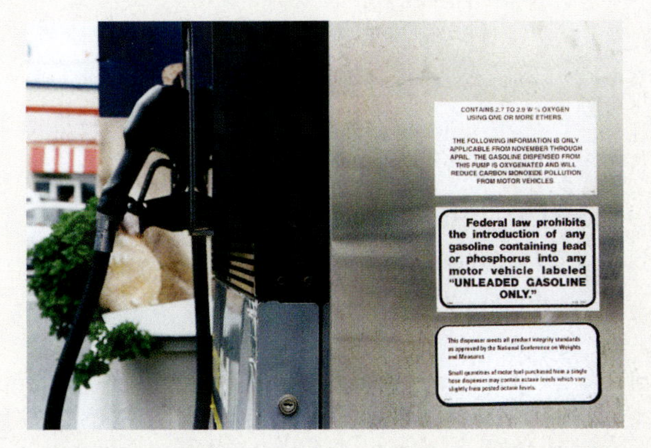

▲ The 1990 amendments to the Clean Air Act required oil companies to put additives in gasoline that increased its oxygen content.

Over time, however, MTBE—a compound that does not readily biodegrade—began to appear in drinking-water supplies across the nation. MTBE got into drinking water through gasoline spills at gas stations, from boat motors, and from leaking underground storage tanks. The consequences have been significant. MTBE, even at low levels, imparts a turpentine-like odor and foul taste to drinking water. It is also a suspected carcinogen.

Public response has been swift and dramatic. Several class-action lawsuits have been filed against the manufacturers of MTBE, against gas stations suspected of leaking it, and against the oil companies that put MTBE into gasoline. Several states have moved to ban MTBE from gasoline, and the U.S. Congress is in the process of imposing a federal ban on MTBE. This does raise a problem, however. MTBE was added to gasoline as a way to meet the requirements of the 1990 Clean Air Act amendments. Will the federal government temporarily suspend these requirements until a substitute for MTBE is found? Or should the government completely remove the requirements, weakening the Clean Air Act? The answers are not simple. However, ethanol, made from the fermentation of grains, is a ready substitute for MTBE that has many of the same pollution-reducing effects without the associated health hazards. Oil companies did not use ethanol originally because it was more expensive than MTBE. Now they are paying the price.

CAN YOU ANSWER THIS? *How many moles of oxygen (O_2) are required to completely react with 425 mol of octane (approximate capacity of a 15-gal automobile gasoline tank)?*

8.4 Making Molecules: Mass-to-Mass Conversions

In Chapter 6, we learned how a chemical *formula* contains conversion factors for converting between moles of a compound and moles of its constituent elements. In this chapter, we have seen how a chemical *equation* contains conversion factors between moles of reactants and moles of products. However, we are often interested in relationships between *mass* of reactant and *mass* of product. For example, we might want to know the mass of carbon dioxide emitted by an automobile that uses a kilogram of gasoline. Or we might want to know the mass of each reactant required to obtain a certain mass of product in a synthesis reaction. These calculations are similar to calculations we learned in Section 6.5, where we converted between mass of a compound and mass of a constituent element. The general outline for these types of calculations is:

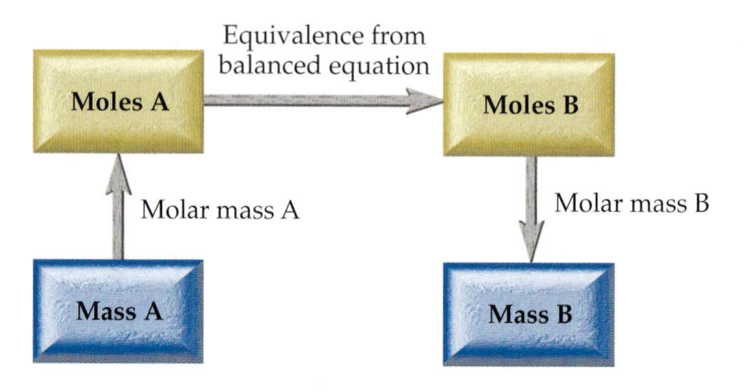

where A and B are two different substances involved in the reaction. The conversion from mass of A to moles of A is achieved using the molar mass of A. The conversion from moles of A to moles of B is achieved using conversion factors from the chemical equation, and the conversion from moles of B to mass of B is achieved using the molar mass of B. For example, suppose we want to calculate the mass of CO_2 emitted upon the combustion of 5.0×10^2 g of pure octane. The balanced chemical equation for octane combustion is:

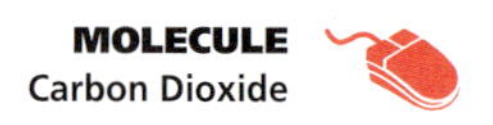

$$2\,C_8H_{18}(l) + 25\,O_2(g) \longrightarrow 16\,CO_2(g) + 18\,H_2O(g)$$

We set up the problem in standard format.

Given: 5.0×10^2 g C_8H_{18}

Find: g CO_2

Conversion Factors:

2 mol $C_8H_{18} \equiv 16$ mol CO_2 (from chemical equation)

Molar mass $C_8H_{18} = 114.2$ g/mol

Molar mass $CO_2 = 44.01$ g/mol

Notice that we are given g C_8H_{18} and asked to find g CO_2. The balanced chemical equation, however, gives us a relationship between moles of C_8H_{18} and moles of CO_2. Consequently, before using that relationship, we must convert from grams to moles.

The solution map uses the general outline

$$\text{Mass A} \longrightarrow \text{Moles A} \longrightarrow \text{Moles B} \longrightarrow \text{Mass B}$$

where A is octane and B is carbon dioxide.

Solution Map:

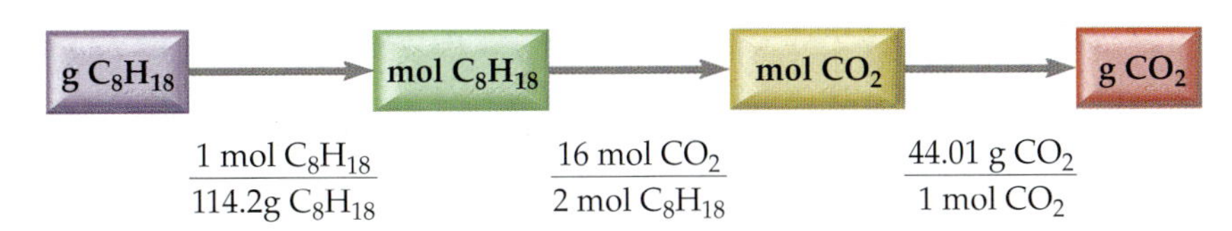

$$\frac{1 \text{ mol } C_8H_{18}}{114.2\text{g } C_8H_{18}} \qquad \frac{16 \text{ mol } CO_2}{2 \text{ mol } C_8H_{18}} \qquad \frac{44.01 \text{ g } CO_2}{1 \text{ mol } CO_2}$$

Solution: We then follow the solution map to solve the problem, beginning with g C_8H_{18} and canceling units to arrive at g CO_2.

$$5.0 \times 10^2 \text{ g } C_8H_{18} \times \frac{1 \text{ mol } C_8H_{18}}{114.2 \text{ g } C_8H_{18}} \times \frac{16 \text{ mol } CO_2}{2 \text{ mol } C_8H_{18}} \times \frac{44.01 \text{ g } CO_2}{1 \text{ mol } CO_2} = 1.5 \times 10^3 \text{ g } CO_2$$

So, upon combustion, 5.0×10^2 g of octane produces 1.5×10^3 g of carbon dioxide.

EXAMPLE 8.2 Mass-to-Mass Conversions

In photosynthesis, plants convert carbon dioxide and water into glucose ($C_6H_{12}O_6$) according to the following reaction.

$$6\, CO_2(g) + 6\, H_2O(l) \xrightarrow[\text{sunlight}]{} 6\, O_2(g) + C_6H_{12}O_6(aq)$$

How many grams of glucose can be synthesized from 58.5 g of CO_2? Assume that there is more than enough water present to react with all of the CO_2.

Set up the problem in the normal way.

Given: 58.5 g CO_2

Find: g $C_6H_{12}O_6$

The main conversion factor is the stoichiometric relationship between moles of carbon dioxide and moles of glucose. This conversion factor comes from the balanced equation. The other conversion factors are simply the molar masses of carbon dioxide and glucose.

Conversion Factors:

6 mol CO_2 ≡ 1 mol $C_6H_{12}O_6$ (from balanced chemical equation)

Molar mass CO_2 = 44.01 g/mol

Molar mass $C_6H_{12}O_6$ = 180.2 g/mol

The solution map uses the general outline

Mass A ⟶ Moles A ⟶ Moles B ⟶ Mass B

where A is carbon dioxide and B is glucose.

Solution Map:

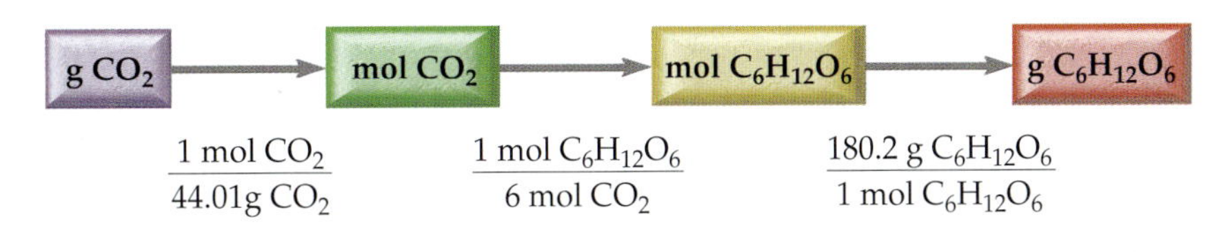

$$\frac{1 \text{ mol } CO_2}{44.01\text{g } CO_2} \qquad \frac{1 \text{ mol } C_6H_{12}O_6}{6 \text{ mol } CO_2} \qquad \frac{180.2 \text{ g } C_6H_{12}O_6}{1 \text{ mol } C_6H_{12}O_6}$$

Follow the solution map to solve the problem. Begin with grams of carbon dioxide and multiply by the appropriate factors to arrive at grams of glucose.

Solution:

$$58.5 \text{ g } CO_2 \times \frac{1 \text{ mol } CO_2}{44.01 \text{ g } CO_2} \times \frac{1 \text{ mol } C_6H_{12}O_6}{6 \text{ mol } CO_2} \times \frac{180.2 \text{ g } C_6H_{12}O_6}{1 \text{ mol } C_6H_{12}O_6} = 39.9 \text{ g } C_6H_{12}O_6$$

SKILLBUILDER 8.2 Mass-to-Mass Conversions

Magnesium hydroxide, the active ingredient in milk of magnesia, neutralizes stomach acid, primarily HCl, according to the following reaction.

$$Mg(OH)_2(aq) + 2\, HCl(aq) \longrightarrow 2\, H_2O(l) + MgCl_2(aq)$$

How much HCl in grams can be neutralized by 5.50 g of $Mg(OH)_2$?

EXAMPLE 8.3 Mass-to-Mass Conversions

One of the components of acid rain is nitric acid, which forms when NO_2, a pollutant, reacts with oxygen and rainwater according to the following reaction.

$$4\,NO_2(g) + O_2(g) + 2\,H_2O(l) \longrightarrow 4\,HNO_3(aq)$$

Assuming that there is more than enough O_2 and H_2O, how much HNO_3 in kilograms forms from 1.5×10^3 kg of NO_2 pollutant?

Set up the problem in the normal way.

The main conversion factor is the stoichiometric relationship between moles of nitrogen dioxide and moles of nitric acid. This conversion factor comes from the balanced equation. The other conversion factors are simply the molar masses of nitrogen dioxide and nitric acid and the relationship between kilograms and grams.

Given: 1.5×10^3 kg NO_2

Find: kg HNO_3

Conversion Factors:

4 mol $NO_2 \equiv 4$ mol HNO_3 (from balanced chemical equation)

Molar mass $NO_2 = 46.01$ g/mol

Molar mass $HNO_3 = 63.02$ g/ mol

1 kg $= 1000$ g

The solution map follows the general format of

Mass $\longrightarrow$ Moles $\longrightarrow$ Moles $\longrightarrow$ Mass.

However, since the original quantity of NO_2 is given in kilograms, you must first convert to grams. Since the final quantity is requested in kilograms, you must convert back to kilograms at the end.

Solution Map:

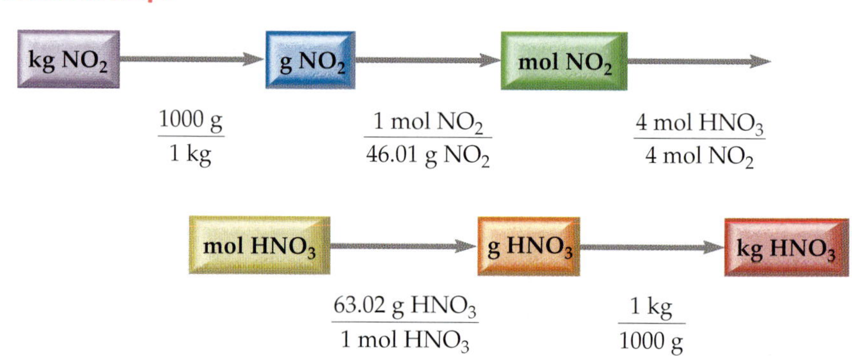

Follow the solution map to solve the problem. Begin with kilograms of nitrogen dioxide and multiply by the appropriate conversion factors to arrive at kilograms of nitric acid.

Solution:

$$1.5 \times 10^3\ \text{kg NO}_2 \times \frac{1000\ \text{g}}{1\ \text{kg}} \times \frac{1\ \text{mol NO}_2}{46.01\ \text{g NO}_2} \times \frac{4\ \text{mol HNO}_3}{4\ \text{mol NO}_2} \times \frac{63.02\ \text{g HNO}_3}{1\ \text{mol HNO}_3} \times \frac{1\ \text{kg}}{1000\ \text{g}} = 2.1 \times 10^3\ \text{kg HNO}_3$$

SKILLBUILDER 8.3 Mass-to-Mass Conversions

Another component of acid rain is sulfuric acid, which forms when SO_2, also a pollutant, reacts with oxygen and rainwater according to the following reaction.

$$2\,SO_2(g) + O_2(g) + 2\,H_2O(l) \longrightarrow 2\,H_2SO_4(aq)$$

Assuming that there is plenty of O_2 and H_2O, how much H_2SO_4 in kilograms forms from 2.6×10^3 kg of SO_2?

8.5 More Pancakes: Limiting Reactant, Theoretical Yield, and Percent Yield

Let's return to our pancake analogy to understand two more concepts important in reaction stoichiometry: limiting reactant and percent yield. Recall our pancake recipe:

$$1 \text{ cup flour} + 2 \text{ eggs} + \tfrac{1}{2} \text{ tsp baking powder} \longrightarrow 5 \text{ pancakes}$$

Suppose we have 3 cups flour, 10 eggs, and 4 tsp baking powder. How many pancakes can we make? We have enough flour to make:

$$3 \text{ cups flour} \times \frac{5 \text{ pancakes}}{1 \text{ cup flour}} = 15 \text{ pancakes}$$

We have enough eggs to make:

$$10 \text{ eggs} \times \frac{5 \text{ pancakes}}{2 \text{ eggs}} = 25 \text{ pancakes}$$

We have enough baking powder to make:

$$4 \text{ tsp baking powder} \times \frac{5 \text{ pancakes}}{\frac{1}{2} \text{ tsp baking powder}} = 40 \text{ pancakes}$$

We have enough flour for 15 pancakes, enough eggs for 25 pancakes, and enough baking powder for 40 pancakes. Consequently, unless we get more ingredients, we can make only 15 pancakes. The flour *limits* the number of pancakes we can make. If this were a chemical reaction, the flour would be the **limiting reactant**, the reactant that limits the amount of product in a chemical reaction. Notice that the limiting reactant is simply the reactant that makes *the least amount of product*. If this were a chemical reaction, 15 pancakes would be the **theoretical yield**, the amount of product that can be made in a chemical reaction based on the amount of limiting reactant.

The term *limiting reagent* is sometimes used in place of *limiting reactant*.

▶ If this were a chemical reaction, the flour would be the limiting reactant and 15 pancakes would be the theoretical yield.

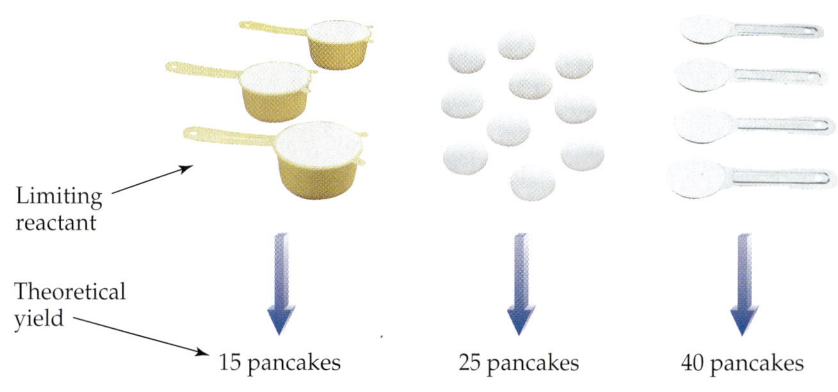

Let us carry this analogy one step further. Suppose we go on to cook our pancakes. We accidentally burn three of them and one falls on the floor. So even though we had enough flour for 15 pancakes, we finished with only 11 pancakes. If this were a chemical reaction, the 11 pancakes would be our **actual yield**, the amount of product actually produced by a chemical reaction. Finally, our **percent yield**, the percentage of the theoretical yield that was actually attained, is:

$$\text{Percent yield} = \frac{11 \text{ pancakes}}{15 \text{ pancakes}} \times 100\% = 73\%$$

The actual yield of a chemical reaction, which must be determined experimentally, often depends on the reaction conditions in various ways. We will explore some of the factors involved in Chapter 15.

Since four of the pancakes were ruined, we got only 73% of our theoretical yield. In a chemical reaction, the actual yield is almost always less than 100% because at least some of the product does not form or is lost in the process of recovering it (in analogy to some of the pancakes being burned).

To summarize:

- Limiting reactant (or limiting reagent)—the reactant that is completely consumed in a chemical reaction.
- Theoretical yield—the amount of product that can be made in a chemical reaction based on the amount of limiting reactant.
- Actual yield—the amount of product actually produced by a chemical reaction.
- Percent yield $= \dfrac{\textbf{Actual yield}}{\textbf{Theoretical yield}} \times 100\%$

Now consider the following reaction.

$$Ti(s) + 2\,Cl_2(g) \longrightarrow TiCl_4(s)$$

If we begin with 1.8 mol of titanium and 3.2 mol of chlorine, what is the limiting reactant and theoretical yield of $TiCl_4$ in moles? We can set up the problem according to our standard problem-solving procedure.

Given:

1.8 mol Ti

3.2 mol Cl_2

Find:

limiting reactant
theoretical yield

Conversion Factors: The conversion factors come from the balanced chemical equation and give the relationships between moles of each of the reactants and moles of product.

$$1 \text{ mol Ti} \equiv 1 \text{ mol } TiCl_4$$

$$2 \text{ mol } Cl_2 \equiv 1 \text{ mol } TiCl_4$$

Like our pancake analogy, we find the limiting reactant by calculating how much product can be made from each reactant. The reactant that makes the *least amount of product* is the limiting reactant.

Solution Map:

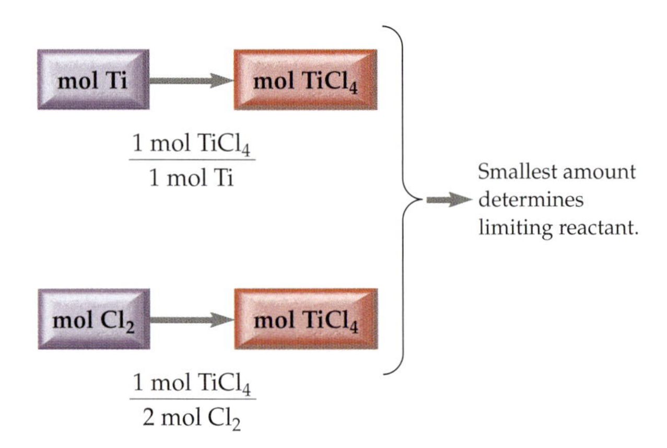

Solution:

$$1.8 \text{ mol Ti} \times \frac{1 \text{ mol TiCl}_4}{1 \text{ mol Ti}} = 1.8 \text{ mol TiCl}_4$$

$$3.2 \text{ mol Cl}_2 \times \frac{1 \text{ mol TiCl}_4}{2 \text{ mol Cl}_2} = 1.6 \text{ mol TiCl}_4$$

Limiting reactant — Least amount of product

In many industrial applications, the more costly reactant or the reactant that is most difficult to remove from the product mixture is chosen to be the limiting reactant.

Since the 3.2 mol of Cl_2 make the least amount of $TiCl_4$, Cl_2 is the limiting reactant. Notice that we began with more moles of Cl_2 than Ti, but since the reaction requires 2 Cl_2 for each Ti, Cl_2 is still the limiting reactant. The theoretical yield is 1.6 mol of $TiCl_4$.

EXAMPLE 8.4 Limiting Reactant and Theoretical Yield from Initial Moles of Reactants

Consider the following reaction.

$$2\,Al(s) + 3\,Cl_2(g) \longrightarrow 2\,AlCl_3(s)$$

If we begin with 0.552 mol of aluminum and 0.887 mol of chlorine, what is the limiting reactant and theoretical yield of $AlCl_3$ in moles?

Set up the problem in the normal way.

Given:

0.552 mol Al

0.887 mol Cl_2

Find:

limiting reactant

theoretical yield

The conversion factors are the stoichiometric relationships (from the balanced equation) between the moles of each reactant and the moles of product.

Conversion Factors:

2 mol Al $\equiv$ 2 mol $AlCl_3$

3 mol Cl_2 $\equiv$ 2 mol $AlCl_3$

The solution map shows how to get from moles of each reactant to moles of $AlCl_3$. The reactant that makes the *least amount of* $AlCl_3$ is the limiting reactant.

Solution Map:

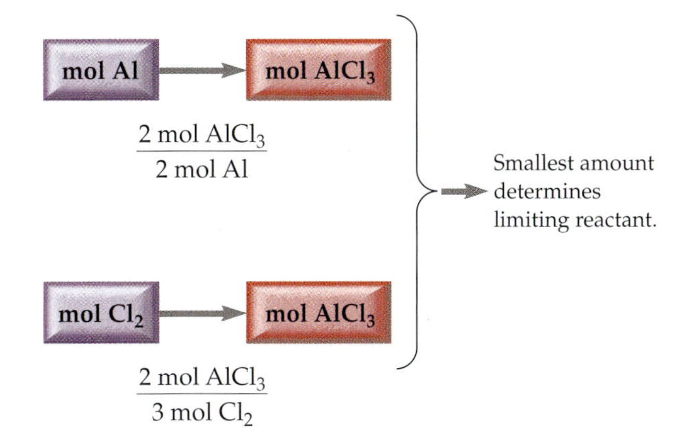

Solution:

$$0.552 \text{ mol Al} \times \frac{2 \text{ mol AlCl}_3}{2 \text{ mol Al}} = 0.552 \text{ mol AlCl}_3$$

Limiting reactant — Least amount of product

$$0.887 \text{ mol Cl}_2 \times \frac{2 \text{ mol AlCl}_3}{3 \text{ mol Cl}_2} = 0.591 \text{ mol AlCl}_3$$

Since the 0.552 mol of Al makes the least amount of $AlCl_3$, Al is the limiting reactant. The theoretical yield is 0.552 mol of $AlCl_3$.

SKILLBUILDER 8.4 Limiting Reactant and Theoretical Yield from Initial Moles of Reactants

Consider the following reaction.

$$2\,\text{Na}(s) + \text{F}_2(g) \longrightarrow 2\,\text{NaF}(s)$$

If you begin with 4.8 mol of sodium and 2.6 mol of fluorine, what is the limiting reactant and theoretical yield of NaF in moles?

CONCEPTUAL CHECKPOINT 8.1

A maker of tricycles has 17 seats, 19 bells, 37 pedals, and 49 wheels in his warehouse. The limiting component for the manufacture of new tricycles would be: (Assume that all the wheels on each tricycle are identical.)

(a) seats
(b) bells
(c) pedals
(d) wheels

8.6 Limiting Reactant, Theoretical Yield, and Percent Yield from Initial Masses of Reactants

When working in the laboratory, we normally measure the initial amounts of reactants in grams. To find limiting reactants and theoretical yields from initial masses, we must add two steps to our calculations. Consider, for example, the following synthesis reaction.

$$2\,\text{Na}(s) + \text{Cl}_2(g) \longrightarrow 2\,\text{NaCl}(s)$$

If we have 53.2 g of Na and 65.8 g of Cl_2, what is the limiting reactant and theoretical yield? We will approach this problem in our standard way.

Given:

53.2 g Na
65.8 g Cl_2

Find:

limiting reactant
theoretical yield

Conversion Factors: From the balanced chemical equation, we get:

$$2 \text{ mol Na} \equiv 2 \text{ mol NaCl}$$
$$1 \text{ mol Cl}_2 \equiv 2 \text{ mol NaCl}$$

We will also have to convert between grams and moles for each of the reactants and the product, so we also need the following molar masses.

$$\text{Molar mass Na} = \frac{22.99 \text{ g Na}}{1 \text{ mol Na}}$$

$$\text{Molar mass } Cl_2 = \frac{70.91 \text{ g } Cl_2}{1 \text{ mol } Cl_2}$$

$$\text{Molar mass NaCl} = \frac{58.44 \text{ g NaCl}}{1 \text{ mol NaCl}}$$

We again find the limiting reactant by calculating how much product can be made from each reactant. Since we are given the initial amounts in grams, we must first convert to moles. After we convert to moles of product, we convert back to grams of product. The reactant that makes the *least amount of product* is the limiting reactant.

Solution Map:

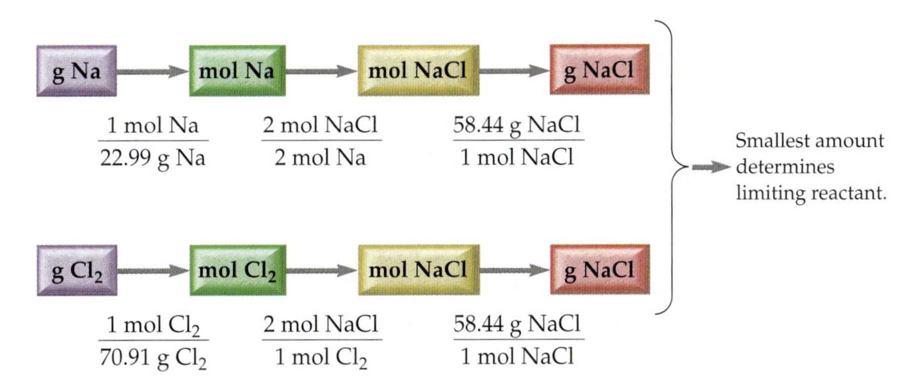

Solution: Beginning with the actual amounts of each reactant, we follow the solution map to calculate how much product can be made from each.

The limiting reactant can also be found by calculating the number of moles of NaCl (rather than grams) that can be made from each reactant. However, since theoretical yields are normally calculated in grams, we take the calculation all the way to grams to determine limiting reactant.

$$53.2 \text{ g Na} \times \frac{1 \text{ mol Na}}{22.99 \text{ g Na}} \times \frac{2 \text{ mol NaCl}}{2 \text{ mol Na}} \times \frac{58.44 \text{ g NaCl}}{1 \text{ mol NaCl}} = 135 \text{ g NaCl}$$

$$65.8 \text{ g } Cl_2 \times \frac{1 \text{ mol } Cl_2}{70.91 \text{ g } Cl_2} \times \frac{2 \text{ mol NaCl}}{1 \text{ mol } Cl_2} \times \frac{58.44 \text{ g NaCl}}{1 \text{ mol NaCl}} = 108 \text{ g NaCl}$$

Limiting reactant (65.8 g Cl_2) **Least amount of product** (108 g NaCl)

Since Cl_2 makes the least amount of product (108 g NaCl is less than 135 g NaCl), it is the limiting reactant. Notice that the limiting reactant is not necessarily the reactant with the least mass. In this case, we had fewer grams of Na than Cl_2, yet Cl_2 was the limiting reactant because it made less NaCl. The theoretical yield is therefore 108 g of NaCl, the amount of product possible based on the limiting reactant.

The limiting reactant is *not necessarily* the reactant with the least mass.

Now suppose that when the synthesis was carried out, the actual yield of NaCl was 86.4 g. What is the percent yield? The percent yield is simply:

The actual yield is always less than the theoretical yield because at least a small amount of product is usually lost or does not form during a reaction.

$$\text{Percent yield} = \frac{\text{Actual yield}}{\text{Theoretical yield}} \times 100\% = \frac{86.4 \text{ g}}{108 \text{ g}} \times 100\% = 80.0\%$$

EXAMPLE 8.5 Finding Limiting Reactant and Theoretical Yield

Ammonia, NH_3, can be synthesized by the following reaction.

$$2\,NO(g) + 5\,H_2(g) \longrightarrow 2\,NH_3(g) + 2\,H_2O(g)$$

What is the maximum amount of ammonia in grams that can be synthesized from 45.8 g of NO and 12.4 g of H_2?

Although this problem does not specifically ask for the limiting reactant, it must be found to determine the theoretical yield, which is the maximum amount of ammonia that can be synthesized. Begin by setting up the problem in the normal way.

Given: 45.8 g NO, 12.4 g H_2

Find: maximum amount of NH_3 (theoretical yield)

The main conversion factors are the stoichiometric relationship between moles of each reactant and moles of ammonia. The other conversion factors are simply the molar masses of nitrogen monoxide, hydrogen gas, and ammonia.

Conversion Factors:

$$2 \text{ mol NO} \equiv 2 \text{ mol } NH_3$$

$$5 \text{ mol } H_2 \equiv 2 \text{ mol } NH_3$$

$$\text{Molar mass NO} = \frac{30.01 \text{ g NO}}{1 \text{ mol NO}}$$

$$\text{Molar mass } H_2 = \frac{2.02 \text{ g } H_2}{1 \text{ mol } H_2}$$

$$\text{Molar mass } NH_3 = \frac{17.04 \text{ g } NH_3}{1 \text{ mol } NH_3}$$

Find the limiting reactant by calculating how much product can be made from each reactant. The reactant that makes the *least amount of product* is the limiting reactant.

Solution Map:

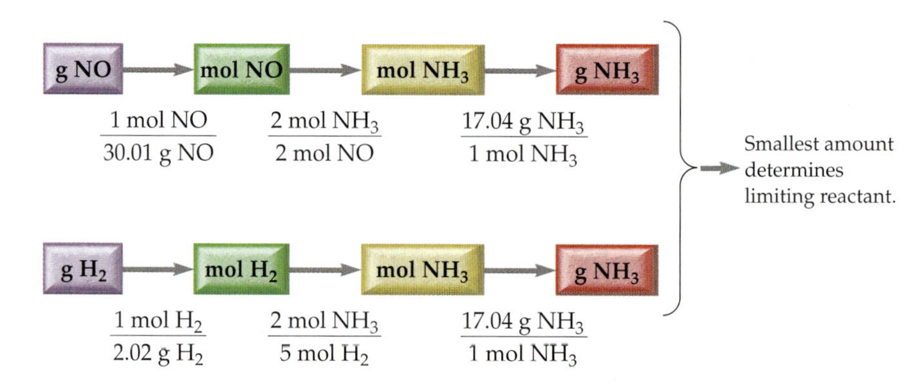

Follow the solution map, beginning with the actual amount of each reactant given, to calculate the amount of product that can be made from each reactant.

Solution:

$$45.8 \text{ g NO} \times \frac{1 \text{ mol NO}}{30.01 \text{ g NO}} \times \frac{2 \text{ mol } NH_3}{2 \text{ mol NO}} \times \frac{17.04 \text{ g } NH_3}{1 \text{ mol } NH_3} = 26.0 \text{ g } NH_3$$

Limiting reactant (NO) — **Least amount of product** (26.0 g NH_3)

$$12.4 \text{ g } H_2 \times \frac{1 \text{ mol } H_2}{2.02 \text{ g } H_2} \times \frac{2 \text{ mol } NH_3}{5 \text{ mol } H_2} \times \frac{17.04 \text{ g } NH_3}{1 \text{ mol } NH_3} = 41.8 \text{ g } NH_3$$

There is enough NO to make 26.0 g of NH_3 and enough H_2 to make 41.8 g of NH_3. Therefore, NO is the limiting reactant, and the maximum amount of ammonia that can possibly be made is 26.0 g, the theoretical yield.

SKILLBUILDER 8.5 Finding Limiting Reactant and Theoretical Yield

Ammonia can also be synthesized by the following reaction.

$$3\,H_2(g) + N_2(g) \longrightarrow 2\,NH_3(g)$$

What is the maximum amount of ammonia in grams that can be synthesized from 25.2 g of N_2 and 8.42 g of H_2?

SKILLBUILDER PLUS

What is the maximum amount of ammonia in kilograms that can be synthesized from 5.22 kg of H_2 and 31.5 kg of N_2?

EXAMPLE 8.6 Finding Limiting Reactant, Theoretical Yield, and Percent Yield

Consider the following reaction.

$$Cu_2O(s) + C(s) \longrightarrow 2\,Cu(s) + CO(g)$$

When 11.5 g of C are allowed to react with 114.5 g of Cu_2O, 87.4 g of Cu are obtained. Find the limiting reactant, theoretical yield, and percent yield.

Set up the problem in the normal way.

Given:

11.5 g C
114.5 g Cu_2O
87.4 g Cu produced

Find:

limiting reactant
theoretical yield
percent yield

The main conversion factors are the stoichiometric relationships between moles of each reactant and moles of copper. The other conversion factors are simply the molar masses of copper(I) oxide, carbon, and copper.

Conversion Factors:

$1 \text{ mol } Cu_2O \equiv 2 \text{ mol Cu}$
$1 \text{ mol C} \equiv 2 \text{ mol Cu}$

Molar mass Cu_2O = 143.08 g/mol
Molar mass C = 12.01 g/mol
Molar mass Cu = 63.54 g/mol

The solution map shows how to find the mass of Cu formed by the initial masses of Cu_2O and C. The reactant that makes the *least amount of product* is the limiting reactant and determines the theoretical yield.

Solution Map:

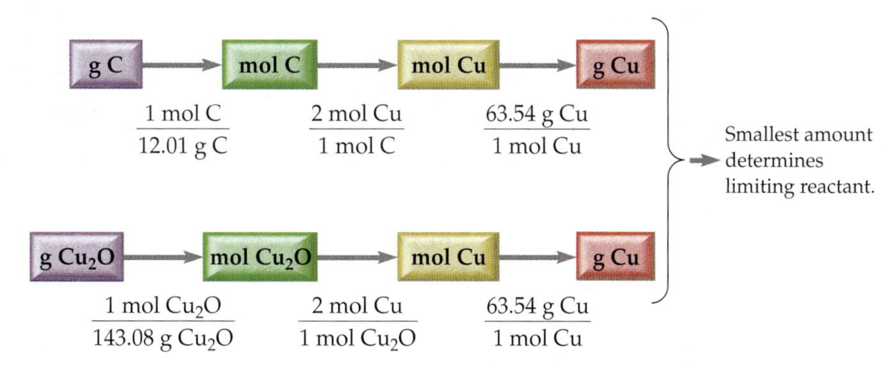

Follow the solution map, beginning with the actual amount of each reactant given, to calculate the amount of product that can be made from each reactant.

Solution:

$$11.5 \text{ g C} \times \frac{1 \text{ mol C}}{12.01 \text{ g C}} \times \frac{2 \text{ mol Cu}}{1 \text{ mol C}} \times \frac{63.54 \text{ g Cu}}{1 \text{ mol Cu}} = 122 \text{ g Cu}$$

$$114.5 \text{ g } Cu_2O \times \frac{1 \text{ mol } Cu_2O}{143.08 \text{ g } Cu_2O} \times \frac{2 \text{ mol Cu}}{1 \text{ mol } Cu_2O} \times \frac{63.54 \text{ g Cu}}{1 \text{ mol Cu}} = 101.7 \text{ g Cu}$$

Limiting reactant (Cu_2O) — Least amount of product (101.7 g Cu)

Since Cu_2O makes the least amount of product, Cu_2O is the limiting reactant. The theoretical yield is simply the amount of product made by the limiting reactant. The percent yield is the actual yield (87.4 g Cu) divided by the theoretical yield (101.7 g Cu) multiplied by 100%.

Theoretical yield = 101.7 g Cu

$$\text{Percent yield} = \frac{\text{Actual yield}}{\text{Theoretical yield}} \times 100\%$$

$$= \frac{87.4 \text{ g}}{101.7 \text{ g}} \times 100\% = 85.9\%$$

SKILLBUILDER 8.6 Finding Limiting Reactant, Theoretical Yield, and Percent Yield

The following reaction is used to obtain iron from iron ore:

$$Fe_2O_3(s) + 3\,CO(g) \longrightarrow 2\,Fe(s) + 3\,CO_2(g)$$

The reaction of 185 g of Fe_2O_3 with 95.3 g of CO produces 87.4 g of Fe. Find the limiting reactant, theoretical yield, and percent yield.

EVERYDAY Chemistry

Bunsen Burners

In the laboratory, we often use Bunsen burners as heat sources. These burners are normally fueled by methane. The balanced equation for methane (CH_4) combustion is:

$$CH_4(g) + 2\,O_2(g) \longrightarrow CO_2(g) + 2\,H_2O(g)$$

Most Bunsen burners have a mechanism to adjust the amount of air (and therefore of oxygen) that is mixed with the methane. If you light the burner with the air completely closed off, you get a yellow, smoky flame that is not very hot. As you increase the amount of air going into the burner, the flame becomes bluer, less smoky, and hotter. When you reach the optimum adjustment, the flame has a sharp, inner blue triangle, no smoke, and is hot enough to melt glass easily. Continuing to increase the air beyond this point causes the flame to become cooler again and may actually extinguish it.

(a) No air

(b) Small amount of air

(c) Optimum

(d) Too much air

▲ Bunsen burner at various stages of air intake adjustment.

CAN YOU ANSWER THIS? *Can you use the concepts from this chapter to explain the changes in the Bunsen burner as the air intake is adjusted?*

CHAPTER IN REVIEW

Chemical Principles

Stoichiometry: A balanced chemical equation gives quantitative relationships between the amounts of reactants and products. For example, the reaction $2\,H_2 + O_2 \longrightarrow 2\,H_2O$ says that 2 mol of H_2 reacts with 1 mol of O_2 to form 2 mol of H_2O. These relationships can be used to calculate quantities such as the amount of product possible with a certain amount of reactant, or the amount of one reactant required to completely react with a certain amount of another reactant. The quantitative relationships between reactants and products in a chemical reaction are called reaction stoichiometry.

Relevance

Stoichiometry: Reaction stoichiometry is important because we often want to know numerical relationships between the reactants and products in a chemical reaction. For example, we might want to know how much carbon dioxide, a greenhouse gas, is formed upon burning a certain amount of a particular fossil fuel.

Limiting Reactant, Theoretical Yield, and Percent Yield: The limiting reactant in a chemical reaction is the reactant that, for the amounts of each reactant used in a particular instance, limits the amount of product that can be made. The theoretical yield in a chemical reaction is the amount of product that can be made based on the amount of the limiting reactant. The actual yield in a chemical reaction is the amount of product actually produced. The percent yield in a chemical reaction is the actual yield divided by theoretical yield times 100%.

Limiting Reactant, Theoretical Yield, and Percent Yield: Calculations of limiting reactant, theoretical yield, and percent yield are central to chemistry because they bring quantitative understanding to chemical reactions. Just as you need to know relationships between ingredients to follow a recipe, so you must know relationships between reactants and products to carry out a chemical reaction. The percent yield in a chemical reaction is often used as a measure of the success of the reaction. Imagine following a recipe and making only 1% of the final product—your cooking would be a failure. Similarly, low percent yields in chemical reactions are usually considered poor, and high percent yields are considered good.

Chemical Skills

Examples

Mole-to-Mole Conversions (Section 8.3)

For mole-to-mole conversions, we follow our standard problem-solving procedure, writing down the given quantities, find quantities, and conversion factors. The conversion factor comes from the balanced chemical equation.

The solution map begins with the number of moles of the given substance and then uses the conversion factor from the balanced chemical equation to get to the number of moles of the substance you are trying to find.

To compute the answer, begin with the number of moles of the given substance and follow the solution map to get to the number of moles of the substance you are trying to find.

EXAMPLE 8.7 Mole-to-Mole Conversions

How many moles of sodium oxide can be synthesized from 4.8 mol of sodium? Assume that there is more than enough oxygen present. The balanced equation is:

$$4\ Na(s) + O_2(g) \longrightarrow 2\ Na_2O(s)$$

Given: 4.8 mol Na

Find: mol Na_2O

Conversion Factor: 4 mol Na $\equiv$ 2 mol Na_2O

Solution Map:

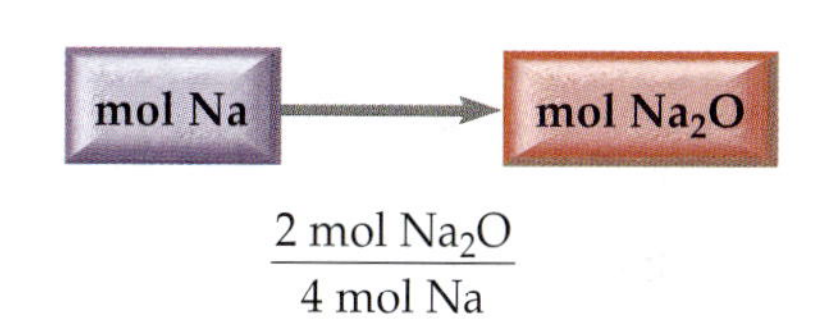

$$\frac{2\ \text{mol}\ Na_2O}{4\ \text{mol Na}}$$

Solution:

$$4.8\ \cancel{\text{mol Na}} \times \frac{2\ \text{mol}\ Na_2O}{4\ \cancel{\text{mol Na}}} = 2.4\ \text{mol}\ Na_2O$$

Mole-to-Mole Conversions (Section 8.4)

For mass-to-mass conversions, we also follow our standard problem-solving procedure, writing down the given quantities, find quantities, and conversion factors. The required conversion factors include those from the balanced chemical equation as well as the molar masses of the substances involved.

LIVE EXAMPLE

EXAMPLE 8.8 Mass-to-Mass Conversions

How many grams of sodium oxide can be synthesized from 17.4 g of sodium? Assume that there is more than enough oxygen present. The balanced equation is:

$$4\ Na(s) + O_2(g) \longrightarrow 2\ Na_2O(s)$$

Given: 17.4 g Na

Find: g Na_2O

Conversion Factors:

4 mol Na = 2 mol Na_2O

Molar mass Na = 22.99 g/mol

Molar mass Na_2O = 61.98 g/mol

Draw the solution map by beginning with the mass of the given substance. Convert to moles using the molar mass and then convert to moles of the substance you are trying to find, using the conversion factor obtained from the balanced chemical equation. Finally, convert to mass of the substance you are trying to find, using its molar mass.

Solution Map:

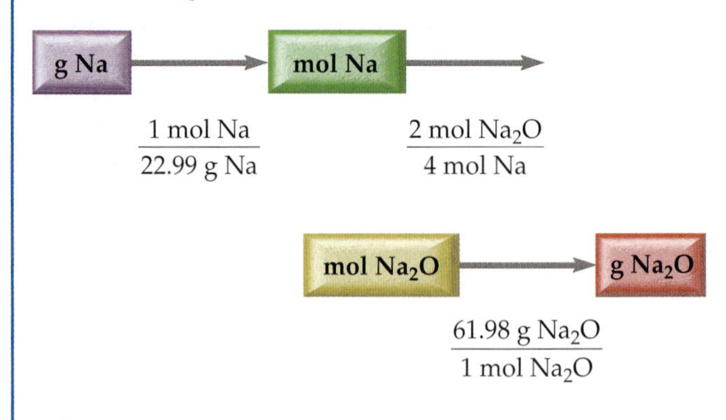

Compute the answer by beginning with the amount of the given substance and multiplying by the appropriate conversion factors to get to the mass of the substance you are trying to find.

Solution:

$$17.4\ \text{g Na} \times \frac{1\ \text{mol Na}}{22.99\ \text{g Na}} \times \frac{2\ \text{mol Na}_2\text{O}}{4\ \text{mol Na}} \times \frac{61.98\ \text{g Na}_2\text{O}}{1\ \text{mol Na}_2\text{O}} = 23.5\ \text{g Na}_2\text{O}$$

Limiting Reactant, Theoretical Yield, and Percent Yield (Sections 8.5, 8.6)

LIVE EXAMPLE

EXAMPLE 8.9 Limiting Reactant, Theoretical Yield, and Percent Yield

10.4 g of Fe reacts with 11.8 g of S to produce 14.2 g of Fe_2S_3. Find the limiting reactant, theoretical yield, and percent yield for this reaction. The balanced chemical equation is:

$$2\ \text{Fe}(s) + 3\ \text{S}(l) \longrightarrow \text{Fe}_2\text{S}_3(s)$$

For limiting-reactant problems, we also follow our standard problem-solving procedure, writing down the given quantities, find quantities, and conversion factors. The required conversion factors include those from the balanced chemical equation as well as the molar masses of the substances involved.

Given:

10.4 g Fe
11.8 g S
14.2 g Fe_2S_3

Find:

limiting reactant
theoretical yield
percent yield

Conversion Factors:

2 mol Fe = 1 mol Fe_2S_3
3 mol S = 1 mol Fe_2S_3
Molar mass Fe = 55.85 g/mol
Molar mass S = 32.06 g/mol
Molar mass Fe_2S_3 = 207.9 g/mol

The solution map for limiting-reactant problems shows how to convert from mass of each of the reactants to mass of the product for each reactant. These are mass-to-mass conversions involving the basic outline of Mass ⟶ Moles ⟶ Moles ⟶ Mass. The reactant that forms the least amount of product is the limiting reactant.

Solution Map:

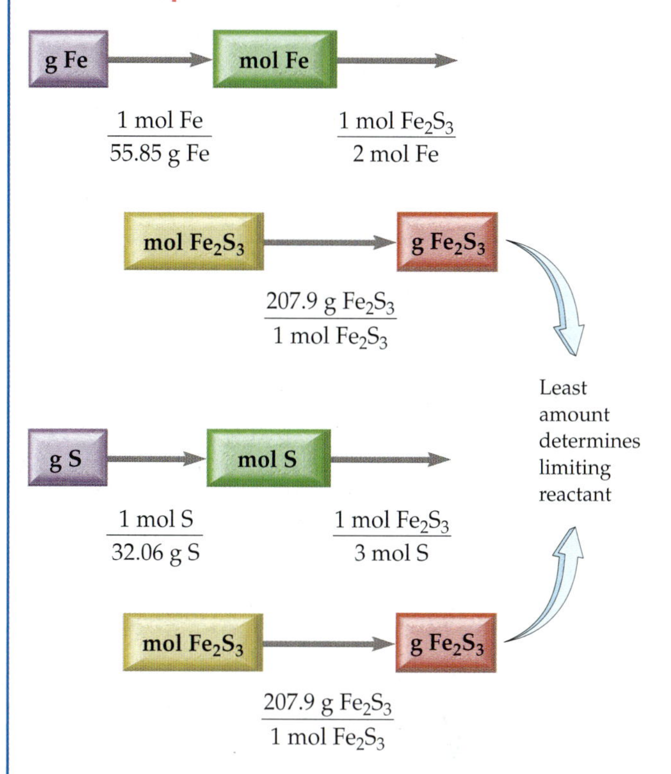

To compute the amount of product formed by each reactant, begin with the given amount of each reactant and multiply by the appropriate conversion factors, as shown in the solution map, to arrive at the mass of product for each reactant. The reactant that forms the least amount of product is the limiting reactant.

Solution:

$$10.4\ \text{g Fe} \times \frac{1\ \text{mol Fe}}{55.85\ \text{g Fe}} \times \frac{1\ \text{mol Fe}_2\text{S}_3}{2\ \text{mol Fe}} \times \frac{207.9\ \text{g Fe}_2\text{S}_3}{1\ \text{mol Fe}_2\text{S}_3} = 19.4\ \text{g Fe}_2\text{S}_3$$

Limiting reactant (Fe) — **Least amount of product** (19.4 g Fe_2S_3)

$$11.8\ \text{g S} \times \frac{1\ \text{mol S}}{32.06\ \text{g S}} \times \frac{1\ \text{mol Fe}_2\text{S}_3}{3\ \text{mol S}} \times \frac{207.9\ \text{g Fe}_2\text{S}_3}{1\ \text{mol Fe}_2\text{S}_3} = 25.5\ \text{g Fe}_2\text{S}_3$$

The theoretical yield is the amount of product formed by the limiting reactant.

The limiting reactant is Fe.
The theoretical yield is 19.4 g of Fe_2S_3.

The percent yield is the actual yield divided by the theoretical yield times 100%.

$$\text{Percent yield} = \frac{\text{Actual yield}}{\text{Theoretical yield}} \times 100\% = \frac{14.2\ \text{g}}{19.4\ \text{g}} \times 100\% = 73.2\%$$

The percent yield is 73.2%.

Key Terms

actual yield [8.5]
equivalent [8.2]
global warming [8.1]
greenhouse gases [8.1]
limiting reactant [8.5]
percent yield [8.5]
stoichiometry [8.1]
theoretical yield [8.5]

Exercises

Questions

1. Why is reaction stoichiometry important? Can you give some examples?
2. What does it mean to say that two quantities in a chemical reaction are equivalent?
3. Write conversion factors showing the relationships between moles of each of the reactants and products in the following reaction.

$$N_2(g) + 3\,H_2(g) \longrightarrow 2\,NH_3(g)$$

4. For the reaction in Problem 3, how many molecules of H_2 are required to completely react with two molecules of N_2? How many moles of H_2 are required to completely react with 2 mol of N_2?
5. Write the conversion factor that you would use to convert from moles of Cl_2 to moles of NaCl in the following reaction.

$$2\,Na(s) + Cl_2(g) \longrightarrow 2\,NaCl$$

6. What is wrong with this statement in reference to the reaction in Problem 5?
"Two grams of Na react with 1 g of Cl_2 to form 2 g of NaCl."
Correct the statement to make it true.
7. What is the general form of the solution map for problems in which you are given the mass of a reactant in a chemical reaction and asked to find the mass of the product that can be made from the given amount of reactant?
8. Consider the following recipe for making tomato and garlic pasta.

$$2 \text{ cups noodles} + 12 \text{ tomatoes} + 3 \text{ cloves garlic} \longrightarrow 4 \text{ servings pasta}$$

If you have 7 cups of noodles, 27 tomatoes, and 9 cloves of garlic, how many servings of pasta can you make? Which ingredient limits the amount of pasta that is possible?
9. In a chemical reaction, what is the limiting reactant?
10. In a chemical reaction, what is the theoretical yield?
11. In a chemical reaction, what are the actual yield and percent yield?
12. If you are given a chemical equation and specific amounts of each reactant in grams, how would you determine how much product can possibly be made?
13. Consider the following generic chemical reaction.

$$A + 2\,B \longrightarrow C + D$$

Suppose you have 12 g of A and 24 g of B. Which of the following statements are true?
(a) A will definitely be the limiting reactant.
(b) B will definitely be the limiting reactant.
(c) A will be the limiting reactant if its molar mass is less than B.
(d) A will be the limiting reactant if its molar mass is greater than B.
14. Consider the following generic chemical equation.

$$A + B \longrightarrow C$$

Suppose 25 g of A were allowed to react with 8 g of B. Analysis of the final mixture showed that A was completely used up and 4 g of B remained. What was the limiting reactant?

Problems

Mole-to-Mole Conversions

15. Consider the following generic chemical reaction.

$$A + 2\,B \longrightarrow C$$

How many moles of C are formed upon complete reaction of:
(a) 1 mol of A
(b) 1 mol of B
(c) 2 mol of A
(d) 2 mol of B
16. Consider the following generic chemical reaction.

$$2\,A + 3\,B \longrightarrow 3\,C$$

How many moles of B are required to completely react with:
(a) 2 mol of A
(b) 4 mol of A
(c) 5 mol of A
(d) 17 mol of A

17. For the reaction shown, calculate how many moles of NO_2 form when each of the following completely reacts.

$$2\,N_2O_5(g) \longrightarrow 4\,NO_2(g) + O_2(g)$$

(a) 1.3 mol N_2O_5
(b) 5.8 mol N_2O_5
(c) 4.45×10^3 mol N_2O_5
(d) 1.006×10^{-3} mol N_2O_5

18. For the reaction shown, calculate how many moles of NH_3 form when each of the following completely reacts.

$$3\,N_2H_4(l) \longrightarrow 4\,NH_3(g) + N_2(g)$$

(a) 5.3 mol N_2H_4
(b) 2.28 mol N_2H_4
(c) 5.8×10^{-2} mol N_2H_4
(d) 9.76×10^7 mol N_2H_4

19. For each reaction, calculate how many moles of the product form when 1.48 mol of the reactant in color completely reacts. Assume that there is more than enough of the other reactant.
(a) $H_2(g) + Cl_2(g) \longrightarrow 2\,HCl(g)$
(b) $2\,H_2(g) + O_2(g) \longrightarrow 2\,H_2O(l)$
(c) $2\,Na(s) + O_2(g) \longrightarrow Na_2O_2(s)$
(d) $2\,S(s) + 3\,O_2(g) \longrightarrow 2\,SO_3(g)$

20. For each reaction, calculate how many moles of the product form when 0.036 mol of the reactant in color completely reacts. Assume that there is more than enough of the other reactant.
(a) $2\,Ca(s) + O_2(g) \longrightarrow 2\,CaO(s)$
(b) $4\,Fe(s) + 3\,O_2(g) \longrightarrow 2\,Fe_2O_3(s)$
(c) $4\,K(s) + O_2(g) \longrightarrow 2\,K_2O(s)$
(d) $4\,Al(s) + 3\,O_2(g) \longrightarrow 2\,Al_2O_3(s)$

21. For the reaction shown, calculate how many moles of each product form when the following amounts of each reactant completely react to form products. Assume that there is more than enough of the other reactant.

$$2\,PbS(s) + 3\,O_2(g) \longrightarrow 2\,PbO(s) + 2\,SO_2(g)$$

(a) 1.2 mol PbS
(b) 1.2 mol O_2
(c) 7.9 mol PbS
(d) 7.9 mol O_2

22. For the reaction shown, calculate how many moles of each product form when the following amounts of each reactant completely react to form products. Assume that there is more than enough of the other reactant.

$$C_3H_8(g) + 5\,O_2(g) \longrightarrow 3\,CO_2(g) + 4\,H_2O(g)$$

(a) 8.7 mol C_3H_8
(b) 8.7 mol O_2
(c) 0.0918 mol C_3H_8
(d) 0.0918 mol O_2

23. Consider the following balanced equation.

$$2\,N_2H_4(g) + N_2O_4(g) \longrightarrow 3\,N_2(g) + 4\,H_2O(g)$$

Complete the following table, showing the appropriate number of moles of reactants and products. If the number of moles of a reactant is provided, fill in the required amount of the other reactant, as well as the moles of each product formed. If the number of moles of a product is provided, fill in the required amount of each reactant to make that amount of product, as well as the amount of the other product that is made.

mol N_2H_4	*mol N_2O_4*	*mol N_2*	*mol H_2O*
_____	2	_____	_____
6	_____	_____	_____
_____	_____	_____	8
_____	5.5	_____	_____
3	_____	_____	_____
_____	_____	12.4	_____

24. Consider the following balanced equation.

$$SiO_2(s) + 3\,C(s) \longrightarrow SiC(s) + 2\,CO(g)$$

Complete the following table, showing the appropriate number of moles of reactants and products. If the number of moles of a reactant is provided, fill in the required amount of the other reactant, as well as the moles of each product formed. If the number of moles of a product is provided, fill in the required amount of each reactant to make that amount of product, as well as the amount of the other product that is made.

mol SiO_2	*mol C*	*mol SiC*	*mol CO*
_____	6	_____	_____
3	_____	_____	_____
_____	_____	_____	10
_____	9.5	_____	_____
3.2	_____	_____	_____

25. Consider the following unbalanced equation for the combustion of butane.

$$C_4H_{10}(g) + O_2(g) \longrightarrow CO_2(g) + H_2O(g)$$

Balance the equation and determine how many moles of O_2 are required to react completely with 4.9 mol of C_4H_{10}.

26. Consider the following unbalanced equation for the neutralization of acetic acid.

$$HC_2H_3O_2(aq) + Ca(OH)_2(aq) \longrightarrow H_2O(l) + Ca(C_2H_3O_2)_2(aq)$$

Balance the equation and determine how many moles of $Ca(OH)_2$ are required to completely neutralize 1.07 mol of $HC_2H_3O_2$.

27. Consider the following unbalanced equation for the reaction of solid lead with silver nitrate.

$$Pb(s) + AgNO_3(aq) \longrightarrow Pb(NO_3)_2(aq) + Ag(s)$$

(a) Balance the equation.
(b) How many moles of silver nitrate are required to completely react with 9.3 mol of lead?
(c) How many moles of Ag(s) are formed by the complete reaction of 28.4 mol of Pb?

28. Consider the following unbalanced equation for the reaction of aluminum with sulfuric acid.

$$Al(s) + H_2SO_4(aq) \longrightarrow Al_2(SO_4)_3(aq) + H_2(g)$$

(a) Balance the equation.
(b) How many moles of H_2SO_4 are required to completely react with 8.3 mol of Al?
(c) How many moles of H_2 are formed by the complete reaction of 0.341 mol of Al?

Mass-to-Mass Conversions

29. For the reaction shown, calculate how many grams of oxygen form when each of the following completely reacts.

$$2\,HgO(s) \longrightarrow 2\,Hg(l) + O_2(g)$$

(a) 2.88 g HgO
(b) 5.21 g HgO
(c) 1.3×10^3 g HgO
(d) 4.6×10^{-3} g HgO

30. For the reaction shown, calculate how many grams of oxygen form when each of the following completely reacts.

$$2\,KClO_3(s) \longrightarrow 2\,KCl(s) + 3\,O_2(g)$$

(a) 8.91 g $KClO_3$
(b) 0.549 g $KClO_3$
(c) 8.4×10^6 g $KClO_3$
(d) 3.3×10^{-4} g $KClO_3$

31. For each of the reactions shown, calculate how many grams of the product form when 1.8 g of the reactant in color completely reacts. Assume that there is more than enough of the other reactant.

(a) $2\,Na(s) + Cl_2(g) \longrightarrow 2\,NaCl(s)$
(b) $CaO(s) + CO_2(g) \longrightarrow CaCO_3(s)$
(c) $2\,Mg(s) + O_2(g) \longrightarrow 2\,MgO(s)$
(d) $Na_2O(s) + H_2O(l) \longrightarrow 2\,NaOH(aq)$

32. For each of the reactions shown, calculate how many grams of the product form when 14.4 g of the reactant in color completely reacts. Assume that there is more than enough of the other reactant.

(a) $Ca(s) + Cl_2(g) \longrightarrow CaCl_2(s)$
(b) $2\,K(s) + Br_2(l) \longrightarrow 2\,KBr(s)$
(c) $4\,Cr(s) + 3\,O_2(g) \longrightarrow 2\,Cr_2O_3(s)$
(d) $2\,Sr(s) + O_2(g) \longrightarrow 2\,SrO(s)$

33. For the reaction shown, calculate how many grams of each product form when the following amounts of reactant completely react to form products. Assume that there is more than enough of the other reactant.

$$2\,Al(s) + Fe_2O_3(s) \longrightarrow Al_2O_3(s) + 2\,Fe(l)$$

(a) 4.7 g Al
(b) 4.7 g Fe_2O_3

34. For the reaction shown, calculate how many grams of each product form when the following amounts of reactant completely react to form products. Assume that there is more than enough of the other reactant.

$$2\,HCl(aq) + Na_2CO_3(aq) \longrightarrow 2\,NaCl(aq) + H_2O(l) + CO_2(g)$$

(a) 10.8 g HCl
(b) 10.8 g Na_2CO_3

35. Consider the following balanced equation for the combustion of methane, a component of natural gas.

$$CH_4(g) + 2\,O_2(g) \longrightarrow CO_2(g) + 2\,H_2O(g)$$

Complete the following table, showing the appropriate masses of reactants and products. If the mass of a reactant is provided, fill in the mass of other reactant required to completely react with the given mass, as well as the mass of each product formed. If the mass of a product is provided, fill in the required masses of each reactant to make that amount of product, as well as the mass of the other product that is formed.

Mass CH_4	*Mass O_2*	*Mass CO_2*	*Mass H_2O*
_____	2.57 g	_____	_____
22.32 g	_____	_____	_____
_____	_____	_____	11.32 g
_____	_____	2.94 g	_____
3.18 kg	_____	_____	_____
_____	_____	2.35×10^3 kg	_____

36. Consider the following balanced equation for the combustion of butane, a fuel often used in lighters.

$$2\,C_4H_{10}(g) + 13\,O_2(g) \longrightarrow 8\,CO_2(g) + 10\,H_2O(g)$$

Complete the following table, showing the appropriate masses of reactants and products. If the mass of a reactant is provided, fill in the mass of other reactant required to completely react with the given mass, as well as the mass of each product formed. If the mass of a product is provided, fill in the required masses of each reactant to make that amount of product, as well as the mass of the other product that is formed.

Mass C_4H_{10}	*Mass O_2*	*Mass CO_2*	*Mass H_2O*
_____	1.11 g	_____	_____
5.22 g	_____	_____	_____
_____	_____	10.12 g	_____
_____	_____	_____	9.04 g
232 mg	_____	_____	_____
_____	_____	118 mg	_____

37. For each of the following acid–base reactions, calculate how many grams of each acid are necessary to completely react with and neutralize 2.5 g of the base.
(a) $HCl(aq) + NaOH(aq) \longrightarrow H_2O(l) + NaCl(aq)$
(b) $2\,HNO_3(aq) + Ca(OH)_2(aq) \longrightarrow 2\,H_2O(l) + Ca(NO_3)_2(aq)$
(c) $H_2SO_4(aq) + 2\,KOH(aq) \longrightarrow 2\,H_2O(l) + K_2SO_4(aq)$

38. For each of the following precipitation reactions, calculate how many grams of the first reactant are necessary to completely react with 17.3 g of the second reactant.
(a) $2\,KI(aq) + Pb(NO_3)_2(aq) \longrightarrow PbI_2(s) + 2\,KNO_3(aq)$
(b) $Na_2CO_3(aq) + CuCl_2(aq) \longrightarrow CuCO_3(s) + 2\,NaCl(aq)$
(c) $K_2SO_4(aq) + Sr(NO_3)_2(aq) \longrightarrow SrSO_4(s) + 2\,KNO_3(aq)$

39. Sulfuric acid can dissolve aluminum metal according to the following reaction.

$$2\,Al(s) + 3\,H_2SO_4(aq) \longrightarrow Al_2(SO_4)_3(aq) + 3\,H_2(g)$$

Suppose you wanted to dissolve an aluminum block with a mass of 22.5 g. What minimum amount of H_2SO_4 in grams would you need? How many grams of H_2 gas would be produced by the complete reaction of the aluminum block?

40. Hydrochloric acid can dissolve solid iron according to the following reaction.

$$Fe(s) + 2\,HCl(aq) \longrightarrow FeCl_2(aq) + H_2(g)$$

How much HCl in grams would you need to dissolve a 2.8-g iron bar on a padlock? How much H_2 would be produced by the complete reaction of the iron bar?

Limiting Reactant, Theoretical Yield, and Percent Yield

41. Consider the following generic chemical equation.

$$2\,A + 4\,B \longrightarrow 3\,C$$

What is the limiting reactant when each of the following amounts of A and B are allowed to react?
(a) 2 mol A; 5 mol B
(b) 1.8 mol A; 4 mol B
(c) 3 mol A; 4 mol B
(d) 22 mol A; 40 mol B

42. Consider the following generic chemical equation.

$$A + 3\,B \longrightarrow C$$

What is the limiting reactant when each of the following amounts of A and B are allowed to react?
(a) 1 mol A; 4 mol B
(b) 2 mol A; 3 mol B
(c) 0.5 mol A; 1.6 mol B
(d) 24 mol A; 75 mol B

43. Determine the theoretical yield of C when each of the following amounts of A and B are allowed to react in the generic reaction:

$$A + 2\,B \longrightarrow 3\,C$$

(a) 1 mol A; 1 mol B
(b) 2 mol A; 2 mol B
(c) 1 mol A; 3 mol B
(d) 32 mol A; 68 mol B

44. Determine the theoretical yield of C when each of the following amounts of A and B are allowed to react in the generic reaction:

$$2\,A + 3\,B \longrightarrow 2\,C$$

(a) 2 mol A; 4 mol B
(b) 3 mol A; 3 mol B
(c) 5 mol A; 6 mol B
(d) 4 mol A; 5 mol B

45. For the reaction shown, find the limiting reactant for each of the following initial amounts of reactants.

$$2\,K(s) + Cl_2(g) \longrightarrow 2\,KCl(s)$$

(a) 1 mol K; 1 mol Cl_2
(b) 1.8 mol K; 1 mol Cl_2
(c) 2.2 mol K; 1 mol Cl_2
(d) 14.6 mol K; 7.8 mol Cl_2

46. For the reaction shown, find the limiting reactant for each of the following initial amounts of reactants.

$$4\,Cr(s) + 3\,O_2(g) \longrightarrow 2\,Cr_2O_3(s)$$

(a) 1 mol Cr; 1 mol O_2
(b) 4 mol Cr; 2.5 mol O_2
(c) 12 mol Cr; 10 mol O_2
(d) 14.8 mol Cr; 10.3 mol O_2

47. For the reaction shown, compute the theoretical yield of product in moles for each of the following initial amounts of reactants.

$$2\,Mn(s) + 3\,O_2(g) \longrightarrow 2\,MnO_3(s)$$

(a) 2 mol Mn; 2 mol O_2
(b) 4.8 mol Mn; 8.5 mol O_2
(c) 0.114 mol Mn; 0.161 mol O_2
(d) 27.5 mol Mn; 43.8 mol O_2

48. For the reaction shown, compute the theoretical yield of the product in moles for each of the following initial amounts of reactants.

$$Ti(s) + 2\,Cl_2(g) \longrightarrow TiCl_4(s)$$

(a) 2 mol Ti; 2 mol Cl_2
(b) 5 mol Ti; 9 mol Cl_2
(c) 0.483 mol Ti; 0.911 mol Cl_2
(d) 12.4 mol Ti; 15.8 mol Cl_2

49. Consider the following reaction:

$$4\,HCl(g) + O_2(g) \longrightarrow 2\,H_2O(g) + 2\,Cl_2(g)$$

Each of the following molecular diagrams represents an initial mixture of the reactants. How many molecules of Cl_2 would be formed by complete reaction in each case? (Assume 100% actual yield.)

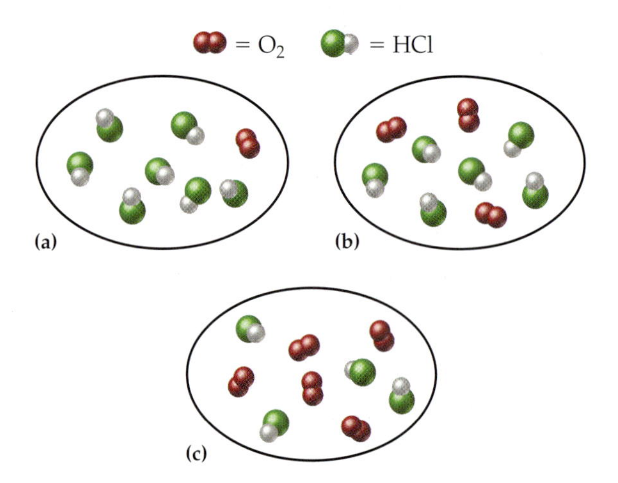

50. Consider the following reaction:

$$2\,CH_3OH(g) + 3\,O_2(g) \longrightarrow 2\,CO_2(g) + 4\,H_2O(g)$$

Each of the following molecular diagrams represents an initial mixture of the reactants. How many CO_2 molecules would be formed by complete reaction in each case? (Assume 100% actual yield.)

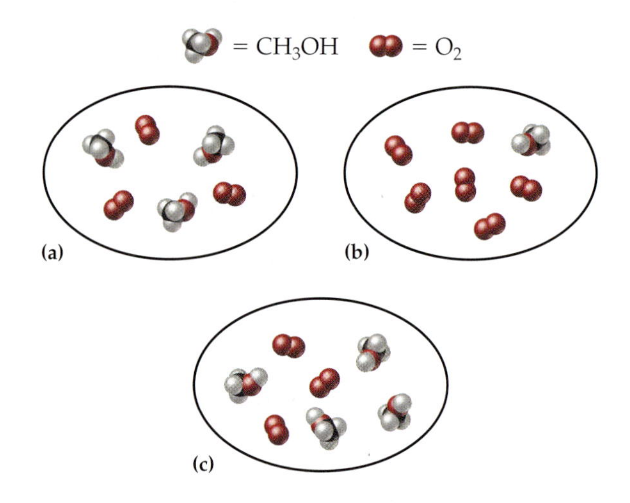

51. For the reaction shown, find the limiting reactant for each of the following initial amounts of reactants.

$$2\ Li(s) + F_2(g) \longrightarrow 2\ LiF(s)$$

(a) 1.0 g Li; 1.0 g F_2
(b) 10.5 g Li; 37.2 g F_2
(c) 2.85×10^3 g Li; 6.79×10^3 g F_2

52. For the reaction shown, find the limiting reactant for each of the following initial amounts of reactants.

$$4\ Al(s) + 3\ O_2(g) \longrightarrow 2\ Al_2O_3(s)$$

(a) 1.0 g Al; 1.0 g O_2
(b) 2.2 g Al; 1.8 g O_2
(c) 0.353 g Al; 0.482 g O_2

53. For the reaction shown, compute the theoretical yield of the product in grams for each of the following initial amounts of reactants.

$$2\ Al(s) + 3\ Cl_2(g) \longrightarrow 2\ AlCl_3(s)$$

(a) 1.0 g Al; 1.0 g Cl_2
(b) 5.5 g Al; 19.8 g Cl_2
(c) 0.439 g Al; 2.29 g Cl_2

54. For the reaction shown, compute the theoretical yield of the product in grams for each of the following initial amounts of reactants.

$$Ti(s) + 2\ F_2(g) \longrightarrow TiF_4(s)$$

(a) 1.0 g Ti; 1.0 g F_2
(b) 4.8 g Ti; 3.2 g F_2
(c) 0.388 g Ti; 0.341 g F_2

55. If the theoretical yield of a reaction is 24.8 g and the actual yield is 18.5 g, what is the percent yield?

56. If the theoretical yield of a reaction is 0.118 g and the actual yield is 0.104 g, what is the percent yield?

57. Consider the following reaction.

$$CaO(s) + CO_2(g) \longrightarrow CaCO_3(s)$$

A chemist allows 14.4 g of CaO and 13.8 g of CO_2 to react. When the reaction is finished, the chemist collects 19.4 g of $CaCO_3$. Determine the limiting reactant, theoretical yield, and percent yield for the reaction.

58. Consider the following reaction.

$$SO_3(g) + H_2O(l) \longrightarrow H_2SO_4(aq)$$

A chemist allows 61.5 g of SO_3 and 11.2 g of H_2O to react. When the reaction is finished, the chemist collects 54.9 g of H_2SO_4. Determine the limiting reactant, theoretical yield, and percent yield for the reaction.

59. Consider the following reaction.

$$2\ NiS_2(s) + 5\ O_2(g) \longrightarrow 2\ NiO(s) + 4\ SO_2(g)$$

When 11.2 g of NiS_2 are allowed to react with 5.43 g of O_2, 4.86 g of NiO are obtained. Determine the limiting reactant, theoretical yield of NiO, and percent yield for the reaction.

60. Consider the following reaction.

$$4\ HCl(g) + O_2(g) \longrightarrow 2\ H_2O(l) + 2\ Cl_2(g)$$

When 63.1 g of HCl are allowed to react with 17.2 g of O_2, 49.3 g of Cl_2 are collected. Determine the limiting reactant, theoretical yield of Cl_2, and percent yield for the reaction.

61. Lead ions can be precipitated from solution with NaCl according to the following reaction.

$$Pb^{2+}(aq) + 2\ NaCl(aq) \longrightarrow PbCl_2(s) + 2\ Na^+(aq)$$

When 135.8 g of NaCl are added to a solution containing 195.7 g of Pb^{2+}, a $PbCl_2$ precipitate forms. The precipitate is filtered and dried and found to have a mass of 252.4 g. Determine the limiting reactant, theoretical yield of $PbCl_2$, and percent yield for the reaction.

62. Magnesium oxide can be made by heating magnesium metal in the presence of oxygen. The balanced equation for the reaction is:

$$2\ Mg(s) + O_2(g) \longrightarrow 2\ MgO(s)$$

When 10.1 g of Mg are allowed to react with 10.5 g of O_2, 11.9 g of MgO are collected. Determine the limiting reactant, theoretical yield, and percent yield for the reaction.

Cumulative Problems

63. A solution contains an unknown mass of dissolved barium ions. When sodium sulfate is added to the solution, a white precipitate forms. The precipitate is filtered and dried and then found to have a mass of 258 mg. What mass of barium was in the original solution? (Assume that all of the barium was precipitated out of solution by the reaction.)

64. A solution contains an unknown mass of dissolved silver ions. When potassium chloride is added to the solution, a white precipitate forms. The precipitate is filtered and dried and then found to have a mass of 212 mg. What mass of silver was in the original solution? (Assume that all of the silver was precipitated out of solution by the reaction.)

65. Sodium bicarbonate is often used as an antacid to neutralize excess hydrochloric acid in an upset stomach. How much hydrochloric acid in grams can be neutralized by 3.5 g of sodium bicarbonate? (Hint: Begin by writing a balanced equation for the reaction between aqueous sodium bicarbonate and aqueous hydrochloric acid.)

66. Toilet bowl cleaners often contain hydrochloric acid to dissolve the calcium carbonate deposits that accumulate within a toilet bowl. How much calcium carbonate in grams can be dissolved by 5.8 g of HCl? (Hint: Begin by writing a balanced equation for the reaction between hydrochloric acid and calcium carbonate.)

67. The combustion of gasoline produces carbon dioxide and water. Assume gasoline to be pure octane (C_8H_{18}) and calculate how many kilograms of carbon dioxide are added to the atmosphere per 1.0 kg of octane burned. (Hint: Begin by writing a balanced equation for the combustion reaction.)

68. Many home barbecues are fueled with propane gas (C_3H_8). How much carbon dioxide in kilograms is produced upon the complete combustion of 18.9 L of propane (approximate contents of one 5-gal tank)? Assume that the density of the liquid propane in the tank is 0.621 g/mL. (Hint: Begin by writing a balanced equation for the combustion reaction.)

69. A hard water solution contains 4.8 g of calcium chloride. How much sodium phosphate in grams should be added to the solution to completely precipitate all of the calcium?

70. Magnesium ions can be precipitated from seawater by the addition of sodium hydroxide. How much sodium hydroxide in grams must be added to a sample of seawater to completely precipitate the 88.4 mg of magnesium present?

71. Hydrogen gas can be prepared in the laboratory by a single displacement reaction in which solid zinc reacts with hydrochloric acid. How much zinc in grams is required to make 14.5 g of hydrogen gas through this reaction?

72. Sodium peroxide (Na_2O_2) reacts with water to form sodium hydroxide and oxygen gas. Write a balanced equation for the reaction and determine how much oxygen in grams is formed by the complete reaction of 35.23 g of Na_2O_2.

73. Ammonium nitrate reacts explosively upon heating to form nitrogen gas, oxygen gas, and gaseous water. Write a balanced equation for this reaction and determine how much oxygen in grams is produced by the complete reaction of 1.00 kg of ammonium nitrate.

74. Pure oxygen gas can be prepared in the laboratory by the decomposition of solid potassium chlorate to form solid potassium chloride and oxygen gas. How much oxygen gas in grams can be prepared from 45.8 g of potassium chlorate?

75. Aspirin can be made in the laboratory by reacting acetic anhydride ($C_4H_6O_3$) with salicylic acid ($C_7H_6O_3$) to form aspirin ($C_9H_8O_4$) and acetic acid ($C_2H_4O_2$). The balanced equation is:

$$C_4H_6O_3 + C_7H_6O_3 \longrightarrow C_9H_8O_4 + C_2H_4O_2.$$

In a laboratory synthesis, a student begins with 5.00 mL of acetic anhydride (density = 1.08 g/mL) and 2.08 g of salicylic acid. Once the reaction is complete, the student collects 2.01 g of aspirin. Determine the limiting reactant, theoretical yield of aspirin, and percent yield for the reaction.

76. The combustion of liquid ethanol (C_2H_5OH) produces carbon dioxide and water. After 3.8 mL of ethanol (density = 0.789 g/mL) was allowed to burn in the presence of 12.5 g of oxygen gas, 3.10 mL of water (density = 1.00 g/mL) was collected. Determine the limiting reactant, theoretical yield of H_2O, and percent yield for the reaction. (Hint: Write a balanced equation for the combustion of ethanol.)

77. Urea (CH_4N_2O), a common fertilizer, can be synthesized by the reaction of ammonia (NH_3) with carbon dioxide:

$$2\,NH_3(aq) + CO_2(aq) \longrightarrow CH_4N_2O(aq) + H_2O(l)$$

An industrial synthesis of urea obtains 87.5 kg of urea upon reaction of 68.2 kg of ammonia with 105 kg of carbon dioxide. Determine the limiting reactant, theoretical yield of urea, and percent yield for the reaction.

78. Silicon, which occurs in nature as SiO_2, is the material from which most computer chips are made. If SiO_2 is heated until it melts into a liquid, it reacts with solid carbon to form liquid silicon and carbon monoxide gas. In an industrial preparation of silicon, 52.8 kg of SiO_2 reacted with 25.8 kg of carbon to produce 22.4 kg of silicon. Determine the limiting reactant, theoretical yield, and percent yield for the reaction.

79. The ingestion of lead from food, water, or other environmental sources can cause lead poisoning, a serious condition that affects the central nervous system, causing symptoms such as distractibility, lethargy, and loss of motor function. Lead poisoning is treated with chelating agents, substances that bind to lead and allow it to be eliminated in the urine. A modern chelating agent used for this purpose is succimer ($C_4H_6O_4S_2$). Suppose you are trying to determine the appropriate dose for succimer treatment of lead poisoning. What minimum mass of succimer in milligrams is needed to bind all of the lead in a patient's bloodstream? Assume that patient blood lead levels are 0.550 mg/L, that total blood volume is 5.0 L, and that 1 mol of succimer binds 1 mol of lead.

80. An emergency breathing apparatus often placed in mines or caves works via the following chemical reaction:

$$4\,KO_2(s) + 2\,CO_2(g) \longrightarrow 2\,K_2CO_3(s) + O_2(g)$$

If the oxygen supply becomes limited or if the air becomes poisoned, a worker can use the apparatus to breathe while exiting the mine. Notice that the reaction produces O_2, which can be breathed, and absorbs CO_2, a product of respiration. What minimum amount of KO_2 is required for the apparatus to produce enough oxygen to allow the user 15 minutes to exit in an emergency? Assume that an adult consumes approximately 4.4 g of oxygen in 15 minutes of normal breathing.

Highlight Problems

81. A loud classroom demonstration involves igniting a hydrogen-filled balloon. The hydrogen within the balloon explosively reacts with oxygen in the air to form water.

$$2\,H_2(g) + O_2(g) \longrightarrow 2\,H_2O(g)$$

If the balloon is filled with a mixture of hydrogen and oxygen, the explosion is even louder than if the balloon is filled with only hydrogen, and the intensity of the explosion depends on the relative amounts of oxygen and hydrogen within the balloon. Look at the following molecular views representing different amounts of hydrogen and oxygen in four different balloons. Based on the balanced chemical equation, which balloon will make the loudest explosion?

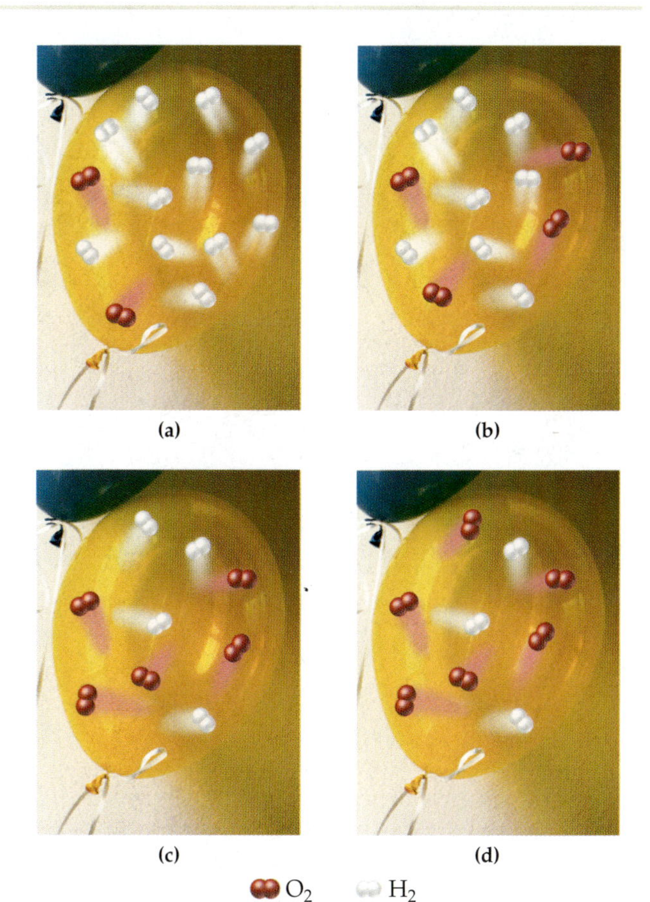

82. A hydrochloric acid solution will neutralize a sodium hydroxide solution. Look at the following molecular views showing one beaker of HCl and four beakers of NaOH. Which NaOH beaker will just neutralize the HCl beaker? Begin by writing a balanced chemical equation for the neutralization reaction.

83. As we have seen, scientists have grown increasingly worried about the potential for global warming caused by increasing atmospheric carbon dioxide levels. The world burns the fossil fuel equivalent of 7.0×10^{12} kg of petroleum per year. Assume that all of this petroleum is in the form of octane (C_8H_{18}) and calculate how much CO_2 in kilograms is produced by world fossil fuel combustion per year. (Hint: Begin by writing a balanced equation for the combustion of octane.) If the atmosphere currently contains approximately 3.0×10^{15} kg of CO_2, how long will it take for the world's fossil fuel combustion to double the amount of atmospheric carbon dioxide?

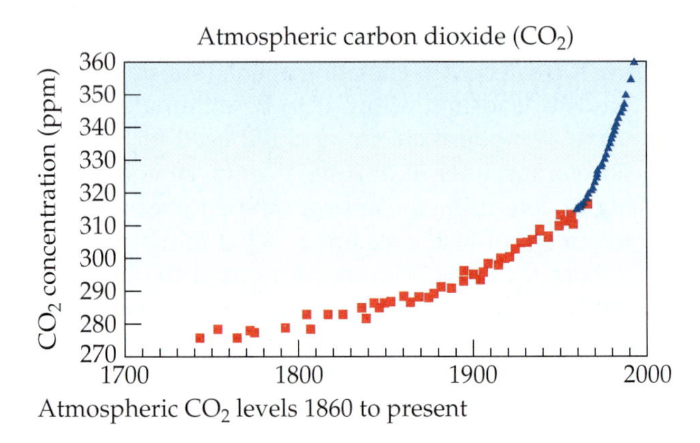

▲ Atmospheric CO_2 levels, 1860 to present.

84. Lakes that have been acidified by acid rain can be neutralized by the addition of limestone ($CaCO_3$). How much limestone in kilograms would be required to completely neutralize a 5.2×10^9–L lake containing 5.0×10^{-3} g of H_2SO_4 per liter?

Answers to Skillbuilder Exercises

Skillbuilder 8.1 49.2 mol H_2O

Skillbuilder 8.2 6.88 g HCl

Skillbuilder 8.3 4.0×10^3 kg H_2SO_4

Skillbuilder 8.4 Limiting reactant is Na; theoretical yield is 4.8 mol of NaF

Skillbuilder 8.5 30.7 g NH_3

Skillbuilder Plus, p. 246 29.4 kg NH_3

Skillbuilder 8.6 Limiting reactant is CO; theoretical yield = 127 g Fe; percent yield = 69.0%

Answers to Conceptual Checkpoints

8.1 **(d)** The manufacturer has enough seats for 17 tricycles, enough bells for 19 tricycles, enough pedals for 18 tricycles (with one left over), but only enough wheels for 16 tricycles (with one left over).

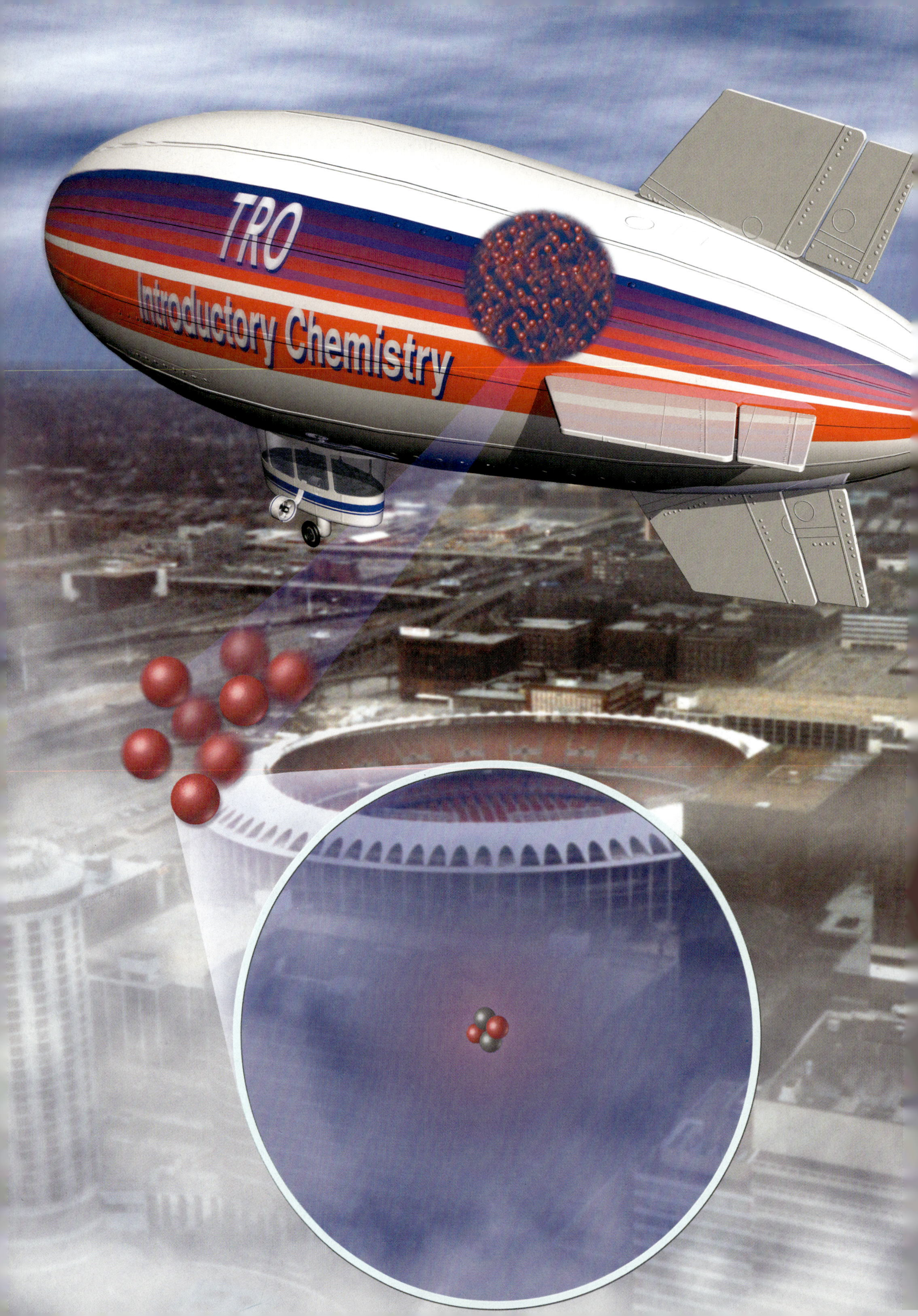
TRO
Introductory Chemistry

CHAPTER 9

Electrons in Atoms and the Periodic Table

"Anyone who is not shocked by quantum mechanics has not understood it."

Niels Bohr (1885–1962)

9.1 Blimps, Balloons, and Models of the Atom

▲ The Hindenburg was filled with hydrogen, a reactive and flammable gas. **Question:** What makes hydrogen reactive?

◀ Modern blimps are filled with helium, an inert gas. The nucleus of the helium atom (inset) has two protons, so the neutral helium atom has two electrons—a highly stable configuration. In this chapter we learn about models that explain the inertness of helium and the reactivity of other elements.

You have probably seen one of the Goodyear blimps floating in the sky. A Goodyear blimp is often present at championship sporting events such as the Rose Bowl, the Indy 500, and the U.S. Open golf tournament. It was present at the Statue of Liberty's 100th birthday party and has made appearances in countless movies and television shows. The blimp's inherent stability allows it to provide spectacular views of the world below for television and film.

The Goodyear blimp is similar to a large balloon. Unlike airplanes, which must be moving fast to stay in flight, a blimp or *airship* floats in air because it is filled with a gas that is lighter than air. The Goodyear blimp is filled with helium. Other airships in history, however, have used hydrogen for buoyancy. For example, the Hindenburg—the largest airship ever constructed—was filled with hydrogen, which turned out to be a poor choice. Hydrogen is a reactive and flammable gas. On May 6, 1937, while landing in New Jersey on its first transatlantic crossing, the Hindenburg burst into flames, destroying the airship and killing thirty-six of the ninety-seven passengers. Apparently, as the Hindenburg was landing, a leak in the hydrogen gas ignited, resulting in an explosion that destroyed the ship. (The skin of the Hindenburg, which was constructed of a flammable material, may has also been partially to blame for its demise.)

A similar accident cannot happen to the Goodyear blimp because it is filled with helium, an inert and therefore nonflammable gas. A spark or even a flame would actually be *extinguished* by helium.

Why is helium inert? What is it about helium *atoms* that makes helium *gas* inert? On the other hand, why is hydrogen reactive? Recall from Chapter 5 that elemental hydrogen exists as a diatomic element. Hydrogen atoms are so reactive that they react with each other to form hydrogen molecules. What is it about hydrogen atoms that make them so reactive? What is the difference between hydrogen and helium that accounts for their different reactivities?

Alkali metals 1 1A	Noble gases 18 8A
1 **H**	2 **He**
3 **Li**	10 **Ne**
11 **Na**	18 **Ar**
19 **K**	36 **Kr**
37 **Rb**	54 **Xe**
55 **Cs**	86 **Rn**
87 **Fr**	

The noble gases are chemically inert and the alkali metals are chemically reactive. Why?

When we examine the properties of hydrogen and helium, we make observations about nature. Mendeleev's periodic law, first discussed in Chapter 4, summarizes the results of many similar observations on the properties of elements:

> When the elements are arranged in order of increasing atomic number, certain sets of properties recur periodically.

We know that hydrogen's reactivity recurs with lithium, sodium, and the other Group I metals. We also know that helium's inertness recurs with neon, argon, and the other noble gases. These are critical parts of the scientific method: observing something about nature and formulating a law that summarizes a large number of observations. Now, we need a model or theory that gives the underlying reasons for these observations and laws.

The periodic law stated here is a modification of Mendeleev's original formulation. Mendeleev listed elements in order of increasing *mass*; today we list them in order of increasing *atomic number*.

In this chapter, we examine two important models—the **Bohr model** and the **quantum-mechanical model**—that propose explanations for the inertness of helium, the reactivity of hydrogen, and the periodic law. These models explain how electrons exist in atoms and how those electrons affect the chemical and physical properties of elements. We have already learned much about the behavior of elements. We know, for example, that sodium tends to form 1+ ions and that fluorine tends to form 1− ions. We know that some elements are metals and that others are nonmetals. We know that the noble gases are chemically inert and that the alkali metals are chemically reactive. But we do not know *why*. The models in this chapter explain why.

These models were developed in the early 1900s and caused a revolution in the physical sciences, changing our fundamental view of matter at its most basic level. The scientists who devised these models—including Niels Bohr, Erwin Schrödinger, and Albert Einstein—were bewildered by their discoveries. Bohr claimed, "Anyone who is not shocked by quantum mechanics has not understood it." Schrödinger lamented, "I don't like it, and I am sorry I ever had anything to do with it." Einstein disbelieved it, insisting that, "God does not play dice with the universe." However, the quantum-mechanical model has such explanatory power that it is rarely questioned today. It forms the basis of the modern periodic table and our understanding of chemical bonding. Its applications include lasers, computers, and semiconductor devices, and it has given us new ways to design drugs that cure disease. The quantum-mechanical model for the atom is, in many ways, the foundation of modern chemistry.

▲ Niels Bohr (left) and Erwin Schrödinger (right), along with Albert Einstein, played a role in the development of quantum mechanics, yet they were bewildered by their own theory.

9.2 Light: Electromagnetic Radiation

Before we explore models of the atom, we must understand a few things about light, because the interaction of light with atoms helped to shape these models. Light is familiar to all of us—we see the world by it—but we may not know much about what light is. Unlike most of what we have encountered so far in this book, light is not matter—it has no mass. Light is a form of **electromagnetic radiation**, a type of energy that travels through space at a constant speed of 3.0×10^8 m/s (186,000 mi/s). At this speed, a flash of light generated at the equator would travel around the world in one-seventh of a second. This extremely fast speed is part of the reason that you see a firework in the sky before you hear the sound of its explosion. The light from the exploding firework reaches your eye almost instantaneously. The sound, traveling much slower, takes longer.

▲ When a water surface is disturbed, waves are created that radiate outward from the site.

Before the advent of quantum mechanics, light was described exclusively as a wave of electromagnetic energy traveling through space. You are probably familiar with water waves (think of the waves created by a rock dropped into a still pond) or you may have created a wave on a rope by moving the end of the rope up and down in a quick motion. In either case, the wave carries energy as it moves through the water or along the rope.

The Greek letter *lambda* (λ) is pronounced "lam-duh."

Waves are generally characterized by their **wavelength (λ),** the distance between adjacent wave crests (▼ Figure 9.1). For visible light, wavelength determines color. For example, orange light has a longer wavelength than blue light. White light, as produced by the sun or by a lightbulb, contains a spectrum of wavelengths and therefore a spectrum of color. We can see these colors—red, orange, yellow, green, blue, indigo, and violet—in a rainbow or when white light is passed through a prism (▼ Figure 9.2). Red light, with a wavelength of 750 nm (nanometers), has the longest wavelength of visible light.

Helpful mnemonic: ROY G BIV—Red, Orange, Yellow, Green, Blue, Indigo, Violet

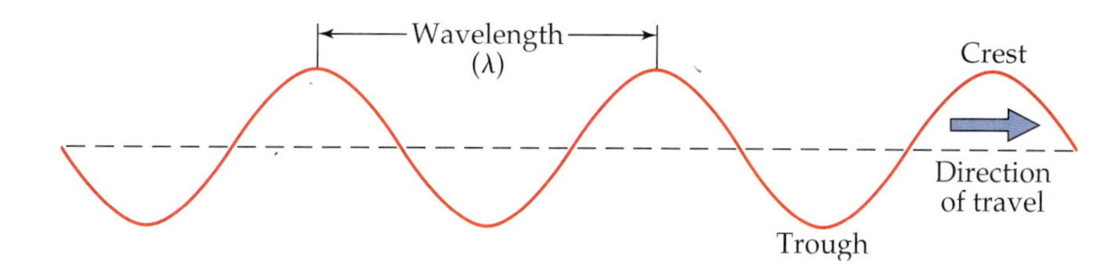

▲ **Figure 9.1 Wavelength** The wavelength of light (λ) is defined as the distance between adjacent wave crests.

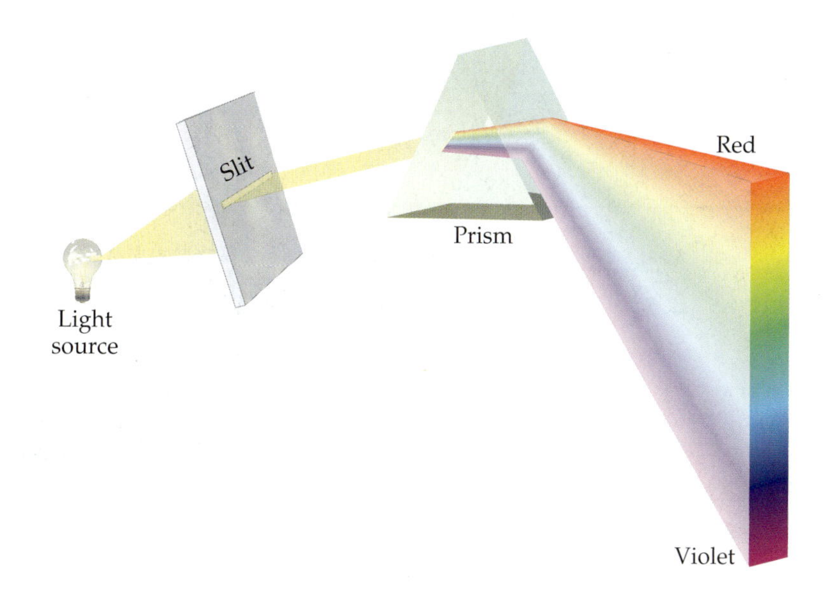

► **Figure 9.2 Components of white light** Light is separated into its constituent colors—red, orange, yellow, green, blue, indigo, and violet—when it is passed through a prism.

▲ **Figure 9.3 Color in objects**
A red shirt appears red because it absorbs all colors except red, which it reflects.

Violet light, with a wavelength of 400 nm, has the shortest (1 nm $= 10^{-9}$ m). The presence of color in white light is responsible for the colors we see in our everyday vision. For example, a red shirt is red because it reflects red light (◄ Figure 9.3). Our eyes see only the reflected light, making the shirt appear red.

Light waves are also often characterized by their **frequency (ν)**, the number of cycles or crests that pass through a stationary point in one second. Wavelength and frequency are inversely related—the shorter the wavelength, the higher the frequency. Blue light, for example, has a higher frequency than red light.

The Greek letter *nu* (ν) is pronounced "noo."

In the early twentieth century, scientists such as Albert Einstein and others discovered that the results of certain experiments could be explained only by describing light not as waves, but as particles. In this description, the light leaving a flashlight, for example, is a stream of particles. A particle of light is called a **photon**, and we can think of a photon as a single packet of light energy. The amount of energy carried in the packet depends on the wavelength of the light—the shorter the wavelength, the greater the energy. Just as water waves carry more energy if their crests are closer together—think about surf pounding a beach—light waves carry more energy if their crests are closer together. Therefore violet light (shorter wavelength) carries more energy per photon than red light (longer wavelength).

To summarize:

- Electromagnetic radiation is a form of energy that travels through space at a constant speed of 3.0×10^8 m/s and exhibits both wavelike and particle-like properties.
- The wavelength of electromagnetic radiation determines the amount of energy carried by one of its photons. The shorter the wavelength, the greater the energy of each photon.
- The frequency and energy of electromagnetic radiation are inversely related to its wavelength.

9.3 The Electromagnetic Spectrum

Electromagnetic radiation ranges in wavelength from 10^{-16} m (gamma rays) to 10^6 m (radio waves). Visible light makes up only a tiny portion of the **electromagnetic spectrum**, which includes all wavelengths of electromagnetic radiation. ▼ Figure 9.4 shows the entire electromagnetic spectrum, with short-wavelength, high-frequency radiation on the right, and long-wavelength, low-frequency radiation on the left. Visible light is only a small sliver in the middle.

▼ **Figure 9.4 The electromagnetic spectrum**

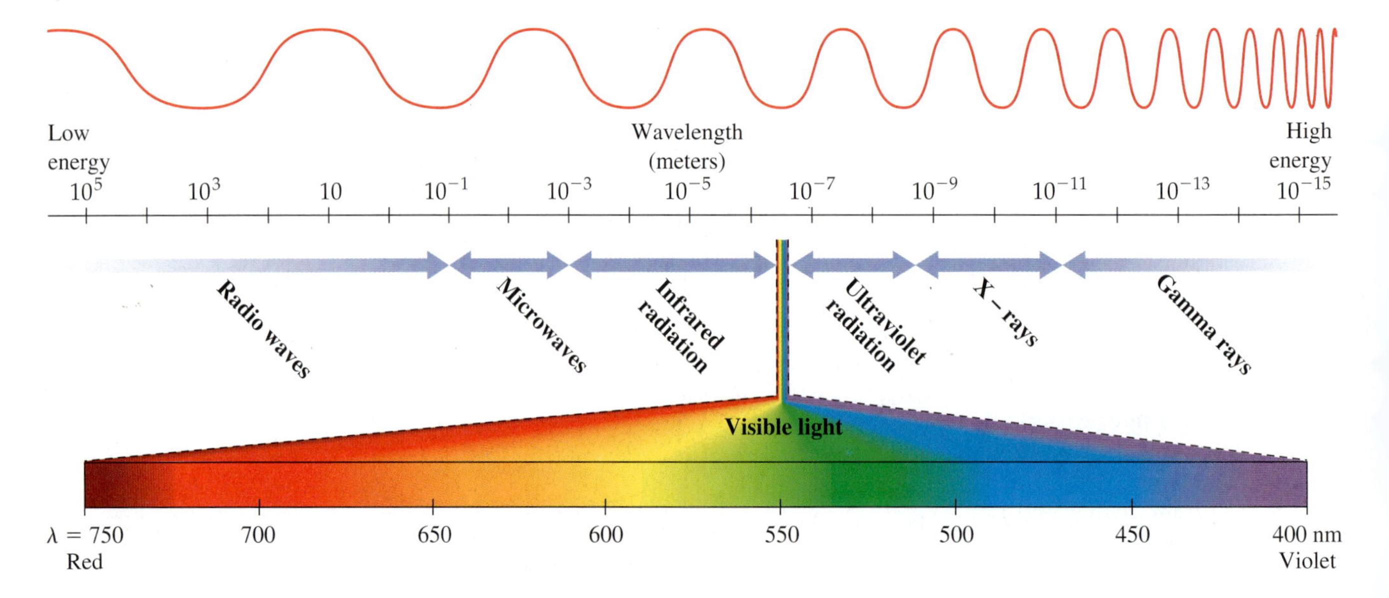

Remember that the energy carried per photon is greater for short wavelengths than for long wavelengths. Therefore, the most energetic photons are those of **gamma rays**, the form of electromagnetic radiation with the shortest wavelength. Gamma rays are produced by the sun, by stars, and by certain unstable atomic nuclei on earth. Human exposure to gamma rays is dangerous because the high energy of gamma-ray photons can damage biological molecules.

Next on the electromagnetic spectrum, at longer wavelengths than gamma rays, are **X-rays**, familiar to us from their medical use. X-rays pass through many substances that block visible light and are therefore used to image internal bones and organs. Like gamma-ray photons, X-ray photons carry enough energy to damage biological molecules. While several yearly exposures to X-rays are relatively harmless, excessive exposure to X-rays increases cancer risk.

Sandwiched between X-rays and visible light in the electromagnetic spectrum is **ultraviolet** or **UV light**, most familiar to us as the component of sunlight that produces a sunburn or suntan. While not as energetic as gamma-ray or X-ray photons, ultraviolet photons still carry enough energy to damage biological molecules. Excessive exposure to ultraviolet light increases the risk of skin cancer and cataracts and causes premature wrinkling of the skin. Next on the spectrum is **visible light**, ranging from violet (shorter wavelength, higher energy) to red (longer wavelength, lower energy). Visible photons do not damage biological molecules. They do, however, cause molecules in our eyes to rearrange and send a signal to our brain that results in vision.

TUTORIAL
The Electromagnetic Spectrum

Beyond visible light lies **infrared light**. The heat you feel when you place your hand near a hot object is infrared light. All warm objects, including human bodies, emit infrared light. While infrared light is invisible to our eyes, infrared sensors can detect it and are often used in night-vision technology to "see" in the dark. In the infrared region of the spectrum, warm objects—such as human bodies, for example—glow, much as a lightbulb glows in the visible region of the spectrum.

Beyond infrared light, at longer wavelengths still, are **microwaves**, used for radar and in microwave ovens. Although microwave light has longer wavelengths—and therefore lower energy per photon—than visible or infrared light, it is efficiently absorbed by water and can therefore heat substances that contain water. For this reason substances that contain water, such as food, are warmed when placed in a microwave oven, but substances that do not contain water, such as a plate, are not.

Some types of dishes contain substances that absorb microwave radiation, but most do not.

The longest wavelengths are **radio waves**, which are used to transmit the signals used by AM and FM radio, cellular telephones, television, and other forms of communication.

Normal photograph

Infrared photograph

▲ Warm objects, such as human or animal bodies, give off infrared light that is easily detected with an infrared camera. In the infrared photograph, the warmest areas appear as red and the coolest as dark blue. (Note that the photo confirms the familiar idea that healthy dogs have cold noses.) (*Source:* Sierra Pacific Innovations. All rights reserved. SPI CORP, www.x20.org)

EXAMPLE 9.1 Wavelength, Energy, and Frequency

Arrange the following three types of electromagnetic radiation—visible light, X-rays, and microwaves—in order of increasing:

(a) wavelength
(b) frequency
(c) energy per photon

	Solution:
(a) wavelength	
Examine Figure 9.4 to see that X-rays have the shortest wavelength, followed by visible light and then microwaves.	X-rays, visible light, microwaves
(b) frequency	
Since frequency and wavelength are inversely proportional—the longer the wavelength, the shorter the frequency—the ordering with respect to frequency is exactly the reverse of the ordering with respect to wavelength.	microwaves, visible light, X-rays
(c) energy per photon	
Energy per photon decreases with increasing wavelength but increases with increasing frequency; therefore the ordering with respect to energy per photon is the same as frequency.	microwaves, visible light, X-rays

SKILLBUILDER 9.1 Wavelength, Energy, and Frequency

Arrange the following colors of visible light—green, red, and blue—in order of increasing:

(a) wavelength
(b) frequency
(c) energy per photon

Chemistry AND HEALTH

Radiation Treatment for Cancer

X-rays and gamma rays are sometimes called ionizing radiation because the high energy in their photons can ionize atoms and molecules. When ionizing radiation interacts with biological molecules, it can permanently change or even destroy them. Consequently, we normally try to limit our exposure to ionizing radiation. However, doctors can use ionizing radiation to destroy molecules within unwanted cells such as cancer cells.

In radiation therapy (or radiotherapy), doctors aim X-ray or gamma-ray beams at cancerous tumors. The ionizing radiation damages the molecules within the tumor's cells that carry genetic information—information necessary for the cell to grow and divide—and the cell dies or stops dividing. Ionizing radiation also damages molecules within healthy cells; however, cancerous cells divide more quickly than healthy cells, making them more

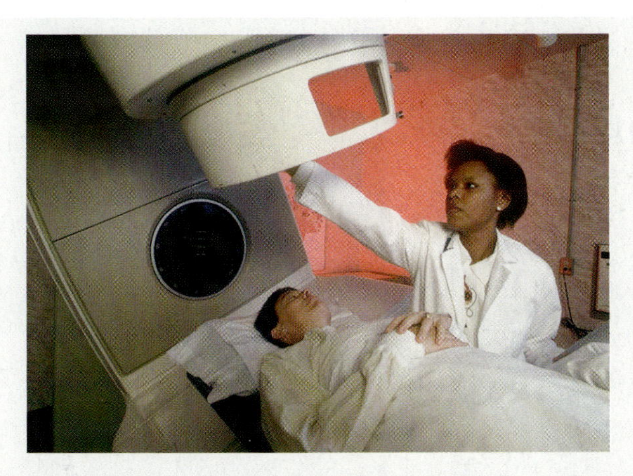

▲ Cancer patient undergoing radiation therapy.

susceptible to genetic damage. Nonetheless, healthy cells are damaged during treatments, resulting in side effects such as fatigue, skin lesions, and hair loss. Doctors try to minimize the exposure of healthy cells by appropriate shielding and by targeting the tumor from multiple directions, minimizing the exposure of healthy cells while maximizing the exposure of cancerous cells (▶ Figure 9.5).

Another side effect of exposing healthy cells to radiation is that they too may become cancerous. So a treatment for cancer may cause cancer. Why do we continue to use it? In radiation therapy, as in most other disease therapies, there is an associated risk. However, we take risks all the time, many for lesser reasons. For example, every time we drive a car, we risk injury or even death. Why? Because we perceive the benefit—such as getting to the grocery store to buy food—to be worth the risk. The situation is similar in cancer therapy or any other therapy for that matter. The benefit of cancer therapy (possibly curing a cancer that will certainly kill you) is worth the risk (a slight increase in the chance of developing a future cancer).

CAN YOU ANSWER THIS? *Why would visible light not work to destroy cancerous tumors?*

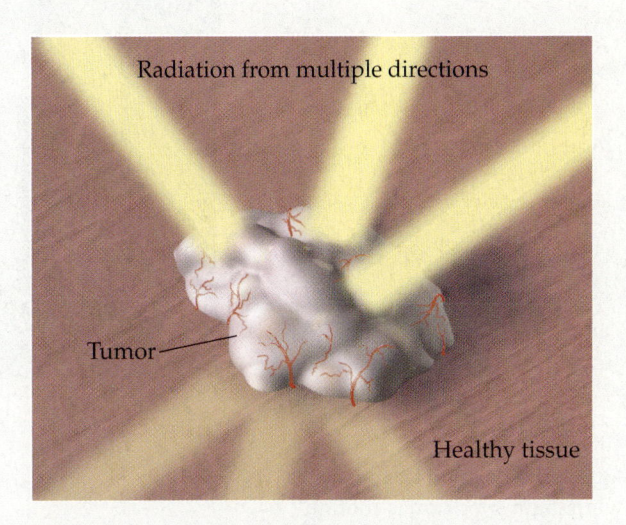

▲ **Figure 9.5 Radiation therapy** By targeting the tumor from various different directions, radiologists minimize damage to healthy tissue.

9.4 The Bohr Model: Atoms with Orbits

When an atom absorbs energy—in the form of heat, light, or electricity—it often re-emits that energy as light. For example, a neon sign is composed of one or more glass tubes filled with gaseous neon atoms. When an electrical current is passed through the tube, the neon atoms absorb some of the electrical energy and re-emit it as the familiar red light of a neon sign (◀ Figure 9.6). If the atoms in the tube are different, the emitted light is a different color. In other words, atoms of a given element emit light of unique colors (or unique wavelengths). Mercury atoms, for example, emit light that appears blue, and hydrogen atoms emit light that appears pink (▶ Figure 9.7), and helium atoms emit light that appears yellow-orange.

▲ **Figure 9.6 A neon sign** Neon atoms inside a glass tube absorb electrical energy and then re-emit the energy as light.

Closer inspection of the light emitted by hydrogen, helium, and neon atoms reveals that each contains several distinct colors or wavelengths. Just as the white light from a lightbulb can be separated into its constituent wavelengths by passing it through a prism, so the light emitted by glowing hydrogen, helium, or neon can also be separated into its constituent wavelengths (▶ Figure 9.8) by passing it through a prism. The result is called an **emission spectrum**. Notice the differences between a white-light spectrum and the emission spectra of hydrogen, helium, and neon. The white-light spectrum is *continuous*, meaning that the light intensity is uninterrupted or smooth across the entire visible range—there is some radiation at all wavelengths, with no gaps. The emission spectra of hydrogen, helium, and neon, however, are not continuous. They consist of bright spots or lines at specific wavelengths with complete darkness in between. Since the emission of light in atoms is related to the motions of electrons within the atoms, a model for how electrons exist in atoms must account for these spectra.

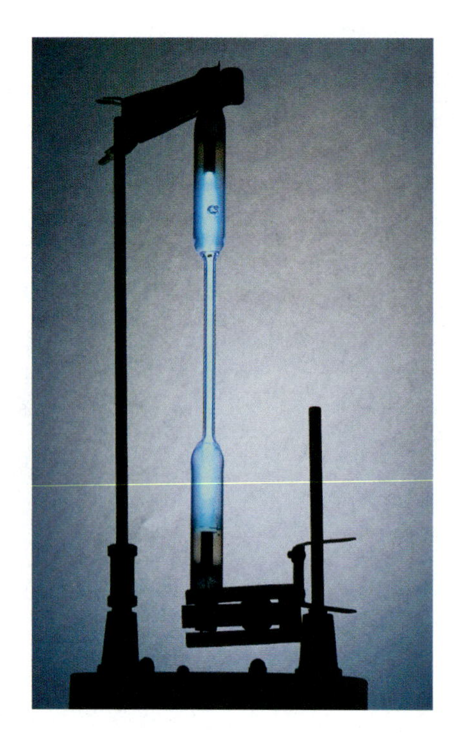

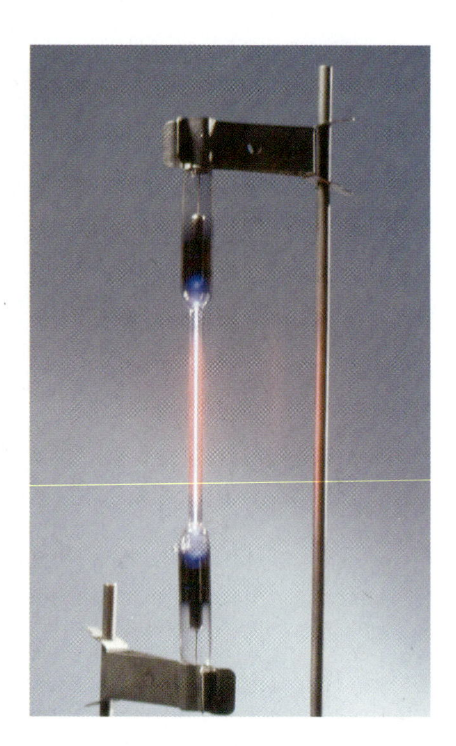

▶ **Figure 9.7 Light emission by different elements** Light emitted from a mercury lamp (left) appears blue, and light emitted from a hydrogen lamp (right) appears pink.

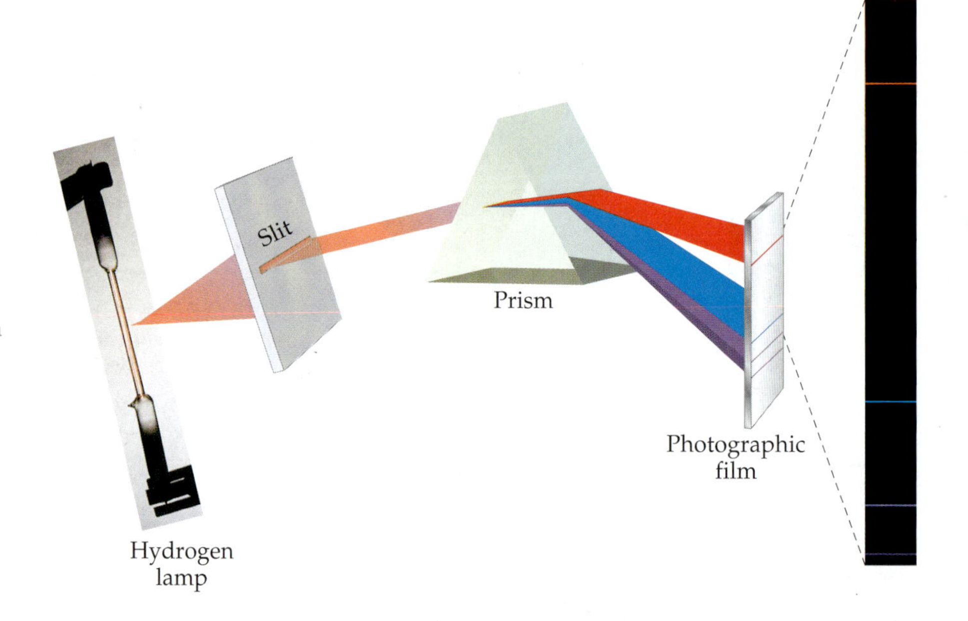

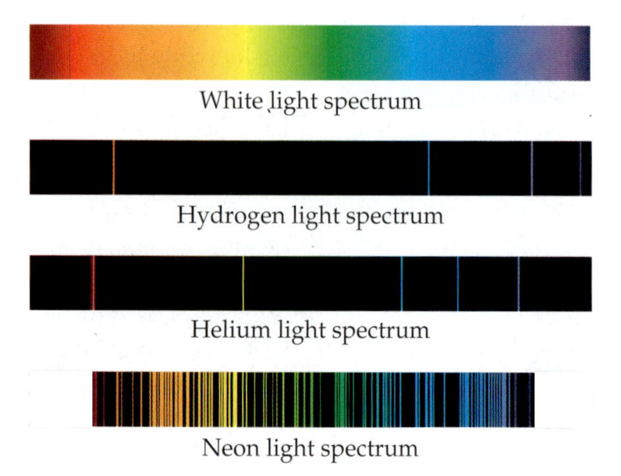

▶ **Figure 9.8 Emission spectra** A white-light spectrum is continuous, with some radiation emitted at every wavelength. The emission spectrum of an individual element, however, includes only certain specific wavelengths. (The different wavelengths appear as *lines* because the light from the source passes through a slit before entering the prism.) Each element produces its own unique and distinctive emission spectrum.

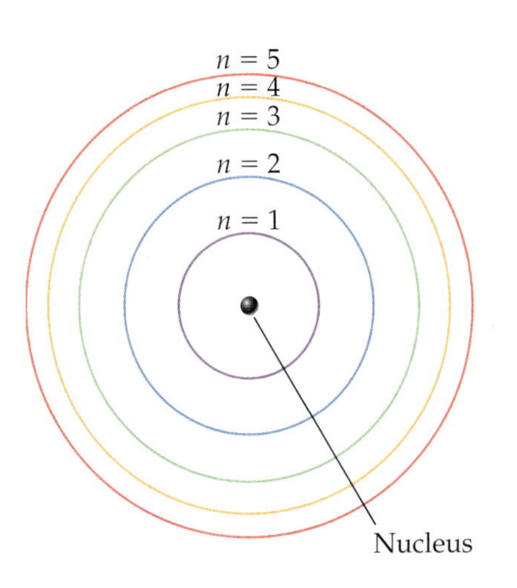

◀ **Figure 9.9 Bohr orbits**

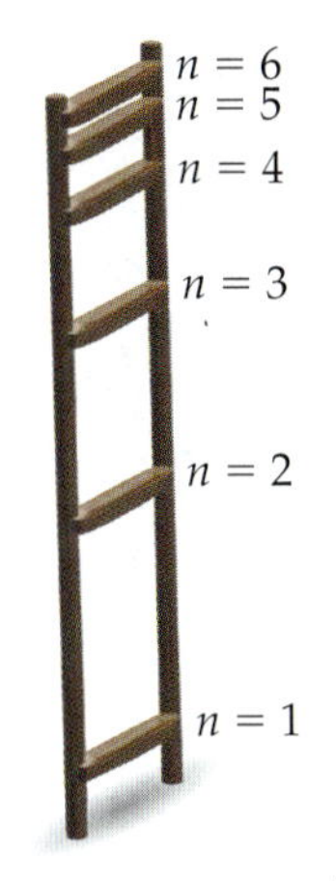

▶ **Figure 9.10 The Bohr energy ladder** Bohr orbits are like steps on a ladder. It is possible to stand on one step or another, but impossible to stand between steps.

A major challenge in developing a model for electrons in atoms was the discrete or bright-line nature of the emission spectra. Why did atoms, researchers wondered, emit light only at particular wavelengths when excited with energy? Why did they *not* emit a continuous spectrum? Niels Bohr developed a simple model to explain these results. In his model, now called the Bohr model, electrons travel around the nucleus in circular orbits that are similar to planetary orbits around the sun. However, unlike planets revolving around the sun—which can theoretically orbit at any distance whatsoever from the sun—electrons in the Bohr model can orbit only at *specific, fixed* distances from the nucleus (▲ Figure 9.9).

The *energy* of each Bohr orbit, specified by **quantum numbers** $n = 1, 2, 3 \ldots$, was also fixed, or **quantized**. Bohr orbits are like steps of a ladder (▲ Figure 9.10), each at a specific distance from the nucleus and each at a specific energy. Just as it is impossible to stand *between steps* on a ladder, so it is impossible for an electron to exist *between orbits* in the Bohr model. An electron in an $n = 3$ orbit, for example, is farther from the nucleus and has more energy than an electron in an $n = 2$ orbit. However, an electron cannot exist at an intermediate distance or energy between the two orbits—the orbits are quantized. As long as an electron remains in a given orbit, it does not absorb or emit light, and its energy remains fixed and constant.

When an atom absorbs energy, an electron in one of these fixed orbits is *excited* or promoted to an orbit that is farther away from the nucleus (◀ Figure 9.11) and therefore higher in energy (this is analogous to moving up a step on the ladder). However, in this new configuration, the atom is less stable, and the electron quickly falls back or *relaxes* to a lower-energy orbit (this is analogous to moving down a step on the ladder). As it does so, it releases a photon of light containing the precise amount of energy—called a **quantum** of energy—that corresponds to the energy difference between the two orbits.

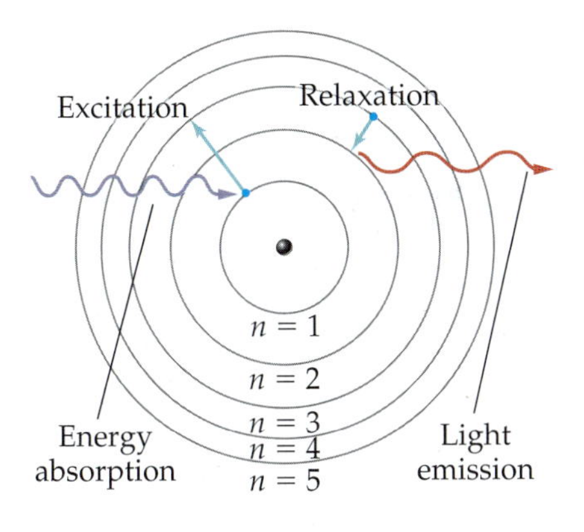

▲ **Figure 9.11 Excitation and emission** When a hydrogen atom absorbs energy, an electron is excited to a higher-energy orbit. The electron then relaxes back to a lower-energy orbit, emitting a photon of light.

Since the amount of energy in a photon is directly related to its wavelength, the photon has a specific wavelength. Consequently, the light emitted by excited atoms consists of specific lines at certain wavelengths, each corresponding to a specific transition between two orbits. For example, the line at 486 nm in the hydrogen emission spectrum corresponds to an electron relaxing from the $n = 4$ orbit to the $n = 2$ orbit (▶ Figure 9.12). In the same way, the line at 657 nm (longer wavelength and therefore lower energy) corresponds to an electron relaxing from the $n = 3$ orbit to the $n = 2$ orbit. Notice that transitions between orbits that are closer together produce lower-energy (and therefore longer-wavelength) light than transitions between orbits that are farther apart.

The great success of the Bohr model of the atom was that it predicted the lines of the hydrogen emission spectrum. However, it failed to predict the emission spectra of other elements that contained more than one electron. For this, and other reasons, the Bohr model was replaced with a more sophisticated model called the quantum-mechanical or wave-mechanical model.

The Bohr model is still important because it provides a logical foundation to the quantum-mechanical model and reveals the historical development of scientific understanding.

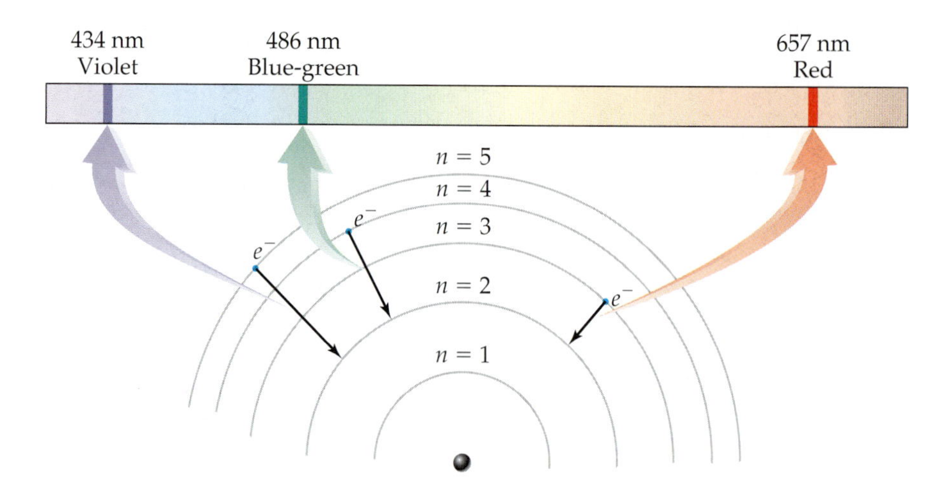

▲ **Figure 9.12 Hydrogen emission lines** The 657-nm line of the hydrogen emission spectrum corresponds to an electron relaxing from the $n = 3$ orbit to the $n = 2$ orbit. The 486-nm line corresponds to an electron relaxing from the $n = 4$ orbit to the $n = 2$ orbit, and the 434-nm line corresponds to an electron relaxing from $n = 5$ to $n = 2$.

To summarize the Bohr model:

- Electrons exist in quantized orbits at specific, fixed energies and specific, fixed distances from the nucleus.
- When energy is put into an atom, electrons are excited to higher-energy orbits.
- When an atom emits light, electrons fall from higher-energy orbits to lower-energy orbits.
- The energy (and therefore the wavelength) of the emitted light corresponds to the difference in energy between the two orbits in the transition. Since these energies are fixed and discrete, the energy (and therefore the wavelength) of the emitted light is fixed and discrete.

CONCEPTUAL CHECKPOINT 9.1

In one transition, an electron in a hydrogen atom falls from the $n = 3$ level to the $n = 2$ level. In a second transition, an electron in a hydrogen atom falls from the $n = 2$ level to the $n = 1$ level. Compared to the radiation emitted by the first of these transitions, the radiation emitted by the second will have:

(a) a lower frequency
(b) a smaller energy per photon
(c) a shorter wavelength
(d) a longer wavelength

9.5 The Quantum-Mechanical Model: Atoms with Orbitals

In the quantum-mechanical model of the atom that has supplanted the Bohr model, orbits are replaced with quantum-mechanical **orbitals**. Orbitals are different from orbits in that they represent, not specific paths that electrons follow, but probability maps that show a statistical distribution of where the electron is likely to be found. This is a nonintuitive property—one that is difficult to visualize—of electrons. A revolutionary concept in quantum mechanics is that electrons *do not* behave like particles flying through space. We cannot, in general, describe their exact paths. An orbital does not represent exactly how

▲ **Figure 9.13 Baseballs follow predictable paths** A baseball follows a well-defined path as it travels from the pitcher to the catcher.

an electron moves—instead, it is a probability map showing where the electron is likely to be found when the atom is probed.

Baseball Paths and Electron Probability Maps

To understand orbitals, let's contrast the behavior of a baseball with that of an electron. Imagine a baseball thrown from the pitcher's mound to a catcher at home plate (▲ Figure 9.13). The baseball's path can easily be traced as it travels from the pitcher to the catcher. The catcher can watch the baseball as it travels through the air, and she can predict exactly where the baseball will cross over home plate. She can even place her mitt in the correct place to catch it. This would be impossible for an electron. Like photons, electrons exhibit wave–particle duality. This duality leads to behavior that makes it impossible to trace an electron's path. If an electron were "thrown" from the pitcher's mound to home plate, it would land in a different place every time, *even if it were thrown in exactly the same way*. Baseballs have predictable paths—electrons do not.

In the quantum-mechanical world of the electron, the catcher could not know exactly where the electron would cross the plate for any given throw. She would have no way of putting her mitt in the right place to catch it. However, if the catcher kept track of hundreds of electron throws, she could observe a reproducible, statistical pattern of where the electron crosses the plate. She could even draw maps in the strike zone showing the probability of an electron crossing a certain area (◀ Figure 9.14). These maps are called *probability maps*.

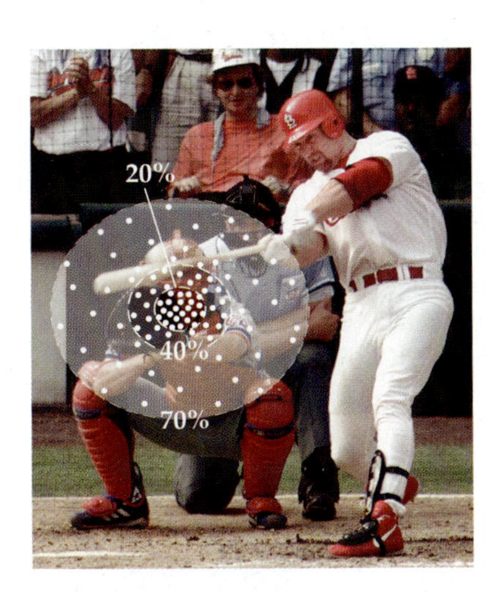

▲ **Figure 9.14 Electrons are unpredictable** To describe the behavior of a "pitched" electron, you would have to construct a probability map of where it would cross home plate.

From Orbits to Orbitals

In the Bohr model, an *orbit* is a circular path—analogous to a baseball's path—that shows the electron's motion around an atomic nucleus. In the quantum-mechanical model, an *orbital* is a probability map, analogous to the probability map drawn by our catcher. It shows the relative likelihood of the electron being found at various locations when the atom is probed. Just as the Bohr model has different orbits with different radii, the quantum-mechanical model has different orbitals with different shapes.

9.6 Quantum-Mechanical Orbitals

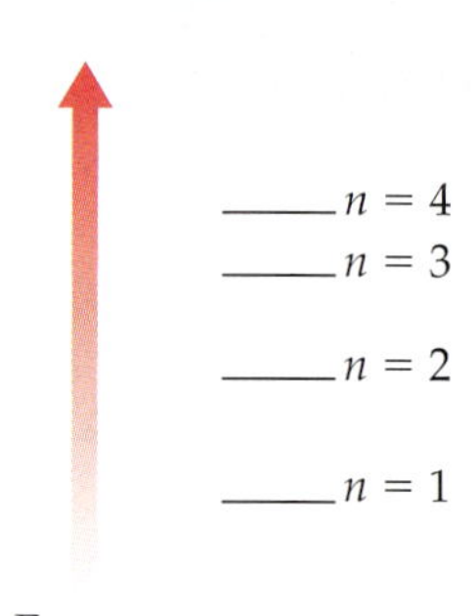

▲ **Figure 9.15 Principal quantum numbers** The principal quantum numbers ($n = 1, 2, 3, \ldots$) determine the energy of the hydrogen quantum-mechanical orbitals.

In the Bohr model of the atom, a single quantum number (n) specifies each orbit. In the quantum-mechanical model, a number and a letter are required to specify an orbital. For example, the lowest-energy orbital in the quantum-mechanical model—analogous to the $n = 1$ orbit in the Bohr model—is called the *1s orbital*. It is specified by the number 1 and the letter *s*. The number is called the **principal quantum number** (n) and specifies the **principal shell** of the orbital. The higher the principal quantum number, the higher the energy of the orbital. The possible principal quantum numbers are $n = 1, 2, 3\ldots$, with energy increasing as n increases (◀ Figure 9.15). Since the 1*s* orbital has the lowest possible principal quantum number, it is in the lowest-energy shell and has the lowest possible energy.

The letter indicates the **subshell** of the orbital and specifies its shape. The possible letters are *s*, *p*, *d*, and *f*, each with a different shape. Orbitals within the *s* subshell have a spherical shape. Unlike the $n = 1$ Bohr orbit, which shows the electron's path, the 1*s* quantum-mechanical orbital is a three-dimensional probability map. These probability maps are best represented by dots (◀ Figure 9.16), where the dot density is proportional to the probability of finding the electron.

This analogy is purely hypothetical. It is impossible to photograph electrons in this way.

We can understand these probability maps and dots better with another analogy. Imagine the electron moving randomly around the nucleus. Imagine also taking a photograph of the electron every second for ten or fifteen minutes. One second the electron is very close to the nucleus; the next second it is farther away and so on. Each photo shows a dot representing the electron's position relative to the nucleus at that time. If you took hundreds of photos and superimposed all of them, you would have a probability map like Figure 9.16—a statistical representation of where the electron spends its time. Notice that the dot density for the 1*s* orbital is greatest near the nucleus and decreases farther away from the nucleus. This means that the electron is more likely to be found close to the nucleus than far away from it.

Orbitals can also be represented as geometric shapes that encompass most of the volume where the electron is likely to be found. For example, the 1*s* orbital can be represented as a sphere (▼ Figure 9.17) that encompasses the vol-

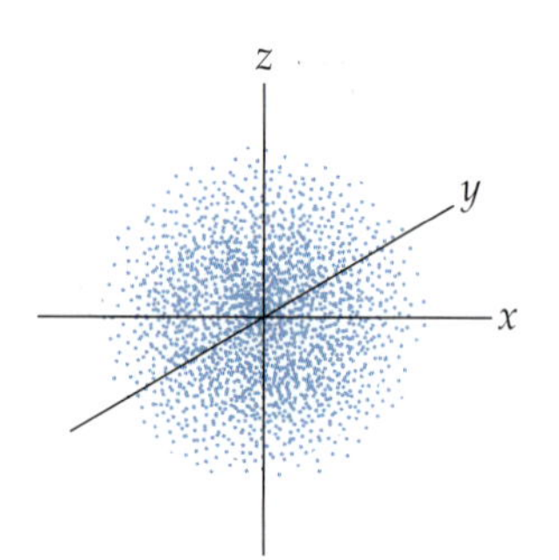

▲ **Figure 9.16 Probability map of the 1*s* orbital** The dot density in this plot is proportional to the probability of finding the electron. The greater dot density near the middle represents a higher probability of finding the electron near the nucleus.

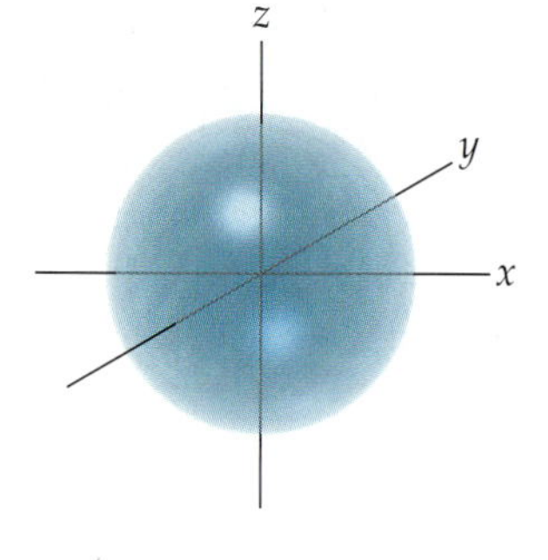

▶ **Figure 9.17 Shape representation of the 1*s* orbital** Because the distribution of electron density around the nucleus in Figure 9.16 is symmetrical—the same in all directions—we can represent the 1*s* orbital as a sphere.

▲ **Figure 9.18 Orbital shape and electron probability for the 1*s* orbital** The shape representation of the 1*s* orbital superimposed on the dot density representation. We can see that when the electron is in the 1*s* orbital, it is most likely to be found within the sphere.

Shell	# of subshells	Letters specifying subshells			
$n = 4$	4	*s*	*p*	*d*	*f*
$n = 3$	3	*s*	*p*	*d*	
$n = 2$	2	*s*	*p*		
$n = 1$	1	*s*			

▲ **Figure 9.19 Subshells** The number of subshells in a given principal shell is equal to the value of *n*.

ume within which the electron is found 90% of the time. If we superimpose the dot density representation of the 1*s* orbital on the shape representation (◀ Figure 9.18), we can see that most of the dots are within the sphere, meaning that the electron is most likely to be found within the sphere when it is in the 1*s* orbital.

The single electron of an undisturbed hydrogen atom at room temperature is in the 1*s* orbital. This is called the **ground state**, or lowest energy state, of the hydrogen atom. However, like the Bohr model, the quantum-mechanical model allows transitions to higher-energy orbitals upon the absorption of energy. What are these higher-energy orbitals? What do they look like?

The next orbitals are those with principal quantum number $n = 2$. Unlike the $n = 1$ principal shell, which contains only one subshell (specified by *s*), the $n = 2$ principal shell contains two subshells, specified by *s* and *p*.

The number of subshells in a given principal shell is equal to the value of *n*.

Therefore the $n = 1$ principal shell has one subshell, the $n = 2$ principal shell has two subshells, and so on (▲ Figure 9.19). The *s* subshell contains the 2*s* orbital, higher in energy than the 1*s* orbital and slightly larger (◀ Figure 9.20), but otherwise similar in shape. The *p* subshell contains three 2*p* orbitals (▼ Figure 9.21), all with the same dumbbell-like shape but with different orientations.

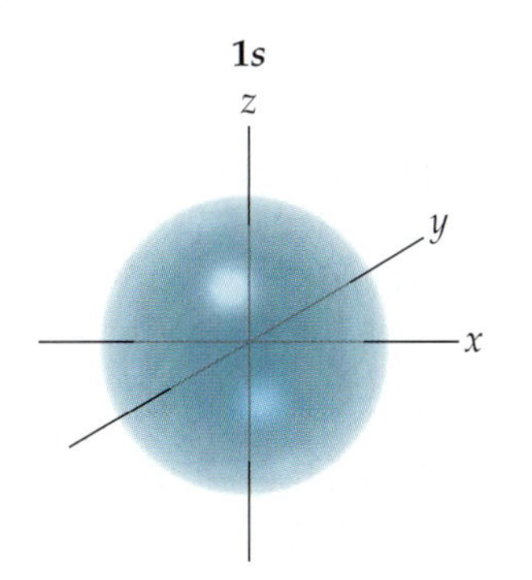

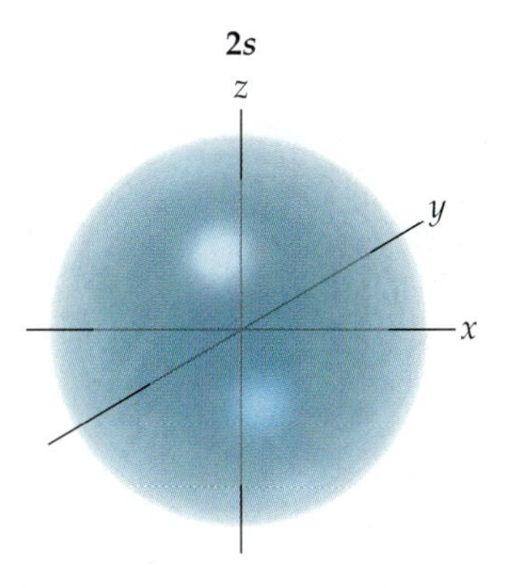

▲ **Figure 9.20 The 2*s* orbital** The 2*s* orbital is similar to the 1*s* orbital, but larger in size.

The next principal shell, $n = 3$, contains three subshells specified by *s*, *p*, and *d*. The *s* and *p* subshells contain the 3*s* and 3*p* orbitals, similar in shape to the 2*s* and 2*p* orbitals, but slightly larger and higher in energy. The *d* subshell contains the five *d* orbitals shown in ▶ Figure 9.22. The next principal shell, $n = 4$, contains four subshells specified by *s*, *p*, *d*, and *f*. The *s*, *p*, and *d* subshells are similar to those in $n = 3$. The *f* subshell contains seven orbitals (called the 4*f* orbitals), whose shape we do not consider in this book.

As we have already discussed, hydrogen's single electron is usually in the 1*s* orbital because electrons seek out the lowest-energy orbital available. In hydrogen, the rest of the orbitals are normally empty. However, the absorption of

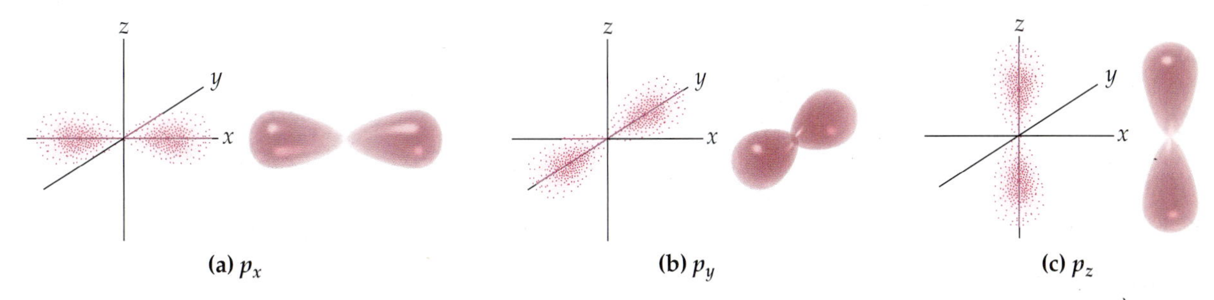

▲ **Figure 9.21 The 2*p* orbitals**

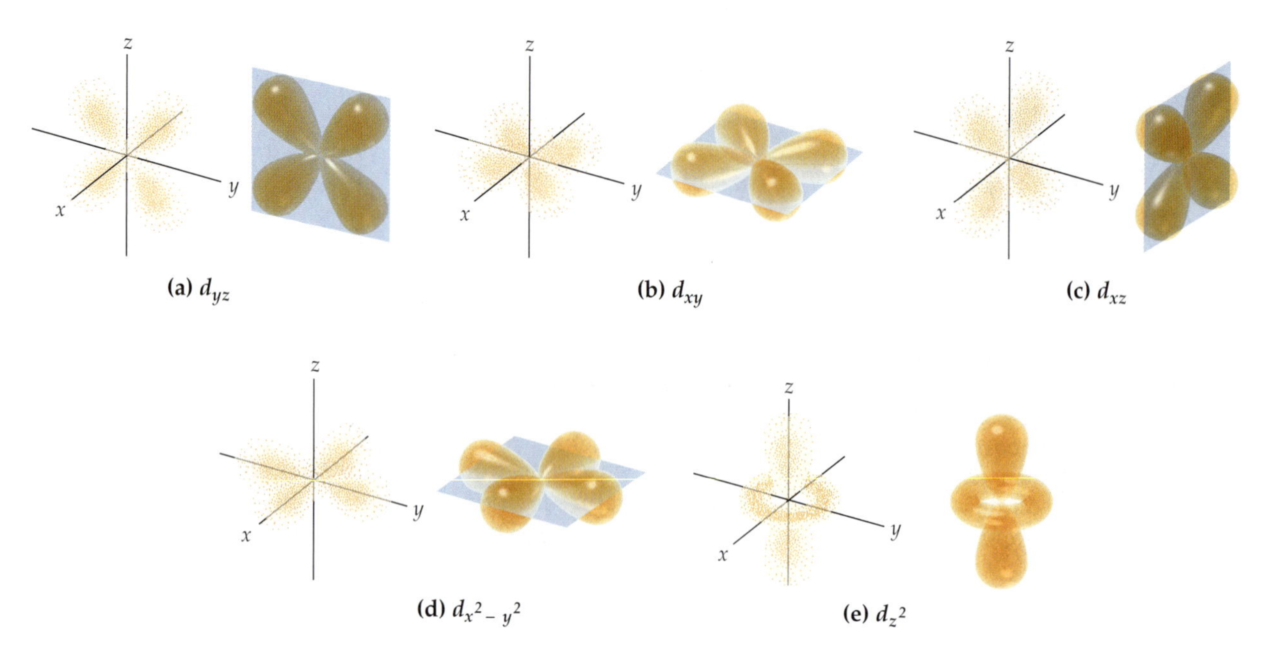

▲ **Figure 9.22 The 3*d* orbitals**

energy by a hydrogen atom can cause the electron to jump (or make a transition) from the 1*s* orbital to a higher-energy orbital. When the electron is in a higher-energy orbital, the hydrogen atom is said to be in an **excited state**.

Because of their higher energy, excited states are unstable, and the electron will usually fall (or relax) back to a lower-energy orbital. In the process the electron emits energy, often in the form of light. As in the Bohr model, the energy difference between the two orbitals involved in the transition determines the wavelength of the emitted light (the greater the energy difference, the shorter the wavelength). The quantum-mechanical model predicts the bright-line spectrum of hydrogen as well as the Bohr model. However, it can also predict the bright-line spectra of other elements as well.

Chemistry AND HEALTH

Magnetic Resonance Imaging

We have just learned that atoms emit radiation of a particular wavelength when one of their electrons makes a transition from a higher-energy orbital to a lower-energy one. The study of these transitions and the associated radiation is called *spectroscopy*. Spectroscopy is an important tool for chemists, allowing them to analyze atoms and molecules by observing how they interact with radiation. Spectroscopy, especially in magnetic resonance imaging or *MRI*, has also become an important tool in medicine.

MRI is based on a type of spectroscopy called *nuclear magnetic resonance* or *NMR spectroscopy*. Unlike the emission spectra discussed earlier, which involve electrons making transitions between energy levels, NMR involves atomic nuclei making transitions between energy levels.

The best way to understand NMR is to think of atomic nuclei as tiny magnets (▶ Figure 9.23). When these magnets are placed in an external magnetic field, they align themselves with this external field to minimize

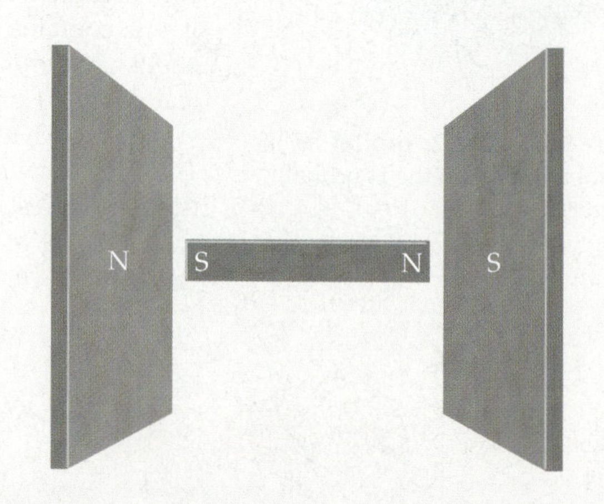

▲ **Figure 9.23 The nucleus as a magnet** A nucleus is like a small magnet that aligns itself with an external magnetic field.

their energy (just as the needle of a compass aligns itself with Earth's magnetic field). However, just as the energies of an electron in an atom are quantized, so the energies of nuclei in an external field are quantized—only certain, fixed energies are allowed. These fixed energies correspond to fixed orientations relative to the external magnetic field. In the simplest case, only two orientations are allowed, one lower in energy than the other (▼ Figure 9.24).

Electromagnetic radiation of the correct energy will cause a transition between the two orientations of the nucleus. The energy of the radiation that causes the transition depends on the energy separation between the two orientations, which in turn depends on the strength of the external magnetic field. In normal NMR spectroscopy, a sample is placed in a uniform magnetic field. The wavelength of electromagnetic radiation striking the sample is then varied to find the wavelength or frequency that causes the transition between the two allowed orientations. This frequency is called the *resonance frequency*.

In MRI, the sample is the patient. The nuclei are those of hydrogen atoms within water molecules contained in the patient's tissues. However, instead of putting the patient in a uniform magnetic field, he or she is put in a magnetic field that varies in space. For example, the magnetic field may be strongest on the left side of the patient and weaker on the right. Then the nuclei on the left side of the patient will have a resonance frequency of higher energy than those on the right side. By mapping these resonance frequencies—each resonance frequency corresponding to a different position in space—MRI can obtain a remarkably clear and detailed image of the patient's internal tissues (▼ Figure 9.25).

CAN YOU ANSWER THIS? *Will the wavelength of electromagnetic radiation required to cause a transition in the preceding example (stronger magnetic field on the left side of the person than the right side) be longer or shorter on the left side versus the right? Explain.*

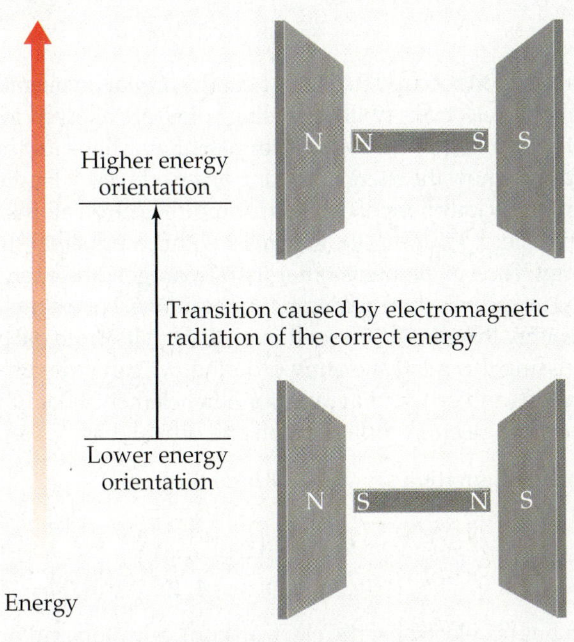

▲ **Figure 9.24 Nuclear energy levels** The energies of a nucleus in an external magnetic field, which correspond to different orientations, are quantized. Light of the correct energy causes a transition from one orientation to the other.

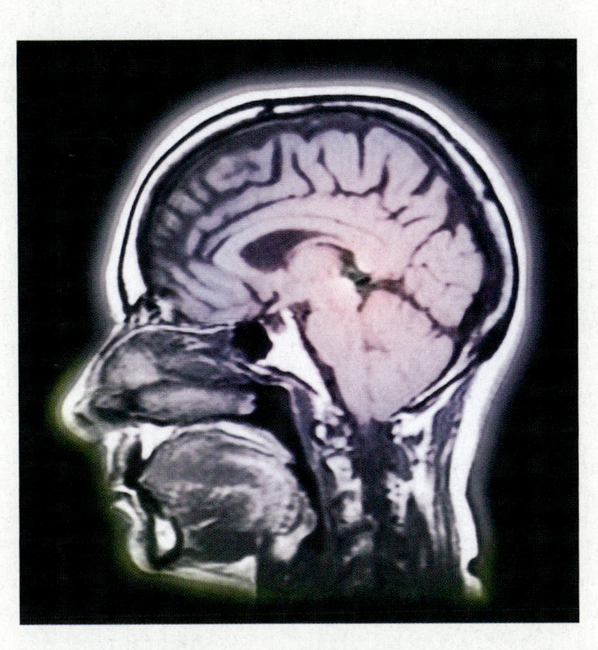

▲ **Figure 9.25 Magnetic resonance imaging** MRI produces remarkably clear images of a patient's internal tissues.

Electron Configurations: How Electrons Occupy Orbitals

An **electron configuration** simply shows the occupation of orbitals by electrons for a particular atom. For example, the electron configuration for a ground-state hydrogen atom is:

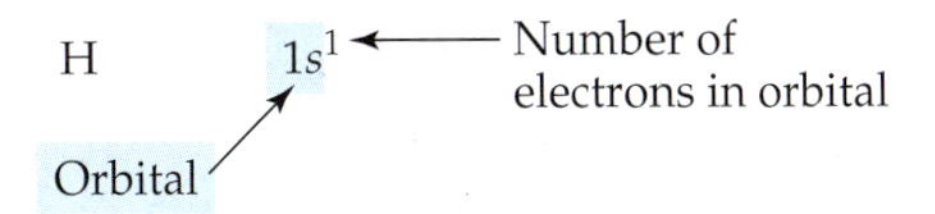

The electron configuration tells us that hydrogen's single electron is in the 1*s* orbital.

Another way to represent this information is with an **orbital diagram**, which gives similar information but shows the electrons as arrows in a box representing the orbital. The orbital diagram for a ground-state hydrogen atom is:

H [↑]
1s

The box represents the 1*s* orbital and the arrow within the box represents the electron in the 1*s* orbital. In orbital diagrams, the direction of the arrow (pointing up or pointing down) represents **electron spin**, a fundamental property of electrons. All electrons have spin. The **Pauli exclusion principle** states that orbitals may hold no more than two electrons with opposing spins. We symbolize this as two arrows pointing in opposite directions ↑↓. Helium atoms, for example, have two electrons. The electron configuration and orbital diagram for helium are:

	Electron configuration	Orbital diagram
He	$1s^2$	[↑↓] 1s

Since we know that electrons occupy the lowest-energy orbitals available, and since we know that only two electrons (with opposing spins) are allowed in each orbital, we can continue to build ground-state electron configurations for the rest of the elements as long as we know the energy ordering of the orbitals. ▼ Figure 9.26 shows the energy ordering of a number of orbitals for multi-electron atoms.

The subshells within a principal shell do not have the same energy because of electron–electron interactions.

Notice that, for multi-electron atoms, the subshells within a principal shell *do not* have the same energy. Thus in elements other than hydrogen, the energy ordering is not determined by the principal quantum number alone. For example, the 4*s* subshell is lower in energy than the 3*d* subshell, even though its principal quantum number is higher. Using this relative energy ordering, we can write ground-state electron configurations and orbital diagrams for other elements. For lithium, which has three electrons, the electron configuration and orbital diagram are:

Remember that the number of electrons in an atom is equal to its atomic number.

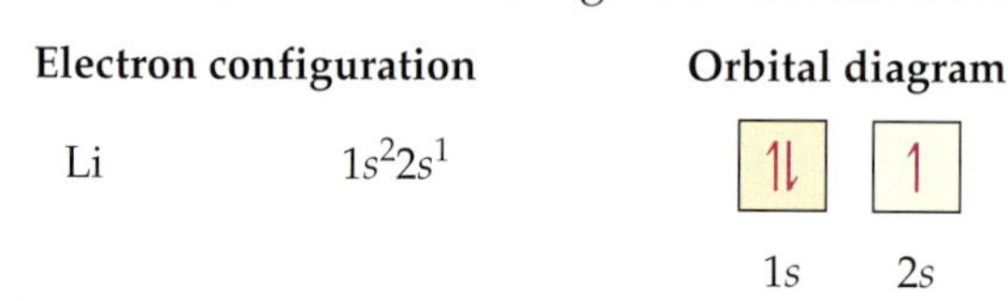

For carbon, which has six electrons, the electron configuration and orbital diagram are:

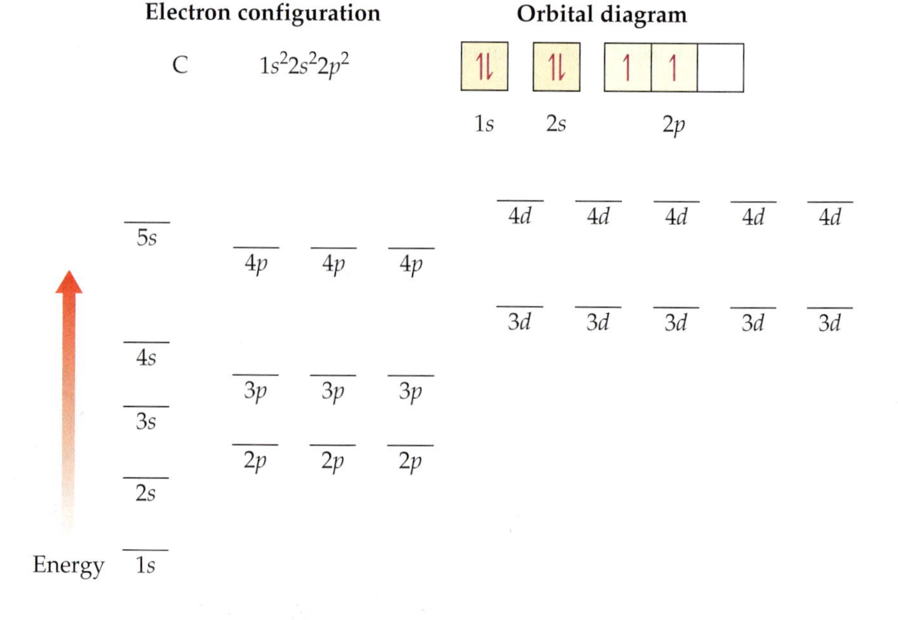

▶ **Figure 9.26 Energy ordering of orbitals for multi-electron atoms** Different subshells within the same principal shell have different energies.

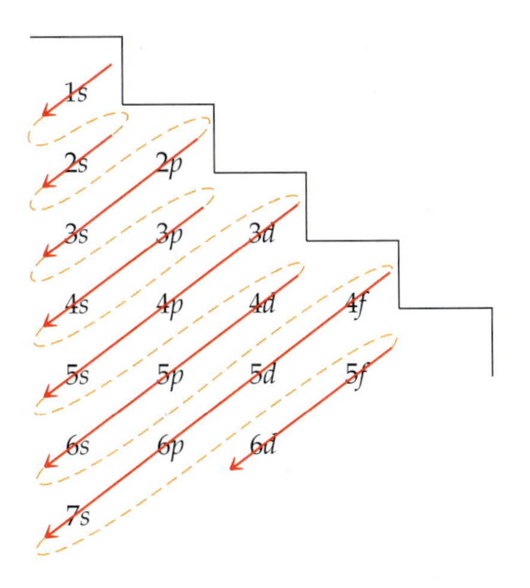

▲ **Figure 9.27 Orbital filling order** The arrows indicate the order in which orbitals fill.

Notice that the 2*p* electrons occupy the *p* orbitals (of equal energy) singly, rather than pairing in one orbital. This is a result of **Hund's rule**, which states that when filling orbitals of equal energy, electrons fill them singly first, with parallel spins.

Before we write electron configurations for other elements, let us summarize what we have learned so far:

- Electrons occupy orbitals so as to minimize the energy of the atom; therefore, lower-energy orbitals fill before higher-energy orbitals. Orbitals fill in the following order: 1*s* 2*s* 2*p* 3*s* 3*p* 4*s* 3*d* 4*p* 5*s* 4*d* 5*p* 6*s* (◀ Figure 9.27).
- Orbitals can hold no more than two electrons each. When two electrons occupy the same orbital, they must have opposing spins. This is known as the Pauli exclusion principle.
- When orbitals of identical energy are available, these are first occupied singly with parallel spins rather than in pairs. This is known as Hund's rule.

Consider the electron configurations and orbital diagrams for elements with atomic numbers 3 through 10.

Symbol (#e^-)	Electron configuration	Orbital diagram 1s	2s	2p		
Li (3)	$1s^22s^1$	↑↓	↑			
Be (4)	$1s^22s^2$	↑↓	↑↓			
B (5)	$1s^22s^22p^1$	↑↓	↑↓	↑		
C (6)	$1s^22s^22p^2$	↑↓	↑↓	↑	↑	
N (7)	$1s^22s^22p^3$	↑↓	↑↓	↑	↑	↑
O (8)	$1s^22s^22p^4$	↑↓	↑↓	↑↓	↑	↑
F (9)	$1s^22s^22p^5$	↑↓	↑↓	↑↓	↑↓	↑
Ne (10)	$1s^22s^22p^6$	↑↓	↑↓	↑↓	↑↓	↑↓

Notice how the p orbitals fill. As a result of Hund's rule, the p orbitals fill with single electrons before they fill with paired electrons. The electron configuration of neon represents the complete filling of the $n = 2$ principal shell. When writing electron configurations for elements beyond neon—or beyond any other noble gas—the electron configuration of the previous noble gas is often abbreviated by the symbol for the noble gas in brackets. For example, the electron configuration of sodium is:

Na $1s^2 2s^2 2p^6 3s^1$

This can also be written as:

Na $[Ne]3s^1$

where [Ne] represents $1s^2 2s^2 2p^6$, the electron configuration for neon.

To write an electron configuration for an element, first find its atomic number from the periodic table—this number equals the number of electrons in the neutral atom. Then use the order of filling from Figure 9.26 or 9.27 to distribute the electrons in the appropriate orbitals. Remember that each orbital can hold a maximum of 2 electrons. Consequently:

- the s subshell has only 1 orbital and therefore can hold only 2 electrons.
- the p subshell has 3 orbitals and therefore can hold 6 electrons.
- the d subshell has 5 orbitals and therefore can hold 10 electrons.
- the f subshell has 7 orbitals and therefore can hold 14 electrons.

EXAMPLE 9.2 Electron Configurations

Write electron configurations for each of the following elements.

(a) Mg
(b) S
(c) Ga

Solution:

(a) Magnesium has 12 electrons. Distribute two of these into the 1s orbital, two into the 2s orbital, six into the 2p orbitals, and two into the 3s orbital.

Mg $1s^2 2s^2 2p^6 3s^2$

or

You can also write the electron configuration more compactly using the noble gas core notation. For magnesium, we use [Ne] to represent $1s^2 2s^2 2p^6$.

Mg $[Ne]3s^2$

(b) Sulfur has 16 electrons. Distribute two of these into the 1s orbital, two into the 2s orbital, six into the 2p orbitals, two into the 3s orbital, and four into the 3p orbitals.

S $1s^2 2s^2 2p^6 3s^2 3p^4$

or

You can write the electron configuration more compactly by using [Ne] to represent $1s^2 2s^2 2p^6$.

S $[Ne]3s^2 3p^4$

(c) Gallium has 31 electrons. Distribute two of these into the 1*s* orbital, two into the 2*s* orbital, six into the 2*p* orbitals, two into the 3*s* orbital, six into the 3*p* orbitals, two into the 4*s* orbital, ten into the 3*d* orbitals, and one into the 4*p* orbitals. Notice that the *d* subshell has five orbitals and can therefore hold 10 electrons.

Ga $1s^22s^22p^63s^23p^64s^23d^{10}4p^1$

or

You can write the electron configuration more compactly by using [Ar] to represent $1s^22s^22p^63s^23p^6$.

Ga $[Ar]4s^23d^{10}4p^1$

SKILLBUILDER 9.2 Electron Configurations

Write electron configurations for each of the following elements.

(a) Al
(b) Br
(c) Sr

SKILLBUILDER PLUS

Write electron configurations for each of the following ions.

(a) Al^{3+}
(b) Cl^-
(c) O^{2-}

EXAMPLE 9.3 Writing Orbital Diagrams

Write an orbital diagram for silicon.

Solution:

Since silicon is atomic number 14, it has 14 electrons. Draw a box for each orbital, putting the lowest-energy orbital (1*s*) on the far left and proceeding to orbitals of higher energy to the right.

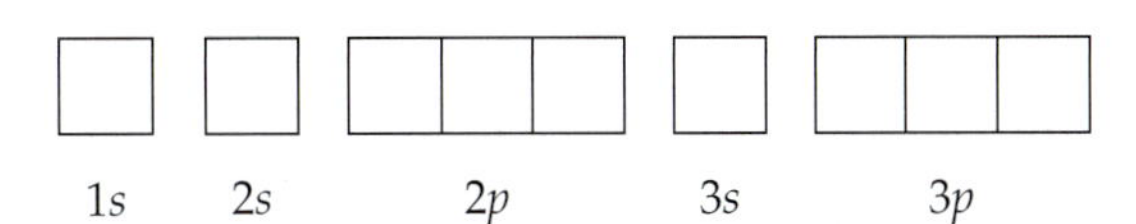

Distribute the 14 electrons into the orbitals, allowing a maximum of 2 electrons per orbital and remembering Hund's rule. The complete orbital diagram is:

Si [↑↓] [↑↓] [↑↓|↑↓|↑↓] [↑↓] [↑|↑|]
1*s* 2*s* 2*p* 3*s* 3*p*

SKILLBUILDER 9.3 Writing Orbital Diagrams

Write an orbital diagram for argon.

CONCEPTUAL CHECKPOINT 9.2

Which of the following pairs of elements have the same *total* number of electrons in *p* orbitals?

(a) Na and K
(b) K and Kr
(c) P and N
(d) Ar and Ca

9.7 Electron Configurations and the Periodic Table

Valence Electrons

Valence electrons are the electrons in the outermost principal shell (the principal shell with the highest principal quantum number, *n*). These electrons are important because, as we will see in the next chapter, they are involved in chemical bonding. Electrons that are *not* in the outermost principal shell are called **core electrons**. For example, silicon, with the electron configuration of $1s^2 2s^2 2p^6 3s^2 3p^2$, has 4 valence electrons (those in the $n = 3$ principal shell) and 10 core electrons.

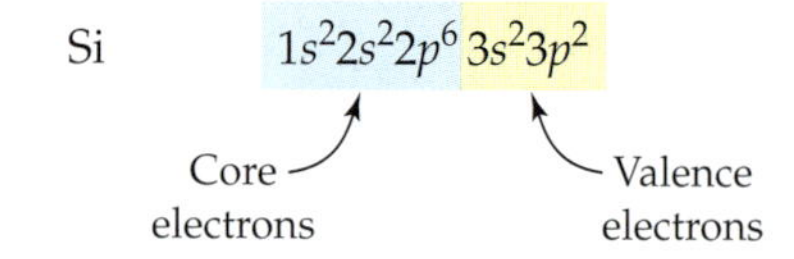

EXAMPLE 9.4 Valence Electrons and Core Electrons

Write an electron configuraton for selenium and identify the valence electrons and the core electrons.

Solution:

Write the electron configuration for selenium by determining the total number of electrons from selenium's atomic number (34) and then distributing them into the appropriate orbitals.

Se $1s^2 2s^2 2p^6 3s^2 3p^6 4s^2 3d^{10} 4p^4$

The valence electrons are those in the outermost principal shell. For selenium, the outermost principal shell is the $n = 4$ shell, which contains 6 electrons (2 in the 4*s* orbital and 4 in the three 4*p* orbitals). All other electrons, including those in the 3*d* orbitals, are core electrons.

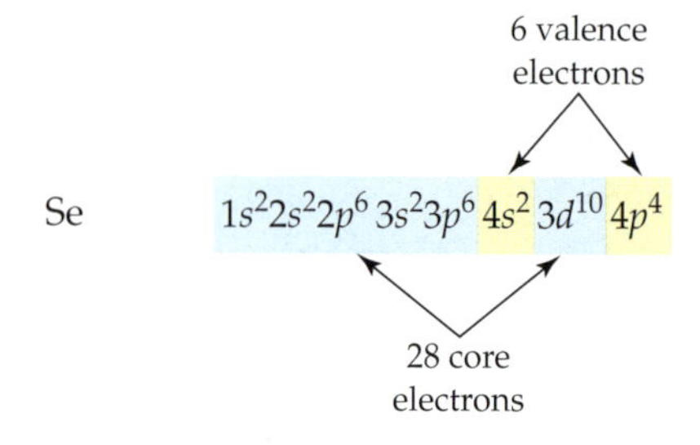

SKILLBUILDER 9.4 Valence Electrons and Core Electrons

Write an electron configuration for chlorine and identify the valence electrons and core electrons.

▶ **Figure 9.28 Outer electron configurations of the first eighteen elements**

1A	2A	3A	4A	5A	6A	7A	8A
1 **H** $1s^1$							2 **He** $1s^2$
3 **Li** $2s^1$	4 **Be** $2s^2$	5 **B** $2s^22p^1$	6 **C** $2s^22p^2$	7 **N** $2s^22p^3$	8 **O** $2s^22p^4$	9 **F** $2s^22p^5$	10 **Ne** $2s^22p^6$
11 **Na** $3s^1$	12 **Mg** $3s^2$	13 **Al** $3s^23p^1$	14 **Si** $3s^23p^2$	15 **P** $3s^23p^3$	16 **S** $3s^23p^4$	17 **Cl** $3s^23p^5$	18 **Ar** $3s^23p^6$

▲ Figure 9.28 shows the first eighteen elements in the periodic table with their outer electron configurations listed below each one. As you move across a row, the orbitals are simply filling in the correct order. As you move down a column, the highest principal quantum number increases, but the number of electrons in each subshell remains the same. Consequently, the elements within a column (or family) all have the same number of valence electrons and similar outer electron configurations.

A similar pattern exists for the entire periodic table (▼ Figure 9.29). Notice that, because of the filling order of orbitals, the periodic table can be divided into blocks representing the filling of particular subshells.

- The first two columns on the left side of the periodic table are the *s* block with outer electron configurations of ns^1 (first column) and ns^2 (second column).
- The six columns on the right side of the periodic table are the *p* block with outer electron configurations of: ns^2np^1, ns^2np^2, ns^2np^3, ns^2np^4, ns^2np^5 (halogens), and ns^2np^6 (noble gases).
- The transition metals are the *d* block.
- The lanthanides and actinides (also called the inner transition metals) are the *f* block.

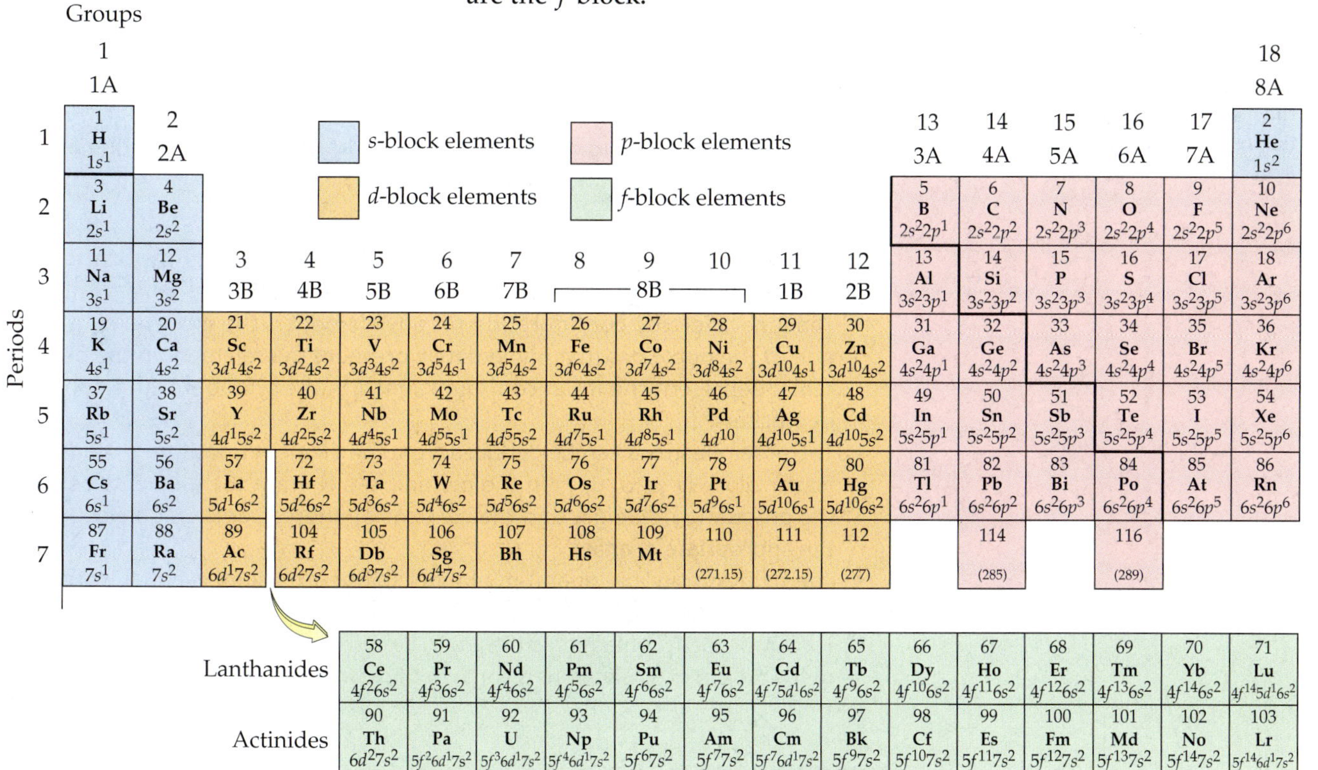

▲ **Figure 9.29 Outer electron configurations of the elements**

Remember that main-group elements are those in the two far left columns (1A, 2A) and the six far right columns (3A–8A) of the periodic table.

Notice that, except for helium, the number of valence electrons for any main-group element is equal to the group number of its column. For example, we can tell that chlorine has 7 valence electrons because it is in the column with group number 7A. The row number in the periodic table is equal to the number of the highest principal shell (n value). For example, since chlorine is in row 3, its highest principal shell is the $n = 3$ shell.

The transition metals have electron configurations with trends that differ somewhat from main-group elements. As you move across a row in the d block, the d orbitals are filling (Figure 9.29). However, the principal quantum number of the d orbital being filled across each row in the transition series is equal to the row number minus one (in the fourth row, the $3d$ orbitals fill; in the fifth row, the $4d$ orbitals fill; and so on). For the first transition series, the outer configuration is $4s^23d^x$ (x = number of d electrons) with two exceptions: Cr is $4s^13d^5$ and Cu is $4s^13d^{10}$. There is a special stability associated with a half-filled d subshell and a completely filled d subshell that causes these exceptions. Otherwise, the number of outer-shell electrons in a transition series does not change as you move across a period. In other words, *the transition series represents the filling of core orbitals and the number of valence electrons is mostly constant.*

We can now see that the organization of the periodic table allows us to write the electron configuration for any element simply based on its position in the periodic table. For example, suppose we want to write an electron configuration for P. The inner electrons of P are simply those of the noble gas that precedes P in the periodic table, Ne. So we can represent the inner electrons with [Ne]. We obtain the outer electron configuration by tracing the elements between Ne and P and assigning electrons to the appropriate orbitals (◀ Figure 9.30). Remember that the highest n value is given by the row number (3 for phosphorus). So we begin with [Ne], then add in the two $3s$ electrons as we trace across the s block, followed by three $3p$ electrons as we trace across the p block to P, which is in the third column of the p block. The electron configuration is:

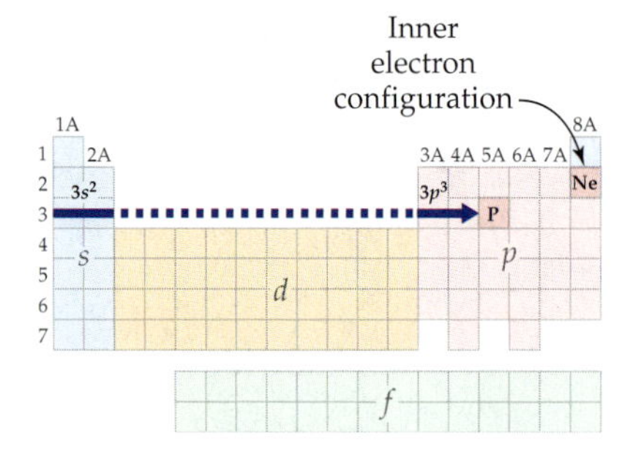

▲ **Figure 9.30 Electron configuration of phosphorus** Determining the electron configuration for P from its position in the periodic table.

P $[Ne]3s^23p^3$

Notice that P is in column 5A and therefore has 5 valence electrons and an outer electron configuration of ns^2np^3.

To summarize, follow these steps to write an electron configuration for an element based on its position in the periodic table.

- The inner electron configuration for any element is the electron configuration of the noble gas that immediately precedes that element in the periodic table. Represent the inner configuration with the symbol for the noble gas in brackets.
- The outer electrons can be deduced from the element's position within a particular block (s, p, d, or f) in the periodic table. Trace the elements between the preceding noble gas and the element of interest, and assign electrons to the appropriate orbitals.
- The highest principal quantum number (highest n value) is equal to the row number of the element in the periodic table.
- For any element containing d electrons, the principal quantum number (n value) of the outermost d electrons is equal to the row number of the element minus 1.

EXAMPLE 9.5 Writing Electron Configurations from the Periodic Table

Use the periodic table to write an electron configuration for arsenic.

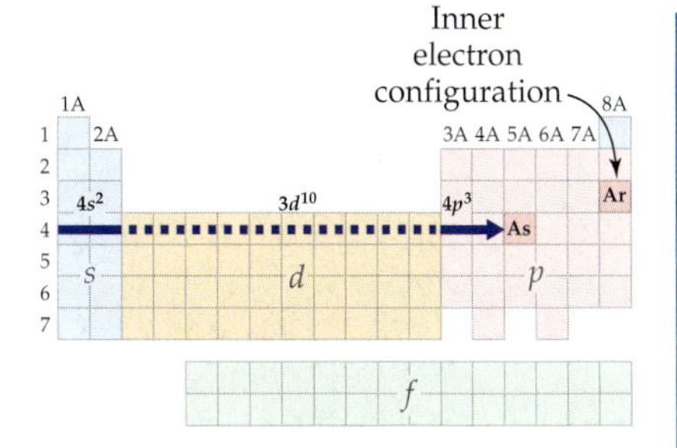

▲ **Figure 9.31 Electron configuration of arsenic** Determining the electron configuration for As from its position in the periodic table.

Solution:

The noble gas that precedes arsenic in the periodic table is argon, so the inner electron configuration is [Ar]. Obtain the outer electron configuration by tracing the elements between Ar and As and assigning electrons to the appropriate orbitals. Remember that the highest n value is given by the row number (4 for arsenic). So, we begin with [Ar], then add in the two $4s$ electrons as we trace across the s block, followed by ten $3d$ electrons as we trace across the d block (the n value for d subshells is equal to the row number minus one), and finally the three $4p$ electrons as we trace across the p block to As, which is in the third column of the p block (◄ Figure 9.31).

The electron configuration is:

$$\text{As} \quad [\text{Ar}]4s^2 3d^{10} 4p^3$$

SKILLBUILDER 9.5 Writing Electron Configurations from the Periodic Table

Use the periodic table to determine the electron configuration for tin.

CONCEPTUAL CHECKPOINT 9.3

Which element has the *fewest* valence electrons?

(a) B
(b) Ca
(c) V
(d) Cr
(e) Ga

9.8 The Explanatory Power of the Quantum-Mechanical Model

Noble gases 18 8A
2 **He** $1s^2$
10 **Ne** $2s^2 2p^6$
18 **Ar** $3s^2 3p^6$
36 **Kr** $4s^2 4p^6$
54 **Xe** $5s^2 5p^6$
86 **Rn** $6s^2 6p^6$

▲ **Figure 9.32 Electron configurations of the noble gases** The noble gases (except for helium) all have 8 valence electrons and completely full outer principal shells.

At the beginning of this chapter, we learned that the quantum-mechanical model explained the chemical properties of the elements such as the inertness of helium, the reactivity of hydrogen, and the periodic law. We can now see how. *The chemical properties of elements are largely determined by the number of valence electrons they contain.* Their properties vary in a periodic fashion because the number of valence electrons is periodic.

Since elements within a column in the periodic table have the same number of valence electrons, they also have similar chemical properties. The noble gases, for example, all have 8 valence electrons, except for helium, which has 2 (◄ Figure 9.32). Although we don't get into the quantitative (or numerical) aspects of the quantum-mechanical model in this book, calculations show that atoms with 8 valence electrons (or 2 for helium) are particularly low in energy, and therefore stable. Consequently, the noble gases are chemically stable, and thus relatively inert or nonreactive.

Elements with electron configurations close to the noble gases are the most reactive because they can attain noble gas electron configurations by losing or gaining a small number of electrons. Alkali metals (Group 1) are among the most reactive metals since their outer electron configuration (ns^1) is 1 electron beyond a noble gas configuration (► Figure 9.33). If they can react to lose the ns^1 electron, they attain a noble gas configuration. This explains why—as we learned in Chapter 4—the Group 1 metals tend to form 1+ cations. As an example, consider the electron configuration of sodium:

$$\text{Na} \quad 1s^2 2s^2 2p^6 3s^1$$

Alkali metals 1 1A
3 **Li** $2s^1$
11 **Na** $3s^1$
19 **K** $4s^1$
37 **Rb** $5s^1$
55 **Cs** $6s^1$
87 **Fr** $7s^1$

▲ **Figure 9.33 Electron configurations of the alkali metals** The alkali metals all have ns^1 electron configurations and are therefore 1 electron beyond a noble gas configuration. In their reactions, they tend to lose that electron, forming 1+ ions and attaining a noble gas configuration.

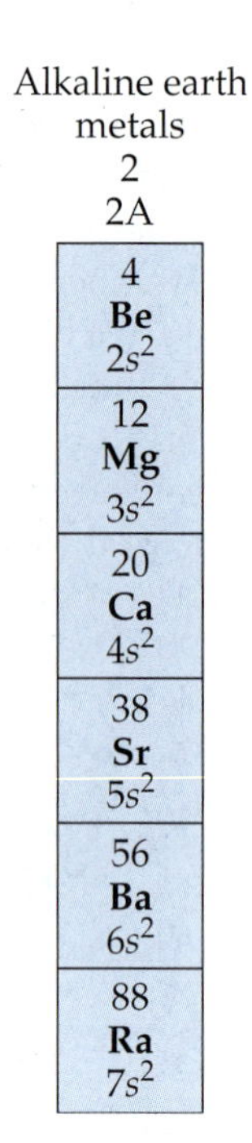

Alkaline earth
metals
2
2A
4
Be
$2s^2$
12
Mg
$3s^2$
20
Ca
$4s^2$
38
Sr
$5s^2$
56
Ba
$6s^2$
88
Ra
$7s^2$

▲ **Figure 9.34 Electron configurations of the alkaline earth metals** The alkaline earth metals all have ns^2 electron configurations and are therefore 2 electrons beyond a noble gas configuration. In their reactions, they tend to lose 2 electrons, forming 2+ ions and attaining a noble gas configuration.

Halogens 17 7A
9 **F** $2s^22p^5$
17 **Cl** $3s^23p^5$
35 **Br** $4s^24p^5$
53 **I** $5s^25p^5$
85 **At** $6s^26p^5$

▲ **Figure 9.35 Electron configurations of the halogens** The halogens all have ns^2np^5 electron configurations and are therefore 1 electron short of a noble gas configuration. In their reactions, they tend to gain 1 electron, forming 1− ions and attaining a noble gas configuration.

Atoms and/or ions that have the same electron configuration are termed *isoelectronic*.

In reactions, sodium loses its 3*s* electron, forming a 1+ ion with the electron configuration of neon.

Na^+ $1s^22s^22p^6$
Ne $1s^22s^22p^6$

Similarly, alkaline earth metals, with an outer electron configuration of ns^2, also tend to be reactive metals, losing their ns^2 electrons to form 2+ cations (▲ Figure 9.34). For example, consider magnesium:

Mg $1s^22s^22p^63s^2$

In reactions, magnesium loses its two 3*s* electrons, forming a 2+ ion with the electron configuration of neon.

Mg^{2+} $1s^22s^22p^6$

On the other side of the periodic table, halogens are among the most reactive nonmetals because of their ns^2np^5 electron configurations (▲ Figure 9.35). They are only one electron away from a noble gas configuration and tend to react to gain that one electron, forming 1− ions. For example, consider fluorine:

F $1s^22s^22p^5$

In reactions, fluorine gains an additional 2*p* electron, forming a 1− ion with the electron configuration of neon.

F^- $1s^22s^22p^6$

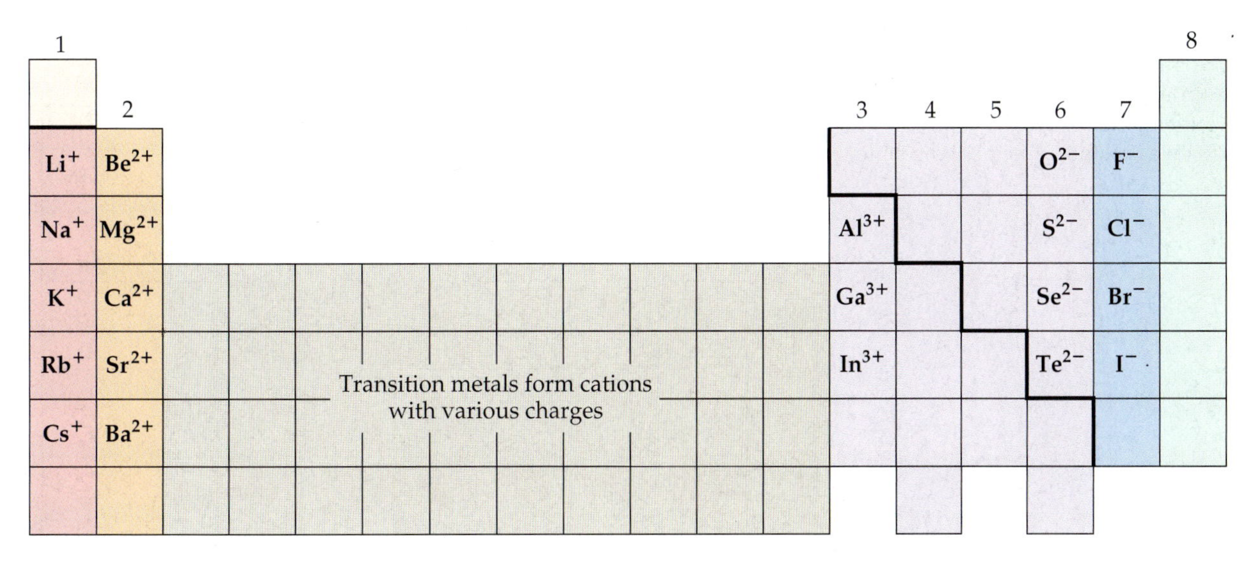

▲ **Figure 9.36** **Elements that form predictable ions**

The elements that form predictable ions are shown in ▲ Figure 9.36 (first introduced in Chapter 4). Notice how the charge of these ions reflects their electron configuration—these elements form ions with noble gas electron configurations.

9.9 Periodic Trends: Atomic Size, Ionization Energy, and Metallic Character

The quantum-mechanical model also explains other periodic trends such as atomic size, ionization energy, and metallic character. We will examine these one at a time.

Atomic Size

The **atomic size** of an atom is determined by how far the outermost electrons are from the nucleus. As we move across a period in the periodic table, we know that electrons are going into orbitals with the same principal quantum number, n. Since the principal quantum number largely determines the size of an orbital, electrons are therefore filling orbitals of approximately the same size, and we might expect atomic size to remain constant across a period. However, with each step across a period, the number of protons in the nucleus is also increasing. This increase in the number of protons results in a greater pull on the electrons, causing atomic size to actually decrease. Therefore:

As you move across a period, or row, to the right in the periodic table, atomic size decreases, as shown in ▶ Figure 9.37.

As you move down a column in the periodic table, the highest principal quantum number, n, increases. Since the size of an orbital increases with increasing principal quantum number, the electrons that occupy the outermost orbitals are farther from the nucleus as you move down a column. Therefore:

TUTORIAL
Atomic Radii Movie

As you move down a column, or family, in the periodic table, atomic size increases, as shown in Figure 9.37.

Figure 9.37 Periodic properties: Atomic size Atomic size decreases as you move to the right across a period and increases as you move down a column in the periodic table.

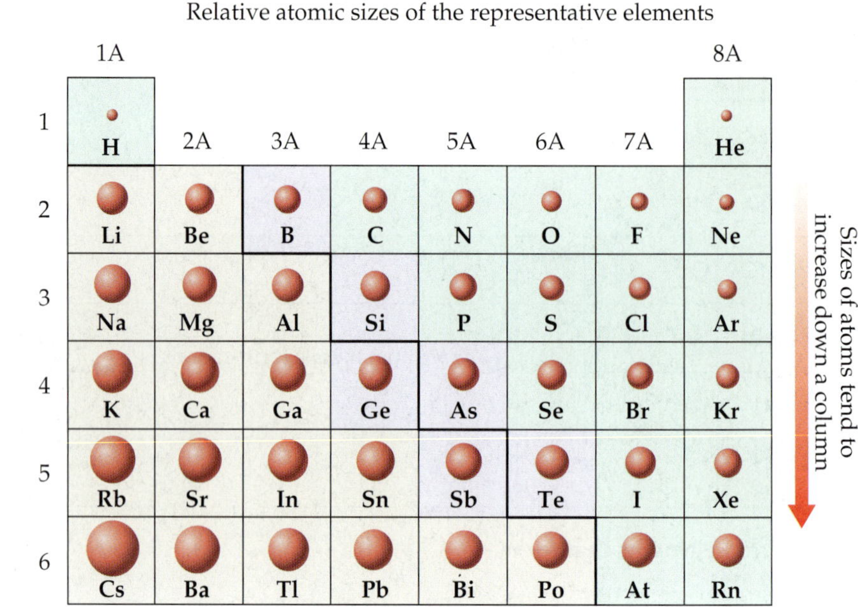

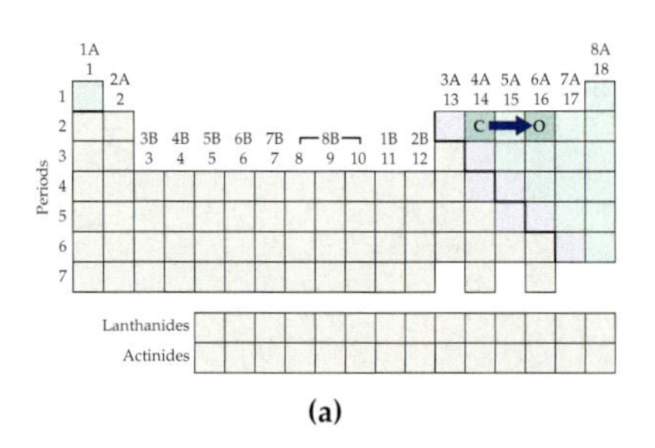

(a)

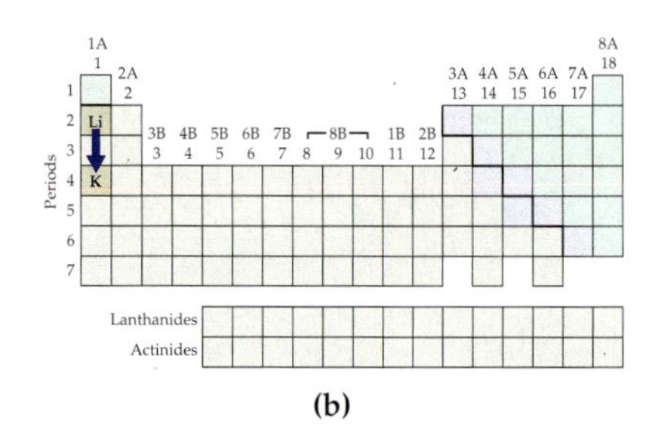

(b)

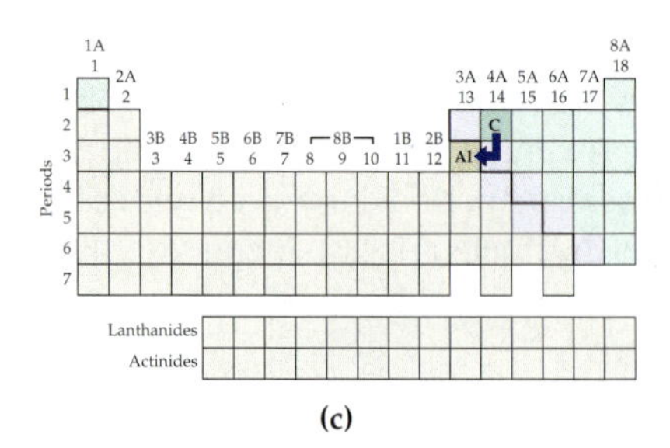

(c)

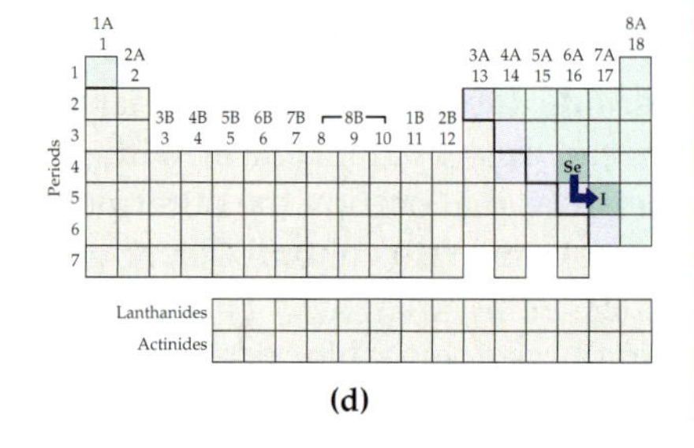

(d)

EXAMPLE 9.6 Atomic Size

Choose the larger atom from each of the following pairs.

(a) C or O
(b) Li or K
(c) C or Al
(d) Se or I

Solution

(a) C or O
Carbon atoms are larger than O atoms because, as we trace the path between C and O on the periodic table (see margin), we move to the right within the same period. Atomic size decreases as you go to the right.

(b) Li or K
Potassium atoms are larger than Li atoms because, as we trace the path between Li and K on the periodic table (see margin), we move down a column. Atomic size increases as you go down a column.

(c) C or Al
Aluminum atoms are larger than C atoms because, as we trace the path between C and Al on the periodic table (see margin), we move down a column (atomic size increases) and then to the left across a period (atomic size increases). These effects add together for an overall increase.

(d) Se or I
Based on periodic properties alone, we cannot tell which atom is larger, because as we trace the path between Se and I (see margin) we go down a column (atomic size increases) and then to the right across a period (atomic size decreases). These effects tend to cancel one another.

Chemistry AND HEALTH

Pumping Ions: Atomic Size and Nerve Impulses

No matter what you are doing at this moment, tiny pumps in each of the trillions of cells that make up your body are hard at work. These pumps, located in the cell membrane, move a number of different ions into and out of the cell. The most important such ions are sodium (Na^+) and potassium (K^+). However, these ions are pumped in opposite directions. Sodium ions are pumped *out of cells*, while potassium ions are pumped *into cells*. The result is a *chemical gradient* for each ion: The concentration of sodium is higher outside the cell than within, while exactly the opposite is true for potassium.

The ion pumps within the cell membrane are analogous to water pumps in a high-rise building that pump water against the force of gravity to a tank on the roof. Other structures within the membrane, called ion channels, are like the building's faucets. When they open momentarily, bursts of sodium and potassium ions, driven by their concentration gradients, flow back across the membrane—sodium flowing in and potassium flowing out. These ion pulses are the basis for the transmission of nerve signals in the brain, heart, and throughout the body. Consequently, every move you make or every thought you have is mediated by the flow of these ions.

How do the pumps and channels differentiate between sodium and potassium ions? How do the ion pumps selectively move sodium out of the cell and potassium into the cell? To answer this question, we must examine the sodium and potassium ions more closely. In what way do they differ? Both are cations of group I metals. All group I metals tend to lose one electron to form cations with 1+ charge, so that cannot be the decisive factor. But potassium (atomic number 19) lies directly below sodium in the periodic table (atomic number 11), and based on periodic properties is therefore larger than sodium. The potassium ion has a radius of 133 pm while the sodium ion has a radius of 95 pm. (Recall from Chapter 2 that 1 pm = 10^{-12} m.) The pumps and channels within cell membranes are so sensitive that they can distinguish between the sizes of these two ions and selectively allow only one or the other to pass. The result is the transmission of nerve signals that allows you to read this page.

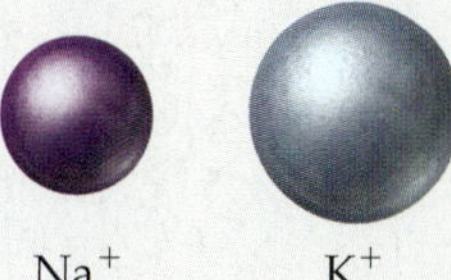

CAN YOU ANSWER THIS? *Other ions, including calcium and magnesium, are also important to nerve signal transmission. Arrange the following four ions in order of increasing size:* K^+, Na^+, Mg^{2+}, *and* Ca^{2+}.

SKILLBUILDER 9.6 **Atomic Size**

Choose the larger atom from each of the following pairs.

(a) Pb or Po
(b) Rb or Na
(c) Sn or Bi
(d) F or Se

Ionization Energy

The **ionization energy** of an atom is the energy required to remove an electron from the atom in the gaseous state. For example, the ionization of sodium can be represented with the following equation.

$$Na + \text{Ionization energy} \longrightarrow Na^+ + 1e^-$$

Based on what you know about electron configurations, what would you predict about ionization energy trends? Would it take more or less energy to remove an electron from Na than from Cl? We know that Na has an outer electron configuration of $3s^1$ and Cl has an outer electron configuration of $3s^23p^5$. Since removing an electron from sodium gives it a noble gas configuration—and removing an electron from Cl does not—we would expect

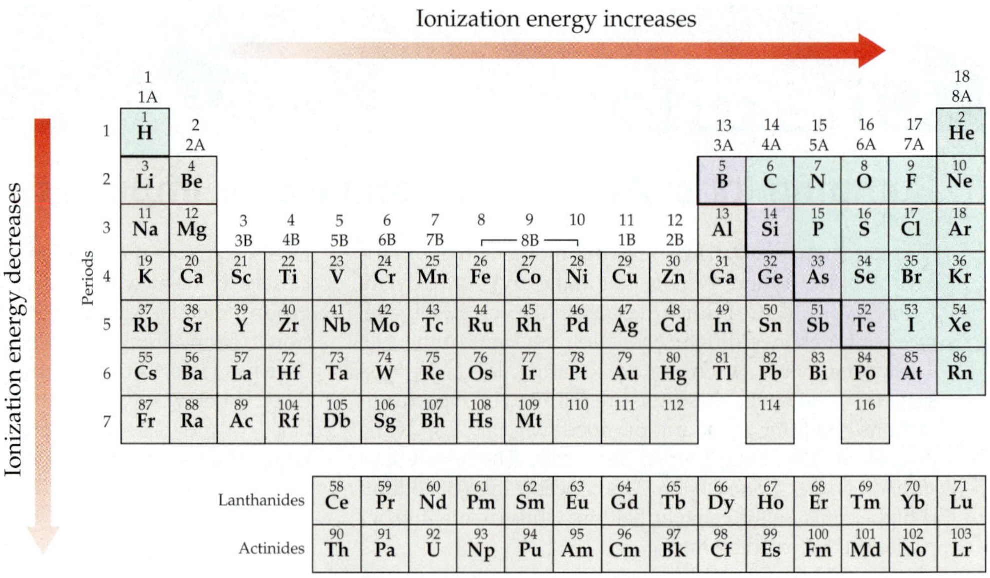

Ionization energy increases

Ionization energy decreases

Periods	1 1A	2 2A	3 3B	4 4B	5 5B	6 6B	7 7B	8 8B	9 8B	10 8B	11 1B	12 2B	13 3A	14 4A	15 5A	16 6A	17 7A	18 8A
1	1 H																	2 He
2	3 Li	4 Be											5 B	6 C	7 N	8 O	9 F	10 Ne
3	11 Na	12 Mg											13 Al	14 Si	15 P	16 S	17 Cl	18 Ar
4	19 K	20 Ca	21 Sc	22 Ti	23 V	24 Cr	25 Mn	26 Fe	27 Co	28 Ni	29 Cu	30 Zn	31 Ga	32 Ge	33 As	34 Se	35 Br	36 Kr
5	37 Rb	38 Sr	39 Y	40 Zr	41 Nb	42 Mo	43 Tc	44 Ru	45 Rh	46 Pd	47 Ag	48 Cd	49 In	50 Sn	51 Sb	52 Te	53 I	54 Xe
6	55 Cs	56 Ba	57 La	72 Hf	73 Ta	74 W	75 Re	76 Os	77 Ir	78 Pt	79 Au	80 Hg	81 Tl	82 Pb	83 Bi	84 Po	85 At	86 Rn
7	87 Fr	88 Ra	89 Ac	104 Rf	105 Db	106 Sg	107 Bh	108 Hs	109 Mt	110	111	112		114		116		

Lanthanides	58 Ce	59 Pr	60 Nd	61 Pm	62 Sm	63 Eu	64 Gd	65 Tb	66 Dy	67 Ho	68 Er	69 Tm	70 Yb	71 Lu
Actinides	90 Th	91 Pa	92 U	93 Np	94 Pu	95 Am	96 Cm	97 Bk	98 Cf	99 Es	100 Fm	101 Md	102 No	103 Lr

▲ **Figure 9.38 Periodic properties: Ionization energy** Ionization energy increases as you move to the right across a period and decreases as you move down a column in the periodic table.

sodium to have a lower ionization energy, and that is the case. We can generalize this idea in the following statement:

> As you move across a period, or row, to the right in the periodic table, ionization energy increases (▲ Figure 9.38).

What happens to ionization energy as you move down a column? As we have learned, the principal quantum number, n, increases as you move down a column. Within a given subshell, orbitals with higher principal quantum numbers are larger than orbitals with smaller principal quantum numbers. Consequently, electrons in the outermost principal shell are farther away from the positively charged nucleus—and therefore held less tightly—as you move down a column. This results in a lower ionization energy (if the electron is held less tightly, it is easier to pull away) as you move down a column. Therefore:

> As you move down a column, or family, in the periodic table, ionization energy decreases (Figure 9.38).

Notice that the trends in ionization energy are consistent with the trends in atomic size. As an atom gets smaller, it is more difficult to ionize because the electrons are held more tightly. Therefore, as you go across a period, atomic size decreases and ionization energy increases. Similarly, as you go down a column, atomic size increases and ionization energy decreases since electrons are farther from the nucleus and therefore less tightly held.

EXAMPLE 9.7 Ionization Energy

Choose the element with the higher ionization energy from each of the following pairs.

(a) Mg or P
(b) As or Sb
(c) N or Si
(d) O or Cl

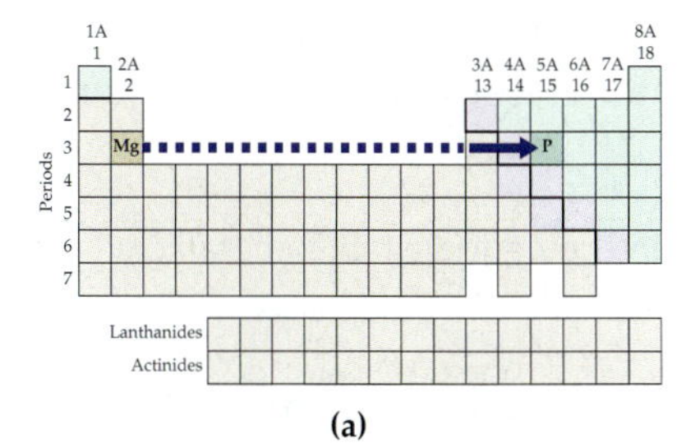

(a)

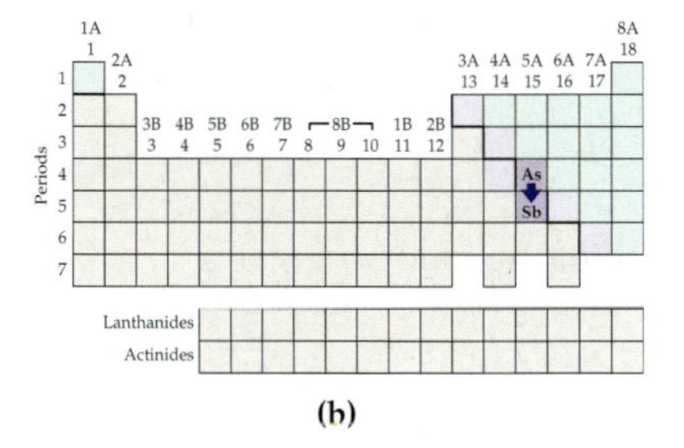

(b)

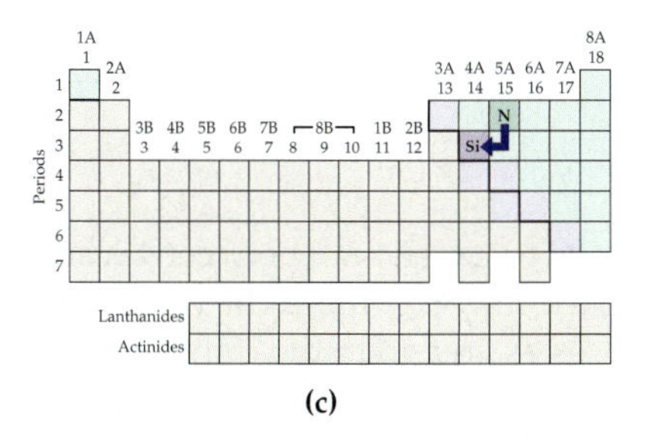

(c)

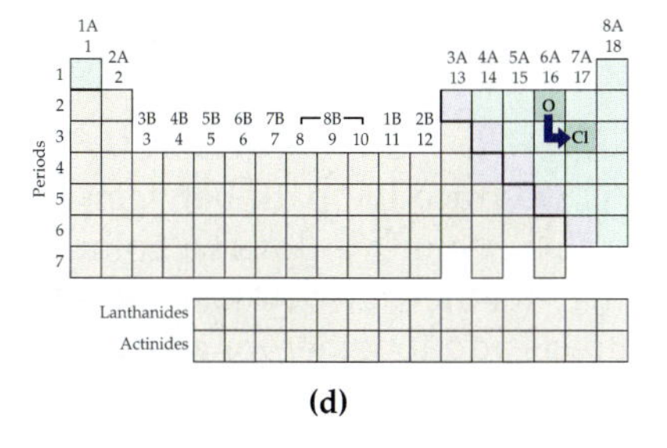

(d)

Solution:

(a) Mg or P
P has a higher ionization than Mg because, as we trace the path between Mg and P on the periodic table (see margin), we move to the right within the same period. Ionization energy increases as you go to the right.

(b) As or Sb
As has a higher ionization energy than Sb because, as we trace the path between As and Sb on the periodic table (see margin), we move down a column. Ionization energy decreases as you go down a column.

(c) N or Si
N has a higher ionization energy than Si because, as we trace the path between N and Si on the periodic table (see margin), we move down a column (ionization energy decreases) and then to the left across a period (ionization energy decreases). These effects sum together for an overall decrease.

(d) O or Cl
Based on periodic properties alone, we cannot tell which has a higher ionization energy because, as we trace the path between O and Cl (see margin) we go down a column (ionization energy decreases) and then to the right across a period (ionization energy increases). These effects tend to cancel.

SKILLBUILDER 9.7 Ionization Energy

Choose the element with the higher ionization energy from each of the following pairs.

(a) Mg or Sr
(b) In or Te
(c) C or P
(d) F or S

Metallic Character

As we learned in Chapter 4, metals tend to lose electrons in their chemical reactions, while nonmetals tend to gain electrons. As you move across a period in the periodic table, ionization energy increases, which means that electrons are less likely to be lost in chemical reactions. Consequently:

> As you move across a period, or row, to the right in the periodic table, **metallic character** decreases (▼ Figure 9.39).

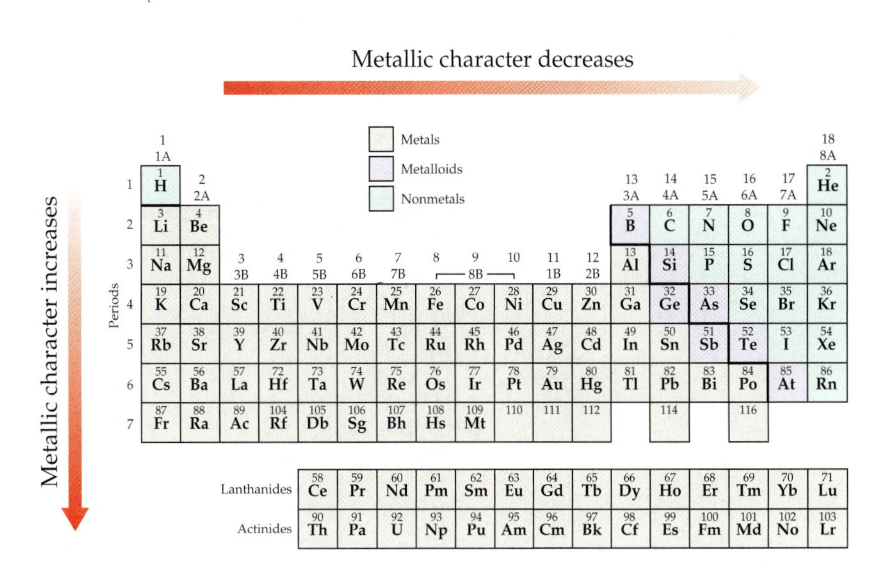

▶ **Figure 9.39 Periodic properties: metallic character** Metallic character decreases as you move to the right across a period and increases as you move down a column in the periodic table.

As you move down a column in the periodic table, ionization energy decreases, making electrons more likely to be lost in chemical reactions. Consequently:

As you move down a column, or family, in the periodic table, metallic character increases (Figure 9.39).

TUTORIAL
Interactive Periodic Table

These trends, based on the quantum-mechanical model, explain the distribution of metals and nonmetals that we learned in Chapter 4. Metals are found toward the left side of the periodic table and nonmetals (with the exception of hydrogen) toward the upper right.

EXAMPLE 9.8 Metallic Character

Choose the more metallic element from each of the following pairs.

(a) Sn or Te
(b) Si or Sn
(c) Br or Te
(d) Se or I

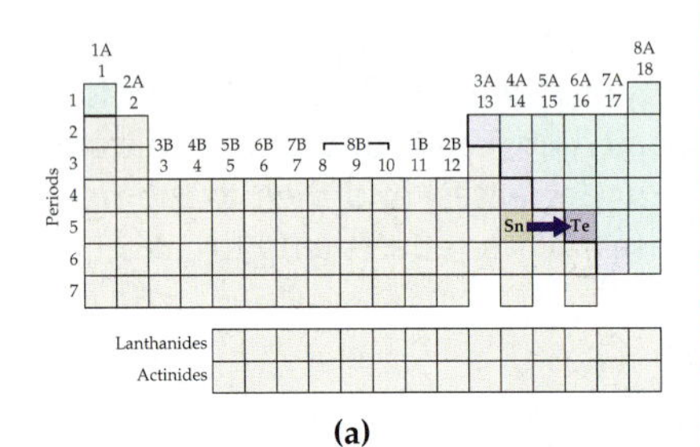

(a)

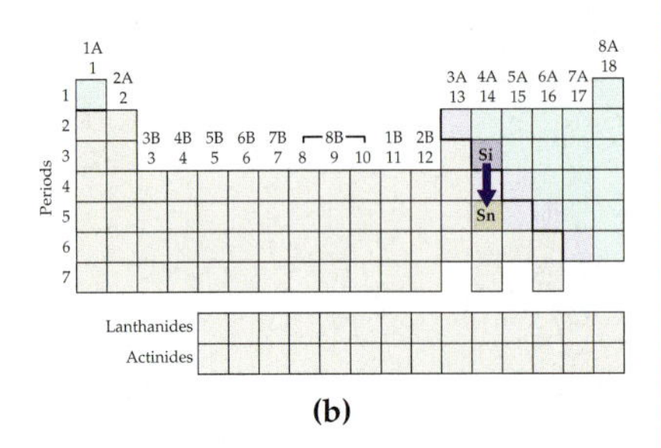

(b)

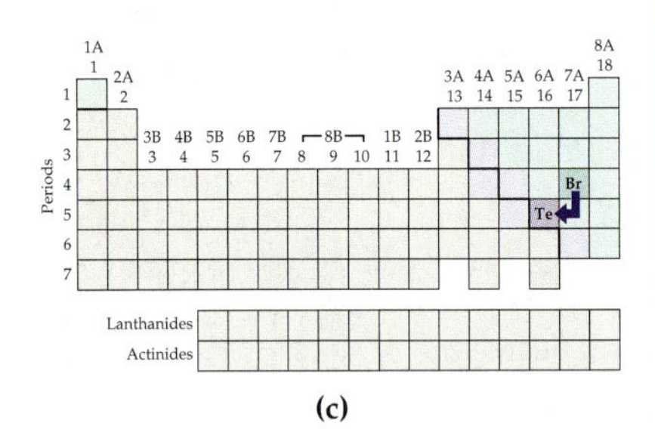

(c)

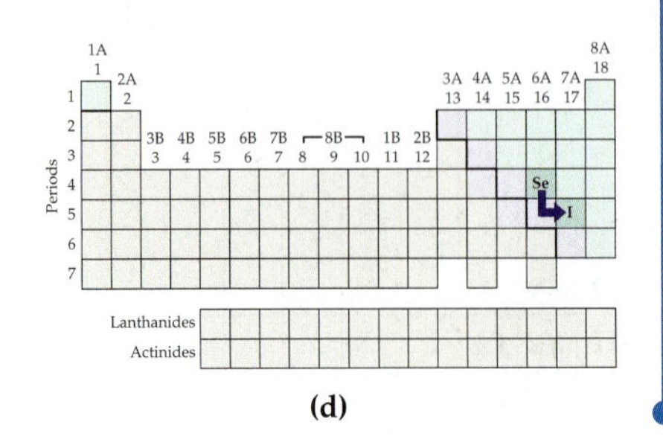

(d)

Solution:

(a) Sn or Te
Sn is more metallic than Te because, as we trace the path between Sn and Te on the periodic table (see margin), we move to the right within the same period. Metallic character decreases as you go to the right.

(b) Si or Sn
Sn is more metallic than Si because, as we trace the path between Si and Sn on the periodic table (see margin), we move down a column. Metallic character increases as you go down a column.

(c) Br or Te
Te is more metallic than Br because, as we trace the path between Br and Te on the periodic table (see margin), we move down a column (metallic character increases) and then to the left across a period (metallic character increases). These effects add together for an overall increase.

(d) Se or I
Based on periodic properties alone, we cannot tell which is more metallic because as we trace the path between Se and I (see margin), we go down a column (metallic character increases) and then to the right across a period (metallic character decreases). These effects tend to cancel.

SKILLBUILDER 9.8 Metallic Character

Choose the more metallic element from each of the following pairs.

(a) Ge or In
(b) Ga or Sn
(c) P or Bi
(d) B or N

CHAPTER IN REVIEW

Chemical Principles

Light: Light is electromagnetic radiation, a kind of energy that travels through space at a constant speed of 3.0×10^8 m/s (186,000 mi/s) and exhibits both wavelike and particle-like behavior. Particles of light are called photons. The wave nature of light is characterized by its wavelength, the distance between adjacent crests in the wave. The wavelength of light is inversely proportional to both the frequency—the number of cycles that pass a stationary point in one second—and the energy of a photon. Electromagnetic radiation ranges in wavelength from 10^{-16} m (gamma rays) to 10^6 m (radio waves). In between these lie X-rays, ultraviolet light, visible light, infrared light, and microwaves.

The Bohr Model: The emission spectrum of hydrogen, consisting of bright lines at specific wavelengths, can be explained by the Bohr model for the hydrogen atom. In this model, electrons occupy circular orbits at specific fixed distances from the nucleus. Each orbit is specified by a quantum number (n), which also specifies the orbit's energy. While an electron is in a given orbit, its energy remains constant. When it jumps between orbits, a quantum of energy is absorbed or emitted. Since the difference in energy between orbits is fixed, the energy emitted or absorbed is also fixed. Emitted energy is carried away by a photon of specific wavelength.

The Quantum-Mechanical Model: The quantum-mechanical model for the atom replaces orbits with orbitals, which are electron probability maps. Orbitals show the relative probability of the electron's being found in various places when the atom is probed. Orbitals are specified with a number (n), called the principal quantum number, and a letter. The principal quantum number, which must be an integer (1, 2, 3 . . .), specifies the principal shell, and the letter (s, p, d, or f) specifies the subshell of the orbital. In the hydrogen atom, the energy of orbitals depends only on n. In multi-electron atoms, the energy ordering is 1s 2s 2p 3s 3p 4s 3d 4p 5s 4d 5p 6s.

An electron configuration shows the occupation of orbitals for a particular atom. These are written by filling the orbitals in order of increasing energy while keeping in mind the Pauli exclusion principle (each orbital can hold a maximum of two electrons with opposing spins) and Hund's rule (electrons occupy orbitals of identical energy singly before pairing).

Relevance

Light: Light enables us to see the world. However, we see only visible light, a small sliver in the center of the electromagnetic spectrum. Other forms of electromagnetic radiation are used for cancer therapy, X-ray imaging, night vision, microwave cooking, and communications. Light is also important to many chemical problems. We can learn about the electronic structure of atoms, for example, by examining their interaction with light.

The Bohr Model: The Bohr model was a first attempt to explain the bright-line spectra of atoms. While it did predict the spectrum of the hydrogen atom, it failed at predicting the spectra of other atoms. Consequently, it was replaced by the quantum-mechanical model.

The Quantum-Mechanical Model: The quantum-mechanical model changed the way we view nature. Before the quantum-mechanical model, electrons were viewed as small particles, much like any other particle. Electrons were supposed to follow the normal laws of motion, just as a baseball does. However, the electron, with its wavelike properties, does not follow these laws. Instead, electron motion is describable only through probabilistic predictions. Quantum theory single-handedly changed the predictability of nature at its most fundamental level.

The quantum-mechanical model of the atom predicts and explains many of the chemical properties we have learned in earlier chapters.

The Periodic Table: Elements within the same column of the periodic table have similar outer electron configurations and the same number of valence electrons (electrons in the outermost principal shell). Consequently, they have similar chemical properties. The periodic table is divisible into blocks (*s* block, *p* block, *d* block, and *f* block) in which particular sublevels are filled. As you move across a period to the right in the periodic table, atomic size decreases, ionization energy increases, and metallic character decreases. As you move down a column in the periodic table, atomic size increases, ionization energy decreases, and metallic character increases.

The Periodic Table: The periodic law exists because the number of valence electrons is periodic, and valence electrons determine chemical properties. Quantum theory also predicts that atoms with 8 outer shell electrons (or 2 for helium) are particularly stable, thus explaining the inertness of the noble gases. Atoms without noble gas configurations undergo chemical reactions to attain them, explaining the reactivity of the alkali metals and the halogens as well as the tendency of several families to form ions with certain charges.

Chemical Skills

Predicting Relative Wavelength, Energy, and Frequency of Light (Section 9.3)

- Relative wavelengths can be obtained from Figure 9.4.
- Energy per photon increases with decreasing (shorter) wavelength.
- Frequency increases with decreasing (shorter) wavelength.

Writing Electron Configurations and Orbital Diagrams (Sections 9.6, 9.7)

To write electron configurations, determine the number of electrons in the atom from the element's atomic number and then follow these rules:

- Electrons occupy orbitals so as to minimize the energy of the atom; therefore, lower-energy orbitals fill before higher-energy orbitals. Orbitals fill in the following order: 1*s* 2*s* 2*p* 3*s* 3*p* 4*s* 3*d* 4*p* 5*s* 4*d* 5*p* 6*s* (Figure 9.27). The *s* subshells hold up to 2 electrons, *p* subshells hold up to 6, *d* subshells hold up to 10, and *f* subshells hold up to 14.
- Orbitals can hold no more than 2 electrons each. When 2 electrons occupy the same orbital, they must have opposing spins.
- When orbitals of identical energy are available, these are first occupied singly with parallel spins rather than in pairs.

Examples

EXAMPLE 9.9 Predicting Relative Wavelength, Energy, and Frequency of Light

Which of the following types of light—infrared or ultraviolet—has the longer wavelength? Higher frequency? Higher energy per photon?

Solution:

Infrared light has the longer wavelength (Figure 9.4). Ultraviolet light has the higher frequency and the higher energy per photon.

LIVE EXAMPLE

EXAMPLE 9.10 Writing Electron Configurations and Orbital Diagrams

Write an electron configuration and orbital diagram (outer electrons only) for germanium.

Solution:

Germanium is atomic number number 32; therefore, it has 32 electrons.

Electron Configuration

Ge $1s^22s^22p^63s^23p^64s^23d^{10}4p^2$

or

Ge $[Ar]4s^23d^{10}4p^2$

Orbital Diagram (Outer Electrons)

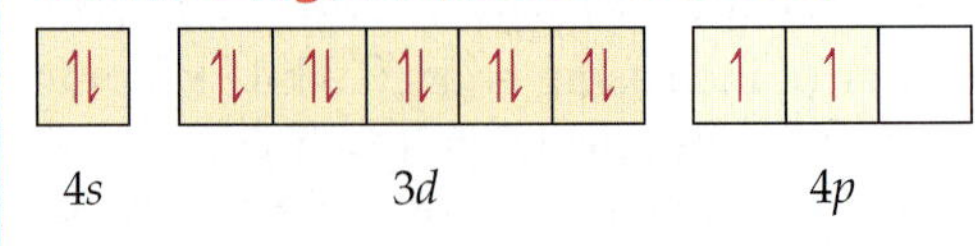

Identifying Valence Electrons and Core Electrons (Section 9.7)

- Valence electrons are the electrons in the outermost principal energy shell (the principal shell with the highest principal quantum number).
- Core electrons are those that are not in the outermost principal shell.

EXAMPLE 9.11 Identifying Valence Electrons and Core Electrons

Identify the valence electrons and core electrons in the electron configuration of germanium (given in Example 9.10).

Solution:

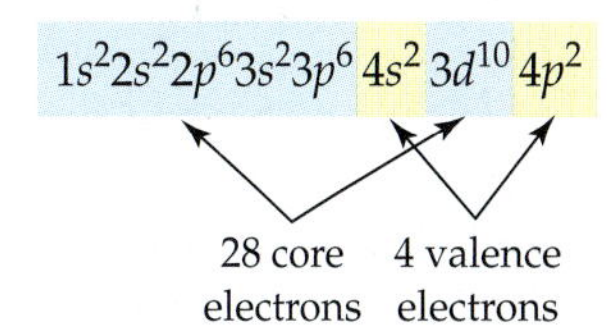

Writing Electron Configurations for an Element Based on Its Position in the Periodic Table (Section 9.7)

1. The inner electron configuration for any element is the electron configuration of the noble gas that immediately precedes that element in the periodic table. Represent the inner configuration with the symbol for the noble gas in brackets.
2. The outer electrons can be deduced from the element's position within a particular block (*s*, *p*, *d*, or *f*) in the periodic table. Trace the elements between the preceding noble gas and the element of interest and assign electrons to the appropriate orbitals. Figure 9.29 shows the outer electron configuration based on the position of an element in the periodic table.
3. The highest principal quantum number (highest *n* value) is equal to the row number of the element in the periodic table.
4. The principal quantum number (*n* value) of the outermost *d* electrons for any element containing *d* electrons is equal to the row number of the element minus 1.

EXAMPLE 9.12 Writing Electron Configurations for an Element Based on Its Position in the Periodic Table

Write an electron configuration for iodine based on its position in the periodic table.

Solution:

The inner configuration for I is [Kr].

We begin with the [Kr] inner electron configuration. As we trace from Kr to I, we add 2 5*s* electrons, 10 4*d* electrons, and 5 5*p* electrons. The overall configuration is:

I [Kr]$5s^24d^{10}5p^5$

Periodic Trends: Atomic Size, Ionization Energy, and Metallic Character (Section 9.9)

On the periodic table:

- Atomic size decreases as you move to the right and increases as you move down.
- Ionization energy increases as you move to the right and decreases as you move down.
- Metallic character decreases as you move to the right and increases as you move down.

EXAMPLE 9.13 Periodic Trends: Atomic Size, Ionization Energy, and Metallic Character

Arrange Si, In, and S in order of **(a)** increasing atomic size, **(b)** increasing ionization energy, and **(c)** increasing metallic character.

Solution:

(a) S, Si, In
(b) In, Si, S
(c) S, Si, In

Key Terms

atomic size [9.9]
Bohr model [9.1]
core electrons [9.7]
electromagnetic radiation [9.2]
electromagnetic spectrum [9.3]
electron configuration [9.6]
electron spin [9.6]
emission spectrum [9.4]
excited state [9.4]
frequency (ν) [9.2]
gamma rays [9.3]
ground state [9.6]
Hund's rule [9.6]
infrared light [9.3]
ionization energy [9.9]
metallic character [9.9]
microwaves [9.3]
orbital diagram [9.6]
orbitals [9.5]
Pauli exclusion principle [9.6]
photon [9.2]
principal quantum number [9.6]
principal shell [9.6]
quantized [9.4]
quantum (plural, *quanta*) [9.4]
quantum-mechanical model [9.1]
quantum numbers [9.4]
radio waves [9.3]
subshell [9.6]
ultraviolet (UV) light [9.3]
valence electrons [9.7]
visible light [9.3]
wavelength (λ) [9.2]
X-rays [9.3]

Exercises

Questions

1. When were the Bohr model and the quantum-mechanical model for the atom developed? What is their purpose?
2. What is light?
3. How fast does light travel? What are some examples of its fast speed?
4. What is white light? Colored light?
5. Explain, in terms of absorbed and reflected light, why a blue object appears blue.
6. What is the relationship between the wavelength of light and the amount of energy carried by its photons?
7. How are wavelength and frequency related?
8. What produces gamma rays?
9. How are X-rays used?
10. Why should excess exposure to gamma rays and X-rays be avoided?
11. Why should excess exposure to ultraviolet light be avoided?
12. What objects emit infrared light? What technology exploits this?
13. Why do microwave ovens heat food but seldom heat the dish the food is on?
14. What type of electromagnetic radiation is used in communications devices such as cellular telephones?
15. Describe the Bohr model for the hydrogen atom.
16. What is an emission spectrum? Use the Bohr model to explain why the emission spectra of atoms consist of distinct lines at specific wavelengths.
17. Explain the difference between a Bohr orbit and a quantum-mechanical orbital.
18. What is the difference between the ground state of an atom and an excited state of an atom?
19. Explain how the motion of an electron is different from the motion of a baseball. What is a probability map?
20. Explain why quantum-mechanical orbitals have "fuzzy" boundaries.
21. List the four possible subshells in the quantum-mechanical model, the number of orbitals in each subshell, and the maximum number of electrons that can be contained in each subshell.
22. List all of the quantum-mechanical orbitals, in the correct energy order for atoms other than hydrogen, through 5*s*.
23. What is the Pauli exclusion principle? Why is it important when writing electron configurations?
24. What is Hund's rule? Why is it important when writing orbital diagrams?
25. Within an electron configuration, what do symbols such as [Ne] and [Kr] represent?
26. Explain the difference between valence electrons and core electrons.
27. Identify each block in the following blank periodic table.
    (a) *s* block
    (b) *p* block
    (c) *d* block
    (d) *f* block

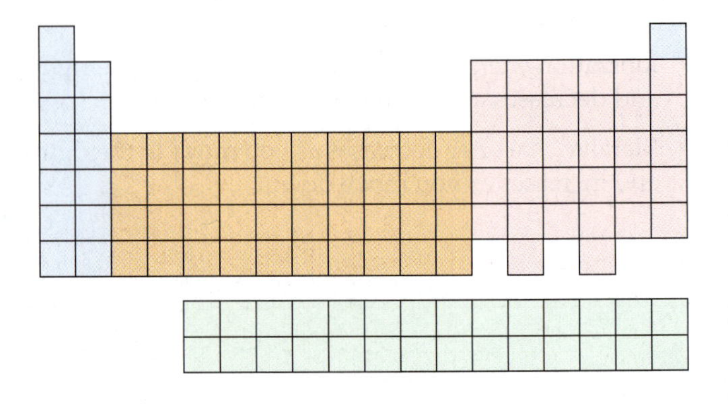

28. Give some examples of the explanatory power of the quantum-mechanical model.
29. Explain why Group 1 elements tend to form 1+ ions and Group 7 elements tend to form 1− ions.
30. Explain the periodic trends in each of the following:
    (a) ionization energy
    (b) atomic size
    (c) metallic character

Problems

Wavelength, Energy, and Frequency of Electromagnetic Radiation

31. Which one of these types of electromagnetic radiation has the longest wavelength?
    (a) visible
    (b) ultraviolet
    (c) infrared
    (d) X-ray
32. Which one of these types of electromagnetic radiation has the shortest wavelength?
    (a) radio waves
    (b) microwaves
    (c) infrared
    (d) ultraviolet
33. Write the following types of electromagnetic radiation in order of increasing energy per photon.
    (a) radio waves
    (b) microwaves
    (c) infrared
    (d) ultraviolet
34. Write the following types of electromagnetic radiation in order of decreasing energy per photon.
    (a) gamma rays
    (b) radio waves
    (c) microwaves
    (d) visible light
35. List two types of electromagnetic radiation with frequencies higher than visible light.
36. List two types of electromagnetic radiation with frequencies lower than infrared light.
37. List the following three types of radiation—X-rays, gamma rays, and microwaves—in order of:
    (a) increasing energy per photon
    (b) increasing frequency
    (c) increasing wavelength
38. List the following three types of electromagnetic radiation—visible, infrared, and radio waves—in order of:
    (a) decreasing energy per photon
    (b) decreasing frequency
    (c) decreasing wavelength

The Bohr Model

39. Bohr orbits have fixed _____ and fixed _____.
40. In the Bohr model, what happens when an electron makes transitions between orbits?
41. Two of the emission wavelengths in the hydrogen emission spectrum are 410 nm and 434 nm. One of these is due to the $n = 6$ to $n = 2$ transition and the other is due to the $n = 5$ to $n = 2$ transition. Which wavelength goes with which transition?
42. Two of the emission wavelengths in the hydrogen emission spectrum are 656 nm and 486 nm. One of these is due to the $n = 4$ to $n = 2$ transition and the other is due to the $n = 3$ to $n = 2$ transition. Which wavelength goes with which transition?

The Quantum-Mechanical Model

43. Make a sketch of the 1*s* and 2*p* orbitals. How would the 2*s* and 3*p* orbitals differ from the 1*s* and 2*p* orbitals?
44. Make a sketch of the 3*d* orbitals. How would the 4*d* orbitals differ from the 3*d* orbitals?
45. Which electron is, on average, closer to the nucleus: an electron in a 2*s* orbital or an electron in a 3*s* orbital?
46. Which electron is, on average, farther from the nucleus: an electron in a 3*p* orbital or an electron in a 4*p* orbital?
47. According to the quantum-mechanical model for the hydrogen atom, which of the following electron transitions would produce light with longer wavelength: 2*p* to 1*s* or 3*p* to 1*s*?
48. According to the quantum-mechanical model for the hydrogen atom, which of the following transitions would produce light with longer wavelength: 3*p* to 2*s* or 4*p* to 2*s*?

Electron Configurations

49. Write full electron configurations for each of the following elements.
    (a) C
    (b) Na
    (c) Ar
    (d) Si

50. Write full electron configurations for each of the following elements.
    (a) N
    (b) S
    (c) Ne
    (d) K

51. Write full orbital diagrams for each of the following elements and indicate the number of unpaired electrons in each.
    (a) Be
    (b) C
    (c) F
    (d) Ne

52. Write full orbital diagrams for each of the following elements and indicate the number of unpaired electrons in each.
    (a) B
    (b) N
    (c) Li
    (d) He

53. Write electron configurations for each of the following elements. Use the symbol of the previous noble gas in brackets to represent the core electrons.
    (a) Ga
    (b) As
    (c) Rb
    (d) Sn

54. Write electron configurations for each of the following elements. Use the symbol of the previous noble gas in brackets to represent the core electrons.
    (a) Te
    (b) Br
    (c) I
    (d) Cs

55. Write electron configurations for each of the following transition metals.
    (a) Ti
    (b) V
    (c) Cr
    (d) Mn

56. Write electron configurations for each of the following transition metals.
    (a) Co
    (b) Ni
    (c) Cu
    (d) Zn

Valence Electrons and Core Electrons

57. Write full electron configurations for each of the following elements and indicate the valence electrons and the core electrons.
    (a) B
    (b) N
    (c) Sb
    (d) K

58. Write full electron configurations for each of the following elements and indicate the valence electrons and the core electrons.
    (a) Sr
    (b) Cl
    (c) Kr
    (d) Ge

59. Write orbital diagrams for the valence electrons of each of the following elements and indicate the number of unpaired electrons in each.
    (a) Si
    (b) Ca
    (c) Cl
    (d) Ar

60. Write orbital diagrams for the valence electrons of each of the following elements and indicate the number of unpaired electrons in each.
    (a) Ga
    (b) K
    (c) Kr
    (d) Br

61. How many valence electrons are in each of the following:
    (a) O
    (b) S
    (c) Br
    (d) Rb

62. How many valence electrons are in each of the following:
    (a) Ba
    (b) Al
    (c) Be
    (d) Se

Electron Configurations and the Periodic Table

63. Give the outer electron configuration for each of the following columns in the periodic table.
(a) 1A
(b) 2A
(c) 5A
(d) 7A

64. Give the outer electron configuration for each of the following columns in the periodic table.
(a) 3A
(b) 4A
(c) 6A
(d) 8A

65. Use the periodic table to write electron configurations for each of the following elements.
(a) Al
(b) Be
(c) In
(d) Zr

66. Use the periodic table to write electron configurations for each of the following elements.
(a) Xe
(b) Sc
(c) Zr
(d) Ba

67. Use the periodic table to write electron configurations for each of the following elements.
(a) As
(b) Ba
(c) Ni
(d) Bi

68. Use the periodic table to write electron configurations for each of the following elements.
(a) Se
(b) Sn
(c) Pb
(d) Cd

69. How many $2p$ electrons are in each of the following elements?
(a) C
(b) N
(c) F
(d) P

70. How many $3d$ electrons are in each of the following elements?
(a) Fe
(b) Zn
(c) K
(d) As

71. Give the number of elements in periods 1 and 2 of the periodic table. Explain why they are different.

72. Give the number of elements in periods 3 and 4 of the periodic table. Explain why they are different.

73. Name an element in the third period (row) of the periodic table with:
(a) 3 valence electrons
(b) a total of 4 $3p$ electrons
(c) 6 $3p$ electrons
(d) 2 $3s$ electrons and no $3p$ electrons

74. Name an element in the fourth period of the periodic table with:
(a) 5 valence electrons
(b) a total of 4 $4p$ electrons
(c) a total of 3 $3d$ electrons
(d) a complete outer shell

75. Use the periodic table to identify the element with the following electron configuration.
(a) $[Ne]3s^2 3p^5$
(b) $[Ar]4s^2 3d^{10} 4p^1$
(c) $[Ar]4s^2 3d^6$
(d) $[Kr]5s^1$

76. Use the periodic table to identify the element with the following electron configuration.
(a) $[Ne]3s^1$
(b) $[Kr]5s^2 4d^{10}$
(c) $[Xe]6s^2$
(d) $[Kr]5s^2 4d^{10} 5p^3$

Periodic Trends

77. Choose the element with the higher ionization energy from each of the following pairs.
(a) Na or Rb
(b) Ga or Ge
(c) P or I
(d) P or Sn

78. Choose the element with the higher ionization energy from each of the following pairs.
(a) As or At
(b) Br or Bi
(c) Si or Cl
(d) P or Sb

79. Arrange the following elements in order of increasing ionization energy: Te, Pb, Cl, S, Sn

80. Arrange the following elements in order of increasing ionization energy: Ga, In, F, Si, N

81. Choose the element with the larger atoms from each of the following pairs.
(a) Al or In
(b) Si or N
(c) P or Pb
(d) C or F

82. Choose the element with the larger atoms from each of the following pairs.
(a) Sn or Si
(b) Br or Ga
(c) Sn or Bi
(d) Se or Sn

83. Arrange the following elements in order of increasing atomic size: Ca, Rb, S, Si, Ge, F

84. Arrange the following elements in order of increasing atomic size: Cs, Sb, S, Pb, Se

85. Choose the more metallic element from each of the following pairs.
(a) Sr or Sb
(b) As or Bi
(c) Cl or O
(d) S or As

86. Choose the more metallic element from each of the following pairs.
(a) Sb or Pb
(b) K or Ge
(c) Ge or Sb
(d) As or Sn

87. Arrange the following elements in order of increasing metallic character: Fr, Sb, In, S, Ba, Se

88. Arrange the following elements in order of increasing metallic character: Sr, N, Si, P, Ga, Al

Cumulative Problems

89. What is the maximum number of electrons that can occupy the $n = 3$ quantum shell?

90. What is the maximum number of electrons that can occupy the $n = 4$ quantum shell?

91. Use the electron configurations of the alkali metals to explain why they tend to form 1+ ions.

92. Use the electron configurations of the halogens to explain why they tend to form 1− ions.

93. Write electron configurations for each of the following ions.
(a) Ca^{2+}
(b) K^{+}
(c) S^{2-}
(d) Br^{-}

94. Write electron configurations for each of the following ions.
(a) F^{-}
(b) P^{3-}
(c) Li^{+}
(d) Al^{3+}

95. Examine Figure 4.12, which shows the division of the periodic table into metals, nonmetals, and metalloids. Use what you know about electron configurations to explain these divisions.

96. Examine Figure 4.14, which shows the elements that form predictable ions. Use what you know about electron configurations to explain these trends.

97. Explain what is wrong with each of the following electron configurations and write the correct configuration based on the number of electrons.
(a) $1s^3 2s^3 2p^9$
(b) $1s^2 2s^2 2p^6 2d^4$
(c) $1s^2 1p^5$
(d) $1s^2 2s^2 2p^8 3s^2 3p^1$

98. Explain what is wrong with each of the following electron configurations and write the correct configuration based on the number of electrons.
(a) $1s^4 2s^4 2p^{12}$
(b) $1s^2 2s^2 2p^6 3s^2 3p^6 3d^{10}$
(c) $1s^2 2p^6 3s^2$
(d) $1s^2 2s^2 2p^6 3s^2 3p^6 4s^2 4d^{10} 4p^3$

99. Bromine is a highly reactive liquid while krypton is an inert gas. Explain the difference based on their electron configurations.

100. Potassium is a highly reactive metal while argon is an inert gas. Explain the difference based on their electron configurations.

101. Based on periodic trends, which one of the following elements do you expect to be most easily oxidized: Ge, K, S, or N?

102. Based on periodic trends, which one of the following elements do you expect to be most easily reduced: Ca, Sr, P, or Cl?

103. When an electron makes a transition from the $n = 3$ to the $n = 2$ hydrogen atom Bohr orbit, the energy difference between these two orbits (3.0×10^{-19} J) is given off in a photon of light. The relationship between the energy of a photon and its wavelength is given by $E = hc/\lambda$, where E is the energy of the photon in J, h is Planck's constant (6.626×10^{-34} J·s), and c is the speed of light (3.00×10^8 m/s). Find the wavelength of light emitted by hydrogen atoms when an electron makes this transition.

104. When an electron makes a transition from the $n = 4$ to the $n = 2$ hydrogen atom Bohr orbit, the energy difference between these two orbits (4.1×10^{-19} J) is given off in a photon of light. The relationship between the energy of a photon and its wavelength is given by $E = hc/\lambda$, where E is the energy of the photon in J, h is Planck's constant (6.626×10^{-34} J·s), and c is the speed of light (3.00×10^8 m/s). Find the wavelength of light emitted by hydrogen atoms when an electron makes this transition.

105. The distance from the sun to Earth is 1.496×10^8 km. How long does it take light to travel from the sun to Earth?

106. The nearest star is Alpha Centauri, at a distance of 4.3 light-years from Earth. A light-year is the distance that light travels in one year (365 days). How far away, in kilometers, is Alpha Centauri from Earth?

107. In the beginning of this chapter, we learned that the quantum-mechanical model for the atom is the foundation for modern chemical understanding. Explain why this is so.

108. Niels Bohr said, "Anyone who is not shocked by quantum mechanics has not understood it." What did he mean by this?

109. The wave nature of matter was first proposed by Louis de Broglie, who suggested that the wavelength (λ) of a particle was related to its mass (m) and its velocity (v) by the following equation: $\lambda = h/mv$, where h is Planck's constant (6.626×10^{-34} J·s). Calculate the de Broglie wavelength of each of the following: (a) a 0.0459-kg golf ball traveling at 95 m/s; (b) an electron traveling at 3.88×10^6 m/s. Can you explain why the wave nature of matter is significant for the electron, but not for the golf ball? (Hint: Express mass in kilograms.)

110. The particle nature of light was first proposed by Albert Einstein, who suggested that light could be described as a stream of particles called photons. A photon of wavelength λ has an energy (E) given by the following equation: $E = hc/\lambda$, where E is the energy of the photon in J, h is Planck's constant (6.626×10^{-34} J·s), and c is the speed of light (3.00×10^8 m/s). Calculate the energy of 1 mol of photons with a wavelength of 632 nm.

Highlight Problems

111. Excessive exposure to sunlight increases the risk of skin cancer because some of the photons have enough energy to break chemical bonds in biological molecules. These bonds require approximately 250–800 kJ/mol of energy to break. The energy of a single photon is given by $E = hc/\lambda$, where E is the energy of the photon in J, h is Planck's constant ($6.626 \times 10^{-34}\ \text{J} \cdot \text{s}$), and c is the speed of light ($3.00 \times 10^{8}\ \text{m/s}$). Determine which of the following kinds of light contain enough energy to break chemical bonds in biological molecules by calculating the total energy in 1 mol of photons for light of each wavelength.
   (a) infrared light (1500 nm)
   (b) visible light (500 nm)
   (c) ultraviolet light (150 nm)

112. The quantum-mechanical model, besides revolutionizing chemistry, shook the philosophical world because of its implications regarding determinism. Determinism is the idea that the outcomes of future events are determined by preceding events. The trajectory of a baseball, for example, is deterministic; that is, its trajectory—and therefore its landing place—is determined by its position, speed, and direction of travel. Before quantum mechanics, most scientists thought that fundamental particles—such as electrons and protons—also behaved deterministically. The implication of this was that the entire universe must behave deterministically—its future must be determined by preceding events. Quantum mechanics challenged this reasoning because fundamental particles did not behave deterministically—their future paths were not determined by preceding events. Some scientists struggled with this idea. Einstein himself refused to believe it, stating, "God does not play dice with the universe." Explain what Einstein meant by this statement.

▲ "God does not play dice with the universe."

Answers to Skillbuilder Exercises

Skillbuilder 9.1 **(a)** blue, green, red **(b)** red, green, blue **(c)** red, green, blue

Skillbuilder 9.2 **(a)** Al $1s^2 2s^2 2p^6 3s^2 3p^1$ or $[Ne]3s^2 3p^1$

(b) Br $1s^2 2s^2 2p^6 3s^2 3p^6 4s^2 3d^{10} 4p^5$ or $[Ar]4s^2 3d^{10} 4p^5$

(c) Sr $1s^2 2s^2 2p^6 3s^2 3p^6 4s^2 3d^{10} 4p^6 5s^2$ or $[Kr]5s^2$

Skillbuilder Plus, p. 281 Subtract 1 electron for each unit of positive charge. Add 1 electron for each unit of negative charge.

(a) Al^{3+} $1s^2 2s^2 2p^6$

(b) Cl^- $1s^2 2s^2 2p^6 3s^2 3p^6$

(c) O^{2-} $1s^2 2s^2 2p^6$

Skillbuilder 9.3

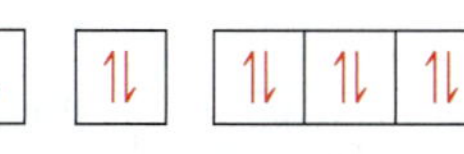

Skillbuilder 9.4

Cl $1s^2 2s^2 2p^6 \, 3s^2 3p^5$

10 core electrons — 7 valence electrons

Skillbuilder 9.5 $[Kr]5s^2 4d^{10} 5p^2$

Skillbuilder 9.6 **(a)** Pb **(b)** Rb **(c)** cannot tell based on periodic properties **(d)** Se

Skillbuilder 9.7 **(a)** Mg **(b)** Te **(c)** cannot tell based on periodic properties **(d)** F

Skillbuilder 9.8 **(a)** In **(b)** cannot tell based on periodic properties **(c)** Bi **(d)** B

Answers to Conceptual Checkpoints

9.1 (c) The higher energy levels are more closely spaced then the lower ones, so the difference in energy between $n = 2$ and $n = 1$ is greater than the difference in energy between $n = 3$ and $n = 2$. The photon emitted when an electron falls from $n = 2$ to $n = 1$ therefore carries more energy, corresponding to radiation with a shorter wavelength and higher frequency.

9.2 (d) Both have six electrons in $2p$ orbitals and six electrons in $3p$ orbitals.

9.3 (d) The outermost principal shell for Cr is $n = 4$, which contains only a single valence electron, $4s^1$. (The five d electrons are in the $n = 3$ shell, and so are core electrons.)

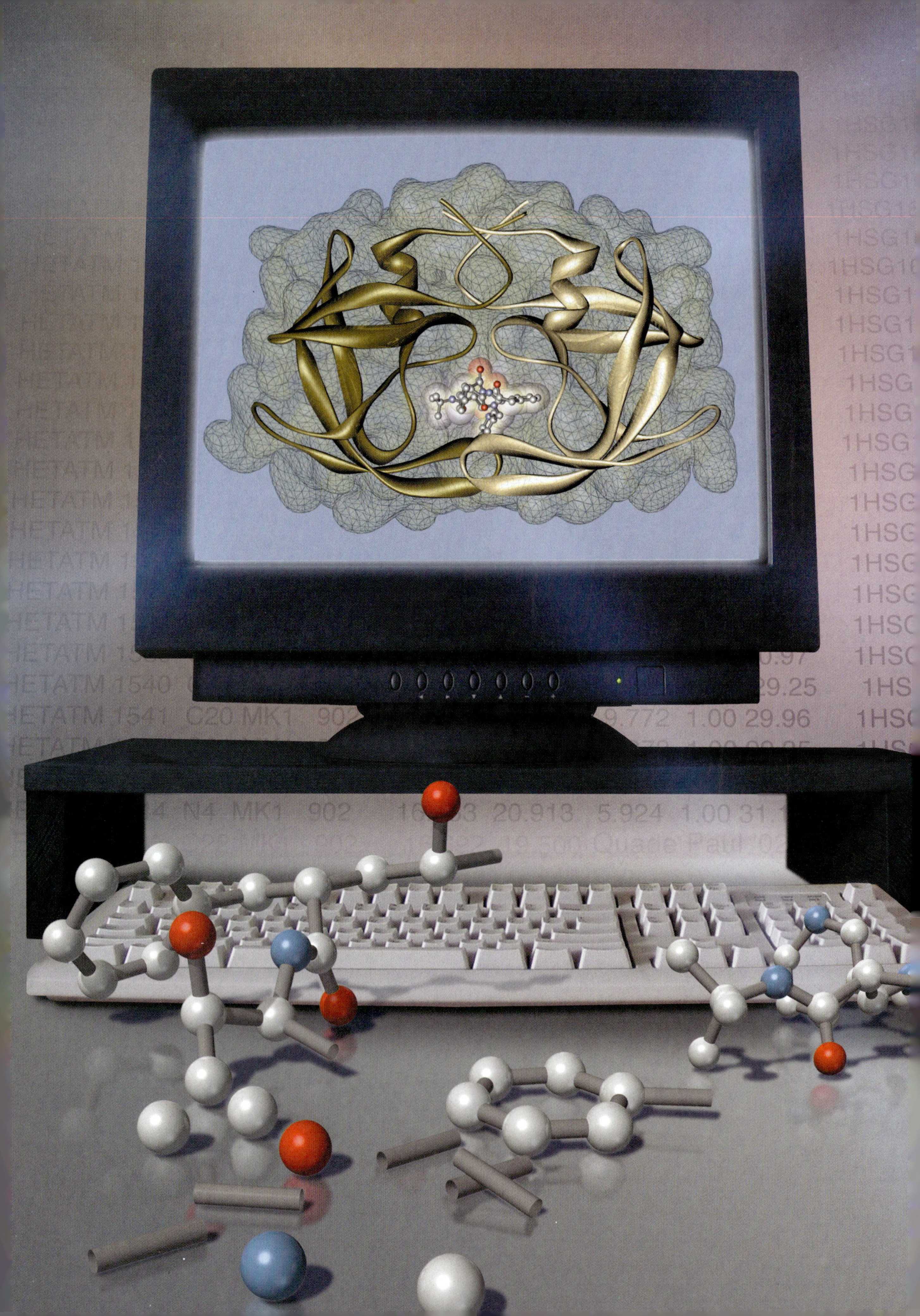
HETATM 1540
HETATM 1541 C20 MK1 902 9.772 1.00 29.96
N4 MK1 902 20.913 5.924 1.00
Quade Paul '02

CHAPTER 10

Chemical Bonding

"The fascination of a growing science lies in the work of the pioneers at the very borderland of the unknown, but to reach this frontier one must pass over well traveled roads . . ."

Gilbert N. Lewis (1875–1946)

10.1 Bonding Models and AIDS Drugs

Proteins are discussed in more detail in Chapter 19.

In 1989, researchers discovered the structure of a molecule called HIV-protease. HIV-protease is a protein (a class of biological molecules) synthesized by the human immunodeficiency virus (HIV), which causes AIDS. HIV-protease is crucial to the virus's ability to replicate itself. Without HIV-protease, HIV could not spread in the human body because the virus could not copy itself, and AIDS would not develop.

With knowledge of the HIV-protease structure, drug companies set out to design a molecule that would disable protease by sticking to the working part of the molecule (called the *active site*). To design such a molecule, researchers used **bonding theories**—models that predict how atoms bond together to form molecules—to simulate how potential drug molecules would interact with the protease molecule. By the early 1990s, these companies had developed several drug molecules that seemed to work. Since these molecules inhibit the action of HIV-protease, they are called *protease inhibitors.* In human trials, protease inhibitors in combination with other drugs have decreased the viral count in HIV-infected individuals to undetectable levels. Many AIDS patients are still alive today because of the development of these drugs.

Bonding theories are central to chemistry because they predict how atoms bond together to form compounds. They predict what combinations of atoms form compounds and what combinations do not. For example, bonding theories predict why salt is NaCl and not $NaCl_2$ and why water is H_2O and not H_3O. Bonding theories also explain the shapes of molecules, which in turn determine many of their physical and chemical properties. The bonding theory you will learn in this chapter is called **Lewis theory,** named after the American chemist who developed it, G. N. Lewis (1875–1946). It involves representing electrons as dots and drawing what are called *dot structures* or *Lewis structures* to represent molecules. These structures, which are fairly simple to draw, have tremendous predictive power. It takes just a few minutes to

◀ The gold-colored structure on the computer screen is a representation of HIV-protease. The molecule shown in the center is Indinavir, a protease inhibitor.

use Lewis theory to determine whether a particular set of atoms will form a stable molecule and what that molecule might look like. Although modern chemists also use more advanced bonding theories to better predict molecular properties, Lewis theory remains the simplest method for making quick, everyday predictions about molecules.

10.2 Representing Valence Electrons with Dots

In the previous chapter, we learned that valence electrons are the electrons in the outermost principal shell. Since valence electrons are most important in bonding, Lewis theory focuses on these. In Lewis theory, the valence electrons of an element are represented as dots surrounding the symbol of the element. The result is called a **Lewis structure**, or **dot structure**. For example, the electron configuration of O is:

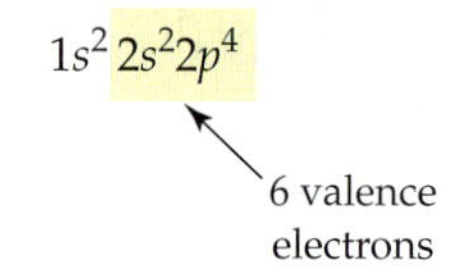

and the Lewis structure is:

·Ö: ← 6 dots representing valence electrons

Each dot represents a valence electron. The dots are placed around the element's symbol with a maximum of two dots per side. While the exact location of dots is not critical, in this book we fill in the dots singly first and then pair them.

The Lewis structures for all of the period 2 elements are:

Li· ·Be· ·Ḃ· ·Ċ· ·N̈: ·Ö: :F̈: :N̈e:

Remember, the number of valence electrons for any main-group element (except helium, which has 2 valence electrons but is in Group 8A) is equal to the group number of the element.

Lewis structures allow us to easily see the number of valence electrons in an atom. Atoms with 8 valence electrons—which are particularly stable—are easily identified because they have eight dots, an **octet**.

Helium is somewhat of an exception. Its electron configuration and Lewis structure are:

$1s^2$ He:

The Lewis structure of helium contains only two dots (a **duet**). For helium, a duet represents a stable electron configuration.

In Lewis theory, a **chemical bond** involves the sharing or transfer of electrons to attain stable electron configurations for the bonding atoms. If the electrons are transferred, the bond is an **ionic bond**. If the electrons are shared, the bond is a **covalent bond**. In either case, the bonding atoms get stable electron configurations. As we have seen, a stable configuration usually consists of eight electrons in the outermost or valence shell. This observation leads to the **octet rule**:

> In chemical bonding, atoms transfer or share electrons so that all obtain outer shells with eight electrons.

Hydrogen, lithium, and beryllium are exceptions: Each of these achieves stability when it has two electrons in its outermost shell.

EXAMPLE 10.1 Writing Lewis Structures for Elements

Write a Lewis structure for phosphorus.

Since phosphorus is in Group 5A in the periodic table, it has 5 valence electrons. Represent these as five dots surrounding the symbol for phosphorus.

Solution:

$$\cdot\dot{\underset{\cdot}{\mathrm{P}}}:$$

SKILLBUILDER 10.1 Writing Lewis Structures for Elements

Write a Lewis structure for Mg.

10.3 Lewis Structures for Ionic Compounds: Electrons Transferred

Recall from Chapter 5 that when metals bond with nonmetals, electrons are transferred from the metal to the nonmetal. The metal becomes a cation and the nonmetal becomes an anion. The attraction between the cation and the anion results in an ionic compound. In Lewis theory, we represent this by moving electron dots from the metal to the nonmetal. For example, potassium and chlorine have the following Lewis structures.

$$\mathrm{K}\cdot \quad :\ddot{\underset{\cdot}{\mathrm{Cl}}}:$$

When potassium and chlorine bond, potassium transfers its valence electron to chlorine.

$$\mathrm{K}\cdot \quad :\ddot{\underset{\cdot}{\mathrm{Cl}}}: \longrightarrow \mathrm{K}^+ \ [:\ddot{\underset{\cdot\cdot}{\mathrm{Cl}}}:]^-$$

Recall from Section 4.7 that atoms that lose electrons become positively charged and atoms that gain electrons become negatively charged.

The transfer of the electron gives chlorine an octet (shown as eight dots around chlorine), and leaves potassium with an octet in the previous principal shell, which is now the valence shell. The potassium, because it lost an electron, becomes positively charged, while the chlorine, which gained an electron, becomes negatively charged. The Lewis structure of an anion is usually written within brackets with the charge in the upper right corner (outside the brackets). The positive and negative charges attract one another, which then results in the compound KCl.

EXAMPLE 10.2 Writing Ionic Lewis Structures

Write a Lewis structure for the compound MgO.

Draw the Lewis structures of magnesium and oxygen by drawing two dots around the symbol for magnesium and six dots around the symbol for oxygen.

Solution:

$$\cdot\mathrm{Mg}\cdot \quad \cdot\ddot{\underset{\cdot}{\mathrm{O}}}:$$

In MgO, magnesium loses its 2 valence electrons, forming a 2+ charge, and oxygen gains two electrons, forming a 2− charge and acquiring an octet.

$$\mathrm{Mg}^{2+} \ [:\ddot{\underset{\cdot\cdot}{\mathrm{O}}}:]^{2-}$$

SKILLBUILDER 10.2 Writing Ionic Lewis Structures

Write a Lewis structure for the compound NaBr.

Recall from Section 5.4 that ionic compounds do not exist as distinct molecules, but rather as part of a large lattice of alternating cations and anions.

Lewis theory predicts the correct chemical formulas for ionic compounds. For the compound that forms between K and Cl, for example, Lewis theory predicts one potassium cation to every chlorine anion, KCl. As another example, consider the ionic compound formed between sodium and sulfur. The Lewis structures for sodium and sulfur are:

Na· ·S̤̈:

Notice that sodium must lose its 1 valence electron to get an octet (in the previous principal shell), while sulfur must gain 2 electrons to get an octet. Consequently, the compound that forms between sodium and sulfur requires two sodium atoms to every one sulfur atom. The Lewis structure is:

Na^{+} [:S̤̈:]$^{2-}$ Na^{+}

The two sodium atoms each lose their 1 valence electron, while the sulfur atom gains 2 electrons and gets an octet. The correct chemical formula is Na_2S.

EXAMPLE 10.3 Using Lewis Theory to Predict the Chemical Formula of an Ionic Compound

Use Lewis theory to predict the formula for the compound that forms between calcium and chlorine.

Draw the Lewis structures of calcium and chlorine by drawing two dots around the symbol for calcium and seven dots around the symbol for chlorine.

Solution:

·Ca· :C̤̈l:

Calcium must lose its 2 valence electrons (to effectively get an octet in its previous principal shell), while chlorine needs to gain only 1 electron to get an octet. Consequently, the compound that forms between Ca and Cl must have two chlorine atoms to every one calcium atom.

[:C̤̈l:]$^{-}$ Ca^{2+} [:C̤̈l:]$^{-}$

The formula is therefore $CaCl_2$.

SKILLBUILDER 10.3 Using Lewis Theory to Predict the Chemical Formula of an Ionic Compound

Use Lewis theory to predict the formula for the compound that forms between magnesium and nitrogen.

CONCEPTUAL CHECKPOINT 10.1

Which of the following nonmetals forms an ionic compound with aluminum with the formula Al_2X_3 (where X represents the nonmetal)?

(a) Cl
(b) S
(c) N
(d) C

10.4 Covalent Lewis Structures: Electrons Shared

Recall from Chapter 5 that when nonmetals bond with other nonmetals, a molecular compound results. Molecular compounds contain covalent bonds, in which electrons are shared between atoms rather than transferred. In Lewis theory, we represent covalent bonding by allowing neighboring atoms to share some of their valence electrons in order to attain octets (or duets for hydrogen). For example, hydrogen and oxygen have the following Lewis structures.

H· ·Ö̤:

In water, hydrogen and oxygen share their electrons so that each hydrogen atom gets a duet and the oxygen atom gets an octet.

H:Ö:H

The shared electrons—those that appear in the space between the two atoms—count toward the octets (or duets) of *both of the atoms.*

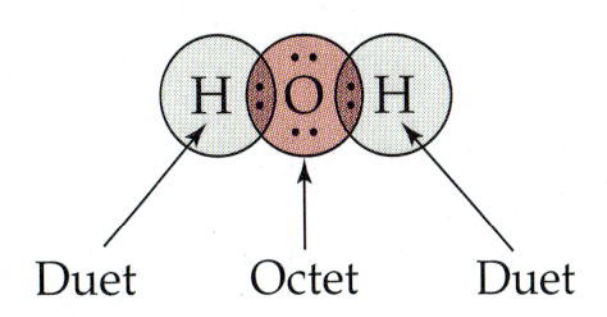

Sometimes lone pair electrons are also called nonbonding electrons.

Electrons that are shared between two atoms are called **bonding pair** electrons, while those that are only on one atom are called **lone pair** electrons.

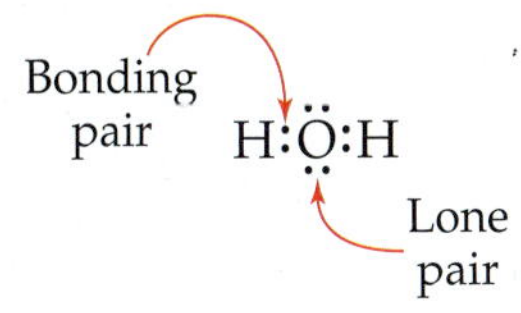

Remember that each dash represents a *pair* of shared electrons.

Bonding pair electrons are often represented by dashes to emphasize that they are a chemical bond.

H—Ö—H

Lewis theory also explains why the halogens form diatomic molecules. Consider the Lewis structure of chlorine.

:C̤̈l:

If two Cl atoms pair together, they can each get an octet.

:C̤̈l:C̤̈l: *or* :C̤̈l—C̤̈l:

When we examine elemental chlorine, it indeed exists as a diatomic molecule, just as Lewis theory predicts. The same is true for the other halogens.

Similarly, Lewis theory predicts that hydrogen, which has the following Lewis structure:

H·

should exist as H_2. When two hydrogen atoms share their valence electrons, they each get a duet, a stable configuration for hydrogen.

H:H *or* H—H

TUTORIAL
H_2 Bond Formation

Again, Lewis theory is correct. In nature, elemental hydrogen exists as H_2 molecules.

Double and Triple Bonds

In Lewis theory, two atoms may share more than one electron pair to get octets. For example, we know from Chapter 5 that oxygen exists as the diatomic molecule, O_2. The Lewis structure of an oxygen atom is:

·Ö:

If we pair two oxygen atoms together and then try to write a Lewis structure, we do not have enough electrons to give each O atom an octet.

:Ö:Ö:

However, we can convert a lone pair into an additional bonding pair by moving it into the bonding region.

$$:\ddot{\text{O}}:\ddot{\text{O}}: \longrightarrow :\ddot{\text{O}}::\ddot{\text{O}}: \quad or \quad :\ddot{\text{O}}=\ddot{\text{O}}:$$

Each oxygen atom now has an octet because the additional bonding pair counts toward the octet of both oxygen atoms.

Octet → $:\ddot{\text{O}}::\ddot{\text{O}}:$ ← Octet

When two electron pairs are shared between two atoms, the resulting bond is a **double bond**. In general, double bonds are shorter and stronger than single bonds. For example, the distance between oxygen nuclei in an oxygen–oxygen double bond is 121 pm. In a single bond, it is 148 pm.

1 pm = 10^{-12} m

Atoms can also share three electron pairs. Consider the Lewis structure of N_2. Since each N atom has 5 valence electrons, the Lewis structure for N_2 has 10 electrons. A first attempt at writing the Lewis structure gives:

$$:\ddot{\text{N}}:\ddot{\text{N}}:$$

As with O_2, we do not have enough electrons to satisfy the octet rule for both N atoms. However, if we convert two additional lone pairs into bonding pairs, each nitrogen atom can get an octet.

$$:\ddot{\text{N}}:\ddot{\text{N}}: \longrightarrow :\text{N}:::\text{N}: \quad or \quad :\text{N}\equiv\text{N}:$$

The resulting bond is called a **triple bond**. Triple bonds are even shorter and stronger than double bonds. For example, the distance between nitrogen nuclei in a nitrogen–nitrogen triple bond is 110 pm. In a double bond, the distance is 124 pm. When we examine nitrogen in nature, we find that it indeed exists as a diatomic molecule with a very strong short bond between the two nitrogen atoms. The bond is so strong that it is difficult to break, making N_2 a relatively unreactive molecule.

10.5 Writing Lewis Structures for Covalent Compounds

To write a Lewis structure for a covalent compound, follow these steps:

1. **Write the correct skeletal structure for the molecule.** The Lewis structure of a molecule must have the atoms in the correct positions. For example, you could *not* write a Lewis structure for water if you started with the hydrogen atoms next to each other and the oxygen atom at the end (H H O). In nature, oxygen is the central atom and the hydrogens are **terminal atoms** (at the ends). The correct skeletal structure is H O H.

   The only way to absolutely know the correct skeletal structure for any molecule is by examining its structure in nature. However, we can write likely skeletal structures by remembering two guidelines. First, *hydrogen atoms will always be terminal.* Since hydrogen requires only a duet, it will never be a central atom because central atoms must form at least two bonds and hydrogen can form only one. Second, *many molecules tend to be symmetrical*, so when a molecule contains several atoms of the same type, these tend to be in terminal positions. This second guideline, however, has many exceptions. In cases where the skeletal structure is unclear, this text will provide you with the correct skeletal structure.

When guessing at skeletal structures, put the less metallic elements in terminal positions and the more metallic elements in central positions. Halogens, being among the least metallic elements, are nearly always terminal.

2. **Calculate the total number of electrons for the Lewis structure by summing the valence electrons of each atom in the molecule.** Remember that the number of valence electrons for any main-group element is equal to its group number in the periodic table. **If you are writing a Lewis structure for a polyatomic ion, the charge of the ion must be considered when calculating the total number of electrons.** Add 1 electron for each negative charge and subtract 1 electron for each positive charge.

3. **Distribute the electrons among the atoms, giving octets (or duets for hydrogen) to as many atoms as possible.** Begin by placing 2 electrons between each pair of atoms. These are the minimal number of bonding electrons. Then distribute the remaining electrons, first to terminal atoms, and then to the central atom, giving octets to as many atoms as possible.

4. **If any atoms lack an octet, form double or triple bonds as necessary to give them octets.** Do this by moving lone electron pairs from terminal atoms into the bonding region with the central atom.

TUTORIAL
Learning Lewis Dot Structures

A brief version of this procedure is shown in the left column below. Two examples of applying it are shown in the center and right columns.

Writing Lewis Structures for Covalent Compounds	EXAMPLE 10.4 Write a Lewis structure for CO_2.	EXAMPLE 10.5 Write a Lewis structure for CCl_4.
1. Write the correct skeletal structure for the molecule.	**Solution:** Following the symmetry guideline, we write: O C O	**Solution:** Following the symmetry guideline, we write: Cl Cl C Cl Cl
2. Calculate the total number of electrons for the Lewis structure by summing the valence electrons of each atom in the molecule.	Total number of electrons for Lewis structure = $\begin{pmatrix}\text{\# valence}\\ e^- \text{ for C}\end{pmatrix} + 2\begin{pmatrix}\text{\# valence}\\ e^- \text{ for O}\end{pmatrix}$ $= 4 + 2(6)$ $= 16$	Total number of electrons for Lewis structure = $\begin{pmatrix}\text{\# valence}\\ e^- \text{ for C}\end{pmatrix} + 4\begin{pmatrix}\text{\# valence}\\ e^- \text{ for Cl}\end{pmatrix}$ $= 4 + 4(7)$ $= 32$
3. Distribute the electrons among the atoms, giving octets (or duets for hydrogen) to as many atoms as possible. Begin with the bonding electrons, and then proceed to lone pairs on terminal atoms, and finally to lone pairs on the central atom.	Bonding electrons first. O:C:O (4 of 16 electrons used) Lone pairs on terminal atoms next. :Ö:C:Ö: (16 of 16 electrons used)	Bonding electrons first. Cl Cl:C:Cl Cl (8 of 32 electrons used) Lone pairs on terminal atoms next. :Cl: :Cl:C:Cl: :Cl: (32 of 32 electrons used)
4. If any atoms lack an octet, form double or triple bonds as necessary to give them octets.	Move lone pairs from the oxygen atoms to bonding regions to form double bonds. :Ö:C:Ö: ⟶ :O::C::O:	Since all of the atoms have octets, the Lewis structure is complete.
	SKILLBUILDER 10.4 Write a Lewis structure for CO.	**SKILLBUILDER 10.5** Write a Lewis structure for H_2CO.

Writing Lewis Structures for Polyatomic Ions

We write Lewis structures for polyatomic ions by following the same procedure, but we pay special attention to the charge of the ion when calculating the number of electrons for the Lewis structure. Add 1 electron for each negative charge and subtract 1 electron for each positive charge. We normally show the Lewis structure for a polyatomic ion within brackets and write the charge of the ion in the upper right corner. For example, suppose we want to write a Lewis structure for the CN^- ion. We begin by writing the skeletal structure.

CN

We next calculate the total number of electrons for the Lewis structure by summing the number of valence electrons for each atom and adding one for the negative charge.

Total number of electrons for Lewis structure = (# valence e^- in C) + (# valence e^- in N) + 1

= 4 + 5 + 1

= 10

Add one e^- to account for −1 charge of ion.

We then place two electrons between each pair of atoms

C:N (2 of 10 electrons used)

and then distribute the remaining electrons.

:C̈:N̈: (10 of 10 electrons used)

Since neither of the atoms has octets, we move two lone pairs into the bonding region to form a triple bond, giving both atoms octets. We also enclose the Lewis structure in brackets and write the charge of the ion in the upper right corner.

$[:C:::N:]^-$ *or* $[:C{\equiv}N:]^-$

EXAMPLE 10.6 Writing Lewis Structures for Polyatomic Ions

Write the Lewis structure for the NH_4^+ ion.

Solution:

Begin by writing the skeletal structure. Since hydrogen atoms must be terminal, and following the guideline of symmetry, put the nitrogen atom in the middle surrounded by four hydrogen atoms.

```
   H
H  N  H
   H
```

Calculate the total number of electrons for the Lewis structure by summing the number of valence electrons for each atom and subtracting 1 for the positive charge.

Total number of electrons for Lewis structure = 5 + 4 − 1 = 8

(5 = # valence e^- in N; 4 = 4 × (# valence e^- in H); Subtract 1 e^- to account for +1 charge of ion.)

Next, place 2 electrons between each pair of atoms.

H:N:H with H above and below (8 of 8 electrons used)

Since the nitrogen atom has an octet and since all of the hydrogen atoms have duets, the placement of electrons is complete. Write the entire Lewis structure in brackets and write the charge of the ion in the upper right corner.

$[H:\ddot{N}:H]^+$ (with H above and below N) *or* $[H—N—H]^+$ (with H bonded above and below N)

SKILLBUILDER 10.6 **Writing Lewis Structures for Polyatomic Ions**

Write a Lewis structure for the ClO^- ion.

Exceptions to the Octet Rule

Lewis theory is often correct in its predictions; however, there are some exceptions. For example, if we try to write a Lewis structure for NO, which has 11 electrons, the best we can do is:

:N::Ö: *or* :N=Ö:

The nitrogen atom does not have an octet, so this is not a great Lewis structure. However, NO exists in nature. Why? As with any simple theory, Lewis theory is not sophisticated enough to be correct every time. It is impossible to write good Lewis structures for molecules with odd numbers of electrons, yet some of these molecules exist in nature. In such cases, we simply write the best Lewis structure that we can. Another significant exception to the octet rule is boron, which tends to form compounds with only 6 electrons around B, rather than 8. For example, BF_3 and BH_3—both of which exist in nature—lack an octet for B.

:F:B:F: (with :F: above B)   H:B:H (with H above B)

A third type of exception to the octet rule is also common. A number of molecules, such as SF_6 and PCl_5, have more than 8 electrons around a central atom in their Lewis structure.

[Lewis structures of SF_6 (S bonded to six F atoms) and PCl_5 (P bonded to five Cl atoms)]

These are often referred to as expanded octets. Beyond mentioning them, we do not cover expanded octets in this book. In spite of these exceptions, which can largely be accommodated within the theory, Lewis theory remains a powerful and simple way to understand chemical bonding.

CONCEPTUAL CHECKPOINT 10.2

Which two species have the same number of lone electron pairs in their Lewis structures?

(a) H_2O and H_3O^+
(b) NH_3 and H_3O^+
(c) NH_3 and CH_4
(d) NH_3 and NH_4^+

10.6 Resonance: Equivalent Lewis Structures for the Same Molecule

When writing Lewis structures, we may find that, for some molecules, we can write more than one good Lewis structure. For example, consider writing a Lewis structure for SO_2. We begin with the skeletal structure:

O S O

We then sum the valence electrons.

Total number of electrons for Lewis structure
$= (\text{\# valence e}^- \text{ in S}) + 2(\text{\# valence e}^- \text{ in O})$
$= 6 + 2(6)$
$= 18$

We next place two electrons between each pair of atoms

O:S:O (4 of 18 electrons used)

and then distribute the remaining electrons, first to terminal atoms

:Ö:S:Ö: (16 of 18 electrons used)

and finally to the central atom.

:Ö:S̈:Ö: (18 of 18 electrons used)

Since the central atom lacks an octet, we move one lone pair from an oxygen atom into the bonding region to form a double bond, giving all of the atoms octets.

:Ö::S̈:Ö: *or* :Ö=S̈—Ö:

However, we could have formed the double bond with the other oxygen atom.

:Ö—S̈=Ö:

These two Lewis structures are equally correct. In cases such as this—where we can write two or more equivalent (or nearly equivalent) Lewis structures for the same molecule—we find that the molecule exists in nature as an average or intermediate between the two Lewis structures. Any *one* of the two Lewis structures for SO_2 would predict that SO_2 should contain two different kinds of bonds (one double bond and one single bond). However, when we examine SO_2 in nature, we find that both of the bonds are equivalent and intermediate in strength and length between a double bond and single bond. We account for this in Lewis theory by representing the molecule with both structures, called **resonance structures**, with a double-headed arrow between them.

$$:\ddot{O}=\ddot{S}-\ddot{\underset{..}{O}}: \longleftrightarrow :\ddot{\underset{..}{O}}-\ddot{S}=\ddot{O}:$$

The true structure of SO_2 is intermediate between these two resonance structures.

EXAMPLE 10.7 Writing Resonance Structures

Write a Lewis structure for the NO_3^- ion. Include resonance structures.

Solution:

Begin by writing the skeletal structure. Using the guideline of symmetry, make the three oxygen atoms terminal.

O
O N O

Sum the valence electrons (adding 1 electron to account for the −1 charge) to determine the total number of electrons in the Lewis structure.

Total number of electrons for Lewis structure = 5 + 3(6) + 1 = 24

(5: # valence e⁻ in N; 3(6): 3 × (# valence e⁻ in O); 1: Add one e⁻ to account for negative charge of ion.)

Place 2 electrons between each pair of atoms.

O:N:O with O above N (6 of 24 electrons used)

Distribute the remaining electrons, first to terminal atoms.

(24 of 24 electrons used)

Since there are no electrons remaining to complete the octet of the central atom, form a double bond by moving a lone pair from one of the oxygen atoms into the bonding region with nitrogen. Enclose the structure in brackets and write the charge at the upper right.

$[:\ddot{O}-N=\ddot{O}:]^-$ (with $:\ddot{O}:$ bonded above N) *or* dot structure

Notice that you could have formed the double bond with either of the other two oxygen atoms.

$[:\ddot{O}=N-\ddot{O}:]^-$ *or* $[:\ddot{O}-N-\ddot{O}:]^-$ (with top O double-bonded)

Since the three Lewis structures are equally correct, write the three structures as resonance structures.

$$[:\ddot{O}=N-\ddot{O}:]^- \longleftrightarrow [:\ddot{O}-N(=O)-\ddot{O}:]^- \longleftrightarrow [:\ddot{O}-N=\ddot{O}:]^-$$

SKILLBUILDER 10.7 Writing Resonance Structures

Write a Lewis structure for the NO_2^- ion. Include resonance structures.

Chemistry IN THE ENVIRONMENT

The Lewis Structure of Ozone

Ozone is a form of oxygen in which three oxygen atoms bond together. Its Lewis structure consists of the following two resonance structures:

:Ö=Ö−Ö̤: ⟷ :Ö̤−Ö=Ö:

Compare the Lewis structure of ozone to the Lewis structure of O_2:

:Ö=Ö:

Which molecule, O_3 or O_2, do you think has the stronger oxygen–oxygen bond? If you deduced O_2, you are correct. O_2 has a stronger bond because it is a pure double bond. Ozone, on the other hand, has bonds that are intermediate between single and double, so O_3 has weaker bonds. The effects of this are significant. As we learned in the *Chemistry in the Environment* box in Chapter 6, O_3 shields us from harmful ultraviolet light entering Earth's atmosphere. O_3 is ideally suited to do this because photons at wavelengths of 280–320 nm (the most dangerous components of sunlight to humans) are just strong enough to break ozone's bonds. In the process, the photons are absorbed.

:Ö̤−Ö=Ö: + UV light ⟶ :Ö=Ö: + ·Ö̤·

The same wavelengths of UV light, however, do not have sufficient energy to break the stronger double bond of O_2, which is therefore transparent to UV. As the ozone layer continues to thin, no other molecules in our atmosphere can do the job that ozone does. Consequently, it is important that we continue, and even strengthen, the ban on ozone-depleting compounds.

CAN YOU ANSWER THIS? *Why are the following Lewis structures for ozone incorrect?*

:Ö̤−Ö̤−Ö̤: :Ö̤=O=Ö̤:

10.7 Predicting the Shapes of Molecules

Lewis theory, in combination with **valence shell electron pair repulsion (VSEPR) theory**, can be used to predict the shapes of molecules. VSEPR is based on the simple idea that **electron groups**—lone pairs, single bonds, or multiple bonds—repel each other. This repulsion between the negative charge of electron groups on the central atom determines the geometry of the molecule. For example, consider CO_2, which has the following Lewis structure:

:Ö=C=Ö:

The geometry of CO_2 is determined by the repulsion between the two electron groups (or double bonds) on the central carbon atom. These two electron groups get as far away from each other as possible, resulting in a bond angle of 180° and a **linear** geometry for CO_2.

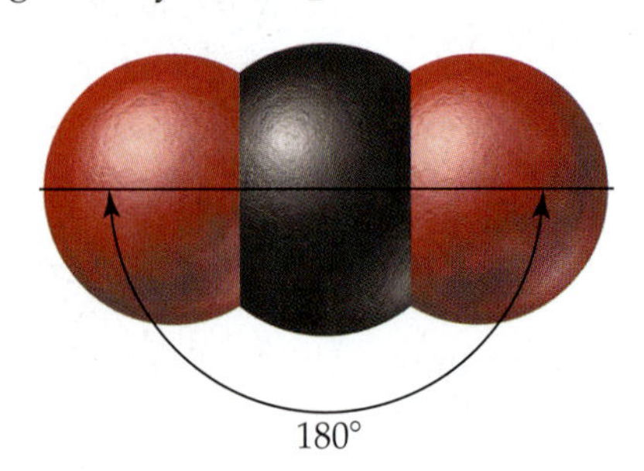

As another example, consider the molecule H_2CO. Its Lewis structure is:

```
   :O:
    ‖
H — C — H
```

This molecule has three electron groups around the central atom. These three electron groups get as far away from each other as possible, resulting in a bond angle of 120° and a **trigonal planar** geometry.

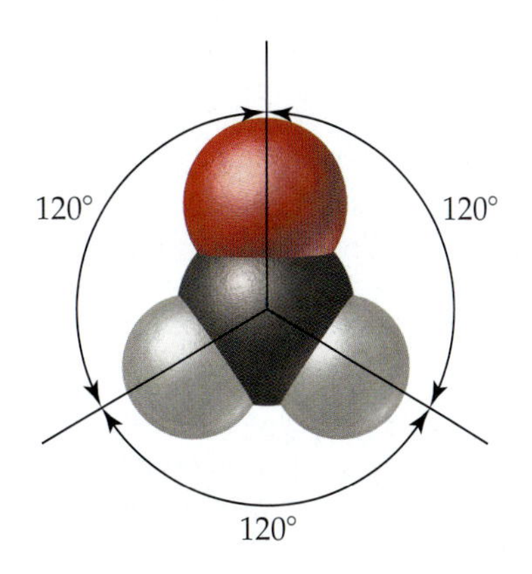

A tetrahedron is a geometric shape with four triangular faces.

If a molecule has four electron groups around the central atom, as in CH_4, it has a **tetrahedral** geometry with bond angles of 109.5°.

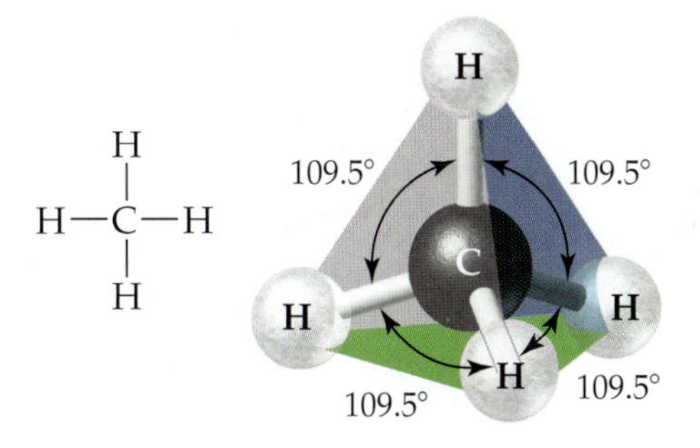

CH_4 is shown here with both a ball-and-stick model (above) and a space-filling model (below). Although space-filling models more closely portray molecules, ball-and-stick models are often used to clearly illustrate molecular geometries.

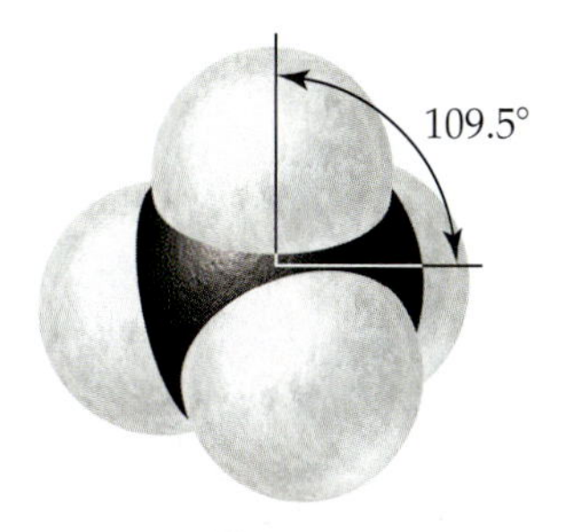

MOLECULE Methane

The electron groups repelling each other cause the tetrahedral shape—the tetrahedron allows the maximum separation among the four groups. When we write the structure of CH_4 on paper, it may seem that the molecule should be square planar, with bond angles of 90°. However, in three dimensions the electron groups can get farther away from each other by forming the tetrahedral geometry.

Each of the preceding examples has only bonding groups of electrons around the central atom. What happens in molecules with lone pairs around the central atom? These lone pairs also repel other electron groups. For example, consider NH_3:

H
|
H—N̤—H

The four electron groups (one lone pair and three bonding pairs) get as far away from each other possible. If we look only at the electrons, we find that the **electron geometry**—the geometrical arrangement of the electron groups—is tetrahedral.

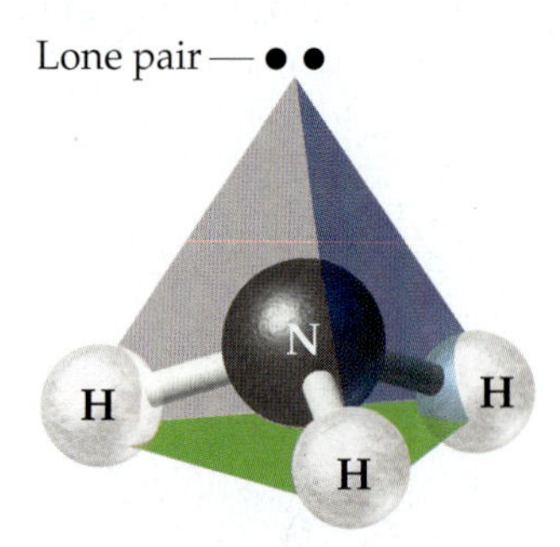

However, the **molecular geometry**—the geometrical arrangement of the atoms—is **trigonal pyramidal**.

Trigonal pyramidal structure

Notice that, although the electron geometry and the molecular geometry are different, the electron geometry is relevant to the molecular geometry. In other words, the lone pair exerts its influence on the bonding pairs.

Consider one last example, H_2O. Its Lewis structure is:

$$\mathrm{H{-}\ddot{\underset{\cdot\cdot}{O}}{-}H}$$

Since it has four electron groups, its electron geometry is also tetrahedral.

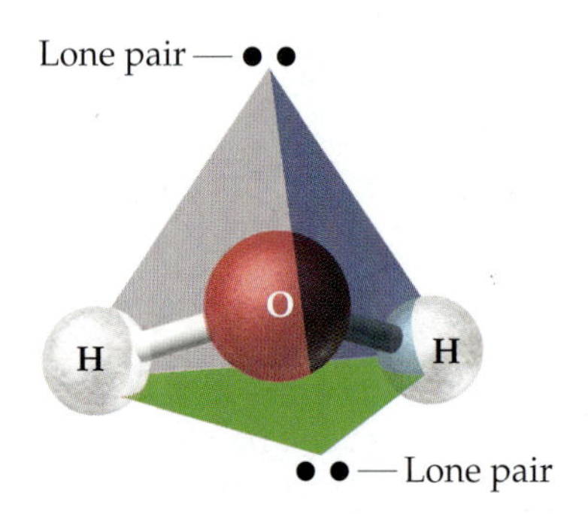

However, its molecular geometry is **bent.**

For reasons we don't cover here, the bond angles in NH_3 and H_2O are actually a few degrees smaller than the ideal tetrahedral angles.

Bent structure

Table 10.1 summarizes the electron and molecular geometry of a molecule based on the total number of electron groups, the number of bonding groups, and the number of lone pairs.

To determine the geometry of any molecule, use the following procedure. As usual, we give the steps in the left column and two examples of applying the steps in the center and right columns.

TUTORIAL
VSEPR Movie

TABLE 10.1 Electron and Molecular Geometries

Electron Groups*	Bonding Groups	Lone Pairs	Electron Geometry	Angle between Electron Groups**	Molecular Geometry	Example
2	2	0	linear	180°	linear	:Ö=C=Ö:
3	3	0	trigonal planar	120°	trigonal planar	H—C(=Ö:)—H
3	2	1	trigonal planar	120°	bent	:Ö=S̈—Ö:
4	4	0	tetrahedral	109.5°	tetrahedral	CH_4 (H—C—H with H above and below)
4	3	1	tetrahedral	109.5°	trigonal pyramidal	H—N̈—H (H below)
4	2	2	tetrahedral	109.5°	bent	H—Ö—H

*Count only electron groups around the *central* atom. Each of the following is considered one electron group: a lone pair, a single bond, a double bond, and a triple bond.
**Angles listed here are idealized. Actual angles in specific molecules may vary by several degrees.

LIVE EXAMPLE

Predicting Geometry Using VSEPR

	EXAMPLE 10.8	EXAMPLE 10.9
	Predict the electron and molecular geometry of PCl_3.	Predict the electron and molecular geometry of the $[NO_3]^-$ ion.
1. *Draw a Lewis structure for the molecule.*	**Solution:** PCl_3 has 26 electrons. :C̈l:P̈:C̈l: (with :C̈l: above P)	**Solution:** $[NO_3]^-$ has 24 electrons. [:Ö:N::Ö: (with :Ö: above N)]$^-$
2. *Determine the total number of electron groups around the central atom.* Lone pairs, single bonds, double bonds, and triple bonds each count as one group.	The central atom (P) has four electron groups.	The central atom (N) has three electron groups (the double bond counts as one group).
3. *Determine the number of bonding groups and the number of lone pairs around the central atom.* These should sum to the result from Step 2. Bonding groups include single bonds, double bonds, and triple bonds.	:C̈l:P̈:C̈l: (with :C̈l: above P) — Lone pair. Three of the four electron groups around P are bonding groups and one is a lone pair.	[:Ö:N::Ö: (with :Ö: above N)]$^-$ — No lone pairs. All three of the electron groups around N are bonding groups.

4. *Use Table 10.1 to determine the electron geometry and molecular geometry.*

The electron geometry is tetrahedral (four electron groups) and the molecular geometry—the shape of the molecule—is trigonal pyramidal (four electron groups, three bonding groups, and one lone pair).

SKILLBUILDER 10.8

Predict the molecular geometry of ClNO (N is the central atom).

The electron geometry is trigonal planar (three electron groups) and the molecular geometry—the shape of the molecule—is trigonal planar (three electron groups, three bonding groups, and no lone pairs).

SKILLBUILDER 10.9

Predict the molecular geometry of the SO_3^{2-} ion.

CONCEPTUAL CHECKPOINT 10.3

Which of the following conditions necessarily leads to a molecular geometry that is identical to the electron geometry?

(a) The presence of a double bond between the central atom and a terminal atom.
(b) The presence of two or more identical terminal atoms bonded to the central atom.
(c) The presence of one or more lone pairs on the central atom.
(d) The absence of any lone pairs on the central atom.

Representing Molecular Geometries on Paper

Since molecular geometries are three-dimensional, they are often difficult to represent on two-dimensional paper. Many chemists use the following notation for bonds to indicate three-dimensional structures on two-dimensional paper.

—	||||··	◄
Straight line	*Hashed lines*	*Wedge*
Bond in plane of paper	Bond going into the paper	Bond coming out of the paper

The major molecular geometries used in this book are shown here using this notation:

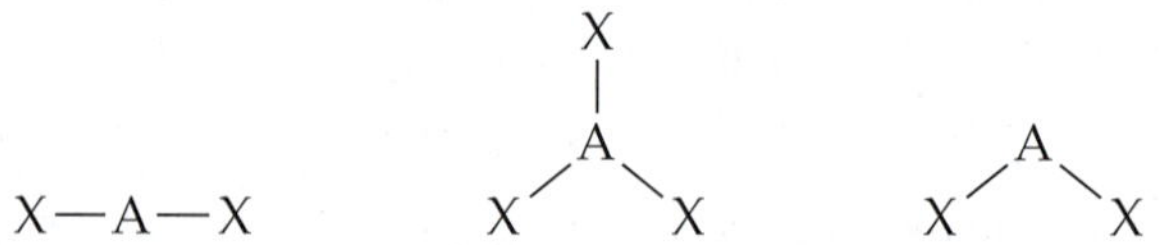

X—A—X
Linear

Trigonal planar

Bent

Tetrahedral

Trigonal pyramidal

Chemistry AND HEALTH

Fooled by Molecular Shape

Artificial sweeteners, such as aspartame (Nutrasweet™), taste sweet but have few or no calories. Why? Because taste and caloric value are entirely separate properties of foods. The caloric value of a food depends on the amount of energy released when the food is metabolized. For example, sucrose (table sugar) is metabolized by oxidation to carbon dioxide and water:

$$C_{12}H_{22}O_{11} + 6\,O_2 \longrightarrow 12\,CO_2 + 11\,H_2O + 5644\text{ kJ}$$

When your body metabolizes one mole of sucrose, it obtains 5644 kJ of energy. Some artificial sweeteners, such as saccharin, are not metabolized at all—they just pass through the body unchanged—and therefore have no caloric value. Other artificial sweeteners, such as aspartame, are metabolized, but have a much lower caloric content (for a given amount of sweetness) than sucrose.

The *taste* of a food, however, is independent of its metabolism. The sensation of taste originates in the tongue, where specialized cells called taste cells act as highly sensitive and specific molecular detectors. These cells can detect sugar molecules out of the thousands of different types of molecules present in a mouthful of food. The main basis for this discrimination is the molecule's *shape*.

The surface of a taste cell contains specialized protein molecules called taste receptors. A particular *tastant*—a molecule that we can taste—fits snugly into a special pocket on the taste receptor protein called the *active site*, just as a key fits into a lock (see Section 15.12). For example, a sugar molecule just fits into the active site of the sugar receptor protein called Tlr3. When the sugar molecule (the key) enters the active site (the lock), the different subunits of the Tlr3 protein split apart. This split causes a series of events that result in transmission of a nerve signal, which reaches the brain and registers a sweet taste.

Artificial sweeteners taste sweet because they fit into the receptor pocket that normally binds sucrose. In fact, both aspartame and saccharin actually bind to the active site in the Tlr3 protein more strongly than sugar does! For this reason, artificial sweeteners are "sweeter than sugar." It takes 200 times as much sucrose as aspartame to trigger the same amount of nerve signal transmission from taste cells.

This type of lock-and-key fit between the active site of a protein and a particular molecule is important not only to taste but to many other biological functions as well. For example, immune response, the sense of smell, and many types of drug action all depend on shape-specific interactions between molecules and proteins. In fact, the ability of scientists to determine the shapes of key biological molecules is largely responsible for the revolution in biology that has occurred over the last 50 years.

CAN YOU ANSWER THIS? *Proteins are long-chain molecules in which each link is an amino acid. The simplest amino acid is glycine, which has the following structure:*

```
   H  :O:
   |   ||
H—N̈—C—C—Ö—H
   |   |
   H   H
```

Determine the geometry about each interior atom in the glycine structure and make a three-dimensional sketch of the molecule.

10.8 Electronegativity and Polarity: Why Oil and Water Don't Mix

If you combine oil and water in a container, they separate into distinct regions (▶ Figure 10.1). Why? Something about water molecules causes them to bunch together into one region, expelling the oil molecules into a separate region. What is that something? We can begin to understand the answer by examining the Lewis structure of water.

H—Ö—H

▲ **Figure 10.1 Oil and water don't mix Question:** Why not?

The two bonds between O and H each consist of an electron pair—2 electrons shared between the oxygen atom and the hydrogen atom. The oxygen and hydrogen atoms each donate 1 electron to this electron pair; however, like unruly children, they don't share them equally. The oxygen atom takes more than its fair share of the electron pair.

Electronegativity

The ability of an element to attract electrons within a covalent bond is called **electronegativity**. Oxygen is more electronegative than hydrogen, which means that, on average, the shared electrons are more likely to be found near the oxygen atom than near the hydrogen atom. Consider one of the two OH bonds:

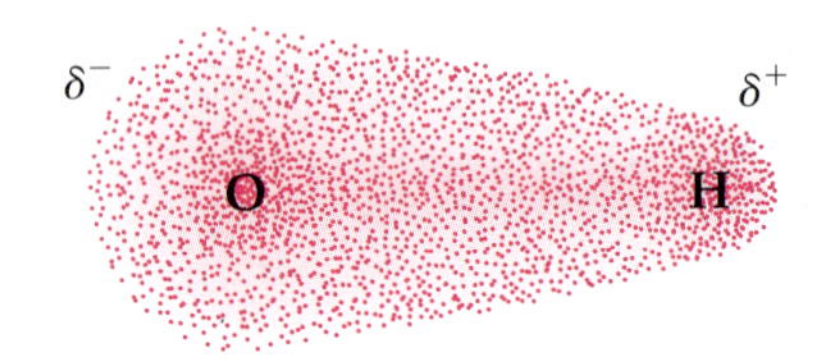

Dipole moment

The representation for depicting electron density in this figure is introduced in Section 9.6.

Since the electron pair is unequally shared (with oxygen getting the larger share), the oxygen atom has a partial negative charge, symbolized by $\delta-$ (delta minus). The hydrogen atom (getting the smaller share) has a partial positive charge, symbolized by $\delta+$ (delta plus). The result of this is uneven electron sharing is a **dipole moment**, a separation of charge within the bond. Covalent bonds that have a dipole moment are called **polar covalent bonds.** The magnitude of the dipole moment, and therefore the degree of polarity of the bond, depends on the electronegativity difference between the two elements in the bond and the length of the bond. For a fixed bond length, the greater the electronegativity difference, the greater the dipole moment and the more polar the bond.

The value of electronegativity is assigned using a relative scale on which fluorine, the most electronegative element, has an electronegativity of 4.0. All other electronegativities are defined relative to fluorine.

▼ Figure 10.2 shows the relative electronegativities of the elements. Notice that electronegativity increases as you go toward the right across a period in the periodic table and decreases as you go down a column in the periodic table. If two elements with identical electronegativities form a covalent bond,

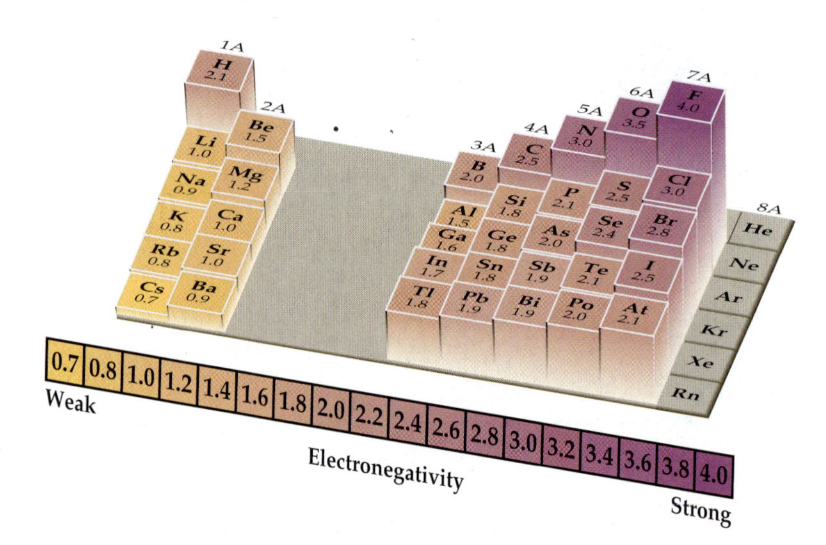

▲ **Figure 10.2 Electronegativity of the elements** Linus Pauling introduced the scale shown here. He arbitrarily set the electronegativity of fluorine at 4.0 and computed all other values relative to fluorine.

TUTORIAL
Periodic Trends: Electronegativity Movie

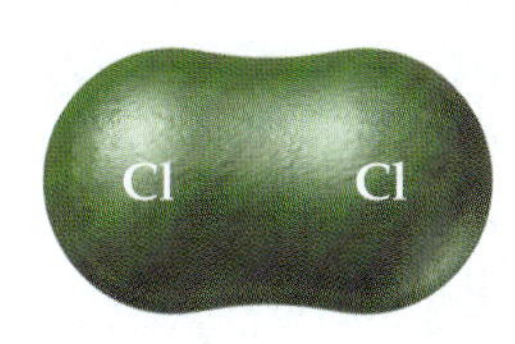

◀ **Figure 10.3 Pure covalent bonding** In Cl_2, the two Cl atoms share the electrons evenly. This is a pure covalent bond.

they equally share the electrons, and there is no dipole moment. For example, the chlorine molecule, composed of two chlorine atoms (which of course have identical electronegativities), has a covalent bond in which electrons are evenly shared (▲ Figure 10.3). The bond has no dipole moment and the molecule is **nonpolar**.

If there is a large electronegativity difference between the two elements in a bond, such as normally occurs between a metal and a nonmetal, the electron is completely transferred and the bond is ionic. For example, sodium and chlorine form an ionic bond (▼ Figure 10.4).

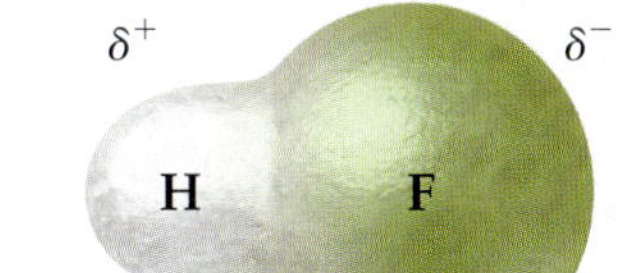

◀ **Figure 10.4 Ionic bonding** In NaCl, Na completely transfers an electron to Cl. This is an ionic bond.

If there is an intermediate electronegativity difference between the two elements, such as between two different nonmetals, then the bond is polar covalent. For example, HF forms a polar covalent bond (▼ Figure 10.5).

TUTORIAL Electronegativity and Bonding

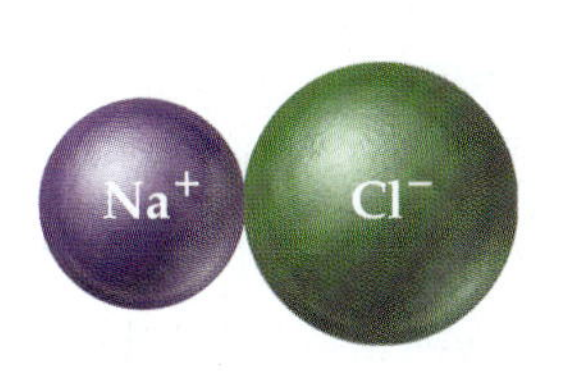

◀ **Figure 10.5 Polar covalent bonding** In HF, the electrons are shared, but the shared electrons are more likely to be found on F than on H. The bond is polar covalent.

These concepts are summarized in Table 10.2 and ▼ Figure 10.6.

TABLE 10.2 The Effect of Electronegativity Difference on Bond Type

Electronegativity Difference (ΔEN)	Bond Type	Example
zero (0–0.4)	pure covalent	Cl_2
intermediate (0.4–2.0)	polar covalent	HF
large (2.0+)	ionic	NaCl

The degree of bond polarity is a continuous function. The guidelines given here are only approximate.

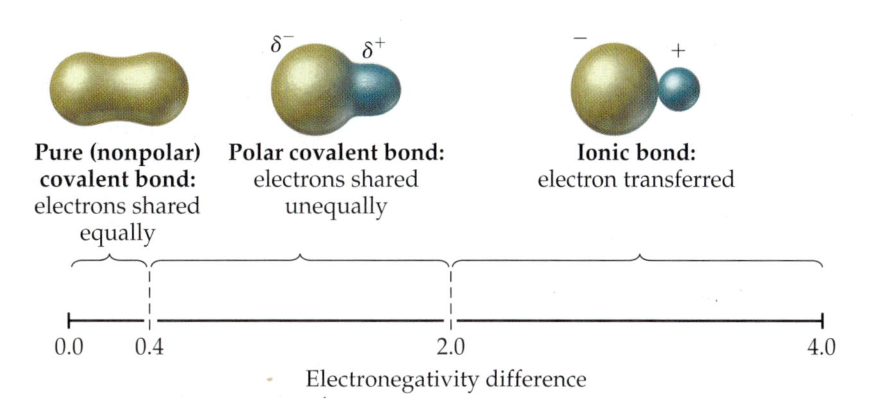

▶ **Figure 10.6 The continuum of bond types** The type of bond (pure covalent, polar covalent, or ionic) is related to the electronegativity difference between the bonded atoms.

EXAMPLE 10.10 Classifying Bonds as Pure Covalent, Polar Covalent, or Ionic

Determine whether the bond formed between each of the following pairs of atoms is pure covalent, polar covalent, or ionic.

(a) Sr and F

(b) N and Cl

(c) N and O

Solution:

(a) From Figure 10.2, we find the electronegativity of Sr (1.0) and of F (4.0). The electronegativity difference (ΔEN) is:

$$\Delta EN = 4.0 - 1.0 = 3.0$$

Using Table 10.2, we classify this bond as ionic.

(b) From Figure 10.2, we find the electronegativity of N (3.0) and of Cl (3.0). The electronegativity difference (ΔEN) is:

$$\Delta EN = 3.0 - 3.0 = 0$$

Using Table 10.2, we classify this bond as pure covalent.

(c) From Figure 10.2, we find the electronegativity of N (3.0) and of O (3.5). The electronegativity difference (ΔEN) is:

$$\Delta EN = 3.5 - 3.0 = 0.5$$

Using Table 10.2, we classify this bond as polar covalent.

SKILLBUILDER 10.10 Classifying Bonds as Pure Covalent, Polar Covalent, or Ionic

Determine whether the bond formed between each of the following pairs of atoms is pure covalent, polar covalent, or ionic.

(a) I and I

(b) Cs and Br

(c) P and O

Polar Bonds and Polar Molecules

We just learned how to identify polar bonds. Does the presence of one or more polar bonds in a molecule always result in a polar molecule? The answer is no. A **polar molecule** is one with polar bonds that add together—they do not cancel each other—to form a net dipole moment. For diatomic molecules, you can easily tell polar molecules from nonpolar ones. If a diatomic molecule contains a polar bond, then the molecule is polar. However, for molecules with more than two atoms, it is more difficult to tell polar molecules from nonpolar ones because two or more polar bonds may cancel one another. For example, consider carbon dioxide:

$$:\ddot{O}=C=\ddot{O}:$$

Each C=O *bond* is polar because oxygen and carbon have different electronegativities (3.5 and 2.5). However, since CO_2 has a linear geometry, the dipole moment of one bond completely cancels the dipole moment of the other and the *molecule* is nonpolar. We can understand this with a simple analogy. Consider each polar bond as a rope pulling on the central atom. In CO_2, we can see how the two ropes pulling in opposing directions cancel each other:

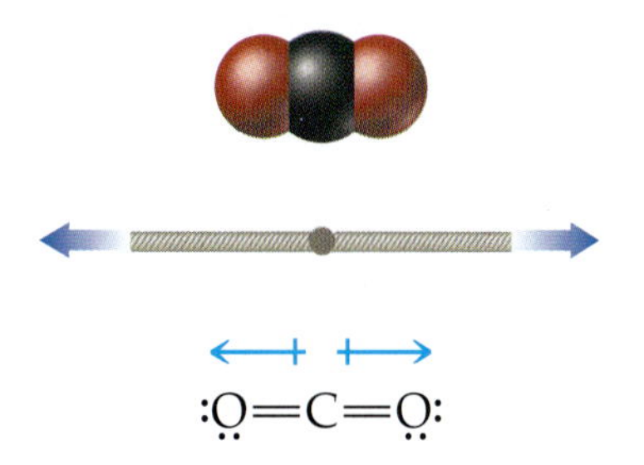

In the vector representation of a dipole moment, the vector points in the direction of the atom with the partial negative charge.

We can also represent polar bonds with arrows (or vectors) that point in the direction of the negative pole and have a plus sign at the positive pole (as shown above for carbon dioxide.) If the arrows (or vectors) point in exactly opposing directions as in carbon dioxide, the dipole moments cancel.

Water, on the other hand, has two dipole moments that do not cancel. If we imagine each bond as a rope pulling on oxygen, we see that, because of the angle between the bonds, the pulls of the two ropes do not cancel:

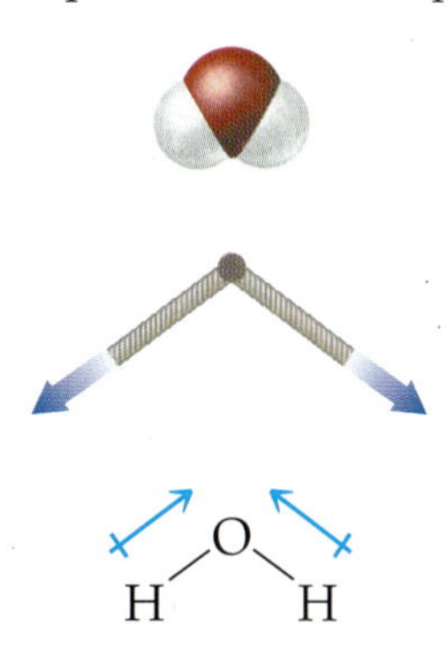

Consequently, water is a polar molecule. We can use symmetry as a guide to determine whether a molecule containing polar bonds is indeed polar. Highly symmetric molecules will tend to be nonpolar even if they have polar bonds because the bond dipole moments (or the pulls of the ropes) tend to cancel. Asymmetric molecules that contain polar bonds will tend to be polar because the bond dipole moments (or the pulls of the ropes) tend not to cancel. Table 10.3 summarizes various common cases.

TUTORIAL
Bond Polarity and Geometry

In summary, to determine whether a molecule is polar:

- **Determine whether the molecule contains polar bonds.** A bond is polar if the two bonding atoms have different electronegativities. If there are no polar bonds, the molecule is nonpolar.
- **Determine whether the polar bonds add together to form a net dipole moment.** You must first use VSEPR to determine the geometry of the molecule. Then visualize each bond as a rope pulling on the central atom. Is the molecule highly symmetrical? Do the pulls of the ropes cancel? If so, there is no net dipole moment and the molecule is nonpolar. If the molecule is asymmetrical and the pulls of the rope do not cancel, the molecule is polar.

TABLE 10.3 Common Cases of Adding Dipole Moments to Determine whether a Molecule Is Polar

Nonpolar

Two identical polar bonds pointing in opposite directions will cancel. The molecule is nonpolar.

Polar

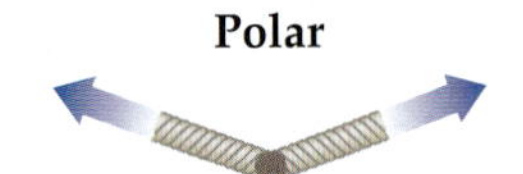

Two polar bonds with an angle of less than 180° between them will not cancel. The molecule is polar.

Nonpolar

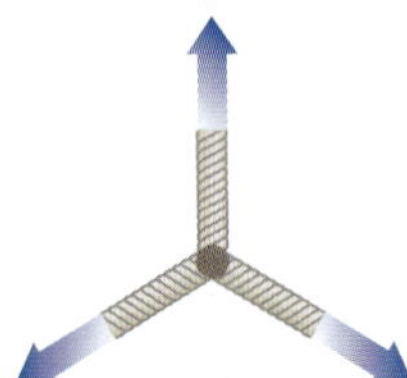

Three identical polar bonds at 120° from each other will cancel. The molecule is nonpolar.

Nonpolar

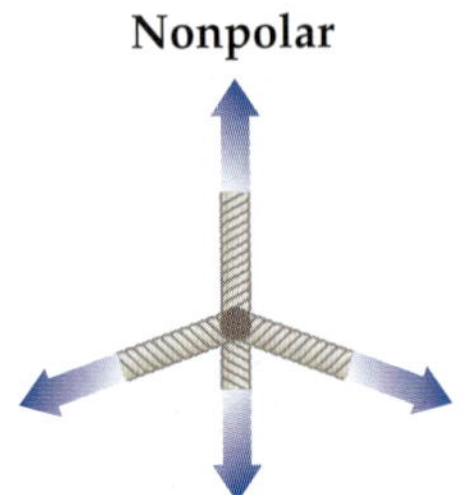

Four identical polar bonds in a tetrahedral arrangement (109.5° from each other) will cancel. The molecule is nonpolar.

Polar

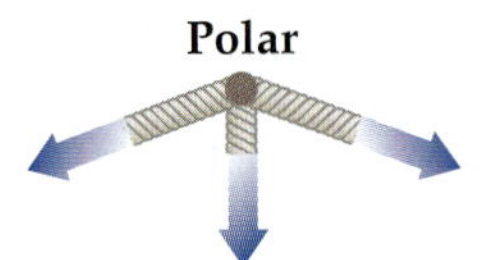

Three polar bonds in a trigonal pyramidal arrangement (109.5°) will not cancel. The molecule is polar.

Note: In all cases where the polar bonds cancel, the bonds are assumed to be identical. If one or more of the bonds are different than the other(s), the bonds will not cancel and the molecule is polar.

EXAMPLE 10.11 Determining whether a Molecule Is Polar

Determine whether NH_3 is polar.

Solution:

Begin by drawing the Lewis structure of NH_3. Since N and H have different electronegativities, the bonds are polar.

H
|
H—N—H

The geometry of NH_3 is trigonal pyramidal (four electron groups, three bonding groups, one lone pair). Draw a three-dimensional picture of NH_3 and imagine each bond as a rope that is being pulled. The pulls of the ropes do not cancel and the molecule is polar.

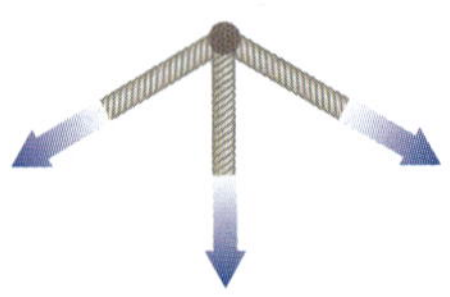

NH_3 is polar

SKILLBUILDER 10.11 Determining whether a Molecule Is Polar

Determine whether CH_4 is polar.

Whether or not a molecule is polar is important because polar molecules tend to behave differently than nonpolar molecules. Water and oil do not mix, for example, because water molecules are polar and the molecules that compose oil are generally nonpolar. Polar molecules interact strongly with other polar molecules because the positive end of one molecule is attracted to the negative end of another, just as the south pole of a magnet is attracted to the north pole of another magnet (▼ Figure 10.7). A mixture of polar and nonpolar molecules is similar to a mixture of small magnetic and nonmagnetic particles. The magnetic particles clump together, excluding the nonmagnetic ones and separating into distinct regions (▼ Figure 10.8). Similarly, the polar molecules attract one another, forming regions from which the nonpolar molecules are excluded (▼ Figure 10.9).

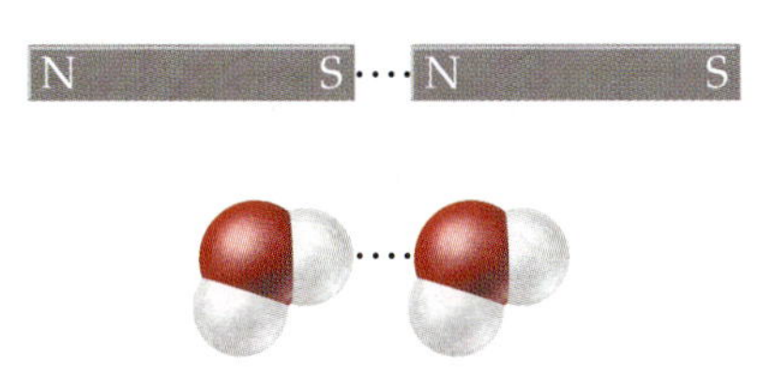

▲ **Figure 10.7 Dipole–dipole attraction** Just as the north pole of one magnet is attracted to the south pole of another, so the positive end of one molecule with a dipole is attracted to the negative end of another molecule with a dipole.

▲ **Figure 10.8 Magnetic and nonmagnetic particles** Magnetic particles attract one another, excluding nonmagnetic particles. This behavior is analogous to that of polar and nonpolar molecules.

▶ **Figure 10.9 Polar and nonpolar molecules** A mixture of polar and nonpolar molecules, like a mixture of magnetic and nonmagnetic particles, separates into distinct regions because the polar molecules attract one another, excluding the nonpolar ones. **Question:** Can you think of some examples of this behavior?

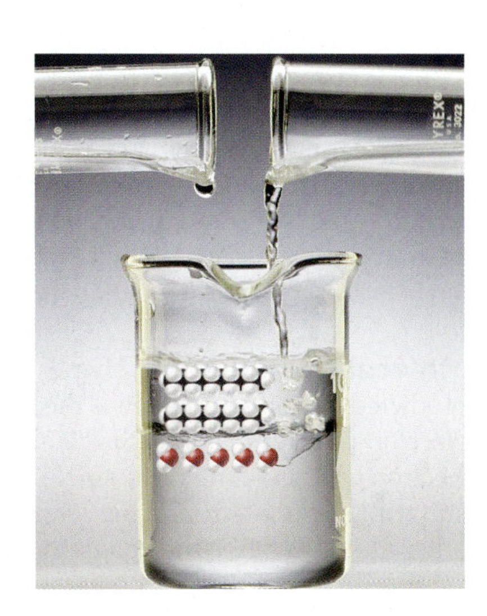

EVERYDAY Chemistry

How Soap Works

Imagine eating a greasy cheeseburger with both hands and no napkins. By the end of the meal, your hands are coated with grease and oil. If you try to wash them with only water, they remain greasy. However, if you add a little soap, the grease washes away. Why? As we learned previously, water molecules are polar and the molecules that compose grease and oil are nonpolar. As a result, water and grease repel each other.

The molecules that compose soap, however, have a special structure that allows them to interact strongly with both water and grease. One end of a soap molecule is polar while the other end is nonpolar.

Soap molecule

Polar head attracts water

Nonpolar tail attracts grease

The polar head of a soap molecule strongly attracts water molecules, while the nonpolar tail strongly attracts grease and oil molecules. Soap is a sort of molecular liaison, one end interacting with water and the other end interacting with grease. Soap therefore allows water and grease to mix, removing the grease from your hands and washing it down the drain.

CAN YOU ANSWER THIS? *Consider the following detergent molecule. Which end do you think is polar? Which end is nonpolar?*

$CH_3(CH_2)_{11}OCH_2CH_2OH$

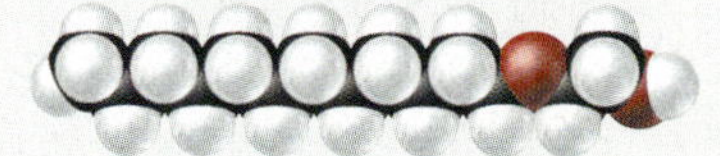

CHAPTER IN REVIEW

Chemical Principles

Lewis Theory: Lewis theory is a model for chemical bonding. In Lewis theory, chemical bonds are formed when atoms transfer valence electrons (ionic bonding) or share valence electrons (covalent bonding) to attain noble gas electron configurations. In Lewis theory, valence electrons are represented as dots surrounding the symbol for an element. When two or more elements bond together, the dots are transferred or shared so that every atom gets eight dots (an octet), or two dots (a duet) in the case of hydrogen.

Relevance

Lewis Theory: Bonding theories predict what combinations of elements will form stable compounds. They can also be used to predict the properties of those compounds. For example, bonding theories are used by pharmaceutical companies when they are designing drug molecules that must interact with a specific part of a protein molecule.

Molecular Shapes: The shapes of molecules can be predicted by combining Lewis theory with valence shell electron pair repulsion (VSEPR) theory. In this model, electron groups—lone pairs, single bonds, double bonds, and triple bonds—around the central atom repel one another and determine the geometry of the molecule.

Molecular Shapes: Molecular shapes determine many of the properties of compounds. Water's bent geometry, for example, causes it to be a liquid at room temperature instead of a gas. It also causes ice to float on water and is responsible for the hexagonal patterns in snowflakes.

Electronegativity and Polarity: Electronegativity refers to the relative ability of elements to attract electrons within a chemical bond. Electronegativity increases as you move to the right across a period in the periodic table and decreases as you move down a column. When two nonmetal atoms of different electronegativities form a covalent bond, the electrons in the bond are not evenly shared and the bond is polar. In diatomic molecules, a polar bond results in a polar molecule. In multi-atom molecules, polar bonds may cancel, forming a nonpolar molecule, or they may sum, forming a polar molecule.

Electronegativity and Polarity: The polarity of a molecule influences many of its properties such as whether it will be a solid, liquid, or gas at room temperature and whether it will mix with other compounds. Oil and water, for example, do not mix because water is polar while oil is nonpolar.

Chemical Skills

Examples

Lewis Structures for Elements (Section 10.2)

The Lewis structure of any element is simply the symbol for the element with the valence electrons represented as dots drawn around the element. The number of valence electrons is equal to the group number of the element (for main-group elements).

EXAMPLE 10.12 Lewis Structures for Elements

What is the Lewis structure of sulfur?

Solution:

Since S is in Group 6, it has 6 valence electrons. We draw these as dots surrounding its symbol, S.

Writing Lewis Structures for Ionic Compounds (Section 10.3)

In an ionic Lewis structure, the metal loses all of its valence electrons to the nonmetal, which gets an octet. The nonmetal, with its octet, is normally written in brackets with the charge in the upper right corner.

EXAMPLE 10.13 Writing Lewis Structures for Ionic Compounds

Write a Lewis structure for lithium bromide.

Solution:

$$Li^+ [:\ddot{\underset{..}{Br}}:]^-$$

Using Lewis Theory to Predict the Chemical Formula of an Ionic Compound (Section 10.3)

To determine the chemical formula of an ionic compound, write the Lewis structures of each of the elements. Then choose the correct number of each type of atom so that the metal atom(s) lose all of their valence electrons and the nonmetal atom(s) get an octet.

EXAMPLE 10.14 Using Lewis Theory to Predict the Chemical Formula of an Ionic Compound

Use Lewis theory to predict the formula for the compound that forms between potassium and sulfur.

Solution:

The Lewis structures of K and S are:

$$K\cdot \qquad \cdot\ddot{\underset{\cdot}{S}}:$$

Potassium must lose 1 electron and sulfur must gain 2. Consequently, we need two potassium atoms to every sulfur atom. The Lewis structure is:

$$K^+[:\ddot{\underset{..}{S}}:]^{2-}K^+$$

The correct formula is K_2S.

Writing Lewis Structures for Covalent Compounds (Sections 10.4, 10.5)

To write covalent Lewis structures, follow these steps:

1. *Write the correct skeletal structure for the molecule.* Hydrogen atoms will always be terminal, halogens will usually be terminal, and many molecules tend to be symmetrical.
2. *Calculate the total number of electrons for the Lewis structure by summing the valence electrons of each atom in the molecule.* Remember that the number of valence electrons for any main-group element is equal to its group number in the periodic table. For polyatomic ions, add 1 electron for each negative charge and subtract 1 electron for each positive charge.
3. *Distribute the electrons among the atoms, giving octets (or duets for hydrogen) to as many atoms as possible.* Begin by placing 2 electrons between each pair of atoms. These are the bonding electrons. Then distribute the remaining electrons, first to terminal atoms and then to the central atom.
4. *If any atoms lack an octet, form double or triple bonds as necessary to give them octets.* Do this by moving lone electron pairs from terminal atoms into the bonding region with the central atom.

LIVE EXAMPLE

EXAMPLE 10.15 Writing Lewis Structures for Covalent Compounds

Write a Lewis structure for CS_2.

Solution:

S C S

$$\begin{aligned}\text{Total \# e}^- &= 1 \times (\text{\# valence e}^- \text{ in C}) \\ &\quad + 2 \times (\text{\# valence e}^- \text{ in S}) \\ &= 4 + 2(6) \\ &= 16\end{aligned}$$

S:C:S (4 of 16 e⁻ used)

:S̤̈:C:S̤̈: (16 of 16 e⁻ used)

:S̈::C::S̈: or :S̈=C=S̈:

Writing Resonance Structures (Section 10.6)

When two or more equivalent (or nearly equivalent) Lewis structures can be written for a molecule, the true structure is an average between these. Represent this by writing all of the correct structures (called resonance structures) with double-headed arrows between them.

EXAMPLE 10.16 Writing Resonance Structures

Write resonance structures for SeO_2.

Solution:

We can write a Lewis structure for SeO_2 by following the steps for writing covalent Lewis structures. We find that we can write two equally correct structures, so we draw them both as resonance structures.

:Ö̤—S̈e=Ö: ⟷ :Ö=S̈e—Ö̤:

Predicting the Shapes of Molecules (Section 10.7)

To determine the shape of a molecule, follow these steps:

1. *Draw a Lewis structure for the molecule.*
2. *Determine the total number of electron groups around the central atom.* Lone pairs, single bonds, double bonds, and triple bonds each count as one group.
3. *Determine the number of bonding groups and the number of lone pairs around the central atom.* These should sum to the result from Step 2. Bonding groups include single bonds, double bonds, and triple bonds.
4. *Use Table* 10.1 *to determine the electron geometry and molecular geometry.*

EXAMPLE 10.17 Predicting the Shapes of Molecules

Predict the geometry of SeO_2.

Solution:

The Lewis structure for SeO_2 (as we saw in Example 10.16) is composed of the following two resonance structures.

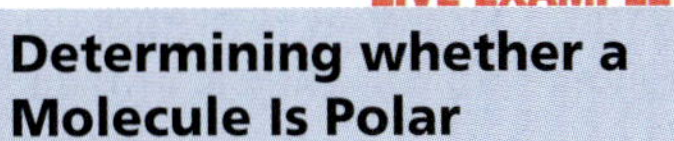

Either of the resonance structures will give the same geometry.

Total number of electron groups = 3

Number of bonding groups = 2

Number of lone pairs = 1

Electron geometry = Trigonal planar

Molecular geometry = Bent

Determining whether a Molecule Is Polar (Section 10.8)

- *Determine whether the molecule contains polar bonds.* A bond is polar if the two bonding atoms have different electronegativities. If there are no polar bonds, the molecule is nonpolar.
- *Determine whether the polar bonds add together to form a net dipole moment.* Use VSEPR to determine the geometry of the molecule. Then visualize each bond as a rope pulling on the central atom. Is the molecule highly symmetrical? Do the pulls of the ropes cancel? If so, there is no net dipole moment and the molecule is nonpolar. If the molecule is asymmetrical and the pulls of the rope do not cancel, the molecule is polar.

LIVE EXAMPLE

EXAMPLE 10.18 Determining whether a Molecule Is Polar

Determine whether SeO_2 is polar.

Solution:

Se and O are nonmetals with different electronegativities (2.4 for Se and 3.5 for O). Therefore, the Se–O bonds are polar.

As we saw in Example 10.17, the geometry of SeO_2 is bent.

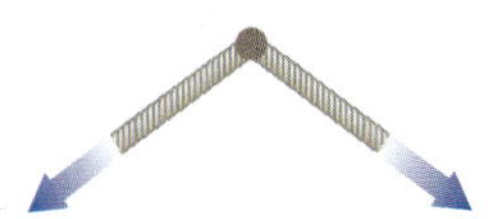

The polar bonds do not cancel but rather sum to give a net dipole moment. Therefore the molecule is polar.

Key Terms

bent [10.7]
bonding pair [10.4]
bonding theory [10.1]
chemical bond [10.2]
covalent bond [10.2]
dipole moment [10.8]
dot structure [10.2]
double bond [10.4]
duet [10.2]
electron geometry [10.7]
electron group [10.7]
electronegativity [10.8]
ionic bond [10.2]
Lewis structure [10.2]
Lewis theory [10.1]
linear [10.7]
lone pair [10.4]
molecular geometry [10.7]
nonpolar [10.8]
octet [10.2]
octet rule [10.2]
polar covalent bond [10.8]
polar molecule [10.8]
resonance structures [10.6]
terminal atom [10.5]
tetrahedral [10.7]
trigonal planar [10.7]
trigonal pyramidal [10.7]
triple bond [10.4]
valence shell electron pair repulsion theory (VSEPR) [10.7]

Exercises

Questions

1. Why are bonding theories important?
2. Give some examples of what bonding theories can predict.
3. Write the electron configurations for Ne and Ar. How many valence electrons do they each have?
4. In Lewis theory, what is an octet? What is a duet?
5. According to Lewis theory, what is a chemical bond?
6. What is the difference between ionic bonding and covalent bonding?
7. How can Lewis theory be used to determine the formula of ionic compounds? You may explain this with an example.
8. What is the difference between lone pair and bonding pair electrons?
9. How are double and triple bonds physically different from single bonds?
10. What is the procedure for writing a covalent Lewis structure?
11. How do you determine the number of electrons that go into the Lewis structure of a molecule?
12. How do you determine the number of electrons that go into the Lewis structure of a polyatomic ion?
13. Why does the octet rule have exceptions? Give some examples.
14. What are resonance structures? Why are they necessary?
15. Explain how VSEPR predicts the shapes of molecules.
16. If all of the electron groups around a central atom are bonding groups (that is, there are no lone pairs), what is the molecular geometry for:
    (a) two electron groups
    (b) three electron groups
    (c) four electron groups
17. Give the bond angles for each of the geometries in Question 16.
18. What is the difference between electron geometry and molecular geometry in VSEPR?
19. What is electronegativity?
20. What is the most electronegative element on the periodic table?
21. What is a polar covalent bond?
22. What is a dipole moment?
23. Why is it important to know if a molecule is polar or nonpolar?
24. If a molecule has polar bonds, will the molecule itself be polar? Why or why not?

Problems

Writing Lewis Structures for Elements

25. Write an electron configuration for each of the following elements and the corresponding Lewis structure. Indicate which electrons in the electron configuration are included in the Lewis structure.
    (a) N
    (b) C
    (c) Cl
    (d) Ar
26. Write an electron configuration for each of the following elements and the corresponding Lewis structure. Indicate which electrons in the electron configuration are included in the Lewis structure.
    (a) Li
    (b) P
    (c) F
    (d) Ne

27. Write Lewis structures for each of the following elements.
(a) K
(b) Al
(c) P
(d) Ar

28. Write Lewis structures for each of the following elements.
(a) Mg
(b) Si
(c) S
(d) Br

29. Write a generic Lewis structure for the halogens. Do the halogens tend to gain or lose electrons in chemical reactions? How many?

30. Write a generic Lewis structure for the alkali metals. Do the alkali metals tend to gain or lose electrons in chemical reactions? How many?

31. Write a generic Lewis structure for the alkaline earth metals. Do the alkaline earth metals tend to gain or lose electrons in chemical reactions? How many?

32. Write a generic Lewis structure for the elements in the oxygen family (Group 6). Do the elements in the oxygen family tend to gain or lose electrons in chemical reactions? How many?

33. Write a Lewis structure for each of the following ions.
(a) Cl^-
(b) Se^{2-}
(c) Na^+
(d) Mg^{2+}

34. Write a Lewis structure for each of the following ions.
(a) N^{3-}
(b) S^{2-}
(c) Ca^{2+}
(d) Al^{3+}

35. For each of the following ions, indicate the noble gas that has the same Lewis structure as the ion.
(a) Br^-
(b) O^{2-}
(c) Rb^+
(d) Ba^{2+}

36. For each of the following ions, indicate the noble gas that has the same Lewis structure as the ion.
(a) Se^{2-}
(b) I^-
(c) Sr^{2+}
(d) F^-

Lewis Structures for Ionic Compounds

37. Determine whether each of the following compounds would be best represented by an ionic or a covalent Lewis structure.
(a) Rb_2O
(b) CO_2
(c) Al_2S_3
(d) NO

38. Determine whether each of the following compounds would be best represented by an ionic or a covalent Lewis structure.
(a) $MgCl_2$
(b) K_2S
(c) BrCl
(d) SF_6

39. Write a Lewis structure for each of the following ionic compounds.
(a) NaF
(b) CaO
(c) $SrBr_2$
(d) K_2O

40. Write a Lewis structure for each of the following ionic compounds.
(a) SrO
(b) Li_2S
(c) CaI_2
(d) RbF

41. Use Lewis theory to determine the formula for the compound that forms from:
(a) Sr and Se
(b) Ba and Cl
(c) Na and S
(d) Al and O

42. Use Lewis theory to determine the formula for the compound that forms from:
(a) Ca and N
(b) Mg and I
(c) Ca and S
(d) Cs and F

43. Draw the Lewis structure for the ionic compound that forms from Li and:
(a) F
(b) O
(c) N

44. Draw the Lewis structure for the ionic compound that forms from Ba and:
(a) F
(b) O
(c) N

45. Determine what is wrong with each of the following ionic Lewis structures and write the correct structure.

(a) $[Cs:]^+ [:\ddot{\underset{..}{Cl}}:]^-$

(b) $Ba^+ [:\ddot{\underset{..}{O}}:]^-$

(c) $Ca^{2+} [:\ddot{\underset{..}{I}}:]^-$

46. Determine what is wrong with each of the following ionic Lewis structures and write the correct structure.

(a) $[:\ddot{\underset{..}{O}}:]^{2-} Na^+ [:\ddot{\underset{..}{O}}:]^{2-}$

(b) $Mg:\ddot{\underset{..}{O}}:$

(c) $[Li:]^+ [:\ddot{\underset{..}{S}}:]^-$

Lewis Structures for Covalent Compounds

47. Write a Lewis structure for each of the following molecules.
(a) PH_3
(b) SCl_2
(c) F_2
(d) HI

48. Write a Lewis structure for each of the following molecules.
(a) CH_4
(b) NF_3
(c) OF_2
(d) H_2O

49. Write a Lewis structure for each of the following molecules.
(a) O_2
(b) CO
(c) HONO (N is central)
(d) SO_2

50. Write a Lewis structure for each of the following molecules.
(a) N_2O (oxygen is terminal)
(b) SiH_4
(c) CI_4
(d) Cl_2CO (carbon is central)

51. Write a Lewis structure for each of the following molecules.
(a) C_2H_2
(b) C_2H_4
(c) N_2H_2
(d) N_2H_4

52. Write a Lewis structure for each of the following molecules.
(a) H_2CO (carbon is central)
(b) H_3COH (carbon and oxygen are both central)
(c) H_3COCH_3 (oxygen is between the two carbon atoms)
(d) H_2O_2

53. Determine what is wrong with each of the following Lewis structures and write the correct structure.

(a) $:\ddot{N}{=}\ddot{N}:$

(b) $:\ddot{\underset{..}{S}}{-}Si{-}\ddot{\underset{..}{S}}:$

(c) $H{-}H{-}\ddot{\underset{..}{O}}:$

(d) $:\ddot{\underset{..}{I}}{-}N{-}\ddot{\underset{..}{I}}:$ with $:\underset{..}{I}:$ bonded below N

54. Determine what is wrong with each of the following Lewis structures and write the correct structure.

(a) $H{-}H{-}H{-}\ddot{\underset{..}{N}}:$

(b) $:\ddot{\underset{..}{Cl}}{=}O{=}\ddot{\underset{..}{Cl}}:$

(c) $H{-}C{-}\ddot{\underset{..}{O}}{-}H$ with $:\ddot{O}:$ bonded above C

(d) $H{=}\ddot{\underset{..}{Br}}:$

55. Write a Lewis structure for each of the following molecules or ions. Include resonance structures if necessary.
(a) SeO_2
(b) CO_3^{2-}
(c) ClO^-
(d) ClO_2^-

56. Write a Lewis structure for each of the following molecules or ions. Include resonance structures if necessary.
(a) ClO_3^-
(b) ClO_4^-
(c) NO_3^-
(d) SO_3

57. Write a Lewis structure for each of the following ions. Include resonance structures if necessary.
(a) PO_4^{3-}
(b) CN^-
(c) NO_2^-
(d) SO_3^{2-}

58. Write a Lewis structure for each of the following ions. Include resonance structures if necessary.
(a) SO_4^{2-}
(b) HSO_4^- (S is central; H is attached to one of the O atoms)
(c) NH_4^+
(d) BrO_2^- (Br is central)

59. Write a Lewis structure for each of the following molecules that are exceptions to the octet rule.
(a) BCl_3
(b) NO_2
(c) BH_3

60. Write a Lewis structure for each of the following molecules that are exceptions to the octet rule.
(a) BBr_3
(b) NO

Predicting the Shapes of Molecules

61. Determine the number of electron groups around the central atom for each of the following molecules.
(a) PCl_3
(b) SBr_2
(c) CH_2Cl_2
(d) CS_2

62. Determine the number of electron groups around the central atom for each of the following molecules.
(a) CH_4
(b) NF_3
(c) OF_2
(d) H_2S

63. Determine the number of bonding groups and the number of lone pairs for each of the molecules in Problem 61. The sum of these should equal your answer to Problem 61.

64. Determine the number of bonding groups and the number of lone pairs for each of the molecules in Problem 62. The sum of these should equal your answer to Problem 62.

65. Determine the molecular geometry of each of the following molecules.
(a) CBr_4
(b) H_2CO
(c) CS_2
(d) BH_3

66. Determine the molecular geometry of each of the following molecules.
(a) SiO_2
(b) BF_3
(c) $CFCl_3$ (carbon is central)
(d) H_2CS (carbon is central)

67. Determine the bond angles for each molecule in Problem 65.

68. Determine the bond angles for each molecule in Problem 66.

69. Determine the electron and molecular geometries of each of the following molecules.
(a) N_2O (oxygen is terminal)
(b) SO_2
(c) H_2S
(d) PF_3

70. Determine the electron and molecular geometries of each of the following molecules. (Hint: Determine the geometry around each of the two central atoms.)
(a) C_2H_2 (skeletal structure HCCH)
(b) C_2H_4 (skeletal structure H_2CCH_2)
(c) C_2H_6 (skeletal structure H_3CCH_3)

71. Determine the bond angles for each molecule in Problem 69.

72. Determine the bond angles for each molecule in Problem 70.

73. Determine the electron and molecular geometries of each of the following molecules. For those with two central atoms, indicate the geometry about each central atom.
(a) N_2
(b) N_2H_2 (skeletal structure HNNH)
(c) N_2H_4 (skeletal structure H_2NNH_2)

74. Determine the electron and molecular geometries of each of the following molecules. For those with more than one central atom, indicate the geometry about each central atom.
(a) CH_3OH (skeletal structure H_3COH)
(b) H_3COCH_3 (skeletal structure H_3COCH_3)
(c) H_2O_2 (skeletal structure HOOH)

75. Determine the molecular geometry of each of the following polyatomic ions.
(a) CO_3^{2-}
(b) ClO_2^-
(c) NO_3^-
(d) NH_4^+

76. Determine the molecular geometry of each of the following polyatomic ions.
(a) ClO_4^-
(b) BrO_2^-
(c) NO_2^-
(d) SO_4^{2-}

Electronegativity and Polarity

77. Use Figure 10.2 to determine the electronegativity of each of the following elements.
(a) Mg
(b) Si
(c) Br

78. Use Figure 10.2 to determine the electronegativity of each of the following elements.
(a) F
(b) C
(c) S

79. List the following elements in order of increasing electronegativity: Rb, Si, Cl, Ca, Ga

80. List the following elements in order of decreasing electronegativity: Ba, N, F, Si, Cs

81. Use Figure 10.2 to find the electronegativity difference between each of the following pairs of elements and then use Table 10.2 to classify the bonds that occur between them as pure covalent, polar covalent, or ionic.
(a) Mg and Br
(b) Cr and F
(c) Br and Br
(d) Si and O

82. Use Figure 10.2 to find the electronegativity difference between each of the following pairs of elements and then use Table 10.2 to classify the bonds that occur between them as pure covalent, polar covalent, or ionic.
(a) K and Cl
(b) N and N
(c) C and S
(d) C and Cl

83. Arrange the following diatomic molecules in order of increasing bond polarity:

ICl, HBr, H_2, CO

84. Arrange the following diatomic molecules in order of decreasing bond polarity:

HCl, NO, F_2, HI

85. Classify each of the following diatomic molecules as polar or nonpolar.
(a) CO
(b) O_2
(c) F_2
(d) HBr

86. Classify each of the following diatomic molecules as polar or nonpolar.
(a) I_2
(b) NO
(c) HCl
(d) N_2

87. For each polar molecule in Problem 85, draw the molecule and indicate the positive and negative ends of the dipole moment.

88. For each polar molecule in Problem 86, draw the molecule and indicate the positive and negative ends of the dipole moment.

89. Classify each of the following molecules as polar or nonpolar.
(a) CS_2
(b) SO_2
(c) CH_4
(d) CH_3Cl

90. Classify each of the following molecules as polar or nonpolar.
(a) H_2CO
(b) CH_3OH
(c) CH_2Cl_2
(d) SiO_2

91. Classify each of the following molecules as polar or nonpolar.
(a) BH_3
(b) $CHCl_3$
(c) C_2H_2
(d) NH_3

92. Classify each of the following molecules as polar or nonpolar.
(a) N_2H_2
(b) H_2O_2
(c) CF_4
(d) NO_2

Cumulative Problems

93. Write electron configurations and Lewis structures for each of the following elements. Indicate which of the electrons in the electron configuration are shown in the Lewis structure.
(a) Ca
(b) Ga
(c) As
(d) I

94. Write electron configurations and Lewis structures for each of the following elements. Indicate which of the electrons in the electron configuration are shown in the Lewis structure.
(a) Rb
(b) Ge
(c) Kr
(d) Se

95. Determine whether each of the following compounds is ionic or covalent and write an appropriate Lewis structure.
(a) K_2S
(b) CHFO (carbon is central)
(c) MgSe
(d) PBr_3

96. Determine whether each of the following compounds is ionic or covalent and write an appropriate Lewis structure.
(a) HCN
(b) ClF
(c) MgI_2
(d) CaS

97. Write a Lewis structure for $OCCl_2$ (carbon is central) and determine whether the molecule is polar. Draw a three-dimensional structure for the molecule.

98. Write a Lewis structure for CH_3COH and determine whether the molecule is polar. Draw a three-dimensional structure for the molecule. The skeletal structure is:

```
  H O
H C C H
  H
```

99. Write a Lewis structure for acetic acid (a component of vinegar), CH_3COOH, and draw a three-dimensional sketch of the molecule. The skeletal structure is:

```
  H O
H C C O H
  H
```

100. Write a Lewis structure for benzene, C_6H_6, and draw a three-dimensional sketch of the molecule. The skeletal structure is the ring shown here. (Hint: Include two resonance structures.)

```
    H
    C
HC     CH
HC     CH
    C
    H
```

101. Consider the following neutralization reaction.

$$HCl(aq) + NaOH(aq) \longrightarrow H_2O(l) + NaCl(aq)$$

Write the reaction showing the Lewis structures of each of the reactants and products.

102. Consider the following precipitation reaction.

$$Ba(NO_3)_2(aq) + 2\,LiCl(aq) \longrightarrow BaCl_2(s) + 2\,LiNO_3(aq)$$

Write the reaction showing the Lewis structures of each of the reactants and products.

103. Consider the following redox reaction:

$$2\,K(s) + Cl_2(g) \longrightarrow 2\,KCl(s)$$

Draw Lewis structures for each reactant and product and determine which reactant was oxidized and which one was reduced.

104. Consider the following redox reaction:

$$Ca(s) + Br_2(g) \longrightarrow CaBr_2(s)$$

Draw Lewis structures for each reactant and product and determine which reactant was oxidized and which one was reduced.

105. Each of the following compounds contains both ionic and covalent bonds. Write an ionic Lewis structure for each one, including the covalent structure for the ion in brackets. Write resonance structures if necessary.
(a) KOH
(b) KNO_3
(c) LiIO
(d) $BaCO_3$

106. Each of the following compounds contains both ionic and covalent bonds. Write an ionic Lewis structure for each one, including the covalent structure for the ion in brackets. Write resonance structures if necessary.
(a) $RbIO_2$
(b) $Ca(OH)_2$
(c) NH_4Cl
(d) $Sr(CN)_2$

107. Each of the following molecules contains an expanded octet (10 or 12 electrons) around the central atom. Write a Lewis structure for each molecule.
(a) PF_5
(b) SF_4
(c) SeF_4

108. Each of the following molecules contains an expanded octet (10 or 12 electrons) around the central atom. Write a Lewis structure for each molecule.
(a) ClF_5
(b) SF_6
(c) IF_5

109. Formic acid is responsible for the sting in biting ants. By mass, formic acid is 26.10% C, 4.38% H, and 69.52% O. The molar mass of formic acid is 46.02 g/mol. Find the molecular formula of formic acid and draw its Lewis structure.

110. Diazomethane has the following composition by mass: 28.57% C, 4.80% H, and 66.64% N. The molar mass of diazomethane is 42.04 g/mol. Find the molecular formula of diazomethane and draw its Lewis structure.

Highlight Problems

111. Some theories on aging suggest that a class of molecules and ions called *free radicals* cause some diseases and aging. Free radicals are molecules or ions containing an unpaired electron. As you know from Lewis theory, such molecules are not chemically stable and quickly react with other molecules. Free radicals may attack molecules within the cell, such as DNA, changing them and causing cancer or other diseases. Free radicals may also attack molecules on the surfaces of cells, making them appear foreign to the body's immune system. The immune system then attacks the cell and destroys it, weakening the body. Draw a Lewis structure for each of the following free radicals implicated in this theory of aging.
(a) O_2^-
(b) O^-
(c) OH
(d) CH_3OO (unpaired electron on terminal oxygen)

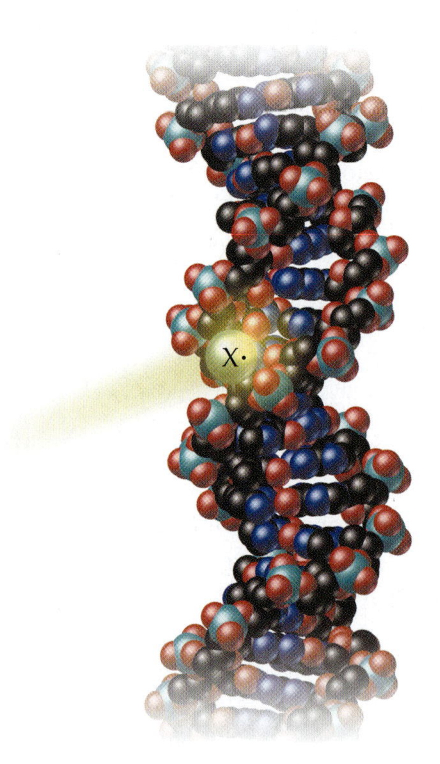

▲ Free radicals, molecules containing unpaired electrons (represented here as X·), may attack biological molecules such as the DNA molecule depicted here.

112. Free radicals (see Problem 111) are also important in many environmentally significant reactions. For example, photochemical smog, which forms as a result of the action of sunlight on air pollutants, is formed in part by the following two steps.

$$NO_2 \xrightarrow{\text{UV light}} NO + O$$

$$O + O_2 \longrightarrow O_3$$

The product of this reaction, ozone, is a pollutant in the lower atmosphere. Ozone is an eye and lung irritant and also accelerates the weathering of rubber products. Write Lewis structures for each of the reactants and products in the preceding reactions.

▲ Ozone damages rubber products.

113. The following are formulas of several molecules along with a space-filling model of their structure. Determine whether the structure is correct and if not, make a sketch of the correct structure.

(a) H_2Se

(b) CSe_2

(c) PCl_3

(d) CF_2Cl_2

Answers to Skillbuilder Exercises

Skillbuilder 10.1 ·Mg·

Skillbuilder 10.2

Na^+ [:B̤̈r:]$^-$

Skillbuilder 10.3 Mg_3N_2

Skillbuilder 10.4

:C≡O:

Skillbuilder 10.5

```
    Ö:
    ||
 H—C—H
```

Skillbuilder 10.6

[:C̤̈l—Ö̤:]$^-$

Skillbuilder 10.7

[:Ö=N̈—Ö̤:]$^-$ ⟷ [:Ö̤—N̈=Ö:]$^-$

Skillbuilder 10.8 bent

Skillbuilder 10.9 trigonal pyramidal

Skillbuilder 10.10 **(a)** pure covalent **(b)** ionic **(c)** polar covalent

Skillbuilder 10.11 CH_4 is nonpolar

Answers to Conceptual Checkpoints

10.1 (b) Aluminum must lose its 3 valence electrons to get an octet. Sulfur must gain 2 electrons to get an octet. Therefore, 2 Al atoms are required for every 3 S atoms.

10.2 (b) Both NH_3 and H_3O^+ have 1 lone electron pair.

10.3 (d) If there are no lone pairs on the central atom, all of its valence electrons are involved in bonds, so the molecular geometry must be the same as the electron geometry.

CHAPTER 11

Gases

"We live immersed at the bottom of a sea of elemental air."

Evangelista Torricelli (1608–1647)

11.1 Extralong Straws

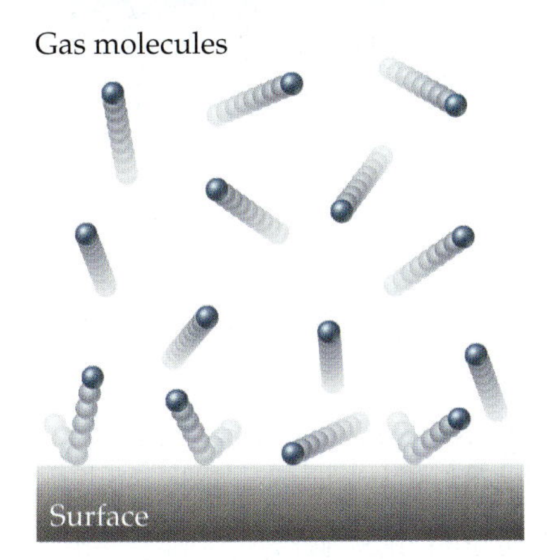

▲ **Figure 11.1 Gas pressure** Pressure is the force exerted by gas molecules as they collide with the surfaces around them.

◀ When you drink from a straw, you remove some of the molecules from inside the straw. This creates a pressure difference between the inside of the straw and the outside of the straw that results in the liquid being pushed up the straw. The pushing is done by molecules in the atmosphere—primarily nitrogen and oxygen—as shown here.

Like most kids, I grew up preferring fast-food restaurants to home cooking. My favorite stunt at the burger restaurant was drinking my orange soda from an extralong straw that I pieced together from several smaller straws. I would pinch the end of one straw and fit it into the end of another. By attaching several straws together, I could put my orange soda on the floor and drink it while standing on my chair (for some reason, my parents did not appreciate my scientific curiosity). I sometimes planned ahead and brought duct tape with me to form extratight seals between adjacent straws. My brother and I would have contests to make the longest working straw. Because I was older, I usually won.

Sometimes I wondered how long the straw could be if I made perfect seals between the straws. Could I drink my orange soda from a cup on the ground while I sat in my tree house? Could I drink it from the top of a ten-story building? It seemed to me that I could. I was wrong. Even if the extended straw had perfect seals and rigid walls and I sucked hard enough to create a perfect vacuum (the absence of all air), I could never suck my orange soda from a straw longer than about 10.3 m (34 ft). Why?

Straws work because sucking creates a pressure difference between the inside of the straw and the outside. We define pressure more thoroughly later. For now think of pressure as the force exerted per unit area by gas molecules as they collide with the surfaces around them (◀ Figure 11.1). Just as a ball exerts a force when it bounces against a wall, so a molecule exerts a force when it collides with a surface. The result of many of these collisions is pressure. The total amount of pressure exerted by a gas sample depends on several factors, including the concentration of gas molecules in the sample. On Earth at sea level, the gas molecules in our atmosphere exert an average pressure of 101,325 N/m^2 or, in English units, 14.7 lb/in^2.

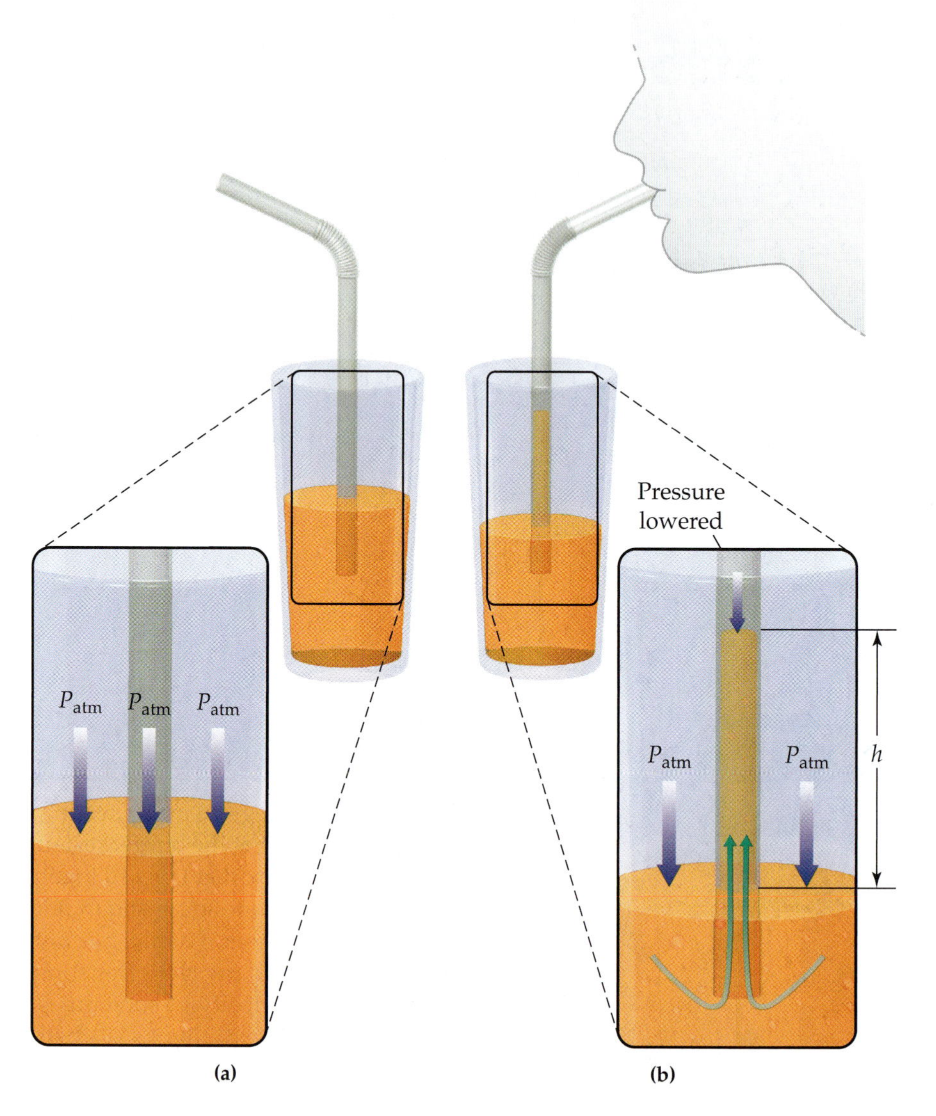

▲ **Figure 11.2 Sipping soda** **(a)** When a straw is put into a glass of orange soda, the pressure inside and outside the straw are the same, so the liquid levels inside and outside the straw are the same. **(b)** When a person sucks on the straw, the pressure inside the straw is lowered. The greater pressure on the surface of the liquid outside the straw pushes the liquid up the straw.

When you put a straw in orange soda, the pressure inside and outside the straw are the same, so the soda does not rise within the straw (▲ Figure 11.2a). When you suck on the straw, however, you remove some of the air molecules, lowering the number of collisions that occur inside the straw and therefore lowering the pressure (▲ Figure 11.2b). However, the pressure outside the straw remains the same. The result is a pressure differential—the pressure outside the straw becomes greater than the pressure inside of the straw. This greater external pressure pushes the liquid up the straw and into your mouth.

How high can this greater external pressure push the liquid up the straw? If you formed a perfect vacuum within the straw, the pressure out-

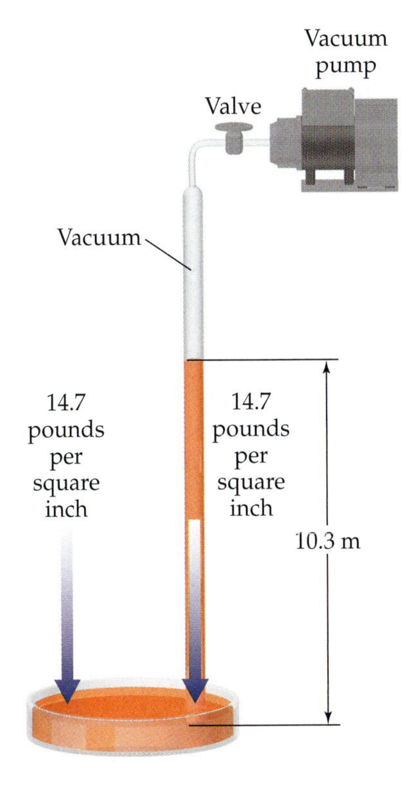

▶ **Figure 11.3 Atmospheric pressure** Even if you formed a perfect vacuum, atmospheric pressure could only push orange soda to a total height of about 10 m. This is because a column of water 10.3 m high exerts the same pressure (14.7 lb/in.^{2}) as the gas molecules in our atmosphere.

side of the straw at sea level would be enough to push the orange soda (which is mostly water) to a total height of about 10.3 m (▲ Figure 11.3). This is because a 10.3-m column of water exerts the same pressure—101,325 N/m^2 or 14.7 lb/in.^{2}—as the gas molecules in our atmosphere. In other words, the orange soda would rise up the straw until the pressure exerted by its weight equaled the pressure exerted by the molecules in our atmosphere.

11.2 Kinetic Molecular Theory: A Model for Gases

In past chapters, we have learned the importance of models or theories in understanding nature. A simple model for gases is called the **kinetic molecular theory.** In this chapter, we examine this model and how it predicts the correct behavior for most gases under many conditions. Like other models, the kinetic molecular theory is not perfect and breaks down under certain conditions. In this book, however, we focus on conditions where it works well.

Kinetic molecular theory has the following assumptions (▶ Figure 11.4):

1. A gas is a collection of particles (molecules or atoms) in constant, straight-line motion.
2. The particles do not attract or repel each other—they do not interact. The particles collide with each other and with the surfaces around them, but they bounce back from these collisions like perfect billiard balls.
3. There is a lot of space between the particles compared with the size of the particles themselves.
4. The average kinetic energy—energy due to motion—of the particles is proportional to the temperature of the gas in kelvins. This means that the higher the temperature, the more energy the particles have and the faster they move.

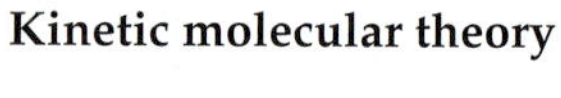

1. Collection of particles in constant motion
2. No attractions or repulsions between particles; collisions like billiard ball collisions
3. A lot of space between the particles compared to the size of the particles themselves
4. The speed of the particles increases with increasing temperature

▶ **Figure 11.4 Simplified representation of an ideal gas** In reality, the spaces between the gas molecules would be larger in relation to the size of the molecules than shown here.

See Table 3.1.

Kinetic molecular theory is consistent with, and indeed predicts, the properties of gases. As we learned in Section 3.3, gases:

- are compressible;
- assume the shape and volume of their container; and
- have low densities in comparison with liquids and solids.

Gases are compressible because the atoms or molecules that compose them have a lot of space between them. By applying external pressure to a gas sample, the atoms or molecules are forced closer together, compressing the gas. Liquids and solids, in contrast, are not compressible because the atoms or molecules composing them are already in close contact—they cannot be forced any closer together. The compressibility of a gas can be seen, for example, by pushing a piston into a cylinder containing a gas. The piston goes down (▼ Figure 11.5) in response to the external pressure. If the cylinder were filled with a liquid or a solid, however, the piston would not move when pushed (▼ Figure 11.6).

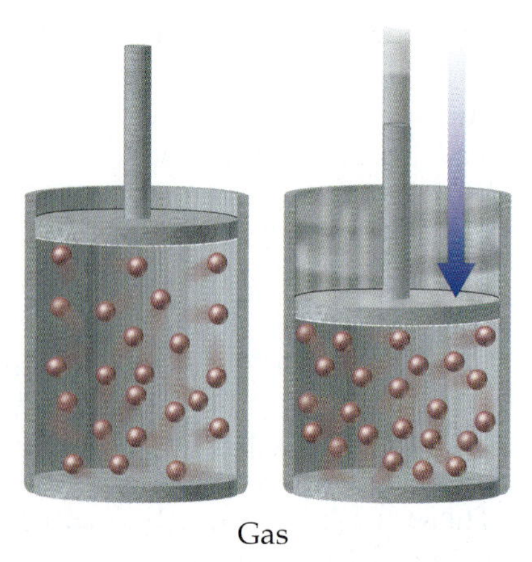

▲ **Figure 11.5 Compressibility of gases** Gases are compressible because there is so much empty space between gas particles.

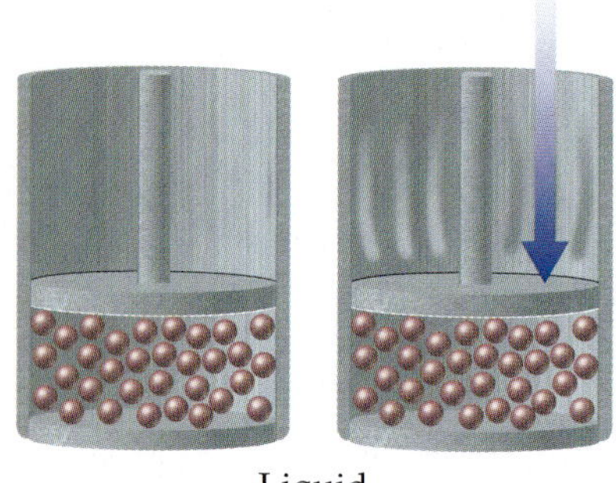

▲ **Figure 11.6 Incompressibility of liquids** Liquids are not compressible because there is so little space between the liquid particles.

▲ **Figure 11.7** **A gas assumes the shape of its container**

Gases assume the shape and volume of their container because gaseous atoms or molecules are in constant, straight-line motion. In contrast to a solid or liquid, whose atoms or molecules interact with one another, the atoms or molecules in a gas do not interact with one another (or more precisely, their interactions are negligible). They simply move in straight lines, colliding with each other and with the walls of their container. As a result, they fill the entire container, collectively assuming its shape (◀ Figure 11.7).

Gases have a low density in comparison with solids and liquids because there is so much empty space between the atoms or molecules in a gas. For example, if the water in a 350-mL (12-oz) can of soda were converted to steam (gaseous water), the steam would occupy a volume of 595 L (the equivalent of 1700 soda cans).

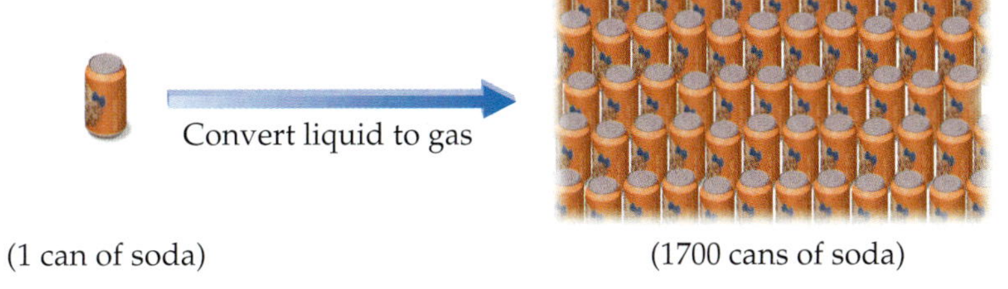

▲ If all of the water in a 12-oz (350-mL) can of orange soda were converted to gaseous steam (at 1 atm pressure and 100 °C), the steam would occupy a volume equal to 1700 soda cans.

11.3 Pressure: The Result of Constant Molecular Collisions

A prediction of kinetic molecular theory—which we already encountered in explaining how straws work—is pressure. **Pressure** is the result of the constant collisions between the atoms or molecules in a gas and the surfaces around them. Because of pressure, we can drink from straws, inflate basketballs, and move air into and out of our lungs. Variation in pressure in Earth's atmosphere creates wind, and changes in pressure help predict weather. Pressure is all around us and even inside us. The pressure exerted by a gas sample is defined as the force per unit area that results from the collisions of gas particles with surrounding surfaces.

$$\text{Pressure} = \frac{\text{Force}}{\text{Area}}$$

The pressure exerted by a gas depends on several factors. One of these factors is the number of gas particles in a given volume (◀ Figure 11.8). The fewer the gas particles, the lower the pressure. The pressure decreases, for example, with increasing altitude. As we climb a mountain or ascend in an airplane, there are fewer molecules per unit volume in air and the pressure consequently drops. For this reason, most airplane cabins are artificially pressurized (see the *Everyday Chemistry* box on page 348.)

You can often feel the effect of a drop in pressure as a pain in your ears. This pain is caused by air-containing cavities within your ear (▶ Figure 11.9). When you climb a mountain, for example, the external pressure (that pressure that surrounds you) drops while the pressure within your ear cavities (the internal pressure) remains the same. This creates an imbalance—the greater internal pressure causes your eardrum to bulge outward, causing pain. With time and a yawn or two, the excess air within your ears' cavities escapes, equalizing the internal and external pressure and relieving the pain.

▲ **Figure 11.8** **Pressure** Since pressure is a result of collisions between gas particles and the surfaces around them, the amount of pressure increases when the number of particles in a given volume increases.

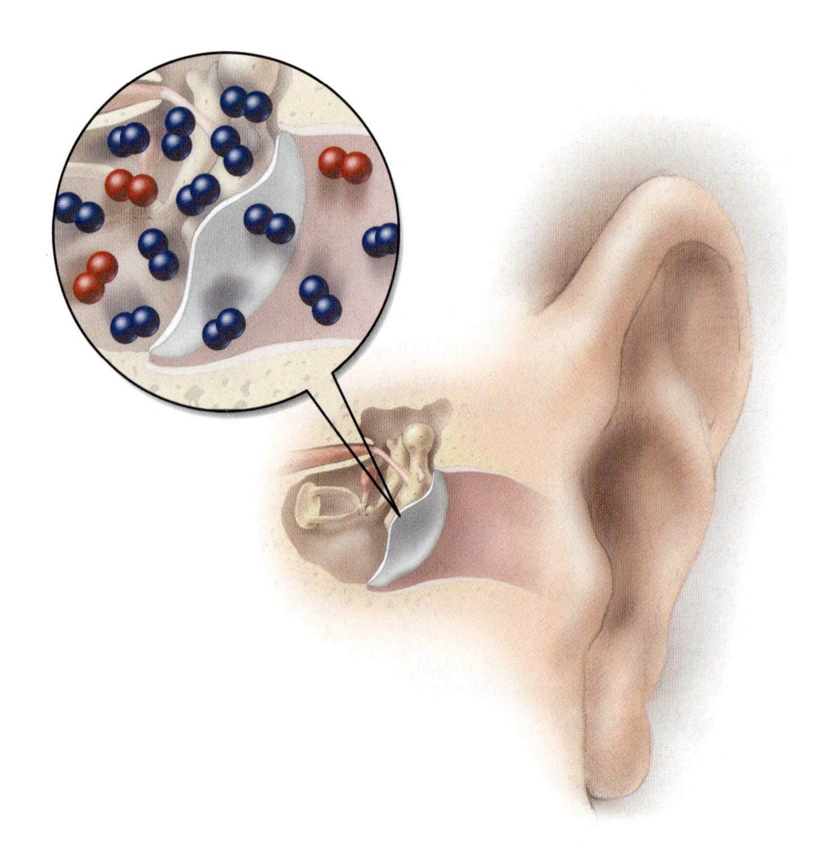

▶ **Figure 11.9 Pressure imbalance** The pain you feel in your ears upon climbing a mountain or ascending in an airplane is caused by an imbalance of pressure between the cavities inside your ear and the outside air.

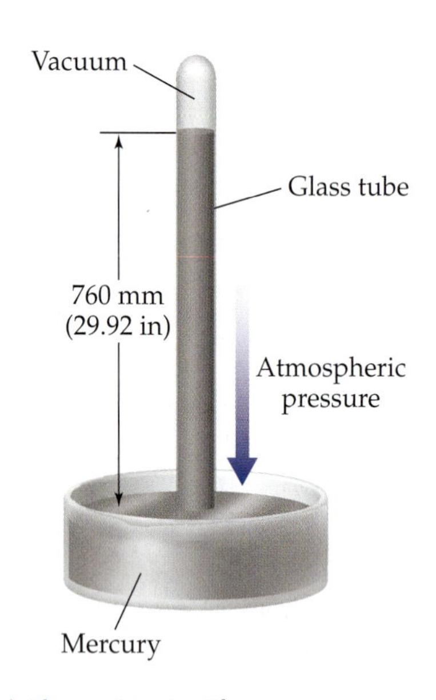

▲ **Figure 11.10 The mercury barometer** Average atmospheric pressure at sea level pushes a column of mercury to a height of 760 mm (29.92 in.). **Question:** What would happen to the height of the mercury column if the external pressure became lower? Higher?

Since mercury is 13.5 times as dense as water, it is pushed up 1/13.5 times as high as water by atmospheric pressure.

Pressure Units

The simplest unit of pressure is the **atmosphere (atm),** the average pressure at sea level.

$$1\text{ atm} = \text{Average pressure of air at sea level}$$

In this unit, the pressure inside a fully inflated mountain bike tire is about 6 atm, and the pressure on top of Mt. Everest is about 0.311 atm.

The SI unit of pressure is the **pascal (Pa)**, defined as 1 newton (N) per square meter.

$$1\text{ Pa} = 1\text{ N/m}^2$$

The pascal is a much smaller unit of pressure, with 1 atm being equal to 101,325 Pa.

$$1\text{ atm} = 101{,}325\text{ Pa}$$

A third unit of pressure, **millimeter of mercury (mm Hg)**, originates from how pressure is measured with a barometer (◀ Figure 11.10). A barometer is an evacuated glass tube whose tip is submerged in a pool of mercury. As we learned in Section 11.1, a liquid is pushed up an evacuated tube by atmospheric gas pressure on the liquid's surface. We learned that water is pushed up to a height of 10.3 m by the average pressure at sea level. Mercury, however, with its higher density, is pushed up to a height of only 0.760 m, or 760 mm, by the average pressure at sea level. This shorter length—0.760 m instead of 10.3 m—makes a column of mercury a convenient way to measure pressure.

In a barometer, the mercury column rises or falls with changes in pressure. If the pressure increases, the level of mercury within the column rises. If the pressure decreases, the level of mercury within the column falls. Since

TABLE 11.1 Common Units of Pressure

Unit	Average Air Pressure at Sea Level
pascal (Pa)	101,325
atmosphere (atm)	1
millimeter of mercury (mm Hg)	760
torr (torr)	760
pounds per square inch (psi)	14.7
inches of mercury (in. Hg)	29.92

1 atm of pressure pushes a column of mercury to a height of 760 mm, 1 atm and 760 mm Hg are equal.

$$1 \text{ atm} = 760 \text{ mm Hg}$$

A millimeter of mercury is also called a **torr** after Italian physicist Evangelista Torricelli (1608–1647), who invented the barometer.

$$1 \text{ mm Hg} = 1 \text{ torr}$$

Inches of mercury is still a widely used unit in weather reports. You have probably heard a weather forecaster say, "The barometer is 30.07 and rising," meaning that the atmospheric pressure is currently 30.07 in. Hg.

Other common units of pressure include inches of mercury (in. Hg) and **pounds per square inch (psi)**.

$$1 \text{ atm} = 14.7 \text{ psi} \qquad 1 \text{ atm} = 29.92 \text{ in. Hg}$$

These units are all summarized in Table 11.1.

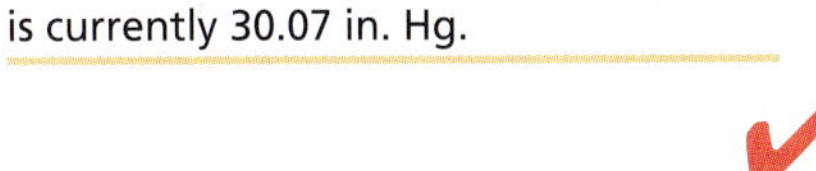

CONCEPTUAL CHECKPOINT 11.1

A liquid that is about twice as dense as water is used in a barometer. With this barometer, normal atmospheric pressure would be about:

(a) 0.38 m **(b)** 1.52 m **(c)** 5.15 m **(d)** 20.6 m

Pressure Unit Conversion

See Section 2.6.

We convert one pressure unit to another in the same way that we converted between other units in Chapter 2. For example, suppose we want to convert 0.311 atm (the approximate average pressure at the top of Mount Everest) to millimeters of mercury. We set up the problem in the normal way.

Given: 0.311 atm

Find: mm Hg

Conversion Factor:

$$1 \text{ atm} = 760 \text{ mm Hg (from Table 11.1)}$$

Solution Map: The solution map shows how to convert from atm to mm Hg.

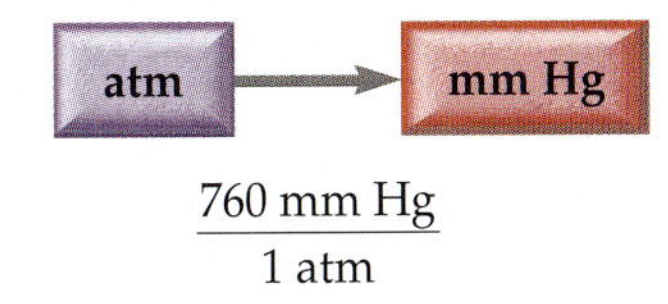

$$\frac{760 \text{ mm Hg}}{1 \text{ atm}}$$

Solution: The solution begins with the given value (0.311 atm) and converts it to mm Hg.

$$0.311 \text{ atm} \times \frac{760 \text{ mm Hg}}{1 \text{ atm}} = 236 \text{ mm Hg}$$

EVERYDAY Chemistry

Airplane Cabin Pressurization

Most commercial airplanes fly at elevations between 25,000 and 40,000 ft. At these elevations, atmospheric pressure is below 0.50 atm, much less than the 1.0 atm of pressure to which our bodies are accustomed. The physiological effects of these lowered pressures—and the correspondingly lowered oxygen levels—include dizziness, headache, shortness of breath, and even unconsciousness. Consequently, commercial airplanes pressurize the air in their cabins. If, for some reason, an airplane cabin should lose its pressurization, passengers are directed to breathe oxygen through an oxygen mask.

▲ Commercial airplane cabins must be pressurized to a pressure greater than the equivalent atmospheric pressure at an elevation of 8,000 ft.

Cabin air pressurization is accomplished as part of the cabin's overall air circulation system. As air flows into the plane's jet engines, the large turbines at the front of the engines compress it. Most of this compressed (or pressurized) air exits out the back of the engines, creating the thrust that drives the plane forward. However, some of the pressurized air is directed into the cabin, where it is cooled and mixed with existing cabin air. This air is then circulated through the cabin through the overhead vents. The air leaves the cabin through ducts that direct it into the lower portion of the airplane. About half of this exiting air is mixed with incoming, pressurized air to circulate again. The other half is vented out of the plane through an outflow valve. This valve is adjusted to maintain the desired cabin pressure. Federal regulations require that cabin pressures in commercial airliners be greater than the equivalent of outside air pressure at 8,000 ft.

CAN YOU ANSWER THIS? *Atmospheric pressure at elevations of 8,000 ft average about 0.72 atm. Convert this pressure to millimeters of mercury, inches of mercury, and pounds per square inch. Would a cabin pressurized at 500 mm Hg meet federal standards?*

EXAMPLE 11.1 Converting between Pressure Units

A high-performance road bicycle tire is inflated to a *total* pressure of 125 psi. What is this pressure in millimeters of mercury?

You are given a pressure in psi and asked to convert it to mm Hg. Find the required conversion factors in Table 11.1.

Given: 125 psi

Find: mm Hg

Conversion Factors:

$$1 \text{ atm} = 14.7 \text{ psi}$$
$$760 \text{ mm Hg} = 1 \text{ atm}$$

Begin the solution map with the given units of psi. Use the conversion factors to convert first to atm and then to mm Hg.

Solution Map:

psi → atm → mm Hg

$$\frac{1 \text{ atm}}{14.7 \text{ psi}} \qquad \frac{760 \text{ mm Hg}}{1 \text{ atm}}$$

Follow the solution map to solve the problem.

Solution:

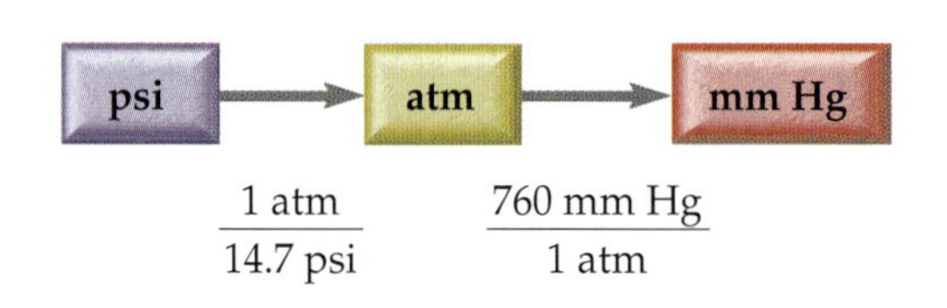

$$125 \text{ psi} \times \frac{1 \text{ atm}}{14.7 \text{ psi}} \times \frac{760 \text{ mm Hg}}{1 \text{ atm}} = 6.46 \times 10^3 \text{ mm Hg}$$

SKILLBUILDER 11.1 **Converting between Pressure Units**

Convert a pressure of 173 in. Hg into pounds per square inch.

SKILLBUILDER PLUS

Convert a pressure of 23.8 in. Hg into kilopascals.

11.4 Boyle's Law: Pressure and Volume

The pressure of a gas sample changes when its volume changes. If the temperature and the amount of gas are constant, the pressure of a gas sample *increases* for a *decrease* in volume and *decreases* for an *increase* in volume. A simple hand pump, for example, works on this principle. A hand pump is basically a cylinder equipped with a moveable piston (▼ Figure 11.11). The volume in the cylinder increases when you pull the handle up (the upstroke) and decreases when you push the handle down (the downstroke). On the upstroke, the *increasing* volume causes a *decrease* in the internal pressure (the pressure within the pump's cylinder). This, in turn, draws air into the pump's cylinder through a one-way valve. On the downstroke, the *decreasing* volume causes an *increase* in the internal pressure. This increase forces the air out of the pump, through a different one-way valve, and into the tire or whatever else is being inflated.

The relationships between gas properties—such as pressure and volume—are described by gas laws. These laws show how a change in one of these properties affects one or more of the others. The relationship between volume and pressure was discovered by Robert Boyle (1627—1691) and is called **Boyle's law**.

Boyle's law assumes constant temperature and a constant number of gas particles.

Boyle's law: The volume of a gas and its pressure are inversely proportional.

$$V \propto \frac{1}{P} (\propto \text{ means "proportional to"})$$

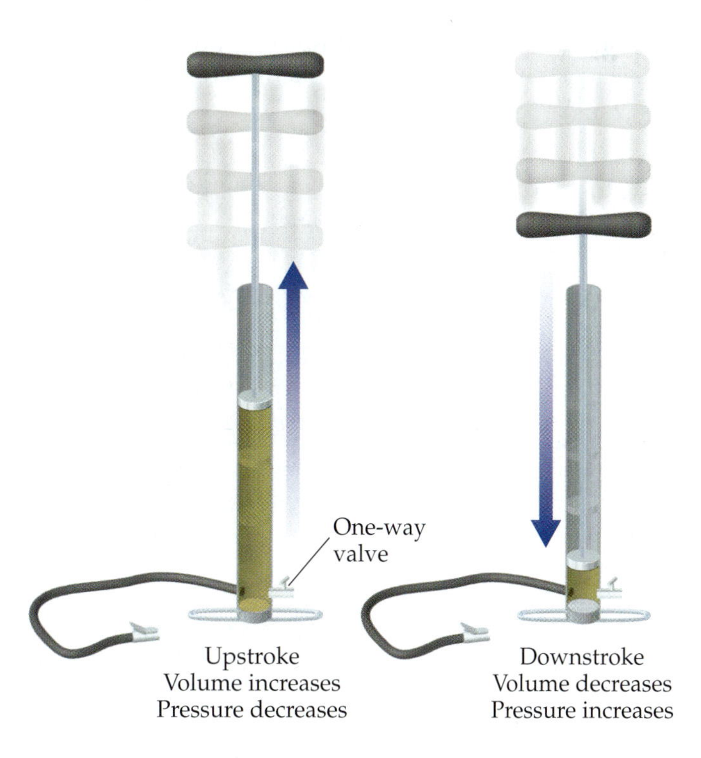

▲ **Figure 11.11** **Operation of a hand pump**

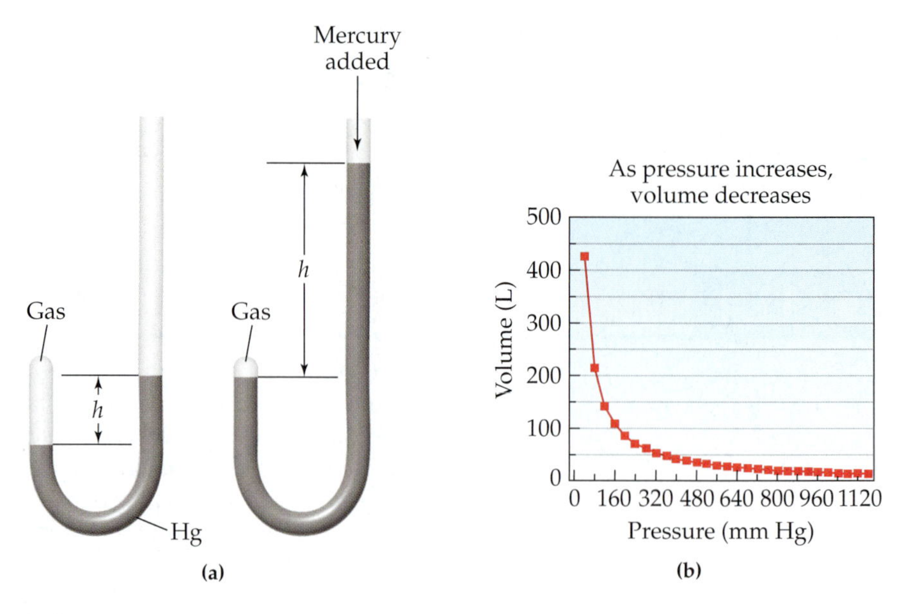

▲ **Figure 11.12 Volume versus pressure** **(a)** A J-tube, such as the one shown here, can be used to measure the volume of a gas at different pressures. Simply adding mercury to the J-tube causes the pressure on the gas sample to increase and its volume to decrease. **(b)** A plot of the volume of a gas as a function of pressure.

If two quantities are inversely proportional, then increasing one decreases the other (▲ Figure 11.12). As we saw for our hand pump, when the volume of a sample of gas is decreased its pressure increases and vice versa. This follows from kinetic molecular theory. If the volume of a gas sample is decreased, the same number of gas particles is crowded into a smaller volume, causing more collisions with the walls of the container and therefore increasing the pressure (▼ Figure 11.13).

Scuba divers learn about Boyle's law during certification because it explains why ascending quickly toward the surface is dangerous. For every 10 m of depth that a diver descends in water, he experiences an additional 1 atm of pressure due to the weight of the water above him (▶ Figure 11.14). The pressure regulator used in scuba diving delivers air

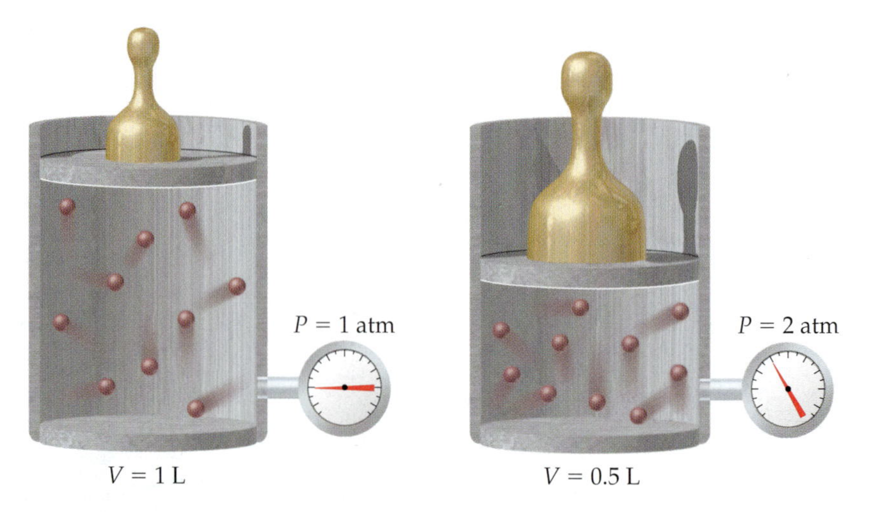

▲ **Figure 11.13 Volume versus pressure: a molecular view** As the volume of a sample of gas is decreased, the number of collisions between the gas molecules and each square meter of the container increases. This raises the pressure exerted by the gas.

▶ **Figure 11.14 Pressure at depth** For every 10 m of depth, a diver experiences an additional 1 atm of pressure due to the weight of the water surrounding him. At 20 m, the diver experiences a total pressure of 3 atm (1 atm from atmospheric pressure plus an additional 2 atm from the weight of the water).

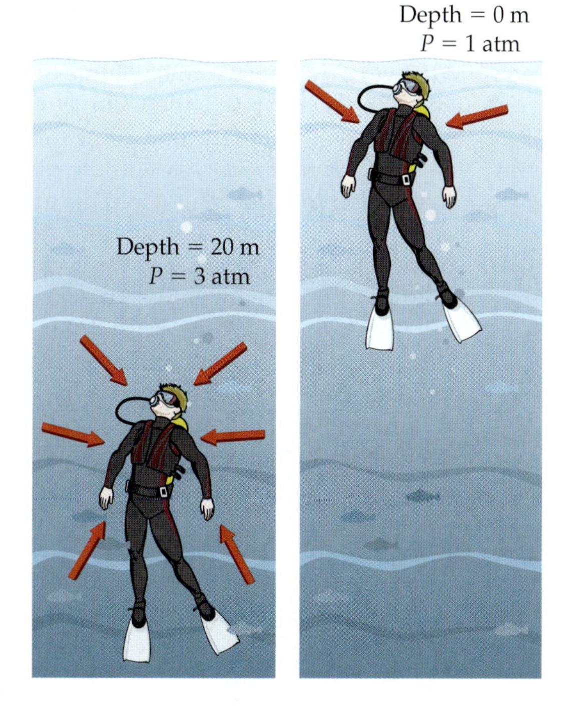

at a pressure that matches the external pressure; otherwise the diver could not inhale the air (see the *Everyday Chemistry* box on page 354). For example, when a diver is at 20 m of depth, the regulator delivers air at a pressure of 3 atm to match the 3 atm of pressure around the diver—1 atm due to normal atmospheric pressure and 2 additional atmospheres due to the weight of the water at 20 m (▼ Figure 11.15).

▶ **Figure 11.15 The dangers of decompression** **(a)** A diver at 20 m experiences an external pressure of 3 atm and breathes air pressurized at 3 atm. **(b)** If the diver shoots toward the surface with lungs full of 3 atm air, his lungs will expand as the external pressure drops to 1 atm.

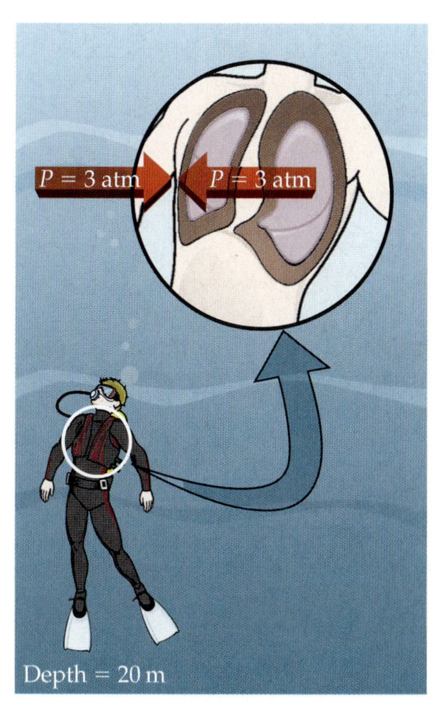

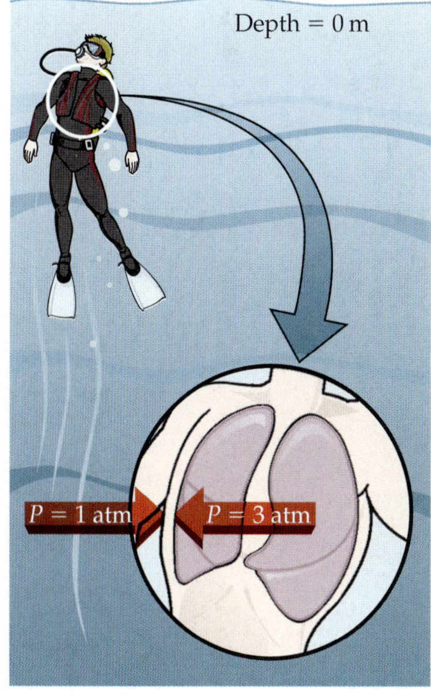

Suppose that a diver inhaled a lungful of 3-atm air and swam quickly to the surface (where the pressure drops to 1 atm) while holding his breath. What would happen to the volume of air in his lungs? Since the pressure decreases by a factor of 3, the volume of the air in his lungs would increase by a factor of 3, severely damaging his lungs and possibly killing him. Of course, the volume increase in the diver's lungs would be so great that the diver would not be able to hold his breath all the way to the surface—the air would force itself out of his mouth. Nonetheless, the most important rule in diving is *never hold your breath.* Divers must ascend slowly and breathe continuously, allowing the regulator to bring the air pressure in their lungs back to 1 atm by the time they reach the surface.

Boyle's law can be used to compute the volume of a gas following a pressure change or the pressure of a gas following a volume change *as long as the temperature and the amount of gas remain constant.* For these calculations, we must write Boyle's law in a slightly different way.

$$\text{Since } V \propto \frac{1}{P}, \text{ then, } V = \frac{\text{Constant}}{P}$$

If we multiply both sides by P, we get:

$$PV = \text{Constant}$$

This relationship is true because if the pressure increases, the volume decreases, but the product $P \times V$ is always equal to the same constant. For two different sets of conditions, we can say that

$$P_1V_1 = \text{Constant} = P_2V_2, \text{ or}$$

$$P_1V_1 = P_2V_2$$

where P_1 and V_1 are the initial pressure and volume of the gas, and P_2 and V_2 are the final volume and pressure. For example, suppose we want to calculate the pressure of a gas that was initially at 765 mm Hg and 1.78 L and later compressed to 1.25 L. We set up the problem as follows:

Given:

$$P_1 = 765 \text{ mm Hg}$$
$$V_1 = 1.78 \text{ L}$$
$$V_2 = 1.25 \text{ L}$$

Based on Boyle's law, and before doing any calculations, do you expect P_2 to be greater than or less than P_1?

Find: P_2

Equation: Since this problem involves an equation, we write the appropriate equation here.

$$P_1V_1 = P_2V_2$$

Solution Map:

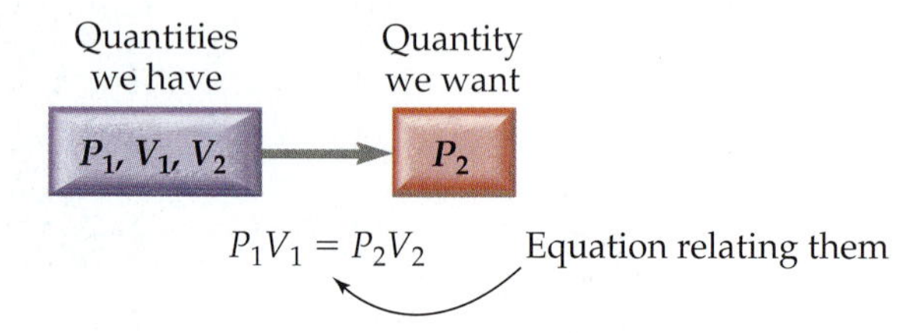

The solution map shows how the equation takes us from the given quantities (what we have) to the find quantity (what we want).

Solution: We then solve the equation for the quantity we are trying to find (P_2).

$$P_1V_1 = P_2V_2$$

$$P_2 = \frac{P_1V_1}{V_2}$$

Lastly, we substitute the numerical values into the equation and compute the answer.

$$P_2 = \frac{P_1V_1}{V_2}$$

$$= \frac{(765 \text{ mm Hg})(1.78 \cancel{\text{L}})}{1.25 \cancel{\text{L}}}$$

$$= 1.09 \times 10^3 \text{ mm Hg}$$

EXAMPLE 11.2 Boyle's Law

A cylinder equipped with a moveable piston has an applied pressure of 4.0 atm and a volume of 6.0 L. What is the volume of the cylinder if the applied pressure is decreased to 1.0 atm?

You are given an initial pressure, an initial volume, and final pressure. You are asked to find the final volume.

Given:

$$P_1 = 4.0 \text{ atm}$$

$$V_1 = 6.0 \text{ L}$$

$$P_2 = 1.0 \text{ atm}$$

Find: V_2

This problem requires the use of Boyle's law.

Equation: $P_1V_1 = P_2V_2$

Draw a solution map beginning with the given quantities. Boyle's law shows the relationship necessary to get to the find quantity.

Solution Map:

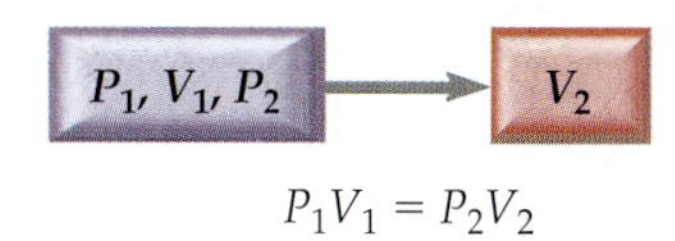

$$P_1V_1 = P_2V_2$$

Solve the equation for the quantity you are trying to find (V_2), and then substitute the numerical quantities into the equation to compute the answer.

Solution:

$$P_1V_1 = P_2V_2$$

$$V_2 = \frac{V_1P_1}{P_2}$$

$$= \frac{(6.0 \text{ L})(4.0 \cancel{\text{atm}})}{1.0 \cancel{\text{atm}}}$$

$$= 24 \text{ L}$$

SKILLBUILDER 11.2 Boyle's Law

A snorkeler takes a syringe filled with 16 mL of air from the surface, where the pressure is 1.0 atm, to an unknown depth. The volume of the air in the syringe at this depth is 7.5 mL. What is the pressure at this depth? If the pressure increases by an additional 1 atm for every 10 m of depth, how deep is the snorkeler?

EVERYDAY Chemistry

Extralong Snorkels

Several episodes of *The Flintstones* featured Fred Flintstone and Barney Rubble snorkeling. Their snorkels, however, were not the modern kind, but long reeds that stretched from the surface of the water down to many meters of depth. Fred and Barney swam around in deep water while breathing air provided to them by these extralong snorkels. Would this work? Why do people bother with scuba diving equipment if they could simply use 10-m snorkels the way that Fred and Barney did?

When we breathe, we expand the volume of our lungs, lowering the pressure within them (Boyle's law). Air from outside our lungs then flows into them. Extralong snorkels, such as those used by Fred and Barney, do not work because of the pressure caused by water at depth. A diver at 10 m experiences a pressure of 2 atm that compresses the air in his lungs to a pressure of 2 atm. If the diver had a snorkel that went to the surface—where the air pressure is 1 atm—air would flow out of his lungs, not into them. It would be impossible to breathe.

CAN YOU ANSWER THIS? *Suppose a diver takes a balloon with a volume of 2.5 L from the surface, where the pressure is 1.0 atm, to a depth of 20 m, where the pressure is 3.0 atm. What would happen to the volume of the balloon? What if the end of the balloon was on a long tube that went to the surface and was attached to another balloon, as shown in the drawing? Which way would air flow as the diver descends?*

▲ Fred and Barney used reeds to breathe air from the surface, even when they were at depth. This would not work because the pressure at depth would push air out of their lungs, preventing them from breathing.

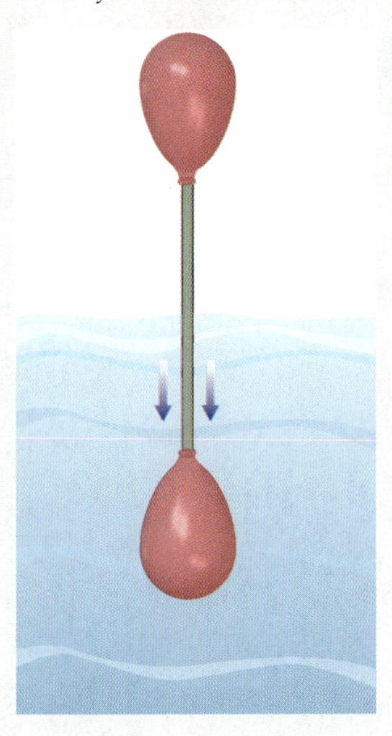

▲ If one end of a long tube with balloons tied on both ends were submerged in water, in which direction would air flow?

11.5 Charles's Law: Volume and Temperature

Have you ever noticed that hot air rises? You may have walked upstairs in your house and noticed it getting warmer. Or you may have witnessed a hot-air balloon take flight. The air that fills a hot-air balloon is warmed with a burner, which then causes the balloon to rise in the cooler air around it. Why does hot air rise? Hot air rises because the volume of a gas sample at constant pressure increases with increasing temperature. As long as the amount of gas (and therefore its mass) remains constant, warming it decreases its density because density is mass divided by volume. A lower-density gas floats in a higher-density gas just as wood floats in water.

Recall from Section 2.9 that Density = Mass/Volume. If the volume increases and the mass remains constant, the density must decrease.

▶ Heating the air in a balloon makes it expand (Charles's law). As the volume occupied by the hot air increases, its density decreases, allowing the balloon to float in the cooler, denser air that surrounds it.

Suppose you keep the pressure of a gas sample constant and measure its volume at a number of different temperatures. The results of a number of such measurements are shown in ▼ Figure 11.16. From the plot we can see the relationship between volume and temperature: The volume of a gas increases with increasing temperature. Looking at the plot, however, reveals more; temperature and volume are *linearly related*. If two variables are linearly related, then plotting one against the other produces a straight line.

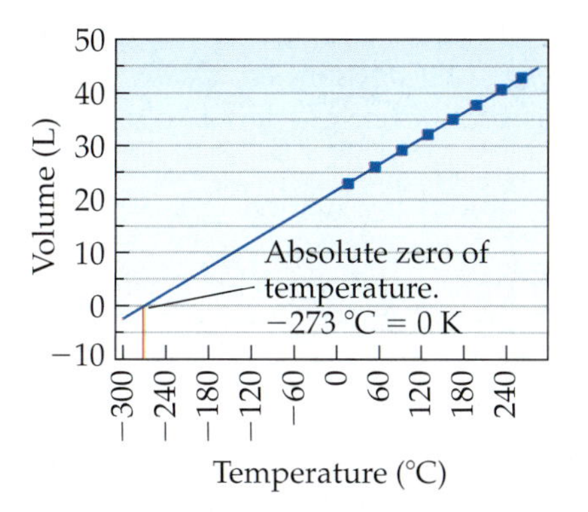

▶ **Figure 11.16 Volume versus temperature** The volume of a gas increases linearly with increasing temperature. **Question:** How does this graph demonstrate that −273 °C is the coldest possible temperature?

Another interesting feature arises if we extend the line on our plot backward from the lowest measured point—a process called *extrapolation*. Our extrapolated line shows that the gas should have a zero volume at −273 °C. Recall from Chapter 3 that −273 °C corresponds to 0 K, the coldest possible temperature. Our extrapolated line shows that below −273 °C, our gas would have a negative volume, which is physically impossible. For this reason, we refer to 0 K as **absolute zero**—colder temperatures do not exist.

The extrapolated line could not be measured experimentally because all gases would condense into liquids before −273 °C is reached.

Section 3.9 summarizes the three different temperature scales.

The first person to carefully quantify the relationship between the volume of a gas and its temperature was J. A. C. Charles (1746–1823), a French mathematician and physicist. Charles was interested in gases and was among the first people to ascend in a hydrogen-filled balloon. The law he formulated is called **Charles's law**.

Charles's law assumes constant pressure and a constant amount of gas.

Charles's law: The volume (V) of a gas and its Kelvin temperature (T) are directly proportional.

$$V \propto T$$

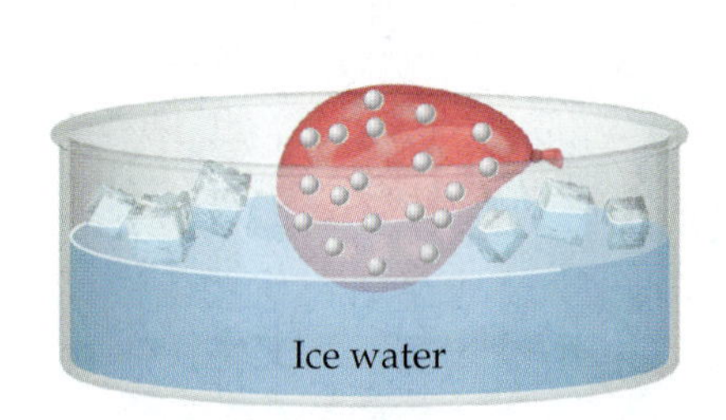

▲ **Figure 11.17** **Volume versus temperature: a molecular view** If a balloon is moved from an ice-water bath into a boiling-water bath, the gas molecules inside it move faster due to the increased temperature. If the external pressure remains constant, the molecules will expand the balloon and collectively occupy a larger volume.

If two variables are directly proportional, then increasing one by some factor increases the other by the same factor. For example, when the temperature of a gas sample (in kelvins) is doubled, its volume doubles; when the temperature is tripled, its volume triples; and so on. This also follows from kinetic molecular theory. If the temperature of a gas sample is increased, the gas particles move faster and, if the pressure is to remain constant, the volume must increase (▲ Figure 11.17).

You can experience Charles's law directly by holding a partially inflated balloon over a warm toaster. As the air in the balloon warms, you can feel the balloon expanding. Alternatively, you can a put an inflated balloon in the freezer or take it outside on a very cold day (below freezing) and see that it becomes smaller as it cools.

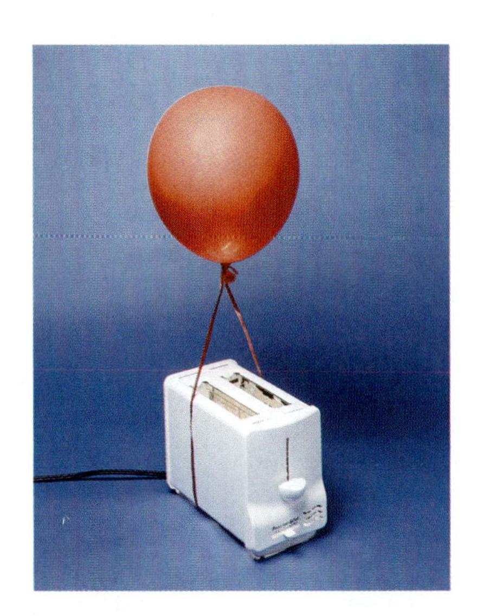

▲ If you hold a partially inflated balloon over a warm toaster, the balloon will expand as the air within the balloon warms.

Charles's law can be used to compute the volume of a gas following a temperature change or the temperature of a gas following a volume change *as long as the pressure and the amount of gas are constant*. For these calculations, we express Charles's law in a different way as follows:

Since $V \propto T$, then, $V = \text{Constant} \times T$

If we divide both sides by T, we get:

$V/T = \text{Constant}$

If the temperature increases, the volume increases in direct proportion so that the quotient, V/T, is always equal to the same constant. So, for two different measurements, we can say that

$V_1/T_1 = \text{Constant} = V_2/T_2$, or

$$\frac{V_1}{T_1} = \frac{V_2}{T_2}$$

where V_1 and T_1 are the initial volume and temperature of the gas and V_2 and T_2 are the final volume and temperature. *All temperatures must be expressed in kelvins.*

For example, suppose we want to calculate the volume of a gas that was initially at 298 K and 2.37 L and later heated to 354 K with no change in pressure. We set up the problem in the normal way.

Given: $T_1 = 298\text{ K}$
$V_1 = 2.37\text{ L}$
$T_2 = 354\text{ K}$

Based on Charles's law, and before doing any calculations, do you expect V_2 to be greater than or less than V_1?

Find: V_2

Equation: $\frac{V_1}{T_1} = \frac{V_2}{T_2}$

Solution Map: The solution map shows how the equation takes us from the given quantities to the unknown quantity.

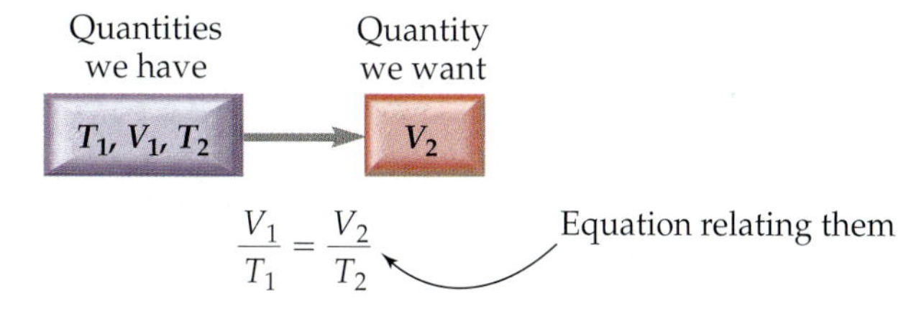

Solution: We then solve the equation for the quantity we are trying to find (V_2).

$$\frac{V_1}{T_1} = \frac{V_2}{T_2}$$

$$V_2 = \frac{V_1}{T_1}T_2$$

Lastly, we substitute the numerical values into the equation and compute the answer.

$$\begin{aligned} V_2 &= \frac{V_1}{T_1}T_2 \\ &= \frac{2.37 \text{ L}}{298 \cancel{\text{K}}} 354 \cancel{\text{K}} \\ &= 2.82 \text{ L} \end{aligned}$$

EXAMPLE 11.3 Charles's Law

A sample of gas has a volume of 2.80 L at an unknown temperature. When the sample is submerged in ice water at $t = 0$ °C, its volume decreases to 2.57 L. What was its initial temperature (in kelvins and in Celsius)? Assume a constant pressure. (To distinguish between the two temperature scales, use t for temperature in °C and T for temperature in K.)

You are given an initial volume, a final volume, and a final temperature. You are asked to find the initial temperature in both kelvins (T_1) and degrees Celcius (t_1).

Given:

$$V_1 = 2.80 \text{ L}$$
$$V_2 = 2.57$$
$$t_2 = 0 \text{ °C}$$

Find: T_1 and t_1

This problem requires the use of Charles's law.

Equation:

$$\frac{V_1}{T_1} = \frac{V_2}{T_2}$$

Draw a solution map beginning with the given quantities. Charles's law shows the relationship necessary to get to the find quantity.

Solution Map:

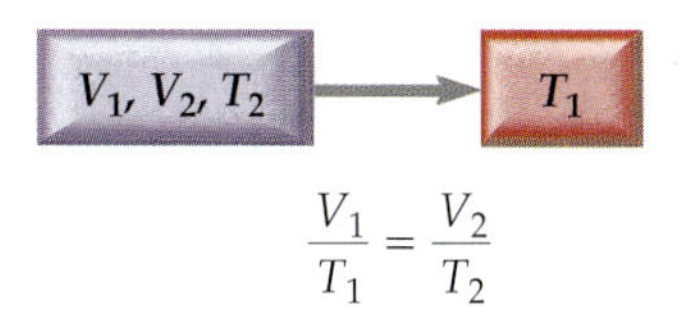

Solve the equation for the quantity you are trying to find (T_1).

Solution:

$$\frac{V_1}{T_1} = \frac{V_2}{T_2}$$

$$T_1 = \frac{V_1}{V_2} T_2$$

Before you substitute in the numerical values, you must convert the temperature to kelvins. *Remember, gas law problems must always be worked using Kelvin temperatures.* Once you have converted the temperature to kelvins, substitute into the equation to find T_1. Convert the temperature to degrees Celsius to find t_1.

$$T_2 = 0 + 273 = 273 \text{ K}$$

$$T_1 = \frac{V_1}{V_2} T_2$$

$$= \frac{2.80 \text{ L}}{2.57 \text{ L}} 273 \text{ K}$$

$$= 297 \text{ K}$$

$$t_1 = 297 - 273 = 24 \text{ °C}$$

SKILLBUILDER 11.3 Charles's Law

A gas in a cylinder with a moveable piston with an initial volume of 88.2 mL is heated from 35 °C to 155 °C. What is the final volume of the gas in milliliters?

11.6 The Combined Gas Law: Pressure, Volume, and Temperature

Boyle's law shows how P and V are related at constant temperature, and Charles's law shows how V and T are related at constant pressure. But what if two of these variables change at once? For example, what happens to the volume of a gas if both its pressure and its temperature are changed? Since volume is inversely proportional to pressure ($V \propto 1/P$) and directly proportional to temperature ($V \propto T$), we can write:

$$V \propto \frac{T}{P} \quad \text{or} \quad \frac{PV}{T} = \text{Constant}$$

The combined gas law includes both Boyle's law and Charles's law within it and can be used in place of them. If one physical property (P, V, or T) is constant, it will cancel out of your calculations when you use the combined gas law.

For a sample of gases under two different sets of conditions, therefore, we can write the **combined gas law:**

$$\frac{P_1V_1}{T_1} = \frac{P_2V_2}{T_2}$$

The combined gas law only applies when the amount of gas is constant. The temperature (as with Charles's law) must be expressed in kelvins.

Suppose you carry a cylinder with a moveable piston that has an initial volume of 3.65 L up a mountain. The pressure at the bottom of the mountain is 755 mm Hg and the temperature is 302 K. The pressure at the top of the mountain is 687 mm Hg and the temperature is 291 K. What is the volume of the cylinder at the top of the mountain? We set up the problem as follows:

Given:

$P_1 = 755 \text{ mm Hg}$ $\quad T_1 = 302 \text{ K}$

$V_1 = 3.65 \text{ L}$ $\quad P_2 = 687 \text{ mm Hg}$

$T_2 = 291 \text{ K}$

Find: V_2

Equation:

$$\frac{P_1V_1}{T_1} = \frac{P_2V_2}{T_2}$$

Solution Map: The solution map shows how the equation takes us from the given quantities to the find quantity.

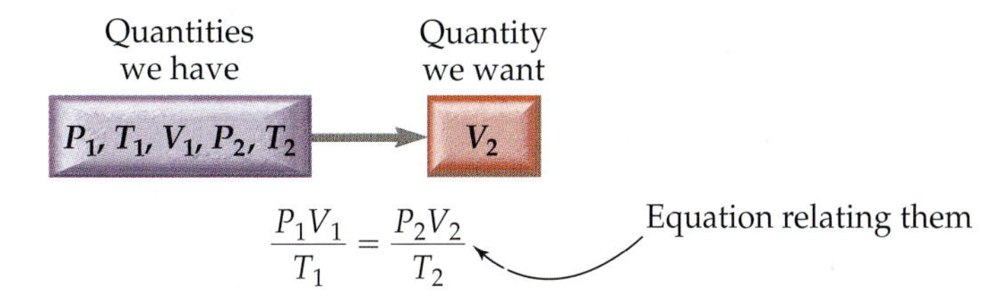

Solution: We then solve the equation for the quantity we are trying to find (V_2).

$$\frac{P_1V_1}{T_1} = \frac{P_2V_2}{T_2}$$

$$V_2 = \frac{P_1V_1T_2}{T_1P_2}$$

Lastly, we substitute in the appropriate values and compute the answer.

$$\begin{aligned} V_2 &= \frac{P_1V_1T_2}{T_1P_2} \\ &= \frac{755 \cancel{\text{mm Hg}} \times 3.65 \text{ L} \times 291 \cancel{\text{K}}}{302 \cancel{\text{K}} \times 687 \cancel{\text{mm Hg}}} \\ &= 3.87 \text{ L} \end{aligned}$$

LIVE EXAMPLE

EXAMPLE 11.4 The Combined Gas Law

A sample of gas has an initial volume of 158 mL at a pressure of 735 mm Hg and a temperature of 34 °C. If the gas is compressed to a volume of 108 mL and heated to a temperature of 85 °C, what is its final pressure in millimeters of mercury?

You are given an initial pressure, temperature, and volume as well as a final temperature and volume. You are asked to find the final pressure.

Given:

$P_1 = 735$ mm Hg

$t_1 = 34$ °C $\quad t_2 = 85$ °C

$V_1 = 158$ mL $\quad V_2 = 108$ mL

Find: P_2

This problem requires the use of the combined gas law.

Equation:

$$\frac{P_1V_1}{T_1} = \frac{P_2V_2}{T_2}$$

Draw a solution map beginning with the given quantities. The combined gas law shows the relationship necessary to get to the find quantity.

Solution Map:

P_1, T_1, V_1, T_2, V_2 → P_2

$$\frac{P_1V_1}{T_1} = \frac{P_2V_2}{T_2}$$

Solve the equation for the quantity you are trying to find (P_2).

Solution:

$$\frac{P_1V_1}{T_1} = \frac{P_2V_2}{T_2}$$

$$P_2 = \frac{P_1V_1T_2}{T_1V_2}$$

Before you substitute in the numerical values, you must convert the temperatures to kelvins.

$$T_1 = 34 + 273 = 307 \text{ K}$$

$$T_2 = 85 + 273 = 358 \text{ K}$$

Once you have converted the temperature to kelvins, substitute into the equation to find P_2.

$$P_2 = \frac{735 \text{ mm Hg} \times 158 \text{ mL} \times 358 \text{ K}}{307 \text{ K} \times 108 \text{ mL}}$$

$$= 1.25 \times 10^3 \text{ mm Hg}$$

SKILLBUILDER 11.4 The Combined Gas Law

A balloon has a volume of 3.7 L at a pressure of 1.1 atm and a temperature of 30 °C. If the balloon is submerged in water to a depth where the pressure is 4.7 atm and the temperature is 15 °C, what will its volume be (assume that any changes in pressure caused by the skin of the balloon are negligible)?

CONCEPTUAL CHECKPOINT 11.2

A volume of gas is confined to a cylinder with a moveable piston at one end. If you apply enough heat to double the Kelvin temperature of the gas,

(a) the pressure and volume will both double.
(b) the pressure will double but the volume will remain the same.
(c) the volume will double but the pressure will remain the same.
(d) the volume will double but the pressure will be halved.

11.7 Avogadro's Law: Volume and Moles

So far, we have learned how V, P, and T are interrelated, but we have considered only a constant amount of a gas. What happens when the amount of gas changes? If we make several measurements of the volume of a gas sample (at constant temperature and pressure) while varying the number of moles in the sample, the results look like ▶ Figure 11.18. We can see that the relationship between volume and number of moles is linear. An extrapolation to zero moles shows a zero volume, as we might expect. This relationship was first stated formally by Amadeo Avogadro (1776–1856) and is called **Avogadro's law**.

Avogadro's law assumes constant temperature and pressure.

Avogadro's law: The volume of a gas and the amount of the gas in moles (n) are directly proportional.

$$V \propto n$$

When the amount of gas in a sample is increased, its volume increases in direct proportion. This also follows from kinetic molecular theory. If the number of gas particles increases at constant pressure and temperature, the particles occupy more volume.

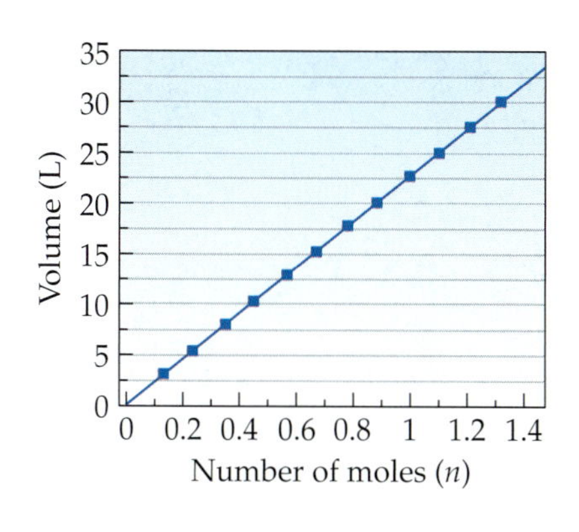

◀ **Figure 11.18 Volume versus number of moles** The volume of a gas sample increases linearly with the number of moles in the sample.

▶ **Figure 11.19 Blow-up** As you exhale into a balloon, you add gas molecules to the inside of the balloon, increasing its volume.

You experience Avogadro's law when you inflate a balloon, for example. With each exhaled breath, you add more gas particles to the inside of the balloon, increasing its volume (▲ Figure 11.19). Avogadro's law can be used to compute the volume of a gas following a change in the amount of the gas *as long as the pressure and temperature of the gas are constant*. For these calculations, Avogadro's law is expressed as

$$\frac{V_1}{n_1} = \frac{V_2}{n_2}$$

Since $V \propto n$, then V/n = Constant. If the number of moles increases, then the volume increases in direct proportion so that the quotient, V/n is always equal to the same constant. Thus, for two different measurements, we can say that $\frac{V_1}{n_1} = \text{Constant} = \frac{V_2}{n_2}$ or $\frac{V_1}{n_1} = \frac{V_2}{n_2}$.

where V_1 and n_1 are the initial volume and number of moles of the gas and V_2 and n_2 are the final volume and number of moles. In calculations, Avogadro's law is used in a manner similar to the other gas laws, as shown in the following example.

EXAMPLE 11.5 Avogadro's Law

A 4.8-L sample of helium gas contains 0.22 mol of helium. How many additional moles of helium gas must be added to the sample to obtain a volume of 6.4 L? Assume constant temperature and pressure.

You are given an initial volume, an initial number of moles, and a final volume. You are (essentially) asked to find the final number of moles.

Given:

$V_1 = 4.8 \text{ L}$

$n_1 = 0.22 \text{ mol}$

$V_2 = 6.4 \text{ L}$

Find: n_2

This problem requires the use of Avogadro's law.

Equation:

$$\frac{V_1}{n_1} = \frac{V_2}{n_2}$$

Draw a solution map beginning with the given quantities. Avogadro's law shows the relationship necessary to get to the find quantity.

Solution Map:

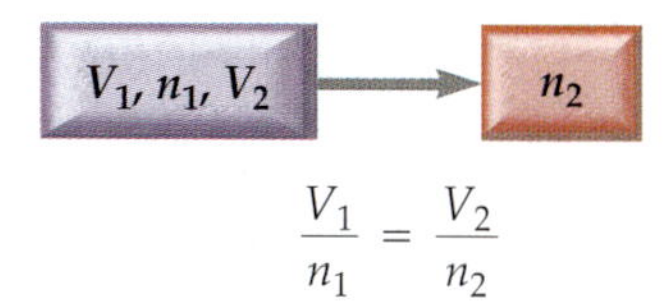

$$\frac{V_1}{n_1} = \frac{V_2}{n_2}$$

Solve the equation for the quantity you are trying to find (n_2) and substitute the appropriate quantities to compute n_2.

Solution:

$$\frac{V_1}{n_1} = \frac{V_2}{n_2}$$

$$n_2 = \frac{V_2}{V_1} n_1$$

$$= \frac{6.4 \cancel{L}}{4.8 \cancel{L}} 0.22 \text{ mol}$$

$$= 0.29 \text{ mol}$$

Since the balloon already contains 0.22 mol, subtract this quantity from the final number of moles to determine how much must be added.

mol to add = 0.29 − 0.22 = 0.07 mol

SKILLBUILDER 11.5 **Avogadro's Law**

A chemical reaction occurring in a cylinder equipped with a moveable piston produces 0.58 mol of a gaseous product. If the cylinder contained 0.11 mol of gas before the reaction and had an initial volume of 2.1 L, what was its volume after the reaction?

11.8 The Ideal Gas Law: Pressure, Volume, Temperature, and Moles

The relationships that we have learned so far can be combined into a single law that encompasses all of them. So far, we know that:

$$V \propto \frac{1}{P} \quad \text{(Boyle's law)}$$

$$V \propto T \quad \text{(Charles's law)}$$

$$V \propto n \quad \text{(Avogadro's law)}$$

Combining these three expressions, we get:

$$V \propto \frac{nT}{P}$$

The volume of a gas is directly proportional to the number of moles of gas and the temperature of the gas and is inversely proportional to the pressure of the gas. We can replace the proportional sign with an equal sign by adding R, a proportionality constant called the **ideal gas constant**.

$$V = \frac{RnT}{P}$$

Rearranging, we get:

$$PV = nRT$$

The preceding equation is called the **ideal gas law**. The value of R, the ideal gas constant, is:

$$R = 0.0821 \frac{\text{L} \cdot \text{atm}}{\text{mol} \cdot \text{K}}$$

R can also be expressed in other units, but its numerical value will be different.

The ideal gas law contains within it the simple gas laws we have learned. For example, recall that Boyle's law states that $V \propto 1/P$ when the amount of gas (n) and the temperature of the gas (T) are kept constant. To derive Boyle's law, we can rearrange the ideal gas law as follows:

$$PV = nRT$$

First, divide both sides by P.

$$V = \frac{nRT}{P}$$

Then put the variables that are constant in parentheses.

$$V = (nRT)\frac{1}{P}$$

Since n and T are constant in this case and since R is always a constant,

$$V = (\text{Constant}) \times \frac{1}{P}$$

which gives us Boyle's law $\left(V \propto \frac{1}{P}\right)$.

The ideal gas law also shows how other pairs of variables are related. For example, from Charles's law we know that volume is proportional to temperature at constant pressure and a constant number of moles. But what if we heat a sample of gas at constant *volume* and a constant number of moles? This question applies to the warning labels on aerosol cans such as hair spray or deodorants. These labels warn the user against excessive heating or incineration of the can, even after the contents are used up. Why? A seemingly empty aerosol can is not really empty but contains a fixed amount of gas trapped in a fixed volume. What would happen if you heated the can? Let's rearrange the ideal gas law to clearly see the relationship between pressure and temperature at constant volume and a constant number of moles.

$$PV = nRT$$

If we divide both sides by V, we get:

$$P = \frac{nRT}{V}$$

$$P = \left(\frac{nR}{V}\right)T$$

Since n and V are constant and since R is always a constant:

The relationship between pressure and temperature is also known as Gay-Lussac's law.

$$P = \text{Constant} \times T$$

As the temperature of a fixed amount of gas in a fixed volume increases, the pressure increases. In an aerosol can, this pressure increase can cause the can to explode, which is why aerosol cans should not be heated or incinerated. The relationships between all of the simple gas laws and the ideal gas law are summarized in Table 11.2.

The ideal gas law can also be used to determine the value of any one of the four variables (P, V, n, or T) given the other three. However, each of the quantities in the ideal gas law *must be expressed* in the units within R.

TABLE 11.2 Relationship between Simple Gas Laws and Ideal Gas Law

Variable Quantities	Constant Quantities	Ideal Gas Law in Form of Variables-constant	Simple Gas Law	Name of Simple Law
V and P	n and T	$PV = nRT$	$P_1V_1 = P_2V_2$	Boyle's law
V and T	n and P	$\frac{V}{T} = \frac{nR}{P}$	$\frac{V_1}{T_1} = \frac{V_2}{T_2}$	Charles's law
P and T	n and V	$\frac{P}{T} = \frac{nR}{V}$	$\frac{P_1}{T_1} = \frac{P_2}{T_2}$	Gay-Lussac's law
P and n	V and T	$\frac{P}{n} = \frac{RT}{V}$	$\frac{P_1}{n_1} = \frac{P_2}{n_2}$	
V and n	T and P	$\frac{V}{n} = \frac{RT}{P}$	$\frac{V_1}{n_1} = \frac{V_2}{n_2}$	Avogadro's law

- Pressure *(P)* must be expressed in atmospheres.
- Volume *(V)* must be expressed in liters.
- Amount of gas *(n)* must be expressed in moles.
- Temperature *(T)* must be expressed in kelvins.

For example, suppose we want to know the pressure of 0.18 mol of a gas in a 1.2-L flask at 298 K.

Given:

$$n = 0.18 \text{ mol}$$
$$V = 1.2 \text{ L}$$
$$T = 298 \text{ K}$$

Find: P

Equation: The equation that relates these is the ideal gas law.

$$PV = nRT$$

Solution Map: The solution map shows how the ideal gas law takes us from the given quantities to the find quantity.

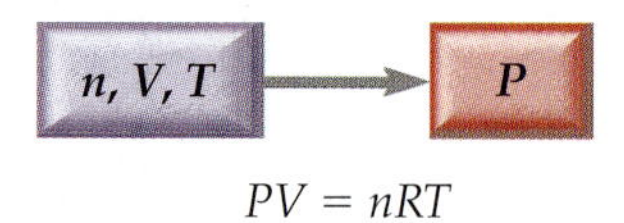

$PV = nRT$

Solution: We then solve the equation for the quantity we are trying to find (in this case, P).

$$PV = nRT$$
$$P = \frac{nRT}{V}$$

We then substitute in the numerical values and compute the answer.

$$P = \frac{0.18 \cancel{\text{mol}} \times 0.0821 \frac{\cancel{\text{L}} \cdot \text{atm}}{\cancel{\text{mol}} \cdot \cancel{\text{K}}} \times 298 \cancel{\text{K}}}{1.2 \cancel{\text{L}}}$$
$$= 3.7 \text{ atm}$$

Notice that all units cancel except the units of the quantity we need (atm).

LIVE EXAMPLE

EXAMPLE 11.6 The Ideal Gas Law

Calculate the volume occupied by 0.845 mol of nitrogen gas at a pressure of 1.37 atm and a temperature of 315 K.

You are given the number of moles, the pressure, and the temperature of a gas sample. You are asked to find the volume.

Given:

$n = 0.845\text{ mol}$
$P = 1.37\text{ atm}$
$T = 315\text{ K}$

Find: V

This problem requires the use of the ideal gas law.

Equation:

$$PV = nRT$$

Draw a solution map beginning with the given quantities. The ideal gas law shows the relationship necessary to get to the find quantity.

Solution Map:

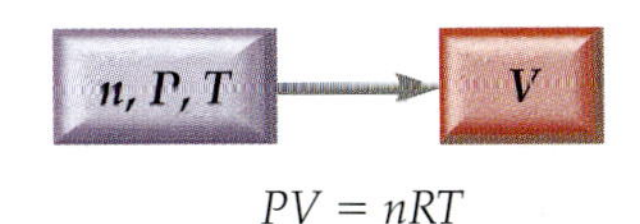

$PV = nRT$

Solve the equation for the quantity you are trying to find (V) and substitute the appropriate quantities to compute V.

Solution:

$$PV = nRT$$

$$V = \frac{nRT}{P}$$

$$V = \frac{0.845\ \cancel{\text{mol}} \times 0.0821 \frac{\text{L} \cdot \cancel{\text{atm}}}{\cancel{\text{mol}} \cdot \cancel{\text{K}}} \times 315\ \cancel{\text{K}}}{1.37\ \cancel{\text{atm}}}$$

$$= 16.0\text{ L}$$

SKILLBUILDER 11.6 The Ideal Gas Law

An 8.5-L tire is filled with 0.55 mol of gas at a temperature of 305 K. What is the pressure of the gas in the tire?

If the units given in an ideal gas law problem are different from those of the ideal gas constant (atm, L, mol, and K), you must convert to the correct units before you substitute into the ideal gas equation, as demonstrated in the following example.

EXAMPLE 11.7 The Ideal Gas Law Requiring Unit Conversion

Calculate the number of moles of gas in a basketball inflated to a total pressure of 24.3 psi with a volume of 3.2 L at 25 °C.

You are given the pressure, the volume and the temperature of a gas sample. You are asked to find the number of moles.

Given:

$P = 24.3\text{ psi}$
$V = 3.2\text{ L}$
$t = 25\text{ °C}$

Find: n

This problem requires the use of the ideal gas law.

Equation: $PV = nRT$

Note: The total pressure is not the same as the *gauge pressure*, the pressure read on a pressure gauge. Gauge pressure is excess pressure—the *difference* between the total pressure and atmospheric pressure. In this case, if atmospheric pressure is 14.7 psi, the gauge pressure would be 9.6 psi. However, for calculations involving the ideal gas law, you must use the total pressure of 24.3 psi.

Draw a solution map beginning with the given quantities. The ideal gas law shows the relationship necessary to get to the find quantity.

Solution Map:

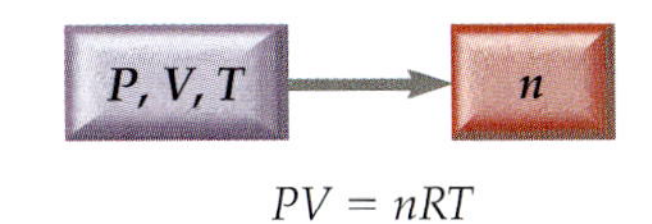

$PV = nRT$

Solve the equation for the quantity you are trying to find (n).

Solution:

$$PV = nRT$$

$$n = \frac{PV}{RT}$$

Before substituting into the equation, you must convert P and t into the correct units. (Since 1.6462 is an intermediate answer, mark the least significant digit, but don't round until the end.)

$$P = 24.2\ \text{psi} \times \frac{1\ \text{atm}}{14.7\ \text{psi}} = 1.6\underline{4}62\ \text{atm}$$

$$T = t + 273$$
$$= 25 + 273 = 298\ \text{K}$$

Finally, substitute into the equation to compute n.

$$n = \frac{1.6\underline{4}62\ \text{atm} \times 3.2\ \text{L}}{0.0821 \frac{\text{L} \cdot \text{atm}}{\text{mol} \cdot \text{K}} \times 298\ \text{K}}$$
$$= 0.22\ \text{mol}$$

SKILLBUILDER 11.7 The Ideal Gas Law Requiring Unit Conversion

How much volume does 0.556 mol of gas occupy when its pressure is 715 mm Hg and its temperature is 58 °C?

SKILLBUILDER PLUS

Find the pressure in millimeters of mercury of a 0.133-g sample of helium gas at 32 °C and contained in a 648-mL container.

Molar Mass of a Gas from the Ideal Gas Law

The ideal gas law can be used in combination with mass measurements to calculate the molar mass of a gas. For example, a sample of gas has a mass of 0.136 g. Its volume is 0.112 L at a temperature of 298 K and a pressure of 1.06 atm. Find its molar mass.

We first set up the problem.

Given: $m = 0.136\ \text{g}$ $V = 0.112\ \text{L}$

$T = 298\ \text{K}$ $P = 1.06\ \text{atm}$

Find: molar mass (g/mol)

Equations: We need two equations to solve this problem, the ideal gas law and the definition of molar mass.

$$PV = nRT$$

$$\text{Molar mass} = \frac{\text{Mass}}{\text{Moles}}$$

Solution Map: The solution map has two parts. In the first part, we use P, V, and T to find the number of moles of gas. In the second part, we use the number of moles of gas and the given mass to find the molar mass.

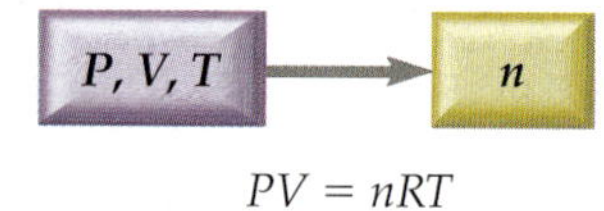

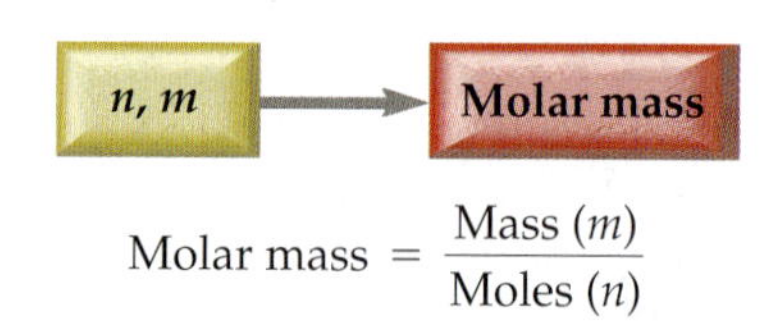

Solution:

$$PV = nRT$$

$$n = \frac{PV}{RT}$$

$$= \frac{1.06 \cancel{\text{atm}} \times 0.112 \cancel{\text{L}}}{0.0821 \dfrac{\cancel{\text{L}} \cdot \cancel{\text{atm}}}{\text{mol} \cdot \cancel{\text{K}}} \times 298 \cancel{\text{K}}}$$

$$= 4.8525 \times 10^{-3}\ \text{mol}$$

$$\text{Molar mass} = \frac{\text{Mass } (m)}{\text{Moles } (n)}$$

$$= \frac{0.136\ \text{g}}{4.8525 \times 10^{-3}\ \text{mol}}$$

$$= 28.0\ \text{g/mol}$$

Therefore the gas is N_2.

EXAMPLE 11.8 Molar Mass Using the Ideal Gas Law and Mass Measurement

A sample of gas has a mass of 0.311 g. Its volume is 0.225 L at a temperature of 55 °C and a pressure of 886 mm Hg. Find its molar mass.

You are given the mass, the volume, the temperature, and the pressure of a gas sample. You are asked to find the molar mass of the gas.

Given:

$m = 0.311\ \text{g}$

$V = 0.225\ \text{L}$

$t = 55\ °\text{C}$

$P = 886\ \text{mm Hg}$

Find: molar mass (g/mol)

This problem requires the use of the ideal gas law and the definition of molar mass.

Equations:

$$PV = nRT$$

$$\text{Molar mass} = \frac{\text{Mass } (m)}{\text{Moles } (n)}$$

In the first part of the solution map, use the ideal gas law to find the number of moles of gas from the other given quantities. In the second part, use the number of moles from the first part, as well as the given mass, to find the molar mass.

Solution Map:

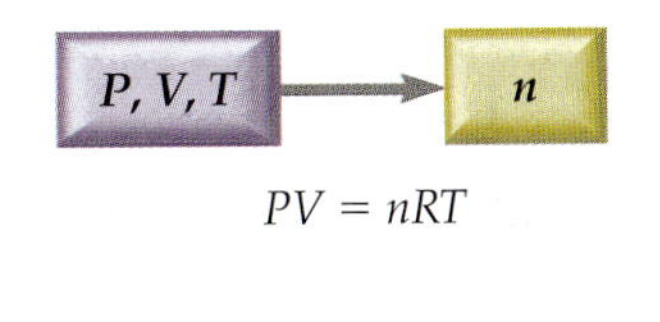

$$PV = nRT$$

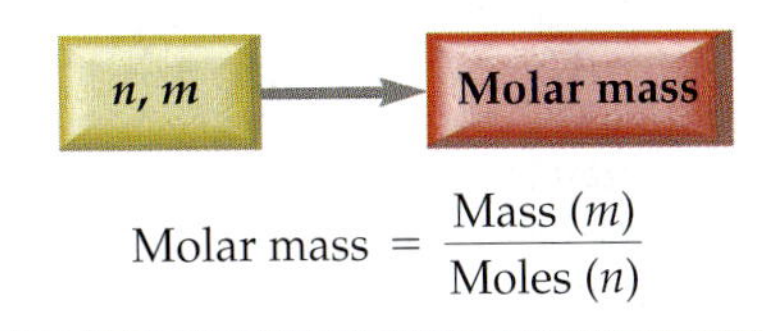

$$\text{Molar mass} = \frac{\text{Mass } (m)}{\text{Moles } (n)}$$

First, solve the ideal gas law for n.

Solution:

$$PV = nRT$$

$$n = \frac{PV}{RT}$$

Before substituting into the equation, you must convert the pressure to atm and temperature to K.

$$P = 886 \text{ mm Hg} \times \frac{1 \text{ atm}}{760 \text{ mm Hg}} = 1.1\underline{6}58 \text{ atm}$$

$$T = 55\ ^\circ\text{C} + 273 = 328 \text{ K}$$

Now, substitute into the equation to compute n, the number of moles.

$$n = \frac{1.1\underline{6}58 \text{ atm} \times 0.225 \text{ L}}{0.0821 \frac{\text{L} \cdot \text{atm}}{\text{mol} \cdot \text{K}} \times 328 \text{ K}}$$

$$= 9.7\underline{4}06 \times 10^{-3} \text{ mol}$$

Finally, use the number of moles just found and the given mass (m) to find the molar mass.

$$\text{Molar mass} = \frac{\text{Mass } (m)}{\text{Moles } (n)}$$

$$= \frac{0.311 \text{ g}}{9.7\underline{4}06 \times 10^{-3} \text{ mol}}$$

$$= 31.9 \text{ g/mol}$$

SKILLBUILDER 11.8 Molar Mass Using the Ideal Gas Law and Mass Measurement

A sample of gas has a mass of 827 mg. Its volume is 0.270 L at a temperature of 88 °C and a pressure of 975 mm Hg. Find its molar mass.

Although a complete derivation is beyond the scope of this text, the ideal gas law follows directly from the kinetic molecular theory of gases. Consequently, the ideal gas law holds only under conditions where the kinetic molecular theory holds. The ideal gas law works exactly only for gases that are acting ideally (▼ Figure 11.20), which means that (a) the volume of the gas particles is small compared to the space between them and (b) the forces between the gas particles are not significant. These assumptions break down (▼ Figure 11.21)

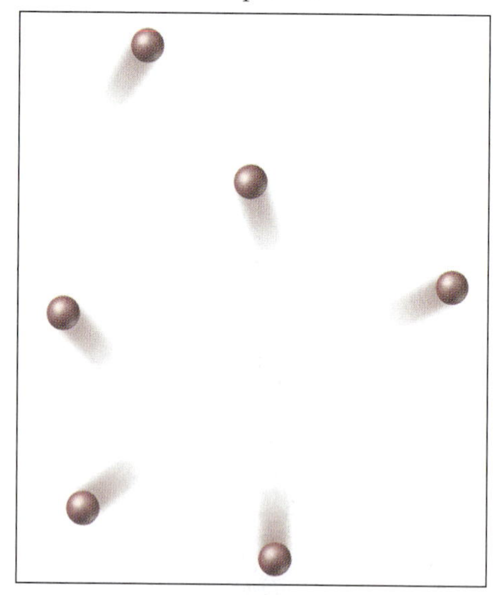

Ideal gas conditions
•High temperature
•Low pressure
•Particle size small compared to space between particles.
•Interactions between particles are insignificant.

▲ **Figure 11.20 Conditions for ideal gas behavior** At high temperatures and low pressures, the assumptions of the kinetic molecular theory apply.

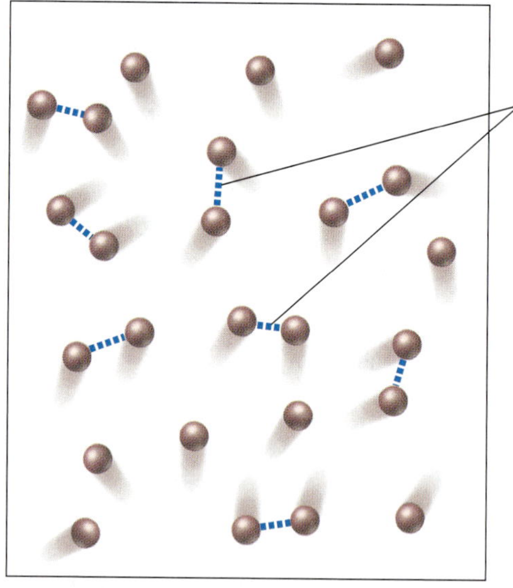

▲ **Figure 11.21 Conditions for non-ideal gas behavior** At low temperatures and high pressures, the assumptions of the kinetic molecular theory are not valid.

under conditions of high pressure (when the space between the gas particles becomes smaller) or low temperatures (when the gas particles move so slowly that their interactions become significant). For all of the problems encountered in this book, you may assume ideal gas behavior.

11.9 Mixtures of Gases: Why Deep-Sea Divers Breathe a Mixture of Helium and Oxygen

TABLE 11.3 Composition of Dry Air

Gas	Percent by Volume (%)
nitrogen (N_2)	78
oxygen (O_2)	21
argon (Ar)	0.9
carbon dioxide (CO_2)	0.03

Many gas samples are not pure but consist of mixtures of gases. The air we breathe, for example, is a mixture containing 78% nitrogen, 21% oxygen, 0.9% argon, 0.03% carbon dioxide (Table 11.3), and a few other gases in smaller amounts.

According to the kinetic molecular theory, each of the components in a gas mixture acts independently of the others. For example, the nitrogen molecules in air exert a certain pressure—78% of the total pressure—that is independent of the presence of the other gases in the mixture. Likewise, the oxygen molecules in air exert a certain pressure—21% of the total pressure—that is also independent of the presence of the other gases in the mixture. The pressure due to any individual component in a gas mixture is called the **partial pressure** of that component. The partial pressure of any component is that component's fractional composition times the total pressure of the mixture (▼ Figure 11.22).

The fractional composition is the percent composition divided by 100.

Partial pressure of component
= Fractional composition of component × Total pressure

For example, the partial pressure of nitrogen (P_{N_2}) in air at 1.0 atm is:

MOLECULE Oxygen 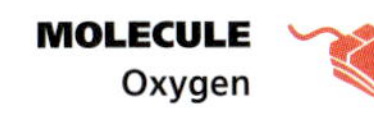

$$P_{N_2} = 0.78 \times 1 \text{ atm} = 0.78 \text{ atm}$$

Similarly, the partial pressure of oxygen in air at 1.0 atm is:

MOLECULE Nitrogen

$$P_{O_2} = 0.21 \times 1 \text{ atm} = 0.21 \text{ atm}$$

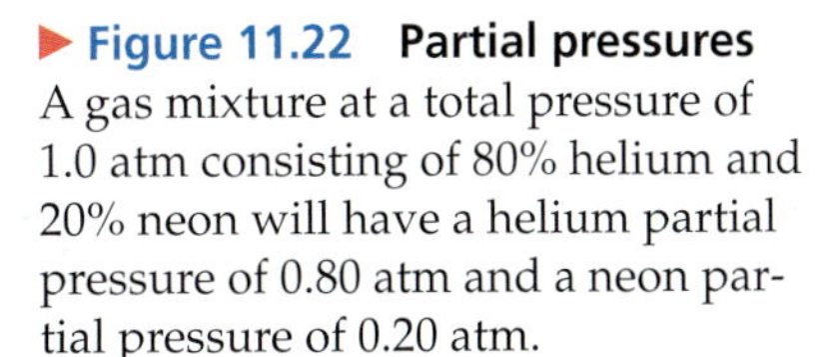

▶ **Figure 11.22 Partial pressures** A gas mixture at a total pressure of 1.0 atm consisting of 80% helium and 20% neon will have a helium partial pressure of 0.80 atm and a neon partial pressure of 0.20 atm.

Gas mixture (80% He, 20% Ne)
$P_{tot} = 1.0$ atm
$P_{He} = 0.80$ atm
$P_{Ne} = 0.20$ atm

The sum of the partial pressures of each of the components in a gas mixture must equal the total pressure

$$P_{tot} = P_a + P_b + P_c + \ldots$$

where P_{tot} is the total pressure and $P_a, P_b, P_c, \ldots$ are the partial pressures of the components. This is known as **Dalton's law of partial pressures**.

For 1 atm air:

We can ignore the contribution of CO_2 and other trace gases because they are present in very small amounts.

$$P_{tot} = P_{N_2} + P_{O_2} + P_{Ar}$$
$$P_{tot} = 0.78 \text{ atm} + 0.21 \text{ atm} + 0.01 \text{ atm}$$
$$= 1.0 \text{ atm}$$

EXAMPLE 11.9 Total Pressure and Partial Pressure

A mixture of helium, neon, and argon has a total pressure of 558 mm Hg. If the partial pressure of helium is 341 mm Hg and the partial pressure of neon is 112 mm Hg, what is the partial pressure of argon?

You are given the total pressure of a gas mixture and the partial pressures of two (of its three) components. You are asked to find the partial pressure of the third component.

Given:

$$P_{tot} = 558 \text{ mm Hg}$$
$$P_{He} = 341 \text{ mm Hg}$$
$$P_{Ne} = 112 \text{ mm Hg}$$

Find: P_{Ar}

This problem requires the use of Dalton's law of partial pressures.

Equation:

$$P_{tot} = P_a + P_b + P_c + \ldots$$

To solve this problem, simply solve Dalton's law for the partial pressure of argon and substitute the correct values to compute it.

Solution:

$$P_{tot} = P_{He} + P_{Ne} + P_{Ar}$$
$$P_{Ar} = P_{tot} - P_{He} - P_{Ne}$$
$$= 558 \text{ mm Hg} - 341 \text{ mm Hg} - 112 \text{ mm Hg}$$
$$= 105 \text{ mm Hg}$$

SKILLBUILDER 11.9 Total Pressure and Partial Pressure

A sample of hydrogen gas is mixed with water vapor. The mixture has a total pressure of 745 torr and the water vapor has a partial pressure of 24 torr. What is the partial pressure of the hydrogen gas?

Deep-Sea Diving and Partial Pressure

TUTORIAL
Diving and Dalton's Law of Partial Pressures

Our lungs have evolved to breathe oxygen at a partial pressure of $P_{O_2} = 0.21$ atm. If the total pressure decreases—when climbing a mountain, for example—the partial pressure of oxygen also decreases. For example, on top of Mt. Everest, where the total pressure is only 0.311 atm, the partial pressure of oxygen is only 0.065 atm. As we learned earlier, low oxygen levels have physiological effects and result in a condition called **hypoxia** or oxygen starvation. Mild hypoxia causes dizziness, headache, and shortness of breath. Severe hypoxia, which occurs when P_{O_2} drops below 0.1 atm, may cause unconsciousness or even death. For this reason, climbers hoping to make the summit of Mt. Everest usually carry oxygen to breathe.

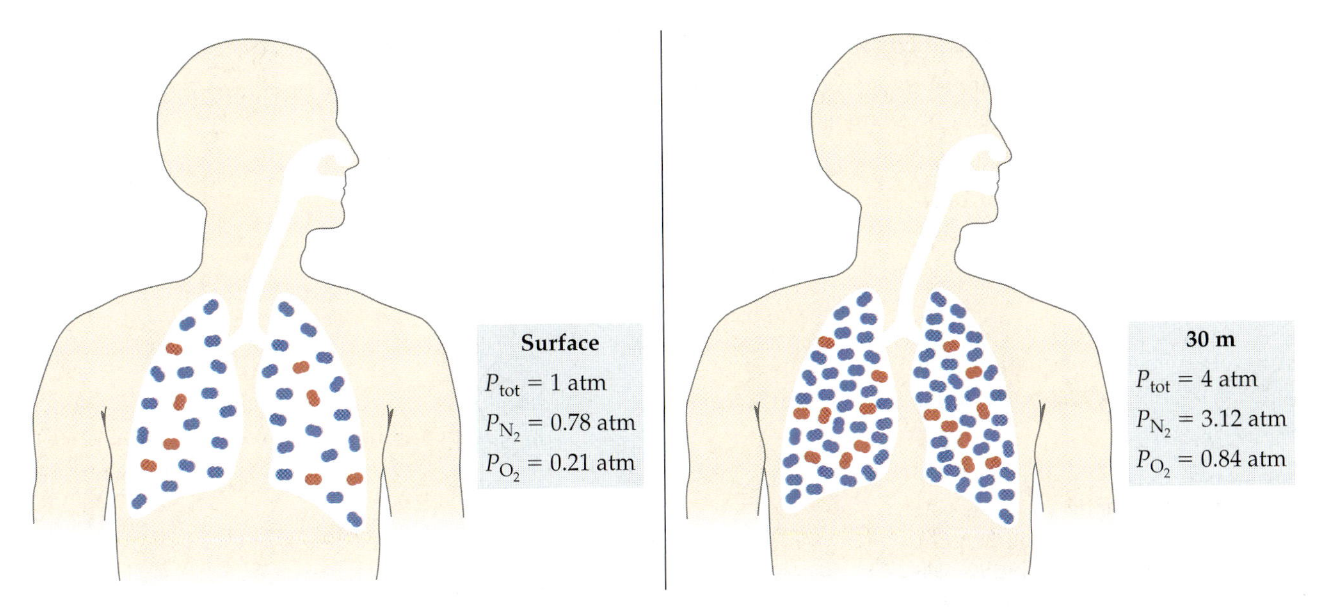

▲ **Figure 11.23 Too much of a good thing** When a person is breathing compressed air, there is a larger partial pressure of oxygen in the lungs. A large oxygen partial pressure in the lungs results in a larger amount of oxygen in bodily tissues. When the oxygen partial pressure increases beyond 1.4 atm, oxygen toxicity results. (In this figure, the red molecules are oxygen and the blue ones are nitrogen.)

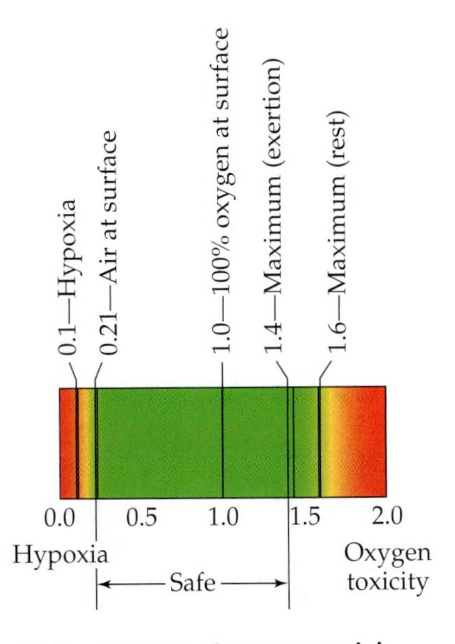

▲ **Figure 11.24 Oxygen partial pressure limits** The partial pressure of oxygen in air at sea level is 0.21 atm. If this pressure drops by 50%, fatal hypoxia can result. High oxygen levels can also be harmful, but only if the partial pressure of oxygen increases by a factor of 7 or more.

High oxygen levels can also have physiological effects. Scuba divers, as we have learned, breathe pressurized air. At 30 m, a scuba diver breathes air at a total pressure of 4.0 atm, making P_{O_2} about 0.84 atm. This increased partial pressure of oxygen results in a higher density of oxygen molecules in the lungs (▲ Figure 11.23), which results in a higher concentration of oxygen in body tissues. When P_{O_2} increases beyond 1.4 atm, the increased oxygen concentration in body tissues causes a condition called **oxygen toxicity**, which results in muscle twitching, tunnel vision, and convulsions (◄ Figure 11.24). Divers who venture too deep without proper precautions have drowned because of oxygen toxicity.

A second problem associated with breathing pressurized air is the increase in nitrogen in the lungs. At 30 m a scuba diver breathes nitrogen at P_{N_2} = 3.1 atm, which causes an increase in nitrogen concentration in bodily tissues and fluids. When P_{N_2} increases beyond about 4 atm, a condition called **nitrogen narcosis** or *rapture of the deep* results. Divers describe this condition as a feeling of being tipsy. A diver breathing compressed air at 60 m feels as if he has had too much wine.

To avoid oxygen toxicity and nitrogen narcosis, deep-sea divers—those venturing beyond 50 m—breathe specialized mixtures of gases. One common mixture is called heliox, a mixture of helium and oxygen. These mixtures usually contain a smaller percentage of oxygen than would be found in air, thereby lowering the risk of oxygen toxicity. Heliox also contains helium instead of nitrogen, eliminating the risk of nitrogen narcosis.

EXAMPLE 11.10 Partial Pressure, Total Pressure, and Percent Composition

Calculate the partial pressure of oxygen that a diver breathes with a heliox mixture containing 2.0% oxygen at a depth of 100 m where the total pressure is 10.0 atm.

You are given the percent oxygen in the mixture and the total pressure. You are asked to find the partial pressure of oxygen.

Given:

O_2 percent = 2.0%

P_{tot} = 10.0 atm

Find: P_{O_2}

You will need the equation that relates partial pressure to total pressure.

Equation:

Partial pressure of component = Fractional composition of component × Total pressure

Calculate the fractional composition of O_2 by dividing the percent composition by 100. Calculate the partial pressure of O_2 by multiplying the fractional composition by the total pressure.

Solution:

$$\text{Fractional composition of } O_2 = \frac{2.0}{100} = 0.020$$

$$P_{O_2} = 0.020 \times 10.0 \text{ atm} = 0.20 \text{ atm}$$

SKILLBUILDER 11.10 Partial Pressure, Total Pressure, and Percent Composition

What must the total pressure be for a diver breathing heliox with an oxygen composition of 5.0% to breathe $P_{O_2} = 0.21$ atm?

Collecting Gases over Water

When the product of a chemical reaction is gaseous, it is often collected by the displacement of water. For example, suppose the following reaction is used as a source of hydrogen gas.

$$Zn(s) + 2\,HCl(aq) \longrightarrow ZnCl_2(aq) + H_2(g)$$

As the hydrogen gas is formed, it bubbles through the water and gathers in the collection flask. However, the hydrogen gas collected in this way is not pure but is mixed with water vapor because some water molecules evaporate and become mixed with the hydrogen molecules (▼ Figure 11.25).

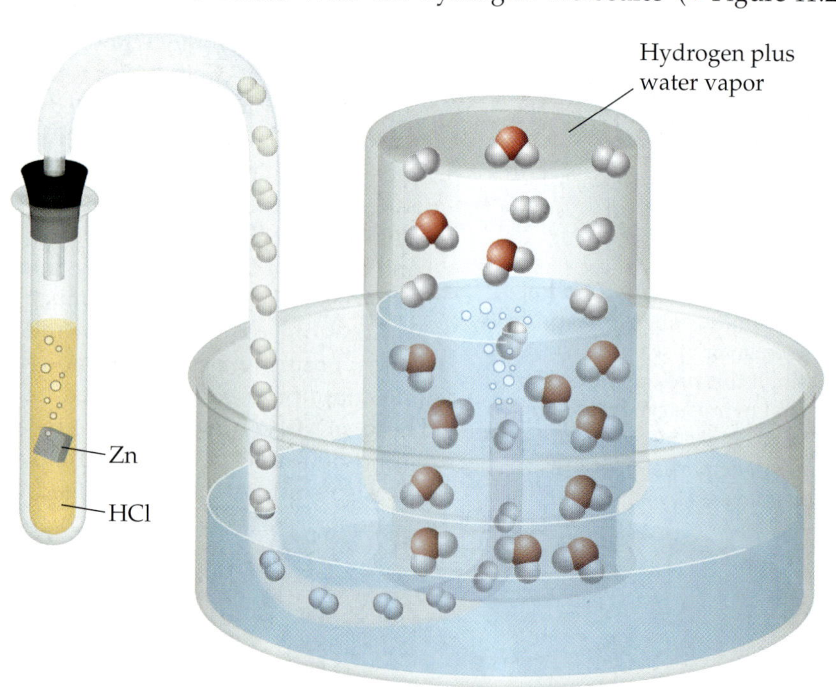

▶ **Figure 11.25 Vapor pressure** When a gas from a chemical reaction is collected through water, water molecules become mixed with the gas molecules. The pressure of water vapor in the final mixture is the vapor pressure of water at the temperature at which the gas is collected.

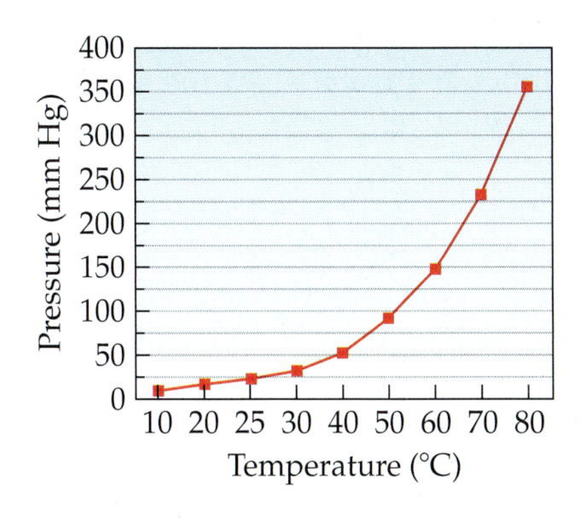

▲ **Figure 11.26 Vapor pressure of water as a function of temperature** Vapor pressure increases with increasing temperature.

Vapor pressure is covered in detail in Chapter 12.

TABLE 11.4 Vapor Pressure of Water versus Temperature

Temperature (°C)	Pressure (mm Hg)
10 °C	9.2
20 °C	17.5
25 °C	23.8
30 °C	31.8
40 °C	55.3
50 °C	92.5
60 °C	149.4
70 °C	233.7
80 °C	355.1

The partial pressure of water in the mixture depends on the temperature and is called its **vapor pressure** (Table 11.4 and ◄ Figure 11.26). Vapor pressure increases with increasing temperature because the higher temperatures cause more water molecules to evaporate.

Suppose we collect the hydrogen gas over water at a total pressure of 758 mm Hg and a temperature of 25 °C. What is the partial pressure of the hydrogen gas? We know that the total pressure is 758 mm Hg and that the partial pressure of water is 23.8 mm Hg (its vapor pressure at 25 °C).

$$P_{tot} = P_{H_2} + P_{H_2O}$$
$$758 \text{ mm Hg} = P_{H_2} + 23.8 \text{ mm Hg}$$

Therefore,

$$P_{H_2} = 758 \text{ mm Hg} - 23.8 \text{ mm Hg}$$
$$= 734 \text{ mm Hg}$$

The partial pressure of the hydrogen in the mixture will be 734 mm Hg.

11.10 Gases in Chemical Reactions

See Section 8.3.

In Chapter 8, we learned how the coefficients in chemical equations can be used as conversion factors between moles of reactants and moles of products in a chemical reaction. These conversion factors could be used to determine, for example, the amount of product obtained in a chemical reaction based on a given amount of reactant or the amount of one reactant needed to completely react with a given amount of another reactant. The general solution map for these kinds of calculations is

$$\text{Moles A} \longrightarrow \text{Moles B}$$

where A and B are two different substances involved in the reaction and the conversion factor between them comes from the stoichiometric coefficients in the balanced chemical equation.

In reactions involving gaseous reactant or products, the amount of gas is often specified in terms of its volume at a given temperature and pressure. In these cases, we can use the ideal gas law to convert pressure, volume, and temperature to moles.

$$n = \frac{PV}{RT}$$

We can then use the stoichiometric coefficients to convert to other quantities in the reaction. For example, consider the following reaction for the synthesis of ammonia.

$$3\,H_2(g) + N_2(g) \longrightarrow 2\,NH_3(g)$$

How many moles of NH_3 are formed by the complete reaction of 2.5 L of hydrogen at 381 K and 1.32 atm? Assume that there is more than enough N_2.

We set up the problem in the normal way.

Given:

$V = 2.5\text{ L}$

$T = 381\text{ K}$

$P = 1.32\text{ atm (of } H_2)$

Find: mol NH_3

Equation and Conversion Factor:

$PV = nRT$

$3\text{ mol } H_2 = 2\text{ mol } NH_3$

Solution Map: The solution map for this problem is similar to the solution maps for other stoichiometric problems. We first use the ideal gas law to find mol H_2 from P, V, and T. Then we use the stoichiometric coefficients from the equation to convert mol H_2 to mol NH_3.

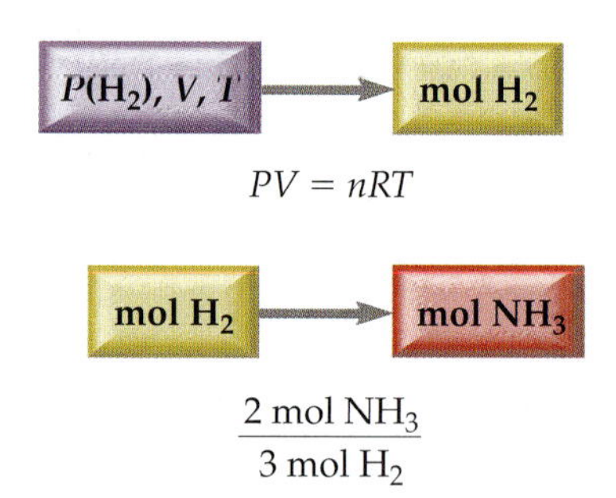

Solution: We first solve the ideal gas equation for n.

$PV = nRT$

$$n = \frac{PV}{RT}$$

Then we substitute in the appropriate values.

$$n = \frac{1.32\text{ atm} \times 2.5\text{ L}}{0.0821\dfrac{\text{L}\cdot\text{atm}}{\text{mol}\cdot\text{K}} \times 381\text{ K}}$$

$$= 0.1\underline{0}55\text{ mol } H_2$$

Next, we convert mol H_2 to mol NH_3.

$$0.1055\text{ mol } H_2 \times \frac{2\text{ mol } NH_3}{3\text{ mol } H_2} = 0.070\text{ mol } NH_3$$

There is enough H_2 to form 0.070 mol NH_3.

LIVE EXAMPLE 

EXAMPLE 11.11 Gases in Chemical Reactions

How many liters of oxygen gas form when 294 g of $KClO_3$ completely react in the following reaction (which is used in the ignition of fireworks)?

$$2\,KClO_3(s) \longrightarrow 2\,KCl(s) + 3\,O_2(g)$$

Assume that the oxygen gas is collected at $P = 755$ mm Hg and $T = 305$ K.

You are given the mass of a reactant in a chemical reaction. You are asked to find the volume of a gaseous product at a given pressure and temperature.

Given:

294 g $KClO_3$

$P = 755$ mm Hg (of oxygen gas)

$T = 305$ K

Find: Volume of O_2 in liters

You will need the molar mass of $KClO_3$ and the stoichiometric relationship between $KClO_3$ and O_2 (from the balanced chemical equation.) You will also need the ideal gas law.

Equations and Conversion Factors:

1 mol $KClO_3$ = 122.5 g

2 mol $KClO_3$ = 3 mol O_2

$PV = nRT$

The solution map has two parts. In the first part, convert from g $KClO_3$ to mol $KClO_3$ and then to mol O_2.

In the second part, use mol O_2 as n in the ideal gas law to find the volume of O_2.

Solution Map:

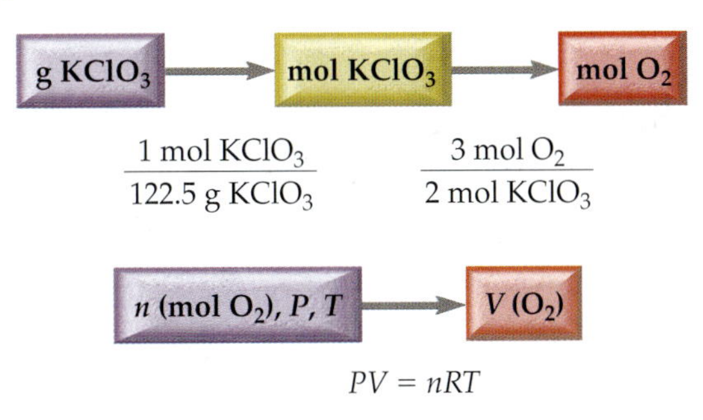

Begin by converting mass $KClO_3$ to mol $KClO_3$ and then to mol O_2.

Solution:

$$294\ \text{g KClO}_3 \times \frac{1\ \text{mol KClO}_3}{122.5\ \text{g KClO}_3} \times \frac{3\ \text{mol O}_2}{2\ \text{mol KClO}_3} = 3.60\ \text{mol O}_2$$

Then solve the ideal gas equation for V.

$$PV = nRT$$

$$V = \frac{nRT}{P}$$

Before substituting the values into this equation, you must convert the pressure to atm.

$$P = 755\ \text{mm Hg} \times \frac{1\ \text{atm}}{760\ \text{mm Hg}} = 0.99342\ \text{atm}$$

Finally, substitute the given quantities along with the number of moles just calculated to compute the volume.

$$V = \frac{3.60\ \text{mol} \times 0.0821\ \frac{\text{L} \cdot \text{atm}}{\text{mol} \cdot \text{K}} \times 305\ \text{K}}{0.99342\ \text{atm}}$$

$$= 90.7\ \text{L}$$

SKILLBUILDER 11.11 Gases in Chemical Reactions

In the following reaction, 4.58 L of O_2 were formed at 745 mm Hg and 308 K. How many grams of Ag_2O must have decomposed?

$$2\,Ag_2O(s) \longrightarrow 4\,Ag(s) + O_2(g)$$

Molar Volume at Standard Temperature and Pressure

The volume occupied by 1 mol of gas at 0 °C (273 K) and 1 atm can be easily calculated using the ideal gas law. These conditions are called **standard temperature and pressure (STP)** and the volume occupied by 1 mol of gas under these conditions is called the **molar volume** of an ideal gas. Using the ideal gas law, molar volume is:

$$V = \frac{nRT}{P}$$

$$= \frac{1.00\ \text{mol} \times 0.0821 \frac{\text{L} \cdot \text{atm}}{\text{mol} \cdot \text{K}} \times 273\ \text{K}}{1.00\ \text{atm}}$$

$$= 22.4\ \text{L}$$

Under standard conditions, therefore, we can use the following equivalence as a conversion factor.

The molar volume of 22.4 L applies only at STP.

$$1\ \text{mol} \equiv 22.4\ \text{L}$$

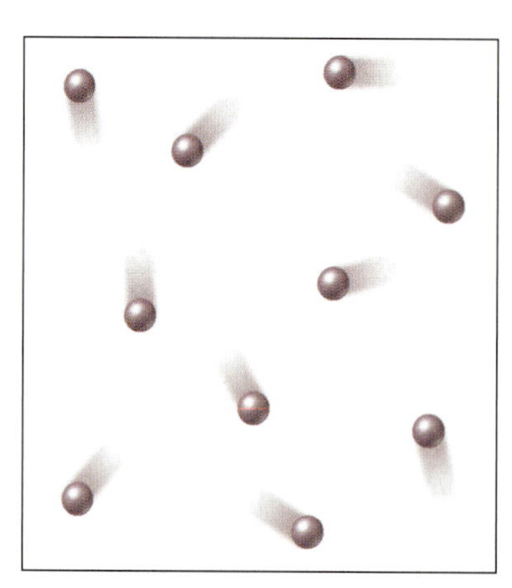

1 mol helium at STP
Volume = 22.4 L
Mass = 4.00 g

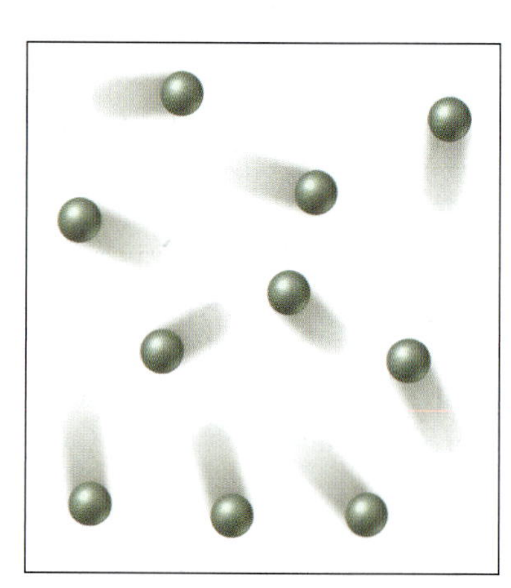

1 mol xenon at STP
Volume = 22.4 L
Mass = 131.3 g

▲ One mole of any gas at standard temperature and pressure (STP) occupies 22.4 L.

For example, suppose we wanted to calculate the number of liters of CO_2 gas that forms at STP when 0.879 moles of $CaCO_3$ undergoes the following reaction.

$$CaCO_3(s) \longrightarrow CaO(s) + CO_2(g)$$

We set up the problem in the normal way.

Given: 0.879 mol $CaCO_3$

Find: $CO_2(g)$ in liters

Conversion Factors:

$1\ \text{mol} = 22.4\ \text{L (at STP)}$

$1\ \text{mol}\ CaCO_3 = 1\ \text{mol}\ CO_2$

Solution Map: The solution map shows how to convert from mol $CaCO_3$ to mol CO_2 to L CO_2 using the conversion factor of 1 mol = 22.4 L.

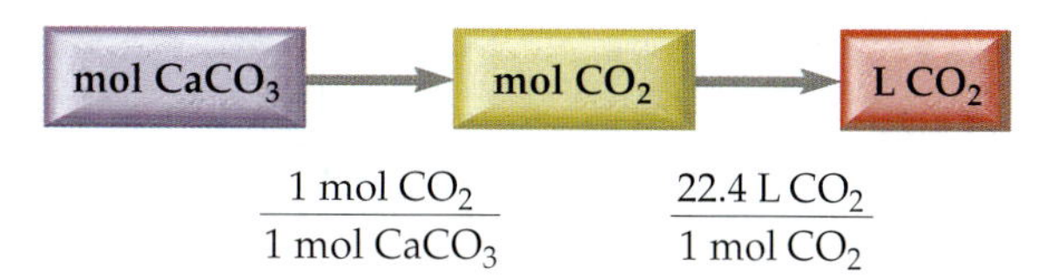

Solution:

$$0.879\ \text{mol}\ CaCO_3 \times \frac{1\ \text{mol}\ CO_2}{1\ \text{mol}\ CaCO_3} \times \frac{22.4\ \text{L}\ CO_2}{1\ \text{mol}\ CO_2} = 19.7\ \text{L}\ CO_2$$

EXAMPLE 11.12 Using Molar Volume in Calculations

How many grams of water form when 1.24 L of H_2 gas at STP completely reacts with O_2?

$$2\ H_2(g) + O_2(g) \longrightarrow 2\ H_2O(g)$$

You are given the volume of a reactant at STP and asked to find the mass of the product formed.

Given: 1.24 L H_2

You will need the molar volume at STP, the stoichiometric relationship between H_2 and H_2O (from the balanced chemical equation), and the molar mass of H_2O.

Find: g H_2O

Conversion Factors:

$1\ \text{mol} = 22.4\ \text{L (at STP)}$

$2\ \text{mol}\ H_2 \equiv 2\ \text{mol}\ H_2O$

$18.02\ \text{g}\ H_2O = 1\ \text{mol}\ H_2O$

In the solution map, use the molar volume to convert from volume H_2 to mol H_2. Then use the stoichiometric relationship to convert to mol H_2O and finally the molar mass of H_2O to get to mass H_2O.

Solution Map:

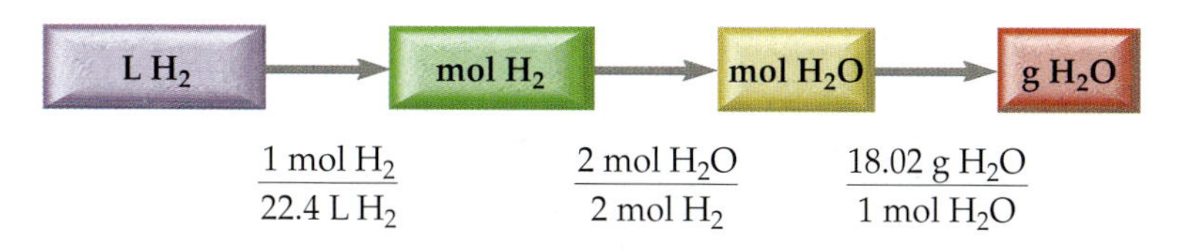

Begin with the volume of H_2 and follow the solution map to arrive at mass H_2O in grams.

Solution:

$$1.24\ \text{L}\ H_2 \times \frac{1\ \text{mol}\ H_2}{22.4\ \text{L}\ H_2} \times \frac{2\ \text{mol}\ H_2O}{2\ \text{mol}\ H_2} \times \frac{18.02\ \text{g}\ H_2O}{1\ \text{mol}\ H_2O} = 0.998\ \text{g}\ H_2O$$

SKILLBUILDER 11.12 Using Molar Volume in Calculations

How many liters of oxygen (at STP) are required to form 10.5 g of H_2O?

$$2\ H_2(g) + O_2(g) \longrightarrow 2\ H_2O(g)$$

Chemistry IN THE ENVIRONMENT

Air Pollution

All major cities in the world have polluted air. This pollution comes from a number of sources, including electricity generation, motor vehicles, and industrial waste. While there are many different kinds of air pollutants, the major gaseous air pollutants include the following:

Sulfur dioxide (SO_2)—Sulfur dioxide is emitted primarily as a by-product of electricity generation and industrial metal refining. SO_2 is a lung and eye irritant that affects the respiratory system. SO_2 is also one of the main precursors of acid rain.

Carbon monoxide (CO)—Carbon monoxide is formed by the incomplete combustion of fossil fuels (petroleum, natural gas, and coal). It is emitted mainly by motor vehicles. CO displaces oxygen in the blood and causes the heart and lungs to work harder. At high levels, CO can cause sensory impairment, decreased thinking ability, unconsciousness, and even death.

Ozone (O_3)—Ozone in the upper atmosphere is a normal part of our environment. Upper atmospheric ozone filters out part of the harmful UV light contained in sunlight. Lower-atmospheric or *ground-level* ozone, on the other hand, is a pollutant that results from the action of sunlight on motor vehicle emissions. Ground-level ozone is an eye and lung irritant. Prolonged exposure to ozone has been shown to permanently damage the lungs.

▲ Air pollution plagues most large cities.

Nitrogen dioxide (NO_2)—Nitrogen dioxide is emitted by motor vehicles and by electricity generation plants. It is an orange-brown gas that causes the dark haze often seen over polluted cities. NO_2 is an eye and lung irritant and a precursor of acid rain.

In the United States, the U.S. Environmental Protection Agency (EPA) has set standards for these pollutants. Beginning in the 1970s, the U.S. Congress passed the Clean Air Act and its amendments, requiring U.S. cities to reduce their pollution and maintain levels below the standards set by the EPA. As a result of this legislation, pollutant levels in U.S. cities have decreased significantly over the last twenty years, even as the number of vehicles has increased. For example, according to the EPA's 1998 National Air Quality and Emissions Trends Report, the levels of all four of the previously mentioned pollutants in major U.S cities decreased during the period 1979–1998. The amounts of these decreases are shown in Table 11.5.

TABLE 11.5 Changes in Pollutant Levels for Major U.S. Cities, 1979–1998

Pollutant	Change, 1979–1998
SO_2	−40%
CO	−55%
O_3	−15%
NO_2	−25%

Although the levels of pollutants (especially ozone) in many cities are still above what the EPA considers safe, much progress has been made. These trends demonstrate that good legislation can clean up our environment.

CAN YOU ANSWER THIS? *Calculate the amount (in grams) of SO_2 emitted when 1.0 kg of coal containing 4.0% S by mass is completely burned. Under standard conditions, what volume in liters would this SO_2 occupy?*

MOLECULE Sulfur Dioxide 

MOLECULE Ozone 

CHAPTER IN REVIEW

Chemical Principles

Kinetic Molecular Theory: The kinetic molecular theory is a model for gases. In this model, gases are composed of widely spaced, noninteracting particles whose average kinetic energy depends on temperature.

Pressure: Pressure is the force per unit area that results from the collision of gas particles with surfaces. The SI unit of pressure is the pascal, but pressure is often expressed in other units such as atmospheres, millimeters of mercury, torr, pounds per square inch, and inches of mercury.

Simple Gas Laws: The simple gas laws show how one of the properties of a gas varies with another. They are:

Volume *(V)* and Pressure *(P)*

$$V \propto \frac{1}{P} \qquad \text{(Boyle's law)}$$

Volume *(V)* and Temperature *(T)*

$$V \propto T \qquad \text{(Charles's law)}$$

Volume *(V)* and Moles *(n)*

$$V \propto n \qquad \text{(Avogadro's law)}$$

The Combined Gas Law: The combined gas law joins Boyle's law and Charles's law.

$$\frac{P_1V_1}{T_1} = \frac{P_2V_2}{T_2}$$

The Ideal Gas Law: The ideal gas law combines the four properties of a gas—pressure, volume, temperature, and number of moles—in a single equation showing their interrelatedness.

$$PV = nRT$$

Relevance

Kinetic Molecular Theory: The kinetic molecular theory predicts many of the properties of gases, including their low density in comparison to solids, their compressibility, and their tendency to assume the shape and volume of their container. The kinetic molecular theory also predicts the ideal gas law.

Pressure: Pressure is a fundamental property of a gas. It allows tires to be inflated and makes it possible to drink from straws.

Simple Gas Laws: Each of the simple gas laws allows us to see how two properties of a gas are interrelated. They are also useful in calculating how one of the properties of a gas changes when another does. Boyle's law, for example, can be used to compute how the volume of a gas will change in response to a change in pressure, or vice versa.

The Combined Gas Law: The combined gas law is used to calculate how a property of a gas (pressure, volume, or temperature) changes when two other properties are changed at the same time.

The Ideal Gas Law: The ideal gas law lets you find any one of the four properties of a gas if you know the other three.

Mixtures of Gases: The pressure due to an individual component in a mixture of gases is called its partial pressure and is defined as the fractional composition of the component multiplied by the total pressure:

Partial pressure of component = Fractional composition of component × Total pressure

Dalton's law states that the total pressure of a mixture of gases is equal to the sum of the partial pressures of its components:

$$P_{tot} = P_a + P_b + P_c + \ldots$$

Mixtures of Gases: Since many gases are not pure but mixtures of several components, it is useful to know how each component contributes to the properties of the entire mixture. The concepts of partial pressure are relevant to deep-sea diving, for example, and to collecting gases over water, where water vapor mixes with the gas being collected.

Gases in Chemical Reactions: Stoichiometric calculations involving gases are similar to those that do not involve gases in that the coefficients in a balanced chemical equation provide conversion factors among moles of reactants and products in the reaction. For gases, the amount of a reactant or product is often specified by the volume of reactant or product at a given temperature and pressure. The ideal gas law is then used to convert from these quantities to moles of reactant or product. Alternatively, at standard temperature and pressure, volume can be converted directly to moles with the following equality:

1 mol = 22.4 L (at STP)

Gases in Chemical Reactions: Reactions involving gases are common in chemistry. For example, many atmospheric reactions—some of which are important to the environment—occur as gaseous reactions.

Chemical Skills

Pressure Unit Conversion (Section 11.3)

Perform pressure unit conversions by following the normal procedure for unit conversion problems. Begin with the quantity you are given and multiply by the appropriate conversion factor(s) to get to the quantity you are trying to find. Find the necessary conversion factors in Table 11.1.

Examples

EXAMPLE 11.13 Pressure Unit Conversion

Convert 18.4 in. Hg to torr.

Given: 18.4 in. Hg

Find: torr

Conversion Factors:

1 atm = 29.92 in. Hg
760 torr = 1 atm

Solution Map:

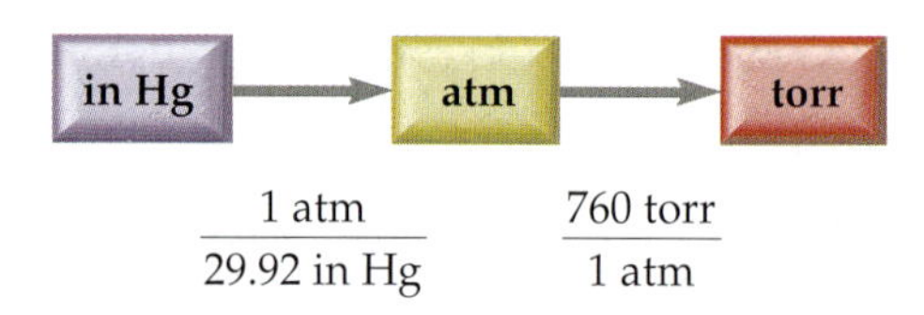

$$\frac{1 \text{ atm}}{29.92 \text{ in Hg}} \qquad \frac{760 \text{ torr}}{1 \text{ atm}}$$

Solution:

$$18.4 \text{ in. Hg} \times \frac{1 \text{ atm}}{29.92 \text{ in. Hg}} \times \frac{760 \text{ torr}}{1 \text{ atm}} = 467 \text{ torr}$$

Simple Gas Laws (Sections 11.4, 11.5, 11.7)

Calculations involving the simple gas laws usually consist of finding one of the initial or final conditions given the other initial and final conditions. Use the relevant formula (Boyle's law, Charles's law, or Avogadro's law) and solve it for the quantity you are trying to find. Then substitute in the appropriate given values and compute the quantity you are trying to find.

$$P_1V_1 = P_2V_2 \qquad \text{(Boyle's law)}$$

$$\frac{V_1}{T_1} = \frac{V_2}{T_2} \qquad \text{(Charles's law)}$$

$$\frac{V_1}{n_1} = \frac{V_2}{n_2} \qquad \text{(Avogadro's law)}$$

EXAMPLE 11.14 Simple Gas Laws

A gas has a volume of 5.7 L at a pressure of 3.2 atm. What is its volume at 4.7 atm?

Given:

$$P_1 = 3.2 \text{ atm}$$
$$V_1 = 5.7 \text{ L}$$
$$P_2 = 4.7 \text{ atm}$$

Find: V_2

Equation:

$$P_1V_1 = P_2V_2$$

Solution Map:

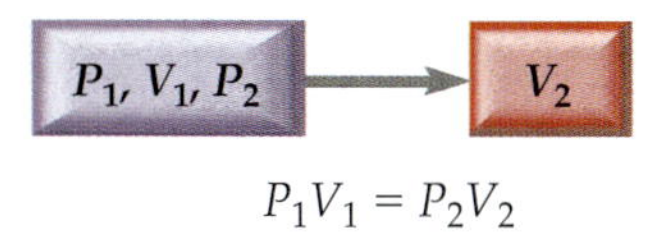

$P_1V_1 = P_2V_2$

Solution:

$$P_1V_1 = P_2V_2$$

$$V_2 = \frac{P_1}{P_2}V_1$$

$$= \frac{3.2 \text{ atm}}{4.7 \text{ atm}} 5.7 \text{ L}$$

$$= 3.9 \text{ L}$$

The Combined Gas Law (Section 11.6)

Calculations involving the combined gas laws usually consist of finding one of the initial or final conditions given the other initial and final conditions. Solve the formula for the quantity you are trying to find. Then substitute in the appropriate given values and compute the quantity you are trying to find.

$$\frac{P_1V_1}{T_1} = \frac{P_2V_2}{T_2} \quad \text{(combined gas law)}$$

EXAMPLE 11.15 The Combined Gas Law

A sample of gas has an initial volume of 2.4 L at a pressure of 855 mm Hg and a temperature of 298 K. If the gas is heated to a temperature of 387 K and expanded to a volume of 4.1 L, what is its final pressure in millimeters of mercury?

Given:

$P_1 = 855$ mm Hg
$V_1 = 2.4$ L
$T_1 = 298$ K
$V_2 = 4.1$ L
$T_2 = 387$ K

Find: P_2

Equation:

$$\frac{P_1V_1}{T_1} = \frac{P_2V_2}{T_2}$$

Solution Map:

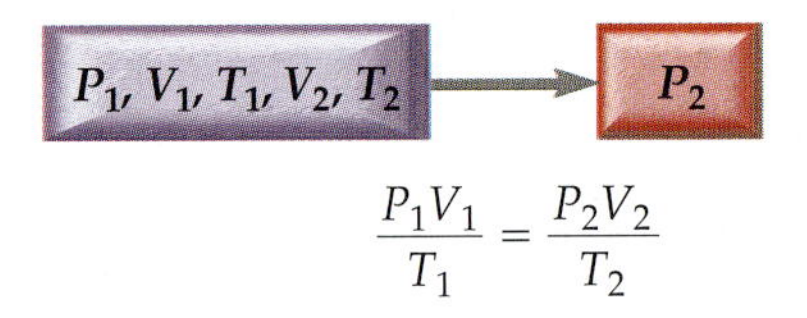

$$\frac{P_1V_1}{T_1} = \frac{P_2V_2}{T_2}$$

Solution:

$$\frac{P_1V_1}{T_1} = \frac{P_2V_2}{T_2}$$

$$P_2 = \frac{P_1V_1T_2}{T_1V_2}$$

$$= \frac{855 \text{ mm Hg} \times 2.4 \cancel{\text{L}} \times 387 \cancel{\text{K}}}{298 \cancel{\text{K}} \times 4.1 \cancel{\text{L}}}$$

$$= 6.5 \times 10^2 \text{ mm Hg}$$

The Ideal Gas Law (Section 11.8)

Calculations involving the ideal gas law often involve finding one of the four quantities (P, V, n, or T) given the other three. Do this by solving the equation for the quantity you are trying to find. Then convert each given quantity to the correct units (if necessary) and substitute into the equation to compute the answer. The correct units for each quantity are:

Pressure (P)—atm
Volume (V)—L
Moles (n)—mol
Temperature (T)—K

$$PV = nRT \quad \text{(ideal gas law)}$$

$$R = 0.0821 \frac{\text{L} \cdot \text{atm}}{\text{mol} \cdot \text{K}}$$

EXAMPLE 11.16 The Ideal Gas Law

Calculate the pressure exerted by 1.2 mol of gas in a volume of 28.2 L and at a temperature of 334 K.

Given:

$n = 1.2$ mol
$V = 28.2$ L
$T = 334$ K

Find: P

Equation:

$$PV = nRT$$

Solution Map:

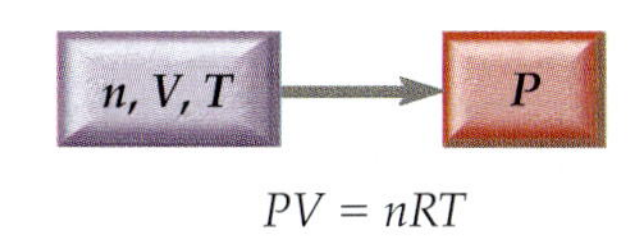

$$PV = nRT$$

Solution:

$$PV = nRT$$

$$P = \frac{nRT}{V}$$

$$= \frac{1.2 \text{ mol} \times 0.0821 \frac{\text{L} \cdot \text{atm}}{\text{mol} \cdot \text{K}} \times 334 \text{ K}}{28.2 \text{ L}}$$

$$= 1.2 \text{ atm}$$

Total Pressure and Partial Pressure (Section 11.9)

The total pressure of a mixture of gases is the sum of the partial pressures of each of the components.

$$P_{\text{tot}} = P_a + P_b + P_c + \ldots$$

EXAMPLE 11.17 Total Pressure and Partial Pressure

A mixture of three gases has the following partial pressures.

$P_{CO_2} = 289$ mm Hg
$P_{O_2} = 342$ mm Hg
$P_{N_2} = 122$ mm Hg

What is the total pressure of the mixture?

Given:

$P_{CO_2} = 289$ mm Hg
$P_{O_2} = 342$ mm Hg
$P_{N_2} = 122$ mm Hg

Find: P_{tot}

Equation:

$$P_{\text{tot}} = P_a + P_b + P_c + \ldots$$

Solution:

$$P_{\text{tot}} = P_{CO_2} + P_{O_2} + P_{N_2}$$

$$= 289 \text{ mm Hg} + 342 \text{ mm Hg} + 122 \text{ mm Hg}$$

$$= 753 \text{ mm Hg}$$

Key Terms

absolute zero [11.5]
atmosphere (atm) [11.3]
Avogadro's law [11.7]
Boyle's law [11.4]
Charles's law [11.5]
combined gas law [11.6]
Dalton's law of partial pressures [11.9]
hypoxia [11.9]
ideal gas constant (R) [11.8]
ideal gas law [11.8]
kinetic molecular theory [11.2]
millimeter of mercury (mm Hg) [11.3]
molar volume [11.10]
nitrogen narcosis [11.9]
oxygen toxicity [11.9]
partial pressure [11.9]
pascal (Pa) [11.3]
pounds per square inch (psi) [11.3]
pressure [11.3]
standard temperature and pressure (STP) [11.10]
torr [11.3]
vapor pressure [11.9]

Exercises

Questions

1. What is pressure?
2. Explain how drinking from a straw works. What causes the drink to go up the straw?
3. Is there an upper limit to how long a straw can theoretically be and still work as a drinking straw?
4. What are the main assumptions of kinetic molecular theory?
5. Describe the main properties of a gas. How are these predicted by kinetic molecular theory?
6. Why do we experience pain in our ears during changes in altitude?
7. What are the main units used to measure pressure?
8. What is Boyle's law?
9. Explain Boyle's law from the perspective of kinetic molecular theory.
10. Explain why scuba divers should not hold their breath as they ascend to the surface.
11. Why would it be impossible to breathe air through an extralong snorkel (greater than a couple of meters) while swimming underwater?
12. What is Charles's law?
13. Explain Charles's law from the perspective of kinetic molecular theory.
14. Explain why hot-air balloons float above the ground.
15. What is the combined gas law? When is it useful?
16. What is Avogadro's law?
17. Explain Avogadro's law from the perspective of kinetic molecular theory.
18. What is the ideal gas law? When is it useful?
19. Under what conditions is the ideal gas law most accurate? Under what conditions does the ideal gas law break down? Why?
20. What is partial pressure?
21. What is Dalton's law?
22. Describe hypoxia and oxygen toxicity.
23. Why do deep-sea divers breathe a mixture of helium and oxygen?
24. When a gas is collected over water, is the gas pure? Why or why not?
25. What is vapor pressure?
26. What is standard temperature and pressure (STP)? What is the molar volume of a gas at STP?

Problems

Converting between Pressure Units

27. Convert each of the following pressure measurements to atm.
    - **(a)** 879 torr
    - **(b)** 19.5 psi
    - **(c)** 30.07 in. Hg
    - **(d)** 98.4×10^3 Pa

28. Convert each of the following pressure measurements to atm.
    - **(a)** 577 mm Hg
    - **(b)** 958 torr
    - **(c)** 115 psi
    - **(d)** 1.78×10^5 Pa

29. Perform each of the following pressure conversions.
    - **(a)** 2.3 atm to torr
    - **(b)** 4.7×10^{-2} atm to millimeters of mercury
    - **(c)** 24.8 psi to millimeters of mercury
    - **(d)** 32.84 in. Hg to torr

30. Perform each of the following pressure conversions.
    - **(a)** 1.06 atm to millimeters of mercury
    - **(b)** 95,422 Pa to millimeters of mercury
    - **(c)** 22.3 psi to torr
    - **(d)** 35.78 in. Hg to millimeters of mercury

31. Complete the following table:

Pascals	*Atmospheres*	*Millimeters of mercury*	*Torr*	*Pounds per square inch*
882 Pa	_____	6.62 mm Hg	_____	_____
_____	0.558 atm	_____	_____	_____
_____	_____	_____	_____	24.8 psi
_____	_____	_____	764 torr	_____
_____	_____	249 mm Hg	_____	_____

32. Complete the following table:

Pascals	*Atmospheres*	*Millimeters of mercury*	*Torr*	*Pounds per square inch*
_____	1.91 atm	_____	1.45×10^3 torr	_____
1.15×10^4 Pa	_____	_____	_____	_____
_____	_____	_____	721 torr	_____
_____	_____	109 mm Hg	_____	_____
_____	_____	_____	_____	38.9 psi

33. The pressure in Denver, Colorado (5280 ft elevation), averages about 24.9 in. Hg. Convert this pressure to:
(a) atmospheres
(b) millimeters of mercury
(c) pounds per square inch
(d) pascals

34. The pressure on top of Mt. Everest averages about 235 mm Hg. Convert this pressure to:
(a) torr
(b) pounds per square inch
(c) inches of mercury
(d) atmospheres

35. The North American record for highest recorded barometric pressure is 31.85 in. Hg, set in 1989 in Northway, Alaska. Convert this pressure to:
(a) millimeters of mercury
(b) atmospheres
(c) torr
(d) kilopascals

36. The world record for lowest pressure (at sea level) was 658 mm Hg, recorded inside Typhoon Ida on September 24, 1958, in the Philippine Sea. Convert this pressure to:
(a) torr
(b) atmospheres
(c) inches of mercury
(d) pounds per square inch

Simple Gas Laws

37. A sample of gas has an initial volume of 2.8 L at a pressure of 755 mm Hg. If the volume of the gas is increased to 3.7 L, what will the pressure be?

38. A sample of gas has an initial volume of 32.6 L at a pressure of 1.3 atm. If the sample is compressed to a volume of 13.8 L, what will its pressure be?

39. A snorkeler with a lung capacity of 6.3 L inhales a lungful of air at the surface, where the pressure is 1.0 atm. The snorkeler then descends to a depth of 25 m, where the pressure increases to 3.5 atm. What is the volume of the snorkeler's lungs at this depth?

40. A scuba diver with a lung capacity of 5.2 L inhales a lungful of air at a depth of 45 m and a pressure of 5.5 atm. If the diver were to ascend to the surface (where the pressure is 1.0 atm) while holding her breath, to what volume would the air in her lungs expand?

41. Use Boyle's law to complete the following table (assume temperature and number of moles of gas to be constant):

P_1	V_1	P_2	V_2
755 mm Hg	2.85 L	885 mm Hg	_____
_____	1.33 L	4.32 atm	2.88 L
192 mm Hg	382 mL	_____	482 mm Hg
2.11 atm	_____	3.82 atm	125 mL

42. Use Boyle's law to complete the following table (assume temperature and number of moles of gas to be constant):

P_1	V_1	P_2	V_2
_____	1.90 L	4.19 atm	1.09 L
755 mm Hg	118 mL	709 mm Hg	_____
2.75 atm	6.75 mL	_____	49.8 mL
343 torr	_____	683 torr	8.79 L

43. A balloon with an initial volume of 3.8 L at a temperature of 305 K is warmed to 385 K. What is its volume at the final temperature?

44. A dramatic classroom demonstration involves cooling a balloon from room temperature (298 K) to liquid nitrogen temperature (77 K). If the initial volume of the balloon is 2.4 L, what will its volume be after it cools?

45. A 48.3-mL sample of gas in a cylinder is warmed from 22 °C to 87 °C. What is its volume at the final temperature?

46. A syringe containing 1.55 mL of oxygen gas is cooled from 95.3 °C to 0.0 °C. What is the final volume of oxygen gas?

47. Use Charles's law to complete the following table (assume pressure and number of moles of gas to be constant):

V_1	T_1	V_2	T_2
1.08 L	25.4 °C	1.33 L	_____
_____	77 K	228 mL	298 K
115 cm^3	_____	119 cm^3	22.4 °C
232 L	18.5 °C	_____	96.2 °C

48. Use Charles's law to complete the following table (assume pressure and number of moles of gas to be constant):

V_1	T_1	V_2	T_2
119 L	10.5 °C	_____	112.3 °C
_____	135 K	176 mL	315 K
2.11 L	15.4 °C	2.33 L	_____
15.4 cm^3	_____	19.2 cm^3	10.4 °C

49. A volume of 3.42 L is occupied by 0.15 mol of nitrogen gas. What is the volume of 0.25 mol of nitrogen gas under the same conditions?

50. A volume of 12.8 L is occupied by 0.57 mol of helium gas. What is the volume of 0.78 mol of helium gas under the same conditions?

51. A balloon contains 0.128 mol of gas and has a volume of 2.76 L. If an additional 0.073 mol of gas is added to the balloon, what will its final volume be?

52. A cylinder with a moveable piston contains 0.87 mol of gas and has a volume of 334 mL. What will its volume be if an additional 0.22 mol of gas is added to the cylinder?

53. Use Avogadro's law to complete the following table (assume pressure and temperature to be constant):

V_1	n_1	V_2	n_2
38.5 mL	1.55×10^{-3} mol	49.4 mL	_____
_____	1.37 mol	26.8 L	4.57 mol
11.2 L	0.628 mol	_____	0.881 mol
422 mL	_____	671 mL	0.0174 mol

54. Use Avogadro's law to complete the following table (assume pressure and temperature to be constant):

V_1	n_1	V_2	n_2
25.2 L	5.05 mol	_____	3.03 mol
_____	1.10 mol	414 mL	0.913 mol
8.63 L	0.0018 mol	10.9 L	_____
53 mL	_____	13 mL	2.61×10^{-4} mol

The Combined Gas Law

55. A sample of gas with an initial volume of 32.5 L at a pressure of 755 mm Hg and a temperature of 315 K is compressed to a volume of 15.8 L and warmed to a temperature of 395 K. What is the final pressure of the gas?

56. A cylinder with a moveable piston contains 188 mL of nitrogen gas at a pressure of 1.12 atm and a temperature of 298 K. What must the final volume be for the pressure of the gas to be 1.42 atm at a temperature of 345 K?

57. A scuba diver takes a 2.8-L balloon from the surface, where the pressure is 1.0 atm and the temperature is 34 °C, to a depth of 25 m, where the pressure is 3.5 atm and the temperature is 18 °C. What is the volume of the balloon at this depth?

58. A bag of potato chips contains 585 mL of air at 25 °C and a pressure of 765 mm Hg. Assuming the bag does not break, what will be its volume at the top of a mountain where the pressure is 442 mm Hg and the temperature is 5.0 °C?

59. A gas sample with a volume of 5.3 L has a pressure of 735 mm Hg at 28 °C. What is the pressure of the sample if the volume remains at 5.3 L but the temperature rises to 86 °C?

60. The total pressure in a 11.7-L automobile tire is 44 psi at 11 °C. By how much does the pressure in the tire rise if it warms to a temperature of 37 °C and the volume remains at 11.7 L?

61. Use the combined gas law to complete the following table (assume the number of moles of gas to be constant):

P_1	V_1	T_1	P_2	V_2	T_2
1.21 atm	1.58 L	12.2 °C	1.54 atm	_____	32.3 °C
721 torr	141 mL	135 K	801 torr	152 mL	_____
5.51 atm	0.879 L	22.1 °C	_____	1.05 L	38.3 °C

62. Use the combined gas law to complete the following table (assume the number of moles of gas to be constant):

P_1	V_1	T_1	P_2	V_2	T_2
1.01 atm	_____	2.7 °C	0.54 atm	0.58 L	42.3 °C
123 torr	41.5 mL	_____	626 torr	36.5 mL	205 K
_____	1.879 L	20.8 °C	0.412 atm	2.05 L	48.1 °C

The Ideal Gas Law

63. What is the volume occupied by 0.118 mol of helium gas at a pressure of 0.97 atm and a temperature of 305 K?

64. What is the pressure in a 10.0-L cylinder filled with 0.448 mol of nitrogen gas at a temperature of 315 K?

65. A cylinder contains 28.5 L of oxygen gas at a pressure of 1.8 atm and a temperature of 298 K. How many moles of gas are in the cylinder?

66. What is the temperature of 0.52 mol of gas at a pressure of 1.3 atm and a volume of 11.8 L?

67. A cylinder contains 11.8 L of air at a total pressure of 43.2 psi and a temperature of 25 °C. How many moles of gas does the cylinder contain?

68 What is the pressure in millimeters of mercury of 0.0115 mol of helium gas with a volume of 214 mL at a temperature of 45 °C?

69. Use the ideal gas law to complete the following table:

P	*V*	*n*	*T*
1.05 atm	1.19 L	0.112 mol	_____
112 torr	_____	0.241 mol	304 K
_____	28.5 mL	1.74×10^{-3} mol	25.4 °C
0.559 atm	0.439 L	_____	255 K

70. Use the ideal gas law to complete the following table:

P	*V*	*n*	*T*
2.39 atm	1.21 L	_____	205 K
512 torr	_____	0.741 mol	298 K
0.433 atm	0.192 L	0.0131 mol	_____
_____	20.2 mL	5.71×10^{-3} mol	20.4 °C

71. How many moles of gas must be forced into a 3.5-L ball to give it a gauge pressure of 9.4 psi at 25 °C? The gauge pressure is relative to atmospheric pressure. Assume that atmospheric pressure is 14.7 psi so that the total pressure in the ball is 24.1 psi.

72. How many moles of gas must be forced into a 4.8-L tire to give it a gauge pressure of 32.4 psi at 25 °C? The gauge pressure is relative to atmospheric pressure. Assume that atmospheric pressure is 14.7 psi so that the total pressure in the tire is 47.1 psi.

73. An experiment shows that a 248-mL gas sample has a mass of 0.433 g at a pressure of 745 mm Hg and a temperature of 28 °C. What is the molar mass of the gas?

74. An experiment shows that a 113-mL gas sample has a mass of 0.171 g at a pressure of 721 mm Hg and a temperature of 32 °C. What is the molar mass of the gas?

75. A sample of gas has a mass of 38.8 mg. Its volume is 224 mL at a temperature of 55 °C and a pressure of 886 torr. Find the molar mass of the gas.

76. A sample of gas has a mass of 0.555 g. Its volume is 117 mL at a temperature of 85 °C and a pressure of 753 mm Hg. Find the molar mass of the gas.

Partial Pressure

77. A gas mixture contains each of the following gases at the indicated partial pressure.

N_2	355 torr
O_2	128 torr
He	229 torr

What is the total pressure of the mixture?

78. A gas mixture contains each of the following gases at the indicated partial pressure.

CO_2	455 mm Hg
Ar	124 mm Hg
O_2	167 mm Hg
H_2	92 mm Hg

What is the total pressure of the mixture?

79. A heliox deep-sea diving mixture delivers an oxygen partial pressure of 0.30 atm when the total pressure is 11.0 atm. What is the partial pressure of helium in this mixture?

80. A mixture of helium, nitrogen, and oxygen has a total pressure of 752 mm Hg. The partial pressures of helium and nitrogen are 234 mm Hg and 197 mm Hg, respectively. What is the partial pressure of oxygen in the mixture?

81. The hydrogen gas formed in a chemical reaction is collected over water at 30 °C at a total pressure of 732 mm Hg. What is the partial pressure of the hydrogen gas collected in this way?

82. The oxygen gas emitted from an aquatic plant during photosynthesis is collected over water at a temperature of 25 °C and a total pressure of 753 torr. What is the partial pressure of the oxygen gas?

83. A gas mixture contains 78% nitrogen and 22% oxygen. If the total pressure is 1.12 atm, what are the partial pressures of each component?

84. An air sample contains 0.038% CO_2. If the total pressure is 758 mm Hg, what is the partial pressure of CO_2?

85. A heliox deep-sea diving mixture contains 4.0% oxygen and 96.0% helium. What is the partial pressure of oxygen when this mixture is delivered at a total pressure of 8.5 atm?

86. A scuba diver breathing normal air descends to 100 m of depth, where the total pressure is 11 atm. What is the partial pressure of oxygen that the diver experiences at this depth? Is the diver in danger of experiencing oxygen toxicity?

Molar Volume

87. Calculate the volume of each of the following gas samples at STP.
(a) 2.5 mol helium
(b) 5.9 mol nitrogen
(c) 32.7 mol Cl_2
(d) 41 mol CH_4

88. Calculate the volume of each of the following gas samples at STP.
(a) 7.8 mol C_2H_6
(b) 0.435 mol CO
(c) 0.298 mol CO_2
(d) 38.9 mol N_2O

89. Calculate the volume of each of the following gas samples at STP.
(a) 73.9 g N_2
(b) 42.9 g O_2
(c) 148 g NO_2
(d) 245 mg CO_2

90. Calculate the volume of each of the following gas samples at STP.
(a) 48.9 g He
(b) 45.2 g Xe
(c) 48.2 mg Cl_2
(d) 3.83 kg SO_2

91. Calculate the mass of each of the following gas samples at STP.
(a) 178 mL CO_2
(b) 155 mL O_2
(c) 1.25 L SF_6

92. Calculate the mass of each of the following gas samples at STP.
(a) 5.82 L NO
(b) 0.324 L N_2
(c) 139 cm^3 Ar

Gases in Chemical Reactions

93. Consider the following chemical reaction:

$$C(s) + H_2O(g) \longrightarrow CO(g) + H_2(g)$$

How many liters of hydrogen gas are formed from the complete reaction of 1.45 mol of C? Assume that the hydrogen gas is collected at a pressure of 1.0 atm and temperature of 355 K.

94. Consider the following chemical reaction:

$$2\,H_2O(l) \longrightarrow 2\,H_2(g) + O_2(g)$$

How many moles of H_2O are required to form 1.8 L of O_2 at a temperature of 315 K and a pressure of 0.957 atm?

95. CH_3OH can be synthesized by the following reaction:

$$CO(g) + 2\,H_2(g) \longrightarrow CH_3OH(g)$$

How many liters of H_2 gas, measured at 748 mm Hg and 86 °C, are required to synthesize 0.55 mol of CH_3OH? How many liters of CO gas, measured under the same conditions, are required?

96. Oxygen gas reacts with powdered aluminum according to the following reaction:

$$4\,Al(s) + 3\,O_2(g) \longrightarrow 2\,Al_2O_3(s)$$

How many liters of O_2 gas, measured at 782 mm Hg and 25 °C, are required to completely react with 2.4 mol of Al?

97. Nitrogen reacts with powdered aluminum according to the following reaction:

$$2\,Al(s) + N_2(g) \longrightarrow 2\,AlN(s)$$

How many liters of N_2 gas, measured at 892 torr and 95 °C, are required to completely react with 18.5 g of Al?

98. Sodium reacts with chlorine gas according to the following reaction:

$$2\,Na(s) + Cl_2(g) \longrightarrow 2\,NaCl(s)$$

What volume of Cl_2 gas, measured at 687 torr and 35 °C, is required to form 28 g of NaCl?

99. How many grams of NH_3 form when 24.8 L of $H_2(g)$ (measured at STP) reacts with N_2 to form NH_3 according to the following reaction?

$$N_2(g) + 3\,H_2(g) \longrightarrow 2\,NH_3(g)$$

100. Lithium reacts with nitrogen gas according to the following reaction:

$$6\,Li(s) + N_2(g) \longrightarrow 2\,Li_3N(s)$$

How many grams of lithium are required to completely react with 58.5 mL of N_2 gas at STP?

101. How many grams of calcium are consumed when 156.8 mL of oxygen gas, measured at STP, reacts with calcium according to the following reaction?

$$2\,Ca(s) + O_2(g) \longrightarrow 2\,CaO(s)$$

102. How many grams of magnesium oxide are formed when 14.8 L of oxygen gas, measured at STP, completely reacts with magnesium metal according to the following reaction?

$$2\,Mg(s) + O_2(g) \longrightarrow 2\,MgO(s)$$

Cumulative Problems

103. Use the ideal gas law to show that the molar volume of a gas at STP is 22.4 L.

104. Use the ideal gas law to show that 28.0 g of nitrogen gas and 4.00 g of helium gas occupy the same volume at any temperature and pressure.

105. The mass of an evacuated 255-mL flask is 143.187 g. The mass of the flask filled with 267 torr of an unknown gas at 25 °C is 143.289 g. Calculate the molar mass of the unknown gas.

106. A 118-mL flask is evacuated and its mass is measured as 97.129 g. When the flask is filled with 768 torr of helium gas at 35 °C, it is found to have a mass of 97.171 g. Is the gas pure helium?

107. A gaseous compound containing hydrogen and carbon is decomposed and found to contain 82.66% carbon and 17.34% hydrogen by mass. The mass of 158 mL of the gas, measured at 556 mm Hg and 25 °C, is found to be 0.275 g. What is the molecular formula of the compound?

108. A gaseous compound containing hydrogen and carbon is decomposed and found to contain 85.63% C and 14.37% H by mass. The mass of 258 mL of the gas, measured at STP, is found to be 0.646 g. What is the molecular formula of the compound?

109. The following reaction is carried out as a source of hydrogen gas in the laboratory:

$$Zn(s) + 2\,HCl(aq) \longrightarrow ZnCl_2(aq) + H_2(g)$$

If 325 mL of hydrogen gas is collected over water at 25 °C at a total pressure of 748 mm Hg, how many grams of Zn reacted?

110. Consider the following reaction:

$$2\,NiO(s) \longrightarrow 2\,Ni(s) + O_2(g)$$

If O_2 is collected over water at 40 °C and a total pressure of 745 mm Hg, what volume of gas will be collected for the complete reaction of 24.78 g of NiO?

111. How many grams of hydrogen are collected in a reaction where 1.78 L of hydrogen gas is collected over water at a temperature of 40 °C and a total pressure of 748 torr?

112. How many grams of oxygen are collected in a reaction where 235 mL of oxygen gas is collected over water at a temperature of 25 °C and a total pressure of 697 torr?

113. The following reaction forms 15.8 g of Ag(*s*):

$$2\,Ag_2O(s) \longrightarrow 4\,Ag(s) + O_2(g)$$

What total volume of gas forms if it is collected over water at a temperature of 25 °C and a total pressure of 752 mm Hg?

114. The following reaction consumes 2.45 kg of CO(*g*):

$$CO(g) + H_2O(g) \longrightarrow CO_2(g) + H_2(g)$$

How many total liters of gas are formed if the products are collected at STP?

115. When hydrochloric acid is poured over a sample of sodium bicarbonate, 28.2 mL of carbon dioxide gas is produced at a pressure of 0.954 atm and a temperature of 22.7 °C. Write an equation for the gas evolution reaction and determine how much sodium bicarbonate reacted.

116. When hydrochloric acid is poured over potassium sulfide, 42.9 mL of hydrogen sulfide gas is produced at a pressure of 752 torr and a temperature of 25.8 °C. Write an equation for the gas evolution reaction and determine how much potassium sulfide (in grams) reacted.

117. Consider the following reaction:

$$2\,SO_2(g) + O_2(g) \longrightarrow 2\,SO_3(g)$$

(a) If 285.5 mL of SO_2 is allowed to react with 158.9 mL of O_2 (both measured at STP), what is the limiting reactant and the theoretical yield of SO_3?
(b) If 187.2 mL of SO_3 is collected (measured at STP), what is the percent yield for the reaction?

118. Consider the following reaction:

$$P_4(s) + 6\,H_2(g) \longrightarrow 4\,PH_3(g)$$

(a) If 88.6 L of $H_2(g)$, measured at STP, is allowed to react with 158.3 g of P_4, what is the limiting reactant?
(b) If 48.3 L of PH_3, measured at STP, forms, what is the percent yield?

119. Consider the following equation for the synthesis of nitric acid:

$$3\,NO_2(g) + H_2O(l) \longrightarrow 2\,HNO_3(aq) + NO(g)$$

(a) If 12.8 L of $NO_2(g)$, measured at STP, is allowed to react with 14.9 g of water, find the limiting reagent and the theoretical yield of HNO_3 in grams.
(b) If 14.8 g of HNO_3 forms, what is the percent yield?

120. Consider the following equation for the production of NO_2 from NO:

$$2\,NO(g) + O_2(g) \longrightarrow 2\,NO_2(g)$$

(a) If 84.8 L of $O_2(g)$, measured at 35 °C and 632 mm Hg, is allowed to react with 158.2 g of NO, find the limiting reagent.
(b) If 97.3 L of NO_2 forms, measured at 35 °C and 632 mm Hg, what is the percent yield?

121. Ammonium carbonate decomposes upon heating according to the following balanced equation:

$$(NH_4)_2CO_3(s) \longrightarrow 2\,NH_3(g) + CO_2(g) + H_2O(g)$$

Calculate the total volume of gas produced at 22 °C and 1.02 atm by the complete decomposition of 11.83 g of ammonium carbonate.

122. Ammonium nitrate decomposes explosively upon heating according to the following balanced equation:

$$2\,NH_4NO_3(s) \longrightarrow 2\,N_2(g) + O_2(g) + 4\,H_2O(g)$$

Calculate the total volume of gas (at 25 °C and 748 mm Hg) produced by the complete decomposition of 1.55 kg of ammonium nitrate.

Highlight Problems

123. Which of the following gas samples, all at the same temperature, will have the greatest pressure? Explain.

(a)

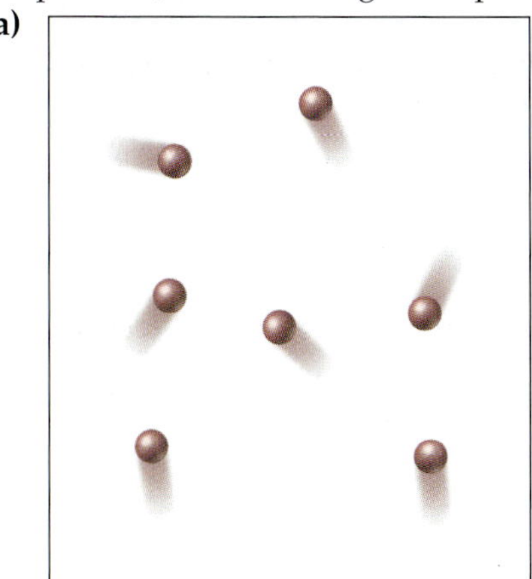

(b)

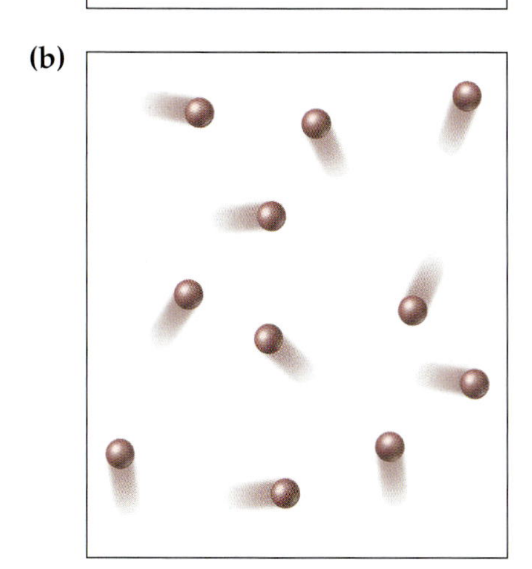

(c)

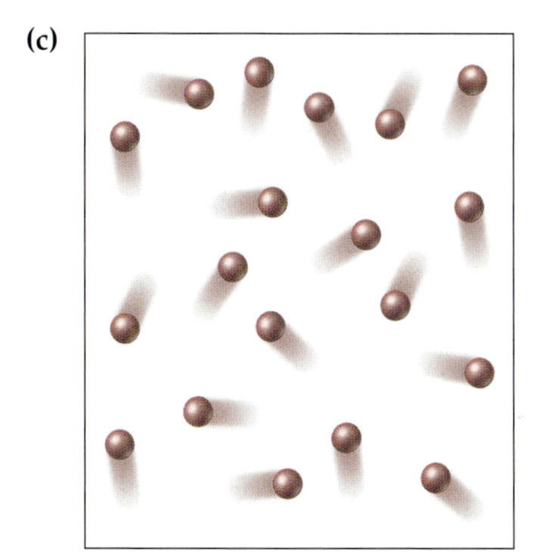

124. The following picture represents a sample of gas at a pressure of 1 atm, a volume of 1 L, and a temperature of 25 °C. Draw a similar picture showing what happens if the volume were reduced to 0.5 L and the temperature increased to 250 °C. What happens to the pressure?

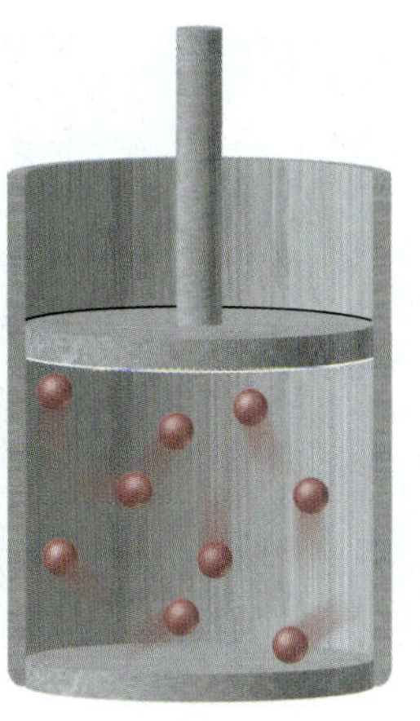

$V = 1.0$ L
$T = 25$ °C
$P = 1.0$ atm

125. Automobile air bags inflate following a serious impact. The impact triggers the following chemical reaction:

$$2\,NaN_3(s) \longrightarrow 2\,Na(s) + 3\,N_2(g)$$

If an automobile air bag has a volume of 11.8 L, how much NaN_3 in grams is required to fully inflate the air bag upon impact? Assume STP conditions.

126. Olympic cyclists fill their tires with helium to make them lighter. Calculate the mass of air in an air-filled tire and the mass of helium in a helium-filled tire. What is the mass difference between the two? Assume that the volume of the tire is 855 mL, that it is filled with a total pressure of 125 psi, and that the temperature is 25 °C. Also, assume an average molar mass for air of 28.8 g/mol.

127. In a common classroom demonstration, a balloon is filled with air and submerged into liquid nitrogen. The balloon contracts as the gases within the balloon cool. Suppose the balloon initially contains 2.95 L of air at a temperature of 25.0 °C and a pressure of 0.998 atm. Calculate the expected volume of the balloon upon cooling to −196 °C (the boiling point of liquid nitrogen). When the demonstration is carried out, the actual volume of the balloon decreases to 0.61 L. How well does the observed volume of the balloon compare to your calculated value? Can you explain the difference?

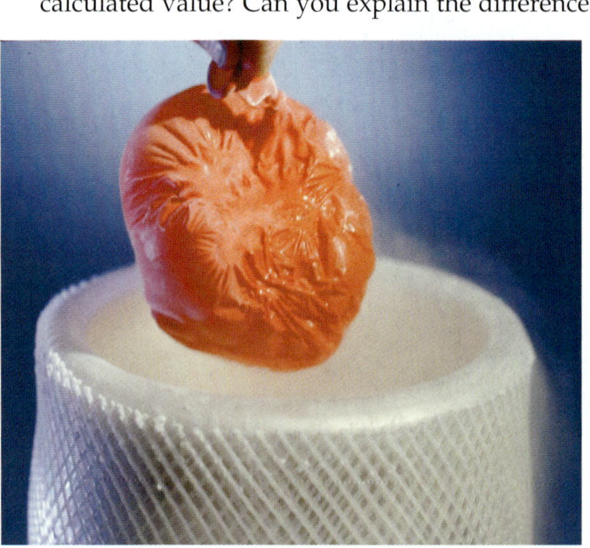

128. Aerosol cans carry clear warnings against incineration because of the high pressures that can develop upon heating. Suppose a can contains a residual amount of gas at a pressure of 755 mm Hg and a temperature of 25 °C. What would the pressure be if the can were heated to 1155 °C?

Answers to Skillbuilder Exercises

Skillbuilder 11.1 85.0 psi

Skillbuilder Plus, p. 349 80.6 kPa

Skillbuilder 11.2 $P_2 = 2.1$ atm; depth is approximately 11 m

Skillbuilder 11.3 123 mL

Skillbuilder 11.4 0.82 L

Skillbuilder 11.5 13 L

Skillbuilder 11.6 1.6 atm

Skillbuilder 11.7 16.1 L

Skillbuilder Plus, p. 366 977 mm Hg

Skillbuilder 11.8 70.8 g/mol

Skillbuilder 11.9 721 torr

Skillbuilder 11.10 $P_{tot} = 4.2$ atm

Skillbuilder 11.11 82.3 g

Skillbuilder 11.12 6.53 L O_2

Answers to Conceptual Checkpoints

11.1 (c) Atmospheric pressure will support a column of water 10.3 m in height. If the liquid in a barometer were twice as dense as water, a column of it would be twice as heavy and the pressure it exerted at its base would be twice as great. Therefore, atmospheric pressure would be able to support a column only half as high.

11.2 (c) The piston will move in response to any pressure difference, keeping the pressure of the gas in the cylinder equal to that of the surrounding atmosphere. Thus, we know that the final pressure of the gas will be the same as it was initially. At constant pressure, the volume of the gas will be proportional to the temperature—if the Kelvin temperature doubles, the volume will double.

CHAPTER 12

Liquids, Solids, and Intermolecular Forces

"It will be found that everything depends on the composition of the forces with which the particles of matter act upon one another; and from these forces . . . all phenomena of nature take their origin."

Roger Joseph Boscovich (1711–1787)

12.1 Interactions between Molecules

◀ Flavors are caused by the interactions of molecules in foods or drinks with molecular receptors on the surface of the tongue. This image shows a caffeine molecule, one of the substances responsible for the sometimes bitter flavors in coffee.

Bite into a candy bar and taste its sweetness. Drink a cup of strong coffee and experience its bitterness. What causes these flavors? Most tastes depend on interactions between molecules. Certain molecules in coffee, for example, interact with molecular receptors on the surface of specialized cells on the tongue. The receptors are highly specific, some recognizing only bitter molecules, others recognizing only sweet. The interaction triggers a signal that goes to the brain, which we interpret as a bitter taste. Bitter tastes are usually unpleasant because many of the molecules that cause them are poisons. The sensation of bitterness is probably an evolutionary adaptation that helps us avoid these poisons.

The interaction between the bitter molecules in coffee and taste receptors on the tongue is caused by **intermolecular forces**—attractive forces that exist *between* molecules. Living organisms depend on intermolecular forces not only for taste but also for many other physiological processes. For example, in Chapter 19, we will learn how intermolecular forces help determine the shapes of protein molecules—the workhorse molecules in living organisms. Later in this chapter—in the *Chemistry and Health* box in Section 12.6—we learn how intermolecular forces are central to DNA, the inheritable molecules that serve as blueprints for life.

The interactions between bitter molecules in coffee and molecular receptors on the tongue are highly specific. However, less-specific intermolecular forces exist between all molecules and atoms. These intermolecular forces are responsible for the very existence of liquids and solids. The state of a sample of matter—solid, liquid, or gas—depends on the magnitude of intermolecular forces relative to the amount of thermal energy in the sample.

Recall from Section 3.9 that the molecules and atoms that compose matter are in constant random motion that increases with increasing temperature. The energy associated with this motion is called *thermal energy*. The weaker the intermolecular forces relative to thermal energy, the more likely the sample will be gaseous. The stronger the intermolecular forces relative to thermal energy, the more likely the sample will be liquid or solid.

12.2 Properties of Liquids and Solids

We are all familiar with solids and liquids. Water, gasoline, rubbing alcohol, and fingernail-polish remover are all common liquids that you have probably encountered. Ice, dry ice, and diamond are familiar solids. In contrast to gases—in which molecules or atoms are separated by large distances—the molecules or atoms that compose liquids and solids are in close contact with one another (▼ Figure 12.1).

The difference between solids and liquids is in the freedom of movement of the constituent molecules or atoms. In liquids, even though the atoms or molecules are in close contact, they are still free to move around each other. In solids, the atoms or molecules are fixed in their positions, although thermal energy causes them to vibrate about a fixed point. These molecular properties of solids and liquids result in the following macroscopic properties.

Properties of Liquids

- High densities in comparison to gases.
- Indefinite shape; they assume the shape of their container.
- Definite volume; they are not easily compressed.

Properties of Solids

- High densities in comparison to gases.
- Definite shape; they do not assume the shape of their container.
- Definite volume; they are not easily compressed.
- May be crystalline (ordered) or amorphous (disordered).

These properties, as well as the properties of gases for comparison, are summarized in Table 12.1.

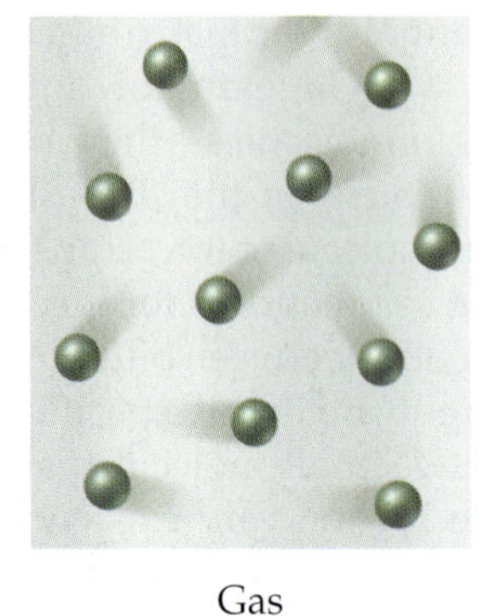
Gas

Liquid

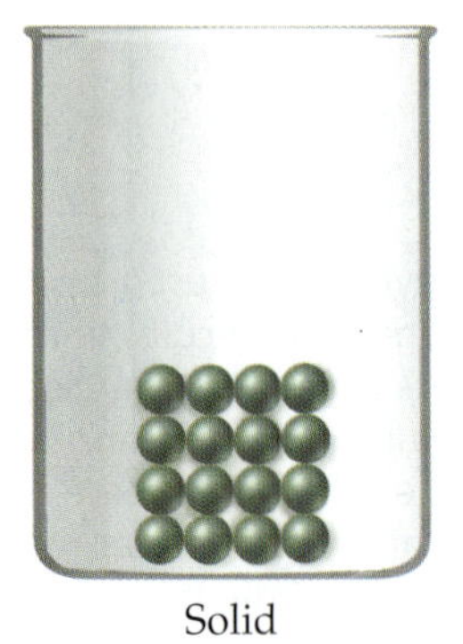
Solid

▲ Figure 12.1 **Gas, liquid, and solid states**

TABLE 12.1 Properties of the Phases of Matter

Phase	Density	Shape	Volume	Strength of Intermolecular Forces*	Example
gas	low	indefinite	indefinite	weak	carbon dioxide gas (CO_2)
liquid	high	indefinite	definite	moderate	liquid water (H_2O)
solid	high	definite	definite	strong	sugar ($C_{12}H_{22}O_{11}$)

*Relative to thermal energy.

▲ **Figure 12.2 A liquid assumes the shape of its container** Because the molecules in liquid water are free to move around each other, they flow and assume the shape of their container.

In comparison to gases, liquids have high densities because the atoms or molecules that compose liquids are much closer together. The density of liquid water, for example, is 1.0 g/cm^3 (at 25 °C), while the density of gaseous water at 100 °C and 1 atm is 0.59 g/L, or 5.9×10^{-4} g/cm^3. Liquids assume the shape of their containers because the atoms or molecules that compose them are free to flow. When you pour water into a flask, the water flows and assumes the shape of the flask (◀ Figure 12.2). Liquids are not easily compressed because the molecules or atoms that compose them are in close contact—they cannot be pushed closer together.

Like liquids, solids have high densities in comparison to gases because the atoms or molecules that compose solids are also close together. The densities of solids are usually just slightly greater than those of the corresponding liquids. A major exception is water, whose solid (ice) is slightly less dense than liquid water. Solids have a definite shape because, in contrast to liquids or gases, the molecules or atoms that compose solids are fixed in place—each molecule or atom only vibrates about a fixed point (▼ Figure 12.3). Like liquids, solids have a definite volume and cannot be compressed because the molecules or atoms composing them are in close contact. As we saw in Section 3.3, solids may be *crystalline*, in which case the atoms or molecules that compose them arrange themselves in a well-ordered, three-dimensional array, or they may be *amorphous*, in which case the atoms or molecules that compose them have no long-range order.

As we will see in Section 12.8, ice is less dense than liquid water because water expands when it freezes due to its unique crystalline structure.

▲ **Figure 12.3 Solids have a definite shape** In a solid such as ice, the molecules are fixed in place. However, they vibrate about fixed points.

12.3 Intermolecular Forces in Action: Surface Tension and Viscosity

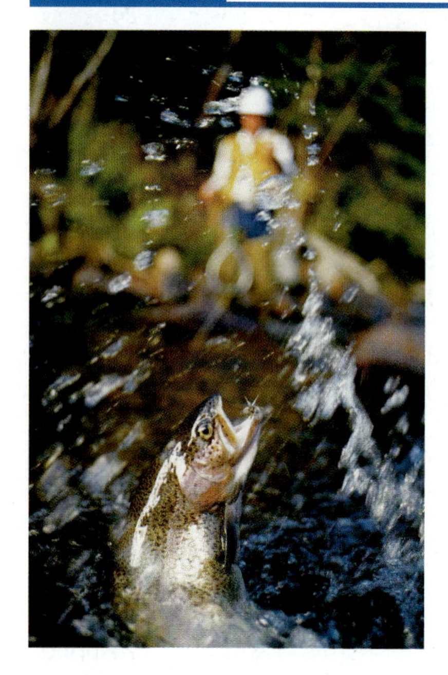

▲ **Figure 12.4 Floating flies** Even though they are denser than water, fly-fishing lures float on the surface of a stream or lake because of surface tension.

The most important manifestation of intermolecular forces is the very existence of liquids and solids. Without intermolecular forces, solids and liquids would not exist. In liquids, we also observe several other manifestations of intermolecular forces.

Surface Tension

A fly fisherman delicately casts a small metal hook (with a few feathers and strings attached to make it look like an insect) onto the surface of a moving stream. The hook floats on the surface of the water and attracts trout (◀ Figure 12.4). Why? The hook floats because of **surface tension**, the tendency of liquids to minimize their surface area. This tendency causes liquids to have a sort of skin that resists penetration. For the fisherman's hook to sink into the water, the water's surface area must increase slightly. This is resisted because molecules at the surface interact only with molecules on the interior of the liquid (▼ Figure 12.5). The surface molecules are therefore subjected to a net inward force (not balanced by any force from the external side), so the surface is under tension that tends to minimize its area. You can observe surface tension by carefully placing a paper clip on the surface of water (▼ Figure 12.6). The paper clip, even though it is denser than water, will float on the surface of the water. A slight tap on the clip will overcome surface tension and cause the clip to sink. Surface tension increases with increasing intermolecular forces. If you try to float a paper clip on gasoline, for example, you can't, because the intermolecular forces among the molecules composing gasoline are weaker than the intermolecular forces among water molecules.

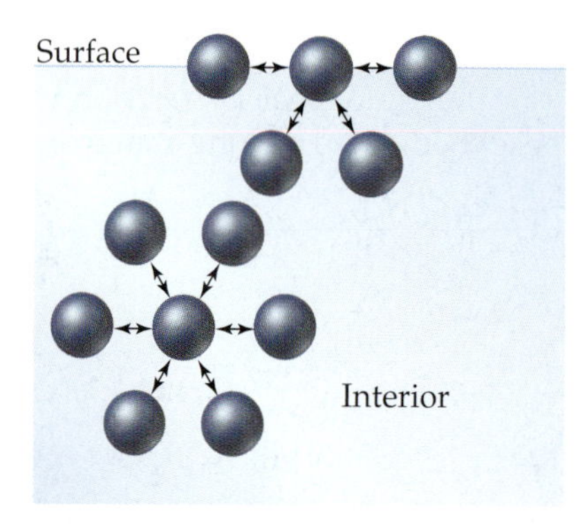

▲ **Figure 12.5 Origin of surface tension** Molecules at the surface of a liquid interact with neighbors on only one side, resulting in a net inward force that creates tension at the surface. This tension tends to minimize the area of the surface and makes it resist penetration.

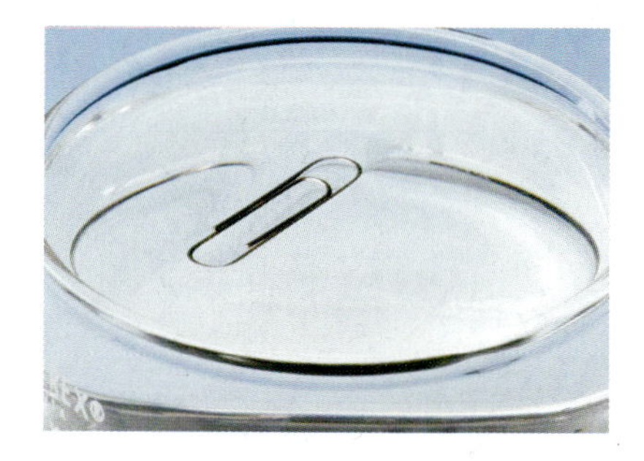

▲ **Figure 12.6 Surface tension at work** A paper clip will float on water if it is carefully placed on the surface of the water. It is held up by surface tension.

Viscosity

▲ **Figure 12.7 Viscosity** Maple syrup is more viscous than water because its molecules interact strongly, and so cannot flow past one another easily.

Another manifestation of intermolecular forces is **viscosity**, the resistance of a liquid to flow. Motor oil, for example, is more viscous than gasoline, and maple syrup is more viscous than water (◀ Figure 12.7). Viscosity is greater in substances with stronger intermolecular forces because molecules cannot move around each other as freely, hindering flow.

EVERYDAY Chemistry

Why Are Water Drops Spherical?

Have you ever seen a close-up photograph of tiny water droplets (▼ Figure 12.8) or carefully watched water in free fall? In both cases, the distorting effects of gravity are diminished, and the water forms nearly perfect spheres. On the space shuttle, the complete absence of gravity results in floating spheres of water (▼ Figure 12.9). Why?

▲ **Figure 12.8 An almost perfect sphere** If a water droplet is small enough, it will largely be free of the distorting effects of gravity and be almost perfectly spherical.

▲ **Figure 12.9 A perfect sphere** In the absence of gravity, as in this picture taken on the space shuttle, water assumes the shape of a sphere.

Water drops are spherical because of the surface tension caused by the attractive forces between water molecules. Just as gravity pulls matter within a planet or star inward to form a sphere, so intermolecular forces pull water molecules inward to form a sphere. The sphere minimizes the surface-area-to-volume ratio, thereby minimizing the number of molecules at the surface.

A collection of magnetic marbles provides a good physical model of a water drop. Each magnetic marble is like a water molecule, attracted to the marbles around it. If you agitate these marbles slightly, so that they can find their preferred configuration, they tend toward a spherical shape (▼ Figure 12.10) because the attractions between the marbles cause them to minimize the number of marbles at the surface.

▲ **Figure 12.10 An analogy for surface tension** Magnetic marbles tend to arrange themselves in a spherical shape.

CAN YOU ANSWER THIS? *How would the tendency of a liquid to form spherical drops depend on the strength of intermolecular forces? Would liquids with weaker intermolecular forces have a higher or lower tendency to form spherical drops?*

12.4 Evaporation and Condensation

Leave a glass of water in the open for several days and the water level within the glass slowly drops. Why? The first reason is that water molecules at the surface of the water—which experience fewer attractions to neighboring molecules and are therefore held less tightly—break away from the rest of the liquid more easily. The second reason is that all of the molecules in the liquid have

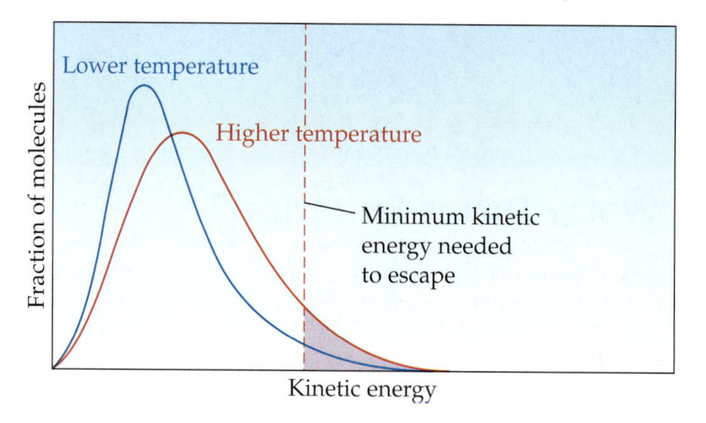

▶ **Figure 12.11 Energy distribution** At a given temperature, a sample of molecules or atoms will have a distribution of kinetic energies, as shown here. Only a small fraction of molecules have enough energy to escape. At a higher temperature, the fraction of molecules with enough energy to escape increases.

a *distribution of energy* at any given temperature (▲ Figure 12.11). At any given moment, some molecules in the liquid are moving faster than the average and others are moving more slowly. Those molecules that are moving much faster than the average have enough energy to break free from the surface, resulting in **evaporation** or **vaporization**, a physical change in which a substance is converted from its liquid form to its gaseous form (◀ Figure 12.12). If you spill the same water on the table, it evaporates faster, probably within a few hours. The surface area of the spilled water is greater, leaving more molecules susceptible to evaporation. If you warm the water, it also evaporates faster, because the greater thermal energy causes more molecules to have higher energy, allowing them to break away from the surface. If you fill the glass with rubbing alcohol instead of water, the liquid again evaporates faster because the intermolecular forces between the alcohol molecules are weaker than the intermolecular forces between water molecules. In general, the rate of vaporization increases with:

In evaporation or vaporization, a substance is converted from its liquid form into its gaseous form.

- Increasing surface area
- Increasing temperature
- Decreasing strength of intermolecular forces

▲ **Figure 12.12 Evaporation** Because molecules on the surface of a liquid are held less tightly than those in the interior, the most energetic among them can break away into the gas phase. This is called *evaporation*.

Liquids that evaporate easily are termed **volatile**, while those that do not vaporize easily are termed **nonvolatile**. Rubbing alcohol, for example, is more volatile than water. Motor oil is virtually nonvolatile.

If you leave water in a *closed* container, it will not evaporate away because the molecules that leave the liquid are trapped in the air space above the water. These gaseous molecules bounce off of the walls of the container and eventually hit the surface of the water again and recondense. **Condensation** is a physical change in which a substance is converted from its gaseous form to its liquid form.

Evaporation and condensation are opposites: Evaporation is a liquid turning into a gas, and condensation is a gas turning into a liquid. When liquid water is first put into a closed container, more evaporation happens than condensation, because there are so few gaseous water molecules in the space above the water (▶ Figure 12.13a). However, as the number of gaseous water molecules increases, the rate of condensation also increases (Figure 12.13b). At the point where the rates of condensation and evaporation become equal (Figure 12.13c), **dynamic equilibrium** is reached and the number of gaseous water molecules above the liquid remains constant. **Vapor pressure** is the partial pressure of a gas in dynamic equilibrium with its liquid. For water at 25 °C, the vapor pressure is 23.8 mm Hg. Vapor pressure increases with:

We call this *dynamic* equilibrium because both condensation and evaporation of individual molecules are still occurring, but at the same rate.

- Increasing temperature
- Decreasing strength of intermolecular forces

Vapor pressure is independent of surface area because an increase in surface area at equilibrium equally affects both the rate of evaporation and the rate of condensation.

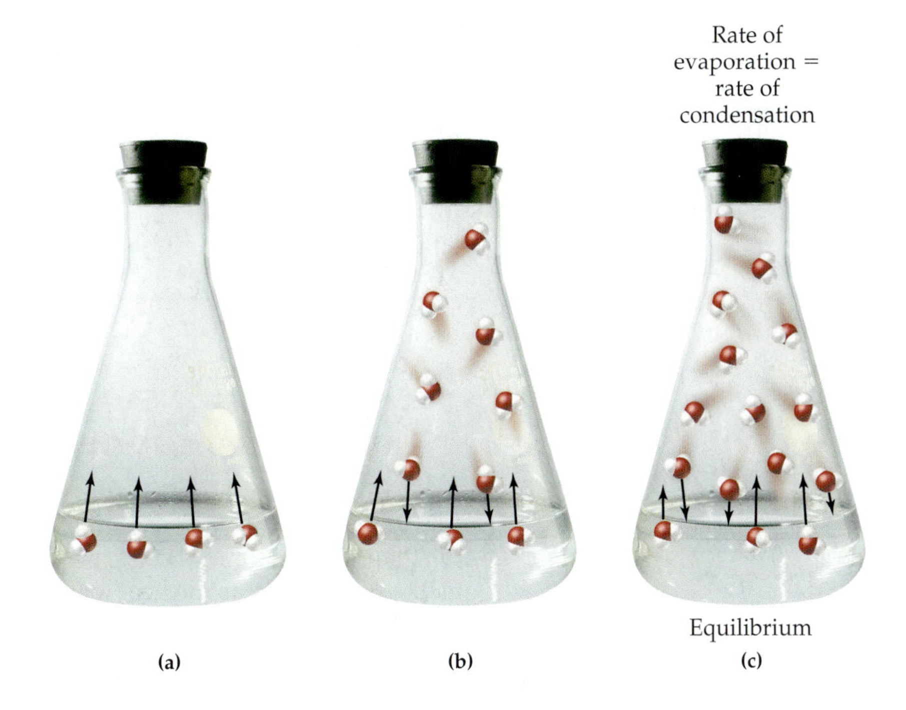

▶ **Figure 12.13 Evaporation and condensation** **(a)** When water is first put into a closed container, water molecules begin to evaporate. **(b)** As the number of gaseous molecules increases, some of the molecules begin to collide with the liquid and are recaptured. That is, they recondense into liquid. **(c)** When the rate of evaporation equals the rate of condensation, dynamic equilibrium occurs, and the number of gaseous molecules remains constant.

Boiling

As you increase the temperature of water in an open container, the thermal energy causes molecules to leave the surface and vaporize at a faster and faster rate. At the **boiling point**—the temperature at which the vapor pressure of a liquid is equal to the pressure above it—the thermal energy is enough for molecules within the interior of the liquid (not just those at the surface) to break free into the gas phase (▼ Figure 12.14). Water's **normal boiling point**—its boiling point at a pressure of 1 atmosphere—is 100 °C. When a sample of water reaches 100 °C, you see bubbles form. These bubbles are pockets of gaseous water that have formed within the liquid water. The bubbles float to the surface and leave as gaseous water, or steam. Once the boiling point of a liquid is reached, additional heating only causes more rapid boiling; it does not raise the temperature of the liquid above its boiling point (▼ Figure 12.15). Therefore, boiling water at 1 atm will always have a temperature of 100 °C. After all the water has been converted to steam, the temperature of the steam can continue to rise beyond 100 °C.

Sometimes you see bubbles begin to form in hot water below 100 °C. These bubbles are dissolved air—not gaseous water—leaving the liquid. Dissolved air comes out of water as you heat it because the solubility of a gas in a liquid decreases with increasing temperature (Section 13.4).

▼ **Figure 12.14 Boiling** During boiling, thermal energy is enough to cause water molecules in the interior of the liquid to become gaseous, forming bubbles containing gaseous water molecules.

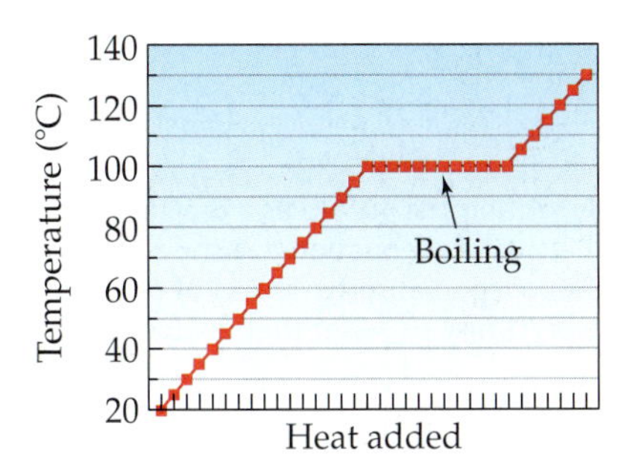

▲ **Figure 12.15 Heating curve during boiling** The temperature of water as it is heated from room temperature through boiling. During boiling, the temperature remains at 100 °C until all the liquid is evaporated.

CONCEPTUAL CHECKPOINT 12.1

The gas over a rapidly boiling pot of water is sampled and analyzed. Which of the following substances do you expect to compose a large fraction of the gas sample?

(a) H_2
(b) H_2O
(c) O_2
(d) H_2O_2

Energetics of Evaporation and Condensation

In an endothermic process, heat is absorbed. In an exothermic process, heat is released.

Evaporation is **endothermic**—heat is absorbed when a liquid is converted into a gas because energy is required to break a molecule away from the rest of the liquid. If you turn off the heat beneath a boiling pot of water, it will quickly stop boiling as the heat lost due to vaporization causes the water to cool below its boiling point. Our bodies use evaporation as a cooling mechanism. When we get overheated, we sweat, causing our skin to be covered with liquid water. As this water evaporates, the fastest-moving molecules break away, leaving the slower-moving molecules behind. This process lowers the overall energy of the water molecules remaining on the skin, resulting in a cooling effect. A fan intensifies the cooling effect because it blows newly vaporized water away from the skin, allowing more sweat to vaporize and causing even more cooling. High humidity, on the other hand, slows down evaporation, preventing cooling. When the air already contains high amounts of water vapor, sweat does not evaporate easily, making our cooling system less efficient.

Condensation, the opposite of evaporation, is **exothermic**—heat is released when a gas condenses to a liquid. If you have ever accidentally put your hand above a steaming kettle, you may have experienced a *steam burn*. As the steam condenses to a liquid on your skin, it releases heat, causing a severe burn. The condensation of water vapor is also the reason that winter overnight temperatures in coastal cities, which tend to have water vapor in the air, do not get as low as in deserts, which tend to have dry air. As the air temperature in a coastal city drops, water condenses out of the air, releasing heat and preventing the temperature from dropping further. In deserts, there is little moisture in the air to condense, so the temperature drop is greater.

Heat of Vaporization

The amount of heat required to vaporize one mole of liquid is called the **heat of vaporization** (ΔH_{vap}). The heat of vaporization of water at its normal boiling point (100 °C) is 40.7 kJ/mole.

By writing the 40.7 kJ as a reactant in the chemical equation, we show that the reaction absorbs 40.7 kJ of heat for every 1 mol of water that is vaporized.

$$40.7 \text{ kJ} + H_2O(l) \longrightarrow H_2O(g) \qquad \text{(at 100 °C)}$$

The same amount of heat is involved when 1 mol of gas condenses, but the heat is emitted rather than absorbed.

By writing the 40.7 kJ as a product in the chemical equation, we show that the reaction releases 40.7 kJ of heat for every 1 mol of water that condenses.

$$H_2O(g) \longrightarrow H_2O(l) + 40.7 \text{ kJ} \qquad \text{(at 100 °C)}$$

Different liquids have different heats of vaporization (Table 12.2). Heats of vaporization are also *temperature dependent* (they change with temperature). The higher the temperature, the easier it is to vaporize a given liquid and therefore the lower the heat of vaporization.

TABLE 12.2 Heats of Vaporization of Several Liquids at Their Boiling Points and at 25 °C

Liquid	Chemical Formula	Normal Boiling Point (°C)	Heat of Vaporization (kJ/mol) at Boiling Point	Heat of Vaporization (kJ/mol) at 25 °C
water	H_2O	100	40.7	44.0
isopropyl alcohol (rubbing alcohol)	C_3H_8O	82.3	39.9	45.4
acetone	C_3H_6O	56.1	29.1	31.0
diethyl ether	$C_4H_{10}O$	34.5	26.5	27.1

The heat of vaporization of a liquid can be used to calculate the amount of heat energy required to vaporize a given amount of the liquid. The heat of vaporization can be viewed as a conversion factor between moles of a liquid and the amount of heat required to vaporize it. For example, suppose we want to calculate the amount of heat required to vaporize 25.0 g of water at its boiling point. We set up the problem in the normal way.

Given: 25.0 g H_2O

Find: heat (kJ)

Conversion Factors:

$$\Delta H_{vap} = 40.7 \text{ kJ/mol} \qquad \text{(at 100 °C)}$$
$$1 \text{ mol } H_2O = 18.02 \text{ g } H_2O$$

Solution Map:

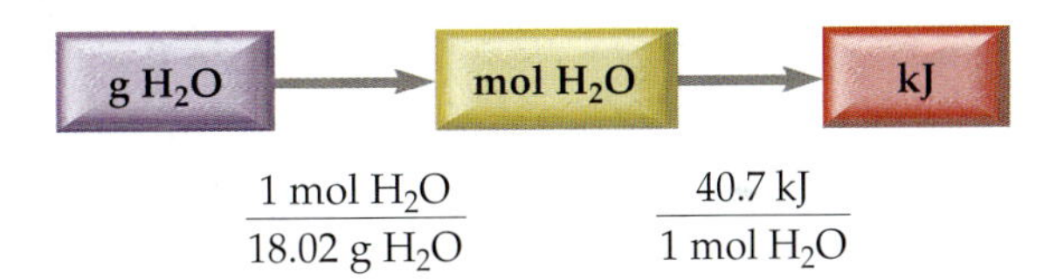

$$\frac{1 \text{ mol } H_2O}{18.02 \text{ g } H_2O} \qquad \frac{40.7 \text{ kJ}}{1 \text{ mol } H_2O}$$

The solution map begins with g of water, shows the conversion to mol of water using the molar mass of water, and then shows the conversion to kJ using ΔH_{vap}.

Solution:

$$25.0 \text{ g } H_2O \times \frac{1 \text{ mol } H_2O}{18.02 \text{ g } H_2O} \times \frac{40.7 \text{ kJ}}{1 \text{ mol } H_2O} = 56.5 \text{ kJ}$$

EXAMPLE 12.1 Using the Heat of Vaporization in Calculations

Calculate the amount of water in grams that can be vaporized at its boiling point with 155 kJ of heat.

Set up the problem in the normal way. You are given the number of kilojoules of heat energy and asked to find the mass of water that can be vaporized with the given amount of energy. The required conversion factors are the heat of vaporization and the molar mass of water.

Given: 155 kJ

Find: g H_2O

Conversion Factors:

$$\Delta H_{vap} = 40.7 \text{ kJ/mol} \qquad \text{(at 100 °C)}$$
$$18.02 \text{ g } H_2O = 1 \text{ mol } H_2O$$

Draw the solution map beginning with the energy in kilojoules and converting to moles of water and then to grams of water.

Solution Map:

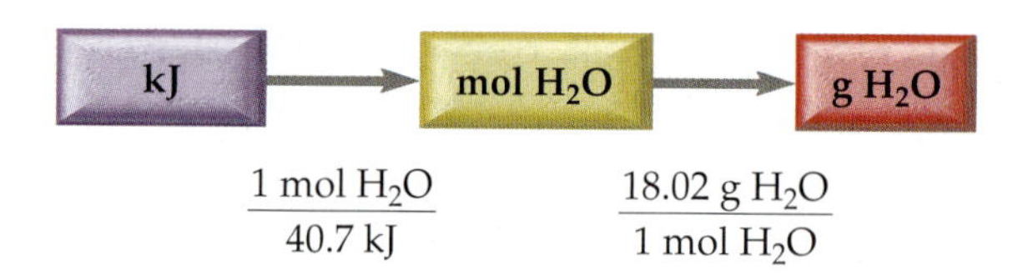

$$\frac{1 \text{ mol H}_2\text{O}}{40.7 \text{ kJ}} \qquad \frac{18.02 \text{ g H}_2\text{O}}{1 \text{ mol H}_2\text{O}}$$

Follow the solution map to solve the problem.

Solution:

$$155 \cancel{\text{kJ}} \times \frac{1 \cancel{\text{mol H}_2\text{O}}}{40.7 \cancel{\text{kJ}}} \times \frac{18.02 \text{ g}}{1 \cancel{\text{mol H}_2\text{O}}} = 68.6 \text{ g}$$

SKILLBUILDER 12.1 Using the Heat of Vaporization in Calculations

Calculate the amount of heat in kilojoules required to vaporize 2.58 kg of water at its boiling point.

SKILLBUILDER PLUS

A drop of water weighing 0.48 g condenses on the surface of a 55-g block of aluminum that is initially at 25 °C. If the heat released during condensation goes only toward heating the metal, what is the final temperature in Celsius of the metal block? (The specific heat capacity of aluminum is 0.903 J/g °C.)

12.5 Melting, Freezing, and Sublimation

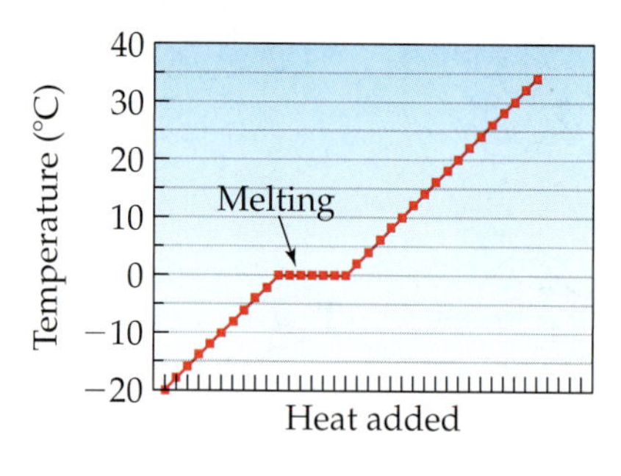

▲ **Figure 12.16 Heating curve during melting** A graph of the temperature of ice as it is heated from −20 °C to 35 °C. During melting, the temperature remains at 0 °C until the entire solid is melted.

As you increase the temperature of a solid, thermal energy causes the molecules and atoms composing the solid to vibrate faster. At the **melting point**, atoms and molecules have enough thermal energy to overcome the intermolecular forces that hold them at their stationary points, and the solid turns into a liquid. The melting point of ice, for example, is 0 °C. Once the melting point of a solid is reached, additional heating only causes more rapid melting; it does not raise the temperature of the solid above its melting point (◄ Figure 12.16). Only after all of the ice has melted will additional heating raise the temperature of the liquid water past 0 °C. A mixture of water *and* ice will always have a temperature of 0 °C (at 1 atm pressure).

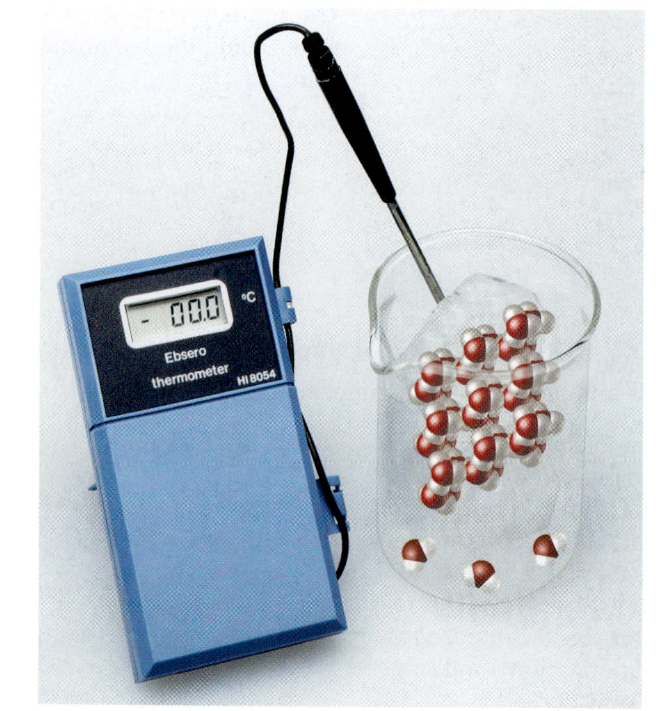

► When ice melts, water molecules break free from the solid structure and become a liquid. As long as ice and water are both present, the temperature will be 0.0 °C.

Energetics of Melting and Freezing

The most common way to cool down a drink is to drop several ice cubes into it. As the ice melts, the drink cools because melting is endothermic—heat is absorbed when a solid is converted into a liquid. The melting ice absorbs heat from the liquid in the drink and cools the liquid.

TUTORIAL Changes of State 

Freezing, the opposite of melting, is exothermic—heat is released when a liquid freezes into a solid. For example, as water in your freezer turns into ice, it releases heat, which must be removed by the refrigeration system of the freezer. If the refrigeration system did not remove the heat, the water would not completely freeze into ice. The heat released as it began to freeze would warm the freezer, preventing further freezing.

Heat of Fusion

The amount of heat required to melt 1 mol of a solid is called the **heat of fusion** (ΔH_{fus}). The heat of fusion for water is 6.02 kJ/mol.

$$6.02 \text{ kJ} + H_2O(s) \longrightarrow H_2O(l)$$

The same amount of heat is involved when 1 mol of liquid water freezes, but the heat is emitted rather than absorbed.

$$H_2O(l) \longrightarrow H_2O(s) + 6.02 \text{ kJ}$$

Different substances have different heats of fusion (Table 12.3).

TABLE 12.3 Heats of Fusion of Several Substances

Liquid	Chemical Formula	Melting Point (°C)	Heat of Fusion (kJ/mol)
water	H_2O	0.00	6.02
isopropyl alcohol (rubbing alcohol)	C_3H_8O	−89.5	5.37
acetone	C_3H_6O	−94.8	5.69
diethyl ether	$C_4H_{10}O$	−116.3	7.27

Notice that, in general, the heat of fusion is significantly less than the heat of vaporization. It takes less energy to melt 1 mol of ice than it does to vaporize 1 mol of liquid water. Why? Vaporization requires complete separation of one molecule from another, so the intermolecular forces must be completely overcome. Melting, on the other hand, requires that intermolecular forces be only partially overcome, allowing molecules to move around one another while still remaining in contact.

The heat of fusion can be used to calculate the amount of heat energy required to melt a given amount of a solid. The heat of fusion can be viewed as a conversion factor between moles of a solid and the amount of heat required to melt them. For example, suppose we want to calculate the amount of heat required to melt 25.0 g of ice. We set up the problem in the normal way.

Given: 25.0 g H_2O

Find: heat (kJ)

Conversion Factor:

$\Delta H_{fus} = 6.02$ kJ/mol

1 mol H_2O = 18.02 g H_2O

Solution Map:

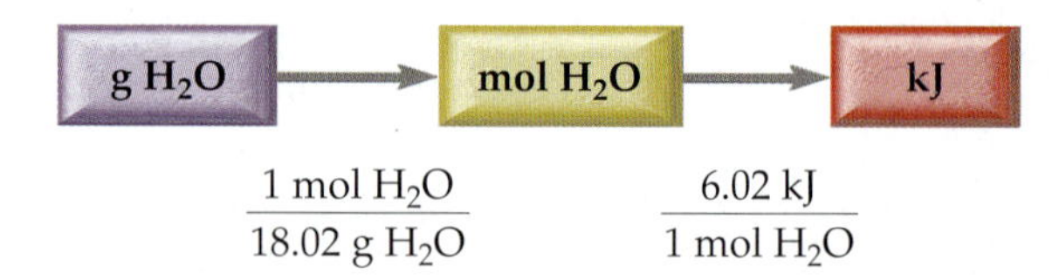

$$\frac{1 \text{ mol } H_2O}{18.02 \text{ g } H_2O} \qquad \frac{6.02 \text{ kJ}}{1 \text{ mol } H_2O}$$

The solution map begins with g of ice, shows the conversion to mol of ice using the molar mass of water, and then shows the conversion to kJ using ΔH_{fus}.

Solution:

$$25.0 \text{ g } H_2O \times \frac{1 \text{ mol } H_2O}{18.02 \text{ g } H_2O} \times \frac{6.02 \text{ kJ}}{1 \text{ mol } H_2O} = 8.35 \text{ kJ}$$

EXAMPLE 12.2 Using the Heat of Fusion in Calculations

Calculate the amount of ice in grams that, upon melting, absorbs 237 kJ of heat.

Set up the problem in the normal way. You are given the number of kJ of heat energy and asked to find the mass of ice that absorbs the given amount of energy upon melting. The required conversion factors are the heat of fusion and the molar mass of water.

Given: 237 kJ

Find: g H_2O (ice)

Conversion Factors:

$\Delta H_{fus} = 6.02$ kJ/mol

1 mol H_2O = 18.02 g H_2O

Draw the solution map beginning with the energy in kilojoules and converting to moles of water and then to grams of water.

Solution Map:

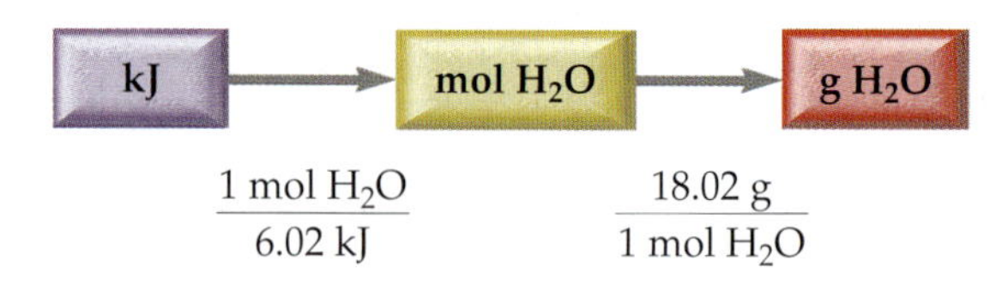

$$\frac{1 \text{ mol } H_2O}{6.02 \text{ kJ}} \qquad \frac{18.02 \text{ g}}{1 \text{ mol } H_2O}$$

Follow the solution map to solve the problem.

Solution:

$$237 \text{ kJ} \times \frac{1 \text{ mol } H_2O}{6.02 \text{ kJ}} \times \frac{18.02 \text{ g}}{1 \text{ mol } H_2O} = 709 \text{ g}$$

SKILLBUILDER 12.2 Using the Heat of Fusion in Calculations

Calculate the amount of heat absorbed when a 15.5-g ice cube melts.

SKILLBUILDER PLUS

A 5.6-g ice cube is placed into 195 g of water initially at 25 °C. If the heat absorbed for melting the ice comes only from the 195 g of water, what is the temperature change of the 195 g of water?

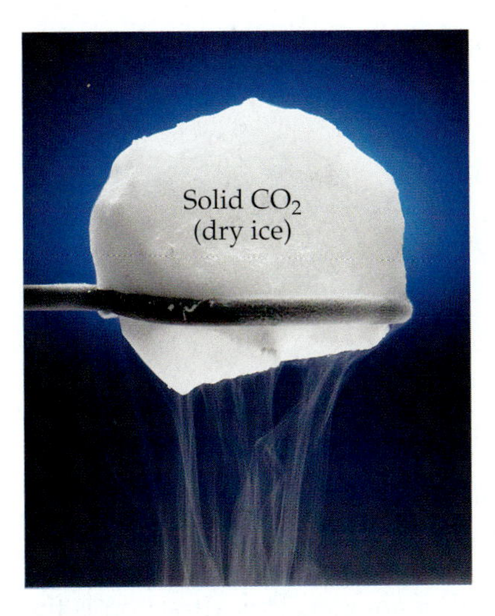

▲ Dry ice is solid carbon dioxide. The solid does not melt but rather sublimes. It goes directly from solid carbon dioxide to gaseous carbon dioxide.

Sublimation

Sublimation is a physical change in which a substance is converted from its solid form directly to its gaseous form. When a substance sublimes, molecules leave the surface of the solid, where they are held less tightly than in the interior, and become gaseous. For example, dry ice, which is solid carbon dioxide, does not melt under atmospheric pressure. At −78 °C the CO_2 molecules have enough energy to leave the surface of the dry ice and become gaseous. Regular ice will slowly sublime at temperatures below 0 °C. In cold climates, ice and snow will slowly disappear, even if the temperature remains below 0 °C. Similarly, ice cubes left in the freezer for a long time slowly become smaller, even though the freezer is always below 0 °C. In both cases, the ice is subliming, turning directly into water vapor.

Ice also sublimes out of frozen foods. This can be seen, for example, when food is frozen in an airtight plastic bag for a long time. The ice crystals that form in the bag are water that has sublimed out of the food and redeposited on the surface of the bag. For this reason, food that remains frozen for too long becomes dried out. This can be avoided to some degree by freezing foods to colder temperatures, a process called deep-freezing. The colder temperature lowers the rate of sublimation and preserves the food longer.

12.6 Types of Intermolecular Forces: Dispersion, Dipole–Dipole, and Hydrogen Bonding

The strength of the intermolecular forces between the molecules or atoms that compose a substance determines the state—solid, liquid, or gas—of the substance at room temperature. Strong intermolecular forces tend to result in liquids and solids (high melting and boiling points). Weak intermolecular forces tend to result in gases (low melting and boiling points). In this book, we focus on three fundamental types of intermolecular forces. In order of increasing strength, they are the dispersion force, the dipole–dipole force, and the hydrogen bond.

Dispersion Force

The nature of dispersion forces was first recognized by Fritz W. London (1900–1954), a German-American physicist.

The default intermolecular force, present in all molecules and atoms, is the **dispersion force** (also called the *London force*). Dispersion forces are caused by fluctuations in the electron distribution within molecules or atoms. Since all atoms and molecules have electrons, they all have dispersion forces. The electrons in an atom or molecule may, at any one instant, be unevenly distributed. For example, imagine a frame-by-frame movie of a helium atom in which each "frame" captures the position of the helium atom's two electrons (▼ Figure 12.17). In any one frame, the electrons are not symmetrically arranged around the nucleus. In frame 3, for example, helium's two electrons are on the left side of the helium atom. The left side then acquires a slightly negative charge (δ^-). The right side of the atom, which is void of electrons, acquires a slightly positive charge (δ^+).

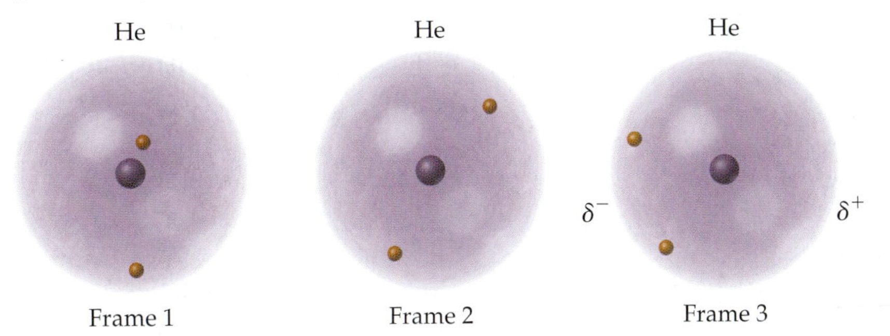

▶ **Figure 12.17 Instantaneous dipoles** Random fluctuations in the electron distribution of a helium atom cause instantaneous dipoles to form.

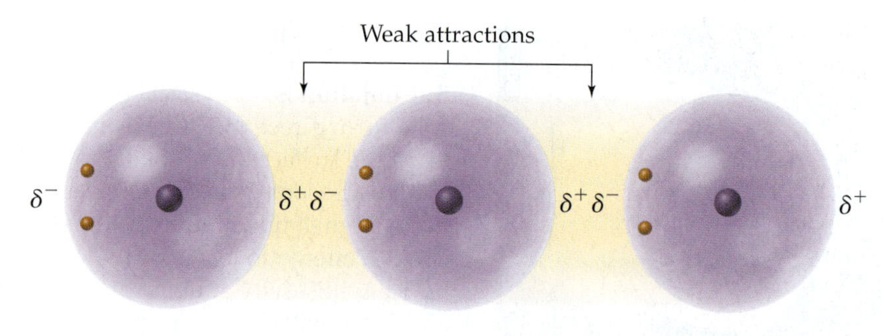

▶ **Figure 12.18 Dispersion force** An instantaneous dipole on any one helium atom induces instantaneous dipoles on neighboring atoms. The neighboring atoms then attract one another. This attraction is called the dispersion force.

This fleeting charge separation is called an **instantaneous dipole** (or *temporary dipole*). An instantaneous dipole on one helium atom induces an instantaneous dipole on its neighboring atoms because the positive end of the instantaneous dipole attracts electrons in the neighboring atoms (▲ Figure 12.18). The neighboring atoms then attract one another—the positive end of one instantaneous dipole attracts the negative end of another. This attraction is the dispersion force. The dipoles responsible for the dispersion force are transient, constantly appearing and disappearing in response to fluctuations in the electron clouds.

To *polarize* means to form a dipole moment.

The magnitude of the dispersion force depends on how easily the electrons in the atom or molecule can move or *polarize* in response to an instantaneous dipole, which in turn depends on the size of the electron cloud. A larger electron cloud results in a greater dispersion force because the electrons are held less tightly by the nucleus and therefore can polarize more easily. If all other variables are constant, the dispersion force increases with increasing molar mass. For example, consider the boiling points of the noble gases displayed in Table 12.4. As the molar mass of the noble gas increases, its boiling point increases. While molar mass alone does not determine the magnitude of the dispersion force, it can be used as a guide when comparing dispersion forces within a family of similar elements or compounds.

TABLE 12.4 Noble Gas Boiling Points

Noble Gas	Molar Mass (g/mol)	Boiling Point (K)
He	4.00	4.2 K
Ne	20.18	27 K
Ar	39.95	87 K
Kr	83.80	120 K
Xe	131.29	165 K

EXAMPLE 12.3 Dispersion Forces

Which halogen, Cl_2 or I_2, has the higher boiling point?

Solution:

The molar mass of Cl_2 is 70.90 g/mol and the molar mass of I_2 is 253.81 g/mol. Since I_2 has the higher molar mass, it has stronger dispersion forces and therefore the higher boiling point.

SKILLBUILDER 12.3 Dispersion Forces

Which hydrocarbon, CH_4 or C_2H_6, has the higher boiling point?

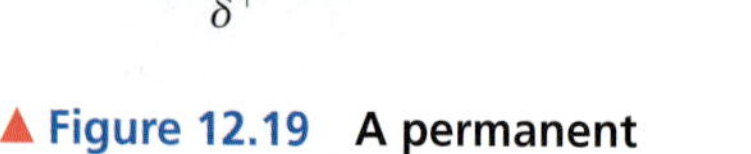

▲ **Figure 12.19 A permanent dipole** Molecules such as formaldehyde are polar and therefore have a permanent dipole.

Dipole–Dipole Force

The **dipole–dipole force** exists in all molecules that are polar. Polar molecules have **permanent dipoles** (Section 10.8) that interact with the permanent dipoles of neighboring molecules (◀ Figure 12.19). The positive end of one

See Section 10.8 to review how to determine whether a molecule is polar.

permanent dipole is attracted to the negative end of another; this attraction is the dipole–dipole force (◀ Figure 12.20). Polar molecules, therefore, have higher melting and boiling points than nonpolar molecules of similar molar mass. Remember that all molecules (including polar ones) have dispersion forces. In addition, polar molecules have dipole–dipole forces. This additional attractive force raises their melting and boiling points relative to nonpolar molecules of similar molar mass. For example, consider the following two compounds:

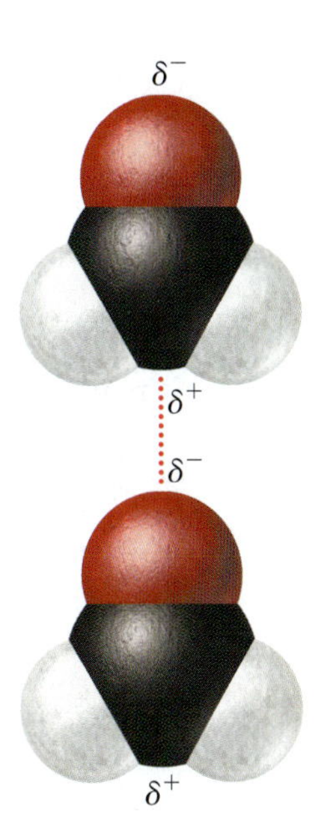

▲ **Figure 12.20 Dipole–dipole attraction** The positive end of a polar molecule is attracted to the negative end of its neighbor, giving rise to the dipole–dipole force.

Name	Formula	Molar mass (g/mol)	Structure	bp (°C)	mp (°C)
Formaldehyde	CH_2O	30.0	H—C(=O)—H	−19.5	−92
Ethane	C_2H_6	30.0	$H_3C—CH_3$ (H—C(H)(H)—C(H)(H)—H)	−88	−172

Formaldehyde is polar, and therefore has a higher melting point and boiling point than nonpolar ethane, even though the two compounds have the same molar mass.

The polarity of molecules composing liquids is also important in determining the **miscibility**—the ability to mix without separating into two phases—of liquids. In general, polar liquids are miscible with other polar liquids but are not miscible with nonpolar liquids. For example, water, a polar liquid, is not miscible with pentane (C_5H_{12}), a nonpolar liquid (▼ Figure 12.21). Similarly, water and oil (also nonpolar) do not mix. Consequently, oily hands or oily stains on clothes cannot be washed away with plain water (see Chapter 10, *Everyday Chemistry: How Soap Works*).

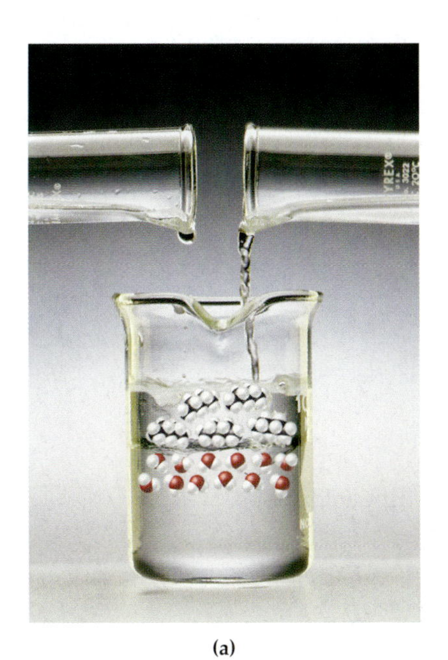

(a)

(b)

(c)

▲ **Figure 12.21 Polar and nonpolar compounds** **(a)** Pentane, a nonpolar compound, does not mix with water, a polar compound. **(b)** For the same reason, the oil and vinegar (largely a water solution of acetic acid) in salad dressing tend to separate into distinct layers. **(c)** An oil spill from a tanker demonstrates dramatically that petroleum and seawater are not miscible.

EXAMPLE 12.4 Dipole–Dipole Forces

Which of the following molecules have dipole–dipole forces?

(a) CO_2 **(b)** CH_2Cl_2 **(c)** CH_4

Solution:

A molecule will have dipole–dipole forces if it is polar. To determine whether a molecule is polar, you must:

1. determine whether the molecule contains polar bonds, and
2. determine whether the polar bonds add together to form a net dipole moment (Section 10.8).

(a) Since the electronegativities of carbon and oxygen are 2.5 and 3.5, respectively (Figure 10.2), CO_2 has polar bonds. The geometry of CO_2 is linear. Consequently, the polar bonds cancel; the molecule is not polar and does not have dipole–dipole forces.

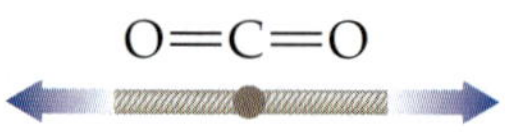

O=C=O

Nonpolar; no dipole–dipole forces

(b) The electronegativities of C, H, and Cl are 2.5, 2.1, and 3.5, respectively. Consequently, CH_2Cl_2 has two polar bonds (C—Cl) and two bonds that are nearly nonpolar (C—H). The geometry of CH_2Cl_2 is tetrahedral. Since the C—Cl bonds and the C—H bonds are different, they do not cancel, but sum to a net dipole moment. Therefore the molecule is polar and has dipole–dipole forces.

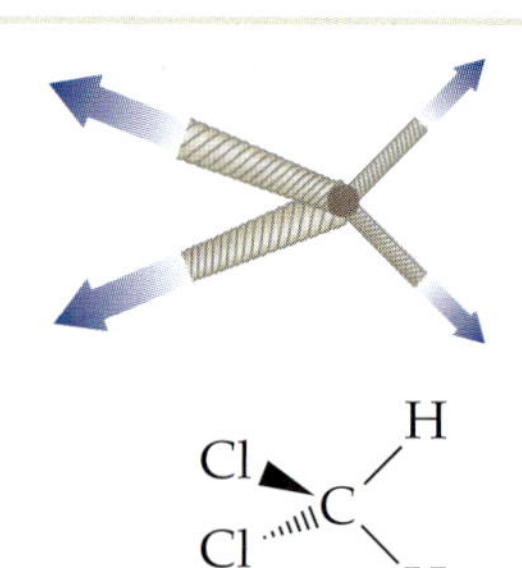

Polar; dipole–dipole forces

(c) Since the electronegativities of C and H are 2.5 and 2.1, respectively, the C—H bonds are nearly nonpolar. In addition, since the geometry of the molecule is tetrahedral, any slight polarities that the bonds might have will cancel. CH_4 is therefore nonpolar and does not have dipole–dipole forces.

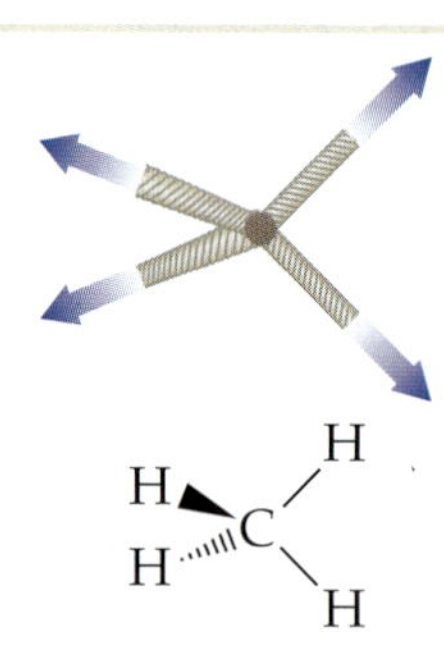

Nonpolar; no dipole–dipole forces

SKILLBUILDER 12.4 Dipole–Dipole Forces

Which of the following molecules have dipole–dipole forces?

(a) CI_4
(b) CH_3Cl
(c) HCl

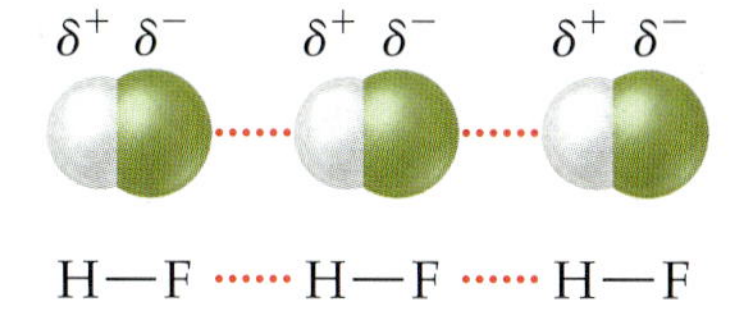

▶ **Figure 12.22 The hydrogen bond** In HF, the hydrogen on one molecule is strongly attracted to the fluorine on its neighbors. This attraction is called a *hydrogen bond*.

Hydrogen Bonding

Polar molecules containing hydrogen atoms bonded directly to fluorine, oxygen, or nitrogen exhibit an additional intermolecular force called a **hydrogen bond**. HF, NH_3 and H_2O, for example, all undergo hydrogen bonding. The hydrogen bond is a sort of *super* dipole–dipole force. The large electronegativity difference between hydrogen and these electronegative elements, as well as the small size of these atoms (which allows neighboring molecules to get very close to each other), gives rise to a strong attraction between the hydrogen in each of these molecules and the F, O, or N on its neighbors. This attraction is the hydrogen bond. For example, in HF the hydrogen is strongly attracted to the fluorine on neighboring molecules (▲ Figure 12.22).

Hydrogen bonds should not be confused with chemical bonds. Chemical bonds occur between *individual atoms within a molecule* and are generally much stronger than hydrogen bonds. A hydrogen bond has only 2% to 5% the strength of a typical covalent chemical bond. Hydrogen bonds—like dispersion forces and dipole–dipole forces—are intermolecular forces that occur *between molecules*. In liquid water, for example, the hydrogen bonds are transient, constantly forming, breaking, and reforming as water molecules move within the liquid. Hydrogen bonds are, however, the strongest of the three intermolecular forces. Substances composed of molecules that form hydrogen bonds have much higher melting and boiling points than you would predict based on molar mass. For example, consider the following two compounds.

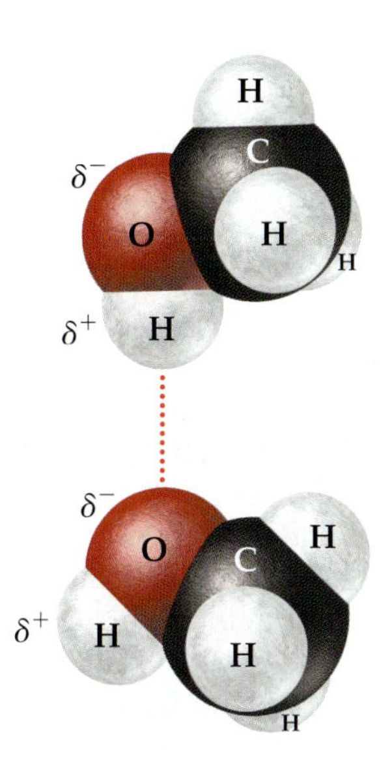

▲ **Figure 12.23 Hydrogen bonding in methanol** Since methanol contains hydrogen atoms directly bonded to oxygen, methanol molecules form hydrogen bonds to one another. The hydrogen atom on one methanol molecule is attracted to the oxygen atom of its neighbor.

Name	Formula	Molar mass (g/mol)	Structure	bp (°C)	mp (°C)
Methanol	CH_3OH	32.0	H | H—C—O—H | H	64.7	−97.8
Ethane	C_2H_6	30.0	H H | | H—C—C—H | | H H	−88	−172

Since methanol contains hydrogen directly bonded to oxygen, its molecules have hydrogen bonding as an intermolecular force. The hydrogen that is directly bonded to oxygen is strongly attracted to the oxygen on neighboring molecules (◀ Figure 12.23). This strong attraction makes the boiling point of methanol 64.7 °C. Consequently, methanol is a liquid at room temperature. Water is another good example of a molecule with hydrogen bonding as an intermolecular force (◀ Figure 12.24). The boiling point of water (100 °C) is remarkably high for a molecule with such a low molar mass (18.0 g/mol). Hydrogen bonding is also important in biological molecules. For example, the shapes of proteins are largely influenced by hydrogen bonding, and the two halves of DNA are held together by hydrogen bonds (see the *Chemistry and Health* box in this section).

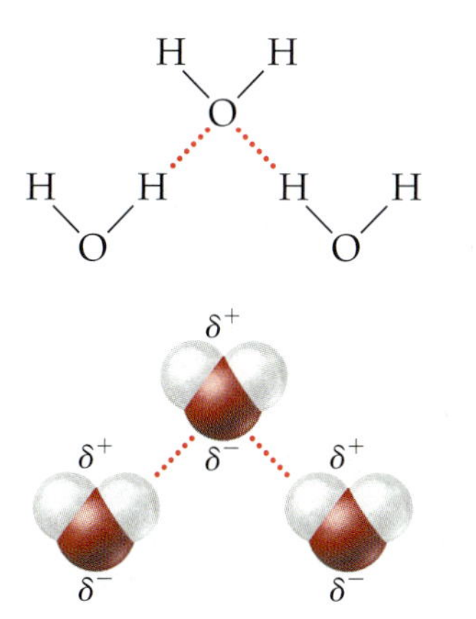

▲ **Figure 12.24 Hydrogen bonding in water** Water molecules form strong hydrogen bonds with one another.

TUTORIAL Hydrogen Bonding

EXAMPLE 12.5 Hydrogen Bonding

One of the following compounds is a liquid at room temperature. Which one and why?

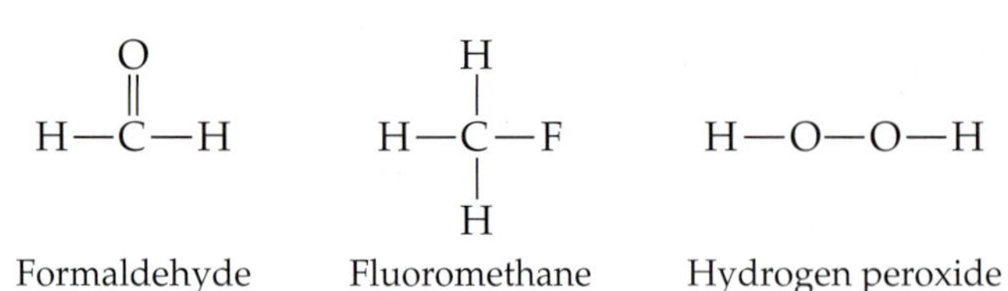

TUTORIAL Forces Between Molecules and Ions

Solution:

The three compounds have similar molar masses.

formaldehyde	30.03 g/mol
fluoromethane	34.03 g/mol
hydrogen peroxide	34.02 g/mol

Therefore, the strengths of their dispersion forces are similar. All three compounds are also polar, so they have dipole–dipole forces. Hydrogen peroxide, however, is the only compound to also contain H bonded directly to F, O, or N. Therefore it also has hydrogen bonding and is most likely to have the highest boiling point of the three. Since the problem stated that only one of the compounds was a liquid, we can safely assume that hydrogen peroxide is the liquid. Note that although fluoromethane *contains* both H and F, H is not *directly bonded* to F, so fluoromethane does not have hydrogen bonding as an intermolecular force. Similarly, although formaldehyde *contains* both H and O, H is not *directly bonded* to O, so formaldehyde does not have hydrogen bonding either.

SKILLBUILDER 12.5 Hydrogen Bonding

Which has the higher boiling point, HF or HCl? Why?

TABLE 12.5 Types of Intermolecular Forces

Type of Force	Relative Strength	Present in	Example
dispersion force (or London force)	weak, but increases with increasing molar mass	all atoms and molecules	H_2 (figure: H_2 ⋯ H_2)
dipole–dipole force	moderate	only polar molecules	HCl (figure: δ^+ HCl δ^- ⋯ δ^+ HCl δ^-)
hydrogen bond	strong	molecules containing H bonded directly to F, O, or N	HF (figure: δ^+ HF δ^- ⋯ δ^+ HF δ^-)

In some cases, hydrogen bonding can occur between one molecule in which H is directly bonded to F, O, or N and another molecule containing an electronegative atom. (See the Box, *Hydrogen Bonding in DNA*, in this section for an example.)

The different types of intermolecular forces are summarized in Table 12.5. Remember that dispersion forces, the weakest kind of intermolecular force, are present in all molecules and atoms and increase with increasing molar mass. These forces are always weak in small molecules, but they become substantial in molecules with high molar masses. Dipole–dipole forces are present in polar molecules. Hydrogen bonds, the strongest kind of intermolecular force, are present in molecules containing hydrogen bonded directly to fluorine, oxygen, or nitrogen.

CONCEPTUAL CHECKPOINT 12.2

When dry ice sublimes, what kind of forces are overcome?

(a) chemical bonds between carbon atoms and oxygen atoms
(b) hydrogen bonds between carbon dioxide molecules
(c) dispersion forces between carbon dioxide molecules
(d) dipole–dipole forces between carbon dioxide molecules

12.7 Types of Crystalline Solids: Molecular, Ionic, and Atomic

As we learned in Section 12.2, solids may be crystalline (a well-ordered array of atoms or molecules) or amorphous (having no long-range order). Crystalline solids can be divided into three categories—molecular, ionic, and atomic—based on the individual units that compose the solid (▼ Figure 12.25).

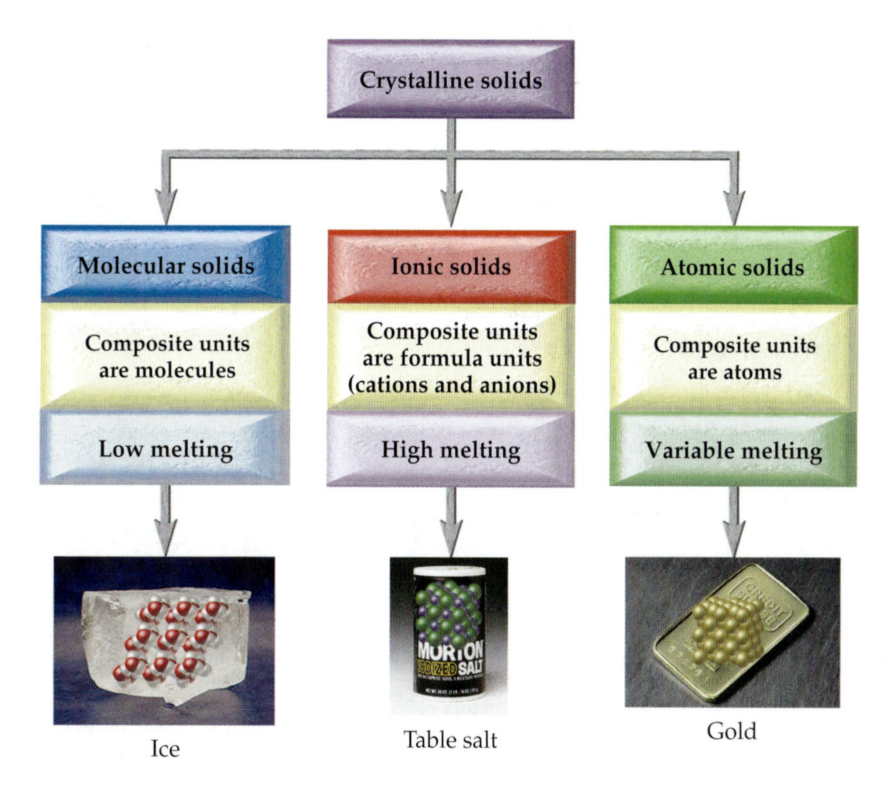

▲ **Figure 12.25 A classification scheme for crystalline solids**

Chemistry AND HEALTH

Hydrogen Bonding in DNA

DNA is a long chainlike molecule that acts as a blueprint for living organisms. Copies of DNA are passed from parent to offspring, which is why we inherit traits from our parents. A DNA molecule is composed of thousands of repeating units called *nucleotides* (▶ Figure 12.26). Each nucleotide contains one of four different bases: adenine, thymine, cytosine, and guanine (abbreviated A, T, C, and G). The order of these bases along DNA encodes the instructions that specify how proteins—the workhorse molecules in living organisms—are made in each cell of the body. Proteins determine virtually all human characteristics, including how we look, how we fight infections, and even how we behave. Consequently, human DNA is a blueprint for how humans are made.

Each time a human cell divides, it must copy the genetic instructions—which means replicating its DNA. The replicating mechanism is related to the structure of DNA, discovered in 1953 by James Watson and Francis Crick. DNA consists of two complementary strands wrapped around each other in the now famous double helix. Each strand is held to the other by hydrogen bonds that occur between the bases on each strand. DNA replicates because each base (A, T, C, and G) has a complementary partner with which it hydrogen bonds (▶ Figure 12.27). Adenine (A) hydrogen bonds with thymine (T) and cytosine (C) hydrogen bonds with guanine (G). The hydrogen bonds are so specific that each base will pair only with its complementary partner. When a cell is going to divide, the DNA unzips across the hydrogen bonds that run along its length. Then new bases, complementary to the bases in each half, add along each of the halves, forming hydrogen bonds with their complement. The result is two identical copies of the original DNA (see Chapter 19).

CAN YOU ANSWER THIS? *Why would dispersion forces not work as a way to hold the two halves of DNA together? Why would covalent bonds not work?*

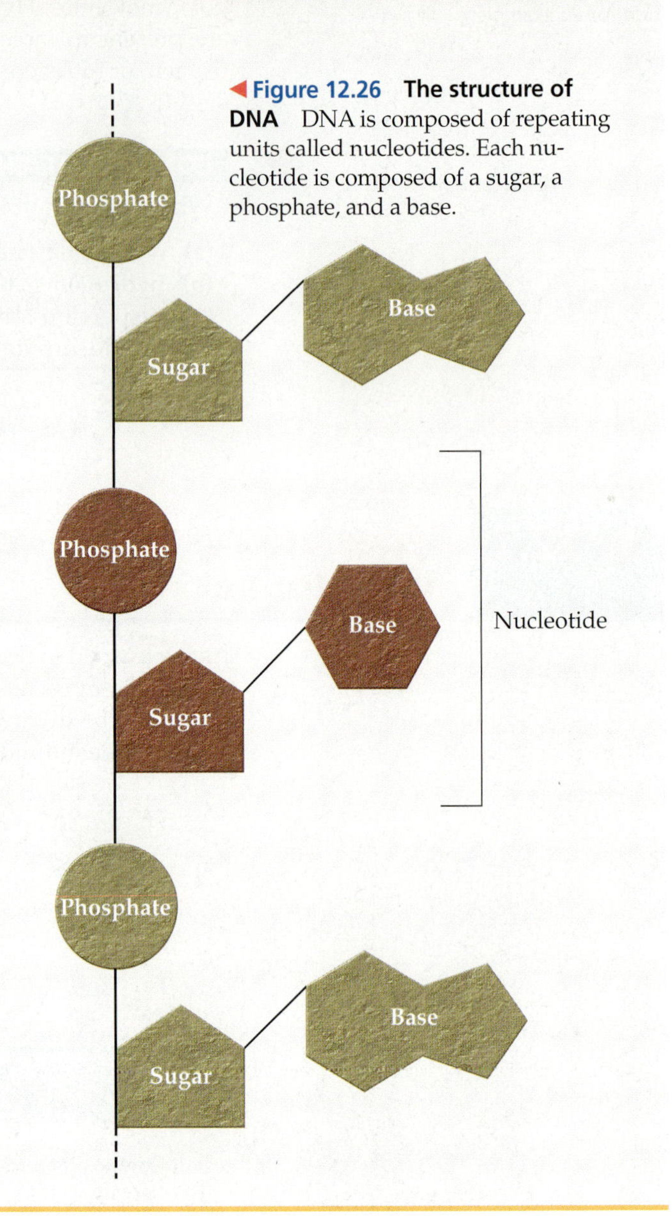

◀ **Figure 12.26 The structure of DNA** DNA is composed of repeating units called nucleotides. Each nucleotide is composed of a sugar, a phosphate, and a base.

Molecular Solids

Molecular solids are solids whose composite units are *molecules*. Ice (solid H_2O) and dry ice (solid CO_2) are examples of molecular solids. Molecular solids are held together by the kinds of intermolecular forces—dispersion forces, dipole–dipole forces, and hydrogen bonding—that we have just discussed. For example, ice is held together by hydrogen bonds and dry ice is held together by dispersion forces. Molecular solids as a whole tend to have low to moderately low melting points; ice melts at 0 °C and dry ice sublimes at −78.5 °C. However, strong intermolecular forces can increase their melting points relative to one another.

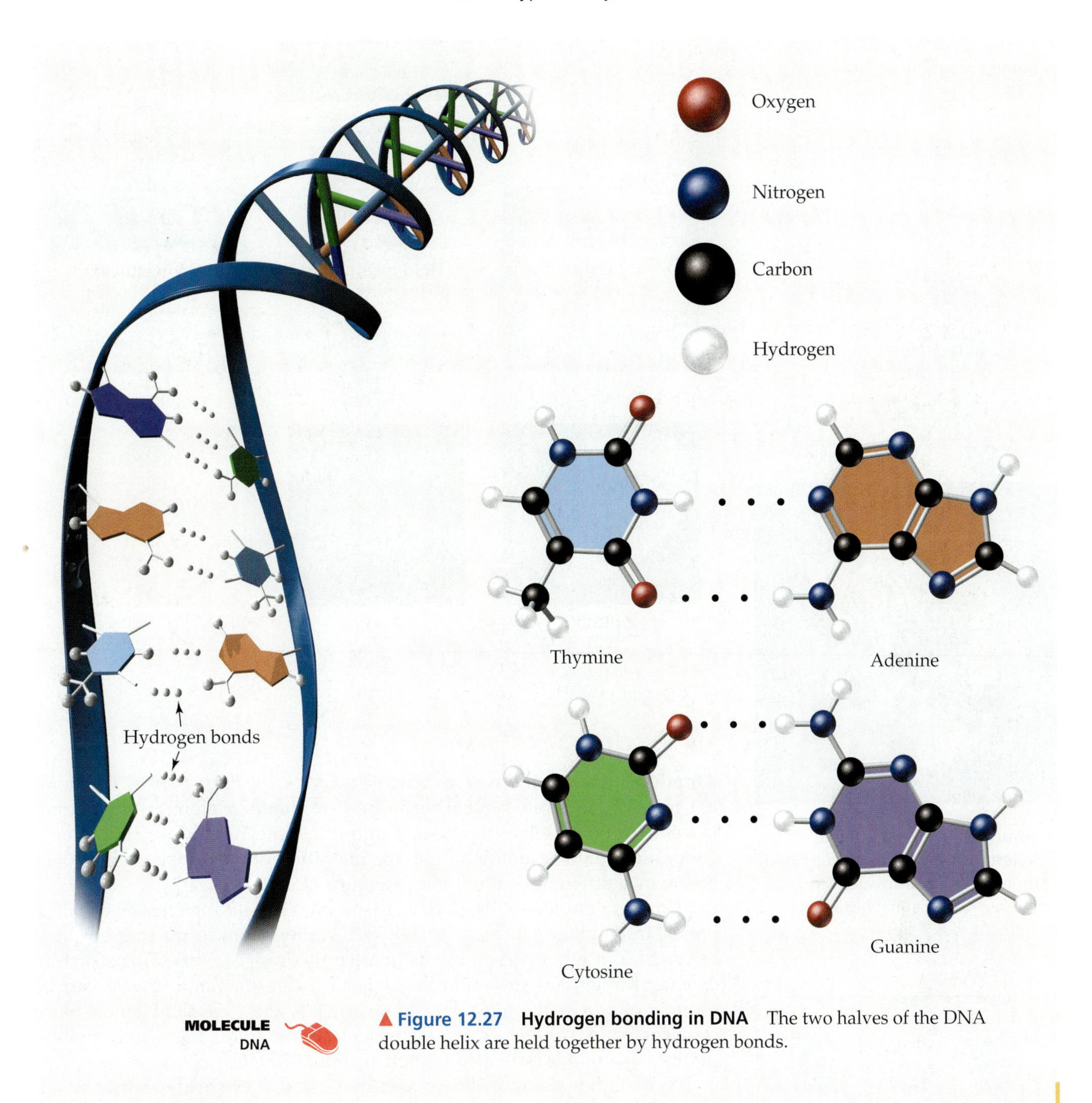

▲ **Figure 12.27 Hydrogen bonding in DNA** The two halves of the DNA double helix are held together by hydrogen bonds.

Ionic Solids

See Section 5.4 for a complete description of the formula unit.

Ionic solids are solids whose composite units are *formula units*, the smallest electrically neutral collection of cations and anions that compose the compound. Table salt (NaCl) and calcium fluoride (CaF_2) are good examples of ionic solids. Ionic solids are held together by the electrostatic attractions that occur between cations and anions. For example, in NaCl, the same force that attracts a Na^+ cation to a Cl^- anion holds the entire solid lattice together because the lattice is composed of alternating Na^+ cations and Cl^- anions in a three-dimensional array. In other words, the forces that hold ionic solids together are actual ionic bonds. Since ionic bonds are much stronger than any of the intermolecular forces discussed previously, ionic solids tend to have much higher melting points than molecular solids. For example, sodium chloride melts at 801 °C, while carbon disulfide (CS_2)—a molecular solid with a higher molar mass—melts at −110 °C.

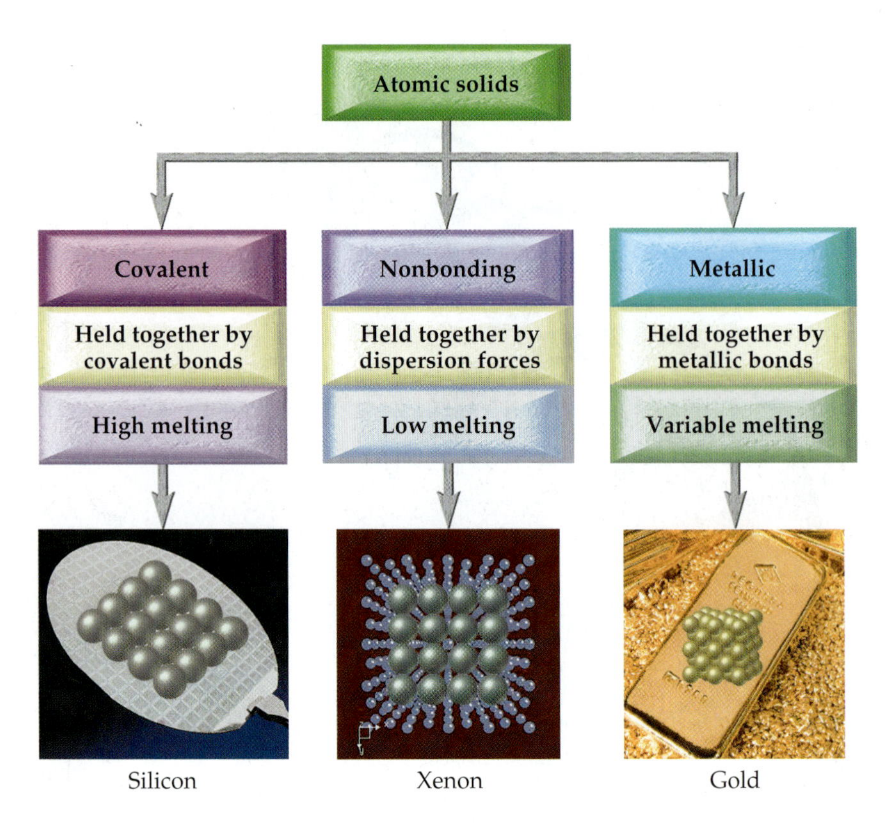

▶ **Figure 12.28 A classification scheme for atomic solids**

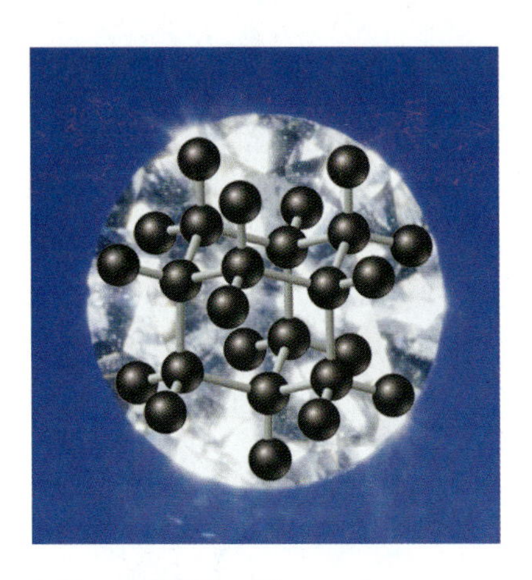

▲ **Figure 12.29 Diamond: a covalent atomic solid** In diamond, carbon atoms form covalent bonds in a three-dimensional hexagonal pattern.

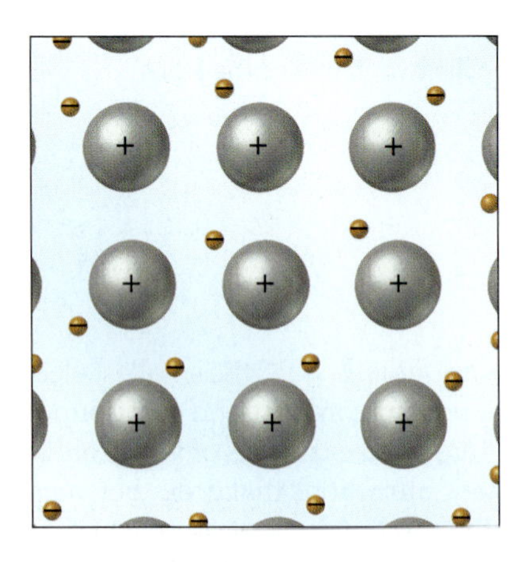

▲ **Figure 12.30 Structure of a metallic atomic solid** In the simplest model of a metal, each atom donates one or more electrons to an "electron sea." The metal then consists of the metal cations in a negatively charged electron sea.

Atomic Solids

Atomic solids are solids whose composite units are *individual atoms*. Diamond (C), iron (Fe), and solid xenon (Xe) are good examples of atomic solids. Atomic solids can themselves be divided into three categories—**covalent atomic solids**, **nonbonding atomic solids**, and **metallic atomic solids**—each held together by a different kind of force (▲ Figure 12.28).

Covalent atomic solids, such as diamond, are held together by covalent bonds. In diamond (◀ Figure 12.29), each carbon atom forms four covalent bonds to four other carbon atoms in a tetrahedral geometry. This structure extends throughout the entire crystal, so that a diamond crystal can be thought of as a giant molecule held together by these covalent bonds. Since covalent bonds are very strong, covalent atomic solids have high melting points. Diamond is estimated to melt at about 3800 °C.

Nonbonding atomic solids, such as solid xenon, are held together by relatively weak dispersion forces. Xenon atoms have stable electron configurations and therefore do not form covalent bonds with each other. Consequently, solid xenon, like other nonbonding atomic solids, has a very low melting point (about −112 °C).

Metallic atomic solids, such as iron, have variable melting points. Metals are held together by metallic bonds that, in the simplest model, consist of positively charged ions in a sea of electrons (◀ Figure 12.30). Metallic bonds are of varying strengths, with some metals, such as mercury, having melting points below room temperature, and other metals, such as iron, having relatively high melting points (iron melts at 1809 °C).

EXAMPLE 12.6 Identifying Types of Crystalline Solids

Identify each of the following solids as molecular, ionic, or atomic.

(a) $CaCl_2(s)$ **(b)** $Co(s)$ **(c)** $CS_2(s)$

Solution:

(a) $CaCl_2$ is an ionic compound (metal and nonmetal) and therefore forms an ionic solid ($CaCl_2$ melts at 772 °C).

(b) Co is a metal and therefore forms a metallic atomic solid (Co melts at 1768 °C).

(c) CS_2 is a molecular compound (nonmetal bonded to a nonmetal) and therefore forms a molecular solid (CS_2 melts at −110 °C).

SKILLBUILDER 12.6 Identifying Types of Crystalline Solids

Identify each of the following solids as molecular, ionic, or atomic.

(a) $NH_3(s)$ **(b)** $CaO(s)$ **(c)** $Kr(s)$

12.8 Water: A Remarkable Molecule

Water is easily the most common and important liquid on Earth. It fills our oceans, lakes, and streams. In its solid form, it covers nearly an entire continent (Antarctica), as well as large regions around the North Pole, and caps our tallest mountains. In its gaseous form, it humidifies our air. We drink water, we sweat water, and we excrete bodily wastes dissolved in water. Indeed, the majority of our body mass *is* water. Life is impossible without water, and in most places on Earth where liquid water exists, life exists. Recent evidence of water on Mars—either in the past or in the present—has fueled hopes of finding life or evidence of past life there. Water is remarkable.

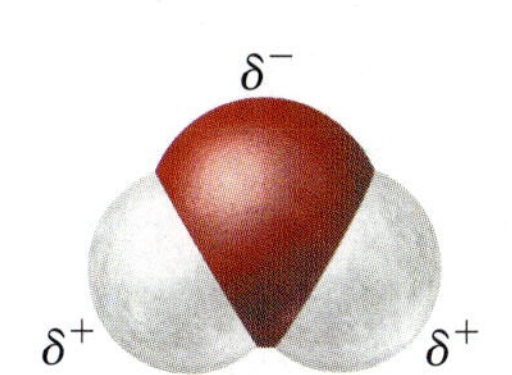

▲ **Figure 12.31 The water molecule**

Among liquids, water is unique. It has a low molar mass (18.02 g/mol), yet it is a liquid at room temperature. No other compound of similar molar mass even comes close to being a liquid at room temperature. For example, nitrogen (28.02 g/mol) and carbon dioxide (44.01 g/mol) are both gases at room temperature. Water's relatively high boiling point (for its low molar mass) can be understood by examining the structure of the water molecule (◀ Figure 12.31). The bent geometry of the water molecule and the highly polar nature of the O—H bonds result in a molecule with a significant dipole moment. Water's two O—H bonds (hydrogen directly bonded to oxygen) allow water molecules to form strong hydrogen bonds with other water molecules, resulting in a relatively high boiling point. Water's high polarity also allows it to dissolve many other polar and ionic compounds. Consequently, water is the main solvent of living organisms, transporting nutrients and other important compounds throughout the body.

Water reaches its maximum density at 4.0 °C.

The way water freezes is also unique. Unlike other substances, which contract upon freezing, water expands upon freezing. This seemingly trivial property has significant consequences. For example, because liquid water expands when it freezes, ice is less dense than liquid water. Consequently, ice cubes and icebergs both float. The frozen layer of ice at the surface of a winter lake insulates the water in the lake from further freezing. If this ice layer sank, it would kill bottom-dwelling aquatic life and possibly allow the lake to freeze solid, eliminating virtually all aquatic life in the lake.

The expansion of water upon freezing, however, is one reason that most organisms do not survive freezing. When the water within a cell freezes, it expands and often ruptures the cell, just as water freezing within a pipe bursts the pipe. Many foods, especially those with high water content, do not survive freezing very well either. Have you ever tried, for example, to freeze your own vegetables? Try putting lettuce or spinach in the freezer. When you defrost it, it will be limp and damaged. The frozen food industry gets around this problem by *flash-freezing* vegetables and other foods. In this process, foods are frozen instantaneously, not allowing water molecules to settle into their preferred crystalline structure. Consequently, the water does not expand very much, and the food remains largely undamaged.

▲ Lettuce does not survive freezing because the expansion of water upon freezing ruptures the cells within the lettuce leaf.

Chemistry IN THE ENVIRONMENT

Water Pollution

Water quality is critical to human health. Many human diseases—especially in developing nations—are caused by poor water quality. Several kinds of pollutants, including biological contaminants and chemical contaminants, can get into water supplies. Biological contaminants are microorganisms that cause diseases such as hepatitis, cholera, dysentery, and typhoid. Drinking water in developed nations is usually treated to kill microorganisms. Most biological contaminants can be eliminated from untreated water by boiling. Water containing biological contaminants poses an immediate danger to human health and should not be consumed.

Chemical contaminants get into drinking water from sources such as industrial dumping, pesticide and fertilizer use, and household dumping. These contaminants include organic compounds, such as carbon tetrachloride and dioxin, and inorganic elements and compounds such as mercury, lead, and nitrates. Since many chemical contaminants are neither volatile nor alive like biological contaminants, they are *not* eliminated through boiling.

The Environmental Protection Agency (EPA), under the Safe Drinking Water Act of 1974 and its amendments, sets standards that specify the maximum contamination level (MCL) for nearly one hundred biological and chemical contaminants in water. Water providers that serve more than twenty-five people must periodically test the water they deliver to their consumers for these contaminants. If levels exceed the standards set by the EPA, the water provider must notify the consumer and take appropriate measures to remove the contaminant from the water. According to the EPA, if water comes from a provider that serves more than twenty-five people, it should be safe to consume over a lifetime. If it is not safe to drink for a short period of time, you will be notified.

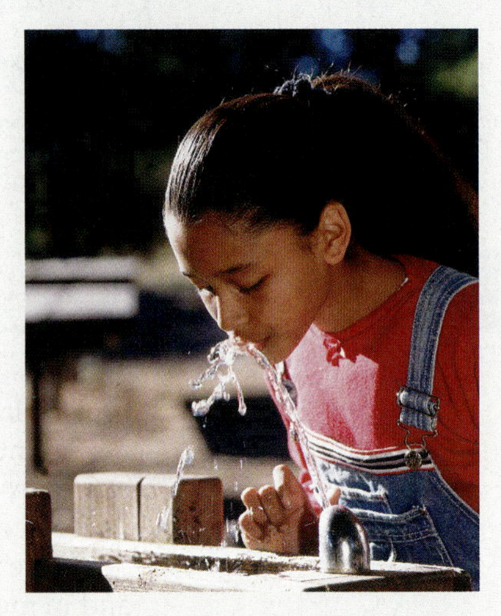

▲ Safe drinking water has a major effect on public health and the spread of disease. In many parts of the world, the water supply is unsafe to drink. In the United States the Environmental Protection Agency (EPA) is charged with maintaining water safety.

CAN YOU ANSWER THIS? *Suppose a sample of water is contaminated by a nonvolatile contaminant such as lead. Why does boiling not eliminate the contaminant?*

CHAPTER IN REVIEW

Chemical Principles	Relevance
Properties of Liquids: • High densities in comparison to gases. • Indefinite shape; they assume the shape of their container. • Definite volume; they are not easily compressed.	**Properties of Liquids:** Common liquids include water, acetone (fingernail-polish remover), and rubbing alcohol. Water is the most common and most important liquid on Earth. It is difficult to imagine life without water.
Properties of Solids: • High densities in comparison to gases. • Definite shape; they do not assume the shape of their container. • Definite volume; they are not easily compressed. • May be crystalline (ordered) or amorphous (disordered).	**Properties of Solids:** Much of the matter we encounter is solid. Common solids include ice, dry ice, and diamond. Understanding the properties of solids involves understanding the particles that compose them and how they interact.

Manifestations of Intermolecular Forces: Surface tension—the tendency for liquids to minimize their surface area—is a direct result of intermolecular forces. Viscosity—the resistance of liquids to flow—is another result of intermolecular forces. Both surface tension and viscosity increase with greater intermolecular forces.

Manifestations of Intermolecular Forces: Many insects can walk on water due to surface tension. Water is drawn from roots of trees and up into the leaves because of capillary action, a direct result of intermolecular forces.

Evaporation and Condensation: Evaporation or vaporization—an endothermic physical change—is the conversion of a liquid to a gas. Condensation—an exothermic physical change—is the conversion of a gas to a liquid. When the rate of evaporation and condensation in a liquid/gas sample are equal, dynamic equilibrium is reached and the partial pressure of the gas at that point is called its vapor pressure. When the vapor pressure equals the external pressure, the boiling point is reached. At the boiling point, thermal energy causes molecules in the interior of the liquid, as well as those at the surface, to convert to gas, resulting in the familiar bubbling. The heat absorbed or emitted during evaporation and condensation (respectively) can be calculated using the heat of vaporization.

Evaporation and Condensation: Evaporation is the body's natural cooling system. When we get overheated, we sweat; the sweat then evaporates and cools us. Evaporation and condensation both play roles in moderating climate. Humid areas, for example, cool less at night because as the temperature drops, water condenses out of the air, releasing heat and preventing a further temperature drop.

Melting and Freezing: Melting—an endothermic physical change—is the conversion of a solid to a liquid, and freezing—an exothermic physical change—is the conversion of liquid to a solid. The heat absorbed or emitted during melting and freezing (respectively) can be calculated using the heat of fusion.

Melting and Freezing: The melting of solid ice is used, for example, to cool drinks when we place ice cubes in them. Since melting is endothermic, it absorbs heat from the liquid and cools it.

Types of Intermolecular Forces: The three main types of intermolecular forces are:

Dispersion forces—Dispersion forces occur between all molecules and atoms due to instantaneous fluctuations in electron charge distribution. The strength of the dispersion force increases with increasing molar mass.

Dipole–dipole forces—Dipole–dipole forces exist between molecules that are polar. Consequently, polar molecules have higher melting and boiling points than nonpolar molecules of similar molar mass.

Hydrogen bonding—Hydrogen bonding exists between molecules that have H bonded directly to F, O, or N. Hydrogen bonds are the strongest of the three intermolecular forces.

Types of Intermolecular Forces: The type of intermolecular force present in a substance determines many of the properties of the substance. The stronger the intermolecular force, for example, the greater the melting and boiling point of the substance. In addition, the miscibility of liquids—their ability to mix without separating—depends on the relative kinds of intermolecular forces present within them. In general, polar liquids are miscible with other polar liquids, but not with nonpolar liquids. Hydrogen bonding is important in many biological molecules such as proteins and DNA.

Types of Crystalline Solids: Crystalline solids can be divided into three categories based on the individual units composing the solid:

> **Molecular solids**—Molecular solids have molecules as their composite units and are held together by dispersion forces, dipole–dipole forces, or hydrogen bonding.
> **Ionic solids**—Ionic solids have formula units (the smallest electrically neutral collection of cations and anions) as their composite units. They are held together by the electrostatic attractions that occur between cations and anions.
> **Atomic solids**—Atomic solids have atoms as their composite units. They are held together by different forces depending on the particular solid.

Types of Crystalline Solids: Solids have different properties depending on their individual units and the forces that hold those units together. Molecular solids tend to have low melting points. Ionic solids tend to have intermediate to high melting points. Atomic solids have varied melting points, depending on the particular solid.

Water: Water is a unique molecule. Because of its strong hydrogen bonding, water is a liquid at room temperature. Unlike most liquids, water expands when it freezes. In addition, water is highly polar, making it a good solvent for many polar substances.

Water: Water is critical to life. On Earth, wherever there is water, there is life. Water acts as a solvent and transport medium, and virtually all the chemical reactions on which life depends take place in aqueous solution. The expansion of water upon freezing allows life within frozen lakes to survive the winter. The ice on top of the lake acts as insulation, protecting the rest of the lake (and the life within it) from freezing.

Chemical Skills

Using Heat of Vaporization in Calculations (Section 12.4)

The heat of vaporization can be used as a conversion factor between moles of a substance and the amount of heat required to vaporize it. To calculate the amount of heat required to vaporize a given amount of a substance, first convert the given amount of the substance to moles and then use the heat of vaporization as a conversion factor to get to kilojoules. For vaporization, the heat is always absorbed. For condensation, follow the same procedure, but the heat is always emitted.

Examples

LIVE EXAMPLE

EXAMPLE 12.7 Using Heat of Vaporization in Calculations

Calculate the amount of heat required to vaporize 84.8 g of water at its boiling point.

Given: 84.8 g H_2O

Find: heat (kJ)

Conversion Factors:

$$\Delta H_{vap} = 40.7 \text{ kJ/mol} \quad (\text{at } 100\,°\text{C})$$
$$1 \text{ mol } H_2O = 18.02 \text{ g } H_2O$$

Solution Map:

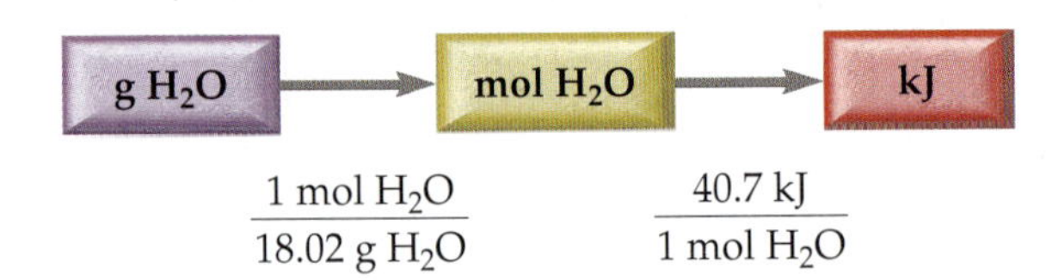

$$\frac{1 \text{ mol } H_2O}{18.02 \text{ g } H_2O} \qquad \frac{40.7 \text{ kJ}}{1 \text{ mol } H_2O}$$

Solution:

$$84.8 \text{ g } H_2O \times \frac{1 \text{ mol } H_2O}{18.02 \text{ g } H_2O} \times \frac{40.7 \text{ kJ}}{1 \text{ mol } H_2O} = 192 \text{ kJ}$$

Using Heat of Fusion in Calculations (Section 12.5)

The heat of fusion can be used as a conversion factor between moles of a substance and the amount of heat required to melt it. To calculate the amount of heat required to melt a given amount of a substance, first convert the given amount of the substance to moles and then use the heat of fusion as a conversion factor to get to kilojoules. For melting, the heat is always absorbed. For freezing, follow the same procedure, but the heat is always emitted.

EXAMPLE 12.8 Using Heat of Fusion in Calculations

Calculate the amount of heat emitted when 12.4 g of water freezes to solid ice.

Given: 12.4 g H_2O

Find: heat (kJ)

Conversion Factors:

$$\Delta H_{fus} = 6.02 \text{ kJ/mol}$$
$$1 \text{ mol } H_2O = 18.02 \text{ g } H_2O$$

Solution Map:

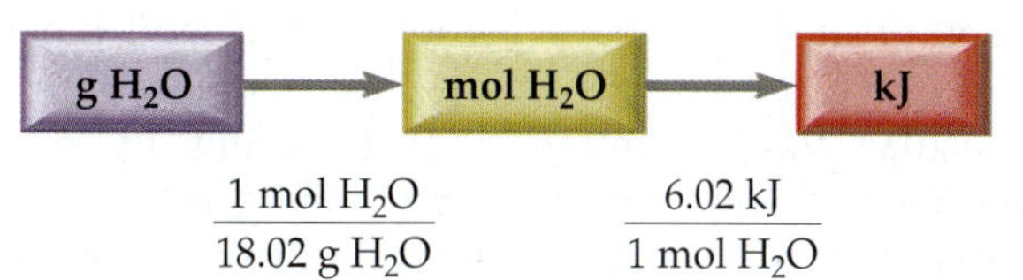

$$\frac{1 \text{ mol } H_2O}{18.02 \text{ g } H_2O} \qquad \frac{6.02 \text{ kJ}}{1 \text{ mol } H_2O}$$

Solution:

$$12.4 \text{ g } H_2O \times \frac{1 \text{ mol } H_2O}{18.02 \text{ g } H_2O} \times \frac{6.02 \text{ kJ}}{1 \text{ mol } H_2O} = 4.14 \text{ kJ}$$

The heat emitted is 4.14 kJ.

Determining the Types of Intermolecular Forces in a Compound (Section 12.6)

All substances exhibit dispersion forces. Polar substances—those whose molecules have polar bonds that add to net dipole moment—also exhibit dipole–dipole forces. Substances whose molecules contain H bonded directly to F, O, or N exhibit hydrogen bonding as well.

EXAMPLE 12.9 Determining the Types of Intermolecular Forces in a Compound

Determine the types of intermolecular forces present in each of the following substances.

(a) N_2
(b) CO
(c) NH_3

Solution:

(a) N_2 is nonpolar and therefore has only dispersion forces.
(b) CO is polar and therefore has dipole–dipole forces (in addition to dispersion forces).
(c) NH_3 has hydrogen bonding (in addition to dispersion forces and dipole–dipole forces).

Using Intermolecular Forces to Determine Melting and/or Boiling Points (Section 12.6)

To determine relative boiling points and melting points among compounds you must evaluate the types of intermolecular forces that each compound exhibits. Dispersion forces are the weakest kind of intermolecular force, but they increase with increasing molar mass. Dipole–dipole forces are stronger than dispersion forces. If two compounds have similar molar mass, but one is polar, it will have higher melting and boiling points. Hydrogen bonds are the strongest type of intermolecular force. Substances that exhibit hydrogen bonding will have much higher boiling and melting points than substances without hydrogen bonding, even if the substance without hydrogen bonding is of higher molar mass.

EXAMPLE 12.10 Using Intermolecular Forces to Determine Melting and/or Boiling Points

Arrange each of the following in order of increasing boiling point.

(a) F_2, Cl_2, Br_2
(b) HF, HCl, HBr

Solution:

(a) Since these all have only dispersion forces, and since they are similar substances (all halogens), the strength of the dispersion force will increase with increasing molar mass. Therefore, the correct order is $F_2 < Cl_2 < Br_2$.

(b) Since HF has hydrogen bonding, it has the highest boiling point. Between HCl and HBr, HBr (because of its higher molar mass) has a higher boiling point. Therefore the correct order is $HCl < HBr < HF$.

Key Terms

atomic solid [12.7]
boiling point [12.4]
condensation [12.4]
covalent atomic solid [12.7]
dipole–dipole force [12.6]
dispersion force [12.6]
dynamic equilibrium [12.4]
endothermic [12.4]
evaporation [12.4]
exothermic [12.4]
heat of fusion (ΔH_{fus}) [12.5]
heat of vaporization (ΔH_{vap}) [12.4]
hydrogen bond [12.6]
instantaneous (temporary) dipole [12.6]
intermolecular forces [12.1]
ionic solid [12.7]
melting point [12.5]
metallic atomic solid [12.7]
miscibility [12.6]
molecular solid [12.7]
nonbonding atomic solid [12.7]
nonvolatile [12.4]
normal boiling point [12.4]
permanent dipole [12.6]
sublimation [12.5]
surface tension [12.3]
vaporization [12.4]
vapor pressure [12.4]
viscosity [12.3]
volatile [12.4]

Exercises

Questions

1. Explain what causes a substance to taste bitter.
2. What are intermolecular forces?
3. Why are intermolecular forces important?
4. Why are water droplets spherical?
5. What determines whether a substance is a solid, liquid, or gas?
6. What are the properties of liquids?
7. What are the properties of solids?
8. Explain the properties of liquids in terms of the molecules or atoms that compose them.
9. Explain the properties of solids in terms of the molecules or atoms that compose them.
10. What is the difference between a crystalline solid and an amorphous solid?
11. What is surface tension? How does it depend on intermolecular forces?
12. What is viscosity? How does it depend on intermolecular forces?
13. What is evaporation? Condensation?
14. Why does a glass of water evaporate more slowly in the glass than if you spilled the same amount of water on a table?
15. Explain the difference between evaporation below the boiling point of a liquid and evaporation at the boiling point of a liquid.
16. What is the boiling point of a liquid? What is the normal boiling point?
17. Acetone evaporates more quickly than water at room temperature. What can you say about the relative strength of the intermolecular forces in the two compounds? Which is more volatile?
18. Explain condensation and dynamic equilibrium.

19. How is the vapor pressure of a substance defined? How does it depend on temperature and strength of intermolecular forces?
20. Explain why you feel cooler when you sweat.
21. Explain why a steam burn from gaseous water at 100 °C is worse than a water burn involving the same amount of liquid water at 100 °C.
22. Explain what happens when a liquid boils.
23. Explain why simply placing a cup of water in a small ice chest (without a refrigeration mechanism) initially at −5 °C will *not* result in the freezing of the cup of water.
24. Explain how ice cubes cool down beverages.
25. What are dispersion forces? What can you say about the strength of dispersion forces as a function of molar mass?
26. What are dipole–dipole forces? How can you tell whether a compound has dipole–dipole forces?
27. What is hydrogen bonding? How can you tell whether a compound has hydrogen bonding?
28. What is a molecular solid? What kinds of forces hold molecular solids together?
29. What can you say about the melting points of molecular solids relative to those of other types of solids?
30. What is an ionic solid? What kinds of forces hold ionic solids together?
31. What can you say about the melting points of ionic solids relative to those of other types of solids?
32. What is an atomic solid? What can you say about the properties of atomic solids?
33. In what ways is water unique?
34. How would ice be different if it were denser than water? How would that affect aquatic life in cold-climate lakes?

Problems

Evaporation, Condensation, Melting, and Freezing

35. Which will evaporate more quickly: 55 mL of water in a beaker with a diameter of 4.5 cm or 55 mL of water in a dish with a diameter of 12 cm? Why?
36. Two samples of water of equal volume are put into two dishes and kept at room temperature for several days. The water in the first dish is completely vaporized after 2.8 days while the water in the second dish takes 8.3 days to completely evaporate. What can you conclude about the two dishes?
37. One milliliter of water is poured onto one hand and one milliliter of acetone (fingernail-polish remover) is poured onto the other. As they evaporate, they both feel cool. Which one feels cooler and why? (Hint: Which substance is more volatile?)
38. Spilling water over your skin on a hot day will cool you down. Spilling vegetable oil over your skin on a hot day will not. Explain the difference.
39. Several ice cubes are placed in a beaker on a lab bench and their temperature, initially at −5.0 °C, is monitored. Explain what happens to the temperature as a function of time. Make a sketch of how the temperature might change with time. (Assume that the lab is at 25 °C.)
40. Water is put into a beaker and heated with a Bunsen burner. The temperature of the water, initially at 25 °C, is monitored. Explain what happens to the temperature as a function of time. Make a sketch of how the temperature might change with time. (Assume that the Bunsen burner is hot enough to heat the water to its boiling point.)
41. Which of the following would cause a more severe burn: spilling 0.50 g of 100 °C water on your hand, or allowing 0.50 g of 100 °C steam to condense on your hand? Why?
42. The nightly winter temperature drop in a seaside town is usually less than that in nearby towns that are farther inland. Explain.
43. A watery bag of ice is placed in an ice chest initially at −8 °C and the temperature of the ice chest goes up. Why?
44. The refrigeration mechanism in a freezer with an automatic ice maker usually has to run extensively each time ice is forming from liquid water in the freezer. Why?
45. An ice chest is filled with 3.5 kg of ice at 0 °C. A second ice chest is filled with 3.5 kg of water at 0 °C. After several hours, which ice chest is colder? Why?
46. Why will 50 g of water initially at 0 °C warm more quickly than 50 g of an ice/water mixture initially at 0 °C?
47. In Denver, Colorado, water boils at 95 °C. Explain why.
48. At the top of Mt. Everest, water boils at 70 °C. Explain why.

Heat of Vaporization and Heat of Fusion

49. How much heat is required to vaporize 34.8 g of water at 100 °C?

50. How much heat is required to vaporize 78.2 g of acetone at its boiling point?

51. How much heat does your body lose when 2.8 g of sweat evaporates from your skin at 25 °C? (Assume that the sweat is only water.)

52. How much heat does your body lose when 2.56 g of sweat evaporates from your skin at 25 °C? (Assume that the sweat is only water.)

53. How much heat is emitted when 2.56 g of water condenses at 25 °C?

54. How much heat is emitted when 58.2 g of isopropyl alcohol condenses at 25 °C?

55. The human body obtains 835 kJ of energy from a chocolate chip cookie. If this energy were used to vaporize water at 100 °C, how many grams of water could be vaporized?

56. The human body obtains 1078 kJ from a candy bar. If this energy were used to vaporize water at 100 °C, how much water in liters could be vaporized? (Assume that the density of water is 1.0 g/mL.)

57. How much heat is required to melt 28.9 g of ice?

58. How much heat is required to melt 43.8 g of solid diethyl ether?

59. How much energy is released when 47.5 g of water freezes?

60. How much energy is released when 1.8 kg of diethyl ether freezes?

Intermolecular Forces

61. Determine the kinds of intermolecular forces that are present in each of the following:
    (a) Kr
    (b) N_2
    (c) CO
    (d) HF

62. Determine the kinds of intermolecular forces that are present in each of the following:
    (a) HCl
    (b) H_2O
    (c) Br_2
    (d) He

63. Determine the kinds of intermolecular forces that are present in each of the following:
    (a) NCl_3 (trigonal pyramidal)
    (b) NH_3 (trigonal pyramidal)
    (c) SiH_4 (tetrahedral)
    (d) CCl_4 (tetrahedral)

64. Determine the kinds of intermolecular forces present in each of the following:
    (a) O_3
    (b) HBr
    (c) CH_3OH
    (d) I_2

65. Which of the following has the highest boiling point? Why? Hint: They are all nonpolar.
    (a) CH_4
    (b) CH_3CH_3
    (c) $CH_3CH_2CH_3$
    (d) $CH_3CH_2CH_2CH_3$

66. Which of the following has the highest boiling point? Why?
    (a) Kr
    (b) Xe
    (c) Rn

67. One of the following two substances is a liquid at room temperature. Which one and why?

    CH_3OH CH_3SH

68. One of the following two substances is a liquid at room temperature. Which one and why?

    CH_3OCH_3 CH_3CH_2OH

69. A flask containing a mixture of $NH_3(g)$ and $CH_4(g)$ is cooled. At −33.3 °C a liquid begins to form in the flask. What is the liquid?

70. Explain why CS_2 is a liquid at room temperature while CO_2 is a gas.

71. Are $CH_3CH_2CH_2CH_2CH_3$ and H_2O miscible?

72. Are CH_3OH and H_2O miscible?

73. Which of the following pairs of substances will form homogeneous solutions when combined?
    (a) CCl_4 and H_2O
    (b) Br_2 and CCl_4
    (c) CH_3CH_2OH and H_2O

74. Which of the following pairs of compounds will form homogeneous solutions when combined?
    (a) $CH_3CH_2CH_2CH_2CH_3$ and $CH_3CH_2CH_2CH_2CH_2CH_3$
    (b) CBr_4 and H_2O
    (c) Cl_2 and H_2O

Types of Solids

75. Identify each of the following solids as molecular, ionic, or atomic.
   (a) $Ar(s)$
   (b) $H_2O(s)$
   (c) $K_2O(s)$
   (d) $Fe(s)$

76. Identify each of the following solids as molecular, ionic, or atomic.
   (a) $CaCl_2(s)$
   (b) $CO_2(s)$
   (c) $Ni(s)$
   (d) $I_2(s)$

77. Identify each of the following solids as molecular, ionic, or atomic.
   (a) $H_2S(s)$
   (b) $KCl(s)$
   (c) $N_2(s)$
   (d) $NI_3(s)$

78. Identify each of the following solids as molecular, ionic, or atomic.
   (a) $SF_6(s)$
   (b) $C(s)$
   (c) $MgCl_2(s)$
   (d) $Ti(s)$

79. Which of the following solids has the highest melting point? Why?
   (a) $Ar(s)$
   (b) $CCl_4(s)$
   (c) $LiCl(s)$
   (d) $CH_3OH(s)$

80. Which of the following solids has the highest melting point? Why?
   (a) C (*s*, diamond)
   (b) $Kr(s)$
   (c) $NaCl(s)$
   (d) $H_2O(s)$

81. Of each pair of solids, which one has the higher melting point and why?
   (a) $Ti(s)$ and $Ne(s)$
   (b) $H_2O(s)$ and $H_2S(s)$
   (c) $Kr(s)$ and $Xe(s)$
   (d) $NaCl(s)$ and $CH_4(s)$

82. Of each pair of solids, which one has the higher melting point and why?
   (a) $Fe(s)$ and $CCl_4(s)$
   (b) $KCl(s)$ or $HCl(s)$
   (c) $TiO_2(s)$ or $HOOH(s)$
   (d) $CCl_4(s)$ or $SiCl_4(s)$

83. List the following substances in order of increasing boiling point:

   H_2O, Ne, NH_3, NaF, SO_2

84. List the following substances in order of decreasing boiling point.

   CO_2, Ne, CH_3OH, KF

Cumulative Problems

85. Ice actually has negative caloric content. How much energy, in each of the following units, does your body lose from eating (and therefore melting) 58 g of ice?
   (a) joules
   (b) kilojoules
   (c) calories (1 cal = 4.18 J)
   (d) nutritional Calories or capital "C" Calories (1000 cal = 1 Cal)

86. As mentioned in the previous problem, ice has negative caloric content. How much energy, in each of the following units, does your body lose from eating (and therefore melting) 135 g of ice?
   (a) joules
   (b) kilojoules
   (c) calories (1 cal = 4.18 J)
   (d) nutritional Calories or capital "C" calories (1000 cal = 1 Cal)

87. An 8.5-g ice cube is placed into 255 g of water. Calculate the temperature change in the water upon the complete melting of the ice. Hint: Determine how much heat is absorbed by the melting ice and then use $q = mC\,\Delta T$ to calculate the temperature change of the 255 g of water.

88. A 14.7-g ice cube is placed into 324 g of water. Calculate the temperature change in the water upon complete melting of the ice. Hint: Determine how much heat is absorbed by the melting ice and then use $q = mC\,\Delta T$ to calculate the temperature change of the 324 g of water.

89. How much ice in grams would have to melt to lower the temperature of 352 mL of water from 25 °C to 5 °C? (Assume that the density of water is 1.0 g/mL.)

90. How much ice in grams would have to melt to lower the temperature of 55.8 g of water from 55.0 °C to 47.0 °C? (Assume that the density of water is 1.0 g/mL.)

91. How much heat in kilojoules is evolved in converting 1.00 mol of steam at 145 °C to ice at −50.0 °C? The heat capacity of steam is 1.84 J/g °C and that of ice is 2.09 J/g °C.

92. How much heat in kilojoules is required to warm 10.0 g of ice, initially at −10.0 °C, to steam at 110.0 °C. The heat capacity of ice is 2.09 J/g °C and that of steam is 1.84 J/g °C.

93. Draw a Lewis structure for each of the following molecules and determine its molecular geometry. What kind of intermolecular forces are present in each substance?
(a) H_2Se
(b) SO_2
(c) $CHCl_3$
(d) CO_2

94. Draw a Lewis structure for each of the following molecules and determine its molecular geometry. What kind of intermolecular forces are present in each substance?
(a) BCl_3 (remember that B is a frequent exception to the octet rule)
(b) HCOH (carbon is central; each H and O bonded directly to C)
(c) CS_2
(d) NCl_3

95. The melting point of ionic solids depends on the magnitude of the electrostatic attractions that hold the solid together. Draw ionic Lewis structures for NaF and MgO. Which do you think has the higher melting point?

96. Draw ionic Lewis structures for KF and CaO. Use the information and the method in the previous problem to predict which of these two ionic solids has the higher melting point.

97. Explain the observed trend in the melting points of the alkyl halides. Why is HF atypical?

Compound	*Melting Point*
HI	−50.8 °C
HBr	−88.5 °C
HCl	−114.8 °C
HF	−83.1 °C

98. Explain the following trend in the boiling points of the compounds listed. Why is H_2O atypical?

Compound	*Boiling Point*
H_2Te	−2 °C
H_2Se	−41.5 °C
H_2S	−60.7 °C
H_2O	+100 °C

Highlight Problems

99. Look at the following molecular view of water. Pick a molecule in the interior and draw a line to each of its direct neighbors. Pick a molecule near the edge (analogous to a molecule on the surface in three dimensions) and do the same. Which molecule has the most neighbors? Which molecule is more likely to evaporate?

100. Water does not easily remove grease from dirty hands because grease is nonpolar and water is polar; therefore they are immiscible. The addition of soap, however, results in the removal of the grease. Examine the following structure of soap and explain why soap works.

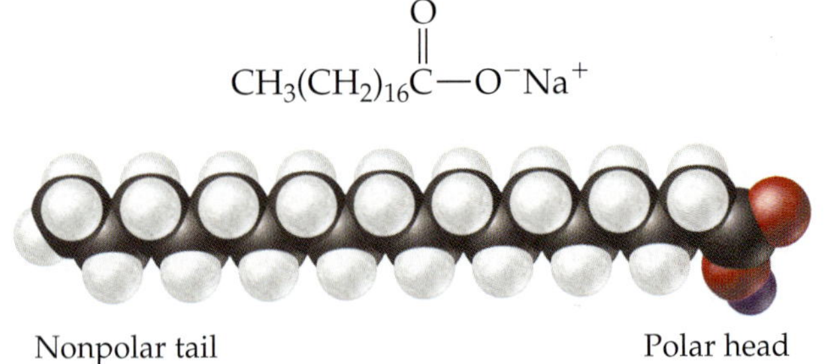

Sodium stearate
a soap

101. One prediction of global warming is the melting of global ice, which may result in coastal flooding. A criticism of this prediction is that the melting of icebergs does not increase ocean levels any more than the melting of ice in a glass of water increases the level of liquid in the glass.
 (a) Is this a valid criticism? Does the melting of an ice cube in a cup of water raise the level of the liquid in the cup? Why or why not?
 A response to this criticism is that the criticism is incorrect. Scientists are not worried about rising ocean levels due to melting icebergs, but rather, scientists are worried about rising ocean levels due to melting ice sheets that sit on the continent of Antarctica.
 (b) Would the melting of this ice increase ocean levels? Why or why not?

102. Explain why rubbing alcohol feels cold when applied to the skin.

Answers to Skillbuilder Exercises

Skillbuilder 12.1 5.83×10^3 kJ

Skillbuilder Plus, p. 404 47 °C

Skillbuilder 12.2 5.18 kJ

Skillbuilder Plus, p. 406 2.3 °C

Skillbuilder 12.3 C_2H_6

Skillbuilder 12.4 **(a)** no dipole–dipole forces **(b)** yes, it has dipole–dipole forces **(c)** yes, it has dipole–dipole forces

Skillbuilder 12.5 HF, because it has hydrogen bonding as an intermolecular force

Skillbuilder 12.6 **(a)** molecular **(b)** ionic **(c)** atomic

Answers to Conceptual Checkpoints

12.1 (b) Since boiling is a physical rather than a chemical change, the water molecules (H_2O) undergo no chemical alteration—they merely change from the liquid to the gaseous state.

12.2 (c) The chemical bonds between carbon and oxygen atoms are not affected by changes of state such as sublimation. Because carbon dioxide contains no hydrogen atoms, it cannot undergo hydrogen bonding, and because the molecule is nonpolar, it does not experience dipole–dipole interactions.

CHAPTER 13

Solutions

"Life can be thought of as water kept at the right temperature in the right atmosphere in the right light for a long enough period of time."

N. J. Berrill (1903–1996)

13.1 Tragedy in Cameroon

Carbon dioxide, a colorless and odorless gas, displaced the air in low-lying regions surrounding the lake, leaving no oxygen for the inhabitants to breathe. The rotten-egg smell indicates the presence of some additional sulfur-containing gases.

Most people living near Lake Nyos in Cameroon, West Africa, had an ordinary day on August 22, 1986. Unfortunately, the day ended in tragedy. On that evening, a large cloud of carbon dioxide gas, burped up from the depths of Lake Nyos, killed more than 1700 people and about 3000 head of cattle. Survivors tell of smelling rotten eggs, feeling a warm sensation, and then losing consciousness. Two years before that, a similar tragedy had occurred in Lake Monoun, just 60 miles away, killing thirty-seven people. Today, scientists are taking steps to prevent these lakes, both of which are in danger of burping again, from accumulating the carbon dioxide that caused the disaster.

▲ Cameroon is in West Africa.

◀ Carbon dioxide bubbled out of Lake Nyos and flowed into the adjacent valley. The carbon dioxide came from the bottom of the lake, where it was held in solution by the pressure of the water above it. When the layers in the lake were disturbed, the carbon dioxide came out of solution due to the decrease in pressure—with lethal consequences.

▲ Engineers watch as carbon dioxide vented from the bottom of Lake Nyos creates a geyser. The controlled release of carbon dioxide from the lake bed is designed to prevent future catastrophes like the one that killed more than 1700 people in 1986.

Lake Nyos is a water-filled volcanic crater. Some 50 miles beneath the surface of the lake, molten volcanic rock (magma) produces carbon dioxide gas that seeps into the lake through the volcano's plumbing system. The carbon dioxide mixes with the lake water, and the great pressure at the bottom of the deep lake allows the concentration of carbon dioxide to become very high (just as the pressure in a beer can allows the concentration of carbon dioxide in beer to be very high). Over time, the carbon dioxide and water mixture at the bottom of the lake became so concentrated that—either because of the high concentration itself or because of some other natural trigger such as a landslide—some gaseous carbon dioxide escaped. The rising bubbles disrupted the stratified layers of lake water, causing the highly concentrated carbon dioxide and water mixture at the bottom of the lake to rise, lowering the pressure on it. The drop in pressure on the mixture released more carbon dioxide bubbles just as the drop in pressure upon opening a beer can releases carbon dioxide bubbles. This in turn caused more churning and more carbon dioxide release. Since carbon dioxide is heavier than air, it traveled down the sides of the volcano and into the nearby valley, displacing air and asphyxiating many of the local residents.

In efforts to prevent these events from occurring again—by 2001, carbon dioxide concentrations had already returned to dangerously high levels—scientists built a piping system that slowly vents carbon dioxide from the lake bottom. This system gradually releases the carbon dioxide into the atmosphere, preventing a repeat of the tragedy.

13.2 Solutions: Homogenous Mixtures

The carbon dioxide and water mixture at the bottom of Lake Nyos is an example of a **solution**, a homogenous mixture of two or more substances. Solutions are all around us—most of the liquids and gases that we encounter every day are actually solutions. When most people think of a solution, they think of a solid dissolved in water. The ocean, for example, is a solution of salt and other solids dissolved in water, and our blood is a solution of several solids (as well as some gases) dissolved in water. However, many other kinds of solutions exist. A solution may be composed of a gas and a liquid (like the carbon dioxide and water of Lake Nyos), a liquid and another liquid, a solid and a gas, or other combinations (see Table 13.1).

Aqueous comes from the Latin word *aqua*, meaning "water."

The most common solutions, however, are those containing a solid, a liquid, or a gas and water. These are called *aqueous solutions*—they are critical to life and are the main focus of this chapter. Common examples of aqueous solutions include sugar water and salt water, both solutions of solids and water. Similarly, ethyl alcohol—the alcohol in alcoholic beverages—readily

TABLE 13.1 Common Types of Solutions

Solution Phase	Solute Phase	Solvent Phase	Example
gaseous solutions	gas	gas	air (mainly oxygen and nitrogen)
liquid solutions	gas	liquid	soda water (CO_2 and water)
	liquid	liquid	vodka (ethanol and water)
	solid	liquid	seawater (salt and water)
solid solutions	solid	solid	brass (copper and zinc) and other alloys

TABLE 13.2 Common Laboratory Solvents

Common Polar Solvents	Common Nonpolar Solvents
water (H_2O)	hexane (C_6H_{12})
acetone (CH_3COCH_3)	ethyl ether ($CH_3CH_2OCH_2CH_3$)
methyl alcohol (CH_3OH)	toluene (C_7H_8)

mixes with water to form a solution of a liquid with water, and we have already seen an example of a gas-and-water solution in Lake Nyos.

In a solid/liquid solution, the liquid is usually considered the solvent, regardless of the relative proportions of the components.

A solution has at least two components. The majority component is usually called the **solvent**, and the minority component is called the **solute**. In our carbon-dioxide-and-water solution, carbon dioxide is the solute and water is the solvent. In a salt-and-water solution, salt is the solute and water is the solvent. Because water is so abundant on Earth, it is a common solvent. However, other solvents are often used in the laboratory and even in the home, especially to form solutions with nonpolar solutes. For example, you may use paint thinner, a nonpolar solvent, to remove grease from a dirty bicycle chain or from ball bearings. The paint thinner dissolves (or forms a solution with) the grease, removing it from the metal. In general, polar solvents dissolve polar or ionic solutes and nonpolar solvents dissolve nonpolar solutes. This tendency is described by the rule *like dissolves like*. Thus, similar kinds of solvents dissolve similar kinds of solutes. Table 13.2 lists some common polar and nonpolar laboratory solvents.

See Sections 10.8 and 12.6 to review the concept of polarity.

Like dissolves like—polar solvents dissolve polar solutes and nonpolar solvents dissolve nonpolar solutes.

CONCEPTUAL CHECKPOINT 13.1

Which of these compounds would you expect to be *least* soluble in water?

(a) CCl_4
(b) CH_3Cl
(c) H_2S
(d) KF

13.3 Solutions of Solids Dissolved in Water: How to Make Rock Candy

We have already seen several examples of solutions of a solid dissolved in water. The ocean, for example, is a solution of salt and other solids dissolved in water. A sweetened cup of coffee is a solution of sugar and other solids dissolved in water. Our blood is a solution of several solids (and some gases) dissolved in water. Not all solids, however, dissolve in water. We already know that nonpolar solids—such as lard and shortening, for example—do not dissolve in water. However, solids such as calcium carbonate and sand do not dissolve either.

When a solid is put into water, there is a competition between the attractive forces that hold the solid together (the solute–solute interactions) and the attractive forces occurring between the water molecules and the particles that compose the solid (the solvent–solute interactions). For example, when sodium chloride is put into water, there is a competition between the attraction of Na^+ cations and Cl^- anions to each other and the attraction of Na^+

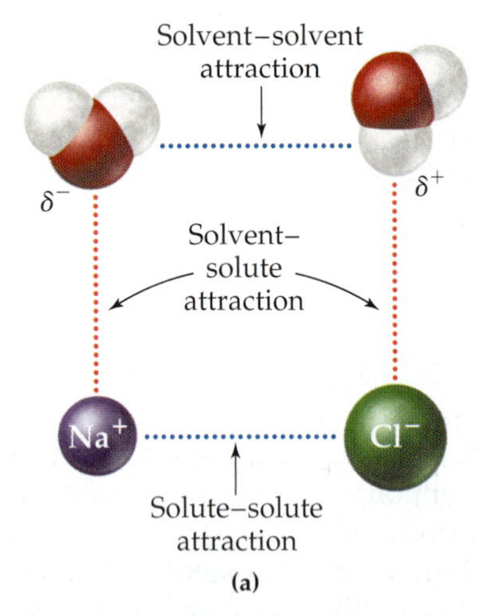

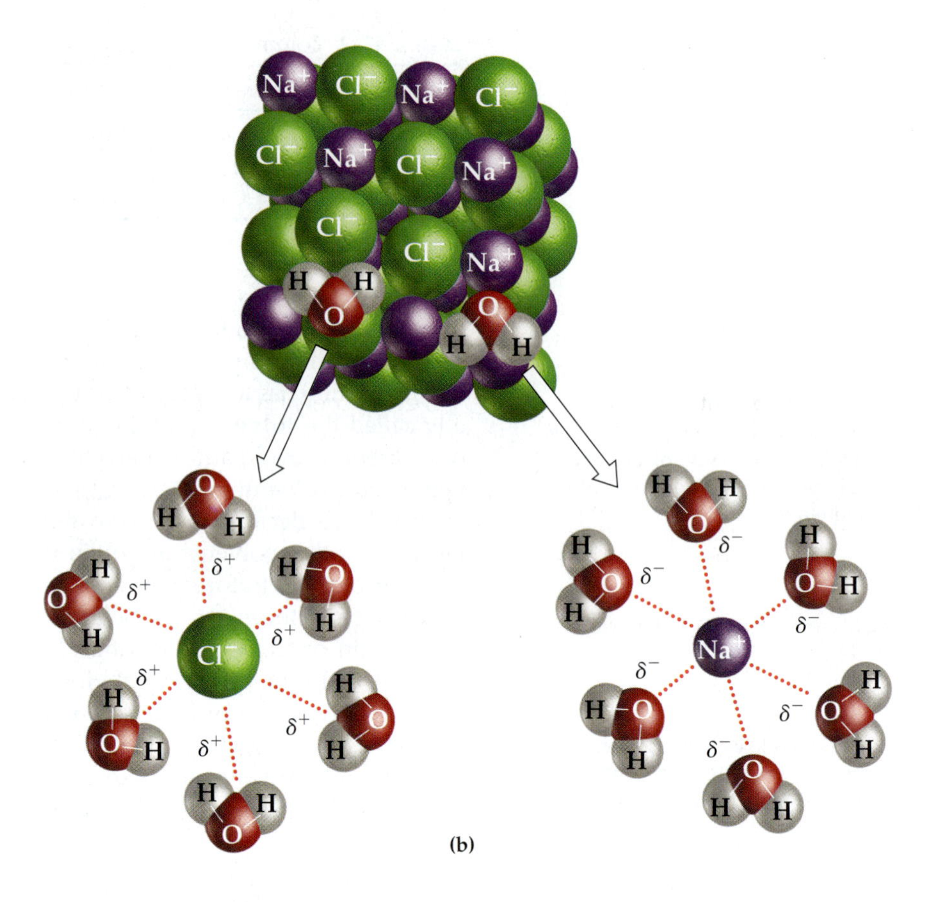

▲ **Figure 13.1 How a solid dissolves in water** **(a)** When NaCl is put into water, the attraction between water molecules and Na^+ and Cl^- ions (solvent–solute attraction) overcomes the attraction between Na^+ and Cl^- ions (solute–solute attraction). **(b)** The positive ends of the water dipoles are attracted to the negatively charged Cl^- ions and the negative ends of the water dipoles are attracted to the positively charged Na^+ ions. The result is that water molecules surround the ions of NaCl and disperse them in the solution.

and Cl^- to water molecules (▲ Figure 13.1a). For sodium ions, the attraction is between the positive charge of the sodium ion and the negative side of water's dipole moment (▲ Figure 13.1b). For chloride ions, the attraction is between the negative charge of the chloride ion and the positive side of water's dipole moment. In the case of NaCl, the attraction to water wins, and sodium chloride dissolves (▼ Figure 13.2). In the case of calcium carbonate ($CaCO_3$), the attraction between Ca^{2+} ions and $CO_3{}^{2-}$ ions wins and calcium carbonate does not dissolve in water.

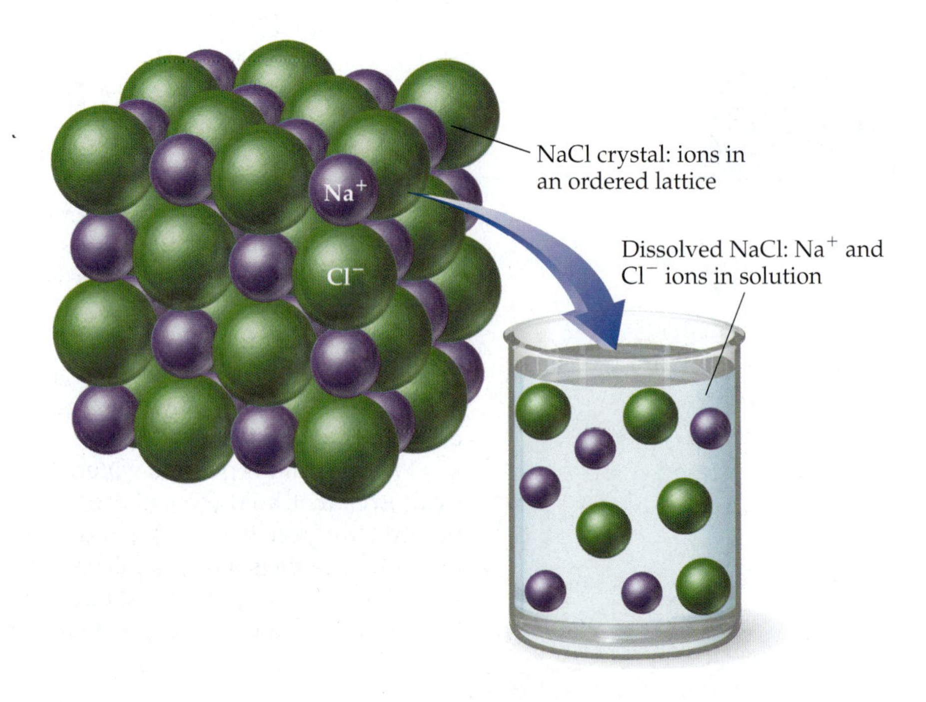

► **Figure 13.2 A sodium chloride solution** In a solution of NaCl, the Na^+ and Cl^- ions are dispersed in the water.

Solubility and Saturation

The **solubility** of a compound is defined as the amount of the compound, usually in grams, that will dissolve in a certain amount of liquid. For example, the solubility of sodium chloride at 25 °C is 36 g NaCl per 100 g water, while the solubility of calcium carbonate is close to zero. A solution that has 36 g of NaCl per 100 g water is called a saturated sodium chloride solution. A **saturated solution** is one that holds the maximum amount of solute under the solution conditions. If additional solute is added to a saturated solution, it will not dissolve. An **unsaturated solution** is one holding less than the maximum amount of solute. If additional solute is added to an unsaturated solution, it will dissolve. A **supersaturated solution** is one holding more than the normal maximum amount of solute. The solute will normally precipitate from (or come out of) a supersaturated solution. As the carbon dioxide and water solution rose from the bottom of Lake Nyos, for example, it became supersaturated because of the drop in pressure. The excess gas came out of the solution and rose to the surface of the lake, where it was emitted into the surrounding air.

Supersaturated solutions can form under special circumstances, such as the sudden release in pressure that occurs in a soda can when it is opened.

(a)

(b)

(c)

▲ A supersaturated solution holds more than the normal maximum amount of solute. In some cases, such as the sodium acetate solution pictured here, a supersaturated solution may be temporarily stable. Any disturbance however, such as dropping in a small piece of solid sodium acetate **(a)**, will cause the solid to come out of solution **(b, c)**.

Table 7.2 lists the solubility rules for ionic compounds.

In Chapter 7 we learned the solubility rules, which give us a qualitative description of the solubility of ionic solids. Molecular solids may also be soluble in water depending on whether the solid is polar. Table sugar ($C_{12}H_{22}O_{11}$), for example, is polar and soluble in water. Nonpolar solids, such as lard and vegetable shortening, are usually insoluble in water.

Electrolyte Solutions: Dissolved Ionic Solids

There is an important difference, however, between a sugar solution (containing a molecular solid) and a salt solution (containing an ionic solid) (▶ Figure 13.3). In a salt solution the dissolved particles are ions, while in a sugar solution the dissolved particles are molecules. The ions in the salt solution are mobile charged particles and can therefore conduct electricity. As we learned in Section 7.5, a solution containing a solute that dissociates into ions is called an **electrolyte solution**. The sugar solution contains dissolved sugar molecules and cannot conduct electricity; it is called a **nonelectrolyte solution**. In general, soluble ionic solids form electrolyte solutions, while soluble molecular solids form nonelectrolyte solutions.

NaCl forms a strong electrolyte solution (Section 7.5). Weak electrolyte solutions will be covered in Chapter 14.

▶ **Figure 13.3 Electrolyte and nonelectrolyte solutions** Electrolyte solutions contain dissolved ions (charged particles) and therefore conduct electricity. Nonelectrolyte solutions contain dissolved molecules (neutral particles) and so do not conduct electricity.

How Solubility Varies with Temperature

The solubility of solids in water can be highly dependent on temperature. Have you ever noticed how much easier it is to dissolve sugar in hot tea than in cold tea? In general, the solubility of *solids* in water increases with increasing temperature (▼ Figure 13.4). For example, the solubility of potassium nitrate (KNO_3) at 20 °C is about 37 g KNO_3 per 100 g of water. However, at 50 °C, the solubility rises to 88 g KNO_3 per 100 g of water. A common way to purify a solid is a technique called **recrystallization**. In this technique, the solid is put into water (or some other solvent) at an elevated temperature. Enough solid is added to the solvent to create a saturated solution at the elevated temperature. As the solution cools, the solubility decreases, causing some of the solid to precipitate from solution. If the solution cools slowly, the solid will form crystals as it comes out. The crystalline structure tends to reject impurities, resulting in a purer solid.

▲ Rock candy is composed of sugar crystals that have been grown through recrystallization.

Rock Candy

A similar effect can be seen if you make rock candy. To make rock candy, a saturated sucrose (table sugar) solution is prepared at an elevated temperature. A string is left to dangle in the solution, and the solution is allowed to cool and stand for several days. As the solution cools, it becomes supersaturated and sugar crystals grow on the string. After several days, beautiful and delicious crystals, or "rocks," of sugar cover the string, ready to be admired and eaten.

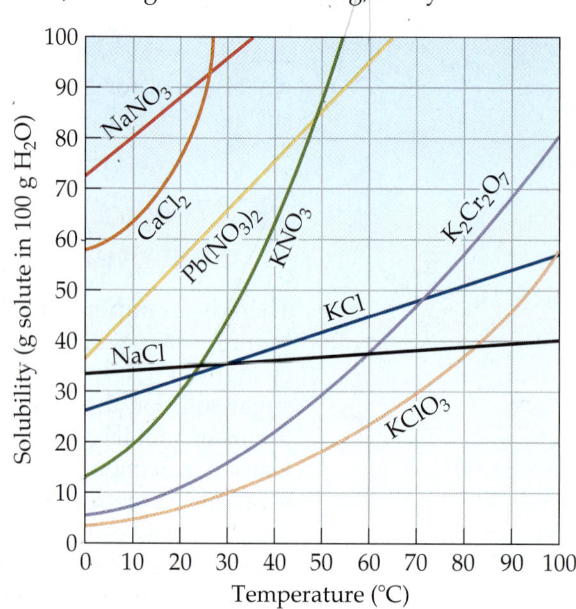

▶ **Figure 13.4 Solubility of some ionic solids as a function of temperature**

13.4 Solutions of Gases in Water: How Soda Pop Gets Its Fizz

The water at the bottom of Lake Nyos and a can of soda pop are both examples of a gas dissolved in a liquid. They are solutions of carbon dioxide and water. Most liquids exposed to air contain some dissolved gases. Lake water and seawater, for example, contain dissolved oxygen necessary for the survival of fish. Our blood contains dissolved nitrogen, oxygen, and carbon dioxide. Even tap water contains dissolved nitrogen and oxygen.

You can see the dissolved gases in ordinary tap water by heating it on a stove. Before the water reaches its boiling point, you will see small bubbles develop in the water. These bubbles are dissolved air (mostly nitrogen and oxygen) coming out of solution. Once the water boils, the bubbling becomes more vigorous—these larger bubbles are composed of water vapor. The dissolved air comes out of solution because—unlike solids, whose solubility *increases* with increasing temperature—the solubility of gases in water *decreases* with increasing temperature. As the temperature of the water rises, the solubility of the dissolved nitrogen and oxygen decreases and these gases come out of solution, forming small bubbles around the bottom of the pot. The decrease in the solubility of gases with increasing temperature is the reason that warm soda pop bubbles more than cold soda pop and also the reason that warm beer goes flat faster than cold beer. The carbon dioxide comes out of solution faster (bubbles more) at room temperature than at lower temperature because it is less soluble at room temperature.

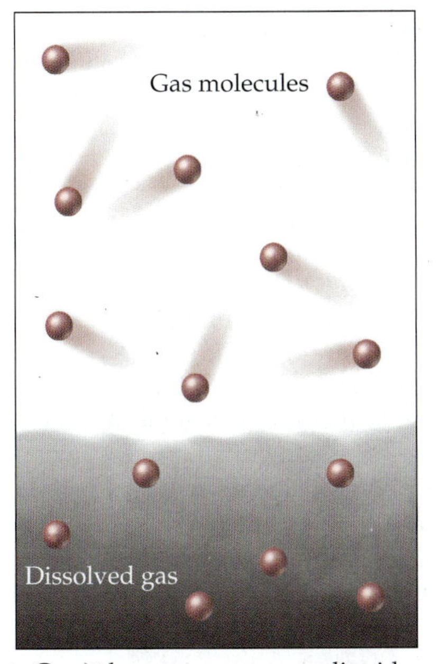

Gas at low pressure over a liquid

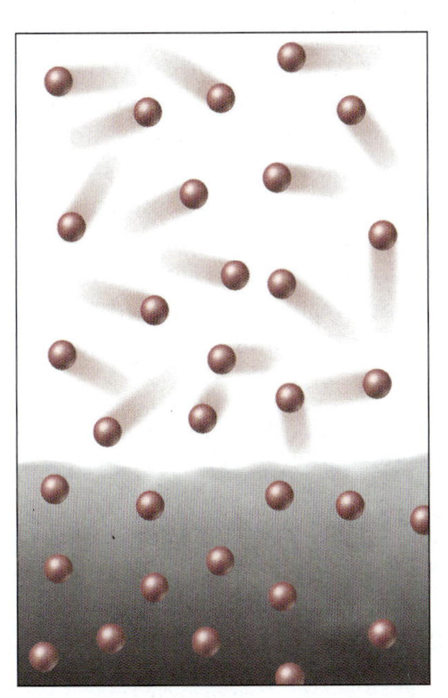
Gas at high pressure over a liquid

▲ **Figure 13.5 Pressure and solubility** The higher the pressure above a liquid, the more soluble the gas is in the liquid.

Cold soda pop Warm soda pop

▲ Warm soda pop fizzes more than cold soda pop because the solubility of the dissolved carbon dioxide decreases with increasing temperature.

The solubility of gases also depends on pressure. The higher the pressure above a liquid, the more soluble the gas is in the liquid (◀ Figure 13.5). In a can of soda pop and in Lake Nyos, carbon dioxide is maintained in solution by high pressure. In soda pop, the pressure is provided by a large amount of carbon dioxide gas that is pumped into the can before sealing it.

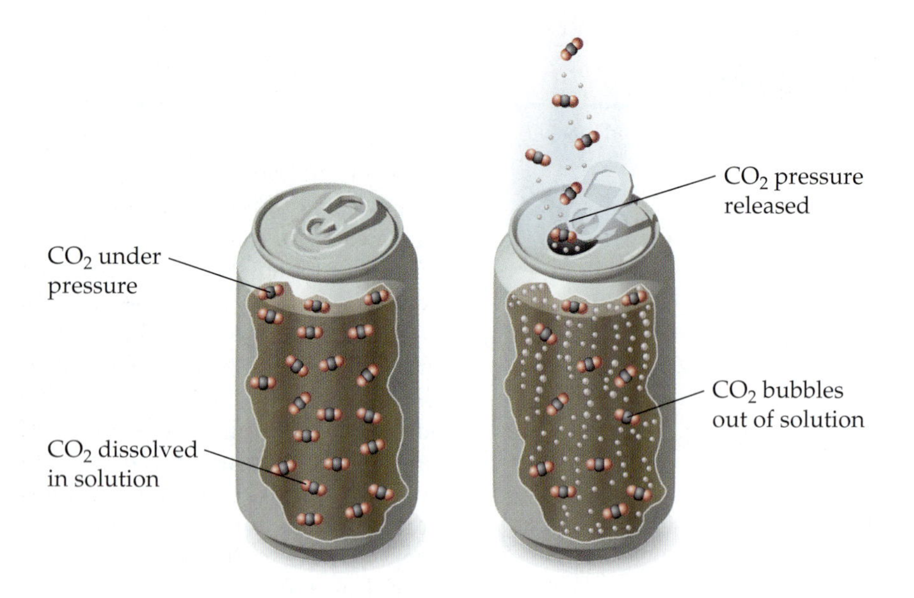

▶ **Figure 13.6 Pop! Fizz!** A can of soda pop is pressurized with carbon dioxide. When the can is opened the pressure is released, lowering the solubility of carbon dioxide in the solution and causing it to come out of solution as bubbles.

The increase in solubility of a gas with increasing pressure is known as *Henry's law*.

When the can is opened, the pressure is released and the solubility of carbon dioxide decreases, resulting in bubbling (▲ Figure 13.6). In Lake Nyos, the pressure is provided by the mass of the lake water itself pushing down on the carbon-dioxide-rich water at the bottom of the lake. When the stratification (or layering) of the lake is disturbed, the pressure on the carbon dioxide solution is lowered and the solubility of carbon dioxide decreases, resulting in the release of excess carbon dioxide gas.

13.5 Specifying Solution Concentration: Mass Percent

As we have seen, the amount of solute in a solution is an important property of the solution. For example, the amount of carbon dioxide in the water at the bottom of Lake Nyos is an important predictor of when the deadly event may repeat itself. A **dilute solution** is one containing small amounts of solute relative to solvent. If the water at the bottom of Lake Nyos were a dilute carbon dioxide solution, it would pose little threat. A **concentrated solution** is one containing large amounts of solute relative to solvent. If the water at the bottom of Lake Nyos becomes concentrated in carbon dioxide (through the continual feeding of carbon dioxide from magma into the lake), it becomes a large threat. A common way to report solution concentration is *mass percent*.

Mass Percent

Also in common use are *parts per million (ppm)*, the number of grams of solute per 1 million g of solution, and *parts per billion (ppb)*, the number of grams of solute per 1 billion g of solution.

Mass percent is simply the number of grams of solute per 100 g of solution. So a solution with a concentration of 14% by mass, for example, contains 14 g of solute per 100 g of solution. To calculate mass percent, simply divide the mass of the solute by the mass of the solution (solute *and* solvent) and multiply by 100%.

Note that the denominator is the mass of *solution*, not the mass of solvent.

$$\text{Mass percent} = \frac{\text{Mass solute}}{\text{Mass solution}} \times 100\%$$

For example, suppose you wanted to calculate the mass percent of NaCl in a solution containing 15.3 g of NaCl and 155.0 g of water. Set up the problem as you normally would.

Given:

15.3 g NaCl

155.0 g H_2O

Find: mass percent

Equation: This problem requires the use of the equation that defines mass percent.

$$\text{Mass percent} = \frac{\text{Mass solute}}{\text{Mass solution}} \times 100\%$$

Solution: The mass of solution is simply the mass of NaCl plus the mass of H_2O.

$$\begin{aligned}\text{Mass solution} &= \text{Mass NaCl} + \text{Mass } H_2O \\ &= 15.3\text{ g} + 155.0\text{ g} \\ &= 170.3\text{ g}\end{aligned}$$

We then substitute the correct quantities into the equation.

$$\begin{aligned}\text{Mass percent} &= \frac{\text{Mass solute}}{\text{Mass solution}} \times 100\% \\ &= \frac{15.3\text{ g}}{170.3\text{ g}} \times 100\% \\ &= 8.98\%\end{aligned}$$

The solution is 8.98% NaCl by mass.

EXAMPLE 13.1 Calculating Mass Percent

Calculate the mass percent of a solution containing 27.5 g of ethanol (C_2H_6O) and 175 mL of H_2O. (Assume that the density of water is 1.00 g/mL.)

Begin by setting up the problem. You are given the mass of ethanol and the volume of water and asked to find the mass percent of the solution.

Given:

27.5 g C_2H_6O
175 mL H_2O

Find: mass percent

You will need the equation that defines mass percent and the density of water

Equation and Conversion Factor:

$$\text{Mass percent} = \frac{\text{Mass solute}}{\text{Mass solution}} \times 100\%$$

$$d(H_2O) = \frac{1.00\text{ g}}{\text{mL}}$$

To find the mass percent, substitute into the equation for mass percent. You need the mass of the solution, which is simply the mass of ethanol plus the mass of water. The mass of water is obtained from the volume of water by using the density as a conversion factor.

Solution:

$$\text{Mass } H_2O = 175\text{ mL } H_2O \times \frac{1.00\text{ g}}{\text{mL}} = 175\text{ g}$$

$$\begin{aligned}\text{Mass solution} &= \text{Mass } C_2H_6O + \text{Mass } H_2O \\ &= 27.5\text{ g} + 175\text{ g} \\ &= 20\underline{2}.5\text{ g}\end{aligned}$$

Finally, substitute the correct quantities into the equation and calculate the mass percent.

$$\text{Mass percent} = \frac{\text{Mass solute}}{\text{Mass solution}} \times 100\%$$
$$= \frac{27.5 \text{ g}}{202.5 \text{ g}} \times 100\%$$
$$= 13.6\%$$

SKILLBUILDER 13.1 Calculating Mass Percent

Calculate the mass percent of a sucrose solution containing 11.3 g of sucrose and 412.1 mL of water. (Assume that the density of water is 1.00 g/mL.)

Using Mass Percent in Calculations

The mass percent of a solution can be used as a conversion factor between mass of the solute and mass of the solution. The key to using mass percent in this way is to write it as a fraction.

$$\text{Mass percent} = \frac{\text{g solute}}{100 \text{ g solution}}$$

A solution containing 3.5% sodium chloride, for example, has the following conversion factor.

$$\frac{3.5 \text{ g NaCl}}{100 \text{ g solution}} \quad \text{converts g solution} \longrightarrow \text{g NaCl}$$

This conversion factor converts from grams of solution to grams of NaCl. If you want to go the other way, simply invert the conversion factor.

$$\frac{100 \text{ g solution}}{3.5 \text{ g NaCl}} \quad \text{converts g NaCl} \longrightarrow \text{g solution}$$

As an example of using mass percent as a conversion factor, consider a water sample from the bottom of Lake Nyos containing 8.5% carbon dioxide by mass. How much carbon dioxide in grams is contained in 28.6 L of the water solution? (Assume that the density of the solution is 1.03 g/mL.) We set up the problem in the normal way.

Given:

8.5% CO_2 by mass

28.6 L solution

Find: g CO_2

Conversion Factors: The main conversion factor is the mass percent of the solution. We write it as g CO_2 per 100 g solution.

$$\frac{8.5 \text{ g } CO_2}{100 \text{ g solution}}$$

We will also need the density of the solution and the conversion factor between L and mL.

$$\frac{1.03 \text{ g}}{\text{mL}}$$

$$1000 \text{ mL} = 1 \text{ L}$$

Solution Map: The solution map begins with L solution and shows the conversion to mL solution and then to g solution using the density. The map then proceeds from g solution to g CO_2, using the mass percent (expressed as a fraction) as a conversion factor.

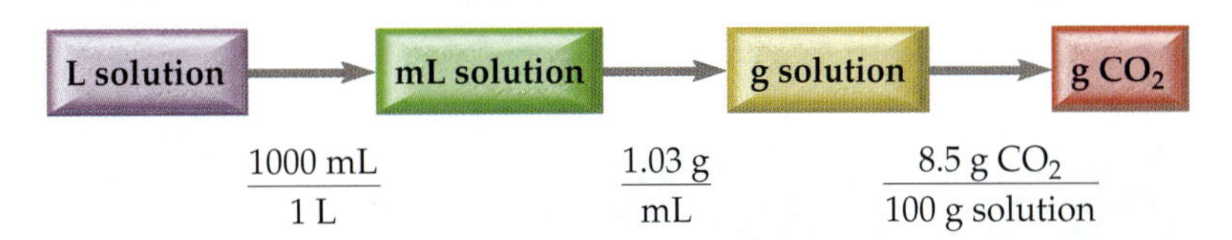

$$\frac{1000 \text{ mL}}{1 \text{ L}} \qquad \frac{1.03 \text{ g}}{\text{mL}} \qquad \frac{8.5 \text{ g } CO_2}{100 \text{ g solution}}$$

Solution: Finally, we follow the solution map to compute the answer.

$$28.6 \cancel{\text{L}} \text{ solution} \times \frac{1000 \cancel{\text{mL}}}{\cancel{\text{L}}} \times \frac{1.03 \cancel{\text{g}}}{\cancel{\text{mL}}} \times \frac{8.5 \text{ g } CO_2}{100 \cancel{\text{g}} \text{ solution}} = 2.5 \times 10^3 \text{ g } CO_2$$

In this example, we used mass percent to convert from a given amount of *solution* to the amount of *solute* present in the solution. In Example 13.2, we use mass percent to convert from a given amount of *solute* to the amount of *solution* containing that solute.

EXAMPLE 13.2 Using Mass Percent in Calculations

A soft drink contains 11.5% sucrose ($C_{12}H_{22}O_{11}$) by mass. What volume of the soft drink solution in milliliters contains 85.2 g of sucrose? (Assume a density of 1.00 g/mL).

You are given the concentration of sucrose in a soft drink and a mass of sucrose. You are asked to find the volume of the soft drink that contains the given mass of sucrose.

Write the mass percent concentration of sucrose as a conversion factor, remembering that percent means *per hundred*. You will also need the density to use as a conversion factor between mass and volume of the soft drink.

Given:

11.5% $C_{12}H_{22}O_{11}$ by mass
85.2 g $C_{12}H_{22}O_{11}$

Find: mL solution (soft drink)

Conversion Factors:

$$\frac{11.5 \text{ g } C_{12}H_{22}O_{11}}{100 \text{ g solution}}$$

$$d = \frac{1.00 \text{ g}}{\text{mL}}$$

Convert from g solute ($C_{12}H_{22}O_{11}$) to g solution using the mass percent in fractional form as the conversion factor. Convert to mL using the density.

Solution Map:

g $C_{12}H_{22}O_{11}$ → g solution → mL solution

$$\frac{100 \text{ g solution}}{11.5 \text{ g } C_{12}H_{22}O_{11}} \qquad \frac{1 \text{ mL}}{1.00 \text{ g}}$$

Follow the solution map to solve the problem.

Solution:

$$85.2 \cancel{\text{g } C_{12}H_{22}O_{11}} \times \frac{100 \cancel{\text{g}} \text{ solution}}{11.5 \cancel{\text{g } C_{12}H_{22}O_{11}}} \times \frac{1 \text{ mL solution}}{1.00 \cancel{\text{g}}} = 741 \text{ mL solution}$$

SKILLBUILDER 13.2 Using Mass Percent in Calculations

How much sucrose ($C_{12}H_{22}O_{11}$) in grams is contained in 355 mL (12 oz) of the soft drink in Example 13.2?

Chemistry IN THE ENVIRONMENT

The Dirty Dozen

▲ Potentially dangerous chemicals can leak into the environment and contaminate water and food supplies.

A number of potentially harmful chemicals—such as DDT, dioxin, and polychlorinated biphenyls (PCBs)—can make their way into our water sources from industrial dumping, atmospheric emissions, agriculture, and household dumping. Since crops, livestock, and fish all rely on water, they too can accumulate these chemicals from water. Human consumption of food or water contaminated with these chemicals can lead to a number of diseases and adverse health effects such as increased cancer risk, liver damage, or central nervous system damage. Governments around the world have joined forces to ban a number of these kinds of chemicals—called persistent organic pollutants or POPs—from production. The original treaty targeted twelve such substances called the dirty dozen (Table 13.3).

One difficult problem with all of these chemicals is their persistence. Once they get into the environment, they stay there for a long time. A second problem is a process called *bioamplification*. Because these chemicals are nonpolar, they are stored and concentrated in the fatty tissues of the organisms that consume them. As larger organisms eat smaller ones they consume more of the stored chemicals. The result is an increase in the concentrations of these chemicals as they move up the food chain. Under the draft treaty, nearly all intentional production of these chemicals will be banned.

TABLE 13.3 The Dirty Dozen

1. aldrin (insecticide)
2. chlordane (insecticide by-product)
3. DDT (insecticide)
4. dieldrin (insecticide)
5. dioxin (industrial by-product)
6. eldrin (insecticide)
7. furan (industrial by-product)
8. heptachlor (insecticide)
9. hexachlorobenzene (fungicide, industrial by-product)
10. mirex (insecticide, fire retardant)
11. polychlorinated biphenyls (PCBs) (electrical insulators)
12. toxaphene (insecticide)

In the United States, the presence of these contaminants in water supplies is monitored under supervision of the Environmental Protection Agency (EPA). The EPA has set limits, called maximum contaminant levels (MCLs), for each of these in food and drinking water. Some MCLs for selected compounds in water supplies are listed in Table 13.4.

TABLE 13.4 EPA Maximum Contaminant Level (MCL) for Several "Dirty Dozen" Chemicals

chlordane	0.002 mg/L
dioxin	0.00000003 mg/L
heptachlor	0.0004 mg/L
hexachlorobenzene	0.001 mg/L

Notice the units that the EPA uses to express the concentration of the contaminants: milligrams per liter. This unit is a conversion factor between liters of water consumed and the mass in milligrams of the pollutant. According to the EPA, as long as the contaminant concentrations are below these levels, the water is safe to drink.

CAN YOU ANSWER THIS? *Using what you know about conversion factors, calculate how much of each of the chemicals in Table 13.4 (at their MCL) would be present in 715 L of water, the approximate amount of water consumed by an adult in one year.*

13.6 Specifying Solution Concentration: Molarity

A second way to express solution concentration is **molarity (M)**, defined as the number of moles of solute per liter of solution. We calculate the molarity of a solution as follows:

$$\text{Molarity (M)} = \frac{\text{Moles solute}}{\text{Liters solution}}$$

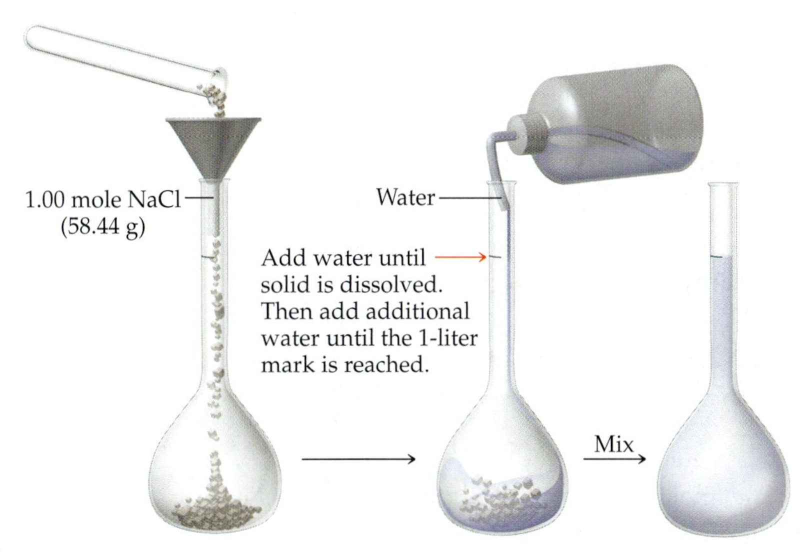
How to prepare a 1.00 molar NaCl solution.
1.00 mole NaCl
(58.44 g)
Water
Add water until solid is dissolved. Then add additional water until the 1-liter mark is reached.
Mix
First add 1.00 mole of NaCl.
A 1.00 molar NaCl solution

▶ **Figure 13.7 Making a solution of specific molarity** To make 1.00 L of a 1.00 M NaCl solution, you add 1.00 mol (58.44 g) of sodium chloride to a flask and then dilute to 1.00 L of total volume. **Question:** What would happen if you added 1 L of water to 1 mol of sodium chloride? Would the resulting solution be 1 M?

Note that molarity is moles of solute per liter of *solution*, not per liter of solvent. To make a solution of a specified molarity, you usually put the solute into a flask and then add water to the desired volume of solution. For example, to make 1.00 L of a 1.00 M NaCl solution, you add 1.00 mol of NaCl to a flask and then add water to make 1.00 L of solution (▲ Figure 13.7). You *do not* combine 1.00 mol of NaCl with 1.00 L of water because that would result in a total volume exceeding 1.00 L and therefore a molarity of less than 1.00 M. Molarity is moles of solute per liter of *solution*.

TUTORIAL
Solution Formation from a Solid

To calculate molarity, simply divide the number of moles of the solute by the volume of the solution (solute *and* solvent) in liters. For example, calculate the molarity of a sucrose ($C_{12}H_{22}O_{11}$) solution made with 1.58 mol of sucrose diluted to a total volume of 5.0 L of solution. Set up the problem in the normal way.

Given:

1.58 mol $C_{12}H_{22}O_{11}$
5.0 L solution

Find: molarity (M)

Equation: This problem requires the use of the equation that defines molarity.

$$\text{Molarity (M)} = \frac{\text{Moles solute}}{\text{Liters solution}}$$

Solution: Simply substitute the correct values into the equation and compute the answer.

$$\begin{aligned}\text{Molarity (M)} &= \frac{\text{Moles solute}}{\text{Liters solution}}\\ &= \frac{1.58 \text{ mol } C_{12}H_{22}O_{11}}{5.0 \text{ L solution}}\\ &= 0.32 \text{ M}\end{aligned}$$

EXAMPLE 13.3 Calculating Molarity

Calculate the molarity of a solution made by putting 15.5 g NaCl into a beaker and adding water to make 1.50 L of NaCl solution.

You are given the mass of sodium chloride (the solute) and the volume of solution. You are asked to find the molarity of the solution.

Given:

15.5 g NaCl

1.50 L solution

Find: molarity (M)

You will need the equation that defines molarity and the molar mass of NaCl.

Equation and Conversion Factor:

$$\text{Molarity (M)} = \frac{\text{Moles solute}}{\text{Liters solution}}$$

$$\text{Molar mass of NaCl} = \frac{58.44\ \text{g}}{1\ \text{mol}}$$

To calculate molarity, substitute the correct values into the equation and compute the answer. However, you must first convert the amount of NaCl from grams to moles using the molar mass of NaCl.

Solution:

$$\text{mol NaCl} = 15.5\ \cancel{\text{g NaCl}} \times \frac{1\ \text{mol NaCl}}{58.44\ \cancel{\text{g NaCl}}} = 0.26\underline{5}2\ \text{mol NaCl}$$

$$\text{Molarity (M)} = \frac{\text{Moles solute}}{\text{Liters solution}}$$

$$= \frac{0.26\underline{5}2\ \text{mol NaCl}}{1.50\ \text{L solution}}$$

$$= 0.177\ \text{M}$$

SKILLBUILDER 13.3 Calculating Molarity

Calculate the molarity of a solution made by putting 55.8 g of $NaNO_3$ into a beaker and diluting to 2.50 L.

Using Molarity in Calculations

The molarity of a solution can be used as a conversion factor between moles of the solute and liters of the solution. For example, a 0.500 M NaCl solution contains 0.500 mol NaCl for every liter of solution.

$$\frac{0.500\ \text{mol NaCl}}{\text{L solution}} \quad \text{converts L solution} \longrightarrow \text{mol NaCl}$$

This conversion factor converts from liters of solution to moles of NaCl. If you want to go the other way, simply invert the conversion factor.

$$\frac{\text{L solution}}{0.500\ \text{mol NaCl}} \quad \text{converts mol NaCl} \longrightarrow \text{L solution}$$

For example, how many grams of sucrose ($C_{12}H_{22}O_{11}$) are contained in 1.72 L of 0.758 M sucrose solution? We set up the problem in the normal way.

Given:

0.758 M $C_{12}H_{22}O_{11}$

1.72 L solution

Find: g $C_{12}H_{22}O_{11}$

Conversion Factors: The main conversion factor is the molarity of the solution. We write it as mol $C_{12}H_{22}O_{11}$ per L solution.

$$\frac{0.758 \text{ mol } C_{12}H_{22}O_{11}}{\text{L solution}}$$

We will also need the molar mass of sucrose.

$$\text{Molar mass of } C_{12}H_{22}O_{11} = \frac{342.34 \text{ g}}{\text{mol}}$$

Solution Map: The solution map begins with L solution and shows the conversion to mol $C_{12}H_{22}O_{11}$ using the molarity and then the conversion to g using the molar mass.

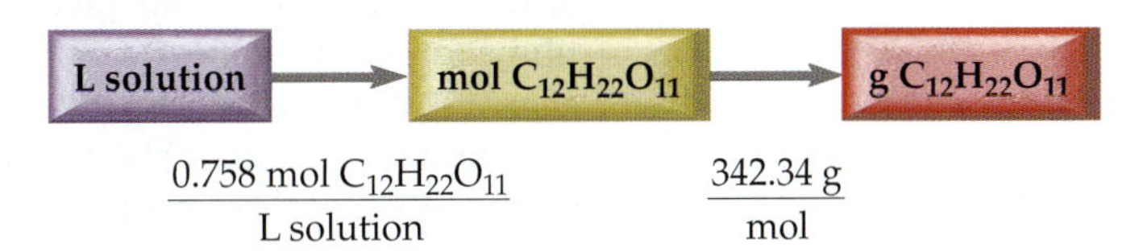

$$\frac{0.758 \text{ mol } C_{12}H_{22}O_{11}}{\text{L solution}} \qquad \frac{342.34 \text{ g}}{\text{mol}}$$

Solution: We then follow the solution map to compute the answer.

$$1.72 \text{ L solution} \times \frac{0.758 \text{ mol } C_{12}H_{22}O_{11}}{\text{L solution}} \times \frac{342.34 \text{ g } C_{12}H_{22}O_{11}}{\text{mol } C_{12}H_{22}O_{11}} = 446 \text{ g } C_{12}H_{22}O_{11}$$

In this example, we used molarity to convert from a given amount of *solution* to the amount of *solute* in that solution. In the example that follows, we use molarity to convert from a given amount of *solute* to the amount of *solution* containing that solute.

EXAMPLE 13.4 Using Molarity in Calculations

How many liters of a 0.114 M NaOH solution contains 1.24 mol of NaOH?

You are given the molarity of an NaOH solution and the number of moles of NaOH. You are asked to find the volume of solution that contains the given number of moles.

You will need to use the molarity of the solution (that is given) as a conversion factor.

Given:

0.114 M NaOH

1.24 mol NaOH

Find: L solution

Conversion Factor:

$$\frac{0.114 \text{ mol NaOH}}{\text{L solution}}$$

The solution map begins with mol NaOH and shows the conversion to liters of solution using the molarity.

Solution Map:

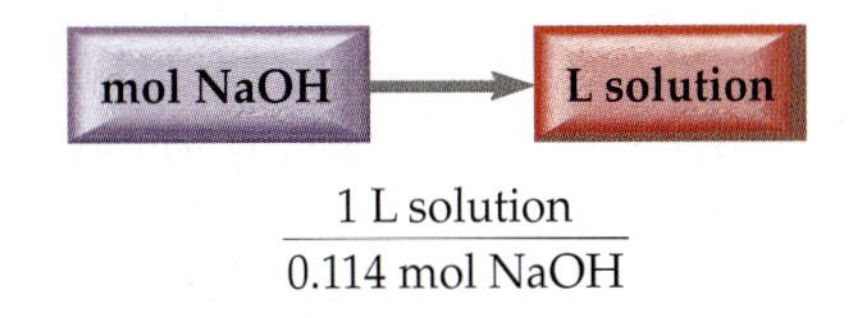

$$\frac{1 \text{ L solution}}{0.114 \text{ mol NaOH}}$$

Solve the problem by following the solution map.

Solution:

$$1.24 \text{ mol NaOH} \times \frac{1 \text{ L solution}}{0.114 \text{ mol NaOH}} = 10.9 \text{ L solution}$$

SKILLBUILDER 13.4 Using Molarity in Calculations

How much of a 0.225 M KCl solution contains 55.8 g of KCl?

Ion Concentrations

The reported concentration of a solution containing a molecular compound usually reflects the concentration of the solute as it actually exists in solution. For example, a 1.0 M glucose ($C_6H_{12}O_6$) solution indicates that the solution contains 1.0 mol of $C_6H_{12}O_6$ per liter of solution. However, the reported concentration of solution containing an ionic compound reflects the concentration of the solute *before it is dissolved in solution*. For example, a 1.0 M $CaCl_2$ solution contains 1.0 mol of Ca^{2+} per liter and 2.0 mol of Cl^- per liter. The concentration of the individual ions present in a solution containing an ionic compound can usually be deduced from the overall concentration, as shown by the following example.

EXAMPLE 13.5 Ion Concentration

Determine the molar concentrations of Na^+ and PO_4^{3-} in a 1.50 M Na_3PO_4 solution.

You are given the concentration of an ionic solution and asked to find the concentrations of the component ions.

Given: 1.50 M Na_3PO_4

Find: molarity (M) of Na^+ and PO_4^{3-}

Since a formula unit of Na_3PO_4 contains 3 Na^+ ions (as indicated by the subscript), the concentration of Na^+ is three times the concentration of Na_3PO_4. Since the same formula unit contains one PO_4^{3-} ion, the concentration of PO_4^{3-} is equal to the concentration of Na_3PO_4.

Solution:

molarity of Na^+ = 3(1.50 M) = 4.50 M

molarity of PO_4^{3-} = 1.50 M

SKILLBUILDER 13.5 Ion Concentration

Determine the molar concentrations of Ca^{2+} and Cl^- in a 0.75 M $CaCl_2$ solution.

13.7 Solution Dilution

When diluting acids, always add the concentrated acid to the water. *Never add water to concentrated acid solutions.*

To save space in laboratory storerooms, solutions are often stored in concentrated forms called **stock solutions**. For example, hydrochloric acid is often stored as a 12 M stock solution. However, many lab procedures call for much less concentrated hydrochloric acid solutions, so chemists must dilute the stock solution to the required concentration. This is normally done by diluting a certain amount of the stock solution with water. How do we know how much of the stock solution to use? The easiest way to solve these problems is to use the following dilution equation:

$$M_1V_1 = M_2V_2$$

where M_1 and V_1 are the molarity and volume of the initial concentrated solution and M_2 and V_2 are the molarity and volume of the final diluted solution. This equation works because the molarity multiplied by the volume gives the number of moles of solute ($M \times V =$ mol), which is the same in both solutions. For example, suppose a laboratory procedure calls for 5.00 L of a 1.50 M KCl solution. How should you prepare this solution from a 12.0 M stock solution? Set up the problem in the normal way.

TUTORIAL
Solution Formation by Dilution

Given: $M_1 = 12.0$ M
$M_2 = 1.50$ M
$V_2 = 5.00$ L

Find: V_1

Equation: $M_1V_1 = M_2V_2$

Solution: We solve the equation for V_1, the volume of the stock solution required for the dilution, and then substitute in the correct values to compute it.

$$M_1V_1 = M_2V_2$$

$$V_1 = \frac{M_2V_2}{M_1}$$

$$= \frac{1.50 \frac{\cancel{\text{mol}}}{\cancel{\text{L}}} \times 5.00 \text{ L}}{12.0 \frac{\cancel{\text{mol}}}{\cancel{\text{L}}}}$$

$$= 0.625 \text{ L}$$

Consequently, we make the solution by diluting 0.625 L of the stock solution to a total volume of 5.00 L (V_2). The resulting solution will be 1.50 M in KCl (▼ Figure 13.8).

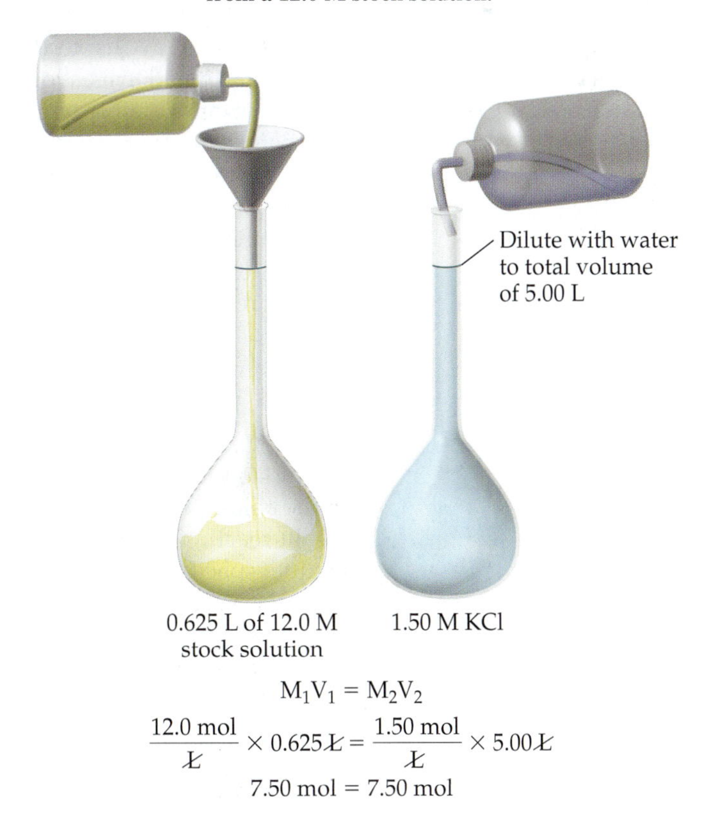
How to make 5.00 L of a 1.50 M KCl solution
from a 12.0 M stock solution.
Dilute with water
to total volume
of 5.00 L
0.625 L of 12.0 M
stock solution
1.50 M KCl
$M_1V_1 = M_2V_2$
$\frac{12.0 \text{ mol}}{\cancel{\text{L}}} \times 0.625 \cancel{\text{L}} = \frac{1.50 \text{ mol}}{\cancel{\text{L}}} \times 5.00 \cancel{\text{L}}$
7.50 mol = 7.50 mol

▶ **Figure 13.8** **Making a solution by dilution of a more concentrated solution**

EXAMPLE 13.6 Solution Dilution

To what volume should you dilute 0.100 L of a 15 M NaOH solution to obtain a 1.0 M NaOH solution?

Your are given the initial volume and concentration of an NaOH solution and a final concentration. You are asked to find the volume required to dilute the initial solution to the given final concentration.

Given:

$$V_1 = 0.100 \text{ L}$$
$$M_1 = 15 \text{ M}$$
$$M_2 = 1.0 \text{ M}$$

Find: V_2

The necessary equation is the solution dilution equation given previously in this section.

Equation:

$$M_1V_1 = M_2V_2$$

Solve the equation for V_2, the volume of the final solution, and substitute the required quantities to compute V_2. You would make the solution by diluting 0.100 L of the stock solution to a total volume of 1.5 L (V_2). The resulting solution will have a concentration of 1.0 M.

Solution:

$$M_1V_1 = M_2V_2$$
$$V_2 = \frac{M_1V_1}{M_2}$$
$$= \frac{15 \frac{\cancel{\text{mol}}}{\cancel{\text{L}}} \times 0.100 \text{ L}}{1.0 \frac{\cancel{\text{mol}}}{\cancel{\text{L}}}}$$
$$= 1.5 \text{ L}$$

SKILLBUILDER 13.6 Solution Dilution

How much 6.0 M $NaNO_3$ solution should be used to make 0.585 L of a 1.2 M $NaNO_3$ solution?

13.8 Solution Stoichiometry

See Sections 8.2 through 8.4 for a review of reaction stoichiometry.

As we discussed in Chapter 7, many chemical reactions take place in aqueous solutions. Precipitation reactions, neutralization reactions, and gas evolution reactions, for example, all occur in aqueous solutions. In Chapter 8, we learned how the coefficients in chemical equations are used as conversion factors between moles of reactants and moles of products in stoichiometric calculations. These conversion factors are often used to determine, for example, the amount of product obtained in a chemical reaction based on a given amount of reactant or the amount of one reactant needed to completely react with a given amount of another reactant. The general solution map for these kinds of calculations is

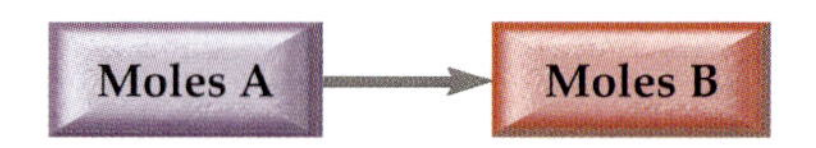

where A and B are two different substances involved in the reaction and the conversion factor between them comes from the stoichiometric coefficients in the balanced chemical equation.

In reactions involving aqueous reactant and products, it is often convenient to specify the amount of reactants or products in terms of their volume and concentration. We can then use the volume and concentration to calculate the number of moles of reactants or products, and then use the stoichiometric coefficients to convert to other quantities in the reaction. The general solution map for these kinds of calculations is

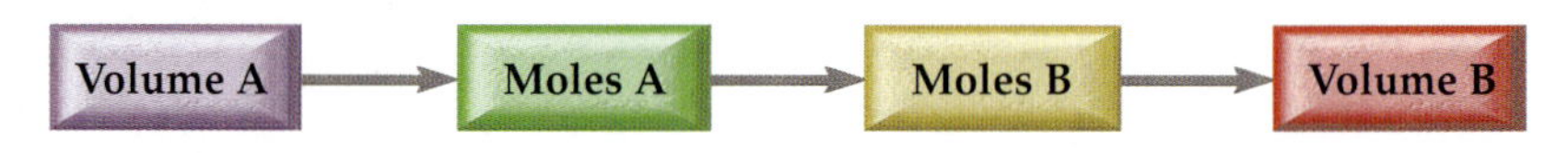

where the conversions between volume and moles are achieved using the molarities of the solutions. For example, consider the following reaction for the neutralization of sulfuric acid.

$$H_2SO_4(aq) + 2\,NaOH(aq) \longrightarrow Na_2SO_4(aq) + 2\,H_2O(l)$$

How much 0.125 M NaOH solution is required to completely neutralize 0.225 L of 0.175 M H_2SO_4 solution?

We set up the problem in the normal way.

Given:

0.225 L H_2SO_4 solution
0.175 M H_2SO_4
0.125 M NaOH

Find: L NaOH solution

Conversion Factors: The conversion factors for this problem are the molarities of the two solutions expressed in mol of solute per L of solution and the stoichiometric relationship (from the balanced equation) between mol of H_2SO_4 and mol of NaOH.

$$M\ (H_2SO_4) = \frac{0.175\ \text{mol}\ H_2SO_4}{\text{L}\ H_2SO_4\ \text{solution}}$$

$$M\ (NaOH) = \frac{0.125\ \text{mol NaOH}}{\text{L NaOH solution}}$$

$$1\ \text{mol}\ H_2SO_4 \equiv 2\ \text{mol NaOH}$$

Solution Map: The solution map for this problem is similar to the solution maps for other stoichiometric problems. We first use the volume and molarity of H_2SO_4 solution to get mol H_2SO_4. Then we use the stoichiometric coefficients from the equation to convert mol H_2SO_4 to mol NaOH. Finally, we use the molarity of NaOH to get to L NaOH solution.

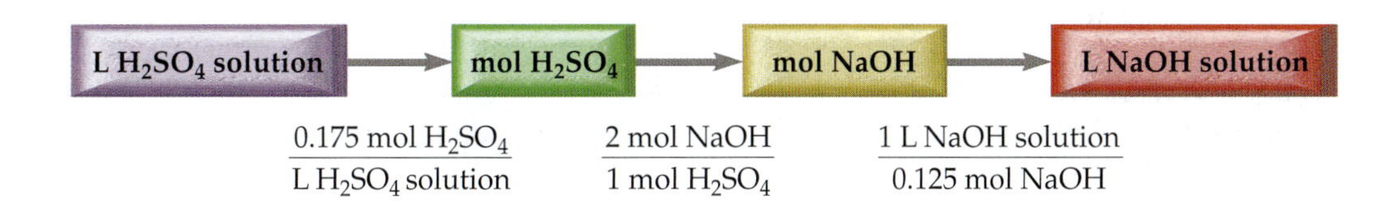

Solution: To solve the problem, we follow the solution map and compute the answer.

$$0.225\ \text{L}\ H_2SO_4\ \text{solution} \times \frac{0.175\ \text{mol}\ H_2SO_4}{\text{L}\ H_2SO_4\ \text{solution}} \times \frac{2\ \text{mol NaOH}}{1\ \text{mol}\ H_2SO_4} \times \frac{1\ \text{L NaOH solution}}{0.125\ \text{mol NaOH}} = 0.630\ \text{L NaOH solution}$$

It will take 0.630 L of the NaOH solution to completely neutralize the H_2SO_4.

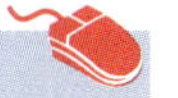
LIVE EXAMPLE

EXAMPLE 13.7 Solution Stoichiometry

Consider the following precipitation reaction:

$$2\,KI(aq) + Pb(NO_3)_2(aq) \longrightarrow PbI_2(s) + 2\,KNO_3(aq)$$

How much 0.115 M KI solution in liters is required to completely precipitate the Pb^{2+} in 0.104 L of 0.225 M $Pb(NO_3)_2$ solution?

You are given the concentration of a reactant, KI, in a chemical reaction. You are also given the volume and concentration of a second reactant, $Pb(NO_3)_2$. You are asked to find the volume of the first reactant that completely reacts with the given amount of the second.

The conversion factors for this problem are the molarities of the two solutions expressed in mol of solute per L of solution and the stoichiometric relationship (from the balanced equation) between mol KI and mol $Pb(NO_3)_2$.

Given:

0.115 M KI

0.104 L $Pb(NO_3)_2$ solution

0.225 M $Pb(NO_3)_2$

Find: L KI solution

Conversion Factors:

$$\text{M KI} = \frac{0.115 \text{ mol KI}}{\text{L KI solution}}$$

$$\text{M Pb(NO}_3)_2 = \frac{0.225 \text{ mol Pb(NO}_3)_2}{\text{L Pb(NO}_3)_2 \text{ solution}}$$

$$2 \text{ mol KI} \equiv 1 \text{ mol Pb(NO}_3)_2$$

The solution map for this problem is similar to the solution maps for other stoichiometric problems. First use the volume and molarity of $Pb(NO_3)_2$ solution to get mol $Pb(NO_3)_2$. Then use the stoichiometric coefficients from the equation to convert mol $Pb(NO_3)_2$ to mol KI. Finally, use mol KI to find L KI solution.

Solution Map:

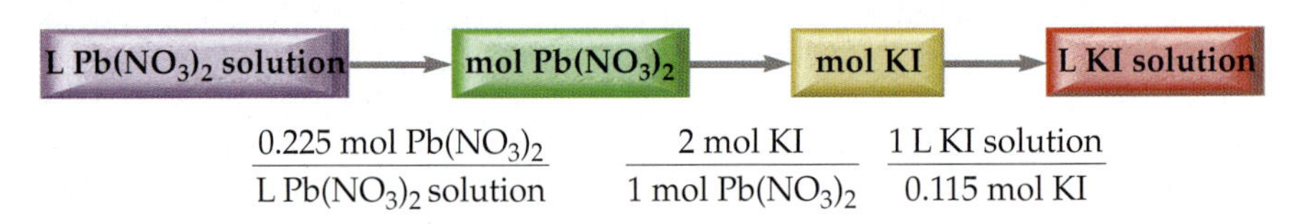

$$\frac{0.225 \text{ mol Pb(NO}_3)_2}{\text{L Pb(NO}_3)_2 \text{ solution}} \qquad \frac{2 \text{ mol KI}}{1 \text{ mol Pb(NO}_3)_2} \qquad \frac{1 \text{ L KI solution}}{0.115 \text{ mol KI}}$$

Follow the solution map to solve the problem. Begin with volume of $Pb(NO_3)_2$ solution and cancel units to arrive at volume of KI solution.

Solution:

$$0.104 \text{ L Pb(NO}_3)_2 \text{ solution} \times \frac{0.225 \text{ mol Pb(NO}_3)_2}{\text{L Pb(NO}_3)_2 \text{ solution}} \times \frac{2 \text{ mol KI}}{\text{mol Pb(NO}_3)_2} \times \frac{\text{L KI solution}}{0.115 \text{ mol KI}} = 0.407 \text{ L KI solution}$$

SKILLBUILDER 13.7 Solution Stoichiometry

How many milliliters of 0.112 M Na_2CO_3 are necessary to completely react with 27.2 mL of 0.135 M HNO_3 according to the following reaction?

$$2\,HNO_3(aq) + Na_2CO_3(aq) \longrightarrow H_2O(l) + CO_2(g) + 2\,NaNO_3(aq)$$

SKILLBUILDER PLUS

A 25.0-mL sample of HNO_3 solution requires 35.7 mL of 0.108 M Na_2CO_3 to completely react with all of the HNO_3 in the solution. What was the concentration of the HNO_3 solution?

13.9 Freezing Point Depression and Boiling Point Elevation: Making Water Freeze Colder and Boil Hotter

Have you ever wondered why salt is added to ice in an ice-cream maker? Or why salt is often scattered on icy roads in cold climates? Salt actually lowers the melting point of ice. A salt-and-water solution will remain a liquid even below 0 °C. By adding salt to ice in the ice-cream maker, you form a mixture of ice, salt, and water that can reach a temperature of about −10 °C, cold enough to freeze the cream. On the road, the salt allows the ice to melt, even if the ambient temperature is below freezing.

▶ Sprinkling salt on icy roads lowers the freezing point of water, allowing ice to melt even if the temperature is below 0 °C.

▲ Ethylene glycol is the chief component of antifreeze, which keeps engine coolant from freezing in winter or boiling over in summer.

Adding a nonvolatile solute to a liquid extends the temperature range over which the liquid remains a liquid. The solution has a lower melting point and a higher boiling point than the pure solvent. These effects are called **freezing point depression** and **boiling point elevation**. Freezing point depression and boiling point elevation depend only on the number of solute particles in solution, not on the type of solute particles. Properties such as these—which depend on the amount of solute and not the type of solute—are called **colligative properties**.

Freezing Point Depression

The freezing point of a solution containing a nonvolatile solute is lower than the freezing point of the pure solvent because the solute molecules interfere with the freezing of the solvent molecules. For example, antifreeze, used to prevent the freezing of engine coolant in cold climates, is an aqueous solution of ethylene glycol ($C_2H_6O_2$). If the temperature drops below zero, the ethylene glycol molecules make it more difficult for water molecules to crystallize into their ice crystal structure. The result is a lowering of the freezing point for the solution. The more concentrated the solution is, the lower the freezing point becomes. For freezing point depression and boiling point elevation, the concentration of the solution is usually expressed in **molality (m)**, the number of moles of solute per kilogram of solvent.

$$\text{Molality (m)} = \frac{\text{Moles solute}}{\text{Kilograms solvent}}$$

Notice that molality is defined with respect to kilograms of *solvent*, not kilograms of *solution*.

Note that molality is abbreviated with a lowercase *m* while molarity is abbreviated with a capital *M*.

EXAMPLE 13.8 Calculating Molality

Calculate the molality of a solution containing 17.2 g of ethylene glycol ($C_2H_6O_2$) dissolved in 0.500 kg of water.

You are given the mass of ethylene glycol in grams and the mass of the solvent in kilograms. You are asked to find the molality of the resulting solution.

Given:

$17.2 \text{ g } C_2H_6O_2$

$0.500 \text{ kg } H_2O$

Find: molality (m)

You will need the equation that defines molality and the molar mass of ethylene glycol.

Equation and Conversion Factor:

$$\text{Molality (m)} = \frac{\text{Moles solute}}{\text{Kilograms solvent}}$$

$$\text{Molar mass of } C_2H_6O_2 = \frac{62.08 \text{ g}}{\text{mol}}$$

To calculate molality, simply substitute the correct values into the equation and compute the answer. However, you must first convert the amount of $C_2H_6O_2$ from grams to moles using the molar mass of $C_2H_6O_2$.

Solution:

$$\text{mol } C_2H_6O_2 = 17.2 \cancel{\text{g } C_2H_6O} \times \frac{1 \text{ mol } C_2H_6O}{62.08 \cancel{\text{g } C_2H_6O}}$$

$$= 0.27\underline{7}1 \text{ mol } C_2H_6O$$

$$\text{Molality (m)} = \frac{\text{Moles solute}}{\text{Kilograms solvent}}$$

$$= \frac{0.27\underline{7}1 \text{ mol } C_2H_6O}{0.500 \text{ kg } H_2O}$$

$$= 0.554 \text{ m}$$

SKILLBUILDER 13.8 Calculating Molality

Calculate the molality (m) of a sucrose ($C_{12}H_{22}O_{11}$) solution containing 50.4 g sucrose and 0.332 kg of water.

CONCEPTUAL CHECKPOINT 13.2

A laboratory procedure calls for a 2.0 molal aqueous solution. A student accidentally makes a 2.0 molar solution. Will the solution made by the student be:

(a) too concentrated
(b) too dilute
(c) just right
(d) it would depend on the molecular weight of the solute

With an understanding of molality, we can now quantify freezing point depression. The amount that the freezing point of a solution is lowered by a particular amount of solute is given by the following equation:

The equations for freezing point depression and boiling point elevation given in this section apply only to nonelectrolyte solutions.

$$\Delta T_f = m \times K_f$$

where

- ΔT_f is the change in temperature of the freezing point in °C (from the freezing point of the pure solvent).
- m is the molality of the solution in $\frac{\text{mol solute}}{\text{kg solvent}}$.
- K_f is the freezing point depression constant for the solvent.

For water:

Different solvents have different values of K_f.

$$K_f = 1.86 \frac{°\text{C kg solvent}}{\text{mol solute}}$$

Calculating the freezing point of a solution involves substituting into the preceding equation, as the following example demonstrates.

EXAMPLE 13.9 Freezing Point Depression

Calculate the freezing point of a 1.7 m ethylene glycol solution.

You are given the molality of an aqueous solution and asked to find the freezing point depression.

You will need the freezing point depression equation given previously in this section.

Given: 1.7 m solution

Find: ΔT_f

Equation:

$$\Delta T_f = m \times K_f$$

To solve this problem, simply substitute the values into the equation for freezing point depression and calculate ΔT_f.

Solution:

$$\begin{aligned}\Delta T_f &= m \times K_f \\ &= 1.7 \frac{\cancel{\text{mol solute}}}{\cancel{\text{kg solvent}}} \times 1.86 \frac{°\text{C}\ \cancel{\text{kg solvent}}}{\cancel{\text{mol solute}}} \\ &= 3.2\ °\text{C}\end{aligned}$$

The actual freezing point will be the freezing point of pure water (0.00 °C) − ΔT_f.

$$\begin{aligned}\text{Freezing point} &= 0.00\ °\text{C} - 3.2\ °\text{C} \\ &= -3.2\ °\text{C}\end{aligned}$$

SKILLBUILDER 13.9 Freezing Point Depression

Calculate the freezing point of a 2.6 m sucrose solution.

Boiling Point Elevation

The boiling point of a solution containing a nonvolatile solute is higher than the boiling point of the pure solvent because the solute molecules interfere with the vaporization of the solvent molecules. In automobiles, antifreeze not only prevents the freezing of coolant within engine blocks in cold climates, it also prevents the boiling of engine coolant in hot climates. The amount that the boiling point is raised for solutions is given by the following equation:

$$\Delta T_b = m \times K_b$$

where

- ΔT_b is change in temperature of the boiling point in °C (from the boiling point of the pure solvent).
- m is the molality of the solution in $\frac{\text{mol solute}}{\text{kg solvent}}$.
- K_b is the boiling point elevation constant for the solvent.

For water:

Different solvents have different values of K_b.

$$K_b = 0.512 \frac{°\text{C kg solvent}}{\text{mol solute}}$$

The boiling point of solutions is calculated by simply substituting into the preceding equation as the following example demonstrates.

EVERYDAY Chemistry

Antifreeze in Frogs

On the outside, wood frogs *(Rana sylvatica)* look like most other frogs. They are only a few inches long and have characteristic greenish-brown skin. However, wood frogs survive cold winters in a remarkable way—they partially freeze. In the frozen state, the frog has no heartbeat, no blood circulation, no breathing, and no brain activity. Within 1 to 2 hours of thawing, however, these vital functions return, and the frog hops off to find food. How is this possible?

Most cold-blooded animals cannot survive freezing temperatures because the water within their cells freezes. As we learned in Section 12.8, when water freezes, it expands, irreversibly damaging cells. When the wood frog hibernates for the winter, however, it secretes large amounts of glucose into its blood and into the interior of its cells. When the temperature drops below freezing, extracellular bodily fluids, such as those in the abdominal cavity, freeze solid. Fluids within cells, however, remain liquid because the high glucose concentration lowers their freezing point. In other words, the concentrated glucose solution within the frog's cells acts as antifreeze, preventing the water within the cells from freezing and allowing the frog to survive.

CAN YOU ANSWER THIS? *The wood frog can survive at body temperatures as low as −8.0 °C. Calculate the molality of a glucose solution ($C_6H_{12}O_6$) required to lower the freezing point of water to −8.0 °C.*

◀ The wood frog survives cold winters by partially freezing. The fluids within the frog's cells, however, remain liquid to temperatures as low as −8 °C. These fluids are protected by a high concentration of glucose that acts as antifreeze, lowering their freezing point.

EXAMPLE 13.10 Boiling Point Elevation

Calculate the boiling point of a 1.7 m ethylene glycol solution.

You are given the molality of an aqueous solution and asked to find the boiling point.

You will need the boiling point elevation equation given previously in this section.

Given: 1.7 m solution

Find: boiling point

Equation:

$$\Delta T_b = m \times K_b$$

To solve this problem, simply substitute the values into the equation for boiling point elevation and calculate ΔT_b.

Solution:

$$\Delta T_b = m \times K_b$$

$$= 1.7 \frac{\cancel{\text{mol solute}}}{\cancel{\text{kg solvent}}} \times 0.512 \frac{°\text{C } \cancel{\text{kg solvent}}}{\cancel{\text{mol solute}}}$$

$$= 0.87\ °\text{C}$$

The actual boiling point of the solution will be the boiling point of pure water (100.00 °C) plus ΔT_b.

$$\text{Boiling point} = 100.00\ °\text{C} + 0.87\ °\text{C} = 100.87\ °\text{C}$$

SKILLBUILDER 13.10 Boiling Point Elevation

Calculate the boiling point of a 3.5 m glucose solution.

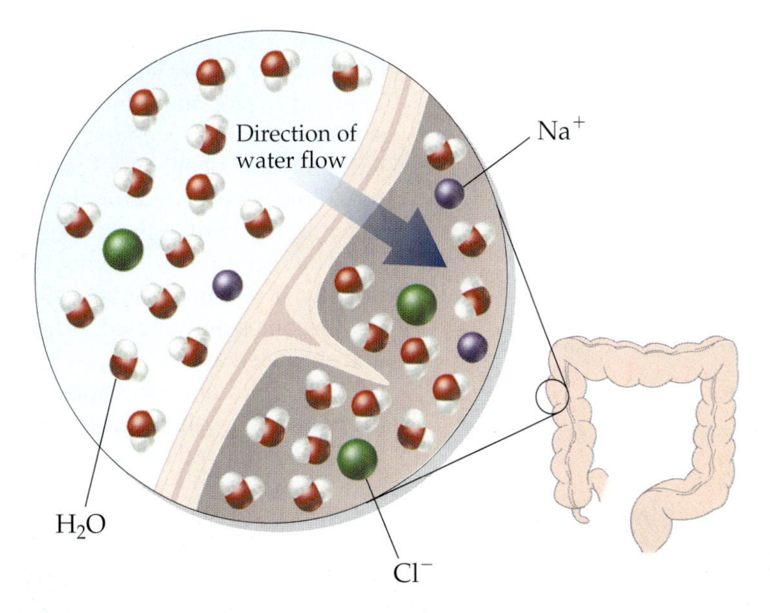

▶ **Figure 13.9 Seawater is a *thirsty* solution** As it flows through the stomach and intestine, seawater draws water *out of* bodily tissues, promoting dehydration.

13.10 Osmosis: Why Drinking Salt Water Causes Dehydration

Humans adrift at sea are surrounded by water, yet drinking that water would only accelerate their dehydration. Why? Salt water causes dehydration because of **osmosis**, the flow of solvent from a less-concentrated solution to a more-concentrated solution. Solutions containing a high concentration of solute draw solvent from solutions containing a lower concentration of solute. In other words, aqueous solutions with high concentrations of solute, such as seawater, are actually *thirsty solutions*—they draw water away from other, less-concentrated solutions, including the human body (▲ Figure 13.9).

▼ Figure 13.10 shows an osmosis cell. The left side of the cell contains a concentrated saltwater solution and the right side of the cell contains pure water. A **semipermeable membrane**—a membrane that selectively allows some substances to pass through but not others—separates the two halves of the cell. Through osmosis, water flows from the pure-water side of the cell through the semipermeable membrane and into the saltwater side. Over time, the water level on the left side of the cell rises while the water level on the right

▼ **Figure 13.10 An osmosis cell** In an osmosis cell, water flows through a semipermeable membrane from a less concentrated solution into a more concentrated solution. As a result, fluid rises in one arm of the tube until the weight of the excess fluid creates enough pressure to stop the flow. This pressure is the osmotic pressure of the solution.

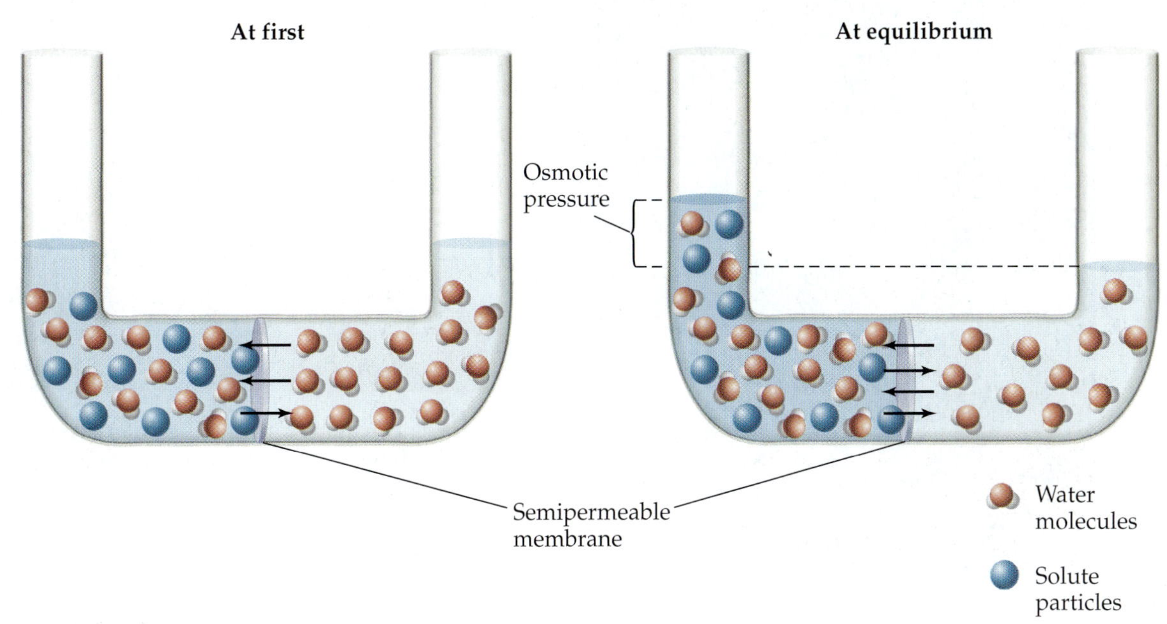

TUTORIAL 
Osmosis and Osmotic Pressure

side of the cell falls. This will continue until the pressure created by the weight of the water on the left side is enough to stop the osmotic flow. The pressure required to stop the osmotic flow is a measure of the **osmotic pressure** of the solution. Osmotic pressure—like freezing point depression and boiling point elevation—is a colligative property; it depends only on the concentration of the solute particles, not on the type of solute. The more concentrated the solution, the greater its osmotic pressure.

The membranes of living cells act as semipermeable membranes. Consequently, if you put a living cell into seawater, it loses water through osmosis and becomes dehydrated. (That is why your skin can start to look wrinkled and shriveled after a long swim in the ocean.) ▶ Figure 13.11 shows red blood cells in solutions of various concentrations. The cell in Figure 13.11a, immersed in a solution with the same solute concentration as the cell interior, has the normal red blood cell shape. The cell in Figure 13.11b, in pure water, is

Chemistry AND HEALTH

Solutions in Medicine

Doctors and others working in health fields must often administer solutions to patients. The osmotic pressure of these solutions is controlled for the desired effect on the patient. Solutions having osmotic pressures less than that of bodily fluids are called *hypoosmotic*. These solutions tend to pump water into cells. When a human cell is placed in a hypoosmotic solution—such as pure water, for example—water enters the cell, sometimes causing it to burst (Figure 13.11b). Solutions having osmotic pressures greater than that of bodily fluids are called *hyperosmotic*. These solutions tend to take water out of cells and tissues. When a human cell is placed in a hyperosmotic solution, it tends to shrivel as it loses water to the surrounding solution (Figure 13.11c).

Intravenous solutions—those that are administered directly into a patient's veins—must have osmotic pressures equal to that of bodily fluids. These solutions are called *isoosmotic*. When a patient is given an IV in a hospital, the majority of the fluid is usually an isoosmotic saline solution—a solution containing 0.9 g NaCl per 100 mL of solution. In medicine and in other health-related fields, solution concentrations are often reported in units that indicate the mass of the solute in a given volume of solution. Also common is *percent mass to volume*—which is simply the mass of the solute in grams divided by volume of the solution in milliliters times 100%. In these units, the concentration of an isoosmotic saline solution is 0.9% mass/volume.

CAN YOU ANSWER THIS? *An isoosmotic sucrose ($C_{12}H_{22}O_{11}$) solution has a concentration of 0.30 M. Calculate its concentration in percent mass to volume.*

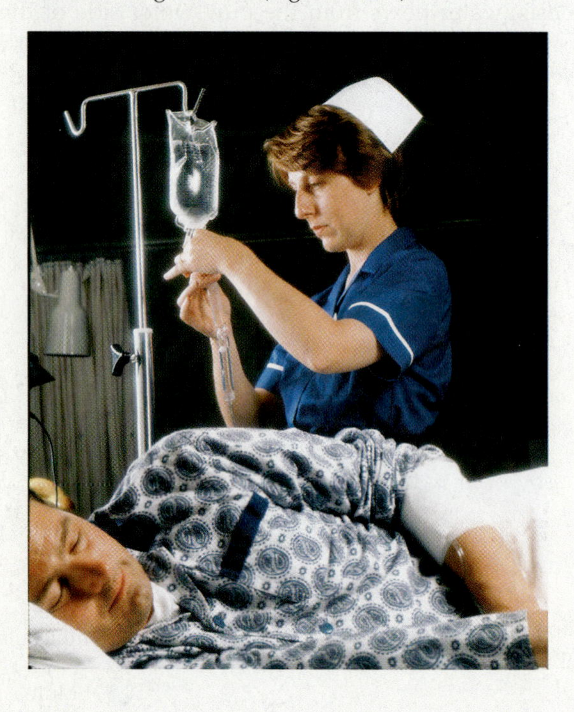

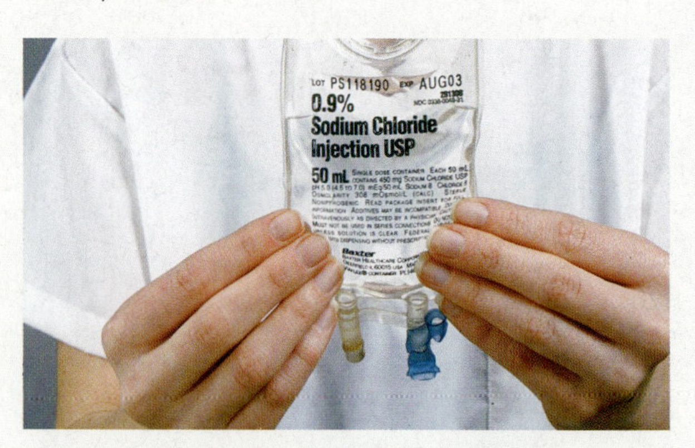

▲ Intravenous fluids consist mostly of saline solutions with an osmotic pressure equal to that of bodily fluids. **Question:** Why would it be dangerous to administer intravenous fluids that do not have an osmotic pressure comparable to that of bodily fluids?

TUTORIAL
Colligative Properties Activity

swollen. Because the solute concentration within the cell is higher than that of the surrounding fluid, osmosis has pulled water across the membrane *into* the cell. The cell in Figure 13.11c, in a solution more concentrated than the cell interior, is starting to shrivel as osmosis draws water *out of* the cell. Similarly, if you drink seawater, the seawater actually draws water out of your body as it passes through your stomach and intestines. All of that extra water in your intestine promotes dehydration of bodily tissues and diarrhea. Consequently, seawater should never be consumed as drinking water.

(a)

(b)

(c)

▶ **Figure 13.11 Red blood cells in solutions of different concentration** **(a)** When the solute concentration of the surrounding fluid is to equal that within the cell, there is no net osmotic flow, and the red blood cell exhibits its typical shape. **(b)** When a cell is placed in pure water, osmotic flow of water into the cell causes it to swell up. Eventually it may burst. **(c)** When a cell is placed in a concentrated solution, osmosis draws water out of the cell, distorting its normal shape.

CHAPTER IN REVIEW

Chemical Principles

Solutions: A solution is a homogeneous mixture with two or more components. The solvent is the majority component and the solute is the minority component. Aqueous solutions are those with water as the solvent.

Solid-and-Liquid Solutions: The solubility—the amount of solute that dissolves in a certain amount of solvent—of solids in liquids increases with increasing temperature. Recrystallization involves dissolving a solid into hot solvent to saturation and then allowing it to cool. As the solution cools, it becomes supersaturated and the solid crystallizes.

Gas-and-Liquid Solutions: The solubility of gases in liquids decreases with increasing temperature but increases with increasing pressure.

Solution Concentration: Solution concentration is used to specify how much of the solute is present in a given amount of solution. Three common ways to express solution concentration are mass percent, molarity, and molality.

$$\text{Mass percent} = \frac{\text{Mass solute}}{\text{Mass solution}} \times 100\%$$

$$\text{Molarity (M)} = \frac{\text{Moles solute}}{\text{Liters solution}}$$

$$\text{Molality (m)} = \frac{\text{Moles solute}}{\text{Kilograms solvent}}$$

Relevance

Solutions: Solutions are all around us—most of the fluids that we encounter every day are solutions. Common solutions include seawater (solid and liquid), soda pop (gas and liquid), alcoholic spirits such as vodka (liquid and liquid), air (gas and gas), and blood (solid, gas, and liquid).

Solid-and-Liquid Solutions: Solutions of solids dissolved in liquids, such as seawater, coffee, and sugar water, are important both in chemistry and in everyday life. Recrystallization is used extensively in the laboratory to purify solids.

Gas-and-Liquid Solutions: The temperature and pressure dependence of gas solubility is the reason that soda pop fizzes when opened and also the reason that warm beer goes flat.

Solution Concentration: Solution concentration is useful in converting between amounts of solute and solution. Mass percent and molarity are the most common concentration units. Molality is useful in quantifying colligative properties such as freezing point depression and boiling point elevation.

Solution Dilution: Solution dilution problems are most conveniently solved using the following equation:

$$M_1V_1 = M_2V_2$$

Solution Dilution: Since many solutions are stored in concentrated form, it is often necessary to dilute them to a desired concentration.

Freezing Point Depression and Boiling Point Elevation: A nonvolatile solute will extend the liquid temperature range of a solution relative to the pure solvent. The freezing point of a solution is lower than the freezing point of the pure solvent and the boiling point of a solution is higher than the boiling point of the pure solvent. These relationships are quantified by the following equations.

Freezing point depression:

$$\Delta T_f = m \times K_f$$

Boiling point elevation:

$$\Delta T_b = m \times K_b$$

Freezing Point Depression and Boiling Point Elevation: Salt is often added to ice in making ice cream, and is used to melt ice on roads in frigid weather. The salt lowers the freezing point of water, allowing the cream within the ice-cream maker to freeze and the ice on icy roads to melt. Antifreeze is used in the cooling systems of cars both to lower the freezing point of the coolant in winter and to raise its boiling point in summer.

Osmosis: Osmosis is the flow of water from a low-concentration solution to a high-concentration solution through a semipermeable membrane.

Osmosis: Osmosis explains why drinking seawater causes dehydration. As seawater goes through the stomach and intestines, it draws water away from the body through osmosis, resulting in diarrhea and dehydration. To avoid damage to body tissues, transfused fluids must always be isoosmotic with body fluids. Most transfused fluids consist in whole or part of 0.9% mass/volume saline solution.

Chemical Skills

Calculating Mass Percent (Section 13.5)

Begin by setting up the problem in the normal way.

To calculate mass percent concentration, divide the mass of the solute by the mass of the solution (solute and solvent) and multiply by 100%.

Examples

LIVE EXAMPLE 

EXAMPLE 13.11 Calculating Mass Percent

Find the mass percent concentration of a solution containing 19 g of solute and 158 g of solvent.

Given: 19 g solute
158 g solvent

Find: mass percent

Equation:

$$\text{Mass percent} = \frac{\text{Mass solute}}{\text{Mass solution}} \times 100\%$$

Solution:

$$\text{Mass solution} = 19 \text{ g} + 158 \text{ g} = 177 \text{ g}$$

$$\text{Mass percent} = \frac{19 \text{ g}}{177 \text{ g}} \times 100\%$$

$$= 11\%$$

Using Mass Percent in Calculations (Section 13.5)

When using mass percent in calculations, set up the problem in the normal way. When writing the conversion factors, express the mass percent as:

$$\frac{\text{g solute}}{\text{100 g solution}}$$

This will be your key conversion factor in converting from the amount of solution to the amount of solute. If you are asked to convert from the amount of solute to the amount of solution, invert the conversion factor.

When writing the solution map, begin with the given quantity and convert to grams of solution (through other units as necessary). Then convert to grams of solute. If you are asked to convert from the amount of solute to the amount of solution, first convert to the amount of solute in grams, then use the concentration (inverted) to convert to the amount of solution in grams, and then to whatever units the problem asks. Finally, follow your solution map to compute the answer.

EXAMPLE 13.12 Using Mass Percent in Calculations

How much KCl in grams is in 0.337 L of a 5.8% mass percent KCl solution? (Assume that the density of the solution is 1.05 g/mL.)

Given: 5.80% KCl by mass
0.337 L solution

Find: g KCl

Conversion Factors:

$$\frac{\text{5.80 g KCl}}{\text{100 g solution}}$$

$$\frac{\text{1.05 g}}{\text{mL}}$$

$$1000 \text{ mL} = 1 \text{ L}$$

Solution Map:

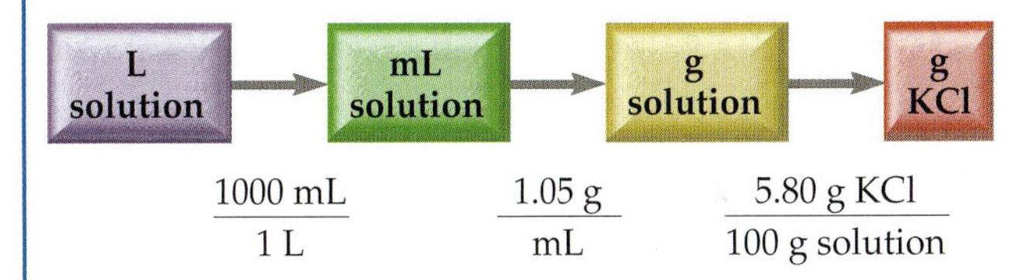

$$\frac{\text{1000 mL}}{\text{1 L}} \qquad \frac{\text{1.05 g}}{\text{mL}} \qquad \frac{\text{5.80 g KCl}}{\text{100 g solution}}$$

Solution:

$$0.337 \text{ L solution} \times \frac{1000 \text{ mL}}{\text{L}} \times \frac{1.05 \text{ g}}{\text{mL}} \times \frac{5.80 \text{ g KCl}}{100 \text{ g solution}} = 20.5 \text{ g KCl}$$

Calculating Molarity (Section 13.6)

Begin by setting up the problem in the normal way.

To calculate molarity, divide the number of moles of solute by the volume of the solution in liters.

EXAMPLE 13.13 Calculating Molarity

Calculate the molarity of a KCl solution containing 0.22 mol of KCl in 0.455 L of solution.

Given:

0.22 mol KCl
0.455 L solution

Find: molarity (M)

Equation:

$$\text{Molarity (M)} = \frac{\text{Moles solute}}{\text{Liters solution}}$$

Solution:

$$\text{Molarity (M)} = \frac{\text{0.22 mol KCl}}{\text{0.455 L solution}} = 0.48 \text{ M}$$

Using Molarity in Calculations (Section 13.6)

When using molarity in calculations, set up the problem in the normal way.

When writing the conversion factors, express the molarity as:

$$\frac{\text{Moles solute}}{\text{Liters solution}}$$

This will be your key conversion factor in converting from the amount of solution to the amount of solute. If you are asked to convert from the amount of solute to the amount of solution, invert the conversion factor. You may also need molar mass of the solute as a conversion factor.

The solution map begins with liters of solution and shows the conversion to moles of solute using the molarity as a conversion factor. You should then show the conversion to grams using the molar mass. If you are asked to convert from the amount of solute to the amount of solution, first convert the amount of solute to moles then use the molarity (inverted) to convert to the amount of solution in liters.

Finally, follow your solution map to compute the answer.

EXAMPLE 13.14 Using Molarity in Calculations

How much KCl in grams is contained in 0.488 L of 1.25 M KCl solution?

Given:

1.25 M KCl
0.488 L solution

Find: g KCl

Conversion Factors:

$$\frac{1.25 \text{ mol KCl}}{\text{L solution}}$$

$$\text{KCl molar mass} = \frac{74.55 \text{ g KCl}}{\text{mol KCl}}$$

Solution Map:

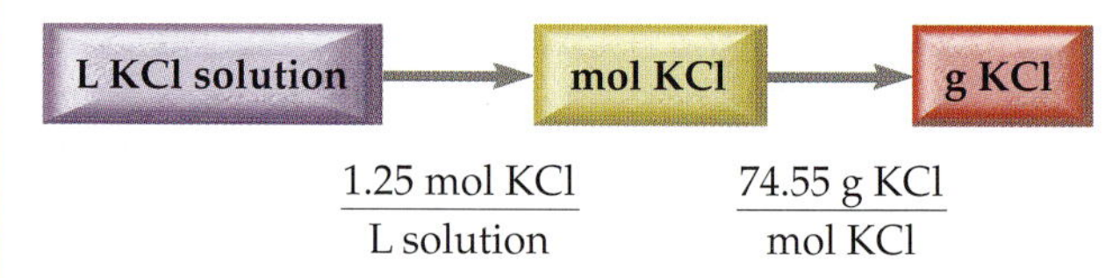

$$\frac{1.25 \text{ mol KCl}}{\text{L solution}} \qquad \frac{74.55 \text{ g KCl}}{\text{mol KCl}}$$

Solution:

$$0.488 \cancel{\text{L solution}} \times \frac{1.25 \cancel{\text{mol KCl}}}{\cancel{\text{L solution}}} \times \frac{74.55 \text{ g KCl}}{\cancel{\text{mol KCl}}}$$

$$= 45.5 \text{ g KCl}$$

Solution Dilution (Section 13.7)

Begin by setting up the problem in the normal way.

Most solution dilution problems will use equation $M_1V_1 = M_2V_2$.

Solve the equation for the quantity you are trying to find (in this case, V_1) and then substitute in the correct values to compute it.

EXAMPLE 13.15 Solution Dilution

How much of an 8.0 M HCl solution should be used to make 0.400 L of a 2.7 M HCl solution?

Given:

$M_1 = 8.0$ M
$M_2 = 2.7$ M
$V_2 = 0.400$ L

Find: V_1

Equation:

$$M_1V_1 = M_2V_2$$

Solution:

$$M_1V_1 = M_2V_2$$

$$V_1 = \frac{M_2V_2}{M_1}$$

$$= \frac{2.7 \frac{\cancel{\text{mol}}}{\cancel{\text{L}}} \times 0.400 \text{ L}}{8.0 \frac{\cancel{\text{mol}}}{\cancel{\text{L}}}}$$

$$= 0.14 \text{ L}$$

Solution Stoichiometry (Section 13.8)

EXAMPLE 13.16 Solution Stoichiometry

Consider the following reaction:

$$HCl(aq) + NaOH(aq) \longrightarrow NaCl(aq) + H_2O(l)$$

How much 0.113 M NaOH solution is required to completely neutralize 1.25 L of 0.228 M HCl solution?

Begin by setting up the problem in the normal way.

Given:

1.25 L HCl solution
0.228 M HCl
0.113 M NaOH

Find: L NaOH solution

The conversion factors for these kinds of problems are the molarities of the solutions expressed in moles of solute per liter of solution and the stoichiometric relationships (from the balanced equation) between the reactants and products of interest.

Conversion Factors:

$$M\ (HCl) = \frac{0.228\ \text{mol HCl}}{\text{L HCl solution}}$$

$$M\ (NaOH) = \frac{0.113\ \text{mol NaOH}}{\text{L NaOH solution}}$$

$$1\ \text{mol HCl} \equiv 1\ \text{mol NaOH}$$

In the solution map, use the volume and molarity of A (in this case, HCl) to get to moles of A. Then use the stoichiometric coefficients to convert to moles of other reactants or products (in this case, NaOH). Finally, convert back to volume of B (in this case, NaOH) using the molarity of B.

Solution Map:

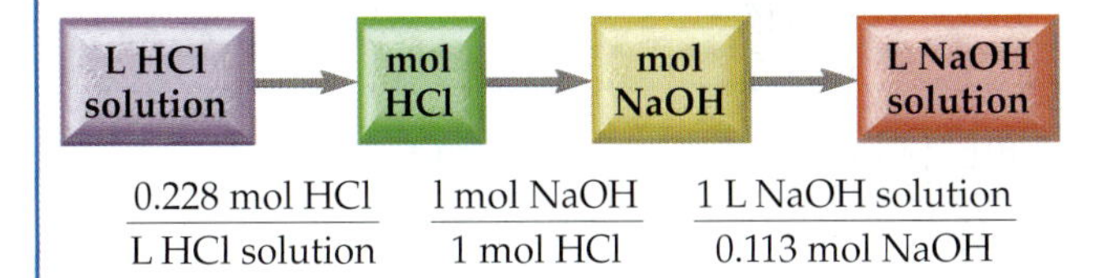

$$\frac{0.228\ \text{mol HCl}}{\text{L HCl solution}} \quad \frac{1\ \text{mol NaOH}}{1\ \text{mol HCl}} \quad \frac{1\ \text{L NaOH solution}}{0.113\ \text{mol NaOH}}$$

Follow the solution map to compute the answer.

Solution:

$$1.25\ \text{L HCl solution} \times \frac{0.228\ \text{mol HCl}}{\text{L HCl solution}} \times \frac{1\ \text{mol NaOH}}{1\ \text{mol HCl}} \times \frac{\text{L NaOH solution}}{0.113\ \text{mol NaOH}} = 2.52\ \text{L NaOH solution}$$

Calculating Molality (Section 13.9)

EXAMPLE 13.17 Calculating Molality

Calculate the molality of a solution containing 0.183 mol of sucrose dissolved in 1.10 kg of water.

Begin by setting up the problem in the normal way.

Given:

0.183 mol of sucrose
1.10 kg H_2O

Find: molality (m)

To calculate molality, you need the definition of molality, which is moles of solute per kilogram of solvent.

Equation:

$$\text{Molality (m)} = \frac{\text{Moles solute}}{\text{Kilograms solvent}}$$

Substitute the correct values into the definition of molality and compute the answer. If any of the quantities are not in the correct units, convert them into the correct units before substituting into the equation.

Solution:

$$\text{Molality (m)} = \frac{0.183\ \text{mol sucrose}}{1.10\ \text{kg } H_2O} = 0.166\ \text{m}$$

Freezing Point Depression and Boiling Point Elevation (Section 13.9)

Begin by setting up the problem in the normal way.

For freezing point depression problems, use $\Delta T_f = m \times K_f$. For boiling point elevation problems, use $\Delta T_b = m \times K_b$.

To compute ΔT_f or ΔT_b, simply substitute the values into the equation and compute.

The freezing point will be the freezing point of pure water (0.00 °C) $- \Delta T_f$.

The boiling point will be the boiling point of pure water (100.00 °C) $+ \Delta T_b$.

EXAMPLE 13.18 Freezing Point Depression and Boiling Point Elevation

Calculate the freezing point of a 2.5 m aqueous sucrose solution.

Given: 2.5 m solution

Find: ΔT_f

Equation:

$$\Delta T_f = m \times K_f$$

Solution:

$$\Delta T_f = m \times K_f$$

$$= 2.5 \frac{\text{mol solute}}{\text{kg solvent}} \times 1.86 \frac{\text{°C kg solvent}}{\text{mol solute}}$$

$$= 4.7 \text{ °C}$$

$$\text{Freezing point} = 0.00 \text{ °C} - 4.7 \text{ °C}$$

$$= -4.7 \text{ °C}$$

Key Terms

boiling point elevation [13.9]
colligative properties [13.9]
concentrated solution [13.5]
dilute solution [13.5]
electrolyte solution [13.3]
freezing point depression [13.9]
mass percent [13.5]
molality (m) [13.9]
molarity (M) [13.6]
nonelectrolyte solution [13.3]
osmosis [13.10]
osmotic pressure [13.10]
recrystallization [13.3]
saturated solution [13.3]
semipermeable membrane [13.10]
solubility [13.3]
solute [13.2]
solution [13.2]
solvent [13.2]
stock solution [13.7]
supersaturated solution [13.3]
unsaturated solution [13.3]

Exercises

Questions

1. What is a solution? Give some examples.
2. What is an aqueous solution?
3. In a solution, what is the solvent? What is the solute? Give some examples.
4. Explain what "like dissolves like" means.
5. What is solubility?
6. Describe what happens when additional solute is added to:
   (a) a saturated solution
   (b) an unsaturated solution
   (c) a supersaturated solution
7. Explain the difference between a strong electrolyte solution and a nonelectrolyte solution. What kinds of solutes form strong electrolyte solutions?
8. How does the solubility of gases depend on temperature?
9. Explain recrystallization.
10. How is rock candy made?
11. When you heat water on a stove, bubbles form on the bottom of the pot *before* the water boils. What are these bubbles? Why do they form?
12. Explain why warm soda pop goes flat faster than cold soda pop.
13. How does the solubility of gases depend on pressure?
14. Explain why a can of soda pop fizzes when opened.
15. What is the difference between a dilute solution and a concentrated solution?

16. What is mass percent?
17. What is molarity?
18. What is a stock solution?
19. How does the presence of a nonvolatile solute affect the boiling point and melting point of a solution relative to the boiling point and melting point of the pure solvent?
20. What are colligative properties?
21. Define molality.
22. What is osmosis?
23. Two shipwreck survivors were rescued from a life raft. One had drunk seawater while the other had not. The one who had drunk the seawater was more severely dehydrated than the one who did not. Explain.
24. Why are intravenous fluids always isoosmotic saline solutions? What would happen if pure water were administered intravenously?

Problems

Solutions

25. Which of the following are solutions?
(a) sand and water mixture
(b) oil and water mixture
(c) salt and water mixture
(d) sterling silver cup

26. Which of the following are solutions?
(a) air
(b) carbon dioxide and water mixture
(c) a blueberry muffin
(d) a brass buckle

27. Identify the solute and solvent in each of the following solutions.
(a) salt water
(b) sugar water
(c) soda water

28. Identify the solute and solvent in each of the following solutions.
(a) 80-proof vodka (40% ethyl alcohol)
(b) oxygenated water
(c) antifreeze (ethylene glycol and water)

29. Pick an appropriate solvent from Table 13.2 to dissolve:
(a) motor oil (nonpolar)
(b) sugar (polar)
(c) lard (nonpolar)
(d) potassium chloride (ionic)

30. Pick an appropriate solvent from Table 13.2 to dissolve:
(a) glucose (polar)
(b) salt (ionic)
(c) vegetable oil (nonpolar)
(d) sodium nitrate (ionic)

Solids Dissolved in Water

31. What are the dissolved particles in a solution containing an ionic solute? What is the name for this kind of solution?

32. What are the dissolved particles in a solution containing a molecular solute? What is the name for this kind of solution?

33. A solution contains 25 g of NaCl per 100 g of water at 25 °C. Is the solution unsaturated, saturated, or supersaturated? (Use Figure 13.4.)

34. A solution contains 32 g of KNO_3 per 100 g of water at 25 °C. Is the solution unsaturated, saturated, or supersaturated? (Use Figure 13.4.)

35. A KNO_3 solution containing 45 g of KNO_3 per 100 g of water is cooled from 40 °C to 0 °C. What will happen during cooling? (Use Figure 13.4.)

36. A KCl solution containing 42 g of KCl per 100 g of water is cooled from 60 °C to 0 °C. What will happen during cooling? (Use Figure 13.4.)

37. Use Figure 13.4 to determine whether each of the given amounts of solid will completely dissolve in the given amount of water at the indicated temperature.
(a) 30.0 g $KClO_3$ in 85.0 g of water at 35 °C
(b) 65.0 g $NaNO_3$ in 125 g of water at 15 °C
(c) 32.0 g KCl in 70.0 g of water at 82 °C

38. Use Figure 13.4 to determine whether each of the given amounts of solid will completely dissolve in the given amount of water at the indicated temperature.
(a) 45.0 g $CaCl_2$ in 105 g of water at 5 °C
(b) 15.0 g $KClO_3$ in 115 g of water at 25 °C
(c) 50.0 g $Pb(NO_3)_2$ in 95.0 g of water at 10 °C

Gases Dissolved in Water

39. Some laboratory procedures involving oxygen-sensitive reactants or products call for using preboiled (and then cooled) water. Explain why this is so.

40. A person preparing a fish tank uses preboiled (and then cooled) water to fill it. When the fish is put into the tank, it dies. Explain.

41. Scuba divers breathing air at increased pressure can suffer from nitrogen narcosis—a condition resembling drunkenness—when the partial pressure of nitrogen exceeds about 4 atm. What property of gas/water solutions causes this to happen? How could the diver reverse this effect?

42. Scuba divers breathing air at increased pressure can suffer from oxygen toxicity—too much oxygen in the bloodstream—when the partial pressure of oxygen exceeds about 1.4 atm. What happens to the amount of oxygen in a diver's bloodstream when he or she breathes oxygen at elevated pressures? How can this be reversed?

Mass Percent

43. Calculate the concentration of each of the following solutions in mass percent.
(a) 12.8 g NaCl in 145 g H_2O
(b) 55.1 g $C_{12}H_{22}O_{11}$ in 478 g H_2O
(c) 355 mg $C_6H_{12}O_6$ in 5.22 g H_2O

44. Calculate the concentration of each of the following solutions in mass percent.
(a) 11.3 g C_2H_6O in 67.4 g H_2O
(b) 104 g KCl in 558 g H_2O
(c) 38.2 mg KNO_3 in 2.58 g H_2O

45. A soft drink contains 45 g of sugar in 309 g of H_2O. What is the concentration of sugar in the soft drink in mass percent?

46. A soft drink contains 35 mg of sodium in 315 g of H_2O. What is the concentration of sodium in the soft drink in mass percent?

47. Complete the following table:

Mass Solute	*Mass Solvent*	*Mass Solution*	*Mass Percent*
15.5 g	238.1 g	____	____
22.8 g	____	____	12.0%
____	183.3 g	212.1 g	____
____	315.2 g	____	15.3%

48. Complete the following table:

Mass Solute	*Mass Solvent*	*Mass Solution*	*Mass Percent*
2.55 g	25.0 g	____	____
____	45.8 g	____	3.8%
1.38 g	____	27.2 g	____
23.7 g	____	____	5.8%

49. Ocean water contains 3.5% NaCl by mass. How much salt can be obtained from 274 g of seawater?

50. A saline solution contains 1.1% NaCl by mass. How much NaCl is present in 87.2 g of this solution?

51. Determine the amount of sucrose in each of the following solutions.
(a) 55 g of a solution containing 4.9% sucrose by mass
(b) 122 mg of a solution containing 11.2% sucrose by mass
(c) 5.8 kg of a solution containing 17.1% sucrose by mass

52. Determine the amount of potassium chloride in each of the following solutions.
(a) 22.9 g of a solution containing 1.33% KCl by mass
(b) 38.2 kg of a solution containing 22.5% KCl by mass
(c) 52 mg of a solution containing 14% KCl by mass

53. Determine how much of each of the following NaCl solutions in grams contains 1.5 g of NaCl.
(a) 0.045% NaCl by mass
(b) 1.94% NaCl by mass
(c) 9.82% NaCl by mass

54. Determine how much of each of the following sucrose solutions in grams contains 12 g of sucrose.
(a) 5.8% sucrose by mass
(b) 2.2% sucrose by mass
(c) 14.9% sucrose by mass

55. $AgNO_3$ solutions are often used to plate silver onto other metals. What is the maximum amount of silver in grams that can be plated out of 4.8 L of an $AgNO_3$ solution containing 3.4% Ag by mass? (Assume that the density of the solution is 1.01 g/mL.)

56. A dioxin-contaminated water source contains 0.085% dioxin by mass. How much dioxin is present in 2.5 L of this water? (Assume that the density of the solution is 1.00 g/mL.)

57. Ocean water contains 3.5% NaCl by mass. How much ocean water in grams contains 45.8 g of NaCl?

58. A hard water sample contains 0.0085% Ca by mass (in the form of Ca^{2+} ions). How much water in grams contains 1.2 g of Ca? (1.2 g of Ca is the recommended daily allowance of calcium for 19- to 24-year-olds.)

59. Lead is a toxic metal that affects the central nervous system. A Pb-contaminated water sample contains 0.0011% Pb by mass. How much of the water in milliliters contains 150 mg of Pb? (Assume that the density of the solution is 1.0 g/mL.)

60. Benzene is a carcinogenic (cancer-causing) compound. A benzene-contaminated water sample contains 0.000037% benzene by mass. How much of the water in liters contains 125 mg of benzene? (Assume that the density of the solution is 1.0 g/mL.)

Molarity

61. Calculate the molarity of each of the following solutions.
(a) 1.3 mol of KCl in 2.5 L of solution
(b) 0.225 mol of KNO_3 in 0.855 L of solution
(c) 0.117 mol of sucrose in 588 mL of solution

62. Calculate the molarity of each of the following solutions.
(a) 0.11 mol of $LiNO_3$ in 5.2 L of solution
(b) 1.77 mol of LiCl in 33.2 L of solution
(c) 0.0441 mol of glucose in 84.2 mL of solution

63. Calculate the molarity of each of the following solutions.
(a) 22.6 g of $C_{12}H_{22}O_{11}$ in 0.442 L of solution
(b) 42.6 g of NaCl in 1.58 L of solution
(c) 315 mg of $C_6H_{12}O_6$ in 58.2 mL of solution

64. Calculate the molarity of each of the following solutions.
(a) 33.2 g of KCl in 0.895 L of solution
(b) 61.3 g of C_2H_6O in 3.4 L of solution
(c) 38.2 mg of KI in 112 mL of solution

65. A 205-mL sample of ocean water is found to contain 7.2 g of NaCl. What is the molarity of the solution with respect to NaCl?

66. A 355-mL can of soda pop is found to contain 45 g of sucrose ($C_{12}H_{22}O_{11}$). What is the molarity of the solution with respect to sucrose?

67. How many moles of NaCl are contained in each of the following?
(a) 1.5 L of a 1.2 M NaCl solution
(b) 0.448 L of a 0.85 M NaCl solution
(c) 144 mL of a 1.65 M NaCl solution

68. How many moles of sucrose are contained in each of the following?
(a) 3.4 L of a 0.100 M sucrose solution
(b) 0.952 L of a 1.88 M sucrose solution
(c) 21.5 mL of a 0.528 M sucrose solution

69. What volume of each of the following solutions contains 0.10 mol of KCl?
(a) 0.255 M KCl
(b) 1.8 M KCl
(c) 0.995 M KCl

70. What volume of each of the following solutions contains 0.225 mol of NaI?
(a) 0.152 M NaI
(b) 0.982 M NaI
(c) 1.76 M NaI

71. Complete the following table:

Solute	*Mass Solute*	*Mol Solute*	*Volume Solution*	*Molarity*
KNO_3	22.5 g	____	125.0 mL	____
$NaHCO_3$	____	____	250.0 mL	0.100 M
$C_{12}H_{22}O_{11}$	55.38 g	____	____	0.150 M

72. Complete the following table:

Solute	*Mass Solute*	*Mol Solute*	*Volume Solution*	*Molarity*
$MgSO_4$	0.588 g	____	25.0 mL	____
NaOH	____	____	100.0 mL	1.75 M
CH_3OH	12.5 g	____	____	0.500 M

73. Calculate the mass of NaCl in a 55-mL sample of a 1.5 M NaCl solution.

74. Calculate the mass of glucose ($C_6H_{12}O_6$) in a 125-mL sample of a 1.15 M glucose solution.

75. A chemist wants to make 2.5 L of a 0.100 M KCl solution. How much KCl in grams should the chemist use?

76. A laboratory procedure calls for making 500.0 mL of a 1.4 M KNO_3 solution. How much KNO_3 in grams is needed?

77. How many liters of a 0.500 M sucrose ($C_{12}H_{22}O_{11}$) solution contain 1.5 kg of sucrose?

78. What volume of a 0.35 M $Mg(NO_3)_2$ solution contains 87 g of $Mg(NO_3)_2$?

Solution Dilution

79. A 158-mL sample of a 1.2 M sucrose solution is diluted to 500.0 mL. What is the molarity of the diluted solution?

80. A 2.5-L sample of a 5.8 M NaCl solution is diluted to 55 L. What is the molarity of the diluted solution?

81. Describe how you would make 2.5 L of a 0.100 M KCl solution from a 5.5 M stock KCl solution.

82. Describe how you would make 500.0 mL of a 0.200 M NaOH solution from a 15.0 M stock NaOH solution.

83. To what volume should you dilute 25 mL of a 12 M stock HCl solution to obtain a 0.500 M HCl solution?

84. To what volume should you dilute 75 mL of a 10.0 M H_2SO_4 solution to obtain a 1.75 M H_2SO_4 solution?

85. How much of a 12.0 M HNO_3 solution should you use to make 850.0 mL of a 0.250 M HNO_3 solution?

86. How much of a 5.0 M sucrose solution should be used to make 85.0 mL of a 0.040 M solution?

Solution Stoichiometry

87. Determine the volume of 0.150 M NaOH solution required to neutralize each of the following samples of hydrochloric acid. The neutralization reaction is:

$$NaOH(aq) + HCl(aq) \longrightarrow H_2O(l) + NaCl(aq)$$

(a) 25 mL of a 0.150 M HCl solution
(b) 55 mL of a 0.055 M HCl solution
(c) 175 mL of a 0.885 M HCl solution

88. Determine the volume of 0.225 M KOH solution required to neutralize each of the following samples of sulfuric acid. The neutralization reaction is:

$$H_2SO_4(aq) + 2\,KOH(aq) \longrightarrow K_2SO_4(aq) + 2\,H_2O(l)$$

(a) 45 mL of 0.225 M H_2SO_4
(b) 185 mL of 0.125 M H_2SO_4
(c) 75 mL of 0.100 M H_2SO_4

89. Consider the following reaction.

$$2\,K_3PO_4(aq) + 3\,NiCl_2(aq) \longrightarrow Ni_3(PO_4)_2(s) + 6\,KCl(aq)$$

What volume of 0.225 M K_3PO_4 solution is necessary to completely react with 134 mL of 0.0112 M $NiCl_2$?

90. Consider the following reaction.

$$K_2S(aq) + Co(NO_3)_2(aq) \longrightarrow 2\,KNO_3(aq) + CoS(s)$$

What volume of 0.225 M K_2S solution is required to completely react with 175 mL of 0.115 M $Co(NO_3)_2$?

91. A 10.0-mL sample of an unknown H_3PO_4 solution requires 112 mL of 0.100 M KOH to completely react with the H_3PO_4 according the following reaction. What was the concentration of the unknown H_3PO_4 solution?

$$H_3PO_4(aq) + 3\,KOH(aq) \longrightarrow 3\,H_2O(l) + K_3PO_4(aq)$$

92. A 25.0-mL sample of an unknown $HClO_4$ solution requires 45.3 mL of 0.101 M NaOH for complete neutralization. What was the concentration of the unknown $HClO_4$ solution? The neutralization reaction is:

$$HClO_4(aq) + NaOH(aq) \longrightarrow H_2O(l) + NaClO_4(aq)$$

93. What is the minimum amount of 6.0 M H_2SO_4 necessary to produce 15.0 g of $H_2(g)$ according to the following reaction?

$$2\,Al(s) + 3\,H_2SO_4(aq) \longrightarrow Al_2(SO_4)_3(aq) + 3\,H_2(g)$$

94. What is the molarity of $ZnCl_2(aq)$ that forms when 15.0 g of zinc completely reacts with $CuCl_2(aq)$ according to the following reaction? (Assume a final volume of 175 mL.)

$$Zn(s) + CuCl_2(aq) \longrightarrow ZnCl_2(aq) + Cu(s)$$

Molality, Freezing Point Depression, and Boiling Point Elevation

95. Compute the molality of each of the following solutions.
    (a) 0.25 mol solute; 0.250 kg solvent
    (b) 0.882 mol solute; 0.225 kg solvent
    (c) 0.012 mol solute; 23.1 g solvent

96. Compute the molality of each of the following solutions.
    (a) 0.455 mol solute; 1.97 kg solvent
    (b) 0.559 mol solute; 1.44 kg solvent
    (c) 0.119 mol solute; 488 g solvent

97. Compute the molality of a solution containing 11.5 g of ethylene glycol ($C_2H_6O_2$) dissolved in 145 g of water. (Assume a density of 1.00 g/mL for water.)

98. Compute the molality of a solution containing 287 g glucose ($C_6H_{12}O_6$) dissolved in 1.52 L of water. (Assume a density of 1.00 g/mL for water.)

99. Calculate the freezing points of water solutions with the following molalities.
    (a) 0.85 m
    (b) 1.45 m
    (c) 4.8 m
    (d) 2.35 m

100. Calculate the freezing points of water solutions with the following molalities.
    (a) 0.100 m
    (b) 0.469 m
    (c) 1.44 m
    (d) 5.89 m

101. Calculate the boiling points of water solutions with the following molalities.
    (a) 0.118 m
    (b) 1.94 m
    (c) 3.88 m
    (d) 2.16 m

102. Calculate the boiling points of water solutions with the following molalities.
    (a) 0.225 m
    (b) 2.58 m
    (c) 4.33 m
    (d) 6.77 m

103. A glucose solution contains 55.8 g of glucose ($C_6H_{12}O_6$) in 455 g of water. Compute the freezing point and boiling point of the solution. (Assume a density of 1.00 g/mL for water.)

104. An ethylene glycol solution contains 21.2 g of ethylene glycol ($C_2H_6O_2$) in 85.4 mL of water. Compute the freezing point and boiling point of the solution. (Assume a density of 1.00 g/mL for water.)

Cumulative Problems

105. An NaCl solution is made using 133 g of NaCl and diluting to a total solution volume of 1.00 L. Calculate the molarity and mass percent of the solution. (Assume a density of 1.08 g/mL for the solution.)

106. A KNO_3 solution is made using 88.4 g of KNO_3 and diluting to a total solution volume of 1.50 L. Calculate the molarity and mass percent of the solution. (Assume a density of 1.05 g/mL for the solution.)

107. A 125-mL sample of an 8.5 M NaCl solution is diluted to 2.5 L. What volume of the diluted solution contains 10.8 g of NaCl?

108. A 45.8-mL sample of a 5.8 M KNO_3 solution is diluted to 1.00 L. What volume of the diluted solution contains 15.0 g of KNO_3?

109. To what final volume should you dilute 50.0 mL of a 5.00 M KI solution so that 25.0 mL of the diluted solution contains 3.25 g of KI?

110. To what volume should you dilute 125 mL of an 8.00 M $CuCl_2$ solution so that 50.0 mL of the diluted solution contains 5.9 g $CuCl_2$?

111. Consider the following reaction.

$$2\,Al(s) + 3\,H_2SO_4(aq) \longrightarrow Al_2(SO_4)_3(aq) + 3\,H_2(g)$$

What minimum volume of 4.0 M H_2SO_4 is required to produce 15.0 L of H_2 at STP?

112. Consider the following reaction.

$$Mg(s) + 2\,HCl(aq) \longrightarrow MgCl_2(aq) + H_2(g)$$

What minimum amount of 1.85 M HCl is necessary to produce 28.5 L of H_2 at STP?

113. How much of a 1.25 M sodium chloride solution in milliliters is required to completely precipitate all of the silver in 25.0 mL of a 0.45 M silver nitrate solution?

114. How much of a 1.50 M sodium sulfate solution in milliliters is required to completely precipitate all of the barium in 150.0 mL of a 0.250 M barium nitrate solution?

115. Nitric acid is usually purchased in concentrated form with a 70.3% HNO_3 concentration by mass and a density of 1.41 g/mL. How much of the concentrated solution in milliliters should you use to make 2.5 L of 0.500 M HNO_3?

116. Hydrochloric acid is usually purchased in concentrated form with a 37.0% HCl concentration by mass and a density of 1.20 g/mL. How much of the concentrated solution in milliliters should you use to make 2.5 L of 0.500 M HCl?

117. An ethylene glycol solution is made using 58.5 g of ethylene glycol ($C_2H_6O_2$) and diluting to a total volume of 500.0 mL. Calculate the freezing point and boiling point of the solution. (Assume a density of 1.09 g/mL for the solution.)

118. A sucrose solution is made using 144 g of sucrose ($C_{12}H_{22}O_{11}$) and diluting to a total volume of 1.00 L. Calculate freezing point and boiling point of the solution. (Assume a density of 1.06 g/mL for the final solution.)

119. A 250.0-mL sample of a 5.00 M glucose ($C_6H_{12}O_6$) solution is diluted to 1.40 L. What are the freezing and boiling points of the final solution? (Assume a density of 1.06 g/mL for the final solution.)

120. A 135-mL sample of a 10.0 M ethylene glycol ($C_2H_6O_2$) solution is diluted to 1.50 L. What are the freezing and boiling points of the final solution? (Assume a density of 1.05 g/mL for the final solution.)

121. An aqueous solution containing 17.5 g of an unknown molecular (nonelectrolyte) compound in 100.0 g of water was found to have a freezing point of −1.8 °C. Calculate the molar mass of the unknown compound.

122. An aqueous solution containing 35.9 g of an unknown molecular (nonelectrolyte) compound in 150.0 g of water was found to have a freezing point of −1.3 °C. Calculate the molar mass of the unknown compound.

Highlight Problems

123. Consider the following molecular views of osmosis cells. For each cell, determine the direction of water flow.

(a)

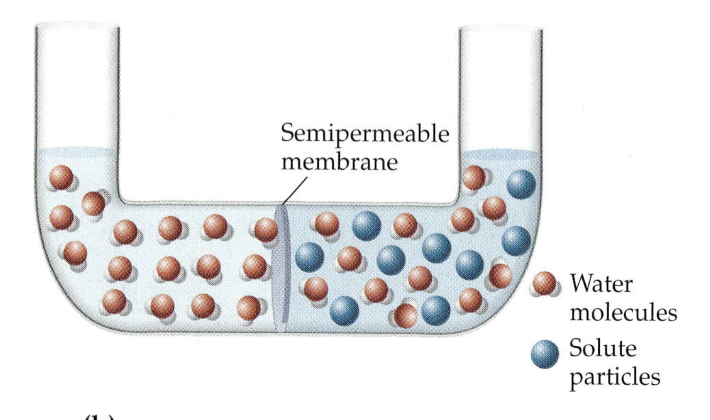

(b)

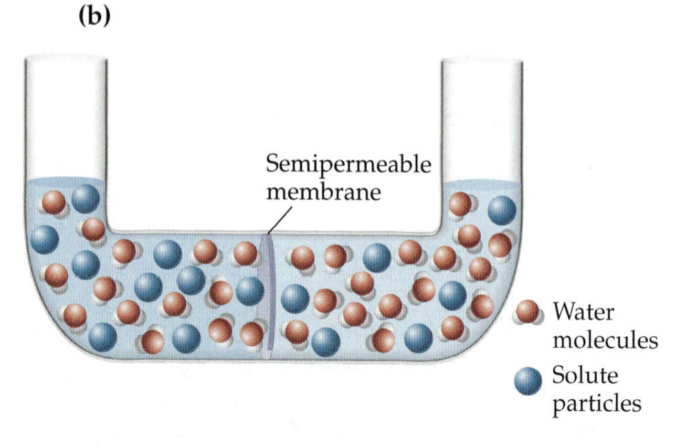

(c)

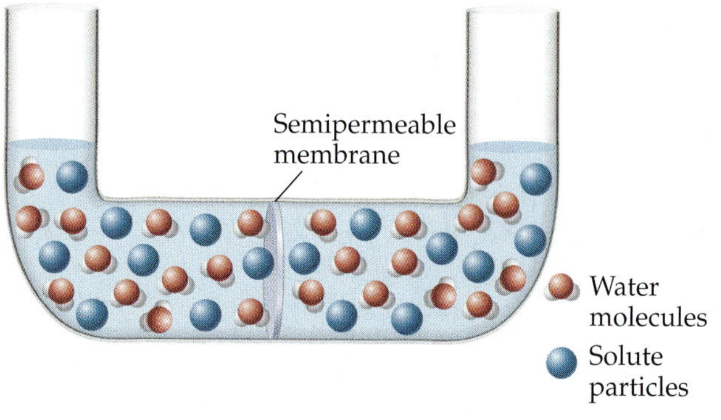

124. What is wrong with the following molecular view of a sodium chloride solution? What would make the picture correct?

125. The Safe Drinking Water Act (SDWA) sets a limit for mercury—a toxin to the central nervous system—at 0.002 mg/L. Water suppliers must periodically test their water to ensure that mercury levels do not exceed 0.002 mg/L. Suppose water becomes contaminated with mercury at twice the legal limit (0.004 mg/L). How much of this water would have to be consumed to ingest 0.100 g of mercury?

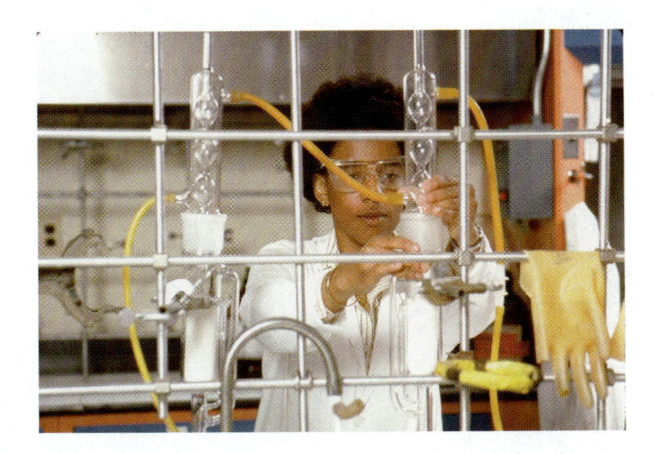

◀ Drinking water must be tested for the presence of various pollutants, including mercury compounds that can damage the nervous system.

126. Water softeners often replace calcium ions in hard water with sodium ions. Since sodium compounds are soluble, the presence of sodium ions in water does not cause the white, scaly residues caused by calcium ions. However, calcium is more beneficial to human health than sodium. Calcium is a necessary part of the human diet, while high levels of sodium intake are linked to increases in blood pressure. The Food and Drug Administration (FDA) recommends that adults ingest less than 2.4 g of sodium per day. How many liters of softened water, containing a sodium concentration of 0.050% sodium by mass, have to be consumed to exceed the FDA recommendation? (Assume a density of 1.0 g/mL for water.)

Answers to Skillbuilder Exercises

Skillbuilder 13.1 2.67%

Skillbuilder 13.2 40.8 g sucrose

Skillbuilder 13.3 0.263 M

Skillbuilder 13.4 3.33 L

Skillbuilder 13.5 0.75 M Ca^{2+} and 1.5 M Cl^-

Skillbuilder 13.6 0.12 L

Skillbuilder 13.7 16.4 mL

Skillbuilder Plus, p. 448 0.308 M

Skillbuilder 13.8 0.443 m

Skillbuilder 13.9 −4.8 °C

Skillbuilder 13.10 101.8 °C

Answers to Conceptual Checkpoints

13.1 (a) CH_3Cl and H_2S are both polar compounds, and KF is ionic. All three would therefore interact more strongly with water molecules (which are polar) than would CCl_4, which is nonpolar.

13.2 (a) A 2.0 m solution would be made by adding 2 mol of solute to 1 kg of solvent. 1 kg of water has a volume of 1 L, but because of the dissolved solute, the final solution would have a volume of slightly *more than* 1 L. A 2.0 M solution, by contrast, would consist of 2 mol of solute in a solution of *exactly* 1 L. Therefore, a 2 M aqueous solution would be slightly more concentrated than a 2 m solution.

REALEMON
LEMON JUICE
FROM CONCENTRATE
BEST YET

CHAPTER 14

Acids and Bases

"The differences between the various acid–base concepts are not concerned with which is 'right,' but which is most convenient to use in a particular situation."

James E. Huheey

14.1 Sour Patch Kids and International Spy Movies

Think about gummy candies, those chewy, sweet little shapes that both children and adults love. From the introduction of the gummy bear to the gummy worm to just about any shape you can imagine, these candies have been incredibly popular. A common variation is the sour gummy candy, whose best-known incarnation is the Sour Patch Kid. Sour Patch Kids are gummy candies shaped like kids and coated with a white powder. When you first put a Sour Patch Kid in your mouth, it tastes incredibly sour. The sour taste is caused by the white powder coating, a mixture of citric acid and tartaric acid. Like all acids, citric and tartaric acid have a sour taste.

A number of other foods contain acids as well. The sour taste of lemons and limes, the bite of sourdough bread, and the tang of a ripe tomato are all caused by acids. Acids are substances that—by one definition that we will elaborate on later—produce H^+ ions in solution. When the citric and tartaric acids from a Sour Patch Kid combine with saliva in your mouth, they produce H^+ ions. Those H^+ ions then react with protein molecules on your tongue. The protein molecules change shape, sending an electrical signal to your brain that you experience as a sour taste (▶ Figure 14.1).

When we say that acids *dissolve* metals, we mean that acids react with metals in a way that causes them to go into solution as metal cations. Bond's pen is made of gold because gold is one of the few metals that is not dissolved by most acids (see Section 16.5).

Acids have also been made famous by their use in spy movies. James Bond, for example, often carries an acid-filled gold pen. When Bond is captured and imprisoned—as inevitably happens at least one time in each movie—he squirts some acid out of his pen and onto the iron bars of his cell. The acid quickly dissolves the metal, providing Bond with a way of escape. While acids do not dissolve iron bars with the ease depicted in the movies, they do dissolve metals. A small piece of aluminum placed in hydrochloric acid, for example, dissolves away in about ten minutes (▶ Figure 14.2). With enough acid, it would be possible to dissolve the iron bars of a prison cell, but it would take more than the amount that fits in a pen.

◀ Acids are found in many common foods. The molecules shown here are citric acid (upper left), the acid found in lemon and limes; acetic acid (upper right), the acid present in vinegar; and tartaric acid (lower left), one of the acids used to coat sour gummy candies.

◀ **Figure 14.1 Acids taste sour** When a person eats a sour food, H^+ ions from the acid in the food react with protein molecules in cells of the taste buds. This interaction causes the protein molecules to change shape, triggering a nerve impulse to the brain that the person experiences as a sour taste.

▶ **Figure 14.2 Acids dissolve many metals** When aluminum is put into hydrochloric acid, the aluminum dissolves. **Question:** What happens to the aluminum atoms? Where do they go?

14.2 Acids: Properties and Examples

NEVER taste or touch laboratory chemicals.

Acids have the following properties:

- Acids have a sour taste.
- Acids dissolve many metals.
- Acids turn litmus paper red.

We have just seen examples of the sour taste of acids (Sour Patch Kids) and their ability to dissolve metals (spy movies). Acids also turn litmus paper red. Litmus paper contains a dye that turns red in acidic solutions (◀ Figure 14.3). In the laboratory, litmus paper is used routinely to test the acidity of solutions.

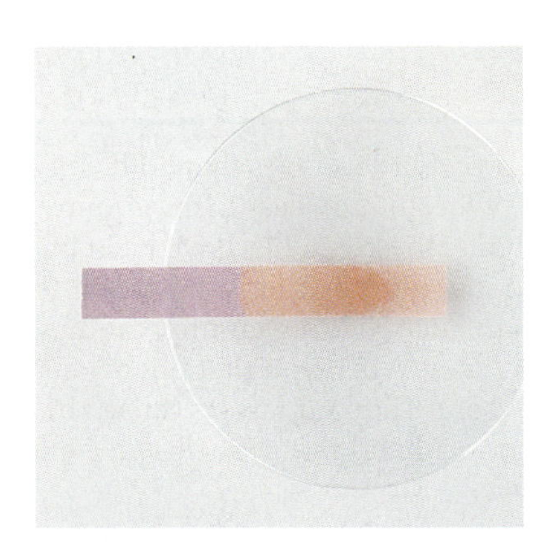

▲ **Figure 14.3 Acids turn litmus paper red**

Some common acids are listed in Table 14.1. Hydrochloric acid is found in most chemistry laboratories. It is used in industry to clean metals, to prepare and process some foods, and to refine metal ores.

For a review of naming acids, see Section 5.9.

TABLE 14.1 Some Common Acids

Name	Uses
hydrochloric acid (HCl)	metal cleaning; food preparation; ore refining; main component of stomach acid
sulfuric acid (H_2SO_4)	fertilizer and explosive manufacturing; dye and glue production; automobile batteries
nitric acid (HNO_3)	fertilizer and explosive manufacturing; dye and glue production
acetic acid ($HC_2H_3O_2$)	plastic and rubber manufacturing; food preservative; active component of vinegar
carbonic acid (H_2CO_3)	found in carbonated beverages due to the reaction of carbon dioxide with water
hydrofluoric acid (HF)	metal cleaning; glass frosting and etching

Hydrochloric acid is also the main component of stomach acid. In the stomach, hydrochloric acid helps break down food and kills harmful bacteria that might enter the body through food. The sour taste sometimes associated with indigestion is caused by the stomach's hydrochloric acid refluxing up into the esophagus (the tube that joins the stomach and the mouth) and throat.

Annual U.S. production of sulfuric acid exceeds 36,000 tons.

Sulfuric acid—the most widely produced chemical in the United States—and nitric acid are also commonly used in the laboratory.

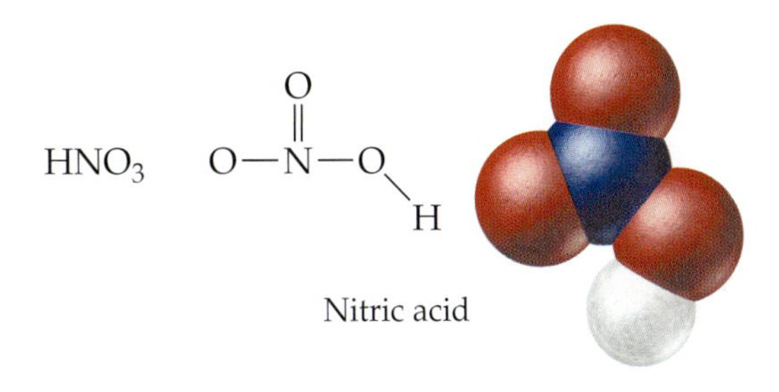

Sulfuric acid

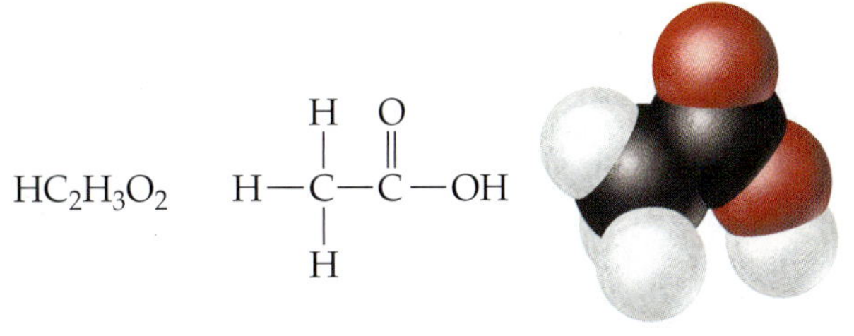

Nitric acid

In addition, they are used in the manufacture of fertilizers, explosives, dyes and glue. Sulfuric acid is contained in most automobile batteries.

MOLECULE Acetic Acid

Acetic acid is found in most people's homes as the active component of vinegar.

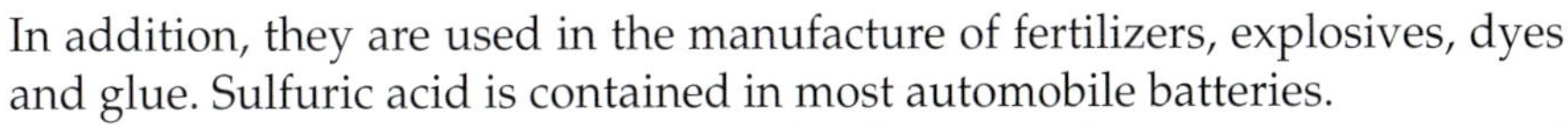

Acetic acid

▲ Acetic acid is the active component in vinegar.

It is also produced in improperly stored wines. The word *vinegar* originates from the French words *vin aigre*, which mean "sour wine." The presence of vinegar in wines is considered a serious fault, making the wine taste like salad dressing.

Acetic acid is an example of a **carboxylic acid**, an acid containing the following grouping of atoms.

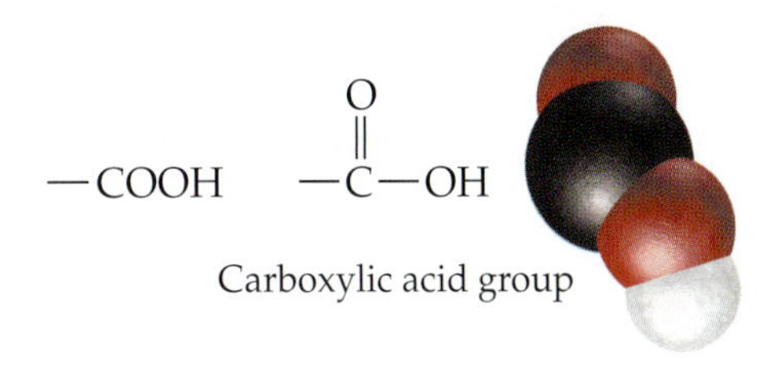

Carboxylic acid group

Carboxylic acids, covered in more detail in Chapter 18 (Section 18.15), are often found in substances derived from living organisms. Other

MOLECULE
Citric Acid

carboxylic acids include citric acid, the main acid in lemons and limes, and malic acid, an acid found in apples, grapes, and wine.

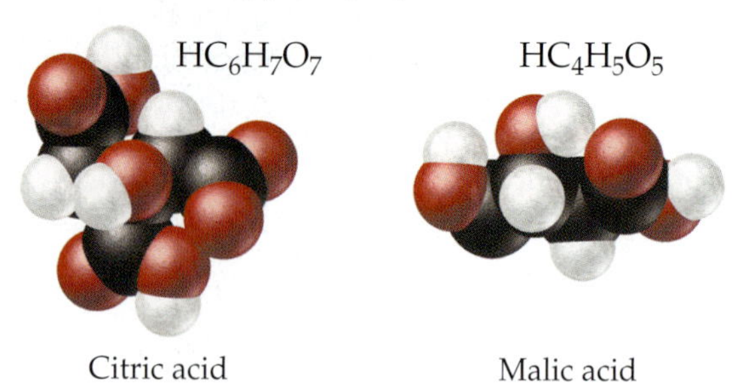

Citric acid Malic acid

14.3 Bases: Properties and Examples

Bases have the following properties:

NEVER taste or touch laboratory chemicals.

- **Bases have a bitter taste.**
- **Bases have a slippery feel.**
- **Bases turn litmus paper blue.**

Bases are less common in foods than acids because of their bitter taste. A Sour Patch Kid coated with a base would never sell. Our aversion to the taste of bases is probably an adaptation to protect us against **alkaloids**, organic bases found in plants. Alkaloids are often poisonous—the active component of hemlock, for example, is the alkaloid coniine—and their bitter taste warns us against eating them. Nonetheless, some foods, such as coffee, contain small amounts of base. Many people enjoy the bitterness, but only after acquiring the taste over time.

Coffee is acidic overall, but bases present in coffee—such as caffeine—impart a bitter flavor.

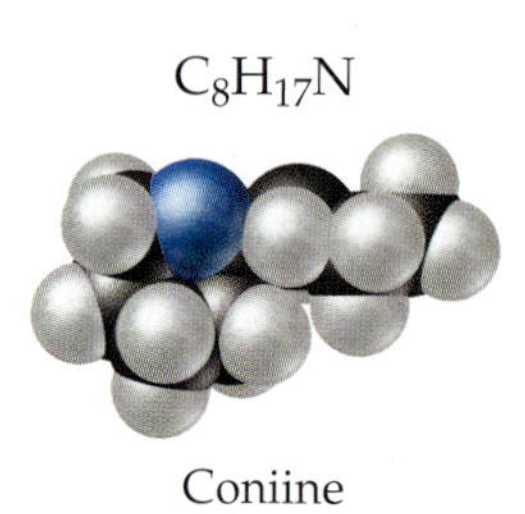

Coniine

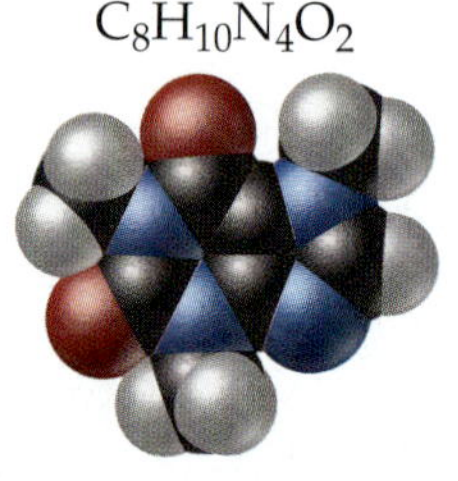

Caffeine

▲ These consumer products all contain bases.

Bases feel slippery because they react with oils on your skin to form soaplike substances. Soap itself is basic and its slippery feel is characteristic of bases. Some household cleaning solutions, such as ammonia, are also basic and have the typical slippery feel of a base. Bases turn litmus paper blue (▶ Figure 14.4). In the laboratory, litmus paper is routinely used to test the basicity of solutions.

Some common bases are listed in Table 14.2. Sodium hydroxide and potassium hydroxide are found in most chemistry laboratories. They are also used in processing petroleum and cotton and in soap and plastic manufacturing. Sodium hydroxide is the active ingredient in products such as Drano that work to unclog drains. Sodium bicarbonate can be found in most homes as baking soda and is also an active ingredient in many antacids. When taken as an antacid, sodium bicarbonate neutralizes stomach acid (see Section 14.5), relieving heartburn and sour stomach.

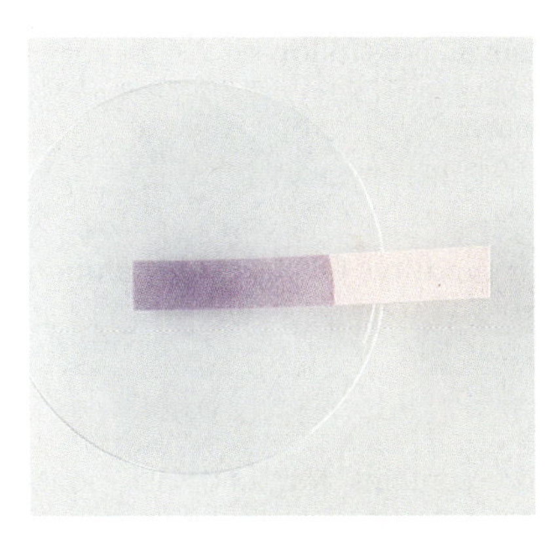

▲ **Figure 14.4 Bases turn litmus paper blue**

TABLE 14.2 Some Common Bases

Name	Uses
sodium hydroxide (NaOH)	petroleum processing; soap and plastic manufacturing
potassium hydroxide (KOH)	cotton processing; electroplating; soap production
sodium bicarbonate ($NaHCO_3$)*	antacid; ingredient of baking soda; source of CO_2
ammonia (NH_3)	detergent; fertilizer and explosive manufacturing; synthetic fiber production

*Sodium bicarbonate is a salt whose anion (HCO_3^-) is the conjugate base of a weak acid (see Section 14.4) and acts as a base.

14.4 Molecular Definitions of Acids and Bases

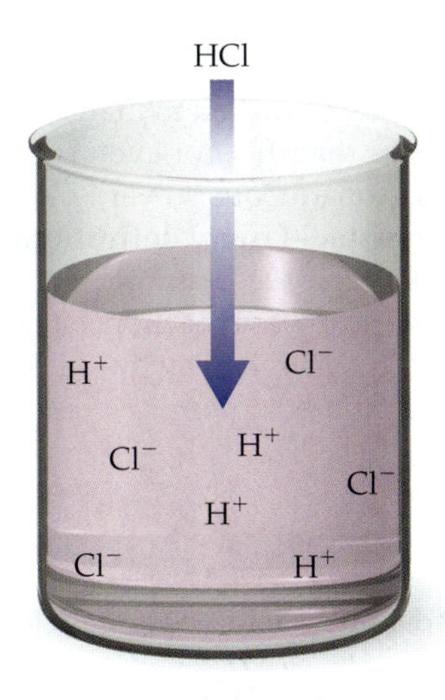

$$HCl(aq) \longrightarrow H^+(aq) + Cl^-(aq)$$

▲ **Figure 14.5 Arrenhius definition of an acid** The Arrhenius definition states that an acid is a substance that produces H^+ ions in solution. These H^+ ions associate with H_2O to form H_3O^+ ions.

We have just seen some of the properties of acids and bases. In this section we examine two different models that explain the molecular basis for acid and base behavior: the Arrhenius model and the Brønsted–Lowry model. The Arrhenius model came first and is more limited in its scope. The Brønsted–Lowry model was developed later and is more broadly applicable.

The Arrhenius Definition

In the 1880s, the Swedish chemist Svante Arrhenius proposed the following molecular definitions of acids and bases.

Acid—An acid produces **H^+** ions in aqueous solution.
Base—A base produces **OH^-** ions in aqueous solution.

For example, under the **Arrhenius definition**, HCl is an **Arrhenius acid** because it produces H^+ ions in solution (◀ Figure 14.5).

$$HCl(aq) \longrightarrow H^+(aq) + Cl^-(aq)$$

HCl is a covalent compound and does not contain ions. However, in water it **ionizes** to form $H^+(aq)$ ions and $Cl^-(aq)$ ions. The H^+ ions are highly reactive. In aqueous solution, they bond to water molecules according to the following reaction.

$$H^+ + H:\ddot{\underset{..}{O}}:H \longrightarrow \left[H:\ddot{\underset{..}{O}}:H \text{ (with H above O)}\right]^+$$

The H_3O^+ ion is called the **hydronium ion**. In water, H^+ ions *always* associate with H_2O molecules to form hydronium ions. Chemists often use $H^+(aq)$ and $H_3O^+(aq)$ interchangeably, to refer to the same thing—a hydronium ion.

In the formula for an acid, the ionizable hydrogen is usually written first. For example, the formula for formic acid is often written as follows:

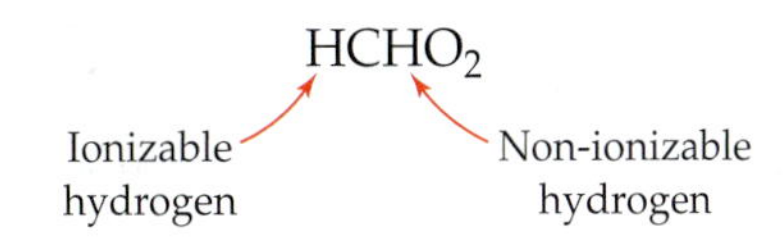

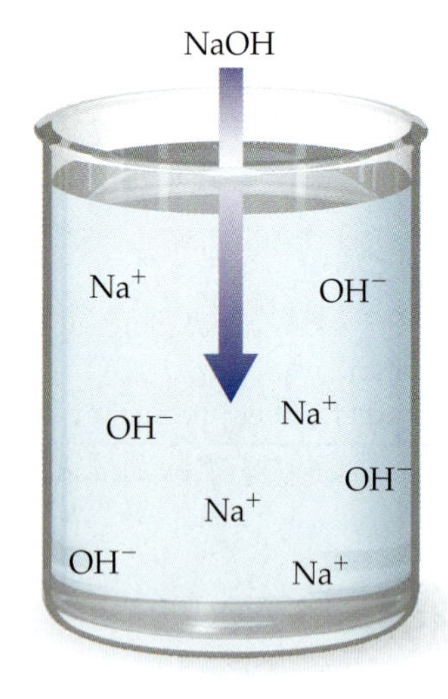

$NaOH(aq) \longrightarrow Na^+(aq) + OH^-(aq)$

▲ **Figure 14.6 Arrenhius definition of a base** The Arrhenius definition states that a base is a substance that produces OH^- ions in solution.

Ionic compounds such as NaOH are composed of positive and negative ions. In solution, soluble ionic compounds dissociate into their component ions. Molecular compounds containing an OH group, such as methanol CH_3OH, *do not* act as bases.

The structural formula for formic acid, however, is as follows:

$$\begin{array}{c} O \\ \| \\ HC{-}OH \end{array}$$

Ionizable hydrogen

NaOH is an **Arrhenius base** because it produces OH^- ions in solution (◀ Figure 14.6).

$$NaOH(aq) \longrightarrow Na^+(aq) + OH^-(aq)$$

NaOH is an ionic compound and therefore contains Na^+ and OH^- ions. When NaOH is added to water, it **dissociates** or breaks apart into its component ions.

Under the Arrhenius definition, acids and bases naturally combine to form water, neutralizing each other in the process.

$$H^+(aq) + OH^-(aq) \longrightarrow H_2O(l)$$

The Brønsted–Lowry Definition

Although the Arrhenius definition of acids and bases worked in many cases, it could not easily explain why some substances that did not even contain OH^- could act as bases and it did not apply to nonaqueous solvents. A second definition of acids and bases, called the **Brønsted–Lowry definition**, introduced in 1923, applies to a wider range of acid–base phenomena. This definition focuses on the transfer of H^+ ions in an acid–base reaction. Since an H^+ ion is a proton—a hydrogen atom with its electron taken away—this definition uses the idea of a proton donor and a proton acceptor.

Acid—An acid is a proton **(H^+ ion) *donor***.
Base—A base is a proton **(H^+ ion) *acceptor***.

Johannes Brønsted, working in Denmark, and Thomas Lowry, working in England, developed the concept of proton transfer in acid–base behavior independently and simultaneously.

According to this definition, HCl is a **Brønsted-Lowry acid** because, in solution, it donates a proton to water.

$$HCl(aq) + H_2O(l) \longrightarrow H_3O^+(aq) + Cl^-(aq)$$

This definition more clearly shows what happens to the H^+ ion from an acid—it associates with a water molecule to form H_3O^+ (a hydronium ion). The Brønsted–Lowry definition also works well with bases (such as NH_3) that do not inherently contain OH^- ions but that still produce OH^- ions in solution. In the Brønsted–Lowry definition, NH_3 is a **Brønsted–Lowry base** because it accepts a proton from water.

The double arrows in this equation indicate that the reaction does not go to completion. We discuss this concept in more detail in Section 14.7.

$$NH_3(aq) + H_2O(l) \rightleftharpoons NH_4^+(aq) + OH^-(aq)$$

In the Brønsted–Lowry definition, acids (proton donors) and bases (proton acceptors) always occur together. In the reaction between HCl and H_2O, HCl is the proton donor (acid) and H_2O is the proton acceptor (base).

$$\underset{\substack{\text{Acid}\\\text{(Proton donor)}}}{HCl(aq)} + \underset{\substack{\text{Base}\\\text{(Proton acceptor)}}}{H_2O(l)} \longrightarrow H_3O^+(aq) + Cl^-(aq)$$

In the reaction between NH_3 and H_2O, H_2O is the proton donor (acid) and NH_3 is the proton acceptor (base).

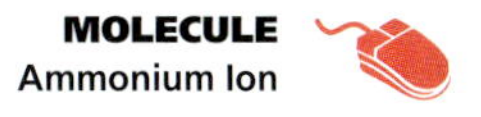

$$\underset{\substack{\text{Base} \\ \text{(Proton acceptor)}}}{NH_3(aq)} + \underset{\substack{\text{Acid} \\ \text{(Proton donor)}}}{H_2O(l)} \rightleftharpoons NH_4^+(aq) + OH^-(aq)$$

Notice that under the Brønsted–Lowry definition, some substances—such as water in the previous two equations—can act as acids *or* bases. Substances that can act as acids or bases are termed **amphoteric**. Notice also what happens when an equation representing Brønsted–Lowry acid–base behavior is reversed.

$$\underset{\substack{\text{Acid} \\ \text{(Proton donor)}}}{NH_4^+(aq)} + \underset{\substack{\text{Base} \\ \text{(Proton acceptor)}}}{OH^-(aq)} \rightleftharpoons NH_3(aq) + H_2O(l)$$

In this reaction, NH_4^+ is the proton donor (acid) and OH^- is the proton acceptor (base). What was the base (NH_3) has become the acid (NH_4^+) and vice versa. NH_4^+ and NH_3 are often referred to as a **conjugate acid–base pair,** two substances related to each other by the transfer of a proton (▼ Figure 14.7). Going back to the original forward reaction, we can identify the conjugate acid–base pairs as follows:

$$\underset{\text{Base}}{NH_3(aq)} + \underset{\text{Acid}}{H_2O(l)} \rightleftharpoons \underset{\substack{\text{Conjugate} \\ \text{acid}}}{NH_4^+(aq)} + \underset{\substack{\text{Conjugate} \\ \text{base}}}{OH^-(aq)}$$

In an acid–base reaction, a base accepts a proton and becomes a conjugate acid. An acid donates a proton and becomes a conjugate base.

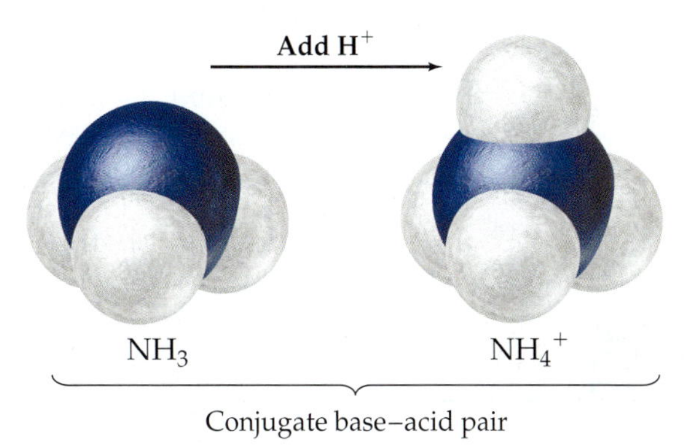

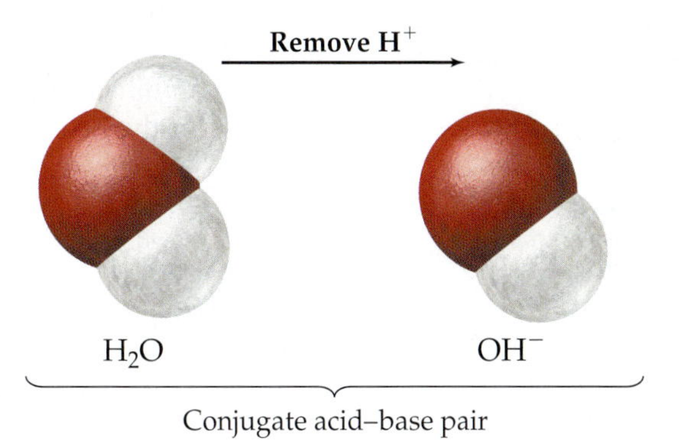

▶ **Figure 14.7 A conjugate acid–base pair** Any two substances related to each other by the transfer of a proton make up a conjugate acid–base pair.

EXAMPLE 14.1 Identifying Brønsted–Lowry Acids and Bases and Their Conjugates

In each of the following reactions, identify the Brønsted–Lowry acid, the Brønsted–Lowry base, the conjugate acid, and the conjugate base.

(a) $H_2SO_4(aq) + H_2O(l) \longrightarrow H_3O^+(aq) + HSO_4^-(aq)$
(b) $HCO_3^-(aq) + H_2O(l) \rightleftharpoons H_2CO_3(aq) + OH^-(aq)$

Solution:

(a) Since H_2SO_4 donates a proton to H_2O in this reaction, it is the acid (proton donor). After H_2SO_4 donates the proton, it becomes HSO_4^-, the conjugate base. Since H_2O accepts a proton, it is the base (proton acceptor). After H_2O accepts the proton it becomes H_3O^+, the conjugate acid.

$$H_2SO_4(aq) + H_2O(l) \longrightarrow HSO_4^-(aq) + H_3O^+(aq)$$

Acid | Base | Conjugate base | Conjugate acid

(b) Since H_2O donates a proton to HCO_3^- in this reaction, it is the acid (proton donor). After H_2O donates the proton, it becomes OH^-, the conjugate base. Since HCO_3^- accepts a proton, it is the base (proton acceptor). After HCO_3^- accepts the proton it becomes H_2CO_3, the conjugate acid.

$$HCO_3^-(aq) + H_2O(l) \longrightarrow H_2CO_3(aq) + OH^-(aq)$$

Base | Acid | Conjugate acid | Conjugate base

SKILLBUILDER 14.1 Identifying Brønsted–Lowry Acids and Bases and Their Conjugates

In each of the following reactions, identify the Brønsted–Lowry acid, the Brønsted–Lowry base, the conjugate acid, and the conjugate base.

(a) $C_5H_5N(aq) + H_2O(l) \rightleftharpoons C_5H_5NH^+(aq) + OH^-(aq)$
(b) $HNO_3(aq) + H_2O(l) \longrightarrow NO_3^-(aq) + H_3O^+(aq)$

14.5 Reactions of Acids and Bases

Neutralization Reactions

Neutralization reactions, also known as acid–base reactions, are covered in Section 7.8.

One of the most important acid–base reactions is **neutralization**, first introduced in Chapter 7. When an acid and a base are mixed, the $H^+(aq)$ from the acid combines with the $OH^-(aq)$ from the base to form $H_2O(l)$. For example, consider the reaction between hydrochloric acid and potassium hydroxide.

The reaction between HCl and KOH is also a double-displacement reaction (see Section 7.10).

$$HCl(aq) + KOH(aq) \longrightarrow H_2O(l) + KCl(aq)$$

Acid | Base | Water | Salt

Acid–base reactions generally form water and a **salt**—an ionic compound—that usually remains dissolved in the solution. The salt contains the cation from the base and the anion from the acid.

$$\text{Acid} + \text{Base} \longrightarrow \text{Water} + \text{Salt}$$

Salt: Ionic compound that contains the cation from the base and the anion from the acid

Net ionic equations are explained in Section 7.7.

The net ionic equation for many neutralization reactions is:

$$H^+(aq) + OH^-(aq) \longrightarrow H_2O(l)$$

A slightly different but common type of neutralization reaction involves an acid reacting with carbonates or bicarbonates (compounds containing CO_3^{2-} or HCO_3^-.) This type of neutralization reaction produces water, gaseous carbon dioxide, and a salt. As an example, consider the reaction of hydrochloric acid and sodium bicarbonate.

$$HCl(aq) + NaHCO_3(aq) \longrightarrow H_2O(l) + CO_2(g) + NaCl(aq)$$

$HCl(aq) + NaHCO_3(aq) \longrightarrow H_2O(l) + CO_2(g) + NaCl(aq)$

▲ The reaction of carbonates or bicarbonates with acids produces water, gaseous carbon dioxide, and a salt.

Since this reaction produces gaseous CO_2, it is also called a *gas evolution reaction* (Section 7.8).

EXAMPLE 14.2 Writing Equations for Neutralization Reactions

Write a molecular equation for the reaction between aqueous HCl and aqueous $Ca(OH)_2$.

Solution:

We must identify the acid and the base and know that they react to form water and a salt. Notice that $Ca(OH)_2$ contains 2 mol of OH^- for every 1 mol of $Ca(OH)_2$ and will therefore require 2 mol of H^+ to neutralize it. We first write the skeletal reaction.

$$HCl(aq) + Ca(OH)_2(aq) \longrightarrow H_2O(l) + CaCl_2(aq)$$

We then balance the equation.

$$2\,HCl(aq) + Ca(OH)_2(aq) \longrightarrow 2\,H_2O(l) + CaCl_2(aq)$$

SKILLBUILDER 14.2 Writing Equations for Neutralization Reactions

Write a molecular equation for the reaction that occurs between aqueous H_3PO_4 and aqueous NaOH. Hint: H_3PO_4 is a triprotic acid, meaning that 1 mol of H_3PO_4 requires 3 mol of OH^- to completely react with it.

Acid Reactions

In Section 14.1, we saw that acids dissolve metals, or more precisely, that acids react with metals in a way that causes them to go into solution. The reaction between an acid and a metal usually produces hydrogen gas and a dissolved salt containing the metal ion as the cation. For example, hydrochloric acid reacts with magnesium metal to form hydrogen gas and magnesium chloride.

The reaction between HCl and Mg is also a single-displacement reaction (see Section 7.10).

$$\underset{\text{Acid}}{2\,HCl(aq)} + \underset{\text{Metal}}{Mg(s)} \longrightarrow \underset{\text{Hydrogen gas}}{H_2(g)} + \underset{\text{Salt}}{MgCl_2(aq)}$$

Similarly, sulfuric acid reacts with zinc to form hydrogen gas and zinc sulfate.

$$\underset{\text{Acid}}{H_2SO_4(aq)} + \underset{\text{Metal}}{Zn(s)} \longrightarrow \underset{\text{Hydrogen gas}}{H_2(g)} + \underset{\text{Salt}}{ZnSO_4(aq)}$$

It is through reactions such as these that the acid in James Bond's pen from our opening example dissolves the metal bars that imprison him. For example, if the bars were made of iron and the acid in the pen were hydrochloric acid, the reaction would be:

$$\underset{\text{Acid}}{2\,HCl(aq)} + \underset{\text{Metal}}{Fe(s)} \longrightarrow \underset{\text{Hydrogen gas}}{H_2(g)} + \underset{\text{Salt}}{FeCl_2(aq)}$$

The way to determine whether a particular metal dissolves in an acid is presented in Section 16.5.

$$2\,HCl(aq) + Mg(s) \longrightarrow H_2(g) + MgCl_2(aq)$$

▲ The reaction between an acid and a metal usually produces hydrogen gas and a dissolved salt containing the metal ion.

We should note, however, that some metals do not react with acids. If the bars that imprisoned James Bond were made of gold, for example, a pen filled with hydrochloric acid would not dissolve the bars.

Acids also react with metal oxides to produce water and a dissolved salt. For example, hydrochloric acid reacts with potassium oxide to form water and potassium chloride.

$$\underset{\text{Acid}}{2\,HCl(aq)} + \underset{\text{Metal oxide}}{K_2O(s)} \longrightarrow \underset{\text{Water}}{H_2O(l)} + \underset{\text{Salt}}{2\,KCl(aq)}$$

Similarly, hydrobromic acid reacts with magnesium oxide to form water and magnesium bromide.

$$\underset{\text{Acid}}{2\,HBr(aq)} + \underset{\text{Metal oxide}}{MgO(s)} \longrightarrow \underset{\text{Water}}{H_2O(l)} + \underset{\text{Salt}}{MgBr_2(aq)}$$

EXAMPLE 14.3 Writing Equations for Acid Reactions

Write an equation for each of the following:

(a) The reaction of hydroiodic acid with potassium metal
(b) The reaction of hydrobromic acid with sodium oxide

Solution:

(a) The reaction of hydroiodic acid with potassium metal forms hydrogen gas and a salt. The salt contains the ionized form of the metal (K^+) as the cation and the anion of the acid (I^-) as the anion. We first write the skeletal equation.

$$HI(aq) + K(s) \longrightarrow H_2(g) + KI(aq)$$

We then balance the equation.

$$2\,HI(aq) + 2\,K(s) \longrightarrow H_2(g) + 2\,KI(aq)$$

(b) The reaction of hydrobromic acid with sodium oxide forms water and a salt. The salt contains the cation from the metal oxide (Na^+) and the anion of the acid (Br^-). We first write the skeletal equation.

$$HBr(aq) + Na_2O(s) \longrightarrow H_2O(l) + NaBr(aq)$$

We then balance the equation.

$$2\,HBr(aq) + Na_2O(s) \longrightarrow H_2O(l) + 2\,NaBr(aq)$$

SKILLBUILDER 14.3 Writing Equations for Acid Reactions

Write an equation for each of the following:

(a) The reaction of hydrochloric acid with strontium metal
(b) The reaction of hydroiodic acid with barium oxide

Base Reactions

The most important base reactions are those in which a base neutralizes an acid (see the beginning of this section). The only other kind of base reaction that we cover in this book is the reaction of sodium hydroxide with aluminum and water.

$$2\,NaOH(aq) + 2\,Al(s) + 6\,H_2O(l) \longrightarrow 2\,NaAl(OH)_4(aq) + 3\,H_2(g)$$

Aluminum is one of the few metals that dissolves in a base. Consequently, it is safe to use NaOH (the main ingredient in many drain-opening products) to unclog your drain as long as your pipes are not made of aluminum, which is against most building codes.

EVERYDAY Chemistry

What Is in My Antacid?

Heartburn, a burning sensation in the lower throat and above the stomach, is caused by the reflux or backflow of stomach acid into the esophagus (the tube that joins the stomach to the throat). In healthy individuals, this occurs only occasionally, especially after large meals. Physical activity—such as bending, stooping, or lifting—after meals also aggravates heartburn. In some people, the flap between the esophagus and the stomach that normally prevents acid reflux becomes damaged, in which case heartburn becomes a regular occurrence.

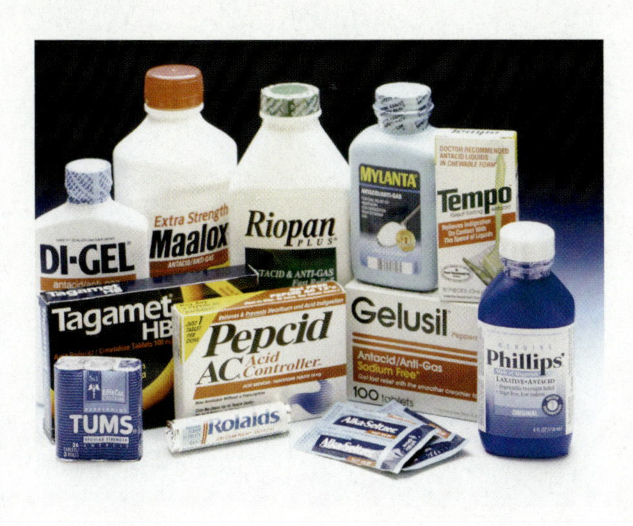

Drugstores carry many products that either reduce the secretion of stomach acid or neutralize the acid that is produced. Antacids such as Mylanta or Phillips' milk of magnesia contain bases that neutralize the refluxed stomach acid, alleviating heartburn.

Drug Facts
Active ingredients (in each 5 mL teaspoon) Purposes
Aluminum hydroxide (equivalent to dried gel, USP) 200 mg..........Antacid
Magnesium hydroxide 200 mg..........Antacid
Simethicone 20 mg..........Antigas
Use relieves:
■ heartburn ■ acid indigestion ■ sour stomach
■ upset stomach due to these symptoms ■ overindulgence in food and drink
Warnings
Ask a doctor before use if you have kidney disease.
Ask a doctor or pharmacist if you are taking a prescription drug. Antacids may interact with certain prescription drugs.
Stop use and ask a doctor if symptoms last more than 2 weeks.
Keep out of reach of children.
Directions ■ shake well ■ take 2-4 teaspoonfuls between meals, at bedtime, or as directed by a doctor ■ do not take more than 24 teaspoonfuls in a 24-hour period, or use the maximum dosage for more than 2 weeks
Other information ■ **does not meet USP requirements for preservative effectiveness** ■ **do not use if breakaway band on plastic cap is broken or missing**

CAN YOU ANSWER THIS? *Look at the label of Mylanta shown in the photograph. Can you identify the bases responsible for the antacid action? Write chemical equations showing the reactions of these bases with stomach acid (HCl).*

14.6 Acid–Base Titration: A Way to Quantify the Amount of Acid or Base in a Solution

The principles we learned in Chapter 13 (Section 13.8) on solution stoichiometry can be applied to a common laboratory procedure called a titration. In a **titration**, a substance in a solution of known concentration is reacted with another substance in a solution of unknown concentration. For example, consider the following acid–base reaction:

$$HCl(aq) + NaOH(aq) \longrightarrow H_2O(l) + NaCl(aq)$$

The net ionic equation for this reaction is as follows:

$$H^+(aq) + OH^-(aq) \longrightarrow H_2O(l)$$

Suppose you have an HCl solution represented by the following molecular diagram. (The Cl^- ions and the H_2O molecules not involved in the reaction have been omitted from this representation for clarity.)

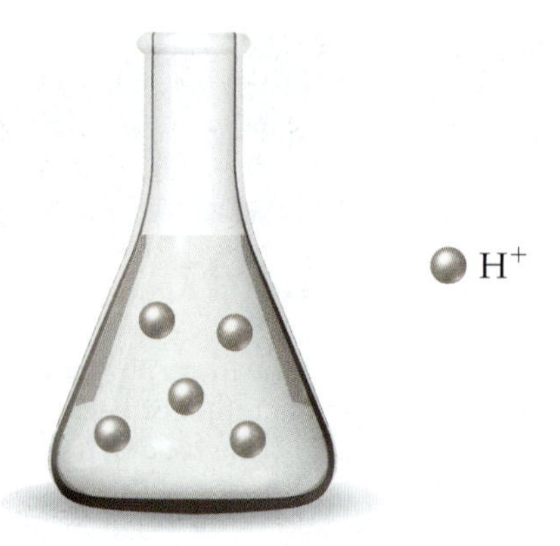

In titrating this sample, we slowly add a solution of known OH^- concentration as represented by the following molecular diagrams.

The OH^- solution also contains Na^+ cations that have been omitted here for clarity.

Beginning of titration

Equivalence point

At the equivalence point, neither reactant is present in excess, and both are limiting. The number of moles of the reactants are related by the reaction stoichiometry (see Chapter 8).

As the OH^- is added, it reacts with and neutralizes the H^+, forming water. At the **equivalence point**—the point in the titration when the number of moles of OH^- added equals the number of moles of H^+ originally in solution—the titration is complete. The equivalence point is usually signaled

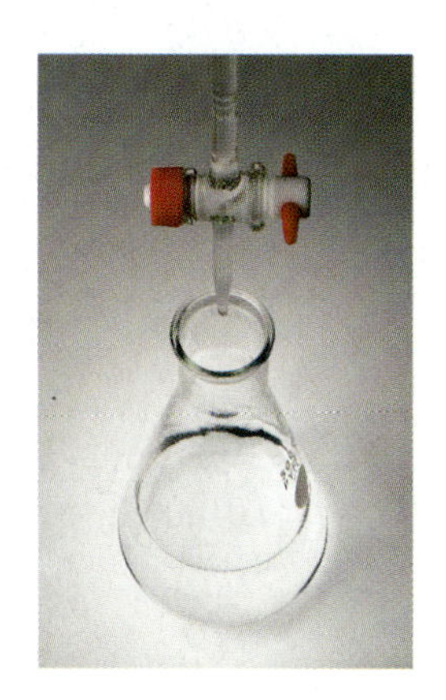

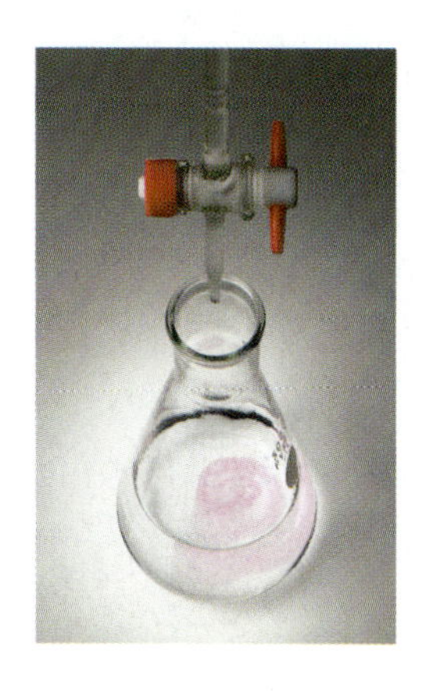

▶ **Figure 14.8 Acid–base titration** In this titration, NaOH is added to an HCl solution. When the NaOH and HCl reach stoichiometric proportions (1 mol of OH^- for every 1 mol of H^+), the indicator (phenolphthalein) changes to pink, signaling the endpoint of the titration. (Phenolphthalein is an indicator that is colorless in acidic solution and pink in basic solution.)

by an **indicator**, a dye whose color depends on the acidity of the solution (▲ Figure 14.8). In most laboratory titrations, the concentration of one of the reactant solutions is unknown and the concentration of the other is precisely known. By carefully measuring the volume of each solution required to reach the equivalence point, the concentration of the unknown solution can be determined, as demonstrated in the following example.

EXAMPLE 14.4 Acid–Base Titration

The titration of 10.00 mL of an HCl solution of unknown concentration requires 12.54 mL of a 0.100 M NaOH solution to reach the endpoint. What is the concentration of the unknown HCl solution?

You are given the volume of an unknown HCl solution and the volume of a known NaOH solution required to titrate the unknown solution. You are asked to find the concentration of the unknown solution.

You will need two equations. The first is the equation for the neutralization reaction of HCl and NaOH (which you should write using your knowledge of acid–base reactions). The second is simply the definition of molarity.

Given:

10.00 mL HCl solution

12.54 mL of a 0.100 M NaOH solution

Find: concentration of HCl solution (mol/L)

Equations:

$$HCl(aq) + NaOH(aq) \longrightarrow H_2O(l) + NaCl(aq)$$

$$\text{Molarity (M)} = \frac{\text{mol solute}}{\text{L solution}}$$

The solution map has two parts. In the first part, use the volume of NaOH required to reach the endpoint to calculate the number of moles of HCl in the solution. The final conversion factor comes from the balanced neutralization equation.

In the second part, use the number of moles of HCl and the volume of HCl solution to determine the molarity of the HCl solution.

Solution Map:

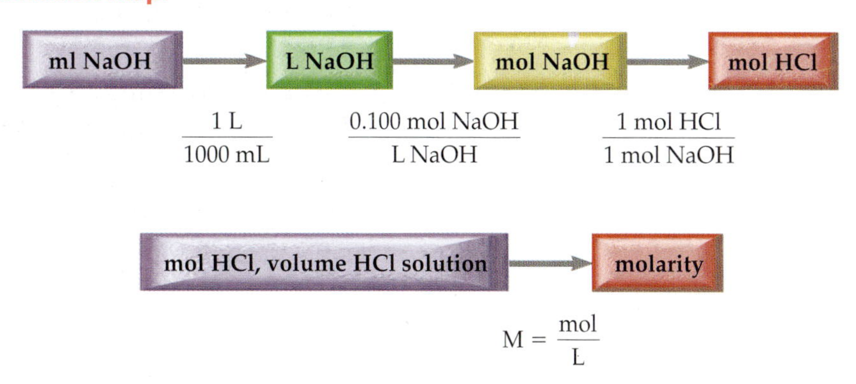

Calculate the moles of HCl in the unknown solution by following the first part of the solution map.

To get the concentration of the solution, divide the number of moles of HCl by the volume of the HCl solution in L. (Note that 10.00 mL is equivalent to 0.01000 L.)

The unknown HCl solution therefore has a concentration of 0.125 M.

Solution:

$$12.54 \text{ mL NaOH} \times \frac{1 \text{ L}}{1000 \text{ mL}} \times \frac{0.100 \text{ mol NaOH}}{\text{L NaOH}} \times \frac{1 \text{ mol HCl}}{1 \text{ mol NaOH}} = 1.25 \times 10^{-3} \text{ mol HCl}$$

$$\text{Molarity} = \frac{1.25 \times 10^{-3} \text{ mol HCl}}{0.01000 \text{ L}} = 0.125 \text{ M}$$

SKILLBUILDER 14.4 Acid–Base Titration

The titration of a 20.0-mL sample of an H_2SO_4 solution of unknown concentration requires 22.87 mL of a 0.158 M KOH solution to reach the endpoint. What is the concentration of the unknown H_2SO_4 solution?

14.7 Strong and Weak Acids and Bases

Strong Acids

Hydrochloric acid (HCl) and hydrofluoric acid (HF) appear similar, but there is an important difference between these two acids. HCl is an example of a **strong acid**, one that completely ionizes in solution.

Single arrow indicates complete ionization

$$HCl(aq) + H_2O(l) \longrightarrow H_3O^+(aq) + Cl^-(aq)$$

We show the *complete* ionization of HCl with a single arrow pointing to the right in the equation. An HCl solution contains almost no intact HCl; virtually all the HCl has reacted with water to form $H_3O^+(aq)$ and $Cl^-(aq)$ (▼ Figure 14.9). A 1.0 M HCl solution will therefore have an H_3O^+ concentration of 1.0 M. The concentration of H_3O^+ is often abbreviated as $[H_3O^+]$. Using this notation, a 1 M HCl solution has $[H_3O^+] = 1.0$ M.

[X] means "molar concentration of X"

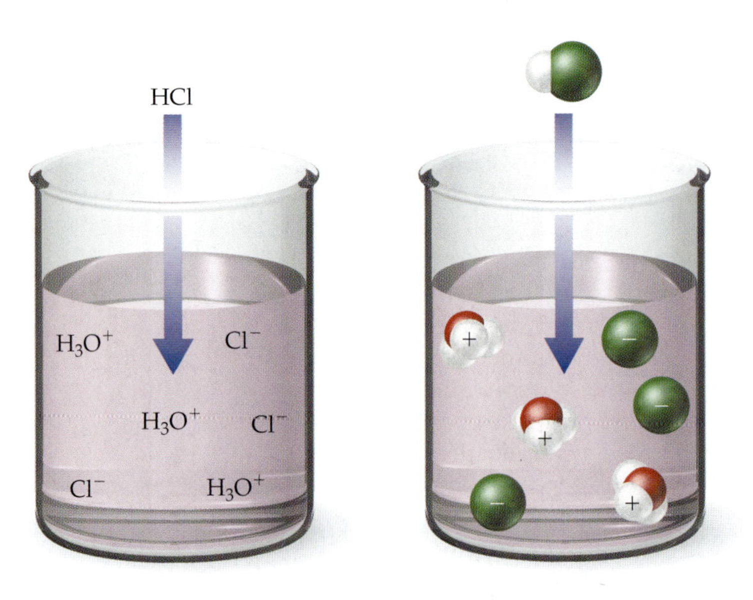

▲ **Figure 14.9 A strong acid** When HCl dissolves in water, it completely ionizes into H_3O^+ and Cl^- ions. The solution contains no intact HCl.

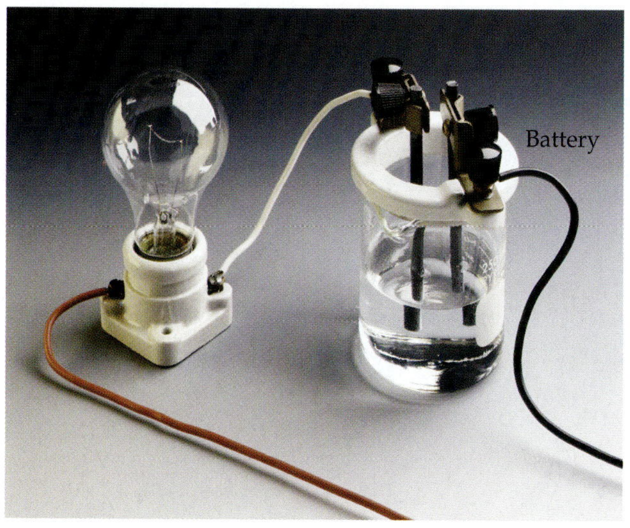

(a) Pure water

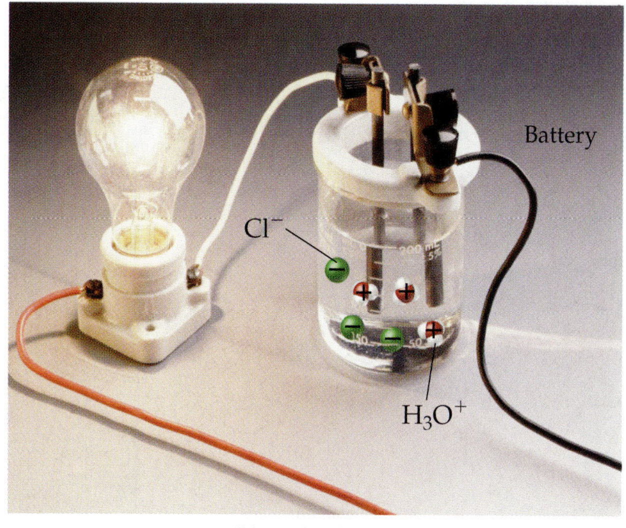

(b) HCl solution

▲ **Figure 14.10 Conductivity of a strong electrolyte solution** **(a)** Pure water will not conduct electricity. **(b)** The presence of ions in an HCl solution results in the conduction of electricity, causing the lightbulb to light. Solutions such as these are called strong electrolyte solutions.

Strong electrolyte solutions were first defined in Section 7.6.

A strong acid is an example of a **strong electrolyte**, a substance whose aqueous solutions are good conductors of electricity (▲ Figure 14.10). Aqueous solutions require the presence of charged particles to conduct electricity. Pure water is not a good conductor of electricity because it has relatively few charged particles. The danger of using electrical devices—such as a hair dryer—while sitting in the bathtub is that water is seldom pure and often contains dissolved ions. If the device were to come in contact with the water, dangerously high levels of electricity could flow through the water and through your body. Strong acid solutions are also strong electrolyte solutions because each acid molecule ionizes into positive and negative ions. These mobile ions are good conductors of electricity.

An ionizable proton is one that becomes an H^+ ion in solution.

Table 14.3 lists the six strong acids. The first five acids in Table 14.3 are **monoprotic acids**, acids containing only one ionizable proton. Sulfuric acid is an example of a **diprotic acid**, an acid that contains two ionizable protons.

TABLE 14.3 Strong Acids

hydrochloric acid (HCl)	nitric acid (HNO_3)
hydrobromic acid (HBr)	perchloric acid ($HClO_4$)
hydroiodic acid (HI)	sulfuric acid (H_2SO_4) *(diprotic)*

Weak Acids

It is a common mistake to confuse the terms *strong* and *weak* acids with the terms *concentrated* and *dilute* acids. Can you state the difference between these terms?

In contrast to HCl, HF is an example of a **weak acid**, one that does not completely ionize in solution.

Double arrow indicates partial ionization

$$HF(aq) + H_2O(l) \rightleftharpoons H_3O^+(aq) + F^-(aq)$$

To show that HF does not completely ionize in solution, the equation for its ionization has two opposing arrows, indicating that the reverse reaction occurs to some degree. An HF solution contains a lot of intact HF; it also

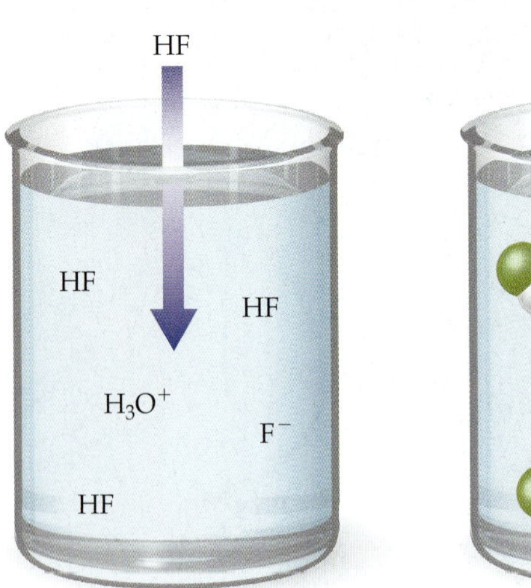

▶ **Figure 14.11 A weak acid** When HF dissolves in water, only a fraction of the dissolved molecules ionize into H_3O^+ and F^- ions. The solution contains many intact HF molecules.

Calculating exact $[H_3O^+]$ for weak acids is beyond the scope of this text.

TUTORIAL
Introduction to Aqueous Acids 

contains some $H_3O^+(aq)$ and $F^-(aq)$ (▲ Figure 14.11). In other words, a 1.0 M HF solution has $[H_3O^+] < 1.0$ M because only some of the HF molecules ionize to form H_3O^+.

A weak acid is an example of a **weak electrolyte**, a substance whose aqueous solutions are poor conductors of electricity (▼ Figure 14.12). Weak acid solutions contain few charged particles because only a small fraction of the acid molecules ionize into positive and negative ions.

The degree to which an acid is strong or weak depends on the attraction between the anion of the acid (the conjugate base) and the hydrogen ion. Suppose H*A* is a generic formula for an acid. Then, the degree to which the following reaction proceeds in the forward direction depends on the strength of the attraction between H^+ and A^-.

$$\underset{\text{Acid}}{HA(aq)} + H_2O(l) \rightleftharpoons H_3O^+(aq) + \underset{\text{Conjugate base}}{A^-(aq)}$$

If the attraction between H^+ and A^- is *weak*, then the reaction favors the forward direction and the acid is *strong* (▶ Figure 14.13). If the attraction

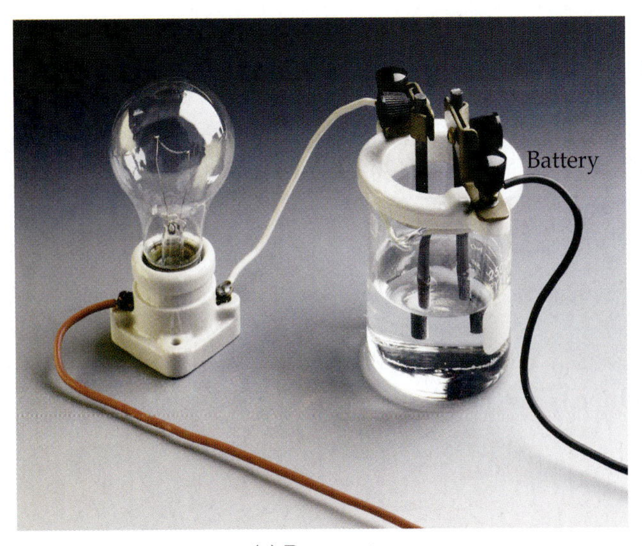

(a) Pure water

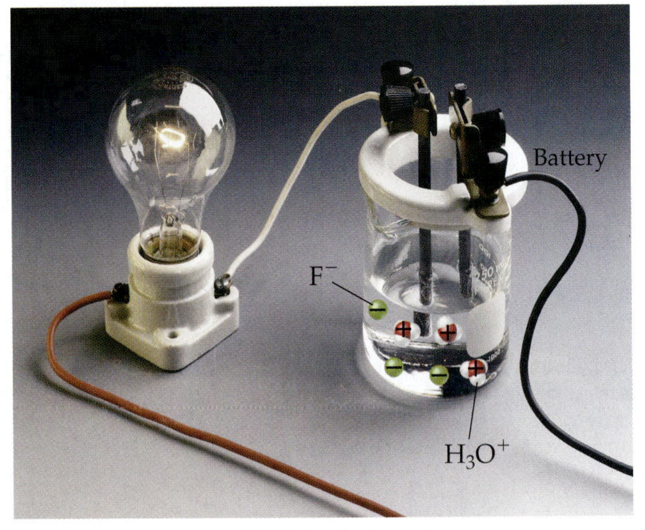

(b) HF solution

▲ **Figure 14.12 Conductivity of a weak electrolyte solution** **(a)** Pure water will not conduct electricity. **(b)** An HF solution contains some ions, but most of the HF is intact. The light glows only dimly. Solutions such as these are called weak electrolyte solutions.

Notice that the strength of a conjugate base is related to its attraction to H^+ in solution.

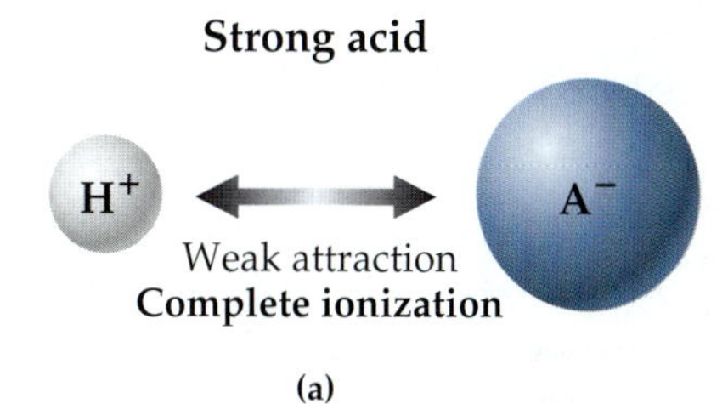

(a)

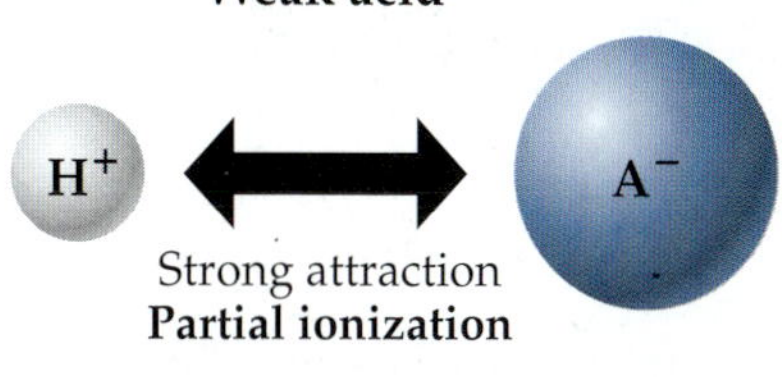

(b)

▲ **Figure 14.13 Strong and weak acids** **(a)** In a strong acid, the attraction between H^+ and A^- is low, resulting in complete ionization. **(b)** In a weak acid, the attraction between H^+ and A^- is high, resulting in partial ionization.

between H^+ and A^- is *strong*, then the reaction favors the reverse direction and the acid is *weak* (Figure 14.13b).

For example, in HCl, the conjugate base (Cl^-) has a relatively weak attraction to H^+, meaning that the reverse reaction does not occur to any significant extent. In HF, on the other hand, the conjugate base (F^-) has a greater attraction to H^+, meaning that the reverse reaction occurs to a significant degree. *In general, the stronger the acid, the weaker the conjugate base and vice versa.* This means that if the forward reaction (that of the acid) has a high tendency to occur, then the reverse reaction (that of the conjugate base) has a low tendency to occur. Table 14.4 lists some common weak acids.

TABLE 14.4 Some Weak Acids

hydrofluoric acid (HF)	sulfurous acid (H_2SO_3) *(diprotic)*
acetic acid ($HC_2H_3O_2$)	carbonic acid (H_2CO_3) *(diprotic)*
formic acid ($HCHO_2$)	phosphoric acid (H_3PO_4) *(triprotic)*

Notice that two of the weak acids in Table 14.4 are diprotic (meaning they have two ionizable protons), and one is triprotic (meaning that it has three ionizable protons). Let us return to sulfuric acid for a moment. Sulfuric acid is a diprotic acid that is strong in its first ionizable proton:

$$H_2SO_4(aq) + H_2O(l) \longrightarrow H_3O^+(aq) + HSO_4^-(aq)$$

but weak in its second ionizable proton.

$$HSO_4^-(aq) + H_2O(l) \rightleftharpoons H_3O^+(aq) + SO_4^{2-}(aq)$$

Sulfurous acid and carbonic acid are weak in both of their ionizable protons, and phosphoric acid is weak in all three of its ionizable protons.

EXAMPLE 14.5 Determining $[H_3O^+]$ in Acid Solutions

What is the H_3O^+ concentration in each of the following solutions?

(a) 1.5 M HCl
(b) 3.0 M $HC_2H_3O_2$
(c) 2.5 M HNO_3

Solution:

(a) Since HCl is a strong acid, it completely ionizes. The concentration of H_3O^+ will be 1.5 M.

$$[H_3O^+] = 1.5 \text{ M}$$

(b) Since $HC_2H_3O_2$ is a weak acid, it partially ionizes. The calculation of the exact concentration of H_3O^+ is beyond the scope of this text, but we know that it will be less than 3.0 M.

$$[H_3O^+] < 3.0 \text{ M}$$

(c) Since HNO_3 is a strong acid, it completely ionizes. The concentration of H_3O^+ will be 2.5 M.

$$[H_3O^+] = 2.5 \text{ M}$$

SKILLBUILDER 14.5 Determining $[H_3O^+]$ in Acid Solutions

What is the H_3O^+ concentration in each of the following solutions?

(a) 0.50 M $HCHO_2$
(b) 1.25 M HI
(c) 0.75 M HF

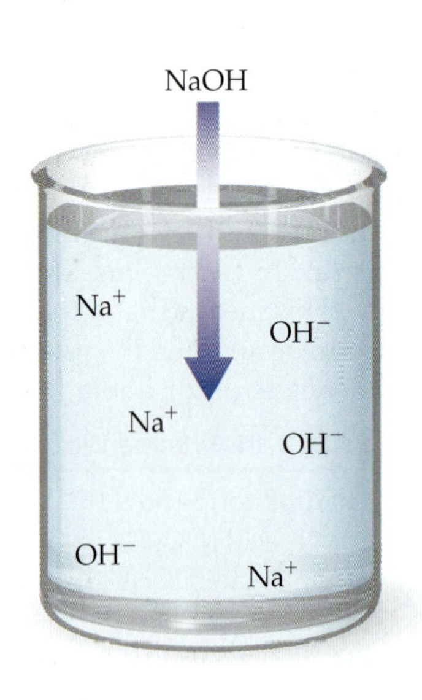

▶ **Figure 14.14 A strong base** When NaOH dissolves in water, it completely dissociates into Na^+ and OH^-. **Question:** The solution contains no intact NaOH. Would NaOH be a strong or weak electrolyte?

Strong Bases

In analogy to the definition of a strong acid, a **strong base** is a base that completely dissociates in solution. NaOH, for example, is a strong base.

$$NaOH(aq) \longrightarrow Na^+(aq) + OH^-(aq)$$

An NaOH solution contains no intact NaOH—it has all dissociated to form $Na^+(aq)$ and $OH^-(aq)$ (▲ Figure 14.14). In other words, a 1.0 M NaOH solution will have $[OH^-] = 1.0\ M$ and $[Na^+] = 1.0\ M$. Some common strong bases are listed in Table 14.5.

TABLE 14.5 Strong Bases

lithium hydroxide (LiOH)	strontium hydroxide ($Sr(OH)_2$)
sodium hydroxide (NaOH)	calcium hydroxide ($Ca(OH)_2$)
potassium hydroxide (KOH)	barium hydroxide ($Ba(OH)_2$)

Unlike diprotic acids, which ionize in two steps, bases containing 2 OH^- ions dissociate in one step.

Some of these strong bases, such as $Sr(OH)_2$, contain two OH^- ions. These bases completely dissociate, producing two moles of OH^- per mole of base. For example, $Sr(OH)_2$ dissociates as follows:

$$Sr(OH)_2(aq) \longrightarrow Sr^{2+}(aq) + 2\ OH^-(aq)$$

Weak Bases

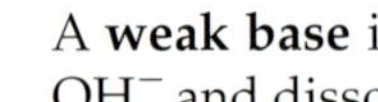
TUTORIAL Introduction to Aqueous Bases

A **weak base** is analogous to a weak acid. Unlike strong bases that contain OH^- and dissociate in water, the most common weak bases produce OH^- by accepting a proton from water, ionizing water to form OH^-.

$$B(aq) + H_2O(l) \rightleftharpoons BH^+(aq) + OH^-(aq)$$

In this equation, *B* is simply generic for a weak base. Ammonia, for example, ionizes water according to the following reaction.

$$NH_3(aq) + H_2O(l) \rightleftharpoons NH_4^+(aq) + OH^-(aq)$$

Calculating exact $[OH^-]$ for weak bases is beyond the scope of this text.

The double arrow indicates that the ionization is not complete. An NH_3 solution contains NH_3, NH_4^+, and OH^- (▶ Figure 14.15). A 1.0 M NH_3 solution will have $[OH^-] < 1.0\ M$. Table 14.6 lists some common weak bases.

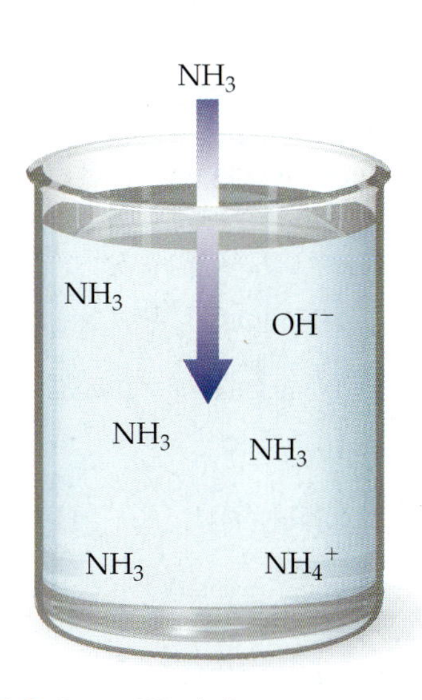

▶ **Figure 14.15 A weak base** When NH_3 dissolves in water, it partially ionizes to form NH_4^+ and OH^-. However, only a fraction of the molecules ionize. Most NH_3 molecules remain as NH_3. **Question:** Would NH_3 be a strong or weak electrolyte?

MOLECULES
Pyridine
Ethylamine

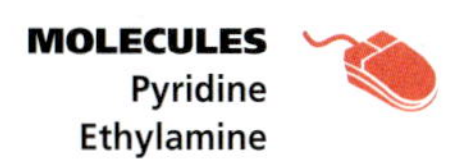

TABLE 14.6 Some Weak Bases

Base	Ionization Reaction
ammonia (NH_3)	$NH_3(aq) + H_2O(l) \rightleftharpoons NH_4^+(aq) + OH^-(aq)$
pyridine (C_5H_5N)	$C_5H_5N(aq) + H_2O(l) \rightleftharpoons C_5H_5NH^+(aq) + OH^-(aq)$
methylamine (CH_3NH_2)	$CH_3NH_2(aq) + H_2O(l) \rightleftharpoons CH_3NH_3^+(aq) + OH^-(aq)$
ethylamine ($C_2H_5NH_2$)	$C_2H_5NH_2(aq) + H_2O(l) \rightleftharpoons C_2H_5NH_3^+(aq) + OH^-(aq)$
bicarbonate ion (HCO_3^-)*	$HCO_3^-(aq) + H_2O(l) \rightleftharpoons H_2CO_3(aq) + OH^-(aq)$

*The bicarbonate ion must occur with a positively charged ion such as Na^+ that serves to balance the charge, but does not have any part in the ionization reaction. It is the bicarbonate ion that makes sodium bicarbonate ($NaHCO_3$) basic.

LIVE EXAMPLE

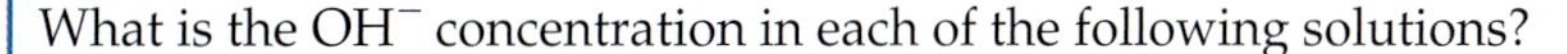

EXAMPLE 14.6 Determining [OH⁻] in Base Solutions

What is the OH^- concentration in each of the following solutions?

(a) 2.25 M KOH
(b) 0.35 M CH_3NH_2
(c) 0.025 M $Sr(OH)_2$

Solution:

(a) Since KOH is a strong base, it completely dissociates into K^+ and OH^- in solution. The concentration of OH^- will be 2.25 M.

$$[OH^-] = 2.25 \text{ M}$$

(b) Since CH_3NH_2 is a weak base, it only partially ionizes water. We cannot calculate the exact concentration of OH^-, but we know it will be less than 0.35 M.

$$[OH^-] < 0.35 \text{ M}$$

(c) Since $Sr(OH)_2$ is a strong base, it completely dissociates into $Sr^{2+}(aq)$ and $2\,OH^-(aq)$. $Sr(OH)_2$ forms 2 mol of OH^- for every 1 mol of $Sr(OH)_2$. Consequently, the concentration of OH^- will be twice the concentration of $Sr(OH)_2$.

$$[OH^-] = 2(0.025 \text{ M}) = 0.050 \text{ M}$$

SKILLBUILDER 14.6 Determining [OH⁻] in Base Solutions

What is the OH^- concentration in each of the following solutions?

(a) 0.055 M $Ba(OH)_2$
(b) 1.05 M C_5H_5N
(c) 0.45 M NaOH

14.8 Water: Acid and Base in One

We saw earlier that water acts as a base when it reacts with HCl and as an acid when it reacts with NH_3.

Water acting as a base

$$\underset{\substack{\text{Acid}\\\text{(Proton donor)}}}{HCl(aq)} + \underset{\substack{\text{Base}\\\text{(Proton acceptor)}}}{H_2O(l)} \longrightarrow H_3O^+(aq) + Cl^-(aq)$$

Water acting as an acid

$$\underset{\substack{\text{Base}\\\text{(Proton acceptor)}}}{NH_3(aq)} + \underset{\substack{\text{Acid}\\\text{(Proton donor)}}}{H_2O(l)} \rightleftharpoons NH_4^+(aq) + OH^-(aq)$$

Water is *amphoteric*; it can act as either an acid or a base. Even in pure water, water acts as an acid and a base with itself, a process called self-ionization.

Water acting as both an acid and a base

$$\underset{\substack{\text{Acid}\\\text{(Proton donor)}}}{H_2O(l)} + \underset{\substack{\text{Base}\\\text{(Proton acceptor)}}}{H_2O(l)} \rightleftharpoons H_3O^+(aq) + OH^-(aq)$$

In pure water, at 25°C, the preceding reaction occurs only to a very small extent, resulting in equal and small concentrations of H_3O^+ and OH^-.

$$[H_3O^+] = [OH^-] = 1.0 \times 10^{-7}\text{ M} \qquad \text{(in pure water at 25 °C)}$$
where $[H_3O^+]$ = the concentration of H_3O^+ in M
and $[OH^-]$ = the concentration of OH^- in M

So all samples of water contain some hydronium and hydroxide ions. The *product* of the concentration of these two ions in aqueous solutions is called the **ion product constant for water (K_w).**

$$K_w = [H_3O^+][OH^-]$$

We can find the value of K_w by simply multiplying the hydronium and hydroxide concentrations for pure water listed earlier.

The units of K_w are normally dropped.

$$\begin{aligned} K_w &= [H_3O^+][OH^-] \\ &= [1.0 \times 10^{-7}][1.0 \times 10^{-7}] \\ &= 1.0 \times 10^{-14} \end{aligned}$$

The preceding equation holds true for all aqueous solutions at room temperature. The concentration of H_3O^+ times the concentration of OH^- will always be 1.0×10^{-14} at room temperature. In pure water, since H_2O is the only source of these ions, there is one H_3O^+ ion for every OH^- ion. Consequently, the concentrations of H_3O^+ and OH^- are equal. Such a solution is a **neutral solution**.

In a neutral solution, $[H_3O^+] = [OH^-]$.

$$[H_3O^+] = [OH^-] = \sqrt{K_w} = 1.0 \times 10^{-7}\text{ M} \qquad \text{(in pure water)}$$

An **acidic solution** contains an acid that creates additional H_3O^+ ions, causing $[H_3O^+]$ to increase. However, the *ion product constant still applies.*

$$[H_3O^+][OH^-] = K_w = 1.0 \times 10^{-14}$$

If $[H_3O^+]$ increases, then $[OH^-]$ must decrease for the ion product to remain 1.0×10^{-14}. For example, suppose $[H_3O^+] = 1.0 \times 10^{-3}$ M; then $[OH^-]$ can be found by solving the ion product expression for $[OH^-]$.

$$[1.0 \times 10^{-3}][OH^-] = 1.0 \times 10^{-14}$$

$$[OH^-] = \frac{1.0 \times 10^{-14}}{1.0 \times 10^{-3}} = 1.0 \times 10^{-11}\ \text{M}$$

In an acidic solution, $[H_3O^+] > [OH^-]$.

In an acidic solution, $[H_3O^+]$ is greater than 1.0×10^{-7} M and $[OH^-]$ is less than 1.0×10^{-7} M.

A **basic solution** contains a base that creates additional OH^- ions, causing the $[OH^-]$ to increase and the $[H_3O^+]$ to decrease. For example, suppose $[OH^-] = 1.0 \times 10^{-2}$ M; then $[H_3O^+]$ can be found by solving the ion product expression for $[H_3O^+]$.

$$[H_3O^+][1.0 \times 10^{-2}] = 1.0 \times 10^{-14}$$

$$[H_3O^+] = \frac{1.0 \times 10^{-14}}{1.0 \times 10^{-2}} = 1.0 \times 10^{-12}\ \text{M}$$

In a basic solution, $[H_3O^+] < [OH^-]$.

In a basic solution, $[OH^-]$ is greater than 1.0×10^{-7} M and $[H_3O^+]$ is less than 1.0×10^{-7} M.

To summarize (see also ▼ Figure 14.16):

- In a *neutral* solution, $[H_3O^+] = [OH^-] = 1.0 \times 10^{-7}$ M
- In an *acidic* solution, $[H_3O^+] > 1.0 \times 10^{-7}$ M $[OH^-] < 1.0 \times 10^{-7}$ M
- In a *basic* solution, $[H_3O^+] < 1.0 \times 10^{-7}$ $[OH^-] > 1.0 \times 10^{-7}$ M
- In *all aqueous* solutions at 25 °C, $[H_3O^+][OH^-] = K_w = 1.0 \times 10^{-14}$

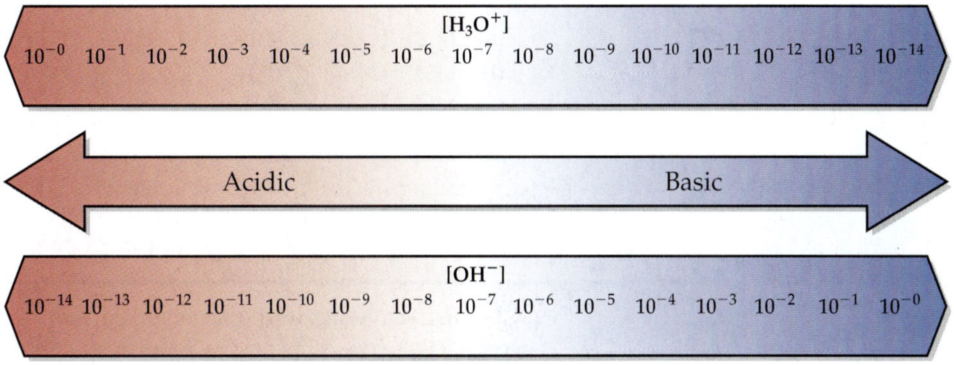

$[H_3O^+]$
10^{-0} 10^{-1} 10^{-2} 10^{-3} 10^{-4} 10^{-5} 10^{-6} 10^{-7} 10^{-8} 10^{-9} 10^{-10} 10^{-11} 10^{-12} 10^{-13} 10^{-14}
Acidic
Basic
$[OH^-]$
10^{-14} 10^{-13} 10^{-12} 10^{-11} 10^{-10} 10^{-9} 10^{-8} 10^{-7} 10^{-6} 10^{-5} 10^{-4} 10^{-3} 10^{-2} 10^{-1} 10^{-0}

▲ **Figure 14.16** **Acidic and basic solutions**

EXAMPLE 14.7 Using K_w in Calculations

Calculate $[OH^-]$ in each of the following solutions and determine whether the solution is acidic, basic, or neutral.

(a) $[H_3O^+] = 7.5 \times 10^{-5}$ M
(b) $[H_3O^+] = 1.5 \times 10^{-9}$ M
(c) $[H_3O^+] = 1.0 \times 10^{-7}$ M

To find $[OH^-]$ use the ion product constant, K_w.

Substitute the given value for $[H_3O^+]$ and solve the equation for $[OH^-]$. Since $[H_3O^+] > 1.0 \times 10^{-7}$ M and $[OH^-] < 1.0 \times 10^{-7}$ M, the solution is acidic.

Solution:

(a) $[H_3O^+][OH^-] = K_w = 1.0 \times 10^{-14}$

$[7.5 \times 10^{-5}][OH^-] = K_w = 1.0 \times 10^{-14}$

$$[OH^-] = \frac{1.0 \times 10^{-14}}{7.5 \times 10^{-5}} = 1.3 \times 10^{-10} \text{ M}$$

acidic solution

Substitute the given value for $[H_3O^+]$ into the ion product constant equation and solve the equation for $[OH^-]$. Since $[H_3O^+] < 1.0 \times 10^{-7}$ M and $[OH^-] > 1.0 \times 10^{-7}$ M, the solution is basic.

(b) $[H_3O^+][OH^-] = K_w = 1.0 \times 10^{-14}$

$[1.5 \times 10^{-9}][OH^-] = 1.0 \times 10^{-14}$

$$[OH^-] = \frac{1.0 \times 10^{-14}}{1.5 \times 10^{-9}} = 6.7 \times 10^{-6} \text{ M}$$

basic solution

Substitute the given value for $[H_3O^+]$ into the ion product constant equation and solve the equation for $[OH^-]$. Since $[H_3O^+] = 1.0 \times 10^{-7}$ M and $[OH^-] = 1.0 \times 10^{-7}$ M, the solution is neutral.

(c) $[H_3O^+][OH^-] = K_w = 1.0 \times 10^{-14}$

$[1.0 \times 10^{-7}][OH^-] = 1.0 \times 10^{-14}$

$$[OH^-] = \frac{1.0 \times 10^{-14}}{1.0 \times 10^{-7}} = 1.0 \times 10^{-7} \text{ M}$$

neutral solution

SKILLBUILDER 14.7 Using K_w in Calculations

Calculate $[H_3O^+]$ in each of the following solutions and determine whether the solution is acidic, basic, or neutral.

(a) $[OH^-] = 1.5 \times 10^{-2}$ M
(b) $[OH^-] = 1.0 \times 10^{-7}$ M
(c) $[OH^-] = 8.2 \times 10^{-10}$ M

CONCEPTUAL CHECKPOINT 14.1

Which of the following would be least likely to act as a base?

(a) H_2O
(b) OH^-
(c) NH_3
(d) NH_4^+

14.9 The pH Scale: A Way to Express Acidity and Basicity

Chemists have devised a scale based on the hydrogen ion concentration to compactly express the acidity or basicity of any solution. The scale is called the **pH** scale and has the following general characteristics:

- pH < 7 *acidic* solution
- pH > 7 *basic* solution
- pH = 7 *neutral* solution

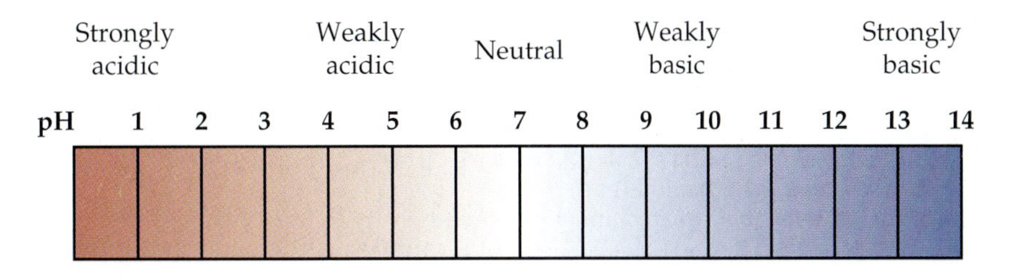

Table 14.7 lists the pH of some common substances. Notice that, as we discussed in Section 14.3, many foods, especially fruits, are acidic and therefore have low pH values. The foods with the lowest pH values are limes and lemons, and they are among the sourest. Relatively few foods, however, are basic.

TABLE 14.7 The pH of Some Common Substances

Substance	pH
gastric (human stomach) acid	1.0–3.0
limes	1.8–2.0
lemons	2.2–2.4
soft drinks	2.0–4.0
plums	2.8–3.0
wine	2.8–3.8
apples	2.9–3.3
peaches	3.4–3.6
cherries	3.2–4.0
beer	4.0–5.0
rainwater (unpolluted)	5.6
human blood	7.3–7.4
egg whites	7.6–8.0
milk of magnesia	10.5
household ammonia	10.5–11.5
4% NaOH solution	14

The pH scale is a **logarithmic scale**; therefore a change of 1 pH unit corresponds to a tenfold change in H_3O^+ concentration. For example, a lime with a pH of 2.0 is 10 times more acidic than a plum with a pH of 3.0 and 100 times more acidic than a cherry with a pH of 4.0. Each change of 1 in pH scale corresponds to a change of 10 in $[H_3O^+]$ (▼ Figure 14.17).

Notice that an *increase* of 1 in pH corresponds to a tenfold *decrease* in $[H_3O^+]$.

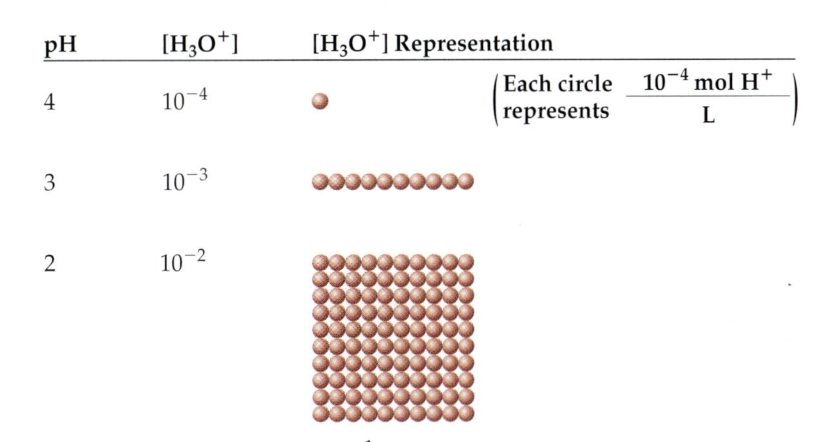

▲ **Figure 14.17 The pH scale is a logarithmic scale** A *decrease* of 1 unit on the pH scale corresponds to an *increase* in H_3O^+ concentration by a factor of 10. Each circle stands for 10^{-4} mol H^+/L, or 6.022×10^{19} H^+ ions per liter. **Question:** How much of an increase in H_3O^+ concentration corresponds to a decrease of 2 pH units?

Calculating pH from $[H_3O^+]$

The pH of a solution is defined as follows:

Note that pH is defined using the log function (base ten), which is different from the natural log (abbreviated ln).

$$pH = -\log[H_3O^+]$$

To calculate pH, you must be able to compute logarithms. Recall that the log of a number is the exponent to which 10 must be raised to obtain that number, as shown in the following examples.

$$\log 10^1 = 1; \log 10^2 = 2; \log 10^3 = 3$$
$$\log 10^{-1} = -1; \log 10^{-2} = -2; \log 10^{-3} = -3$$

In the following example, we need to compute the log of 1.5×10^{-7}. A solution having an $[H_3O^+] = 1.5 \times 10^{-7}$ M (acidic) has a pH of:

When you take the log of a quantity, the result should have the same number of decimal places as the number of significant figures in the original quantity.

$$\begin{aligned} pH &= -\log[H_3O^+] \\ &= -\log(1.5 \times 10^{-7}) \\ &= -(-6.82) \\ &= 6.82 \end{aligned}$$

Notice that the pH is reported to two decimal places here. This is because only the numbers to the right of the decimal place are significant in a log. Since our original value for the concentration had two significant figures, the log of that number has two decimal places.

2 decimal places

$$\log 1.0 \times 10^{-3} = 3.00$$

If the original number had three significant digits, the log would be reported to three decimal places:

3 decimal places

$$-\log 1.00 \times 10^{-3} = 3.000$$

TUTORIAL
pH Prediction

A solution having $[H_3O^+] = 1.0 \times 10^{-7}$ M (neutral) has a pH of:

$$\begin{aligned} pH &= -\log[H_3O^+] \\ &= -\log 1.0 \times 10^{-7} \\ &= -(-7.00) \\ &= 7.00 \end{aligned}$$

EXAMPLE 14.8 Calculating pH from $[H_3O^+]$

Calculate the pH of each of the following solutions and indicate whether the solution is acidic or basic.

(a) $[H_3O^+] = 1.8 \times 10^{-4}$ M
(b) $[H_3O^+] = 7.2 \times 10^{-9}$ M

Solution:
To calculate pH, simply substitute the given $[H_3O^+]$ into the pH equation.

(a)

$$\begin{aligned} pH &= -\log[H_3O^+] \\ &= -\log 1.8 \times 10^{-4} \\ &= -(-3.74) \\ &= 3.74 \end{aligned}$$

Since the pH < 7, this solution is acidic.

(b)

$$\begin{aligned} pH &= -\log[H_3O^+] \\ &= -\log(7.2 \times 10^{-9}) \\ &= -(-8.14) \\ &= 8.14 \end{aligned}$$

Since the pH > 7, this solution is basic.

SKILLBUILDER 14.8 Calculating pH from $[H_3O^+]$

Calculate the pH of each of the following solutions and indicate whether the solution is acidic or basic.

(a) $[H_3O^+] = 9.5 \times 10^{-9}$ M
(b) $[H_3O^+] = 6.1 \times 10^{-3}$ M

SKILLBUILDER PLUS

Calculate the pH of a solution with $[OH^-] = 1.3 \times 10^{-2}$ M and indicate whether the solution is acidic or basic. Hint: Begin by using K_w to find $[H_3O^+]$.

Calculating $[H_3O^+]$ from pH

The inverse log is sometimes called the *antilog*.

Ten raised to the log of a number is equal to that number: $10^{\log x} = x$.

The invlog function "undoes" log: invlog(log x) = x.

To calculate $[H_3O^+]$ from a pH value, you must *undo* the log. The log can be undone using the inverse log function *(Method 1)* on most calculators or it can be undone using the 10^x key *(Method 2)*. Both of these do the same thing; which one you use depends on your calculator.

Method 1: Inverse Log Function	Method 2: 10^x Function
$pH = -\log[H_3O^+]$	$pH = -\log[H_3O^+]$
$-pH = \log[H_3O^+]$	$-pH = \log[H_3O^+]$
$\text{invlog}(-pH) = \text{invlog}(\log[H_3O^+])$	$10^{-pH} = 10^{\log[H_3O^+]}$
$\text{invlog}(-pH) = [H_3O^+]$	$10^{-pH} = [H_3O^+]$

So, to calculate $[H_3O^+]$ from a pH value, simply take the inverse log of the negative of the pH value *(Method 1)* or raise 10 to the negative of the pH value *(Method 2)*.

EXAMPLE 14.9 Calculating $[H_3O^+]$ from pH

Calculate the H_3O^+ concentration for a solution with a pH of 4.80.

Solution:

To find the $[H_3O^+]$ from pH, we must undo the log function. Use either Method 1 or Method 2.

The number of significant figures in the inverse log of a number is equal to the number of decimal places in the number.

Method 1: Inverse Log Function	Method 2: 10^x function
$pH = -\log[H_3O^+]$	$pH = -\log[H_3O^+]$
$4.80 = -\log[H_3O^+]$	$4.80 = -\log[H_3O^+]$
$-4.80 = \log[H_3O^+]$	$-4.80 = \log[H_3O^+]$
$\text{invlog}(-4.80) = \text{invlog}(\log[H_3O^+])$	$10^{-4.80} = 10^{\log[H_3O^+]}$
$\text{invlog}(-4.80) = [H_3O^+]$	$10^{-4.80} = [H_3O^+]$
$[H_3O^+] = 1.6 \times 10^{-5}$ M	$[H_3O^+] = 1.6 \times 10^{-5}$ M

SKILLBUILDER 14.9 Calculating $[H_3O^+]$ from pH

Calculate the H_3O^+ concentration for a solution with a pH of 8.37.

SKILLBUILDER PLUS

Calculate the OH^- concentration for a solution with a pH of 3.66.

Chemistry AND HEALTH

Ulcers

An ulcer is a lesion that forms on the wall of the stomach or small intestine (▼ Figure 14.18). Under normal circumstances, a thick layer of mucus lines the stomach wall and protects it from the hydrochloric acid and other gastric juices in the stomach. When that layer of mucus is damaged, however, gastric juices come into direct contact with the stomach wall and begin to digest it, resulting in an ulcer. The main symptom of an ulcer is a burning or gnawing pain in the stomach.

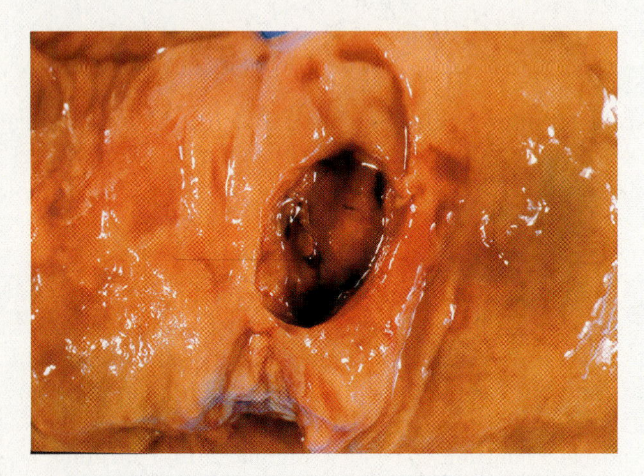

▲ **Figure 14.18 A stomach ulcer** An ulcer is a lesion on the stomach wall caused by the breakdown of the stomach's mucus lining. (A cold sore is a similar lesion.)

Acidic drugs, such as aspirin, and acidic foods, such as citrus fruits and pickling fluids, irritate ulcers. When consumed, these substances increase the acidity of the stomach, worsening the irritation to the stomach wall. On the other hand, antacids—which contain bases—relieve ulcers. Common antacids include Tums and milk of magnesia.

Until relatively recently, the cause of ulcers was unclear. For many years, a stressful lifestyle and a rich diet were blamed. However, it is now accepted that in most cases, a bacterial infection of the stomach lining is largely responsible. Long-term use of some over-the-counter pain relievers, such as aspirin, is also believed to cause ulcers.

CAN YOU ANSWER THIS? *Which of the following are likely to irritate an ulcer? Soothe an ulcer?*

Food	*pH*
limes	1.8–2.0
vinegar	2.0
wine	2.8–3.8
apples	2.9–3.3
egg whites	7.6–8.0
milk of magnesia	10.5

CONCEPTUAL CHECKPOINT 14.2

Solution A has a pH of 13. Solution B has a pH of 10. The concentration of H_3O^+ in solution B is _____ times that in solution A.

(a) 0.001
(b) $\frac{1}{3}$
(c) 3
(d) 1000

14.10 Buffers: Solutions That Resist pH Change

Most solutions will rapidly become more acidic (lower pH) upon addition of an acid or more basic (higher pH) upon addition of a base. A **buffer**, however, resists pH change by neutralizing added acid or added base. Human blood, for example, is a buffer. Any acid or base added to blood is neutralized by components within blood, resulting in a nearly constant pH. In healthy individuals, blood pH is between 7.36 and 7.40. If blood pH were to drop below 7.0 or increase beyond 7.8, death would result.

How does blood maintain such a narrow pH range? Like all buffers, blood contains *significant* amounts of *both a weak acid and its conjugate base*. When additional base is added to blood, the weak acid reacts with the base, neutralizing it. When additional acid is added to blood, the conjugate base reacts with the acid, neutralizing it. In this way, blood maintains a constant pH.

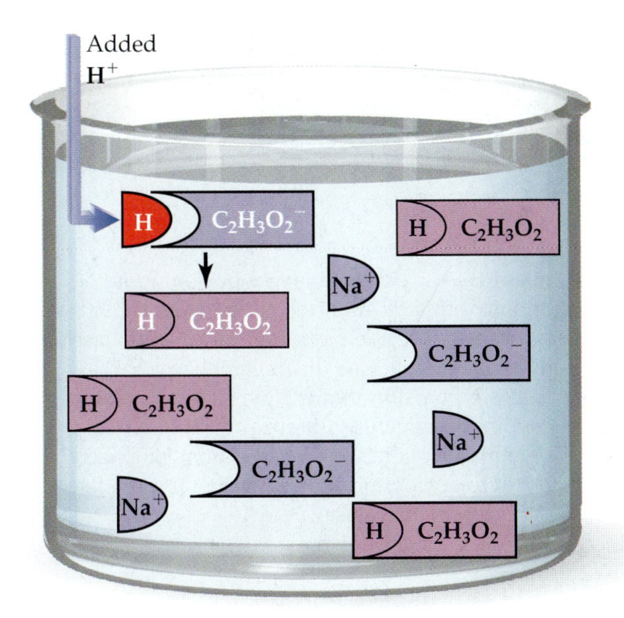

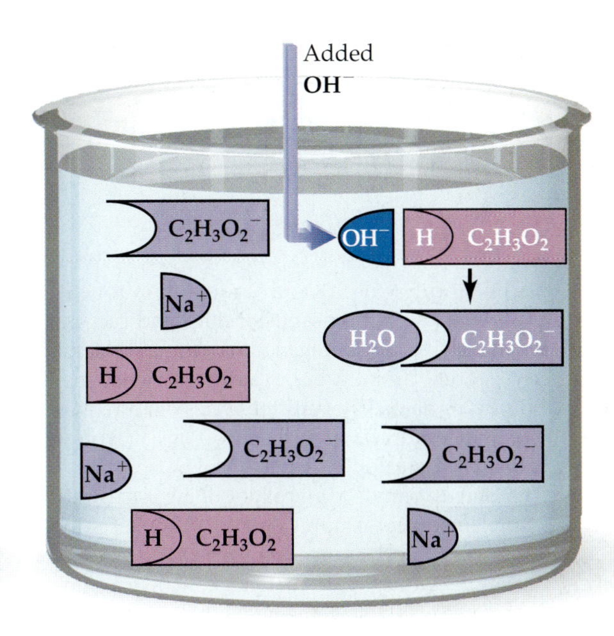

▲ **Figure 14.19 Buffers** A buffer contains significant amounts of a weak acid and its conjugate base. The acid consumes any added base, and the base consumes any added acid. In this way, a buffer resists pH change.

A simple buffer can be made by mixing both acetic acid ($HC_2H_3O_2$) and its conjugate base, sodium acetate ($NaC_2H_3O_2$), in water (▲ Figure 14.19). (The sodium in sodium acetate is just a spectator ion and does not contribute to buffering action.) Since $HC_2H_3O_2$ is a weak acid and since $C_2H_3O_2^-$ is its conjugate base, a solution containing both of these is a buffer. Note that a weak acid by itself, even though it partially ionizes to form some of its conjugate base, does not contain sufficient base to be a buffer. A buffer must contain *significant* amounts of *both* a weak acid and its conjugate base. Suppose that additional base, in the form of NaOH, were added to the buffer solution containing acetic acid and sodium acetate. The acetic acid would neutralize the base according to the following reaction.

$$NaOH(aq) + HC_2H_3O_2(aq) \longrightarrow H_2O(l) + NaC_2H_3O_2(aq)$$

As long as the amount of NaOH added was less than the amount of $HC_2H_3O_2$ in solution, the solution would neutralize the NaOH, and the resulting pH change would be small. Suppose, on the other hand, that additional acid, in the form of HCl, were added to the solution. Then the conjugate base, $NaC_2H_3O_2$, would neutralize the added HCl according to the following reaction.

$$HCl(aq) + NaC_2H_3O_2(aq) \longrightarrow HC_2H_3O_2(aq) + NaCl(aq)$$

As long as the amount of HCl added was less than the amount of $NaC_2H_3O_2$ in solution, the solution would neutralize the HCl and the resulting pH change would be small.

To summarize:

- Buffers resist pH change.
- Buffers contain significant amounts of both a weak acid and its conjugate base.
- The weak acid neutralizes added base.
- The conjugate base neutralizes added acid.

CONCEPTUAL CHECKPOINT 14.3

Which of the following would constitute a buffer solution?

(a) $H_2SO_4(aq)$ and $H_2SO_3(aq)$
(b) $HF(aq)$ and $NaF(aq)$
(c) $HCl(aq)$ and $NaCl(aq)$
(d) $NaCl(aq)$ and $NaOH(aq)$

Chemistry AND HEALTH

The Danger of Antifreeze

Most types of antifreeze used in cars are solutions of ethylene glycol. Every year, thousands of dogs and cats die from ethylene glycol poisoning because they consume improperly stored antifreeze or antifreeze that has leaked out of a radiator. The antifreeze has a somewhat sweet taste, which attracts a curious dog or cat. Young children are also at risk for ethylene glycol poisoning.

The first stage of ethylene glycol poisoning is a drunken state. Ethylene glycol is an alcohol, and it affects the brain of a dog or cat much as an alcoholic beverage would. Once ethylene glycol begins to metabolize, however, the second and more deadly stage begins. Ethylene glycol is metabolized in the liver into glycolic acid ($HC_2H_3O_3$), which enters the bloodstream. If the original quantities of consumed antifreeze are significant, the glycolic acid overwhelms the blood's natural buffering system, causing blood pH to drop to dangerously low levels. At this point, the cat or dog may begin hyperventilating in an effort to overcome the acidic blood's reduced ability to carry oxygen. If no treatment is administered, the animal will eventually go into a coma and die.

One treatment for ethylene glycol poisoning is the administration of ethyl alcohol (the alcohol found in alcoholic beverages). The liver enzyme that metabolizes ethylene glycol is the same one that metabolizes ethyl alcohol, but it has higher affinity for ethyl alcohol than for ethylene glycol. Consequently, the enzyme preferentially metabolizes ethyl alcohol, allowing the unmetabolized ethylene glycol to escape through the urine. If administered early, this treatment can save the life of a dog or cat that has consumed ethylene glycol.

CAN YOU ANSWER THIS? *One of the main buffering systems found in blood consists of carbonic acid (H_2CO_3) and bicarbonate ion (HCO_3^-). Write an equation showing how this buffering system would neutralize glycolic acid ($HC_2H_3O_3$) that might enter the blood from ethylene glycol poisoning. Suppose a cat has 0.15 mol of HCO_3^- and 0.15 mol of H_2CO_3 in its bloodstream. How many grams of $HC_2H_3O_3$ could be neutralized before the buffering system in the blood is overwhelmed?*

14.11 Acid Rain: An Environmental Problem Related to Fossil Fuel Combustion

About 90% of U.S. energy comes from fossil fuel combustion. Fossil fuels include petroleum, natural gas, and coal. Some fossil fuels, especially coal, contain small amounts of sulfur impurities. During combustion, these impurities react with oxygen to form SO_2. In addition, during combustion of any fossil fuel, nitrogen from the air reacts with oxygen to form NO_2. The SO_2 and NO_2 emitted from fossil fuel combustion then react with water in the atmosphere to form sulfuric acid and nitric acid.

These equations represent simplified versions of the reactions that actually occur.

$$2\,SO_2 + O_2 + 2\,H_2O \longrightarrow 2\,H_2SO_4$$
$$4\,NO_2 + O_2 + 2\,H_2O \longrightarrow 4\,HNO_3$$

These acids combine with rain to form **acid rain**. The problem is greatest in the northeastern portion of the United States because many midwestern power plants burn coal. The sulfur and nitrogen oxides produced from coal combustion in the Midwest are carried toward the Northeast by natural air currents, making rain in that portion of the country significantly acidic.

Rain is naturally somewhat acidic because of atmospheric carbon dioxide. Carbon dioxide combines with rainwater to form carbonic acid.

TUTORIAL Carbon Dioxide Behaves as an Acid in Water

$$CO_2 + H_2O \longrightarrow H_2CO_3$$

However, carbonic acid is a relatively weak acid. Even rain that is saturated with CO_2 has a pH of only about 5.6, which is mildly acidic. However, when

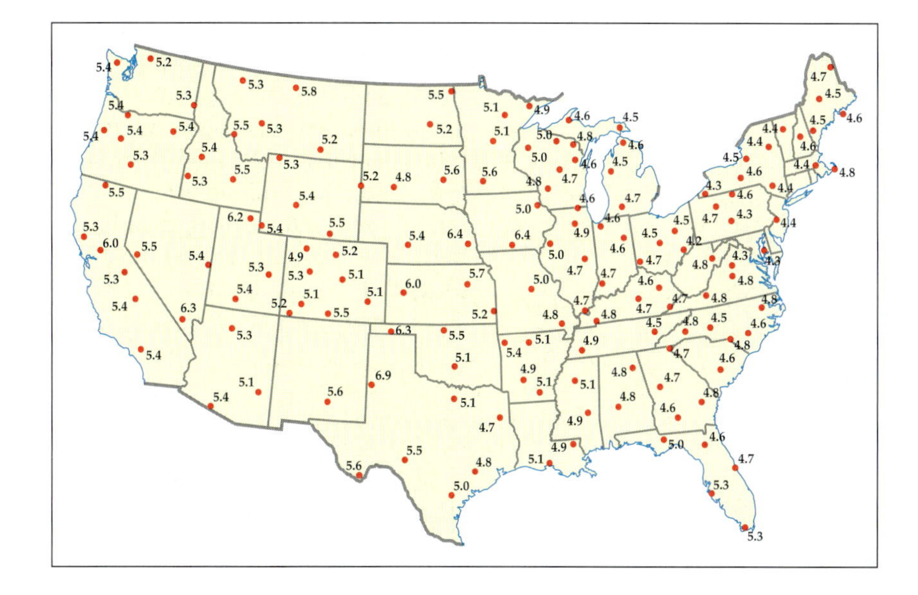

▶ **Figure 14.20 Acid Rain in the United States** Average pH of precipitation in the United States for the period December 25, 2000, to January 1, 2001.

nitric acid and sulfuric acid mix with rain, the pH of the rain can fall as low as 4.3 (▲ Figure 14.20). Remember that, because of the logarithmic nature of the pH scale, rain with a pH of 4.3 has an $[H_3O^+]$ 20 times that of rain with a pH of 5.6. Rain that is this acidic has negative consequences on the environment.

Acid Rain Damage

Since acids dissolve metals, acid rain damages metal structures. Bridges, railroads, and even automobiles can be damaged by acid rain. Since acids react with carbonates (CO_3^{2-}), acid rain also damages building materials that contain carbonates, including marble, cement, and limestone. Statues, buildings, and pathways in the Northeast show significant signs of acid rain damage (▼ Figure 14.21).

See Section 7.8 for the reaction between acids and carbonates.

Acid rain can also accumulate in lakes and rivers and affect aquatic life. In the northeastern United States, more than 2000 lakes and streams have increased acidity levels due to acid rain. Aquatic plants, frogs, salamanders, and some species of fish are sensitive to acid levels and cannot live in the acidified lakes. Trees can also be affected by acid rain because the acid removes nutrients from the soil, making it more difficult for trees to survive.

▶ **Figure 14.21 Acid rain damage** Many monuments and statues, such as this one of George Washington in New York's Washington Square Park, have suffered severe deterioration caused by acid rain. The photo at left was taken in 1935, the one at right some 60 years later. (The statue has recently undergone restoration.)

Acid Rain Legislation

Acid rain has been targeted by legislation. In 1990, the U.S. Congress passed amendments to the Clean Air Act that target acid rain. These amendments force electrical utilities—which are the most significant source of SO_2—to lower their SO_2 emissions gradually over time (▼ Figure 14.22). The acidity of rain in the Northeast has already stabilized and should decrease in the coming years. Scientists expect most lakes, streams, and forests to recover once the pH of the rain returns to normal levels.

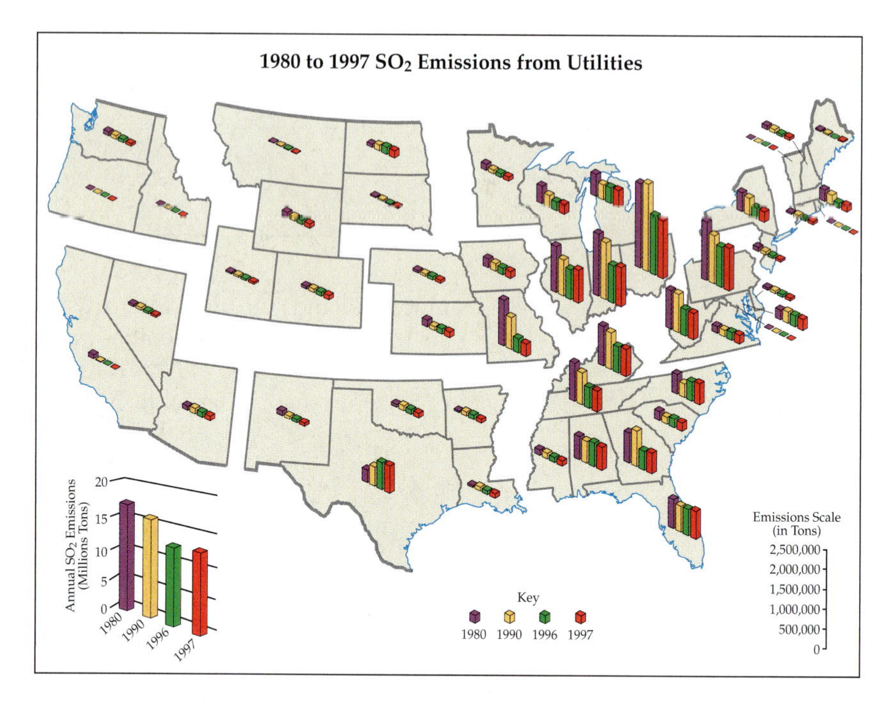

▲ Figure 14.22 **Emissions of SO_2 by utilities from 1980 to 1997** The height of each bar represents annual SO_2 emissions for that region in the year noted. Under the Clean Air Act and its amendments, SO_2 emissions are decreasing in most parts of the country.

CHAPTER IN REVIEW

Chemical Principles

Acid Properties:

- Acids have a sour taste.
- Acids dissolve many metals.
- Acids turn litmus paper red.

Base Properties:

- Bases have a bitter taste.
- Bases have a slippery feel.
- Bases turn litmus paper blue.

Relevance

Acid Properties: Acids are responsible for the sour taste in many foods such as lemons, limes, and vinegar. They also find widespread use in the laboratory and in industry.

Base Properties: Bases are less common in foods, but their presence in some foods—such as coffee and beer—is enjoyed by many as an acquired taste. Bases also have widespread use in the laboratory and in industry.

Molecular Definitions of Acids and Bases:

Arrhenius definition

Acid—substance that produces H^+ ions in solution
Base—substance that produces OH^- ions in solution

Brønsted–Lowry definition

Acid—proton donor
Base—proton acceptor

Molecular Definitions of Acids and Bases: The Arrhenius definition is simpler and easier to use. It also shows how an acid and a base neutralize each other to form water ($H^+ + OH^- \longrightarrow H_2O$). The more generally applicable Brønsted–Lowry definition helps us see that in water, H^+ ions usually associate with water molecules to form H_3O^+. It also easily shows how bases that do not contain OH^- ions can still act as bases by accepting a proton from water.

Reactions of Acids and Bases:

Neutralization Reactions

In a neutralization reaction, an acid and a base react to form water and a salt.

$$\underset{\text{Acid}}{HCl(aq)} + \underset{\text{Base}}{KOH(aq)} \longrightarrow \underset{\text{Water}}{H_2O(l)} + \underset{\text{Salt}}{KCl(aq)}$$

Acid–Metal Reactions

Acids react with many metals to form hydrogen gas and a salt.

$$\underset{\text{Acid}}{2\,HCl(aq)} + \underset{\text{Metal}}{Mg(s)} \longrightarrow \underset{\text{Hydrogen gas}}{H_2(g)} + \underset{\text{Salt}}{MgCl_2(aq)}$$

Acid–Metal Oxide Reactions

Acids react with many metal oxides to form water and a salt.

$$\underset{\text{Acid}}{2\,HCl(aq)} + \underset{\text{Metal oxide}}{K_2O(s)} \longrightarrow \underset{\text{Water}}{H_2O(l)} + \underset{\text{Salt}}{2\,KCl(aq)}$$

Reactions of Acids and Bases: Neutralization reactions have many common uses. Antacids, for example, are bases that react with acids from the stomach to alleviate heartburn and sour stomach.

Acid–metal and acid–metal oxide reactions show the corrosive nature of acids. In both of these reactions, the acid dissolves the metal or the metal oxide. Some of the effects of these kinds of reactions can be seen in the damage to building materials caused by acid rain. Since acids dissolve metals and metal oxides, any building materials composed of these substances are damaged by acid rain.

Acid–Base Titration: In an acid–base titration, an acid (or base) of known concentration is added to a base (or acid) of unknown concentration. The two reactants are combined until they are in exact stoichiometric proportions (moles of H^+ = moles of OH^-), which marks the equivalence point of the titration. Since you know the moles of H^+ (or OH^-) that you added, you can determine the moles of OH^- (or H^+) in the unknown solution.

Acid–Base Titration: An acid–base titration is a laboratory procedure often used to determine the unknown concentration of an acid or base.

Strong and Weak Acids and Bases: Strong acids completely ionize and strong bases completely dissociate in aqueous solutions. For example:

$$HCl(aq) + H_2O(l) \longrightarrow H_3O^+(aq) + Cl^-(aq)$$
$$NaOH(aq) \longrightarrow Na^+(aq) + OH^-(aq)$$

A 1 M HCl solution has $[H_3O^+] = 1$ M and a 1 M NaOH solution has $[OH^-] = 1$ M.

Weak acids only partially ionize in solution. Most weak bases partially ionize water in solution. For example:

$$HF(aq) + H_2O(l) \rightleftharpoons H_3O^+(aq) + F^-(aq)$$
$$NH_3(aq) + H_2O(l) \rightleftharpoons NH_4^+(aq) + OH^-(aq)$$

A 1 M HF solution has $[H_3O^+] < 1$ M and a 1 M NH_3 solution has $[OH^-] < 1$ M.

Strong and Weak Acids and Bases: Whether an acid is strong or weak depends on the conjugate base: The stronger the conjugate base, the weaker the acid. Since the acidity or basicity of a solution depends on $[H_3O^+]$ and $[OH^-]$, we must know whether an acid is strong or weak to know the degree of acidity or basicity.

Self-Ionization of Water: Water can act as both an acid and a base with itself.

$$\underset{\text{Acid}}{H_2O(l)} + \underset{\text{Base}}{H_2O(l)} \rightleftharpoons H_3O^+(aq) + OH^-(aq)$$

The product of $[H_3O^+]$ and $[OH^-]$ in aqueous solutions will always be equal to the ion product constant, $K_w(10^{-14})$.

$$[H_3O^+][OH^-] = K_w = 1.0 \times 10^{-14}$$

Self-Ionization of Water: The self-ionization of water shows us that aqueous solutions always contain some H_3O^+ and some OH^-. In a neutral solution, the concentrations of these are equal (1.0×10^{-7} M). When an acid is added to water, $[H_3O^+]$ increases and $[OH^-]$ decreases. When a base is added to water, the opposite happens. The ion product constant, however, still equals 1.0×10^{-14}, allowing us to calculate $[H_3O^+]$ given $[OH^-]$ and vice versa.

pH Scale: $pH = -\log[H_3O^+]$

$pH > 7$ basic

$pH < 7$ acidic

$pH = 7$ neutral

pH Scale: pH is a convenient way to specify acidity or basicity. Since the pH scale is logarithmic, a change of one on the pH scale corresponds to a tenfold change in the $[H_3O^+]$.

Buffers: Buffers are solutions containing significant amounts of both a weak acid and its conjugate base. Buffers resist pH change by neutralizing added acid or base.

Buffers: Buffers are important in blood chemistry, for example. Blood must maintain a narrow pH range in order to carry oxygen.

Acid Rain: Acid rain is the result of sulfur oxides and nitrogen oxides emitted by fossil fuel combustion. These oxides react with water to form sulfuric acid and nitric acid, which then fall as acid rain.

Acid Rain: Since acids are corrosive, acid rain damages building materials. Since many aquatic plants and animals cannot survive in acidic water, acid rain also affects lakes and rivers, making them too acidic for the survival of some species.

Determining $[OH^-]$ in Base Solutions (Section 14.7)

In a strong base, $[OH^-]$ is equal to the concentration of the base times the number of hydroxide ions in the base. In a weak base, $[OH^-]$ is less than the concentration of the base.

EXAMPLE 14.15 Determining $[OH^-]$ in Base Solutions

What is the OH^- concentration in a 0.25 M NaOH solution, in a 0.25 M $Sr(OH)_2$ solution, and in a 0.25 M NH_3 solution?

Solution:

In the 0.25 M NaOH solution (strong base), $[OH^-] = 0.25$ M. In the 0.25 M $Sr(OH)_2$ solution (strong base), $[OH^-] = 0.50$ M. In the 0.25 M NH_3 solution (weak base), $[OH^-] < 0.25$ M.

Finding the Concentration of $[H_3O^+]$ or $[OH^-]$ from K_w (Section 14.8)

To find $[H_3O^+]$ or $[OH^-]$ use the ion product constant expression.

$$[H_3O^+][OH^-] = 1.0 \times 10^{-14}$$

Substitute the known quantity into the equation ($[H_3O^+]$ or $[OH^-]$) and solve for the unknown quantity.

EXAMPLE 14.16 Finding the Concentration of $[H_3O^+]$ or $[OH^-]$ from K_w

Calculate $[OH^-]$ in a solution with $[H_3O^+] = 1.5 \times 10^{-4}$ M.

Solution:

$$[H_3O^+][OH^-] = 1.0 \times 10^{-14}$$

$$[1.5 \times 10^{-4}][OH^-] = 1.0 \times 10^{-14}$$

$$[OH^-] = \frac{1.0 \times 10^{-14}}{1.5 \times 10^{-4}} = 6.7 \times 10^{-11} \text{ M}$$

Calculating pH from $[H_3O^+]$ (Section 14.9)

To calculate the pH of a solution from $[H_3O^+]$, simply take the negative log of $[H_3O^+]$.

$$pH = -\log[H_3O^+]$$

EXAMPLE 14.17 Calculating pH from $[H_3O^+]$

Calculate the pH of a solution with $[H_3O^+] = 2.4 \times 10^{-5}$ M.

Solution:

$$\begin{aligned} pH &= -\log[H_3O^+] \\ &= -\log 2.4 \times 10^{-5} \\ &= -(-4.62) \\ &= 4.62 \end{aligned}$$

Calculating $[H_3O^+]$ from pH (Section 14.9)

You can calculate $[H_3O^+]$ from pH by taking the inverse log of the negative of the pH value (Method 1):

$$[H_3O^+] = \text{invlog}(-pH)$$

You can also calculate $[H_3O^+]$ from pH by raising 10 to the negative of the pH (Method 2):

$$[H_3O^+] = 10^{-pH}$$

EXAMPLE 14.18 Calculating $[H_3O^+]$ from pH

Calculate the $[H_3O^+]$ concentration for a solution with a pH of 6.22.

Solution:

Method 1: Inverse Log Function

$$\begin{aligned} [H_3O^+] &= \text{invlog}(-pH) \\ &= \text{invlog}(-6.22) \\ &= 6.0 \times 10^{-7} \end{aligned}$$

Method 2: 10^x Function

$$\begin{aligned} [H_3O^+] &= 10^{-pH} \\ &= 10^{-6.22} \\ &= 6.0 \times 10^{-7} \end{aligned}$$

Key Terms

acid [14.2]
acid rain [14.11]
acidic solution [14.8]
alkaloids [14.3]
amphoteric [14.4]
Arrhenius acid [14.4]
Arrhenius base [14.4]
Arrhenius definition [14.4]
base [14.3]
basic solution [14.8]
Brønsted–Lowry acid [14.4]
Brønsted–Lowry base [14.4]
Brønsted–Lowry definition [14.4]
buffer [14.10]
carboxylic acid [14.2]
conjugate acid–base pair [14.4]
diprotic acid [14.7]
dissociation [14.4]
equivalence point [14.6]
hydronium ion [14.4]
indicator [14.6]
ion product constant for water (K_w) [14.8]
ionize [14.4]
logarithmic scale [14.9]
monoprotic acid [14.7]
neutral solution [14.8]
neutralization [14.5]
pH [14.9]
salt [14.5]
strong acid [14.7]
strong base [14.7]
strong electrolyte [14.7]
titration [14.6]
weak acid [14.7]
weak base [14.7]
weak electrolyte [14.7]

Exercises

Questions

1. What makes sour gummy candies, such as Sour Patch Kids, sour?
2. Give some examples of foods that contain acids.
3. What are the properties of acids?
4. What is the main component of stomach acid? Why do we have stomach acid?
5. What are organic acids? Give two examples of organic acids.
6. What are the properties of bases?
7. What are alkaloids?
8. Give some examples of common substances that contain bases.
9. Give the Arrhenius definition of an acid and demonstrate the definition with a chemical equation.
10. Give the Arrhenius definition of a base and demonstrate the definition with a chemical equation.
11. Give the Brønsted–Lowry definitions of acids and bases and demonstrate each with a chemical equation.
12. According to the Brønsted–Lowry definition of acids and bases, what is a conjugate acid–base pair? Give an example.
13. What is an acid–base neutralization reaction? Give an example.
14. Give an example of a reaction between an acid and a metal.
15. Give an example of a reaction between an acid and a metal oxide.
16. Name a metal that dissolves in a base and write an equation for the reaction.
17. What is a titration? What is the equivalence point?
18. If a solution contains 0.85 mol of OH^-, how many moles of H^+ would be required to reach the equivalence point in a titration?
19. What is the difference between a strong acid and a weak acid?
20. How is the strength of an acid related to the strength of its conjugate base?
21. What are monoprotic and diprotic acids?
22. What is the difference between a strong base and a weak base?
23. Does pure water contain any H_3O^+ ions? Explain.
24. What happens to $[OH^-]$ in an aqueous solution when $[H_3O^+]$ increases?
25. Give a possible value of $[OH^-]$ and $[H_3O^+]$ in a solution that is:
    (a) acidic
    (b) basic
    (c) neutral
26. How is pH defined? A change of 1 pH unit corresponds to how much of a change in $[H_3O^+]$?
27. What is a buffer?
28. What are the main components in a buffer?
29. What is the cause of acid rain?
30. Write equations for the chemical reactions by which acid rain forms in the atmosphere.
31. What are the effects of acid rain?
32. How is the problem of acid rain being addressed in the United States?

Problems

Acid and Base Definitions

33. Identify each of the following as an acid or a base and write a chemical equation showing how it is an acid or a base according to the Arrhenius definition.
    (a) $HNO_3(aq)$
    (b) $KOH(aq)$
    (c) $HC_2H_3O_2(aq)$
    (d) $Ca(OH)_2$
34. Identify each of the following as an acid or a base and write a chemical equation showing how it is an acid or a base according to the Arrhenius definition.
    (a) $NaOH(aq)$
    (b) $HBr(aq)$
    (c) $Sr(OH)_2(aq)$
    (d) $H_2SO_4(aq)$

35. For each of the following, identify the Brønsted–Lowry acid, the Brønsted–Lowry base, the conjugate acid, and the conjugate base.
(a) $HBr(aq) + H_2O(l) \longrightarrow H_3O^+(aq) + Br^-(aq)$
(b) $NH_3(aq) + H_2O(l) \rightleftharpoons NH_4^+(aq) + OH^-(aq)$
(c) $HNO_3(aq) + H_2O(l) \longrightarrow H_3O^+(aq) + NO_3^-(aq)$
(d) $C_5H_5N(aq) + H_2O(l) \rightleftharpoons C_5H_5NH^+(aq) + OH^-(aq)$

36. For each of the following, identify the Brønsted–Lowry acid, the Brønsted–Lowry base, the conjugate acid, and the conjugate base.
(a) $HI(aq) + H_2O(l) \longrightarrow H_3O^+(aq) + I^-(aq)$
(b) $CH_3NH_2(aq) + H_2O(l) \rightleftharpoons CH_3NH_3^+(aq) + OH^-(aq)$
(c) $CO_3^{2-}(aq) + H_2O(l) \rightleftharpoons HCO_3^{2-}(aq) + OH^-(aq)$
(d) $H_2CO_3(aq) + H_2O(l) \rightleftharpoons H_3O^+(aq) + HCO_3^-(aq)$

37. Which of the following are conjugate acid–base pairs?
(a) NH_3, NH_4^+
(b) HCl, HBr
(c) $C_2H_3O_2^-, HC_2H_3O_2$
(d) HCO_3^-, NO_3^-

38. Which of the following are conjugate acid–base pairs?
(a) HI, I^-
(b) $HCHO_2, SO_4^{2-}$
(c) PO_4^{3-}, HPO_4^{2-}
(d) CO_3^{2-}, HCl

39. Write the formula for the conjugate base of each of the following acids.
(a) HCl
(b) H_2SO_3
(c) $HCHO_2$
(d) HF

40. Write the formula for the conjugate base of each of the following acids.
(a) HBr
(b) H_2CO_3
(c) $HClO_4$
(d) $HC_2H_3O_2$

41. Write the formula for the conjugate acid of each of the following bases.
(a) NH_3
(b) ClO_4^-
(c) HSO_4^-
(d) CO_3^{2-}

42. Write the formula for the conjugate acid of each of the following bases.
(a) CH_3NH_2
(b) C_5H_5N
(c) Cl^-
(d) F^-

Acid–Base Reactions

43. Write neutralization reactions for each of the following acids and bases.
(a) $HI(aq)$ and $NaOH(aq)$
(b) $HBr(aq)$ and $KOH(aq)$
(c) $HNO_3(aq)$ and $Ba(OH)_2(aq)$
(d) $HClO_4(aq)$ and $Sr(OH)_2(aq)$

44. Write neutralization reactions for each of the following acids and bases.
(a) $HF(aq)$ and $Ba(OH)_2(aq)$
(b) $HClO_4(aq)$ and $NaOH(aq)$
(c) $HBr(aq)$ and $Ca(OH)_2(aq)$
(d) $HCl(aq)$ and $KOH(aq)$

45. Write a balanced chemical equation showing how each of the following metals is dissolved by HBr.
(a) K
(b) Ca
(c) Sr
(d) Rb

46. Write a balanced chemical equation showing how each of the following metals is dissolved by HCl.
(a) Al
(b) Mg
(c) Cs
(d) Ba

47. Write a balanced chemical equation showing how each of the following metal oxides is dissolved by HI.
(a) MgO
(b) K_2O
(c) Rb_2O
(d) CaO

48. Write a balanced chemical equation showing how each of the following metal oxides is dissolved by HCl.
(a) SrO
(b) Na_2O
(c) Li_2O
(d) BaO

49. Predict the product of each of the following reactions:
(a) $HClO_4(aq) + Fe_2O_3(s) \longrightarrow$
(b) $H_2SO_4(aq) + Sr(s) \longrightarrow$
(c) $H_3PO_4(aq) + KOH(aq) \longrightarrow$

50. Predict the product of each of the following reactions:
(a) $HI(aq) + Al(s) \longrightarrow$
(b) $H_2SO_4(aq) + TiO_2(s) \longrightarrow$
(c) $H_2CO_3(aq) + LiOH(aq) \longrightarrow$

Acid–Base Titrations

51. Four solutions of unknown HCl concentration are titrated with solutions of NaOH. The following table lists the volume of each unknown HCl solution, the volume of NaOH solution required to reach the endpoint, and the concentration of each NaOH solution. Calculate the concentration (in M) of the unknown HCl solution in each case.

HCl Volume (mL)	*NaOH Volume (mL)*	*[NaOH] (M)*
(a) 25.00 mL	28.44 mL	0.1231 M
(b) 15.00 mL	21.22 mL	0.0972 M
(c) 20.00 mL	14.88 mL	0.1178 M
(d) 5.00 mL	6.88 mL	0.1325 M

52. Four solutions of unknown NaOH concentration are titrated with solutions of HCl. The following table lists the volume of each unknown NaOH solution, the volume of HCl solution required to reach the endpoint, and the concentration of each HCl solution. Calculate the concentration (in M) of the unknown NaOH solution in each case.

NaOH Volume (mL)	*HCl Volume (mL)*	*[HCl] (M)*
(a) 5.00 mL	9.77 mL	0.1599 M
(b) 15.00 mL	11.34 mL	0.1311 M
(c) 10.00 mL	10.55 mL	0.0889 M
(d) 30.00 mL	36.18 mL	0.1021 M

53. A 25.00-mL sample of an H_2SO_4 solution of unknown concentration is titrated with a 0.1328 M KOH solution. A volume of 38.33 mL of KOH was required to reach the endpoint. What is the concentration of the unknown H_2SO_4 solution?

54. A 5.00-mL sample of an H_3PO_4 solution of unknown concentration is titrated with a 0.1221 M NaOH solution. A volume of 5.99 mL of the NaOH solution was required to reach the endpoint. What is the concentration of the unknown H_3PO_4 solution?

55. What volume in milliliters of a 0.101 M NaOH solution is required to reach the endpoint in the complete titration of a 10.0-mL sample of 0.138 M H_2SO_4?

56. What volume in milliliters of 0.0992 M NaOH solution is required to reach the endpoint in the complete titration of a 15.0-mL sample of 0.107 M H_3PO_4.

Strong and Weak Acids and Bases

57. Classify each of the following acids as strong or weak.
(a) HCl
(b) HF
(c) HBr
(d) H_2SO_3

58. Classify each of the following acids as strong or weak.
(a) $HCHO_2$
(b) H_2SO_4
(c) HNO_3
(d) H_2CO_3

59. Determine $[H_3O^+]$ in each of the following acid solutions. For weak acids, give what $[H_3O^+]$ is less than.
(a) 2.5 M HI
(b) 1.2 M $HClO_2$
(c) 0.25 M H_2CO_3
(d) 2.25 M $HCHO_2$

60. Determine $[H_3O^+]$ in each of the following acid solutions. For weak acids, give what $[H_3O^+]$ is less than.
(a) 0.100 M HNO_3
(b) 0.75 M H_3PO_4
(c) 3.8 M HI
(d) 0.85 M H_2SO_3

61. Classify each of the following bases as strong or weak.
(a) LiOH
(b) NH_4OH
(c) $Ca(OH)_2$
(d) NH_3

62. Classify each of the following bases as strong or weak.
(a) C_5H_5N
(b) NaOH
(c) $Ba(OH)_2$
(d) KOH

63. Determine $[OH^-]$ in each of the following base solutions. For weak bases, give what $[OH^-]$ is less than.
(a) 0.88 M NaOH
(b) 0.88 M NH_3
(c) 0.88 M $Sr(OH)_2$
(d) 1.55 M KOH

64. Determine $[OH^-]$ in each of the following base solutions. For weak bases, give what $[OH^-]$ is less than.
(a) 3.7 M KOH
(b) 1.10 M NH_3
(c) 0.110 M $Ba(OH)_2$
(d) 2.2 M C_5H_5N

Acidity, Basicity, and K_w

65. Determine whether each of the following solutions is acidic, basic, or neutral.
(a) $[H_3O^+] = 1 \times 10^{-5}$ M; $[OH^-] = 1 \times 10^{-9}$ M
(b) $[H_3O^+] = 1 \times 10^{-6}$ M; $[OH^-] = 1 \times 10^{-8}$ M
(c) $[H_3O^+] = 1 \times 10^{-7}$ M; $[OH^-] = 1 \times 10^{-7}$ M
(d) $[H_3O^+] = 1 \times 10^{-8}$ M; $[OH^-] = 1 \times 10^{-6}$ M

66. Determine whether each of the following solutions is acidic, basic, or neutral.
(a) $[H_3O^+] = 1 \times 10^{-9}$ M; $[OH^-] = 1 \times 10^{-5}$ M
(b) $[H_3O^+] = 1 \times 10^{-10}$ M; $[OH^-] = 1 \times 10^{-4}$ M
(c) $[H_3O^+] = 1 \times 10^{-2}$ M; $[OH^-] = 1 \times 10^{-12}$ M
(d) $[H_3O^+] = 1 \times 10^{-13}$ M; $[OH^-] = 1 \times 10^{-1}$ M

67. Calculate $[OH^-]$ given $[H_3O^+]$ in each of the following aqueous solutions and classify the solution as acidic or basic.
(a) $[H_3O^+] = 1.5 \times 10^{-9}$ M
(b) $[H_3O^+] = 9.3 \times 10^{-9}$ M
(c) $[H_3O^+] = 2.2 \times 10^{-6}$ M
(d) $[H_3O^+] = 7.4 \times 10^{-4}$ M

68. Calculate $[OH^-]$ given $[H_3O^+]$ in each of the following aqueous solutions and classify the solution as acidic or basic.
(a) $[H_3O^+] = 1.3 \times 10^{-3}$ M
(b) $[H_3O^+] = 9.1 \times 10^{-12}$ M
(c) $[H_3O^+] = 5.2 \times 10^{-4}$ M
(d) $[H_3O^+] = 6.1 \times 10^{-9}$ M

69. Calculate $[H_3O^+]$ given $[OH^-]$ in each of the following aqueous solutions and classify each solution as acidic or basic.
(a) $[OH^-] = 2.7 \times 10^{-12}$ M
(b) $[OH^-] = 2.5 \times 10^{-2}$ M
(c) $[OH^-] = 1.1 \times 10^{-10}$ M
(d) $[OH^-] = 3.3 \times 10^{-4}$ M

70. Calculate $[H_3O^+]$ given $[OH^-]$ in each of the following aqueous solutions and classify each solution as acidic or basic.
(a) $[OH^-] = 2.1 \times 10^{-11}$ M
(b) $[OH^-] = 7.5 \times 10^{-9}$ M
(c) $[OH^-] = 2.1 \times 10^{-4}$ M
(d) $[OH^-] = 1.0 \times 10^{-2}$ M

pH

71. Classify each of the following solutions as acidic, basic, or neutral based on the pH value.
(a) pH = 9.0
(b) pH = 7.0
(c) pH = 2.0
(d) pH = 6.0

72. Classify each of the following solutions as acidic, basic, or neutral based on the pH value.
(a) pH = 5.0
(b) pH = 4.0
(c) pH = 14.0
(d) pH = 0.5

73. Calculate the pH of each of the following solutions.
(a) $[H_3O^+] = 1.7 \times 10^{-8}$ M
(b) $[H_3O^+] = 1.0 \times 10^{-7}$ M
(c) $[H_3O^+] = 2.2 \times 10^{-6}$ M
(d) $[H_3O^+] = 7.4 \times 10^{-4}$ M

74. Calculate the pH of each of the following solutions.
(a) $[H_3O^+] = 2.4 \times 10^{-10}$ M
(b) $[H_3O^+] = 7.6 \times 10^{-2}$ M
(c) $[H_3O^+] = 9.2 \times 10^{-13}$ M
(d) $[H_3O^+] = 3.4 \times 10^{-5}$ M

75. Calculate $[H_3O^+]$ for each of the following solutions.
(a) pH = 8.55
(b) pH = 11.23
(c) pH = 2.87
(d) pH = 1.22

76. Calculate $[H_3O^+]$ for each of the following solutions.
(a) pH = 1.76
(b) pH = 3.88
(c) pH = 8.43
(d) pH = 12.32

77. Calculate the pH of each of the following solutions.
(a) $[OH^-] = 1.9 \times 10^{-7}$ M
(b) $[OH^-] = 2.6 \times 10^{-8}$ M
(c) $[OH^-] = 7.2 \times 10^{-11}$ M
(d) $[OH^-] = 9.5 \times 10^{-2}$ M

78. Calculate the pH of each of the following solutions.
(a) $[OH^-] = 2.8 \times 10^{-11}$ M
(b) $[OH^-] = 9.6 \times 10^{-3}$ M
(c) $[OH^-] = 3.8 \times 10^{-12}$ M
(d) $[OH^-] = 6.4 \times 10^{-4}$ M

79. Calculate $[OH^-]$ for each of the following solutions.
(a) pH = 4.25
(b) pH = 12.53
(c) pH = 1.50
(d) pH = 8.25

80. Calculate $[OH^-]$ for each of the following solutions.
(a) pH = 1.82
(b) pH = 13.28
(c) pH = 8.29
(d) pH = 2.32

81. Calculate the pH of each of the following solutions:
(a) 0.0155 M HBr
(b) 1.28×10^{-3} M KOH
(c) 1.89×10^{-3} M HNO_3
(d) 1.54×10^{-4} M $Sr(OH)_2$

82. Calculate the pH of each of the following solutions:
(a) 1.34×10^{-3} M $HClO_4$
(b) 0.0211 M NaOH
(c) 0.0109 M HBr
(d) 7.02×10^{-5} M $Ba(OH)_2$

Buffers and Acid Rain

83. Locate where you live on the map in Figure 14.20. What is the pH of rain where you live? What is the $[H_3O^+]$?

84. Locate the area of the United States with the most acidic rainfall on the map in Figure 14.20. What is the pH of the rain? What is the $[H_3O^+]$?

85. Which of the following mixtures are buffers?
(a) HCl and HF
(b) NaOH and NH_3
(c) HF and NaF
(d) $HC_2H_3O_2$ and $KC_2H_3O_2$

86. Which of the following mixtures are buffers?
(a) HBr and NaCl
(b) $HCHO_2$ and $NaCHO_2$
(c) HCl and HBr
(d) KOH and NH_3

87. Write reactions showing how each of the buffers in Problem 85 would neutralize added HCl.

88. Write reactions showing how each of the buffers in Problem 86 would neutralize added NaOH.

89. What substance could you add to each of the following solutions to make it a buffer solution?
(a) 0.100 M $NaC_2H_3O_2$
(b) 0.500 M H_3PO_4
(c) 0.200 M $HCHO_2$

90. What substance could you add to each of the following solutions to make it a buffer solution?
(a) 0.050 M $NaHSO_3$
(b) 0.150 M HF
(c) 0.200 M $KCHO_2$

Cumulative Exercises

91. How much 0.100 M HCl is required to completely neutralize 20.0 mL of 0.250 M NaOH?

92. How much 0.200 M KOH is required to completely neutralize 25.0 mL of 0.150 M $HClO_4$?

93. What is the minimum volume of 5.0 M HCl required to completely dissolve 10.0 g of magnesium metal?

94. What is the minimum volume of 3.0 M HBr required to completely dissolve 15.0 g of potassium metal?

95. When 18.5 g of K_2O is completely dissolved by HI, how many grams of KI(*aq*) are formed in solution?

96. When 5.88 g of CaO is completely dissolved by HBr, how many grams of $CaBr_2$(*aq*) are formed in solution?

97. A 0.125-g sample of a monoprotic acid of unknown molar mass is dissolved in water and titrated with 0.1003 M NaOH. The endpoint is reached after adding 20.77 mL of base. What is the molar mass of the unknown acid?

98. A 0.105-g sample of a diprotic acid of unknown molar mass is dissolved in water and titrated with 0.1288 M NaOH. The endpoint is reached after adding 15.2 mL of base. What is the molar mass of the unknown acid?

99. Antacids, such as milk of magnesia, are often taken to reduce the discomfort of acid stomach or heartburn. The recommended dose of milk of magnesia is 1 teaspoon, which contains 400 mg of $Mg(OH)_2$. What volume of HCl solution with a pH of 1.1 can be neutralized by 1 dose of milk of magnesia? (Assume two significant figures in your calculations.)

100. An antacid tablet requires 25.82 mL of 0.200 M HCl to titrate to its endpoint. What volume in milliliters of stomach acid can be neutralized by the antacid tablet? Assume that stomach acid has a pH of 1.1. (Assume two significant figures in your calculations.)

101. For each of the following $[H_3O^+]$, determine the pH and state whether the solution is acidic or basic.
(a) $[H_3O^+] = 0.0025$ M
(b) $[H_3O^+] = 1.8 \times 10^{-12}$ M
(c) $[H_3O^+] = 9.6 \times 10^{-9}$ M
(d) $[H_3O^+] = 0.0195$ M

102. For each of the following $[OH^-]$, determine the the pH and state whether the solution is acidic or basic.
(a) $[OH^-] = 1.8 \times 10^{-5}$ M
(b) $[OH^-] = 8.9 \times 10^{-12}$ M
(c) $[OH^-] = 3.1 \times 10^{-2}$ M
(d) $[OH^-] = 1.96 \times 10^{-9}$ M

103. Complete the following table. (The first row is completed for you.)

$[H_3O^+]$	$[OH^-]$	*pH*	*Acidic or Basic*
1.0×10^{-4}	1.0×10^{-10}	4.00	acidic
5.5×10^{-3}	____	____	____
____	3.2×10^{-6}	____	____
4.8×10^{-9}	____	____	____
____	____	7.55	____

104. Complete the following table. (The first row is completed for you.)

$[H_3O^+]$	$[OH^-]$	*pH*	*Acidic or Basic*
1.0×10^{-8}	1.0×10^{-6}	8.00	basic
____	____	3.55	____
1.7×10^{-9}	____	____	____
____	____	13.5	____
____	8.6×10^{-11}	____	____

105. For each of the following strong acid solutions, determine $[H_3O^+]$, $[OH^-]$, and pH.
(a) 0.0088 M $HClO_4$
(b) 1.5×10^{-3} M HBr
(c) 9.77×10^{-4} M HI
(d) 0.0878 M HNO_3

106. For each of the following strong acid solutions, determine $[H_3O^+]$, $[OH^-]$, and pH.
(a) 0.0150 M HCl
(b) 1.9×10^{-4} M HI
(c) 0.0226 M HBr
(d) 1.7×10^{-3} M HNO_3

107. For each of the following strong base solutions, determine $[OH^-]$, $[H_3O^+]$, and pH.
(a) 0.15 M NaOH
(b) 1.5×10^{-3} M $Ca(OH)_2$
(c) 4.8×10^{-4} M $Sr(OH)_2$
(d) 8.7×10^{-5} M KOH

108. For each of the following strong base solutions, determine $[OH^-]$, $[H_3O^+]$, and pH.
(a) 8.77×10^{-3} M LiOH
(b) 0.0112 M $Ba(OH)_2$
(c) 1.9×10^{-4} M KOH
(d) 5.0×10^{-4} M $Ca(OH)_2$

109. As described in Section 14.1, jailed spies on the big screen often use acid stored in a pen to dissolve jail bars and escape. What minimum volume of 12.0 M hydrochloric acid would be required to completely dissolve a 500.0-g iron bar? Would this amount of acid fit into a pen?

110. A popular classroom demonstration consists of filing notches into a new penny and soaking the penny in hydrochloric acid overnight. Since new pennies are made of zinc coated with copper, and since hydrochloric acid dissolves zinc and not copper, the inside of the penny is dissolved by the acid, while the outer copper shell remains. Suppose the penny contains 2.5 g of zinc and is soaked in 20.0 mL of 6.0 M HCl. Calculate the concentration of the HCl solution after all of the zinc has dissolved. Hint: The Zn from the penny is oxidized to Zn^{2+}.

111. What is the pH of a solution formed by mixing 125.0 mL of 0.0250 M HCl with 75.0 mL of 0.0500 M NaOH?

112. What is the pH of a solution formed by mixing 175.0 mL of 0.0880 M HI with 125.0 mL of 0.0570 M KOH?

113. How many H^+ (or H_3O^+) ions are present in one drop (0.050 mL) of pure water at 25 °C?

114. Calculate the number of H^+ (or H_3O^+) ions and OH^- ions in 1.0 mL of 0.100 M HCl.

Highlight Problems

115. Consider the following molecular views of acid solutions. Based on the molecular view, determine whether the acid is weak or strong.

(a)

(b)

(c)

(d)

116. Lakes that have been acidified by acid rain can be neutralized by liming, the addition of limestone ($CaCO_3$). How much limestone in kilograms is required to completely neutralize a 3.8×10^9 L lake with a pH of 5.5?

117. Acid rain over the Great Lakes has a pH of about 4.5. Calculate the $[H_3O^+]$ of this rain and compare that value to the $[H_3O^+]$ of rain over the West Coast that has a pH of 5.4. How many times more concentrated is the acid in rain over the Great Lakes?

Answers to Skillbuilder Exercises

Skillbuilder 14.1

(a) $C_5H_5N(aq) + H_2O(l) \rightleftharpoons C_5H_5NH^+(aq) + OH^-(aq)$

Base, Acid, Conjugate acid, Conjugate base

(b) $HNO_3(aq) + H_2O(l) \longrightarrow H_3O^+(aq) + NO_3^-(aq)$

Acid, Base, Conjugate acid, Conjugate base

Skillbuilder 14.2

$H_3PO_4(aq) + 3\ NaOH(aq) \longrightarrow 3\ H_2O(l) + Na_3PO_4(aq)$

Skillbuilder 14.3

(a) $2\ HCl(aq) + Sr(s) \longrightarrow H_2(g) + SrCl_2(aq)$

(b) $2\ HI(aq) + BaO(s) \longrightarrow H_2O(l) + BaI_2(aq)$

Skillbuilder 14.4 9.03×10^{-2} M H_2SO_4

Skillbuilder 14.5

(a) $[H_3O^+] < 0.50$ M

(b) $[H_3O^+] = 1.25$ M

(c) $[H_3O^+] < 0.75$ M

Skillbuilder 14.6

(a) $[OH^-] = 0.11$ M

(b) $[OH^-] < 1.05$ M

(c) $[OH^-] = 0.45$ M

Skillbuilder 14.7

(a) $[H_3O^+] = 6.7 \times 10^{-13}$ M; basic

(b) $[H_3O^+] = 1.0 \times 10^{-7}$ M; neutral

(c) $[H_3O^+] = 1.2 \times 10^{-5}$ M; acidic

Skillbuilder 14.8

(a) pH = 8.02; basic

(b) pH = 2.21; acidic

Skillbuilder Plus, p. 493 pH = 12.11; basic

Skillbuilder 14.9 4.3×10^{-9} M

Skillbuilder Plus, p. 493 4.6×10^{-11} M

Answers to Conceptual Checkpoints

14.1 (d) Each of the others can accept a proton, and thus act as a base. NH_4^+, however, is the conjugate acid of NH_3, and therefore acts as an acid and not as a base.

14.2 (d) Because pH is the *negative* log of the H_3O^+ concentration, a higher pH corresponds to a lower $[H_3O^+]$, and each unit of pH represents a tenfold change in concentration.

14.3 (b) A buffer solution consists of a weak acid and its conjugate base. Of the compounds listed, HF is the only weak acid, and F^- (from NaF in solution) is its conjugate base.

CHAPTER 15

Chemical Equilibrium

"Old chemists never die, they just reach equilibrium."

Anonymous

15.1 Life: Controlled Disequilibrium

◀ Dynamic equilibrium involves two opposing processes occurring at the same rate. This image draws an analogy between a chemical equilibrium ($N_2O_4 \rightleftharpoons 2\ NO_2$) in which the two opposing reactions occur at the same rate and a freeway with traffic moving in opposing directions at the same rate.

Have you ever tried to define life? If you have, you know that defining life is difficult. How are living things different from nonliving things? You may try to define living things as those things that can move. But of course many living things do not move—many plants, for example, do not move very much—and some nonliving things, such as glaciers and Earth itself, do move. So motion is neither unique to nor definitive of life. You may try to define living things as those things that can reproduce. But again, many living things, such as mules or sterile humans, cannot reproduce; yet they are alive. In addition, some nonliving things—such as crystals, for example—reproduce (in some sense). So what is unique about living things?

One definition of life uses the concept of equilibrium. We will define *chemical* equilibrium more carefully soon. For now, we can think more generally of equilibrium as *sameness and constancy*. When an object is in equilibrium with its surroundings, some property of the object has reached sameness with the surroundings and is no longer changing. For example, a cup of hot water is not in equilibrium with its surroundings with respect to temperature. If left undisturbed, the cup of hot water will slowly cool until it reaches equilibrium with its surroundings. At that point, the temperature of the water is the *same as* that of the surroundings (sameness) and *no longer changes* (constancy).

So equilibrium involves sameness and constancy. Part of a definition for living things, then, is that living things *are not* in equilibrium with their surroundings. Our body temperature, for example, is not the same as the temperature of our surroundings. When we jump into a swimming pool, the pH of our blood does not become the same as the pH of the surrounding water. Living things, even the simplest ones, maintain some measure of *disequilibrium* with their environment.

We must add one more concept, however, to complete our definition of life with respect to equilibrium. Our cup of hot water is in disequilibrium with its environment, yet it is not alive. However, the cup of hot water has no control over its disequilibrium and will slowly come to equilibrium with its environment. In contrast, living things—as long as they are alive—maintain and *control* their disequilibrium. Your body temperature, for example, is not only in disequilibrium with your surroundings—it is in controlled disequilibrium. Your body maintains your temperature within a specific range that is not in equilibrium with the surrounding temperature.

So one definition for life is that living things are in *controlled disequilibrium* with their environment. A living thing comes into equilibrium with its surroundings only after it dies. In this chapter, we will examine the concept of equilibrium, especially chemical equilibrium—the state that involves sameness and constancy.

15.2 The Rate of a Chemical Reaction

Reaction rates are related to chemical equilibrium because, as we will see in Section 15.3, a chemical system is at equilibrium when the rate of the forward reaction equals the rate of the reverse reaction.

Before we probe more deeply into the concept of chemical equilibrium, we must first understand something about the rates of chemical reactions. The **rate of a chemical reaction** is the amount of reactant that changes to product in a given period of time. A reaction with a fast rate proceeds quickly, with a large amount of reactant being converted to product in a certain period of time (▶ Figure 15.1a). A reaction with a slow rate proceeds slowly, with only a small amount of reactant being converted to product in the same period of time (Figure 15.1b).

A reaction rate can also be defined as the amount of a product that forms in a given period of time.

Chemists seek to control reaction rates for many chemical reactions. For example, the space shuttle is propelled by the reaction of hydrogen and oxygen to form water. If the reaction proceeds too slowly, the shuttle will not lift off the ground. If, however, the reaction proceeds too quickly, the shuttle can explode. Reaction rates can be controlled if we understand the factors that influence them.

Collision Theory

According to **collision theory**, chemical reactions occur through collisions between molecules or atoms. For example, consider the following gas phase chemical reaction between $H_2(g)$ and $I_2(g)$ to form $HI(g)$.

$$H_2(g) + I_2(g) \longrightarrow 2\,HI(g)$$

The reaction begins when an H_2 molecule collides with an I_2 molecule. If the collision occurs with enough energy—that is, if the colliding molecules are moving fast enough—the product molecules (HI) form. If the collision occurs without enough energy, the reactant molecules (H_2 and I_2) simply bounce off of one another. Since gas-phase molecules have a wide distribution of velocities, collisions occur with a wide distribution of energies. The high-energy collisions lead to products and the low-energy collisions do not.

Whether a collision leads to a reaction also depends on the *orientation* of the colliding molecules, but this topic is beyond our current scope.

The reason that higher-energy collisions are more likely to lead to products is related to a concept called the *activation energy* of a reaction.

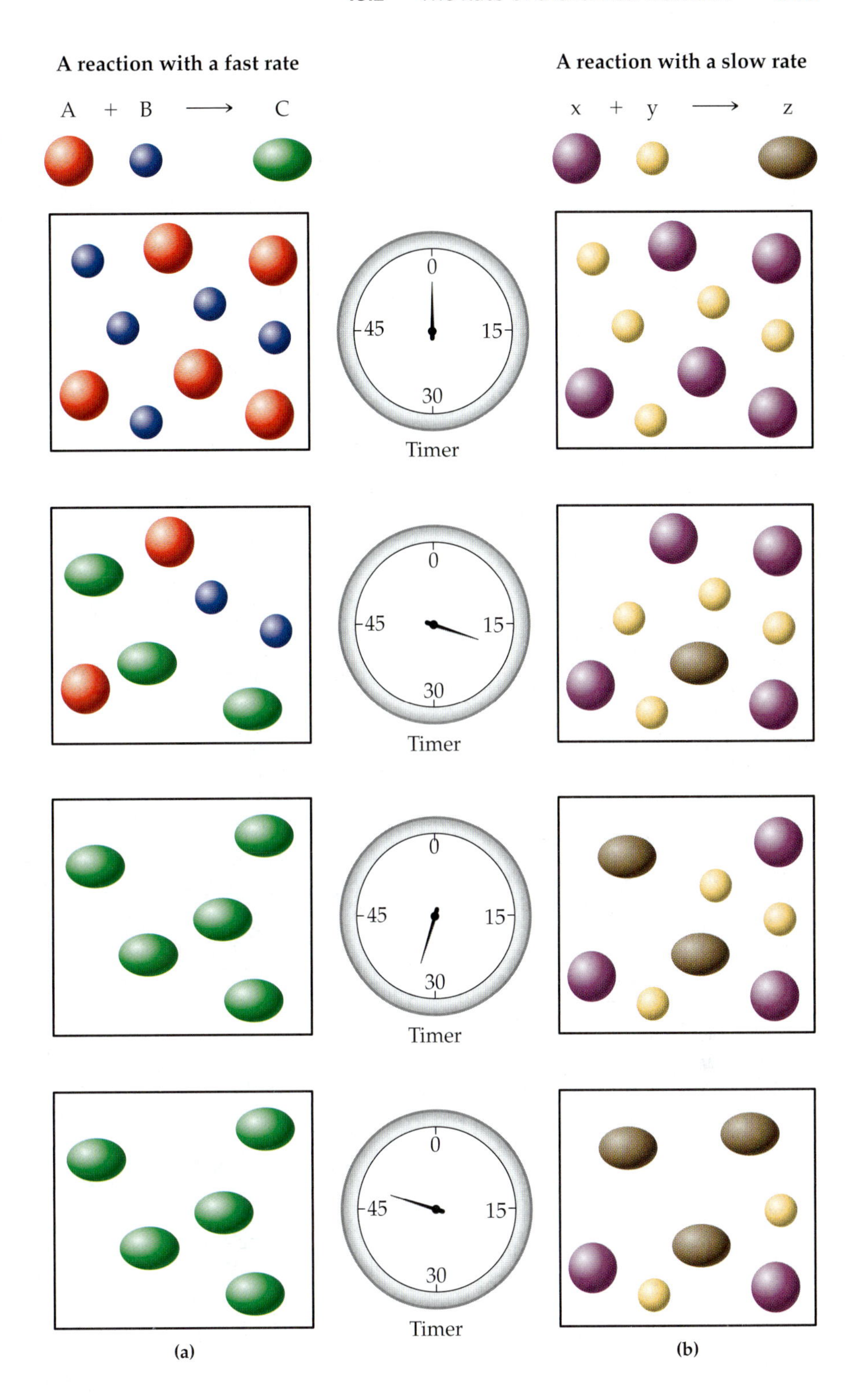

▶ **Figure 15.1 Reaction rates**
(a) In a reaction with a fast rate, the reactants react to form products in a short period of time. **(b)** In a reaction with a slow rate, the reactants react to form products over a long period of time.

The activation energy for chemical reactions is discussed in more detail in Section 15.12. For now, we can think of the activation energy as a barrier that must be overcome for the reaction to proceed. For example, in the case of H_2 reacting with I_2 to form HI, the product (HI) can begin to form only after the H–H bond and the I–I bond each begin to break. The activation energy is the energy required to begin to break these bonds.

If molecules react via high-energy collisions, what factors influence the rate of a reaction? What factors affect the number of high-energy collisions that occur per unit time? There are two important factors: the *concentration* of the reacting molecules and the *temperature* of the reaction mixture.

How Concentration Affects the Rate of a Reaction

Figures 15.2a through c show various mixtures of H_2 and I_2 at the same temperature but different concentrations. If H_2 and I_2 react via collisions to form HI, which mixture do you think will have the highest reaction rate? Since Figure 15.2c has the highest concentration of H_2 and I_2, it will have the most collisions per unit time and therefore the fastest reaction rate. This idea holds true for most chemical reactions.

> The rate of a chemical reaction generally increases with increasing concentration of the reactants.

The exact relationship between increases in concentration and increases in reaction rate varies for different reactions and is beyond our current scope.

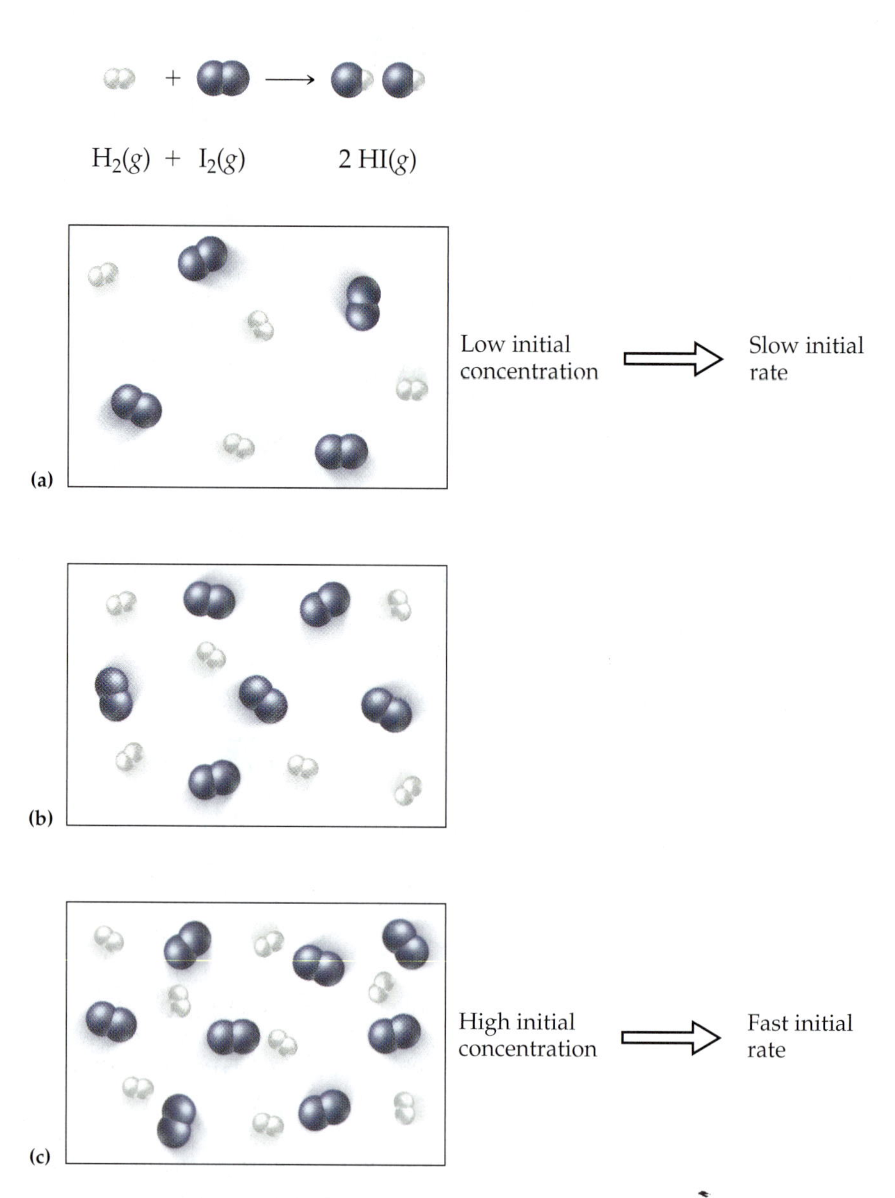

▲ **Figure 15.2 Effect of concentration on reaction rate** **Question:** Which reaction mixture will have the fastest initial rate? The mixture in (c) is fastest because it has the highest concentration of reactants and therefore the highest rate of collisions.

For our purposes, it will suffice to know that for most reactions the reaction rate increases with increasing reactant concentration.

Knowing this, what can we say about the rate of a reaction as the reaction proceeds? Since reactants turn to products in the course of a reaction, their concentration decreases. Consequently, the reaction rate decreases as well. In other words, as a reaction proceeds, there are fewer reactant molecules (because they have turned into products), and the reaction slows down.

How Temperature Affects the Rate of a Reaction

Reaction rates also depend on temperature. ▼ Figures 15.3a through c show various mixtures of H_2 and I_2 at the same concentration, but different temperatures. Which will have the fastest rate? Raising the temperature makes

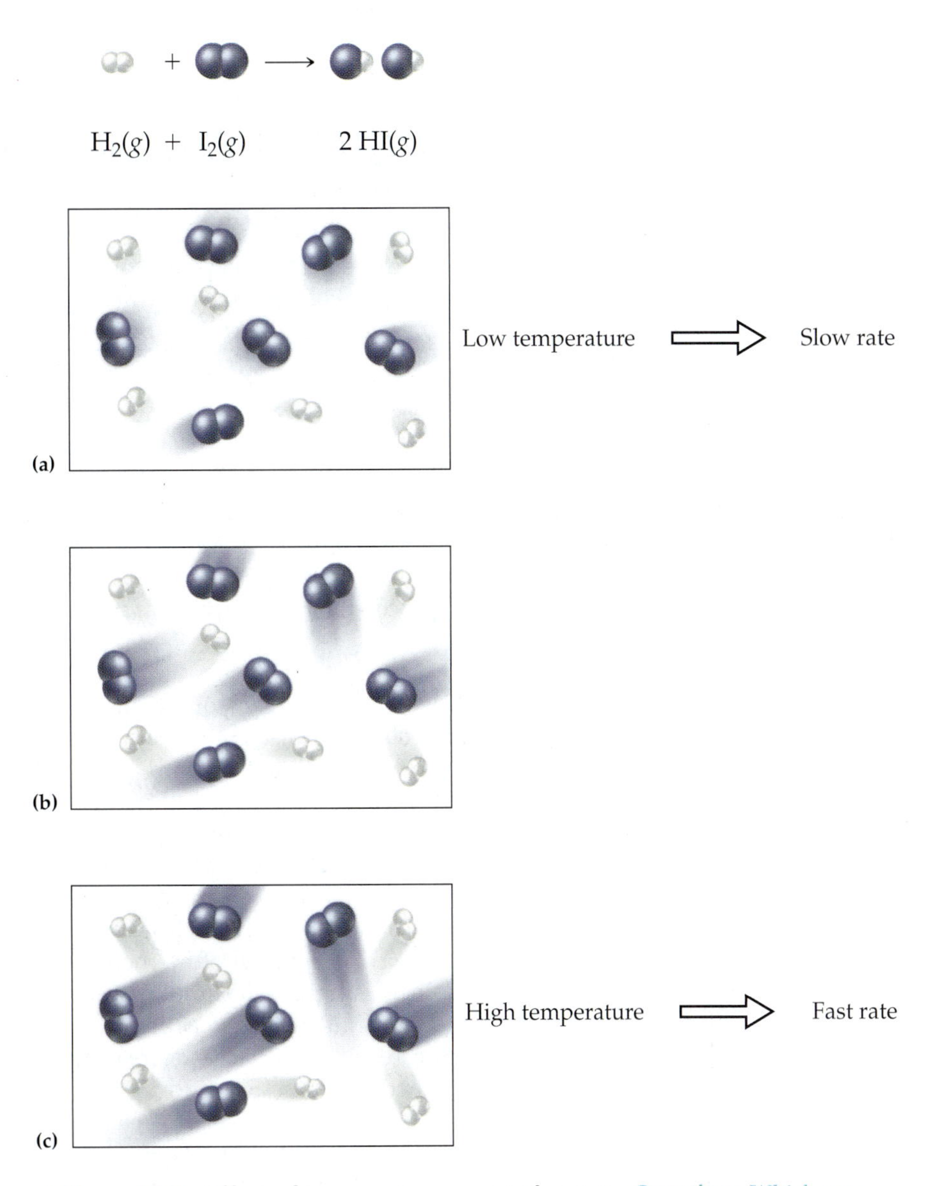

▲ **Figure 15.3 Effect of temperature on reaction rate** **Question:** Which reaction mixture will have the fastest initial rate? The mixture in (c) is fastest because it has the highest temperature.

the molecules move faster (Section 3.9). They therefore experience more collisions per unit time, resulting in a faster reaction rate. In addition, a higher temperature results in more collisions that are (on average) of higher energy. Since it is the high-energy collisions that result in products, this also produces a faster rate. Consequently, Figure 15.3c (which has the highest temperature) has the fastest reaction rate. This relationship holds true for most chemical reactions.

> The rate of a chemical reaction generally increases with increasing temperature of the reaction mixture.

The temperature dependence of reaction rates is the reason that cold-blooded animals become more sluggish at lower temperatures. The reactions required for them to think and move simply become slower, resulting in the sluggish behavior.

To summarize:

- Reaction rates generally increase with increasing reactant concentration.
- Reaction rates generally increase with increasing temperature.
- Reaction rates generally decrease as a reaction proceeds.

CONCEPTUAL CHECKPOINT 15.1

In a chemical reaction between two gases, you would expect that increasing the pressure would probably

(a) increase the reaction rate
(b) decrease the reaction rate
(c) not affect the reaction rate

15.3 The Idea of Dynamic Chemical Equilibrium

What would happen if our reaction between H_2 and I_2 to form HI were able to proceed in both the forward and reverse directions?

$$H_2(g) + I_2(g) \rightleftharpoons 2\,HI(g)$$

Now, H_2 and I_2 can collide and react to form 2 HI molecules, but the 2 HI molecules can also collide and react to reform H_2 and I_2. A reaction that can proceed in both the forward and reverse directions is said to be a **reversible reaction**.

Suppose we begin with only H_2 and I_2 in a container (▶ Figure 15.4a). What happens initially? H_2 and I_2 begin to react to form HI (Figure 15.4b). However, as H_2 and I_2 react their concentration decreases, which in turn decreases the rate of the forward reaction. At the same time, HI begins to form. As the concentration of HI increases, the reverse reaction begins to occur at an increasingly faster rate because there are more HI collisions. Eventually the rate of the reverse reaction (which is increasing) equals the rate of the forward reaction (which is decreasing). At that point, **dynamic equilibrium** is reached (Figures 15.4c and 15.4d).

> **Dynamic equilibrium**—In a chemical reaction, the condition in which the rate of the forward reaction equals the rate of the reverse reaction.

This condition is not static—it is dynamic because the forward and reverse reactions are still occurring, but at a constant rate. When dynamic

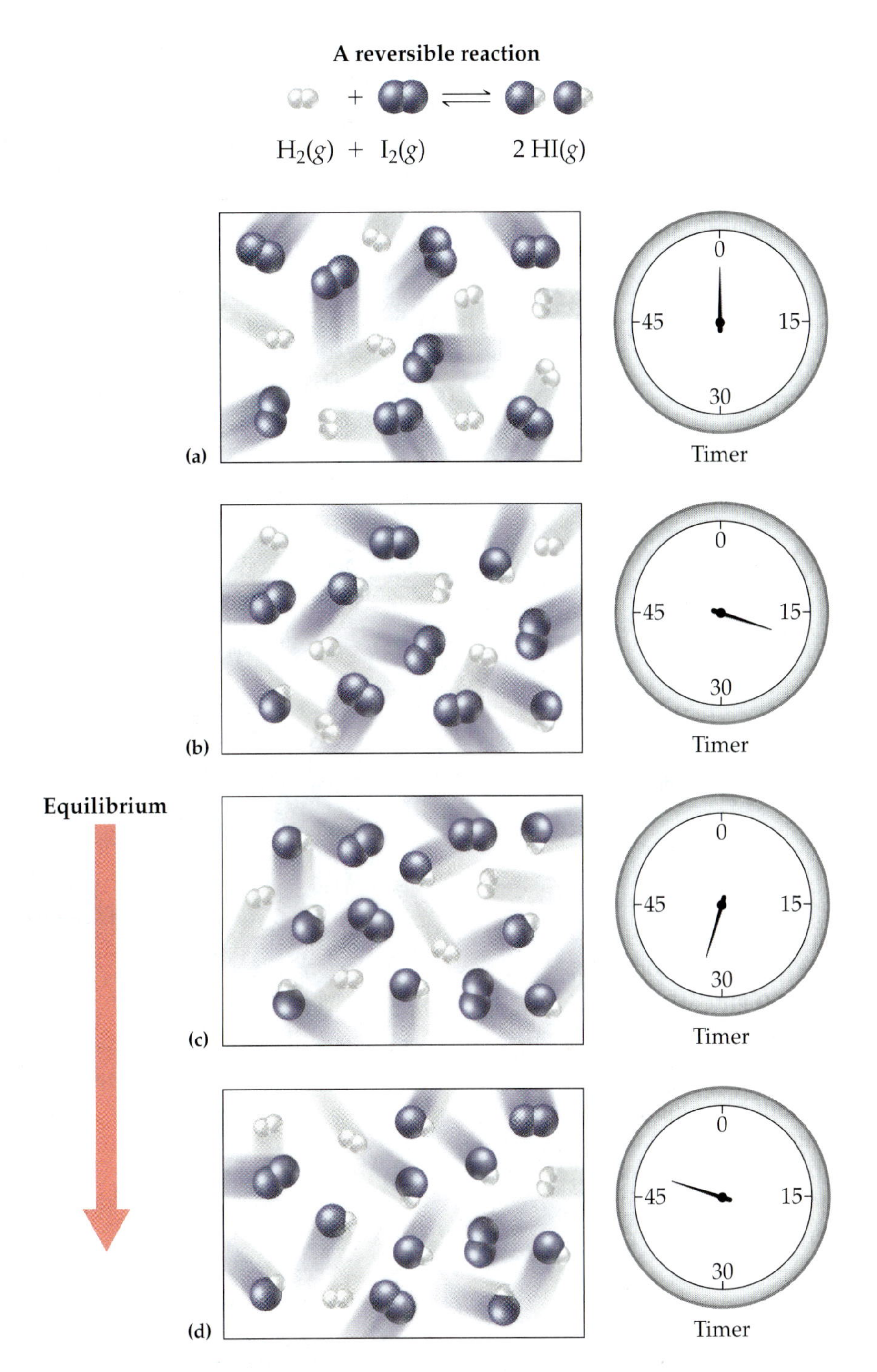

▲ **Figure 15.4 Equilibrium** When the concentrations of the reactants and products no longer change, equilibrium has been reached.

equilibrium is reached, the concentrations of H_2, I_2, and HI no longer change. They remain the same because the reactants and products are being depleted at the same rate at which they are being formed.

Notice that dynamic equilibrium includes the concepts of sameness and constancy that we discussed in Section 15.1. When dynamic equilibrium is reached, the forward reaction rate is the same as the reverse reaction rate (sameness). Because the reaction rates are the same, the concentrations of the reactants and products no longer change (constancy). However, just because the concentrations of reactants and products no longer change at equilibrium does *not* imply that the concentrations of reactants and products are *equal* to one another at equilibrium. Some reactions reach equilibrium only after most of the reactants have formed products. (Recall strong acids from Chapter 14.) Others reach equilibrium when only a small fraction of the reactants have formed products. (Recall weak acids from Chapter 14.) It depends on the reaction.

We can better understand dynamic equilibrium with a simple analogy. Imagine that Narnia and Middle Earth are two neighboring kingdoms (▼Figure 15.5). Narnia is overpopulated and Middle Earth is underpopulated. One day, however, the border between the two kingdoms opens, and people immediately begin to leave Narnia for Middle Earth (call this the forward reaction).

$$\text{Narnia} \longrightarrow \text{Middle Earth} \quad \text{(forward reaction)}$$

The population of Narnia goes down as the population of Middle Earth goes up. As people leave Narnia, however, the *rate* at which they leave begins to slow down (because Narnia becomes less crowded). On the other hand, as

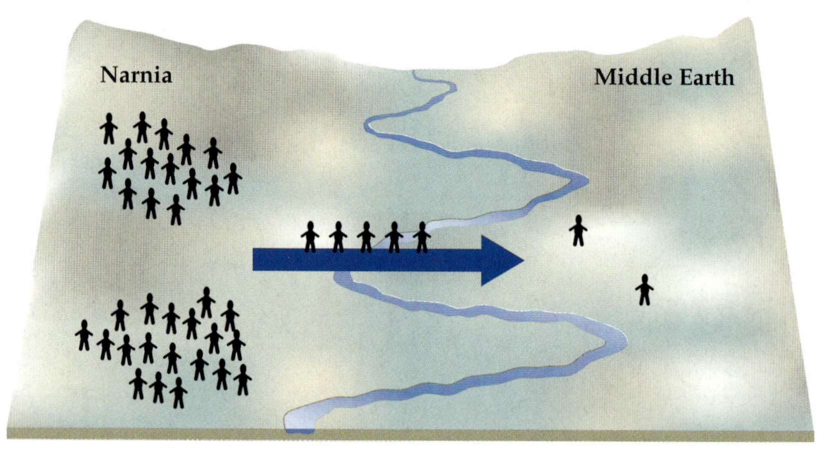

Initial

Represents population

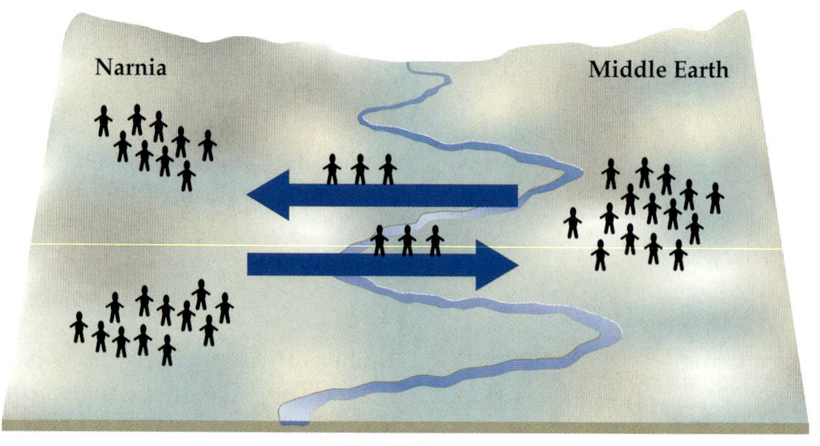

Equilibrium

▲ **Figure 15.5 Population analogy for a chemical reaction proceeding to equilibrium** people move into Middle Earth, some decide it was not for them and begin to move back (call this the reverse reaction).

$$\text{Narnia} \longleftarrow \text{Middle Earth} \quad \text{(reverse reaction)}$$

As Middle Earth fills, the rate of people moving back to Narnia gets faster. Eventually the *rate* of people moving out of Narnia (which has been slowing down as people leave) equals the *rate* of people moving back to Narnia (which has been increasing as Middle Earth gets more crowded). Dynamic equilibrium has been reached.

$$\text{Narnia} \rightleftharpoons \text{Middle Earth}$$

Notice that when the two kingdoms reach dynamic equilibrium, their populations no longer change because the number of people moving out equals the number of people moving in. However, one kingdom—because of its charm, the character of its leader, the availability of better jobs, a lower tax rate, or whatever other reason—may have a higher population than the other kingdom, even when dynamic equilibrium is reached.

Similarly, when a chemical reaction reaches dynamic equilibrium, the rate of the forward reaction (analogous to people moving out of Narnia) equals the rate of the reverse reaction (analogous to people moving back into Narnia), and the relative concentrations of reactants and products (analogous to the relative populations of the two kingdoms) become constant. Also, like our two kingdoms, the concentrations of reactants and products will not necessarily be equal at equilibrium, just as the populations of the two kingdoms are not equal at equilibrium.

15.4 The Equilibrium Constant: A Measure of How Far a Reaction Goes

We have just learned that the *concentrations* of reactants and products are not equal at equilibrium—it is the *rates* of the forward and reverse reactions that are equal. But what about the concentrations? What can we know about them? The equilibrium constant (K_{eq}) is a way to quantify the concentrations of the reactants and products at equilibrium. Consider the following generic chemical reaction

$$a\text{A} + b\text{B} \rightleftharpoons c\text{C} + d\text{D}$$

where A and B are reactants, C and D are products, and a, b, c, and d are the respective stoichiometric coefficients in the chemical equation. The **equilibrium constant (K_{eq})** for the reaction is defined as the ratio—at equilibrium—of the concentrations of the products raised to their stoichiometric coefficients divided by the concentrations of the reactants raised to their stoichiometric coefficients.

$$K_{eq} = \frac{[\text{C}]^c [\text{D}]^d}{[\text{A}]^a [\text{B}]^b} \quad \begin{matrix} \leftarrow \text{Products} \\ \leftarrow \text{Reactants} \end{matrix}$$

The equilibrium constant quantifies the relative concentrations of reactants and products at equilibrium.

Writing Equilibrium Expressions for Chemical Reactions

To write an equilibrium expression for a chemical reaction, simply examine the chemical equation and follow the preceding definition. For example, suppose we want to write an equilibrium expression for the following reaction.

$$2\,N_2O_5(g) \rightleftharpoons 4\,NO_2(g) + O_2(g)$$

The equilibrium constant is $[NO_2]$ raised to the fourth power multiplied by $[O_2]$ raised to the first power divided by $[N_2O_5]$ raised to the second power.

$$K_{eq} = \frac{[NO_2]^4[O_2]}{[N_2O_5]^2}$$

Notice that the *coefficients* in the chemical equation become the *exponents* in the equilibrium expression.

$$2\,N_2O_5(g) \rightleftharpoons 4\,NO_2(g) + O_2(g)$$

$$K_{eq} = \frac{[NO_2]^4\,[O_2]}{[N_2O_5]^2}$$

Implied 1

EXAMPLE 15.1 Writing Equilibrium Expressions for Chemical Reactions

Write an equilibrium expression for the following chemical equation.

$$CO(g) + 2\,H_2(g) \rightleftharpoons CH_3OH(g)$$

Solution:

The equilibrium expression is the concentrations of the products raised to their stoichiometric coefficients divided by the concentrations of the reactants raised to their stoichiometric coefficients. Notice that the expression is a ratio of products over reactants. Notice also that the coefficients in the chemical equation are the exponents in the equilibrium expression.

$$K_{eq} = \frac{[CH_3OH]}{[CO][H_2]^2}$$

Product

Reactants

SKILLBUILDER 15.1 Writing Equilibrium Expressions for Chemical Reactions

Write an equilibrium expression for the following chemical equation.

$$H_2(g) + F_2(g) \rightleftharpoons 2\,HF(g)$$

The Significance of the Equilibrium Constant

Given this definition of an equilibrium constant, what does it mean? What, for example, does a large equilibrium constant ($K_{eq} \gg 1$) imply about a reaction? It means that the forward reaction is largely favored and that there will be more products than reactants when equilibrium is reached. For example, consider the following reaction.

$$H_2(g) + Br_2(g) \rightleftharpoons 2\,HBr(g) \qquad K_{eq} = 1.9 \times 10^{19} \text{ at } 25\,^\circ\text{C}$$

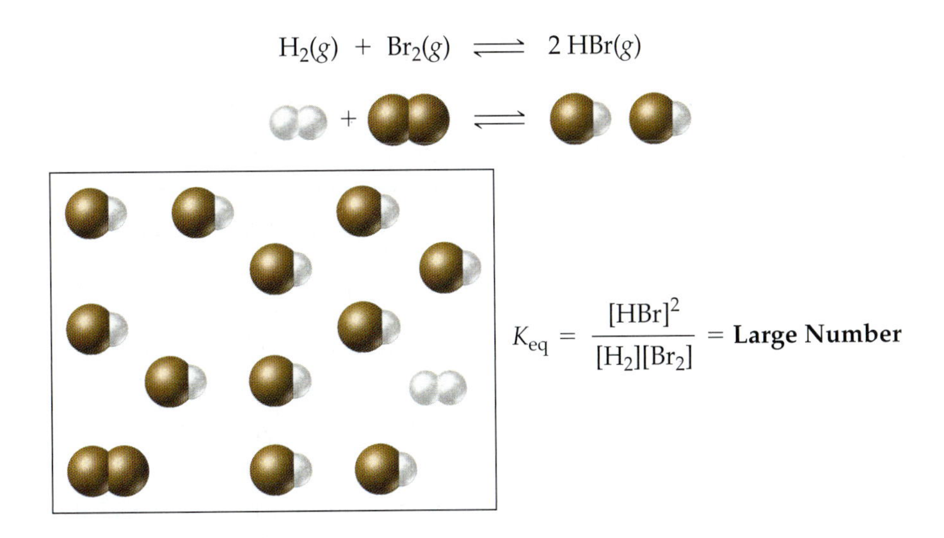

▶ **Figure 15.6 The meaning of a large equilibrium constant** A large equilibrium constant means that there will be a high concentration of products and a low concentration of reactants at equilibrium.

The equilibrium constant is large, meaning that at equilibrium the reaction lies far to the right—high concentrations of products, tiny concentrations of reactants (▲ Figure 15.6).

Conversely, what does a *small* equilibrium constant ($K_{eq} \ll 1$) mean? It means that the reverse reaction is favored and that there will be more reactants than products when equilibrium is reached. For example, consider the following reaction.

$$N_2(g) + O_2(g) \rightleftharpoons 2\,NO(g) \qquad K_{eq} = 4.1 \times 10^{-31} \text{ at } 25\ ^\circ\text{C}$$

The equilibrium constant is very small, meaning that at equilibrium the reaction lies far to the left—high concentrations of reactants, low concentrations of products (▼ Figure 15.7). This is fortunate because N_2 and O_2 are the main components of air. If this equilibrium constant were large, much of the N_2 and O_2 in air would react to form NO, a toxic gas.

The symbol ≈ means "approximately equal to."

To summarize:

- $K_{eq} \ll 1$ Reverse reaction is favored; forward reaction does not proceed very far.
- $K_{eq} \approx 1$ Neither direction is favored; forward reaction proceeds about halfway.
- $K_{eq} \gg 1$ Forward reaction is favored; forward reaction proceeds virtually to completion.

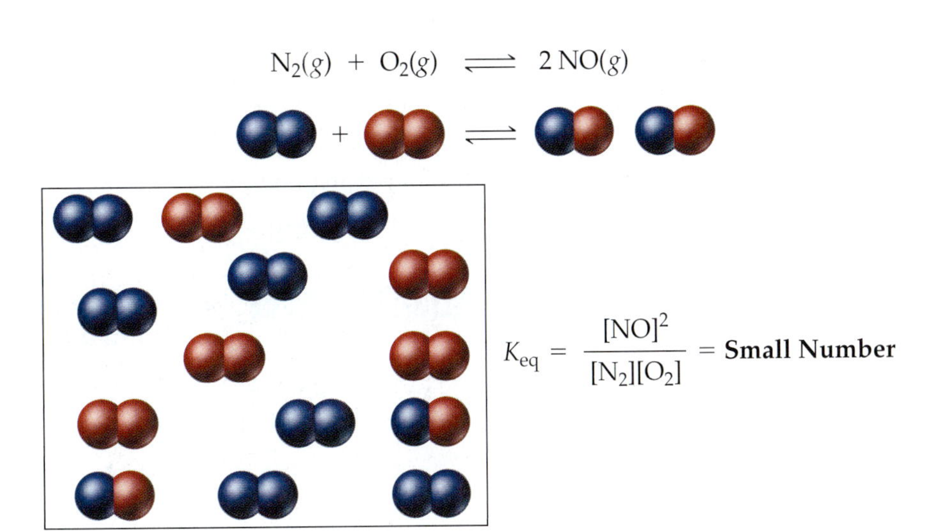

▶ **Figure 15.7 The meaning of a small equilibrium constant** A small equilibrium constant means that there will be a high concentration of reactants and a low concentration of products at equilibrium.

15.5 Heterogeneous Equilibria: The Equilibrium Expression for Reactions Involving a Solid or a Liquid

Consider the following chemical reaction.

$$2\,CO(g) \rightleftharpoons CO_2(g) + C(s)$$

We might expect the expression for the equilibrium constant to be:

$$K_{eq} = \frac{[CO_2][C]}{[CO]^2} \quad \text{(incorrect)}$$

However, since carbon is a solid, its concentration is constant—it does not change. Adding more or less carbon to the reaction mixture does not change the concentration of carbon. The concentration of solids does not change because solids do not expand to fill their container. Their concentration, therefore, depends only on their density, which is constant as long as *some solid is present*. Consequently, pure solids—those reactants or products labeled in the chemical equation with an (*s*)—are not included in the equilibrium expression. The correct equilibrium expression is therefore:

The concentrations of pure solids and pure liquids are excluded from equilibrium expressions because they are constant. Consequently, they simply become incorporated into the value of the equilibrium constant.

$$K_{eq} = \frac{[CO_2]}{[CO]^2} \quad \text{(correct)}$$

Similarly, the concentration of a pure liquid does not change. Consequently, pure liquids—those reactants or products labeled in the chemical equation with an (*l*)—are also excluded from the equilibrium expression. For example, what is the equilibrium expression for the following reaction?

$$CO_2(g) + H_2O(l) \rightleftharpoons H^+(aq) + HCO_3^-(aq)$$

Since $H_2O(l)$ is pure liquid, it is omitted from the equilibrium expression.

$$K_{eq} = \frac{[H^+][HCO_3^-]}{[CO_2]}$$

EXAMPLE 15.2 Writing Equilibrium Expressions for Reactions Involving a Solid or a Liquid

Write an equilibrium expression for the following chemical equation.

$$CaCO_3(s) \rightleftharpoons CaO(s) + CO_2(g)$$

Solution:

Since $CaCO_3(s)$ and $CaO(s)$ are both solids, they are omitted from the equilibrium expression.

$$K_{eq} = [CO_2]$$

SKILLBUILDER 15.2 Writing Equilibrium Expressions for Reactions Involving a Solid or a Liquid

Write an equilibrium expression for the following chemical equation.

$$4\,HCl(g) + O_2(g) \rightleftharpoons 2\,H_2O(l) + 2\,Cl_2(g)$$

15.6 Calculating and Using Equilibrium Constants

Calculating Equilibrium Constants

The most direct way to get a value for the equilibrium constant of a reaction is to measure the concentrations of the reactants and products in a reaction mixture at equilibrium. For example, consider the following reaction.

$$H_2(g) + I_2(g) \rightleftharpoons 2\ HI(g)$$

Equilibrium constants depend on temperature, so temperatures will often be included with equilibrium data. However, the temperature is not a part of the equilibrium expression.

Suppose a mixture of H_2 and I_2 is allowed to come to equilibrium at 445 °C. The measured equilibrium concentrations are $[H_2] = 0.11$ M, $[I_2] = 0.11$ M, and $[HI] = 0.78$ M. What is the value of the equilibrium constant? We begin by setting up the problem in the normal way.

Given:

$$[H_2] = 0.11\ \text{M}$$
$$[I_2] = 0.11\ \text{M}$$
$$[HI] = 0.78\ \text{M}$$

Find: K_{eq}

Solution: The expression for K_{eq} can be written from the balanced equation.

$$K_{eq} = \frac{[HI]^2}{[H_2][I_2]}$$

To calculate the value of K_{eq}, simply substitute the correct equilibrium concentrations into the expression for K_{eq}.

$$\begin{aligned} K_{eq} &= \frac{[HI]^2}{[H_2][I_2]} \\ &= \frac{[0.78]^2}{[0.11][0.11]} \\ &= 5.0 \times 10^1 \end{aligned}$$

The concentrations in an equilibrium expression should always be in units of molarity (M), but the units themselves are normally dropped.

The concentrations within K_{eq} must always be written in moles per liter (M); however, the units are normally dropped in expressing the equilibrium constant so that K_{eq} is unitless.

The particular concentrations of reactants and products for a reaction at equilibrium will *not* always be the same for a given reaction—they will depend on the initial concentrations. However, the *equilibrium constant* will always be the same at a given temperature, regardless of the initial concentrations. For example, Table 15.1 shows several different equilibrium concentrations of H_2, I_2, and HI, each from a different set of initial concentrations. Notice that the equilibrium constant is always the same, regardless of the initial concentrations. In other words, no matter what the initial concentrations are, the reaction will always go in a direction so that the equilibrium concentrations—when substituted into the equilibrium expression—give the same constant, K_{eq}.

A reaction can approach equilibrium from either direction, depending on the initial concentrations, but its K_{eq} at a given temperature will always be the same.

TABLE 15.1 Initial and Equilibrium Concentrations for the Reaction $H_2(g) + I_2(g) \rightleftharpoons 2\ HI(g)$

Initial			Equilibrium			Equilibrium Constant
$[H_2]$	$[I_2]$	$[HI]$	$[H_2]$	$[I_2]$	$[HI]$	$K_{eq} = \frac{[HI]^2}{[H_2][I_2]}$
0.50	0.50	0.0	0.11	0.11	0.78	$\frac{[0.78]^2}{[0.11][0.11]} = 50$
0.0	0.0	0.50	0.055	0.055	0.39	$\frac{[0.39]^2}{[0.055][0.055]} = 50$
0.50	0.50	0.50	0.165	0.165	1.17	$\frac{[1.17]^2}{[0.165][0.165]} = 50$
1.0	0.5	0.0	0.53	0.033	0.934	$\frac{[0.934]^2}{[0.53][0.033]} = 50$

EXAMPLE 15.3 Calculating Equilibrium Constants

Consider the following reaction.

$$2\ CH_4(g) \rightleftharpoons C_2H_2(g) + 3\ H_2(g)$$

A mixture of CH_4, C_2H_2, and H_2 is allowed to come to equilibrium at 1700 °C. The measured equilibrium concentrations are $[CH_4] = 0.0203$ M, $[C_2H_2] = 0.0451$ M, and $[H_2] = 0.112$ M. What is the value of the equilibrium constant at this temperature?

Begin by setting up the problem in the normal format. You are given the concentrations of the reactants and products of a reaction at equilibrium. You are asked to find the equilibrium constant.

Given:

$$[CH_4] = 0.0203 \text{ M}$$
$$[C_2H_2] = 0.0451 \text{ M}$$
$$[H_2] = 0.112 \text{ M}$$

Find: K_{eq}

Write the expression for K_{eq} from the balanced equation. To calculate the value of K_{eq}, simply substitute the correct equilibrium concentrations into the expression for K_{eq}.

Solution:

$$K_{eq} = \frac{[C_2H_2][H_2]^3}{[CH_4]^2}$$

$$K_{eq} = \frac{[0.0451][0.112]^3}{[0.0203]^2}$$
$$= 0.154$$

SKILLBUILDER 15.3 Calculating Equilibrium Constants

Consider the following reaction.

$$CO(g) + 2\ H_2(g) \rightleftharpoons CH_3OH(g)$$

A mixture of CO, H_2, and CH_3OH is allowed to come to equilibrium at 225 °C. The measured equilibrium concentrations are $[CO] = 0.489$ M, $[H_2] = 0.146$ M, and $[CH_3OH] = 0.151$ M. What is the value of the equilibrium constant at this temperature?

SKILLBUILDER PLUS

Suppose that the preceding reaction is carried out at a different temperature and that the initial concentrations of the reactants are $[CO] = 0.500$ M and $[H_2] = 1.00$ M. Assuming that there is no product at the beginning of the reaction, and that at equilibrium $[CO] = 0.15$ M, find the equilibrium constant at this new temperature. Hint: Use the stoichiometric relationships from the balanced equation to find the equilibrium concentrations of H_2 and CH_3OH.

Using Equilibrium Constants in Calculations

The equilibrium constant can also be used to calculate the equilibrium concentration of one of the reactants or products, given the equilibrium concentrations of the others. For example, consider the following reaction.

$$2\,COF_2(g) \rightleftharpoons CO_2(g) + CF_4(g) \qquad K_{eq} = 2.00 \text{ at } 1000\ °C$$

In an equilibrium mixture, the concentration of COF_2 is 0.255 M and the concentration of CF_4 is 0.118 M. What is the equilibrium concentration of CO_2? Again, we set up the problem in the normal way.

Given: $[COF_2] = 0.255$ M

$[CF_4] = 0.118$ M

$K_{eq} = 2.00$

Find: $[CO_2]$

Solution Map:

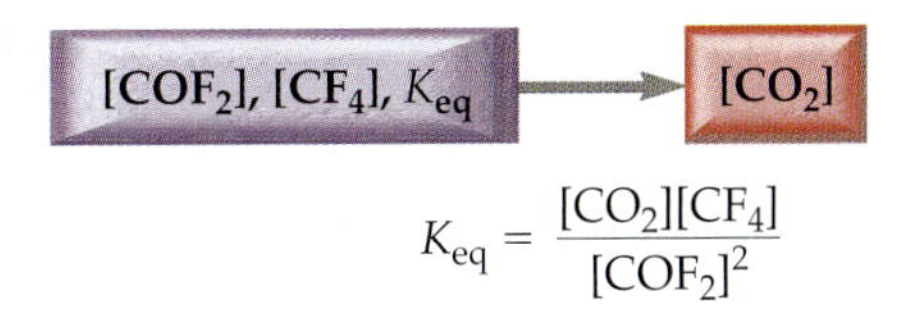

$$K_{eq} = \frac{[CO_2][CF_4]}{[COF_2]^2}$$

In this problem, we are given K_{eq} and the concentrations of one reactant and one product. We are asked to find the concentration of the other product. We can calculate this by using the expression for K_{eq}.

Solution: We first write the equilibrium expression for the reaction, and then solve it for the quantity we are trying to find ($[CO_2]$).

$$K_{eq} = \frac{[CO_2][CF_4]}{[COF_2]^2}$$

$$[CO_2] = K_{eq}\frac{[COF_2]^2}{[CF_4]}$$

Now simply substitute the appropriate values and compute $[CO_2]$.

$$[CO_2] = 2.00\frac{[0.255]^2}{[0.118]}$$

$$= 1.10 \text{ M}$$

EXAMPLE 15.4 Using Equilibrium Constants in Calculations

Consider the following reaction.

$$H_2(g) + I_2(g) \rightleftharpoons 2\,HI(g) \qquad K_{eq} = 69 \text{ at } 340\ °C$$

In an equilibrium mixture, the concentrations of H_2 and I_2 are both 0.020 M. What is the equilibrium concentration of HI?

You are given the equilibrium concentrations of the reactants in a chemical reaction and also given the value of the equilibrium constant. You are asked to find the concentration of the product.

Given:

$[H_2] = [I_2] = 0.020$ M

$K_{eq} = 69$

Find: [HI]

Draw a solution map showing how the equilibrium constant expression gives the relationship between the given concentrations and the concentration you are asked to find.

Solution Map:

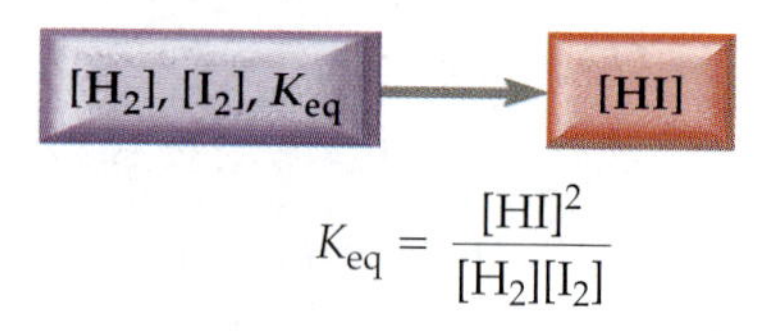

$$K_{eq} = \frac{[HI]^2}{[H_2][I_2]}$$

Solve the equilibrium expression for [HI] and then substitute in the appropriate values to compute it.

Solution:

$$K_{eq} = \frac{[HI]^2}{[H_2][I_2]}$$

$$[HI]^2 = K_{eq}[H_2][I_2]$$

$$[HI] = \sqrt{K_{eq}[H_2][I_2]}$$

$$= \sqrt{69[0.020][0.020]}$$

$$= 0.17 \text{ M}$$

SKILLBUILDER 15.4 Using Equilibrium Constants in Calculations

Diatomic iodine (I_2) decomposes at high temperature to form I atoms according to the following reaction.

$$I_2(g) \rightleftharpoons 2\,I(g) \qquad K_{eq} = 0.011 \text{ at } 1200\ °C$$

In an equilibrium mixture, the concentration of I_2 is 0.10 M. What is the equilibrium concentration of I?

CONCEPTUAL CHECKPOINT 15.2

When the reaction $A(aq) \longrightarrow B(aq) + C(aq)$ is at equilibrium, each of the three compounds has a concentration of 2 M. The equilibrium constant for this reaction is:

(a) 4
(b) 2
(c) 1
(d) 1/2

15.7 Disturbing a Reaction at Equilibrium: Le Châtelier's Principle

We have seen that a chemical system not in equilibrium tends to go toward equilibrium and that the concentrations of the reactants and products at equilibrium correspond to the equilibrium constant, K_{eq}. What happens, however, when a chemical system already at equilibrium is disturbed? **Le Châtelier's principle** states that the chemical system will respond to minimize the disturbance.

Pronounced "le-sha-te-lyay."

Le Châtelier's principle—When a chemical system at equilibrium is disturbed, the system shifts in a direction that minimizes the disturbance.

In other words, a system at equilibrium tries to maintain that equilibrium—it fights back when disturbed.

We can understand Le Châtelier's principle by returning to our Narnia and Middle Earth analogy. Suppose the populations of Narnia and Middle Earth are at equilibrium. This means that the rate of people moving out of Narnia and into Middle Earth is equal to the rate of people moving into Narnia and out of Middle Earth, and the populations of the two kingdoms are

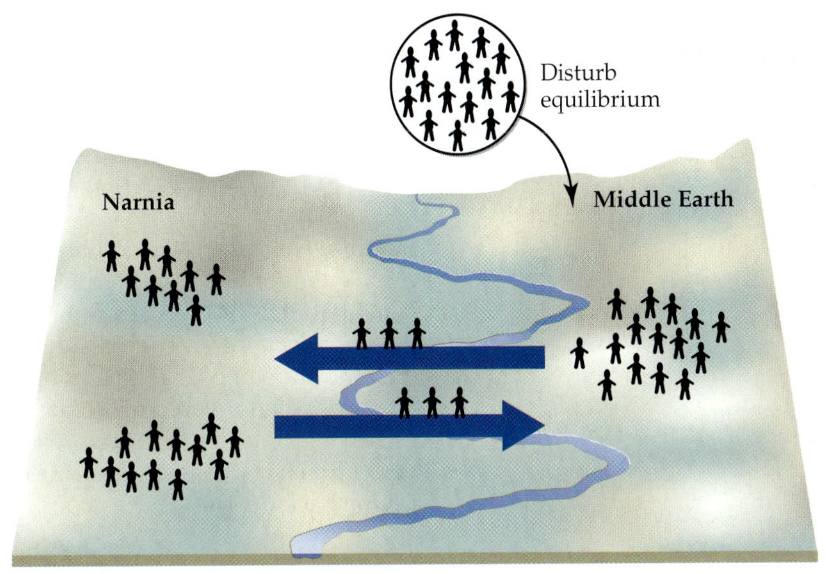

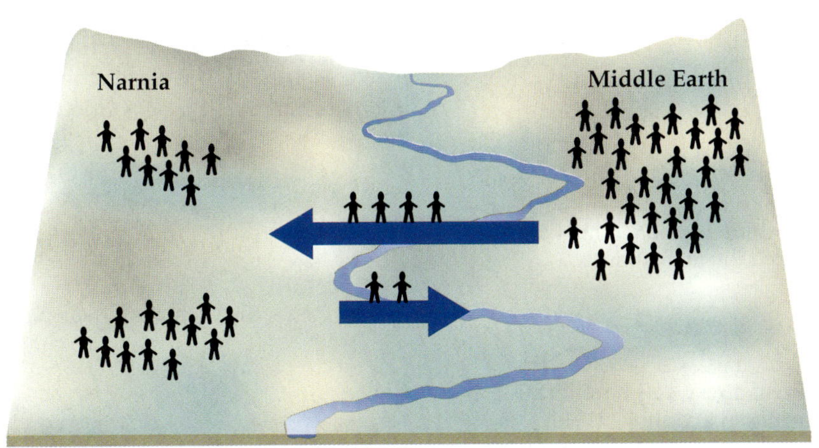

▶ **Figure 15.8 Population analogy for Le Châtelier's principle** When a system at equilibrium is disturbed, it shifts to minimize the disturbance. In this case, adding population to Middle Earth (the disturbance) causes population to move out of Middle Earth (minimizing the disturbance.) **Question:** What would happen if you disturbed the equilibrium by taking population out of Middle Earth? In which direction would the population move to minimize the disturbance?

stable. Now imagine disturbing that balance (▲ Figure 15.8). Suppose we add extra people to Middle Earth. What happens? Since Middle Earth suddenly becomes more crowded, the rate of people leaving Middle Earth increases. The net flow of people is out of Middle Earth and into Narnia. Notice what happened. We disturbed the equilibrium by adding more people to Middle Earth. The system responded by moving people out of Middle Earth—it shifted in the direction that minimized the disturbance.

On the other hand, what happens if we add extra people to Narnia? Since Narnia suddenly gets more crowded, the rate of people leaving Narnia goes up. The net flow of people is out of Narnia and into Middle Earth. We added people to Narnia and the system responded by moving people out of Narnia. When systems at equilibrium are disturbed, they react to counter the disturbance. Chemical systems behave similarly. There are several ways to disturb a system in a chemical equilibrium. We consider each of these separately.

15.8 The Effect of a Concentration Change on Equilibrium

Consider the following reaction at chemical equilibrium.

$$N_2O_4(g) \rightleftharpoons 2\,NO_2(g)$$

Suppose we disturb the equilibrium by adding NO_2 to the equilibrium mixture (▼ Figure 15.9). In other words, we increase the concentration of NO_2. What happens? According to Le Châtelier's principle, the system shifts in a direction to minimize the disturbance. The shift is caused by the increased concentration of NO_2 which in turn increases the rate of the reverse reaction because, as we covered in Section 15.2, reaction rates generally increase with increasing concentration. The reaction goes to the left (it proceeds in the reverse direction), consuming some of the added NO_2 and bringing its concentration back down.

When we say that a reaction *shifts to the left* we mean that it proceeds in the reverse direction, consuming products and forming reactants.

$$N_2O_4(g) \rightleftharpoons 2\,NO_2(g)$$

Reaction shifts left ← Add NO_2

On the other hand, what happens if we add extra N_2O_4, increasing its concentration? In this case, the rate of the forward reaction increases and the reaction shifts to the right, consuming some of the added N_2O_4, and bringing *its* concentration back down (▶ Figure 15.10).

When we say that a reaction *shifts to the right* we mean that it proceeds in the forward direction, consuming reactants and forming products.

$$N_2O_4(g) \rightleftharpoons 2\,NO_2(g)$$

Add N_2O_4 Reaction shifts right →

In both cases, the system shifts in a direction that minimizes the disturbance.

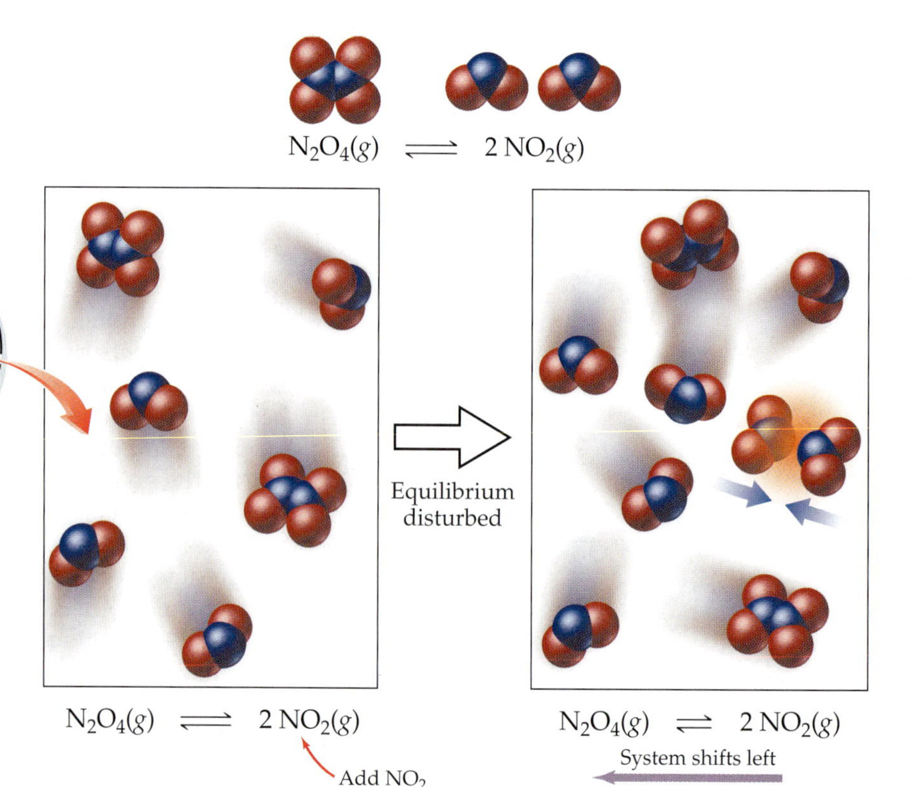

▶ **Figure 15.9 Le Châtelier's principle in action: I** When a system at equilibrium is disturbed, it changes to minimize the disturbance. In this case, adding NO_2 (the disturbance) causes the reaction to shift left, consuming NO_2 by forming more N_2O_4.

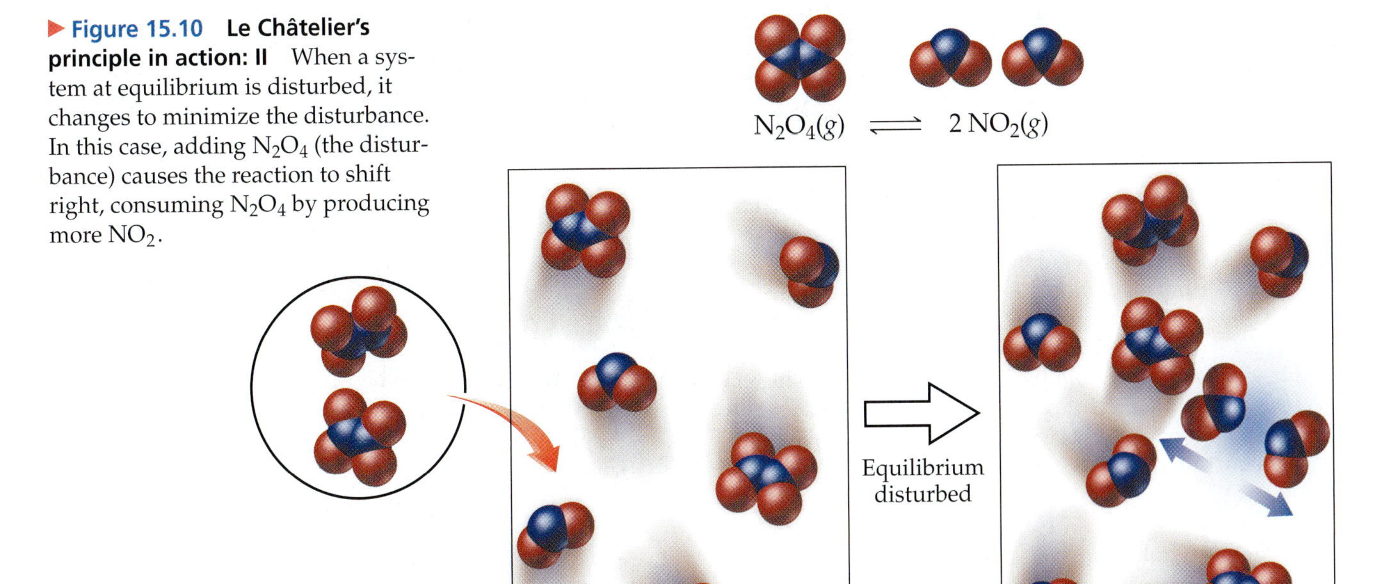

▶ **Figure 15.10 Le Châtelier's principle in action: II** When a system at equilibrium is disturbed, it changes to minimize the disturbance. In this case, adding N_2O_4 (the disturbance) causes the reaction to shift right, consuming N_2O_4 by producing more NO_2.

TUTORIAL
Ammonia Equilibrium

To summarize, if a chemical system is at equilibrium:

- Increasing the concentration of one or more of the reactants causes the reaction to shift to the right (in the direction of the products).
- Increasing the concentration of one or more of the products causes the reaction to shift to the left (in the direction of the reactants).

EXAMPLE 15.5 The Effect of a Concentration Change on Equilibrium

Consider the following reaction at equilibrium.

$$CaCO_3(s) \rightleftharpoons CaO(s) + CO_2(g)$$

What is the effect of adding additional CO_2 to the reaction mixture? What is the effect of adding additional $CaCO_3$?

Solution:

Adding additional CO_2 increases the concentration of CO_2 and causes the reaction to shift to the left. Adding additional $CaCO_3$ does not increase the concentration of $CaCO_3$ because $CaCO_3$ is a solid and thus has a constant concentration. It is therefore not included in the equilibrium expression and has no effect on the position of the equilibrium.

SKILLBUILDER 15.5 The Effect of a Concentration Change on Equilibrium

Consider the following reaction in chemical equilibrium.

$$2\,BrNO(g) \rightleftharpoons 2\,NO(g) + Br_2(g)$$

What is the effect of adding additional Br_2 to the reaction mixture? What is the effect of adding additional BrNO?

SKILLBUILDER PLUS

What is the effect of removing some Br_2 from the preceding reaction mixture?

Chemistry AND HEALTH

How a Developing Fetus Gets Oxygen from Its Mother

Have you ever wondered how a baby in the womb gets oxygen? Unlike you and me, a fetus cannot breathe. Yet like you and me, a fetus needs oxygen. Where does that oxygen come from? In adults, oxygen is carried in the blood by a protein molecule called hemoglobin, which is abundantly present in red blood cells. Hemoglobin (Hb) reacts with oxygen according to the following equilibrium equation.

$$Hb + O_2 \rightleftharpoons HbO_2$$

The equilibrium constant for this reaction is neither large nor small, but intermediate. Consequently, the reaction can shift toward the right or the left, depending on the concentration of oxygen. As blood flows through the lungs, where oxygen concentrations are high, the equilibrium shifts to the right—hemoglobin loads oxygen.

Lung $[O_2]$ high

$$Hb + O_2 \rightleftharpoons HbO_2$$

Reaction shifts right

As blood flows through muscles and organs that are using oxygen (where oxygen concentrations have been depleted) the equilibrium shifts to the left—hemoglobin unloads oxygen.

Muscle $[O_2]$ low

$$Hb + O_2 \rightleftharpoons HbO_2$$

Reaction shifts left

However, a fetus has its own blood circulatory system. The mother's blood never flows into the fetus's body, and the fetus cannot get any air in the womb. So how does the fetus get oxygen?

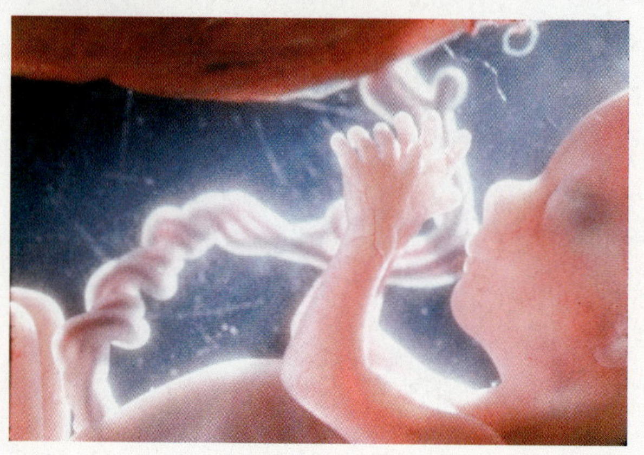

▲ A human fetus. **Question:** How does the fetus get oxygen?

The answer lies in fetal hemoglobin (HbF), which is slightly different from adult hemoglobin. Like adult hemoglobin, fetal hemoglobin is in equilibrium with oxygen.

$$HbF + O_2 \rightleftharpoons HbFO_2$$

However, the equilibrium constant for fetal hemoglobin is larger than the equilibrium constant for adult hemoglobin. In other words, fetal hemoglobin will load oxygen at a lower oxygen concentration than adult hemoglobin. So, when the mother's hemoglobin flows through the placenta, it unloads oxygen into the placenta. The baby's blood also flows into the placenta, and even though the baby's blood never mixes with the mother's blood, the fetal hemoglobin within the baby's blood loads the oxygen (that the mother's hemoglobin unloaded) and carries it to the baby. Nature has thus engineered a chemical system where mother's hemoglobin can in effect *hand off* oxygen to the baby's hemoglobin.

CAN YOU ANSWER THIS? *What would happen if fetal hemoglobin had the same equilibrium constant for the reaction with oxygen as adult hemoglobin?*

15.9 The Effect of a Volume Change on Equilibrium

See Section 11.4 for a complete description of Boyle's law.

How does a system in chemical equilibrium respond to a volume change? Remember from Chapter 11 that changing the volume of a gas (or a gas mixture) results in a change in pressure. Remember also that pressure and volume are inversely related: a *decrease* in volume causes an *increase* in pressure, and an *increase* in volume causes a *decrease* in pressure. So, if the volume of a gaseous reaction mixture at chemical equilibrium is changed, the pressure changes and the system will shift in a direction to minimize that change. For example, consider the following reaction at equilibrium in a cylinder equipped with a moveable piston.

$$N_2(g) + 3\,H_2(g) \rightleftharpoons 2\,NH_3(g)$$

What happens if we push down on the piston, lowering the volume and raising the pressure (▼ Figure 15.11)? How can the chemical system bring the pressure back down? Look carefully at the reaction coefficients. If the reaction shifts to the right, 4 mol of gas particles are converted to 2 mol of gas particles. Fewer gas particles results in lower pressure. So the system shifts to the right, lowering the number of gas molecules and bringing the pressure back down, minimizing the disturbance.

Notice that, from the ideal gas law ($PV = nRT$), we can see that lowering the number of moles of a gas (n) results in a lower pressure (P).

Consider the same reaction mixture at equilibrium again. What happens if, this time, we pull *up* on the piston, *increasing* the volume (▼ Figure 15.12)? The higher volume results in a lower pressure and the system responds by trying to bring the pressure back up. It can do this by shifting to the left, converting 2 mol of gas particles into 4 mol of gas particles, increasing the pressure and minimizing the disturbance.

TUTORIAL NO_2–N_2O_4 Equilibrium

To summarize, if a chemical system is at equilibrium:

- Decreasing the volume causes the reaction to shift in the direction that has fewer moles of gas particles.
- Increasing the volume causes the reaction to shift in the direction that has more moles of gas particles.

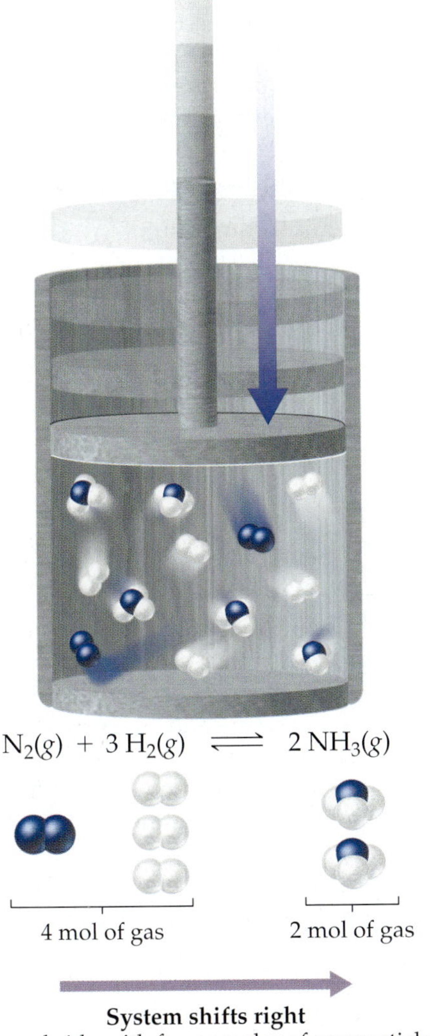

▲ **Figure 15.11 Effect of volume decrease on equilibrium** When the volume of an equilibrium mixture is decreased, the pressure increases. The system responds (to bring the pressure back down) by shifting to the right, the side of the reaction with the fewest moles of gas particles.

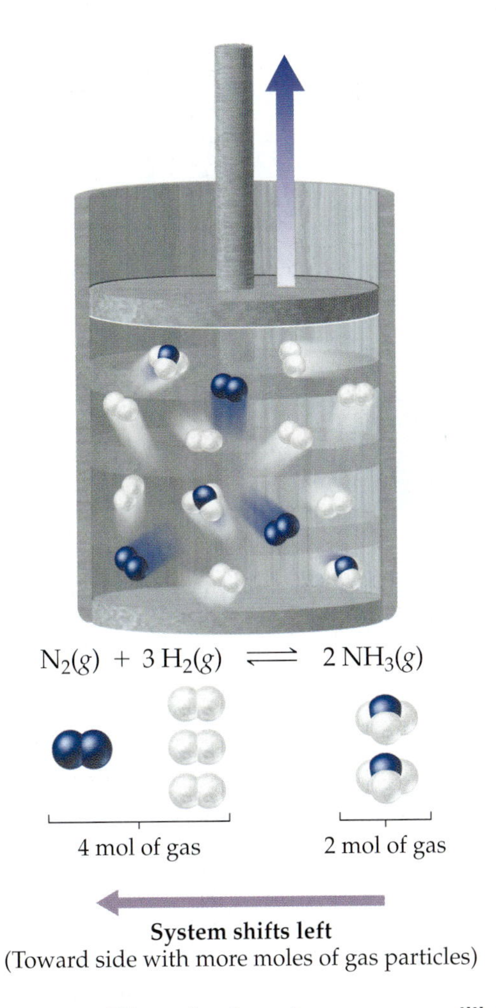

▲ **Figure 15.12 Effect of volume increase on equilibrium** When the volume of an equilibrium mixture is increased, the pressure decreases. The system responds (to raise the pressure) by shifting to the left, the side of the reaction with the most moles of gas particles.

Notice that if a chemical reaction has an equal number of moles of gas particles on both sides of the chemical equation, a change in volume has no effect. For example, consider the following reaction.

$$H_2(g) + I_2(g) \rightleftharpoons 2\,HI(g)$$

Both the left and the right side of the equation contain 2 mol of gas particles, so a change in volume has no effect on this reaction. Similarly, a change in volume has no effect on a reaction that has no gaseous reactants or products.

EXAMPLE 15.6 The Effect of a Volume Change on Equilibrium

Consider the following reaction at chemical equilibrium.

$$2\,KClO_3(s) \rightleftharpoons 2\,KCl(s) + 3\,O_2(g)$$

What is the effect of decreasing the volume of the reaction mixture? Increasing the volume of the reaction mixture?

Solution:

The chemical equation has 3 mol of gas on the right and 0 mol of gas on the left. Decreasing the volume of the reaction mixture increases the pressure and causes the reaction to shift to the left (toward the side with fewer moles of gas particles). Increasing the volume of the reaction mixture decreases the pressure and causes the reaction to shift to the right (toward the side with more moles of gas particles).

SKILLBUILDER 15.6 The Effect of a Volume Change on Equilibrium

Consider the following reaction at chemical equilibrium.

$$2\,SO_2(g) + O_2(g) \rightleftharpoons 2\,SO_3(g)$$

What is the effect of decreasing the volume of the reaction mixture? Increasing the volume of the reaction mixture?

15.10 The Effect of a Temperature Change on Equilibrium

According to Le Châtelier's principle, if the temperature of a system at equilibrium is changed, the system should shift in a direction to counter that change. So if the temperature is increased, the reaction should shift in the direction that attempts to decrease the temperature and vice versa. Recall from Section 3.8 that energy changes are often associated with chemical reactions. If we want to predict the direction in which a reaction will shift upon a temperature change, we must understand how a shift in the reaction affects the temperature.

We can classify chemical reactions according to whether they absorb or emit heat energy in the course of the reaction. An **exothermic reaction** emits heat.

$$\text{Exothermic reaction} \qquad A + B \rightleftharpoons C + D + \text{Heat}$$

In an exothermic reaction, you can think of heat as a product. Consequently, raising the temperature of an exothermic reaction—think of this as adding heat—causes the reaction to shift left. For example, the reaction of nitrogen with hydrogen to form ammonia is exothermic.

$$N_2(g) + 3\,H_2(g) \rightleftharpoons 2\,NH_3 + \text{Heat}$$

Reaction shifts left ← ; Add heat ↑

Raising the temperature of an equilibrium mixture of these three gases causes the reaction to shift left, absorbing some of the added heat. Conversely, lowering the temperature of an equilibrium mixture of these three gases causes the reaction to shift right, releasing heat.

$$N_2(g) + 3\,H_2(g) \rightleftharpoons 2\,NH_3 + \text{Heat}$$

Reaction shifts right → ; Remove heat ↓

In contrast, an **endothermic reaction** absorbs heat.

$$\text{Endothermic reaction:} \quad A + B + \text{Heat} \rightleftharpoons C + D$$

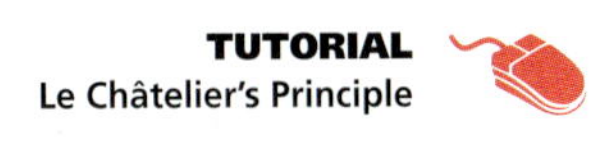

In an endothermic reaction, you can think of heat as a reactant. Consequently, raising the temperature (or adding heat) causes an endothermic reaction to shift right. For example, the following reaction is endothermic.

$$\underset{\text{Colorless}}{N_2O_4(g)} + \text{Heat} \rightleftharpoons \underset{\text{Brown}}{2\,NO_2}$$

Add heat ↑ ; Reaction shifts right →

Raising the temperature of an equilibrium mixture of these two gases causes the reaction to shift right, absorbing some of the added heat. Since N_2O_4 is colorless and NO_2 is brown, the effects of changing the temperature of this reaction are easily seen (▼ Figure 15.13). On the other hand, lowering the temperature of a reaction mixture of these two gases causes the reaction to shift left, releasing heat.

$$\underset{\text{Colorless}}{N_2O_4(g)} + \text{Heat} \rightleftharpoons \underset{\text{Brown}}{2\,NO_2}$$

Remove heat ↓ ; Reaction shifts left ←

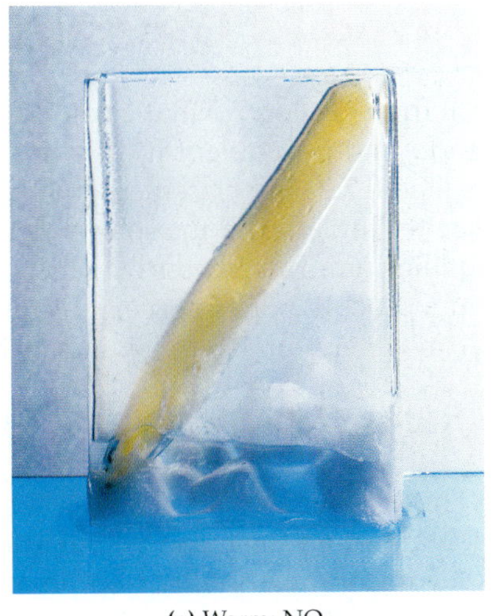
(a) Warm: NO_2

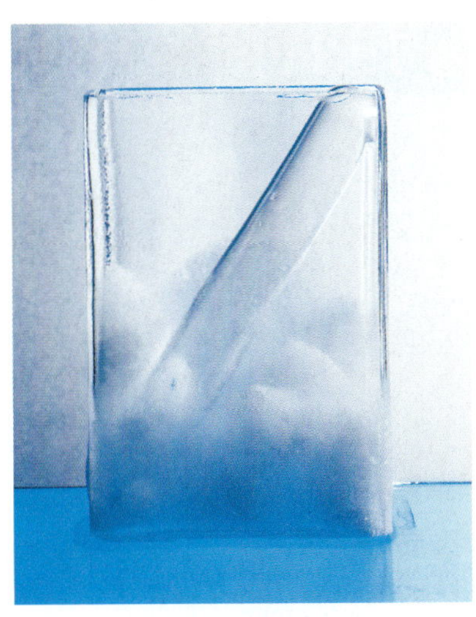
(b) Cool: N_2O_4

▶ **Figure 15.13 Equilibrium as a function of temperature** Since the reaction $N_2O_4(g) \rightleftharpoons 2\,NO_2(g)$ is endothermic, warm temperatures **(a)** cause a shift to the right, toward the production of brown NO_2. Cool temperatures **(b)** cause a shift to the left, to colorless N_2O_4.

To summarize:

In an exothermic chemical reaction, heat is a product and:

- Increasing the temperature causes the reaction to shift left (in the direction of the reactants).
- Decreasing the temperature causes the reaction to shift right (in the direction of the products).

In an endothermic chemical reaction, heat is a reactant and:

- Increasing the temperature causes the reaction to shift right (in the direction of the products).
- Decreasing the temperature causes the reaction to shift left (in the direction of the reactants).

EXAMPLE 15.7 The Effect of a Temperature Change on Equilibrium

The following reaction is endothermic.

$$CaCO_3(s) \rightleftharpoons CaO(s) + CO_2(g)$$

What is the effect of increasing the temperature of the reaction mixture? Decreasing the temperature?

Solution:

Since the reaction is endothermic, we can think of heat as a reactant.

$$\text{Heat} + CaCO_3(s) \rightleftharpoons CaO(s) + CO_2(g)$$

Raising the temperature is adding heat, causing the reaction to shift to the right. Lowering the temperature is removing heat, causing the reaction to shift to the left.

SKILLBUILDER 15.7 The Effect of a Temperature Change on Equilibrium

The following reaction is exothermic.

$$2\,SO_2(g) + O_2(g) \rightleftharpoons 2\,SO_3(g)$$

What is the effect of increasing the temperature of the reaction mixture? Decreasing the temperature?

15.11 The Solubility-Product Constant

Recall from Section 7.7 that a compound is considered soluble if it dissolves in water and insoluble if it does not. Recall also that, through the *solubility rules* (Table 7.2), we classified ionic compounds as soluble or insoluble. We can better understand the solubility of an ionic compound with the concept of equilibrium. The process by which an ionic compound dissolves is an equilibrium process. For example, we can represent the dissolving of calcium fluoride in water with the following chemical equation.

$$CaF_2(s) \rightleftharpoons Ca^{2+}(aq) + 2\,F^-(aq)$$

The equilibrium expression for a chemical equation that represents the dissolving of an ionic compound is called the **solubility-product constant (K_{sp})**. For CaF_2, the solubility-product constant is

$$K_{sp} = [Ca^{2+}][F^-]^2$$

TABLE 15.2 Selected Solubility-Product Constants (K_{sp})

Compound	Formula	K_{sp}
barium sulfate	$BaSO_4$	1.07×10^{-10}
calcium carbonate	$CaCO_3$	4.96×10^{-9}
calcium fluoride	CaF_2	1.46×10^{-10}
calcium hydroxide	$Ca(OH)_2$	4.68×10^{-6}
calcium sulfate	$CaSO_4$	7.10×10^{-5}
copper(II) sulfide	CuS	1.27×10^{-36}
iron(II) carbonate	$FeCO_3$	3.07×10^{-11}
iron(II) hydroxide	$Fe(OH)_2$	4.87×10^{-17}
lead(II) chloride	$PbCl_2$	1.17×10^{-5}
lead(II) sulfate	$PbSO_4$	1.82×10^{-8}
lead(II) sulfide	PbS	9.04×10^{-29}
magnesium carbonate	$MgCO_3$	6.82×10^{-6}
magnesium hydroxide	$Mg(OH)_2$	2.06×10^{-13}
silver chloride	$AgCl$	1.77×10^{-10}
silver chromate	Ag_2CrO_4	1.12×10^{-12}
silver iodide	AgI	8.51×10^{-17}

Notice that, as we discussed in Section 15.5, solids are omitted from the equilibrium expression.

The K_{sp} value is an indicator of the solubility of a compound. A large K_{sp} (forward reaction favored) means that the compound is very soluble. A small K_{sp} (reverse reaction favored) means that the compound is not very soluble. Table 15.2 lists the value of K_{sp} for a number of ionic compounds.

EXAMPLE 15.8 Writing Expressions for K_{sp}

Write expressions for K_{sp} for each of the following ionic compounds.

(a) $BaSO_4$
(b) $Mn(OH)_2$
(c) Ag_2CrO_4

Solution:

To write the expression for K_{sp}, first write the chemical reaction showing the solid compound in equilibrium with its dissolved aqueous ions. Then write the equilibrium expression based on this equation.

(a) $BaSO_4(s) \rightleftharpoons Ba^{2+}(aq) + SO_4^{2-}(aq)$

$K_{sp} = [Ba^{2+}][SO_4^{2-}]$

(b) $Mn(OH)_2(s) \rightleftharpoons Mn^{2+}(aq) + 2\,OH^-(aq)$

$K_{sp} = [Mn^{2+}][OH^-]^2$

(c) $Ag_2CrO_4(s) \rightleftharpoons 2\,Ag^+(aq) + CrO_4^{2-}(aq)$

$K_{sp} = [Ag^+]^2[CrO_4^{2-}]$

SKILLBUILDER 15.8 Writing Expressions for K_{sp}

Write expressions for K_{sp} for each of the following ionic compounds.

(a) AgI
(b) $Ca(OH)_2$

Using K_{sp} to Determine Molar Solubility

Recall from Section 13.3 that the solubility of a compound is the amount of the compound that dissolves in a certain amount of liquid. The **molar solubility** is simply the solubility in units of moles per liter. The molar solubility of a compound can be computed directly from K_{sp}. For example, consider silver chloride.

$$AgCl(s) \rightleftharpoons Ag^+(aq) + Cl^-(aq) \qquad K_{sp} = 1.77 \times 10^{-10}$$

How can we find the molar solubility of AgCl from K_{sp}? First, notice that K_{sp} is *not* the molar solubility; it is the solubility-product constant.

Second, notice that the concentration of either Ag^+ or Cl^- at equilibrium will be equal to the amount of AgCl that dissolved. We know this from the relationship of the stoichiometric coefficients in the balanced equation.

$$1 \text{ mol AgCl} \equiv 1 \text{ mol } Ag^+ \equiv 1 \text{ mol } Cl^-$$

Consequently, to find the solubility, we simply need to find $[Ag^+]$ or $[Cl^-]$ at equilibrium. We can do this by writing the expression for the solubility-product constant.

$$K_{sp} = [Ag^+][Cl^-]$$

EVERYDAY Chemistry

Hard Water

Many parts of the United States obtain their water from lakes or reservoirs that have significant concentrations of $CaCO_3$ and $MgCO_3$. These salts dissolve into rainwater as it flows through soils rich in $CaCO_3$ and $MgCO_3$. Water containing these salts is known as hard water. Hard water is not a health hazard because both calcium and magnesium are part of a healthy diet, but their presence in water can be annoying. For example, because of their relatively low solubility-product constants, water can easily become saturated with $CaCO_3$ and $MgCO_3$. A drop of water, for example, becomes saturated with $CaCO_3$ and $MgCO_3$ as it evaporates. A saturated solution precipitates some of its dissolved ions. These precipitates show up as scaly deposits on faucets, sinks, or cookware. Washing cars or dishes with hard water leaves spots of $CaCO_3$ and $MgCO_3$ as these precipitate out of drying drops of water.

▲ Hard water leaves scaly deposits on plumbing fixtures.

CAN YOU ANSWER THIS? *Is the water in your community hard or soft? Use the solubility-product constants from Table 15.2 to calculate the molar solubility of $CaCO_3$ and $MgCO_3$. How many moles of $CaCO_3$ are in 5 L of water that is saturated with $CaCO_3$? How many grams?*

Since both Ag^+ and Cl^- come from AgCl, their concentrations must be equal. Since the concentration of either one is equal to the solubility, we can write:

$$\text{Solubility} = S = [Ag^+] = [Cl^-]$$

Substituting this into the expression for the solubility constant, we get:

$$\begin{aligned} K_{sp} &= [Ag^+][Cl^-] \\ &= S \times S \\ &= S^2 \end{aligned}$$

Therefore,

$$\begin{aligned} S &= \sqrt{K_{sp}} \\ &= \sqrt{1.77 \times 10^{-10}} \\ &= 1.33 \times 10^{-5}\text{ M} \end{aligned}$$

So the molar solubility of AgCl is 1.33×10^{-5} mol/L.

In this text, we limit the calculation of molar solubility to ionic compounds whose chemical formulas have one cation and one anion.

EXAMPLE 15.9 Calculating Molar Solubility from K_{sp}

Calculate the molar solubility of $BaSO_4$.

Begin by writing the reaction by which solid $BaSO_4$ dissolves into its constituent aqueous ions.

Solution:

$$BaSO_4(s) \rightleftharpoons Ba^{2+}(aq) + SO_4^{2-}(aq)$$

Next, write the expression for K_{sp}.

$$K_{sp} = [Ba^{2+}][SO_4^{2-}]$$

Define the molar solubility (S) as $[Ba^{2+}]$ or $[SO_4^{2-}]$ at equilibrium.

$$S = [Ba^{2+}] = [SO_4^{2-}]$$

Substitute S into the equilibrium expression and solve for it.

$$\begin{aligned} K_{sp} &= [Ba^{2+}][SO_4^{2-}] \\ &= S \times S \\ &= S^2 \end{aligned}$$

Therefore

$$S = \sqrt{K_{sp}}$$

Finally, look up the value of K_{sp} in Table 15.2 and compute S. The molar solubility of $BaSO_4$ is therefore 1.03×10^{-5} mol/L.

$$\begin{aligned} S &= \sqrt{K_{sp}} \\ &= \sqrt{1.07 \times 10^{-10}} \\ &= 1.03 \times 10^{-5}\text{ M} \end{aligned}$$

SKILLBUILDER 15.9 Calculating Molar Solubility from K_{sp}

Calculate the molar solubility of $CaSO_4$.

15.12 The Path of a Reaction and the Effect of a Catalyst

In this chapter, we have learned that the equilibrium constant describes the ultimate fate of a chemical reaction. Large equilibrium constants mean that the reaction favors the products. Small equilibrium constants mean that the reaction favors the reactants. But the equilibrium constant by itself does not tell the whole story. For example, consider the following reaction between hydrogen gas and oxygen gas to form water.

Warning: Hydrogen gas is explosive and should never be handled without proper training.

$$2\,H_2(g) + O_2(g) \rightleftharpoons 2\,H_2O(g) \qquad K_{eq} = 3.2 \times 10^{81} \text{ at } 25\ ^\circ C$$

The equilibrium constant is huge, meaning that the forward reaction is heavily favored. Yet I can mix hydrogen and oxygen in a balloon at room temperature, and no reaction occurs. Hydrogen and oxygen peacefully coexist together inside of the balloon and form virtually no water. Why?

The equilibrium constant describes *how far* a chemical reaction will go. The reaction rate describes *how fast* it will get there.

To answer this question, we must go back to a topic from the beginning of this chapter—the reaction rate. At 25 °C, the reaction rate between hydrogen gas and oxygen gas is virtually zero. Even though the equilibrium constant is large, the reaction rate is small and no reaction occurs. The reaction rate between hydrogen and oxygen is slow because the reaction has a large **activation energy:** energy that must be supplied in order to get a reaction started. Activation energies exist for most chemical reactions because the original bonds must begin to break before new bonds begin to form, and this requires energy. For example, for H_2 and O_2 to react to form H_2O, the H—H and O=O bonds must begin to break before the new bonds can form. The initial weakening of H_2 and O_2 bonds takes energy—this is the activation energy of the reaction.

The activation energy is sometimes called the *activation barrier*.

How Activation Energies Affect Reaction Rates

We can illustrate how activation energies affect reaction rates by means of a graph showing the energy progress of a reaction (▼ Figure 15.14). We can see from the figure that the products have less energy than the reactants, so the reaction is exothermic (it releases energy when it occurs). However, before the reaction can take place, some energy must first be *added*—the energy of the reactants must be raised by an amount that we call the activation energy. The activation energy is thus a kind of "energy hump" that normally exists between the reactants and products.

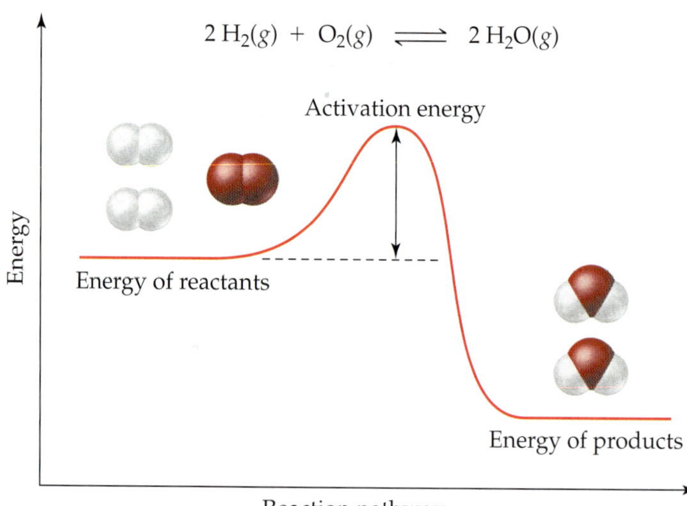

▶ **Figure 15.14 Activation energy** This plot represents the energy of the reactants and products along the reaction pathway (as the reaction occurs). Notice that the energy of the products is lower than the energy of the reactants, so this is an exothermic reaction. However, notice that the reactants must get over an energy hump—called the *activation energy*—to proceed from reactants to products.

We can understand this concept better by means of a simple analogy—getting a chemical reaction to occur is much like trying to push a bunch of boulders over a hill (▼ Figure 15.15a). We can think of each collision that occurs between reactant molecules as an attempt to roll a boulder over the hill. We can think of a successful collision between two molecules (one that leads to product) as a successful attempt to roll a boulder over the hill and down the other side.

For rolling boulders, the higher the hill is, the harder it will be to get the boulders over the hill, and the fewer the number of boulders that make it over the hill in a given period of time. Similarly, for chemical reactions, the higher the activation energy is, the fewer the number of reactant molecules that make it over the barrier, and the slower the reaction rate. In general:

> At a given temperature, the higher the activation energy for a chemical reaction, the slower the reaction rate.

Are there any ways to speed up a slow reaction (one with a high activation barrier)? In Section 15.2 we talked of two ways to increase reaction rates. The first way is to simply increase the concentrations of the reactants, which results in more collisions per unit time. This is analogous to simply pushing more boulders toward the hill in a given period of time. The second way to increase the rate of a reaction is to increase the temperature. This also results in more collisions per unit time, but it also results in collisions with higher energies. Higher-energy collisions are analogous to pushing the boulders harder (with more force), which will result in more boulders making it over the hill per unit time—a faster reaction rate. There is, however, a third way to speed up a slow chemical reaction: by using a *catalyst*.

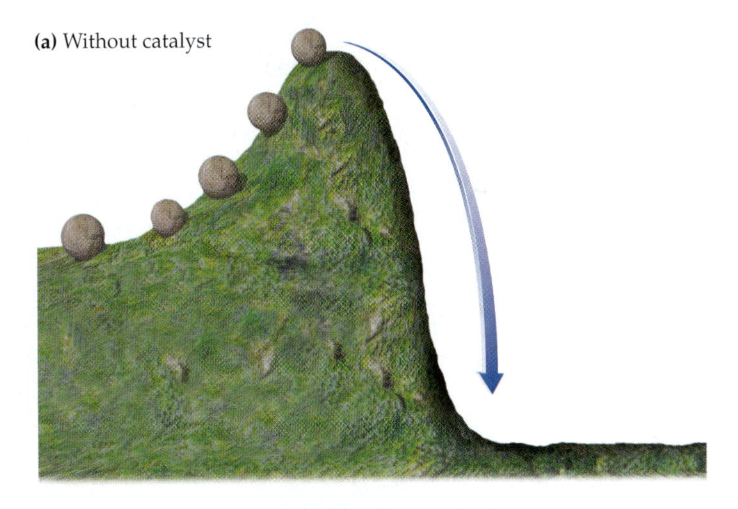

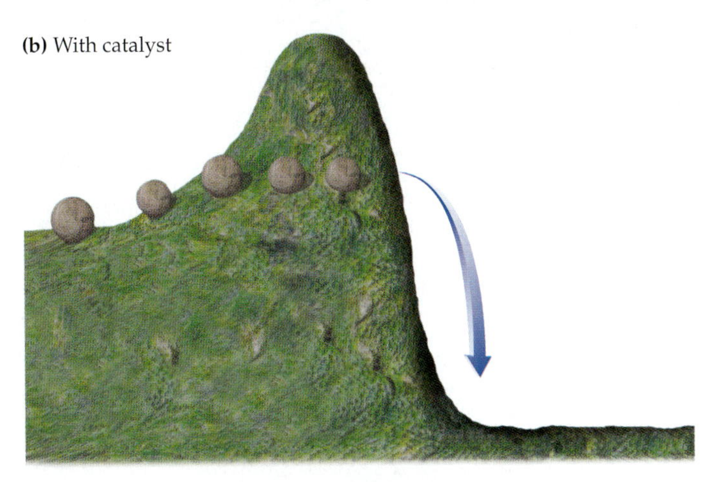

▶ **Figure 15.15 Hill analogy for activation energy** There are several ways to get these boulders over the hill as fast as possible. **(a)** One way is simply to push them harder—this is analogous to an increase in temperature for a chemical reaction. **(b)** The other way is to find a path that goes *around* the hill—this is analogous to the role of a catalyst for a chemical reaction.

Catalysts Lower the Activation Energy

A catalyst does not change the *position* of equilibrium, only *how fast* equilibrium is reached.

A **catalyst** is a substance that increases the rate of a chemical reaction but is not consumed by the reaction. A catalyst works by lowering the activation energy for the reaction, making it easier for reactants to get over the energy hump (▼ Figure 15.16). In our boulder analogy, a catalyst creates another path for the boulders to travel—a path with a smaller hill (see Figure 15.15b). For example, consider the noncatalytic destruction of ozone in the upper atmosphere.

$$O_3 + O \longrightarrow 2\,O_2$$

Upper-atmospheric ozone forms a shield against harmful ultraviolet light that would otherwise enter Earth's atmosphere. See the *Chemistry in the Environment* box in Chapter 6.

The reason that we have a protective ozone layer is that this reaction has a fairly high activation barrier and therefore proceeds at a fairly slow rate. The ozone layer does not rapidly decompose into O_2.

However, the addition of Cl (from synthetic chlorofluorocarbons) to the upper atmosphere has resulted in another pathway by which O_3 can be destroyed. The first step in this pathway—called the catalytic destruction of ozone—is the reaction of Cl with O_3 to form ClO and O_2.

$$Cl + O_3 \longrightarrow ClO + O_2$$

This is followed by a second step in which ClO reacts with O, regenerating Cl.

$$ClO + O \longrightarrow Cl + O_2$$

Notice that, if we add the two reactions, the overall reaction is identical to the noncatalytic reaction.

$$\begin{array}{l} \cancel{Cl} + O_3 \longrightarrow \cancel{ClO} + O_2 \\ \cancel{ClO} + O \longrightarrow \cancel{Cl} + O_2 \\ \hline O_3 + O \longrightarrow 2\,O_2 \end{array}$$

However, the activation energies for the two reactions in this pathway are much smaller than for the first, uncatalyzed pathway, and therefore the reaction occurs at a much faster rate. Note that the Cl is not consumed in the overall reaction—this is characteristic of a catalyst.

In the case of the catalytic destruction of ozone, the catalyst speeds up a reaction that we do *not* want to happen. However, most of the time, catalysts are used to speed up reactions that we *do* want to happen. For example, your car most likely has a catalytic converter in its exhaust system. The catalytic converter contains a catalyst that converts exhaust pollutants (such as carbon monoxide) into less harmful substances (such as carbon dioxide). These reactions occur only with the help of a catalyst because they are too slow to occur otherwise.

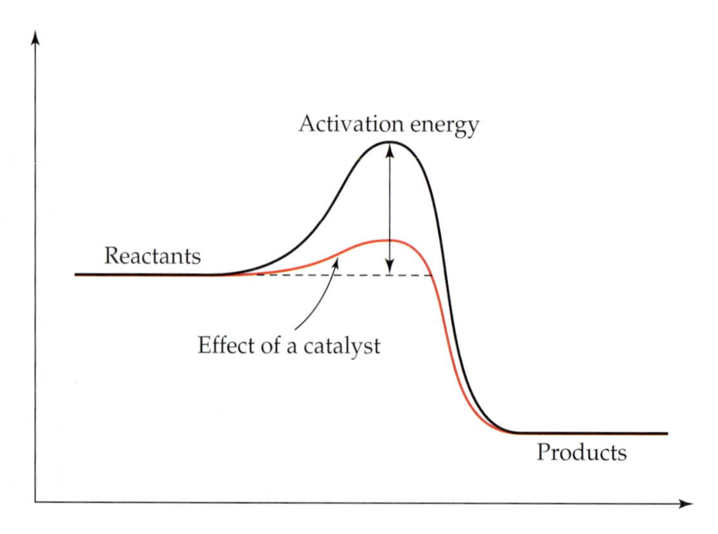

▶ **Figure 15.16 Function of a catalyst** A catalyst provides an alternate pathway with a lower activation energy barrier for the reaction.

A catalyst cannot change the value of K_{eq} for a reaction—it affects only the *rate* of the reaction.

The role of catalysis in chemistry cannot be overstated. Without catalysts, chemistry would be a different field. For many reactions, increasing the reaction rate in other ways—such as increasing the temperature—are simply not feasible. Many reactants are thermally sensitive—increasing the temperature often destroys them. The only way to carry out many reactions is to use catalysts.

Enzymes: Biological Catalysts

Perhaps the best example of chemical catalysis is found in living organisms. Most of the thousands of reactions that must occur for a living organism to survive would be too slow at normal temperatures. So living organisms use **enzymes**, biological catalysts that increase the rates of biochemical reactions. For example, when we eat sucrose (table sugar), our bodies must break it into two smaller molecules called glucose and fructose. The equilibrium constant for this reaction is large, favoring the products. However, at room temperature, or even at body temperature, the sucrose does not break into glucose and fructose because the activation energy is high, resulting in a slow reaction rate. In other words, sugar remains sugar at room temperature even though the equilibrium constant for its reaction to glucose and fructose is high (▼ Figure 15.17). In the body, an enzyme called *sucrase* catalyzes the conversion of sucrose to glucose and fructose. Sucrase has a pocket—called the active site—into which sucrose snugly fits (like a key into a lock). When sucrose is in the active site, the bond between the glucose and fructose units weakens, lowering the activation energy for the reaction

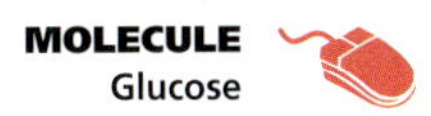

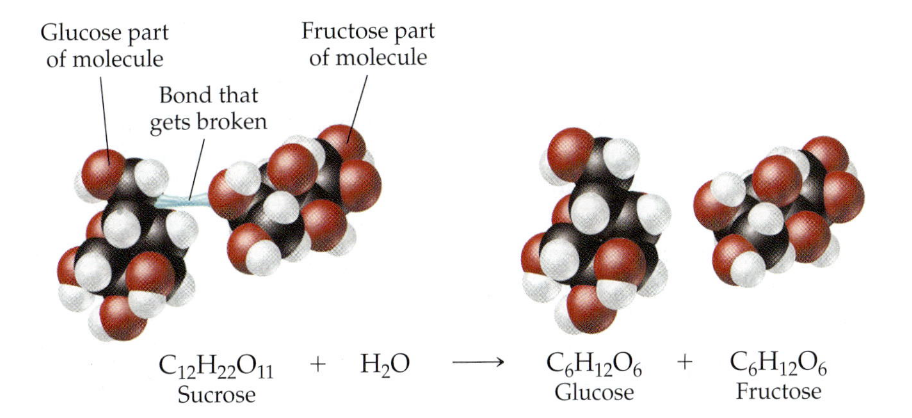

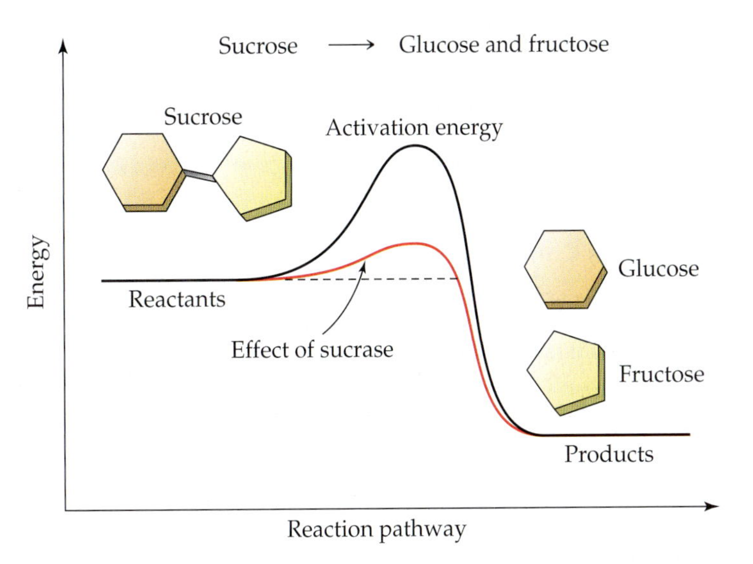

▲ **Figure 15.17 An Enzyme catalyst** The enzyme sucrase creates a pathway with a lower activation energy for the conversion of sucrose to glucose and fructose.

and increasing the reaction rate (▼ Figure 15.18). The reaction can then proceed toward equilibrium—which favors the products—at a much lower temperature.

Not only do enzymes allow otherwise slow reactions to occur at a reasonable rate, they also allow living organisms to have tremendous control over which reactions occur, and when. Enzymes are extremely specific—each enzyme catalyzes only a single reaction. So if a living organism wants to turn a particular reaction on, it simply produces or activates the correct enzyme to catalyze that reaction.

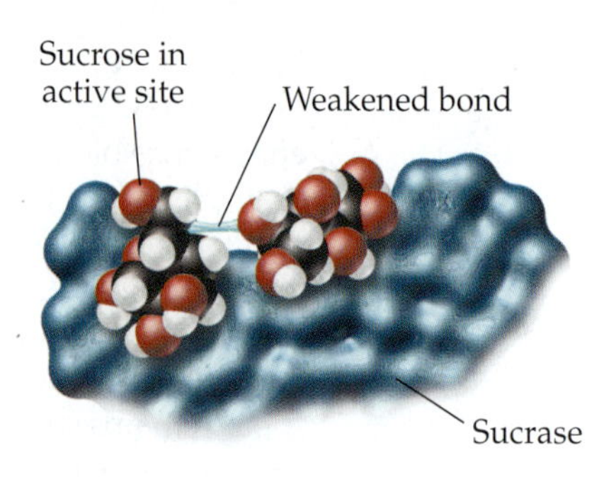

▶ **Figure 15.18 How an enzyme works** Sucrase has a pocket called the active site where sucrose binds. When a molecule of sucrose enters the active site, the bond between glucose and fructose is weakened, lowering the activation energy for the reaction.

CHAPTER IN REVIEW

Chemical Principles	Relevance
The Concept of Equilibrium: Equilibrium involves the ideas of sameness and changelessness. When a system is in equilibrium, there is some property of the system that remains the same and does not change.	**The Concept of Equilibrium:** The equilibrium concept explains many phenomena such as the human body's oxygen delivery system. Life itself can be defined as controlled disequilibrium with the environment.
Rates of Chemical Reactions: The rate of a chemical reaction is the amount of reactant(s) that goes to product(s) in a given period of time. In general, reaction rates increase with increasing reactant concentration and increasing temperature. Since reaction rates depend on the concentration of reactants, and since the concentration of reactants decreases as a reaction proceeds, reaction rates usually slow down as a reaction proceeds.	**Rates of Chemical Reactions:** The rate of a chemical reaction determines how fast a reaction will reach its equilibrium. Chemists want to understand the factors that influence reaction rates so that they can control them.
Dynamic Chemical Equilibrium: Dynamic chemical equilibrium occurs when the rate of the forward reaction equals the rate of the reverse reaction.	**Dynamic Chemical Equilibrium:** When dynamic chemical equilibrium is reached, the concentrations of the reactants and products become constant.
The Equilibrium Constant: For the generic reaction $a\text{A} + b\text{B} \rightleftharpoons c\text{C} + d\text{D}$ the equilibrium constant (K_{eq}) is defined as: $K_{eq} = \frac{[\text{C}]^c[\text{D}]^d}{[\text{A}]^a[\text{B}]^b}$ Only the concentrations of gaseous or aqueous reactants and products are included in the equilibrium constant—the concentrations of solid or liquid reactants or products are omitted.	**The Equilibrium Constant:** The equilibrium constant is a measure of how far a reaction will proceed. A large K_{eq} means the forward reaction is favored (lots of products at equilibrium). A small K_{eq} means the reverse reaction is favored (lots of reactants at equilibrium). An intermediate K_{eq} means that there will be significant amounts of both reactants and products at equilibrium.

Le Châtelier's Principle: Le Châtelier's principle states that when a chemical system at equilibrium is disturbed, the system shifts in a direction that minimizes the disturbance.

Le Châtelier's Principle: Le Châtelier's principle helps us predict what happens to a chemical system at equilibrium when the conditions are changed. This allows chemists to modify the conditions of a chemical reaction to obtain a desired result.

Effect of a Concentration Change on Equilibrium:

- Increasing the concentration of one or more of the *reactants* causes the reaction to shift to the *right*.
- Increasing the concentration of one or more of the products causes the reaction to shift to the *left*.

Effect of a Concentration Change on Equilibrium: There are many cases when a chemist may want to drive a reaction in one direction or another. For example, suppose a chemist is carrying out a reaction to make a desired compound. The reaction can be pushed to the right by continuously removing the product from the reaction mixture as it forms, thus maximizing the amount of product that can be made.

Effect of a Volume Change on Equilibrium:

- Decreasing the volume causes the reaction to shift in the direction that has *fewer* moles of gas particles.
- Increasing the volume causes the reaction to shift in the direction that has more moles of gas particles.

Effect of a Volume Change on Equilibrium: Like the effect of concentration, the effect of pressure on equilibrium allows a chemist to choose the best conditions under which to carry out a chemical reaction. Some reactions are favored in the forward direction by high pressure (those with fewer moles of gas particles in the products) and others (those with fewer moles of gas particles in the reactants) are favored in the forward direction by low pressure.

Effect of a Temperature Change on Equilibrium:
Exothermic **chemical reaction** (heat is a product):

- *Increasing* the temperature causes the reaction to shift *left*.
- *Decreasing* the temperature causes the reaction to shift *right*.

Endothermic **chemical reaction** (heat is a reactant):

- *Increasing* the temperature causes the reaction to shift *right*.
- *Decreasing* the temperature causes the reaction to shift *left*.

Effect of a Temperature Change on Equilibrium: Again, the effect of temperature on a reaction allows chemists to choose conditions that will favor desired reactions. Higher temperatures favor endothermic reactions while lower temperatures favor exothermic reactions. Most reactions will occur *faster* at higher temperature, so the effect of temperature on the rate, not just on the equilibrium constant, must be considered.

The Solubility-Product Constant, K_{sp}: The solubility-product constant of an ionic compound is the equilibrium constant for the chemical equation that describes the dissolving of the compound.

The Solubility-Product Constant, K_{sp}: The solubility-product constant reflects the solubility of a compound. The greater the solubility-product constant, the greater the solubility of the compound.

Reaction Paths and Catalysts: Most chemical reactions must overcome an energy hump, called the *activation energy*, as they proceed from reactants to products. Increasing the temperature of a reaction mixture increases the fraction of reactant molecules that make it over the energy hump, therefore increasing the rate. A catalyst—a substance that increases the rate of the reaction but is not consumed by it—lowers the activation energy so that it is easier to get over the energy hump *without* increasing the temperature.

Reaction Paths and Catalysts: Catalysts are used in many chemical reactions to increase the rates. Without catalysts, many reactions occur too slowly to be of any value. The thousands of reactions that occur in living organisms are controlled by biological catalysts called *enzymes*.

Chemical Skills

Examples

Writing Equilibrium Expressions for Chemical Reactions (Section 15.5)

Examine the definition of the equilibrium constant in the Chemical Principles section (page 544). To write the equilibrium expression for a reaction, write the concentrations of the products raised to their stoichiometric coefficients divided by the concentrations of the reactants raised to their stoichiometric coefficients. Remember that if the reaction contains reactants or products that are liquids or solids, these are omitted from the equilibrium expression.

EXAMPLE 15.10 Writing Equilibrium Expressions for Chemical Reactions

Write an equilibrium expression for the following chemical equation.

$$2\,NO(g) + Br_2(g) \rightleftharpoons 2\,NOBr(g)$$

Solution:

$$K_{eq} = \frac{[NOBr]^2}{[NO]^2[Br_2]}$$

Calculating Equilibrium Constants (Section 15.6)

EXAMPLE 15.11 Calculating Equilibrium Constants

An equilibrium mixture of the following reaction had $[I] = 0.075$ M and $[I_2] = 0.88$ M. What is the value of the equilibrium constant?

$$I_2(g) \rightleftharpoons 2\,I(g)$$

Begin by setting up the problem in the normal way.

Given:

$[I] = 0.075$ M
$[I_2] = 0.88$ M

Find: K_{eq}

Then write the expression for K_{eq} from the balanced equation. To calculate the value of K_{eq}, simply substitute the correct equilibrium concentrations into the expression for K_{eq}. The concentrations within K_{eq} should always be written in moles per liter, M. Units are normally dropped in expressing the equilibrium constant so that K_{eq} is unitless.

Solution:

$$K_{eq} = \frac{[I]^2}{[I_2]} = \frac{[0.075]^2}{[0.88]} = 0.0064$$

Using the Equilibrium Constant to Find the Concentration of a Reactant or Product at Equilibrium (Section 15.6)

LIVE EXAMPLE

EXAMPLE 15.12 Using the Equilibrium Constant to Find the Concentration of a Reactant or Product at Equilibrium

Consider the following reaction.

$$N_2(g) + 3\,H_2(g) \rightleftharpoons 2\,NH_3(g)$$
$$K_{eq} = 152 \text{ at } 225\ °C$$

In an equilibrium mixture, $[N_2] = 0.110$ M and $[H_2] = 0.0935$ M. What is the equilibrium concentration of NH_3?

Begin by setting up the problem in the normal way.

Given:

$[N_2] = 0.110$ M
$[H_2] = 0.0935$ M
$K_{eq} = 152$

Find: $[NH_3]$

Write a solution map that shows how you can use the given concentrations and the equilibrium constant to get to the unknown concentration.

Solution Map:

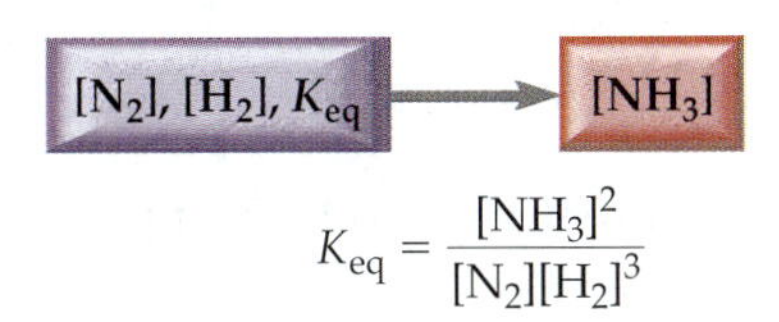

$$K_{eq} = \frac{[NH_3]^2}{[N_2][H_2]^3}$$

Next, solve the equilibrium expression for the quantity you are trying to find and then substitute in the appropriate values to compute the unknown quantity.

Solution:

$$K_{eq} = \frac{[NH_3]^2}{[N_2][H_2]^3}$$

$$[NH_3]^2 = K_{eq}[N_2][H_2]^3$$

$$[NH_3] = \sqrt{K_{eq}[N_2][H_2]^3}$$

$$= \sqrt{(152)[0.110][0.0935]^3}$$

$$= 0.117 \text{ M}$$

Using LeChâtelier's Principle (Sections 15.8, 15.9, 15.10)

EXAMPLE 15.13 Using LeChâtelier's Principle

Consider the following *endothermic* chemical reaction.

$$C(s) + H_2O(g) \rightleftharpoons CO(g) + H_2(g)$$

Predict the effect of:

(a) Increasing [CO]
(b) Increasing $[H_2O]$
(c) Increasing the reaction volume
(d) Increasing the temperature

To apply LeChâtelier's principle, review the effects of concentration, volume, and temperature in the Chemical Principles section (page 545). For each disturbance, predict the direction that the reaction can go to counter the disturbance.

Solution:

(a) Shift left
(b) Shift right
(c) Shift right (more moles of gas on right)
(d) Shift right (heat is a reactant)

Writing an Expression for the Solubility-Product Constant (Section 15.11)

To write the expression for K_{sp}, first write the chemical reaction showing the solid compound in equilibrium with its dissolved aqueous ions. Then write the equilibrium expression as the product of the concentrations of the aqueous ions raised to their stoichiometric coefficients.

EXAMPLE 15.14 Writing an Expression for the Solubility-Product Constant

Write an expression for K_{sp} for $PbCl_2$.

Solution:

$$PbCl_2(s) \rightleftharpoons Pb^{2+}(aq) + 2\,Cl^-(aq)$$
$$K_{sp} = [Pb^{2+}][Cl^-]^2$$

Using K_{sp} to Determine Molar Solubility (Section 15.11)

Begin by writing the reaction by which the solid dissolves into its constituent aqueous ions.

Next, write the expression for K_{sp}.

For the problems assigned in this book, the concentration of individual aqueous ions are equal to the solubility, S.

Substitute S into the equilibrium expression and solve the expression for S.

Finally, look up the value of K_{sp} in Table 15.2 and compute S.

EXAMPLE 15.15 Using K_{sp} to Determine Molar Solubility

Calculate the molar solubility of AgI.

Solution:

$$AgI(s) \rightleftharpoons Ag^+(aq) + I^-(aq)$$
$$K_{sp} = [Ag^+][I^-]$$
$$S = [Ag^+] = [I^-]$$
$$K_{sp} = [Ag^+][I^-]$$
$$= S \times S$$
$$= S^2$$
$$S = \sqrt{K_{sp}}$$
$$S = \sqrt{8.51 \times 10^{-17}}$$
$$= 9.22 \times 10^{-9}\ \text{M}$$

Key Terms

activation energy [15.12]
catalyst [15.12]
collision theory [15.2]
dynamic equilibrium [15.3]
endothermic reaction [15.10]
enzymes [15.12]
equilibrium constant (K_{eq}) [15.4]
exothermic reaction [15.10]
Le Châtelier's principle [15.7]
molar solubility [15.11]
rate of a chemical reaction (reaction rate) [15.2]
reversible reaction [15.3]
solubility-product constant (K_{sp}) [15.11]

Exercises

Questions

1. What are the two *general* concepts involved in equilibrium?
2. How can life be defined with respect to equilibrium?
3. What is the rate of a chemical reaction?
4. What is the difference between a chemical reaction with a fast rate and one with a slow rate?
5. Why do chemists seek to control reaction rates?
6. How do most chemical reactions occur?
7. What factors influence reaction rates? How?
8. What normally happens to the rate of the forward reaction as a reaction proceeds?
9. What is dynamic chemical equilibrium?
10. Explain how dynamic chemical equilibrium involves the concepts of sameness and changelessness.

11. Explain why the concentrations of reactants and products are not necessarily the same at equilibrium.
12. Devise your own analogy—like the Narnia and Middle Earth analogy—to explain chemical equilibrium.
13. What is the equilibrium constant? Why is it significant?
14. Write the expression for the equilibrium constant for the following generic chemical equation.

$$a\text{A} + b\text{B} \rightleftharpoons c\text{C} + d\text{D}$$

15. What does a small equilibrium constant tell you about a reaction? A large equilibrium constant?
16. Why are solids and liquids omitted from the equilibrium expression?
17. Will the concentrations of reactants and products always be the same in every equilibrium mixture of a particular reaction at a given temperature? Explain.
18. What is Le Châtelier's principle?
19. Apply Le Châtelier's principle to your analogy from Question 12.
20. What is the effect of *increasing* the concentration of a reactant in a reaction mixture at equilibrium?
21. What is the effect of *decreasing* the concentration of a reactant in a reaction mixture at equilibrium?
22. What is the effect of *increasing* the concentration of a product in a reaction mixture at equilibrium?
23. What is the effect of *decreasing* the concentration of a product in a reaction mixture at equilibrium?
24. What is the effect of increasing the pressure of a reaction mixture at equilibrium if the reactant side has fewer moles of gas particles than the product side?
25. What is the effect of increasing the pressure of a reaction mixture at equilibrium if the product side has fewer moles of gas particles than the reactant side?
26. What is the effect of decreasing the pressure of a reaction mixture at equilibrium if the reactant side has fewer moles of gas particles than the product side?
27. What is the effect of decreasing the pressure of a reaction mixture at equilibrium if the product side has fewer moles of gas particles than the reactant side?
28. What is the effect of increasing the temperature of an endothermic reaction mixture at equilibrium? Of decreasing the temperature?
29. What is the effect of increasing the temperature of an exothermic reaction mixture at equilibrium? Of decreasing the temperature?
30. What is the solubility-product constant? What does it signify?
31. Write an expression for the solubility-product constant of $AB_2(s)$. Assume that an ion of B has a charge of -1 (that is, B^-).
32. Write an expression for the solubility-product constant of $A_2B(s)$. Assume that an ion of B has charge of -2 (that is, B^{2-}).
33. What are solubility and molar solubility?
34. What is activation energy for a chemical reaction?
35. Explain why two reactants with a large K_{eq} for a particular reaction might not react immediately when combined.
36. What is the effect of a catalyst on reaction? Why are catalysts so important to chemistry?
37. Does a catalyst affect the value of the equilibrium constant?
38. What are enzymes?

Problems

The Rate of Reaction

39. Two gaseous reactants are allowed to react in a 1-L flask and the reaction rate is measured. The experiment is repeated with the same amount of each reactant and at the same temperature, but now in a 2-L flask (so the concentration of each reactant is now less). What happens to the measured reaction rate in the second experiment compared to the first?
40. The rate of phosphorus pentachloride decomposition is measured at a PCl_5 pressure of 0.015 atm and then again at a PCl_5 pressure of 0.30 atm. The temperature is identical in both measurements. Which rate would you predict to be fastest?
41. The body temperature of cold-blooded animals varies with the ambient temperature. From the point of view of reaction rates, explain why cold-blooded animals are more sluggish at cold temperatures.
42. The rate of a reaction is found to double when the temperature is increased from 25 °C to 35 °C. Explain why this is so.

43. The initial rate of a chemical reaction was measured and one of the reactants was found to be reacting at a rate of 0.0011 mol/L/S. The reaction was allowed to proceed for 15 minutes and the rate was measured again. What would you predict about the second measured rate relative to the first?

44. When vinegar is added to a solution of sodium bicarbonate, the mixture immediately begins to bubble furiously. As time passes, however, the bubbling becomes less and less. Explain why this is so.

The Equilibrium Constant

45. Write an equilibrium expression for each of the following chemical equations.
(a) $2\,NO_2(g) \rightleftharpoons N_2O_4(g)$
(b) $2\,BrNO(g) \rightleftharpoons 2\,NO(g) + Br_2(g)$
(c) $H_2O(g) + CO(g) \rightleftharpoons H_2(g) + CO_2(g)$
(d) $CH_4(g) + 2\,H_2S(g) \rightleftharpoons CS_2(g) + 4\,H_2(g)$

46. Write an equilibrium expression for each of the following chemical equations.
(a) $2\,CO(g) + O_2(g) \rightleftharpoons 2\,CO_2(g)$
(b) $N_2(g) + O_2(g) \rightleftharpoons 2\,NO(g)$
(c) $SbCl_5(g) \rightleftharpoons SbCl_3(g) + Cl_2(g)$
(d) $CO(g) + Cl_2(g) \rightleftharpoons COCl_2(g)$

47. Write an equilibrium expression for each of the following chemical equations involving one or more solid or liquid reactants or products.
(a) $PCl_5(g) \rightleftharpoons PCl_3(l) + Cl_2(g)$
(b) $2\,KClO_3(s) \rightleftharpoons 2\,KCl(s) + 3\,O_2(g)$
(c) $HF(aq) + H_2O(l) \rightleftharpoons H_3O^+(aq) + F^-(aq)$
(d) $NH_3(aq) + H_2O(l) \rightleftharpoons NH_4^+(aq) + OH^-(aq)$

48. Write an equilibrium expression for each of the following chemical equations involving one or more solid or liquid reactants or products.
(a) $HCHO_2(aq) + H_2O(l) \rightleftharpoons H_3O^+(aq) + CHO_2^-(aq)$
(b) $CO_3^{2-}(aq) + H_2O(l) \rightleftharpoons HCO_3^-(aq) + OH^-(aq)$
(c) $2\,C(s) + O_2(g) \rightleftharpoons 2\,CO(g)$
(d) $C(s) + CO_2(g) \rightleftharpoons 2\,CO(g)$

49. Consider the following reaction.

$$2\,H_2S(g) \rightleftharpoons 2\,H_2(g) + S_2(g)$$

Find the mistakes in the following equilibrium expression and fix them.

$$K_{eq} = \frac{[H_2][S_2]}{[H_2S]}$$

50. Consider the following reaction.

$$CO(g) + Cl_2(g) \rightleftharpoons COCl_2(g)$$

Find the mistake in the following equilibrium expression and fix it.

$$K_{eq} = \frac{[CO][Cl_2]}{[COCl_2]}$$

51. For each of the following equilibrium constants, indicate whether you would expect an equilibrium reaction mixture to be dominated by reactants, to be dominated by products, or to contain significant amounts of both.
(a) $K_{eq} = 1.8 \times 10^{-5}$
(b) $K_{eq} = 3.8 \times 10^{22}$
(c) $K_{eq} = 9.7 \times 10^{-9}$
(d) $K_{eq} = 0.58$

52. For each of the following equilibrium constants, indicate whether you would expect an equilibrium reaction mixture to be dominated by reactants, to be dominated by products, or to contain significant amounts of both.
(a) $K_{eq} = 6.9 \times 10^{7}$
(b) $K_{eq} = 4.8 \times 10^{-24}$
(c) $K_{eq} = 1.45$
(d) $K_{eq} = 7.7 \times 10^{18}$

Calculating and Using Equilibrium Constants

53. Consider the following reaction.

$$COCl_2(g) \rightleftharpoons CO(g) + Cl_2(g)$$

An equilibrium mixture of this reaction at a certain temperature was found to have $[COCl_2] = 0.225$ M, $[CO] = 0.105$ M, and $[Cl_2] = 0.0844$ M. What is the value of the equilibrium constant at this temperature?

54. Consider the following reaction.

$$CO(g) + 2\,H_2(g) \rightleftharpoons CH_3OH(g)$$

An equilibrium mixture of this reaction at a certain temperature was found to have $[CO] = 0.105$ M, $[H_2] = 0.114$ M, and $[CH_3OH] = 0.185$ M. What is the value of the equilibrium constant at this temperature?

55. Consider the following reaction.

$$2\,H_2S(g) \rightleftharpoons 2\,H_2(g) + S_2(g)$$

An equilibrium mixture of this reaction at a certain temperature was found to have $[H_2S] = 0.562$ M, $[H_2] = 2.74 \times 10^{-2}$ M, and $[S_2] = 7.54 \times 10^{-3}$ M. What is the value of the equilibrium constant at this temperature?

56. Consider the following reaction.

$$CO(g) + H_2O(g) \rightleftharpoons CO_2(g) + H_2(g)$$

An equilibrium mixture of this reaction at a certain temperature was found to have $[CO] = 0.0233$ M, $[H_2O] = 0.0115$ M, $[CO_2] = 0.175$ M, and $[H_2] = 0.0274$ M. What is the value of the equilibrium constant at this temperature?

57. Consider the following reaction.

$$NH_4HS(s) \rightleftharpoons NH_3(g) + H_2S(g)$$

An equilibrium mixture of this reaction at a certain temperature was found to have $[NH_3] = 0.278$ M and $[H_2S] = 0.355$ M. What is the value of the equilibrium constant at this temperature?

58. Consider the following reaction.

$$CaCO_3(s) \rightleftharpoons CaO(s) + CO_2(g)$$

An equilibrium mixture of this reaction at a certain temperature was found to have $[CO_2] = 0.548$ M. What is the value of the equilibrium constant at this temperature?

59. An equilibrium mixture of the following reaction was found to have $[SbCl_3] = 0.0357$ M and $[Cl_2] = 0.0112$ M at 248 °C. What is the concentration of $SbCl_5$?

$$SbCl_5(g) \rightleftharpoons SbCl_3(g) + Cl_2(g)$$
$$K_{eq} = 4.9 \times 10^{-4} \text{ at } 248\,°C$$

60. An equilibrium mixture of the following reaction was found to have $[I_2] = 0.0272$ M at 1200 °C. What is the concentration of I?

$$I_2(g) \rightleftharpoons 2\,I(g)$$
$$K_{eq} = 1.1 \times 10^{-2} \text{ at } 1200\,°C$$

61. An equilibrium mixture of the following reaction was found to have $[I_2] = 0.0112$ M and $[Cl_2] = 0.0155$ M at 25 °C. What is the concentration of ICl?

$$I_2(g) + Cl_2(g) \rightleftharpoons 2\,ICl(g)$$
$$K_{eq} = 81.9 \text{ at } 25\,°C$$

62. An equilibrium mixture of the following reaction was found to have $[SO_3] = 0.391$ M and $[O_2] = 0.125$ M at 600 °C. What is the concentration of SO_2?

$$2\,SO_2(g) + O_2(g) \rightleftharpoons 2\,SO_3(g)$$
$$K_{eq} = 4.34 \text{ at } 600\,°C$$

63. Consider the following reaction.

$$N_2(g) + 3\,H_2(g) \rightleftharpoons 2\,NH_3(g)$$

Complete the following table. Assume that all concentrations are equilibrium concentrations in moles per liter, M.

T (K)	$[N_2]$	$[H_2]$	$[NH_3]$	K_{eq}
500	0.115	0.105	0.439	___
575	0.110	___	0.128	9.6
775	0.120	0.140	___	0.0584

64. Consider the following reaction.

$$H_2(g) + I_2(g) \rightleftharpoons 2\,HI(g)$$

Complete the following table. Assume that all concentrations are equilibrium concentrations in moles per liter, M.

T (°C)	$[H_2]$	$[I_2]$	$[HI]$	K_{eq}
25	0.355	0.388	0.0922	___
340	___	0.0455	0.387	9.6
445	0.0485	0.0468	___	50.2

Le Châtelier's Principle

65. Consider the following reaction at equilibrium.

$$CO(g) + Cl_2(g) \rightleftharpoons COCl_2(g)$$

Predict the effect (shift right, shift left, or no effect) of:
(a) adding Cl_2 to the reaction mixture.
(b) adding $COCl_2$ to the reaction mixture.
(c) adding CO to the reaction mixture.

66. Consider the following reaction at equilibrium.

$$2\,BrNO(g) \rightleftharpoons 2\,NO(g) + Br_2(g)$$

Predict the effect (shift right, shift left, or no effect) of:
(a) adding BrNO to the reaction mixture.
(b) adding NO the reaction mixture.
(c) adding Br_2 to the reaction mixture.

67. Consider the following reaction at equilibrium.

$$C(s) + H_2O(g) \rightleftharpoons CO(g) + H_2(g)$$

Predict the effect (shift right, shift left, or no effect) of:
(a) adding C to the reaction mixture.
(b) condensing H_2O and removing it from the reaction mixture.
(c) adding CO to the reaction mixture.
(d) removing H_2 from the reaction mixture.

68. Consider the following reaction at equilibrium.

$$2\,KClO_3(s) \rightleftharpoons 2\,KCl(s) + 3\,O_2(g)$$

Predict the effect (shift right, shift left, or no effect) of:
(a) adding KCl to the reaction mixture.
(b) adding $KClO_3$ to the reaction mixture.
(c) adding O_2 to the reaction mixture.
(d) removing O_2 from the reaction mixture.

69. Consider the effect of a volume change on the following reaction at equilibrium.

$$I_2(g) \rightleftharpoons 2\,I(g)$$

Predict the effect (shift right, shift left, or no effect) of:
(a) Increasing the reaction volume.
(b) Decreasing the reaction volume.

70. Consider the effect of a volume change on the following reaction at equilibrium.

$$2\,H_2S(g) \rightleftharpoons 2\,H_2(g) + S_2(g)$$

Predict the effect (shift right, shift left, or no effect) of:
(a) Increasing the reaction volume.
(b) Decreasing the reaction volume.

71. Consider the effect of a volume change on the following reaction at equilibrium.

$$I_2(g) + Cl_2(g) \rightleftharpoons 2\,ICl(g)$$

Predict the effect (shift right, shift left, or no effect) of:
(a) Increasing the reaction volume.
(b) Decreasing the reaction volume.

72. Consider the effect of a volume change on the following reaction at equilibrium.

$$CO(g) + H_2O(g) \rightleftharpoons CO_2(g) + H_2(g)$$

Predict the effect (shift right, shift left, or no effect) of:
(a) Increasing the reaction volume.
(b) Decreasing the reaction volume.

73. The following reaction is endothermic.

$$C(s) + CO_2(g) \rightleftharpoons 2\,CO(g)$$

Predict the effect (shift right, shift left, or no effect) of:
(a) Increasing the reaction temperature.
(b) Decreasing the reaction temperature.

74. The following reaction is endothermic.

$$I_2(g) \rightleftharpoons 2\,I(g)$$

Predict the effect (shift right, shift left, or no effect) of:
(a) Increasing the reaction temperature.
(b) Decreasing the reaction temperature.

75. The following reaction is exothermic.

$$C_6H_{12}O_6(s) + 6\,O_2(g) \rightleftharpoons 6\,CO_2(g) + 6\,H_2O(g)$$

Predict the effect (shift right, shift left, or no effect) of:
(a) Increasing the reaction temperature.
(b) Decreasing the reaction temperature.

76. The following reaction is exothermic.

$$C_2H_4(g) + Br_2(g) \rightleftharpoons C_2H_4Br_2(g)$$

Predict the effect (shift right, shift left, or no effect) of:
(a) Increasing the reaction temperature.
(b) Decreasing the reaction temperature.

77. Coal, which is primarily carbon, can be converted to natural gas, primarily CH_4, by the following exothermic reaction.

$$C(s) + 2\,H_2(g) \rightleftharpoons CH_4(g)$$

If this reaction mixture is at equilibrium, predict the effect (shift right, shift left, or no effect) of:
(a) Adding more C to the reaction mixture.
(b) Adding more H_2 to the reaction mixture.
(c) Raising the temperature of the reaction mixture.
(d) Lowering the volume of the reaction mixture.
(e) Adding a catalyst to the reaction mixture.

78. Coal can be used to generate hydrogen gas (a potential fuel) by the following endothermic reaction.

$$C(s) + H_2O(g) \rightleftharpoons CO(g) + H_2(g)$$

If this reaction mixture is at equilibrium, predict the effect (shift right, shift left, or no effect) of:
(a) Adding more C to the reaction mixture.
(b) Adding more $H_2O(g)$ to the reaction mixture.
(c) Raising the temperature of the reaction mixture.
(d) Increasing the volume of the reaction mixture.
(e) Adding a catalyst to the reaction mixture.

The Solubility-Product Constant

79. For each of these compounds, write an equation showing how the compound dissolves in water and write an expression for K_{sp}.
(a) $CaSO_4$
(b) $AgCl$
(c) CuS
(d) $FeCO_3$

80. For each of these compounds, write an equation showing how the compound dissolves in water and write an expression for K_{sp}.
(a) $Mg(OH)_2$
(b) $FeCO_3$
(c) PbS
(d) $PbSO_4$

81. Determine what is wrong with the following K_{sp} expression for $Fe(OH)_2$ and correct it.

$$K_{sp} = [Fe^{2+}][OH^-]$$

82. Determine what is wrong with the following K_{sp} expression for $Ba(OH)_2$ and correct it.

$$K_{sp} = \frac{[Ba(OH)_2]}{[Ba^{2+}][OH^-]^2}$$

83. A saturated solution of MgF_2 has $[Mg^{2+}] = 2.6 \times 10^{-4}$ M and $[F^-] = 5.2 \times 10^{-4}$. What is the value of K_{sp} for MgF_2?

84. A saturated solution of AgI has $[Ag^+] = 9.2 \times 10^{-9}$ M and $[I^-] = 9.2 \times 10^{-9}$ M. What is the value of K_{sp} for AgI?

85. A saturated solution of $PbSO_4$ has $[Pb^{2+}] = 1.35 \times 10^{-4}$ M. What is the concentration of SO_4^{2-}?

86. A saturated solution of $PbCl_2$ has $[Cl^-] = 2.86 \times 10^{-2}$ M. What is the concentration of Pb^{2+}?

87. Calculate the molar solubility of $CaCO_3$.

88. Calculate the molar solubility of PbS.

89. Calculate the molar solubility of $MgCO_3$.

90. Calculate the molar solubility of CuI ($K_{sp} = 1.27 \times 10^{-12}$).

91. Complete the following table. Assume that all concentrations are equilibrium concentrations in moles per liter, M.

Compound	*[Cation]*	*[Anion]*	K_{sp}
$SrCO_3$	2.4×10^{-5}	2.4×10^{-5}	___
SrF_2	1.0×10^{-3}	___	4.0×10^{-9}
Ag_2CO_3	___	1.3×10^{-4}	8.8×10^{-12}

92. Complete the following table. Assume that all concentrations are equilibrium concentrations in moles per liter, M.

Compound	*[Cation]*	*[Anion]*	K_{sp}
CdS	3.7×10^{-15}	3.7×10^{-15}	___
BaF_2	___	7.2×10^{-3}	1.9×10^{-7}
Ag_2SO_4	2.8×10^{-2}	___	1.1×10^{-5}

Cumulative Problems

93. Consider the following reaction.

$$Fe^{3+}(aq) + SCN^-(aq) \rightleftharpoons FeSCN^{2+}(aq)$$

A solution is made containing initial $[Fe^{3+}] = 1.0 \times 10^{-3}$ M and initial $[SCN^-] = 8.0 \times 10^{-4}$ M. At equilibrium, $[FeSCN^{2+}] = 1.7 \times 10^{-4}$. Calculate the value of the equilibrium constant. Hint: Use the chemical reaction stoichiometry to calculate the equilibrium concentrations of Fe^{3+} and SCN^-.

94. Consider the following reaction.

$$SO_2Cl_2(g) \rightleftharpoons SO_2(g) + Cl_2(g)$$

A solution is made containing initial $[SO_2Cl_2] = 0.020$ M. At equilibrium, $[Cl_2] = 1.2 \times 10^{-2}$ M. Calculate the value of the equilibrium constant. Hint: Use the chemical reaction stoichiometry to calculate the equilibrium concentrations of SO_2Cl_2 and SO_2.

95. Consider the following reaction.

$$H_2(g) + I_2(g) \rightleftharpoons 2\,HI(g)$$
$$K_{eq} = 6.17 \times 10^{-2} \text{ at } 25\ °C$$

A 3.67-L flask containing an equilibrium reaction mixture has $[H_2] = 0.104$ M and $[I_2] = 0.0202$ M. How much HI in grams is in the equilibrium mixture?

96. Consider the following reaction.

$$CO(g) + Cl_2(g) \rightleftharpoons COCl_2(g)$$
$$K_{eq} = 2.9 \times 10^{10} \text{ at } 25\ °C$$

A 5.19-L flask containing an equilibrium reaction mixture has $[CO] = 1.8 \times 10^{-6}$ M and $[Cl_2] = 7.3 \times 10^{-7}$ M. How much $COCl_2$ in grams is in the equilibrium mixture?

97. Consider the following exothermic reaction.

$$C_2H_4(g) + Cl_2(g) \rightleftharpoons C_2H_4Cl_2(g)$$

If you were a chemist trying to maximize the amount of $C_2H_4Cl_2$ produced, which of the following might you try? Assume that the reaction mixture reaches equilibrium.

(a) Increasing the reaction volume.
(b) Removing $C_2H_4Cl_2$ from the reaction mixture as it forms.
(c) Lowering the reaction temperature.
(d) Adding Cl_2

98. Consider the following endothermic reaction.

$$C_2H_4(g) + I_2(g) \rightleftharpoons C_2H_4I_2(g)$$

If you were a chemist trying to maximize the amount of $C_2H_4I_2$ produced, which of the following might you try? Assume that the reaction mixture reaches equilibrium.

(a) Decreasing the reaction volume.
(b) Removing I_2 from the reaction mixture.
(c) Raising the reaction temperature.
(d) Adding C_2H_4 to the reaction mixture.

99. Calculate the molar solubility of CuS. How many grams of CuS are present in 15.0 L of a saturated CuS solution?

100. Calculate the molar solubility of $FeCO_3$. How many grams of $FeCO_3$ are present in 15.0 L of a saturated $FeCO_3$ solution?

101. A sample of tap water is found to be 0.025 M in Ca^{2+}. If 105 mg of Na_2SO_4 is added to 100.0 mL of the tap water, will any $CaSO_4$ precipitate out of solution?

102. If 50.0 mg of Na_2CO_3 are added to 150.0 mL of a solution that is 1.5×10^{-3} M in Mg^{2+}, will any $MgCO_3$ precipitate from the solution?

103. The solubility of $CaCrO_4$ at 25 °C is 4.15 g/L. Calculate K_{sp} for $CaCrO_4$.

104. The solubility of nickel(II) carbonate at 25 °C is 0.042 g/L. Calculate K_{sp} for nickel(II) carbonate.

Highlight Problems

105. H_2 and I_2 are combined in a flask and allowed to react according to the following reaction.

$$H_2(g) + I_2(g) \rightleftharpoons 2\,HI(g)$$

Examine the following figures (sequential in time) and determine which figure represents the point where equilibrium is reached.

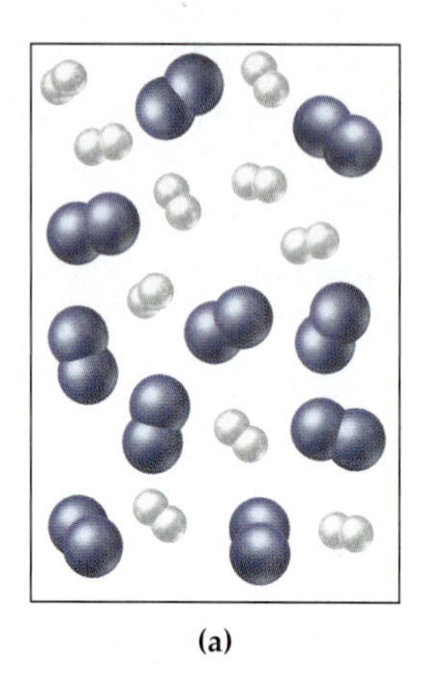
(a)

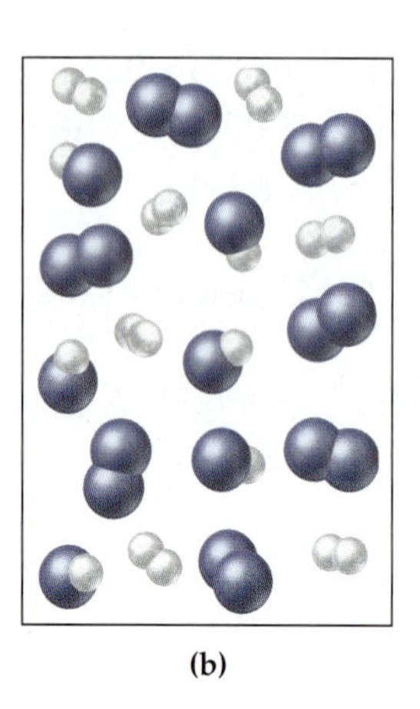
(b)

(c)

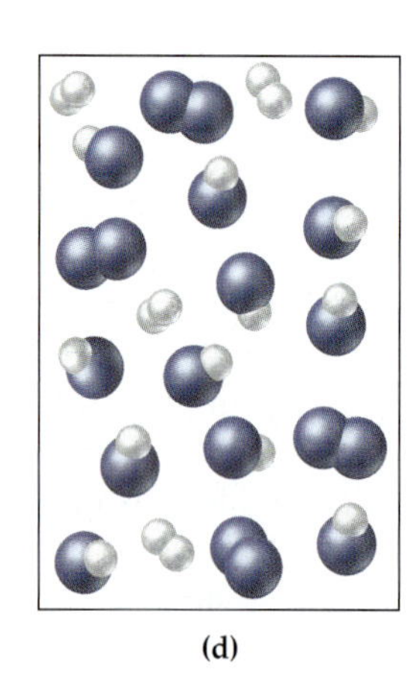
(d)

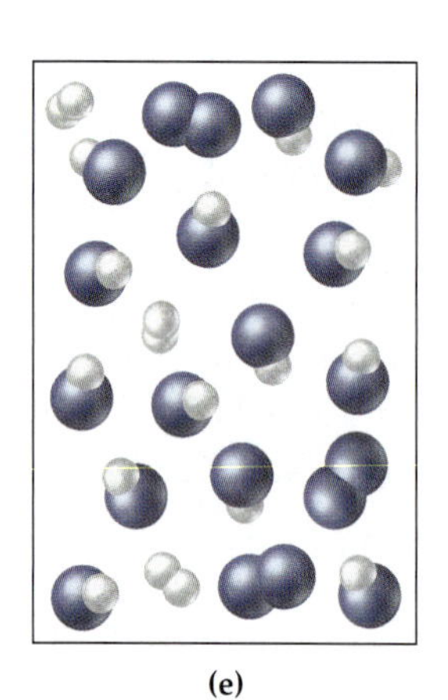
(e)

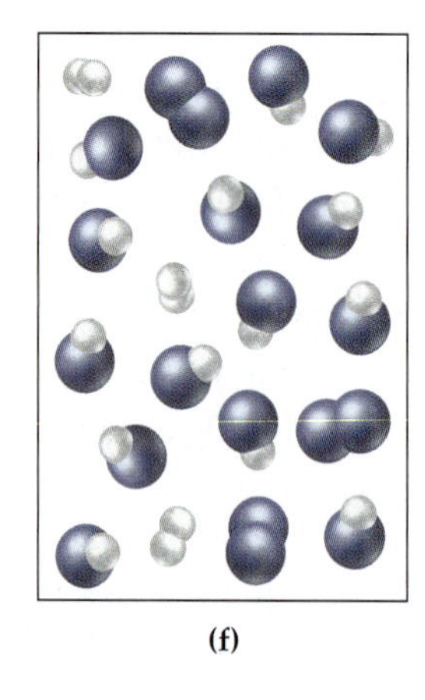
(f)

106. Ethene (C_2H_4) can be halogenated by the following reaction.

$$C_2H_4(g) + X_2(g) \rightleftharpoons C_2H_4X_2(g)$$

where X_2 can be Cl_2, Br_2, or I_2. Examine the following figures representing equilibrium concentrations of this reaction at the same temperature for the three different halogens. Rank the equilibrium constants for these three reactions from largest to smallest.

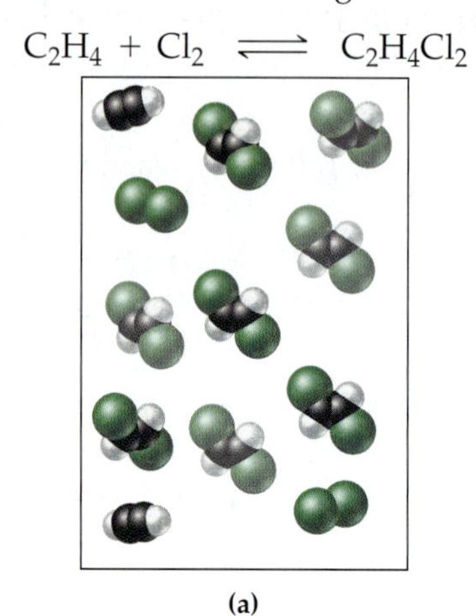

(a)

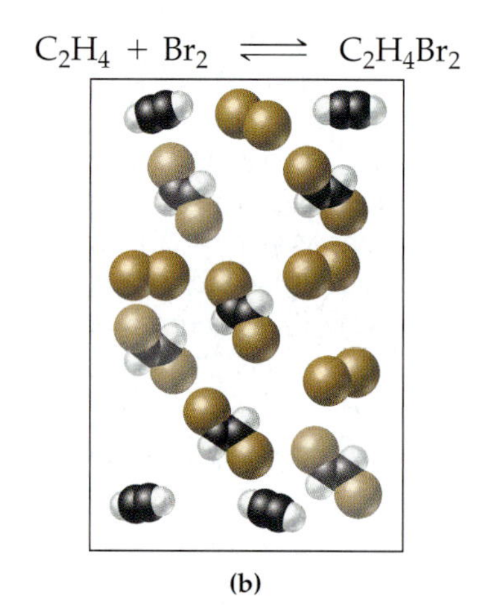

(b)

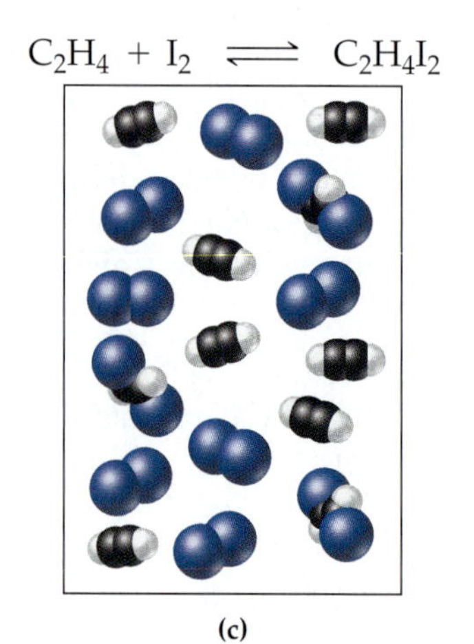

(c)

107. One of the main components of hard water is $CaCO_3$. When hard water evaporates, some of the $CaCO_3$ is left behind as a white mineral deposit. Plumbing fixtures in homes with hard water often acquire these deposits over time. Toilets, for example, often develop these deposits at the water line as the water in the toilet slowly evaporates away. If water is saturated with $CaCO_3$, how much of it has to evaporate to deposit 0.250 g of $CaCO_3$? Hint: Begin by using K_{sp} for $CaCO_3$ to determine its solubility.

Answers to Skillbuilder Exercises

Skillbuilder 15.1 $K_{eq} = \dfrac{[HF]^2}{[H_2][F_2]}$

Skillbuilder 15.2 $K_{eq} = \dfrac{[Cl_2]^2}{[HCl]^4[O_2]}$

Skillbuilder 15.3 $K_{eq} = 14.5$

Skillbuilder Plus, p. 526 $K_{eq} = 26$

Skillbuilder 15.4 0.033 M

Skillbuilder 15.5 Adding Br_2 causes a shift to the left; adding BrNO causes a shift to the right.

Skillbuilder Plus, p. 531 Removing Br_2 causes a shift to the right.

Skillbuilder 15.6 Decreasing volume causes a shift to the right; increasing volume causes a shift to the left.

Skillbuilder 15.7 Increasing the temperature shifts the reaction to the left; decreasing the temperature shifts the reaction to the right.

Skillbuilder 15.8 **(a)** $K_{sp} = [Ag^+][I^-]$
(b) $K_{sp} = [Ca^{2+}][OH^-]^2$

Skillbuilder 15.9 8.43×10^{-3} M

Answers to Conceptual Checkpoints

15.1 (a) In accordance with the gas laws (Chapter 11), increasing the pressure would increase the temperature, decrease the volume, or both. Increasing the temperature would increase the reaction rate. Decreasing the volume would increase the concentration of the reactants, which would also increase the reaction rate. Therefore, we would expect that increasing the pressure would speed up the reaction.

15.2 (b) For this reaction,
$K_{eq} = [B][C]/[A] = (2 \times 2)/2 = 2.$

GAS
BAR CAFE
OUT OF BUS
NO GAS
FCV
TRO-365

CHAPTER 16

Oxidation and Reduction

"In fact, we will have to give up taking things for granted, even the apparently simple things. We have to learn to understand nature and not merely to observe it and endure what it imposes on us."

John Desmond Bernal (1901–1971)

16.1 The End of the Internal Combustion Engine?

▲ **Figure 16.1 A fuel-cell car** The Honda FCX, a hydrogen-powered, fuel-cell automobile. The only emission is water.

It is possible, even likely, that you will see the end of the internal combustion engine within your lifetime. Although it has served us well—powering our airplanes, automobiles, and trains—its time is running out. What will replace it? If our cars don't run on gasoline, what will fuel them? The answers to these questions are not completely settled, but new and better technologies are on the horizon. The most promising of these is the use of **fuel cells** to power electric vehicles. Such whisper-quiet, environmentally friendly supercars are currently available only as prototypes but are slated for production in the next few years (◀ Figure 16.1).

In 2003, General Motors demonstrated its *HydroGen3*—a five-passenger fuel-cell automobile with a top speed of 100 mph and a range of 250 miles on one tank of fuel—to members of the U.S. Congress, who were encouraged to test-drive the vehicle. The electric motor is powered by hydrogen, and the only emission is water—which is so clean you can drink it. Other automakers have similar prototype models in development. In addition, several cities around the world—including Palm Springs, California; Washington, D.C.; and Vancouver, British Columbia—are currently testing the feasibility of fuel-cell buses in a pilot program.

Fuel cells are based on the tendency of some elements to gain electrons from other elements. The most common type of fuel cell—called the hydrogen–oxygen fuel cell—is based on the reaction between hydrogen and oxygen.

$$2\,H_2(g) + O_2(g) \longrightarrow 2\,H_2O(g)$$

◀ Fuel-cell vehicles (FCVs), such as the one shown here, may someday replace vehicles powered by internal combustion engines. As you can see from this image, FCVs produce only water as exhaust.

In this reaction, hydrogen and oxygen form covalent bonds with one another. Recall from Section 10.2 that a single covalent bond is a shared electron pair. However, since oxygen is more electronegative than hydrogen (see Section 10.8), the electron pair in a hydrogen–oxygen bond is not *unequally* shared, with oxygen getting the larger portion. In effect, oxygen has more electrons in H_2O than in elemental O_2—it has *gained electrons* in the reaction.

In a direct reaction between hydrogen and oxygen, oxygen atoms gain the electrons directly from hydrogen atoms as the reaction proceeds. In a hydrogen–oxygen fuel cell, the same reaction occurs, but the hydrogen and oxygen are separated, forcing the electrons to move through an external wire to get from hydrogen to oxygen. These moving electrons constitute an electrical current, which is then used to power the electric motor of a fuel-cell vehicle. In effect, fuel cells use the electron-gaining tendency of oxygen and the electron-losing tendency of hydrogen to force electrons to move through a wire, creating the electricity that powers the car.

Reactions involving the transfer of electrons are called **oxidation–reduction** or **redox reactions**. Besides their application to fuel-cell vehicles, redox reactions are prevalent in nature, in industry, and in many everyday processes. For example, the rusting of iron, the bleaching of hair, and the reactions occurring in batteries all involve redox reactions. Redox reactions are also responsible for providing the energy our bodies need to move, think, and stay alive.

16.2 Oxidation and Reduction: Some Definitions

Consider the following redox reactions.

$$2\,H_2(g) + O_2(g) \longrightarrow 2\,H_2O(g) \qquad \text{(hydrogen–oxygen fuel-cell reaction)}$$

$$4\,Fe(s) + 3\,O_2(g) \longrightarrow 2\,Fe_2O_3(s) \qquad \text{(rusting of iron)}$$

$$CH_4(g) + 2\,O_2(g) \longrightarrow CO_2(g) + 2\,H_2O(g) \qquad \text{(combustion of methane)}$$

What do they all have in common? Each of these reactions involves one or more elements gaining oxygen. In the hydrogen–oxygen fuel-cell reaction, *hydrogen* gains oxygen as it turns into water. In the rusting of iron, *iron* gains oxygen as it turns into iron oxide, the familiar orange substance we call rust (▼ Figure 16.2). In the combustion of methane, *carbon* gains oxygen to form carbon dioxide, producing the brilliant blue flame we see on gas stoves (▼ Figure 16.3). In each

▲ **Figure 16.2 Slow oxidation** Rust is produced by the oxidation of iron to form iron oxide.

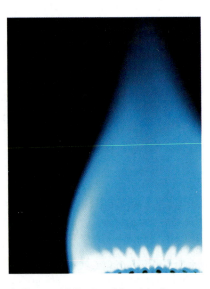

▲ **Figure 16.3 Rapid oxidation** The flame on a gas stove results from the oxidation of carbon in natural gas.

case, the substance that gains oxygen is oxidized in the reaction. One definition of *oxidation*—although not the most fundamental one—is simply the *gaining of oxygen*.

Now consider these same three reactions in reverse.

These definitions of oxidation and reduction are useful because they show the origin of the term *oxidation*, and they allow us to quickly identify reactions involving elemental oxygen as oxidation and reduction reactions. However, as you will see, these definitions are *not* the most fundamental.

$$2\,H_2O(g) \longrightarrow 2\,H_2(g) + O_2(g)$$
$$2\,Fe_2O_3(s) \longrightarrow 4\,Fe(s) + 3\,O_2(g)$$
$$CO_2(g) + 2\,H_2O(g) \longrightarrow CH_4(g) + 2\,O_2(g)$$

Each of these reactions involves loss of oxygen. In the first reaction, hydrogen loses oxygen; in the second reaction, iron loses oxygen; and in the third reaction, carbon loses oxygen. In each case, the substance that loses oxygen is reduced in the reaction. One definition of *reduction* is simply the *loss of oxygen*.

However, redox reactions need not involve oxygen, and a more fundamental definition of oxidation and reduction can be formulated. Consider, for example, the reaction between sodium and oxygen to form sodium oxide. This reaction occurs whenever metallic sodium is exposed to air.

▲ Sodium metal oxidizes immediately upon exposure to air.

$$4\,Na(s) + O_2(g) \longrightarrow 2\,Na_2O(s)$$

Consider also the reaction between sodium and chlorine to form table salt (NaCl).

$$2\,Na(s) + Cl_2(g) \longrightarrow 2\,NaCl(s)$$

Do you notice a similarity between these two reactions? In both cases, sodium (a metal with a strong tendency to lose electrons) reacts with an electronegative nonmetal (which has a tendency to gain electrons). In both cases, sodium atoms lose electrons to become positive ions—sodium is oxidized.

In redox reactions between a metal and a nonmetal, the metal is oxidized and the nonmetal is reduced.

$$Na \longrightarrow Na^+ + e^-$$

The electrons lost by sodium are gained by the nonmetals, which become negative ions—the nonmetals are reduced.

$$O_2 + 4\,e^- \longrightarrow 2\,O^{2-}$$
$$Cl_2 + 2\,e^- \longrightarrow 2\,Cl^-$$

The oxidizing agent may be an element that is itself reduced, or a compound or ion containing an element that is reduced. The reducing agent may be an element that is itself oxidized, or a compound or ion containing an element that is oxidized.

A more fundamental definition of **oxidation**, then, is simply the *loss of electrons*, and a more fundamental definition of **reduction** is the *gain of electrons*.

Notice that *oxidation and reduction must occur together*. If one substance loses electrons (oxidation), then another substance must gain electrons (reduction) (◀ Figure 16.4). The substance that is oxidized is called the **reducing agent** because it causes the reduction of the other substance. Similarly, the substance that is reduced is called the **oxidizing agent** because it causes the oxidation of the other substance. For example, consider our hydrogen–oxgyen fuel-cell reaction.

$$\underset{\text{Reducing agent}}{2\,H_2(g)} + \underset{\text{Oxidizing agent}}{O_2(g)} \longrightarrow 2\,H_2O(g)$$

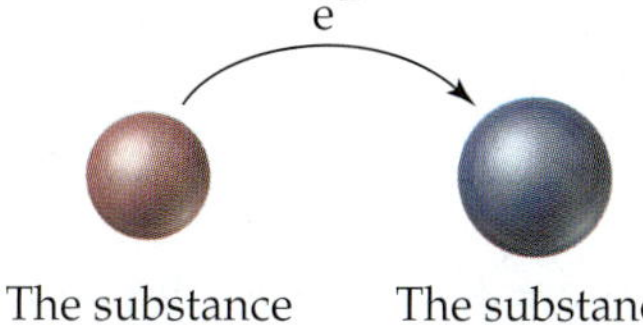
e⁻
The substance losing electrons is oxidized.
The substance gaining electrons is reduced.

▲ **Figure 16.4 Oxidation and reduction** In a redox reaction, one substance loses electrons and another substance gains electrons.

In this reaction, hydrogen is oxidized, making it the reducing agent. Oxygen is reduced, making it the oxidizing agent. Substances such as oxygen, which have a strong tendency to attract electrons, are good oxidizing agents—they tend to cause the oxidation of other substances. Substances such as hydrogen, which have a strong tendency to give up electrons, are good reducing agents—they tend to cause the reduction of other substances.

Helpful mnemonics: OIL RIG—**O**xidation **I**s **L**oss (of electrons); **R**eduction **I**s **G**ain (of electrons). LEO the lion says GER—**L**ose **E**lectrons **O**xidation; **G**ain **E**lectrons **R**eduction.

To summarize:

- Oxidation—the loss of electrons
- Reduction—the gain of electrons
- Oxidizing agent—the substance being reduced
- Reducing agent—the substance being oxidized

TUTORIAL Oxidation–Reduction Reactions, Parts 1 & 2

EXAMPLE 16.1 Identifying Oxidation and Reduction

For each of the following reactions, identify the substance being oxidized and the substance being reduced.

(a) $2\,Mg(s) + O_2(g) \longrightarrow 2\,MgO(s)$
(b) $Fe(s) + Cl_2(g) \longrightarrow FeCl_2(s)$
(c) $Zn(s) + Fe^{2+}(aq) \longrightarrow Zn^{2+}(aq) + Fe(s)$

Solution:

(a) $2\,Mg(s) + O_2(g) \longrightarrow 2\,MgO(s)$
In this reaction, magnesium is gaining oxygen and losing electrons to oxygen. Mg is therefore oxidized and O_2 is reduced.

(b) $Fe(s) + Cl_2(g) \longrightarrow FeCl_2(s)$
In this reaction, a metal (Fe) is reacting with an electronegative nonmetal (Cl_2). Fe loses electrons and is therefore oxidized, while Cl_2 gains electrons and is therefore reduced.

(c) $Zn(s) + Fe^{2+}(aq) \longrightarrow Zn^{2+}(aq) + Fe(s)$
In this reaction, electrons are transferred from the Zn to the Fe^{2+}. Zn loses electrons and is oxidized. Fe^{2+} gains electrons and is reduced.

SKILLBUILDER 16.1 Identifying Oxidation and Reduction

For each of the following reactions, identify the substance being oxidized and the substance being reduced.

(a) $2\,K(s) + Cl_2(g) \longrightarrow 2\,KCl(s)$
(b) $2\,Al(s) + 3\,Sn^{2+}(aq) \longrightarrow 2\,Al^{3+}(aq) + 3\,Sn(s)$
(c) $C(s) + O_2(g) \longrightarrow CO_2(g)$

EXAMPLE 16.2 Identifying Oxidizing and Reducing Agents

For each of the following reactions, identify the oxidizing agent and the reducing agent.

(a) $2\,Mg(s) + O_2(g) \longrightarrow 2\,MgO(s)$
(b) $Fe(s) + Cl_2(g) \longrightarrow FeCl_2(s)$
(c) $Zn(s) + Fe^{2+}(aq) \longrightarrow Zn^{2+}(aq) + Fe(s)$

Solution:
In the previous example, we identified the substance being oxidized and reduced for these reactions. The substance being oxidized is the reducing agent and the substance being reduced is the oxidizing agent.

(a) Mg is oxidized and is therefore the reducing agent; O_2 is reduced and is therefore the oxidizing agent.

(b) Fe is oxidized and is therefore the reducing agent; Cl_2 is reduced and is therefore the oxidizing agent.

(c) Zn is oxidized and is therefore the reducing agent; Fe^{2+} is reduced and is therefore the oxidizing agent.

SKILLBUILDER 16.2 Identifying Oxidizing and Reducing Agents

For each of the following reactions, identify the oxidizing agent and the reducing agent.

(a) $2\,K(s) + Cl_2(g) \longrightarrow 2\,KCl(s)$

(b) $2\,Al(s) + 3\,Sn^{2+}(aq) \longrightarrow 2\,Al^{3+}(aq) + 3\,Sn(s)$

(c) $C(s) + O_2(g) \longrightarrow CO_2(g)$

16.3 Oxidation States: Electron Bookkeeping

For many redox reactions, such as those involving oxygen or other highly electronegative elements, it is easy to identify (by inspection) the substances being oxidized and reduced. For other redox reactions, it is more difficult. For example, consider the redox reaction between carbon and sulfur.

$$C + 2\,S \longrightarrow CS_2$$

Do not confuse oxidation state with ionic charge. A substance need not be ionic to have an assigned oxidation state.

What is oxidized here? What is reduced? In order to easily identify oxidation and reduction, chemists have devised a scheme to track electrons and where they go in chemical reactions. In this scheme—which is like bookkeeping for electrons—all shared electrons are assigned to the most electronegative element. Then a number—called the **oxidation state** or the **oxidation number**—is computed for each element based on the number of electrons assigned to it.

The procedure just described is a bit cumbersome in practice. However, its main results can be summarized in a series of rules. The easiest way to assign oxidation states is to follow these rules.

These rules are hierarchical. If any two rules conflict, follow the rule that is higher on the list.

Rules for Assigning Oxidation States	Examples	
(1) The oxidation state of an atom in a free element is 0.	Cu 0 ox state	Cl_2 0 ox state
(2) The oxidation state of a monoatomic ion is equal to its charge.	Ca^{2+} +2 ox state	Cl^- −1 ox state
(3) The sum of the oxidation states of all atoms in:		
• a neutral molecule or formula unit is 0.	H_2O 2(H ox state) + 1(O ox state) = 0	
• an ion is equal to the charge of the ion.	NO_3^- 1(N ox state) + 3(O ox state) = −1	
(4) In their compounds,		
• Group I metals have an oxidation state of +1.		NaCl +1 ox state
• Group II metals have an oxidation state of +2.		CaF_2 +2 ox state
(5) In their compounds, nonmetals are assigned oxidation states according to the following hierarchical table. Entries at the top of the table have priority over entries at the bottom.		

Nonmetal	Oxidation State	Example
fluorine	−1	MgF_2 −1 ox state
hydrogen	+1	H_2O +1 ox state
oxygen	−2	CO_2 −2 ox state
Group 7A	−1	CCl_4 −1 ox state
Group 6A	−2	H_2S −2 ox state
Group 5A	−3	NH_3 −3 ox state

EXAMPLE 16.3 Assigning Oxidation States

Assign an oxidation state to each atom in each of the following.

(a) Br_2
(b) K^+
(c) LiF
(d) CO_2
(e) SO_4^{2-}
(f) Na_2O_2

Solution:

Since Br_2 is a free element, the oxidation state of both Br atoms is 0 (Rule 1).

(a) Br_2

Br Br
0 0

Since K^+ is a monoatomic ion, the oxidation state of the K^+ ion is +1 (Rule 2).

(b) K^+

K^+
+1

The oxidation state of Li is +1 (Rule 4). The oxidation state of F is −1 (Rule 5). Since this is a neutral compound, the sum of the oxidation states is 0 (Rule 3).

(c) LiF

Li F
+1 −1
sum: $+1 - 1 = 0$

The oxidation state of oxygen is −2 (Rule 5). The oxidation state of carbon must be deduced from Rule 3, which states that the sum of the oxidation states of all the atoms must be 0. Since there are two oxygen atoms, the oxidation state of O must be multiplied by 2 when computing the sum.

(d) CO_2

(C ox state) + 2(O ox state) = 0

(C ox state) + 2(−2) = 0

(C ox state) − 4 = 0

C ox state = +4

C O_2
+4 −2
sum: $+4 + 2(-2) = 0$

The oxidation state of oxygen is −2 (Rule 5). The oxidation state of S is expected to be −2 (Rule 5). However, if that were the case, the sum of the oxidation states would not equal the charge of the ion.

(e) SO_4^{2-}

(S ox state) + 4(O ox state) = −2

(S ox state) + 4(−2) = −2 (Oxygen takes priority over sulfur.)

(S ox state) − 8 = −2

S ox state = −2 + 8

S ox state = +6

Since O is higher on the list, it takes priority and the oxidation state of sulfur is computed by setting the sum of all of the oxidation states equal to −2 (the charge of the ion).

$\underset{+6}{S}\ \underset{-2}{O_4^{2-}}$

sum: $+6 + 4(-2) = -2$

The oxidation state of sodium is +1 (Rule 4). The oxidation state of O is expected to be −2 (Rule 5). However, Na takes priority and we deduce the oxidation state of O by setting the sum of all of the oxidation states equal to 0.

(f) Na_2O_2

2(Na ox state) + 2(O ox state) = 0

2(+1) + 2(O ox state) = 0 (Sodium takes priority over oxygen.)

+2 + 2(O ox state) = 0

O ox state = −1

$\underset{+1}{Na_2}\ \underset{-1}{O_2}$

sum: $2(+1) + 2(-1) = 0$

SKILLBUILDER 16.3 **Assigning Oxidation States**

Assign an oxidation state to each atom in each of the following.

(a) Zn
(b) Cu^{2+}
(c) $CaCl_2$
(d) CF_4
(e) NO_2^-
(f) SO_3

Now let's return to our original question. What is being oxidized and what is being reduced in the following reaction?

$$C + 2\,S \longrightarrow CS_2$$

We use the oxidation state rules to assign oxidation states to all elements on both sides of the equation.

$$\underset{0}{C} + 2\,\underset{0}{S} \longrightarrow \underset{+4\,-2}{CS_2}$$

Carbon went from an oxidation state of 0 to +4. In terms of our electron bookkeeping scheme (the assigned oxidation state), carbon *lost electrons* and was *oxidized*. Sulfur went from an oxidation state of 0 to −2. In terms of our electron bookkeeping scheme, sulfur *gained electrons* and was *reduced*.

$$\underset{0}{C} + 2\,\underset{0}{S} \longrightarrow \underset{+4\,-2}{CS_2}$$

Reduction (S: 0 → −2)
Oxidation (C: 0 → +4)

In terms of oxidation states, oxidation and reduction are defined as follows.

Oxidation—an increase in oxidation state

Reduction—a decrease in oxidation state

EVERYDAY Chemistry

The Bleaching of Hair

College students, both male and female, with bleached hair are a common sight on most campuses. Many students bleach their own hair with home-bleaching kits available at most drugstores and supermarkets. These kits normally contain hydrogen peroxide (H_2O_2), an excellent oxidizing agent. When applied to hair, hydrogen peroxide oxidizes melanin, the dark pigment that gives hair color. Once melanin is oxidized, it no longer imparts a dark color to hair, leaving the hair with the familiar bleached look.

Hydrogen peroxide also oxidizes other components of hair. For example, the protein molecules in hair contain —SH groups called *thiols*. Thiols are normally slippery (they slide across each other). Hydrogen peroxide oxidizes these thiol groups to sulfonic acid groups ($—SO_3H$). Sulfonic acid groups are stickier, causing hair to tangle more easily. Consequently, people with heavily bleached hair often use conditioners. Conditioners contain compounds that form thin, lubricating coatings on individual hair shafts. These coatings prevent tangling and make hair softer and more manageable.

CAN YOU ANSWER THIS? *Assign oxidation states to the atoms of H_2O_2. Which atoms in H_2O_2 do you think change oxidation state when H_2O_2 oxidizes hair?*

▲ Hair is often bleached using hydrogen peroxide, a good oxidizing agent.

EXAMPLE 16.4 Using Oxidation States to Identify Oxidation and Reduction

Use oxidation states to identify the element that is being oxidized and the element that is being reduced in the following redox reaction.

$$Ca(s) + 2\,H_2O(l) \longrightarrow Ca(OH)_2(aq) + H_2(g)$$

Solution:

Assign an oxidation state to each atom in the reaction. Since Ca increased in oxidation state, it was oxidized. Since H decreased in oxidation state, it was reduced. (Note that oxygen has the same oxidation state on both sides of the equation, and was therefore neither oxidized nor reduced.)

	$Ca(s)$	$+\ 2\,H_2O(l)$	$\longrightarrow$	$Ca(OH)_2(aq)$	$+\ H_2(g)$
Oxidation states	0	+1 −2		+2 −2 +1	0

Oxidation: Ca (0 → +2); Reduction: H (+1 → 0)

SKILLBUILDER 16.4 Using Oxidation States to Identify Oxidation and Reduction

Use oxidation states to identify the element that is being oxidized and the element that is being reduced in the following redox reaction.

$$Sn(s) + 4\,HNO_3(aq) \longrightarrow SnO_2(s) + 4\,NO_2(g) + 2\,H_2O(g)$$

CONCEPTUAL CHECKPOINT 16.1

In which of the following does nitrogen have the *lowest* oxidation state?

(a) N_2 **(b)** NO **(c)** NO_2 **(d)** NH_3

16.4 Balancing Redox Equations

In Chapter 7, we learned how to balance chemical equations by inspection. Many redox reactions can be balanced in this way. However, redox reactions occurring in aqueous solutions are usually difficult to balance by inspection and require a special procedure called the *half-reaction method of balancing*. In this procedure, the overall equation is broken down into two **half-reactions**: one for oxidation and one for reduction. The half-reactions are balanced individually and then added together. For example, consider the following redox reaction.

$$Al(s) + Ag^{+}(aq) \longrightarrow Al^{3+}(aq) + Ag(s)$$

We assign oxidation numbers to all atoms to determine what is being oxidized and what is being reduced.

$$\begin{array}{lcccc} & Al(s) & + Ag^{+}(aq) \longrightarrow & Al^{3+}(aq) & + Ag(s) \\ \text{Oxidation states} & 0 & +1 & +3 & 0 \end{array}$$

Reduction
Oxidation

We then divide the reaction into two half-reactions, one for oxidation and one for reduction.

$$\text{Oxidation:} \quad Al(s) \longrightarrow Al^{3+}(aq)$$
$$\text{Reduction:} \quad Ag^{+}(aq) \longrightarrow Ag(s)$$

The two half-reactions are then balanced individually. In this case, the half-reactions are already balanced with respect to mass—the number of each type of atom on both sides of each half-reaction is the same. However, the equations are not balanced with respect to charge—in the oxidation half-reaction, the left side of the equation has 0 charge while the right side has +3 charge, and in the reduction half-reaction, the left side has +1 charge and the right side has 0 charge. We balance the charge of each half-reaction individually by adding the appropriate number of electrons to make the charges on both sides equal.

$$Al(s) \longrightarrow Al^{3+}(aq) + \mathbf{3\,e^{-}} \qquad \text{(zero charge on both sides)}$$
$$\mathbf{1\,e^{-}} + Ag^{+}(aq) \longrightarrow Ag(s) \qquad \text{(zero charge on both sides)}$$

Since these half-reactions must occur together, the number of electrons lost in the oxidation half-reaction must equal the number gained in the reduction half-reaction. We equalize these by multiplying one or both half-reactions by appropriate whole numbers to equalize the electrons lost and gained. In this case, we multiply the reduction half-reaction by 3.

$$Al(s) \longrightarrow Al^{3+}(aq) + 3\,e^{-}$$
$$3 \times (1\,e^{-} + Ag^{+}(aq) \longrightarrow Ag(s))$$

We then add the half-reactions together, canceling electrons and other species as necessary.

$$\begin{array}{r} Al(s) \longrightarrow Al^{3+}(aq) + \cancel{3\,e^{-}} \\ \cancel{3\,e^{-}} + 3\,Ag^{+}(aq) \longrightarrow 3\,Ag(s) \\ \hline Al(s) + 3\,Ag^{+}(aq) \longrightarrow Al^{3+}(aq) + 3\,Ag(s) \end{array}$$

Lastly, we verify that the equation is balanced, both with respect to mass and with respect to charge.

Reactants	Products
1 Al	1 Al
3 Ag	3 Ag
+3 charge	+3 charge

Notice that the charge need not be zero on both sides of the equation—it just has to be *equal* on both sides. The equation is balanced.

A general procedure for balancing redox reactions is given in the following procedure box. Since aqueous solutions are often acidic or basic, the procedure must account for the presence of H^+ ions or OH^- ions. In this book, we cover only the general procedure for acidic solutions.

Balancing Redox Equations Using the Half-Reaction Method	EXAMPLE 16.5	EXAMPLE 16.6
	Balance the following redox reaction.	Balance the following redox reaction.
1. *Assign oxidation states* to all atoms and identify the substances being oxidized and reduced.	$\underset{0}{Al(s)} + \underset{+2}{Cu^{2+}(aq)} \longrightarrow \underset{+3}{Al^{3+}(aq)} + \underset{0}{Cu(s)}$ Oxidation (Al: 0 → +3); Reduction (Cu: +2 → 0)	$\underset{+2}{Fe^{2+}(aq)} + \underset{+7\ -2}{MnO_4^-(aq)} \longrightarrow \underset{+3}{Fe^{3+}(aq)} + \underset{+2}{Mn^{2+}(aq)}$ Oxidation (Fe: +2 → +3); Reduction (Mn: +7 → +2)
2. *Separate the overall reaction into two half-reactions*, one for oxidation and one for reduction.	**Solution:** Oxidation: $Al(s) \longrightarrow Al^{3+}(aq)$ Reduction: $Cu^{2+}(aq) \longrightarrow Cu(s)$	**Solution:** Oxidation: $Fe^{2+}(aq) \longrightarrow Fe^{3+}(aq)$ Reduction: $MnO_4^-(aq) \longrightarrow Mn^{2+}(aq)$
3. *Balance each half-reaction with respect to mass* in the following order.		
• Balance all elements other than H and O.	All elements other than hydrogen and oxygen are balanced, so proceed to next step.	All elements other than hydrogen and oxygen are balanced, so proceed to next step.
• Balance O by adding H_2O.	No oxygen; proceed to next step.	$Fe^{2+}(aq) \longrightarrow Fe^{3+}(aq)$ $MnO_4^-(aq) \longrightarrow Mn^{2+}(aq) + \mathbf{4\ H_2O(\mathit{l})}$
• Balance H by adding H^+.	No hydrogen; proceed to next step.	$Fe^{2+}(aq) \longrightarrow Fe^{3+}(aq)$ $\mathbf{8\ H^+(\mathit{aq})} + MnO_4^-(aq) \longrightarrow Mn^{2+}(aq) + 4\ H_2O(l)$
4. *Balance each half-reaction with respect to charge* by adding electrons to the right side of the oxidation half-reaction and the left side of the reduction half-reaction. (The sum of the charges on both sides of each equation should then be equal.)	$Al(s) \longrightarrow Al^{3+}(aq) + \mathbf{3\ e^-}$ $\mathbf{2\ e^-} + Cu^{2+}(aq) \longrightarrow Cu(s)$	$Fe^{2+}(aq) \longrightarrow Fe^{3+}(aq) + \mathbf{1\ e^-}$ $\mathbf{5\ e^-} + 8\ H^+(aq) + MnO_4^-(aq) \longrightarrow Mn^{2+}(aq) + 4\ H_2O(l)$

5. *Make the number of electrons in both half-reactions equal* by multiplying one or both half-reactions by a small whole number.	$2 \times (Al(s) \longrightarrow Al^{3+}(aq) + 3\,e^-)$ $3 \times [2\,e^- + Cu^{2+}(aq) \longrightarrow Cu(s)]$	$5 \times [Fe^{2+}(aq) \longrightarrow Fe^{3+}(aq) + 1\,e^-]$ $5\,e^- + 8\,H^+(aq) + MnO_4^-(aq) \longrightarrow Mn^{2+}(aq) + 4\,H_2O(l)$
6. *Add the two half-reactions together*, canceling electrons and other species as necessary.	$2\,Al(s) \longrightarrow 2\,Al^{3+}(aq) + \cancel{6\,e^-}$ $\cancel{6\,e^-} + 3\,Cu^{2+}(aq) \longrightarrow 3\,Cu(s)$ $2\,Al(s) + 3\,Cu^{2+}(aq) \longrightarrow 2\,Al^{3+}(aq) + 3Cu(s)$	$5\,Fe^{2+}(aq) \longrightarrow 5\,Fe^{3+}(aq) + \cancel{5\,e^-}$ $\cancel{5\,e^-} + 8\,H^+(aq) + MnO_4^-(aq) \longrightarrow Mn^{2+}(aq) + 4\,H_2O(l)$ $5\,Fe^{2+}(aq) + 8\,H^+(aq) + MnO_4^-(aq) \longrightarrow 5\,Fe^{3+}(aq) + Mn^{2+}(aq) + 4\,H_2O(l)$
7. *Verify that the reaction is balanced* both with respect to mass and with respect to charge.	Reactants / Products: 2 Al / 2 Al 3 Cu / 3 Cu +6 charge / +6 charge	Reactants / Products: 5 Fe / 5 Fe 8 H / 8 H 1 Mn / 1 Mn 4 O / 4 O +17 charge / +17 charge

SKILLBUILDER 16.5

Balance the following redox reaction occurring in acidic solution.

$$H^+(aq) + Cr(s) \longrightarrow H_2(g) + Cr^{3+}(aq)$$

SKILLBUILDER 16.6

Balance the following redox reaction occurring in acidic solution.

$$Cu(s) + NO_3^-(aq) \longrightarrow Cu^{2+}(aq) + NO_2(g)$$

EXAMPLE 16.7 Balancing Redox Reactions

Balance the following redox reaction occurring in acidic solution.

$$I^-(aq) + Cr_2O_7^{2-}(aq) \longrightarrow Cr^{3+}(aq) + I_2(s)$$

1. Assign oxidation states.

Solution:

Follow the half-reaction method for balancing redox reactions.

$$\underset{-1}{I^-(aq)} + \underset{+6\ -2}{Cr_2O_7^{2-}(aq)} \longrightarrow \underset{+3}{Cr^{3+}(aq)} + \underset{0}{I_2(s)}$$

Reduction (Cr: +6 → +3); Oxidation (I: −1 → 0)

2. Separate the overall reaction into two half-reactions.

Oxidation: $I^-(aq) \longrightarrow I_2(s)$
Reduction: $Cr_2O_7^{2-}(aq) \longrightarrow Cr^{3+}(aq)$

3. Balance each half-reaction with respect to mass.

- Balance all elements other than H and O.

$\mathbf{2}\,I^-(aq) \longrightarrow I_2(aq)$
$Cr_2O_7^{2-}(aq) \longrightarrow \mathbf{2}\,Cr^{3+}(s)$

- Balance O by adding H_2O.

$2\,I^-(aq) \longrightarrow I_2(s)$
$Cr_2O_7^{2-}(aq) \longrightarrow 2\,Cr^{3+}(aq) + \mathbf{7\,H_2O(l)}$

- Balance H by adding H^+.

$2\,I^-(aq) \longrightarrow I_2(s)$
$\mathbf{14\,H^+(aq)} + Cr_2O_7^{2-}(aq) \longrightarrow 2\,Cr^{3+}(aq) + 7\,H_2O(l)$

4. Balance each half-reaction with respect to charge.

$2\,I^-(aq) \longrightarrow I_2(s) + \mathbf{2\,e^-}$
$\mathbf{6\,e^-} + 14\,H^+(aq) + Cr_2O_7^{2-}(aq) \longrightarrow 2\,Cr^{3+}(aq) + 7\,H_2O(l)$

5. Make the number of electrons in both half-reactions equal.	$3 \times [2\,I^-(aq) \longrightarrow I_2(s) + 2\,e^-]$ $6\,e^- + 14\,H^+(aq) + Cr_2O_7^{2-}(aq) \longrightarrow 2\,Cr^{3+}(aq) + 7\,H_2O(l)$
6. Add the half-reactions together.	$6\,I^-(aq) \longrightarrow 3\,I_2(s) + \cancel{6\,e^-}$ $\cancel{6\,e^-} + 14\,H^+(aq) + Cr_2O_7^{2-}(aq) \longrightarrow 2\,Cr^{3+}(aq) + 7\,H_2O(l)$ $6\,I^-(aq) + 14\,H^+(aq) + Cr_2O_7^{2-}(aq) \longrightarrow 3\,I_2(s) + 2\,Cr^{3+}(aq) + 7\,H_2O(l)$
7. Verify that the reaction is balanced.	

Reactants	Products
6 I	6 I
14 H	14 H
2 Cr	2 Cr
7 O	7 O
+6 charge	+6 charge

Chemistry IN THE ENVIRONMENT

Photosynthesis and Respiration: Energy for Life

All living things require energy, and most of that energy comes from the Sun. Solar energy reaches Earth in the form of electromagnetic radiation (Chapter 9). This radiation keeps our planet at a temperature that allows life as we know it to flourish. But the wavelengths that make up visible light have an additional and very crucial role to play in the maintenance of life. Plants capture this light and use it to make energy-rich organic molecules such as carbohydrates. (These compounds are discussed more fully in Chapter 19.) Animals get their energy by eating plants, or by eating other animals that have eaten plants. So ultimately, virtually all of the energy for life comes from sunlight.

But in chemical terms, how is this energy captured, transferred from organism to organism, and used? *The key reactions in these processes all involve oxidation and reduction.*

Most living things use chemical energy through a process known as *respiration*. In respiration, energy-rich molecules, typified by the sugar glucose, are "burned" in a reaction that can be summarized as follows:

$$\underset{\text{Glucose}}{C_6H_{12}O_6} + \underset{\text{Oxygen}}{6\,O_2} \longrightarrow \underset{\text{Carbon dioxide}}{6\,CO_2} + \underset{\text{Water}}{6\,H_2O} + \text{energy}$$

You can easily see that respiration is a redox reaction. On the simplest level, it is clear that some of the atoms in glucose are gaining oxygen. More precisely, we can use the rules for assigning oxidation states to show that carbon is oxidized from an oxidation number of 0 in glucose to +4 in carbon dioxide.

Respiration is also an exothermic reaction—it releases energy. If you simply burned glucose in a test tube, the results would be much the same, but the energy would be lost as heat. Living things, however, have devised ways of capturing the energy released and using it to power their life processes, such as movement, growth, and the synthesis of other life-sustaining molecules.

Respiration is one half of a larger cycle. The other half is *photosynthesis*. Photosynthesis is the series of reactions by which green plants capture the energy of sunlight and store it as chemical energy in compounds such as glucose. Photosynthesis can be summarized as follows:

$$\underset{\text{Carbon dioxide}}{6\,CO_2} + \underset{\text{Water}}{6\,H_2O} + \text{energy (sunlight)} \longrightarrow \underset{\text{Glucose}}{C_6H_{12}O_6} + \underset{\text{Oxygen}}{6\,O_2}$$

This reaction—the exact reverse of respiration—is the ultimate source of the molecules that are oxidized in respiration. And just as the key process in respiration is the oxidation of carbon, the key process in photosynthesis is the reduction of carbon. This reduction is driven by solar energy—energy that is stored in the resulting glucose molecule. Living things harvest that energy when they "burn" glucose in the respiration half of the cycle.

Thus, oxidation and reduction reactions are at the very center of all life on Earth!

CAN YOU ANSWER THIS? *What is the oxidation state of the oxygen atoms in CO_2, H_2O, and O_2? What does this information tell you about the reactions of photosynthesis and respiration?*

▲ Sunlight, captured by plants in photosynthesis, is the ultimate source of the chemical energy used by nearly all living things on Earth.

SKILLBUILDER 16.7 **Balancing Redox Reactions**

Balance the following redox reaction occurring in acidic solution.

$$Sn(s) + MnO_4^-(aq) \longrightarrow Sn^{2+}(aq) + Mn^{2+}(aq)$$

16.5 The Activity Series: Predicting Spontaneous Redox Reactions

As we have seen, redox reactions depend on the gaining of electrons by one substance and the losing of electrons by another. Is there a way to predict whether a particular redox reaction will spontaneously occur? Suppose we knew that substance A has a greater tendency to lose electrons than substance B (A is more easily oxidized than B). Then we could predict that if we mix A with cations of B, a redox reaction would occur in which A loses its electrons (A is oxidized) to the cations of B (B cations are reduced). For example, Mg has a greater tendency to lose electrons than Cu. Consequently, if we put solid Mg into a solution containing Cu^{2+} ions, Mg is oxidized and Cu^{2+} is reduced.

$$Mg(s) + Cu^{2+}(aq) \longrightarrow Mg^{2+}(aq) + Cu(s)$$

We see this as the fading of blue (the color of the Cu^{2+} ions in solution), the dissolving of the solid magnesium, and the appearance of solid copper on the remaining magnesium surface (▼ Figure 16.5). This reaction is spontaneous—it occurs on its own when Mg(s) and $Cu^{2+}(aq)$ come into contact. On the other hand, if we put Cu(s) in a solution containing $Mg^{2+}(aq)$ ions, no reaction occurs (▼ Figure 16.6).

$$Cu(s) + Mg^{2+}(aq) \longrightarrow \text{NO REACTION}$$

No reaction occurs because, as we said previously, Mg atoms have a greater tendency to lose electrons than do Cu atoms—Cu atoms will therefore not lose electrons to Mg^{2+} ions.

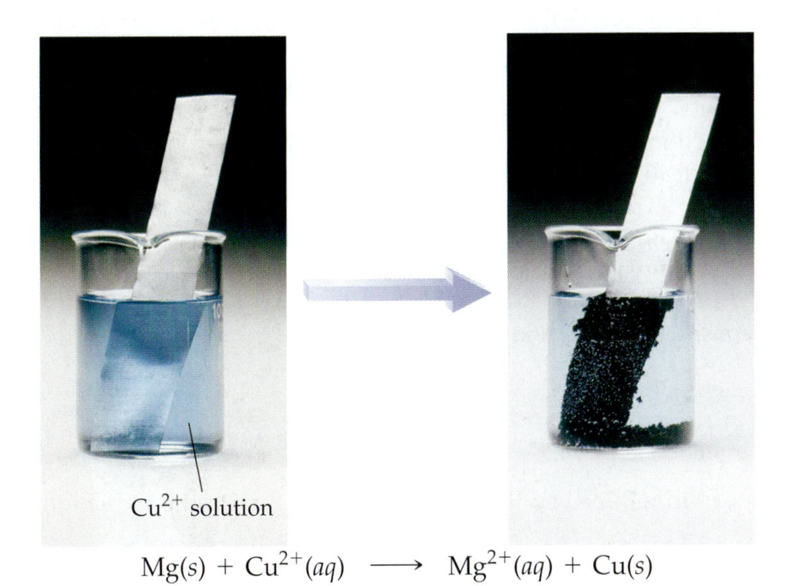

▲ **Figure 16.5** **Cu^{2+} oxidizes magnesium** When a magnesium strip is put into a Cu^{2+} solution, the magnesium is oxidized to Mg^{2+} and the copper ion is reduced to Cu(s). Notice the fading of the blue color (due to Cu^{2+} ions) in solution and the appearance of solid copper on the magnesium strip.

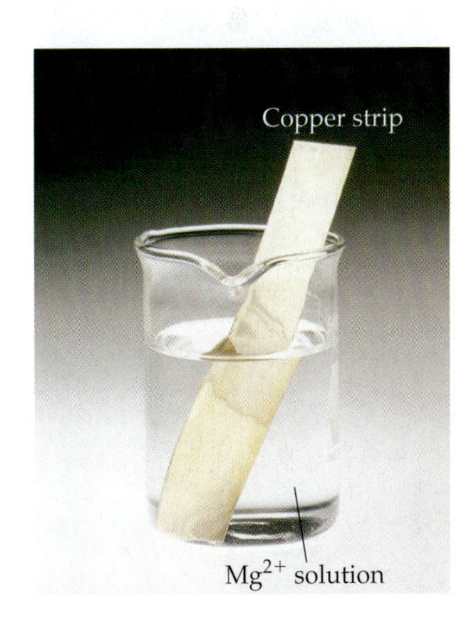

▲ **Figure 16.6** **Mg^{2+} does not oxidize copper** When solid copper is placed in a solution containing Mg^{2+} ions, no reaction occurs.
Question: Why?

TABLE 16.1 Activity Series of Metals

Half-reaction	
$Li(s) \longrightarrow Li^{+}(aq) + e^{-}$	Most reactive
$K(s) \longrightarrow K^{+}(aq) + e^{-}$	Most easily oxidized
$Ca(s) \longrightarrow Ca^{2+}(aq) + 2\,e^{-}$	Strongest tendency to lose electrons
$Na(s) \longrightarrow Na^{+}(aq) + e^{-}$	
$Mg(s) \longrightarrow Mg^{2+}(aq) + 2\,e^{-}$	
$Al(s) \longrightarrow Al^{3+}(aq) + 3\,e^{-}$	
$Mn(s) \longrightarrow Mn^{2+}(aq) + 2\,e^{-}$	
$Zn(s) \longrightarrow Zn^{2+}(aq) + 2\,e^{-}$	
$Cr(s) \longrightarrow Cr^{3+}(aq) + 3\,e^{-}$	
$Fe(s) \longrightarrow Fe^{2+}(aq) + 2\,e^{-}$	
$Ni(s) \longrightarrow Ni^{2+}(aq) + 2\,e^{-}$	
$Sn(s) \longrightarrow Sn^{2+}(aq) + 2\,e^{-}$	
$Pb(s) \longrightarrow Pb^{2+}(aq) + 2\,e^{-}$	
$\mathbf{H_2(g) \longrightarrow 2H^{+}(aq) + 2e^{-}}$	
$Cu(s) \longrightarrow Cu^{2+}(aq) + 2\,e^{-}$	Least reactive
$Ag(s) \longrightarrow Ag^{+}(aq) + e^{-}$	Most difficult to oxidize
$Au(s) \longrightarrow Au^{3+}(aq) + 3\,e^{-}$	Least tendency to lose electrons

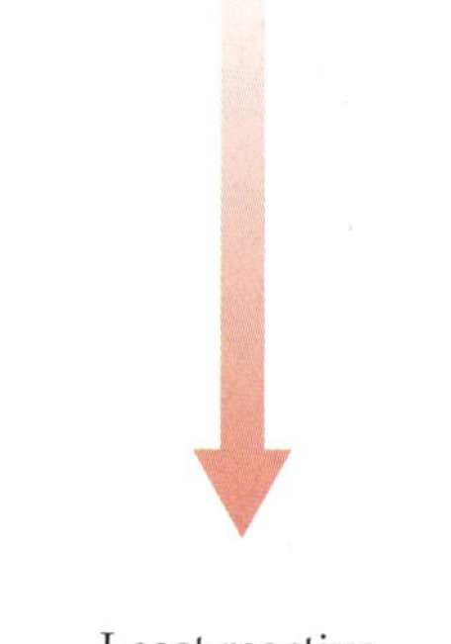

▲ Gold is very low on the activity series. Because it is so difficult to oxidize, it resists the tarnishing and corrosion that more active metals undergo.

Table 16.1 shows the **activity series of metals**. This table lists metals in order of decreasing tendency to lose electrons. The metals at the top of the list have the greatest tendency to lose electrons—they are most easily oxidized and therefore the most reactive. The metals at the bottom of the list have the lowest tendency to lose electrons—they are the most difficult to oxidize and therefore the least reactive. It is not a coincidence that the metals used for jewelry, such as silver and gold, are near the bottom of the list. They are among the least reactive metals and therefore do not form compounds easily. Instead, they tend to remain as solid silver and solid gold rather than being oxidized to silver and gold cations by elements in the environment (such as the oxygen in the air).

Each reaction in the activity series is an oxidation half-reaction. The half-reactions at the top are most likely to occur in the *forward* direction, and the half-reactions at the bottom are most likely to occur in the *reverse* direction. Consequently, if you pair a half-reaction from the top of the list with the reverse of a half-reaction from the bottom of the list, you get a spontaneous reaction. More specifically,

Any half-reaction on the list will be spontaneous when paired with the reverse of any half-reaction below it.

For example, consider the following two half-reactions.

$$Mn(s) \longrightarrow Mn^{2+}(aq) + 2\,e^{-}$$
$$Ni(s) \longrightarrow Ni^{2+}(aq) + 2\,e^{-}$$

The oxidation of Mn will be spontaneous when paired with the reduction of Ni^{2+}.

$$Mn(s) \longrightarrow Mn^{2+}(aq) + \cancel{2\,e^{-}}$$
$$Ni^{2+}(aq) + \cancel{2\,e^{-}} \longrightarrow Ni(s)$$
$$Mn(s) + Ni^{2+}(aq) \longrightarrow Mn^{2+}(aq) + Ni(s) \qquad \text{(spontaneous reaction)}$$

However, if we attempt to pair a half-reaction on the list with the reverse of a half-reaction above it, we get no reaction. For example,

$$\begin{array}{l} Mn(s) \longrightarrow Mn^{2+}(aq) + 2\,e^- \\ \underline{Mg^{2+}(aq) + 2\,e^- \longrightarrow Mg(s)} \\ Mn(s) + Mg^{2+}(aq) \longrightarrow \text{NO REACTION} \end{array}$$

No reaction occurs because Mg has a greater tendency to be oxidized than Mn. Since it is already oxidized in this reaction, nothing else happens.

EXAMPLE 16.8 Predicting Spontaneous Redox Reactions

Will the following redox reactions be spontaneous?

(a) $Fe(s) + Mg^{2+}(aq) \longrightarrow Fe^{2+}(aq) + Mg(s)$

(b) $Fe(s) + Pb^{2+}(aq) \longrightarrow Fe^{2+}(aq) + Pb(s)$

Solution:

(a) $Fe(s) + Mg^{2+}(aq) \longrightarrow Fe^{2+}(aq) + Mg(s)$

This reaction involves the oxidation of Fe

$$Fe(s) \longrightarrow Fe^{2+}(aq) + 2\,e^-$$

with the reverse of a half-reaction *above it* in the activity series.

$$Mg^{2+}(aq) + 2\,e^- \longrightarrow Mg(s)$$

Therefore, the reaction *will not be* spontaneous.

(b) $Fe(s) + Pb^{2+}(aq) \longrightarrow Fe^{2+}(aq) + Pb(s)$

This reaction involves the oxidation of Fe

$$Fe(s) \longrightarrow Fe^{2+}(aq) + 2\,e^-$$

with the reverse of a half-reaction *below it* in the activity series.

$$Pb^{2+}(aq) + 2\,e^- \longrightarrow Pb(s)$$

Therefore, the reaction *will be* spontaneous.

SKILLBUILDER 16.8 Predicting Spontaneous Redox Reactions

Will the following redox reactions be spontaneous?

(a) $Zn(s) + Ni^{2+}(aq) \longrightarrow Zn^{2+}(aq) + Ni(s)$

(b) $Zn(s) + Ca^{2+}(aq) \longrightarrow Zn^{2+}(aq) + Ca(s)$

Predicting Whether a Metal Will Dissolve in Acid

In Chapter 14, we learned that acids dissolve metals. Most acids dissolve metals by the reduction of H^+ ions to hydrogen gas and the corresponding oxidation of the metal to its ion. For example, if solid Zn were dropped into hydrochloric acid, the following reaction would occur.

$$\begin{array}{l} Zn(s) \longrightarrow Zn^{2+}(aq) + 2\,e^- \\ \underline{2\,H^+(aq) + 2\,e^- \longrightarrow H_2(g)} \\ Zn(s) + 2\,H^+(aq) \longrightarrow Zn^{2+}(aq) + H_2(g) \end{array}$$

Unlike most other acids, nitric acid can dissolve some metals below H_2 in the activity series.

$$Zn(s) + 2\,H^+(aq) \longrightarrow Zn^{2+}(aq) + H_2(g)$$

▲ **Figure 16.7 Zinc dissolves in hydrochloric acid** The zinc metal is oxidized to Zn^{2+} ions and the H^+ ions are reduced to hydrogen gas.

We observe the reaction as the dissolving of the zinc and the bubbling of hydrogen gas (◄ Figure 16.7). The zinc is oxidized and the H^+ ions are reduced, dissolving the zinc. Notice that this reaction involves pairing the oxidation half-reaction of Zn with the reverse of a half-reaction below zinc on the activity series (the reduction of H^+). Therefore, this reaction is spontaneous. What would happen, however, if we paired the oxidation of Cu with the reduction of H^+? The reaction would not be spontaneous because it involves pairing the oxidation of copper with the reverse of a half-reaction *above it* in the activity series. Consequently, copper does not react with H^+ and will not dissolve in acids such as HCl. In general,

> Metals above H_2 on the activity series will dissolve in acids, while metals below H_2 will not dissolve in acids.

An important exception to this rule is nitric acid (HNO_3), which through a different reduction half-reaction can dissolve some of the metals below H_2 in the activity series.

EXAMPLE 16.9 Predicting Whether a Metal Will Dissolve in Acid

Will Cr dissolve in hydrochloric acid?

Solution:

Yes. Since Cr is above H_2 in the activity series, it will dissolve in HCl.

SKILLBUILDER 16.9 Predicting Whether a Metal Will Dissolve in Acid

Will Ag dissolve in hydrobromic acid?

CONCEPTUAL CHECKPOINT 16.2

It has been suggested that one cause of the decline of the Roman Empire was widespread chronic poisoning. The suspected source was a metal present in the vessels commonly used to store and serve acidic substances such as wine. Which of the following would you expect might pose such a danger?

(a) silver
(b) gold
(c) lead
(d) copper

16.6 Batteries: Using Chemistry to Generate Electricity

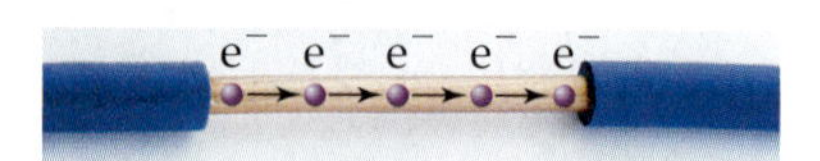

▲ **Figure 16.8 An electrical current** Electrical current is the flow of electrical charge. In this figure, electrons are flowing through a wire.

Electrical current is simply the flow of electric charge (◄ Figure 16.8). Electrons flowing through a wire or ions flowing through a solution are both examples of electrical current. Since redox reactions involve the transfer of electrons from one species to another, they can create electrical current. For example, consider the following spontaneous redox reaction.

$$Zn(s) + Cu^{2+}(aq) \longrightarrow Zn^{2+} + Cu(s)$$

When Zn metal is placed into a Cu^{2+} solution, Zn is oxidized and Cu^{2+} is reduced—electrons are transferred directly from the Zn to the Cu^{2+}. Suppose we could separate the reactants and force the electrons to travel through a wire to get from the Zn to the Cu^{2+}. The flowing electrons would constitute an electrical current and could be used to do electrical work. This is normally done in an **electrochemical cell**, a device that creates electrical current from a redox reaction.

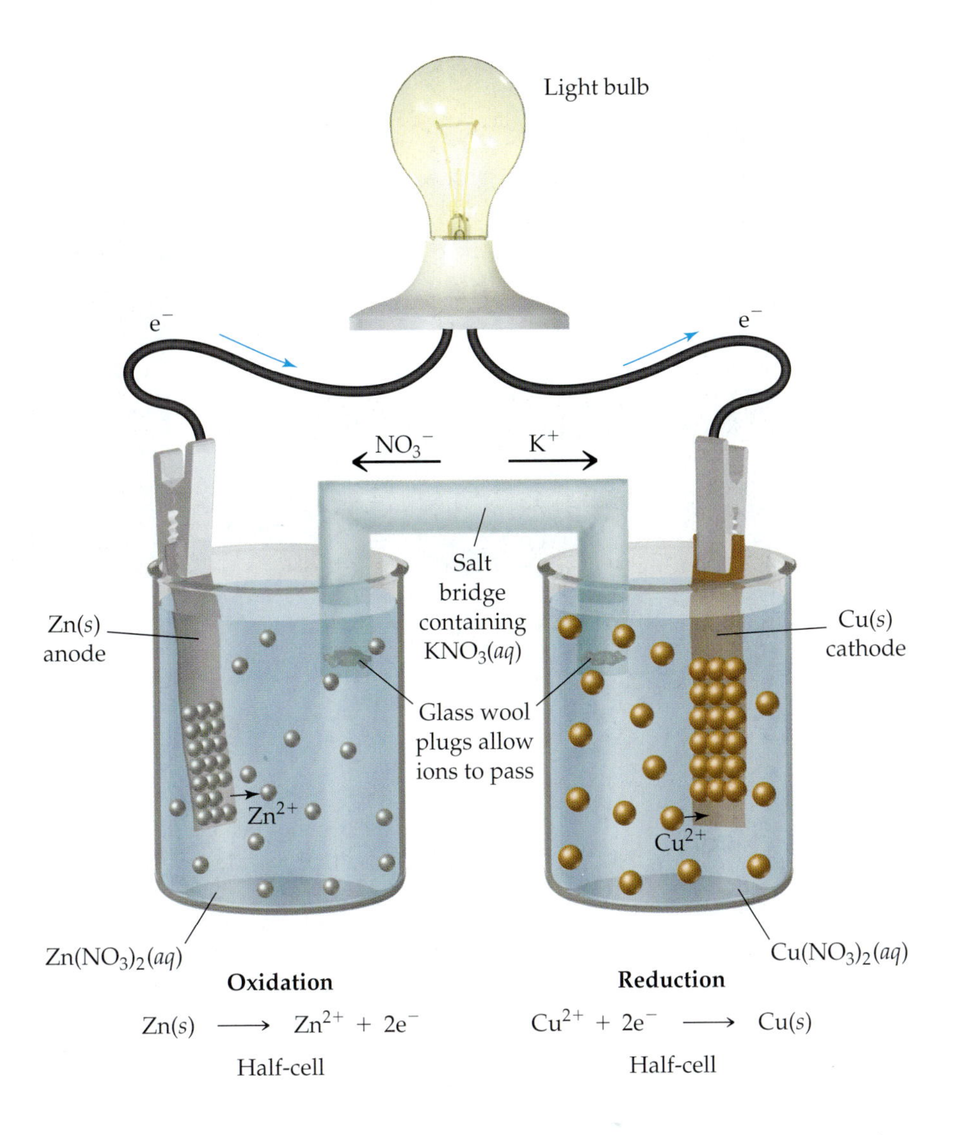

▶ **Figure 16.9 An electrochemical cell** **Question:** Why do electrons flow from left to right in this figure?

The salt bridge completes the circuit—it allows the flow of ions between the two half-cells.

For example, consider the electrochemical cell in ▲ Figure 16.9. In this cell, a solid strip of Zn is placed into a $Zn(NO_3)_2$ solution to form a **half-cell**. Similarly, a solid strip of Cu is placed into a $Cu(NO_3)_2$ solution to form a second half-cell. Then the two half-cells are connected by attaching a wire from the Zn, through a lightbulb or other electrical device, and to the copper. The natural tendency of Zn to oxidize and Cu^{2+} to reduce results in the flow of electrons through the wire. The flowing electrons constitute an electrical current that lights the bulb.

In an electrochemical cell, the metal strip where oxidation occurs is called the **anode** and is labeled with a negative (−) sign. The metal strip where reduction occurs is called the **cathode** and is labeled with a (+) sign. Electrons flow from the anode to the cathode (away from negative and toward positive).

As electrons flow out of the anode, positive ions form in the oxidation half-cell (Zn^{2+} is formed in the preceding example). As electrons flow into the cathode, positive ions deposit as charge-neutral atoms at the reduction half-cell (Cu^{2+} deposits as Cu(s) in the preceding example). However, if this were the only flow of charge, the flow would soon stop as positive charge accumulated at the anode and negative charge at the cathode. The circuit must be completed with a **salt bridge**, an inverted U-shaped tube that joins the two half-cells and contains a strong electrolyte such as KNO_3. The salt bridge allows for the flow of ions that neutralizes the charge imbalance. The negative ions within the salt bridge flow to neutralize the accumulation of positive charge at the anode, and the positive ions flow to neutralize the accumulation of negative charge at the cathode.

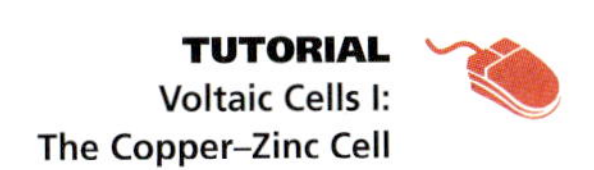

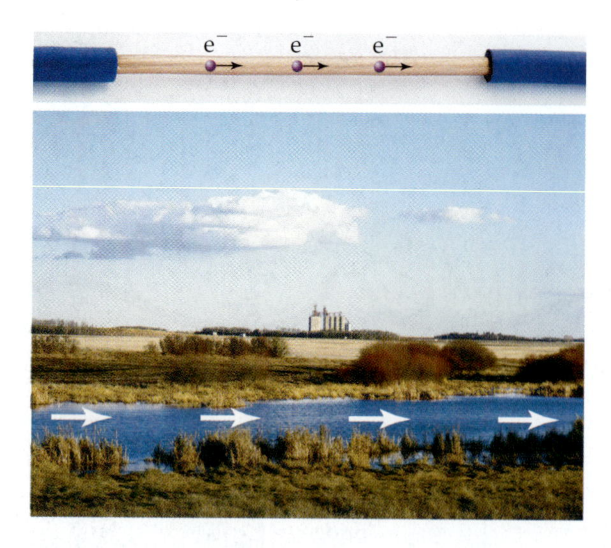

▲ **Figure 16.10 Water analogy for electrical current** The flow of electrons through a wire is similar to the flow of water in a river.

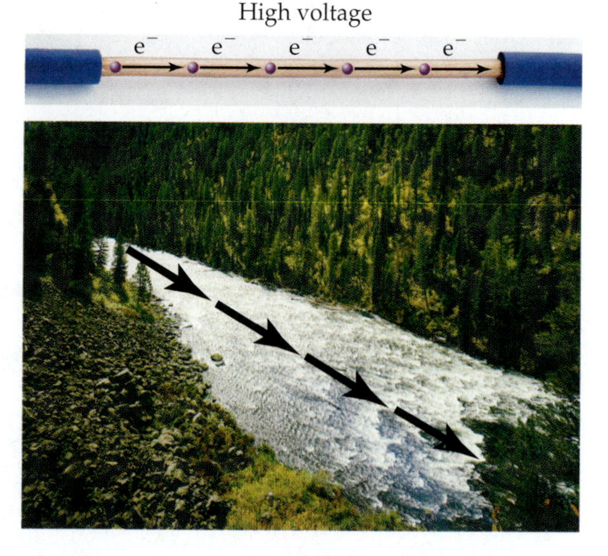

▲ **Figure 16.11 Water analogy for voltage** A high voltage for electricity is analogous to a steep descent for a river.

An electrochemical cell that spontaneously produces electrical current (such as the one just described) is called a **galvanic** or **voltaic cell**. Electrical voltage is a measure of the tendency of electrons to flow. A high voltage corresponds to a high tendency to flow, while a low voltage corresponds to a low tendency. We can understand electrical voltage with an analogy. Electrons flowing through a wire are similar to water flowing in a river (▲ Figure 16.10). The *quantity of electrons* that flows through the wire (electrical current) is analogous to the *amount of water* that flows through the river (the river's current). The driving force that causes the electrons to flow through a wire—called *potential difference* or **voltage**—is analogous to the force of gravity that causes water to flow in a river. A high voltage is analogous to a steeply descending streambed (▲ Figure 16.11).

In a battery, the voltage depends on the relative tendencies of the reactants to undergo oxidation and reduction. Combining the oxidation of a metal high on the activity series with the reduction of a metal ion low on the activity series produces a battery with a relatively high voltage. For example, the oxidation of Li(*s*) combined with the reduction of Cu^{2+}(*aq*) would result in a relatively high voltage. On the other hand, combining the oxidation of a metal on the activity series with the reduction of a metal ion just below it would result in a battery with a relatively low voltage. Combining the oxidation of a metal on the activity series with the reduction of a metal ion above it on the activity series will not produce a battery at all. For example, you could not make a battery by trying to oxidize Cu(*s*) and reduce Li^{+}(*aq*). Such a reaction is not spontaneous and would not produce electrical current.

You can now see why batteries go dead after extended use. As the simple electrochemical cell we have just described is used, the zinc electrode dissolves away as zinc is oxidized to zinc ions. Similarly, the Cu^{2+} solution is depleted of Cu^{2+} ions as they deposit as solid Cu (▶ Figure 16.12). Once the zinc electrode is dissolved and the Cu^{2+} ions are depleted, the electrochemical cell is dead. Some batteries can be recharged by running electrical current—from an external source—in the opposite direction. This causes the regeneration of the reactants, allowing repeated use of the battery.

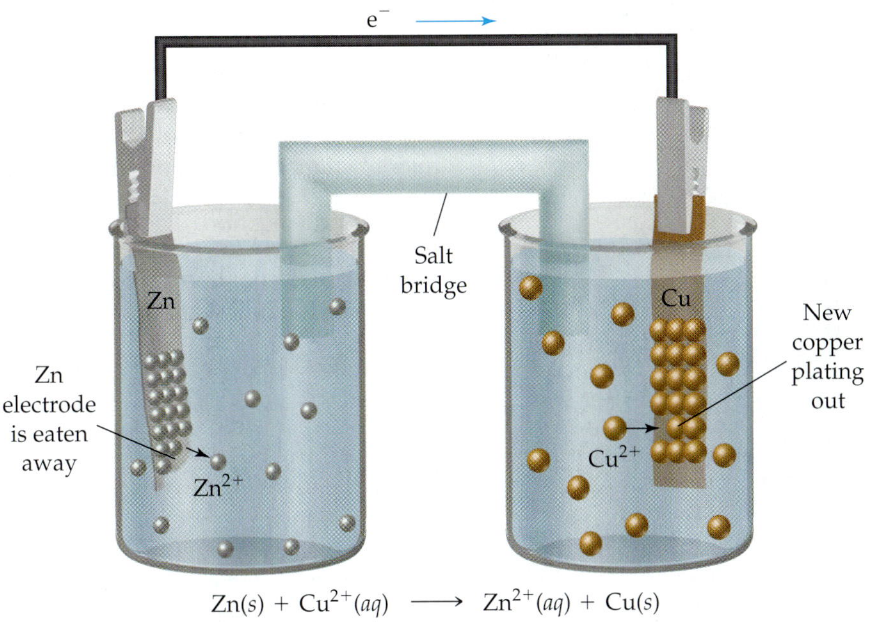

$$Zn(s) + Cu^{2+}(aq) \longrightarrow Zn^{2+}(aq) + Cu(s)$$

▲ **Figure 16.12** **Dead battery** A battery goes dead with extended use because the reactants [in this case $Zn(s)$ and $Cu^{2+}(aq)$] become depleted while the products [in this case $Zn^{2+}(aq)$ and $Cu(s)$] accumulate.

Dry-Cell Batteries

Common flashlight batteries are called **dry cells** because they do not contain large amounts of liquid water. There are several common types of dry-cell batteries. The most inexpensive dry cells are composed of a zinc case that acts as the anode (◀ Figure 16.13). The zinc is oxidized according to the following reaction.

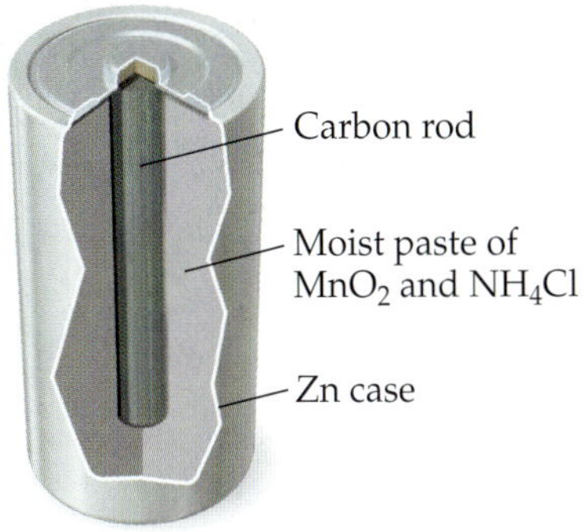

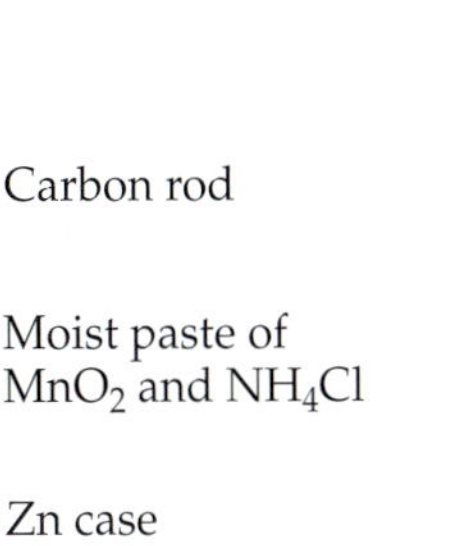

▲ **Figure 16.13** **Common dry-cell battery**

Anode reaction: $$Zn(s) \longrightarrow Zn^{2+}(aq) + 2\,e^- \quad \text{(oxidation)}$$

The cathode is a carbon rod immersed in a moist paste of MnO_2 that also contains NH_4Cl. The MnO_2 is reduced to Mn_2O_3 according to the following reaction.

Cathode reaction: $$2\,MnO_2(s) + 2\,NH_4^+(aq) + 2\,e^- \longrightarrow Mn_2O_3(s) + 2\,NH_3(g) + H_2O(l) \quad \text{(reduction)}$$

These two half-reactions produce a voltage of about 1.5 volts. Two or more of these batteries can be connected in series (cathode-to-anode connection) to produce higher voltages.

The more expensive **alkaline batteries** employ slightly different half-reactions that use a base (therefore the name *alkaline*). In an alkaline battery, the reactions are as follows.

Anode reaction: $$Zn(s) + 2\,OH^-(aq) \longrightarrow Zn(OH)_2(s) + 2\,e^-$$

Cathode reaction: $$2\,MnO_2(s) + 2\,H_2O(l) + 2\,e^- \longrightarrow 2\,MnO(OH)(s) + 2\,OH^-(aq) \quad \text{(reduction)}$$

Alkaline batteries have a longer working life and a longer shelf life than their nonalkaline counterparts.

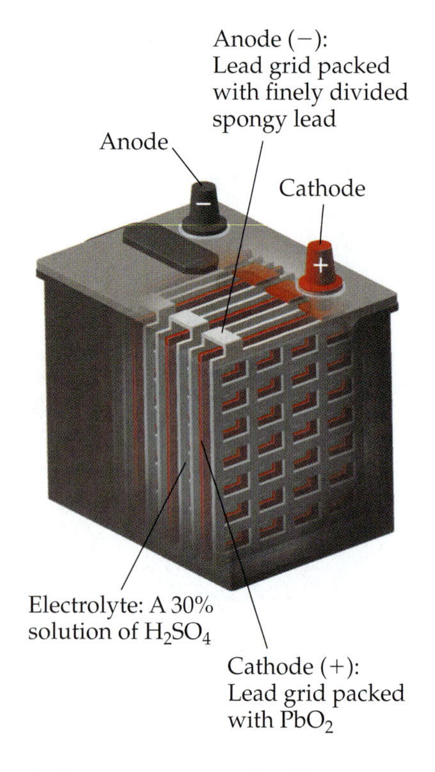

▲ **Figure 16.14 Lead-acid storage battery** **Question:** Why do batteries like this become depleted? How are they recharged?

Lead-Acid Storage Batteries

The batteries in most automobiles are **lead-acid storage batteries**. These batteries consist of six electrochemical cells wired in series (◀ Figure 16.14). Each cell produces 2 volts for a total of 12 volts. Each cell contains a porous lead anode where oxidation occurs according to the following reaction.

$$\text{Anode reaction: } Pb(s) + SO_4^{2-}(aq) \longrightarrow PbSO_4(s) + 2\,e^- \text{ (oxidation)}$$

Each cell also contains a lead (IV) oxide cathode where reduction occurs according to the following reaction.

$$\text{Cathode reaction: } PbO_2(s) + 4\,H^+(aq) + SO_4^{2-}(aq) + 2\,e^- \longrightarrow PbSO_4(s) + 2\,H_2O \text{ (reduction)}$$

Both the anode and the cathode are immersed in sulfuric acid (H_2SO_4). As electrical current is drawn from the battery, both the anode and the cathode become coated with $PbSO_4(s)$. If the battery is run for a long time without recharging, too much $PbSO_4(s)$ develops and the battery goes dead. The lead-acid storage battery can be recharged, however, by running electrical current through it in reverse. The electrical current has to come from an external source, such as an alternator in a car. This causes the preceding reaction to occur in reverse, converting the $PbSO_4(s)$ back to $Pb(s)$ and $PbO_2(s)$, recharging the battery.

Fuel Cells

We discussed the potential for fuel cells in Section 16.1. Electric vehicles powered by fuel cells may one day replace internal combustion vehicles. Fuel cells are like batteries, but the reactants are constantly replenished. Normal batteries lose their voltage with use because the reactants are depleted as electrical current is drawn from the battery. In a fuel cell, the reactants—the fuel—constantly flow through the battery, generating electrical current as they undergo a redox reaction.

The most common fuel cell is the hydrogen–oxygen fuel cell (▼ Figure 16.15). In this cell, hydrogen gas flows past the anode (a screen coated with platinum catalyst) and undergoes oxidation.

▶ **Figure 16.15 Hydrogen–oxygen fuel cell**

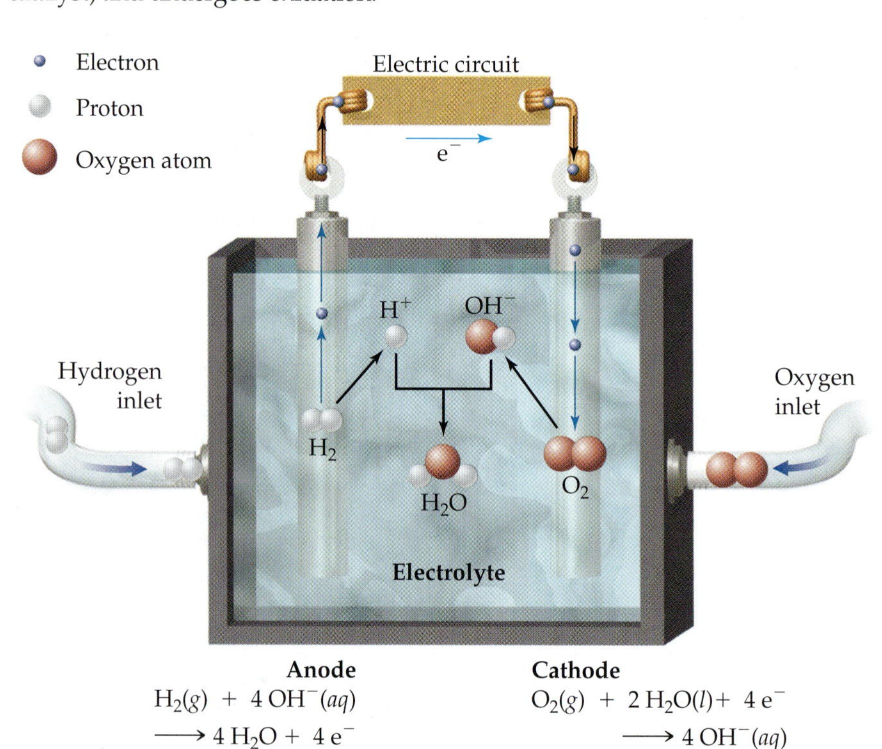

Anode reaction: $2\,H_2(g) + 4\,OH^-(aq) \longrightarrow 4\,H_2O(l) + 4\,e^-$

Oxygen gas flows past the cathode (a similar screen) and undergoes reduction.

Cathode reaction: $O_2(g) + 2\,H_2O(l) + 4\,e^- \longrightarrow 4\,OH^-(aq)$

The half-reactions sum to the following overall reaction.

Overall reaction: $2\,H_2(g) + O_2(g) \longrightarrow 2\,H_2O(l)$

Notice that the only product is water. In the space shuttle program, hydrogen–oxygen fuel cells provide electricity and astronauts drink the resulting water.

16.7 Electrolysis: Using Electricity to Do Chemistry

▲ **Figure 16.16** **Electrolysis of water** As a current passes between the electrodes, liquid water is broken down into hydrogen gas (right tube) and oxygen gas (left tube).

In a battery (or galvanic cell), a spontaneous redox reaction is used to produce electrical current. In **electrolysis**, electrical current is used to drive an otherwise nonspontaneous redox reaction. An electrochemical cell used for electrolysis is called an **electrolytic cell**. For example, we saw that the reaction of hydrogen with oxygen to form water is spontaneous and can be used to produce an electrical current in a fuel cell. However, by providing electrical current, we can cause the reverse reaction to occur, breaking water into hydrogen and oxygen (◀ Figure 16.16).

$2\,H_2(g) + O_2(g) \longrightarrow 2\,H_2O$ (spontaneous—produces electrical current; occurs in a galvanic cell)

$2\,H_2O \longrightarrow 2\,H_2(g) + O_2(g)$ (nonspontaneous—consumes electrical current; occurs in an electrolytic cell)

One of the problems associated with the widespread adoption of fuel cells is the scarcity of hydrogen. Where is the hydrogen to power these fuel cells going to come from? One possible answer is that the hydrogen can come from water through solar-powered electrolysis. In other words, a solar-powered electrolytic cell can be used to make hydrogen from water when the sun is shining. The hydrogen can then be converted back to water to generate electricity when needed. Hydrogen made in this way could also be used to power fuel-cell vehicles.

Electrolysis also has numerous other applications. For example, most metals are found in Earth's crust as metal oxides. Converting them to pure metals requires the reduction of the metal, a nonspontaneous process. Electrolysis can be used to produce these metals. Electrolysis can also be used to plate metals onto other metals. For example, silver can be plated onto an ordinary metal using the electrolytic cell shown in ▶ Figure 16.17. In this cell, a silver electrode is placed in a solution containing silver ions. An electrical current then causes the oxidation of silver at the anode (replenishing the silver ions in solution) and the reduction of silver ions at the cathode (coating the ordinary metal with solid silver).

Anode reaction: $Ag(s) \longrightarrow Ag^+(aq) + e^-$

Cathode reaction: $Ag^+(aq) + e^- \longrightarrow Ag(s)$

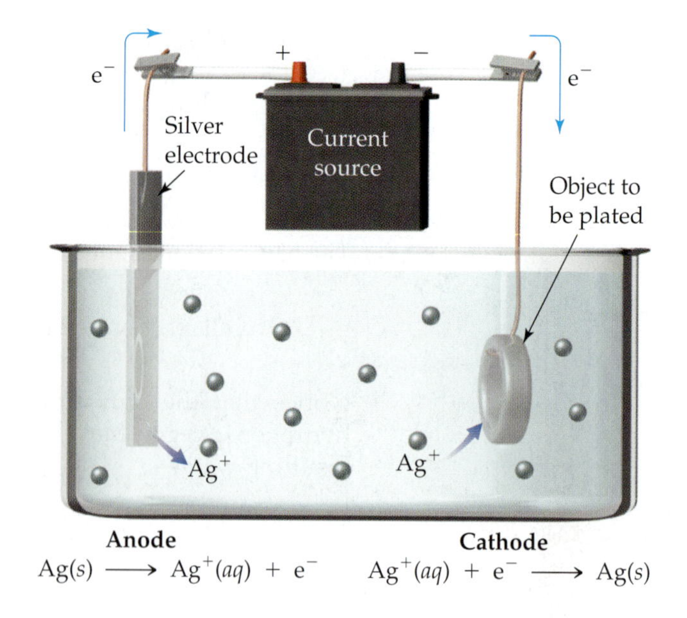

▶ **Figure 16.17 Electrolytic cell for silver plating** Silver is oxidized on the left side of the cell and reduced at the right. As it is reduced, it is deposited on the object to be plated.

16.8 Corrosion: Undesirable Redox Reactions

▲ Paint can prevent underlying iron from rusting. However, if the paint becomes scratched, the iron will rust at the point of the chip. **Question:** Why?

Corrosion is the oxidation of metals. The most common kind of corrosion is the rusting of iron. A significant part of the iron produced each year goes to replace rusted iron. Rusting is a redox reaction in which iron is oxidized and oxygen is reduced.

Oxidation:	$2\,Fe(s) \longrightarrow 2\,Fe^{2+}(aq) + \cancel{4\,e^-}$
Reduction:	$O_2(g) + 2\,H_2O(l) + \cancel{4\,e^-} \longrightarrow 4\,OH^-(aq)$
Overall:	$2\,Fe(s) + O_2(g) + 2\,H_2O(l) \longrightarrow 2\,Fe(OH)_2(s)$

The $Fe(OH)_2$ formed in the overall reaction then undergoes several additional reactions to form Fe_2O_3, the familiar orange substance that we call rust. One of the main problems with Fe_2O_3 is that it crumbles off the solid iron below it, exposing more iron to further rusting. Under the right conditions, an entire piece of iron can rust away.

Iron is not the only metal that undergoes oxidation. Most other metals, such as copper and aluminum, also undergo oxidation. However, the oxides of copper and aluminum do not flake off as iron oxide does. When aluminum oxidizes, the aluminum oxide actually forms a tough clear coating on the underlying metal. This coating protects the underlying metal from further oxidation.

Preventing the rusting of iron is a major industry. The most obvious way to prevent rust is to keep iron dry. Without water, the redox reaction cannot occur. Another way of preventing rust is to coat the iron with a substance that is impervious to water. Cars, for example, are painted and sealed to prevent rust. A scratch in the paint, however, can lead to rusting of the underlying iron.

Rust can also be prevented by placing a *sacrificial electrode* in electrical contact with the iron. The sacrificial electrode must be composed of a metal that is above iron on the activity series. The sacrificial electrode then oxidizes in place of the iron, protecting the iron from oxidation. Another way to protect iron from rusting is to coat it with a metal above it in the activity series. Galvanized nails, for example, are coated with a thin layer of zinc. Since zinc is more active than iron, it will oxidize in place of the underlying iron (just like a sacrificial electrode does). The oxide of zinc is not crumbly and remains on the nail as a protective coating.

EVERYDAY Chemistry

The Fuel-Cell Breathalyzer

Police often use a device called a Breathalyzer to measure the amount of ethyl alcohol (C_2H_5OH) in the bloodstream of a person suspected of driving under the influence of alcohol.

▲ **Fuel-cell Breathalyzer** Simply blowing into the top of this device measures blood alcohol level.

Breathalyzers work because the amount of ethyl alcohol in the breath is proportional to the amount of ethyl alcohol in the bloodstream. One type of Breathalyzer employs a fuel cell to measure the amount of alcohol in the breath. The fuel cell consists of two platinum electrodes (▼ Figure 16.18). When a suspect blows into the Breathalyzer, ethyl alcohol is oxidized to acetic acid at the anode.

Anode:

$$\underset{\text{Ethyl alcohol}}{C_2H_5OH} + 4\,OH^-(aq) \longrightarrow \underset{\text{Acetic acid}}{CH_3COOH(g)} + 3\,H_2O + 4\,e^-$$

At the cathode, oxygen is reduced.

$$\text{Cathode:}\quad O_2(g) + 2\,H_2O(l) + 4\,e^- \longrightarrow 4\,OH^-(aq)$$

The overall reaction is simply the oxidation of ethyl alcohol to acetic acid and water.

$$\text{Overall:}\quad C_2H_5OH(g) + O_2(g) \longrightarrow CH_3COOH(g) + H_2O$$

The amount of electrical current produced depends on the amount of alcohol in the breath. A higher current reveals a higher blood alcohol level. When calibrated correctly, the fuel-cell Breathalyzer can precisely measure the blood alcohol level of a suspected drunk driver.

CAN YOU ANSWER THIS? *Assign oxidation states to each element in the reactants and products in the overall equation for the fuel-cell Breathalyzer. What element is oxidized and what element is reduced in the reaction?*

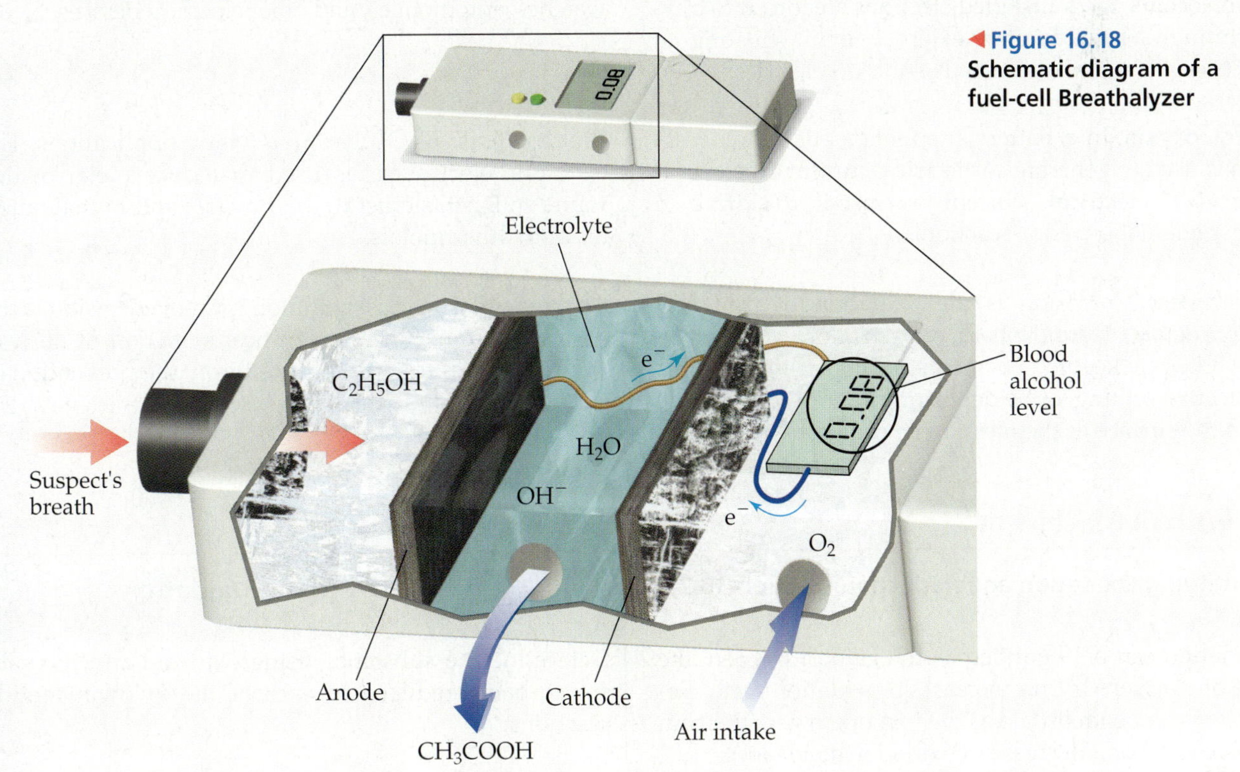

◀ **Figure 16.18** **Schematic diagram of a fuel-cell Breathalyzer**

CHAPTER IN REVIEW

Chemical Principles

Oxidation and Reduction:
Oxidation is defined as:

- The loss of electrons.
- An increase in oxidation state.

Reduction is defined as:

- The gain of electrons.
- A decrease in oxidation state.

Oxidation and reduction reactions always occur together and are sometimes called redox reactions. The substance that is oxidized is called the reducing agent and the substance that is reduced is called the oxidizing agent.

Relevance

Oxidation and Reduction: Redox reactions are common in nature, in industry, and in many everyday processes. Batteries use redox reactions to generate electrical current. Our bodies use redox reactions to obtain energy from glucose. In addition, the bleaching of hair, the rusting of iron, and the electroplating of metals all involve redox reactions.

Good oxidizing agents, such as oxygen, hydrogen peroxide, and chlorine, have a strong tendency to gain electrons. Good reducing agents, such as sodium and hydrogen, have a strong tendency to lose electrons.

Oxidation States: The oxidation state is a fictitious charge assigned to each atom in a compound. It is computed by assigning all bonding electrons in a compound to the most electronegative element.

Oxidation States: Oxidation states can help us more easily identify substances being oxidized and reduced in a chemical reaction.

The Activity Series: The activity series is simply a listing of metals from those that are easiest to oxidize to those that are most difficult to oxidize. Any half-reaction in the activity series will be spontaneous when paired with the reverse of a half-reaction below it on the list.

The Activity Series: The activity series allows us to predict whether a redox reaction (involving half-reactions from the series) will be spontaneous.

Batteries: In a battery, the reactants of a spontaneous redox reaction are physically separated. As the redox reaction occurs, the transferred electrons are forced to travel through a wire or other external circuit, creating an electrical current that can be used to do electrical work.

Batteries: Batteries are common as portable sources of electrical current. They are commonly used in flashlights, watches, automobiles, and other electrical devices.

Electrolysis: In a battery, a spontaneous redox reaction is used to generate an electrical current. In electrolysis, an electrical current is used to drive a nonspontaneous redox reaction.

Electrolysis: Electrolysis has many applications. For example, electrolysis is used to reduce metal oxides found in Earth's crust to their metals and to plate metals onto other metals.

Corrosion: Corrosion is the oxidation of iron and other metals by atmospheric oxygen. Corrosion can be prevented by keeping the metal dry, by sealing it with a protective coating, or by depositing a more active metal onto the surface of the metal to be protected.

Corrosion: The most common form of corrosion is the rusting of iron. Since a significant fraction of all iron produced goes to replace rusted iron, the prevention of rust is a major industry.

Chemical Skills

Identifying Oxidation and Reduction (Section 16.2)

Oxidation can be identified as the gain of oxygen, the loss of electrons, or an increase in oxidation state. Reduction can be identified as the loss of oxygen, the gain of electrons, or a decrease in oxidation state.

Examples

EXAMPLE 16.10 Identifying Oxidation and Reduction

Determine the substance being oxidized and the substance being reduced in each of the following redox reactions.

(a) $Sn(s) + O_2(g) \longrightarrow SnO_2(s)$
(b) $2\,Na(s) + F_2(g) \longrightarrow 2\,NaF(s)$
(c) $Mg(s) + Cu^{2+}(aq) \longrightarrow Mg^{2+}(aq) + Cu(s)$

Solution:

When a substance gains oxygen, the substance is oxidized and the oxygen is reduced.

(a) Sn oxidized; O_2 reduced

When a metal reacts with an electronegative element, the metal is oxidized and the electronegative element is reduced.

(b) Na oxidized; F_2 reduced

When a metal transfers electrons to a metal ion, the metal is oxidized and the metal ion is reduced.

(c) Mg oxidized; Cu^{2+} reduced

Identifying Oxidizing Agents and Reducing Agents (Section 16.2)

EXAMPLE 16.11 Identifying Oxidizing Agents and Reducing Agents

Identify the oxidizing and reducing agents in each of the following redox reactions.

(a) $Sn(s) + O_2(g) \longrightarrow SnO_2(s)$
(b) $2\,Na(s) + F_2(g) \longrightarrow 2\,NaF(s)$
(c) $Mg(s) + Cu^{2+}(aq) \longrightarrow Mg^{2+}(aq) + Cu(s)$

The reducing agent is the substance that is oxidized. The oxidizing agent is the substance that is reduced.

Solution:

(a) Sn is the reducing agent; O_2 is the oxidizing agent.
(b) Na is the reducing agent; F_2 is the oxidizing agent.
(c) Mg is the reducing agent; Cu^{2+} is the oxidizing agent.

Assigning Oxidation States (Section 16.3)

Rules for assigning oxidation states

(These rules are hierarchical. If two rules conflict, follow the rule higher on the list.)

1. The oxidation state of an atom in a free element is 0.
2. The oxidation state of a monoatomic ion is equal to its charge.
3. The sum of the oxidation states of all atoms in:
   - a neutral molecule or formula unit is 0.
   - an ion is equal to the charge of the ion.
4. In their compounds,
   - Group I metals have an oxidation state of +1.
   - Group II metals have an oxidation of state of +2.
5. In their compounds, nonmetals are assigned oxidation states according to the following hierarchical table.

Fluorine	−1
Hydrogen	+1
Oxygen	−2
Group 7A	−1
Group 6A	−2
Group 5A	−3

EXAMPLE 16.12 Assigning Oxidation States

Assign an oxidation state to each atom in each of the following compounds.

(a) Al
(b) Al^{3+}
(c) N_2O
(d) CO_3^{2-}

Solution:

(a) Al (rule 1)
0

(b) Al^{3+} (rule 2)
+3

(c) N_2O (rule 5, O takes priority over N)
+1 −2

(d) CO_3^{2-} (rules 5, 3)
+4 −2

Balancing Redox Reactions (Section 16.4)

To balance redox reactions in aqueous acidic solutions, follow this procedure (brief version).

LIVE EXAMPLE

EXAMPLE 16.13 Balancing Redox Reactions

Balance the following reaction occurring in acidic solution.

$$IO_3^-(aq) + Fe^{2+}(aq) \longrightarrow I_2(s) + Fe^{3+}(aq)$$

Solution:

1. Assign oxidation states.

$$\underset{+5\ -2}{IO_3^-}(aq) + \underset{+2}{Fe^{2+}}(aq) \longrightarrow \underset{0}{I_3}(s) + \underset{+3}{Fe^{3+}}(aq)$$

Reduction (I: +5 → 0); Oxidation (Fe: +2 → +3)

2. Separate the overall reaction into two half-reactions.

Oxidation: $Fe^{2+}(aq) \longrightarrow Fe^{3+}(aq)$

Reduction: $IO_3^-(aq) \longrightarrow I_2(s)$

3. Balance each half-reaction with respect to mass.
   - Balance all elements other than H and O.

$Fe^{2+}(aq) \longrightarrow Fe^{3+}(aq)$

$\mathbf{2}\ IO_3^-(aq) \longrightarrow I_2(s)$

   - Balance O by adding H_2O.

$Fe^{2+}(aq) \longrightarrow Fe^{3+}(aq)$

$2\ IO_3^-(aq) \longrightarrow I_2(s) + \mathbf{6\ H_2O(l)}$

   - Balance H by adding H^+.

$Fe^{2+}(aq) \longrightarrow Fe^{3+}(aq)$

$\mathbf{12\ H^+}(aq) + 2\ IO_3^-(aq) \longrightarrow I_2(s) + 6\ H_2O(l)$

4. Balance each half-reaction with respect to charge by adding electrons.

$Fe^{2+}(aq) \longrightarrow Fe^{3+}(aq) + \mathbf{e^-}$

$\mathbf{10\ e^-} + 12\ H^+(aq) + 2\ IO_3^-(aq) \longrightarrow I_2(s) + 6\ H_2O(l)$

5. Make the number of electrons in both half-reactions equal.

$\mathbf{10} \times [Fe^{2+}(aq) \longrightarrow Fe^{3+}(aq) + e^-]$

$10\ e^- + 12\ H^+(aq) + 2\ IO_3^-(aq) \longrightarrow I_2(s) + 6\ H_2O$

6. Add the two half-reactions together.

$10\ Fe^{2+}(aq) \longrightarrow 10\ Fe^{3+}(aq) + \cancel{10\ e^-}$

$\cancel{10\ e^-} + 12\ H^+(aq) + 2\ IO_3^-(aq) \longrightarrow I_2(s) + 6\ H_2O$

$$10\ Fe^{2+}(aq) + 12\ H^+(aq) + 2\ IO_3^-(aq) \longrightarrow 10\ Fe^{3+}(aq) + I_2(s) + 6\ H_2O$$

7. Verify that the reaction is balanced.

Reactants	Products
10 Fe	10 Fe
12 H	12 H
2 I	2 I
6 O	6 O
+30 charge	+30 charge

Predicting Spontaneous Redox Reactions (Section 16.5)

EXAMPLE 16.14 Predicting Spontaneous Redox Reactions

Predict whether each of the following redox reactions will be spontaneous.

(a) $Cr(s) + 3\ Ag^+(aq) \longrightarrow Cr^{3+}(aq) + 3\ Ag(s)$

(b) $Mn^{2+}(aq) + Fe(s) \longrightarrow Mn(s) + Fe^{2+}(aq)$

Any half-reaction in the activity series will be spontaneous when paired with the reverse of any half-reaction below it.

Solution:

(a) Spontaneous

(b) Nonspontaneous

Key Terms

activity series of metals [16.5]
alkaline battery [16.6]
anode [16.6]
cathode [16.6]
corrosion [16.8]
dry cell [16.6]
electrical current [16.6]
electrochemical cell [16.6]
electrolysis [16.7]
electrolytic cell [16.7]
fuel cell [16.1]
galvanic (voltaic) cell [16.6]
half-cell [16.6]
half-reaction [16.4]
lead-acid storage battery [16.6]
oxidation [16.2]
oxidation state (oxidation number) [16.3]
oxidizing agent [16.2]
redox (oxidation–reduction) reaction [16.1]
reducing agent [16.2]
reduction [16.2]
salt bridge [16.6]
voltage [16.6]

Exercises

Questions

1. What is a fuel-cell electric vehicle?
2. What is an oxidation–reduction or redox reaction?
3. Define oxidation and reduction with respect to:
   (a) oxygen
   (b) electrons
   (c) oxidation state
4. What is an oxidizing agent?
5. What is a reducing agent?
6. Good oxidation agents have a strong tendency to _______ electrons in reactions.
7. Good reducing agents have a strong tendency to _______ electrons in reactions.
8. What is the oxidation state of a free element?
9. What is the oxidation state of a monoatomic ion?
10. For a neutral molecule, the sum of the oxidation states of the individual atoms must add up to _______.
11. For an ion, the sum of the oxidation states of the individual atoms must add up to _______.
12. In their compounds, elements have oxidation states equal to _______. Are there exceptions to this rule? Explain.
13. In a redox reaction, an atom that undergoes an increase in oxidation state is _______. An atom that undergoes a decrease in oxidation state is _______.
14. How does hydrogen peroxide cause hair to change color?
15. When balancing redox equations, the number of electrons lost in the oxidation half-reaction must _______ the number of electrons gained in the reduction half-reaction.
16. When balancing aqueous redox reactions occurring in acidic media, oxygen is balanced using _______.
17. When balancing aqueous redox reactions occurring in acidic media, hydrogen is balanced using _______.
18. When balancing aqueous redox reactions occurring in acidic media, charge is balanced using _______.
19. Are metals at the top of the activity series the most reactive or least reactive?
20. Are metals at the top of the activity series easiest to oxidize or hardest to oxidize?
21. Are metals at the bottom of the activity series most likely or least likely to lose electrons?
22. Any half-reaction in the activity series will be spontaneous when paired with the reverse of any half-reaction _______ it.
23. How can you use the activity series to determine whether a metal will dissolve in acids such as HCl or HBr?
24. What is electrical current? Explain how a simple battery creates electrical current.
25. Oxidation occurs at the _______ of an electrochemical cell.
26. Reduction occurs at the _______ of an electrochemical cell.
27. Explain the role of a salt bridge in an electrochemical cell.
28. A high voltage in an electrochemical cell is analogous to _______ in a river.
29. Describe the common dry-cell battery. Include equations for the anode and cathode reactions.
30. Describe the lead-acid storage battery. Include equations for the anode and cathode reactions.
31. Describe a fuel cell. Include equations for the anode and cathode reactions of the hydrogen–oxygen fuel cell.
32. What is electrolysis? Why is it useful?
33. What is corrosion? Include reactions for the corrosion of iron.
34. How can rust be prevented?

Problems

Oxidation and Reduction

35. What is being oxidized in each of the following reactions?
(a) $2\,H_2(g) + O_2(g) \longrightarrow 2\,H_2O(l)$
(b) $4\,Al(s) + 3\,O_2(g) \longrightarrow 2\,Al_2O_3(s)$
(c) $2\,Al(s) + 3\,Cl_2(g) \longrightarrow 2\,AlCl_3(s)$

36. What is being oxidized in each of the following reactions?
(a) $Zn(s) + O_2(g) \longrightarrow ZnO_2(s)$
(b) $CH_4(g) + 2\,O_2(g) \longrightarrow CO_2(g) + 2\,H_2O(g)$
(c) $Sr(s) + F_2(g) \longrightarrow SrF_2(s)$

37. For each of the following reactions, identify the substance being oxidized and the substance being reduced.
(a) $2\,Sr(s) + O_2(g) \longrightarrow 2\,SrO(s)$
(b) $Ca(s) + Cl_2(g) \longrightarrow CaCl_2(s)$
(c) $Ni^{2+}(aq) + Mg(s) \longrightarrow Mg^{2+}(aq) + Ni(s)$

38. For each of the following reactions, identify the substance being oxidized and the substance being reduced.
(a) $Mg(s) + Br_2(g) \longrightarrow MgBr_2(s)$
(b) $2\,Cr^{3+}(aq) + 3\,Mn(s) \longrightarrow 2\,Cr(s) + 3\,Mn^{2+}(aq)$
(c) $2\,H^+(aq) + Ni(s) \longrightarrow H_2(g) + Ni^{2+}(aq)$

39. For each of the reactions in Problem 37, identify the oxidizing agent and the reducing agent.

40. For each of the reactions in Problem 38, identify the oxidizing agent and the reducing agent.

41. Based on periodic trends, which of the following (in their elemental form) would you expect to be good oxidizing agents?
(a) K
(b) F_2
(c) Fe
(d) Cl_2

42. Based on periodic trends, which of the following (in their elemental form) would you expect to be good oxidizing agents?
(a) O_2
(b) Br_2
(c) Li
(d) Na

43. Based on periodic trends, which of the elements in Problem 41 (in their elemental form) would you expect to be good reducing agents?

44. Based on periodic trends, which of the elements in Problem 42 (in their elemental form) would you expect to be good reducing agents?

45. For each of the following redox reactions, identify the substance being oxidized, the substance being reduced, the oxidizing agent, and the reducing agent.
(a) $N_2(g) + O_2(g) \longrightarrow 2\,NO(g)$
(b) $2\,CO(g) + O_2(g) \longrightarrow 2\,CO_2(g)$
(c) $SbCl_3(g) + Cl_2(g) \longrightarrow SbCl_5(g)$
(d) $2\,K(s) + Pb^{2+}(aq) \longrightarrow 2\,K^+(aq) + Pb(s)$

46. For each of the following redox reactions, identify the substance being oxidized, the substance being reduced, the oxidizing agent, and the reducing agent.
(a) $H_2(g) + I_2(g) \longrightarrow 2\,HI(g)$
(b) $CO(g) + H_2(g) \longrightarrow C(s) + H_2O(g)$
(c) $2\,Al(s) + 6\,H^+(aq) \longrightarrow 2\,Al^{3+}(aq) + 3\,H_2(g)$
(d) $2\,Li(s) + Pb^{2+}(aq) \longrightarrow 2\,Li^+(aq) + Pb(s)$

Oxidation States

47. Assign an oxidation state to each of the following:
(a) Zn
(b) Zn^{2+}
(c) Cl_2
(d) N_2

48. Assign an oxidation state to each of the following:
(a) O_2
(b) Ni^{2+}
(c) Ca^{2+}
(d) Ti

49. Assign an oxidation state to each atom in each of the following compounds.
(a) NaCl
(b) CaF_2
(c) SO_2
(d) H_2S

50. Assign an oxidation state to each atom in each of the following compounds.
(a) CH_4
(b) CH_2Cl_2
(c) $CuCl_2$
(d) HI

51. What is the oxidation state of nitrogen in each of the following compounds?
(a) NO
(b) NO_2
(c) N_2O

52. What is the oxidation state of Cr in each of the following compounds?
(a) CrO
(b) CrO_3
(c) Cr_2O_3

53. Assign an oxidation state to each atom in each of the following polyatomic ions.
(a) CO_3^{2-}
(b) OH^-
(c) NO_3^-
(d) NO_2^-

54. Assign an oxidation state to each atom in each of the following polyatomic ions.
(a) CrO_4^{2-}
(b) $Cr_2O_7^{2-}$
(c) PO_4^{3-}
(d) MnO_4^-

55. What is the oxidation state of Cl in each of the following ions?
(a) ClO^-
(b) ClO_2^-
(c) ClO_3^-
(d) ClO_4^-

56. What is the oxidation state of S in each of the following ions?
(a) SO_4^{2-}
(b) SO_3^{2-}
(c) HSO_3^-
(d) HSO_4^-

57. Assign an oxidation state to each element in each of the following compounds:
(a) $Cu(NO_3)_2$
(b) $Sr(OH)_2$
(c) $K_2Cr_2O_7$
(d) $NaHCO_3$

58. Assign an oxidation state to each element in each of the following compounds:
(a) Na_3PO_4
(b) Hg_2S
(c) $Fe(CN)_3$
(d) NH_4Cl

59. Assign an oxidation state to each element in each of the following reactions and use the change in oxidation state to determine which element is being oxidized and which element is being reduced.
(a) $SbCl_5(g) \longrightarrow SbCl_3(g) + Cl_2(g)$
(b) $CO(g) + Cl_2(g) \longrightarrow COCl_2(g)$
(c) $2\,NO(g) + Br_2(g) \longrightarrow 2\,BrNO(g)$
(d) $H_2(g) + CO_2(g) \longrightarrow H_2O(g) + CO(g)$

60. Assign an oxidation state to each element in each of the following reactions and use the change in oxidation state to determine which element is being oxidized and which element is being reduced.
(a) $CH_4(g) + 2\,H_2S(g) \longrightarrow CS_2(g) + 4\,H_2(g)$
(b) $2\,H_2S(g) \longrightarrow 2\,H_2(g) + S_2(g)$
(c) $C_6H_{12}O_6(s) + 6\,O_2(g) \longrightarrow 6\,CO_2(g) + 6\,H_2O(g)$
(d) $C_2H_4(g) + Cl_2(g) \longrightarrow C_2H_4Cl_2(g)$

61. Use oxidation states to identify the oxidizing agent and the reducing agent in the following redox reaction.

$$2\,Na(s) + 2\,H_2O(l) \longrightarrow 2\,NaOH(aq) + H_2(g)$$

62. Use oxidation states to identify the oxidizing agent and the reducing agent in the following redox reaction.

$$N_2(g) + 3\,H_2(g) \longrightarrow 2\,NH_3$$

Balancing Redox Reactions

63. Balance each of the following redox reactions using the half-reaction method.
(a) $K(s) + Cr^{3+}(aq) \longrightarrow Cr(s) + K^+(aq)$
(b) $Mg(s) + Ag^+(aq) \longrightarrow Mg^{2+}(aq) + Ag(s)$
(c) $Al(s) + Fe^{2+}(aq) \longrightarrow Al^{3+}(aq) + Fe(s)$

64. Balance each of the following redox reactions using the half-reaction method.
(a) $Zn(s) + Sn^{2+}(aq) \longrightarrow Zn^{2+}(aq) + Sn(s)$
(b) $Mg(s) + Cr^{3+}(aq) \longrightarrow Mg^{2+}(aq) + Cr(s)$
(c) $Al(s) + Ag^+(aq) \longrightarrow Al^{3+}(aq) + Ag(s)$

65. Classify each of the following half-reactions occurring in acidic aqueous solution as an oxidation or a reduction and balance the half-reaction.
(a) $MnO_4^-(aq) \longrightarrow Mn^{2+}(aq)$
(b) $Pb^{2+}(aq) \longrightarrow PbO_2(s)$
(c) $IO_3^-(aq) \longrightarrow I_2(s)$
(d) $SO_2(g) \longrightarrow SO_4^{2-}(aq)$

66. Classify each of the following half-reactions occurring in acidic aqueous solution as an oxidation or a reduction and balance the half-reaction.
(a) $S(s) \longrightarrow H_2S(g)$
(b) $S_2O_8^{2-}(aq) \longrightarrow 2\,SO_4^{2-}(aq)$
(c) $Cr_2O_7^{2-}(aq) \longrightarrow Cr^{3+}(aq)$
(d) $NO(g) \longrightarrow NO_3^-(aq)$

67. Balance each of the following redox reactions occurring in acidic aqueous solution. Use the half-reaction method.
(a) $PbO_2(s) + I^-(aq) \longrightarrow Pb^{2+}(aq) + I_2(s)$
(b) $SO_3^{2-}(aq) + MnO_4^-(aq) \longrightarrow SO_4^{2-}(aq) + Mn^{2+}(aq)$
(c) $S_2O_3^{2-}(aq) + Cl_2(g) \longrightarrow SO_4^{2-}(aq) + Cl^-(aq)$

68. Balance each of the following redox reactions occurring in acidic aqueous solution. Use the half-reaction method.
(a) $I^-(aq) + NO_2^-(aq) \longrightarrow I_2(s) + NO(g)$
(b) $BrO_3^-(aq) + N_2H_4(g) \longrightarrow Br^-(aq) + N_2(g)$
(c) $NO_3^-(aq) + Sn^{2+}(aq) \longrightarrow Sn^{4+}(aq) + NO(g)$

69. Balance each of the following redox reactions occurring in acidic aqueous solution. Use the half-reaction method.
(a) $ClO_4^-(aq) + Cl^-(aq) \longrightarrow ClO_3^-(aq) + Cl_2(g)$
(b) $MnO_4^-(aq) + Al(s) \longrightarrow Mn^{2+}(aq) + Al^{3+}(aq)$
(c) $Br_2(aq) + Sn(s) \longrightarrow Sn^{2+}(aq) + Br^-(aq)$

70. Balance each of the following redox reactions occurring in acidic aqueous solution. Use the half-reaction method.
(a) $IO_3^-(aq) + SO_2(g) \longrightarrow I_2(s) + SO_4^{2-}(aq)$
(b) $Sn^{4+}(aq) + H_2(g) \longrightarrow Sn^{2+}(aq) + H^+(aq)$
(c) $Cr_2O_7^{2-}(aq) + Br^-(aq) \longrightarrow Cr^{3+}(aq) + Br_2(aq)$

The Activity Series

71. Which of the following metals has the least tendency to be oxidized?
(a) Mg
(b) Cr
(c) Pb
(d) Fe

72. Which of the following metals has the least tendency to be oxidized?
(a) Li
(b) Zn
(c) K
(d) Au

73. Which of the following metal cations has the greatest tendency to be reduced?
(a) Mn^{2+}
(b) Cu^{2+}
(c) K^+
(d) Ni^{2+}

74. Which of the following metal cations has the greatest tendency to be reduced?
(a) Pb^{2+}
(b) Cr^{3+}
(c) Fe^{2+}
(d) Sn^{2+}

75. Which of the following metals is the best reducing agent?
(a) Mn
(b) Al
(c) Ni
(d) Cr

76. Which of the following metals is the best reducing agent?
(a) Ag
(b) Mg
(c) Fe
(d) Pb

77. Which of the following redox reactions do you expect to occur spontaneously in the forward direction?
(a) $Ni(s) + Zn^{2+}(aq) \longrightarrow Ni^{2+}(aq) + Zn(s)$
(b) $Ni(s) + Pb^{2+}(aq) \longrightarrow Ni^{2+}(aq) + Pb(s)$
(c) $Al(s) + 3\,Ag^+(aq) \longrightarrow 3\,Al^{3+}(aq) + Ag(s)$
(d) $Pb(s) + Mn^{2+}(aq) \longrightarrow Pb^{2+}(aq) + Mn(s)$

78. Which of the following redox reactions do you expect to occur spontaneously in the forward direction?
(a) $Ca^{2+}(aq) + Zn(s) \longrightarrow Ca(s) + Zn^{2+}(aq)$
(b) $2\,Ag^+(aq) + Ni(s) \longrightarrow 2\,Ag(s) + Ni^{2+}(aq)$
(c) $Fe(s) + Mn^{2+}(aq) \longrightarrow Fe^{2+}(aq) + Mn(s)$
(d) $2\,Al(s) + 3\,Pb^{2+}(aq) \longrightarrow 2\,Al^{3+}(aq) + 3\,Pb(s)$

79. Suppose you wanted to cause Ni^{2+} ions to come out of solution as solid Ni. What metal might you use to accomplish this?

80. Suppose you wanted to cause Pb^{2+} ions to come out of solution as solid Pb. What metal might you use to accomplish this?

81. Which metal in the activity series will reduce Mn^{2+} ions but not Mg^{2+} ions?

82. Which metal in the activity series can be oxidized with an Sn^{2+} solution but not with a Fe^{2+} solution?

83. Which of the following metals would dissolve in HCl? For those metals that do dissolve, write a balanced redox reaction showing what happens when the metal dissolves.
(a) Al
(b) Ag
(c) Pb
(d) Cr

84. Which of the following metals would dissolve in HCl? For those metals that do dissolve, write a balanced redox reaction showing what happens when the metal dissolves.
(a) Mn
(b) Cu
(c) Fe
(d) Au

Batteries, Electrochemical Cells, and Electrolysis

85. Make a sketch of an electrochemical cell with the following overall reaction. Label the anode, the cathode, and the salt bridge. Indicate the direction of electron flow. Hint: When drawing electrochemical cells, the anode is usually drawn on the left side.

$$\text{Mn}(s) + \text{Pb}^{2+}(aq) \longrightarrow \text{Mn}^{2+}(aq) + \text{Pb}(s)$$

86. Make a sketch of an electrochemical cell with the following overall reaction. Label the anode, the cathode, and the salt bridge. Indicate the direction of electron flow. Hint: When drawing electrochemical cells, the anode is usually drawn on the left side.

$$\text{Mg}(s) + \text{Ni}^{2+}(aq) \longrightarrow \text{Mg}^{2+}(aq) + \text{Ni}(s)$$

87. An electrochemical cell has the following reaction occurring at the anode.

$$\text{Zn}(s) \longrightarrow \text{Zn}^{2+}(aq) + 2\,\text{e}^-$$

Which of the following cathode reactions would produce a battery with the highest voltage?
(a) $\text{Mg}^{2+}(aq) + 2\,\text{e}^- \longrightarrow \text{Mg}(s)$
(b) $\text{Pb}^{2+}(aq) + 2\,\text{e}^- \longrightarrow \text{Pb}(s)$
(c) $\text{Cr}^{3+}(aq) + 3\,\text{e}^- \longrightarrow \text{Cr}(s)$
(d) $\text{Cu}^{2+}(aq) + 2\,\text{e}^- \longrightarrow \text{Cu}(s)$

88. An electrochemical cell has the following reaction occurring at the cathode.

$$\text{Ni}^{2+}(aq) + 2\,\text{e}^- \longrightarrow \text{Ni}(s)$$

Which of the following anode reactions would produce a battery with the highest voltage?
(a) $\text{Ag}(s) \longrightarrow \text{Ag}^{+}(aq) + \text{e}^-$
(b) $\text{Mg}(s) \longrightarrow \text{Mg}^{2+}(aq) + 2\,\text{e}^-$
(c) $\text{Cr}(s) \longrightarrow \text{Cr}^{3+}(aq) + 3\,\text{e}^-$
(d) $\text{Cu}(s) \longrightarrow \text{Cu}^{2+}(aq) + 2\,\text{e}^-$

89. Use the half-cell reactions for the alkaline battery to determine the overall reaction that occurs in this battery.

90. Use the half-cell reactions for the lead-acid storage battery to determine the overall reaction that occurs in this battery.

91. Make a sketch of an electrolysis cell that might be used to electroplate copper onto other metal surfaces. Label the anode and the cathode and show the reactions that occur at each.

92. Make a sketch of an electrolysis cell that might be used to electroplate nickel onto other metal surfaces. Label the anode and the cathode and show the reactions that occur at each.

Corrosion

93. Which of the following metals, if coated onto iron, would prevent the corrosion of iron?
(a) Zn
(b) Sn
(c) Mn

94. Which of the following metals, if coated onto iron, would prevent the corrosion of iron?
(a) Mg
(b) Cr
(c) Cu

Cumulative Problems

95. Which of the following reactions are redox reactions? For those reactions that are redox reactions, identify the substance being oxidized and the substance being reduced.
(a) $Zn(s) + CoCl_2(aq) \longrightarrow ZnCl_2(aq) + Co(s)$
(b) $HI(aq) + NaOH(aq) \longrightarrow H_2O(l) + NaI(aq)$
(c) $AgNO_3(aq) + NaCl(aq) \longrightarrow AgCl(s) + NaNO_3(aq)$
(d) $2\,K(s) + Br_2(l) \longrightarrow 2\,KBr(s)$

96. Which of the following reactions are redox reactions? For those reactions that are redox reactions, identify the substance being oxidized and the substance being reduced.
(a) $Pb(NO_3)_2(aq) + 2\,LiCl(aq) \longrightarrow PbCl_2(s) + 2\,LiNO_3(aq)$
(b) $2\,HBr(aq) + Ca(OH)_2(aq) \longrightarrow 2\,H_2O(l) + CaBr_2(aq)$
(c) $2\,Al(s) + Fe_2O_3(s) \longrightarrow Al_2O_3(s) + 2\,Fe(l)$
(d) $Na_2O(s) + H_2O(l) \longrightarrow 2\,NaOH(aq)$

97. Consider the following unbalanced redox reaction.

$$MnO_4^-(aq) + Zn(s) \longrightarrow Mn^{2+}(aq) + Zn^{2+}(aq)$$

Balance the equation and determine how much of a 0.500 M $KMnO_4$ solution is required to completely dissolve 2.85 g of Zn.

98. Consider the following unbalanced redox reaction.

$$Cr_2O_7^{2-}(aq) + Cu(s) \longrightarrow Cr^{3+}(aq) + Cu^{2+}$$

Balance the equation and determine how much of a 0.850 M $K_2Cr_2O_7$ solution is required to completely dissolve 5.25 g of Cu.

99. A 10.0-mL sample of a commercial hydrogen peroxide (H_2O_2) solution is titrated with 0.0998 M $KMnO_4$. The endpoint is reached at a volume of 34.81 mL. Find the mass percent of H_2O_2 in the commercial hydrogen peroxide solution. (Assume a density of 1.00 g/mL for the hydrogen peroxide solution.) The unbalanced redox reaction that occurs during the titration is as follows:

$$H_2O_2(aq) + MnO_4^-(aq) \longrightarrow O_2(g) + Mn^{2+}(aq)$$

100. A 1.012-g sample of a salt containing Fe^{2+} is titrated with 0.1201 M $KMnO_4$. The endpoint of the titration is reached at 22.45 mL. Find the mass percent of Fe^{2+} in the sample. The unbalanced redox reaction that occurs during the titration is as follows:

$$Fe^{2+}(aq) + MnO_4^-(aq) \longrightarrow Fe^{3+}(aq) + Mn^{2+}(aq)$$

101. Silver can be electroplated at the cathode of an electrolysis cell by the following half-reaction.

$$Ag^+(aq) + e^- \longrightarrow Ag(s)$$

How many moles of electrons are required to electroplate 5.8 g of Ag?

102. Gold can be electroplated at the cathode of an electrolysis cell by the following half-reaction.

$$Au^{3+}(aq) + 3\,e^- \longrightarrow Au(s)$$

How many moles of electrons are required to electroplate 1.40 g of Au?

103. Determine whether HI can dissolve each of the following metal samples. If so, write a balanced chemical reaction showing how the metal dissolves in HI and determine the minimum amount of 3.5 M HI required to completely dissolve the sample.
(a) 5.95 g Cr
(b) 2.15 g Al
(c) 4.85 g Cu
(d) 2.42 g Au

104. Determine whether HCl can dissolve each of the following metal samples. If so, write a balanced chemical reaction showing how the metal dissolves in HCl and determine the minimum amount of 6.0 M HCl required to completely dissolve the sample.
(a) 5.90 g Ag
(b) 2.55 g Pb
(c) 4.83 g Sn
(d) 1.25 g Mg

105. One drop (assume 0.050 mL) of 6.0 M HCl is placed onto the surface of 0.028-mm-thick aluminum foil. What is the maximum diameter of the hole that will result from the HCl dissolving the aluminum? (density of aluminum = 2.7 g/cm^3)

106. A graduated cylinder containing 1.00 mL of 12.0 M HCl is accidentally tipped over and the contents spill onto a manganese foil with a thickness of 0.055 mm. Calculate the maximum diameter of the hole that will be dissolved in the foil by the reaction between the manganese and hydrochloric acid. (density of manganese = 7.47 g/cm^3)

Highlight Problems

107. Consider the following molecular views of an Al strip and Cu^{2+} solution. Draw a similar sketch showing what happens to the atoms and ions if the Al strip is submerged in the solution for a few minutes.

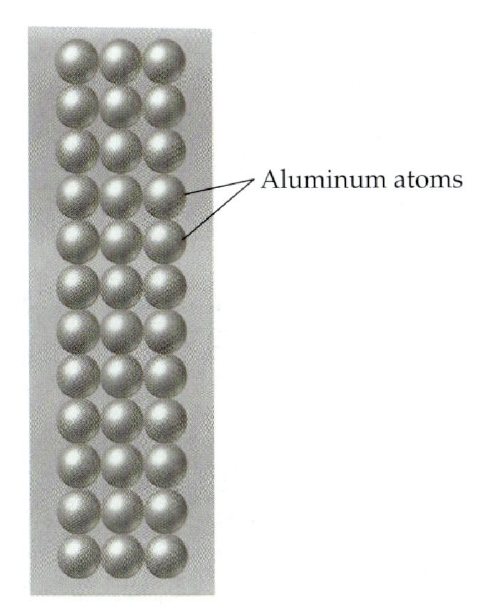

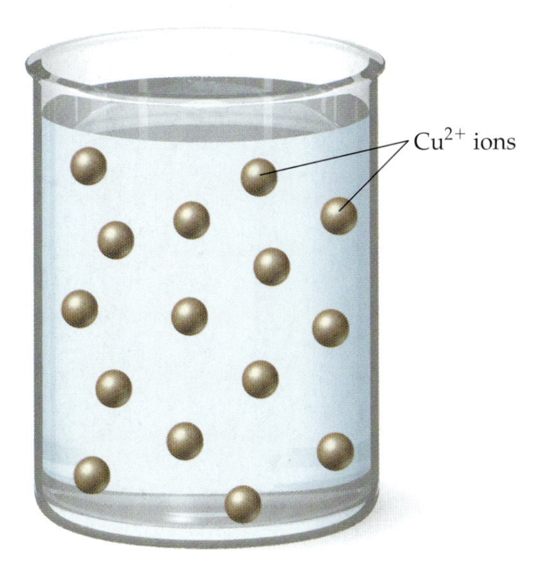

108. Suppose a fuel-cell generator was used to produce electricity for a house. If each H_2 molecule produces $2\,e^-$, how many kilograms of hydrogen would be required to generate the electricity needed for a typical house? Assume the home uses about 850 kWh of electricity per month, which corresponds to approximately 2.65×10^4 mol of electrons at the voltage of a fuel cell.

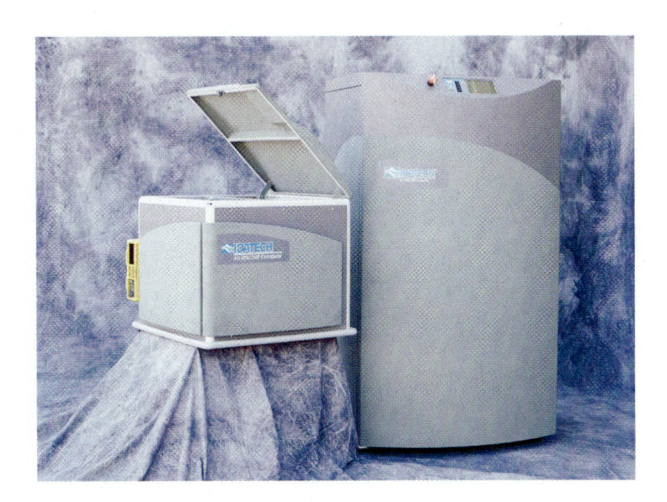

109. Consider the following molecular view of an electrochemical cell involving the following overall reaction.

$$Zn(s) + Ni^{2+}(aq) \longrightarrow Zn^{2+}(aq) + Ni(s)$$

Draw a similar sketch showing how the cell might appear after it has generated a substantial amount of electrical current.

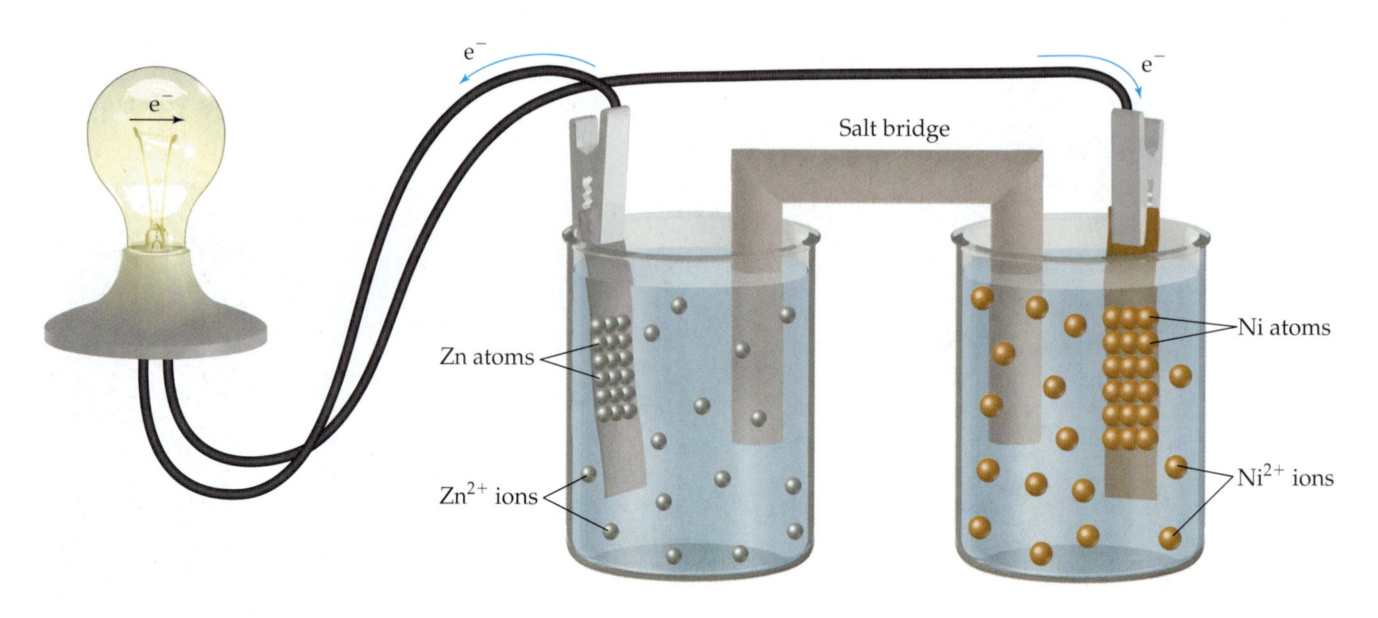

Answers to Skillbuilder Exercises

Skillbuilder 16.1 **(a)** K is oxidized, Cl_2 is reduced **(b)** Al is oxidized, Sn^{2+} is reduced **(c)** C is oxidized, O_2 is reduced

Skillbuilder 16.2 **(a)** K is the reducing agent; Cl_2 is the oxidizing agent **(b)** Al is the reducing agent; Sn^{2+} is the oxidizing agent **(c)** C is the reducing agent; O_2 is the oxidizing agent.

Skillbuilder 16.3

(a) $\underset{0}{\text{Zn}}$

(b) $\underset{+2}{\text{Cu}^{2+}}$

(c) $\underset{+2}{\text{Ca}}\ \underset{-1}{\text{Cl}_2}$

(d) $\underset{+4}{\text{C}}\ \underset{-1}{\text{F}_4}$

(e) $\underset{+3}{\text{N}}\ \underset{-2}{\text{O}_2^-}$

(f) $\underset{+6}{\text{S}}\ \underset{-2}{\text{O}_3}$

Skillbuilder 16.4 Sn oxidized ($0 \longrightarrow +4$); N reduced ($+5 \longrightarrow +4$)

Skillbuilder 16.5 $6\,H^+(aq) + 2\,Cr(s) \longrightarrow 3\,H_2(g) + 2\,Cr^{3+}(aq)$

Skillbuilder 16.6

$Cu(s) + 4\,H^+(aq) + 2\,NO_3^- \longrightarrow Cu^{2+}(aq) + 2\,NO_2(g) + 2\,H_2O(l)$

Skillbuilder 16.7

$5\,Sn(s) + 16\,H^+(aq) + 2\,MnO_4^-(aq) \longrightarrow 5\,Sn^{2+}(aq) + 2\,Mn^{2+}(aq) + 8\,H_2O(l)$

Skillbuilder 16.8 **(a)** Yes **(b)** No

Skillbuilder 16.9 No

Answers to Conceptual Checkpoints

16.1 **(d)** From rule 1, we know that the oxidation state of nitrogen in N_2 is 0. According to rule 3, the sum of the oxidation states of all atoms in a compound = 0. Therefore, by applying rule 5, we can determine that the oxidation state of nitrogen in NO is +2; in NO_2, it is +4; and in NH_3, it is −3.

16.2 **(c)** Lead is the only one of these metals that lies above hydrogen in the activity series, and therefore the only one that would dissolve in an acidic solution.

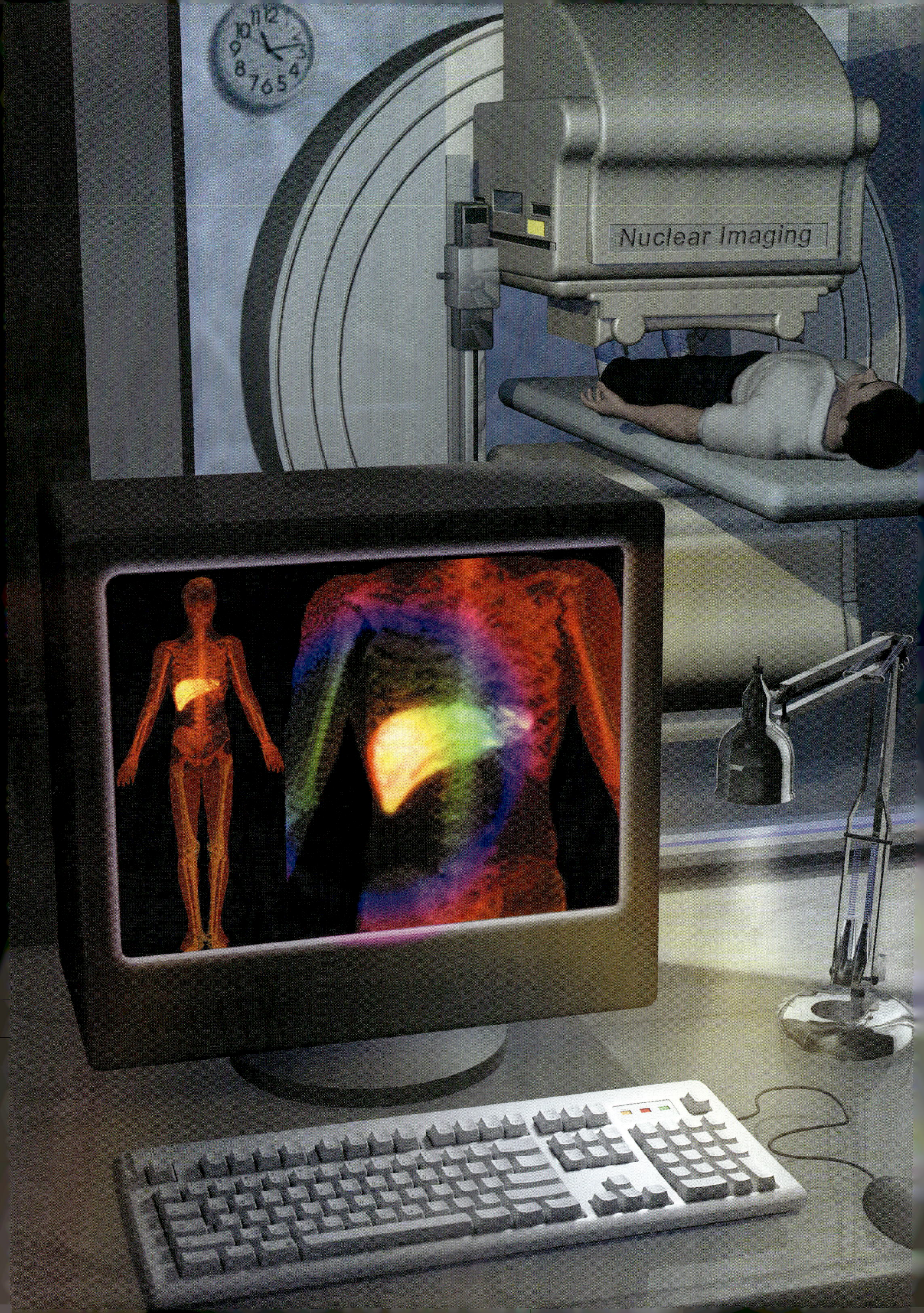
Nuclear Imaging

CHAPTER 17

Radioactivity and Nuclear Chemistry

"Nuclear energy is incomparably greater than the molecular energy which we use today . . . . What is lacking is the match to set the bonfire alight . . . . The scientists are looking for this."

Winston Spencer Churchill (1874–1965), in 1934

17.1 Diagnosing Appendicitis

Several years ago I awoke with a dull pain on the lower right side of my belly. The pain worsened over several hours, so I went to the hospital emergency room for evaluation. I was examined by a doctor who said I might have appendicitis, an inflammation of the appendix. The appendix, which has no known function, is a small pouch that extends from the right side of the large intestine. Occasionally, it becomes diseased and requires surgical removal.

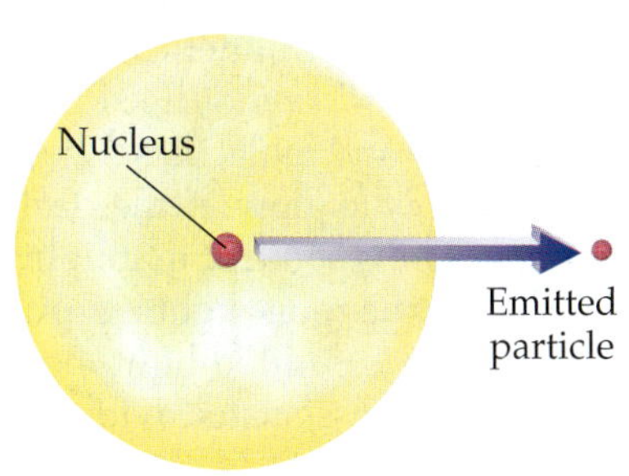

▲ **Figure 17.1 Radioactivity** Radioactivity is the emission of tiny, energetic particles by the nuclei of certain atoms.

Patients with appendicitis usually have a high number of white blood cells, so the hospital performed a blood test to determine my white blood cell count. The test was negative—I had a normal white blood cell count. Although my symptoms were consistent with appendicitis, the negative blood test clouded the diagnosis. The doctor gave me the choice of either having my appendix removed (with the chance of it being healthy) or performing an additional test to confirm appendicitis. I chose the additional test.

The additional test involved nuclear medicine, an area of medical practice that uses radioactivity to diagnose and treat disease. **Radioactivity** is the emission of tiny, invisible particles by the nuclei of certain atoms (◀ Figure 17.1). Many of these particles can pass right through matter. Atoms that emit these particles are said to be **radioactive**.

To perform the test, antibodies—naturally occurring molecules that fight infection—tagged with radioactive atoms were injected into my bloodstream. Since antibodies attack infection, they migrate to areas of the body where infection is present. If my appendix were infected, the antibodies would accumulate there. After waiting about an hour, I was taken to a room and laid on a table. A photographic film was inserted in a panel above me. Although radioactivity is invisible to the eye, it does expose photographic film. If my appendix were indeed infected, it would now contain a high concentration of the radioactively tagged antibodies, and the film would show a

◀ In nuclear medicine, radioactivity is used to obtain clear images of internal organs.

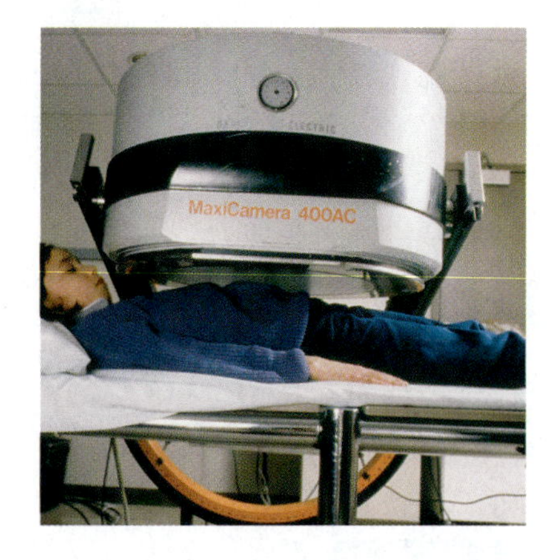

▶ Figure 17.2 **Nuclear medicine** In a test for appendicitis, radioactively tagged antibodies are given to the patient. If the patient has an infection in the appendix, the antibodies accumulate there and the emitted radiation can be detected on photographic film.

bright spot at the location of my appendix (▲ Figure 17.2). In this test, I—or my appendix—was the radiation source that would expose film. The test, however, was negative. No radioactivity was emanating from my appendix. It was healthy. After several hours, the pain in my belly subsided and I went home, appendix and all. I never did find out what caused the pain.

This example from medicine is just one of the many applications of radioactivity. Radioactivity is also used to diagnose and treat many other conditions including cancer, thyroid disease, abnormal kidney and bladder function, and heart disease. Naturally occurring radioactivity allows us to estimate the age of fossils and rocks. Radioactivity also led to the discovery of nuclear fission, used for electricity generation and nuclear weapons. In this chapter, you will learn about radioactivity—how it was discovered, what it is, and how it is used.

17.2 The Discovery of Radioactivity

In 1896, a French scientist named Antoine-Henri Becquerel (1852–1908) discovered radioactivity. Becquerel was interested in the newly discovered X-rays (see Section 9.3), which were the hot topic of physics research in his time. He hypothesized that X-rays were emitted in conjunction with **phosphorescence**. Phosphorescence is the long-lived *emission* of light that sometimes follows the *absorption* of light by some atoms and molecules. Phosphorescence is probably most familiar to you as the glow in glow-in-the-dark toys. After one of these toys is exposed to light, it re-emits some of that light, usually at slightly longer wavelengths. If you turn off the room lights or put the toy in the dark, you can see the greenish glow of the emitted light. Becquerel hypothesized that the visible greenish glow was associated with the emission of X-rays (which are invisible).

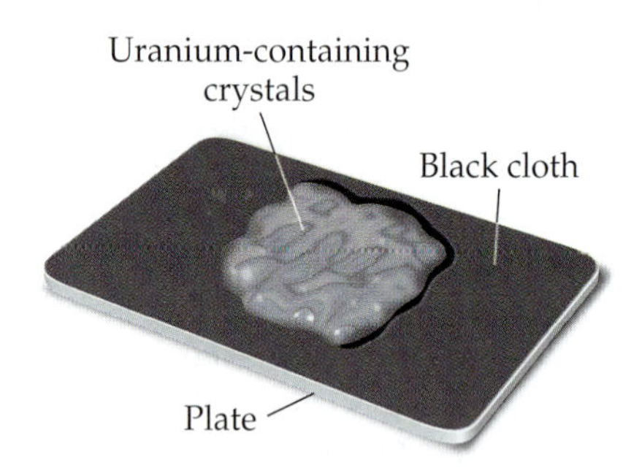

▲ Figure 17.3 **Becquerel's experiment**

To test his hypothesis, Becquerel placed crystals—composed of potassium uranyl sulfate, a compound known to phosphoresce—on top of a photographic plate wrapped in black cloth (◀ Figure 17.3). He then placed the wrapped plate and the crystals outside to expose them to sunlight. He knew that the crystals phosphoresced because he could see the emitted light when he brought them back into the dark. If the crystals also emitted X-rays, the X-rays would pass through the black cloth and expose the underlying photographic plate. Becquerel performed the experiment several times and always got the same result—the photographic plate showed a bright exposure spot where the crystals had been. Becquerel believed his hypothesis was correct and presented the results—that phosphorescence and X-rays were linked—to the French Academy of Sciences.

Becquerel later retracted his results, however, when he discovered that a photographic plate with the same crystals showed a bright exposure spot even when the plate and the crystals were stored in a dark drawer and not exposed to sunlight. Becquerel realized that the crystals themselves were constantly emitting something that exposed the photographic plate, independent of whether they phosphoresced. Becquerel concluded that it was the uranium within the crystals that was the source of the emissions, and he called the emissions *uranic rays*.

Soon after Becquerel's discovery, a young graduate student named Marie Sklodowska Curie (1867–1934), one of the first women in France to attempt doctoral work, decided to pursue the study of uranic rays for her doctoral thesis. Her first task was to determine whether any other substances besides uranium (the heaviest known element at the time) emitted these rays. In her search, Curie discovered two new elements, both of which also emitted uranic rays. Curie named one of her newly discovered elements *polonium* after her home country of Poland. The other element she named *radium* because of the very high amount of radioactivity that it produced. Radium was so radioactive that it gently glowed in the dark and emitted significant amounts of heat. Since it was now clear that these rays were not unique to uranium, Curie changed the name of uranic rays to *radioactivity*. In 1903, Curie received the Nobel Prize in physics—which she shared with Becquerel and her husband, Pierre Curie—for the discovery of radioactivity. In 1911, Curie was awarded a second Nobel Prize, this time in chemistry, for her discovery of the two new elements.

Element 96 (curium) is named in honor of Marie Curie and her contributions to our understanding of radioactivity.

◀ Marie Curie with her two daughters, about 1905. Irene (left) became a distinguished nuclear physicist in her own right, winning a Nobel Prize in 1935. Eve (right) wrote a highly acclaimed biography of her mother.

▶ In the past, radium was added to some paints that were used on watch dials. The radium made the dial glow.

17.3 Types of Radioactivity: Alpha, Beta, and Gamma Decay

While Curie focused her work on discovering the different kinds of radioactive elements, Ernest Rutherford and others focused on characterizing the radioactivity itself. These scientists found that the emissions were produced by the nuclei of radioactive atoms. These nuclei were unstable and would emit small pieces of themselves or electromagnetic radiation to gain stability. These were what Becquerel and Curie detected. There are several different types of radioactive emissions: alpha (α) rays, beta (β) rays, gamma (γ) rays, and positrons.

TUTORIAL
Separation of Alpha, Beta, and Gamma Rays

In order to understand these different types of radioactivity, we must briefly review the notation to symbolize isotopes that was first introduced in Section 4.8. Recall that any isotope can be represented with the following notation.

Remember that the atomic number equals the number of protons and that the mass number equals the number of protons and neutrons.

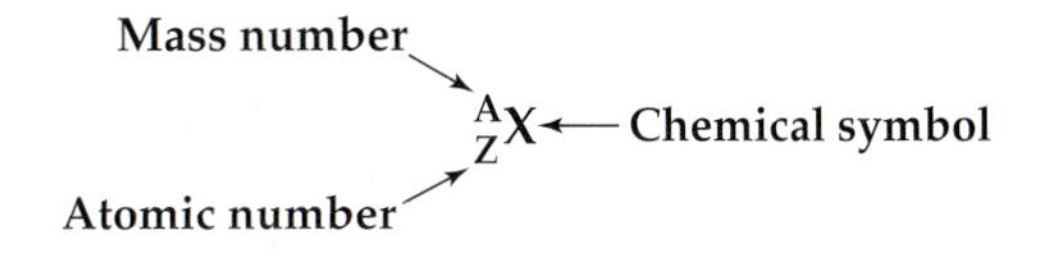

For example, the symbol

$$^{21}_{10}\text{Ne}$$

represents the neon isotope containing 10 protons and 11 neutrons. The symbol

$$^{20}_{10}\text{Ne}$$

represents the neon isotope containing 10 protons and 10 neutrons. Remember that many elements have several different isotopes.

TUTORIAL Particles in Nuclear Reactions

The main subatomic particles—protons, neutrons, and electrons—can all be represented with similar notation.

Proton symbol $^{1}_{1}\text{p}$ Neutron symbol $^{1}_{0}\text{n}$ Electron symbol $^{0}_{-1}\text{e}$

Alpha (α) Radiation

Nuclei are unstable when they are too large or contain an unbalanced ratio of neutrons to protons. Small nuclei need about 1 neutron to every proton to be stable, while larger nuclei need about 1.5 neutrons to every proton.

Alpha (α) radiation occurs when an unstable nucleus emits a small piece of itself consisting of 2 protons and 2 neutrons (▼ Figure 17.4). Since 2 protons and 2 neutrons are identical to a helium-4 nucleus, the symbol for an **alpha (α) particle** is identical to the symbol for helium-4.

alpha (α) particle $^{4}_{2}\text{He}$

An α particle

In nuclear chemistry, we are primarily interested in changes within the nucleus; therefore, the 2+ charge that we would normally write for a helium nucleus is omitted for an alpha particle.

When an atom emits an alpha particle, it becomes a lighter atom. We represent this with a **nuclear equation**, an equation that represents the changes that occur during radioactivity and other nuclear processes. For example, the nuclear equation for the alpha decay of uranium-238 is:

Parent nuclide → $^{238}_{92}\text{U}$; Daughter nuclides → $^{234}_{90}\text{Th}$ and $^{4}_{2}\text{He}$

$$^{238}_{92}\text{U} \longrightarrow {}^{234}_{90}\text{Th} + {}^{4}_{2}\text{He}$$

The term *nuclide* is used in nuclear chemistry to mean a specific isotope.

The original atom is called the **parent nuclide** and the products are called the **daughter nuclides**. Notice that when an element emits an alpha particle, the number of protons in its nucleus changes, transforming it into a different element. In this case, uranium-238 becomes thorium-234. Unlike a chemical reaction, in which elements retain their identity, a nuclear reaction often results in elements changing their identities. Like a chemical equation, however, nuclear equations must be balanced.

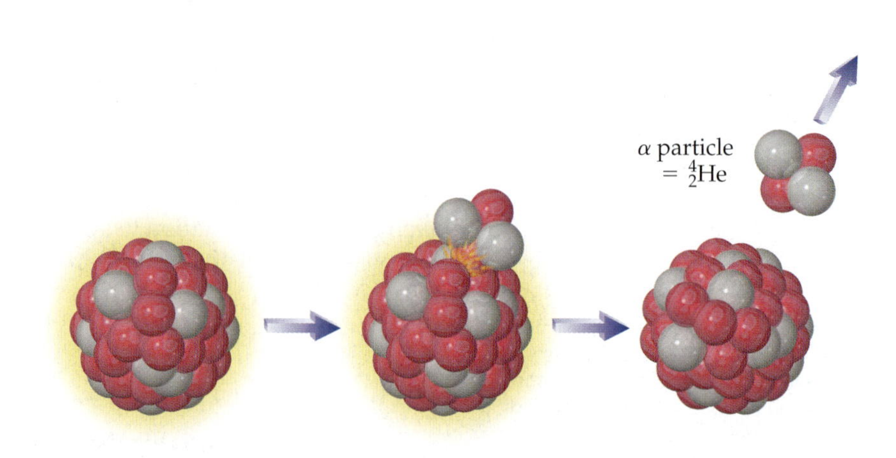

▶ **Figure 17.4 Alpha radiation** Alpha radiation occurs when an unstable nucleus emits a particle composed of 2 protons and 2 neutrons. **Question:** What happens to the atomic number of an element upon emission of an alpha particle?

The sum of the atomic numbers on both sides of a nuclear equation must be equal, and the sum of the mass numbers on both sides must also be equal.

$$^{238}_{92}\text{U} \longrightarrow {}^{234}_{90}\text{Th} + {}^{4}_{2}\text{He}$$

Left Side	Right Side
Sum of mass numbers = 238	Sum of mass numbers = 234 + 4 = 238
Sum of atomic numbers = 92	Sum of atomic numbers = 90 + 2 = 92

The identity and symbol of the daughter nuclide of any alpha decay can be deduced from the mass and atomic numbers of the parent nuclide. During alpha decay, the mass number decreases by 4 and the atomic number decreases by 2. For example, to write a nuclear equation for the alpha decay of Th-232, we begin with the symbol for Th-232 on the left side of the equation and the symbol for an alpha particle on the right side.

$$^{232}_{90}\text{Th} \longrightarrow {}^{x}_{y}? + {}^{4}_{2}\text{He}$$

We can then deduce the mass number and atomic number of the unknown daughter nuclide because the equation must be balanced.

$$^{232}_{90}\text{Th} \longrightarrow \overbrace{\underbrace{{}^{x}_{y}? + {}^{4}_{2}\text{He}}_{y + 2 = 90;\ y = 88}}^{x + 4 = 232;\ x = 228}$$

Therefore,

$$^{232}_{90}\text{Th} \longrightarrow {}^{228}_{88}? + {}^{4}_{2}\text{He}$$

The atomic number must be 88 and the mass number must be 228. Finally, we can deduce the identity of the daughter nuclide and its symbol from its atomic number. Since the atomic number is 88, the unknown daughter nuclide is radium (Ra).

$$^{232}_{90}\text{Th} \longrightarrow {}^{228}_{88}\text{Ra} + {}^{4}_{2}\text{He}$$

EXAMPLE 17.1 Writing Nuclear Equations for Alpha (α) Decay

Write a nuclear equation for the alpha decay of Ra-224.

Solution:

Begin with the symbol for Ra-224 on the left side of the equation and the symbol for an alpha particle on the right side.

$$^{224}_{88}\text{Ra} \longrightarrow {}^{x}_{y}? + {}^{4}_{2}\text{He}$$

Equalize the sum of the mass numbers and the sum of the atomic numbers on both sides of the equation by writing the appropriate mass number and atomic number for the unknown daughter nuclide.

$$^{224}_{88}\text{Ra} \longrightarrow {}^{220}_{86}? + {}^{4}_{2}\text{He}$$

Deduce the identity of the unknown daughter nuclide from the atomic number and write its symbol. Since the atomic number is 86, the daughter nuclide must be radon (Rn).

$$^{224}_{88}\text{Ra} \longrightarrow {}^{220}_{86}\text{Rn} + {}^{4}_{2}\text{He}$$

SKILLBUILDER 17.1 Writing Nuclear Equations for Alpha (α) Decay

Write a nuclear equation for the alpha decay of Po-216.

Recall that Rutherford was using alpha particles to probe the structure of the atom when he discovered the nucleus (Section 4.3).

To ionize means *to create ions* (charged particles)

Alpha radiation is the semi-truck of radioactivity. The alpha particle is by far the most massive of all particles emitted by radioactive nuclei. Consequently, alpha radiation has the most potential to interact with and damage other molecules, including biological ones. Radiation interacts with other molecules and atoms by ionizing them. If radiation ionizes molecules within the cells of living organisms, those molecules become damaged and the cell can die or begin to reproduce abnormally. The ability of radiation to ionize other molecules and atoms is called its **ionizing power**. Of all types of radioactivity, alpha radiation has the highest ionizing power.

However, because of its large size, alpha radiation also has the lowest **penetrating power**—the ability to penetrate matter. (Imagine a semi-truck trying to get through a traffic jam.) In order for radiation to damage important molecules within living cells, it must penetrate into the cell. Alpha radiation does not easily penetrate into cells; it can be stopped by a sheet of paper, by clothing, or even by air. Consequently, a low-level alpha emitter kept outside the body is relatively safe. However, if an alpha emitter is ingested or inhaled, it becomes very dangerous because the alpha particles then have direct access to the molecules that compose organs and tissues.

To summarize:

- Alpha particles are composed of 2 protons and 2 neutrons.
- Alpha particles have the symbol ${}^{4}_{2}He$.
- Alpha particles have a high ionizing power.
- Alpha particles have a low penetrating power.

Beta (β) Radiation

Beta (β) radiation occurs when an unstable nucleus emits an electron (▼ Figure 17.5). How does a nucleus, which contains only protons and neutrons, emit an electron? Where do the electrons come from? They come from neutrons as they convert to protons. In other words, in some unstable nuclei, a neutron will change into a proton and emit an electron in the process.

$$\text{Beta decay} \qquad \text{Neutron} \longrightarrow \text{Proton} + \text{Electron}$$

The symbol for a **beta (β) particle** in a nuclear equation is:

$$\text{Beta } (\beta) \text{ particle} \qquad {}^{0}_{-1}e$$

Beta radiation is also called beta-minus (β^-) radiation because of its negative charge.

The 0 in the upper left corner reflects the mass number of the electron. The −1 in the lower left corner reflects the charge of the electron which is

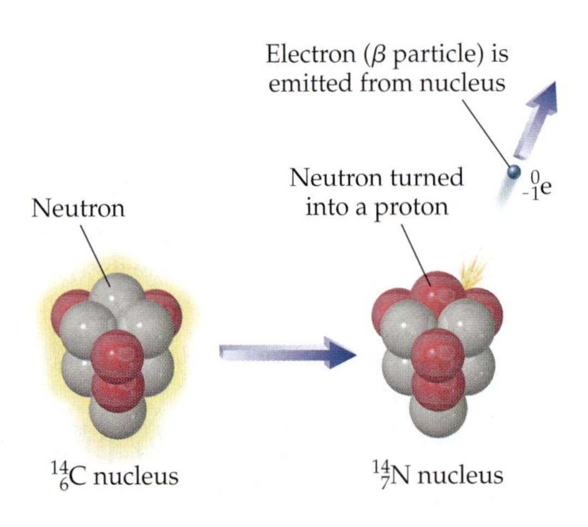

▶ **Figure 17.5 Beta radiation** Beta radiation occurs when an unstable nucleus emits an electron. As the emission occurs, a neutron turns into a proton. **Question:** What happens to the atomic number of an element upon emission of a beta particle?

Remember that the mass number is defined as the sum of the number of protons and neutrons. Since electrons have no protons or neutrons, their mass number is zero.

equivalent to an atomic number of −1 in a nuclear equation. When an atom emits a beta particle, its atomic number increases by one because it now has an additional proton. For example, the nuclear equation for the beta decay of radium-228 is:

$$^{228}_{88}\text{Ra} \longrightarrow {}^{228}_{89}\text{Ac} + {}^{0}_{-1}\text{e}$$

Notice that the nuclear equation is still balanced—the sums of the mass numbers on both sides are equal and the sums of the atomic numbers on both sides are equal.

$$^{228}_{88}\text{Ra} \longrightarrow {}^{228}_{89}\text{Ac} + {}^{0}_{-1}\text{e}$$

Left Side	Right Side
Sum of mass numbers = 228	Sum of mass numbers = 228 + 0 = 228
Sum of atomic numbers = 88	Sum of atomic numbers = 89 − 1 = 88

The identity and symbol of the daughter nuclide of any beta decay can be determined in a manner similar to that for alpha decay, as shown in the following example.

EXAMPLE 17.2 Writing Nuclear Equations for Beta (β) Decay

Write a nuclear equation for the beta decay of Bk-249.

Solution:

Begin with the symbol for Bk-249 on the left side of the equation and the symbol for a beta particle on the right side.

$$^{249}_{97}\text{Bk} \longrightarrow {}^{x}_{y}? + {}^{0}_{-1}\text{e}$$

Equalize the sum of the mass numbers and the sum of the atomic numbers on both sides of the equation by writing the appropriate mass number and atomic number for the unknown daughter nuclide.

$$^{249}_{97}\text{Bk} \longrightarrow {}^{249}_{98}? + {}^{0}_{-1}\text{e}$$

Deduce the identity of the unknown daughter nuclide from the atomic number and write its symbol. Since the atomic number is 98, the daughter nuclide must be californium (Cf).

$$^{249}_{97}\text{Bk} \longrightarrow {}^{249}_{98}\text{Cf} + {}^{0}_{-1}\text{e}$$

SKILLBUILDER 17.2 Writing Nuclear Equations for Beta (β) Decay

Write a nuclear equation for the beta decay of Ac-228.

SKILLBUILDER PLUS

Write three nuclear equations to represent the nuclear decay sequence that begins with the alpha decay of U-235 followed by a beta decay of the daughter nuclide and then another alpha decay.

Beta radiation is the midsized car of radioactivity. Beta particles are much less massive than alpha particles and consequently have a lower ionizing power. However, because of their smaller size, beta particles have a higher penetrating power and require a sheet of metal or a thick piece of wood to stop them. Consequently, a low-level beta emitter outside the body poses a higher risk than an alpha emitter. Inside the body, however, a beta emitter does less damage than an alpha emitter.

To summarize:

- Beta particles are electrons emitted from atomic nuclei when a neutron changes into a proton.
- Beta particles have the symbol ${}_{-1}^{0}e$.
- Beta particles have intermediate ionizing power.
- Beta particles have intermediate penetrating power.

Gamma (γ) Radiation

See Section 9.3 for a review of electromagnetic radiation.

Gamma (γ) radiation is significantly different from alpha or beta radiation. Gamma radiation is not matter but *electromagnetic radiation:* Gamma rays are high-energy (short-wavelength) photons. The symbol for a **gamma ray** is:

Gamma (γ) ray ${}_{0}^{0}\gamma$

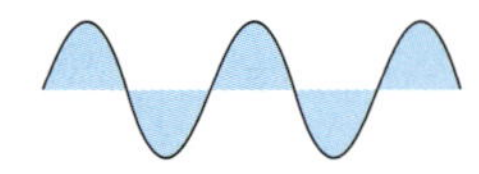

A gamma ray has no charge and no mass. When a gamma ray is emitted from a radioactive atom, it does not change the mass number or the atomic number of the element. Gamma rays, however, are usually emitted in conjunction with other types of radiation. For example, the alpha emission of U-238 (discussed previously) is also accompanied by the emission of a gamma ray.

$$ {}_{92}^{238}U \longrightarrow {}_{90}^{234}Th + {}_{2}^{4}He + {}_{0}^{0}\gamma $$

Gamma rays are the motorbikes of radioactivity. They have the lowest ionizing power, but the highest penetrating power. (Imagine a motorbike zipping through a traffic jam.) Stopping gamma rays requires several inches of lead shielding or thick slabs of concrete.

To summarize:

- Gamma rays are electromagnetic radiation—high-energy, short-wavelength photons.
- Gamma rays have the symbol ${}_{0}^{0}\gamma$.
- Gamma rays have low ionizing power.
- Gamma rays have high penetrating power.

Positron Emission

Positron emission occurs when an unstable nucleus emits a positron (▼ Figure 17.6). A **positron** has the mass of an electron but carries a +1 charge.

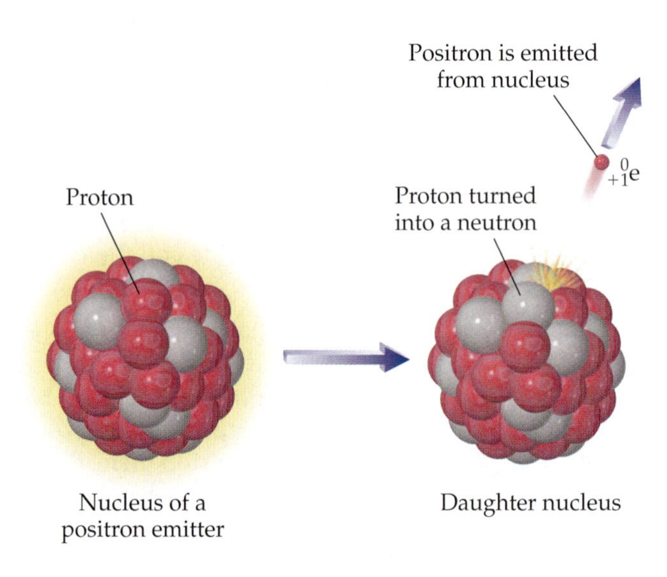

▶ **Figure 17.6 Positron emission** Positron emission occurs when an unstable nucleus emits a positron. As the emission occurs, a proton turns into a neutron. **Question:** What happens to the atomic number of an element upon positron emission?

Positron emission can be thought of as a type of beta emission. It is sometimes referred to as beta-plus (β^{+}) emission.

In some unstable nuclei, a proton will change into a neutron and emit a positron in the process.

$$\text{Positron emission} \qquad \text{Proton} \longrightarrow \text{Neutron} + \text{Positron}$$

The symbol for a positron in a nuclear equation is:

$$\text{Positron} \qquad {}^{0}_{+1}\text{e}$$

The 0 in the upper left corner reflects that a positron has a mass number of 0. The +1 in the lower left corner reflects the charge of the positron, which is equivalent to an atomic number of +1 in a nuclear equation. When an atom emits a positron, its atomic number decreases by 1 because it now has 1 less proton. For example, the nuclear equation for the positron emission of phosphorus-30 is:

$${}^{30}_{15}\text{P} \longrightarrow {}^{30}_{14}\text{Si} + {}^{0}_{+1}\text{e}$$

The identity and symbol of the daughter nuclide of any positron emission can be determined in a manner similar to alpha and beta decay, as shown in the following example. Positron emission is similar to beta emission in its ionizing and penetrating power.

EXAMPLE 17.3 Writing Nuclear Equations for Positron Emission

Write a nuclear equation for the positron emission of potassium-40.

Solution:

Begin with the symbol for K-40 on the left side of the equation and the symbol for a positron on the right side.

$${}^{40}_{19}\text{K} \longrightarrow {}^{x}_{y}? + {}^{0}_{+1}\text{e}$$

Equalize the sum of the mass numbers and the sum of the atomic numbers on both sides of the equation by writing the appropriate mass number and atomic number for the unknown daughter nuclide.

$${}^{40}_{19}\text{K} \longrightarrow {}^{40}_{18}? + {}^{0}_{+1}\text{e}$$

Deduce the identity of the unknown daughter nuclide from the atomic number and write its symbol. Since the atomic number is 18, the daughter nuclide must be argon (Ar).

$${}^{40}_{19}\text{K} \longrightarrow {}^{40}_{18}\text{Ar} + {}^{0}_{+1}\text{e}$$

SKILLBUILDER 17.3 Writing Nuclear Equations for Positron Emission

Write a nuclear equation for the positron emission of sodium-22.

CONCEPTUAL CHECKPOINT 17.1

Which of the following kinds of radioactive decay changes the mass number of the parent element?

(a) alpha (α) decay
(b) beta (β) decay
(c) gamma (γ) decay
(d) positron decay

17.4 Detecting Radioactivity

The particles emitted from radioactive nuclei contain a large amount of energy and can therefore be detected with high sensitivity. Radiation detectors detect such particles through their interactions with atoms or molecules. The simplest radiation detectors are pieces of photographic film that become exposed when radiation passes through them. **Film-badge dosimeters**—which consist of photographic film held in a small case that is pinned to clothing—are standard for most people working with or near radioactive substances (▼ Figure 17.7a). These badges are collected and processed (or developed) regularly as a way to monitor a person's exposure. The more exposed the film has become in a given period of time, the more radioactivity to which the person has been exposed.

Radioactivity can be instantly detected with devices such as a **Geiger-Müller counter** (▼ Figure 17.7b). In such an instrument (commonly referred to simply as a Geiger counter), particles emitted by radioactive nuclei pass through an argon-filled chamber. The energetic particles create a trail of ionized argon atoms as they pass through the chamber. If the applied voltage is high enough, these newly formed ions produce an electrical signal that can be detected on a meter or turned into an audible click. Each click corresponds to a radioactive particle passing through the argon

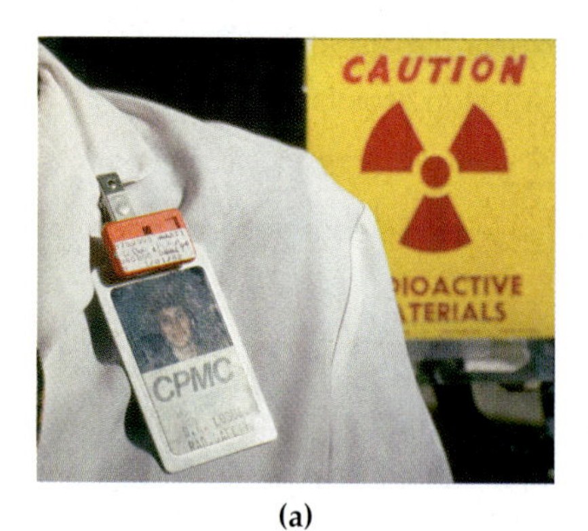

(a)

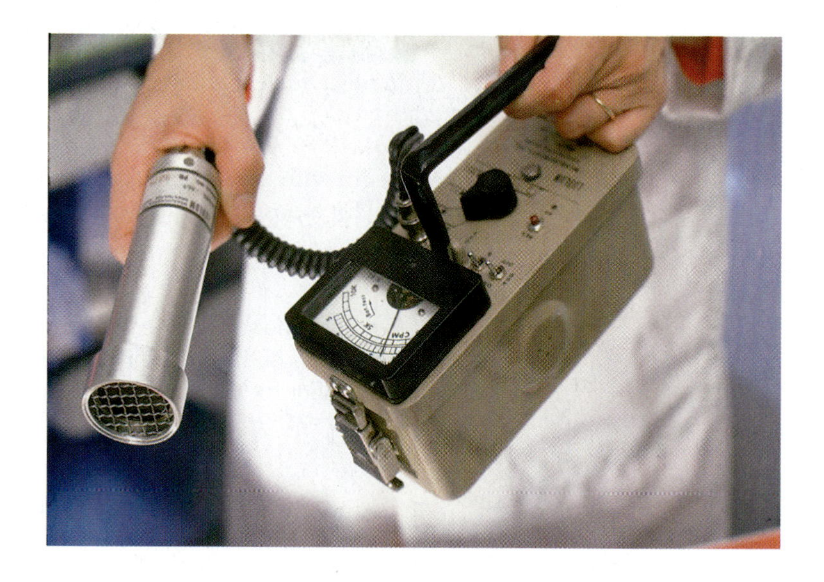

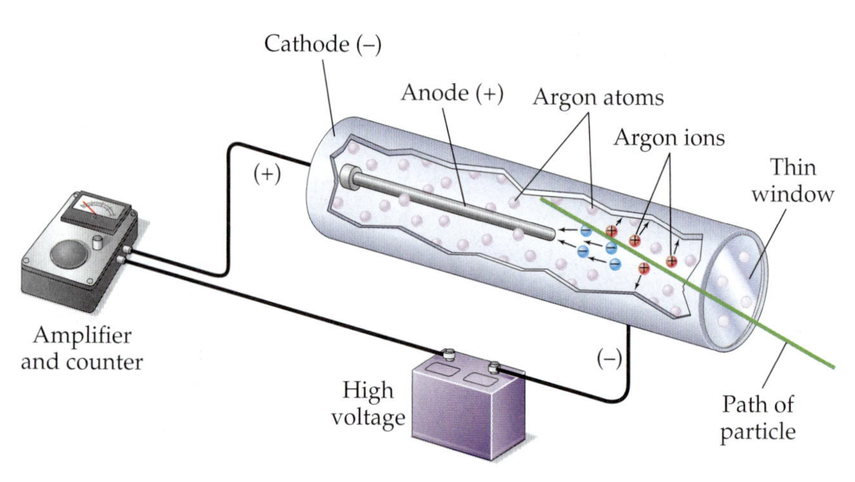

(b)

▲ **Figure 17.7 Detecting radiation** **(a)** A film badge provides a simple and inexpensive way of measuring a person's cumulative exposure to radiation. **(b)** A Geiger counter can record the passage of individual energetic particles as they pass through a chamber filled with argon gas. When an argon atom is ionized, the resulting ion is attracted to the anode and the dislodged electron to the cathode, creating a tiny electrical current that can be recorded in various ways.

gas chamber. This clicking is the stereotypical sound most people associate with a radiation detector.

A second type of device commonly used to detect radiation instantly is called a **scintillation counter**. In a scintillation counter, the radioactive emissions pass through a material (such as NaI or CsI) that emits ultraviolet or visible light in response to excitation by energetic particles. This light is then detected and turned into an electrical signal that can be read on a meter.

17.5 Natural Radioactivity and Half-Life

Radioactivity is a natural component of our environment. The ground beneath you most likely contains radioactive atoms that emit radiation into the air around you. The food you eat contains a residual amount of radioactive atoms that enter into your body fluids and tissues. Small amounts of radiation from space make it through our atmosphere and constantly bombard Earth. Humans and other living organisms have evolved in this environment and have adapted to survive in it.

One reason for the radioactivity in our environment is the instability of all atomic nuclei beyond atomic number 83 (bismuth). In addition, some isotopes of elements with fewer than 83 protons are also unstable and radioactive.

Different radioactive nuclides decay into their daughter nuclides at different rates. Some nuclides decay quickly while others decay slowly. The time it takes for half of the parent nuclides in a radioactive sample to decay to the daughter nuclides is called the **half-life**. Nuclides that decay quickly have short half-lives and are very active (many decay events per unit time), while those that decay slowly have long half-lives and are less active (fewer decay events per unit time). For example, Th-232 is an alpha emitter that decays according to the following nuclear reaction.

A *decay event* is the emission of radiation by a single radioactive nuclide.

$$^{232}_{90}\text{Th} \longrightarrow {}^{228}_{88}\text{Ra} + {}^{4}_{2}\text{He}$$

Th-232 has a half-life of 1.4×10^{10} years or 14 billion years. If we start with a sample of Th-232 containing 1 million atoms, the sample would decay to half a million atoms in 14 billion years and then to a quarter of a million in another 14 billion years and so on (▼ Figure 17.8).

$$\begin{matrix}\text{1 million} \\ \text{Th-232 atoms}\end{matrix} \xrightarrow[\text{14 billion years}]{} \begin{matrix}\tfrac{1}{2}\text{million} \\ \text{Th-232 atoms}\end{matrix} \xrightarrow[\text{14 billion years}]{} \begin{matrix}\tfrac{1}{4}\text{million} \\ \text{Th-232 atoms}\end{matrix}$$

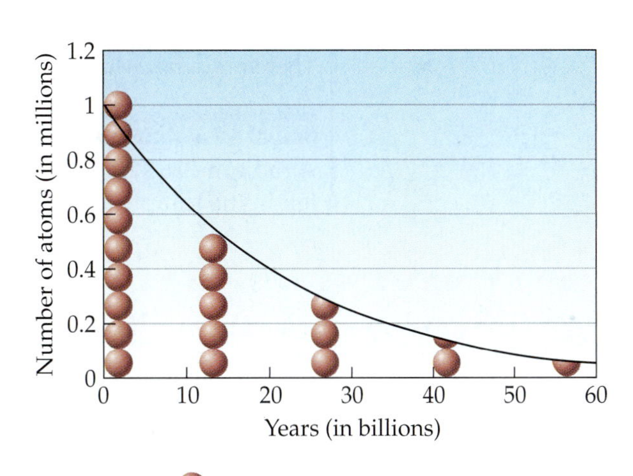

▶ **Figure 17.8 The concept of half-life** A plot of the number of Th-232 atoms in a sample initially containing 1 million atoms as a function of time. Th-232 has a half-life of 14 billion years.

Notice that a radioactive sample *does not* decay to zero atoms in two half-lives—*you can't add two half-lives together to get a "whole" life*. The amount that remains after one half-life is always half of what was present at the start. The amount that remains after two half-lives is a quarter of what was present at the start, and so on.

Some nuclides have short half-lives. Radon-220, for example, has a half-life of approximately 1 minute. If we had a sample of radon-220 that contained 1 million atoms, it would be diminished to $\frac{1}{4}$ million radon-220 atoms in just 2 minutes.

$$\begin{matrix} 1 \text{ million} \\ \text{Rn-220 atoms} \end{matrix} \xrightarrow[\text{1 minute}]{} \begin{matrix} \frac{1}{2}\text{million} \\ \text{Rn-220 atoms} \end{matrix} \xrightarrow[\text{1 minute}]{} \begin{matrix} \frac{1}{4}\text{million} \\ \text{Rn-220 atoms} \end{matrix}$$

Each radioactive nuclide has a unique half-life, which is not affected by physical conditions or chemical environment.

Rn-220 is therefore much more active than Th-232 because it undergoes many more decay events in a given period of time. Some nuclides have even shorter half-lives. Table 17.1 lists several nuclides and their half-lives.

TABLE 17.1 Selected Nuclides and Their Half-Lives

Nuclide	Half-Life	Type of Decay
$^{232}_{90}Th$	1.4×10^{10} yr	alpha
$^{238}_{92}U$	4.5×10^{9} yr	alpha
$^{14}_{6}C$	5730 yr	beta
$^{220}_{86}Rn$	55.6 s	alpha
$^{219}_{90}Th$	1.05×10^{-6} s	alpha

EXAMPLE 17.4 Half-Life

How long does it take for a 1.80-mol sample of Th-228 (which has a half-life of 1.9 years) to decay to 0.225 mol?

Solution:

It is easiest to draw a table showing the amount of Th-228 as a function of number of half-lives. For each half-life, simply divide the amount of Th-228 by 2.

Amount of Th-228	Number of Half-Lives	Time in Years
1.80 mol	0	0
0.900 mol	1	1.9
0.450 mol	2	3.8
0.225 mol	3	5.7

Therefore, it takes three half-lives or 5.7 years for the sample to decay to 0.225 mol.

SKILLBUILDER 17.4 Half-Life

A radium-226 sample initially contains 0.112 mol. How much radium-226 is left in the sample after 6400 years? The half-life of radium-226 is 1600 years.

A Natural Radioactive Decay Series

The radioactive elements in our environment are all undergoing radioactive decay. The reason that they are in our environment at all is that either they have very long half-lives (billions of years) or they are continuously being formed by some other process in the environment. In many cases, the daughter nuclide

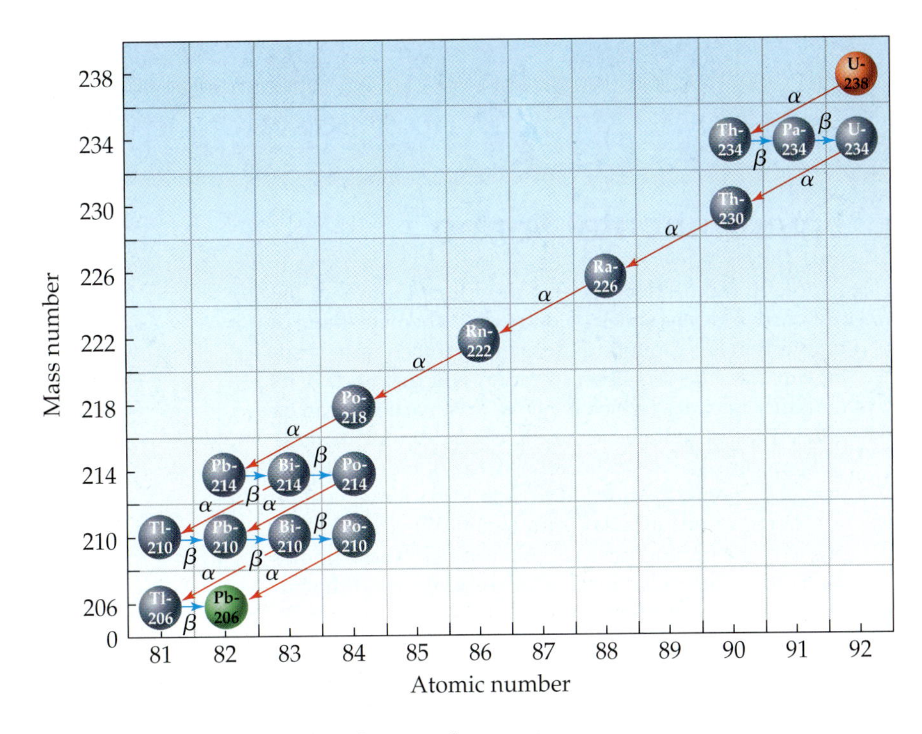

▲ **Figure 17.9 Uranium-238 decay series**

of a radioactive decay is itself radioactive and in turn produces another daughter nuclide that is radioactive and so on, resulting in a radioactive decay series. For example, uranium (atomic number 92) is the heaviest naturally occurring element. It is an alpha emitter that decays to Th-234 with a half-life of 4.47 billion years.

$$^{238}_{92}\text{U} \longrightarrow {}^{234}_{90}\text{Th} + {}^{4}_{2}\text{He}$$

The daughter nuclide, Th-234, is itself radioactive—it is a beta emitter that decays to Pa-234 with a half-life of 24.1 days.

$$^{234}_{90}\text{Th} \longrightarrow {}^{234}_{91}\text{Pa} + {}^{0}_{-1}\text{e}$$

Pa-234 is also radioactive, decaying to U-234 via beta emission with a half-life of 244,500 years. This process continues until it reaches Pb-206, which is stable. The entire uranium-238 decay series is shown in ▲ Figure 17.9. In other words, all of the uranium-238 in the environment is slowly decaying away to lead. Since the half-life for the first step in the series is so long, however, there is still plenty of uranium-238 in the environment. All of the other nuclides in the decay series are also present in the environment in varying amounts, depending on their half-lives.

CONCEPTUAL CHECKPOINT 17.2

Suppose that you start with 1,000,000 atoms of a particular radioactive isotope. How many half-lives would be required to reduce the number of undecayed atoms to fewer than 1000?

(a) 10
(b) 100
(c) 1000
(d) 1001

Chemistry AND HEALTH

Environmental Radon

Radon—a radioactive gas—is one of the products of the radioactive decay series of uranium. Wherever there is uranium in the ground, there is likely to be radon seeping up into the air. Because radon is a gas, it and its daughter nuclides (which attach to dust particles) can be inhaled into the lungs, where they decay and increase lung cancer risk. The radioactive decay of radon is by far the single greatest source of human radiation exposure.

Homes built in areas with significant uranium deposits in the ground pose the greatest risk. These homes can accumulate radon levels that are above what the Environmental Protection Agency (EPA) considers safe. Simple test kits are available to test indoor air and determine radon levels. The higher the level is, the greater the risk. The risk is even higher for smokers who live in these houses. Excessively high indoor radon levels require the installation of a ventilation system to purge radon from the house. Lower levels can be ventilated by keeping windows and doors open.

CAN YOU ANSWER THIS? *Suppose that a house contains 1.80×10^{-3} mol of radon-222 (which has a half-life of 3.8 days). If no new radon entered the house, how long would it take for the radon to decay to 4.50×10^{-4} mol?*

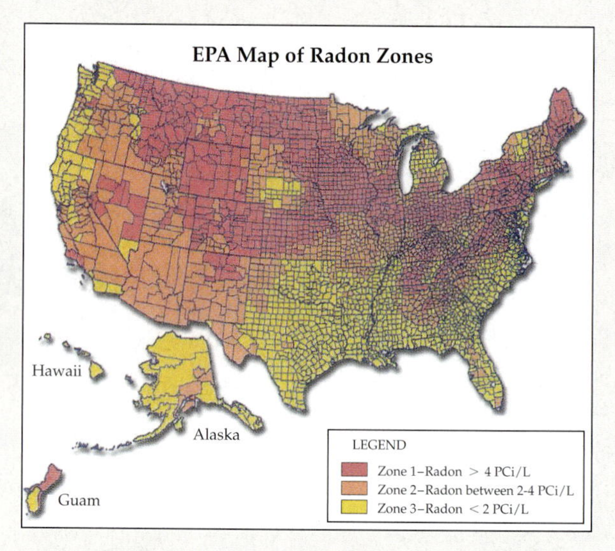

▲ Map of the United States showing radon levels. Zone 1 counties have the highest levels and zone 3 counties have the lowest.

17.6 Radiocarbon Dating: Using Radioactivity to Measure the Age of Fossils and Other Artifacts

Archeologists, geologists, anthropologists, and other scientists take advantage of the presence of natural radioactivity in our environment to estimate the ages of fossils and artifacts through a technique called **radiocarbon dating**. For example, in 1947, young shepherds searching for a stray goat near the Dead Sea (east of Jerusalem) entered a cave and discovered ancient scrolls stuffed into jars. These scrolls—now named the Dead Sea Scrolls—are 2000-year-old biblical manuscripts, predating other existing manuscripts by almost 1000 years.

The Dead Sea Scrolls—like other ancient artifacts—contain a radioactive signature that reveals their age. This signature results from the presence of carbon-14—which is radioactive—in the environment. Carbon-14 is constantly formed in the upper atmosphere by the neutron bombardment of nitrogen.

$$^{14}_{7}\text{N} + ^{1}_{0}\text{n} \longrightarrow ^{14}_{6}\text{C} + ^{1}_{1}\text{H}$$

Carbon-14 then decays back to nitrogen by beta emission, with a half-life of 5730 years.

$$^{14}_{6}\text{C} \longrightarrow ^{14}_{7}\text{N} + ^{0}_{-1}\text{e}$$

TABLE 17.2 Age of Object Based on Its Carbon-14 Content

Concentration of C-14 (% Relative to Living Organisms)	Age of Object in Years
100.0	0
50.0	5,730
25.00	11,460
12.50	17,190
6.250	22,920
3.125	28,650
1.563	34,380

The continuous formation of carbon-14 in the atmosphere and its continuous decay back to nitrogen-14 produces a nearly constant equilibrium concentration of atmospheric carbon-14. That carbon-14 is oxidized to carbon dioxide and then incorporated into plants by photosynthesis. It is also incorporated into animals, because animals ultimately depend on plants for food. Consequently, all living organisms contain residual amounts of carbon-14. When a living organism dies, it stops incorporating new carbon-14 into its tissues. The carbon-14 present at the time of death decays with a half-life of 5730 years. Since many artifacts, such as the Dead Sea Scrolls, are made from materials that were once living—such as papyrus, wood, and other plant and animal derivatives—the amount of carbon-14 in these artifacts indicates their age.

The concentration of carbon-14 in all *living* organisms is the same.

For example, suppose an ancient artifact has a carbon-14 concentration that is 50% of that found in living organisms. How old is the artifact? Since it contains half as much carbon-14 as a living organism, it must be one half-life or 5730 years old. If the artifact has a carbon-14 concentration that is 25% of that found in living organisms, its age is two half-lives or 11,460 years old. Table 17.2 shows the age of an object based on its carbon-14 content. Ages of less than one half-life, or intermediate between a whole number of half-lives, can also be calculated, but the method for doing so is beyond the scope of this book.

We know that carbon-14 dating is accurate because it can be checked against objects whose ages are known from other methods. For example, old trees can be dated by counting the tree rings within their trunks and by carbon-14 dating. Except in cases where the carbon-14 content has been modified by some other source, carbon-14 dating has proven reliable. However, this method is not dependable when attempting to date objects that are more than 50,000 years old; the amount of carbon-14 becomes too low to measure.

EXAMPLE 17.5 Radiocarbon Dating

A skull believed to belong to an early human being is found to have a carbon-14 content of 3.125% of that found in living organisms. How old is the skull?

Solution:

We can examine Table 17.1 and see that a carbon-14 content of 3.125% of that found in living organisms corresponds to an age of 28,650 years.

SKILLBUILDER 17.5 Radiocarbon Dating

An ancient scroll is claimed to have originated from Greek scholars in about 500 B.C. A measure of its carbon-14 content reveals it to contain 100.0% of that found in living organisms. Is the scroll authentic?

The Shroud of Turin

The shroud of Turin—kept in the Cathedral of Turin in Italy—is an old linen cloth that bears the image of a man who appears to have been crucified.

The image becomes clearer if the shroud is photographed and viewed as a negative. Many believe that the shroud is the original burial cloth of Jesus Christ, miraculously imprinted with his image. In 1988, three independent laboratories were chosen by the Roman Catholic Church to perform radiocarbon dating on the shroud. The laboratories took samples from the shroud and measured the carbon-14 content. They all got similar results—the shroud was made from linen originating in about A.D. 1325. Although some have disputed the results, and although no scientific test is 100% reliable, newspapers around the world quickly announced that the shroud was a fake.

CAN YOU ANSWER THIS? *Suppose an artifact is claimed to be from 3000 B.C. Examination of the C-14 content of the artifacts reveals that the concentration of C-14 is 55% of that found in living organisms. Could the artifact be authentic?*

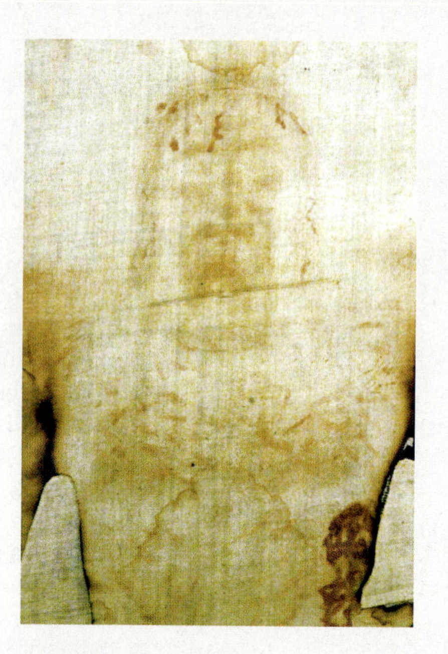

▲ The Shroud of Turin.

17.7 The Discovery of Fission and the Atomic Bomb

The element with atomic number 100 is named *fermium*, in honor of Enrico Fermi.

In the mid-1930s Enrico Fermi (1901–1954), an Italian physicist, tried to synthesize a new element by bombarding uranium—the heaviest known element at that time—with neutrons. Fermi hypothesized that if a neutron were incorporated into the nucleus of a uranium atom, the nucleus might undergo beta decay, converting a neutron into a proton. If that happened, a new element, with atomic number 93, would be synthesized for the first time. The nuclear equation for the process is:

$$^{238}_{92}\text{U} + {}^{1}_{0}\text{n} \longrightarrow {}^{239}_{92}\text{U} \longrightarrow {}^{239}_{93}\text{X} + {}^{0}_{-1}\text{e}$$

Neutron (${}^{1}_{0}\text{n}$); Newly synthesized element (${}^{239}_{93}\text{X}$)

Fermi performed the experiment and detected the emission of beta particles. However, his results were inconclusive. Had he synthesized a new element? Fermi never chemically examined the products to determine their composition and therefore could not say with certainty whether he had.

The element with atomic number 109 is named *meitnerium*, in honor of Lise Meitner.

Three researchers in Germany—Lise Meitner (1878–1968), Fritz Strassmann (1902–1980), and Otto Hahn (1879–1968)—repeated Fermi's experiments but then performed careful chemical analysis of the products. What they found in the products—several elements *lighter* than uranium—would change the world forever. On January 6, 1939, Meitner, Strassmann, and Hahn reported that the neutron bombardment of uranium resulted in **nuclear fission**—the splitting of the atom. The nucleus of the neutron-bombarded uranium atom had fallen apart into barium, krypton, and other

▲ Lise Meitner in Otto Hahn's Berlin laboratory.

smaller products. They also realized that the process emitted enormous amounts of energy. The following is a nuclear equation for a fission reaction, showing how uranium breaks apart into the daughter nuclides.

$$^{235}_{92}\text{U} + {}^{1}_{0}\text{n} \longrightarrow {}^{142}_{56}\text{Ba} + {}^{91}_{36}\text{Kr} + 3\,{}^{1}_{0}\text{n} + \text{Energy}$$

Notice that the initial uranium atom is the U-235 isotope, which comprises less than 1% of all naturally occurring uranium. U-238, the most abundant uranium isotope, does not undergo fission. Notice also that the process produces three neutrons, which have the potential to initiate fission in three other U-235 atoms.

U.S. scientists quickly realized that a sample rich in U-235 could undergo a **chain reaction** in which neutrons produced by the fission of one uranium nucleus would induce fission in other uranium nuclei (▼ Figure 17.10). The result would be a self-amplifying reaction capable of producing an enormous amount of energy—an atomic bomb. However, to make a bomb, a **critical mass** of U-235—enough U-235 to produce a self-sustaining reaction—would be necessary. Fearing that Nazi Germany would develop such a bomb, several U.S. scientists persuaded Albert Einstein, the most well-known scientist of the time, to write a letter to President Franklin Roosevelt warning of this possibility. Einstein wrote, "... and it is conceivable—though much less certain—that extremely powerful bombs of a new type may thus be constructed. A single bomb of this type, carried by boat and exploded in a port, might very well destroy the whole port together with some of the surrounding territory."

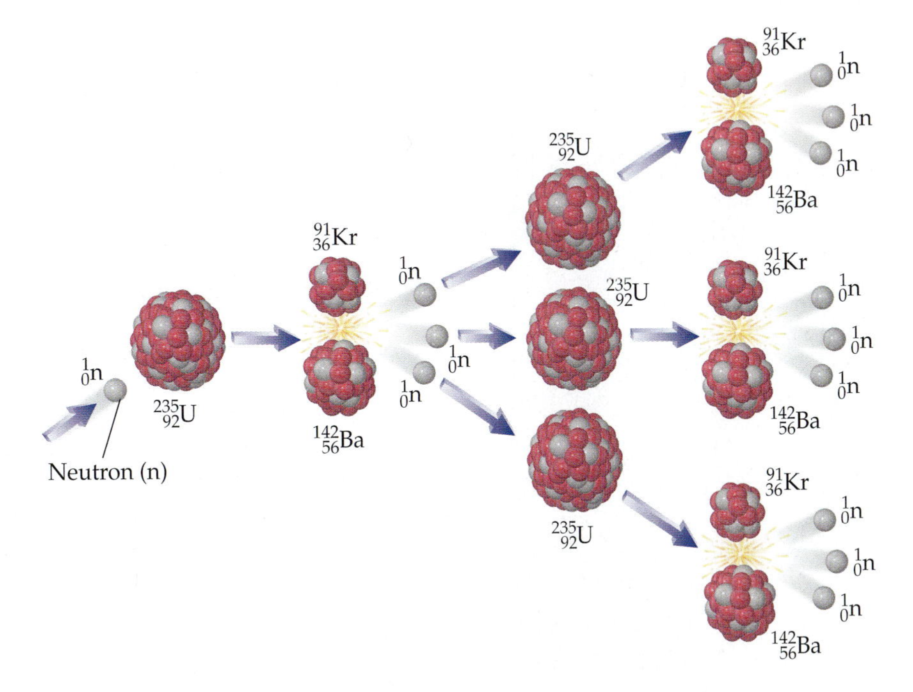

▲ **Figure 17.10 Fission chain reaction** The neutrons produced by the fission of one uranium nucleus induce fission in other uranium nuclei to produce a self-amplifying reaction. **Question:** Why is it necessary that each fission event produce more than one neutron to sustain the chain reaction?

▲ The testing of the world's first nuclear bomb at Alamogordo, New Mexico, in 1945.

Roosevelt was convinced by Einstein's letter, and in 1941 he assembled the resources to begin the costliest scientific project ever attempted. The top-secret endeavor was called the *Manhattan Project*, and its main goal was to build an atomic bomb before the Germans did. The project was led by physicist J. R. Oppenheimer (1904–1967) at a high-security research facility in Los Alamos, New Mexico. Four years later, on July 16, 1945, the world's first nuclear weapon was successfully detonated at a test site in New Mexico. The first atomic bomb exploded with a force equivalent to 18,000 tons of dynamite. Ironically, the Germans—who had *not* made a successful nuclear bomb—had already been defeated by this time. Instead, the atomic bomb was used on Japan. One bomb was dropped on Hiroshima, and a second bomb was dropped on Nagasaki. Together, the bombs killed approximately 200,000 people and forced Japan to surrender. World War II was over. The atomic age had begun.

17.8 Nuclear Power: Using Fission to Generate Electricity

Nuclear reactions, such as fission, generate enormous amounts of energy. In a nuclear bomb, the energy is released all at once. However, the energy can also be released more slowly and used for other purposes such as electricity generation. In the United States, about 20% of electricity is generated by nuclear fission. In some other countries, as much as 70% of electricity is generated by nuclear fission. To get an idea of the amount of energy released during fission, imagine a hypothetical nuclear-powered car. Suppose the fuel for such a car was a uranium cylinder about the size of a pencil. How often would you have to refuel the car? The energy content of the uranium cylinder would be equivalent to about 1000 twenty-gallon tanks of gasoline. If you refuel your gasoline-powered car once a week, your nuclear-powered car could go 1000 weeks—almost 20 years—before refueling. Imagine a pencil-sized fuel rod lasting for twenty years of driving!

A nuclear-powered car really is hypothetical because the amount of uranium-235 needed for a critical mass is beyond the energy needs of a car.

Similarly, a nuclear-powered electrical plant can produce a lot of electricity with a small amount of fuel. Nuclear power plants generate electricity by using fission to generate heat (▶ Figure 17.11). The heat is used to boil water and create steam, which then turns the turbine on a generator to produce electricity. The fission reaction itself occurs in the nuclear core of the power plant, or *reactor*. The core consists of uranium fuel rods—enriched to about 3.5% U-235—interspersed between retractable neutron-absorbing control rods. When the control rods are fully retracted from the fuel rod assembly, the chain reaction can occur. However, when the control rods are fully inserted into the fuel assembly, they absorb the neutrons that would otherwise induce fission, shutting down the chain reaction.

By inserting or retracting the control rods, the operator can control the rate of fission. If more heat is needed, the control rods are retracted slightly. If the fission reaction begins to get too hot, the control rods are inserted a little more. In this way, the fission reaction is controlled to produce the right amount of heat needed for electricity generation. In case of a power failure, the fuel rods automatically drop into the fuel rod assembly, shutting down the fission reaction.

▲ Technicians inspect the core of a nuclear reactor, which will house the fuel rods and control rods.

A typical nuclear power plant generates enough electricity for a city of about 1 million people and uses about 50 kg of fuel per day. In contrast, a coal-burning power plant uses about 2,000,000 kg of fuel to generate the same amount of electricity. Furthermore, a nuclear power plant generates no air pollution and no greenhouse gases. A coal-burning power plant, on the other hand, emits pollutants such as carbon monoxide, nitrogen oxides, and sulfur oxides. Coal-burning power plants also emit carbon dioxide, a greenhouse gas.

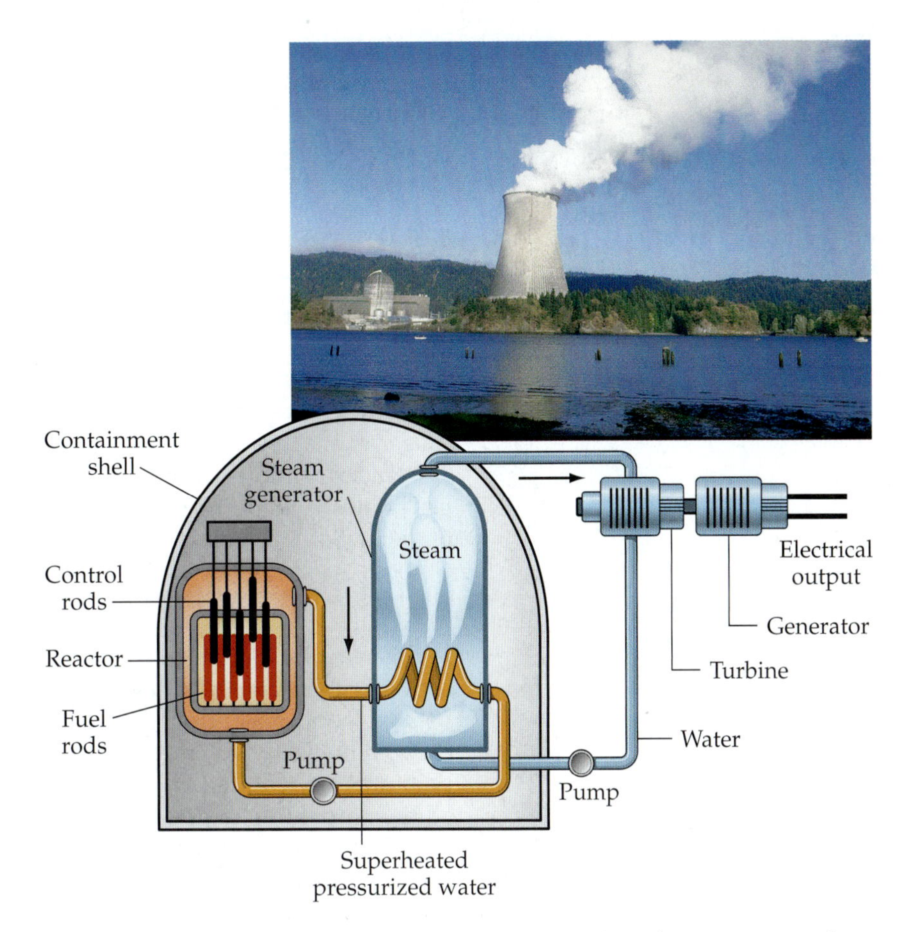

▲ **Figure 17.11 Nuclear power** In a nuclear power plant, fission generates heat that is used to boil water and create steam. The steam then turns a turbine on a generator to produce electricity. Note that the water carrying heat from the reactor core is contained within separate pipes and does not come into direct contact with the steam that drives the turbines.

Reactor cores in the United States are not made of graphite and could not burn in the way that the Chernobyl core did.

Nuclear power generation, however, is not without problems. Foremost among them is the danger of nuclear accidents. In spite of safety precautions, the fission reaction occurring in nuclear power plants can overheat. The most famous example of this occurred in Chernobyl in the former Soviet Union on April 26, 1986. Operators of the plant were performing an experiment designed to reduce maintenance costs. In order to perform the experiment, however, many of the safety features of the reactor core were disabled. The experiment failed, with disastrous results. The nuclear core, composed partly of graphite, overheated and began to burn. The accident directly caused 31 deaths and produced a fire that scattered radioactive debris into the atmosphere, making the surrounding land uninhabitable. The overall death toll from subsequent cancers is undetermined at this time.

It is important to understand, however, that a nuclear power plant *cannot* become a nuclear bomb. The uranium fuel used in electricity generation is not sufficiently enriched in U-235 to produce a nuclear detonation. It is also important to understand that U.S. nuclear power plants have additional safety features designed to prevent similar accidents. For example, U.S. nuclear power plants have large containment structures designed to contain radioactive debris in the event of an accident.

A second problem associated with nuclear power is waste disposal. Although the amount of fuel used in electricity generation is small compared to other fuels, the products of the reaction are radioactive and have very long half-lives (thousands of years or more). What do we do with this waste?

Currently, in the United States, nuclear waste is stored on site at the nuclear power plants. However, a permanent disposal site is being developed in Yucca Mountain, Nevada. The site is scheduled to be operational in 2010, but political resistance may alter current plans.

17.9 Nuclear Fusion: The Power of the Sun

As we have learned, nuclear fission is the splitting of a heavy nucleus to form two or more lighter ones. **Nuclear fusion,** in contrast, is the combination of two light nuclei to form a heavier one. Both fusion and fission emit large amounts of energy. Nuclear fusion is the energy source of stars, including our sun. In stars, hydrogen atoms fuse together to form helium atoms, emitting energy in the process.

Nuclear fusion is also the basis of modern nuclear weapons called hydrogen bombs. A modern hydrogen bomb has up to 1000 times the explosive force of the first atomic bombs. These bombs employ the following fusion reaction.

$$^{2}_{1}H + ^{3}_{1}H \longrightarrow ^{4}_{2}He + ^{1}_{0}n$$

In this reaction, deuterium (the isotope of hydrogen with one neutron) and tritium (the isotope of hydrogen with two neutrons) combine to form helium-4 and a neutron. Because fusion reactions require two positively charged nuclei (which repel each other) to fuse together, extremely high temperatures are required. In a hydrogen bomb, a small fission bomb is detonated first, providing temperatures high enough for fusion to proceed.

Nuclear fusion has been intensely investigated as a way to produce electricity. Because of the higher energy production—fusion provides about ten times more energy per gram of fuel than fission—and because the products of the reaction are less dangerous than those of fission, fusion holds promise as a future energy source. However, in spite of intense efforts, fusion electricity generation remains elusive. One of the main problems is the high temperature required for fusion to occur—no material can withstand these temperatures. After years of pouring billions of dollars into fusion research, the U.S. Congress has reduced funding for these projects. Whether fusion will ever be a viable energy source remains to be seen.

17.10 The Effects of Radiation on Life

As we have seen, radiation can ionize atoms in biological molecules, thereby initiating reactions that can alter the molecules. When radiation damages important molecules in living cells, problems can develop. The ingestion of radioactive materials—especially alpha and beta emitters—is particularly dangerous because the radioactivity is then inside the body and can do more damage. The effects of radiation can be divided into three different types: acute radiation damage, increased cancer risk, and genetic effects.

Acute Radiation Damage

Acute radiation damage results from exposures to large amounts of radiation in a short period of time. The main sources of this kind of exposure are nuclear bombs or exposed nuclear reactor cores. These high levels of radiation kill large numbers of cells. Rapidly dividing cells, such as those in the immune system and the intestinal lining, are most susceptible. Consequently, people exposed to high levels of radiation have weakened immune systems and a lowered ability to absorb nutrients from food. In milder cases, recovery is possible with time. In more extreme cases death results, often from unchecked infection.

Increased Cancer Risk

DNA and its function in the body are explained in more detail in Chapter 19.

Lower doses of radiation over extended periods of time can increase cancer risk. Radiation increases cancer risk because it can damage DNA, the molecules in cells that carry instructions for cell growth and replication. When the DNA within a cell is damaged, the cell normally dies. Occasionally, however, changes in DNA cause cells to grow abnormally and to become cancerous. These cancerous cells grow into tumors that can spread and, in some cases, cause death. Cancer risk increases with increased radiation exposure. However, cancer is so prevalent and has so many convoluted causes that it is difficult to determine an exact threshold for increased cancer risk from radiation exposure.

Genetic Defects

Another possible effect of radiation exposure is genetic defects in offspring. If radiation damages the DNA of reproductive cells—such as eggs or sperm—then the offspring that develop from those cells may have genetic abnormalities. Genetic defects of this type have been observed in laboratory animals exposed to high levels of radiation. However, such genetic defects—with a clear causal connection to radiation exposure—have yet to be observed in humans, even in studies of Hiroshima survivors.

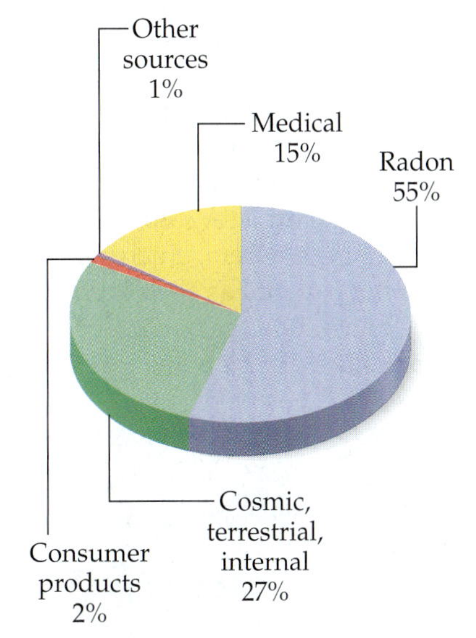

▲ **Figure 17.12 Radiation exposure by source**

Measuring Radiation Exposure

Common units of radioactivity include the *curie*, defined as 3.7×10^{10} decay events per second, and the *roentgen*, defined as the amount of radiation that produces 2.58×10^{-4} C of charge per kilogram of air. Human radiation exposure is often reported in a unit called the **rem**. The rem, which stands for *roentgen equivalent man*, is a weighted measure of radiation exposure that accounts for the ionizing power of the different types of radiation. On average, each of us is exposed to approximately one-third of a rem of radiation per year. This radiation comes primarily from natural sources, especially radon, one of the products in the uranium decay series (◀ Figure 17.12). It takes much more radiation than the natural amount to produce measurable health effects in humans. The first measurable effects, a decreased white blood cell count, occur at instantaneous exposures of approximately 20 rem (Table 17.3). Exposures of 100 rem show a definite increase in cancer risk, and exposures of more than 500 rem often result in death.

TABLE 17.3 Effects of Radiation Exposure

Dose (rem)	Probable Outcome
20–100	decreased white blood cell count; possible increase in cancer risk
100–400 rem	radiation sickness; skin lesions; increase in cancer risk
500 rem	death

17.11 Radioactivity in Medicine

Radioactivity is often perceived as dangerous; however, it is also enormously useful to physicians in the diagnosis and treatment of disease. The use of radioactivity in medicine can be broadly divided into **isotope scanning** and **radiotherapy**.

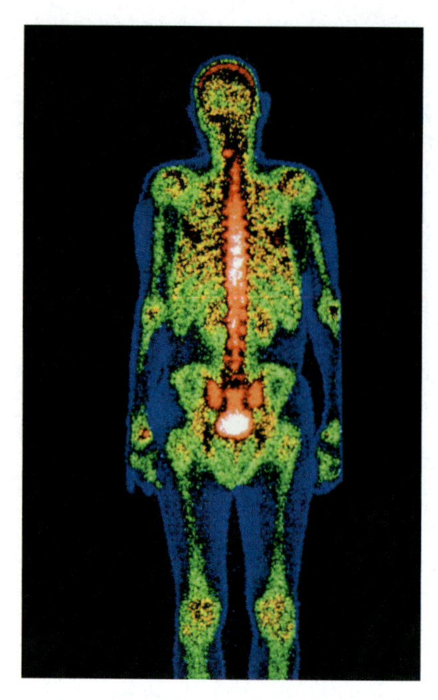

▲ **Figure 17.13 An isotope scan** Technetium-99 is often used as the radiation source for bone scans such as this one.

Isotope Scanning

An example of isotope scanning was presented in the opening section of this chapter. In isotope scanning, a radioactive isotope is introduced into the body. Then the radiation emitted by the isotope is detected using a photographic film or a scintillation counter. Since different isotopes are taken up by different organs or tissues, isotope scanning has a variety of uses. For example, the radioactive isotope phosphorus-32 is preferentially taken up by cancerous tissue. A cancer patient can be given this isotope to find and identify cancerous tumors. Other isotopes commonly used in medicine include iodine-131, used to diagnose thyroid disorders, and technetium-99, which can produce images of several different internal organs (◀ Figure 17.13).

Radiotherapy

Because radiation kills cells, and because it is particularly effective at killing rapidly dividing cells, it is often used as a therapy for cancer (cancer cells divide more quickly than normal cells). Gamma rays are focused on internal tumors to kill them (▼ Figure 17.14). The gamma ray beam is usually aimed at the tumor from a number of different angles, maximizing the exposure of the tumor while minimizing the exposure of the healthy tissue around the tumor. (See *Chemistry and Health: Radiation Treatment for Cancer* in Chapter 9.) Nonetheless, cancer patients undergoing radiation therapy usually develop the symptoms of radiation sickness, which include vomiting, skin burns, and hair loss.

Some people wonder how radiation—which is known to cause cancer—can be used to treat cancer. The answer lies in risk analysis. A cancer patient is normally exposed to radiation doses of about 100 rem. Such a dose increases cancer risk by about 1%. However, if the patient has a 100% chance of dying from the cancer that he or she already has, such a risk becomes acceptable, especially since there is a significant chance of curing the cancer.

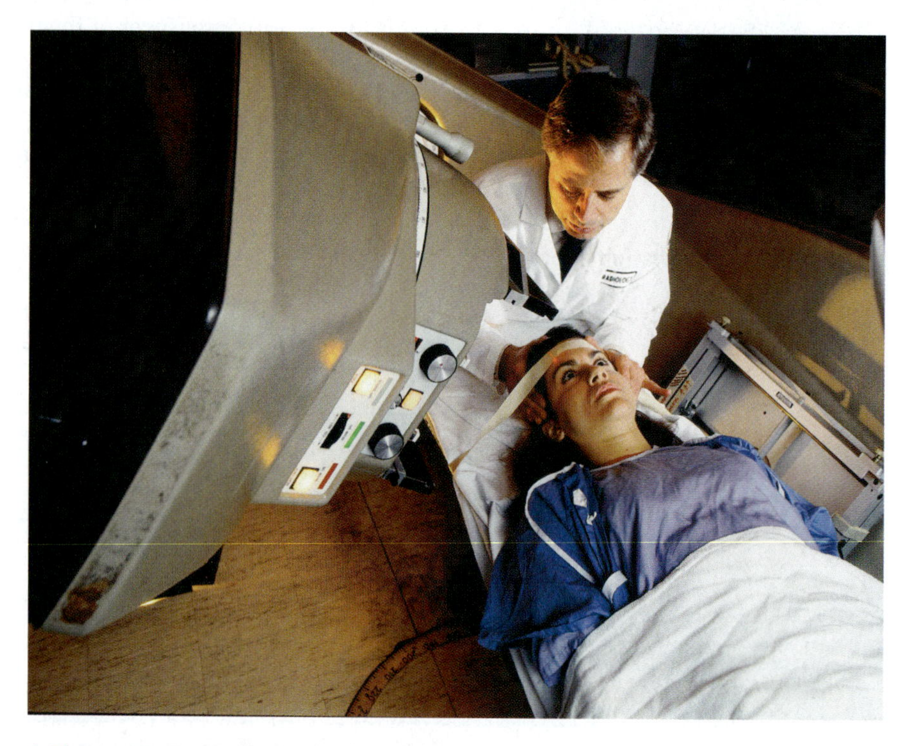

▲ **Figure 17.14 Radiotherapy for cancer** This treatment involves exposing a malignant tumor to gamma rays, typically from radioisotopes such as cobalt-60. The beam is moved in a circular pattern around the tumor to maximize exposure of the cancer cells while minimizing exposure of healthy tissues.

CHAPTER IN REVIEW

Chemical Principles / Relevance

The Nature and Discovery of Radioactivity

Chemical Principles: Radioactivity is the emission of particles from unstable atomic nuclei. It can be divided into four types.

- **Alpha particles,** composed of 2 protons and 2 neutrons and given the symbol ${}^{4}_{2}He$. Alpha particles have a high ionizing power but a low penetrating power.
- **Beta particles,** electrons emitted from atomic nuclei when a neutron changes into a proton. Beta particles have the symbol ${}^{0}_{-1}e$ and have intermediate ionizing power and intermediate penetrating power.
- **Gamma rays,** high-energy, short-wavelength photons. Gamma rays have the symbol ${}^{0}_{0}\gamma$ and have low ionizing power and high penetrating power.
- **Positrons,** emitted from atomic nuclei when protons change into neutrons. Positrons have the same mass as electrons but opposite charge, and so are represented by the symbol ${}^{0}_{+1}e$. They have intermediate ionizing power and intermediate penetrating power.

Relevance: Radioactivity is a fundamental part of the behavior of some atoms, and it also has many applications. For example, radioactivity is used to diagnose and treat diseases, including cancer, thyroid diseases, abnormal kidney and bladder function, and heart disease. Natural radioactivity is part of our environment and can be used to date ancient objects. The discovery of radioactivity led to the discovery of fission, which in turn led to the development of nuclear bombs and nuclear energy.

Detecting Radioactivity

Chemical Principles: Radioactive emissions carry a large amount of energy and are therefore easily detected. The most common and inexpensive way to detect radioactivity is with photographic film, which is used in film-badge dosimeters to monitor exposure in people working with or near radioactive sources. Radioactivity can also be detected with devices such as a Geiger-Müller counter or a scintillation counter, both of which give instantaneous readings of radiation levels.

Relevance: Since radioactivity is invisible, it must be detected using film or instruments. The detection of radioactivity is important as both a scientific tool and a practical one. Our understanding of what radioactivity is and our continuing research to understand it and its effects on living organisms require the ability to detect it. Our need for safety in areas where radioactive substances are used also requires our ability to detect radiation.

Half-Life and Radiocarbon Dating

Chemical Principles: The half-life of a radioactive nuclide is the time it takes for half of the parent nuclides in a radioactive sample to decay. The presence of radioactive carbon-14 (with a half-life of 5730 years) in the environment provides a natural clock by which to estimate the age of many artifacts and fossils. All living things contain carbon-14. When they die, the carbon-14 decays with its characteristic half-life. A measurement of the amount of carbon-14 remaining in a fossil or artifact can therefore reveal its age.

Relevance: The half-life of a radioactive nuclide determines the activity of the nuclide and how long it will be radioactive. Nuclides with short half-lives are very active (many decay events per unit time) but will not be radioactive for long. Nuclides with long half-lives are less active (fewer decays per unit time) but will be radioactive for a long time.

Fission, the Atomic Bomb, and Nuclear Power

Chemical Principles: Fission—the splitting of the atom into smaller fragments—was discovered in 1939. Fission occurs when a neutron is absorbed by a U-235 nucleus. The nucleus becomes unstable, falling apart and producing barium, krypton, neutrons, and a lot of energy.

Relevance: The discovery of fission in 1939 changed the world. Within six years, the United States developed and tested fission nuclear bombs, which ended World War II. Fission can also be used to generate electricity. The fission reaction heats water to create steam, which turns the turbine on an electrical generator. Nuclear reactors generate about 20% of the electricity in the United States and up to 70% in some other nations.

Nuclear Fusion: Nuclear fusion is the combination of two light nuclei to form a heavier one. Nuclear fusion is the energy source of stars, including our sun.

Nuclear Fusion: Modern nuclear weapons are fusion bombs with 1000 times the power of the first fission bombs. Nuclear fusion is being explored as a way to generate electricity but has not yet proven successful.

The Effects of Radiation on Life and Nuclear Medicine: Radiation can damage important molecules within living cells. Instantaneous high exposure to radiation can lead to radiation sickness and even death. Long-term lower exposure levels can increase cancer risk. However, radiation can also be used to attack cancerous tumors and to image internal organs through isotope scanning.

The Effects of Radiation on Life and Nuclear Medicine: Radiation can be used for both good and harm. We have seen how the destructive effects of radiation can be employed in a nuclear bomb. However, we have also seen how radiation can be a precise tool in the physician's arsenal against disease.

Chemical Skills

Examples

Writing Nuclear Equations for Alpha Decay (Section 17.3)

Begin with the symbol for the isotope undergoing decay on the left side of the equation and the symbol for an alpha particle on the right side. Leave a space or a question mark for the unknown daughter nuclide.

Equalize the sum of the mass numbers and the sum of the atomic numbers on both sides of the equation by writing the appropriate mass number and atomic number for the unknown daughter nuclide.

Deduce the identity of the unknown daughter nuclide from the atomic number and write its symbol.

LIVE EXAMPLE

EXAMPLE 17.6 Writing Nuclear Equations for Alpha Decay

Write a nuclear equation for the alpha decay of Po-214.

Solution:

$$^{214}_{84}\text{Po} \longrightarrow {}^{x}_{y}? + {}^{4}_{2}\text{He}$$

$$^{214}_{84}\text{Po} \longrightarrow {}^{210}_{82}? + {}^{4}_{2}\text{He}$$

$$^{214}_{84}\text{Po} \longrightarrow {}^{210}_{82}\text{Pb} + {}^{4}_{2}\text{He}$$

Writing Nuclear Equations for Beta Decay (Section 17.3)

Begin with the symbol for the isotope undergoing decay on the left side of the equation and the symbol for a beta particle on the right side.

Equalize the sum of the mass numbers and the sum of the atomic numbers on both sides of the equation by writing the appropriate mass number and atomic number for the unknown daughter nuclide.

Deduce the identity of the unknown daughter nuclide from the atomic number and write its symbol.

EXAMPLE 17.7 Writing Nuclear Equations for Beta Decay

Write a nuclear equation for the beta decay of Bi-214.

Solution:

$$^{214}_{83}\text{Bi} \longrightarrow {}^{x}_{y}? + {}^{0}_{-1}\text{e}$$

$$^{214}_{83}\text{Bi} \longrightarrow {}^{214}_{84}? + {}^{0}_{-1}\text{e}$$

$$^{214}_{83}\text{Bi} \longrightarrow {}^{214}_{84}\text{Po} + {}^{0}_{-1}\text{e}$$

Writing Nuclear Equations for Positron Decay (Section 17.3)

Begin with the symbol for the isotope undergoing decay on the left side of the equation and the symbol for a positron on the right side.

Equalize the sum of the mass numbers and the sum of the atomic numbers on both sides of the equation by writing the appropriate mass number and atomic number for the unknown daughter nuclide.

Deduce the identity of the unknown daughter nuclide from the atomic number and write its symbol.

EXAMPLE 17.8 Writing Nuclear Equations for Positron Decay

Write a nuclear equation for the positron decay of C-11.

Solution:

$$^{11}_{6}\text{C} \longrightarrow {}^{x}_{y}? + {}^{0}_{+1}\text{e}$$

$$^{11}_{6}\text{C} \longrightarrow {}^{11}_{5}? + {}^{0}_{+1}\text{e}$$

$$^{11}_{6}\text{C} \longrightarrow {}^{11}_{5}\text{B} + {}^{0}_{+1}\text{e}$$

Using Half-Life (Section 17.5)

To use half-life to determine the time it takes for a sample to decay to a specified amount or the amount of a sample left after a specified time, it is easiest to draw a table showing the amount of the nuclide as a function of number of half-lives. For each half-life, simply divide the amount of parent nuclide by 2.

EXAMPLE 17.9 Using Half-Life

Po-210 is an alpha emitter with a half-life of 138 days. How many grams of Po-210 remain after 552 days if the sample initially contained 5.80 g of Po-210?

Solution:

Po-210 (g)	Number of Half-Lives	Time in Days
5.80	0	0
2.90	1	138
1.45	2	276
0.725	3	414
0.363	4	552

The amount of Po-210 left after 552 days is 0.363 g.

Using Carbon-14 Content to Determine the Age of Fossils or Artifacts (Section 17.6)

To determine the age of an artifact or fossil based on its carbon-14 content, you can either examine Table 17.1 or build your own table beginning with 100% carbon-14 (relative to living organisms) and reducing the amount by a factor of one-half for each half-life.

EXAMPLE 17.10 Using Carbon-14 Content to Determine the Age of Fossils or Artifacts

Some wood ashes from a fire pit in the ruins of an ancient village have a carbon-14 content that is 25% of the amount found in living organisms. How old are the ashes and, by implication, the village?

Solution:

C-14(%)*	Number of Half-Lives	Time in Years
100	0	0
50.0	1	5730
25.0	2	11,460

*Percent relative to living organisms

The ashes are from wood that was living 11,460 years ago.

Key Terms

alpha (α) particle [17.3]
alpha (α) radiation [17.3]
beta (β) particle [17.3]
beta (β) radiation [17.3]
chain reaction [17.7]
critical mass [17.7]
daughter nuclide [17.3]
film-badge dosimeter [17.4]
gamma (γ) radiation [17.3]
gamma ray [17.3]
Geiger-Müller counter [17.4]
half-life [17.5]
ionizing power [17.3]
isotope scanning [17.11]
nuclear equation [17.3]
nuclear fission [17.7]
nuclear fusion [17.9]
parent nuclide [17.3]
penetrating power [17.3]
phosphorescence [17.2]
positron [17.3]
positron emission [17.3]
radioactive [17.1]
radioactivity [17.1]
radiocarbon dating [17.6]
radiotherapy [17.11]
rem [17.10]
scintillation counter [17.4]

Exercises

Questions

1. What is radioactivity? What does it mean for an atom to be radioactive?
2. How can radioactivity be used to diagnose appendicitis?
3. How was radioactivity first discovered? By whom?
4. What are uranic rays?
5. What role did Marie Sklodowska Curie play in the discovery of radioactivity?
6. How was Marie Sklodowska Curie acknowledged for her work in radioactivity?
7. Explain what is represented by each symbol in the following notation.

$^{A}_{Z}X$

8. Radioactivity originates from the _______ of radioactive atoms.
9. What is alpha radiation? What is the symbol for an alpha particle?
10. What happens to an atom when it emits an alpha particle?
11. How do the ionizing power and penetrating power of alpha particles compare to other types of radiation?
12. What is beta radiation? What is the symbol for a beta particle?
13. What happens to an atom when it emits a beta particle?
14. How do the ionizing power and penetrating power of beta particles compare to other types of radiation?
15. What is gamma radiation? What is the symbol for a gamma ray?
16. What happens to an atom when it emits a gamma ray?
17. How do the ionizing power and penetrating power of gamma particles compare to other types of radiation?
18. What is positron emission? What is the symbol for a positron?
19. What happens to an atom when it emits a positron?
20. How do the ionizing power and penetrating power of positrons compare to other types of radiation?
21. What is a nuclear equation? What does it mean for a nuclear equation to be balanced?
22. Identify the parent nuclides and daughter nuclides in the following nuclear equation. What kind of radioactive decay is involved?

$$^{231}_{91}\text{Pa} \longrightarrow {}^{227}_{89}\text{Ac} + {}^{4}_{2}\text{He}$$

23. What is a film-badge dosimeter and how does it work?
24. How does a Geiger-Müller counter detect radioactivity?
25. Explain how a scintillation counter works.
26. What are some sources of natural radioactivity?
27. Explain the concept of half-life.
28. What is a radioactive decay series?
29. What is the source of radon in our environment? Why is radon problematic?
30. What is the source of carbon-14 in our environment?
31. Why do all living organisms contain a uniform amount of carbon-14?
32. What happens to the carbon-14 in a living organism when it dies? How can this be used to establish how long ago the organism died?
33. How do we know that carbon-14 (or radiocarbon) dating is accurate? What is the age limit for which carbon-14 dating is useful?
34. Explain Fermi's experiment in which he bombarded uranium with neutrons. Include a nuclear equation in your answer.
35. What is nuclear fission? How and by whom was it discovered?
36. Why can nuclear fission be used in a bomb? Include the concept of a chain reaction in your explanation.
37. What is a critical mass?
38. What was the main goal of the Manhattan Project? Who was the project leader?
39. How can nuclear fission be used to generate electricity?
40. Explain the purpose of the control rods in a nuclear reactor core. How do they work?
41. What are the main advantages of nuclear electricity generation?
42. What are the main problems associated with nuclear electricity generation?
43. Can a nuclear reactor detonate the way a nuclear bomb can? Why or why not?
44. What is nuclear fusion?
45. Do modern nuclear weapons use fission or fusion or both? Explain.
46. Can nuclear fusion be used to generate electricity? What are the advantages of fusion over fission for electricity generation? What are the problems with fusion?
47. How does radiation affect the molecules within living organisms?
48. What is acute radiation damage to living organisms?
49. Explain how radiation can increase cancer risk.
50. Explain how radiation can cause genetic defects. Has this ever been observed in laboratory animals? In humans?
51. What is the main unit of radiation exposure? How much radiation is the average American exposed to per year?
52. Describe the outcomes of radiation exposure at different doses (in rem).
53. Explain the medical use of radioactivity that is known as isotope scanning.
54. How is radioactivity used to treat cancer?

Problems

Isotopic and Nuclear Particle Symbols

55. Provide a symbol for the isotope of lead that contains 125 neutrons.
56. Provide a symbol for the isotope of bismuth that contains 128 neutrons.
57. How many protons and neutrons are in the following nuclide?

$^{207}_{81}\text{Tl}$

58. How many protons and neutrons are in the following nuclide?

$^{219}_{86}\text{Rn}$

59. Identify each of the following as an alpha particle, a beta particle, a gamma ray, a positron, a neutron, or a proton.
(a) ${}^{0}_{-1}\text{e}$ (b) ${}^{1}_{0}\text{n}$ (c) ${}^{0}_{0}\gamma$

60. Identify each of the following as an alpha particle, a beta particle, a gamma ray, a positron, a neutron, or a proton.
(a) ${}^{1}_{1}\text{p}$ (b) ${}^{4}_{2}\text{He}$ (c) ${}^{0}_{+1}\text{e}$

61. Complete the following table:

Chemical Symbol	*Atomic Number (Z)*	*Mass Number (A)*	*# Protons*	*# Neutrons*
Tc	____	95	____	____
____	56	128	____	____
Eu	____	____	____	82
Fr	____	____	____	136

62. Complete the following table:

Chemical Symbol	*Atomic Number (Z)*	*Mass Number (A)*	*# Protons*	*# Neutrons*
Pd	46	____	____	54
Ce	____	136	____	____
____	84	208	____	____
____	____	____	88	138

Radioactive Decay

63. Write a nuclear equation for the alpha decay of each of the following nuclides.
(a) U-234
(b) Th-230
(c) Ra-226
(d) Rn-222

64. Write a nuclear equation for the alpha decay of each of the following nuclides.
(a) Po-218
(b) Po-214
(c) Po-210
(d) Th-227

65. Write a nuclear equation for the beta decay of each of the following nuclides.
(a) Pb-214
(b) Bi-214
(c) Th-231
(d) Ac-227

66. Write a nuclear equation for the beta decay of each of the following nuclides.
(a) Pb-211
(b) Tl-207
(c) Th-234
(d) Pa-234

67. Write a nuclear equation for positron emission by each of the following nuclides.
(a) C-11 (b) N-13 (c) O-15

68. Write a nuclear equation for positron emission by each of the following nuclides.
(a) Co-55 (b) Na-22 (c) F-18

69. Fill in the blanks in the following partial decay series.

$${}^{241}_{94}\text{Pu} \longrightarrow {}^{241}_{95}\text{Am} + \underline{\qquad}$$
$${}^{241}_{95}\text{Am} \longrightarrow {}^{237}_{93}\text{Np} + \underline{\qquad}$$
$${}^{237}_{93}\text{Np} \longrightarrow \underline{\qquad} + {}^{4}_{2}\text{He}$$
$$\underline{\qquad} \longrightarrow {}^{233}_{92}\text{U} + {}^{0}_{-1}\text{e}$$

70. Fill in the blanks in the following partial decay series.

$${}^{225}_{88}\text{Ra} \longrightarrow {}^{225}_{89}\text{Ac} + \underline{\qquad}$$
$${}^{225}_{89}\text{Ac} \longrightarrow \underline{\qquad} + {}^{4}_{2}\text{He}$$
$$\underline{\qquad} \longrightarrow {}^{217}_{85}\text{At} + {}^{4}_{2}\text{He}$$
$${}^{217}_{85}\text{At} \longrightarrow \underline{\qquad} + {}^{4}_{2}\text{He}$$

71. Write a partial decay series for Th-232 undergoing the following sequential decays: α, β, β, α.

72. Write a partial decay series for Rn-220 undergoing the following sequential decays: α, α, β, α.

Half-Life

73. Suppose you a have a 1000-atom sample of a radioactive nuclide that decays with a half-life of 2.0 days. How many radioactive atoms are left after 10 days?

74. Iodine-131 is often used in nuclear medicine to obtain images of the thyroid. If you start with 1.0×10^9 I-131 atoms, how many are left after approximately 1 month? I-131 has a half-life of 8.0 days.

75. A patient is given 0.050 mg of technetium-99m, a radioactive isotope with a half-life of about 6.0 hours. How long until the radioactive isotope decays to 6.3×10^{-3} mg?

76. Radium-223 decays with a half-life of 11.4 days. How long will it take for a 0.240-mol sample of radium to decay to 1.50×10^{-2} mol?

77. One of the nuclides in spent nuclear fuel is U-234, an alpha emitter with a half-life of 2.44×10^5 years. If a spent fuel assembly contains 2.80 kg of U-234, how long will it take for the amount of U-234 to decay to less than 0.10 kg?

78. One of the nuclides in spent nuclear fuel is U-235, an alpha emitter with a half-life of 703 million years. How long would it take for the amount of U-235 to reach one-eighth of its initial amount?

79. A radioactive sample contains 1.55 g of an isotope with a half-life of 3.8 days. How much of the isotope in grams will remain after 11.4 days?

80. A 58-mg sample of a radioactive nuclide is administered to a patient to obtain an image of her thyroid. If the nuclide has a half-life of 12 hours, how much of the nuclide remains in the patient after 4.0 days?

81. Each of the following nuclides is used in nuclear medicine. List them in order of most active (most number of decay events per second) to least active (least number of decay events per second).

Nuclide	*Half-Life*
P-32	14.3 days
Cr-51	27.7 days
Ga-67	78.3 hours
Sr-89	50.5 days

82. Each of the following nuclides is used in nuclear medicine. List them in order of most active (most number of decay events per second) to least active (least number of decay events per second).

Nuclide	*Half-Life*
Y-90	64.1 hours
Tc-99m	6.02 hours
In-111	2.8 days
I-131	8.0 days

Radiocarbon Dating

83. A wooden boat discovered just south of the Great Pyramid in Egypt had a carbon-14 content of approximately 50% of that found in living organisms. How old is the boat?

84. A layer of peat beneath the glacial sediments of the last ice age had a carbon-14 content of 25% of that found in living organisms. How long ago was this ice age?

85. An ancient skull has a carbon-14 content of 1.563% of that found in living organisms. How old is the skull?

86. A mammoth skeleton has a carbon-14 content of 12.50% of that found in living organisms. When did the mammoth live?

Fission and Fusion

87. Write a nuclear reaction for the neutron-induced fission of U-235 to form Xe-144 and Sr-90. How many neutrons are produced in the reaction?

88. Write a nuclear reaction for the neutron-induced fission of U-235 to produce Te-137 and Zr-97. How many neutrons are produced in the reaction?

89. Write a nuclear equation for the fusion of two H-2 atoms to form He-3 and one neutron.

90. Write a nuclear equation for the fusion of H-3 with H-1 to form He-4.

Cumulative Problems

91. Complete each of the following nuclear equations:

(a) ${}^{1}_{1}\text{p} + {}^{9}_{4}\text{Be} \longrightarrow$ ______ $+ {}^{4}_{2}\text{He}$

(b) ${}^{209}_{83}\text{Bi} +$ ______ $\longrightarrow {}^{272}_{111}\text{Uuu} + {}^{1}_{0}\text{n}$

(c) ${}^{179}_{74}\text{W} + {}^{0}_{-1}\text{e} \longrightarrow$ ______

92. Complete each of the following nuclear equations:

(a) ${}^{27}_{13}\text{Al} + {}^{4}_{2}\text{He} \longrightarrow$ ______ $+ {}^{1}_{0}\text{n}$

(b) ______ $+ {}^{1}_{0}\text{n} \longrightarrow {}^{29}_{14}\text{Si} + {}^{4}_{2}\text{He}$

(c) ${}^{241}_{95}\text{Am} \longrightarrow {}^{237}_{93}\text{Np} +$ ______

93. A breeder nuclear reactor is a reactor in which nonfissile U-238 is converted into fissile Pu-239. The process involves bombardment of U-238 by neutrons to form U-239 which then undergoes two sequential beta decays. Write nuclear equations to represent this process.

94. Write a series of nuclear equations in which Al-27 reacts with a neutron and the product undergoes an alpha decay followed by a beta decay.

95. The fission of U-235 produces 3.2×10^{-11} J/atom. How much energy does it produce per mole of U-235? Per kilogram of U-235?

96. The fusion of deuterium and tritium produces 2.8×10^{-12} J for every atom of deuterium and atom of tritium. How much energy is produced per mole of deuterium and mole of tritium?

97. Bi-210 is a beta emitter with a half-life of 5.0 days. If a sample contains 1.2 g of Bi-210, how many beta emissions would occur in 5.0 days?

98. Po-218 is an alpha emitter with a half-life of 3.0 minutes. If a sample contains 55 mg of Po-218, how many alpha emissions would occur in 6.0 minutes?

99. If a person living in a high-radon area is exposed to 0.400 rem of radiation from radon per year, and his total exposure is 0.585 rem, what percentage of his total exposure is due to radon?

100. An X-ray technician is exposed to 0.020 rem of radiation at work. If her total exposure is the national average (0.36 rem), what fraction of her exposure is due to on-the-job exposure?

Highlight Problems

101. Closely examine the following diagram representing the alpha decay of sodium-20 and draw in the missing nucleus.

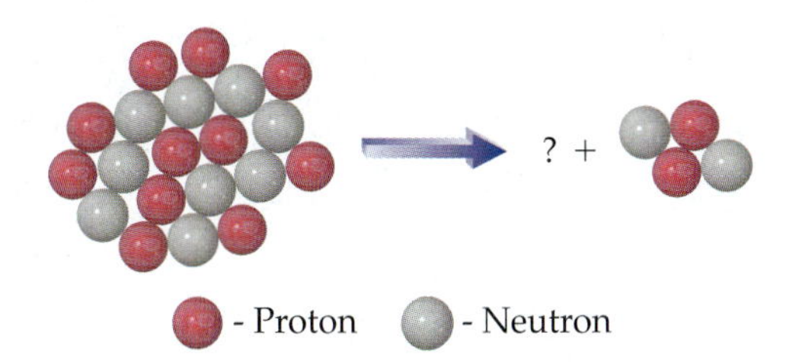

102. Closely examine the following diagram representing the beta decay of fluorine-21 and draw in the missing nucleus.

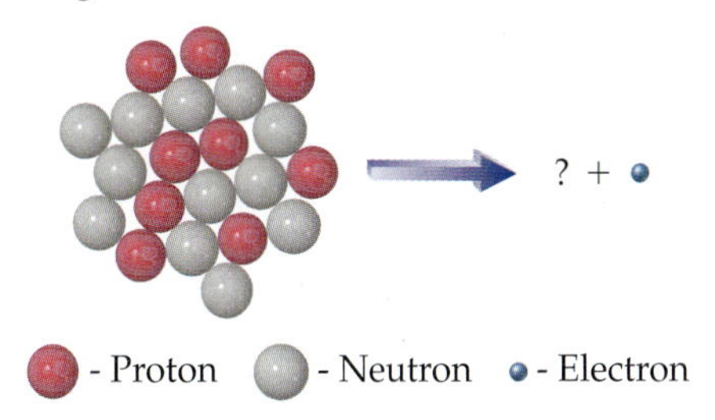

103. Closely examine the following diagram representing the positron emission of carbon-10 and draw the missing nucleus.

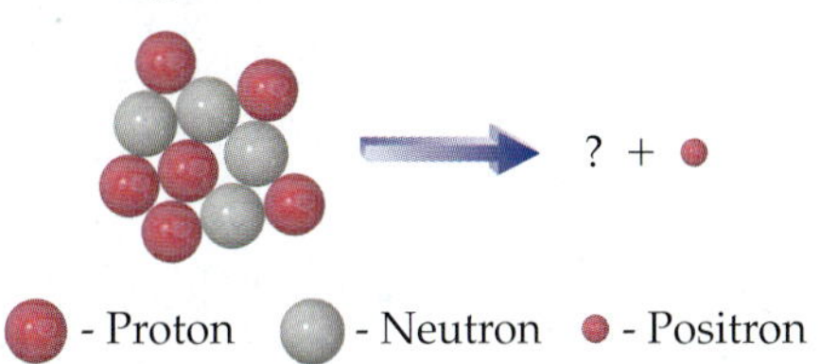

104. A radiometric dating technique uses the decay of U-238 to Pb-206 (the half-life for this process is 4.5 billion years) to determine the age of the oldest rocks on Earth and by implication the age of Earth itself. The oldest uranium-containing rocks on Earth contain approximately equal numbers of uranium atoms and lead atoms. Assuming the rocks were pure uranium when they were formed, how old are the rocks?

Answers to Skillbuilder Exercises

Skillbuilder 17.1 $^{216}_{84}\text{Po} \longrightarrow ^{212}_{82}\text{Pb} + ^{4}_{2}\text{He}$

Skillbuilder 17.2 $^{228}_{89}\text{Ac} \longrightarrow ^{228}_{90}\text{Th} + ^{0}_{-1}\text{e}$

Skillbuilder Plus, p. 601 $^{235}_{92}\text{U} \longrightarrow ^{231}_{90}\text{Th} + ^{4}_{2}\text{He}$; $^{231}_{90}\text{Th} \longrightarrow ^{231}_{91}\text{Pa} + ^{0}_{-1}\text{e}$; $^{231}_{91}\text{Pa} \longrightarrow ^{227}_{89}\text{Ac} + ^{4}_{2}\text{He}$

Skillbuilder 17.3 $^{22}_{11}\text{Na} \longrightarrow ^{22}_{10}\text{Ne} + ^{0}_{+1}\text{e}$

Skillbuilder 17.4 7.00×10^{-3} mol

Skillbuilder 17.5 No, the carbon-14 content suggests that the scroll is from very recent times

Answers to Conceptual Checkpoints

17.1 (a) In alpha decay, the nucleus loses a helium nucleus (2 protons and 2 neutrons), reducing its mass number by 4. The other forms of decay listed involve electrons, which have negligible mass compared to that of nuclear particles, or gamma ray photons, which have no mass.

17.2 (a) If you divide 1,000,000 by 2, then divide the remainder by 2, and repeat this process eight more times, you are left with approximately 977 atoms.

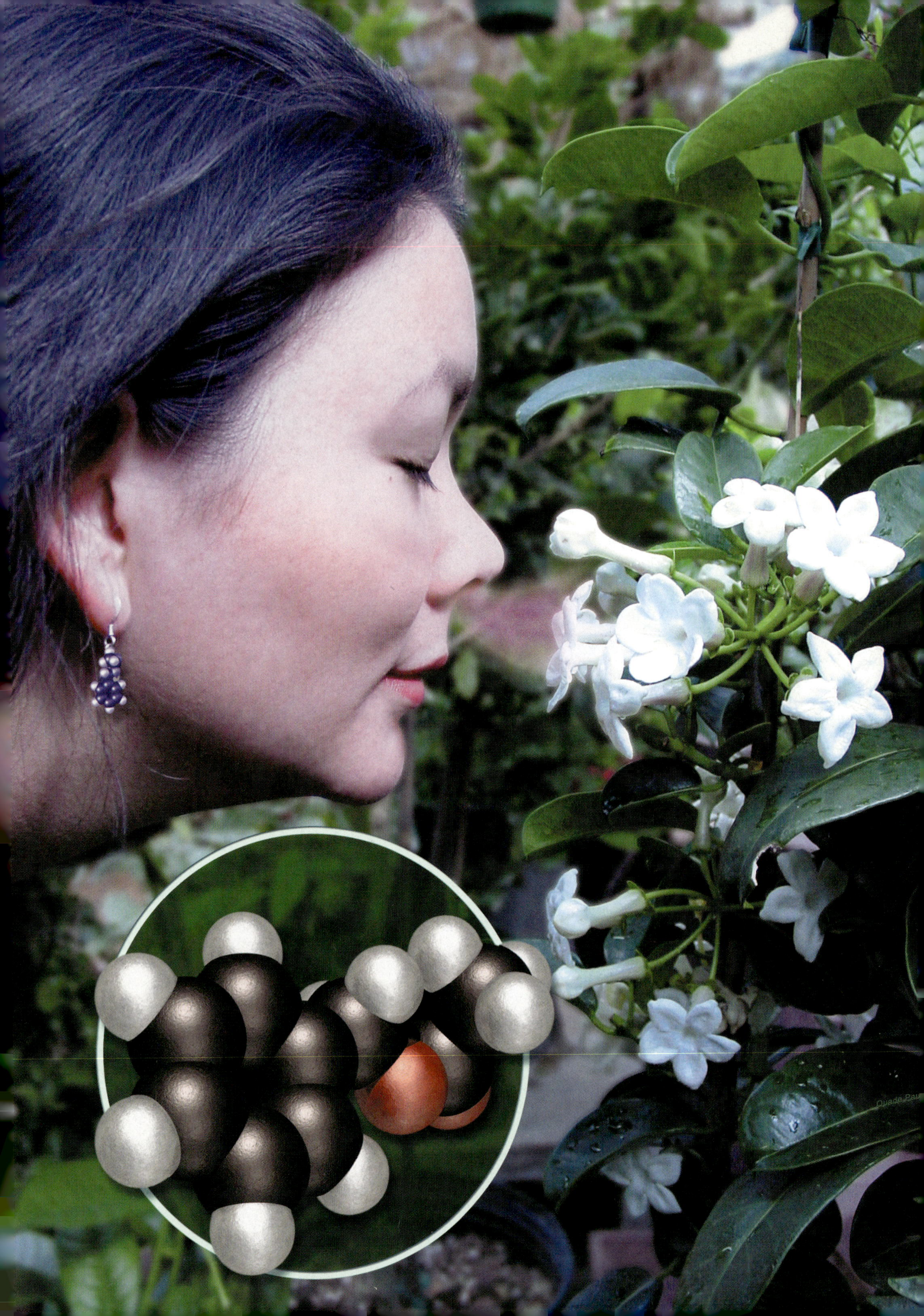

CHAPTER 18

Organic Chemistry

"The atoms come together in different order and position, like the letters, which, though they are few, yet, by being placed together in different ways, produce innumerable words."

Epicurus (341–270 B.C.)

18.1 What Do I Smell?

Perfume companies spend millions of dollars trying to produce the most seductive scents. What causes scent? The answer, of course, is molecules. Certain molecules, when they are inhaled, bind with molecular receptors (called olfactory receptors) in our noses. This interaction sends a nerve signal to the brain that we experience as a smell. Some smells, such as that of a flower, are pleasant. Other smells, such as that of rotten fish, are unpleasant.

What are the molecules that cause smell? Many molecules have no scent at all. Nitrogen, oxygen, water, and carbon dioxide molecules, for example, are constantly passing through our noses, yet they produce no smell. Most of the smells that we experience are caused by **organic molecules**, molecules containing carbon combined with several other elements including hydrogen, nitrogen, oxygen, and sulfur. For example, carbon-containing molecules are responsible for the smells of roses, vanilla, cinnamon, almond, jasmine, body odor, and rotting fish.

The smells of substances are not *always* a reliable guide to what is good to eat.

When you sprinkle cinnamon onto your French toast, some cinnamaldehyde—an organic compound present in cinnamon—evaporates into the air. You inhale some of the cinnamaldehyde molecules and experience the unique smell of cinnamon. When you walk past a rotting fish on a beach, you inhale some triethylamine—an organic compound emitted by the decaying fish—and experience that unique and unpleasant smell.

◀ The sweet smell of jasmine is produced by benzyl acetate, an organic compound. When you smell jasmine, benzyl acetate molecules emitted from the flower bind with molecular receptors in your nose, triggering a nerve signal to your brain that you interpret as a sweet smell.

Our reaction to certain smells, positive or negative, is probably an evolutionary adaptation. The pleasant smell of cinnamon tells you that it is good to eat. The unpleasant smell of rotting fish tells you that it has become spoiled and that you should avoid it.

$CH_3-CH_2-N(CH_2CH_3)-CH_2-CH_3$

Triethylamine

▲ Carbon-containing molecules—especially triethylamine—are responsible for the smell of dead fish.

The study of carbon-containing compounds and their reactions is called **organic chemistry**. Besides composing many of the things we smell, organic compounds are prevalent in foods, drugs, petroleum products, and pesticides. Organic chemistry is also the basis for living organisms. Life has evolved based on carbon-containing compounds, making organic chemistry of utmost importance to any person interested in understanding living organisms.

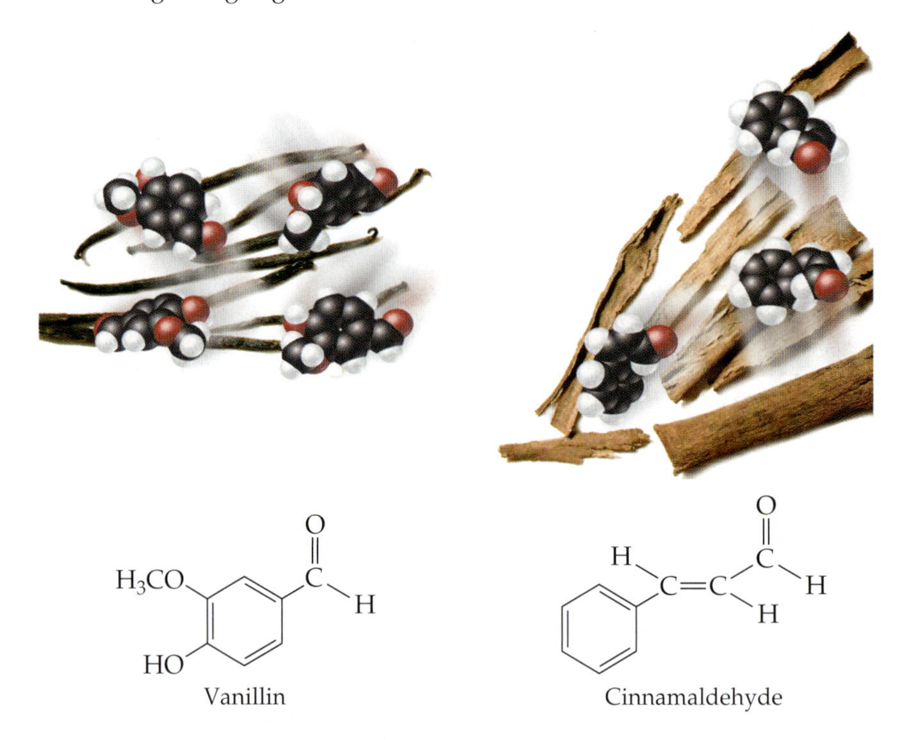

▲ Carbon-containing molecules are responsible for the smell of vanilla beans (vanillin) and cinnamon sticks (cinnamaldehyde).

Organic Inorganic

▶ Sugar, obtained from sugar cane or sugar beets, is an example of an organic compound. Salt, obtained from a salt mine or from the ocean, is an example of an inorganic compound. **Question:** What are the main differences between organic and inorganic compounds?

18.2 Vitalism: The Difference between Organic and Inorganic

By the end of the eighteenth century, chemists had learned that compounds could be broadly divided into two categories: organic and inorganic. Organic compounds came from living things. Sugar—obtained from sugar cane or the sugar beet—is a common example of an organic compound. Inorganic compounds, on the other hand, came from Earth. Salt—mined from the ground or extracted from the ocean—is a common example of an inorganic compound.

Not only were organic and inorganic compounds different in their origin, they were also different in their properties. Organic compounds were easily decomposed. Sugar, for example, easily decomposes into carbon and water when heated. Think of the last time you burned sugar—the apple-pie filling that dripped in the oven or the sweet potatoes that stayed in the pot a little too long. Inorganic compounds, however, were more difficult to decompose. Salt must be heated to very high temperatures before it decomposes. Even more curious to these early chemists was their inability to synthesize a single organic compound in the laboratory. Many inorganic compounds could be easily synthesized, but organic compounds could not.

The origin and properties of organic compounds led early chemists to postulate that organic compounds were unique to living organisms. They postulated that living organisms employed a **vital force**—a mystical or supernatural power—that allowed them to produce organic compounds. They thought that it was impossible to produce an organic compound outside of a living organism because the vital force was not present. This belief—which became known as **vitalism**—explained why no chemist had succeeded in synthesizing an organic compound in the laboratory.

Vitalism is the belief that living things contain a nonphysical "force" that allows them to synthesize organic compounds.

An experiment performed in 1828 by a German chemist named Friedrich Wöhler (1800–1882) proved vitalism wrong. Wöhler heated ammonium cyanate (an inorganic compound) and formed urea (an organic compound).

$$\underset{\text{Ammonium cyanate}}{NH_4OCN} \xrightarrow{\text{Heat}} \underset{\text{Urea}}{H_2NCONH_2}$$

Urea was a known organic compound that had previously been isolated only from urine. Although it was not realized at the time, Wöhler's simple experiment was a key step in opening all of life to scientific investigation. He showed that the compounds that composed living organisms—like all compounds—followed scientific laws and could be studied and understood. Today, known organic compounds number in the millions, and organic chemistry is a vast field that produces substances as diverse as drugs, petroleum, and plastics.

The Origin of Life

The demise of vitalism opened life itself—including its origin—to chemical inquiry. If organic compounds could be made in the laboratory, and if living things were composed of organic compounds, would it be possible to make life in the laboratory? Would it be possible to simulate how life started on Earth?

In 1953, a young scientist named Stanley Miller, working with Harold C. Urey at the University of Chicago, performed an experiment in an attempt to answer this question. Miller recreated the environment of primordial Earth in a flask containing water and certain gases including methane, ammonia, and hydrogen—all believed, at the time, to be components of the early atmosphere. He passed an electrical current through the system to simulate lightning. After several days, Miller analyzed the contents of the flask. What he found made headlines. Not only did the flask contain organic compounds, it contained organic compounds central to life—amino acids. Amino acids, as we will learn in Chapter 19, are the builiding blocks of biological proteins. Apparently, the foundational compounds of life could be synthesized rather simply under the conditions of early Earth.

▲ An artist's conception of the early Earth. It was this environment that Miller's experiment was designed to simulate.

Inspired by Miller's results, a number of other scientists set out to understand, and perhaps recreate, life's origin. Some believed that the creation of life in the laboratory was imminent. It has not turned out that way. More than fifty years have passed since Miller's seminal experiment, yet we are still struggling to understand how life began. In a recent quote, Stanley Miller, now a professor of chemistry at the University of California at San Diego, said, "The problem of the origin of life has turned out to be much more difficult than I and most people envisioned."

Most scientists investigating the origin of life have a basic hypothesis of how life may have started. A group of molecules developed the ability to copy themselves, but not quite perfectly—some of the copies contained inheritable mistakes. In a few cases, these alterations allowed the molecular "offspring" to replicate even more efficiently. Chemical evolution got its start, producing generation after generation of molecules that slowly got better at copying themselves as they assembled into more complex structures. This process eventually produced a living cell, which is (among other things) a very efficient self-replicating machine.

Although this basic hypothesis is still widely accepted, the details are far from clear. What were these early molecules? How did they form? How did they replicate? Earlier origin-of-life theories proposed that the replicating molecules were primitive forms of the molecules that exist in living organisms today—proteins, RNA, and DNA (see Chapter 19). However, the complexity of these molecules and their inability to replicate independently of one another have caused some researchers to look at other materials that may have been involved in the process, such as clays and sulfur-based compounds. In spite of continuing efforts, no single theory has gained widespread acceptance, and the origin of life continues to be an ongoing area of research.

CAN YOU ANSWER THIS? *Why do you think that a belief in vitalism might inhibit research into the origin of life?*

18.3 Carbon: A Versatile Atom

Why did life evolve based on the chemistry of carbon? Why isn't life based on some other element? The answer may not be simple, but we know that life—in order to exist—must have complexity. It is also clear that carbon chemistry is complex. The number of compounds containing carbon is greater than the number of compounds of all the rest of the elements in the periodic table combined. The reason for this is twofold. First, carbon—with its four valence

Lewis structures were first covered in Chapter 10.

electrons—can form four covalent bonds. Recall the Lewis structures of carbon and some carbon compounds.

As you learn to draw structures for organic compounds, remember that carbon always forms four bonds.

Second, carbon, more than any other element, can bond to itself to form chain, branched, and ring structures.

Propane

Isobutane

Cyclohexane

This versatility allows carbon to be the backbone of millions of different chemical compounds—just what is needed for life to exist.

Molecular geometries and VSEPR theory were covered in Chapter 10.

When carbon forms four single bonds, there are four electron groups around it and VSEPR theory predicts a tetrahedral geometry.

Tetrahedral geometry

When carbon forms a double bond and two single bonds, there are three electron groups around each carbon atom and VSEPR theory predicts a trigonal planar geometry.

Trigonal planar geometry

When carbon forms a triple bond and a single bond, there are two electron groups around each carbon atom, resulting in a linear geometry.

H—C≡C—H

Linear geometry

18.4 Hydrocarbons: Compounds Containing Only Carbon and Hydrogen

Hydrocarbons—compounds that contain only carbon and hydrogen—are the simplest organic compounds. However, because carbon atoms are so versatile, many different kinds of hydrocarbons exist. Nature bonds carbon and hydrogen atoms together in different numbers and in different ways to form hundreds of thousands of different compounds. Hydrocarbons are commonly used as fuels. Candle wax, oil, gasoline, liquid propane (LP) gas, and natural gas are all composed of hydrocarbons. Hydrocarbons are also the starting materials in the synthesis of many different consumer products including fabrics, soaps, dyes, cosmetics, drugs, plastic, and rubber.

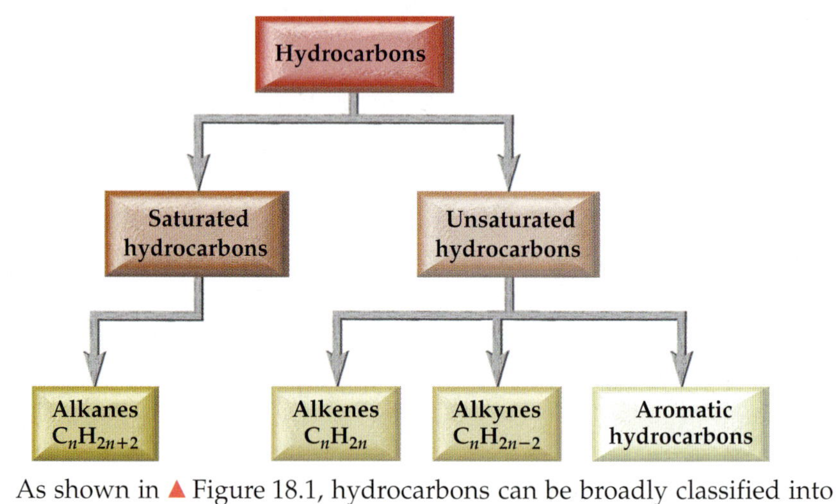

▶ **Figure 18.1 A flow chart for the classification of alkanes** These formulas apply only to open-chain (noncyclical) hydrocarbons.

The difference between saturated and unsaturated hydrocarbons is explained in Sections 18.5 and 18.8.

As shown in ▲ Figure 18.1, hydrocarbons can be broadly classified into four different types: alkanes (which are also called saturated hydrocarbons); and alkenes, alkynes, and aromatic hydrocarbons (all of which are called unsaturated hydrocarbons). Alkanes, alkenes, and alkynes can be differentiated based on their molecular formulas.

These formulas apply only to open-chain (noncyclical) hydrocarbons.

Alkanes C_nH_{2n+2}

Alkenes C_nH_{2n}

Alkynes C_nH_{2n-2}

EXAMPLE 18.1 Differentiating between Alkanes, Alkenes, and Alkynes Based on Their Molecular Formulas

Based on the molecular formula, determine whether the following noncyclical hydrocarbons are alkanes, alkenes, or alkynes.

(a) C_7H_{14}
(b) $C_{10}H_{22}$
(c) C_3H_4

Solution:

(a) C_7H_{14}

The number of carbons is 7; therefore $n = 7$. If $n = 7$, then 14 is $2n$. The molecule must be an alkene.

(b) $C_{10}H_{22}$

The number of carbons is 10; therefore $n = 10$. If $n = 10$, then 22 is $2n + 2$. The molecule must be an alkane.

(c) C_3H_4

The number of carbons is 3; therefore $n = 3$. If $n = 3$, then 4 is $2n - 2$. The molecule must be an alkyne.

SKILLBUILDER 18.1 Differentiating between Alkanes, Alkenes, and Alkynes Based on Their Molecular Formulas

Based on the molecular formula, determine whether the following noncyclical alkanes are alkanes, alkenes, or alkynes.

(a) C_6H_{12}
(b) C_8H_{14}
(c) C_5H_{12}

Environmental Problems Associated with Hydrocarbon Combustion

Hydrocarbon fuels are also called **fossil fuels** because they originate from plant and animal life that existed on Earth in prehistoric times. The main types of fossil fuels are natural gas, petroleum, and coal. Fossil fuels are a convenient form of energy because they are relatively cheap, can be easily transported, and burn easily to release large amounts of energy. However, fossil fuels also have several problems associated with their use, including limited supply, smog, acid rain, and global warming.

▲ Fossil fuels such as gasoline are convenient forms of energy because they are relatively cheap, are easily transported, and release large amounts of energy when burned.

One of the problems with fossil fuels is that they will not last forever. At current rates of consumption, oil and natural gas supplies will be depleted in 50 to 100 years. While there is enough coal to last much longer, it is a dirtier fuel and is less convenient than petroleum and natural gas because it is a solid.

A second problem associated with fossil fuel combustion is smog. Smog is a result of the fossil fuel combustion products that are emitted into the air. As we saw in Section 14.11, these include nitrogen oxides (NO and NO_2), sulfur oxides (SO_2 and SO_3), ozone (O_3), and carbon monoxide (CO). These substances make the air above cities brown and dirty. They also irritate the eyes and lungs and put stress on the heart and lungs. Because of good legislation and catalytic converters, however, the level of these pollutants over most cities is going down. Even so, in many cities, the levels still exceed what the Environmental Protection Agency (EPA) considers safe.

A third problem associated with fossil fuel combustion is acid rain. The nitrogen oxides and sulfur oxides emitted into air make rain acidic. This acidic rain then falls into lakes and streams and makes them acidic as well (see Section 14.11). Some species of aquatic life cannot tolerate the increased acidity and die. Acid rain also affects forests and building materials. Again, good legislation—specifically the Clean Air Amendments of 1990—have addressed acid rain, and sulfur oxide emissions have been decreasing in the United States over the past 10 years. The positive effects of these reductions are expected in lakes and streams in the coming years.

A fourth problem associated with fossil fuel use is global warming, which we discussed in Section 8.1. One of the main products of fossil fuel combustion is carbon dioxide (CO_2). Carbon dioxide is a greenhouse gas, meaning that it allows visible light from the sun to enter Earth's atmosphere, but prevents heat (in the form of infrared light) from escaping. The result is that carbon dioxide acts as a blanket, keeping Earth warm. However, because of fossil fuel combustion, carbon dioxide levels in our atmosphere have been steadily increasing. This increase is expected to raise Earth's average temperature. Current observations suggest that Earth has already warmed by about 0.6 °C in the last century due to an increase in atmospheric carbon dioxide of about 20 percent. Computer models suggest that the warming could worsen if carbon dioxide emissions are not curbed.

CAN YOU ANSWER THIS? *Write a balanced equation for the combustion of octane, a component of gasoline. How many moles of carbon dioxide (the main greenhouse gas) are produced for every mole of octane (C_8H_{18}) burned? How many kilograms of carbon dioxide are produced for every kilogram of octane burned?*

18.5 Alkanes: Saturated Hydrocarbons

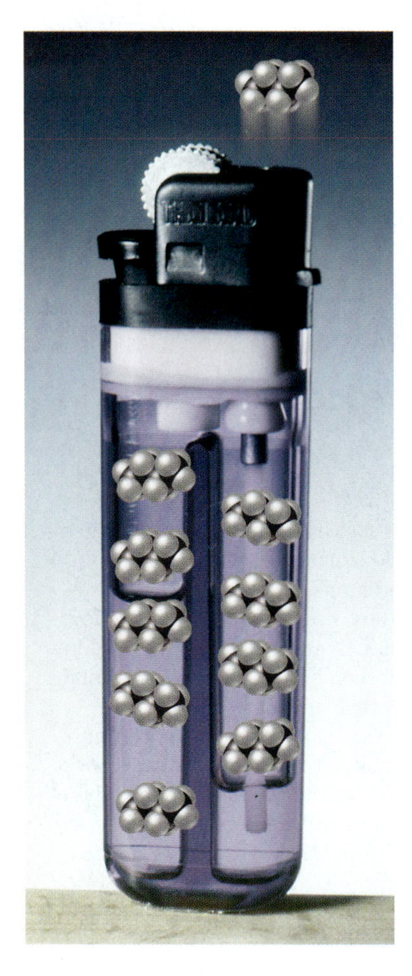

▲ Butane is the main component of lighter fluid.

Alkanes are hydrocarbons containing only single bonds. Alkanes are also called **saturated hydrocarbons** because they are saturated (loaded to capacity) with hydrogen. The simplest hydrocarbon is methane, CH_4, the main component of natural gas.

Methane CH_4

```
    H
    |
H — C — H
    |
    H
```

Formula — Structural formula — Space-filling model

The middle formula above is a **structural formula**, a formula that shows, not only the number and type of each atom in a molecule, but the structure as well. Structural formulas are not three-dimensional representations of the molecule—as space-filling models are—but rather two-dimensional representations that show which atoms are bonded together.

The next simplest hydrocarbon is ethane, C_2H_6. To draw the structural formula of ethane, we remove a hydrogen atom from methane and replace it with a methyl ($-CH_3$) group.

Ethane C_2H_6

```
    H   H
    |   |
H — C — C — H
    |   |
    H   H
```

Formula — Structural formula — Space-filling model

Ethane is a minority component of natural gas.

After ethane, the next simplest hydrocarbon is propane, C_3H_8, the main component of LP (liquid propane) gas.

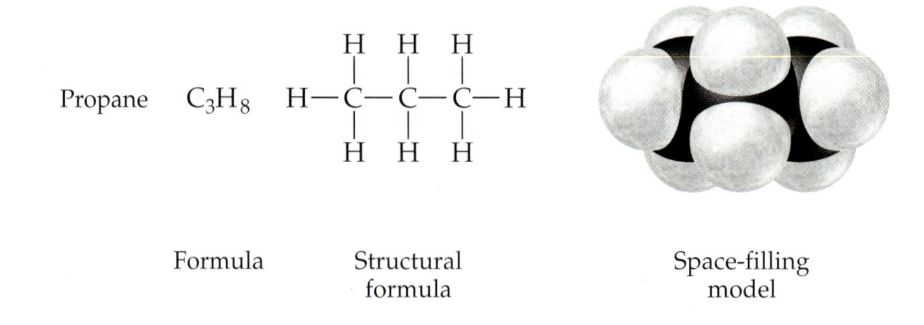

For many organic compounds, it is often useful to write **condensed structural formulas**. A condensed structural formula is a shorthand way to write a structural formula. For example, the condensed structural formula for propane is:

$$CH_3CH_2CH_3$$

This *does not* mean C—H—H—H—C—H—H—C—H—H—H. Such a structure would be absurd because we know that carbon atoms must form four bonds and hydrogen atoms can form only one bond. Rather, the condensed structural formula is simply a shorter way to write the true structural formula of propane shown previously.

Next in the series is butane, C_4H_{10}, the main component in lighter fluid.

Butane C_4H_{10}

$$\begin{array}{ccccccccc} & & H & & H & & H & & H \\ & & | & & | & & | & & | \\ H & - & C & - & C & - & C & - & C & - & H \\ & & | & & | & & | & & | \\ & & H & & H & & H & & H \end{array}$$

Formula — Structural formula — Space-filling model

Alkanes composed of carbon atoms bonded in a straight chain without any branching—like the ones we have just seen—are called **normal alkanes** or ***n*-alkanes**. The *n*-alkanes with three or more carbon atoms have the following general structure:

m = number of —CH_2— groups.

$$CH_3(CH_2)_mCH_3$$

Condensed structural formula

$$\begin{array}{ccccccccc} & & H & & H & & H \\ & & | & & | & & | \\ H & - & C & - & \left[C \right]_m & - & C & - & H \\ & & | & & | & & | \\ & & H & & H & & H \end{array}$$

Structural formula

Alkane	Boiling Point (°C)
methane	−161.5
ethane	−88.6
propane	−42.1
butane	−0.5
pentane	36.0
hexane	68.7
heptane	98.5
octane	125.6

As the number of carbon atoms increases in *n*-alkanes, so does their boiling point. Methane, ethane, propane, and butane are all gases at room temperature, but the next *n*-alkane in the series, pentane, is a liquid at room temperature.

Pentane C_5H_{12}

$$\begin{array}{ccccccccccc} & & H & & H & & H & & H & & H \\ & & | & & | & & | & & | & & | \\ H & - & C & - & C & - & C & - & C & - & C & - & H \\ & & | & & | & & | & & | & & | \\ & & H & & H & & H & & H & & H \end{array}$$

Formula — Structural formula — Space-filling model

Pentane is a component of gasoline. Table 18.1 summarizes the *n*-alkanes through decane, which contains ten carbon atoms. Like pentane, hexane through decane are all components of gasoline. Table 18.2 summarizes the many uses of hydrocarbons.

TABLE 18.1 Alkanes

<table>
<tr><th>n</th><th>Name</th><th>Molecular Formula (C_nH_{2n+2})</th><th>Structural Formula</th><th>Condensed Structural Formula</th></tr>
<tr><td>1</td><td>methane</td><td>CH_4</td><td>H
|
H—C—H
|
H</td><td>CH_4</td></tr>
<tr><td>2</td><td>ethane</td><td>C_2H_6</td><td>H H
| |
H—C—C—H
| |
H H</td><td>CH_3CH_3</td></tr>
<tr><td>3</td><td>propane</td><td>C_3H_8</td><td>H H H
| | |
H—C—C—C—H
| | |
H H H</td><td>$CH_3CH_2CH_3$</td></tr>
<tr><td>4</td><td>*n*-butane</td><td>C_4H_{10}</td><td>H H H H
| | | |
H—C—C—C—C—H
| | | |
H H H H</td><td>$CH_3CH_2CH_2CH_3$</td></tr>
<tr><td>5</td><td>*n*-pentane</td><td>C_5H_{12}</td><td>H H H H H
| | | | |
H—C—C—C—C—C—H
| | | | |
H H H H H</td><td>$CH_3CH_2CH_2CH_2CH_3$</td></tr>
<tr><td>6</td><td>*n*-hexane</td><td>C_6H_{14}</td><td>H H H H H H
| | | | | |
H—C—C—C—C—C—C—H
| | | | | |
H H H H H H</td><td>$CH_3CH_2CH_2CH_2CH_2CH_3$</td></tr>
<tr><td>7</td><td>*n*-heptane</td><td>C_7H_{16}</td><td>H H H H H H H
| | | | | | |
H—C—C—C—C—C—C—C—H
| | | | | | |
H H H H H H H</td><td>$CH_3CH_2CH_2CH_2CH_2CH_2CH_3$</td></tr>
<tr><td>8</td><td>*n*-octane</td><td>C_8H_{18}</td><td>H H H H H H H H
| | | | | | | |
H—C—C—C—C—C—C—C—C—H
| | | | | | | |
H H H H H H H H</td><td>$CH_3CH_2CH_2CH_2CH_2CH_2CH_2CH_3$</td></tr>
<tr><td>9</td><td>*n*-nonane</td><td>C_9H_{20}</td><td>H H H H H H H H H
| | | | | | | | |
H—C—C—C—C—C—C—C—C—C—H
| | | | | | | | |
H H H H H H H H H</td><td>$CH_3CH_2CH_2CH_2CH_2CH_2CH_2CH_2CH_3$</td></tr>
<tr><td>10</td><td>*n*-decane</td><td>$C_{10}H_{22}$</td><td>H H H H H H H H H H
| | | | | | | | | |
H—C—C—C—C—C—C—C—C—C—C—H
| | | | | | | | | |
H H H H H H H H H H</td><td>$CH_3CH_2CH_2CH_2CH_2CH_2CH_2CH_2CH_2CH_3$</td></tr>
</table>

TABLE 18.2 Uses of Hydrocarbons

Number of Atoms	State	Major Uses
1–4	gas	heating fuel, cooking fuel
5–7	low-boiling liquids	solvents, gasoline
6–18	liquids	gasoline
12–24	liquids	jet fuel, portable-stove fuel
18–50	high-boiling liquids	diesel fuel, lubricants, heating oil
50+	solids	petroleum jelly, paraffin wax

EXAMPLE 18.2 Writing Formulas for *n*-Alkanes

Write the structural and condensed structural formula for *n*-octane, C_8H_{18}.

The first step in writing the structural formula is to write out the carbon backbone with eight carbons in it.

Solution:

$$C-C-C-C-C-C-C-C$$

The next step is to add H atoms so that all carbons have four bonds.

$$\begin{array}{c} \text{H}\quad\text{H}\quad\text{H}\quad\text{H}\quad\text{H}\quad\text{H}\quad\text{H}\quad\text{H} \\ |\quad\;\;|\quad\;\;|\quad\;\;|\quad\;\;|\quad\;\;|\quad\;\;|\quad\;\;| \\ \text{H}-\text{C}-\text{C}-\text{C}-\text{C}-\text{C}-\text{C}-\text{C}-\text{C}-\text{H} \\ |\quad\;\;|\quad\;\;|\quad\;\;|\quad\;\;|\quad\;\;|\quad\;\;|\quad\;\;| \\ \text{H}\quad\text{H}\quad\text{H}\quad\text{H}\quad\text{H}\quad\text{H}\quad\text{H}\quad\text{H} \end{array}$$

To write the condensed structural formula, write the hydrogen atoms bonded to each carbon directly to the right of the carbon atom. Use subscripts to indicate the correct number of hydrogen atoms.

$$CH_3CH_2CH_2CH_2CH_2CH_2CH_2CH_3$$

SKILLBUILDER 18.2 Writing Formulas for *n*-Alkanes

Write the structural and condensed structural formula for C_5H_{12}.

18.6 Isomers: Same Formula, Different Structure

In addition to linking together in straight chains to form the *n*-alkanes, carbon atoms can form branched structures called **branched alkanes**. The simplest branched alkane is called isobutane and has the following structure.

Isobutane C_4H_{10}

```
    H   H   H
    |   |   |
H — C — C — C — H
    |   |   |
    H   |   H
    H — C — H
        |
        H
```

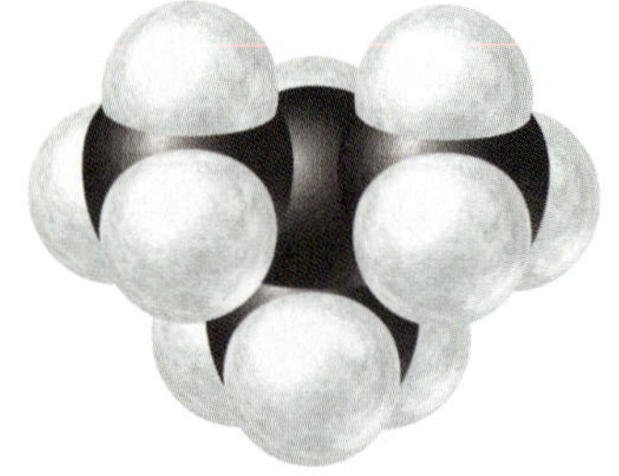

Formula | Structural formula | Space-filling model

TUTORIALS
Isomerism
Isomerism and Boiling Points

Isobutane and butane are **isomers**, molecules with the same molecular formula but different structures. Because of their different structures, they have different properties—indeed, they are different compounds. Isomerism is common in organic compounds. We have seen that butane has two isomers. Pentane (C_5H_{12}) has three isomers, hexane (C_6H_{14}) has five, and decane ($C_{10}H_{22}$) has seventy-five!

EXAMPLE 18.3 Writing Structural Formulas for Isomers

Draw the five isomers for hexane.

Solution:

To start, always draw the carbon backbone. The first isomer is the straight-chain isomer, C—C—C—C—C—C. Then, determine the carbon backbone structure of the other isomers by arranging the carbon atoms in four other unique ways.

```
C — C — C — C — C        C — C — C — C
            |                |   |
            C                C   C

        C
        |
C — C — C — C            C — C — C — C — C
        |                        |
        C                        C
```

Notice that

```
C — C — C — C — C
            |
            C
```

and

```
C — C — C — C — C
    |
    C
```

are identical to each other because the second structure is just the first one flipped around.

Finally, fill in all the hydrogen atoms so that each carbon has four bonds.

```
    H   H   H   H   H   H
    |   |   |   |   |   |
H — C — C — C — C — C — C — H
    |   |   |   |   |   |
    H   H   H   H   H   H

    H   H   H   H   H                  H   H           H   H
    |   |   |   |   |                  |   |           |   |
H — C — C — C — C — C — H          H — C — C ————————— C — C — H
    |   |   |   |   |                  |   |           |   |
    H   H   H   |   H                  H   |           |   H
            H — C — H                  H — C — H   H — C — H
                |                          |           |
                H                          H           H
```

```
            H
            |
          H—C—H
    H   H   |   H
    |   |   |   |
  H—C—C—C—C—H
    |   |   |   |
    H   H   |   H
          H—C—H
            |
            H
```

```
    H   H   H   H   H
    |   |   |   |   |
  H—C—C—C—C—C—H
    |   |   |   |   |
    H   H   |   H   H
          H—C—H
            |
            H
```

SKILLBUILDER 18.3 Writing Structural Formulas for Isomers

Draw the three isomers for pentane.

18.7 Naming Alkanes

TABLE 18.3 Prefixes for Base Names of Alkane Chains

Number of Carbon Atoms	Prefix
1	*meth-*
2	*eth-*
3	*prop-*
4	*but-*
5	*pent-*
6	*hex-*
7	*hept-*
8	*oct-*
9	*non-*
10	*dec-*

Many organic compounds have common names that can be learned only through familiarity. Because there are so many organic compounds, however, a systematic method of nomenclature is required. In this book, we adopt the nomenclature system recommended by IUPAC (International Union of Pure and Applied Chemistry), which is used throughout the world. In this system, the longest continuous chain of carbon atoms—called the **base chain**—determines the base name of the compound. The prefixes for base names depend on the number of carbon atoms in the base chain, as shown in Table 18.3. Base names for alkanes always have the ending *-ane*. Groups of carbon atoms branching off the base chain are called **alkyl groups** and are named as substituents. A **substituent** is simply an atom or group of atoms that has been *substituted* for a hydrogen atom in an organic compound. Common alkyl groups are shown in Table 18.4.

TABLE 18.4 Common Alkyl Groups

Condensed Structural Formula	Name
$—CH_3$	methyl
$—CH_2CH_3$	ethyl
$—CH_2CH_2CH_3$	propyl
$—CH_2CH_2CH_2CH_3$	butyl
$—CHCH_3$ $\vert$ CH_3	isopropyl
$—CH_2CHCH_3$ $\vert$ CH_3	isobutyl
$—CHCH_2CH_3$ $\vert$ CH_3	*sec*-butyl
CH_3 $\vert$ $—CCH_3$ $\vert$ CH_3	*tert*-butyl

The following rules will allow you to name systematically many alkanes. The rules are presented in the left column of the following box and two examples applying the rules are shown in the center and right columns.

Naming Alkanes

EXAMPLE 18.4 Naming Alkanes

Name the following alkane.

$$\begin{array}{l}CH_3CH_2\underset{\begin{array}{c}|\\CH_2\\|\\CH_3\end{array}}{C}HCH_2CH_2CH_3\end{array}$$

1. Count the number of carbon atoms in the longest continuous carbon chain to determine the base name of the compound. Find the prefix corresponding to this number of atoms in Table 18.3 and add the ending *-ane* to form the base name.

Solution:
This compound has six carbon atoms in its longest continuous chain.

$$CH_3CH_2\underset{\begin{array}{c}|\\CH_2\\|\\CH_3\end{array}}{C}HCH_2CH_2CH_3$$

The correct prefix from Table 18.3 is *hex-*. The base name is *hexane.*

2. Consider every branch from the base chain to be a substituent. Name each substituent according to Table 18.4.

This compound has one substituent named *ethyl.*

$$CH_3CH_2\underset{\begin{array}{c}|\\CH_2\\|\\CH_3\end{array}}{C}HCH_2CH_2CH_3$$

Ethyl

3. Beginning with the end closest to the branching, number the base chain and assign a number to each substituent. (If two substituents occur at equal distances from each end, go to the next substituent to determine from which end to start numbering.)

The base chain is numbered as follows:

1 2 3 4 5 6

$$CH_3CH_2\underset{\begin{array}{c}|\\CH_2\\|\\CH_3\end{array}}{C}HCH_2CH_2CH_3$$

The ethyl substituent is assigned the number 3.

4. Write the name of the compound in the following format:

(subst. number)-
(subst. name) (base name)

If there are two or more substituents, give each one a number and list them alphabetically with hyphens between words and numbers.

The name of this compound is:

3-ethylhexane

EXAMPLE 18.5 Naming Alkanes

Name the following alkane.

$$CH_3\underset{\begin{array}{c}|\\CH_3\end{array}}{C}HCH_2\underset{\begin{array}{c}|\\CH_2CH_3\end{array}}{C}HCH_2CH_2CH_3$$

Solution:
This compound has seven carbon atoms in its longest continuous chain.

$$CH_3\underset{\begin{array}{c}|\\CH_3\end{array}}{C}HCH_2\underset{\begin{array}{c}|\\CH_2CH_3\end{array}}{C}HCH_2CH_2CH_3$$

The correct prefix from Table 18.3 is *hept-*. The base name is *heptane.*

This compound has one substituent named *methyl* and one named *ethyl.*

$$CH_3\underset{\begin{array}{c}|\\CH_3\end{array}}{C}HCH_2\underset{\begin{array}{c}|\\CH_2CH_3\end{array}}{C}HCH_2CH_2CH_3$$

Methyl Ethyl

The base chain is numbered as follows:

1 2 3 4 5 6 7

$$CH_3\underset{\begin{array}{c}|\\CH_3\end{array}}{C}HCH_2\underset{\begin{array}{c}|\\CH_2CH_3\end{array}}{C}HCH_2CH_2CH_3$$

The methyl substituent is assigned the number 2 and the ethyl substituent is assigned the number 4.

The name of this compound is:

4-ethyl-2-methylheptane

Ethyl is listed before methyl because substituents are listed in alphabetical order.

5. If a compound has two or more identical substituents, designate the number of identical substituents with the prefix *di-* (2), *tri-* (3), or *tetra-* (4) before the substituent's name. Separate the numbers indicating the positions of the substituents relative to each other with a comma. The prefixes are not taken into account when alphabetizing.

Does not apply to this compound.

Does not apply to this compound.

SKILLBUILDER 18.4 Naming Alkanes

Name the following alkane.

CH_3CHCH_3 with CH_3 on C2

SKILLBUILDER 18.5 Naming Alkanes

Name the following alkane.

$CH_3CHCHCH_2CH_3$ with CH_3 on C2 and CH_2CH_3 on C3

EXAMPLE 18.6 Naming Alkanes

Name the following alkane.

$CH_3CHCH_2CHCH_3$ with CH_3 on C2 and CH_3 on C4

Solution:

1. The longest continuous carbon chain has five atoms. Therefore the base name is *pentane*.

$H_3C-CH-CH_2-CH-CH_3$ with CH_3 on C2 and CH_3 on C4

2. This compound has two substituents, both of which are named *methyl*.

$CH_3CHCH_2CHCH_3$ — Methyl → CH_3, CH_3 ← Methyl

3. Since both substituents are equidistant from the ends, it does not matter which end you start numbering from.

1 2 3 4 5
$CH_3CHCH_2CHCH_3$ with CH_3 on C2 and CH_3 on C4

4. and 5. Since this compound contains two identical substituents, Rule 5 applies and we use the prefix *di-*.

2,4–dimethylpentane

SKILLBUILDER 18.6 Naming Alkanes

Name the following alkane.

$CH_3CHCH_2CHCHCH_3$ with CH_3 on C2, CH_3 on C4 and CH_3 on C5

18.8 Alkenes and Alkynes

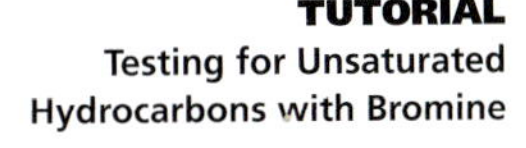

TUTORIAL
Testing for Unsaturated Hydrocarbons with Bromine

Alkenes are hydrocarbons containing at least one double bond between carbon atoms. **Alkynes** are hydrocarbons containing at least one triple bond between carbon atoms. Because of the double or triple bond, alkenes and alkynes have fewer hydrogen atoms than the corresponding alkane and are therefore called **unsaturated hydrocarbons**—they are not loaded to capacity with hydrogen. As we learned earlier, alkenes have the formula C_nH_{2n} and alkynes have the formula C_nH_{2n-2}. The simplest alkene is ethene, C_2H_4, also called *ethylene*.

Ethene or ethylene	C_2H_4	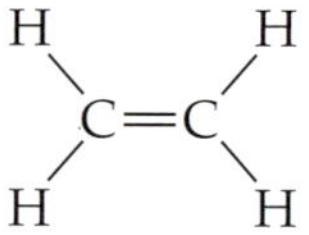 $H_2C{=}CH_2$	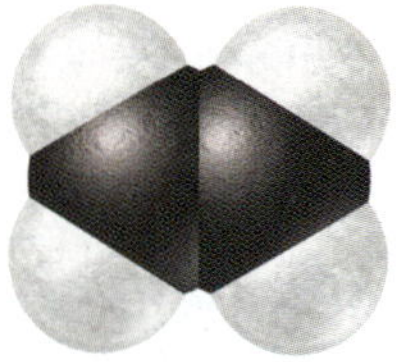
	Formula	Structural formula	Space-filling model

The geometry about each carbon atom in ethene is trigonal planar, making ethene a flat, rigid molecule. Ethene is a ripening agent in fruit. For example, when a banana within a cluster of bananas begins to ripen, it emits ethene. The ethene then causes other bananas in the cluster to ripen. Banana farmers usually pick bananas green for ease of shipping. When the bananas arrive at their destination, they are often "gassed" with ethene to initiate ripening so that they will be ready to sell. The names and structures of several other alkenes are shown in Table 18.5. Most of them do not have familiar uses other than their presence as minority components of fuels.

The simplest alkyne is ethyne, C_2H_2, also called acetylene.

Ethyne or acetylene	C_2H_2	H—C≡C—H	
	Formula	Structural formula	Space-filling model

▲ Ethene is emitted by ripening bananas. It acts as a chemical messenger, inducing bananas in a clump to ripen together.

The geometry about each carbon atom in ethyne is linear, making ethyne a linear molecule. Ethyne (or acetylene) is commonly used as fuel for welding torches. The names and structures of several other alkynes are shown

TABLE 18.5 Alkenes

n	Name	Molecular Formula (C_nH_{2n})	Structural Formula	Condensed Structural Formula
2	ethene	C_2H_4	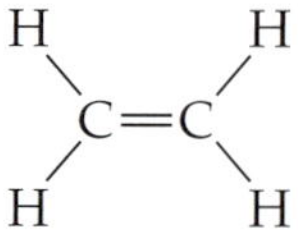 $H_2C{=}CH_2$	$CH_2{=}CH_2$
3	propene	C_3H_6	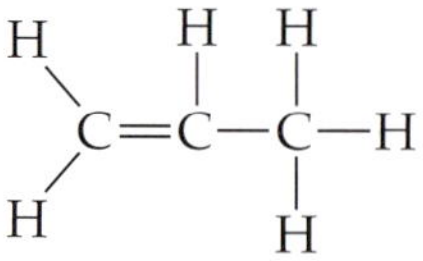 $H_2C{=}CH{-}CH_3$	$CH_2{=}CHCH_3$
4	1-butene*	C_4H_8	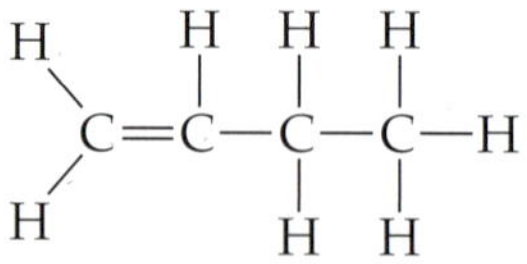 $H_2C{=}CH{-}CH_2{-}CH_3$	$CH_2{=}CHCH_2CH_3$
5	1-pentene*	C_5H_{10}	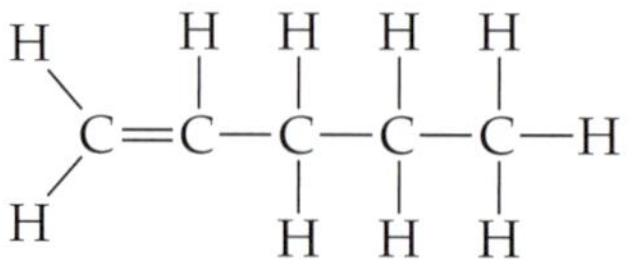 $H_2C{=}CH{-}CH_2{-}CH_2{-}CH_3$	$CH_2{=}CHCH_2CH_2CH_3$
6	1-hexene*	C_6H_{12}	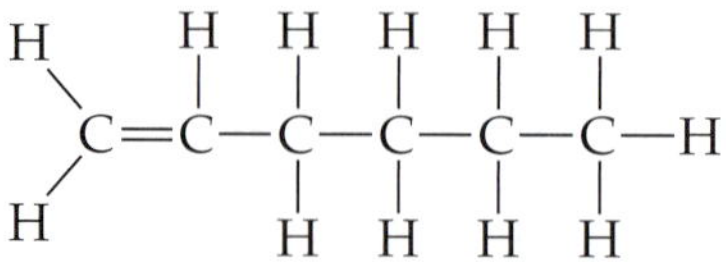 $H_2C{=}CH{-}CH_2{-}CH_2{-}CH_2{-}CH_3$	$CH_2{=}CHCH_2CH_2CH_3$

TABLE 18.6 Alkynes

n	Name	Molecular Formula (C_nH_{2n-2})	Structural Formula	Condensed Structural Formula
2	ethyne	C_2H_2	H—C≡C—H	CH≡CH
3	propyne	C_3H_4	H | H—C≡C—C—H | H	$CH{\equiv}CCH_3$
4	1-butyne*	C_4H_6	H H | | H—C≡C—C—C—H | | H H	$CH{\equiv}CCH_2CH_3$
5	1-pentyne*	C_5H_8	H H H | | | H—C≡C—C—C—C—H | | | H H H	$CH{\equiv}CCH_2CH_2CH_3$
6	1-hexyne*	C_6H_{10}	H H H H | | | | H—C≡C—C—C—C—C—H | | | | H H H H	$CH{\equiv}CCH_2CH_2CH_2CH_3$

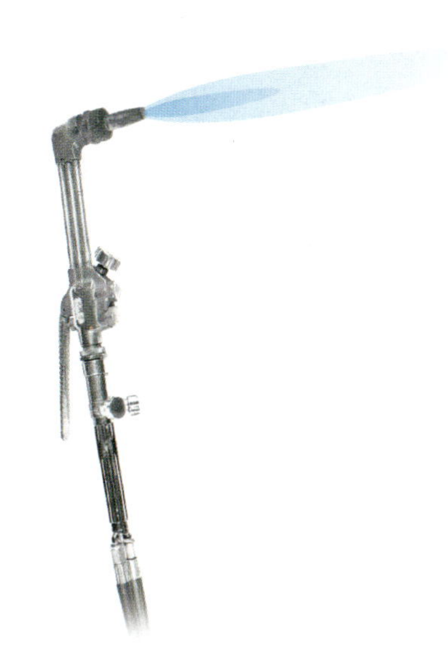

▲ Acetylene is used as a fuel in welding torches.

in Table 18.6. Like alkenes, the alkynes do not have familiar uses other than their presence as minority components of gasoline.

Naming Alkenes and Alkynes

Alkenes and alkynes are named in the same way as alkanes with the following exceptions.

- The base chain is the longest continuous carbon chain that *contains the double or triple bond.*
- The base name has the ending *-ene* for alkenes and *-yne* for alkynes.
- The base chain is numbered to *give the double or triple bond the lowest possible number.*
- A number indicating the position of the double or triple bond (lowest possible number) is inserted just before the base name. For example,

$CH{\equiv}CCH_2CH_3$
1-Butyne

$CH_3CH_2CH{=}CCH_3$ (with CH_3 on the fourth carbon)
2-Methyl-2-pentene

EXAMPLE 18.7 Naming Alkenes and Alkynes

Name the following compounds.

(a) $H_3C{-}C{=}C{-}CH_2{-}CH_3$, with CH_3 on the third carbon and $H_2C{-}CH_3$ on the second carbon

(b) $H_3C{-}HC{-}CH{-}C{\equiv}CH$, with CH_3 on HC and $H_3C{-}CH{-}CH_3$ on CH

(a) Follow the procedure for naming alkanes and consider the preceding exceptions for naming alkenes:

Solution:

1. The longest continuous carbon chain containing the double bond has six carbon atoms. The base name is therefore *hexene.*

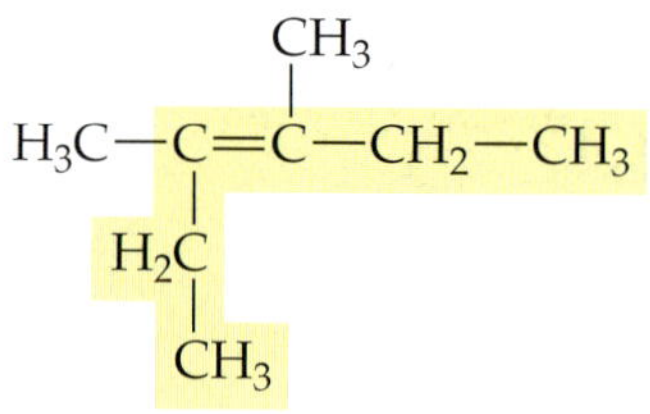

$H_3C-C(CH_2CH_3)=C(CH_3)-CH_2-CH_3$

2. The two substituents are both methyl.

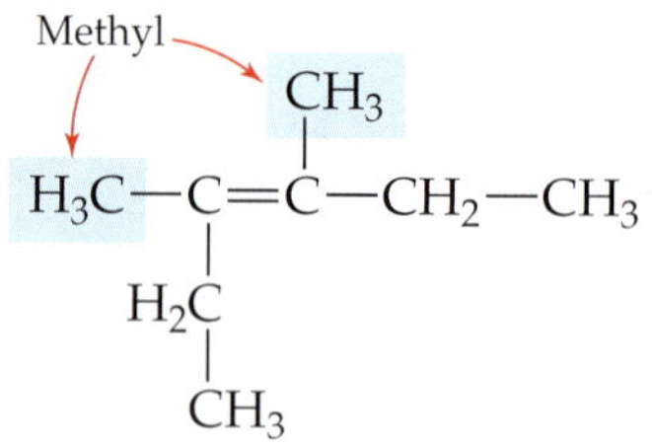

Methyl

3. One of the exceptions for naming alkenes is to number the chain so that the double bond has the lowest number. In this case, the double bond is equidistant from the ends and is assigned the number 3.

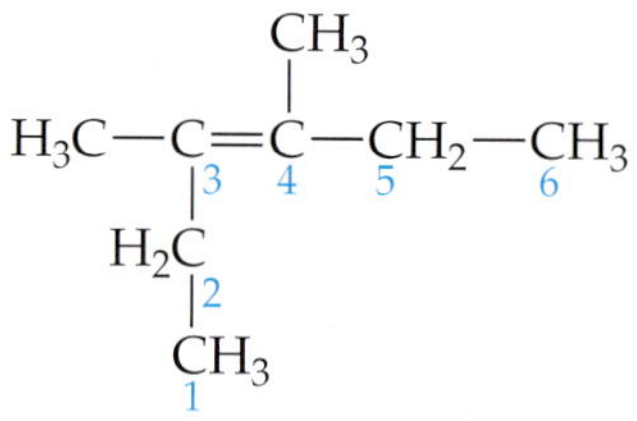

4. Name the compound by assigning numbers to each methyl group and to the double bond. Separate numbers from names using hyphens.

3,4-dimethyl-3-hexene

(b) Follow the procedure for naming alkanes and consider the preceding exceptions for naming alkynes:

1. The longest continuous carbon chain containing the triple bond is five carbons long; therefore the base name is *pentyne.*

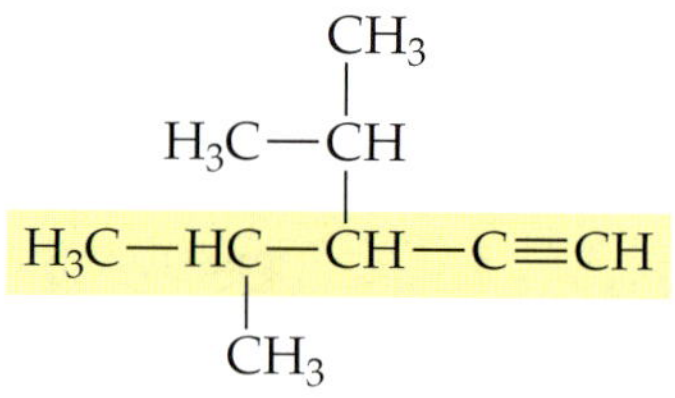

$H_3C-HC(CH_3)-CH(CH(CH_3)_2)-C\equiv CH$

2. There are two substituents: one methyl group and one isopropyl group.

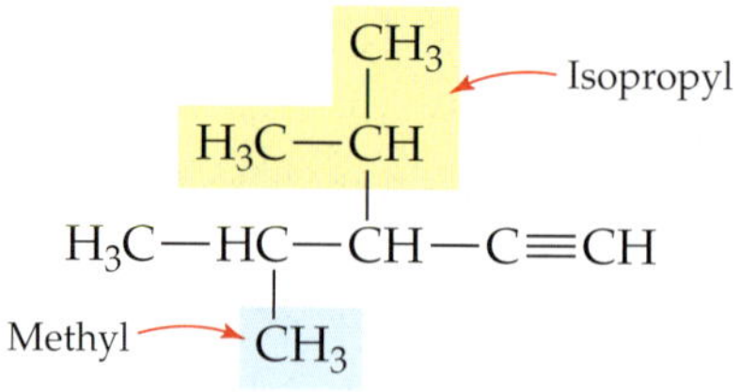

Isopropyl

Methyl

3. Number the base chain, giving the triple bond the lowest number, 1.

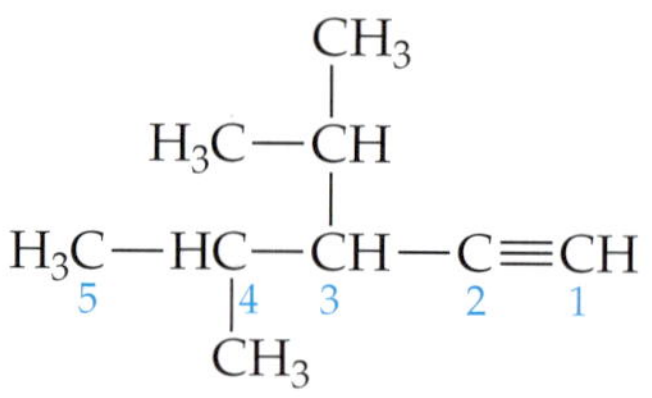

4. Name the compound by assigning numbers to each substituent and to the triple bond. Separate numbers from names using hyphens.

3-isopropyl-4-methyl-1-pentyne

SKILLBUILDER 18.7 Naming Alkenes and Alkynes

Name the following alkene and alkyne.

$$H_3C-C\equiv C-\overset{\displaystyle H_3C}{\underset{\displaystyle H_3C}{C}}-CH_3$$

$$H_3C-\underset{\displaystyle CH_3}{CH}-CH_2-\underset{\displaystyle H_3C}{CH}-\overset{\displaystyle CH_3-H_2C}{CH}-CH=CH_2$$

18.9 Hydrocarbon Reactions

Combustion reactions were first covered in Section 7.9.

One of the most common hydrocarbon reactions is **combustion**, the burning of hydrocarbons in the presence of oxygen. Alkanes, alkenes, and alkynes all undergo combustion. In a combustion reaction, the hydrocarbon reacts with oxygen to form carbon dioxide and water.

$$CH_3CH_2CH_3(g) + 5\,O_2(g) \longrightarrow 3\,CO_2(g) + 4\,H_2O(g) \qquad \text{Alkane combustion}$$
$$CH_2{=}CHCH_2CH_3(g) + 6\,O_2(g) \longrightarrow 4\,CO_2(g) + 4\,H_2O(g) \qquad \text{Alkene combustion}$$
$$CH{\equiv}CCH_3(g) + 4\,O_2(g) \longrightarrow 3\,CO_2(g) + 2\,H_2O(g) \qquad \text{Alkyne combustion}$$

Hydrocarbon combustion reactions are highly exothermic—they emit large amounts of heat. This heat can be used to warm homes and buildings, to generate electricity, or to expand the gas in a cylinder to drive a car forward. Approximately 90% of energy in the United States is generated by hydrocarbon combustion.

Alkanes

As discussed in Chapter 4, the halogens include F, Cl, Br, and I.

In addition to combustion, alkanes also undergo **substitution reactions**, in which one or more hydrogen atoms on an alkane are replaced by one or more other atoms. The most common substitution reaction is halogen substitution. For example, methane reacts with chlorine gas to form chloromethane.

$$H-\overset{\displaystyle H}{\underset{\displaystyle H}{C}}-H + Cl-Cl \xrightarrow{\text{Heat or light}} H-\overset{\displaystyle H}{\underset{\displaystyle H}{C}}-Cl + H-Cl$$

$$CH_4(g) + Cl_2(g) \xrightarrow{\text{Heat or light}} CH_3Cl(g) + HCl(g)$$

Ethane reacts with chlorine gas to form chloroethane.

$$\begin{array}{c} \text{H}\ \ \text{H} \\ |\ \ \ \ | \\ \text{H}-\text{C}-\text{C}-\text{H} \\ |\ \ \ \ | \\ \text{H}\ \ \text{H} \end{array} + \text{Cl}-\text{Cl} \xrightarrow{\text{Heat or light}} \begin{array}{c} \text{H}\ \ \text{H} \\ |\ \ \ \ | \\ \text{H}-\text{C}-\text{C}-\text{Cl} \\ |\ \ \ \ | \\ \text{H}\ \ \text{H} \end{array} + \text{H}-\text{Cl}$$

$$CH_3CH_3(g) + Cl_2(g) \xrightarrow{\text{Heat or light}} CH_3CH_2Cl(g) + HCl(g)$$

The general form for halogen substitution reactions is:

In this equation, R represents a hydrocarbon group.

$$\underset{\text{Alkane}}{\text{R}-\text{H}} + \underset{\text{Halogen}}{X_2} \longrightarrow \underset{\text{Haloalkane}}{\text{R}-\text{X}} + \underset{\text{Hydrogen halide}}{\text{HX}}$$

Multiple halogenation reactions can occur because halogens can replace more than one of the hydrogen atoms on an alkane.

Alkenes and Alkynes

Alkenes and alkynes undergo **addition reactions** in which atoms add across the multiple bond. For example, ethene reacts with chlorine gas to form dichloroethane.

$$\begin{array}{c} \text{H}\ \ \ \ \ \ \ \text{H} \\ \diagdown\ \ \ \diagup \\ \text{C}=\text{C} \\ \diagup\ \ \ \diagdown \\ \text{H}\ \ \ \ \ \ \ \text{H} \end{array} + \text{Cl}-\text{Cl} \longrightarrow \begin{array}{c} \text{H}\ \ \text{H} \\ |\ \ \ \ | \\ \text{H}-\text{C}-\text{C}-\text{H} \\ |\ \ \ \ | \\ \text{Cl}\ \ \text{Cl} \end{array}$$

$$CH_2{=}CH_2(g) + Cl_2(g) \longrightarrow CH_2ClCH_2Cl(g)$$

Notice that the addition of chlorine converts the carbon–carbon double bond into a single bond because each carbon atom now has a new bond to a chlorine atom. Alkenes and alkynes can also add hydrogen in **hydrogenation** reactions. For example, in the presence of an appropriate catalyst, propene reacts with hydrogen gas to form propane.

▲ Many foods contain partially hydrogenated vegetable oil. The name means that some of the double bonds in the carbon chains of these molecules have been converted to single bonds by the addition of hydrogen.

$$\begin{array}{c} \text{H}\ \ \ \ \ \ \ \ \ \ \ \ \ \ \ \text{H} \\ |\ \ \ \ \ \ \ \ \ \ \ \ \diagup \\ \text{H}-\text{C}-\text{C}=\text{C} \\ |\ \ \ \ \ \ |\ \ \ \ \ \ \diagdown \\ \text{H}\ \ \ \text{H}\ \ \ \ \ \ \ \text{H} \end{array} + \text{H}-\text{H} \xrightarrow[\text{Catalyst}]{} \begin{array}{c} \text{H}\ \ \text{H}\ \ \text{H} \\ |\ \ \ \ |\ \ \ \ | \\ \text{H}-\text{C}-\text{C}-\text{C}-\text{H} \\ |\ \ \ \ |\ \ \ \ | \\ \text{H}\ \ \text{H}\ \ \text{H} \end{array}$$

$$CH_3CH{=}CH_2(g) + H_2(g) \xrightarrow[\text{Catalyst}]{} CH_3CH_2CH_3(g)$$

Hydrogenation reactions convert unsaturated hydrocarbons into saturated hydrocarbons. Have you ever read *partially hydrogenated vegetable oil* on a food ingredient label? Vegetable oil is an unsaturated fat—its carbon chains contain double bonds. Unsaturated fats tend to be liquids at room temperature. By means of hydrogenation reactions, hydrogen is added to the double bonds, converting the unsaturated fat into saturated fat, which tends to be solid at room temperature.

More information about fats and oils can be found in Chapter 19.

To summarize:

- All hydrocarbons undergo combustion reactions.
- Alkanes undergo substitution reactions.
- Alkenes and alkynes undergo addition reactions.

18.10 Aromatic Hydrocarbons

As you might imagine, determining the structure of organic compounds has not always been easy. In the mid-1800s chemists were trying to determine the structure of a particularly stable organic compound named benzene that had the formula C_6H_6. In 1865, Friedrich August Kekulé (1829–1896) had a dream in which he envisioned chains of carbon atoms as snakes. The snakes danced before him and one of them twisted around and bit its tail. Based on that vision, Kekulé proposed the following structure for benzene.

```
      H     H
       \   /
        C—C
       //   \\
  H—C         C—H
       \     /
        C=C
       /   \
      H     H
```

This structure shows alternating single and double bonds. When we examine the bond lengths in benzene, however, we find that all of the bonds are of the same length.

The structure of benzene is better represented by the following resonance structures.

```
      H     H                    H     H
       \   /                      \   /
        C—C                        C=C
       //   \\                    /   \
  H—C         C—H   ⟷    H—C         C—H
       \     /                    \\   //
        C=C                        C—C
       /   \                      /   \
      H     H                    H     H
```

The concept of resonance structures was first introduced in Section 10.6.

Remember that the resonance structures mean that the true structure of benzene is an average between the two resonance structures. In other words, all carbon–carbon bonds in benzene are equivalent and are midway between a single and double bond. The space-filling model of benzene is as follows:

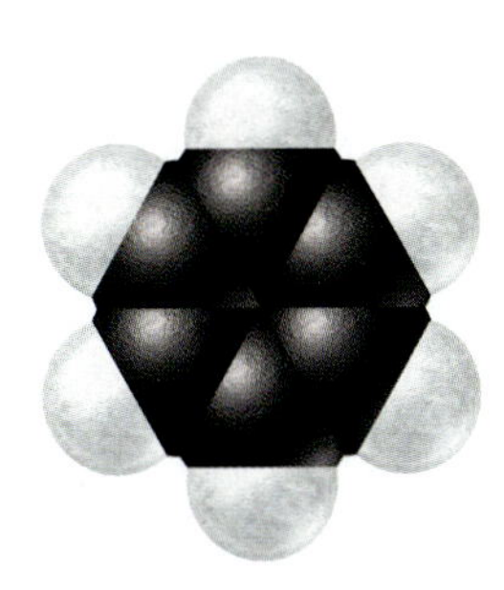

Benzene is often represented with the following shorthand notations.

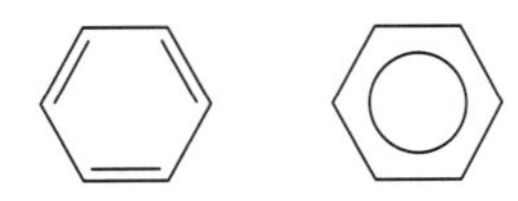

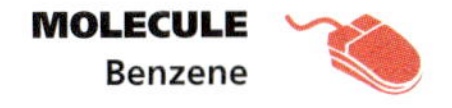

Each point in the hexagon represents a carbon atom with a hydrogen atom attached to it.

The ring structure of benzene occurs in many organic compounds. An atom or group of atoms can be substituted for one or more of the six hydrogen atoms to form substituted benzenes. The following are two examples of substituted benzenes.

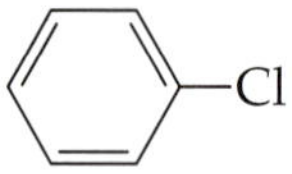

—Cl Chlorobenzene

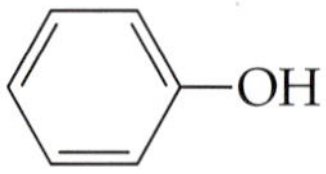

—OH Phenol

Since many compounds containing benzene rings have pleasant aromas, benzene rings are also called **aromatic rings**, and compounds containing them are called *aromatic compounds*. For example, the pleasant smells of cinnamon, vanilla, and jasmine are all caused by aromatic compounds.

Naming Aromatic Hydrocarbons

Monosubstituted benzenes—benzenes in which only one of the hydrogen atoms has been substituted—are often named as derivatives of benzene.

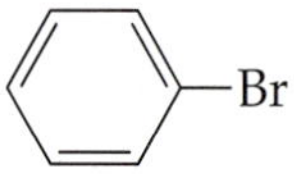

—Br Bromobenzene

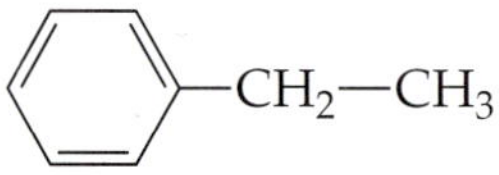

$-CH_2-CH_3$ Ethylbenzene

These names have the following general form.

(name of substituent)*benzene*

However, many monosubstituted benzenes have common names that can be learned only through familiarity.

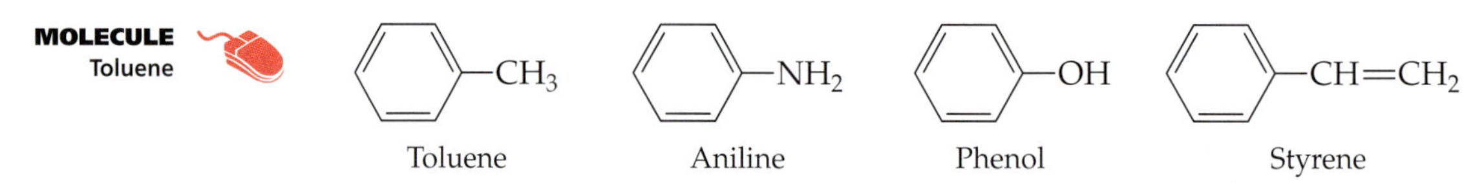

Toluene Aniline Phenol Styrene

Some substituted benzenes, especially those with large substituents, are named by treating the benzene ring as the substituent. In these cases, the benzene substituent is called a **phenyl group**.

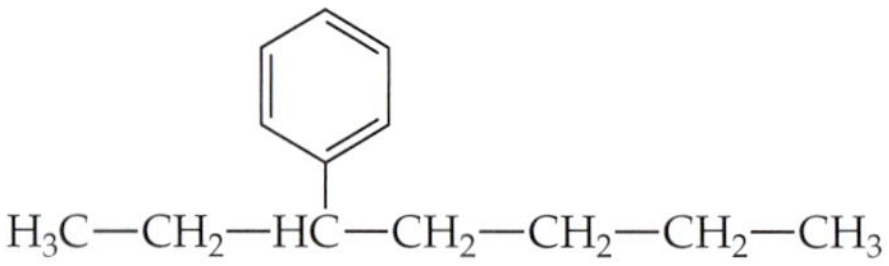

$H_3C-CH_2-HC-CH_2-CH_2-CH_2-CH_3$

3-Phenylheptane

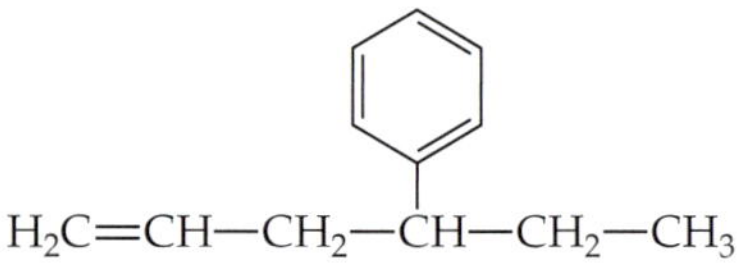

$H_2C=CH-CH_2-CH-CH_2-CH_3$

4-Phenyl-1-hexene

Disubstituted benzenes, benzenes in which two hydrogen atoms have been substituted, are numbered and the substituents are listed alphabetically. The order of numbering within the ring is then determined by the alphabetical order of the substituents.

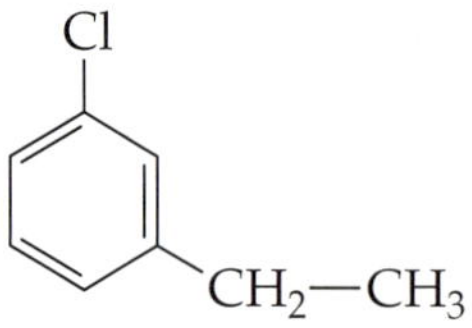

Cl, CH_2-CH_3

1-Chloro-3-ethylbenzene

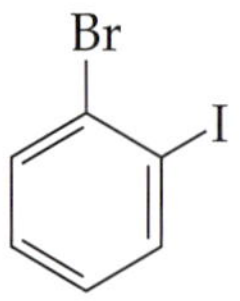

Br, I

1-Bromo-2-iodobenzene

When the two substituents are identical, use the prefix *di-*.

1,2-Dichlorobenzene 1,3-Dichlorobenzene 1,4-Dichlorobenzene

Also in common use—in place of numbering—are the prefixes *ortho-* (1,2 disubstituted), *meta-* (1,3 disubstituted), and *para-* (1,4 disubstituted).

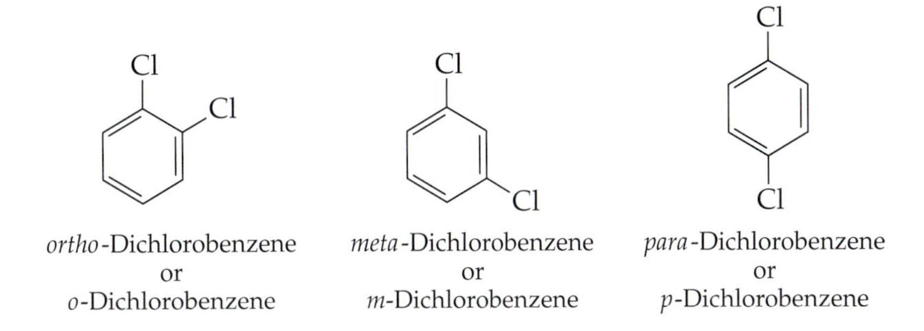

ortho-Dichlorobenzene or *o*-Dichlorobenzene

meta-Dichlorobenzene or *m*-Dichlorobenzene

para-Dichlorobenzene or *p*-Dichlorobenzene

EXAMPLE 18.8 Naming Aromatic Compounds

Name the following compound.

Solution:

Benzene derivatives are named using the general form (name of substituent)*benzene*. Because this derivative has two substituents, the substituents are numbered and listed alphabetically. The two substituents are *bromo-* and *chloro-*.

Because *bromo-* is first alphabetically, it is assigned the number 1; therefore *chloro-* is assigned the number 2. Following the general form, the name of the compound is 1-bromo-2-chlorobenzene.

SKILLBUILDER 18.8 Naming Aromatic Compounds

Name the following compound.

18.11 Functional Groups

Most other families of organic compounds can be thought of as hydrocarbons in which a **functional group**—a characteristic atom or group of atoms—has been inserted into the hydrocarbon. The letter R is often used to represent a hydrocarbon group. If the letter G represents a functional group, then a generic formula for families of organic compounds is:

$$\underbrace{R}_{\text{Hydrocarbon group}}-\underbrace{G}_{\text{Functional group}}$$

A group of organic compounds with the same functional group forms a **family**. The members of the family of **alcohols** have an —OH functional group and the general formula R—OH. Some specific examples of alcohols are methanol and isopropyl alcohol.

$$\underbrace{CH_3}_{\text{R group}}-\underbrace{OH}_{\text{—OH functional group}}$$

Methanol

$$\underbrace{CH_3-\underset{|}{\overset{CH_3}{}}CH}_{\text{R group}}-\underbrace{OH}_{\text{—OH functional group}}$$

2-propanol or isopropyl alcohol

The insertion of a functional group into a hydrocarbon usually alters the properties of the compound significantly. For example, *methanol*—which can be thought of as methane with an —OH group substituted for one of the hydrogen atoms—is a polar, hydrogen-bonded liquid at room temperature. *Methane*, on the other hand, is a nonpolar gas. While each member of a family is unique and different, their common functional group also gives them some similarities in both their physical and chemical properties. Table 18.7 lists some common functional groups, their general formulas, and an example of each.

TABLE 18.7 Functional Groups

Family	General Formula	Condensed General Formula	Example	Name
alcohols	R—OH	ROH	CH_3CH_2—OH	ethanol (ethyl alcohol)
ethers	R—O—R	ROR	CH_3—O—CH_3	dimethyl ether
aldehydes	$R-\overset{O}{\overset{\Vert}{C}}-H$	RCHO	$H_3C-\overset{O}{\overset{\Vert}{C}}-H$	ethanal (acetaldehyde)
ketones	$R-\overset{O}{\overset{\Vert}{C}}-R$	RCOR	$H_3C-\overset{O}{\overset{\Vert}{C}}-CH_3$	propanone (acetone)
carboxylic acids	$R-\overset{O}{\overset{\Vert}{C}}-OH$	RCOOH	$H_3C-\overset{O}{\overset{\Vert}{C}}-OH$	acetic acid
esters	$R-\overset{O}{\overset{\Vert}{C}}-OR$	RCOOR	$H_3C-\overset{O}{\overset{\Vert}{C}}-OCH_3$	methyl acetate
amines	$R-\overset{R}{\overset{\vert}{N}}-R$	R_3N	$H_3CH_2C-\overset{H}{\overset{\vert}{N}}-H$	ethyl amine

18.12 Alcohols

As previously mentioned, alcohols are organic compounds containing the —OH functional group. They have the general formula R—OH. In addition to methanol and isopropyl alcohol (shown previously), other common alcohols include the following:

$H_3C—CH_2—OH$

Ethanol

$H_3C—CH_2—CH_2—CH_2—OH$

1-Butanol

▲ Ethanol is the alcohol in alcoholic beverages.

Naming Alcohols

Alcohols are named similarly to alkanes with the following exceptions.

- The base chain is the longest continuous carbon chain that *contains the —OH functional group.*
- The base name has the ending *-ol.*
- The base chain is numbered to *give the* —OH *group the lowest possible number.*
- A number indicating the position of the —OH group is inserted just before the base name. For example,

$$\underset{\text{OH}}{CH_3CH_2CH_2\overset{|}{C}HCH_3}$$

2-Pentanol

$$CH_2CH_2CHCH_3 \text{ (OH on C1, } CH_3 \text{ on C3)}$$

3-Methyl-1-butanol

About Alcohols

Among the most familiar alcohols is ethanol, the alcohol in alcoholic beverages. Ethanol is most commonly formed by the yeast fermentation of sugars, such as glucose, from fruits and grains.

$$\underset{\text{Glucose}}{C_6H_{12}O_6} \xrightarrow{\text{Yeast}} 2\,\underset{\text{Ethanol}}{CH_3CH_2OH} + 2\,CO_2$$

Alcoholic beverages contain primarily ethanol and water and a few other components that give flavor and color. Beer usually contains 3% to 6% ethanol. Wine contains about 12% ethanol, and spirits—beverages such as whiskey, rum, or tequila—range from 40% to 80% ethanol, depending on their *proof*. The proof of an alcoholic beverage is twice the percentage of its ethanol content, so an 80-proof whiskey contains 40% ethanol. Ethanol is also used as a gasoline additive because it increases the octane rating of gasoline and fosters its complete combustion, eliminating certain pollutants such as carbon monoxide and the precursors of ozone.

Isopropyl alcohol (or 2-propanol) can be purchased at any drugstore as rubbing alcohol. It is commonly used as a disinfectant for wounds and

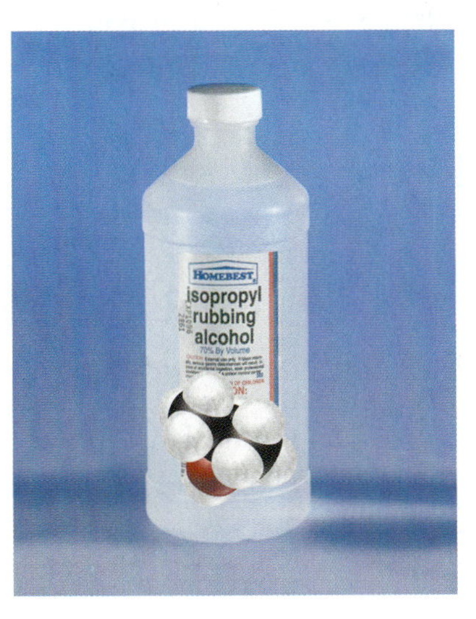

▲ Rubbing alcohol is composed of isopropyl alcohol, or 2-propanol.

Isopropyl alcohol is the common name for 2-propanol.

MOLECULES
1-Propanol
Isopropyl Alcohol

to sterilize medical instruments. Isopropyl alcohol should never be consumed internally, as it is highly toxic. A few ounces of isopropyl alcohol can cause death. A third common alcohol is methanol, also called wood alcohol. Methanol is commonly used as a laboratory solvent and as a fuel additive. Like isopropyl alcohol, methanol is toxic and should never be consumed.

18.13 Ethers

Ethers are organic compounds with the general formula R—O—R. The R groups may be the same or different. Some common ethers include the following:

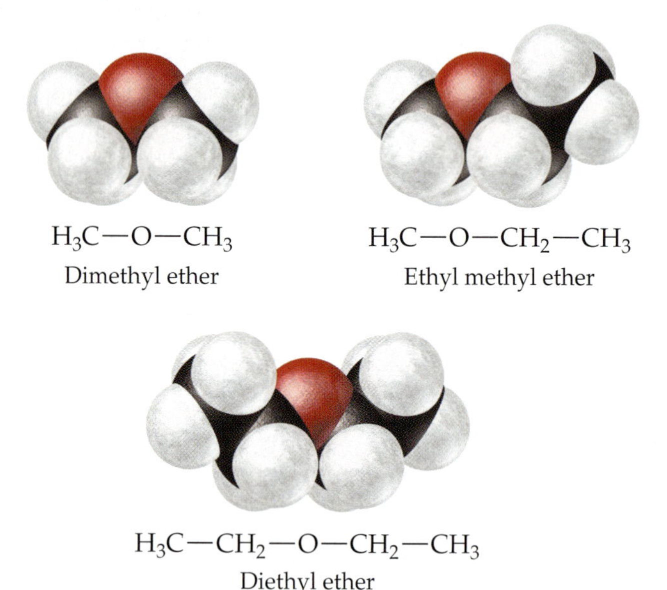

Naming Ethers

The IUPAC names for ethers are beyond the scope of this text. Common names for ethers have the following format.

(R group 1)(R group 2) ether

MOLECULE
Diethyl Ether

If the two R groups are different, use each of their names. If the two R groups are the same, use the prefix *di-*. Some examples include the following:

$H_3C—CH_2—CH_2—O—CH_2—CH_2—CH_3$
Dipropyl ether

$H_3C—CH_2—O—CH_2—CH_2—CH_3$
Ethyl propyl ether

About Ethers

The most common ether is diethyl ether. Diethyl ether is a common laboratory solvent because of its ability to dissolve many organic compounds and because of its low boiling point (34.6 °C). The low boiling point allows for easy removal of the solvent when necessary. Diethyl ether was also used as a general anesthetic for many years. When inhaled, diethyl ether depresses the central nervous system, causing unconsciousness and insensitivity to pain. Its use as an anesthetic, however, has decreased in recent years because other compounds have the same anesthetic effect with fewer side effects (such as nausea).

18.14 Aldehydes and Ketones

> The condensed structural formula for aldehydes is R—CHO and for ketones is R—CO—R. In ketones the R groups may be the same or different.

Aldehydes and **ketones** have the following general formulas.

$$\underset{\text{Aldehyde}}{\mathrm{R{-}\overset{\displaystyle O}{\overset{\|}{C}}{-}H}} \qquad \underset{\text{Ketone}}{\mathrm{R{-}\overset{\displaystyle O}{\overset{\|}{C}}{-}R}}$$

Both aldehydes and ketones contain a **carbonyl group** ($\mathrm{\diagup\overset{\displaystyle O}{\overset{\|}{C}}\diagdown}$). Ketones have an R group attached to both sides of the carbonyl, while aldehydes have one R group and a hydrogen atom. (An exception is formaldehyde, which is an aldehyde with two H atoms attached to the carbonyl group.)

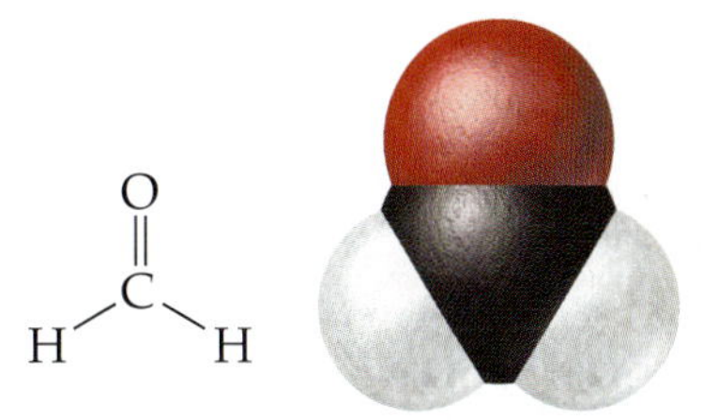

Methanal or formaldehyde

The following are other common aldehydes and ketones.

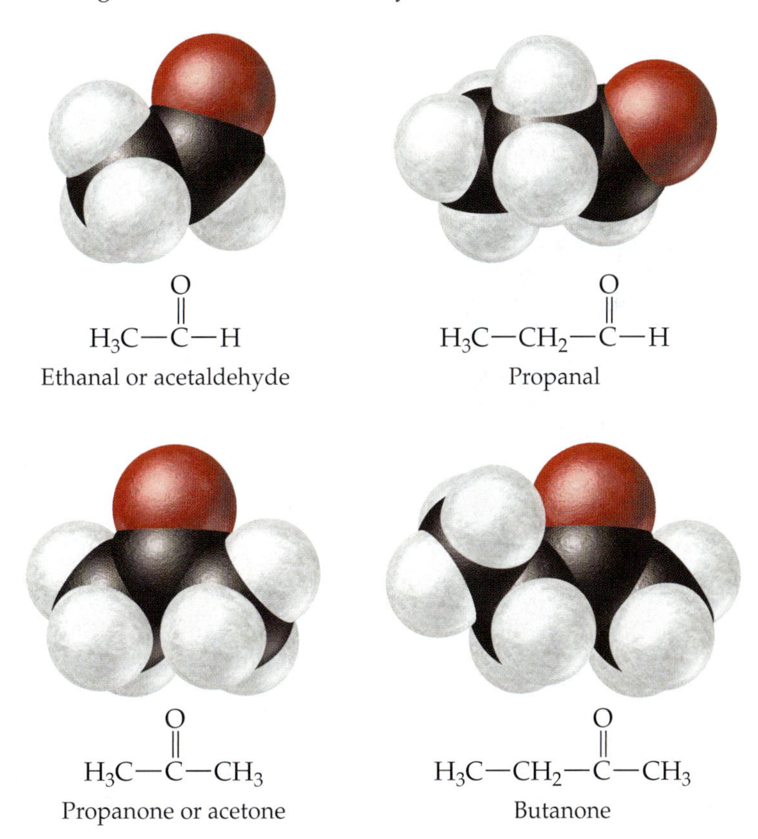

Ethanal or acetaldehyde

Propanal

Propanone or acetone

Butanone

Naming Aldehydes and Ketones

Many aldehydes and ketones have common names that can be learned only by becoming familiar with them. Simple aldehydes are systematically named according to the number of carbon atoms in the longest continuous carbon chain that contains the carbonyl group. Form the base name from the name of the corresponding alkane by dropping the *-e* and adding the ending *-al*.

Simple ketones are systematically named according to the longest continuous carbon chain containing the carbonyl group. Form the base name from the name of the corresponding alkane by dropping the *-e* and adding the ending *-one*. For ketones, number the chain to give the carbonyl group the lowest possible number (when necessary).

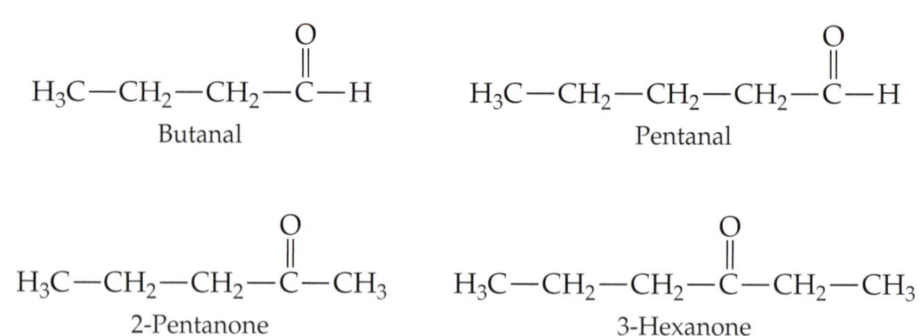

▲ Benzaldehyde is responsible for the smell of almonds.

About Aldehydes and Ketones

The most familiar aldehyde probably is formaldehyde, shown earlier in this section. Formaldehyde is a gas with a pungent odor. It is often mixed with water to make formalin, a preservative and disinfectant. Formaldehyde is also found in wood smoke, which is one reason that smoking foods preserves them—the formaldehyde kills the bacteria. Aromatic aldehydes, those that also contain an aromatic ring, have pleasant aromas. For example, cinnamaldehyde is the sweet-smelling component of cinnamon, benzaldehyde accounts for the smell of almonds, and vanillin is responsible for the smell of vanilla.

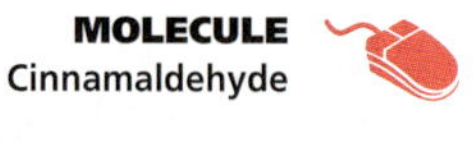

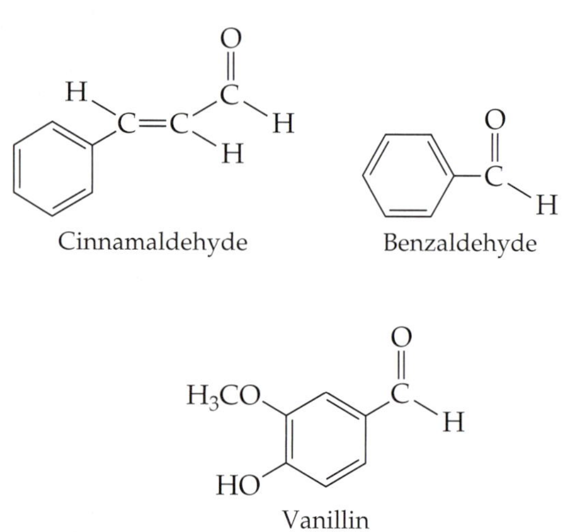

▲ Fingernail-polish remover is composed primarily of acetone, a ketone.

The most familiar ketone is acetone, the main component of fingernail-polish remover. Many ketones also have pleasant aromas. For example, 2-heptanone is responsible for the smell of cloves, carvone for the smell of spearmint, and ionone for the smell of raspberries

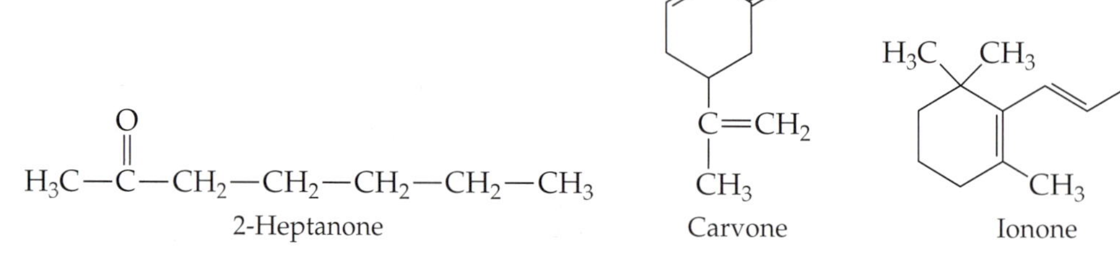

Ionone

▲ Ionone, a ketone, is largely responsible for the smell of raspberries.

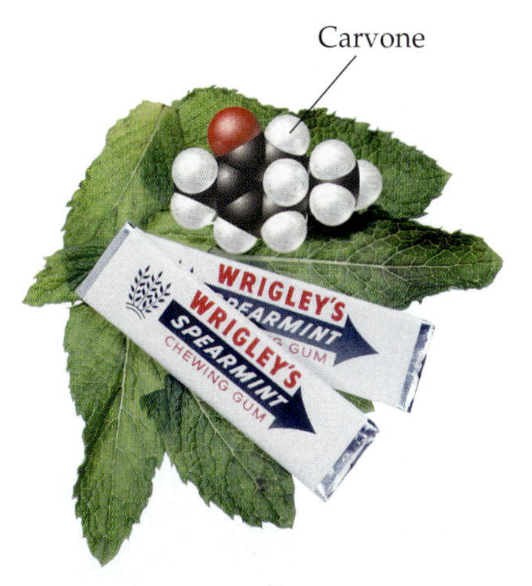

▲ The smell and taste of spearmint are produced by carvone, an aromatic ketone.

18.15 Carboxylic Acids and Esters

> The condensed structural formula for carboxylic acids is R—COOH and for esters is R—COO—R. The R groups in esters may be the same or different.

Carboxylic acids and **esters** have the following general formulas.

$$\underset{\text{Carboxylic acid}}{\mathrm{R{-}\overset{\overset{\large O}{\|}}{C}{-}OH}} \qquad \underset{\text{Ester}}{\mathrm{R{-}\overset{\overset{\large O}{\|}}{C}{-}OR}}$$

The following are common carboxylic acids and esters.

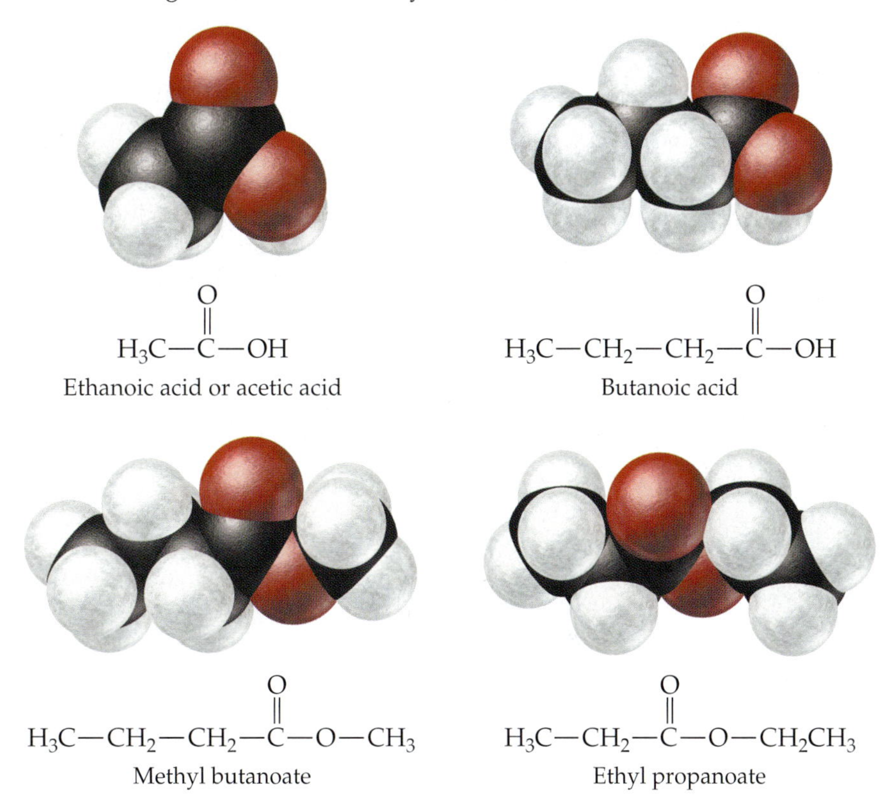

$$\mathrm{H_3C{-}\overset{\overset{O}{\|}}{C}{-}OH}$$

Ethanoic acid or acetic acid

$$\mathrm{H_3C{-}CH_2{-}CH_2{-}\overset{\overset{O}{\|}}{C}{-}OH}$$

Butanoic acid

$$\mathrm{H_3C{-}CH_2{-}CH_2{-}\overset{\overset{O}{\|}}{C}{-}O{-}CH_3}$$

Methyl butanoate

$$\mathrm{H_3C{-}CH_2{-}\overset{\overset{O}{\|}}{C}{-}O{-}CH_2CH_3}$$

Ethyl propanoate

The generic condensed structural formula for carboxylic acids is RCOOH. Carboxylic acids act as weak acids in solution according to the following equation:

$$\mathrm{RCOOH}(aq) + \mathrm{H_2O}(l) \rightleftharpoons \mathrm{H_3O^+}(aq) + \mathrm{RCOO^-}(aq)$$

Naming Carboxylic Acids and Esters

Carboxylic acids are systematically named according to the number of carbon atoms in the longest chain containing the —COOH functional group. Form the base name by dropping the *-e* from the name of the corresponding alkane and adding the ending *-oic acid*.

$$H_3C-CH_2-\overset{\overset{\displaystyle O}{\|}}{C}-OH \qquad H_3C-CH_2-CH_2-CH_2-\overset{\overset{\displaystyle O}{\|}}{C}-OH$$

Propanoic acid   Pentanoic acid

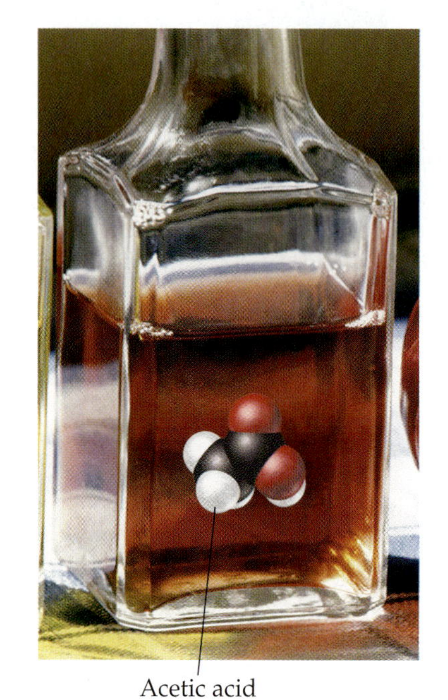

▲ Acetic acid, a carboxylic acid, is the active ingredient in vinegar.

Esters are systematically named as if they were derived from a carboxylic acid by replacing the H on the OH with an alkyl group. The R group from the parent acid forms the base name of the compound. Change the *-ic* on the name of the corresponding carboxylic acid to *-ate*. The R group that replaced the H on the carboxylic acid is named as an alkyl group with the ending *-yl*. For example:

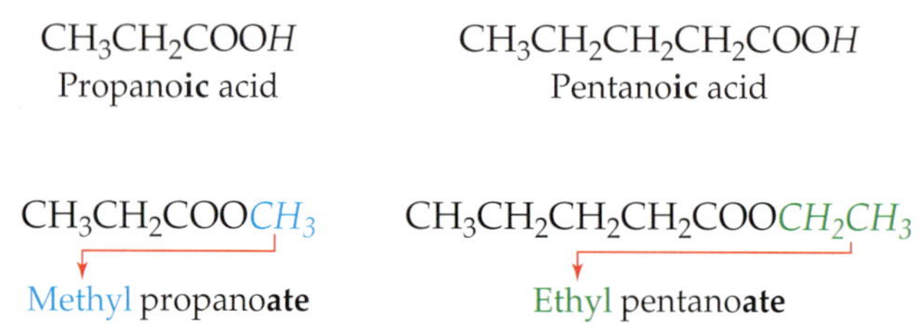

About Carboxylic Acids and Esters

Like all acids, carboxylic acids taste sour. The most familiar carboxylic acid is ethanoic acid, which is known by its common name, *acetic acid*. Acetic acid is the active ingredient in vinegar. It can be formed by the oxidation of ethanol, which is why wines left open to air become sour. Some yeasts and bacteria also form acetic acid when they metabolize sugars in bread dough. These are often added to bread dough to make sourdough bread. Other common carboxylic acids include methanoic acid (usually called formic acid), the acid present in bee stings and ant bites; citric acid, the acid present in limes, lemons, and oranges; and lactic acid, the acid that causes muscle soreness after intense exercise.

MOLECULE
Formic Acid

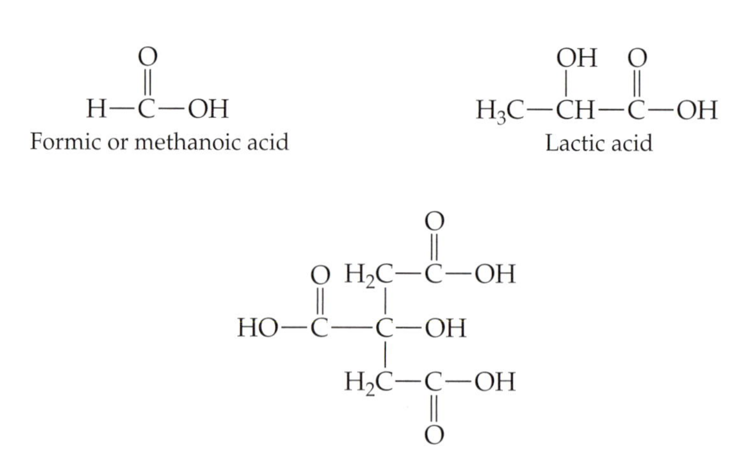

▲ Citric acid is responsible for the sour taste of limes and other citrus fruits. **Question:** Can you think of other fruits that might contain citric acid?

Esters are best known for their sweet smells. For example, ethyl butanoate is responsible for the smell and taste of pineapples, and methyl butanoate is responsible for the smell and taste of apples.

▲ Methyl butanoate is an ester found in apples.

$$H_3C-CH_2-CH_2-\overset{\overset{\Large O}{\|}}{C}-O-CH_2-CH_3$$

Ethyl butanoate

$$H_3C-CH_2-CH_2-\overset{\overset{\Large O}{\|}}{C}-O-CH_3$$

Methyl butanoate

Esters are formed from the reaction of a carboxylic acid and an alcohol as follows:

$$\underset{\text{Acid}}{R\overset{\overset{\Large O}{\|}}{C}-OH} + \underset{\text{Alcohol}}{HO-R'} \longrightarrow \underset{\text{Ester}}{R-\overset{\overset{\Large O}{\|}}{C}-O-R'} + \underset{\text{Water}}{H_2O}$$

An important example of this reaction is the formation of acetylsalicylic acid (aspirin) from acetic acid and salicylic acid (originally obtained from the bark of the willow tree).

$$\underset{\text{Acetic acid}}{CH_3\overset{\overset{\Large O}{\|}}{C}-OH} + \underset{\text{Salicylic acid}}{HO-\overset{\overset{\Large O}{\|}}{C}-C_6H_4-OH} \longrightarrow \underset{\text{Acetylsalicylic acid}}{HO-\overset{\overset{\Large O}{\|}}{C}-C_6H_4-O-\underset{\underset{\Large O}{\|}}{C}-CH_3} + \underset{\text{Water}}{H_2O}$$

CONCEPTUAL CHECKPOINT 18.1

Which of the following terms could *not* be applied to the compound shown?

(a) unsaturated
(b) aromatic
(c) an acid
(d) organic

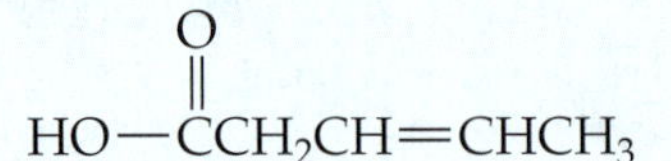

18.16 Amines

Amines are organic compounds containing nitrogen. The simplest nitrogen-containing compound is ammonia (NH_3). All other amines are derivatives of ammonia with one or more of the hydrogen atoms replaced by alkyl groups. They are systematically named according to the hydrocarbon groups attached to the nitrogen and given the ending *-amine*.

MOLECULE Ethylamine

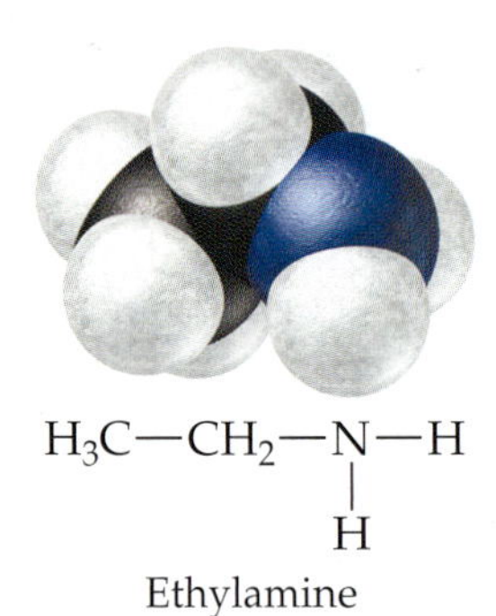

$H_3C-CH_2-N(H)-H$

Ethylamine

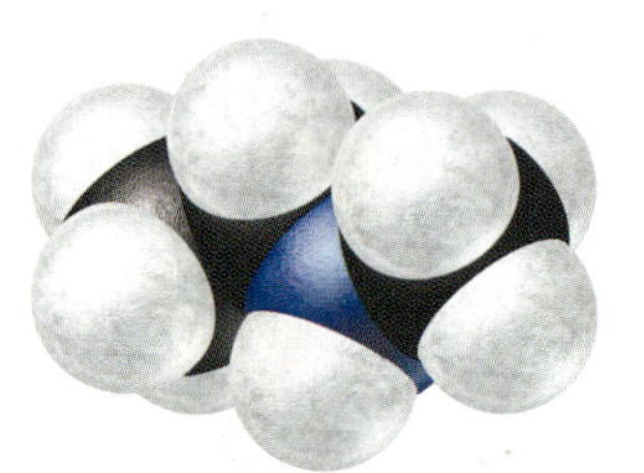

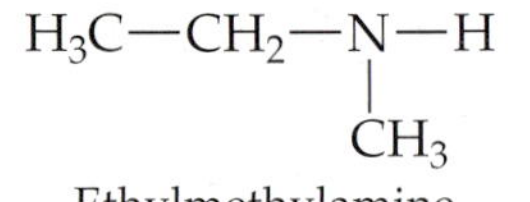

Ethylmethylamine

Amines are best known for their awful odors. When a living organism dies, bacteria that feast on its proteins emit amines. For example, trimethylamine causes the smell of rotten fish and cadaverine causes the smell of decaying animal flesh.

$H_3C-N(CH_3)-CH_3$

Trimethylamine

$H_2N-[CH_2]_5-NH_2$

Cadaverine

18.17 Polymers

Polymers are long chainlike molecules composed of repeating units. The individual repeating units are called **monomers**. In Chapter 19 we will learn about natural polymers such as starches, proteins, and DNA. These natural polymers play important roles in living organisms. In this section, we learn about synthetic polymers. Synthetic polymers compose many frequently encountered plastic products such as PVC tubing, Styrofoam coffee cups, nylon rope, and Plexiglas windows. Polymer materials are common in our everyday lives since they are found in everything from computers to toys to packaging materials. How many things can you think of that are made of plastic?

The simplest synthetic polymer is probably polyethylene. The polyethylene monomer is ethene (also called ethylene).

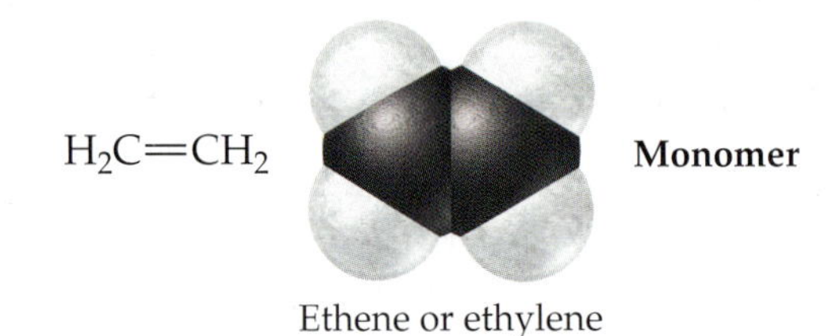

Ethene or ethylene

MOLECULE
Polyethylene

Ethene monomers can be made to react with each other, breaking the double bond between carbons and adding together to make a long polymer chain.

$$\cdots CH_2{-}CH_2{-}CH_2{-}CH_2{-}CH_2{-}CH_2{-}CH_2{-}CH_2{-}CH_2 \cdots$$

Polyethylene

Polyethylene is the plastic that composes milk jugs, juice containers, and garbage bags. It is an example of an **addition polymer**, a polymer in which the monomers simply link together without the elimination of any atoms.

An entire class of polymers can be thought of as substituted polyethylenes. For example, polyvinyl chloride (PVC)—the plastic used to make certain kinds of pipes and plumbing fixtures—is composed of monomers in which a chlorine atom has been substituted for one of the hydrogen atoms in ethene.

▲ Polyethylene is widely used in containers for beverages.

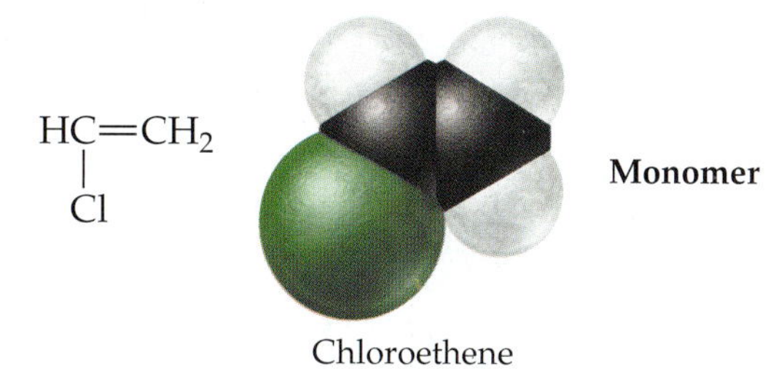

Chloroethene

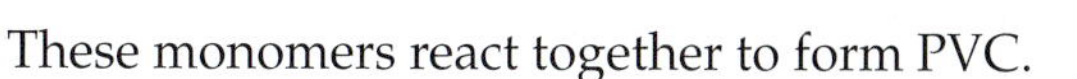

These monomers react together to form PVC.

$$\cdots \underset{\mid \atop Cl}{CH}{-}CH_2{-}\underset{\mid \atop Cl}{CH}{-}CH_2{-}\underset{\mid \atop Cl}{CH}{-}CH_2{-}\underset{\mid \atop Cl}{CH}{-}CH_2{-}\underset{\mid \atop Cl}{CH} \cdots$$

Polyvinyl chloride (PVC)

▶ Polyvinyl chloride (PVC) is used for pipes and plumbing fittings.

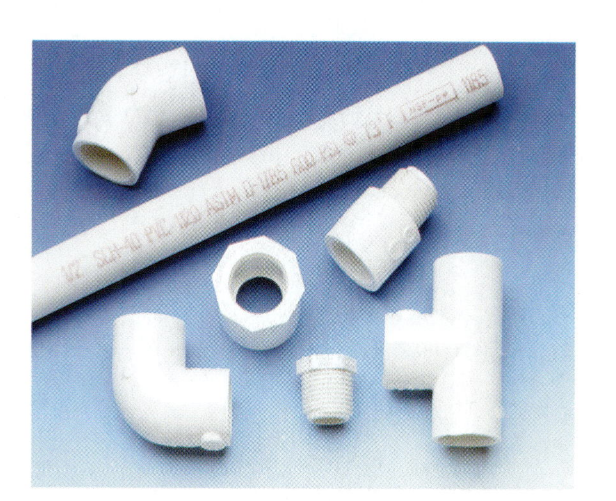

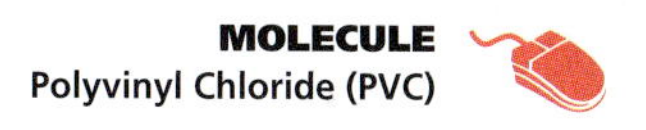

TABLE 18.8 Polymers of Commercial Importance

Polymer	Structure	Uses
Addition polymers		
polyethylene	$-\!\!\left(CH_2-CH_2\right)\!\!-_n$	films, packaging, bottles
polypropylene	$\left[CH_2-CH(CH_3)\right]_n$	kitchenware, fibers, appliances
polystyrene	$\left[CH_2-CH(C_6H_5)\right]_n$	packaging, disposable food containers, insulation
polyvinyl chloride	$\left[CH_2-CH(Cl)\right]_n$	pipe fittings, clear film for meat packaging
Condensation polymers		
polyurethane	$\left[C(=O)-NH-R-NH-C(=O)-O-R'-O\right]_n$ R, R′ = $-CH_2-CH_2-$ (for example)	"foam" furniture stuffing, spray-on insulation, automotive parts, footwear, water-protective coatings
polyethylene terephthalate (a polyester)	$\left[O-CH_2-CH_2-O-C(=O)-C_6H_4-C(=O)\right]_n$	tire cord, magnetic tape, apparel, soft-drink bottles
nylon 6,6	$\left[NH-(CH_2)_6-NH-C(=O)-(CH_2)_4-C(=O)\right]_n$	home furnishings, apparel, carpet fibers, fish line, polymer blends

Table 18.8 shows several other substituted polyethylene polymers.

Some polymers—called **copolymers**—consist of two different kinds of monomers. For example, the monomers that compose nylon 6,6 are hexamethylenediamine and adipic acid. These two monomers add together by eliminating a water molecule for each bond that forms between monomers. Polymers that eliminate an atom or a small group of atoms during polymerization are called **condensation polymers**.

$$H_2N-CH_2CH_2CH_2CH_2CH_2CH_2-NH_2 \quad + \quad HO-\overset{O}{\overset{\|}{C}}-CH_2CH_2-CH_2-CH_2-\overset{O}{\overset{\|}{C}}-OH \qquad \textbf{Monomers}$$

Hexamethylenediamine Adipic acid

$$\downarrow$$

$$H_2N-CH_2CH_2CH_2CH_2CH_2CH_2-\overset{H}{\overset{|}{N}}-\overset{O}{\overset{\|}{C}}-CH_2CH_2-CH_2-CH_2-\overset{O}{\overset{\|}{C}}-OH + H_2O \qquad \textbf{Dimer}$$

The product that forms between the reaction of two monomers is called a **dimer**. The polymer (nylon 6,6) forms as the dimer continues to add more monomers to form a polymer. Nylon 6,6 and other similar nylons can be drawn into fibers and used to make consumer products such as pantyhose, carpet fibers, and fishing line. Table 18.8 shows other condensation polymers.

EVERYDAY Chemistry

Kevlar: Stronger Than Steel

In 1965, Stephanie Kwolek, working for DuPont to develop new polymer fibers, noticed an odd cloudy product from a polymerization reaction. Some researchers might have rejected the product, but Kwolek insisted on examining its properties more carefully. The results were astonishing—when the polymer was spun into a fiber, it was stronger than any other fiber known before. Kwolek had discovered Kevlar, a material that is pound for pound five times stronger than steel.

Kevlar is a condensation polymer containing aromatic rings and amide linkages:

Aromatic ring

Amide linkage

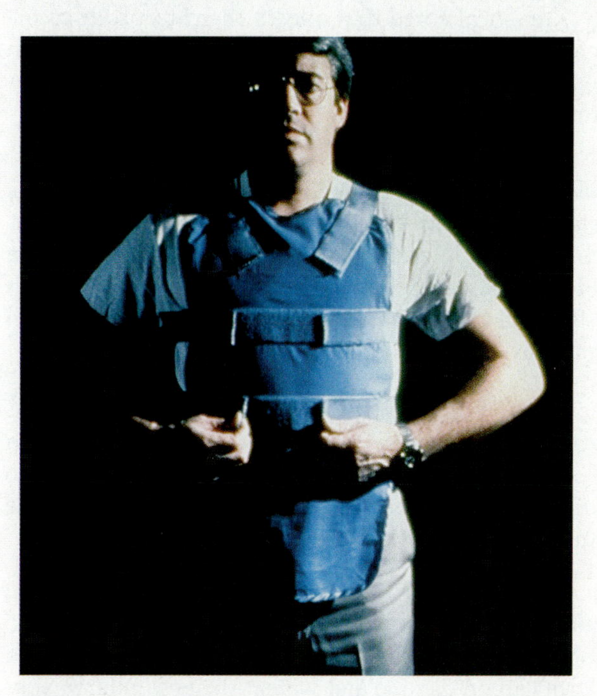

▲ The great strength of Kevlar fibers makes this polymer ideal for use in bullet-proof vests.

The polymeric chains within Kevlar crystallize in a parallel arrangement (like dry spaghetti noodles in a box), with strong cross-linking between neighboring chains due to hydrogen bonding. The hydrogen bonding occurs between the —N—H groups on one chain and the C=O groups on neighboring chains:

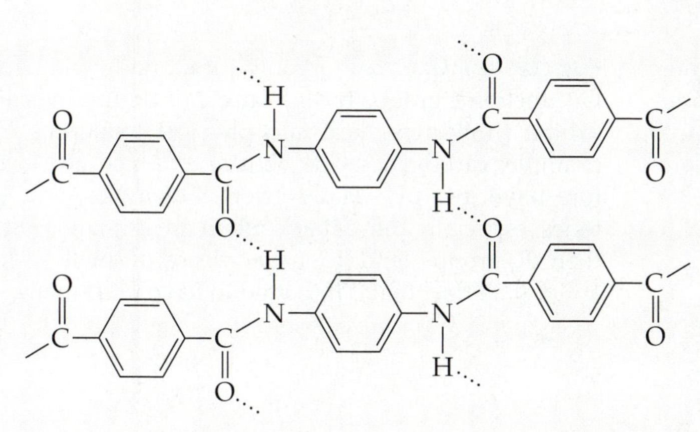

This structure is responsible for Kevlar's high strength and its other properties, including chemical resistance and flame resistance.

Today, DuPont sells hundreds of millions of dollars' worth of Kevlar every year. Kevlar is particularly well known for its use in bulletproof vests. With this application alone, Kwolek's discovery has saved thousands of lives. In addition, Kevlar is used to make helmets, radial tires, brake pads, racing sails, suspension bridge cables, skis, and high-performance hiking and camping gear.

CAN YOU ANSWER THIS? *Examine the structure of the Kevlar polymer. Knowing that the polymer is a condensation polymer, draw the structures of the monomers before the condensation reaction.*

CHAPTER IN REVIEW

Chemical Principles

Organic Chemistry: Organic chemistry is the study of carbon-containing compounds and their reactions. Carbon is unique because it can form four bonds and can bond to itself to form chains, branched structures, and ring structures.

Vitalism: Vitalism is the belief that living things contain a mystical force that allows them to produce organic compounds. Vitalism was overthrown when Friedrich Wöhler synthesized urea, an organic compound, in his laboratory.

Hydrocarbons: Hydrocarbons are organic compounds containing only carbon and hydrogen. They can be classified into four different types: alkanes, with the formula C_nH_{2n+2}; alkenes, with the formula C_nH_{2n}; alkynes, with the formula C_nH_{n-2}; and aromatic hydrocarbons, which contain six-carbon-atom ring structures.

Isomers: Isomers are two or more compounds with the same chemical formula but different structures. The more atoms in an organic compound, the greater the number of isomers that exist.

Hydrocarbon Reactions: All hydrocarbons undergo combustion, the reaction of the hydrocarbon with oxygen to form carbon dioxide and water. Alkanes undergo substitution reactions, in which one or more hydrogen atoms on the alkane are replaced by one or more other atoms, usually a halogen. Alkenes and alkynes undergo addition reactions, in which two atoms—often hydrogen or halogens—add across the multiple bond.

Functional Groups: Most other families of organic compounds can be thought of as substituted hydrocarbons in which a functional group has been inserted into the hydrocarbon. The main families of organic compounds according to their functional groups are:

alcohols	R—OH
ethers	R—O—R
aldehydes	R—COH
ketones	R—CO—R
carboxylic acids	R—COOH
esters	R—COO—R
amines	R_3N

Relevance

Organic Chemistry: Organic compounds are prevalent in food, drugs, petroleum products, and pesticides. Organic chemistry is also the basis for living organisms.

Vitalism: The demise of vitalism opened all of life to scientific inquiry and led to our current understanding of organic chemistry.

Hydrocarbons: One of the main uses of hydrocarbons is as fuels. Petroleum and natural gas are both composed of hydrocarbons. Hydrocarbons are also the starting materials in the synthesis of many consumer products including fabrics, soaps, dyes, cosmetics, drugs, plastics, and rubber.

Isomers: Although isomers may be closely related chemically, they are different compounds with different properties.

Hydrocarbon Reactions: The combustion of hydrocarbons provides about 90% of U.S. energy. We use the combustion of hydrocarbons to heat our homes, to propel our automobiles, and to generate electricity. Substitution reactions, especially hydrogenation, are important in the petroleum industry.

Functional Groups: Organic compounds with similar functional groups form a family of compounds and exhibit similar chemical and physical properties. For example, carboxylic acids behave as acids and therefore have low pH. Many esters, aldehydes, and ketones, especially those that contain an aromatic ring in their R group, tend to have pleasant smells. Most amines, on the other hand, tend to have foul smells.

Polymers: Polymers are long chainlike molecules composed of repeating units called monomers. Polymers that are formed from monomers that simply attach together, without eliminating any atoms or groups of atoms, are called addition polymers. Polymers that eliminate water or other atoms during the polymerization process are called condensation polymers. Copolymers are composed of two different kinds of monomers.

Polymers: Many consumer products are composed of polymers. The plastic you use for trash bags and the plastic container you buy water in are both composed of polyethylene. Nylon is used to make clothing and fishing line and polyvinyl chloride composes plastic pipes and pipe fittings.

Chemical Skills

Differentiating between Alkanes, Alkenes, and Alkynes Based on Their Molecular Formulas (Section 18.4)

Examine the formula for the molecule in question, and, based on the relative number of carbon and hydrogen atoms, determine whether it is an alkane, alkene, or alkyne.

Alkanes	C_nH_{2n+2}
Alkenes	C_nH_{2n}
Alkynes	C_nH_{n-2}

These formulas apply to open-chain (noncyclical) hydrocarbons only.

Writing Structural Formulas for Hydrocarbon Isomers (Section 18.6)

To write structural formulas for a set of hydrocarbon isomers, begin by writing the carbon backbone in as many unique ways as possible. Include double or triple carbon–carbon bonds in your backbone structures.

Once you have determined all of the unique backbone structures, add hydrogen atoms to give each carbon atom four bonds.

Examples

EXAMPLE 18.9 Differentiating between Alkanes, Alkenes, and Alkynes Based on Their Molecular Formulas

Identify each of the following noncyclical hydrocarbons as an alkane, alkene, or alkyne.

(a) C_3H_6
(b) C_6H_{14}
(c) $C_{10}H_{18}$

Solution:

(a) C_3H_6
$n = 3, 6 = 2n$; therefore, this is an alkene.
(b) C_6H_{14}
$n = 6, 14 = 2n + 2$; therefore, this is an alkane.
(c) $C_{10}H_{18}$
$n = 10, 18 = 2n - 2$; therefore, this is an alkyne.

EXAMPLE 18.10 Writing Structural Formulas for Hydrocarbon Isomers

Write structural formulas for all of the isomers of pentyne (C_5H_8).

Solution:
Pentyne has five carbon atoms with one triple bond between two of them. The possible backbone structures are:

C≡C—C—C—C

C—C≡C—C—C

```
C≡C—C—C
    |
    C
```

We then add hydrogen atoms so that each carbon atom forms four bonds.

$HC{\equiv}C{-}CH_2{-}CH_2{-}CH_3$

$H_3C{-}C{\equiv}C{-}CH_2{-}CH_3$

```
HC≡C—CH—CH3
     |
     H3C
```

Naming Alkanes (Section 18.7)

For a full review of naming alkanes, see the rules in Section 18.7. The following is a condensed version.

1. Count the number of carbon atoms in the longest continuous carbon chain to determine the base name of the compound. Use the ending *-ane*.
2. Consider every branch from the base chain to be a substituent. Name each substituent according to the number of carbon atoms in the substituent.
3. Beginning with the end closest to the branching, number the base chain and assign a number to each substituent.
4. Write the name of the compound in the following format.

   (subst. number)-(subst. name)(base name)

   If there are two or more substituents, give each one a number and list them alphabetically. Words and numbers are separated by hyphens.
5. If a compound has two or more identical substituents, use the prefix *di-* (2), *tri-* (3), or *tetra-* (4) before their name. Separate the numbers indicating their positions relative to each other with a comma.

EXAMPLE 18.11 Naming Alkanes

Name the following alkane.

$$\begin{array}{l} CH_3\underset{\displaystyle CH_3}{\underset{|}{C}}HCH_2\underset{\displaystyle CH_3}{\underset{|}{C}}HCH_3 \end{array}$$

Solution:

The longest continuous chain has five carbon atoms. Therefore, the base name is *pentane*.

$$\begin{array}{l} \overset{1}{C}H_3\underset{\displaystyle CH_3}{\underset{|}{\overset{2}{C}}}H\overset{3}{C}H_2\underset{\displaystyle CH_3}{\underset{|}{\overset{4}{C}}}H\overset{5}{C}H_3 \end{array}$$

The substituents are both methyl groups.

The substituents are at the 2 and 4 positions.
The name of this compound is 2,4-dimethylpentane.

Naming Alkenes and Alkynes (Section 18.8)

Alkenes and alkynes are named similarly to alkanes with the following exceptions.

- The base chain must contain the double or triple bond.
- The base name has the ending *-ene* for alkenes and *-yne* for alkynes.
- The base chain is numbered to give the double or triple bond the lowest possible number.
- A number indicating the position of the double bond (lowest possible number) is inserted just before the base name.

EXAMPLE 18.12 Naming Alkenes and Alkynes

Name the following alkyne.

$$\begin{array}{l} CH_3\underset{\displaystyle CH_3}{\underset{|}{C}}H_2C{\equiv}CCH_3 \end{array}$$

Solution:

4-methyl-2-pentyne

Naming Aromatic Compounds (Section 18.10)

The names of substituted benzenes have the following general forms.

Monosubstituted benzenes

(Name of substituent)*benzene*

Compounds in which benzene is better viewed as a substituent

Name the benzene ring as a substituent with the name phenyl.

Disubstituted benzenes

#-(substituent name)-#-(substituent name)benzene

- The substituents are listed in alphabetical order.
- The order of numbering within the ring is also determined by the names of the substituents.
- The prefixes *ortho-* (1,2), *meta-* (1,3), and *para-* (1,4) are often used in place of numbers when the substituents are identical.

EXAMPLE 18.13 Naming Aromatic Compounds

The following are examples of aromatic compounds and their names.

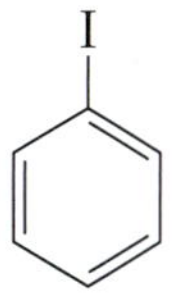

I

Iodobenzene

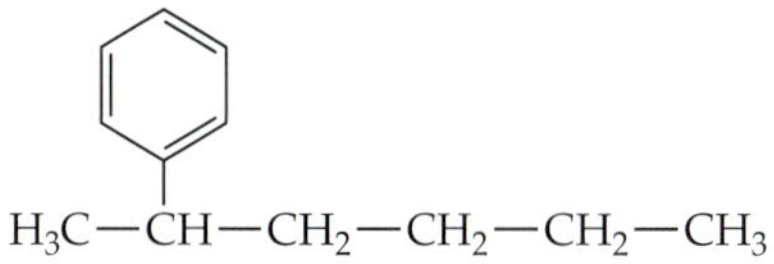

$H_3C-CH-CH_2-CH_2-CH_2-CH_3$

2-Phenylhexane

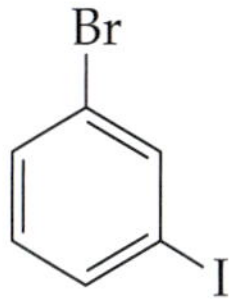

Br

I

1-Bromo-3-iodobenzene

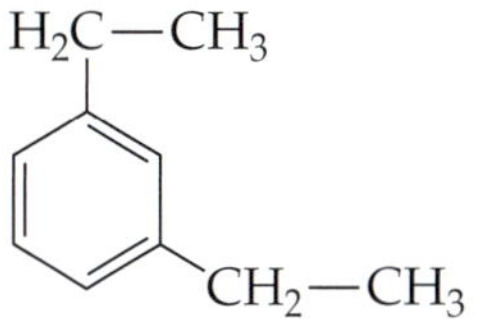

H_2C-CH_3

CH_2-CH_3

meta-Diethylbenzene or *m*-Diethylbenzene

Key Terms

addition polymer [18.17]
addition reaction [18.9]
alcohol [18.11]
aldehyde [18.14]
alkane [18.5]
alkene [18.8]
alkyl group [18.7]
alkyne [18.8]
amine [18.16]
aromatic ring [18.10]
base chain [18.7]
branched alkane [18.6]
carbonyl group [18.14]
carboxylic acid [18.15]
combustion [18.9]
condensation polymer [18.17]
condensed structural formula [18.5]
copolymer [18.17]
dimer [18.17]
disubstituted benzene [18.10]
ester [18.15]
ethers [18.13]
family (of organic compounds) [18.11]
fossil fuels [18.4]
functional group [18.11]
hydrocarbon [18.4]
hydrogenation [18.9]
isomer [18.6]
ketone [18.14]
monomer [18.17]
monosubstituted benzene [18.10]
normal alkane (*n*-alkane) [18.5]
organic chemistry [18.1]
organic molecule [18.1]
phenyl group [18.10]
polymer [18.17]
saturated hydrocarbon [18.5]
structural formula [18.5]
substituent [18.7]
substitution reaction [18.9]
unsaturated hydrocarbon [18.8]
vital force [18.2]
vitalism [18.2]

Exercises

Questions

1. What kinds of molecules are often involved in smell?
2. What is organic chemistry?
3. Explain the difference—as it was viewed at the end of the eighteenth century—between organic and inorganic compounds.
4. What is vitalism?
5. How was vitalism overthrown?
6. What is unique about carbon and carbon-based compounds? Why did life evolve around carbon?
7. Describe the geometry about a carbon atom that forms:
   (a) four single bonds
   (b) two single bonds and one double bond
   (c) one single bond and one triple bond
8. What are hydrocarbons?
9. What are the main uses of hydrocarbons?
10. What are the main classifications of hydrocarbons? What are their generic molecular formulas?
11. What is the difference between saturated and unsaturated hydrocarbons?
12. Explain the difference between a molecular formula, a structural formula, and a condensed structural formula.
13. Explain the difference between *n*-alkanes and branched alkanes.
14. What are isomers? Give some examples.
15. What are alkenes? How are they different from alkanes?
16. What are alkynes? How are they different from alkanes?
17. What are hydrocarbon combustion reactions? Give an example.
18. What are alkane substitution reactions? Give an example.
19. What is an alkene addition reaction? Give an example.
20. What is an alkyne addition reaction? Give an example.
21. What is the structure of benzene? What are the different ways in which this structure is represented?
22. What is a functional group? Give some examples.
23. What is the generic structure of alcohols? Write the structures of two specific alcohols.
24. Give examples of some common alcohols and where you might find them.
25. What is the generic structure of ethers? Write the structures of two specific ethers.
26. Give an example of a common ether and its main uses.
27. What are the generic structures of aldehydes and ketones? Write the structure of a specific aldehyde and a specific ketone.
28. Give some examples of common aldehydes and ketones and where you might find them.
29. What are the generic structures of carboxylic acids and esters? Write the structure of a specific carboxylic acid and a specific ester.
30. Give some examples of common carboxylic acids and esters and where you might find them.
31. What is the generic structure of amines? Write the structures of two specific amines.
32. Give an example of a common amine and where you might find it.
33. Explain what a polymer is and state the difference between a polymer and a copolymer.
34. Explain the difference between an addition polymer and a condensation polymer.

Problems

Hydrocarbons

35. Which of the following are hydrocarbons?
   (a) $C_5H_{12}O$
   (b) NH_3
   (c) C_8H_{16}
   (d) C_2H_6

36. Which of the following are hydrocarbons?
   (a) CH_2O
   (b) CH_4
   (c) C_6H_6
   (d) C_2H_7N

37. Based on the molecular formula, determine whether each of the following is an alkane, alkene, or alkyne.
   (a) C_5H_{12}
   (b) C_2H_4
   (c) C_8H_{14}
   (d) C_4H_8

38. Based on the molecular formula, determine whether each of the following is an alkane, alkene, or alkyne.
   (a) $C_{10}H_{20}$
   (b) C_9H_{16}
   (c) C_7H_{16}
   (d) C_5H_8

Alkanes

39. Write a structural formula and a condensed structural formula for each of the following alkanes.
    (a) ethane
    (b) butane
    (c) pentane
    (d) methane

40. Write a structural formula and a condensed structural formula for each of the following alkanes.
    (a) propane
    (b) hexane
    (c) octane
    (d) heptane

41. Write structural formulas for each of the two isomers of butane.

42. Write structural formulas for each of the three isomers of pentane.

43. Write structural formulas for each of the five isomers of hexane.

44. Write structural formulas for any five of the nine isomers of heptane.

45. Name each of the following alkanes.

(a) $H_3C-CH_2-CH_2-CH_2-CH_3$

(b) 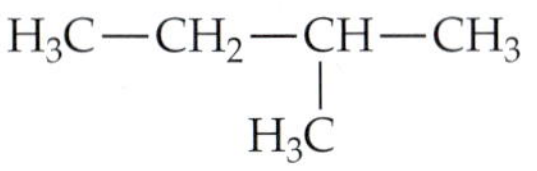 $H_3C-CH_2-CH-CH_3$
$\qquad\qquad\qquad|$
$\qquad\qquad\quad H_3C$

(c) 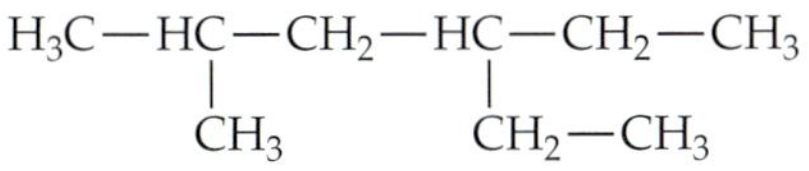 $H_3C-HC-CH_2-HC-CH_2-CH_3$
$\qquad\qquad|\qquad\qquad\quad|$
$\qquad\quad CH_3\qquad\quad CH_2-CH_3$

(d) 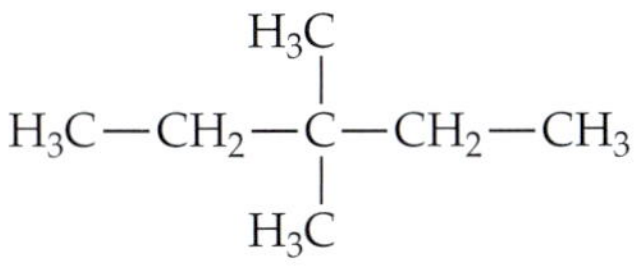
$\qquad\qquad\quad H_3C$
$\qquad\qquad\qquad|$
$H_3C-CH_2-C-CH_2-CH_3$
$\qquad\qquad\qquad|$
$\qquad\qquad\quad H_3C$

46. Name each of the following alkanes.

(a) $H_3C-CH_2-CH_2-CH_3$

(b) 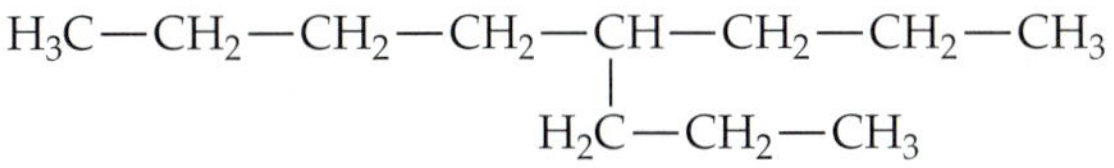 $H_3C-CH_2-CH_2-CH_2-CH-CH_2-CH_2-CH_3$
$\qquad\qquad\qquad\qquad\qquad\qquad|$
$\qquad\qquad\qquad\qquad\qquad H_2C-CH_2-CH_3$

(c) 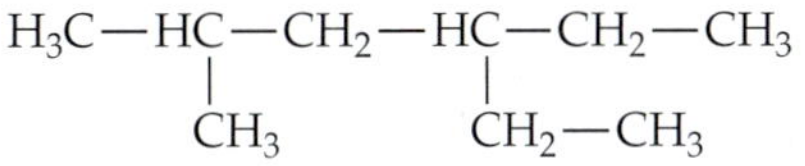 $H_3C-HC-CH_2-HC-CH_2-CH_3$
$\qquad\qquad|\qquad\qquad\quad|$
$\qquad\quad CH_3\qquad\quad CH_2-CH_3$

(d) 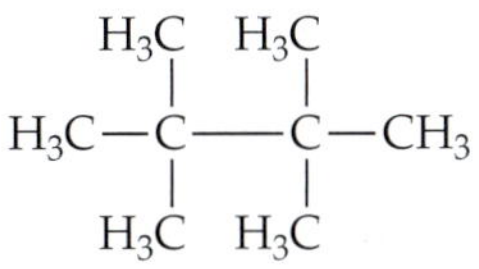
$\qquad\quad H_3C\quad H_3C$
$\qquad\qquad|\qquad\ \ |$
$H_3C-C-\!\!-C-CH_3$
$\qquad\qquad|\qquad\ \ |$
$\qquad\quad H_3C\quad H_3C$

47. Draw a structure for each of the following alkanes.
    (a) 2-methylbutane
    (b) 3-ethyl-2-methylhexane
    (c) 3-isopropylheptane
    (d) 2,5-dimethyloctane

48. Draw a structure for each of the following alkanes.
    (a) 3-ethylhexane
    (b) 3,3-dimethylpentane
    (c) 3-ethyl-3-methylpentane
    (d) 4,4-diethyloctane

49. Determine what is wrong with the names of each of the following alkanes and provide the correct name.

(a) 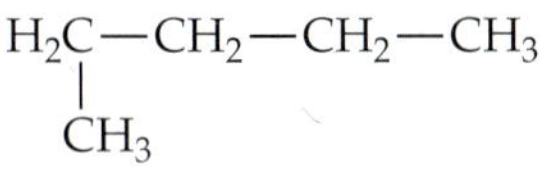 $H_2C-CH_2-CH_2-CH_3$
$\quad\ \ |$
$\quad CH_3$

1-Methylbutane

(b) 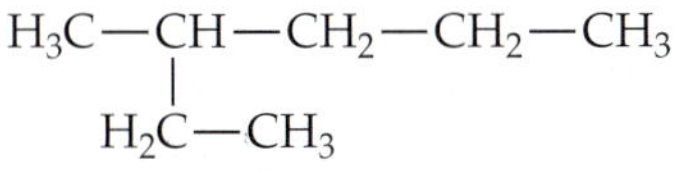 $H_3C-CH-CH_2-CH_2-CH_3$
$\qquad\qquad|$
$\qquad\ \ H_2C-CH_3$

2-Ethylpentane

(c) 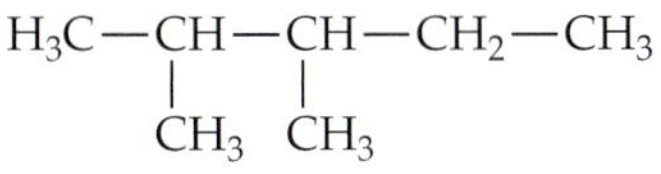 $H_3C-CH-CH-CH_2-CH_3$
$\qquad\qquad|\quad\ \ |$
$\qquad\quad CH_3\ CH_3$

2-Methyl-3-methyl pentane

50. Determine what is wrong with the names of each of the following alkanes and provide the correct name.

(a) 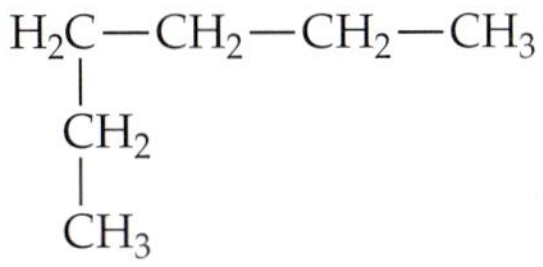 $H_2C-CH_2-CH_2-CH_3$
$\quad\ \ |$
$\quad CH_2$
$\quad\ \ |$
$\quad CH_3$

2-Ethyl-butane

(b) 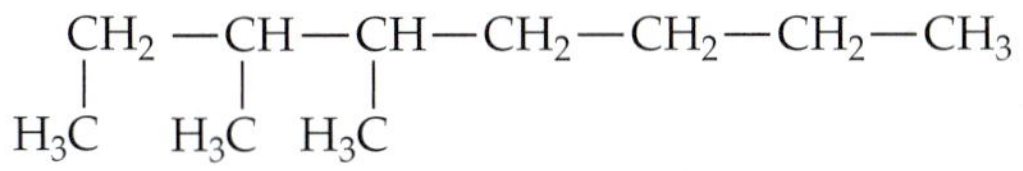 $CH_2-CH-CH-CH_2-CH_2-CH_2-CH_3$
$\ \ \ |\qquad\ \ |\qquad |$
$H_3C\quad H_3C\ H_3C$

1,2,3-Trimethylheptane

(c) 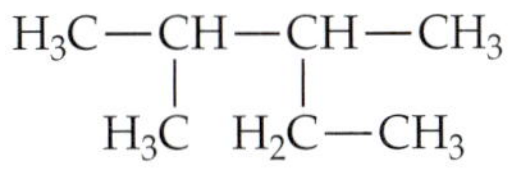 $H_3C-CH-CH-CH_3$
$\qquad\qquad|\quad\ \ |$
$\qquad\quad H_3C\ H_2C-CH_3$

2-Ethyl-3-methylbutane

51. Complete the following table:

Name	*Molecular Formula*	*Structural Formula*	*Condensed Structural Formula*
2,2,3-trimethylpentane	______	______	______
______	______	______	$CH_3CH(CH_3)CH(CH_2CH_2CH_3)CH_2CH_2CH_3$
______	______	CH_3 CH_3 | | $CH_3-C-C-CH_2-CH_2-CH_3$ | | CH_3 CH_3	______
______	______	______	$CH_3CH(CH_3)CH(CH_3)CH(CH_2CH_3)_2CH_2CH_3$

52. Complete the following table:

Name	*Molecular Formula*	*Structural Formula*	*Condensed Structural Formula*
______	______	______	$CH_3C(CH_3)_2C(CH_2CH_3)_2CH_2CH_2CH_2CH_2CH_3$
______	______	CH_3 | CH_2 CH_3 | | $CH_3-CH_2-C-C-CH_2-CH_2-CH_2-CH_3$ | | CH_2 CH_3 | CH_3	______
2,2-dimethyl-3-ethylpentane	______	______	______
______	______	H | $CH_3-CH_2-CH_2-C-CH_2-CH_2-CH_2-CH_2-CH_3$ | CH_2 | CH_2 | CH_3	______

Alkenes and Alkynes

53. Write a structural formula and a condensed structural formula for any two alkenes.

54. Write a structural formula and a condensed structural formula for any two alkynes.

55. Write structural formulas for all of the possible isomers of *n*-pentene that can be formed by moving the position of the double bond.

56. Write structural formulas for all of the possible isomers of *n*-hexyne that can be formed by moving the position of the triple bond.

57. Name each of the following alkenes.

(a) $H_3C-CH=CH-CH_2-CH_3$

(b) $H_3C-HC-CH=CH-CH_3$ (with CH_3 attached to the HC carbon)

58. Name each of the following alkenes

(a) $H_2C=CH-CH_2-CH_3$

(b) $H_3C-CH=CH-CH-CH_2-CH_3$ (with H_2C-CH_3 attached to the fourth carbon)

(c)

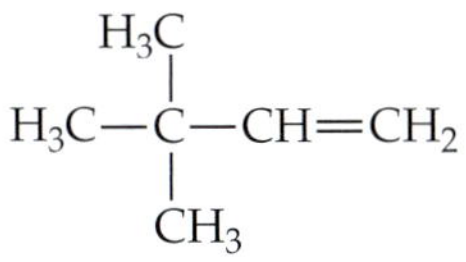

$$\begin{array}{c} \quad H_3C \\ \quad | \\ H_3C{-}C{-}CH{=}CH_2 \\ \quad | \\ \quad CH_3 \end{array}$$

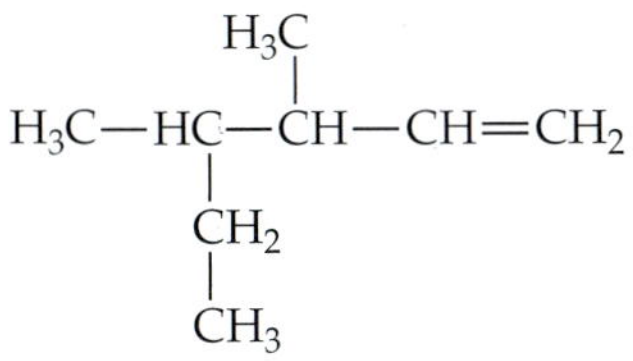

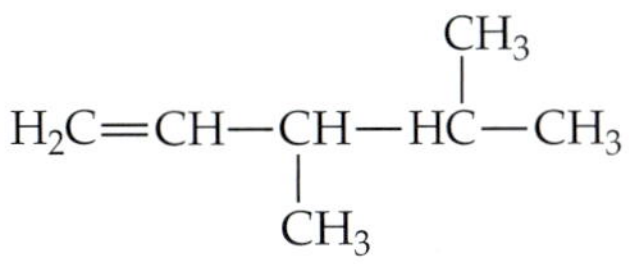

(d)

$$H_3C{-}HC(CH_2CH_3){-}CH(CH_3){-}CH{=}CH_2$$

(the second carbon carries $-CH_2-CH_3$ below; the third carbon carries H_3C above)

(c)

$$H_2C{=}CH{-}CH(CH_3){-}HC(CH_3){-}CH_3$$

(the CH carries CH_3 below; the HC carries CH_3 above)

(d) $H_2C{=}CH{-}CH{-}CH_2{-}CH_2{-}CH_3$

(the third carbon carries $H_3C{-}CH{-}CH_3$ below, i.e. $-CH(CH_3)-CH_3$)

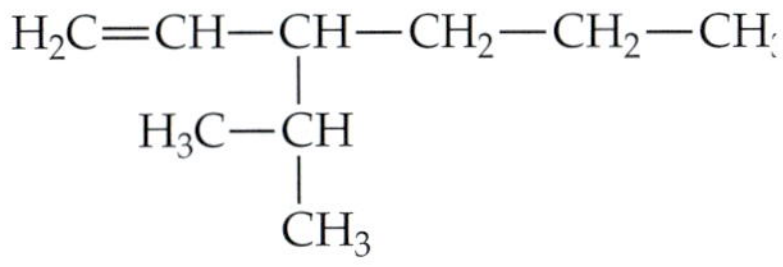

59. Name each of the following alkynes.

(a) $H_3C{-}C{\equiv}C{-}CH_3$

(b) $H_3C{-}C{\equiv}C{-}HC{-}CH_3$ (with CH_3 below HC)

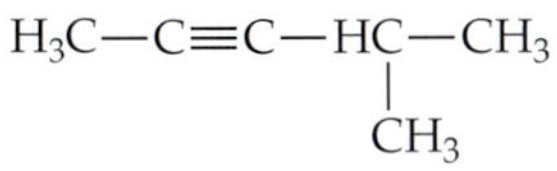

(c) $H_3C{-}C{\equiv}C{-}C{-}CH_2{-}CH_3$ (with CH_3 above and CH_3 below the quaternary C)

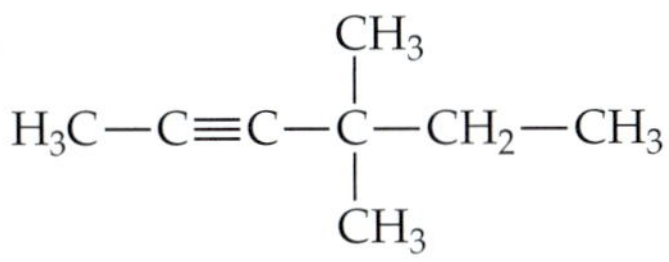

(d) $HC{\equiv}C{-}C{-}CH_2{-}CH_3$ (with CH_3 above and $CH_2{-}CH_3$ below the quaternary C)

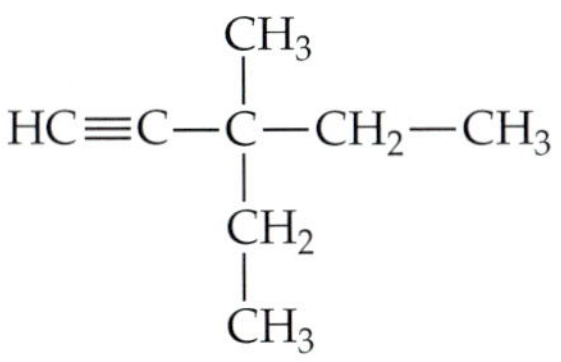

60. Name each of the following alkynes.

(a) $HC{\equiv}C{-}CH_2{-}CH_2{-}CH_3$

(b) $HC{\equiv}C{-}CH{-}CH_2{-}CH_2{-}CH_3$ (with $HC{-}CH_3$ below CH, and CH_3 below that HC)

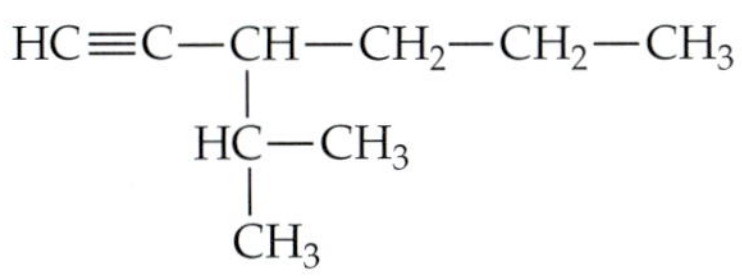

(c) $H_3C{-}HC{-}C{\equiv}C{-}CH{-}CH_3$ (with CH_3 below HC and H_3C below CH)

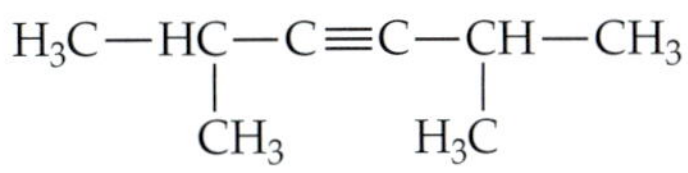

(d) $H_3C{\equiv}CH_2{-}C{\equiv}C{-}HC{-}CH_2{-}CH_3$ (with $CH_2{-}CH_3$ below HC)

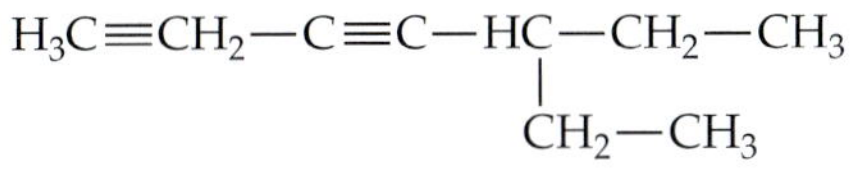

61. Provide correct structures for each of the following:
(a) 2-hexene
(b) 3-heptyne
(c) 3-methyl-1-pentyne
(d) 4,4-dimethyl-2-hexene

62. Provide correct structures for each of the following:
(a) 3-octyne
(b) 1-pentene
(c) 3,3-dimethyl-1-pentyne
(d) 4-ethyl-3-methyl-2-octene

63. Draw and name all of the possible isomers of $CH_2{=}CHCH_2CH_2CH_3$.

64. Draw and name all of the possible alkyne isomers of $CH{\equiv}CCH_2CH_2CH_2CH_3$.

65. Complete the following table:

Name	*Molecular Formula*	*Structural Formula*	*Condensed Structural Formula*
2,2-dimethyl-3-hexene	______	______	______
______	______	______	$CH_3C(CH_3)_2C(CH_2CH_3)_2C{\equiv}CCH_3$
______	______	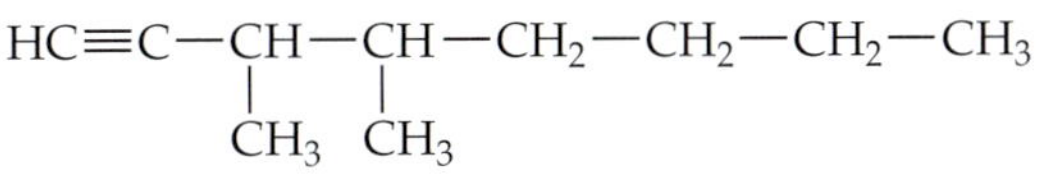 $HC{\equiv}C{-}CH{-}CH{-}CH_2{-}CH_2{-}CH_2{-}CH_3$ (with CH_3 below each CH)	______
______	______	______	$CH_3C(CH_3)_2C(CH_2CH_2)_2CH{=}CHCH_3$

66. Complete the following table:

Name	Molecular Formula	Structural Formula	Condensed Structural Formula
3-ethyl-4-methyl-1-heptene	________	________	________
________	________	________	$CH_3CH{=}C(CH_2CH_3)CH_2CH_3$
________	________	$HC{\equiv}C{-}CH(CH_3){-}CH(CH_2CH_2CH_3){-}CH_2{-}CH_2{-}CH_3$	________
3,3-dimethyl-4-ethyl-1-hexyne	________	________	________

Hydrocarbon Reactions

67. Complete and balance each of the following hydrocarbon combustion reactions.
(a) $CH_3CH_3 + O_2 \longrightarrow$
(b) $CH_2{=}CHCH_3 + O_2 \longrightarrow$
(c) $CH{\equiv}CH + O_2 \longrightarrow$

68. Complete and balance each of the following hydrocarbon combustion reactions.
(a) $CH_3CH_2CH_2CH_3 + O_2 \longrightarrow$
(b) $CH_2{=}CH_2 + O_2 \longrightarrow$
(c) $CH{\equiv}CCH_2CH_3 + O_2 \longrightarrow$

69. What are the products for the following alkane substitution reaction? (Assume monosubstitution.)

$$CH_4 + Br_2 \longrightarrow$$

70. What are the products for the following alkane substitution reaction? (Assume monosubstitution.)

$$CH_3CH_3 + I_2 \longrightarrow$$

71. What are the products of the following alkene addition reaction?

$$CH_3CH{=}CHCH_3 + Cl_2 \longrightarrow$$

72. What are the products of the following alkene addition reaction?

$$CH_3CH(CH_3)CH{=}CH_2 + Cl_2 \longrightarrow$$

73. Complete the following hydrogenation reaction.

$$CH_2{=}CH_2 + H_2 \longrightarrow$$

74. Complete the following hydrogenation reaction.

$$CH_3CH_2CH_2CH{=}CH_2 + H_2 \longrightarrow$$

Aromatic Hydrocarbons

75. Draw the full structural formula that is represented by the following shorthand formulas.

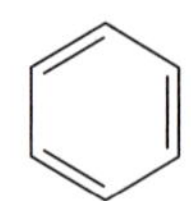

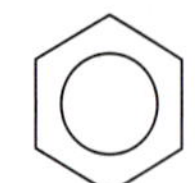

76. Explain how the two resonance structures *together* represent the true structure of benzene.

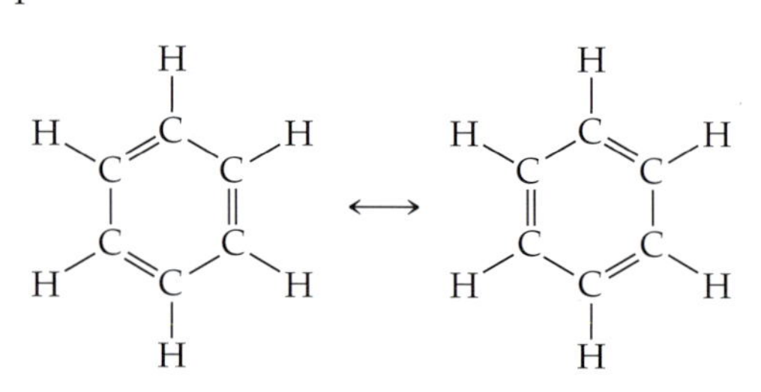

77. Name each of the following monosubstituted benzenes.

(a)
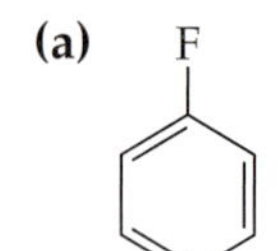

(b)
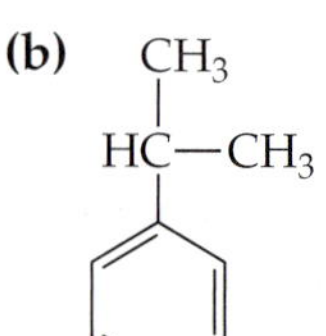

(c)
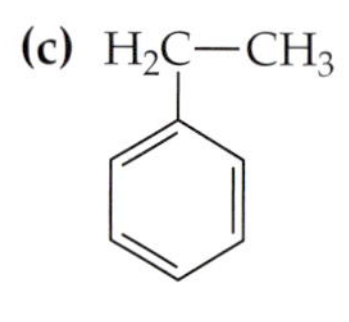

78. Name each of the following monosubstituted benzenes.

(a)
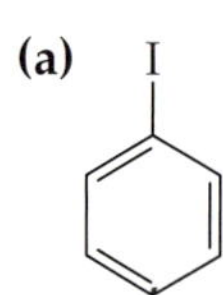

(b)
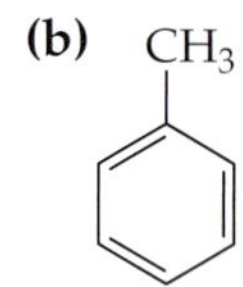

(c) $C_6H_5{-}C(CH_3)_2{-}CH_3$ (drawn as $H_3C{-}C(CH_3){-}CH_3$ attached to a benzene ring)

79. Name each of the following compounds in which the benzene ring is best treated as a substituent.

(a)

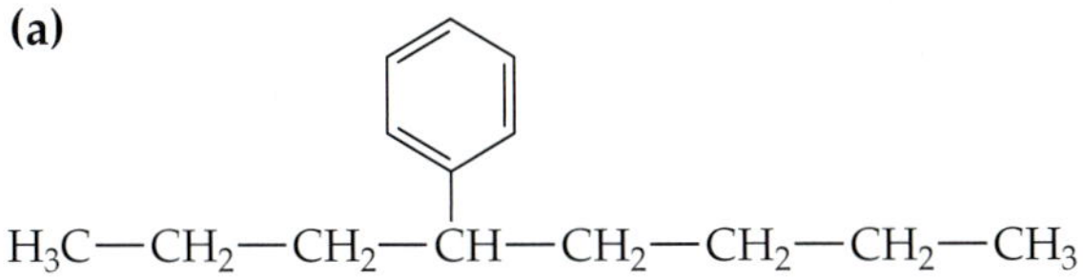

$H_3C{-}CH_2{-}CH_2{-}CH(C_6H_5){-}CH_2{-}CH_2{-}CH_2{-}CH_3$

(b)

$H_3C{-}CH_2{-}CH(C_6H_5){-}CH{=}CH{-}CH_2{-}CH_3$

(c)

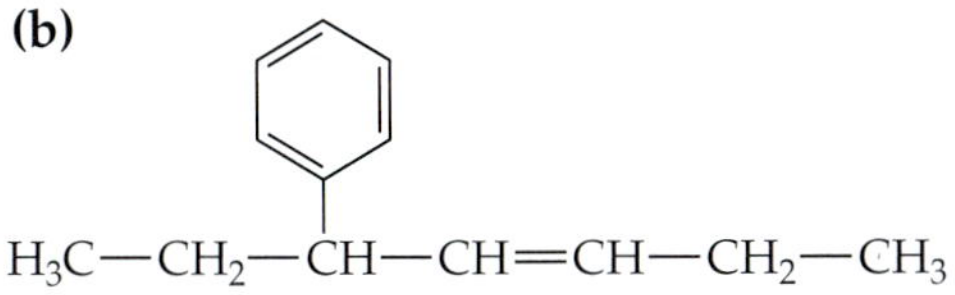

$(C_6H_5)CH_2{-}CH_2{-}CH_2{-}CH_2{-}C{\equiv}C{-}CH_3$

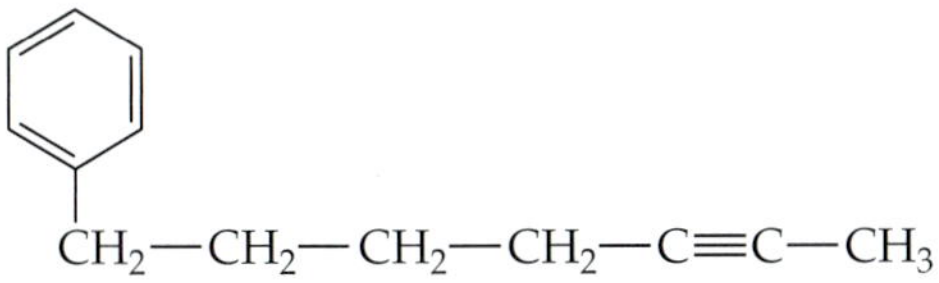

80. Name each of the following compounds in which the benzene ring is best treated as a substituent.

(a)

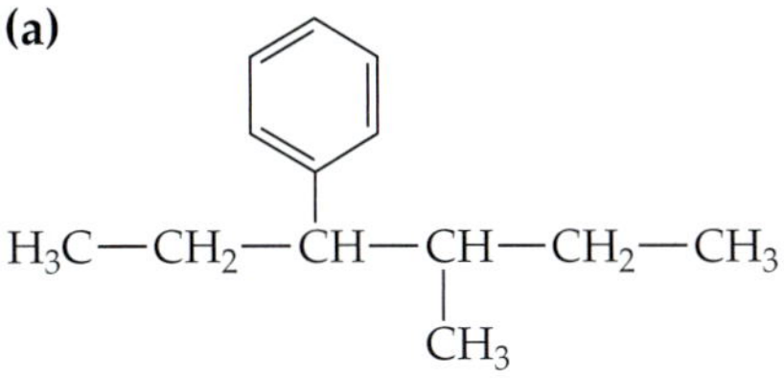

$H_3C{-}CH_2{-}CH(C_6H_5){-}CH(CH_3){-}CH_2{-}CH_3$

(b)

$H_3C{-}CH(C_6H_5){-}CH{=}CH{-}CH_2{-}CH_2{-}CH_2{-}CH_3$

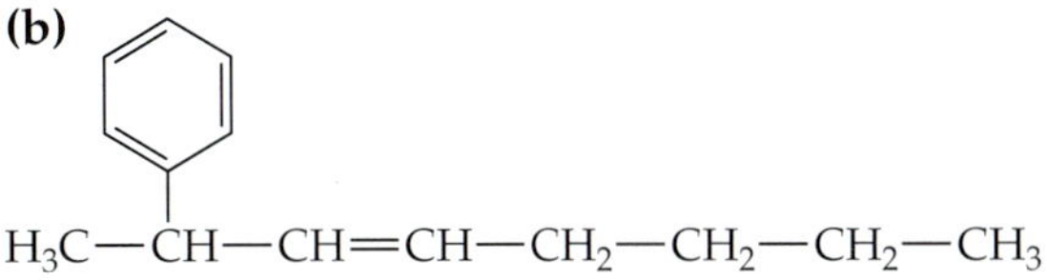

(c)

$H_3C{-}CH(C_6H_5){-}CH_2{-}CH(CH_2{-}CH_3){-}CH_2{-}CH_2{-}CH_3$

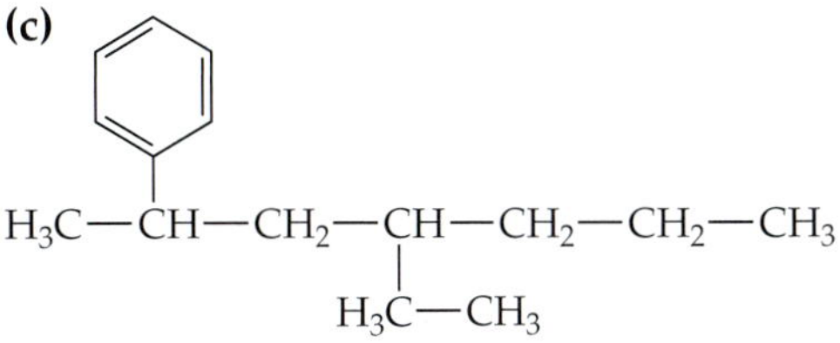

81. Name each of the following disubstituted benzenes.

(a) Br, Cl (benzene ring with adjacent Br and Cl)

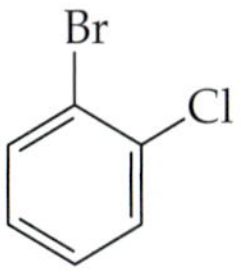

(b) $H_2C{-}CH_3$, $CH_2{-}CH_3$ (benzene ring with adjacent ethyl groups)

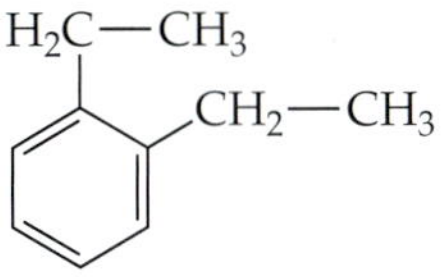

(c) F, F (benzene ring with F substituents at 1 and 3 positions)

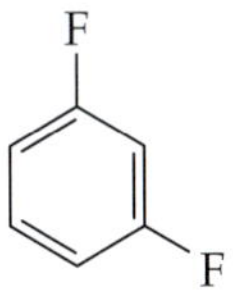

82. Name each of the following disubstituted benzenes.

(a) F, Cl (benzene ring with adjacent F and Cl)

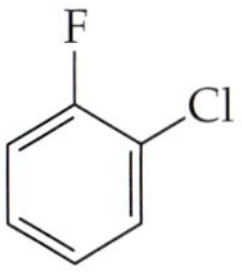

(b) F, $H_2C{-}CH_3$ (benzene ring with F and ethyl at opposite positions)

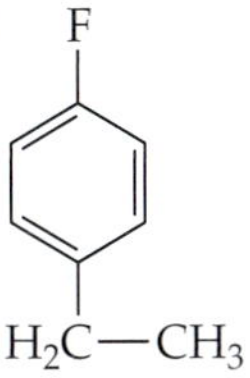

(c) I, I (benzene ring with I at opposite positions)

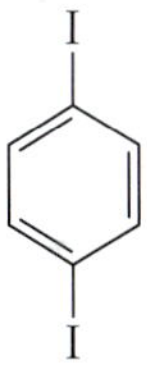

83. Draw structures for each of the following:
(a) butylbenzene
(b) 1-ethyl-2-iodobenzene
(c) paradimethylbenzene

84. Draw structures for each of the following:
(a) isopropylbenzene
(b) metadibromobenzene
(c) 1-bromo-4-ethylbenzene

Functional Groups

85. Based on its functional group, match the structure on the left with the correct name on the right.

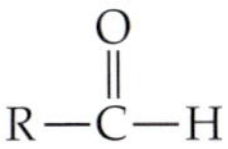

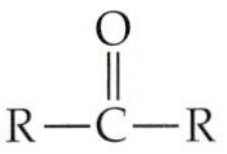

$R{-}C({=}O){-}H$	Ether
$R{-}C({=}O){-}R$	Aldehyde
$R{-}O{-}R$	Amine
$R{-}N(R){-}R$	Ketone

86. Based on its functional group, match the structure on the left with the correct name on the right.

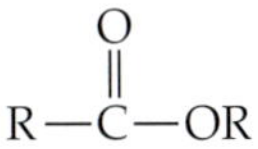

$R{-}C({=}O){-}OR$	Carboxylic acid
$R{-}C({=}O){-}OH$	Alcohol
$R{-}OH$	Ester
$R{-}O{-}R$	Ether

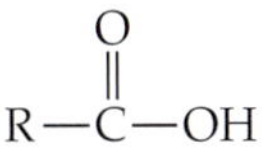

87. For each of the following molecules, identify the functional group and determine the family to which the molecule belongs.

(a)
$$\begin{array}{ccccccc} H_3C & - & CH_2 & - & CH_2 & - & NH \\ &&&&&& | \\ &&&&&& CH_3 \end{array}$$

(b)
$$\begin{array}{ccccccc} &&&&&& O \\ &&&&&& \| \\ H_3C & - & CH_2 & - & CH_2 & - & CH \end{array}$$

(c)
$$\begin{array}{ccccccc} && H_3C && CH_3 && \\ && | && | && \\ H_3C & - & C & - & C & - & OH \\ && | && | && \\ && H_3C && CH_3 && \end{array}$$

(d)
$$\begin{array}{ccccccccc} && CH_3 &&&&&& \\ && | &&&&&& \\ H_3C & - & C & - & O & - & CH_2 & - & CH_3 \\ && | &&&&&& \\ && CH_3 &&&&&& \end{array}$$

88. For each of the following molecules, identify the functional group and determine the family to which the molecule belongs.

(a)
$$\begin{array}{ccccccccc} &&&&&& O && \\ &&&&&& \| && \\ H_3C & - & CH_2 & - & CH_2 & - & C & - & CH_3 \end{array}$$

(b)
$$\begin{array}{ccccccccc} &&&& O &&&& \\ &&&& \| &&&& \\ H_3C & - & CH_2 & - & C & - & O & - & CH_3 \end{array}$$

(c)
$$\begin{array}{ccccc} && OH && \\ && | && \\ H_3C & - & HC & - & CH_3 \end{array}$$

(d)
$$\begin{array}{ccccccc} &&&& O && \\ &&&& \| && \\ H_3C & - & CH_2 & - & C & - & OH \end{array}$$

Alcohols

89. Name each of the following alcohols.

(a)
$$\begin{array}{ccccccc} && OH &&&& \\ && | &&&& \\ H_3C & - & HC & - & CH_2 & - & CH_3 \end{array}$$

(b)
$$\begin{array}{ccccc} OH &&&& \\ | &&&& \\ H_2C & - & CH & - & CH_3 \\ && | && \\ && H_3C && \end{array}$$

(c)
$$\begin{array}{ccccccccccccc} H_3C & - & CH_2 & - & CH_2 & - & CH & - & CH_2 & - & CH_2 & - & OH \\ &&&&&& | &&&&&& \\ &&&&&& H_2C & - & CH_3 &&&& \end{array}$$

(d)
$$\begin{array}{ccccccccc} &&&& OH &&&& \\ &&&& | &&&& \\ H_3C & - & CH_2 & - & C & - & CH_2 & - & CH_3 \\ &&&& | &&&& \\ &&&& CH_3 &&&& \end{array}$$

90. Name each of the following alcohols.

(a)
$$\begin{array}{ccccccccccccccc} CH_3 & - & CH_2 & - & CH & - & CH_2 & - & CH_2 & - & CH_2 & - & CH_2 & - & CH_3 \\ &&&& | &&&&&&&&&& \\ &&&& OH &&&&&&&&&& \end{array}$$

(b)
$$\begin{array}{ccccccccccc} CH_3 & - & CH & - & CH_2 & - & CH & - & CH_2 & - & CH_3 \\ && | &&&& | &&&& \\ && OH &&&& CH_3 &&&& \end{array}$$

(c)
$$\begin{array}{ccccccccc} && CH_3 &&&&&& \\ && | &&&&&& \\ CH_3 & - & C & - & CH_2 & - & CH_2 & - & CH_3 \\ && | &&&&&& \\ && OH &&&&&& \end{array}$$

(d)
$$\begin{array}{ccccccccccccccc} HO & - & CH_2 & - & CH & - & CH & - & CH_2 & - & CH_2 & - & CH_2 & - & CH_3 \\ &&&& | && | &&&&&&&& \\ &&&& CH_3 && CH_3 &&&&&&&& \end{array}$$

91. Draw a structure for each of the following alcohols.
(a) 3-pentanol
(b) 2-methyl-1-butanol
(c) 3-ethyl-2-hexanol
(d) ethanol

92. Draw a structure for each of the following alcohols.
(a) 1-hexanol
(b) 3,4-dimethyl-2-heptanol
(c) 3-propyl-3-octanol
(d) 3,3-diethyl-2,2-dimethyl-1-hexanol

Ethers

93. For each of the following, provide a name if the structure is given, or provide a structure if the name is given.
(a) dibutyl ether
(b) $H_3C—CH_2—O—CH_2—CH_2—CH_3$
(c) $H_3C—CH_2—CH_2—O—CH_2—CH_2—CH_3$
(d) methyl pentyl ether

94. For each of the following, provide a name if the structure is given, or provide a structure if the name is given.
(a) ethyl butyl ether
(b) $H_3C—CH_2—O—CH_2—CH_3$
(c) $H_3C—CH_2—CH_2—O—CH_2—CH_2—CH_2(—CH_2—CH_3)$
(d) ethyl hexyl ether

Aldehydes and Ketones

95. For each of the following, provide a name if the structure is given, or provide a structure if the name is given.
(a) octanal
(b) $H_3C—CH_2—CH_2—CH(=O)$
(c) $H_3C—CH_2—CH_2—C(=O)—CH_2—CH_2—CH_3$
(d) 3-hexanone

96. For each of the following, provide a name if the structure is given, or provide a structure if the name is given.
(a) $H_3C—CH_2—CH_2—C(=O)—CH_3$
(b) 3-pentanone
(c) propanal
(d) $H_3C—CH_2—CH_2—CH_2—CH_2—CH_2—CH(=O)$

Carboxylic Acids and Esters

97. For each of the following, provide a name if the structure is given, or provide a structure if the name is given.
(a) octanoic acid
(b) $H_3C—C(=O)—O—CH_3$
(c) ethyl butanoate
(d) $H_3C—CH_2—CH_2—CH_2—CH_2—CH_2—C(=O)—OH$

98. For each of the following, provide a name if the structure is given, or provide a structure if the name is given.
(a) hexanoic acid
(b) $H_3C—CH_2—CH_2—C(=O)—OH$
(c) $H_3C—CH_2—C(=O)—O—CH_2—CH_2—CH_3$
(d) butyl propanoate

Amines

99. For each of the following, provide a name if the structure is given, or provide a structure if the name is given.
(a) diethylamine
(b) $H_3C—CH_2—N(—CH_2—CH_3)—CH_2—CH_3$
(c) $H_3C—CH_2—CH_2—NH—CH_2—CH_2—CH_2—CH_3$

100. For each of the following, provide a name if the structure is given, or provide a structure if the name is given.
(a) tributylamine
(b) $H_2N—HC(CH_3)—CH_3$
(c) ethylmethylamine

Polymers

101. Polyisobutylene is an addition polymer formed from the following monomer. Draw the structure of the polymer.

$$H_2C{=}C(CH_3)_2$$

102. Teflon is an addition polymer formed from the following monomer. Draw the structure of the polymer.

$$F_2C{=}CF_2$$

103. One kind of polyester is a condensation copolymer formed between terephthalic acid and ethylene glycol. Draw the structure of the dimer and circle the ester functional group. Hint: Water (circled) is eliminated when the bond between the monomers forms.

Terephthalic acid: $HO(O{=})C{-}C_6H_4{-}C({=}O)OH$ Ethylene glycol: $HO{-}CH_2{-}CH_2{-}OH$

104. Lexan, a polycarbonate, is a condensation copolymer formed between carbonic acid and bisphenol A. Draw the structure of the dimer. Hint: Water (circled) is eliminated when the bond between the monomers forms.

Carbonic acid: $HO{-}C({=}O){-}OH$ Bisphenol A: $HO{-}C_6H_4{-}C(CH_3)_2{-}C_6H_4{-}OH$

Cumulative Problems

105. Identify each of the following organic compounds as an alkane, alkene, alkyne, aromatic, alcohol, ether, aldehyde, ketone, carboxylic acid, ester, or amine.

(a) $H_3C{-}CH_2HC(CH_3){-}CH_2{-}OH$

(b) $H_3C{-}HC(CH_3){-}CH_2{-}NH{-}CH_2{-}CH(CH_3){-}CH_3$

(c) $H_3C{-}HC(CH_3){-}C(CH_3)_2{-}CH(CH_3){-}CH_3$

(d) $H_3C{-}HC(CH_3){-}CH_2{-}C({=}O){-}OH$

(e) $H_3C{-}HC(CH_3){-}CH_2{-}O{-}CH_3$

(f) $H_3C{-}HC(CH_3){-}C(CH_3){=}C(CH_3){-}CH_3$

106. Identify each of the following organic compounds as an alkane, alkene, alkyne, aromatic, alcohol, ether, aldehyde, ketone, carboxylic acid, ester, or amine.

(a) $H_2C(CH_3)$ attached to a benzene ring bearing a CH_3 (meta)

(b) $H_3C{-}HC(CH_3){-}C{\equiv}C{-}CH_3$

(c) $H_3C{-}HC(CH_3){-}CH_2{-}C({=}O){-}O{-}CH_3$

(d) $H_3C{-}CH_2{-}CH_2{-}C({=}O){-}CH(CH_3){-}CH_3$

(e) $H_3C{-}CH_2{-}CH_2{-}C({=}O)H$

(f) $H_3C{-}CH(CH_3){-}CH_2(OH)$

107. Name each of the following compounds.

(a)

$$
\begin{array}{c}
H_3C\\
H_3C-CH_2-CH_2-CH-CH-CH_2-CH_3\\
H_3C-C-CH_3\\
CH_3
\end{array}
$$

(b)

$$
\begin{array}{c}
O\\
H_3C-CH-CH_2-CH\\
H_3C
\end{array}
$$

(c)

$$
\begin{array}{c}
H_3C\\
H_3C-CH=C-HC-CH_2-CH_2-CH_3\\
CH-CH_3\\
H_3C
\end{array}
$$

(d)

$$
\begin{array}{c}
O\\
H_3C-CH_2-CH_2-C-O-CH_2-CH_2-CH_3
\end{array}
$$

108. Name each of the following compounds.

(a)

$$
\begin{array}{c}
CH_3\ CH_3\\
H_3C-CH=CH-C-CH-CH_2-CH_3\\
CH_2\\
CH_3
\end{array}
$$

(b) F (benzene ring with F at para positions) F

(c)

$$
\begin{array}{c}
OH\\
H_3C-CH-CH_2\\
CH_3
\end{array}
$$

(d)

$$
\begin{array}{c}
CH_3\\
H_3C-CH_2-CH-CH_2-CH-CH_2-CH_2-CH_2\\
CH-CH_3 \qquad CH_3\\
CH_2\\
CH_3
\end{array}
$$

109. For each of the following, determine whether the two structures are isomers or the same molecule drawn in two different ways.

(a)

$$
\begin{array}{c}
H_3C-CH-CH-CH_3\\
CH_3\ \ CH_3
\end{array}
\qquad
\begin{array}{c}
CH_3\\
H_3C-CH-CH-CH_3\\
CH_3
\end{array}
$$

(b)

$$
\begin{array}{c}
O\\
H_3C-CH_2-CH_2-CH
\end{array}
\qquad
\begin{array}{c}
O\\
H_3C-CH-CH\\
CH_3
\end{array}
$$

(c)

$$
\begin{array}{c}
H_3C-CH_2-CH-CH_2-CH_3\\
CH_3
\end{array}
\qquad
\begin{array}{c}
CH_3\\
CH_2\\
H_3C-CH_2-CH\\
CH_3
\end{array}
$$

110. For each of the following, determine whether the two structures are isomers or the same molecule drawn in two different ways.

(a)

$$
\begin{array}{c}
O\\
H_3C-CH_2-C-O-CH_3
\end{array}
\qquad
\begin{array}{c}
O\\
H_3C-C-O-CH_2-CH_3
\end{array}
$$

(b) $H_3C-C{\equiv}C-CH_2-CH_3$ $\qquad$ $H_3C-CH_2-C{\equiv}C-CH_3$

(c)

$$
\begin{array}{c}
O\\
H_3C-C-OH
\end{array}
\qquad
\begin{array}{c}
O\\
HO-C-CH_3
\end{array}
$$

111. Complete the following equation:

$$CH_2{=}CH_2 + HCl \longrightarrow$$

112. Complete the following equation (assume only one addition):

$$CH{\equiv}CH + HI \longrightarrow$$

113. What is the minimum amount of hydrogen gas in grams required to completely hydrogenate 15.5 kg of 2-butene? Hint: Begin by writing a balanced equation for the hydrogenation reaction.

114. How many kilograms of CO_2 are produced by the complete combustion of 3.8 kg of *n*-octane? Hint: Begin by writing a balanced equation for the combustion of *n*-octane.

Highlight Problems

115. Based on the following space-filling models, identify the family to which each of the following molecules belongs.

Hydrogen

Carbon

Oxygen

Nitrogen

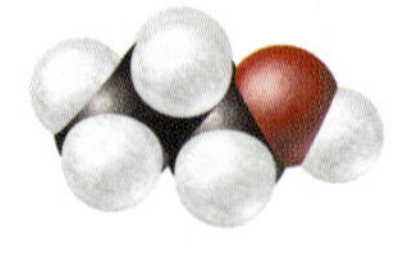
(a)

(b)

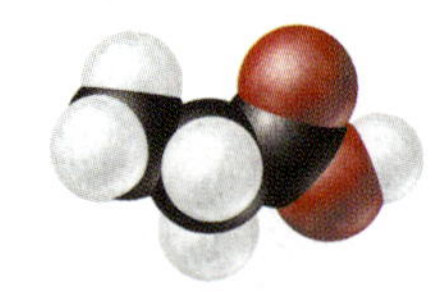
(c)

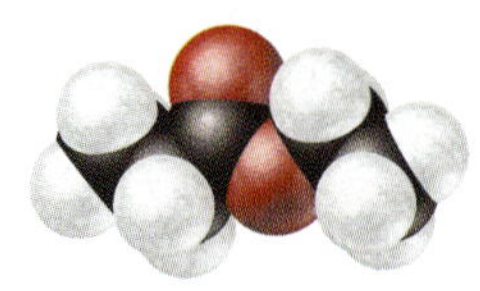
(d)

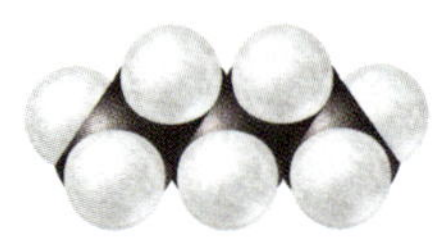
(e)

(f)

116. In Chapter 13, we learned about the "dirty dozen," twelve chemical compounds that have been targeted by governments around the world to be banned. These compounds are known as persistent organic pollutants (POPs) because once they enter the environment, they remain for long periods of time. Examine the structures of the following compounds. What functional groups can you identify within the compounds? (They may have more than one.) What structural features do many of these compounds have in common?

Dioxin—industrial by-product

Furan—industrial by-product

Hexachlorobenzene—fungicide, industrial by-product

DDT—insecticide

Answers to Skillbuilder Exercises

Skillbuilder 18.1 **(a)** alkene **(b)** alkyne **(c)** alkane

Skillbuilder 18.2

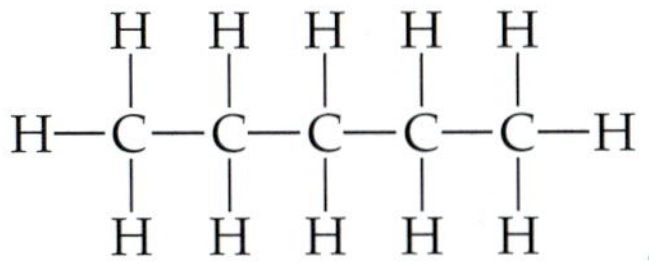

```
    H  H  H  H  H
    |  |  |  |  |
 H—C—C—C—C—C—H
    |  |  |  |  |
    H  H  H  H  H
```

, $CH_3CH_2CH_2CH_2CH_3$

Skillbuilder 18.3

```
    H  H  H  H  H
    |  |  |  |  |
 H—C—C—C—C—C—H
    |  |  |  |  |
    H  H  H  H  H
```

```
    H  H  H  H
    |  |  |  |
 H—C—C—C—C—H
    |  |  |  |
    H  |  H  H
    H—C—H
       |
       H
```

```
       H
       |
    H—C—H
    H  |  H
    |  |  |
 H—C—C—C—H
    |  |  |
    H  |  H
    H—C—H
       |
       H
```

Skillbuilder 18.4 2-methylpropane

Skillbuilder 18.5 3-ethyl-2-methylpentane

Skillbuilder 18.6 2,3,5-trimethylhexane

Skillbuilder 18.7 **(a)** 4,4-dimethyl-2-pentyne
(b) 3-ethyl-4,6-dimethyl-1-heptene

Skillbuilder 18.8 1,3-dibromobenzene, *meta*-dibromobenzene, *m*-dibromobenzene

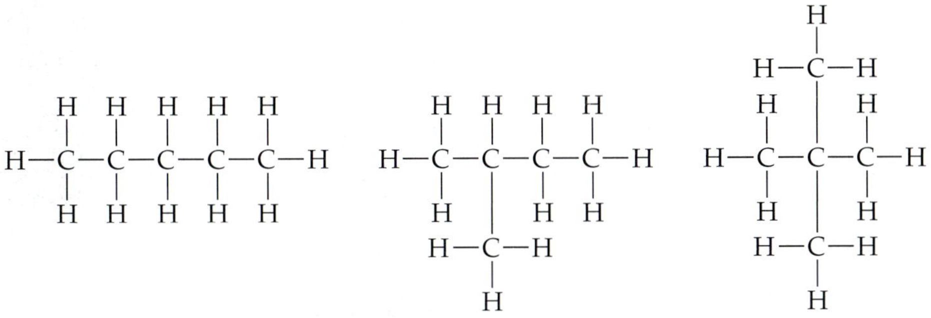

Answers to Conceptual Checkpoints

18.1 **(b)** The compound shown is organic (carbon-based), unsaturated (it contains a double bond in the carbon chain), and an acid (it includes a —COOH group). However, it does not contain a benzene ring, and so is not classified as an aromatic compound.

THE

CHAPTER 19

Biochemistry

"Can—and should—life be described in terms of molecules? For many, such description seems to diminish the beauty of Nature. For others of us, the wonder and beauty of nature are nowhere more manifest than in the submicroscopic plan of life . . . "

Robert A. Weinberg (1943–)

19.1 The Human Genome Project

In 1990, the U.S. Department of Energy (DOE) and the National Institutes of Health (NIH) embarked on a fifteen-year project to map the **human genome**, all of the genetic material of a human being. We will define genetic material—and genes—more carefully later. For now, think of genetic material as the *inheritable blueprint* for making organisms. Think of **genes** as *specific parts* of that blueprint. Each organism has a blueprint unique to itself. The genetic material of humans, for example, is unique to humans and different from that of other organisms. When an organism reproduces, it passes its genetic material to the next generation.

Within the genetic material of a given species of organism, however, there is variation among individuals. For example, whether you have brown eyes or blue eyes depends on the specific genes for eye color that you inherited from your parents. Many traits—such as physical appearance, intelligence, susceptibility to certain diseases, response to certain drug therapies, and even temperament—are at least partially determined by your specific genes. So understanding the human genome is part of understanding ourselves.

In 2001 the early results—a nearly complete rough draft of the human genome—were published in two different scientific journals. These results contained some surprises. For example, the results show that humans have only about 32,000 genes. This may seem like a large number, but scientists initially expected more. The number of genes in humans is not much larger than in many simpler organisms. For example, the number of genes in a roundworm is nearly 20,000. Whatever makes humans unique, it is not the number of genes in our genome.

Nucleotides are defined in Section 19.7. For now think of them as the units that compose genes.

The Human Genome Project is also mapping the specific variations between the DNA of different people. These variations are called single-nucleotide polymorphisms or SNPs. Understanding SNPs can help in identifying individuals who are susceptible to certain diseases. For example, in the future, a genetic test may reveal that you have the SNPs associated with a certain type of cancer. You can then take preventive steps, or even preventive drug therapy, to avoid actually getting the cancer. Knowledge of SNPs may also allow physicians to tailor drug therapies to match individuals. A genetic test may allow a doctor to give you the drug that is most effective for you.

◀ The similarities between parents and their children are caused by genes, inheritable blueprints for making organisms. The structure at the bottom of this image is DNA, the molecular basis of genetic information.

Analysis of the human genome is also expected to lead to the development of new drugs in two ways. First, an understanding of genes can lead to smart drug design. Instead of developing drugs by trial and error (the current procedure for many drugs), knowledge of a specific gene will allow scientists to design drugs to carry out a specific function related to that gene. Second, human genes themselves can provide the blueprint for certain types of drugs. For example, interferon, a drug taken by people with multiple sclerosis, is a complex compound normally found in humans. The blueprint for making interferon is in the human genome. Scientists have been able to take this blueprint out of human cells and put it into bacteria, which then synthesize the needed drug. The drug can be harvested from bacteria, purified, and given to patients.

Interferon is a protein. Proteins are discussed in Section 19.5.

The Human Genome Project was possible because of decades of research in **biochemistry**, the study of the chemical substances and processes that occur in plants, animals, and microorganisms. In this chapter, we examine the chemical substances that make life possible and some of the new technology that has resulted from this understanding.

19.2 The Cell and Its Main Chemical Components

The **cell** is the smallest structural unit of living organisms that has the properties associated with life (◀ Figure 19.1). A cell can be an independent living organism or a building block of a more complex organism. Most cells in higher animals contain a **nucleus**, the control center of the cell and the part of the cell that contains the genetic material. The perimeter of the cell is bound by a **cell membrane** that holds the contents of the cell together. The region between the nucleus and the cell membrane is called the **cytoplasm**. This region contains a number of specialized structures that carry out much of the cell's work. The main *chemical* components of the cell can be divided into four classes: *carbohydrates, lipids, proteins*, and *nucleic acids*.

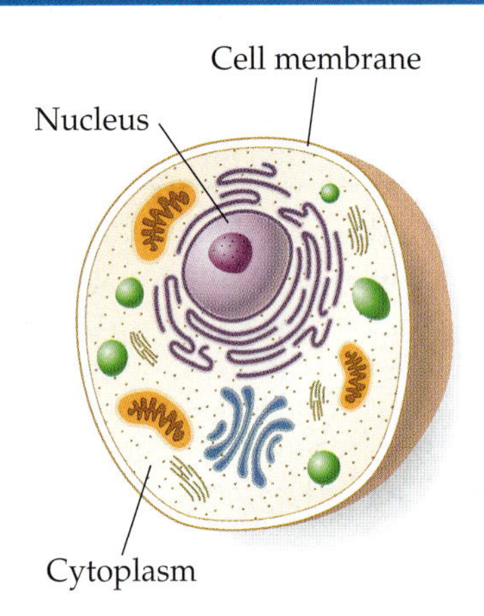

◀ **Figure 19.1 A typical cell** The cell is the smallest structural unit of living organisms. The primary genetic material is stored in the nucleus.

19.3 Carbohydrates: Sugar, Starch, and Fiber

Carbohydrates are the primary molecules responsible for short-term energy storage in living organisms. They also form the main structural components of plants. Carbohydrates—as their name, which means carbon and water, implies—often have the general formula $(CH_2O)_n$. Structurally, we identify carbohydrates as aldehydes or ketones containing multiple —OH groups. For example, glucose, with the formula $C_6H_{12}O_6$, has the following structure.

As we learned in Section 18.14, aldehydes have the general structure R—CHO and ketones have the general structure R—CO—R.

HC=O
H—C—OH
HO—C—H
H—C—OH
H—C—OH
H_2C—OH

Glucose

Notice that glucose is an aldehyde (it contains the —CHO group) with —OH groups on most of the carbon atoms. The many —OH groups make glucose soluble in water (and therefore in blood), which is important in glucose's role as the primary fuel of cells. Glucose is easily transported in the bloodstream and is soluble within the aqueous interior of a cell.

Monosaccharides

Glucose is an example of a **monosaccharide**, a carbohydrate that cannot be broken down into simpler carbohydrates. Monosaccharides such as glucose rearrange in aqueous solution to form ring structures (▼ Figure 19.2).

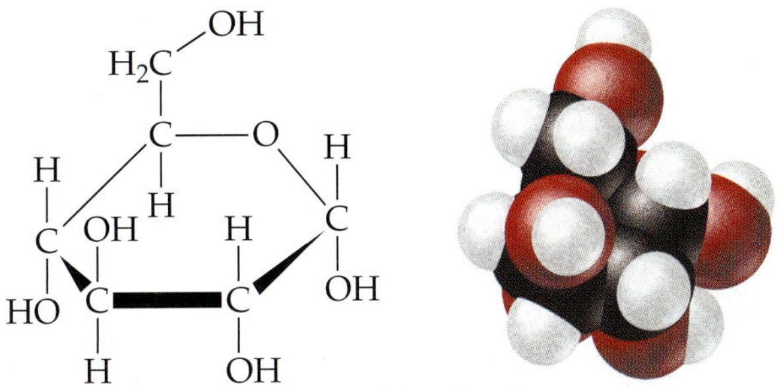

Glucose (ring form)

Three-carbon sugar—triose; four-carbon sugar—tetrose; five-carbon sugar—pentose; six-carbon sugar—hexose; seven-carbon sugar—heptose; eight-carbon sugar—octose.

Glucose is also an example of a *hexose*, a six-carbon sugar. The general names for monosaccharides have a prefix that depends on the number of carbon atoms, followed by the suffix *-ose*. The most common monosaccharides in living organisms are pentoses and hexoses.

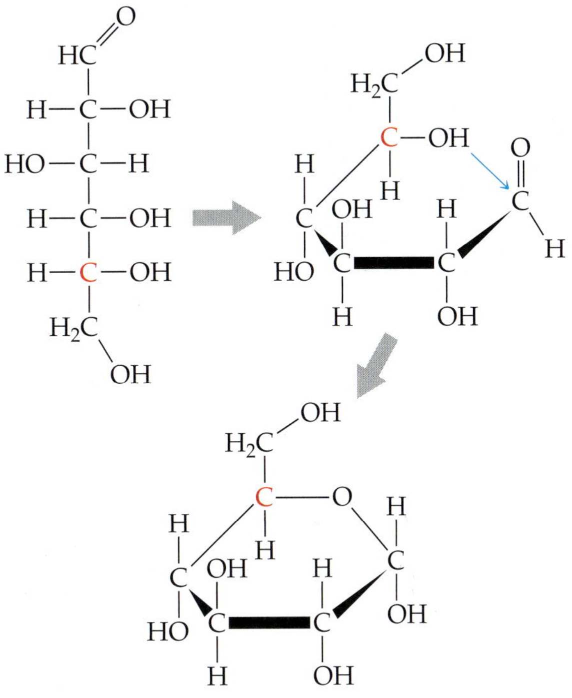

▲ **Figure 19.2 Rearrangement of glucose from straight-chain to ring form**
Question: Can you verify that the straight-chain form and the ring form of glucose are isomers?

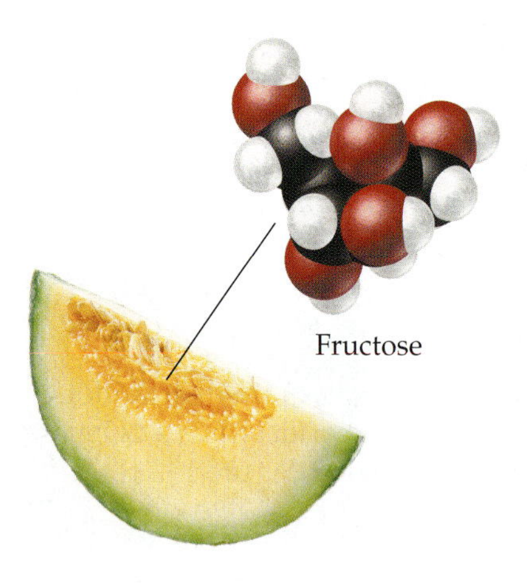

▲ Fructose is the main sugar that occurs in fruit.

The following show other monosaccharides in their ring form.

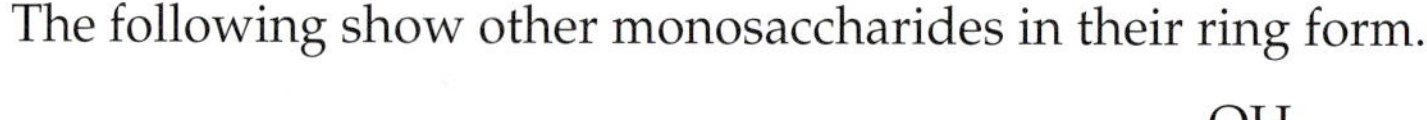

Fructose

Galactose

Fructose, also known as fruit sugar, is a hexose found in many fruits and vegetables and is a major component of honey. Galactose, also known as brain sugar, is a hexose usually found combined with other monosaccharides in substances such as lactose (see next section). Galactose also occurs within the brain and nervous system of most animals.

Disaccharides

Two monosaccharides can react, eliminating water to form a carbon–oxygen–carbon bond called a **glycoside linkage** that connects the two rings. The resulting compound is a **disaccharide**, a carbohydrate that can be decomposed into two simpler carbohydrates. For example, glucose and fructose link together to form sucrose, commonly known as table sugar.

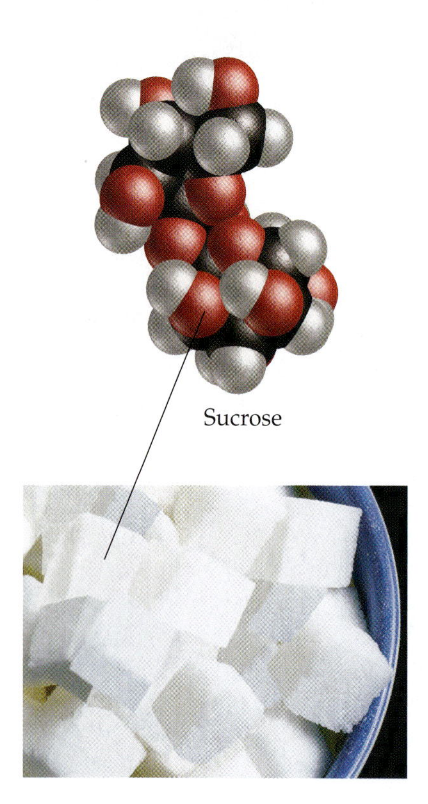

▲ Table sugar is composed of sucrose, a disaccharide.

Glucose + Fructose ⟶

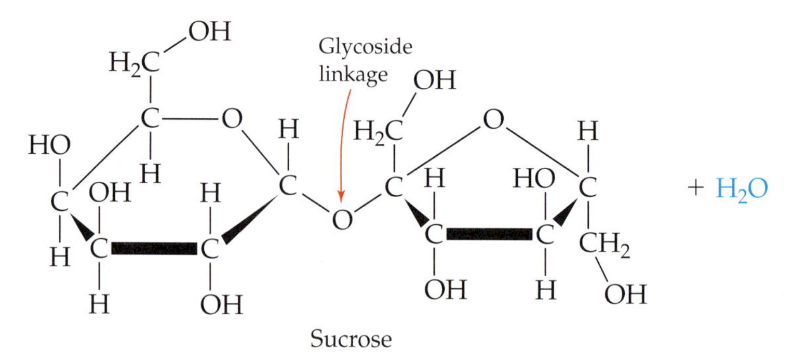

Sucrose

The link between individual monosaccharides is broken during digestion, allowing the individual monosaccharides to pass through the intestinal wall and enter the bloodstream (▼ Figure 19.3).

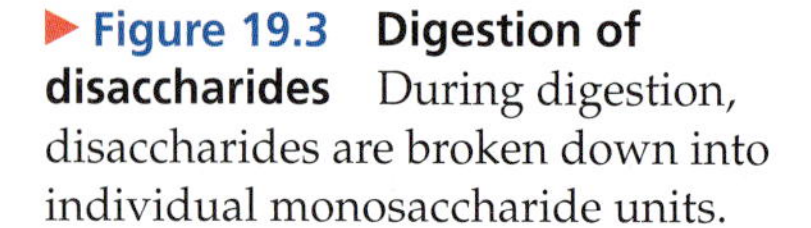

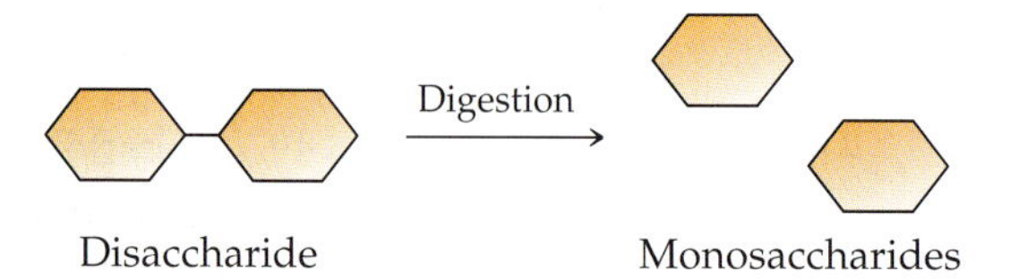

► **Figure 19.3 Digestion of disaccharides** During digestion, disaccharides are broken down into individual monosaccharide units.

Polysaccharides

Monosaccharides can also link together to form **polysaccharides**, long, chain-like molecules composed of many monosaccharide units. Polysaccharides are a type of *polymer*—chemical compounds (introduced in Section 18.17) composed of repeating structural units in a long chain. Monosaccharides and disaccharides are known as **simple sugars** or **simple carbohydrates**. Polysaccharides are known as **complex carbohydrates**. Some common polysacchharides include **starch** and **cellulose**, both composed of repeating glucose units.

Starch

The difference between starch and cellulose is the link between the glucose units. In starch, the oxygen atom joining neighboring glucose units points down (as conventionally drawn) relative to the planes of the rings. This is called an *alpha linkage*. In cellulose, the oxygen atoms are roughly parallel with the planes of the rings but pointing slightly up. This is called a *beta linkage*. This difference in linkage causes the differences in the properties of starch and cellulose.

Starch is common in potatoes and grains. It is a soft and pliable substance that we easily chew and swallow. During digestion, the links between individual glucose units are broken, allowing glucose molecules to pass through the intestinal wall and into the bloodstream (▼ Figure 19.4). On the other hand, cellulose—also known as fiber—is a more stiff and rigid substance. Cellulose is the main structural component of plants. The bonding in cellulose makes it indigestible by humans. When we eat cellulose, it passes right through the intestine, providing bulk to stools and preventing constipation.

A third kind of polysaccharide is **glycogen**. Glycogen has a structure similar to starch, but the chain is highly branched. In animals, excess glucose in the blood is stored as glycogen until it is needed.

One form of starch, amylopectin, is also branched, but less so than glycogen.

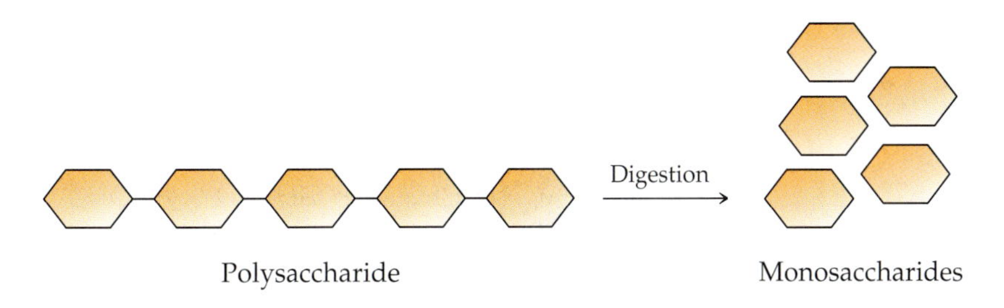

▲ **Figure 19.4 Digestion of polysaccharides** During digestion, polysaccharides are broken down into individual monosaccharide units.

EXAMPLE 19.1 Identifying Carbohydrates

Determine which of the following molecules are carbohydrates. Classify each carbohydrate as a monosaccharide, disaccharide, or polysaccharide.

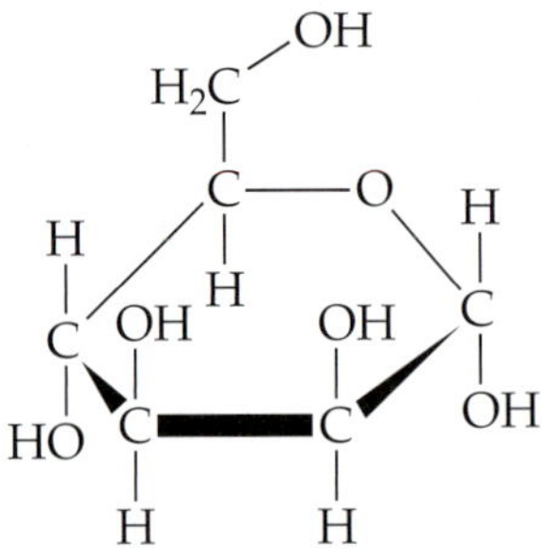

(a)

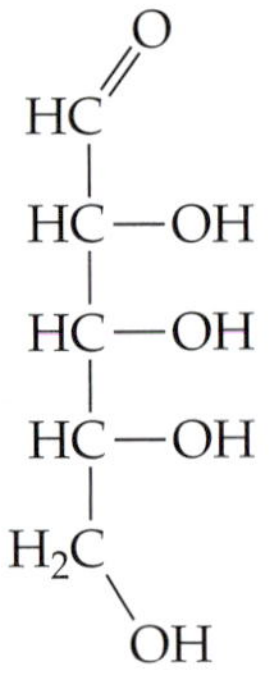

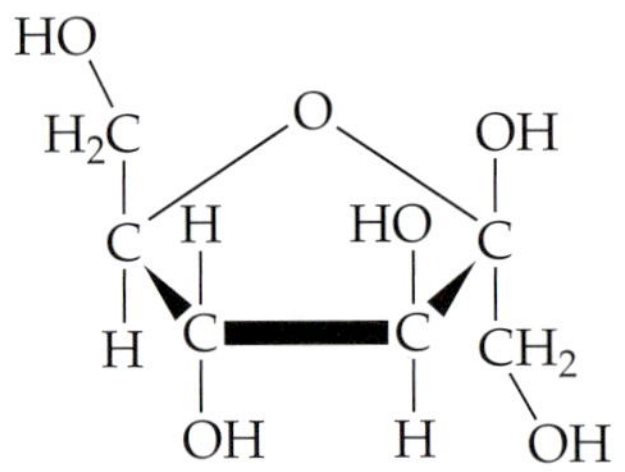

(b)

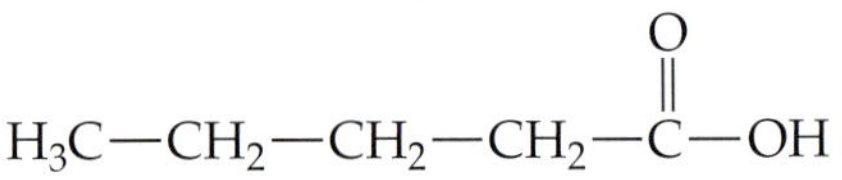

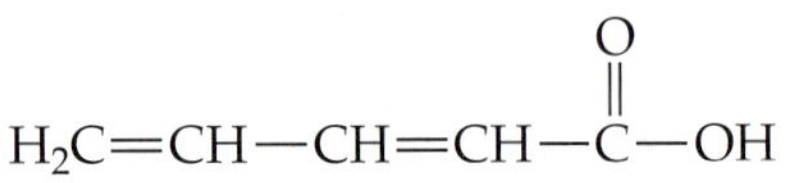

$H_3C-CH_2-CH_2-CH_2-C(=O)-OH$

(c)

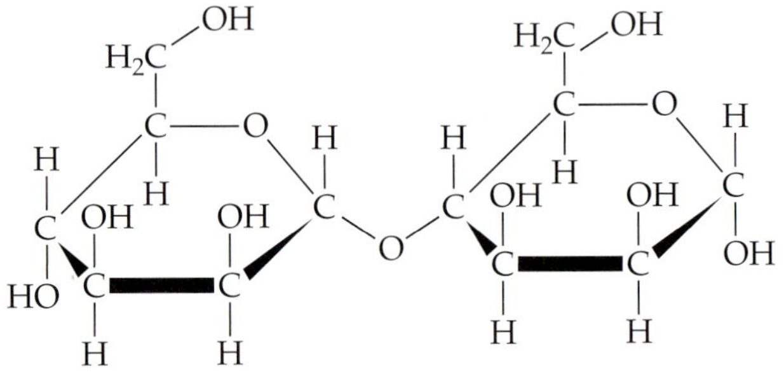

(d)

Solution:

You can identify carbohydrates as either an aldehyde or ketone with multiple —OH groups attached or as one or more rings of carbon atoms that include one oxygen atom and also have —OH groups attached to most of the carbon atoms. (a), (b), and (d) are carbohydrates. (a) and (b) are both monosaccharides and (d) is a disaccharide. (c) is not a carbohydrate because it has only a carboxylic acid group, which is not a characteristic of carbohydrates.

SKILLBUILDER 19.1 Identifying Carbohydrates

Determine which of the following molecules are carbohydrates and classify each carbohydrate as a monosaccharide, disaccharide, or polysaccharide.

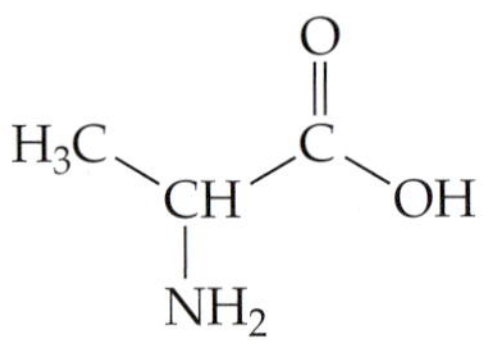

$H_3C-CH(NH_2)-C(=O)-OH$

(a)

(b)

$H_2C{=}CH-CH{=}CH-C(=O)-OH$

(c)

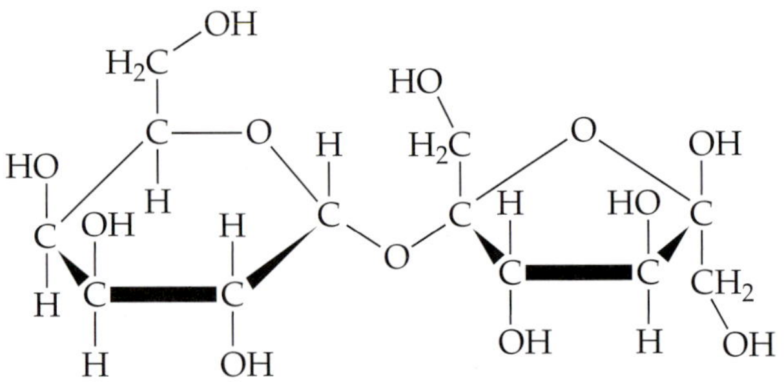

(d)

19.4 Lipids

> Carboxylic acids were first defined in Section 18.15.

Lipids are chemical components of the cell that are insoluble in water but soluble in nonpolar solvents. Lipids include fatty acids, fats, oils, phospholipids, glycolipids, and steroids. Their insolubility in water makes lipids ideal as the structural components of cell membranes. In other words, lipids make up the containers that separate the interior of the cell from its external environment. Lipids are also used for long-term energy storage and for insulation. We all store extra calories from food as lipids, some of us more than others.

Fatty Acids

A class of lipids are the **fatty acids**, carboxylic acids with long hydrocarbon tails. The general structure for a fatty acid is

Hydrocarbon tail → R; Carboxylic acid group → $-\overset{\overset{\large O}{\|}}{C}-OH$

$$R-\overset{\overset{O}{\|}}{C}-OH$$

Fatty acid general structure

where *R* represents a hydrocarbon chain containing three to nineteen carbon atoms. Fatty acids differ only in their *R* group. A common fatty acid is myristic acid, where the *R* group is $CH_3(CH_2)_{12}-$.

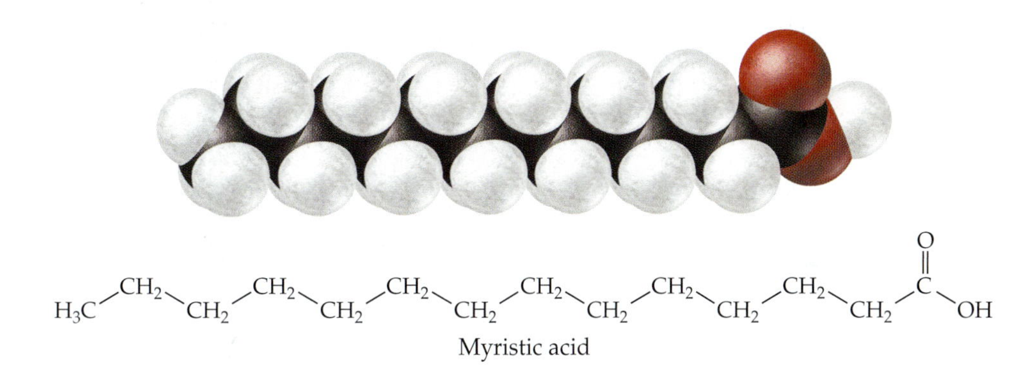

Myristic acid

Myristic acid occurs in butterfat and in coconut oil. Myristic acid is an example of a *saturated* fatty acid—its carbon chain has no double bonds. Other fatty acids—called *monounsaturated* or *polyunsaturated* fatty acids—have one or more double bonds, respectively, in their carbon chains. For example, oleic acid—found in olive oil, peanut oil, and human fat—is an example of a monounsaturated fatty acid.

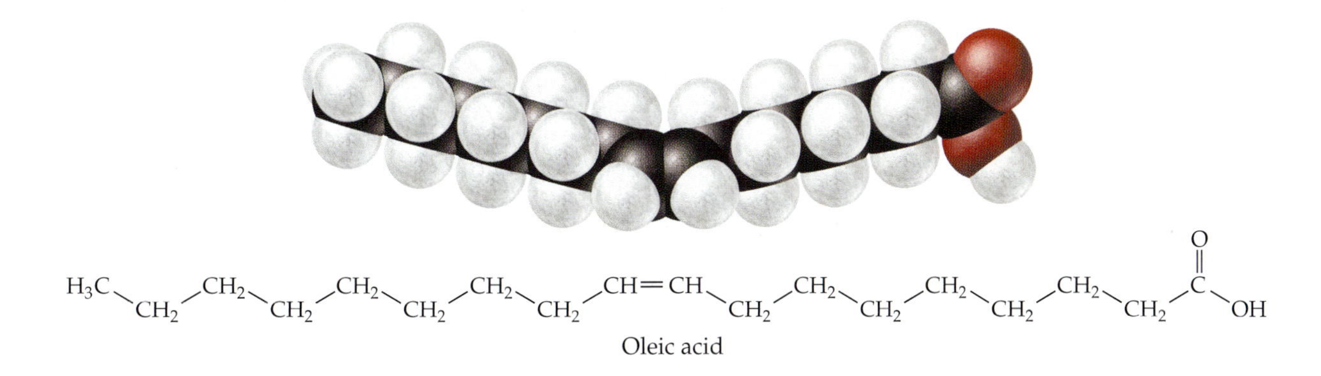

Oleic acid

TABLE 19.1 Fatty Acids

Saturated Fatty Acids			
Name	Number of Carbon Atoms	Structure	Sources
butyric acid	4	$CH_3CH_2CH_2COOH$	milk fat
capric acid	10	$CH_3(CH_2)_8COOH$	milk fat, whale oil
myristic acid	14	$CH_3(CH_2)_{12}COOH$	butterfat, coconut oil
palmitic acid	16	$CH_3(CH_2)_{14}COOH$	beef fat, butterfat
stearic acid	18	$CH_3(CH_2)_{16}COOH$	beef fat, butterfat

Unsaturated Fatty Acids				
Name	Number of Carbon Atoms	Number of Double Bonds	Structure	Sources
oleic acid	18	1	$CH_3(CH_2)_7CH{=}CH(CH_2)_7COOH$	olive oil, peanut oil
linoleic acid	18	2	$CH_3(CH_2)_4(CH{=}CHCH_2)_2(CH_2)_6COOH$	linseed oil, corn oil
linolenic acid	18	3	$CH_3CH_2(CH{=}CHCH_2)_3(CH_2)_6COOH$	linseed oil, corn oil

The long hydrocarbon tails of fatty acids make them insoluble in water. Table 19.1 contains a list of several different fatty acids and some common sources for each.

Fats and Oils

Fats and oils are **triglycerides**, triesters composed of glycerol linked to three fatty acids, as shown in the following block diagram.

Esters, first defined in Section 18.15, have the general structure R—COO—R. Glycerol has the following structure.

$$\begin{array}{l} H_2C{-}OH \\ \quad | \\ HO{-}CH \\ \quad | \\ H_2C{-}OH \end{array}$$

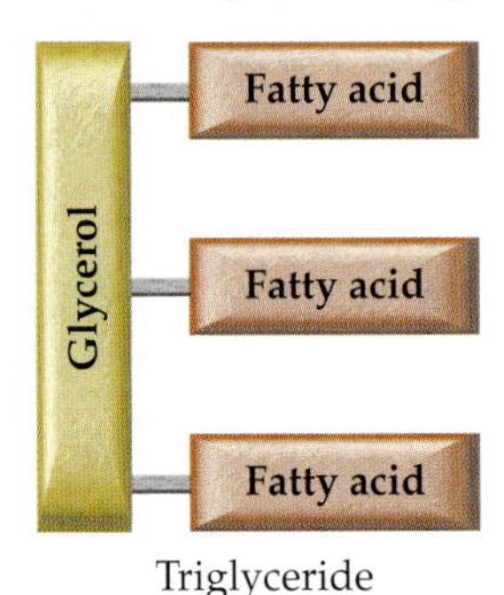

Triglyceride

Triglycerides form by the reaction of glycerol with three fatty acids.

MOLECULE Triglyceride

$$\text{Glycerol } (H_2C{-}OH,\ HO{-}CH,\ H_2C{-}OH) + \text{3 fatty acids } (3\ R{-}\overset{O}{\overset{\|}{C}}{-}OH) \longrightarrow \text{Triglyceride } (H_2C{-}O{-}\overset{O}{\overset{\|}{C}}{-}R,\ R{-}\overset{O}{\overset{\|}{C}}{-}O{-}CH,\ H_2C{-}O{-}\overset{O}{\overset{\|}{C}}{-}R) + 3\,H_2O$$

Ester linkages

The bonds that join the glycerol to the fatty acids are called **ester linkages**. For example, tristearin—the main component of beef fat—is formed from the reaction of glycerol and three stearic acid molecules.

$$\underset{\text{Glycerol}}{\begin{array}{l} H_2C{-}OH \\ HO{-}CH \\ H_2C{-}OH \end{array}} + 3\Big[\underset{\text{3 stearic acid}}{H_3C{-}CH_2{-}CH_2{-}CH_2{-}CH_2{-}CH_2{-}CH_2{-}CH_2{-}CH_2{-}CH_2{-}CH_2{-}CH_2{-}CH_2{-}CH_2{-}CH_2{-}CH_2{-}CH_2{-}\overset{O}{\overset{\|}{C}}{-}OH}\Big] \longrightarrow$$

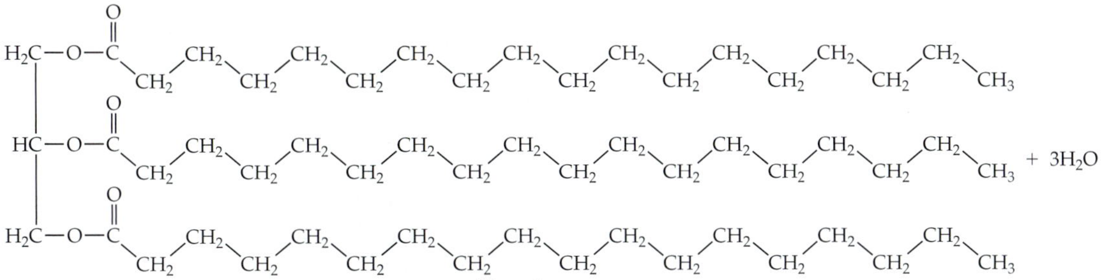

$$
\begin{array}{l}
H_2C-O-\overset{\displaystyle O}{\overset{\|}{C}}-CH_2-CH_2-CH_2-CH_2-CH_2-CH_2-CH_2-CH_2-CH_2-CH_2-CH_2-CH_2-CH_2-CH_2-CH_2-CH_2-CH_3 \\
\quad | \\
HC-O-\overset{\displaystyle O}{\overset{\|}{C}}-CH_2-CH_2-CH_2-CH_2-CH_2-CH_2-CH_2-CH_2-CH_2-CH_2-CH_2-CH_2-CH_2-CH_2-CH_2-CH_2-CH_3 \quad + \; 3H_2O \\
\quad | \\
H_2C-O-\overset{\displaystyle O}{\overset{\|}{C}}-CH_2-CH_2-CH_2-CH_2-CH_2-CH_2-CH_2-CH_2-CH_2-CH_2-CH_2-CH_2-CH_2-CH_2-CH_2-CH_2-CH_3
\end{array}
$$

Tristearin

If the fatty acids in a triglyceride are saturated, the triglyceride is called a **saturated fat** and tends to be solid at room temperature. Lard and many animal fats are examples of saturated fat. On the other hand, if the fatty acids in a triglyceride are unsaturated, the triglyceride is called an **unsaturated fat** or an *oil* and tends to be liquid at room temperature. Canola oil, olive oil, and most other vegetable oils are examples of unsaturated fats.

EXAMPLE 19.2 Identifying Triglycerides

Identify the triglycerides among the following molecules and classify each as saturated or unsaturated.

(a) 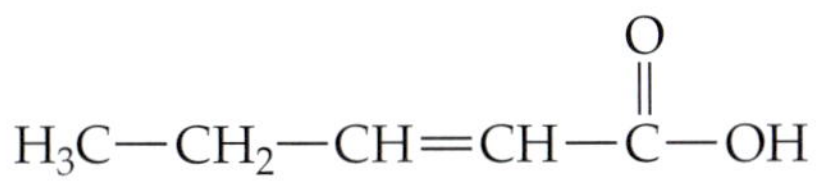 $H_3C-CH_2-CH=CH-\overset{\displaystyle O}{\overset{\|}{C}}-OH$

(b)

$$
\begin{array}{l}
H_2C-O-\overset{\displaystyle O}{\overset{\|}{C}}-CH_2-CH_2-CH_2-CH_2-CH_2-CH_2-CH_2-CH_2-CH_2-CH_2-CH_3 \\
\quad | \\
HC-O-\overset{\displaystyle O}{\overset{\|}{C}}-CH_2-CH_2-CH_2-CH_2-CH_2-CH_2-CH_2-CH_2-CH_2-CH_2-CH_3 \\
\quad | \\
H_2C-O-\overset{\displaystyle O}{\overset{\|}{C}}-CH_2-CH_2-CH_2-CH_2-CH_2-CH_2-CH_2-CH_2-CH_2-CH_2-CH_3
\end{array}
$$

(c)

$$
\begin{array}{l}
H_2C-O-\overset{\displaystyle O}{\overset{\|}{C}}-CH_2-CH_2-CH_2-CH_2-CH_2-CH_2-CH_2-CH=CH-CH_2-CH_2-CH_2-CH_2-CH_2-CH_2 \\
\quad | \\
HC-O-\overset{\displaystyle O}{\overset{\|}{C}}-CH_2-CH_2-CH_2-CH_2-CH_2-CH_2-CH_2-CH=CH-CH_2-CH_2-CH_2-CH_2-CH_2-CH_2 \\
\quad | \\
H_2C-O-\overset{\displaystyle O}{\overset{\|}{C}}-CH_2-CH_2-CH_2-CH_2-CH_2-CH_2-CH_2-CH=CH-CH_2-CH_2-CH_2-CH_2-CH_2-CH_2
\end{array}
$$

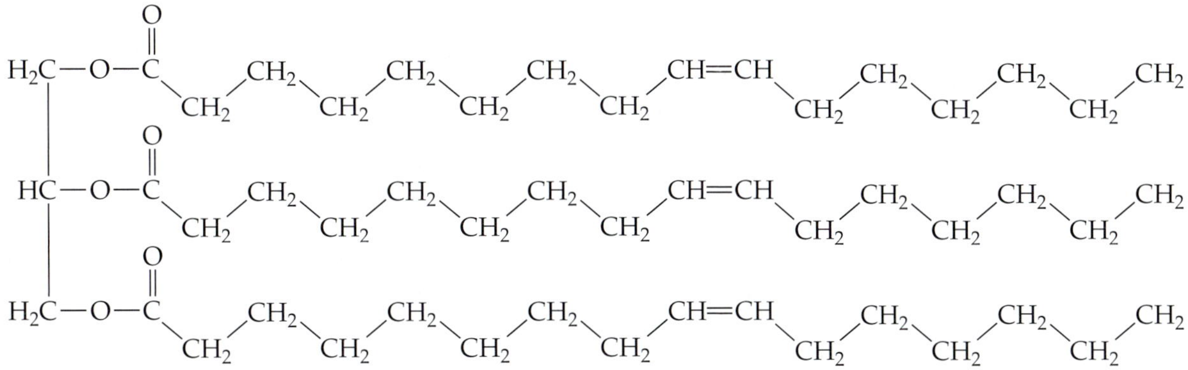

(d)

OH
H_2C
C—O
H H
C OH H C
HO C C OH
H OH

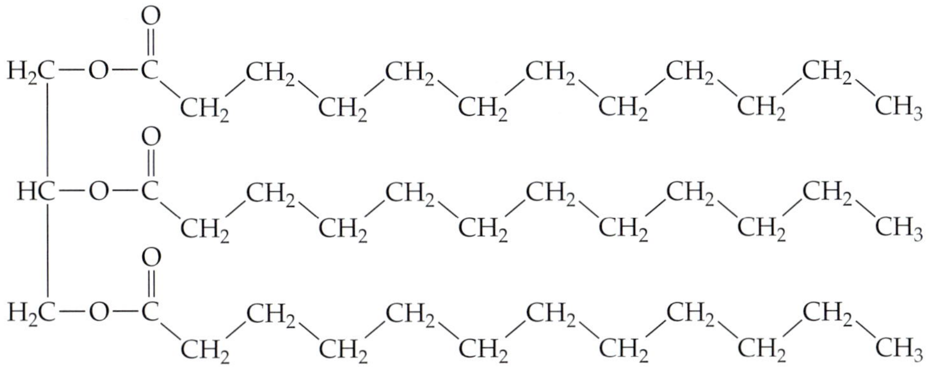

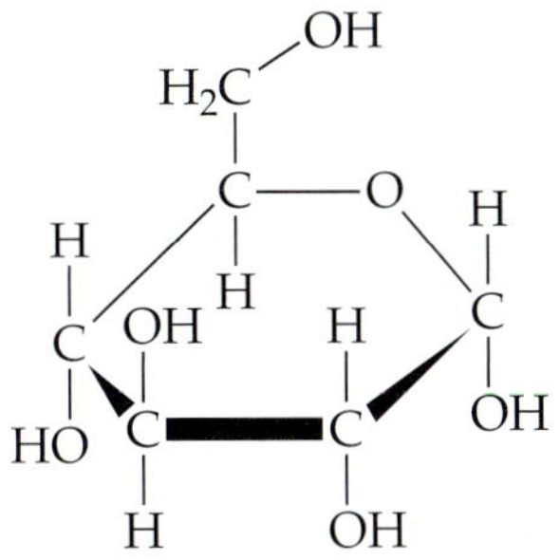

Solution:

Triglycerides are easily identified by the three-carbon backbone with long fatty acid tails. Both (b) and (c) are triglycerides. (b) is a saturated fat because it does not have any double bonds in its carbon chains. (c) is an unsaturated fat because it contains double bonds in its carbon chains.

SKILLBUILDER 19.2 Identifying Triglycerides

Identify the triglycerides among the following molecules and classify each as a saturated or unsaturated fat.

(a) O, NH_2

(b)
$H_2C{-}O{-}C(=O){-}(CH_2)_7{-}CH{=}CH{-}(CH_2)_5{-}CH_3$
$HC{-}O{-}C(=O){-}(CH_2)_7{-}CH{-}CH_2{-}CH_2{-}CH_3$
$H_2C{-}O{-}C(=O){-}(CH_2)_7{-}CH{=}CH{-}(CH_2)_7{-}CH_3$

(c) HO, H_2C, O, H, C, H, HO, C, HO, C, C, CH_2, OH, H, OH

(d)
$H_2C{-}O{-}C(=O){-}(CH_2)_{12}{-}CH_3$
$HC{-}O{-}C(=O){-}(CH_2)_{10}{-}CH_3$
$H_2C{-}O{-}C(=O){-}(CH_2)_{12}{-}CH_3$

Other Lipids

Other lipids found in cells include phospholipids, glycolipids, and steroids. **Phospholipids** have the same basic structure as triglycerides, except that one of the fatty acid groups is replaced with a phosphate group.

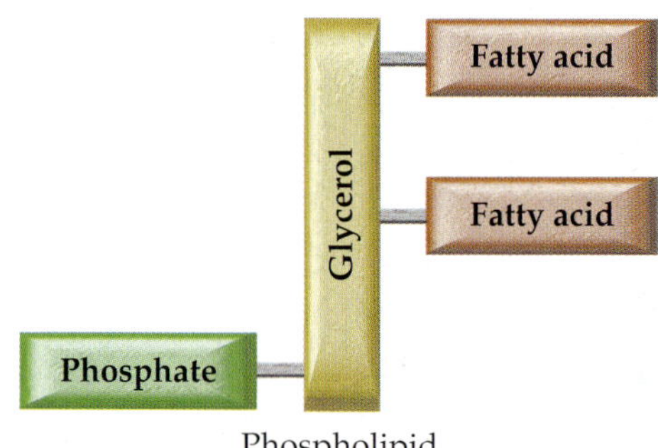

Phospholipid

Unlike a fatty acid, which is nonpolar, the phosphate group is polar, and often has another polar group attached to it. The phospholipid molecule therefore has a polar section and a nonpolar section. For example, consider the structure of a phosphatidyl choline, a phospholipid found in the cell membranes of higher animals.

Polar section

Nonpolar section

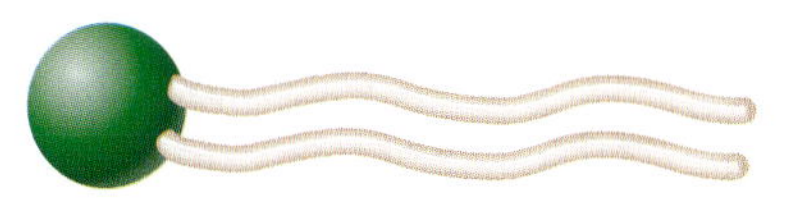

▲ **Figure 19.5 Schematic representation of phospholipids and glycolipids** The green circle represents the polar part of the molecule and the tails represent the nonpolar hydrocarbon chains. **Question:** If this molecule were placed in water, how do you think it might orient itself at the surface?

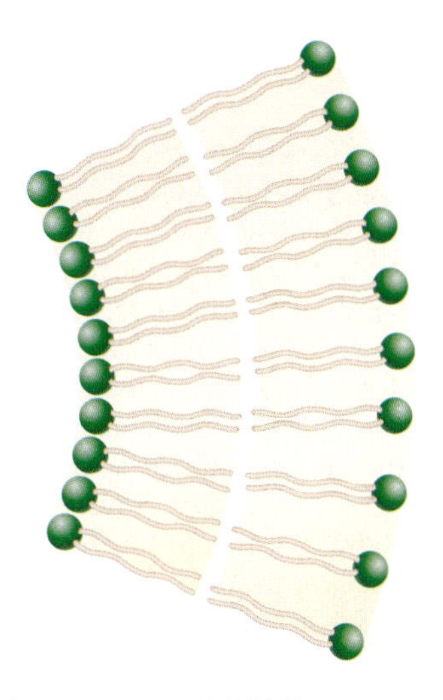

▲ **Figure 19.6 Lipid bilayer membrane** Cell membranes are composed of lipid bilayers, in which phospholipids or glycolipids form a double layer. In this bilayer, the polar heads of the molecules point outward and the nonpolar tails point inward.

The polar part of the molecule is *hydrophilic* (has a strong affinity for water) while the nonpolar part is *hydrophobic* (is repelled by water). **Glycolipids** have similar structures and properties. The nonpolar section of a glycolipid is composed of a fatty acid chain and a hydrocarbon chain. The polar section is a sugar molecule such as glucose. A schematic way to represent both phospholipids and glycolipids is as a circle with two long tails (◀ Figure 19.5). The circle represents the polar hydrophilic part of the molecule and the tails represent the nonpolar hydrophobic parts. The structure of phospholipids and glycolipids make them ideal as cell membranes, where the polar parts can interact with the aqueous environments of the cell and the nonpolar parts interact with each other. In a membrane, these lipids form a structure called a **lipid bilayer** (◀ Figure 19.6). Lipid bilayer membranes encapsulate cells and many cellular structures.

Steroids are lipids that contain the following four-ring structure.

Some common steroids include cholesterol, testosterone, and estrogen.

Cholesterol

Testosterone

Estrogen

Although cholesterol has a bad reputation, it serves many important functions in the body. Like phospholipids and glycolipids, cholesterol is part of cell membranes. Cholesterol also serves as a starting material (or precursor) for the body to synthesize other steroids such as testosterone, a principal male hormone, and estrogen, a principal female hormone. Hormones are chemical messengers that regulate many body processes, such as growth and metabolism. They are secreted by specialized tissues and transported in the blood.

Chemistry AND HEALTH

Dietary Fats

Most of the fats and oils in our diet are triglycerides. During digestion, triglycerides are broken down into fatty acids, glycerol, monoglycerides, and diglycerides. These products pass through the intestinal wall and then reassemble into triglycerides before they are absorbed into the blood. This process, however, is slower than the digestion of other food types, and therefore eating fats and oils gives a lasting feeling of fullness.

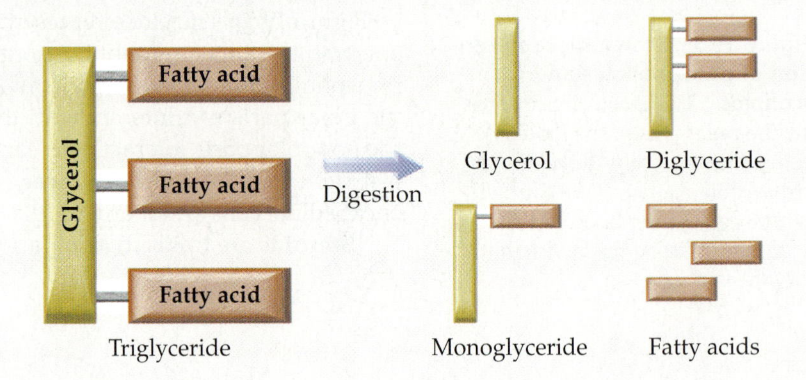

The effect of fats and oils on health has been widely debated. Some diets call for drastic reduction of daily intake of fats and oils. Other diets actually call for an increase in fats and oils. The Food and Drug Administration (FDA) recommends that fats and oils compose less than 30% of total caloric intake. However, because fats and oils have a higher caloric content per gram than other food types, it is easy to eat too much of them. The FDA also recommends that of those fats that are consumed, no more than one-third (10% of total caloric intake) be saturated fats. This is because a diet high in saturated fats increases the risk of artery blockages that can lead to stroke and heart attack. Monounsaturated fats, by contrast, may help protect against these threats.

Nutrition Facts
Serving Size 1 oz (28g/about 18 chips)
Servings Per Container 7

Amount Per Serving	
Calories 150	Calories from Fat 80
	% Daily Value*
Total Fat 9g	14%
Saturated Fat 1g	5%
Polyunsaturated Fat 1g	
Monounsaturated Fat 7g	
Cholesterol 0mg	0%
Sodium 160mg	7%
Total Carbohydrate 16g	5%
Dietary Fiber 1g	4%
Sugars 1g	
Protein 2g	
Vitamin A 0% •	Vitamin C 15%

◀ The Food and Drug Administration (FDA) recommends that fats and oils provide less than 30% of total caloric intake.

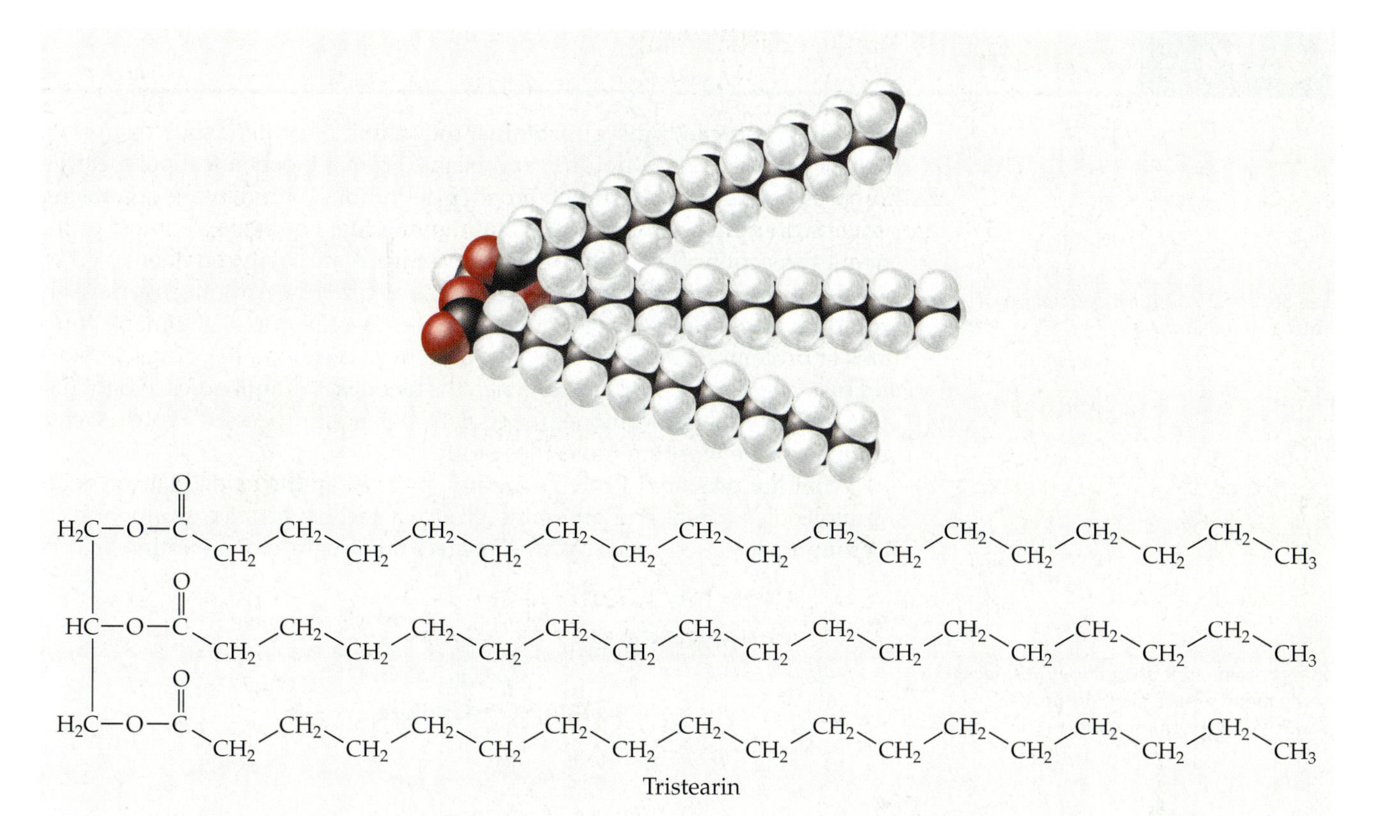

Tristearin

CAN YOU ANSWER THIS? *Saturated fats tend to be solid at room temperature, while unsaturated fats tend to be liquid. One reason saturated fats taste good is that they tend to melt in your mouth. Since unsaturated fats are liquid at room temperature, they don't have the same effect. Examine the structures of tristearin, a saturated triglyceride, and trilinolenin, an unsaturated tryglyceride.*

From the structures, give a reason for why tristearin has a greater tendency to be a solid at room temperature, while trilinolenin has a greater tendency to be a liquid. Hint: Think of the interactions between molecules. Which molecules do you think can interact with neighboring molecules better?

O
H2C—O—C—CH2—CH2—CH2—CH2—CH2—CH2—CH2—CH=CH—CH2—CH=CH—CH2—CH=CH—CH2—CH3
O
HC—O—C—CH2—CH2—CH2—CH2—CH2—CH2—CH2—CH=CH—CH2—CH=CH—CH2—CH=CH—CH2—CH3
O
H2C—O—C—CH2—CH2—CH2—CH2—CH2—CH2—CH2—CH=CH—CH2—CH=CH—CH2—CH=CH—CH2—CH3

Trilinolenin

19.5 Proteins

When most people think of **proteins**, they think of protein sources in their diet such as beef, eggs, poultry, and beans. From a biochemical perspective, however, proteins have a much broader definition. Within living organisms, proteins do much of the work of maintaining life. For example, most of the chemical reactions that occur in living organisms are catalyzed or enabled by proteins. Proteins that act as catalysts are called *enzymes*. Without enzymes, life would be impossible. But acting as enzymes is only one of the many functions of proteins. Proteins are also the structural components of muscle, skin, and cartilage. They transport oxygen in the blood, act as antibodies to fight disease, and function as hormones to regulate metabolic processes. Proteins reign supreme as the working molecules of life.

See Section 15.12 for a description of catalysts and enzymes.

What are proteins? Proteins are polymers of amino acids. **Amino acids** are molecules containing an amine group, a carboxylic acid group, and an **R group** (also called a *side chain*). The general structure of an amino acid is:

In a protein, an R group does not necessarily mean a pure alkyl group. See Table 19.2 for common R groups.

Amine group — Carboxylic acid group — R group

$$H_2N-\overset{\displaystyle H}{\underset{\displaystyle R}{C}}-\overset{\displaystyle O}{\overset{\|}{C}}-OH$$

Amino acid general stucture

Amino acids differ from each other only in their R groups. A simple amino acid is alanine, in which the R group is simply a methyl ($-CH_3$) group.

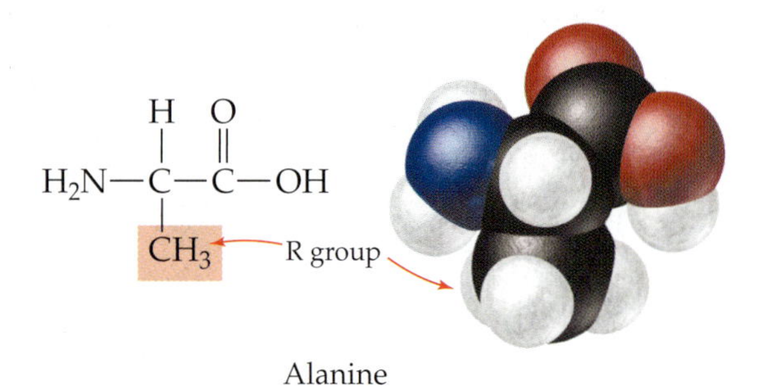

Alanine

Other amino acids, include serine, R = $-CH_2OH$; aspartic acid, R = $-CH_2COOH$; and lysine, R = $-CH_2(CH_2)_3NH_2$.

$$H_2N-\overset{\displaystyle H}{\underset{\displaystyle CH_2-OH}{C}}-\overset{\displaystyle O}{\overset{\|}{C}}-OH$$

Serine

$$H_2N-\underset{\underset{\underset{\underset{OH}{|}}{C=O}}{\underset{|}{CH_2}}}{\underset{|}{\overset{\overset{H}{|}}{C}}}-\overset{\overset{O}{\|}}{C}-OH$$

Aspartic acid

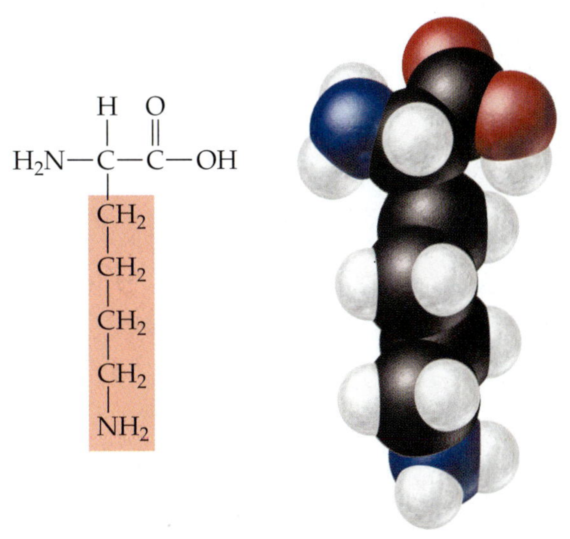

Lysine

Notice that the R groups, or side chains, of different amino acids can be very different chemically. Alanine, for example, has a nonpolar side chain while serine has a polar one. Aspartic acid has an acidic side chain while lysine, since it contains nitrogen, has a basic one. When amino acids are strung together to make a protein, these differences determine the structure and properties of the protein. Table 19.2 shows the most common amino acids in proteins.

Amino acids link together because the amine end of one amino acid reacts with the carboxylic end of another amino acid.

$$H_2N-\underset{\underset{R_1}{|}}{\overset{\overset{H}{|}}{C}}-\overset{\overset{O}{\|}}{C}-OH \;+\; H_2N-\underset{\underset{R_2}{|}}{\overset{\overset{H}{|}}{C}}-\overset{\overset{O}{\|}}{C}-OH \longrightarrow H_2N-\underset{\underset{R_1}{|}}{\overset{\overset{H}{|}}{C}}-\overset{\overset{O}{\|}}{C}-NH-\underset{\underset{R_2}{|}}{\overset{\overset{H}{|}}{C}}-\overset{\overset{O}{\|}}{C}-OH \;+\; H_2O$$

Peptide bond

The resulting bond is called a **peptide bond** and the resulting molecule—two amino acids linked together—is called a **dipeptide**. A dipeptide can link to a third amino acid to form a tripeptide, and so on. Short chains of amino acids are generally called **polypeptides**. Functional proteins usually contain hundreds or even thousands of amino acids joined by peptide bonds.

TABLE 19.2 Common Amino Acids

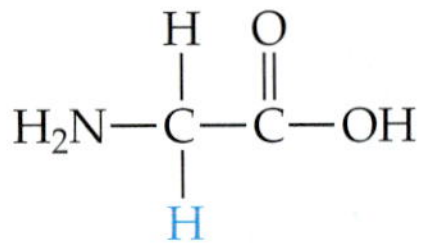

H O
$H_2N-C-C-OH$
H
Glycine (Gly)

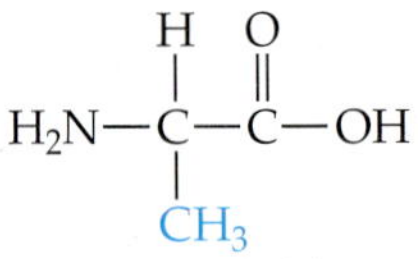

H O
$H_2N-C-C-OH$
CH_3
Alanine (Ala)

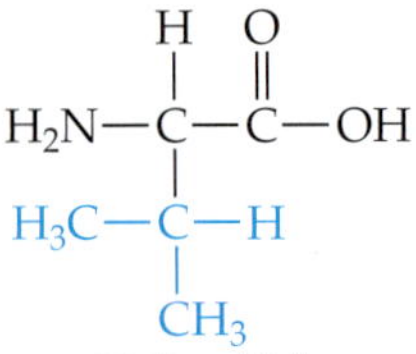

H O
$H_2N-C-C-OH$
H_3C-C-H
CH_3
Valine (Val)

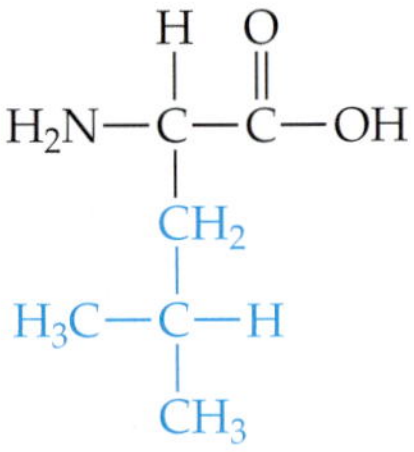

H O
$H_2N-C-C-OH$
CH_2
H_3C-C-H
CH_3
Leucine (Leu)

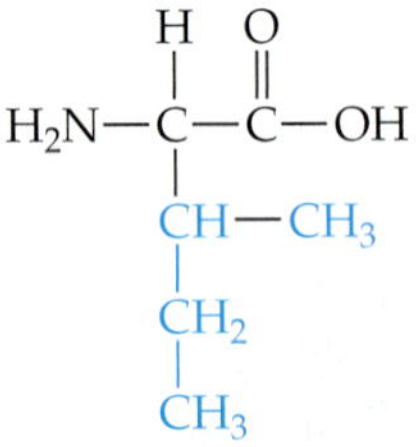

H O
$H_2N-C-C-OH$
$CH-CH_3$
CH_2
CH_3
Isoleucine (Ile)

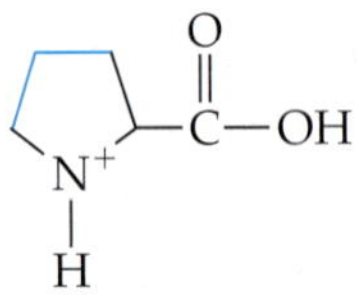

O
$C-OH$
N^+
H
Proline (Pro)

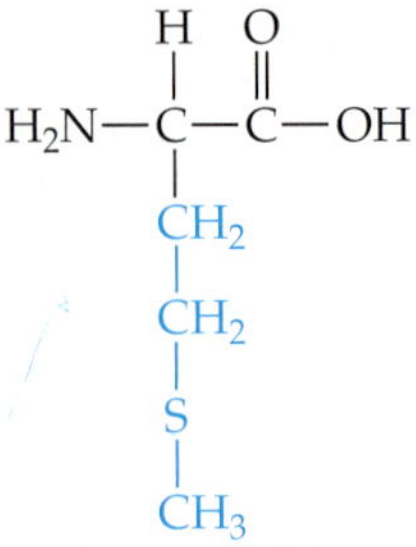

H O
$H_2N-C-C-OH$
CH_2
CH_2
S
CH_3
Methionine (Met)

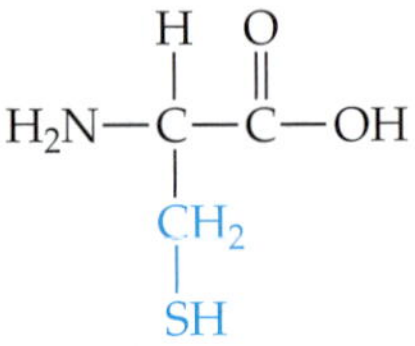

H O
$H_2N-C-C-OH$
CH_2
SH
Cysteine (Cys)

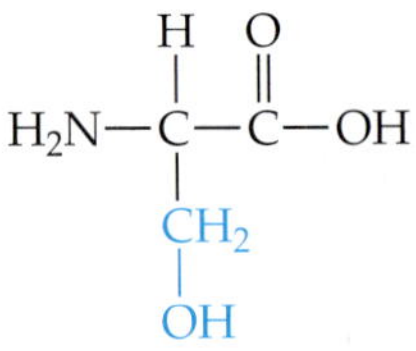

H O
$H_2N-C-C-OH$
CH_2
OH
Serine (Ser)

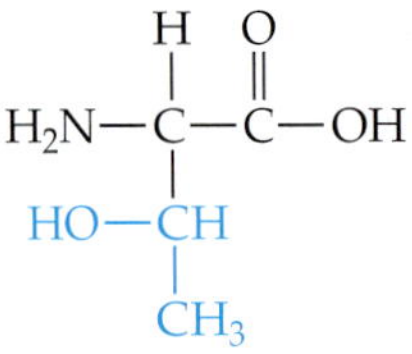

H O
$H_2N-C-C-OH$
$HO-CH$
CH_3
Threonine (Thr)

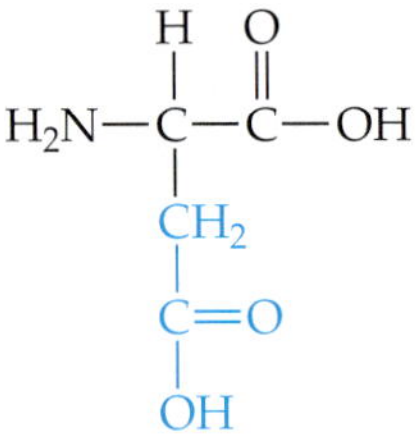

H O
$H_2N-C-C-OH$
CH_2
$C=O$
OH
Aspartic acid (Asp)

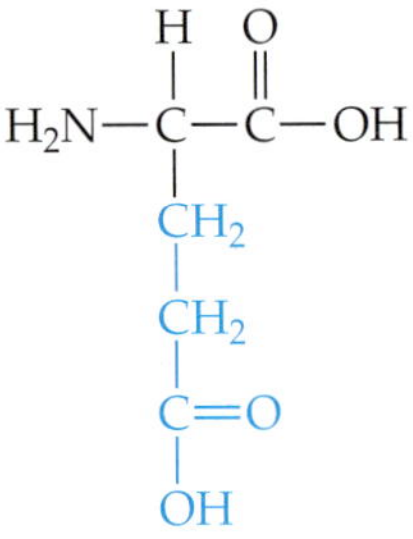

H O
$H_2N-C-C-OH$
CH_2
CH_2
$C=O$
OH
Glutamic acid (Glu)

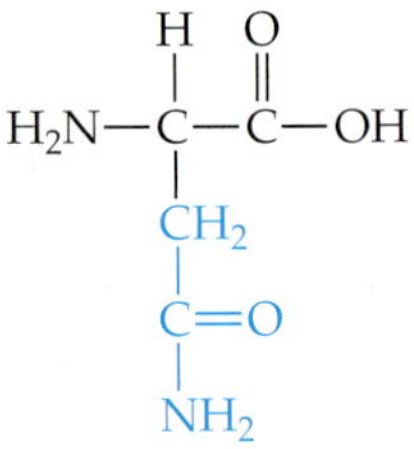

H O
$H_2N-C-C-OH$
CH_2
$C=O$
NH_2
Asparagine (Asn)

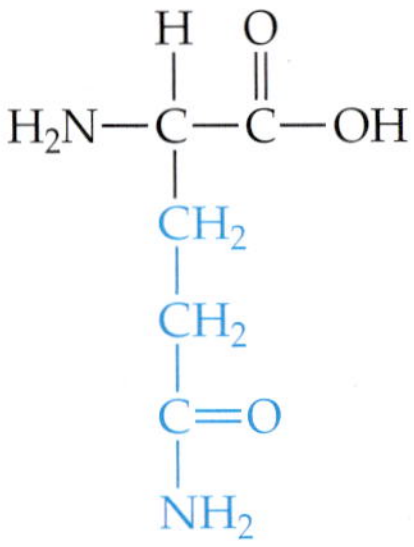

H O
$H_2N-C-C-OH$
CH_2
CH_2
$C=O$
NH_2
Glutamine (Glu)

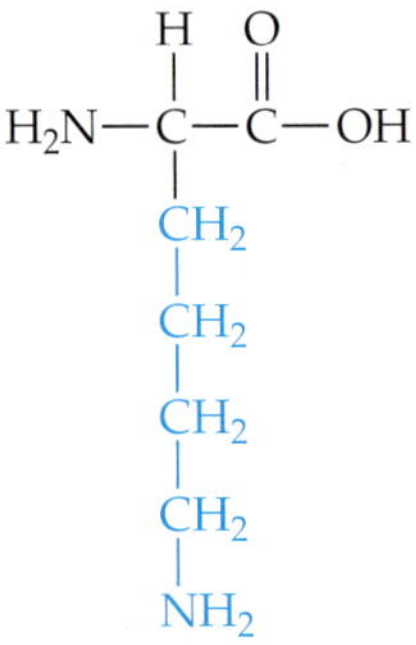

H O
$H_2N-C-C-OH$
CH_2
CH_2
CH_2
CH_2
NH_2
Lysine (Lys)

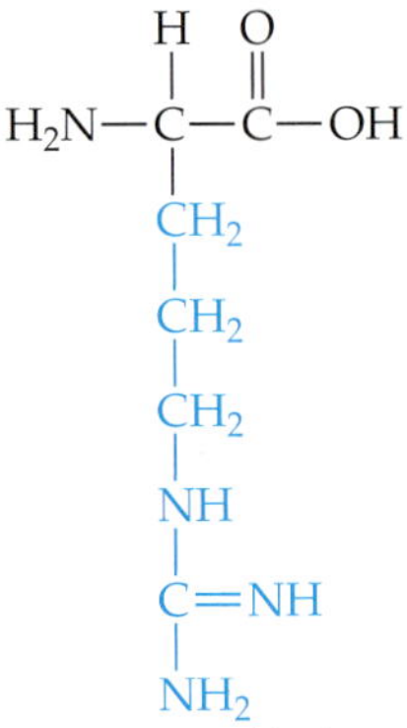

H O
$H_2N-C-C-OH$
CH_2
CH_2
CH_2
NH
$C=NH$
NH_2
Arginine (Arg)

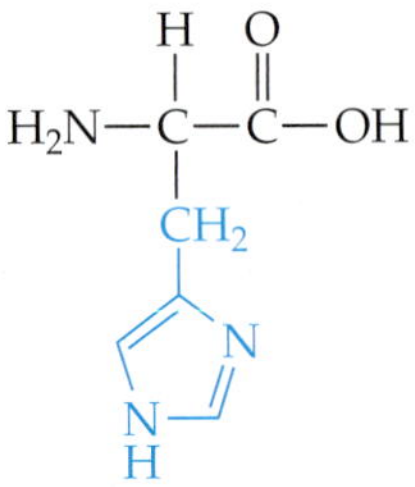

H O
$H_2N-C-C-OH$
CH_2
N
N
H
Histidine (His)

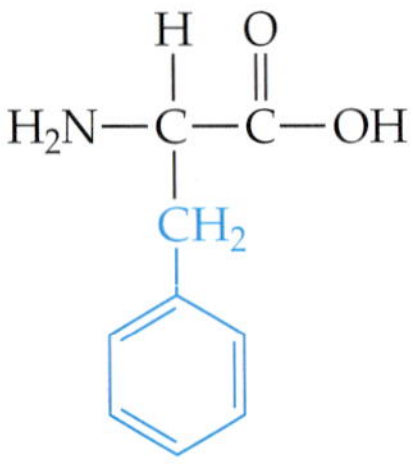

H O
$H_2N-C-C-OH$
CH_2
Phenylalanine (Phe)

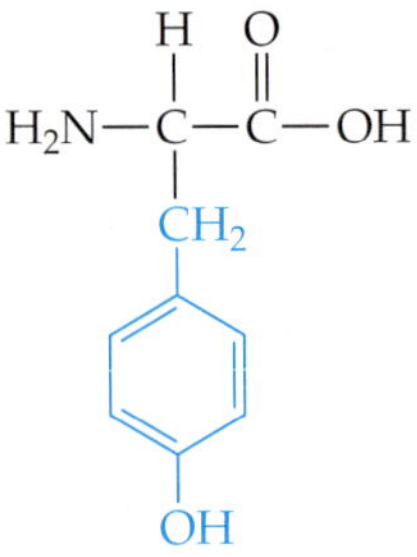

H O
$H_2N-C-C-OH$
CH_2
OH
Tyrosine (Tyr)

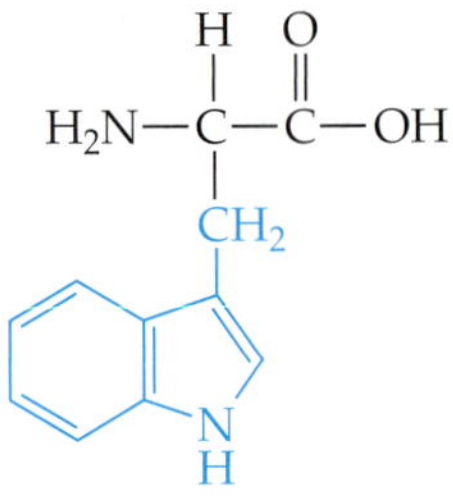

H O
$H_2N-C-C-OH$
CH_2
N
H
Tryptophan (Trp)

EXAMPLE 19.3 Peptide Bonds

Show the reaction by which glycine and alanine form a peptide bond.

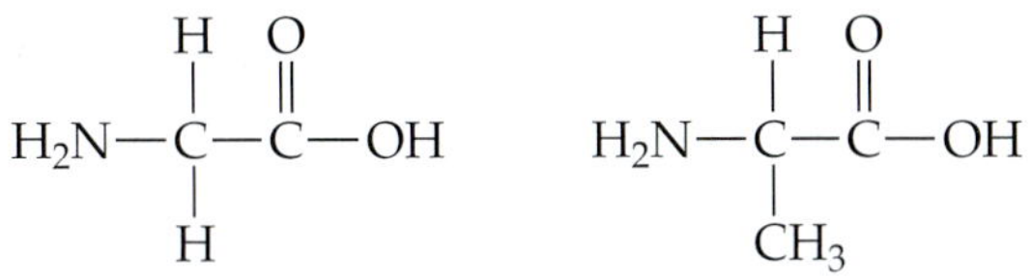

$$\underset{\text{Glycine}}{\mathrm{H_2N{-}CH(H){-}C({=}O){-}OH}} \qquad \underset{\text{Alanine}}{\mathrm{H_2N{-}CH(CH_3){-}C({=}O){-}OH}}$$

Solution:

Peptide bonds are formed when the carboxylic end of one amino acid reacts with the amine end of a second amino acid to form a dipeptide and water.

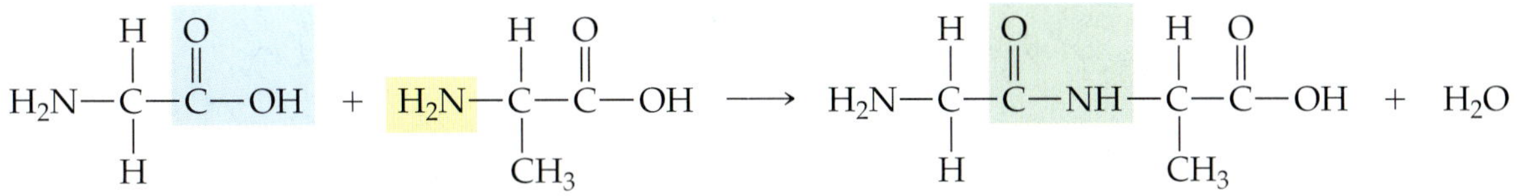

$$\mathrm{H_2N{-}CH(H){-}C({=}O){-}OH} + \mathrm{H_2N{-}CH(CH_3){-}C({=}O){-}OH} \longrightarrow \mathrm{H_2N{-}CH(H){-}C({=}O){-}NH{-}CH(CH_3){-}C({=}O){-}OH} + \mathrm{H_2O}$$

Note that this reaction can also take place between the $-NH_2$ end of glycine and the $-COOH$ end of alanine, producing a slightly different dipeptide.

SKILLBUILDER 19.3 Peptide Bonds

Show the reaction by which valine and leucine form a peptide bond.

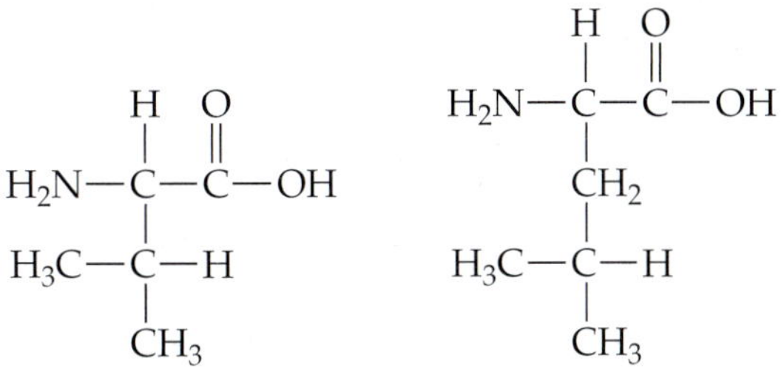

$$\underset{\text{Valine}}{\mathrm{H_2N{-}CH(CH(CH_3)CH_3){-}C({=}O){-}OH}} \qquad \underset{\text{Leucine}}{\mathrm{H_2N{-}CH(CH_2{-}CH(CH_3)CH_3){-}C({=}O){-}OH}}$$

19.6 Protein Structure

When amino acids link together to form proteins, the amino acids interact with one another, causing the protein chain to twist and fold in a very specific way. The exact shape that a protein takes depends on the types of amino acids and their sequence in the protein chain. Different amino acids and different sequences result in different shapes, and these shapes are extremely important.

For example, insulin is a protein that promotes the absorption of glucose out of the blood and into muscle cells where glucose is needed for energy. Insulin recognizes muscle cells because their surfaces contain insulin receptors, molecules that fit a specific portion of the insulin protein. If insulin were a different shape, it would not latch onto insulin receptors on muscle cells and therefore would not do its job. So the shape, or *conformation*, of proteins is crucial to their function. We can understand protein structure by dividing it into four categories: primary structure, secondary structure, tertiary structure, and quaternary structure (▶ Figure 19.7).

▶ **Figure 19.7 Protein structure**
(a) Primary structure is the amino acid sequence. **(b)** Secondary structure refers to small-scale repeating patterns such as the helix or the pleated sheet. **(c)** Tertiary structure refers to the large-scale bends and folds of the protein. **(d)** Quaternary structure is the arrangement of individual polypeptide chains.

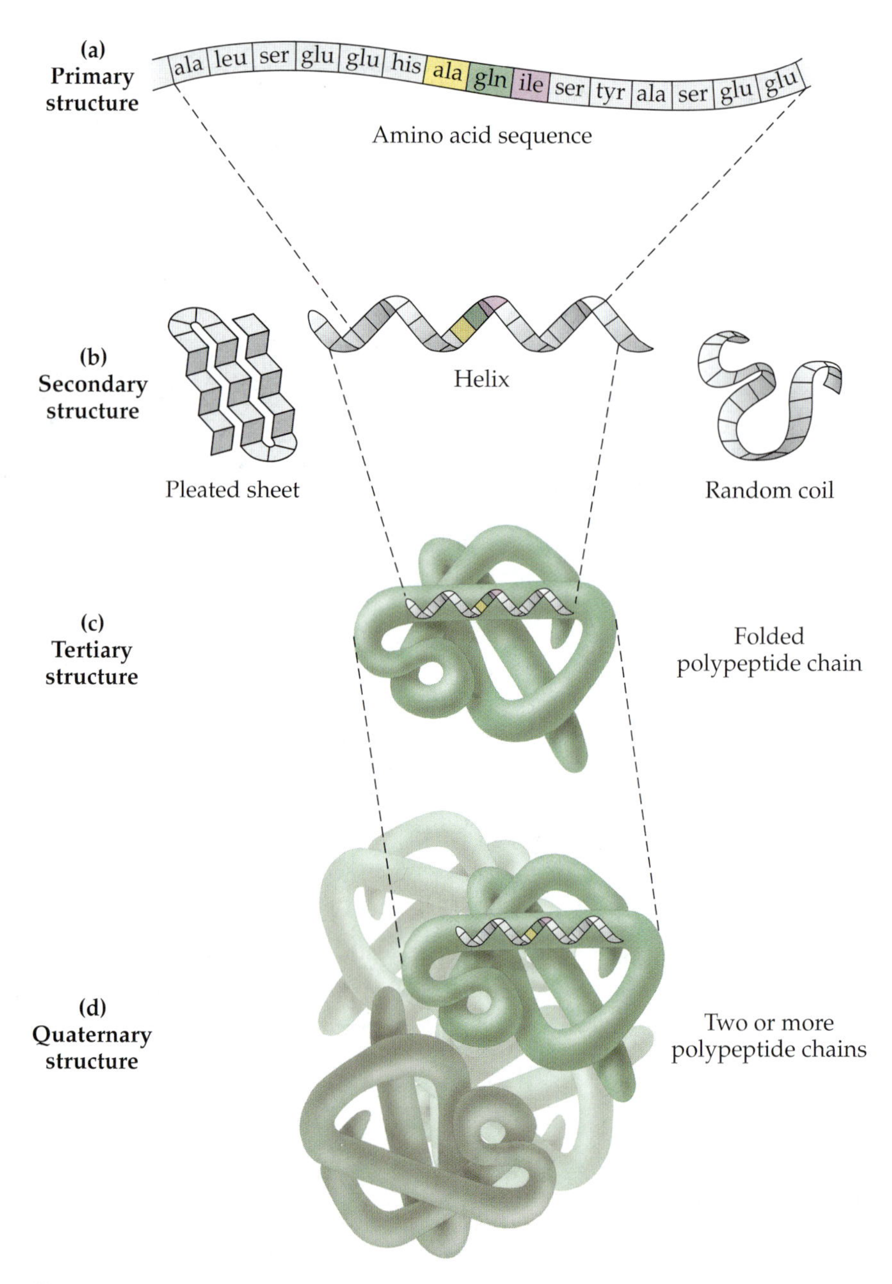

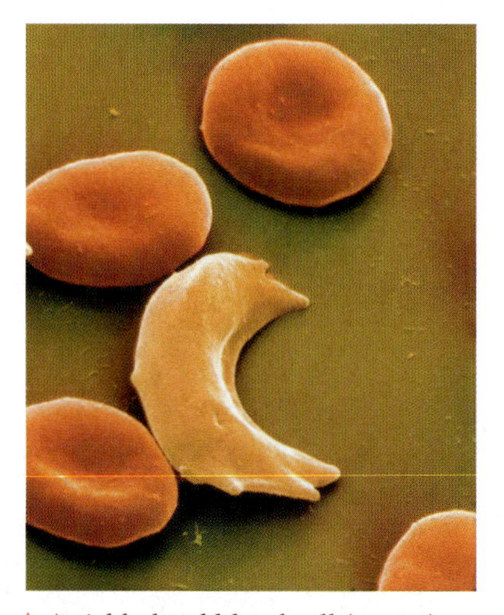

▲ A sickled red blood cell (center) surrounded by normal red blood cells. The sickled cells, characteristic of sickle-cell anemia, are fragile and easily damaged. They are also more rigid and so tend to become stuck in tiny capillaries, interfering with the flow of blood to tissues and organs.

Primary Structure

The **primary structure** of a protein is simply the sequence of amino acids in its chain. Primary structure is maintained by the covalent peptide bonds between individual amino acids. For example, one section of the insulin protein has the following sequence.

Gly-Ile-Val-Glu-Gln-Cys-Cys-Ala-Ser-Val-Cys

Each three-letter abbreviation represents an amino acid (see Table 19.2). The first amino acid sequences for proteins were determined in the 1950s. Today, the amino acid sequences for thousands of proteins are known.

Changes in the amino acid sequence of a protein, even minor ones, can have devastating effects on the function of a protein. Hemoglobin, for example, is a protein that transports oxygen in the blood. It is composed of four protein chains, each containing 146 amino acid units (▶ Figure 19.8) for a total of 584 amino acid units. The substitution of glutamic acid for valine in just one position on two of these chains results in the disease known as sickle-cell anemia, in which red blood cells take on a sickle shape that ultimately leads to damage of major organs. In the past, sickle-cell anemia has been fatal, often before age 30—all because of a change in 2 amino acids out of 584.

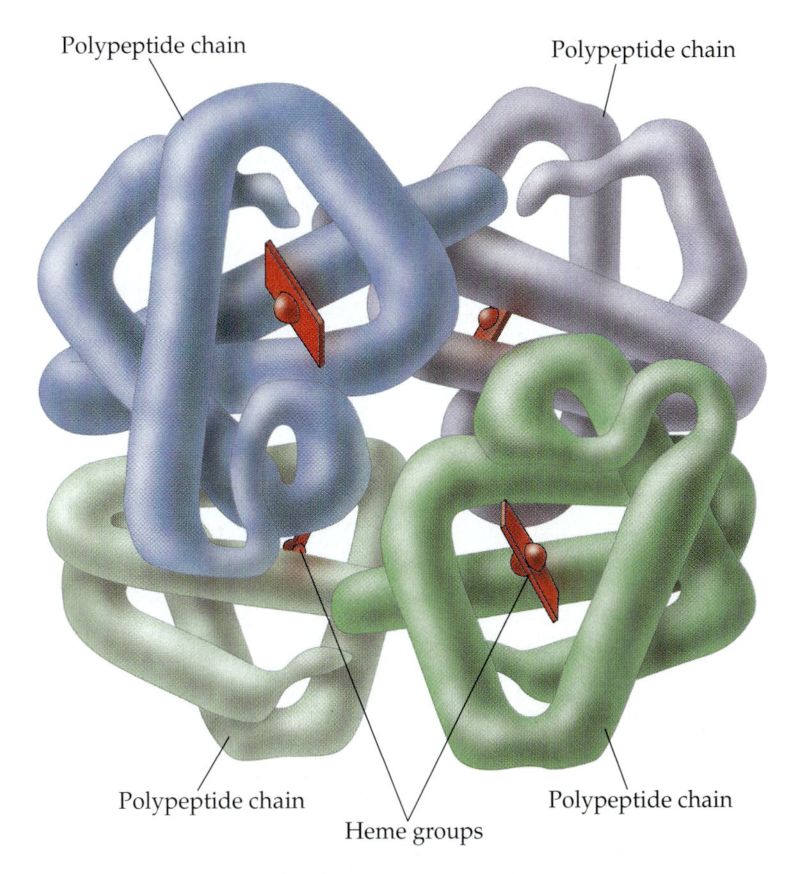

▶ **Figure 19.8 Hemoglobin** Hemoglobin is a protein composed of four chains, each containing 146 amino acid units. Each chain holds a molecule called a *heme,* which contains an iron atom in its center. Oxygen binds at the iron atom.

Secondary Structure

The **secondary structure** of a protein refers to certain short-range periodic or repeating patterns often found along protein chains. Secondary structure is maintained by interactions between amino acids that are fairly close together in the linear sequence of the protein chain or adjacent to each other on neighboring chains. The most common of these patterns is called the **alpha (α)-helix,** shown in ▼ Figure 19.9. In the α-helix structure, the amino acid chain is wrapped into a tight coil in which the side chains extend outward from the coil. The structure is maintained by hydrogen-bonding interactions between NH and CO groups along the peptide backbone of the protein. Some proteins—such as keratin, which composes hair—have the α-helix pattern throughout their entire chain. Other proteins have very little or no α-helix pattern in their chain. It depends on the particular protein.

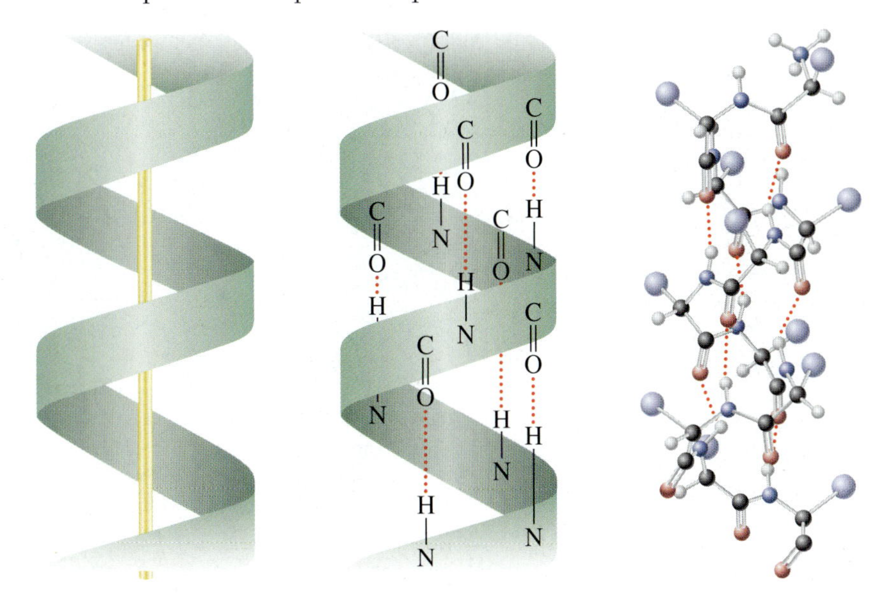

▶ **Figure 19.9 Alpha-helix protein structure** The α-helix is maintained by interactions between the peptide backbones of amino acids that are close to each other in the linear sequence of the protein chain.

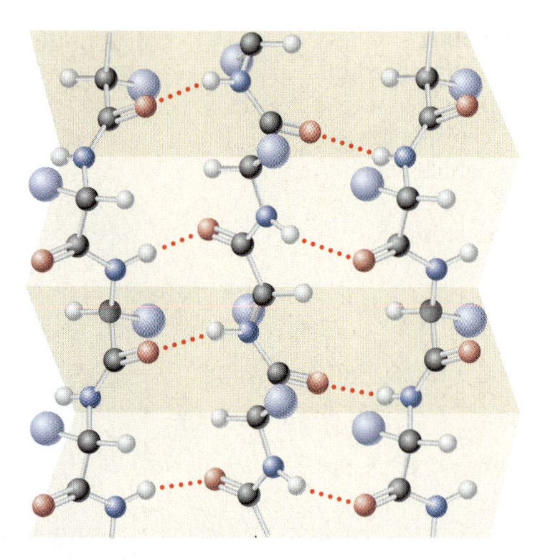

▶ **Figure 19.10 Beta-pleated sheet protein structure** The β-pleated sheet is maintained by interactions between the peptide backbones of neighboring protein chains.

A second common pattern in the secondary structure of proteins is called the **beta (β)-pleated sheet** (▲ Figure 19.10). In this structure, the chain is extended (as opposed to coiled) and forms a zigzag pattern. The peptide backbones of the chains interact with one another through hydrogen bonding to form zigzagged sheets. Some proteins—such as silk—have the β-pleated sheet structure throughout their entire chain. Since the protein chains in the β-pleated sheet are fully extended, silk is inelastic. Many proteins have some sections that are β-pleated sheet, other sections that are α-helix, and still other sections that have less regular patterns called **random coils**.

EVERYDAY **Chemistry**

Why Hair Gets Longer When It Is Wet

Have you ever noticed that your hair gets longer when it is wet? Why does this happen?

Hair is composed of a protein called keratin. The secondary structure of keratin is α-helix throughout, meaning that the protein has a wound-up helical structure. As we learned earlier, this structure is maintained by hydrogen bonding.

Individual hair fibers are composed of several strands of keratin coiled around each other. When hair is dry, the keratin protein is tightly coiled, resulting in the normal length of dry hair. However, when hair becomes wet, water molecules interfere with the hydrogen bonding that maintains the α-helix structure. The result is the relaxation of the α-helix structure and the lengthening of the hair fiber. Completely wet hair is 10% to 12% longer than dry hair.

CAN YOU ANSWER THIS? *When curlers are put into wet hair and the wet hair is allowed to dry, the hair tends to retain the shape of the curler. Can you explain why this happens?*

▲ Wet hair is 10% to 12% longer than dry hair.

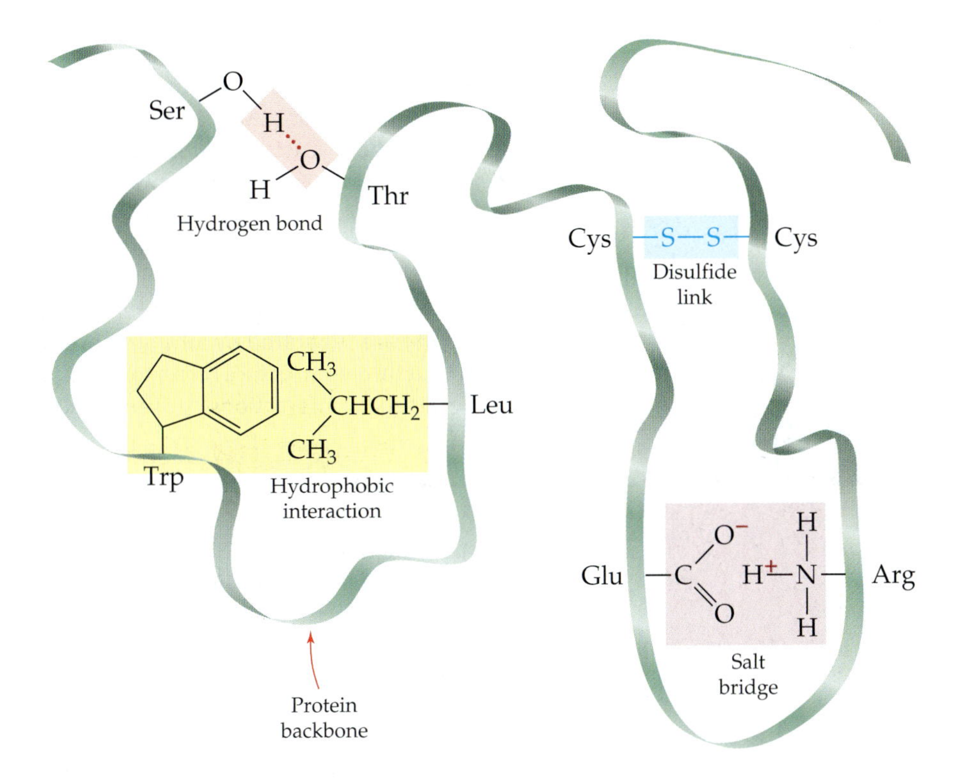

▶ **Figure 19.11 Interactions that create tertiary and quaternary structure** The tertiary structure of a protein is maintained by interactions between the R groups of amino acids that are separated by large distances in the linear sequence of the protein chain. These interactions include hydrogen bonds, disulfide linkages, hydrophobic interactions, and salt bridges. (The figure shows a typical example of each kind of interaction.) The same interactions can also hold different amino acid chains together (quaternary structure).

Tertiary Structure

The **tertiary structure** of a protein consists of the large-scale bends and folds due to interactions between the R groups of amino acids that are separated by large distances in the linear sequence of the protein chain. These interactions, as shown in ▲ Figure 19.11, include the following:

- hydrogen bonds
- disulfide linkages (covalent bonds between sulfur atoms on different R groups)
- hydrophobic interactions (attractions between large nonpolar groups)
- salt bridges (acid–base interactions between acidic and basic groups)

Proteins with structural functions—such as keratin, which composes hair, or collagen, which composes tendons and much of the skin—tend to have tertiary structures in which coiled amino acid chains align roughly parallel to one another, forming long, water-insoluble fibers. These kinds of proteins are called **fibrous proteins**. Proteins with nonstructural functions—such as hemoglobin, which carries oxygen, or lysozyme, which fights infections—tend to have tertiary structures in which amino acid chains fold in on themselves, forming water-soluble globules that can travel through the bloodstream. These kinds of proteins are called **globular proteins**. The overall shape of a protein may seem random, but it is not. It is determined by the amino acid sequence and, as we have seen, is critical to its function.

Quaternary Structure

Many proteins are composed of more than one amino acid chain. As we have seen, for example, hemoglobin is composed of four amino acid chains. The way that these chains fit together is called **quaternary structure**. Quaternary structure is maintained by the same kinds of interactions between amino acids as those that maintain tertiary structure.

To summarize protein structure:

- Primary structure is simply the amino acid sequence. It is maintained by the peptide bonds that hold amino acids together.
- Secondary structure refers to the small-scale repeating patterns often found in proteins. These are maintained by interactions between the peptide backbones of amino acids that are close together in the chain sequence or adjacent to each other on neighboring chains.
- Tertiary structure refers to the large-scale twists and folds within the protein. These are maintained by interactions between the R groups of amino acids that are separated by long distances in the chain sequence.
- Quaternary structure refers to the arrangement of chains in proteins. It is maintained by interactions between amino acids on the individual chains.

19.7 Nucleic Acids: Molecular Blueprints

We have seen the importance of amino acid sequence in determining protein structure and function. If the amino acid sequence in a protein is incorrect, the protein is unlikely to function properly. How do our bodies constantly synthesize the many thousands of different proteins—each with the correct amino acid sequence—that we need to survive? What ensures that proteins have the correct amino acid sequence? The answer to this question lies in **nucleic acids**. Nucleic acids contain a chemical code that specifies the correct amino acid sequences for proteins. Nucleic acids can be divided into two types: deoxyribonucleic acid or **DNA**, which exists primarily in the nucleus of the cell; and ribonucleic acid or **RNA**, which exists through the entire interior of the cell.

Like proteins, nucleic acids are polymers. The individual units composing nucleic acids are called **nucleotides**. Each nucleotide has three parts: a phosphate, a sugar, and a base (▼ Figure 19.12). In DNA, the sugar is deoxyribose, while in RNA the sugar is ribose.

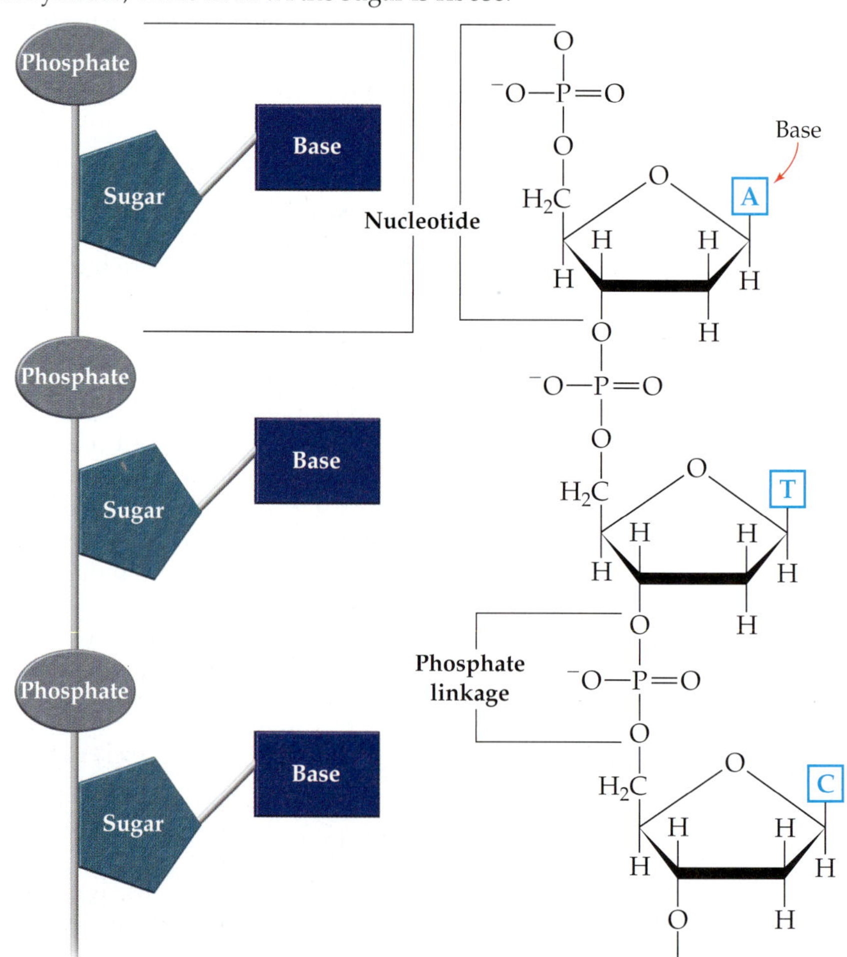

▶ **Figure 19.12 Components of DNA** DNA is a polymer of nucleotides. Each nucleotide has three parts: a sugar group, a phosphate group, and a base. Nucleotides are joined by phosphate linkages.

Deoxyribose

Ribose

Nucleotides link together via phosphate linkages to form nucleic acids. Every nucleotide in DNA has the same phosphate and sugar, but can have one of four different bases. In DNA, the four bases are adenine (A), cytosine (C), guanine (G), and thymine (T).

Adenine

Cytosine

Guanine

Thymine

In RNA, the base uracil (U) replaces thymine.

Uracil

The order of bases in a nucleic acid chain specifies the order of amino acids in a protein. However, since there are only four bases and about twenty different amino acids that must be specified, a single base cannot code for a single amino acid. It takes a sequence of three bases—called a **codon**—to code for one amino acid (▼ Figure 19.13). The genetic code—the code that links a specific codon to an amino acid—was discovered in 1961. It is nearly universal, meaning that the same codons specify the same amino acids in nearly all organisms. For example, in DNA the sequence AGT codes for the amino acid serine and the sequence TGA codes for the amino acid threonine. It doesn't matter whether you are a rat, a bacterium, or a human—the code is the same.

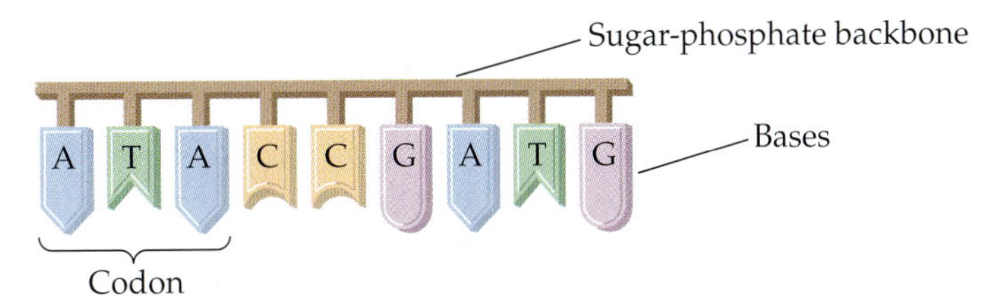

▶ **Figure 19.13 Codons** A sequence of three nucleotides with their associated bases is called a *codon*. Each codon codes for one amino acid.

A gene is a sequence of codons within a DNA molecule that codes for a single protein. Since proteins vary in size from fifty to thousands of amino acids, so genes vary in length from fifty to thousands of codons. For example, insulin is composed of fifty-one amino acids. The insulin gene, then, must contain fifty-one codons—one for each amino acid in the insulin protein. Each codon is like a three-letter word that specifies one amino acid. String the correct number of codons together in the correct sequence, and you have a gene, the instructions for the amino acid sequence in a protein. Genes are contained in structures called **chromosomes**—forty-six in humans—within the nuclei of cells (▶ Figure 19.14).

▶ **Figure 19.14 Organization of the genetic material**

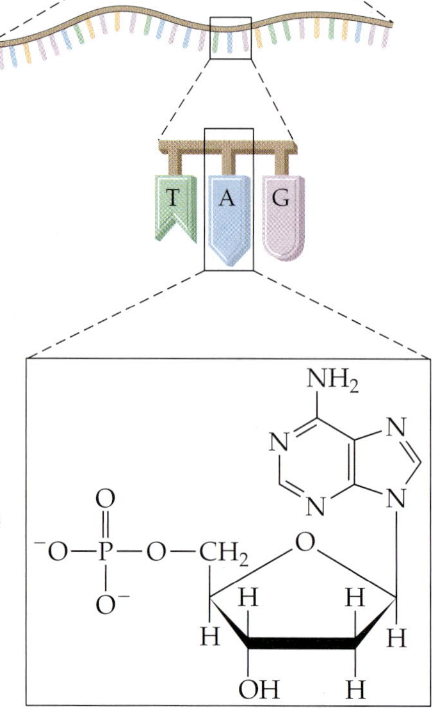
Chromosome—structure within cell nucleus that houses DNA.
Gene—portion of DNA that codes for a single protein.
Codon—sequence of three nucleotides and their associated bases. A codon codes for one amino acid.
T
A
G
Nucleotide—individual links in the nucleic acid chain. Nucleotides are composed of a sugar group, a phosphate group, and a base.
NH_2
N
N
N
N
O
$^-O—P—O—CH_2$
O
O^-
H
H
H
H
OH
H

CONCEPTUAL CHECKPOINT 19.1

The number of DNA bases needed to code for insulin (51 amino acids) is:

(a) 17
(b) 20
(c) 51
(d) 153

19.8 DNA Structure, DNA Replication, and Protein Synthesis

DNA Structure

Most of the cells in our bodies contain a complete set of instructions—within the DNA in the nucleus—to make all of the proteins that we need. All cells, however, do not synthesize every protein specified by the genes of its DNA. Cells synthesize only those proteins that are important to their function. For example, pancreatic cells synthesize insulin, and therefore use the insulin gene within their nucleus for the instructions. Pancreatic cells do not, however, synthesize keratin (hair protein), even though the keratin gene is also contained within their nucleus. The cells in the scalp, on the other hand (which also have both insulin and keratin genes in their nuclei) synthesize keratin but not insulin. Cells synthesize only the proteins that are specific to their function.

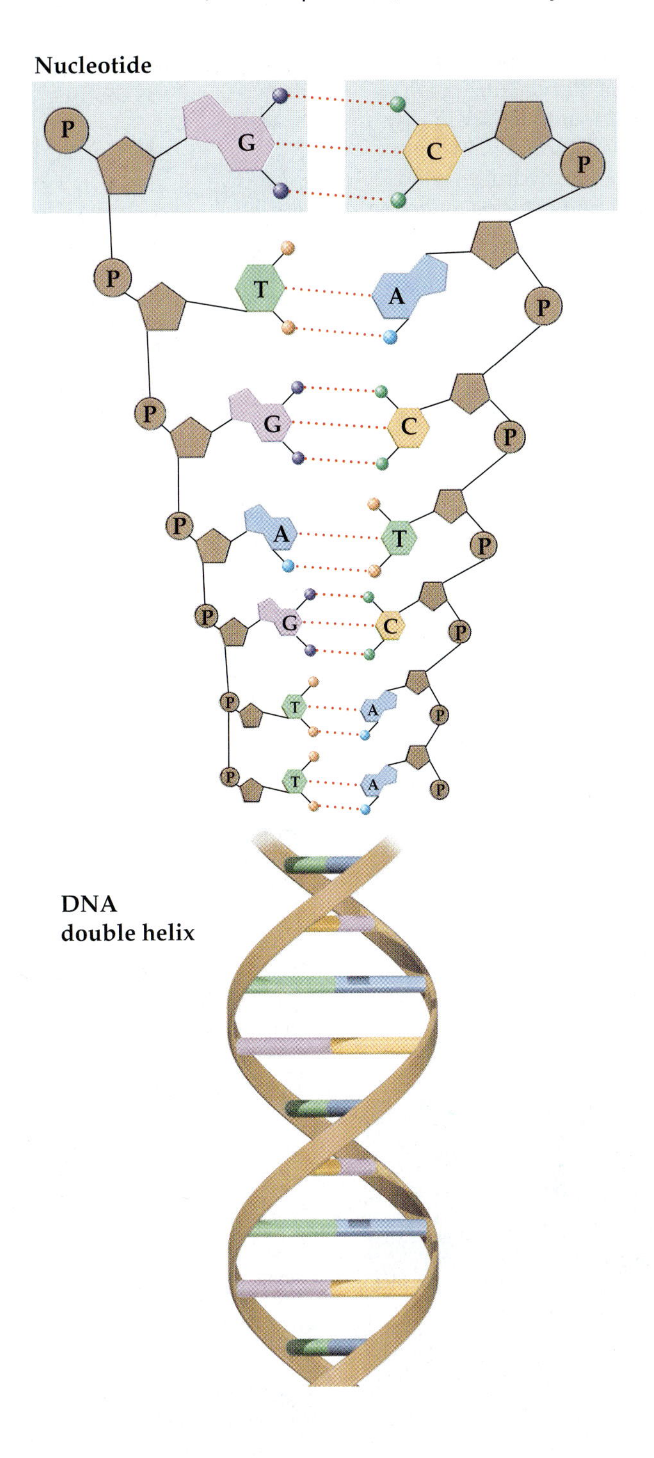

▶ **Figure 19.15** **Structure of the DNA molecule** DNA has a double-stranded helical structure. Each strand is complementary to the other.

How did most of the cells in our bodies get a complete copy of our entire DNA? The answer lies in DNA replication. Cells reproduce by dividing—a parent cell divides into two daughter cells. As it divides, it makes complete copies of its DNA for each daughter cell. The ability of DNA to copy itself is related to its structure. DNA exists as a double-stranded helix (▲ Figure 19.15). The bases on each DNA strand are directed toward the interior of the helix, where they hydrogen bond to bases on the other strand. However, the hydrogen bonding between bases is not random. Each base is **complementary**—capable of precise pairing—with only one other base. Adenine (A) hydrogen bonds only with thymine (T), and

▶ **Figure 19.16 Complementarity** The complementary nature of DNA is related to the unique way in which the bases interact through hydrogen bonding. Adenine hydrogen bonds with thymine and cytosine hydrogen bonds with guanine.

Thymine Adenine

(a)

Cytosine Guanine

(b)

cytosine (C) hydrogen bonds only with guanine (G) (▲ Figure 19.16). For example, consider a section of DNA containing the following bases.

The complementary strand would then have the following sequence.

The two complementary strands are tightly wrapped into a helical coil, the famous DNA double helix structure.

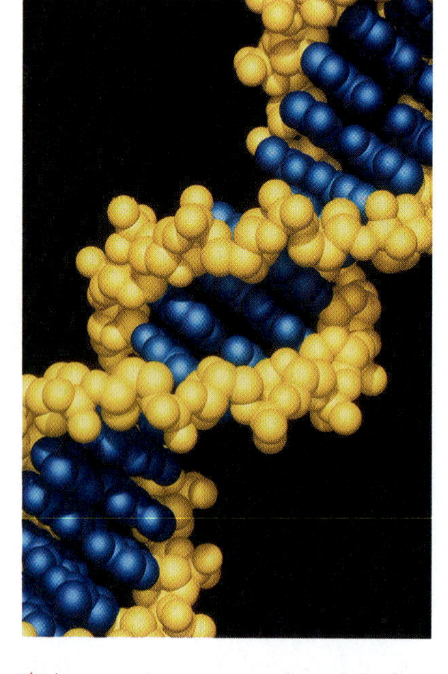

▲ A computer-generated model of the DNA double-helix structure. The yellow atoms are the sugar-phosphate chains while the blue atoms make up the paired complementary bases.

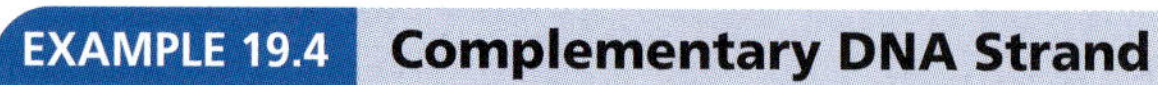

EXAMPLE 19.4 Complementary DNA Strand

Show the sequence of the complementary strand for the following DNA strand.

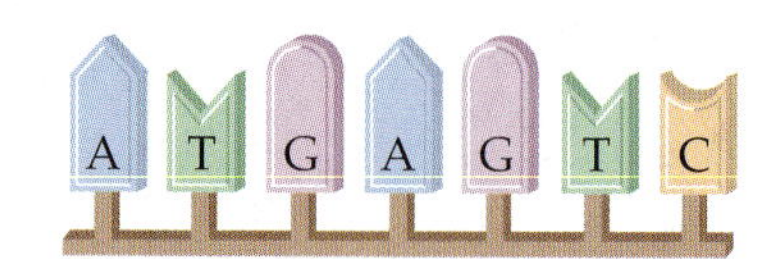

Solution:

Draw the complementary strand, remembering that A pairs with T and C pairs with G.

SKILLBUILDER 19.4 Complementary DNA Strand

Show the sequence of the complementary strand for the following DNA strand.

DNA Replication

TUTORIAL
DNA Replication Movie

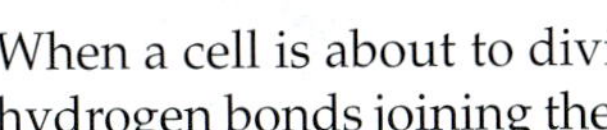

When a cell is about to divide, the DNA within its nucleus unwinds and the hydrogen bonds joining the complementary bases break, forming two daughter strands (▼ Figure 19.17). With the help of enzymes, a complement to each daughter strand—with the correct complementary bases in the correct order—

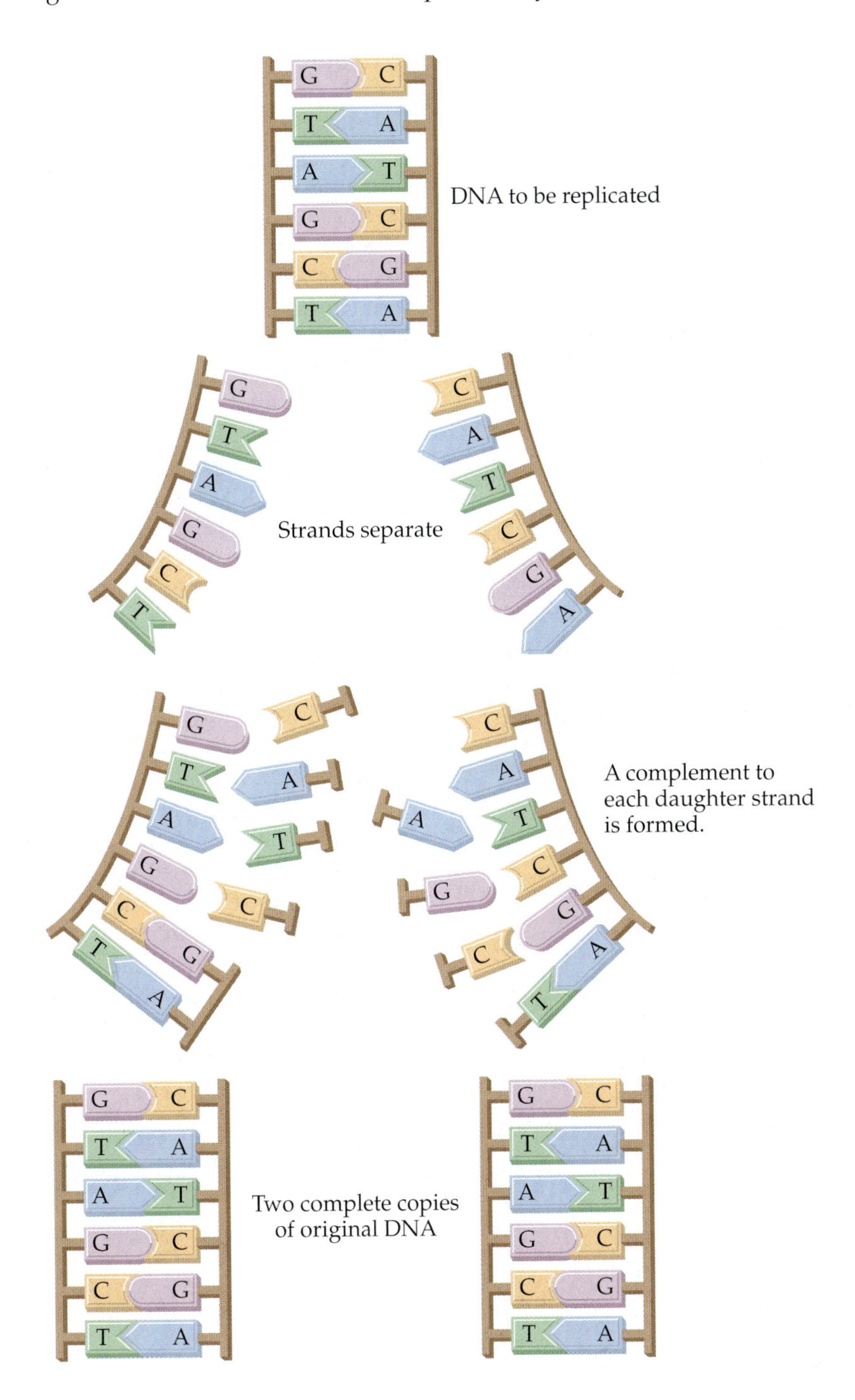

▶ Figure 19.17 DNA replication

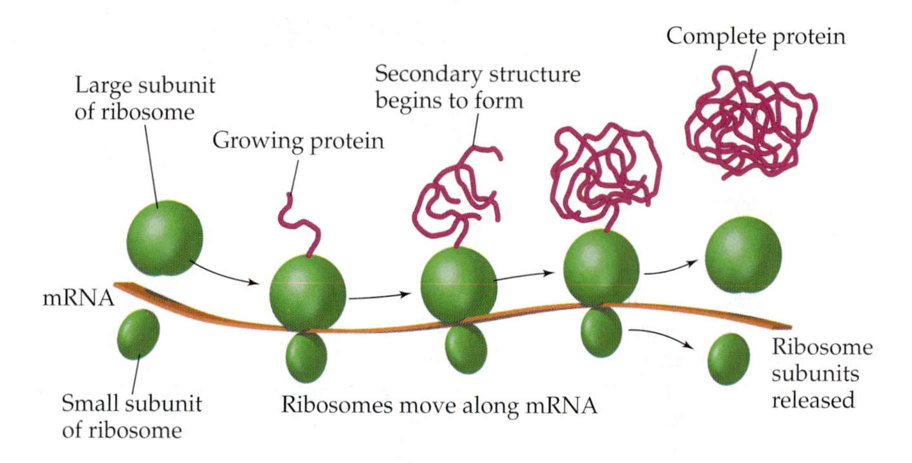

▶ **Figure 19.18 Protein synthesis** The mRNA strand that codes for a protein moves through the ribosome. At each codon, the correct amino acid is brought into place and bonds with the previous amino acid.

is formed. The hydrogen bonds between the strands then reform, resulting in two complete copies of the original DNA, one for each daughter cell.

Protein Synthesis

Living organisms must synthesize the proteins they need to survive. Dietary protein is not used in the same form as it was consumed. Rather, dietary protein is split into its constituent amino acids during digestion. These amino acids are then reconstructed into the correct proteins—those needed by the particular organism—in the organism's cells. Nucleic acids, of course, direct the process. When a cell needs to make a particular protein, the gene—the section of the DNA that codes for that particular protein—unravels. A complementary copy of that gene is then synthesized using **messenger RNA** (or **mRNA**). The mRNA then moves out of the cell's nucleus to a cell structure within the cytoplasm called a ribosome. At the ribosome, protein synthesis occurs. The mRNA chain that codes for the protein moves through the ribosome. At each codon, the correct amino acid is brought into place and a peptide forms with the previous amino acid (▲ Figure 19.18). As the mRNA moves through the ribosome, the protein is formed.

To summarize:

- DNA contains the code for the sequence of amino acids in proteins.
- A codon—three nucleotides with their bases—codes for one amino acid.
- Each base is complementary—capable of precise pairing—with only one other base.
- A gene—a sequence of codons—codes for one protein.
- Genes are contained in structures called chromosomes that occur in the nuclei of cells. Humans have forty-six chromosomes.
- When a cell divides, each daughter cell receives a complete copy of the DNA—all forty-six chromosomes—within the cell's nucleus.
- When a cell synthesizes a protein, the base sequence of the gene that codes for that protein is transferred to mRNA. The mRNA then moves out to a ribosome, where the amino acids are linked in the correct sequence to synthesize the protein. The general sequence is

$$\text{DNA} \longrightarrow \text{RNA} \longrightarrow \text{Protein}$$

CONCEPTUAL CHECKPOINT 19.2

Which of the following biological molecules are *not* polymers?

(a) proteins
(b) steroids
(c) nucleic acids
(d) polysaccharides

Chemistry AND HEALTH

Drugs for Diabetes

Diabetes is a disease in which a person's body does not make enough insulin, the substance that promotes the absorption of sugar from the blood. Consequently, diabetics have high blood sugar levels, which can—over time—lead to a number of complications, including kidney failure, heart attacks, strokes, blindness, and nerve damage. One treatment for diabetes is the injection of insulin, which can help manage blood sugar levels and reduce the risk of these complications. However, insulin is a human protein and cannot be easily synthesized in the laboratory. So where can diabetics get life-saving insulin? For many years, the primary source was animals, particularly pigs and cattle. Although animal insulin worked to lower blood sugar levels, some patients did not tolerate it.

Today, diabetics inject *human* insulin. Where does it come from? Its source is one of the success stories of biotechnology. Scientists were able to remove the gene for insulin from a sample of healthy human cells. They then inserted that gene into bacteria, which incorporated the gene into their genome. When the bacteria reproduced, they passed on exact copies of the gene to their offspring. The result was a colony of bacteria that all contained the human insulin gene. Even more amazing, the chemical machinery within the bacteria expressed the gene—the bacteria synthesized the human insulin that the gene codes for. The insulin is harvested from the cell cultures and bottled for distribution to diabetics. Today, millions of diabetics manage their disease with human insulin made in this way.

CAN YOU ANSWER THIS? *Can all drugs be made in this way? What kinds of drugs can be made with these techniques?*

CHAPTER IN REVIEW

Chemical Principles	Relevance
The Cell: The cell is the smallest structural unit of a living organism that has the properties normally associated with life. The main chemical components of the cell can be divided into four categories: • Carbohydrates • Lipids • Proteins • Nucleic acids	**The Cell:** Some living organisms, such as bacteria, are composed of a single cell. In more complex organisms, cells are the building blocks that compose organs, which together compose the organism.
Carbohydrates: Carbohydrates are aldehydes or ketones containing multiple —OH groups. Monosaccharides, such as glucose and fructose, are carbohydrates that cannot be broken down into simpler carbohydrates. Disaccharides, such as sucrose and lactose, are two monosaccharides linked together by glycoside linkages. Polysaccharides, such as starch and cellulose, consist of many monosaccharides linked together. Polysaccharides are also called complex carbohydrates.	**Carbohydrates:** Living organisms use carbohydrates for short-term energy storage. Complex carbohydrates also form the main structural components of plants. Carbohydrates are a major part of our diet. Table sugar (or sucrose), for example, is a disaccharide. Starch and fiber (also called cellulose) are polysaccharides.

Lipids: Lipids are chemical components of the cell that are insoluble in water but soluble in nonpolar solvents. Important lipids include fatty acids, triglycerides, phospholipids, glycolipids, and steroids.

Lipids: Lipids compose many of the structural components of cells. Lipids are also used for long-term energy storage and for insulation. In the diet, saturated fats—triglycerides containing no double bonds in their carbon chains—are more likely to increase the risk of stroke and heart attack than unsaturated fats.

Proteins: Proteins are polymers of amino acids. Amino acids are molecules composed of an amine group on one end and a carboxylic acid on the other. Between these two groups is a central carbon atom that has an R group (also called a *side chain*) attached. Amino acids link together by means of peptide bonds, formed by the reaction between the amine end of one amino acid and the carboxylic acid of another. Functional proteins are composed of hundreds or thousands of amino acids.

Proteins: Proteins are the working molecules of life. They act as biological catalysts (called *enzymes*) to enable thousands of chemical reactions. They compose the structural elements of muscle, skin, and cartilage. They transport oxygen in the blood. They act as antibodies to fight disease and they function as hormones to regulate metabolic processes.

Protein Structure: Primary protein structure is the amino acid sequence in the protein chain. It is maintained by the peptide bonds that hold amino acids together.

Secondary structure refers to the small-scale repeating patterns often found in proteins. These are maintained by interactions between the peptide backbones of amino acids that are close together in the chain sequence or on neighboring chains.

Tertiary structure refers to the large-scale twists and folds within the protein. These are maintained by interactions between R groups of amino acids that are separated by long distances in the chain sequence.

Quaternary structure refers to the arrangement of chains in proteins. Quaternary structure is maintained by interactions between amino acids on the individual chains.

Protein Structure: The structure of proteins is critical to their function. The shapes of proteins largely determine how they interact with other molecular structures to do their job. That structure depends on the sequence of amino acids within the protein chain and how those amino acids interact with one another.

Nucleic Acids, DNA Replication, and Protein Synthesis: Nucleic acids, including DNA and RNA, are polymers of nucleotides. In DNA, each nucleotide contains one of four bases: adenine (A), cytosine (C), thymine (T), and guanine (G). The order of these bases contains a code that specifies the amino acid sequence in proteins. A codon, a sequence of three bases, codes for an amino acid. A gene, a sequence of hundreds to thousands of codons, codes for a protein. Genes are held in cellular structures called chromosomes.

Complete copies of DNA are transferred from parent cells to daughter cells via DNA replication. In this process, the two complementary strands of DNA within a cell unravel and two new strands that complement the original strands are synthesized. In this way, two complete copies of the DNA are made, one for each daughter cell.

When a cell synthesizes a protein, the base sequence of the gene that codes for that protein is transferred to mRNA. The mRNA then moves out to a ribosome, where the amino acids are linked in the correct sequence to synthesize the protein. The general sequence is:

$$\text{DNA} \longrightarrow \text{RNA} \longrightarrow \text{Protein}$$

Nucleic Acids and DNA Replication: Since DNA contains the instructions for making proteins and since proteins are the working molecules of life, our DNA determines a great deal of who and what we are. Humans have basically the same body parts, organs, and metabolic processes because most of our DNA is the same. The differences between humans are at least partly caused by slight differences in their DNA. In recent years, scientists have learned how to manipulate and change DNA, and in this way they can manipulate and change the organisms that result from that DNA.

Key Terms

alpha (α)-helix [19.6]
amino acid [19.5]
beta (β)-pleated sheet [19.6]
biochemistry [19.1]
carbohydrate [19.3]
cell [19.2]
cell membrane [19.2]
cellulose [19.3]
chromosome [19.7]
codon [19.7]
complementary base [19.8]
complex carbohydrate [19.3]
cytoplasm [19.2]
dipeptide [19.5]
disaccharide [19.3]
DNA [19.7]
ester linkages [19.4]
fatty acid [19.4]
fibrous proteins [19.6]
gene [19.1]
globular proteins [19.6]
glycogen [19.3]
glycolipid [19.4]
glycoside linkage [19.3]
human genome [19.1]
lipid [19.4]
lipid bilayer [19.4]
messenger RNA (mRNA) [19.8]
monosaccharide [19.3]
nucleic acid [19.7]
nucleotide [19.7]
nucleus (of a cell) [19.2]
peptide bond [19.5]
phospholipid [19.4]
polypeptide [19.5]
polysaccharide [19.3]
primary protein structure [19.6]
protein [19.5]
quaternary protein structure [19.6]
random coil [19.6]
R group (side chain) [19.5]
RNA [19.7]
saturated fat [19.4]
secondary protein structure [19.6]
simple carbohydrate [19.3]
simple sugar [19.3]
starch [19.3]
steroid [19.4]
tertiary protein structure [19.6]
triglyceride [19.4]
unsaturated fat [19.4]

Exercises

Questions

1. What is the Human Genome Project?
2. What has been one of the surprises of the Human Genome Project?
3. What are some of the expected benefits of the Human Genome Project?
4. Explain what a cell is and list its main chemical components.
5. What are carbohydrates?
6. What functions do carbohydrates serve in living organisms?
7. Is glucose soluble in water? How can you tell? Why is this important?
8. Explain the differences between a monosaccharide, a disaccharide, and a polysaccharide.
9. What happens to disaccharides and polysaccharides during digestion?
10. What is the difference between a simple sugar and a complex carbohydrate?
11. What is the difference between starch and cellulose? What is the result of this difference with respect to digestion?
12. What are lipids?
13. What are the main functions of lipids?
14. What are fatty acids?
15. Draw the general structure of a fatty acid.
16. What is the difference between a saturated fatty acid and an unsaturated fatty acid?
17. What is a triglyceride?
18. Draw the general structure of a triglyceride.
19. What is the difference between a saturated fat and an unsaturated fat, both in terms of structure and in terms of properties?
20. What are phospholipids and glycolipids? What properties do they have in common?
21. What are the main functions of phospholipids and glycolipids in the body?
22. What are steroids?
23. What are some of the functions of steroids in the body?
24. What are proteins?
25. What are the main functions of proteins within living organisms?
26. What are amino acids? Draw a general structure for amino acids.
27. How do amino acids differ from one another?
28. What is a peptide bond?
29. Use two generic amino acids to show how a peptide bond forms.
30. What determines the shapes of proteins? Why is the shape of a protein so important?
31. What does primary protein structure refer to? What kinds of interactions maintain primary protein structure?
32. What does secondary protein structure refer to? What kinds of interactions maintain secondary protein structure?
33. What does tertiary protein structure refer to? What kinds of interactions maintain tertiary protein structure?
34. What does quaternary protein structure refer to? What kinds of interactions maintain quaternary protein structure?
35. Explain the α-helix structure and the β-pleated sheet structure.
36. What are nucleic acids?
37. What is the main function of nucleic acids?
38. What are the two main types of nucleic acids?
39. What are the four different bases that occur within DNA?

40. What is a codon?
41. What is the genetic code?
42. Is the genetic code different for different organisms?
43. What is a gene?
44. If a protein contains 300 amino acids, about how many nucleotides are in the gene that codes for it?
45. What are chromosomes?
46. Do most cells in the human body contain genes for all of the proteins that a human needs?
47. Do most cells in the human body synthesize all the proteins that they have genes for?
48. Explain how DNA replication occurs.
49. Give the complementary base of:
    (a) adenine (A)
    (b) thymine (T)
    (c) cytosine (C)
    (d) guanine (G)
50. Explain how protein synthesis occurs.

Problems

Carbohydrates

51. Determine whether each of the following molecules is a carbohydrate. If it is, classify it as a monosaccharide, a disaccharide, or a trisaccharide.

(a)
O
HC
H—C—OH
HO—C—H
H—C—OH
H—C—OH
H_2C
OH

(b)
CH_3
CH_2
CH_2
CH_2
CH_2
CH_2
CH_2
CH_2
CH_2
CH_2
CH_2
CH_2
CH_2
O=C
OH

(c)
H O
H_2N—C—C—OH
CH_2
OH

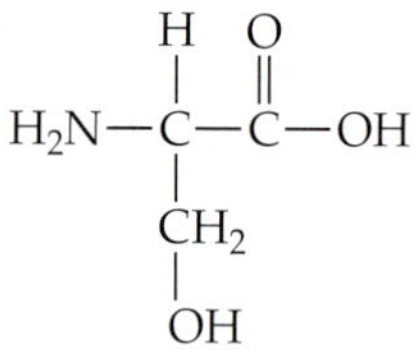

(d) HO
H_2C
H O H OH OH
H OH O OH H CH_2
OH
HO H CH_2 O H
H H HO

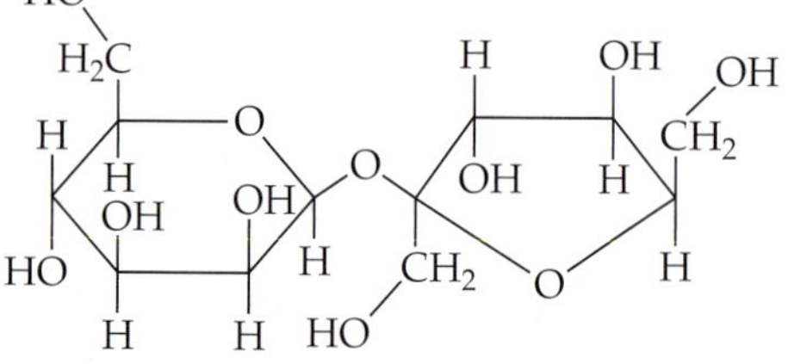

52. Determine whether each of the following molecules is a carbohydrate. If it is, classify it as a monosaccharide, a disaccharide, or a trisaccharide.

(a)
CH_2—CH—CH_2
O O O
C=O C=O C=O
CH_2 CH_2 CH_2
CH_2 CH_2 CH_2
CH_2 CH_2 CH_2
CH_2 CH_2 CH_2
CH_2 CH_2 CH_2
CH_2 CH_2 CH_2
CH_2 CH_2 CH_2
CH_2 CH_2 CH_2
CH_2 CH_2 CH_2
CH_2 CH_2 CH_2
CH_2 CH_3 CH_2
CH_2 CH_2
CH_3 CH_3

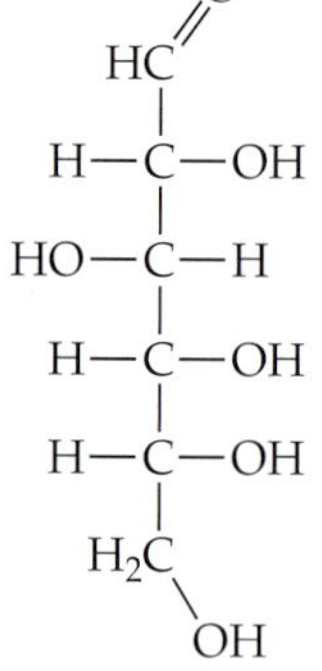

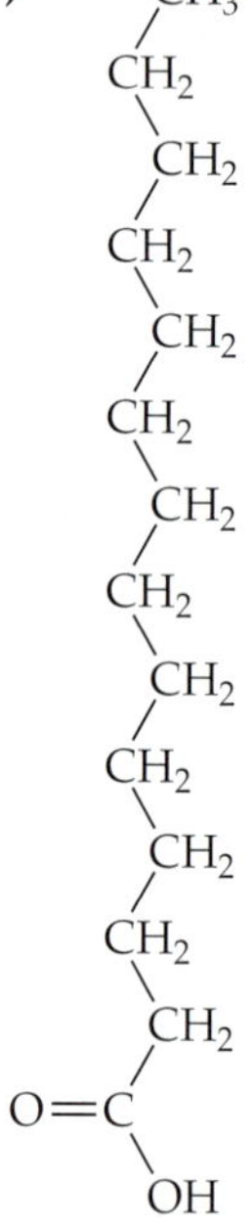

(b)
O
HC
HC—OH
CH_2—C—OH
HO CH_2
HO

(c)
HO
H_2C
HO
CH—O
HC OH CH
CH—CH
O—CH_2
OH
OH
CH—O H
H H_2C O
HC OH C C HO CH
CH—CH O CH—CH CH_2
HO
OH OH OH

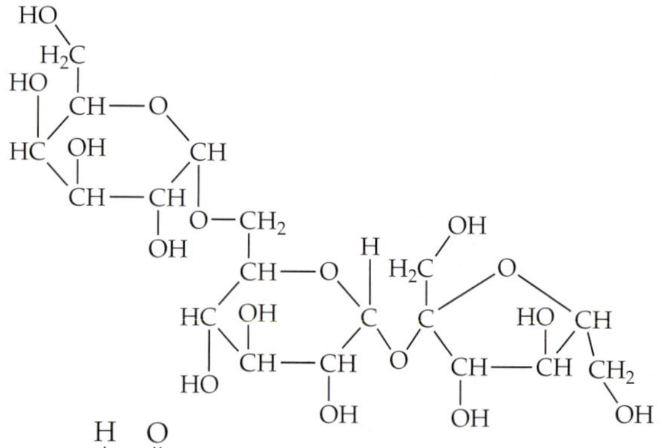

(d)
H O
H_2N—C—C—OH
CH_2
C=O
H_2N

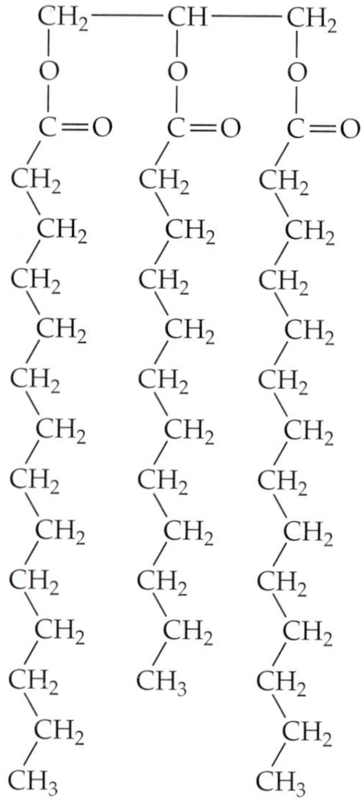

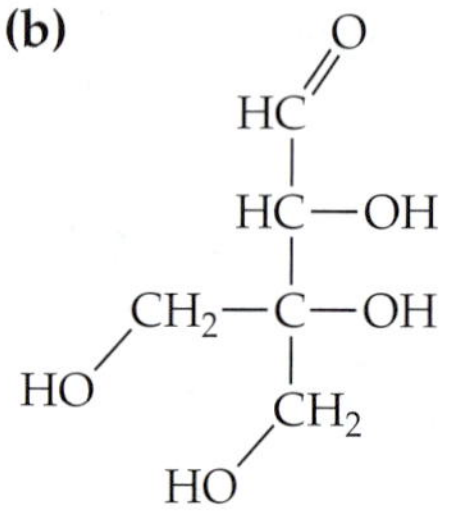

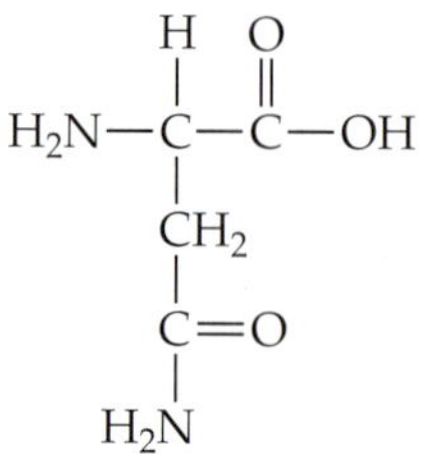

53. Classify each of the following carbohydrates as a triose, tetrose, pentose, and so on.

(a)
HO
CH_2
HO CH—O H
C OH H C
H C—C OH
H OH

(b)
O
HC
HO—C—OH
H—C—OH
H_2C
OH

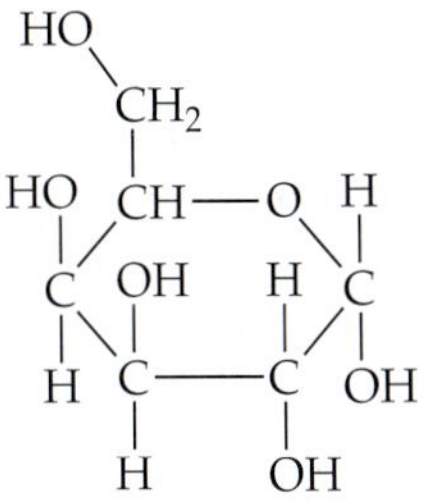

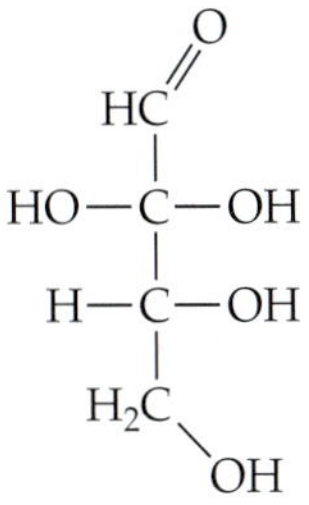

(c)
O
HC
H—C—OH
H—C—OH
H—C—OH
H_2C
OH

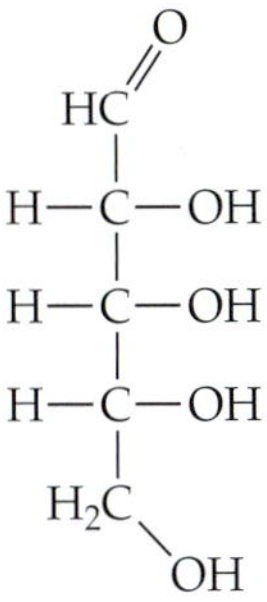

(d)
H O H
C OH H C
H C—C OH
H OH

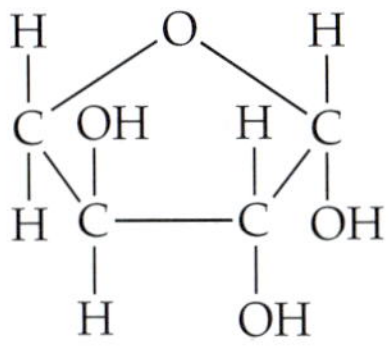

54. Classify each of the following carbohydrates as a triose, tetrose, pentose, and so on.

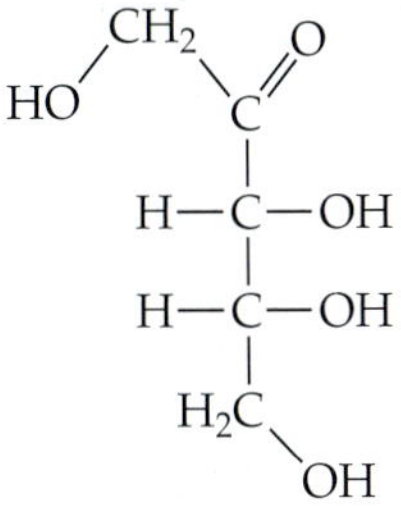

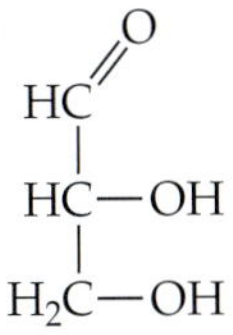

(a)
CH_2 O
HO C
H—C—OH
H—C—OH
H_2C
OH

(b)
O
HC
HC—OH
H_2C—OH

(c) HO
CH_2 O H
C H H C
H C—C OH
OH OH

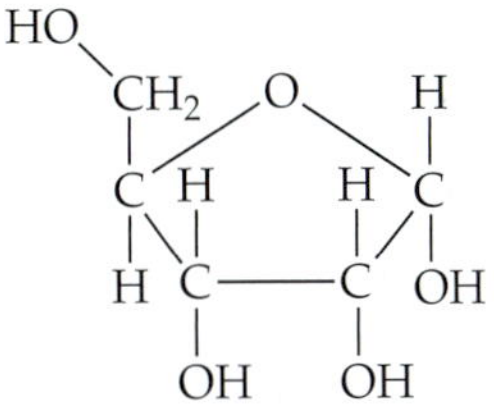

(d)
H H
H C—O H
C OH OH C
HO C—C OH
H H

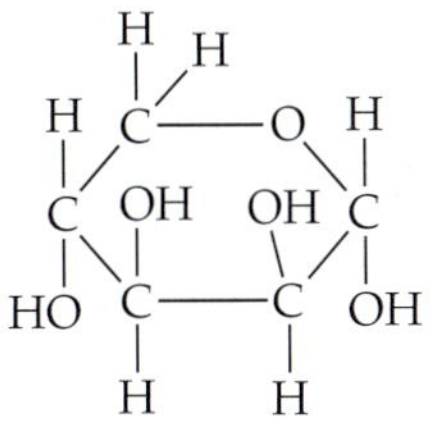

55. Draw the structure of glucose in both its straight chain and its ring form.

56. Draw the structure of fructose in its ring form. Are fructose and glucose isomers?

57. Draw the structure of sucrose. Label the glucose and fructose rings in this disaccharide.

58. Draw the structure of lactose. Label the glucose and galactose rings in this disaccharide.

Lipids

59. Determine whether each of the following molecules is a lipid. If the molecule is a lipid, determine the kind of lipid. If it is a fatty acid or a triglyceride, classify it as saturated or unsaturated.

(a) 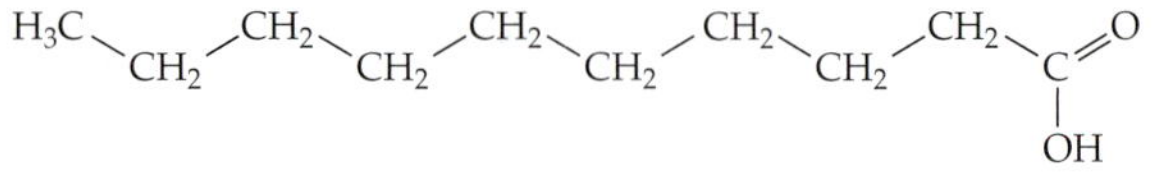 $H_3C-CH_2-CH_2-CH_2-CH_2-CH_2-CH_2-CH_2-CH_2-CH_2-C(=O)-OH$

(b) [steroid ring structure with labels: CH_3, OH, CH_3, O]

(c) $CH_2-CH-CH_2$, each carbon bonded to $O-C(=O)-CH_2-CH_2-CH_2-CH_2-CH_2-CH_2-CH_2-CH=CH-CH_2-CH=CH-CH_2-CH=CH-CH_2-CH_3$ (three identical chains)

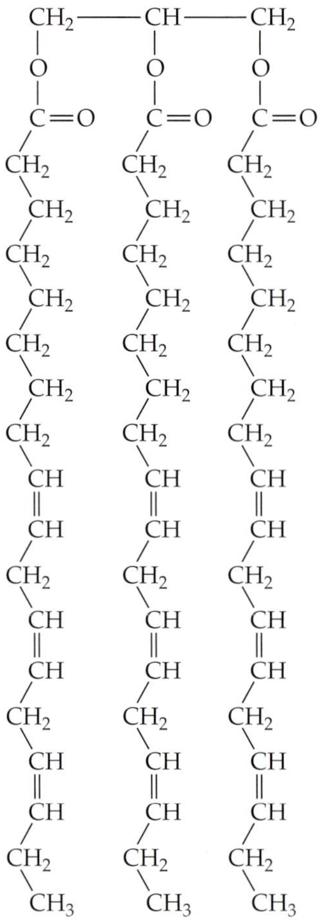

(d) 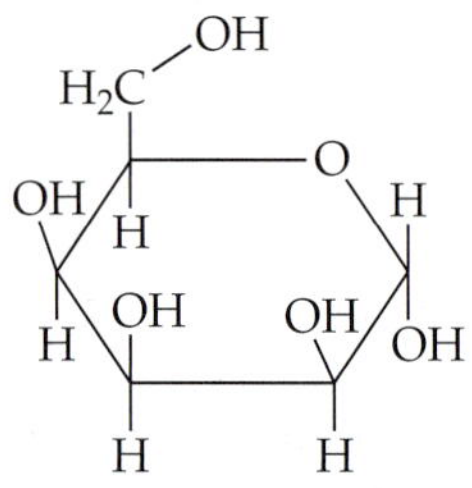 [ring structure with labels: H_2C, OH, O, OH, H, H, OH, H, OH, OH, H, H]

60. Determine whether each of the following molecules is a lipid. If the molecule is a lipid, determine the kind of lipid. If it is a fatty acid or a triglyceride, classify it as saturated or unsaturated.

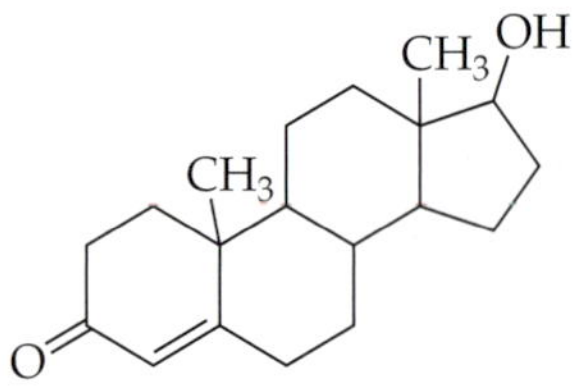

(a) 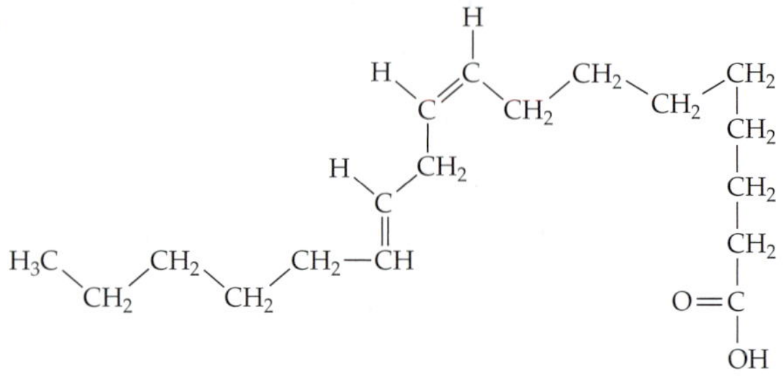 $H_3C-CH_2-CH_2-CH_2-CH_2-CH=CH-CH_2-CH=CH-CH_2-CH_2-CH_2-CH_2-CH_2-CH_2-CH_2-C(=O)-OH$

(b) $O=C(NH_2)-CH_2-CH_2-CH(NH_2)-C(=O)-OH$

(c) [steroid ring structure with labels: OH, O, CH_3, HO, OH, CH_3, O]

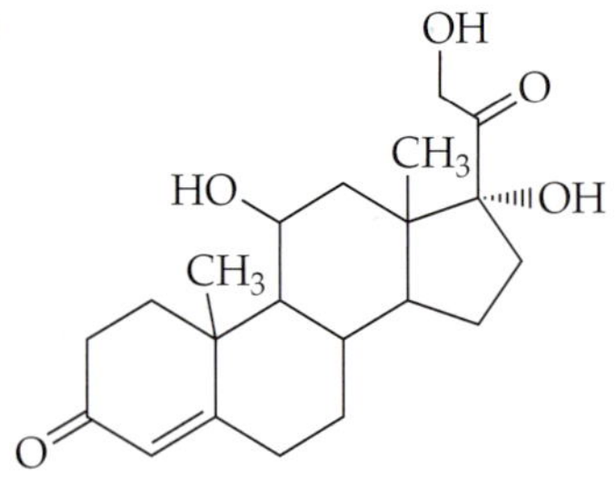

(d) 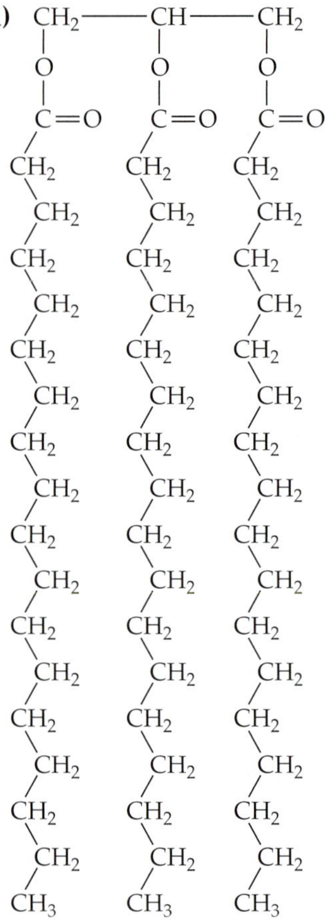 $CH_2-CH-CH_2$, each carbon bonded to $O-C(=O)-CH_2-CH_2-CH_2-CH_2-CH_2-CH_2-CH_2-CH_2-CH_2-CH_2-CH_2-CH_2-CH_2-CH_2-CH_2-CH_2-CH_3$ (three identical chains)

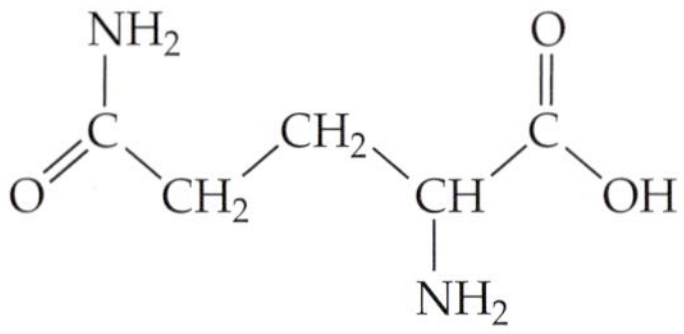

61. Sketch the block diagram for a triglyceride.

62. Sketch the block diagram for a phospholipid. How are phospholipids different from triglycerides?

63. Draw the structure of the triglyceride that would form from the reaction of capric acid with glycerol. Would you expect this triglyceride to be a fat or an oil?

64. Draw the structure of the triglyceride that would form from the reaction of linolenic acid with glycerol. Would you expect this triglyceride to be a fat or an oil?

Amino Acids and Proteins

65. Determine whether each of the following molecules is an amino acid.

(a) $H_3C{-}CH_2{-}CH_2{-}C({=}O){-}OH$

(b) $H_2N{-}CH(CH_2{-}OH){-}C({=}O){-}OH$

(c) $H_2N{-}C({=}O){-}CH_2{-}CH_3$

(d) $H_2N{-}CH(H_2C{-}C_6H_4{-}OH){-}C({=}O){-}OH$

66. Determine whether each of the following molecules is an amino acid.

(a) $H_2N{-}CH_2{-}C({=}O){-}OH$

(b) $H_2C{-}OH$ / $HO{-}CH$ / $H_2C{-}OH$ (i.e., $HOCH_2{-}CH(OH){-}CH_2OH$)

(c) HO–CH_2, O, H, H, OH, H, CH_2–OH, OH, H, OH (five-membered ring: $HOCH_2$ and H on one ring carbon; ring O; H and CH_2OH on the next; OH and H on the next; H and OH on the last)

(d) $H_2N{-}CH(H_2C{-}C({=}O){-}OH){-}C({=}O){-}OH$

67. Show the reaction by which valine and leucine form a peptide bond.

68. Show the reaction by which cysteine and serine form a peptide bond.

69. Draw structures for each of the following tripeptides.
(a) thr-ala-leu
(b) asn-ser-gly
(c) val-phe-ala

70. Draw structures for each of the following tripeptides.
(a) asp-gln-tyr
(b) met-ile-lys
(c) ser-val-gly

71. A cysteine on a protein strand forms a covalent bond with another cysteine that is thirty-five amino acid units away. The resulting fold in the protein is an example of what kind of structure? (primary, secondary, tertiary, or quaternary)

72. An amino acid on a protein strand hydrogen bonds to another amino acid that is four amino acid units away. The next amino acid on the chain does the same, hydrogen bonding to an amino acid that is four amino acids away from it. This pattern repeats itself over a significant part of the protein chain. The resulting pattern in the protein is an example of what kind of structure? (primary, secondary, tertiary, or quaternary)

73. The following is the amino acid sequence in one section of a protein. It represents what kind of structure? (primary, secondary, tertiary, or quaternary)

-lys-glu-thr-ala-ala-ala-lys-phe-glu-

74. A particular protein is composed of two individual chains of amino acids. The way these two chains fit together is an example of what kind of structure? (primary, secondary, tertiary, or quaternary)

Nucleic Acids

75. Determine whether each of the following is a nucleotide. For each nucleotide, identify A, T, C, or G.

(a)

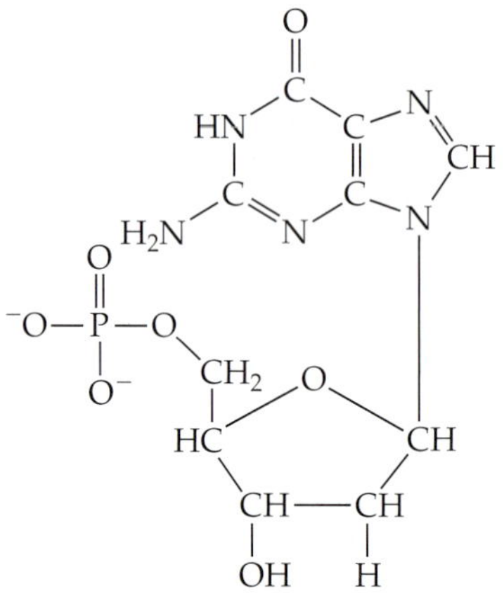

(b)

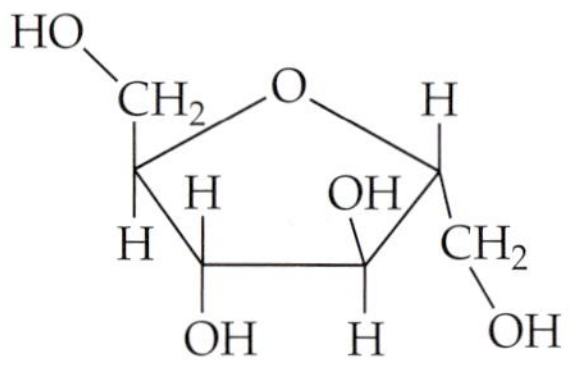

(c) $H_2N-CH(CH_2C(=O)NH_2)-C(=O)-OH$

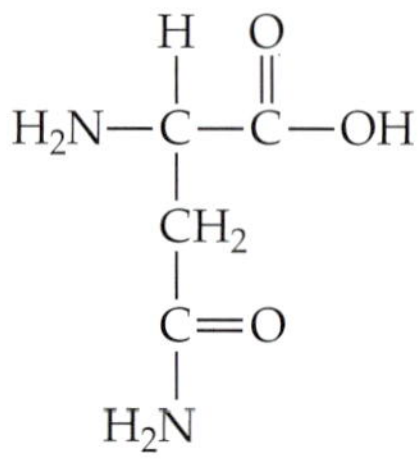

(d)

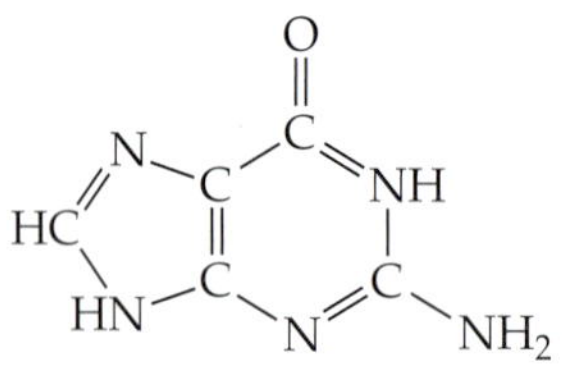

76. Determine whether each of the following molecules is a nucleotide. For each nucleotide, identify the base as A, T, C, or G.

(a) $H_2N-CH_2-C(=O)-OH$

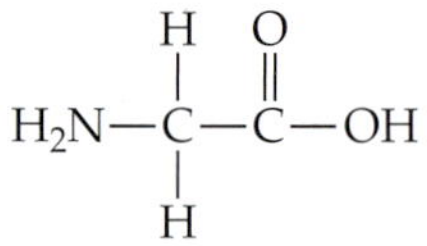

(b)

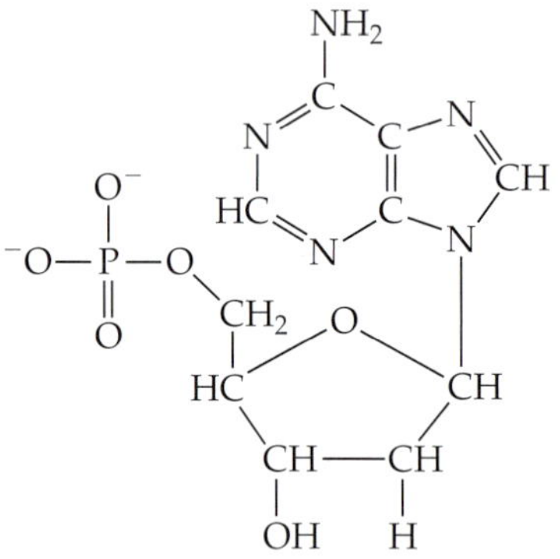

(c)

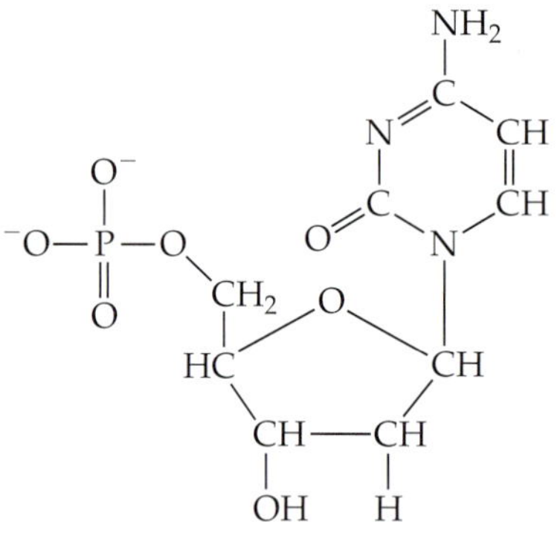

(d)

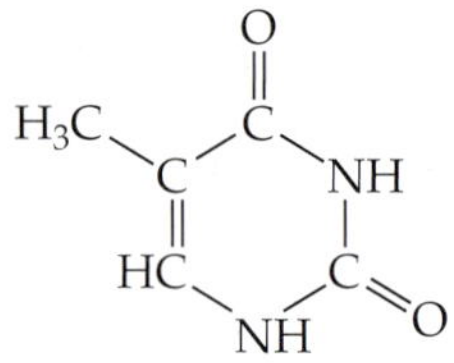

77. Draw the complementary strand of the following DNA strand.

78. Draw the complementary strand of the following DNA strand.

79. In a step-by-step fashion, show how the following section of DNA would replicate to form two copies.

80. In a step-by-step fashion, show how the following section of DNA would replicate to form two copies.

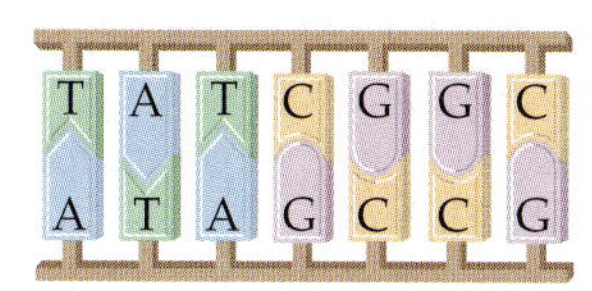

Cumulative Problems

81. Match each of the following linkages with the correct class of biochemicals.
    (a) glycoside linkage
    (b) peptide bonds
    (c) ester linkage

- proteins
- triglycerides
- carbohydrates

82. Match each of the following monomers with the correct class of biopolymers.
    (a) nucleotide
    (b) saccharide
    (c) amino acid

- protein
- DNA
- starch

83. Match each of the following biochemicals with the correct function in living organisms.
    (a) glucose
    (b) DNA
    (c) phospholipids
    (d) triglycerides

- compose cell membranes
- long-term energy storage
- short-term energy storage
- blueprint for proteins

84. Match each of the following biochemicals with the correct function in living organisms.
    (a) proteins
    (b) cellulose
    (c) RNA

- act as enzymes (among other things)
- involved in protein synthesis
- structural components of plants

85. Match each of the following terms with the correct meaning.
    (a) codon
    (b) gene
    (c) human genome
    (d) chromosome

- codes for a single protein
- codes for a single amino acid
- all of the genetic material of a human
- structure that contains genes

86. Match each of the following terms with the correct meaning.
    (a) pentose
    (b) dipeptide
    (c) diglyceride
    (d) fatty acid

- a carboxylic acid with a long hydrocarbon R group
- two amino acids joined by a peptide bond
- a glycerol molecule with two fatty acids attached
- a five-carbon sugar

87. The amino acid glycine has the following condensed structural formula:

NH_2CH_2COOH

Determine the VSEPR geometry about each internal atom and make a three-dimensional sketch of the molecule.

88. The amino acid serine has the following condensed structural formula:

$NH_2CH(CH_2OH)COOH$

Determine the VSEPR geometry about each internal atom and make a three-dimensional sketch of the molecule.

89. Since amino acids are asymetrical, a peptide with amino acids in a certain order is different from a peptide with the amino acids in the reverse order. For example, ser-ala is different from ala-ser. Draw the structures of these two dipeptides and show how they are different.

90. Use the abbreviations to write the sequences for all the possible polypeptides that can be made from the following three amino acids: ala, ser, val.

91. Determining the amino acid sequence in a protein usually involves treating the protein with various reagents that break up the protein into smaller fragments that can be individually sequenced. Treating a particular eleven-amino-acid polypeptide with one reagent produced the following fragments:

trp-glu-val, gly-arg, ala-ser-phe-gly-asn-lys

Treating the same polypeptide with a different reagent produced the following fragments:

gly-asn-lys-trp, glu-val, gly-arg-ala-ser-phe

What is the amino acid sequence of the polypeptide?

92. Treating a particular polypeptide with one reagent (as described in the previous problem) produced the following fragments:

asp-thr-ala-trp, gly-glu-ser-lys, trp-arg

Treating the same polypeptide with a different reagent produced the following fragments:

thr-ala-trp, gly-glu, ser-lys-trp-arg-asp

What is the amino acid sequence of the polypeptide?

93. The insulin protein contains fifty-one amino acids. How many DNA base pairs are required to code for all the amino acids in insulin?

94. A hemoglobin subchain contains 146 amino acids. How many DNA base pairs are required to code for all the amino acids in the subchain?

Highlight Problems

95. One way to fight viral infections is to prevent viruses from replicating their DNA. Without DNA replication, the virus cannot multiply. Some viral drug therapies cause the introduction of *fake* nucleotides into cells. When the virus uses one of these fake nucleotides in an attempt to replicate its DNA, the fake nucleotide doesn't work and DNA replication is halted. For example, AZT, a drug used to fight the human immunodeficiency virus (HIV) that causes AIDS, results in the introduction of the following fake thymine-containing nucleotide into cells.

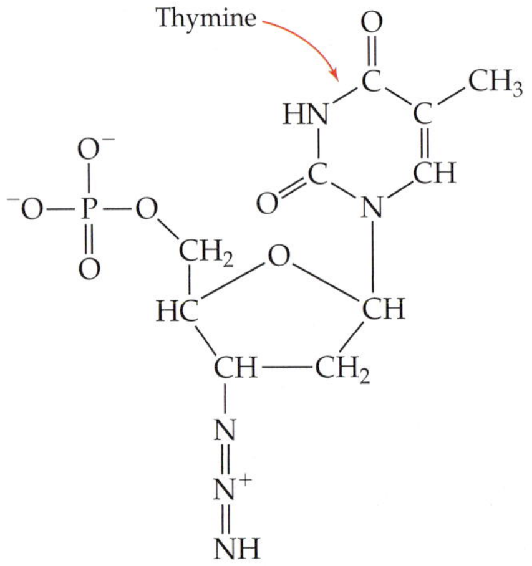

Thymine

Fake nucleotide that results from taking azidothymidine (AZT)

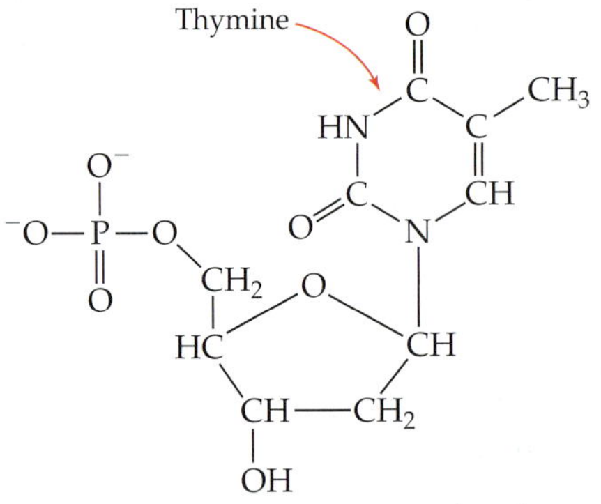

Thymine

Actual thymine-containing nucleotide

Examine the structures of the real nucleotide and the AZT fake nucleotide. Can you propose a mechanism for how this fake nucleotide might halt DNA replication?

96. Sickle-cell anemia is caused by a genetic defect that substitutes valine for glutamic acid at one position in two of the four chains of the hemoglobin protein. The result is a decrease in the water solubility of hemoglobin. Examine the structures of valine and glutamic acid and explain why this might be so.

Answers to Skillbuilder Exercises

Skillbuilder 19.1 (b) monosaccharide; (d) disaccharide

Skillbuilder 19.2 (b) unsaturated fat; (d) saturated fat

Skillbuilder 19.3

$$H_2N-CH(CH(CH_3)_2)-C(=O)-OH + H_2N-CH(CH_2CH(CH_3)_2)-C(=O)-OH \longrightarrow H_2N-CH(CH(CH_3)_2)-C(=O)-NH-CH(CH_2CH(CH_3)_2)-C(=O)-OH + H_2O$$

Skillbuilder 19.4

Answers to Conceptual Checkpoints

19.1 (d) Each of the 51 amino acids is coded for by a single codon. A codon consists of three nucleotides, each containing one base.

19.2 (b) Proteins are polymers of amino acids; nucleic acids are polymers of nucleotides; polysaccharides are polymers of monosaccharides. Steroids, however, are not chains of any repeating unit.

APPENDIX: MATHEMATICS REVIEW

Basic Algebra

In chemistry, you often have to solve an equation for a particular variable. For example, suppose you want to solve the following equation for V.

$$PV = nRT$$

To solve an equation for a particular variable, you must isolate that variable on one side of the equation. The rest of the variables or numbers will then be on the other side of the equation. To solve the above equation for V, simply divide both sides by P.

$$\frac{\cancel{P}V}{\cancel{P}} = \frac{nRT}{P}$$

$$V = \frac{nRT}{P}$$

The Ps cancel and we are left with an expression for V. For another example, consider solving the following equation for °F.

$$°C = \frac{(°F - 32)}{1.8}$$

First, eliminate the 1.8 in the denominator of the right side by multiplying both sides by 1.8.

$$(1.8)\ °C = \frac{(°F - 32)}{\cancel{1.8}}(\cancel{1.8})$$

$$(1.8)\ °C = (°F - 32)$$

Then eliminate the -32 on the right by adding 32 to both sides.

$$(1.8)\ °C + 32 = (°F - 32) + 32$$

$$(1.8)\ °C + 32 = °F$$

We are now left with an expression for °F.

In general, solve equations by following these guidelines:

- Cancel numbers or symbols in the denominator (bottom part of a fraction) by multiplying by the number or symbol to be canceled.
- Cancel numbers or symbols in the numerator (upper part of a fraction) by dividing by the number or symbol to be canceled.
- Eliminate numbers or symbols that are added by subtracting the same number or symbol.
- Eliminate numbers or symbols that are subtracted by adding the same number or symbol.

- Whether you add, subtract, multiply, or divide, **always perform the same operation for both sides of a mathematical equation**. (Otherwise, the two sides will no longer be equal.)

For a final example, solve the following equation for x.

$$\frac{67x - y + 3}{6} = 2z$$

Cancel the 6 in the denominator by multiplying both sides by 6.

$$(\not{6})\frac{67x - y + 3}{\not{6}} = (6)2z$$

$$67x - y + 3 = 12z$$

Eliminate the +3 by subtracting 3 from both sides.

$$67x - y + 3 - 3 = 12z - 3$$

$$67x - y = 12z - 3$$

Eliminate the $-y$ by adding y to both sides.

$$67x - y + y = 12z - 3 + y$$

$$67x = 12z - 3 + y$$

Cancel the 67 by dividing both sides by 67.

$$\frac{\not{67}x}{\not{67}} = \frac{12z - 3 + y}{67}$$

$$x = \frac{12z - 3 + y}{67}$$

FOR PRACTICE Using Algebra to Solve Equations

Solve each of the following for the indicated variable.

(a) $P_1V_1 = P_2V_2$; solve for V_2

(b) $\frac{V_1}{T_1} = \frac{V_2}{T_2}$; solve for T_1

(c) $PV = nRT$; solve for n

(d) K = °C + 273; solve for °C

(e) $\frac{3x + 7}{2} = y$; solve for x

(f) $\frac{32}{y + 3} = 8$; solve for y

Answers:

(a) $V_2 = \frac{P_1V_1}{P_2}$

(b) $T_1 = \frac{V_1T_2}{V_2}$

(c) $n = \frac{PV}{RT}$

(d) °C = K − 273

(e) $x = \frac{2y - 7}{3}$

(f) $y = 1$

 Mathematical Operations with Scientific Notation

Writing numbers in scientific notation is covered in detail in Section 2.2. Briefly, a number written in scientific notation consists of a **decimal part**, a number that is usually between 1 and 10, and an **exponential part**, 10 raised to an **exponent**, n.

Each of the following numbers is written in both scientific and decimal notation.

$$1.0 \times 10^5 = 100{,}000 \qquad 1.0 \times 10^{-6} = 0.000001$$
$$6.7 \times 10^3 = 6700 \qquad 6.7 \times 10^{-3} = 0.0067$$

Multiplication and Division

To multiply numbers expressed in scientific notation, multiply the decimal parts and add the exponents.

$$(A \times 10^m)(B \times 10^n) = (A \times B) \times 10^{m+n}$$

To divide numbers expressed in scientific notation, divide the decimal parts and subtract the exponent in the denominator from the exponent in the numerator.

$$\frac{(A \times 10^m)}{(B \times 10^n)} = \left(\frac{A}{B}\right) \times 10^{m-n}$$

Consider the following example involving multiplication.

$$\begin{aligned}(3.5 \times 10^4)(1.8 \times 10^6) &= (3.5 \times 1.8) \times 10^{4+6} \\ &= 6.3 \times 10^{10}\end{aligned}$$

Consider the following example involving division.

$$\begin{aligned}\frac{(5.6 \times 10^7)}{(1.4 \times 10^3)} &= \left(\frac{5.6}{1.4}\right) \times 10^{7-3} \\ &= 4.0 \times 10^4\end{aligned}$$

Addition and Subtraction

To add or subtract numbers expressed in scientific notation, rewrite all the numbers so that they have the same exponent, then add or subtract the decimal parts of the numbers. The exponents remained unchanged.

$$\begin{array}{r} A \times 10^n \\ \pm B \times 10^n \\ \hline (A \pm B) \times 10^n \end{array}$$

Notice that the numbers *must have* the same exponent.

Consider the following example involving addition.

$$\begin{array}{r} 4.82 \times 10^7 \\ +3.4 \times 10^6 \\ \hline \end{array}$$

First, express both numbers with the same exponent. In this case, we rewrite the lower number and perform the addition as follows:

$$\begin{array}{r} 4.82 \times 10^7 \\ +0.34 \times 10^7 \\ \hline 5.16 \times 10^7 \end{array}$$

Consider the following example involving subtraction.

$$\begin{array}{r} 7.33 \times 10^5 \\ -1.9 \times 10^4 \\ \hline \end{array}$$

First, express both numbers with the same exponent. In this case, we rewrite the lower number and perform the subtraction as follows:

$$\begin{array}{r} 7.33 \times 10^5 \\ -0.19 \times 10^5 \\ \hline 7.14 \times 10^5 \end{array}$$

FOR PRACTICE Mathematical Operations with Scientific Notation

Perform each of the following operations.

(a) $(2.1 \times 10^7)(9.3 \times 10^5)$

(b) $(5.58 \times 10^{12})(7.84 \times 10^{-8})$

(c) $\dfrac{(1.5 \times 10^{14})}{(5.9 \times 10^8)}$

(d) $\dfrac{(2.69 \times 10^7)}{(8.44 \times 10^{11})}$

(e) $\begin{array}{r} 1.823 \times 10^9 \\ +1.11 \times 10^7 \\ \hline \end{array}$

(f) $\begin{array}{r} 3.32 \times 10^{-5} \\ +3.400 \times 10^{-7} \\ \hline \end{array}$

(g) $\begin{array}{r} 6.893 \times 10^9 \\ -2.44 \times 10^8 \\ \hline \end{array}$

(h) $\begin{array}{r} 1.74 \times 10^4 \\ -2.9 \times 10^3 \\ \hline \end{array}$

Answers:

(a) 2.0×10^{13}

(b) 4.37×10^5

(c) 2.5×10^5

(d) 3.19×10^{-5}

(e) 1.834×10^9

(f) 3.35×10^{-5}

(g) 6.649×10^9

(h) 1.45×10^4

Logarithms

The log of a number is the exponent to which 10 must be raised to obtain that number. For example, the log of 100 is 2 because 10 must be raised to the 2nd power to get 100. Similarly, the log of 1000 is 3 because 10 must be raised to the 3rd power to get 1000. The logs of several multiples of 10 are shown on the following page.

$\log 10 = 1$
$\log 100 = 2$
$\log 1000 = 3$
$\log 10{,}000 = 4$

Because $10^0 = 1$ by definition, $\log 1 = 0$.

The log of a number smaller than 1 is negative because 10 must be raised to a negative exponent to get a number smaller than 1. For example, the log of 0.01 is -2 because 10 must be raised to the power of -2 to get 0.01. Similarly, the log of 0.001 is -3 because 10 must be raised to the power of -3 to get 0.001. The logs of several fractional numbers are shown below.

$\log 0.1 = -1$
$\log 0.01 = -2$
$\log 0.001 = -3$
$\log 0.0001 = -4$

The logs of numbers that are not multiples of 10 can be computed on your calculator. See your calculator manual for specific instructions.

Inverse Logarithms

The inverse logarithm or invlog function (sometimes called antilog) is exactly the opposite of the log function. For example, the log of 100 is 2 and the inverse log of 2 is 100. The log function and the invlog function undo one another.

$\log 1000 = 3$
$\text{invlog } 3 = 1000$
$\text{invlog } (\log 1000) = 1000$

The inverse log of a number is simply 10 raised to that number.

$\text{invlog } x = 10^x$
$\text{invlog } 3 = 10^3 = 1000$

The inverse logs of numbers can be computed on your calculator. See your calculator manual for specific instructions.

FOR PRACTICE Logarithms and Inverse Logarithms

Perform each of the following operations.

(a) $\log 1.0 \times 10^5$
(b) $\log 59$
(c) $\log 1.0 \times 10^{-5}$
(d) $\log 0.068$
(e) $\text{invlog } 7.0$
(f) $\text{invlog } 1.44$
(g) $\text{invlog } -6.0$
(h) $\text{invlog } -0.250$
(i) $\text{invlog } (\log 88)$

Answers:

(a) 5.00
(b) 1.77
(c) -5.00
(d) -1.17
(e) 1×10^7
(f) 28
(g) 1×10^{-6}
(h) 0.56
(i) 88

GLOSSARY

absolute zero The coldest temperature possible. Absolute zero (0 K or −273 °C or −459 °F) is the temperature at which molecular motion stops. Lower temperatures do not exist.

acid A molecular compound that dissolves in solution to form H^+ ions. Acids have the ability to dissolve some metals and will turn litmus paper red.

acid rain Acidic precipitation in the form of rain; created when fossil fuels are burned, releasing SO_2 and NO_2, which then react with water in the atmosphere to form sulfuric acid and nitric acid.

acid–base reaction A reaction that forms water and typically a salt.

acidic solution A solution containing a concentration of H_3O^+ ions greater than 1.0×10^{-7} (pH < 7).

activation energy The amount of energy that must be absorbed by reactants before a reaction can occur; an energy hump that normally exists between the reactants and products.

activity series of metals A listing of metals (and hydrogen) in order of decreasing activity, decreasing ability to oxidize, and decreasing tendency to lose electrons.

actual yield The amount of product actually produced by a chemical reaction.

addition polymer A polymer formed by addition of monomers to one another without elimination of any atoms.

alcohol An organic compound containing an —OH functional group bonded to a carbon atom and having the general formula *R*OH.

aldehyde An organic compound with the general formula *R*CHO.

alkali metals The Group 1A elements, which are highly reactive metals.

alkaline battery A dry cell employing half-reactions that use a base.

alkaline earth metals The Group 2A elements, which are fairly reactive metals.

alkaloids Organic bases found in plants.

alkanes Hydrocarbons in which all carbon atoms are connected by single bonds. Noncyclic alkanes have the general formula C_nH_{2n+2}.

alkene A hydrocarbon that contains at least one double bond between carbon atoms. Noncyclic alkenes have the general formula C_nH_{2n}.

alkyl group In an organic molecule, any group containing only singly bonded carbon atoms and hydrogen atoms.

alkyne A hydrocarbon that contains at least one triple bond between carbon atoms. Noncylclic alkynes have the general formula C_nH_{2n-2}.

alpha particle A particle consisting of two protons and two neutrons (a helium nucleus), represented by the symbol ${}^4_2He$.

alpha (α) radiation Radiation emitted by an unstable nucleus, consisting of alpha particles.

alpha (α)-helix The most common secondary protein structure. The amino acid chain is wrapped into a tight coil from which the side chains extend outward. The structure is maintained by hydrogen bonding interactions between NH and CO groups along the peptide backbone of the protein.

amine An organic compound that contains nitrogen and has the general formula NR_3, where *R* may be an alkyl group or a hydrogen atom.

amino acid A molecule containing an amine group, a carboxylic acid group, and an *R* group (also called a side chain). Amino acids are the building blocks of proteins.

amorphous A type of solid matter in which atoms or molecules do not have long-range order (e.g., glass and plastic).

amphoteric In Brønsted–Lowry terminology, able to act as either an acid or a base.

anion A negatively charged ion.

anode The electrode where oxidation occurs in an electrochemical cell.

aqueous solution A homogeneous mixture of a substance with water.

aromatic ring A ring of carbon atoms containing alternating single and double bonds. Another name for the benzene ring.

Arrhenius acid A substance that produces H^+ ions in aqueous solution.

Arrhenius base A substance that produces OH^- ions in aqueous solution.

atmosphere (atm) The average pressure at sea level, 101,325 Pa.

atom The smallest identifiable unit of an element.

atomic element An element that exists in nature with single atoms as the basic unit.

atomic mass A weighted average of the masses of each naturally occurring isotope of an element; atomic mass is the average mass of the atoms of an element.

atomic mass unit (amu) The unit commonly used to express the masses of protons, neutrons, and nuclei. 1 amu = 1.66×10^{-24} g.

atomic number (Z) The number of protons in the nucleus of an atom.

atomic size The atomic size of an atom is determined by how far the outermost electrons are from the nucleus.

atomic solid A solid whose component units are individual atoms (e.g., diamond, C; iron, Fe).

atomic theory A theory that states that all matter is composed of tiny particles called atoms.

Avogadro's law The volume (V) of a gas and the amount of the gas in moles (n) are directly proportional.

Avogadro's number The number of entities in a mole, 6.022×10^{23}.

balanced equation A chemical equation in which the numbers of each type of atom on both sides of the equation are equal.

base A molecular compound that dissolves in solution to form OH^- ions. Bases have a slippery feel and turn litmus paper blue.

base chain The longest continuous chain of carbon atoms in an organic compound.

basic solution A solution containing a concentration of OH^- ions greater than 1.0×10^{-7} (pH > 7).

bent The molecular geometry in which 3 atoms are not in a straight line. This geometry occurs when the central atoms contain 4 electron groups (2 bonding and 2 nonbonding) or 3 electron groups (2 bonding and 1 nonbonding).

benzene (C_6H_6) A particularly stable organic compound consisting of six carbon atoms joined by alternating single and double bonds in a ring structure.

beta particle A form of radiation consisting of an energetic electron and represented by the symbol ${}_{-1}^{0}e$.

beta (β) radiation Energetic electrons emitted by an unstable nucleus.

beta (β)-pleated sheet A common pattern in the secondary structure of proteins. The protein chain is extended in a zigzag pattern, and the peptide backbones of neighboring chains interact with one another through hydrogen bonding to form sheets.

binary acid An acid containing only hydrogen and a nonmetal.

binary compound A compound containing only two different kinds of elements.

biochemistry The study of the chemical substances and processes that occur in living organisms.

Bohr model A model for the atom in which electrons travel around the nucleus in circular orbits at specific, fixed distances from the nucleus.

boiling point The temperature at which the vapor pressure of a liquid is equal to the pressure above it.

boiling point elevation The increase in the boiling point of a solution caused by the presence of the solute.

bonding pair Electrons that are shared between two atoms in a chemical bond.

bonding theory A model that predicts how atoms bond together to form molecules.

Boyle's law The volume (V) of a gas and its pressure (P) are inversely proportional.

branched alkane An alkane composed of carbon atoms bonded in chains containing branches.

Brønsted–Lowry acid A proton (H^+ ion) donor.

Brønsted–Lowry base A proton (H^+ ion) acceptor.

buffer A solution that resists pH change by neutralizing added acid or added base.

Calorie (Cal) An energy unit equivalent to 1000 little-*c* calories.

calorie (cal) The amount of energy required to raise the temperature of 1 g of water by 1 °C.

carbohydrates Polyhydroxyl aldehydes or ketones or their derivatives, containing multiple —OH groups and often having the general formula $(CH_2O)_n$.

carbonyl group A carbon atom double bonded to an oxygen atom.

carboxylic acid An organic compound with the general formula *RCOOH*.

catalyst A substance that increases the rate of a chemical reaction but is not consumed by the reaction.

cathode The electrode where reduction occurs in an electrochemical cell.

cation A positively charged ion.

cell The smallest structural unit of living organisms that has the properties associated with life.

cell membrane The structure that bounds the cell and holds the contents of the cell together.

cellulose A common polysaccharide composed of repeating glucose units linked together.

Celsius (°C) scale A temperature scale often used by scientists. On this scale, water freezes at 0 °C and boils at 100 °C. Room temperature is approximately 25 °C.

chain reaction A self-sustaining chemical or nuclear reaction yielding energy or products that cause further reactions of the same kind.

charge A fundamental property of protons and electrons. Charged particles experience forces such that like charges repel and unlike charges attract.

Charles's law The volume (V) of a gas and its temperature (T) expressed in kelvins are directly proportional.

chemical bond The sharing or transfer of electrons to attain stable electron configurations among the bonding atoms.

chemical change A change in which matter changes its composition.

chemical energy Energy associated with chemical changes.

chemical equation An equation that represents a chemical reaction; the reactants are on the left side of the equation and the products are on the right side.

chemical formula A way to represent a compound. At a minimum, the chemical formula indicates the elements present in the compound and the relative number of atoms of each element.

chemical properties Properties that a substance can display only through changing its composition.

chemical reaction The process by which one or more substances transform into different substances via a chemical change. Chemical reactions often emit or absorb energy.

chemical symbol A one- or two-letter abbreviation for an element. Chemical symbols are listed directly below the atomic number in the periodic table.

chemistry The science that seeks to understand what matter does by studying what atoms and molecules do.

chromosome A biological structure containing genes, located within the nucleus of a cell.

codon A sequence of three bases in a nucleic acid that codes for one amino acid.

colligative properties Physical properties of solutions that depend on the number of solute particles present but not the type of solute particles.

collision theory A theory of reaction rates stating that effective collisions between reactant molecules must take place in order for the reaction to occur.

color change One type of evidence of a chemical reaction, involving the change in color of a substance after a reaction.

combined gas law A law that combines Boyle's law and Charles's law; it is used to calculate how a property of a gas (P, V, or T) changes when two other properties are changed at the same time.

combustion reaction A reaction in which a substance reacts with oxygen, emitting heat and forming one or more oxygen-containing compound.

complementary base In DNA, a base capable of precise pairing with a specific other DNA base.

complete ionic equation A chemical equation showing all the species as they are actually present in solution.

complex carbohydrate A carbohydrate composed of many repeating saccharide units.

compound A substance composed of two or more elements in fixed, definite proportions.

compressible Able to occupy a smaller volume when subjected to increased pressure. Gases are compressible because, in the gas phase, atoms or molecules are widely separated.

concentrated solution A solution containing large amounts of solute.

condensation A physical change in which a substance is converted from its gaseous form to its liquid form.

condensation polymer A class of polymers that expel atoms, usually water, during their formation or polymerization.

condensed structural formula A shorthand way of writing a structural formula.

conjugate acid–base pair In Brønsted–Lowry terminology, two substances related to each other by the transfer of a proton.

conservation of energy, law of A law stating that energy can be neither created nor destroyed. The total amount of energy is constant and cannot change; it can only be transferred from one object to another or converted from one form to another.

conservation of mass, law of A law stating that in a chemical reaction, matter is neither created nor destroyed.

constant composition, law of A law stating that all samples of a given compound have the same proportions of their constituent elements.

conversion factor A factor used to convert between two separate units; a conversion factor is constructed from any two quantities known to be equivalent.

copolymers Polymers that are composed of two different kinds of monomers and result in chains composed of alternating units rather than a single repeating unit.

core electrons The electrons that are not in the outermost principal shell of an atom.

corrosion The oxidation of metals (e.g., rusting of iron).

covalent atomic solid An atomic solid, such as diamond, that is held together by covalent bonds.

covalent bond The bond that results when two nonmetals combine in a chemical reaction. In a covalent bond, the atoms share their electrons.

critical mass The mass of uranium or plutonium required for a nuclear reaction to be self-sustaining.

crystalline A type of solid matter with atoms or molecules arranged in a well-ordered, three-dimensional array with long-range, repeating order (e.g., salt and diamond).

cytoplasm In a cell, the region between the nucleus and the cell membrane.

Dalton's law of partial pressure A law stating that the sum of the partial pressures of each of the components in a gas mixture equals the total pressure.

daughter nuclide The nuclide product of a nuclear decay.

decimal part One part of a number expressed in scientific notation.

decomposition A reaction in which a complex substance decomposes to form simpler substances; $AB \longrightarrow A + B$.

density (d) A fundamental property of materials that differs from one substance to another. The units of density are those of mass divided by volume, most commonly expressed in g/cm^3, g/mL, or g/L.

derived unit A unit formed from the combination of other units.

dilute solution A solution containing small amounts of solute.

dimer A molecule formed by the joining together of two smaller molecules.

dipeptide Two amino acids linked together via a peptide bond.

dipole moment A measure of the separation of charge in a bond or in a molecule.

diprotic acid An acid containing two ionizable protons.

disaccharide A carbohydrate that can be decomposed into two simpler carbohydrates.

dispersion force The intermolecular force present in all molecules and atoms. Dispersion forces are caused by fluctuations in the electron distribution within molecules or atoms.

displacement A reaction in which one element displaces another in a compound; $A + BC \longrightarrow AC + B$.

dissociation In aqueous solution, the process by which a solid ionic compound separates into its ions.

disubstituted benzene A benzene in which two hydrogen atoms have been replaced by an atom or group of atoms.

DNA (deoxyribonucleic acid) Long chainlike molecules that occur in the nucleus of cells and act as blueprints for the construction of proteins.

dot structure A drawing that represents the valence electrons in atoms as dots; it shows a chemical bond as the sharing or transfer of electron dots.

double bond The bond that exists when two electron pairs are shared between two atoms. In general, double bonds are shorter and stronger than single bonds.

double displacement A reaction in which two elements or groups of elements in two different compounds exchange places to form two new compounds; $AB + CD \longrightarrow AD + CB$.

dry cell An ordinary battery (voltaic cell); it does not contain large amounts of liquid water.

duet The name for the two electrons corresponding to a stable Lewis structure in hydrogen and helium.

dynamic equilibrium In a chemical reaction, the condition in which the rate of the forward reaction equals the rate of the reverse reaction.

electrical current The flow of electric charge — for example, electrons flowing through a wire or ions through a solution.

electrical energy Energy associated with the flow of electric charge.

electrochemical cell A device that creates electrical current from a redox reaction.

electrolysis A process in which electrical current is used to drive an otherwise nonspontaneous redox reaction.

electrolytic cell An electrochemical cell used for electrolysis.

electromagnetic radiation A type of energy that travels through space at a constant speed of 3.0×10^8 m/s (186,000 miles/s) and exhibits both wavelike and particlelike behavior. Light is a form of electromagnetic radiation.

electromagnetic spectrum A spectrum that includes all wavelengths of electromagnetic radiation.

electron A negatively charged particle that occupies most of the atom's volume but contributes almost none of its mass.

electron configuration A representation that shows the occupation of orbitals by electrons for a particular element.

electron geometry The geometrical arrangement of the electron groups in a molecule.

electron group A general term for a lone pair, single bond, or multiple bond in a molecule.

electron spin A fundamental property of all electrons that causes them to have magnetic fields associated with them. The spin of an electron can either be oriented up $\left(+\frac{1}{2}\right)$ or down $\left(-\frac{1}{2}\right)$.

electronegativity The ability of an element to attract electrons within a covalent bond.

element A substance that cannot be broken down into simpler substances.

emission spectrum A spectrum associated with the emission of electromagnetic radiation by elements or compounds.

empirical formula A formula for a compound that gives the smallest whole-number ratio of each type of atom.

empirical formula molar mass The sum of the molar masses of all the atoms in an empirical formula.

endothermic Describes a process that absorbs heat energy.

endothermic reaction A chemical reaction that absorbs energy from the surroundings.

endpoint The point in a reaction at which the reactants are in exact stoichiometric proportions.

energy The capacity to do work.

English system A unit system commonly used in the United States.

enzymes Biological catalysts that increase the rates of biochemical reactions; enzymes are abundant in living organisms.

equilibrium constant (K_{eq}) The ratio, at equilibrium, of the concentrations of the products raised to their stoichiometric coefficients divided by the concentrations of the reactants raised to their stoichiometric coefficients.

equivalent The stoichiometric proportions of elements and compounds in a chemical equation.

ester An organic compound with the general formula *RCOOR*.

ester linkage A type of bond with the general structure —COO—. Ester linkages join glycerol to fatty acids.

ether An organic compound with the general formula *ROR*.

evaporation A process in which molecules of a liquid, undergoing constant random motion, acquire enough energy to overcome attractions to neighbors and enter the gas phase.

excited state An unstable state for an atom or molecule in which energy has been absorbed but not reemitted, raising an electron from the ground state into a higher energy orbital.

exothermic Describes a process that releases heat energy.

exothermic reaction A chemical reaction that releases energy to the surroundings.

experiment A procedure that makes use of observable predictions to test a theory.

exponent A number that represents the number of times a term is multiplied by itself. For example, in 2^4 the exponent is 4 and represents $2 \times 2 \times 2 \times 2$.

exponential part One part of a number expressed in scientific notation; it represents the number of places the decimal point has moved.

Fahrenheit (°F) scale The temperature scale that is most familiar in the United States; water freezes at 32 °F and boils at 212 °F.

family (of elements) A group elements that have similar outer electron configurations and therefore similar properties. Families occur in vertical columns in the periodic table.

family (of organic compounds) A group of organic compounds with the same functional group.

fatty acid A type of lipid consisting of a carboxylic acid with a long hydrocarbon tail.

film badge dosimeter Badges used to measure radiation exposure, consisting of photographic film held in a small case that is pinned to clothing.

fission, nuclear The process by which a heavy nucleus is split into nuclei of smaller masses and energy is emitted.

formula mass The average mass of the molecules (or formula units) that compose a compound.

formula unit The basic unit of ionic compounds; the smallest electrically neutral collection of cations and anions that compose the compound.

freezing point depression The decrease in the freezing point of a solvent caused by the presence of a solute.

frequency The number of wave cycles or crests that pass through a stationary point in one second.

fuel cell A voltaic cell in which the reactants are constantly replenished.

functional group A set of atoms that characterizes a family of organic compounds.

fusion, nuclear The combination of light atomic nuclei to form heavier ones with emission of large amounts of energy.

galvanic (voltaic) cell An electrochemical cell that spontaneously produces electrical current.

gamma radiation High-energy, short-wavelength electromagnetic radiation emitted by an atomic nucleus.

gamma rays The shortest-wavelength, most energetic form of electromagnetic radiation. Gamma ray photons are represented by the symbol ${}^{0}_{0}\gamma$.

gas A state of matter in which atoms or molecules are widely separated and free to move relative to one another.

gas evolution reaction A reaction that occurs in liquids and forms a gas as one of the products.

gas formation One type of evidence of a chemical reaction, a gas forms when two substances are mixed together.

Geiger-Müller counter A radioactivity detector consisting of a chamber filled with argon gas that discharges electrical signals when high-energy particles pass through it.

gene A sequence of codons within a DNA molecule that codes for a single protein. Genes vary in length from hundreds to thousands of codons.

genetic material The inheritable blueprint for making organisms.

glycogen A type of polysaccharide; it has a structure similar to that of starch, but the chain is highly branched.

glycolipid A biological molecule composed of a nonpolar fatty acid and hydrocarbon chain and a polar section composed of a sugar molecule such as glucose.

glycoside linkage The link between monosaccharides in a polysaccharide.

ground state The state of an atom or molecule in which the electrons occupy the lowest possible energy orbitals available.

group (of elements) Elements that have similar outer electron configurations and therefore similar properties. Groups occur in vertical columns in the periodic table.

half-cell A compartment in which the oxidation or reduction half-reaction occurs in a galvanic or voltaic cell.

half-life The time it takes for one-half of the parent nuclides in a radioactive sample to decay to the daughter nuclides.

half-reaction Either the oxidation part or the reduction part of a redox reaction.

halogens The Group 7A elements, which are very reactive nonmetals.

heat absorption One type of evidence of a chemical reaction, involving the intake of energy.

heat capacity The quantity of heat energy required to change the temperature of a given amount of a substance by 1 °C.

heat emission One type of evidence of a chemical reaction, involving the evolution of thermal energy.

heat of fusion The amount of heat required to melt one mole of a solid at its melting point with no change in temperature.

heat of vaporization The amount of heat required to vaporize one mole of a liquid at its boiling point with no change in temperature.

heterogeneous mixture A mixture, such as oil and water, that has two or more regions with different compositions.

homogeneous mixture A mixture, such as salt water, that has the same composition throughout.

human genome All of the genetic material of a human being; the total DNA of a human cell.

Hund's rule A rule stating that when filling orbitals of equal energy, electrons will occupy empty orbitals singly before pairing with other electrons.

hydrocarbon A compound that contains only carbon and hydrogen atoms.

hydrogen bond A strong dipole–dipole interaction between molecules containing hydrogen directly bonded to a small, highly electronegative atom, such as N, O, or F.

hydrogenation The chemical addition of hydrogen to a compound.

hydronium ion The H_3O^+ ion. Chemists often use $H^+(aq)$ and $H_3O^+(aq)$ interchangeably to mean the same thing—a hydronium ion.

hypothesis A theory before it has become well established; a tentative explanation for an observation or scientific problem that can be tested by further investigation.

hypoxia A shortage of oxygen in the tissues of the body.

ideal gas law A law that combines the four properties of a gas—pressure (P), volume (V), temperature (T), and number of moles (n) in a single equation showing their interrelatedness: $PV = nRT$ (R = ideal gas constant).

indicator A substance that changes color with acidity level, often used to detect the endpoint of a titration.

infrared (IR) light The fraction of the electromagnetic spectrum between visible light and microwaves. Infrared light is invisible to the human eye.

insoluble Not soluble in water.

instantaneous dipole A type of intermolecular force resulting from transient shifts in electron density within an atom or molecule.

intermolecular forces Attractive forces that exist between molecules.

International System (SI) The standard set of units for science measurements, based on the metric system.

ion An atom (or group of atoms) that has gained or lost one or more electrons, so that it has an electric charge.

ion product constant (K_w) The product of the H_3O^+ ion concentration and the OH^- ion concentration in an aqueous solution. At room temperature, $K_w = 1.0 \times 10^{-14}$.

ionic bond The bond that results when a metal and a nonmetal combine in a chemical reaction. In an ionic bond, the metal transfers one or more electrons to the nonmetal.

ionic compound A compound formed between a metal and one or more nonmetals.

ionic solid A solid compound composed of metals and nonmetals joined by ionic bonds.

ionization The forming of ions.

ionization energy The energy required to remove an electron from an atom in the gaseous state.

ionizing power The ability of radiation to ionize other molecules and atoms.

isomers Molecules with the same molecular formula but different structures.

isoosmotic Describes solutions having osmotic pressure equal to that of bodily fluids.

isotope scanning The use of radioactive isotopes to identify disease in the body.

isotopes Atoms with the same number of protons but different numbers of neutrons.

Kelvin (K) scale The temperature scale that assigns 0 K to the coldest temperature possible, absolute zero (−273 °C or −459 °F), the temperature at which molecular motion stops. The size of the kelvin is identical to that of the Celsius degree.

ketone An organic compound with the general formula *RCOR*.

kilogram (kg) The SI standard unit of mass.

kilowatt-hour (kWh) A unit of energy equal to 3.6 million joules.

kinetic energy Energy associated with the motion of an object.

kinetic molecular theory A simple model for gases that predicts the behavior of most gases under many conditions.

Le Châtelier's principle A principle stating that when a chemical system at equilibrium is disturbed, the system shifts in a direction that minimizes the disturbance.

lead-acid storage battery An automobile battery consisting of six electrochemical cells wired in series. Each cell produces 2 volts for a total of 12 volts.

Lewis structure A drawing that represents chemical bonds between atoms as shared or transferred electrons; the valence electrons of atoms are represented as dots.

Lewis theory A simple theory for chemical bonding involving diagrams showing bonds between atoms as lines or dots. In this theory, atoms bond together to obtain stable octets (8 valence electrons).

light emission One type of evidence of a chemical reaction, involving the giving off of electromagnetic radiation.

limiting reactant The reactant that limits the amount of product formed in a chemical reaction.

linear Describes the molecular geometry of a molecule containing 2 electron groups (2 bonding groups and no lone pairs).

linearly related A relationship between two variables such that, when they are plotted one against the other, the graph produced is a straight line.

lipid A cellular component that is insoluble in water, but soluble in nonpolar solvents.

lipid bilayer A structure formed by lipids in the cell membrane.

liquid A state of matter in which atoms or molecules are packed close to each other (about as closely as in a solid) but are free to move around and by each other.

logarithmic scale A scale involving logarithms. A logarithm entails an exponent that indicates the power to which a number is raised to produce a given number (e.g., the logarithm of 100 to the base 10 is 2).

lone pair Electrons that are only on one atom in a Lewis structure.

main-group elements Groups 1A–8A on the periodic table. These groups have properties that tend to be predictable based on their position in the periodic table.

mass A measure of the quantity of matter within an object.

mass number (A) The sum of the number of neutrons and protons in an atom.

mass percent composition (or mass percent) The percentage, by mass, of each element in a compound.

matter Anything that occupies space and has mass. Matter exists in three different states: solid, liquid, and gas.

melting point The temperature at which a solid turns into a liquid.

messenger RNA (mRNA) Long chainlike molecules that act as blueprints for the construction of proteins.

metallic atomic solid An atomic solid, such as iron, which is held together by metallic bonds that, in the simplest model, consist of positively charged ions in a sea of electrons.

metallic character The properties typical of a metal, especially the tendency to lose electrons in chemical reactions. Elements become more metallic as you move from right to left across the periodic table.

metalloids Those elements that fall along the boundary between the metals and the nonmetals in the periodic table; their properties are intermediate between those of metals and those of nonmetals.

metals Elements that tend to lose electrons in chemical reactions. They are found at the left side and in the center of the periodic table.

meter (m) The SI standard unit of length.

metric system The unit system commonly used throughout most of the world.

microwaves The part of the electromagnetic spectrum between the infrared region and the radio wave region. Microwaves are efficiently absorbed by water molecules and can therefore be used to heat water-containing substances.

millimeter of mercury (mm Hg) A unit of pressure that originates from the method used to measure pressure with a barometer. Also called a *torr*.

miscibility The ability of two liquids to mix without separating into two phases, or the ability of one liquid to mix with (dissolve in) another liquid.

mixture A substance composed of two or more different types of atoms or molecules combined in variable proportions.

molality (m) A common unit of solution concentration, defined as the number of moles of solute per kilogram of solvent.

molar mass The mass of one mole of atoms of an element. An element's molar mass in grams per mole is numerically equivalent to the element's atomic mass in amu.

molar solubility The solubility of a substance in units of moles per liter (mol/L).

molar volume The volume occupied by one mole of gas under standard temperature and pressure conditions (22.4 L).

molarity (M) A common unit of solution concentration, defined as the number of moles of solute per liter of solution.

mole Avogadro's number (6.022×10^{23}) of particles—especially, of atoms, ions, or molecules. A mole of any element has a mass in grams that is numerically equivalent to its atomic mass in amu.

molecular compound A compound formed from two or more nonmetals. Molecular compounds have distinct molecules as their simplest identifiable units.

molecular element An element that does not normally exist in nature with single atoms as the basic unit. These elements usually exist as diatomic molecules—two atoms of that element bonded together—as their basic units.

molecular equation A chemical equation showing the complete, neutral formulas for every compound in a reaction.

molecular formula A formula for a compound that gives the specific number of each type of atom in a molecule.

molecular geometry The geometrical arrangement of the atoms in a molecule.

molecular solid A solid whose composite units are molecules.

molecule Two or more atoms joined in a specific arrangement by chemical bounds. A molecule is the smallest identifiable unit of a molecular compound.

monomer An individual repeating unit that makes up a polymer.

monoprotic acid An acid containing only one ionizable proton.

monosaccharide A carbohydrate that cannot be decomposed into simpler carbohydrates.

monosubstituted benzene A benzene in which one of the hydrogen atoms has been replaced by another atom or group of atoms.

net ionic equation An equation that shows only the species that actually participate in a reaction.

neutral solution A solution in which the concentrations of H_3O^+ and OH^- are equal (pH = 7).

neutralization A reaction that takes place when an acid and a base are mixed; the $H^+(aq)$ from the acid combines with the $OH^-(aq)$ from the base to form $H_2O(l)$.

neutron A nuclear particle with no electrical charge and nearly the same mass as a proton.

nitrogen narcosis An increase in nitrogen concentration in bodily tissues and fluids that results in feelings of drunkenness.

noble gases The Group 8A elements, which are chemically inert gases.

nonbonding atomic solid An atomic solid that is held together by relatively weak dispersion forces.

nonelectrolyte solution A solution containing a solute that dissolves as molecules; therefore, the solution does not conduct electricity.

nonmetals Elements that tend to gain electrons in chemical reactions. They are found at the upper right side of the periodic table.

nonpolar molecule A molecule that does not have a net dipole moment.

nonvolatile Describes a compound that does not vaporize easily.

normal boiling point The boiling point of a liquid at a pressure of 1 atmosphere.

normal alkane (or *n*-alkane) An alkane composed of carbon atoms bonded in a straight chain with no branches.

nuclear equation An equation that represents the changes that occur during radioactivity and other nuclear processes.

nuclear radiation The energetic particles emitted from the nucleus of an atom when it is undergoing a nuclear process.

nuclear theory of the atom A theory stating that most of the atom's mass and all of its positive charge is contained in a small, dense nucleus. Most of the volume of the atom is empty space occupied by negatively charged electrons.

nucleic acids Biological molecules, such as deoxyribonucleic acid (DNA) and ribonucleic acid (RNA), that store and transmit genetic information.

nucleotide An individual unit of a nucleic acid. Nucleic acids are polymers of nucleotides.

nucleus (of a cell) The part of the cell that contains the genetic material.

nucleus (of an atom) The small core containing most of the atom's mass and all of its positive charge. The nucleus is made up of protons and neutrons.

observation The first step in the scientific method. An observation must measure or describe something about the physical world.

octet The number of electrons, eight, around atoms with stable Lewis structures.

octet rule A rule that states that an atom will give up, accept, or share electrons in order to achieve a filled outer electron shell, which usually consists of 8 electrons.

orbital diagram An electron configuration in which electrons are represented as arrows in boxes corresponding to orbitals of a particular atom.

orbital The region around the nucleus of an atom where an electron is most likely to be found.

organic chemistry The study of carbon-containing compounds and their reactions.

organic molecule A molecule whose main structural component is carbon.

osmosis The flow of solvent from a lower-concentration solution through a semipermeable membrane to a higher-concentration solution.

osmotic pressure The pressure produced on the surface of a semipermeable membrane by osmosis or the pressure required to stop osmotic flow.

oxidation The gain of oxygen, the loss of hydrogen, or the loss of electrons (the most fundamental definition).

oxidation state (or oxidation number) A number that can be used as an aid in writing formulas and balancing equations. It is computed for each element based on the number of electrons assigned to it in a scheme where the most electronegative element is assigned all of the bonding electrons.

oxidation–reduction (redox) reaction A reaction in which electrons are transferred from one substance to another.

oxidizing agent In a redox reaction, the substance being reduced. Oxidizing agents tend to gain electrons easily.

oxyacid An acid containing hydrogen, a nonmetal, and oxygen.

oxyanion An anion containing oxygen. Most polyatomic ions are oxyanions.

oxygen toxicity The result of increased oxygen concentration in bodily tissues.

parent nuclide The original nuclide in a nuclear decay.

partial pressure The pressure due to any individual component in a gas mixture.

pascal (Pa) The SI unit of pressure, defined as 1 newton per square meter.

Pauli exclusion principle A principle stating that no more than two electrons can occupy an orbital and that the two electrons must have opposite spins.

penetrating power The ability of a radioactive particle to penetrate matter.

peptide bond The bond between the amine end of one amino acid and the carboxylic acid end of another. Amino acids link together via peptide bonds to form proteins.

percent natural abundance The percentage amount of each isotope of an element in a naturally occurring sample of the element.

percent yield In a chemical reaction, the percentage of the theoretical yield that was actually attained.

period A horizontal row of the periodic table.

periodic law A law that states that when the elements are arranged in order of increasing relative mass, certain sets of properties recur periodically.

periodic table An arrangement of the elements in which atomic number increases from left to right and elements with similar properties fall in columns called families or groups.

permanent dipole A separation of charge resulting from the unequal sharing of electrons between atoms.

pH scale A scale used to quantify acidity or basicity. A pH of 7 is neutral; a pH lower than 7 is acidic; and a pH greater than 7 is basic. The pH is defined as follows: $pH = -\log[H_3O^+]$.

phenyl group The term for a benzene ring when other substituents are attached to it.

phospholipid A lipid with the same basic structure as a triglyceride, except that one of the fatty acid groups is replaced with a phosphate group.

phosphorescence The slow, long-lived emission of light that sometimes follows the absorption of light by some atoms and molecules.

photon A particle of light or a packet of light energy.

physical change A change in which matter does not change its composition even though its appearance might change.

physical properties Those properties that a substance displays without changing its composition.

polar covalent bond A covalent bond between atoms of different electronegativities. Polar covalent bonds have a dipole moment.

polar molecule A molecule with polar bonds that add together to create a net dipole moment.

polyatomic ion An ion composed of a group of atoms with an overall charge.

polymer A molecule with many similar units, called monomers, bonded together in a long chain.

polypeptide A short chain of amino acids joined by peptide bonds.

polysaccharide A long, chainlike molecule composed of many linked monosaccharide units. Polysaccharides are polymers of monosaccharides

positron A nuclear particle that has the mass of an electron but carries a +1 charge.

positron emission Expulsion of a positron from an unstable atomic nucleus. In positron emission, a proton is transformed into a neutron.

potential energy The energy of a body that is associated with its position or the arrangement of its parts.

precipitate An insoluble product formed through the reaction of two solutions containing soluble compounds.

precipitation reaction A reaction that forms a solid or precipitate when two aqueous solutions are mixed.

prefix multipliers Prefixes used by the SI system with the standard units. These multipliers change the value of the unit by powers of 10.

pressure The force exerted per unit area by gaseous molecules as they collide with the surfaces around them.

primary protein structure The sequence of amino acids in a protein's chain. Primary protein structure is maintained by the covalent peptide bonds between individual amino acids.

principal quantum number A number that indicates the shell that an electron occupies.

principal shell The shell indicated by the principal quantum number.

products The final substances produced in a chemical reaction; represented on the right side of a chemical equation.

properties The characteristics we use to distinguish one substance from another.

protein A biological molecule composed of a long chain of amino acids joined by peptide bonds. In living organisms, proteins serve many varied and important functions.

proton A positively charged nuclear particle. A proton's mass is approximately 1 amu.

pure substance A substance composed of only one type of atom or molecule.

quantification The assigning of a number to an observation so as to specify a quantity or property precisely.

quantum (*pl.* quanta) The precise amount of energy possessed by a photon; the difference in energy between two atomic orbitals.

quantum number (n) An integer that specifies the energy of an orbital. The higher the quantum number n, the greater the distance between the electron and the nucleus and the higher its energy.

quantum-mechanical model The foundation of modern chemistry; explains how electrons exist in atoms, and how they affect the chemical and physical properties of elements.

quaternary structure In a protein, the way that individual chains fit together to compose the protein. Quaternary structure is maintained by interactions between the R groups of amino acids on the different chains.

R group (side chain) An organic group attached to the central carbon atom of an amino acid.

radio waves The longest-wavelength and least energetic form of electromagnetic radiation.

radioactive Describes a substance that emits tiny, invisible, energetic particles from the nuclei of its component atoms.

radioactivity The emission of tiny, invisible, energetic particles from the unstable nuclei of atoms. Many of these particles can penetrate matter.

radiocarbon dating A technique used to estimate the age of fossils and artifacts through the measurement of natural radioactivity of carbon atoms in the environment.

radiotherapy Treatment of disease with radiation, such as the use of gamma rays to kill rapidly dividing cancer cells.

random coil The name given to an irregular pattern of a secondary protein structure.

rate of a chemical reaction (reaction rate) The amount of reactant that changes to product in a given period of time. Also defined as the amount of a product that forms in a given period of time.

reactants The initial substances in a chemical reaction, represented on the left side of a chemical equation.

recrystallization A technique used to purify a solid; involves dissolving the solid in a solvent at high temperature, creating a saturated solution, then cooling the solution to cause the crystallization of the solid.

reducing agent In a redox reaction, the substance being oxidized. Reducing agents tend to lose electrons easily.

reduction The loss of oxygen, the gain of hydrogen, or the gain of electrons (the most fundamental definition).

rem Stands for *roentgen equivalent man;* a weighted measure of radiation exposure that accounts for the ionizing power of the different types of radiation.

resonance structures Two or more Lewis structures that are necessary to describe the bonding in a molecule or ion.

reversible reaction A reaction that is able to proceed in both the forward and reverse directions.

RNA (ribonucleic acid) Long chainlike molecules that occur throughout cells and take part in the construction of proteins.

salt An ionic compound that usually remains dissolved in a solution after an acid–base reaction has occurred.

salt bridge An inverted, U-shaped tube containing a strong electrolyte; completes the circuit in an electrochemical cell by allowing the flow of ions between the two half-cells.

saturated fat A triglyceride composed of saturated fatty acids. Saturated fat tends to be solid at room temperature.

saturated hydrocarbon A hydrocarbon that contains no double or triple bonds between the carbon atoms.

saturated solution A solution that holds the maximum amount of solute under the solution conditions. If additional solute is added to a saturated solution, it will not dissolve.

scientific law A statement that summarizes past observations and predicts future ones. Scientific laws are usually formulated from a series of related observations.

scientific method The way that scientists learn about the natural world. The scientific method involves observations, laws, hypotheses, theories, and experimentation.

scientific notation A system used to write very big or very small numbers, often containing many zeros, more compactly and precisely. A number written in scientific notation consists of a decimal part and an exponential part (10 raised to a particular exponent).

scintillation counter A device used to detect radioactivity in which energetic particles traverse a material that emits ultraviolet or visible light when excited by their passage. The light is detected and turned into an electrical signal.

second (s) The SI standard unit of time.

secondary protein structure Short-range periodic or repeating patterns often found in proteins. Secondary protein structure is maintained by interactions between amino acids that are fairly close together in the linear sequence of the protein chain or adjacent to each other on neighboring chains.

semiconductor A compound or element exhibiting intermediate electrical conductivity that can be changed and controlled.

semipermeable membrane A membrane that selectively allows some substances to pass through but not others.

SI units The most convenient system of units for science measurements, based on the metric system. The set of standard units agreed on by scientists throughout the world.

significant digits (figures) The non-place-holding digits in a reported measurement; they represent the precision of a measured quantity.

simple carbohydrate (simple sugar) A monosaccharide or disaccharide.

single bond A chemical bond in which one electron pair is shared between two atoms.

solid A state of matter in which atoms or molecules are packed close to each other in fixed locations.

solid formation One type of evidence of a chemical reaction, involving the formation of a solid.

solubility The amount of a compound, usually in grams, that will dissolve in a certain amount of solvent.

solubility rules A set of empirical rules used to determine whether an ionic compound is soluble.

solubility-product constant (K_{sp}) The equilibrium expression for a chemical equation that represents the dissolving of an ionic compound in solution.

soluble Dissolves in solution.

solute The minority component of a solution.

solution A homogenous mixture of two or more substances.

solvent The majority component of a solution.

specific heat capacity (or specific heat) The heat capacity of a substance in joules per gram degree celsius (J/g °C).

spectator ions Ions that do not participate in a reaction; they appear unchanged on both sides of a chemical equation.

standard temperature and pressure (STP) Conditions often assumed in calculations involving gases: $T = 0$ °C (273 K) and $P = 1$ atm.

starch A common polysaccharide composed of repeating glucose units.

states of matter The three forms in which matter can exist: solid, liquid, and gas.

steroid A biological compound containing a 17-carbon 4-ring system.

stock solution A concentrated form in which solutions are often stored.

stoichiometry The numerical relationships among chemical quantities in a balanced chemical equation. Stoichiometry allows us to predict the amounts of products that form in a chemical reaction based on the amounts of reactants.

strong acid An acid that completely ionizes in solution.

strong base A base that completely dissociates in solution.

strong electrolyte A substance whose aqueous solutions are good conductors of electricity.

strong electrolyte solution A solution containing a solute that dissociates into ions; therefore, a solution that conducts electricity well.

structural formula A two-dimensional representation of molecules that not only shows the number and type of atoms, but also how the atoms are bonded together.

sublimation A physical change in which a substance is converted from its solid form directly into its gaseous form.

subshell In quantum mechanics, specifies the shape of the orbital and is represented by different letters (*s*, *p*, *d*, *f*).

substituent An atom or group of atoms that has been substituted for a hydrogen atom in an organic compound.

substitution reaction A reaction in which one or more atoms are replaced by one or more different atoms.

supersaturated solution A solution holding more than the normal maximum amount of solute.

surface tension The tendency of liquids to minimize their surface area, resulting in a "skin" on the surface of the liquid.

synthesis A reaction in which simpler substances combine to form more complex substances; $A + B \longrightarrow AB$.

temporary dipole A type of intermolecular force resulting from transient shifts in electron density within an atom or molecule.

terminal atom An atom that is located at the end of a molecule or chain.

tertiary structure A protein's structure that consists of the large-scale bends and folds due to interactions between the *R* groups of amino acids that are separated by large distances in the linear sequence of the protein chain.

tetrahedral The molecular geometry of a molecule containing 4 electron groups (4 bonding groups and no lone pairs).

theoretical yield The amount of product that can be made in a chemical reaction based on the amount of limiting reactant.

theory A proposed explanation for observations and laws. A theory presents a model of the way nature works and predicts behavior that extends well beyond the observations and laws from which it was formed.

titration A laboratory procedure used to determine the amount of a substance in solution. In a titration, a reactant in a solution of known concentration is reacted with another reactant in a solution of unknown concentration until the reaction reaches the endpoint.

torr A unit of pressure named after the Italian physicist Evangelista Torricelli. Also called a millimeter of mercury.

transition metals The elements in the middle of the periodic table whose properties tend to be less predictable based simply on their position in the periodic table. Transition metals lose electrons in their chemical reactions, but do not necessarily acquire noble gas configurations.

triglyceride A fat or oil; a tryglyceride is a tri-ester composed of glycerol with three fatty acids attached.

trigonal planar The molecular geometry of a molecule containing 3 electron groups, 3 bonding groups, and no lone pairs.

trigonal pyramidal The molecular geometry of a molecule containing 4 electron groups, 3 bonding groups, and 1 lone pair.

triple bond A chemical bond consisting of three electron pairs shared between two atoms. In general, triple bonds are shorter and stronger than double bonds.

Type I compounds Compounds containing metals that always form cations with the same charge.

Type II compounds Compounds containing metals that can form cations with different charges.

ultraviolet (UV) light The fraction of the electromagnetic spectrum between the visible region and the X-ray region. UV light is invisible to the human eye.

units Previously agreed-on quantities used to report experimental measurements. Units are vital in chemistry.

unsaturated fat (or oil) A triglyceride composed of unsaturated fatty acids. Unsaturated fats tend to be liquids at room temperature.

unsaturated hydrocarbon A hydrocarbon that contains one or more double or triple bonds between its carbon atoms.

unsaturated solution A solution holding less than the maximum possible amount of solute under the solution conditions.

valence electrons The electrons in the outermost principal shell of an atom; they are involved in chemical bonding.

valence shell electron pair repulsion (VSEPR) A theory that allows prediction of the shapes of molecules based on the idea that electrons—either as lone pairs or as bonding pairs—repel one another.

vapor pressure The partial pressure of a vapor in dynamic equilibrium with its liquid.

vaporization The phase transition between a liquid and a gas.

viscosity The resistance of a liquid to flow; manifestation of intermolecular forces.

visible light The fraction of the electromagnetic spectrum that is visible to the human eye, bounded by wavelengths of 400 nm (violet) and 780 nm (red).

vital force A mystical or supernatural power that, it was once believed, was possessed only by living organisms and allowed them to produce organic compounds.

vitalism The belief that living things contain a nonphysical "force" that allows them to synthesize organic compounds.

volatile Tending to vaporize easily.

voltage The potential difference between two electrodes; the driving force that causes electrons to flow.

volume A measure of space. Any unit of length, when cubed, becomes a unit of volume.

wavelength The distance between adjacent wave crests in a wave.

weak acid An acid that does not completely ionize in solution.

weak base A base that does not completely dissociate in solution.

weak electrolyte A substance whose aqueous solutions are poor conductors of electricity.

X-rays The portion of the electromagnetic spectrum between the ultraviolet (UV) region and the gamma-ray region.

ANSWERS TO ODD-NUMBERED EXERCISES

Note: Answers in the Questions section are written as briefly as possible. Student answers may vary and still be correct.

Chapter 1

Questions

1. Soda fizzes due to the interactions between carbon dioxide and water under high pressures. At room temperature, carbon dioxide is a gas and water is a liquid. Through the use of pressure, the makers of soda force the carbon dioxide gas to dissolve in the water. When the can is sealed, the solution remains mixed. When the can is opened, the pressure is released and the carbon dioxide molecules escape in bubbles of gas.
3. Chemists study molecules and interactions at the molecular level to learn about and explain macroscopic events. Chemists attempt to explain why ordinary things are as they are.
5. Chemistry is the science that seeks to understand what matter does by studying what atoms and molecules do.
7. The scientific method is the way chemists investigate the chemical world. The first step consists of observing the natural world. Later observations can be combined to create a scientific law, which summarizes and predicts behavior. Theories are models that strive to explain the cause of the observed phenomenon. Theories are tested through experiment. When a theory is not well established, it is sometimes referred to as a hypothesis.
9. A law is simply a general statement that summarizes and predicts observed behavior. Theories seek to explain the causes of observed behavior.
11. To say "It is just a theory" makes it seem as if theories are easily discardable. However, many theories are very well established and are as close to truth as we get in science. Established theories are backed up with years of experimental evidence, and they are the pinnacle of scientific understanding.
13. The atomic theory states that all matter is composed of small, indestructible particles called atoms. John Dalton formulated this theory.

Problems

15. Carbon dioxide contains one carbon atom and two oxygen atoms. Water contains one oxygen atom and two hydrogen atoms.
17. **a.** observation **b.** theory
    c. law **d.** observation
19. **a.** All atoms contain a degree of chemical reactivity. The larger the size of an atom, the higher the chemical reactivity of that atom.

    b. There are many correct answers. One example is: Conceivably, when the size of an atom is increased, the surface area of the atom is also increased; an atom with a greater surface area is more likely to react chemically.

Chapter 2

Questions

1. Without units, the results are unclear and it is hard to keep track of what each separate measurement entails.
3. Often scientists work with very large or very small numbers that contain a lot of zeros. Scientific notation allows these numbers to be written more compactly, and the information is more organized.
5. Zeros count as significant digits when they are interior zeros (zeros between two numbers) and when they are trailing zeros (zeros after a decimal point). Zeros are **not** significant digits when they are leading zeros, which are zeros to the left of the first nonzero number.
7. For calculations involving only multiplication and division, the result carries the same number of significant figures as the factor with the fewest significant figures.
9. When rounding numbers, one must round down if the last (or left-most) digit dropped is four or less, or one must round up if the last (or left-most) digit is five or more.
11. The basic SI unit of length is the meter. The kilogram is the SI unit of mass. Lastly, the second is the SI unit of time.
13. For measuring a Frisbee, the unit would be the meter and the prefix multiplier would be *centi-*. The final measurement would be in centimeters.
15. **a.** 2.42 cm **b.** 1.69 cm
    c. 21.58 cm **d.** 20.85 cm
17. Units act as a guide in the calculation and are able to show if the calculation is off track. The units must be followed in the calculation, so that the answer is correctly written and understood.
19. A conversion factor is a quantity used to relate two separate units. They are constructed from any two quantities known to be equivalent.

21. The solution map for converting grams to pounds is:

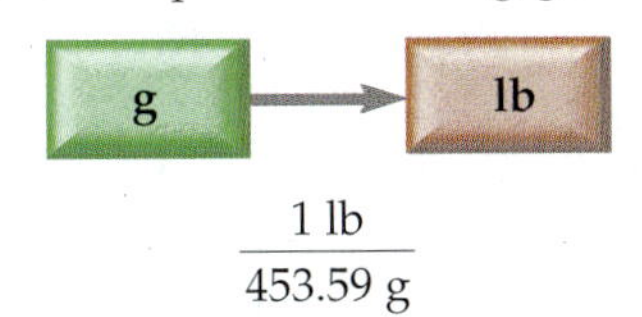

$\frac{1 \text{ lb}}{453.59 \text{ g}}$

23. The solution map for converting meters to feet is:

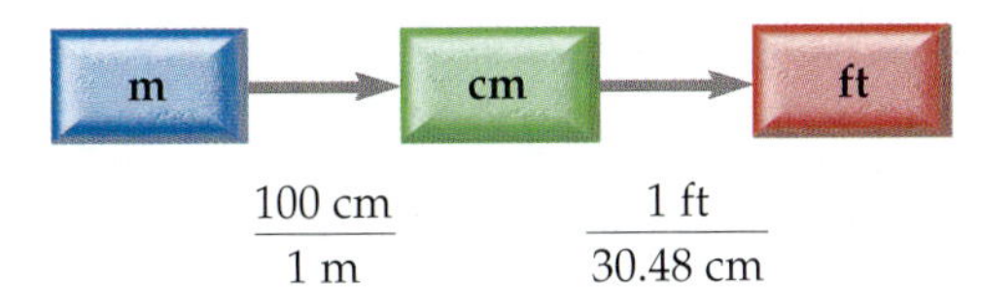

$\frac{100 \text{ cm}}{1 \text{ m}}$ $\frac{1 \text{ ft}}{30.48 \text{ cm}}$

25. The density of a substance is the ratio of its mass to its volume. Density is a fundamental property of materials and differs from one substance to another. Density can be used to relate two separate units, thus working as a conversion factor. Density is a conversion factor between mass and volume.

Problems

27. **a.** 3.2667×10^7 **b.** 1.193×10^6 **c.** 1.8175×10^7 **d.** 4.81×10^5

29. **a.** 7.461×10^{-11} m **b.** 1.58×10^{-5} mi **c.** 6.32×10^{-7} m **d.** 1.5×10^{-5} m

31. **a.** 602,200,000,000,000,000,000,000
b. 0.00000000000000000016 C
c. 299,000,000 m/s
d. 344 m/s

33. **a.** 3,890,000,000 **b.** 0.00059 **c.** 8,680,000,000,000 **d.** 0.0000000786

35.

2,000,000,000	2×10^9
1,211,000,000	1.211×10^9
0.000874	8.74×10^{-4}
320,000,000,000	3.2×10^{11}

37. **a.** 54.9 mL **b.** 48.7 °C **c.** 5.550 mL **d.** 46.83 °C

39. **a.** 0.005050 **b.** 0.0000000000000000060 **c.** 220,103 **d.** 0.00108

41. **a.** 4 **b.** 4 **c.** 6 **d.** 5

43. **a.** correct **b.** 3 **c.** 7 **d.** correct

45. **a.** 343.0 **b.** 0.009651 **c.** 3.526×10^{-8} **d.** 1.127×10^9

47. **a.** 2.3 **b.** 2.4 **c.** 2.3 **d.** 2.4

49. **a.** 42.3 **b.** correct **c.** correct **d.** 0.0456

51. **a.** 0.054 **b.** 0.619 **c.** 1.2×10^8 **d.** 6.6

53. **a.** 4.22×10^3 **b.** correct **c.** 3.9969 **d.** correct

55. **a.** 200.6 **b.** 41.4 **c.** 183.3 **d.** 1.22

57. **a.** correct **b.** 1.0982 **c.** correct **d.** 3.53

59. **a.** 3.9×10^3 **b.** 632 **c.** 8.93×10^4 **d.** 6.34

61. **a.** 3.15×10^3 **b.** correct **c.** correct **d.** correct

63. **a.** 2.14×10^3 g **b.** 6.172 m **c.** 1.316×10^{-3} kg **d.** 25.6 mL

65. **a.** 0.588 L **b.** 34.1 μg **c.** 10.1 ns **d.** 2.19×10^{-12} m

67. **a.** 57.2 cm **b.** 38.4 m **c.** 0.754 km **d.** 61 mm

69. **a.** 15.7 in **b.** 91.2 ft **c.** 6.21 mi **d.** 8478 lb

71.

5.08×10^8 m	5.08×10^5 km	508 Mm
5.08×10^{-1} Gm	5.08×10^{-4} Tm	
2.7976×10^{10} m	2.7976×10^7 km	27,976 Mm
2.7976×10^1 Gm	2.7976×10^{-2} Tm	
1.77×10^{12} m	1.77×10^9 km	1.77×10^6 Mm
1.77×10^3 Gm	1.77 Tm	
1.5×10^8 m	1.5×10^5 km	1.5×10^2 Mm
0.15 Gm	1.5×10^{-4} Tm	
4.23×10^{11} m	4.23×10^8 km	4.23×10^5 Mm
423 Gm	0.423 Tm	

73. **a.** 2.255×10^7 kg **b.** 2.255×10^4 Mg **c.** 2.255×10^{13} mg **d.** 2.255×10^4 metric tons

75. 1.1×10^3 g

77. 5.0×10^1 min

79. 4.7×10^3 cm^3

81. **a.** 1.0×10^6 m^2 **b.** 1.0×10^{-6} m^3 **c.** 1.0×10^{-9} m^3

83. **a.** 6.2×10^5 pm^3 **b.** 6.2×10^{-4} nm^3 **c.** 6.2×10^{-1} Å^3

85. **a.** 2.15×10^{-4} km^2 **b.** 2.15×10^4 dm^2 **c.** 2.15×10^6 cm^2

87. 1.49×10^6 mi^2

89. 11.4 g/cm^3, lead

91. 1.26 g/cm^3

93. Yes, the density of the crown is 19.3 g/cm^3

95. **a.** 463 g **b.** 3.7 L

97. **a.** 3.38×10^4 g (gold); 5.25×10^3 g (sand)
b. Yes, the mass of the bag of sand is different from the mass of the gold vase; thus, the weight-sensitive pedestal will sound the alarm.

99. 10.6 g/cm^3

101. $2.7 \times 10^3 \frac{\text{kg}}{\text{m}^3}$

103. 2.5×10^5 lbs

105. 1.19×10^5 kg

107. 3.0×10^1 km/L

109. 108 km; 47.2 km

111. 9.1×10^{10} g/cm^3

Chapter 3

Questions

1. Matter is defined as anything that occupies space and possesses mass. It can be thought of as the physical material that makes up the universe.
3. The three states of matter are solid, liquid, and gas.
5. In a crystalline solid, the atoms/molecules are arranged in geometric patterns with repeating order. In amorphous solids, the atoms/molecules do not have long-range order.
7. The atoms/molecules in gases are not in contact with each other and are free to move relative to one another. The spacing between separate atoms/molecules is very far apart. A gas has no fixed volume or shape; rather it assumes both the shape and the volume of the container it occupies.
9. A mixture is two or more pure substances combined in variable proportions.
11. Pure substances are those composed of only one type of atom or molecule.
13. A mixture is formed when two or more pure substances are mixed together; however a new substance is not formed. A compound is formed when two or more elements are bonded together and form a new substance.
15. In a physical change the composition of the substance does not change, even though its appearance might change. However, in a chemical change, the substance undergoes a change in its composition.
17. Energy is defined as the capacity to do work.
19. Some different kinds of energy are kinetic, potential, electrical, chemical, and thermal.
21. Three common units for temperature are kelvin, Celsius, and Fahrenheit. They differ in how they define zero and in the size of the respective degrees.
23. Heat capacity is the quantity of heat energy required to change the temperature of a given amount of the substance by 1 °C.
25. $°F = \frac{9}{5}(°C) + 32$

Problems

27. **a.** element **b.** element **c.** compound **d.** compound
29. **a.** homogeneous **b.** heterogeneous **c.** homogeneous **d.** homogeneous
31. **a.** pure substance-element
    b. mixture-homogeneous
    c. mixture-heterogeneous
    d. mixture-heterogeneous
33. **a.** chemical **b.** physical **c.** physical **d.** chemical
35. physical–colorless; odorless; gas at room temperature; one liter has a mass of 1.260 g under standard conditions; mixes with acetone chemical–flammable; polymerizes to form polyethylene
37. **a.** chemical **b.** physical **c.** chemical **d.** chemical
39. **a.** physical **b.** chemical
41. 2.10×10^2 kg
43. **a.** Yes **b.** No
45. 15.1 g of water
47. **a.** 7.77 cal **b.** 2.35×10^3 J **c.** 8.53×10^3 cal **d.** 1.20 kJ
49. **a.** 9.0×10^7 J **b.** 0.249 Cal **c.** 1.31×10^{-4} kWh **d.** 1.1×10^4 cal
51.

J	*cal*	*Cal*	*kWh*
225	53.8	5.38×10^{-2}	6.25×10^{-5}
3.44×10^6	8.21×10^5	8.21×10^2	9.54×10^{-1}
1.06×10^9	2.54×10^8	2.54×10^5	295
6.49×10^5	1.55×10^5	155	1.80×10^{-1}

53. 3.44×10^9 J
55. 8×10^2 kJ; 17 days
57. **a.** 1.00×10^2 °C **b.** -3.2×10^2 °F **c.** 298 K **d.** 3.10×10^2 K
59. −62 °C, 211 K
61. 159 K, −173 °F
63.

0.0 K	−459.4 °F	−273.0 °C
301 K	82.5 °F	28.1 °C
282 K	47 °F	8.5 °C

65. 1.0×10^4 J
67. 8.7×10^5 J
69. 58 °
71. 31 °
73. 1.0×10^1 °C
75. 0.24 J/g °C; silver
77. 2.2 J/g °C;
79. When warm drinks are placed into the ice, they release heat, which then melts the ice. The prechilled drinks, on the other hand, are already cold, so they do not release much heat.
81. 49 °C
83. 70.2 J
85. 1.7×10^4 kJ
87. 67 °C
89. 6.0 kWh
91. 22 g of fuel
93. 78 g
95. 27.2 °C
97. **a.** pure substance **b.** pure substance **c.** pure substance **d.** mixture
99. physical change
101. Small temperature changes in the ocean have a great impact on global weather because of the high heat capacity of water.
103. **a.** Sacramento is farther inland than San Francisco, so Sacramento is not as close to the ocean. The ocean water has a high heat capacity and will be able to keep San Francisco cooler in hot days of summer. However, Sacramento is away from the ocean in a valley, so it will experience high temperatures in the summer.
     b. San Francisco is located right next to the ocean, so the high heat capacity of the seawater keeps the temperature in the city from dropping. In the winter, the ocean actually helps to keep the city warmer, compared to an inland city like Sacramento.

Chapter 4

Questions

1. It is important to understand atoms because they compose matter and the properties of atoms determine matter's properties. The atom is the key to connecting the microscopic world with the macroscopic world.
3. Democritus theorized that matter was ultimately composed of small, indivisible particles called atoms. Upon dividing matter, one would find tiny, indestructible atoms.
5. Rutherford's gold foil experiment involved sending positively charged alpha particles through a thin sheet of gold foil and detecting if there was any deflection of the particles. He found that most passed straight through, yet some particles showed some deflection. This result contradicts the plum-pudding model of the atom because the plum-pudding model does not explain the deflection of the alpha-particles.
7.

Particle	*Mass (kg)*	*Mass (amu)*	*Charge*
Proton	1.67262×10^{-27}	1	+1
Neutron	1.67493×10^{-27}	1	0
Electron	0.00091×10^{-27}	0.00055	−1

9. Matter is usually charge neutral due to protons and electrons having opposite charges. If matter were not charge neutral, many unnatural things would occur, such as objects repelling or attracting each other.
11. A chemical symbol is a unique one- or two-letter abbreviation for an element. It is listed below the atomic number for that element on the periodic table.
13. Mendeleev noticed that many patterns were evident when elements were organized by increasing mass; from this he formulated the periodic law. He also organized the elements based on this law and created the basis for the periodic table being used today.
15. The periodic table is organized by listing the elements in order of increasing atomic number.
17. Nonmetals have varied properties (solid, liquid, or gas at room temperature); however, as a whole they tend to be poor conductors of heat and electricity and they all tend to gain electrons when they undergo chemical changes. They are located toward the upper right side of the periodic table.
19. Each column within the main group elements in the periodic table is labeled as a family or group of elements. The elements within a group usually have similar chemical properties.
21. An ion is an atom or group of atoms that has lost or gained electrons and has become charged.
23. **a.** ion charge = +1 **b.** ion charge = +2 **c.** ion charge = +3 **d.** ion charge = −2 **e.** ion charge = −1
25. The percent natural abundance of isotopes is the relative amount of each different isotope in a naturally occurring sample of a given element.
27. Isotopes are noted in this manner, ${}^{A}_{Z}X$. X represents the chemical symbol, A represents the mass number, and Z represents the atomic number.

Problems

29. **a.** Correct.
b. False; different elements contain different types of atoms according to Dalton.
c. False; one cannot have 1.5 hydrogen atoms; combinations must be in simple, whole-number ratios.
d. Correct.
31. **a.** Correct.
b. False; most of the volume of the atom is empty space occupied by tiny, negatively charged electrons.
c. False, the number of negatively charged particles outside the nucleus equals the number of positively charged particles inside the nucleus.
d. False, the majority of the mass of an atom is found in the nucleus.
33. a, b, d
35. b, d
37. approximately 1.8×10^{3} electrons
39. 5.4×10^{-4} g
41. **a.** 27 **b.** 77 **c.** 92 **d.** 14 **e.** 4
43. **a.** 25 **b.** 47 **c.** 79 **d.** 82 **e.** 16
45. **a.** C, 6 **b.** N, 7 **c.** Na, 11 **d.** K, 19 **e.** Cu, 29
47. **a.** gold, 79 **b.** silicon, 14 **c.** nickel, 28 **d.** zinc, 30 **e.** tungsten, 74
49.

Element Name	*Element Symbol*	*Atomic Number*
Gold	Au	79
Tin	Sn	50
Arsenic	As	33
Copper	Cu	29
Iron	Fe	26
Mercury	Hg	80

51. **a.** metal **b.** metal **c.** nonmetal **d.** nonmetal **e.** metalloid
53. a, d, e
55. a, b
57. c, d
59. a, b, e
61. **a.** halogen **b.** noble gas **c.** halogen **d.** neither **e.** noble gas
63. **a.** 6A **b.** 3A **c.** 4A **d.** 4A **e.** 5A
65. b, oxygen; it is in the same group or family.
67. b, chlorine and fluorine; they are in the same family or group.
69.

Chemical Symbol	*Group Number*	*Group Name*	*Metal or Nonmetal*
K	1A	Alkali Metals	Metals
Br	7A	Halogens	Nonmetal
Sr	2A	Alkaline Earth	Metal
He	8A	Noble Gas	Nonmetal
Ar	8A	Noble Gas	Nonmetal

71. **a.** e^- **b.** O^{2-}
c. $2e^-$ **d.** Cl^-

73. **a.** −2 **b.** +3
c. +4 **d.** −1

75. **a.** 19 protons, 18 electrons
b. 16 protons, 18 electrons
c. 38 protons, 36 electrons
d. 24 protons, 21 electrons

77. **a.** False; Ti^{2+} has 22 protons and 20 electrons.
b. True
c. False; Mg^{2+} has 12 protons and 10 electrons
d. True

79. **a.** Rb^+ **b.** K^+
c. Al^{3+} **d.** O^{2-}

81. **a.** 3 electrons lost **b.** 1 electron lost
c. 1 electron gained **d.** 2 electrons gained

83.

Symbol	*Ion Commonly Formed*	*Number of Electrons in Ion*	*Number of Protons in Ion*
Te	Te^{2-}	54	52
In	In^{3+}	46	49
Sr	Sr^{2+}	36	38
Mg	Mg^{2+}	10	12
Cl	Cl^-	18	17

85. **a.** Z = 1, A = 2 **b.** Z = 24, A = 51
c. Z = 20, A = 40 **d.** Z = 73, A = 181

87. **a.** $^{16}_{8}O$ **b.** $^{19}_{9}F$
c. $^{23}_{11}Na$ **d.** $^{27}_{13}Al$

89. **a.** $^{60}_{27}Co$ **b.** $^{22}_{10}Ne$
c. $^{131}_{53}I$ **d.** $^{244}_{94}Pu$

91. **a.** 11 protons, 12 neutrons
b. 88 protons, 178 neutrons
c. 82 protons, 126 neutrons
d. 7 protons, 7 neutrons

93. 6 protons, 8 neutrons, $^{14}_{6}C$

95. 85.47 amu

97. **a.** 49.31% **b.** 78.92 amu

99. 121.8 amu, Sb

101. 7.8×10^{17} electrons

103. $6.8 \times 10^{-13}\%$

105.

Number Symbol	*Number of Protons*	*Number of Neutrons*	*A (Mass Number)*	*Natural Abundance*
Sr-84 or $^{84}_{38}Sr$	38	46	84	0.56%
Sr-86 or $^{86}_{38}Sr$	38	48	86	9.86%
Sr-87 or $^{87}_{38}Sr$	38	49	87	7.00%
Sr-88 or $^{88}_{38}Sr$	38	50	88	82.58%

Atomic mass of Sr = 87.62 amu

107.

Symbol	*Z*	*A*	*Number of Protons*	*Number of Electrons*	*Number of Neutrons*	*Charge*
Zn^+	30	64	30	29	34	1+
Mn^{3+}	25	55	25	22	30	3+
P	15	31	15	15	16	0
O^{2-}	8	16	8	10	8	2−
S^{2-}	16	34	16	18	18	2−

109. 153 amu, 52.2%

111. The atomic theory and nuclear model of the atom are both theories because they attempt to provide a broader understanding and model behavior of chemical systems.

113. **a.** Nt-304 = 72%; Nt-305 = 4%; Nt-306 = 24%
b.

120
Nt
304.5

Chapter 5

Questions

1. Yes; when elements combine with other elements, a compound is created. Each compound is unique and contains properties different from those of the elements that compose it.

3. The law of constant composition states that all samples of a given compound have the same proportions of their constituent elements. Joseph Proust formulated this law.

5. The more metallic element is generally listed first in a chemical formula.

7. An atomic element is one that exists in nature with a single atom as the basic unit. A molecular element is one that exists as a diatomic molecule as the basis unit. Molecular elements include H_2, N_2, O_2, F_2, Cl_2, Br_2, and I_2.

9. The systematic name can be directly derived by looking at the compound's formula. The common name for a compound acts like a nickname and can only be learned through familiarity.

11. The block that contains the elements for Type II compounds is known as the transition metals.

13. The basic form for the names of Type II ionic compounds is to have the name of the metal cation first, followed by the charge of the metal cation (in parentheses, using Roman numerals), and finally the base name of the nonmetal anion with *-ide* attached to the end.

15. For compounds containing a polyatomic anion, the name of the cation is first, followed by the name for the polyatomic anion. Also, if the compound contains both a polyatomic cation and a polyatomic anion, one would just use the names of both polyatomic ions.

17. The form for naming molecular compounds is to have the first element preceded by a prefix to indicate the number of atoms present. This is then followed by the second element with its corresponding prefix and *-ide* placed on the end of the second element.

19. To correctly name a binary acid one must begin the first word with *hydro-*, which is followed by the base name of the nonmetal plus *-ic* added on the end. Finally, the word *acid* follows the first word.

21. To name an acid with oxyanions ending with *-ite*, one must take the base name of the oxyanion and attach *-ous* to it; the word *acid* follows this.

Problems

23. Yes; the ratios of sodium to chlorine in both samples were equal.

25. 2.06×10^3 g

27.

	Mass N_2O	*Mass N*	*Mass O*
Sample A	2.85	1.82	1.03
Sample B	4.55	2.91	1.64
Sample C	3.74	2.39	1.35
Sample D	1.74	1.11	0.63

29. NBr_3

31. **a.** Fe_2O_3 **b.** PCl_3 **c.** PCl_5 **d.** Ag_2O

33. **a.** 4 **b.** 4 **c.** 6 **d.** 4

35. **a.** magnesium, 1; chlorine, 2
b. sodium, 1; nitrogen, 1; oxygen, 3
c. calcium, 1; nitrogen, 2; oxygen, 4
d. strontium, 1; oxygen, 2; hydrogen, 2

37. **a.** atomic **b.** molecular **c.** molecular **d.** atomic

39. **a.** molecular **b.** ionic **c.** ionic **d.** molecular

41. helium → single atoms
CCl_4 → molecules
K_2SO_4 → formula units
bromine → diatomic molecules

43. **a.** formula units **b.** single atoms **c.** molecules **d.** molecules

45. **a.** ionic, Type I **b.** molecular **c.** molecular **d.** ionic, Type II

47. **a.** SrO **b.** Na_2O **c.** Al_2S_3 **d.** $MgBr_2$

49. **a.** $KC_2H_3O_2$ **b.** K_2CrO_4 **c.** K_3PO_4 **d.** KCN

51. **a.** K_3N, K_2O, KF **b.** Ba_3N_2, BaO, BaF_2 **c.** AlN, Al_2O_3, AlF_3

53. **a.** cesium chloride **b.** strontium bromide **c.** potassium oxide **d.** lithium fluoride

55. **a.** chromium(II) chloride
b. chromium(III) chloride
c. tin(IV) oxide
d. lead(II) iodide

57. **a.** Type II, chromium(III) oxide
b. Type I, sodium iodide
c. Type I, calcium bromide
d. Type II, tin(II) oxide

59. **a.** barium nitrate **b.** lead(II) acetate **c.** ammonium iodide **d.** potassium chlorate **e.** cobalt(II) sulfate **f.** sodium perchlorate

61. **a.** hypobromite ion **b.** bromite ion **c.** bromate ion **d.** perbromate ion

63. **a.** $CuBr_2$ **b.** $AgNO_3$ **c.** KOH **d.** Na_2SO_4 **e.** $KHSO_4$ **f.** $NaHCO_3$

65. **a.** sulfur dioxide
b. nitrogen triiodide
c. bromine pentafluoride
d. nitrogen monoxide
e. tetranitrogen tetraselenide

67. **a.** CO **b.** S_2F_4 **c.** Cl_2O **d.** PF_5 **e.** BBr_3 **f.** P_2S_5

69. **a.** PBr_5 phosphorus pentabromide
b. P_2O_3 diphosphorus trioxide
c. SF_4 sulfur tetraflouride
d. correct

71. **a.** chlorous acid **b.** hydroiodic acid **c.** sulfuric acid **d.** nitric acid

73. **a.** H_3PO_4 **b.** HBr **c.** H_2SO_3

75. **a.** 47.02 amu **b.** 110.98 amu **c.** 153.81 amu **d.** 148.33 amu

77. PBr_3, Ag_2O, PtO_2, $Al(NO_3)_3$

79. **a.** CH_4 **b.** SO_3 **c.** NO_2

81. **a.** 12 **b.** 4 **c.** 12 **d.** 7

83. **a.** 8 **b.** 12 **c.** 12

85.

Formula	*Type*	*Name*
N_2H_4	molecular	dinitrogen tetrahydride
KCl	ionic	potassium chloride
H_2CrO_4	acid	chromic acid
$Co(CN)_3$	ionic	cobalt(III) cyanide

87. **a.** calcium nitrite **b.** potassium oxide **c.** phosphorus trichloride **d.** correct **e.** potassium iodite

89. **a.** $Sn(SO_4)_2$ 310.8 amu **b.** HNO_2 47.02 amu **c.** $NaHCO_3$ 84.01 amu **d.** PF_5 125.97 amu

91. **a.** platinum(IV) oxide 227.08 amu
b. dinitrogen pentoxide 108.02 amu
c. aluminum chlorate 277.33 amu
d. phosphorous pentabromide 430.47 amu

93. **a.** molecular element **b.** atomic element **c.** ionic compound **d.** molecular compound

95. **a.** NaOCl; NaOH **b.** $CaCO_3$; $NaHCO_3$;
c. $Al(OH)_3$; $Mg(OH)_2$
d. $NaHCO_3$, $Ca_3(PO_4)_2$, $NaAl(SO_4)_2$

Chapter 6

Questions

1. Chemical composition lets us determine how much of a particular element is contained within a particular compound.
3. There are 6.022×10^{23} atoms in 1 mole of atoms.
5. One mole of any element has a mass equal to its atomic mass in grams.
7. **a.** 30.97 g **b.** 195.08 g **c.** 12.01 g **d.** 52.00 g
9. Each element has a different atomic mass number. So, the subscripts that represent mole ratios cannot be used to represent the ratios of grams of a compound. The grams per mole of one element always differs from the grams per mole of a different element.
11. **a.** 11.19 g H ≡ 100 g H_2O
b. 53.29 g O ≡ 100 g fructose
c. 84.12 g C ≡ 100 g octane
d. 52.14 g C ≡ 100 g ethanol

13. The empirical formula gives the smallest whole number ratio of each type of atom. The molecular formula gives the specific number of each type of atom in the molecule. The molecular formula is always a multiple of the empirical formula.

15. The empirical formula mass of a compound is the sum of the masses of all the atoms in the empirical formula.

Problems

17. 3.6×10^{24} atoms

19. **a.** 2.0×10^{24} atoms **b.** 5.8×10^{21} atoms **c.** 1.38×10^{25} atoms **d.** 1.29×10^{23} atoms

21.

Element	*Moles*	*Number of Atoms*
Ne	0.552	3.32×10^{23}
Ar	5.40	3.25×10^{24}
Xe	1.78	1.07×10^{24}
He	1.79×10^{-4}	1.08×10^{20}

23. **a.** 72.7 dozen **b.** 6.06 gross **c.** 1.74 reams **d.** 1.45×10^{-21} moles

25. 0.356 mol

27. 28.6g

29. **a.** 2.36×10^{-2} mol **b.** 0.571 mol **c.** 0.476 mol **d.** 4.9×10^{-3} mol

31.

Element	*Moles*	*Mass*
Ne	1.11	22.5 g
Ar	0.117	4.67 g
Xe	7.62	1.00 kg
He	1.44×10^{-4}	5.76×10^{-4}g

33. 8.07×10^{18} atoms

35. 8.44×10^{22} atoms

37. **a.** 8.80×10^{22} atoms **b.** 4.87×10^{23} atoms **c.** 2.84×10^{22} atoms **d.** 7.02×10^{23} atoms

39. 2.6×10^{21} atoms

41. 1.61×10^{25} atoms

43.

Element	*Mass*	*Moles*	*Number of Atoms*
Na	38.5 mg	1.67×10^{-3}	1.01×10^{21}
C	13.5 g	1.12	6.74×10^{23}
V	1.81×10^{-20}g	3.55×10^{-22}	214
Hg	1.44 kg	7.18	4.32×10^{24}

45. **a.** 0.945 mol **b.** 0.903 mol **c.** 109 mol **d.** 8.65×10^{-6} mol

47.

Compound	*Mass*	*Moles*
H_2O	112 kg	6.22×10^{3}
N_2O	6.33 g	0.144
SO_2	156	2.44
CH_2Cl_2	5.46	0.0643

49. 6.20×10^{21} molecules

51. **a.** 1.2×10^{23} molecules
b. 1.21×10^{24} molecules
c. 3.5×10^{23} molecules
d. 6.4×10^{22} molecules

53. 0.10 mg

55. 9.4 mol Cl

57. d, 3 mol O

59. **a.** 3.8 mol C **b.** 0.546 mol C **c.** 19.6 mol C **d.** 183 mol C

61. **a.** 2 moles H per mole of molecules; 8 H atoms present
b. 4 moles H per mole of molecules; 20 H atoms present
c. 3 moles H per mole of molecules; 9 H atoms present

63. **a.** 32 g **b.** 43 g **c.** 31 g **d.** 19 g

65. **a.** 1.4×10^{3} kg **b.** 1.4×10^{3} kg **c.** 2.1×10^{3} kg

67. 84.8% Sr

69. 36.1% Ca; 63.9% Cl

71. 10.7 g

73. 6.6 mg

75. **a.** 63.65% **b.** 46.68% **c.** 30.45% **d.** 25.94%

77. **a.** 39.99% C; 6.73% H; 53.28% O
b. 26.09% C; 4.39% H; 69.52% O
c. 60.93% C; 15.37% H; 23.69% N
d. 54.48% C; 13.74% H; 31.78% N

79. Fe_3O_4, 72.36% Fe; Fe_2O_3, 69.94% Fe; $FeCO_3$, 48.20% Fe; magnetite

81. NO_2

83. **a.** NiI_2 **b.** $SeBr_4$ **c.** $BeSO_4$

85. C_2H_6N

87. **a.** C_3H_6O **b.** $C_5H_{10}O_2$ **c.** $C_9H_{10}O_2$

89. P_2O_3

91. NCl_3

93. C_4H_8

95. **a.** C_6Cl_6 **b.** C_2HCl_3 **c.** $C_6H_3Cl_3$

97. 2.43×10^{23} atoms

99. 2×10^{21} molecules

101.

Substance	*Mass*	*Moles*	*Number of Particles*
Ar	0.018 g	4.5×10^{-4}	2.7×10^{20}
NO_2	8.33×10^{-3} g	1.81×10^{-4}	1.09×10^{20}
K	22.4 mg	5.73×10^{-4}	3.45×10^{20}
C_8H_{18}	3.76 kg	32.9	1.98×10^{25}

103. **a.** CuI_2: 20.03% Cu; 79.97% I
b. $NaNO_3$: 27.05% Na; 16.48% N; 56.47% O
c. $PbSO_4$: 68.32% Pb; 10.57% S; 21.10% O
d. CaF_2: 51.33% Ca; 48.67% F

105. 1.8×10^{3} kg rock

107. 59 kg Cl

109. 1.1×10^{2} g H

111.

Formula	*Molar Mass*	*%C (by mass)*	*%H (by mass)*
C_2H_4	28.06	85.60%	14.40%
C_4H_{10}	58.12	82.66%	17.34%
C_4H_8	56.12	85.60%	14.40%
C_3H_8	44.09	81.71%	18.29

113. $C_4H_6O_2$

115. $C_{10}H_{14}N_2$

117. **a.** 1×10^{57} atoms per star
b. 1×10^{68} atoms per galaxy
c. 1×10^{79} atoms in the universe

119. $C_{16}H_{10}$

Chapter 7

Questions

1. A chemical reaction is the change of one or more substances into different substances, for example burning wood, rusting iron, and protein synthesis.
3. The main evidences of a chemical reaction are a color change, the formation of a solid, the formation of a gas, the emission of light, and the emission or absorption of heat.
5. **a.** gas **b.** liquid **c.** solid **d.** aqueous
7. **a.** reactants: 4 Ag, 2 O, 1 C products: 4 Ag, 2 O, 1 C balanced: yes
   b. reactants: 1 Pb, 2 N, 6 O, 2 Na, 2 Cl products: 1 Pb, 2 N, 6 O, 2 Na, 2 Cl balanced: yes
   c. reactants: 3 C, 8 H, 2 O products: 3 C, 8 H, 10 O balanced: no
9. If a compound dissolves in water than it is soluble. If it does not dissolve in water, it is insoluble.
11. When ionic compounds containing polyatomic ions dissolve in water, the polyatomic ions usually dissolve as intact units.
13. The solubility rules are a set of empirical rules for ionic compounds that were deduced from observations on many compounds. The rules help us determine whether particular compounds will be soluble or insoluble.
15. The precipitate will always be insoluble; it is the solid that forms upon mixing two aqueous solutions.
17. A complete ionic equation shows the reactants and products as they are actually present in solution. The net ionic equation shows only the species that actually change during the reaction. Spectator ions are shown in the complete ionic equation but are not shown in the net ionic equation. An example of each:

    Complete ionic equation:

    $$Na^+(aq) + OH^-(aq) + H^+(aq) + NO_3^-(aq) \longrightarrow H_2O(l) + Na^+(aq) + NO_3^-(aq)$$

    Net ionic equation:

    $$H^+(aq) + OH^-(aq) \longrightarrow H_2O(l)$$
19. An acid is a compound characterized by its sour taste, its ability to dissolve some metals, and its tendency to form H^+ ions in solution. A base is a compound characterized by its bitter taste, its slippery feel, and its tendency to form OH^- ions in solution.
21. A redox reaction is a reaction involving the transfer of electrons. An example of a redox reaction is the rusting of iron: $4\,Fe(s) + 3\,O_2(g) \longrightarrow 2\,Fe_2O_3(s)$
23. You can classify chemical reactions by either 1) the type of chemistry occurring during the reaction, such as acid–base chemistry verses precipitation chemistry, or 2) by studying what happens to atoms during the reaction. We use both methods for classifying reactions so we can study the similarities and differences between reactions.
25. Two examples of synthesis reactions:

    $2\,Na(s) + Cl_2(g) \longrightarrow 2\,NaCl(s)$

    $CaO(s) + CO_2(g) \longrightarrow CaCO_3(s)$
27. Two examples of single-displacement reactions:

    $2\,Na(s) + 2\,HOH(l) \longrightarrow 2\,NaOH(aq) + H_2(g)$

    $Zn(s) + CuCl_2(aq) \longrightarrow ZnCl_2(aq) + Cu(s)$

Problems:

29. **a.** Yes; there is a color change showing a chemical reaction.
    b. No; the state of the compound changes, but no chemical reaction takes place.
    c. Yes; there is a formation of a solid in a previously clear solution.
    d. Yes; there is a formation of a gas when the yeast was added to the solution.
31. Yes; a chemical reaction has occurred; for the presence of the bubbles is evidence for the formation of a gas.
33. Yes; a chemical reaction has occurred. We know this due to the color change of the hair.
35. **a.** $2\,Cu(s) + S(s) \longrightarrow Cu_2S(s)$
    b. $2\,SO_2(g) + O_2(g) \longrightarrow 2\,SO_3(g)$
    c. $4\,HCl(aq) + MnO_2(s) \longrightarrow 2\,H_2O(l) + Cl_2(g) + MnCl_2(aq)$
    d. $2\,C_6H_6(l) + 15\,O_2(g) \longrightarrow 12\,CO_2(g) + 6\,H_2O(l)$
37. **a.** $Mg(s) + 2\,CuNO_3(aq) \longrightarrow 2\,Cu(s) + Mg(NO_3)_2(aq)$
    b. $2\,N_2O_5(g) \longrightarrow 4\,NO_2(g) + O_2(g)$
    c. $Ca(s) + 2\,HNO_3(aq) \longrightarrow H_2(g) + Ca(NO_3)_2(aq)$
    d. $2\,CH_3OH(l) + 3\,O_2(g) \longrightarrow 2\,CO_2(g) + 4\,H_2O(g)$
39. $2\,H_2(g) + O_2(g) \longrightarrow 2\,H_2O(l)$; $Cl(g) + O_3(g) \longrightarrow ClO(g) + O_2(g)$
41. $2\,Na(s) + 2\,H_2O(l) \longrightarrow H_2(g) + 2\,NaOH(aq)$
43. $2\,SO_2(g) + O_2(g) + 2\,H_2O(l) \longrightarrow 2\,H_2SO_4(aq)$
45. $V_2O_5(s) + 2\,H_2(g) \longrightarrow V_2O_3(s) + 2\,H_2O(l)$
47. $C_{12}H_{22}O_{11}(aq) + H_2O(l) \longrightarrow 4\,CO_2(g) + 4\,C_2H_5OH(aq)$
49. **a.** $3\,N_2H_4(l) \longrightarrow 4\,NH_3(g) + N_2(g)$
    b. $3\,H_2(g) + N_2(g) \longrightarrow 2\,NH_3(g)$
    c. $Cu_2O(s) + C(s) \longrightarrow 2\,Cu(s) + CO(g)$
    d. $H_2(g) + Cl_2(g) \longrightarrow 2\,HCl(g)$
51. **a.** $BaO_2(s) + H_2SO_4(aq) \longrightarrow BaSO_4(s) + H_2O_2(aq)$
    b. $2\,Co(NO_3)_3(aq) + 3\,(NH_4)_2S(aq) \longrightarrow Co_2S_3(s) + 6\,NH_4NO_3(aq)$
    c. $Li_2O(s) + H_2O(l) \longrightarrow 2\,LiOH(aq)$
    d. $Hg_2(C_2H_3O_2)_2(aq) + 2\,KCl(aq) \longrightarrow Hg_2Cl_2(s) + 2\,KC_2H_3O_2(aq)$
53. **a.** $2\,Rb(s) + 2\,H_2O(l) \longrightarrow 2\,RbOH(aq) + H_2(g)$
    b. Ok
    c. $2\,NiS(s) + 3\,O_2(g) \longrightarrow 2\,NiO(s) + 2\,SO_2(g)$
    d. $3\,PbO(s) + 2\,NH_3(g) \longrightarrow 3\,Pb(s) + N_2(g) + 3\,H_2O(l)$
55. $C_6H_{12}O_6(aq) + 6\,O_2(g) \longrightarrow 6\,CO_2(g) + 6\,H_2O(l)$
57. $2\,NO(g) + 2\,CO(g) \longrightarrow N_2(g) + 2\,CO_2(g)$
59. **a.** soluble; Na^+ and NO_3^-
    b. soluble; Pb^{2+} and $C_2H_3O_2^-$
    c. insoluble
    d. soluble; NH_4^+ and S^{2-}
61. $AgCl$; $BaSO_4$; $CuCO_3$; Fe_2S_3

63.

Soluble	*Insoluble*
K_2S	Hg_2I_2
BaS	$Cu_3(PO_4)_2$
NH_4Cl	MgS
Na_2CO_3	$CaSO_4$
K_2SO_4	$PbSO_4$
SrS	$PbCl_2$
Li_2S	Hg_2Cl_2

65. **a.** $NH_4Cl(aq) + AgNO_3(aq) \longrightarrow AgCl(s) + NH_4NO_3(aq)$
b. NO REACTION
c. $CrCl_2(aq) + Li_2CO_3(aq) \longrightarrow CrCO_3(s) + 2\,LiCl(aq)$
d. $3\,KOH(aq) + FeCl_3(aq) \longrightarrow Fe(OH)_3(s) + 3\,KCl(aq)$

67. **a.** $Na_2CO_3(aq) + Pb(NO_3)_2(aq) \longrightarrow PbCO_3(s) + 2\,NaNO_3(aq)$
b. $K_2SO_4(aq) + Pb(CH_3CO_2)_2(aq) \longrightarrow PbSO_4(s) + 2\,KCH_3CO_2(aq)$
c. $Cu(NO_3)_2(aq) + BaS(aq) \longrightarrow CuS(s) + Ba(NO_3)_2(aq)$
d. NO REACTION

69. **a.** correct **b.** NO REACTION
c. correct
d. $Pb(NO_3)_2(aq) + 2\,LiCl(aq) \longrightarrow PbCl_2(s) + 2\,LiNO_3(aq)$

71. K^+, $C_2H_3O_2^-$

73. **a.** $Ag^+(aq) + NO_3^-(aq) + K^+(aq) + Cl^-(aq) \longrightarrow AgCl(s) + K^+(aq) + NO_3^-(aq)$
$Ag^+(aq) + Cl^-(aq) \longrightarrow AgCl(s)$
b. $Ca^{2+}(aq) + S^{2-}(aq) + Cu^{2+}(aq) + 2\,Cl^-(aq) \longrightarrow CuS(s) + Ca^{2+}(aq) + 2\,Cl^-(aq)$
$Cu^{2+}(aq) + S^{2-}(aq) \longrightarrow CuS(s)$
c. $Na^+(aq) + OH^-(aq) + H^+(aq) + NO_3^-(aq) \longrightarrow H_2O(l) + Na^+(aq) + NO_3^-(aq)$
$H^+(aq) + OH^-(aq) \longrightarrow H_2O(l)$
d. $6\,K^+(aq) + 2\,PO_4^{3-}(aq) + 3\,Ni^{2+}(aq) + 6\,Cl^-(aq) \longrightarrow Ni_3(PO_4)_2(s) + 6\,K^+(aq) + 6\,Cl^-(aq)$
$3\,Ni^{2+}(aq) + 2\,PO_4^{3-}(aq) \longrightarrow Ni_3(PO_4)_2(s)$

75. $Hg_2^{2+}(aq) + 2\,NO_3^-(aq) + 2\,Na^+(aq) + 2\,Cl^-(aq) \longrightarrow Hg_2Cl_2(s) + 2\,Na^+(aq) + 2\,NO_3^-(aq)$
$Hg_2^{2+}(aq) + 2\,Cl^-(aq) \longrightarrow Hg_2Cl_2(s)$

77. **a.** $2\,Na^+(aq) + CO_3^{2-}(aq) + Pb^{2+}(aq) + 2\,NO_3^-(aq) \longrightarrow PbCO_3(s) + 2\,Na^+(aq) + 2\,NO_3^-(aq)$
$Pb^{2+}(aq) + CO_3^{2-}(aq) \longrightarrow PbCO_3(s)$
b. $2\,K^+(aq) + SO_4^{2-}(aq) + Pb^{2+}(aq) + 2\,CH_3CO_2^-(aq) \longrightarrow PbSO_4(s) + 2\,K^+(aq) + 2\,CH_3CO_2^-(aq)$
$Pb^{2+}(aq) + SO_4^{2-}(aq) \longrightarrow PbSO_4(s)$
c. $Cu^{2+}(aq) + 2\,NO_3^-(aq) + Ba^{2+}(aq) + S^{2-}(aq) \longrightarrow CuS(s) + Ba^{2+}(aq) + 2\,NO_3^-(aq)$
$Cu^{2+}(aq) + S^{2-}(aq) \longrightarrow CuS(s)$
d. NO REACTION

79. $HCl(aq) + KOH(aq) \longrightarrow H_2O(l) + KCl(aq)$
$H^+(aq) + OH^-(aq) \longrightarrow H_2O(l)$

81. **a.** $2\,HCl(aq) + Ba(OH)_2(aq) \longrightarrow 2\,H_2O(l) + BaCl_2(aq)$
b. $H_2SO_4(aq) + 2\,KOH(aq) \longrightarrow 2\,H_2O(l) + K_2SO_4(aq)$
c. $HClO_4(aq) + NaOH(aq) \longrightarrow H_2O(l) + NaClO_4(aq)$

83. **a.** $HBr(aq) + NaHCO_3(aq) \longrightarrow H_2O(l) + CO_2(g) + NaBr(aq)$
b. $NH_4I(aq) + KOH(aq) \longrightarrow H_2O(l) + NH_3(g) + KI(aq)$
c. $2\,HNO_3(aq) + K_2SO_3(aq) \longrightarrow H_2O(l) + SO_2(g) + 2\,KNO_3(aq)$
d. $2\,HI(aq) + Li_2S(aq) \longrightarrow H_2S(g) + 2\,LiI(aq)$

85. b and d are redox reactions; a and c are not.

87. **a.** $2\,C_2H_6(g) + 7\,O_2(g) \longrightarrow 4\,CO_2(g) + 6\,H_2O(g)$
b. $2\,Ca(s) + O_2(g) \longrightarrow 2\,CaO(s)$
c. $2\,C_3H_8O(l) + 9\,O_2(g) \longrightarrow 6\,CO_2(g) + 8\,H_2O(g)$
d. $2\,C_4H_{10}S(l) + 15\,O_2(g) \longrightarrow 8\,CO_2(g) + 10\,H_2O(g) + 2\,SO_2(g)$

89. **a.** $2\,K(s) + Cl_2(g) \longrightarrow 2\,KCl(s)$
b. $Ca(s) + Cl_2(g) \longrightarrow CaCl_2(s)$
c. $2\,Al(s) + 3\,Cl_2(g) \longrightarrow 2\,AlCl_3(s)$
d. $Zn(s) + Cl_2(g) \longrightarrow ZnCl_2(s)$

91. **a.** double displacement
b. synthesis or combination
c. single displacement
d. decomposition

93. **a.** synthesis
b. decomposition
c. synthesis

95. **a.** $2\,Na^+(aq) + 2\,I^-(aq) + Hg_2^+(aq) + 2\,NO_3^-(aq) \longrightarrow Hg_2I_2(s) + 2\,Na^+(aq) + 2\,NO_3^-(aq)$
$Hg_2^+(aq) + 2\,I^-(aq) \longrightarrow Hg_2I_2(s)$
b. $2\,H^+(aq) + ClO_4^-(aq) + Ba^{2+}(aq) + 2\,OH^-(aq) \longrightarrow 2\,H_2O(l) + Ba^{2+}(aq) + ClO_4^-(aq)$
$H^+(aq) + OH^-(aq) \longrightarrow H_2O(s)$
c. NO REACTION
d. $2\,H^+(aq) + 2\,Cl^-(aq) + 2\,Li^+(aq) + CO_3^{2-}(aq) \longrightarrow H_2O(l) + CO_2(g) + 2\,Li^+(aq) + 2\,Cl^-(aq)$
$2\,H^+(aq) + CO_3^{2-}(aq) \longrightarrow H_2O(l) + CO_2(g)$

97. **a.** NO REACTION
b. NO REACTION
c. $K^+(aq) + HSO_3^-(aq) + H^+(aq) + NO_3^-(aq) \longrightarrow H_2O(l) + SO_2(g) + K^+(aq) + NO_3^-(aq)$
$H^+(aq) + HSO_3^-(aq) \longrightarrow H_2O(l) + SO_2(g)$
d. $Mn^{3+}(aq) + 3\,Cl^-(aq) + 3\,K^+(aq) + PO_4^{3-}(aq) \longrightarrow MnPO_4(s) + 3\,K^+(aq) + 3\,Cl^-(aq)$
$Mn^{3+}(aq) + PO_4^{3-}(aq) \longrightarrow MnPO_4(s)$

99. **a.** acid–base; $KOH(aq) + HC_2H_3O_2(aq) \longrightarrow H_2O(l) + KC_2H_3O_2(aq)$
b. gas evolution; $2\,HBr(aq) + K_2CO_3(aq) \longrightarrow H_2O(l) + CO_2(g) + 2\,KBr(aq)$
c. synthesis; $2\,H_2(g) + O_2(g) \longrightarrow 2\,H_2O(l)$
d. precipitation; $2\,NH_4Cl(aq) + Pb(NO_3)_2(aq) \longrightarrow PbCl_2(s) + 2\,NH_4NO_3(aq)$

101. **a.** oxidation–reduction; single displacement
b. gas evolution; acid–base
c. gas evolution; double displacement
d. precipitation; double displacement

103. $3\,CaCl_2(aq) + 2\,Na_3PO_4(aq) \longrightarrow Ca_3(PO_4)_2(s) + 6\,NaCl(aq)$

$3\,Ca^{2+}(aq) + 6\,Cl^-(aq) + 6\,Na^+(aq) + 2\,PO_4^{3-}(aq) \longrightarrow Ca_3(PO_4)_2(s) + 6\,Na^+(aq) + 6\,Cl^-(aq)$

$3\,Ca^{2+}(aq) + 2\,PO_4^{3-}(aq) \longrightarrow Ca_3(PO_4)_2(s)$

$3\,Mg(NO_3)_2(aq) + 2\,Na_3PO_4(aq) \longrightarrow Mg_3(PO_4)_2(s) + 6\,NaNO_3(aq)$

$3\,Mg^{2+}(aq) + 6\,NO_3^-(aq) + 6\,Na^+(aq) + 2\,PO_4^{3-}(aq) \longrightarrow Mg_3(PO_4)_2(s) + 6\,Na^+(aq) + 6\,NO_3^-(aq)$

$3\,Mg^{2+}(aq) + 2\,PO_4^{3-}(aq) \longrightarrow Mg_3(PO_4)_2(s)$

105. *Correct answers may vary, representative correct answers are:

a. addition of a solution containing SO_4^{2-}; $Pb^{2+}(aq) + SO_4^{2-}(aq) \longrightarrow PbSO_4(s)$

b. addition of a solution containing SO_4^{2-}; $Ca^{2+}(aq) + SO_4^{2-}(aq) \longrightarrow CaSO_4(s)$

c. addition of a solution containing SO_4^{2-}; $Ba^{2+}(aq) + SO_4^{2-}(aq) \longrightarrow BaSO_4(s)$

d. addition of a solution containing Cl^-; $Hg_2^{2+}(aq) + 2\,Cl^-(aq) \longrightarrow Hg_2Cl_2(s)$

107. Ca^{2+} and Cu^{2+} were present in the original solution.

1st: $Ca^{2+}(aq) + SO_4^{2-}(aq) \longrightarrow CaSO_4(s)$

2nd: $Cu^{2+}(aq) + CO_3^{2-}(aq) \longrightarrow CuCO_3(s)$

109. **a.** chemical; **b.** physical

Chapter 8

Questions

1. Reaction stoichiometry is very important to chemistry. It gives us a numerical relationship between the reactants and products that allows chemists to plan and carry out chemical reactions to obtain products in the desired quantities,

e.g., How much CO_2 is produced when a given amount of C_8H_{10} is burned?

How much $H_2(g)$ is produced when a given amount of water decomposes?

3. $1\text{ mol } N_2 \equiv 2\text{ mol } NH_3$

$3\text{ mol } H_2 \equiv 2\text{ mol } NH_3$

5. $1\text{ mol } Cl_2 \equiv 2\text{ mol } NaCl$

7. mass A $\longrightarrow$ moles A $\longrightarrow$ moles B $\longrightarrow$ mass B (A = reactant, B = product)

9. The limiting reactant is the reactant that limits the amount of product in a chemical reaction.

11. The actual yield is the amount of product actually produced by a chemical reaction. The percent yield is the percentage of the theoretical yield that was actually attained.

13. d

Problems

15. **a.** 1 mol C **b.** 0.5 mol C **c.** 2 mol C **d.** 1 mol C

17. **a.** 2.6 mol NO_2 **b.** 11.6 mol NO_2 **c.** 8.90×10^3 mol NO_2 **d.** 2.012×10^{-3} mol NO_2

19. **a.** 2.96 mol HCl **b.** 2.96 mol H_2O **c.** 0.740 mol Na_2O_2 **d.** 0.987 mol SO_3

21. **a.** 1.2 mol PbO(*s*), 1.2 mol $SO_2(g)$ **b.** 0.80 mol PbO(*s*), 0.80 mol $SO_2(g)$ **c.** 7.9 mol PbO(*s*), 7.9 mol $SO_2(g)$ **d.** 5.3 mol PbO(*s*), 5.3 mol $SO_2(g)$

23.

mol N_2H_4	*mol* N_2O_4	*mol* N_2	*mol* H_2O
4	2	6	8
6	3	9	12
4	2	6	8
11	5.5	16.5	22
3	1.5	4.5	6
8.26	4.13	12.4	16.5

25. $2\,C_4H_{10}(g) + 13\,O_2(g) \longrightarrow 8\,CO_2(g) + 10\,H_2O(g)$; 32 mol O_2

27. **a.** $Pb(s) + 2\,AgNO_3(aq) \longrightarrow Pb(NO_3)_2(aq) + 2\,Ag(s)$ **b.** 19 mol $AgNO_3$ **c.** 56.8 mol Ag

29. **a.** 0.213 g O_2 **b.** 0.385 g O_2 **c.** 96 g O_2 **d.** 3.4×10^{-4} g O_2

31. **a.** 3.0 g NaCl **b.** 3.2 g $CaCO_3$ **c.** 3.0 g MgO **d.** 2.3 g NaOH

33. **a.** 8.9 g Al_2O_3, 9.7 g Fe **b.** 3.0 g Al_2O_3, 3.3 g Fe

35.

Mass CH_4	*Mass* O_2	*Mass* CO_2	*Mass* H_2O
0.645 g	2.57 g	1.77 g	1.45 g
22.32 g	89.00 g	61.20 g	50.09 g
5.044 g	20.11 g	13.83 g	11.32 g
1.07 g	4.28 g	2.94 g	2.41 g
3.18 kg	12.7 kg	8.72 kg	7.14 kg
8.57×10^2 kg	3.42×10^3 kg	2.35×10^3 kg	1.92×10^3 kg

37. **a.** 2.3 g HCl **b.** 4.3 g HNO_3 **c.** 2.2 g H_2SO_4

39. 123 g H_2SO_4, 2.53 g H_2

41. **a.** 2 mol A **b.** 1.8 mol A **c.** 4 mol B **d.** 40 mol B

43. **a.** 1.5 mol C **b.** 3 mol C **c.** 3 mol C **d.** 96 mol C

45. **a.** 1 mol K **b.** 1.8 mol K **c.** 1 mol Cl_2 **d.** 14.6 mol K

47. **a.** 1.3 mol MnO_3 **b.** 4.8 mol MnO_3 **c.** 0.107 mol MnO_3 **d.** 27.5 mol MnO_3

49. **a.** 2 Cl_2 **b.** 3 Cl_2 **c.** 2 Cl_2

51. **a.** 1.0 g F_2 **b.** 10.5 g Li **c.** 6.79×10^3 g F_2

53. **a.** 1.3 g $AlCl_3$ **b.** 24.8 g $AlCl_3$ **c.** 2.17 g $AlCl_3$

55. 74.6%

57. CaO; 25.7 g $CaCO_3$; 75.4%

59. O_2; 5.07 g NiO; 95.9%

61. Pb^{2+}; 262.7 g $PbCl_2$; 96.09%

63. 0.152 g Ba^{2+}

65. 1.5 g HCl

67. 3.1 kg CO_2

69. 4.7 g Na_3PO_4
71. 469 g Zn
73. $2\ NH_4NO_3(s) \longrightarrow 2\ N_2(g) + O_2(g) + 4\ H_2O(l)$; 2.00×10^2 g O_2
75. salicylic acid ($C_7H_6O_3$); 2.71 g $C_9H_8O_4$; 74.1%
77. NH_3; 120 kg CH_4N_2O; 72.9%
79. 2.4 mg $C_4H_6O_4S_2$
81. b; the loudest explosion will occur when the ratio is 2 hydrogen to 1 oxygen, for that is the ratio that occurs in water.
83. 2.2×10^{13} kg CO_2 per year; 1.4×10^2 years

Chapter 9

Questions

1. Both the Bohr model and the quantum mechanical model for the atom were developed in the early 1900s. These models serve to explain how electrons are arranged within the atomic structure and how the electrons affect the chemical and physical properties of each element.
3. Light travels at a speed of 3.0×10^8 m/s. The speed of light is extremely fast, for a flash of light from the equator could circle our earth 7 times in one second.
5. A blue object appears blue because when white light is present the object absorbs all wavelengths of light except blue, which it reflects.
7. Wavelength and frequency are inversely related; the longer the wavelength, the lower the frequency, and vice versa.
9. X-rays pass through many substances that block visible light and are therefore used to image bones and organs.
11. Ultraviolet light contains enough energy to damage biological molecules, and excessive exposure increases the risk of skin cancer and cataracts.
13. Microwaves can only heat things containing water, and therefore the food, which contains water, becomes hot, but the plate does not.
15. The Bohr model is a representation for the atom in which electrons travel around the nucleus in circular orbits with a fixed energy at specific, fixed distances from the nucleus.
17. The Bohr orbit describes the path of an electron as an orbit or trajectory (a specified path). A quantum mechanical orbital describes the path of an electron using a probability map.
19. The e^- has wave particle duality, which means the path of an electron is not predictable. The motion of a baseball is predictable. A probability map shows a statistical, reproducible pattern of where the electron is located.
21. The subshells are *s* (1 orbital which contains a maximum of 2 electrons); *p* (3 orbitals which contain a maximum of 6 electrons); *d* (5 orbitals which contain a maximum of 10 electrons); and *f* (7 orbitals which contain a maximum of 14 electrons).
23. The Pauli exclusion principle states that separate orbitals may hold no more than 2 electrons, and when 2 electrons are present in a single orbital, they must have opposite spins. When writing electron configurations, the principle means that no box can have more than 2 arrows, and the arrows will point in opposite directions.
25. [Ne] represents $1s^22s^22p^6$
[Kr] represents $1s^22s^22p^63s^23p^64s^23d^{10}4p^6$
27.

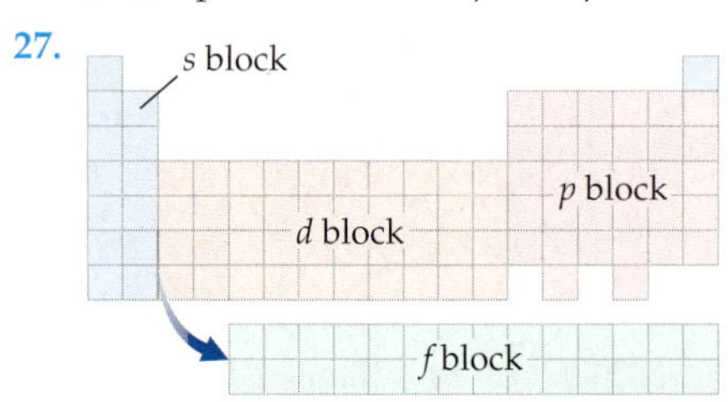

29. Group 1 elements form 1+ ions because they lose one valence electron in the outer *s* shell to obtain a noble gas configuration. Group 7 elements form 1− ions because they gain an electron to fill their outer *p* orbital to obtain a noble gas configuration.

Problems

31. infrared
33. radiowaves < microwaves < infrared < ultraviolet
35. gamma, ultraviolet, or X-rays
37. **a.** microwaves < X-rays < gamma rays
b. microwaves < X-rays < gamma rays
c. gamma rays < X-ray < microwaves
39. energies, distances
41. $n = 6 \longrightarrow n = 2$: 410 nm
$n = 5 \longrightarrow n = 2$: 434 nm
43. z, y, x — 1s; 2p

The 2*s* and 3*p* orbitals are bigger than the 1*s* and 2*p* orbitals.

45. Electron in the 2*s* orbital
47. $2p \longrightarrow 1s$
49. **a.** $1s^22s^22p^2$ **b.** $1s^22s^22p^63s^1$
c. $1s^22s^22p^63s^23p^6$ **d.** $1s^22s^22p^63s^23p^2$
51. **a.** Be [↑↓] [↑↓]
1s 2s

b. C

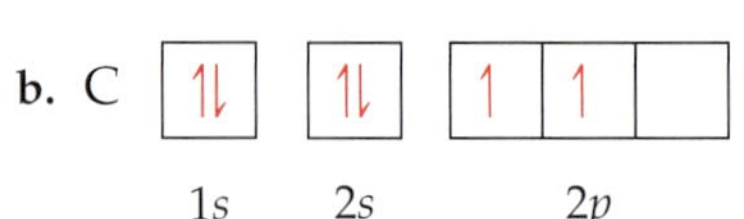

c. F [↑↓] [↑↓] [↑↓|↑↓|↑]
1s 2s 2p

d. Ne [↑↓] [↑↓] [↑↓|↑↓|↑↓]
1s 2s 2p

53. **a.** [Ar] $4s^2 3d^{10} 4p^1$
b. [Ar] $4s^2 3d^{10} 4p^3$
c. [Kr] $5s^1$
d. [Kr] $5s^2 4d^{10} 5p^2$

55. **a.** [Ar] $4s^2 3d^2$
b. [Ar] $4s^2 3d^3$
c. [Ar] $4s^1 3d^5$
d. [Ar] $4s^2 3d^5$

57. Valence electrons are underlined
a. $1s^2 \underline{2s^2 2p^1}$
b. $1s^2 \underline{2s^2 2p^3}$
c. $1s^2 2s^2 2p^6 3s^2 3p^6 4s^2 3d^{10} 4p^6 \underline{5s^2} 4d^{10} \underline{5p^3}$
d. $1s^2 2s^2 2p^6 3s^2 3p^6 \underline{4s^1}$

59. **a.** Si [↑↓] [↑][↑][ ] 2 unpaired electrons
$3s$ $3p$

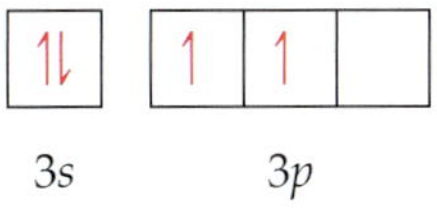

b. Ca [↑↓] 0 unpaired electrons
$4s$

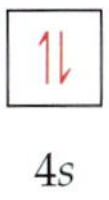

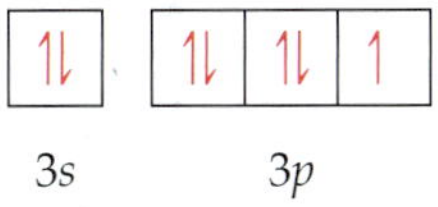

c. Cl [↑↓] [↑↓][↑↓][↑] 1 unpaired electron
$3s$ $3p$

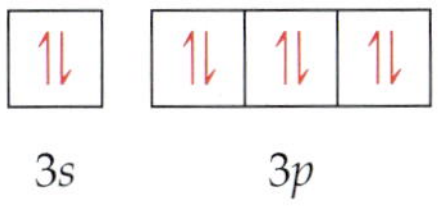

d. Ar [↑↓] [↑↓][↑↓][↑↓] 0 unpaired electrons
$3s$ $3p$

61. **a.** 6 **b.** 6
c. 7 **d.** 1

63. **a.** ns^1 **b.** ns^2
c. $ns^2 np^3$ **d.** $ns^2 np^5$

65. **a.** [Ne] $3s^2 3p^1$ **b.** [He] $2s^2$
c. [Kr] $5s^2 4d^{10} 5p^1$ **d.** [Kr] $5s^2 4d^2$

67. **a.** [Ar] $4s^2 3d^{10} 4p^3$ **b.** [Xe] $6s^2$
c. [Ar] $4s^2 3d^8$ **d.** [Xe] $6s^2 4f^{14} 5d^{10} 6p^3$

69. **a.** 2 **b.** 3
c. 5 **d.** 6

71. Period 1 has two elements. Period 2 has eight elements. The number of subshells is equal to the principle quantum number. For Period 1, $n = 1$ and the s subshell contains only two elements. For Period 2, $n = 2$ and contains s and p subshells that have a total of 8 elements.

73. **a.** Al **b.** S
c. Ar **d.** Mg

75. **a.** Cl **b.** Ga
c. Fe **d.** Rb

77. **a.** Na **b.** Ge
c. cannot tell **d.** P

79. Pb < Sn < Te < S < Cl

81. **a.** In **b.** Si
c. Pb **d.** C

83. F < S < Si < Ge < Ca < Rb

85. **a.** Sr **b.** Bi
c. cannot tell **d.** As

87. S < Se < Sb < In < Ba < Fr

89. 18 e^-

91. Alkali metals have the general electron configuration of ns^1. If they lose their one s electron, they will obtain the electron configuration of a noble gas. This loss of an electron will give the alkali metal a +1 charge.

93. **a.** $1s^2 2s^2 2p^6 3s^2 3p^6$
b. $1s^2 2s^2 2p^6 3s^2 3p^6$
c. $1s^2 2s^2 2p^6 3s^2 3p^6$
d. $1s^2 2s^2 2p^6 3s^2 3p^6 4s^2 3d^{10} 4p^6$

95. Metals tend to form positive ions because they tend to lose electrons. Elements on the left side of the periodic table have only a few extra electrons, which they will lose to gain a noble gas configuration. Metalloids tend to be elements with 3 to 5 valence electrons; they could lose or gain electrons to obtain a noble gas configuration. Nonmetals tend to gain electrons to fill their almost full valence shell, so they tend to form negative ions and are on the right side of the table.

97. **a.** Can only have 2 in the s shell and 6 in the p shell: $1s^2 2s^2 2p^6 3s^2 3p^3$.
b. There is no $2d$ subshell: $1s^2 2s^2 2p^6 3s^2 3p^2$.
c. There is no $1p$ subshell: $1s^2 2s^2 2p^3$.
d. Can only have 6 in the p shell: $1s^2 2s^2 2p^6 3s^2 3p^3$.

99. Bromine is highly reactive because it reacts quickly to gain an electron and obtain a stable valence shell. Krypton is a noble gas because it already has a stable valence shell.

101. K

103. 660 nm

105. 8 min, 19 sec

107. The quantum-mechanical model provided the ability to understand and predict chemical bonding, which is the basic level of understanding of matter and how it interacts. This model was critical in the areas of lasers, computers, semiconductors, and drug design. The quantum-mechanical model for the atom is considered the foundation of modern chemistry.

109. **a.** 1.5×10^{-34} m
b. 1.88×10^{-10} m
electrons have particle-wave duality whereas golf balls do not.

111. Ultraviolet light is the only one of these three types of light that contains enough energy to break chemical bonds in biological molecules.

Chapter 10

Questions

1. Bonding theories predict how atoms bond together to form molecules and also predict what combinations of atoms form molecules and what combinations do not. Likewise, bonding theories explain the shapes of molecules, which in turn determine many of their physical and chemical properties.

3. Ne: $1s^2 2s^2 2p^6$, 8 valence electrons
Ar: $1s^2 2s^2 2p^6 3s^2 3p^6$, 8 valence electrons

5. According to Lewis theory, a chemical bond is the sharing or transfer of electrons to attain stable electron configurations among the bonding atoms.

7. The Lewis structure for potassium has 1 valence electron while the Lewis structure for monatomic chlorine has 7 valence electrons. From these structures we can determine that if potassium gives up its one valence electron to chlorine, K^+ and Cl^- are formed; therefore the formula must be KCl.

9. Double and triple bonds are shorter and stronger than single bonds.

11. You determine the number of electrons that go into the Lewis structure of a molecule by summing the valence electrons of each atom in the molecule.

13. The octet rule is not sophisticated enough to be correct every time. For example, some molecules that exist in nature have an odd number of valence electrons, and thus will not have octets on all their constituent atoms. Some elements tend to form compounds in nature in which they have more (sulfur) or less (boron) than 8 valence electrons.

15. VSEPR theory predicts the shape of molecules using the idea that electron groups repel each other.

17. **a.** 180 ° **b.** 120 ° **c.** 109.5 °

19. Electronegativity is the ability of an element to attract electrons within a covalent bond.

21. A polar covalent bond is a covalent bond that has a dipole moment.

23. It is important to know if a molecule is polar or nonpolar in order to determine how it will react with other molecules.

Problems

25. **a.** $1s^22s^22p^3$, ·N̈:

b. $1s^22s^22p^2$, ·Ċ·

c. $1s^22s^22p^63s^23p^5$, :C̤̈l·

d. $1s^22s^22p^63s^23p^6$, :Ä̤r:

27. **a.** K·

b. Ȧl:

c. ·P̈:

d. :Ä̤r:

29. :Ẍ̤: Halogens tend to gain one electron in a chemical reaction.

31. M: Alkaline earth metals tend to lose two electrons in a chemical reaction.

33. **a.** $[:\ddot{Cl}:]^-$ **b.** $[:\ddot{Se}:]^{2-}$

c. Na^+ **d.** Mg^{2+}

35. **a.** Kr **b.** Ne

c. Kr **d.** Xe

37. **a.** Ionic **b.** Covalent

c. Ionic **d.** Covalent

39. **a.** $Na^+[:\ddot{F}:]^-$ **b.** $Ca^{2+}[:\ddot{O}:]^{2-}$

c. $[:\ddot{Br}:]^-Sr^{2+}[:\ddot{Br}:]^-$ **d.** $K^+[:\ddot{O}:]^{2-}K^+$

41. **a.** SrSe **b.** $BaCl_2$

c. Na_2S **d.** Al_2O_3

43. **a.** $Li^+[:\ddot{F}:]^-$ **b.** $Li^+[:\ddot{O}:]^{2-}Li^+$

c. $Li^+[:\ddot{N}:]^{3-}Li^+$
Li^+

45. **a.** $Cs^+[:\ddot{Cl}:]^-$

b. $Ba^{2+}[:\ddot{O}:]^{2-}$

c. $[:\ddot{I}:]^-Ca^{2+}[:\ddot{I}:]^-$

47. **a.** 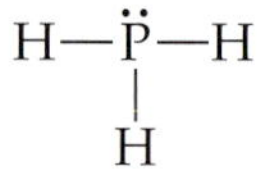 **b.** :C̤̈l—S̤̈—C̤̈l:

c. :F̤̈—F̤̈: **d.** H—Ï̤:

49. **a.** Ö=Ö **b.** :C≡O:

c. H—Ö̤—N̈—Ö̤: **d.** :Ö=S̈—Ö̤:

51. **a.** H—C≡C—H **b.** 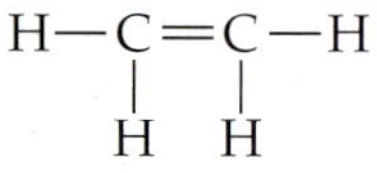

c. H—N̈=N̈—H **d.** 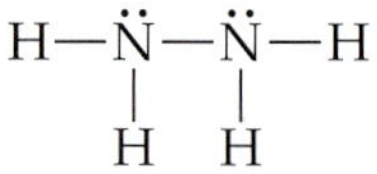

53. **a.** :N≡N:

b. S̤̈=Si=S̤̈

c. H—Ö̤—H

d. 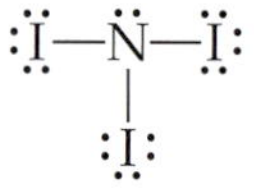

55. **a.** Ö=S̈e—Ö̤: ⟷ :Ö̤—S̈e=Ö

b.

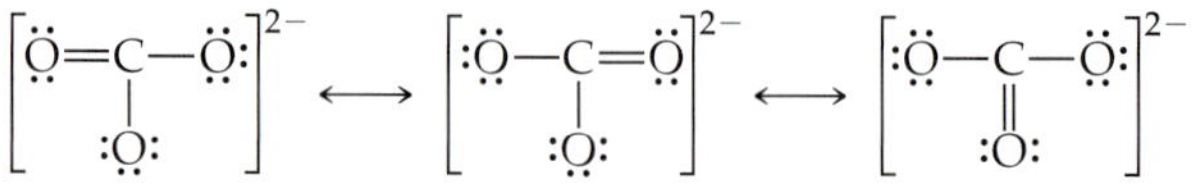

c. $[:\ddot{Cl}—\ddot{O}:]^-$

d. $[:\ddot{O}—\ddot{Cl}—\ddot{O}:]^-$

57. **a.** 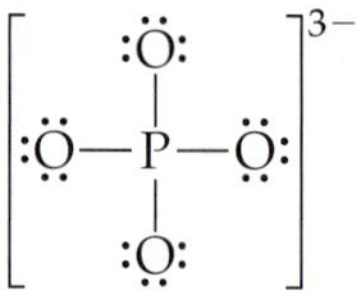

b. [:C≡N:]

c. $[:\ddot{O}=\ddot{N}—\ddot{O}:]^-$ ⟷ $[:\ddot{O}—\ddot{N}=\ddot{O}:]^-$

d. 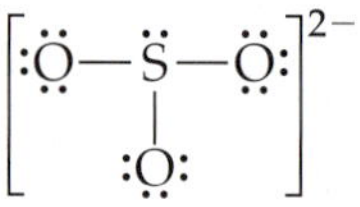

59. **a.** :Cl—B—Cl: with :Cl: bonded to B

b. O=N—O: ⟷ :O—N=O

c. H—B—H with H bonded to B

61. **a.** 4 **b.** 4 **c.** 4 **d.** 2

63. **a.** 3 bonding groups, 1 lone pair
b. 2 bonding groups, 2 lone pairs
c. 4 bonding groups, 0 lone pairs
d. 2 bonding groups, 0 lone pairs

65. **a.** tetrahedral
b. trigonal planar
c. linear
d. trigonal planar

67. **a.** 109.5 ° **b.** 120 ° **c.** 180 ° **d.** 120 °

69. **a.** linear, linear
b. trigonal planar, bent
c. tetrahedral, bent
d. tetrahedral, trigonal pyramidal

71. **a.** 180 °
b. 120 °
c. 109.5 °
d. 109.5 °

73. **a.** linear, linear
b. trigonal planar, bent
c. tetrahedral, trigonal pyramidal

75. **a.** trigonal planar **b.** bent **c.** trigonal planar **d.** tetrahedral

77. **a.** 1.2
b. 1.8
c. 2.8

79. Rb < Ca < Ga < Si < Cl

81. **a.** polar covalent **b.** ionic **c.** pure covalent **d.** polar covalent

83. H_2 < ICl < HBr < CO

85. **a.** polar **b.** nonpolar **c.** nonpolar **d.** polar

87. **a.** (+):C≡O:(−)
b. nonpolar
c. nonpolar
d. (+) H—Br:(−)

89. **a.** nonpolar **b.** polar **c.** nonpolar **d.** polar

91. **a.** nonpolar **b.** polar **c.** nonpolar **d.** polar

93. **a.** $1s^22s^22p^63s^23p^6\underline{4s^2}$, Ca: (underlined electrons are the ones included)
b. $1s^22s^22p^33s^23p^6\underline{4s^23d^{10}4p^1}$, Ga:
c. [Ar] $\underline{4s^23d^{10}4p^3}$, ·As: **d.** [Kr] $\underline{5s^24d^{10}5p^5}$, :I:

95. **a.** ionic, $K^+[:S:]^{2-}K^+$ **b.** covalent, H—C(=O)—F:
c. ionic, $Mg^{2+}[:Se:]^{2-}$ **d.** covalent, :Br—P(—Br:)—Br:

97. :Cl—C(=O)—Cl: polar,

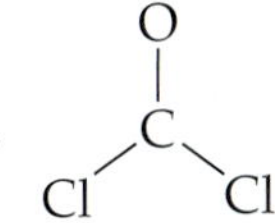

99. H—C(H)(H)—C(=O)—OH H····C(H)(H)—C(=O)OH

101.

H:Cl: + $Na^+[:O:H]^-$ ⟶ H:O:H + $Na^+[:Cl:]^-$

103. K·, :Cl—Cl:, $K^+[:Cl:]^-$, Cl reduced, K oxidized

105. **a.** $K^+[:O:H]^-$
b.

$K^+[:O=N:O:]^-$ ⟷ $K^+[:O:N:O:]^-$ ⟷ $K^+[:O:N=O:]^-$

c. $Li^+[:I:O:]^-$
d.

$Ba^{2+}[CO_3]^{2-}$ ⟷ $Ba^{2+}[CO_3]^{2-}$ ⟷ $Ba^{2+}[CO_3]^{2-}$

107. **a.** PF_5 **b.** SF_4 **c.** SeF_4

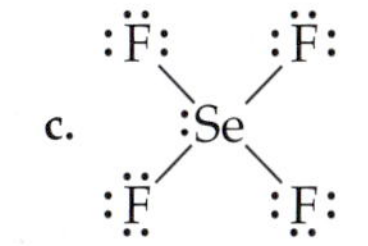

109. CH_2O_2 or H—C(=O)—O—H

111. **a.** $[:O—O:]^-$ **b.** $[:O·]^-$
c. ·O—H **d.** H—C(H)(H)—O—O·

113. **a.** The structure has 2 bonding electron pairs and 2 lone pairs. The Lewis structure is analogous to that of water, and the molecular geometry is bent. **b.** Correct **c.** The structure has 3 bonding electron pairs and 1 lone pair. The Lewis structure is analogous to that of NH_3, and the geometry is trigonal pyramidal.

Chapter 11

Questions

1. Pressure is the push (or force) exerted per unit area by gaseous molecules as they collide with the surfaces around them.
3. The longest straw that could work correctly is about 10.3 m long.
5. Gases are compressible due to the large amount of empty space between their molecules. Gases assume the shape and volume of their container because the atoms or molecules are in constant motion. Gases have low densities because they are mostly empty space.
7. The main units used to measure pressure are Pa, atm, mm Hg, torr.
9. Boyle's law from the perspective of kinetic molecular theory states that if the volume of a gas container is decreased, the same number of gas particles is crowded into a smaller volume, causing more collisions with the walls of the container and therefore increasing the pressure.
11. When an individual is more than a couple of meters underwater, the air pressure in the lungs is greater than the air pressure at the water's surface. If a snorkel were used, it would move the air from the lungs to the surface, making it very difficult to breath.
13. Charles's law from the perspective of kinetic molecular theory states that if the temperature of a gas is increased, the particles move faster and will collectively occupy more space.
15. The combined gas law is $\frac{P_1V_1}{T_1} = \frac{P_2V_2}{T_2}$. It is useful when more than one variable of a gas changes at the same time.
17. Avogadro's law from the perspective of kinetic molecular theory states that if the numbers of particles increase, they occupy more volume.
19. The ideal gas law is most accurate when the volume of gas particles is small compared to the space between them. It is also accurate when the forces between particles are not important. The ideal gas law breaks down at high pressures and low temperatures. This breakdown occurs because the gases are no longer acting according to the kinetic molecular theory.
21. Dalton's law states that the sum of the partial pressures in a gas mixture must equal the total pressure. $P_{tot} = P_A + P_B + P_C + \cdots$
23. Deep-sea divers breathe helium with oxygen because helium, unlike nitrogen, does not have physiological effects under high-pressure conditions. The oxygen concentration in the mixture is low to avoid oxygen toxicity.
25. Vapor pressure is the partial pressure of a gas in with its liquid. Partial pressure increases with increasing temperature.

Problems

27. **a.** 1.16 atm **b.** 1.33 atm **c.** 1.005 atm **d.** 0.971 atm

29. **a.** 1.7×10^3 torr **b.** 36 mm Hg **c.** 1.28×10^3 mm Hg **d.** 834.2 torr

31.

Pascals	*Atmospheres*	*mm Hg*	*Torr*	*PSI*
882	0.00871	6.62	6.62	0.128
5.65×10^4	0.558	424	424	8.20
1.71×10^5	1.69	1.28×10^3	1.28×10^3	24.8
1.02×10^5	1.01	764	764	14.8
3.32×10^4	0.328	249	249	4.82

33. **a.** 0.832 atm **b.** 632 mm Hg **c.** 12.2 psi **d.** 8.43×10^4 Pa

35. **a.** 809.0 mm Hg **b.** 1.065 atm **c.** 809.0 torr **d.** 107.9 kPa

37. 5.7×10^2 mm Hg

39. 1.8 L

41.

P_1	V_1	P_2	V_2
755 mm Hg	2.85 L	885 mm Hg	2.43 L
9.35 atm	1.33 L	4.32 atm	2.88 L
192 mm Hg	382 mL	152 mm Hg	482 mm Hg
2.11 atm	226 mL	3.82 atm	125

43. 4.8 L

45. 58.9 mL

47.

V_1	T_1	V_2	T_2
1.08 L	25.4 °C	1.33 L	94.5 °C
58.9 mL	77 K	228 mL	298 K
115 cm^3	12.5 °C	119 cm^3	22.4 °C
232 L	18.5 °C	294 L	96.2 °C

49. 5.7 L

51. 4.33 L

53.

V_1	n_1	V_2	n_2
38.5 mL	1.55×10^{-3}	49.4 mL	1.99×10^{-3}
8.03 L	1.37	26.8 L	4.57
11.2 L	0.628	15.7 L	0.881
422 mL	0.0109	671 mL	0.0174

55. 1.95×10^3 mm Hg

57. 0.76 L

59. 877 mm Hg

61.

P_1	V_1	T_1	P_2	V_2	T_2
121 atm	1.58 L	12.2 °C	1.54 torr	1.33 L	32.3 °C
721 torr	141 mL	135 K	801 torr	152	162 K
5.51 atm	0.879 L	22.1 °C	4.87 atm	1.05 L	38.3 °C

63. 3.0 L

65. 2.1 mol

67. 1.42 mol

69.

P	V	n	T
1.05 atm	1.19 L	0.112 mol	136 K
112 torr	40.8 L	0.241 mol	304 K
1.50 atm	28.5 mL	1.74×10^{-3} mol	25.4 °C
0.559 atm	0.439 L	0.0117 mol	255 K

71. 0.23 mol

73. 44.0 g/mol

75. 4.00 g/mol

77. 712 torr

79. 10.7 atm

81. 7.00×10^2 mm Hg
83. 0.87 atm N; 0.25 atm O
85. 0.34 atm
87. a. 56 L b. 1.3×10^2 L
c. 732 L d. 9.2×10^2 L
89. a. 59.1 L b. 30.0 L
c. 72.1 L d. 0.125 L
91. a. 0.350 g b. 0.221 g
c. 8.15 g
93. 42 L
95. 33 L H_2; 16 L CO
97. 8.83 L
99. 12.6 g
101. 0.5611 g
103. $V = \frac{nRT}{P}$

$$= \frac{1.00\ \text{mol}\left(0.0821 \frac{\text{L}\cdot\text{atm}}{\text{mol}\cdot\text{K}}\right)(273\ \text{K})}{1.00\ \text{atm}} = 22.4\ \text{L}$$

105. 27.9 g/mol
107. C_4H_{10}
109. 0.828 g
111. 0.128 g
113. 0.935 L
115. $HCl + NaHCO_3 \longrightarrow CO_2 + H_2O + NaCl$; 1.11×10^{-3} mol
117. a. SO_2, 0.0127 mol b. 65.6%
119. a. NO_2, 24.0 g b. 61.7%
121. 11.7 L
123. c. From the ideal gas law, we see that pressure is directly proportional to the number of moles of gas per unit volume (n/V). The gas in (c) contains the greatest concentration of particles, and thus has the highest pressure.
125. 22.8 g
127. $V_2 = 0.76$ L, actual volume is 0.61 L. Difference is due to the fact that the ideal gas law is not ideal, especially at low temperatures.

Chapter 12

Questions

1. The bitter taste is due to the interaction of molecules with receptors on the surface of specialized cells on the tongue.
3. Intermolecular forces are what living organisms depend on for many physiological processes. Intermolecular forces are also responsible for the existence of liquids and solids.
5. The magnitude of intermolecular forces relative to the amount of thermal energy in the sample determines the state of the matter.
7. Properties of solids:
   a. Solids have high densities in comparison to gases.
   b. Solids have a definite shape.
   c. Solids have a definite volume.
   d. Solids may be crystalline or amorphous.
9. Properties of solids explained
   a. Solids have high densities in comparison to gases. The solid particles are in close contact, whereas gas particles are not.
   b. Solids have a definite shape.
   The particles of a solid are in fixed positions.
   c. Solids have a definite volume.
   The particles of a solid are in close contact.
   d. Solids may be crystalline or amorphous.
   A crystalline solid is a well-ordered three-dimensional array of solid particles.
11. Surface tension is the tendency of liquids to minimize their surface area. Molecules at the surface have few neighbors to interact with via intermolecular forces.
13. Evaporation is a physical change in which a substance is converted from its liquid form to its gaseous form. Condensation is a physical change in which a substance is converted from its gaseous form to its liquid form.
15. Evaporation below the boiling point occurs because molecules on the surface of the liquid experience fewer attractions to the neighboring molecules and can therefore break away. At the boiling point, evaporation occurs faster due to more of the molecules having sufficient thermal energy to break away (including internal molecules).
17. Acetone has weaker intermolecular forces than water. Acetone is more volatile than water.
19. Vapor pressure is the partial pressure of a gas in dynamic equilibrium with its liquid. It increases with increasing temperature and also increases with decreasing strength of intermolecular forces.
21. A steam burn is worse than a water burn at the same temperature (100 °C), because when the steam condenses on the skin, it releases large amounts of additional heat.
23. As the first molecules freeze, they release heat, making it harder for other molecules to freeze without the aid of a refrigeration mechanism, which would draw heat out.
25. Dispersion forces are the default intermolecular force present in all molecules and atoms. Dispersion forces are caused by fluctuations in the electron distribution within molecules or atoms. Dispersion forces are the weakest type of intermolecular force and increase with increasing molar mass.
27. Hydrogen bonding is an intermolecular force and is sort of a super dipole–dipole force. Hydrogen bonding occurs in compounds containing hydrogen atoms bonded directly to fluorine, oxygen, or nitrogen.
29. Molecular solids as a whole tend to have low to moderately low melting points relative to other types of solids; however, strong molecular forces can increase their melting points relative to each other.
31. Ionic solids tend to have much higher melting points relative to the melting points of other types of solids.
33. Water is unique for a couple reasons. Water has a low molar mass, yet it is still liquid at room temperature and has a relatively high boiling point. Unlike other substances, which contract upon freezing, water expands upon freezing.

Problems

35. The 55 mL of water in a dish with a diameter of 12 cm will evaporate more quickly because it has a larger surface area.

37. Acetone feels cooler while evaporating from one's hand, for it is more volatile than water and evaporates much faster.

39. The ice's temperature will increase from −5 °C to 0 °C, where it will then stay constant while the ice completely melts. After the melting process is complete, the water will continue to steadily rise in temperature until it reaches room temperature (25 °C)

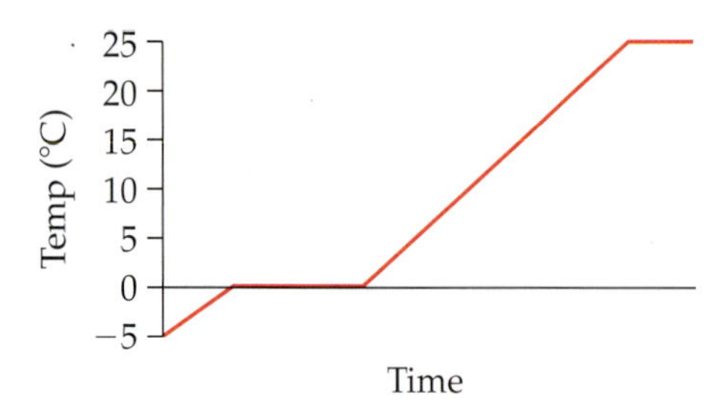

41. Allowing 0.50 g of 100 °C steam to condense on your hand would cause a more severe burn than spilling 0.50 g of 100 °C water on your hand. When the steam condenses on the skin, it releases large amounts of additional heat.

43. The −8 °C ice chest will cause the water in the watery bag of ice to freeze, which is an exothermic process. The freezing process of the water will release heat, and the temperature of the ice chest will increase.

45. The ice chest filled with ice at 0 °C will be colder after a couple hours than the ice chest filled with water at 0 °C. This is because the ice can absorb additional heat as it melts.

47. Denver is a mile above sea level and thus has a lower air pressure. Due to this low pressure, the point at which the vapor pressure of water equals the external pressure will occur at a lower temperature.

49. 78.6 kJ

51. 6.8 kJ

53. 6.25 kJ

55. 371 g

57. 9.65 kJ

59. 15.9 kJ

61. **a.** dispersion
b. dispersion
c. dispersion, dipole–dipole
d. dispersion, dipole–dipole, hydrogen bonding

63. **a.** dispersion, dipole–dipole
b. dispersion, dipole–dipole, hydrogen bonding
c. dispersion
d. dispersion

65. d, because it has the highest molecular weight

67. CH_3OH, due to strong hydrogen bonding

69. NH_3, because it has the ability to form hydrogen bonds

71. These two substances are not miscible, for H_2O is polar and $CH_3CH_2CH_2CH_2CH_3$ is nonpolar.

73. **a.** no **b.** yes
c. yes

75. **a.** atomic **b.** molecular
c. ionic **d.** atomic

77. **a.** molecular **b.** ionic
c. molecular **d.** molecular

79. **c.** LiCl(*s*); it is an ionic solid and possesses ionic bonds resulting in a higher melting point.

81. **a.** Ti(*s*); Ti is a covalent atomic solid and Ne is a nonbonding atomic solid.
b. H_2O(*s*); while both are molecular solids, water has strong hydrogen bonding.
c. Xe(*s*); both are nonbonding atomic solids, but Xe has a higher molar mass.
d. NaCl(*s*); NaCl(*s*) is an ionic solid and CH_4(*s*) is molecular solid.

83. Ne < SO_2 < NH_3 < H_2O < NaF

85. **a.** 1.9×10^4 J **b.** 19 kJ
c. 4.6×10^3 cal **d.** 4.6 Cal

87. 2.7 °C

89. 88.1 g

91. 57.5 kJ

93. **a.** H—S̈e—H Bent: Dispersion and dipole–dipole
b. Ö=S̈—Ö: Bent: Dispersion and dipole–dipole
c. H—C(:C̈l:)(:C̈l:)—C̈l: Tetrahedral: Dispersion and dipole–dipole
d. :Ö=C=Ö: Linear: Dispersion

95. $Na^+[:\ddot{F}:]^-$; $Mg^{2+}[:\ddot{O}:]^{2-}$; MgO has a higher melting point, because the magnitude of the charges on the ions is greater.

97. As the molecular weight increases from Cl to I, the greater the London dispersion forces present, which will increase the boiling point as observed. However, HF is the only compound listed that has the ability to form hydrogen bonds, which explains the anomaly in the trend.

99. The molecule in the interior has the most neighbors. The molecule on the surface is more likely to evaporate. In three dimensions, the molecules that make up the surface area have the least amount of neighbors and are more likely to evaporate than the molecules in the interior.

101. **a.** Yes, if the focus is on the melting of icebergs. No, the melting of an ice cube in a cup of water will *not* raise the level of the liquid in the cup; when the ice cube melts, the volume of the water created will be less than the volume of the initial ice cube.
b. Yes, for the ice sheets that sit on the continent of Antarctica are above sea level and if some melted the water created would be added to the ocean without decreasing the amount of ice below the ocean's surface.

Chapter 13

Questions

1. A solution is a homogeneous mixture of two or more substances. Some examples are air, seawater, soda water, and brass.
3. In a solution, the solvent is the majority component of the mixture and the solute is the minority component. For example, in a seawater solution, the water is the solvent and the salt content is the solute.
5. Solubility is the amount of the compound, usually in grams, that will dissolve in a specified amount of solvent.
7. In solutions with solids, soluble ionic solids form strong electrolyte solutions, while soluble molecular solids form nonelectrolyte solutions. Strong electrolyte solutions are solutions containing solutes that dissociate into ions, for example, $BaCl_2$ and NaOH.
9. Recrystallization is a common way to purify a solid. In recrystallization, enough solid is put into high-temperature water until a saturated solution is created. Then the solution cools slowly, and crystals result from the solution. The crystalline structure tends to reject impurities, resulting in a purer solid.
11. The bubbles formed on the bottom of a pot of heated water (before boiling) are dissolved air coming out of the solution. These gases come out of solution because the solubility of the dissolved nitrogen and oxygen decreases as the temperature of the water rises.
13. The higher the pressure above a liquid, the more soluble the gas is in the liquid.
15. A dilute solution is one containing small amounts of a solute relative to solvent and a concentrated solution is one containing large amounts of solute relative to solvent.
17. Molarity is a common unit of concentration of a solution defined as moles of solute per liter of solution.
19. The boiling point of a solution containing a nonvolatile solute is higher than the boiling point of the pure solvent. The melting point of the solution, however is lower.
21. Molality is a common unit of concentration of a solution expressed as number of moles of solute per kilogram of solvent.
23. Water tends to move from lower concentrations to higher concentrations and when the salt water is being passed through the human body, the salt content draws the water out of the body, causing dehydration.

Problems

25. c, d
27. **a.** solute: salt, solvent: water
b. solute: sugar, solvent: water
c. solute: CO_2, solvent: water
29. **a.** hexane **b.** water
c. ethyl ether **d.** water
31. ions, strong electrolyte solution
33. unsaturated
35. recrystallization
37. **a.** no **b.** yes **c.** yes
39. At room temperature water contains some dissolved oxygen gas; however, the boiling of the water will remove dissolved gases.
41. Under higher pressure the gas (nitrogen) will be more easily dissolved in the blood. To reverse this process, the diver should ascend to relieve the pressure.
43. **a.** 8.11% **b.** 10.3% **c.** 6.37%
45. 13%
47.

Mass Solute	*Mass Solvent*	*Mass Solution*	*Mass%*
15.5	238.1	<u>253.6</u>	<u>6.11%</u>
22.8	<u>167.2</u>	<u>190.0</u>	12.0%
<u>28.8</u>	183.3	212.1	<u>13.6%</u>
<u>56.9</u>	315.2	<u>372.1</u>	15.3%

49. 9.6 g
51. **a.** 2.7 g **b.** 13.7 mg **c.** 0.99 kg
53. **a.** 3.3×10^3 g
b. 77 g
c. 15 g
55. 1.6×10^2 g
57. 1.3×10^3 g
59. 1.4×10^4 mL
61. **a.** 0.52 M **b.** 0.263 M **c.** 0.199 M
63. **a.** 0.149 M **b.** 0.461 M **c.** 3.00×10^{-2} M
65. 0.60 M
67. **a.** 1.8 mol **b.** 0.38 mol **c.** 0.238 mol
69. **a.** 0.39 L **b.** 0.056 L **c.** 0.10 L
71.

Solute	*Mass Solute*	*Mol Solute*	*Volume Solution*	*Molartiy*
KNO_3	22.5 g	<u>0.223</u>	125 mL	<u>1.78 M</u>
$NaHCO_3$	<u>2.10 g</u>	<u>0.0250</u>	250.0 mL	0.100 M
$C_{12}H_{22}O_{11}$	55.38 g	<u>0.162</u>	<u>1.08 L</u>	0.150 M

73. 4.8 g
75. 19 g
77. 8.8 L
79. 0.38 M
81. Dilute 0.045 L of the stock solution to 2.5 L.
83. 6.0×10^2 mL
85. 17.7 mL
87. **a.** 0.025 L **b.** 0.020 L **c.** 1.03 L
89. 4.45 mL
91. 0.373 M
93. 1.2 L
95. **a.** 1.0 m **b.** 3.92 m **c.** 0.52 m
97. 1.28 m
99. **a.** −1.6 °C **b.** −2.70 °C
c. −8.9 °C **d.** −4.37 °C
101. **a.** 100.060 °C **b.** 100.993 °C
c. 101.99 °C **d.** 101.11 °C
103. −1.27 °C, 100.348 °C
105. 2.28 M, 12.3%
107. 0.43 L
109. 319 mL
111. 0.17 L
113. 9.00 mL
115. 79.5 mL

117. -3.22 °C, 100.885 °C
119. -1.56 °C, 100.431 °C
121. 1.8×10^2 g/mol
123. **a.** Water will flow from left to right.
 b. Water will flow from right to left.
 c. Water won't flow between the two.
125. 3×10^4 L

Chapter 14

Questions

1. Sour gummy candies are coated with a white powder that is a mixture of citric acid and tartaric acid. The combination of these two acids creates the sour taste.
3. Acids have a sour taste; acids dissolve many metals; and acids turn litmus paper red.
5. Carboxylic acids are organic compounds with the general formula R—COOH. Some examples are acetic acid ($HC_2H_3O_2$) and formic acid ($HCHO_2$)
7. Alkaloids are organic bases found in plants.
9. The Arrhenius definition of an acid is a substance that produces H^+ ions in aqueous solution. An example:

 $HCl(aq) \longrightarrow H^+(aq) + Cl^-(ag)$

11. The Brønsted–Lowery definition states that an acid is a proton donor and a base is a proton acceptor. The following is an example of a chemical equation demonstrating this definition:

 $HCl(aq) + H_2O(l) \longrightarrow H_3O^+(aq) + Cl^-(aq)$

13. An acid–base neutralization reaction occurs when an acid and a base are mixed and the $H^+(aq)$ from the acid combines with the $OH^-(aq)$ from the base to form $H_2O(l)$ An example follows.

 $HCl(aq) + KOH(aq) \longrightarrow H_2O(l) + KCl(aq)$

15. $2\,HCl(aq) + K_2O(s) \longrightarrow H_2O(l) + 2\,KCl(aq)$
17. A titration is a laboratory procedure where a reactant in a solution of known concentration is reacted with another reactant in a solution of unknown concentration until the reaction has reached the equivalence point. The equivalence point is the point at which the reactants are in exact stoichiometric proportions.
19. A strong acid is one that will completely dissociate in solution, while a weak acid does not completely dissociate in solution.
21. Monoprotic acids (such as HCl) contain only one hydrogen ion that will dissociate in solution, while diprotic acids (such as H_2SO_4) contain two hydrogen ions that will dissociate in solution.
23. Yes, pure water contains H_3O^+ ions. Through self-ionization, water acts as an acid and a base with itself; water is amphoteric.
25. **a.** $[H_3O^+] > 1.0 \times 10^{-7}$ M; $[OH^-] < 1.0 \times 10^{-7}$ M
 b. $[H_3O^+] < 1.0 \times 10^{-7}$ M; $[OH^-] > 1.0 \times 10^{-7}$ M
 c. $[H_3O^+] = 1.0 \times 10^{-7}$ M; $[OH^-] = 1.0 \times 10^{-7}$ M
27. A buffer is a solution that resists pH change by neutralizing added acid or added base.
29. The cause of acid rain is the formation of SO_2, NO, and NO_2 during the combustion of fossil fuels.
31. Acid rain damages structures made out of metal, marble, cement, and limestone, as well as harming and possibly killing aquatic life and trees.

Problems

33. **a.** acid; $HNO_3(aq) \longrightarrow H^+(aq) + NO_3^-(aq)$
 b. base; $KOH(aq) \longrightarrow K^+(aq) + OH^-(aq)$
 c. acid; $HC_2H_3O_2(aq) \longrightarrow H^+(aq) + C_2H_3O_2^-(aq)$
 d. base; $Ca(OH)_2(aq) \longrightarrow Ca^{2+}(aq) + 2\,OH^-(aq)$
35.

	B-L Acid	*B-L Base*	*Conjugate Acid*	*Conjugate Base*
a.	HBr	H_2O	H_3O^+	Br^-
b.	H_2O	NH_3	NH_4^+	OH^-
c.	HNO_3	H_2O	H_3O^+	NO_3^-
d.	H_2O	C_5H_5N	$C_5H_5NH^+$	OH^-

37. a, c
39. **a.** Cl^- **b.** HSO_3^- **c.** CHO_2^- **d.** F^-
41. **a.** NH_4^+ **b.** $HClO_4$ **c.** H_2SO_4 **d.** HCO_3^-
43. **a.** $HI(aq) + NaOH(aq) \longrightarrow H_2O(l) + NaI(aq)$
 b. $HBr(aq) + KOH(aq) \longrightarrow H_2O(l) + KBr(aq)$
 c. $2\,HNO_3(aq) + Ba(OH)_2(aq) \longrightarrow 2\,H_2O(l) + Ba(NO_3)_2(aq)$
 d. $2\,HClO_4(aq) + Sr(OH)_2(aq) \longrightarrow 2\,H_2O(l) + Sr(ClO_4)_2(aq)$
45. **a.** $2\,K(aq) + 2\,HBr(aq) \longrightarrow H_2(g) + 2\,KBr(aq)$
 b. $Ca(aq) + 2\,HBr(aq) \longrightarrow H_2(g) + CaBr_2(aq)$
 c. $Sr(aq) + 2\,HBr(aq) \longrightarrow H_2(g) + SrBr_2(aq)$
 d. $2\,Rb(aq) + 2\,HBr(aq) \longrightarrow H_2(g) + 2\,RbBr(aq)$
47. **a.** $MgO(aq) + 2\,HI(aq) \longrightarrow H_2O(l) + MgI_2(aq)$
 b. $K_2O(aq) + 2\,HI(aq) \longrightarrow H_2O(l) + 2\,KI(aq)$
 c. $Rb_2O(aq) + 2\,HI(aq) \longrightarrow H_2O(l) + 2\,RbI(aq)$
 d. $CaO(aq) + 2\,HI(aq) \longrightarrow H_2O(l) + CaI_2(aq)$
49. **a.** $6\,HClO_4(aq) + Fe_2O_3(s) \longrightarrow 2\,Fe(ClO_4)_3(aq) + 3\,H_2O(l)$
 b. $H_2SO_4(aq) + Sr(s) \longrightarrow SrSO_4(aq) + H_2(g)$
 c. $H_3PO_4(aq) + 3\,KOH(aq) \longrightarrow 3\,H_2O(l) + K_3PO_4(aq)$
51. **a.** 0.1400 M **b.** 0.138 M **c.** 0.08764 M **d.** 0.182 M
53. 0.1018 M
55. 27.3 mL
57. **a.** strong **b.** weak **c.** strong **d.** weak
59. **a.** $[H_3O^+] = 2.5$ M **b.** $[H_3O^+] < 1.2$ M **c.** $[H_3O^+] < 0.25$ M **d.** $[H_3O^+] < 2.25$ M
61. **a.** strong **b.** weak **c.** strong **d.** weak
63. **a.** $[OH^-] = 0.88$ M **b.** $[OH^-] < 0.88$ M **c.** $[OH^-] = 1.76$ M **d.** $[OH^-] = 1.55$ M
65. **a.** acidic **b.** acidic **c.** neutral **d.** basic
67. **a.** 6.7×10^{-6} M, basic **b.** 1.1×10^{-6} M, basic **c.** 4.5×10^{-9} M, acidic **d.** 1.4×10^{-11} M, acidic
69. **a.** 3.7×10^{-3} M, acidic **b.** 4.0×10^{-13} M, basic **c.** 9.1×10^{-5} M, acidic **d.** 3.0×10^{-11} M, basic

71. **a.** basic **b.** neutral
c. acidic **d.** acidic

73. **a.** 7.77 **b.** 7.0
c. 5.66 **d.** 3.13

75. **a.** 2.8×10^{-9} M **b.** 5.9×10^{-12} M
c. 1.3×10^{-3} M **d.** 6.0×10^{-2} M

77. **a.** 7.28 **b.** 6.42
c. 3.86 **d.** 12.98

79. **a.** 1.8×10^{-10} M **b.** 3.4×10^{-2} M
c. 3.2×10^{-13} M **d.** 1.8×10^{-6} M

81. **a.** 1.810 **b.** 11.107
c. 2.724 **d.** 10.489

83. various answers

85. c, d

87. $HCl(aq) + NaF(aq) \longrightarrow HF(aq) + NaCl(aq)$
$HCl(aq) + KC_2H_3O_2(aq) \longrightarrow HC_2H_3O_2(aq) + KCl(aq)$

89. **a.** $HC_2H_3O_2$ **b.** NaH_2PO_4
c. NaCHOO

91. 50.0 mL

93. 0.16 L

95. 65.2 g

97. 60.0 g/mol

99. 0.17 L

101. **a.** 2.60, acidic **b.** 11.75, basic
c. 8.02, basic **d.** 1.710, acidic

103.

$[H_3O^+]$	$[OH^-]$	*pH*	*Acidic or Basic*
1.0×10^{-4}	1.0×10^{-10}	4.00	acidic
5.5×10^{-3}	1.8×10^{-12}	2.26	acidic
3.1×10^{-9}	3.2×10^{-6}	8.50	basic
4.8×10^{-9}	2.1×10^{-6}	8.32	basic
2.8×10^{-8}	3.5×10^{-7}	7.55	basic

105. **a.** $[H_3O^+] = 0.0088$ M
$[OH^-] = 1.1 \times 10^{-12}$ M
pH = 2.06
b. $[H_3O^+] = 1.5 \times 10^{-3}$ M
$[OH^-] = 6.7 \times 10^{-12}$M
pH = 2.82
c. $[H_3O^+] = 9.77 \times 10^{-4}$ M
$[OH^-] = 1.02 \times 10^{-11}$ M
pH = 3.010
d. $[H_3O^+] = 0.0878$ M
$[OH^-] = 1.14 \times 10^{-13}$ M
pH = 1.057

107. **a.** $[OH^-] = 0.15$ M
$[H_3O^+] = 6.7 \times 10^{-14}$ M
pH = 13.18
b. $[OH^-] = 3.0 \times 10^{-3}$ M
$[H_3O^+] = 3.3 \times 10^{-12}$ M
pH = 11.48
c. $[OH^-] = 9.6 \times 10^{-4}$ M
$[H_3O^+] = 1.0 \times 10^{-11}$ M
pH = 10.98
d. $[OH^-] = 8.7 \times 10^{-5}$ M
$[H_3O^+] = 1.1 \times 10^{-10}$ M
pH = 9.94

109. 1.49L

111. 11.495

113. 3.0×10^{12} H^+ ions

115. **a.** weak **b.** strong
c. weak **d.** strong

117. approximately 8 times more concentrated

Chapter 15

Questions

1. The two general concepts involved in equilibrium are sameness and changelessness.
3. The rate of a chemical reaction is the amount of reactant that changes to product in a given period of time.
5. By controlling reaction rates, chemists can control the amount of a product that forms in a given period of time and have control of the outcome.
7. The two factors that influence reaction rates are concentration and temperature. The rate of a reaction increases with increasing concentration. The rate of a reaction increases with increasing temperature.
9. In a chemical reaction, dynamic equilibrium is the condition in which the rate of the forward reaction equals the rate of the reverse reaction.
11. Because the rate of the forward and reverse reactions is the same at equilibrium, the relative concentrations of reactants and products becomes constant.
13. The equilibrium constant is a measure of how far a reaction goes; it is significant because it is a way to quantify the concentrations of the reactants and products at equilibrium.
15. A small equilibrium constant shows that a reverse reaction is favored and that when equilibrium is reached, there will be more reactants than products. A large equilibrium constant shows that a forward reaction is favored and that when equilibrium is reached, there will be more products than reactants.
17. No, the particular concentrations of reactants and products at equilibrium will not always be the same for a given reaction—they will depend on the initial concentrations.
19. Various answers depending on the answer for Question 12.
21. Decreasing the concentration of a reactant in a reaction mixture at equilibrium causes the reaction to shift to the left.
23. Decreasing the concentration of a product in a reaction mixture at equilibrium causes the reaction to shift to the right.
25. Increasing the pressure of a reaction mixture at equilibrium if the product side has fewer moles of gas particles than the reactant side causes the reaction to shift to the right.
27. Decreasing the pressure of a reaction mixture at equilibrium if the product side has fewer moles of gas particles than the reactant side causes the reaction to shift to the left.
29. Increasing the temperature of an exothermic reaction mixture at equilibrium causes the reaction to shift left, absorbing some of the added heat. Decreasing the temperature of an exothermic reaction mixture at equilibrium causes the reaction to shift right, releasing heat.

31. $K_{sp} = [A^{2+}][B^-]^2$
33. The solubility of a compound is the amount of the compound that dissolves in a certain amount of liquid, and the molar solubility is the solubility in units of moles per liter.
35. Two reactants with a large K_{eq} for a particular reaction might not react immediately when combined because of a large activation energy, which is an energy hump that normally exists between the reactants and products. The activation energy must be overcome before the system will undergo a reaction.
37. No, a catalyst does not affect the value of the equilibrium constant; it simply lowers the activation energy and increases the rate of a chemical reaction.

Problems

39. Rate would decrease because the effective concentration of the reactants has been decreased which lowers the rate of a reaction.
41. Reaction rates tend to decrease with decreasing temperature so all life processes (chemical reactions) would have decreased rates.
43. The rate would be lower because the concentration of reactants decreases as they are consumed in the reaction.
45. **a.** $K_{eq} = \dfrac{[N_2O_4]}{[NO_2]^2}$ **b.** $K_{eq} = \dfrac{[NO]^2[Br_2]}{[BrNO]^2}$
c. $K_{eq} = \dfrac{[H_2][CO_2]}{[H_2O][CO]}$ **d.** $K_{eq} = \dfrac{[CS_2][H_2]^4}{[CH_4][H_2S]^2}$
47. **a.** $K_{eq} = \dfrac{[Cl_2]}{[PCl_5]}$ **b.** $K_{eq} = [O_2]^3$
c. $K_{eq} = \dfrac{[H_3O^+][F^-]}{[HF]}$ **d.** $K_{eq} = \dfrac{[NH_4^+][OH^-]}{[NH_3]}$
49. $K_{eq} = \dfrac{[H_2]^2[S_2]}{[H_2S]^2}$
51. **a.** reactants **b.** products
c. reactants **d.** both
53. 0.0394
55. 1.79×10^{-5}
57. 0.0987
59. 0.82 M
61. 0.119 M
63.

T(K)	$[N_2]$	$[H_2]$	$[NH_3]$	K_{eq}
500	0.115	0.105	0.439	1.45×10^3
575	0.110	0.25	0.128	9.6
775	0.120	0.140	0.00439	0.0584

65. **a.** shift right **b.** shift left
c. shift right
67. **a.** no effect **b.** shift left
c. shift left **d.** shift right
69. **a.** shift right **b.** shift left
71. **a.** no effect **b.** no effect
73. **a.** shift right **b.** shift left
75. **a.** shift left **b.** shift right
77. **a.** no effect **b.** shift right
c. shift left **d.** shift right
e. no effect
79. **a.** $CaSO_4(s) \longleftrightarrow Ca^{2+}(aq) + SO_4^{2-}(aq)$
$K_{sp} = [Ca^{2+}][SO_4^{2-}]$
b. $AgCl(s) \longleftrightarrow Ag^+(aq) + Cl^-(aq)$
$K_{sp} = [Ag^+][Cl^-]$
c. $CuS(s) \longleftrightarrow Cu^{2+}(aq) + S^{2-}(aq)$
$K_{sp} = [Cu^{2+}][S^{2-}]$
d. $FeCO_3(s) \longleftrightarrow Fe^{2+}(aq) + CO_3^{2-}(aq)$
$K_{sp} = [Fe^{2+}][CO_3^{2-}]$
81. $K_{sp} = [Fe^{2+}][OH^-]^2$
83. 7.0×10^{-11}
85. 1.35×10^{-4}
87. 7.04×10^{-5} M
89. 2.61×10^{-3} M
91.

Compound	*[Cation]*	*[Anion]*	K_{sp}
$SrCO_3$	2.4×10^{-5}	2.4×10^{-5}	5.8×10^{-10}
SrF_2	1.0×10^{-3}	2.0×10^{-3}	4.0×10^{-9}
Ag_2CO_3	2.6×10^{-4}	1.3×10^{-4}	8.8×10^{-12}

93. 3.3×10^2
95. 5.34 g
97. b, c, d
99. 1.13×10^{-18} M, 1.62×10^{-15} g
101. Yes
103. 7.07×10^{-4} g
105. e
107. 35.5 L

Chapter 16

Questions

1. A fuel-cell electric vehicle is an automobile running on an electric motor that is powered by hydrogen. The fuel cells use the electron-gaining tendency of oxygen and the electron-losing tendency of hydrogen to force electrons to move through a wire, creating the electricity that powers the car.
3. **a.** Oxidation is the gaining of oxygen and reduction is the losing of oxygen.
b. Oxidation is the loss of electrons and reduction is the gain of electrons.
c. Oxidation is an increase in oxidation state and reduction is a decrease in oxidation state.
5. A reducing agent is the substance being oxidized that causes the reduction of the other substance.
7. Good reducing agents have a strong tendency to *lose* electrons in reactions.
9. The oxidation state of a monotomic ion is equal to its charge.
11. For an ion, the sum of the oxidation states of the individual atoms must add up to *the charge of the ion.*
13. In a redox reaction, an atom that undergoes an increase in oxidation state is *oxidized*. An atom that undergoes a decrease in oxidation state is *reduced*.
15. When balancing redox equations, the number of electrons lost in the oxidation half-reaction must *equal* the number of electrons gained in the reduction half-reaction.
17. When balancing aqueous redox reactions occurring in acidic media, hydrogen is balanced using *H^+ ions*.

19. The metals at the top of the activity series are the most reactive.
21. The metals at the bottom of the activity series are least likely to lose electrons.
23. If the metal is listed above H_2 on the activity series, it will dissolve in acids such as HCl or HBr.
25. Oxidation occurs at the *anode* of an electrochemical cell.
27. The salt bridge joins the two half-cells or completes the circuit—it allows the flow of ions between the two half-cells.
29. The common dry cell battery does not contain large amounts of liquid water and is composed of a zinc case that acts as the anode. The cathode is a carbon rod immersed in a moist paste of MnO_2 that also contains NH_4Cl. The anode and cathode reactions occur to produce a voltage of about 1.5 volts.
    anode reaction:
    $Zn(s) \longrightarrow Zn^{2+}(aq) + 2\,e^-$
    cathode reaction:
    $2\,MnO_2(s) + 2\,NH_2^+(aq) + 2\,e^- \longrightarrow 2\,Mn_2O_3(s) + 2\,NH_3(g) + H_2O(l)$
31. Fuel cells are like batteries, but the reactants are constantly replenished. The reactants constantly flow through the battery, generating electrical current as they undergo a redox reaction.
    anode reaction:
    $2\,H_2(g) + 4\,OH^-(aq) \longrightarrow 4\,H_2O(l) + 4\,e^-$
    cathode reaction:
    $O_2(g) + 2\,H_2O(l) + 4\,e^- \longrightarrow 4\,OH^-(aq)$
33. Corrosion is the oxidation of metals; the most common is rusting of iron.
    oxidation:
    $2\,Fe(s) \longrightarrow 2\,Fe^{2+}(aq) + 4\,e^-$
    reduction:
    $O_2(g) + 2\,H_2O(l) + 4\,e^- \longrightarrow 4\,OH^-(aq)$
    overall:
    $2\,Fe(s) + O_2(g) + 2\,H_2O(l) \longrightarrow 2\,Fe(OH)_2(s)$

Problems

35. **a.** H_2 **b.** Al **c.** Al
37. **a.** Sr is oxidized, O_2 is reduced.
    b. Ca is oxidized, Cl_2 is reduced.
    c. Mg is oxidized, Ni^{2+} is reduced.
39. **a.** Sr is the reducing agent, O_2 is the oxidizing agent.
    b. Ca is the reducing agent, Cl_2 is the oxidizing agent.
    c. Mg is the reducing agent, Ni^{2+} is the oxidizing agent.
41. b (F_2), d (Cl_2)
43. a (K), c (Fe)
45. **a.** N_2 is oxidized and is the reducing agent. O_2 is reduced and is the oxidizing agent.
    b. C is oxidized and is the reducing agent. O_2 is reduced and is the oxidizing agent.
    c. Sb is oxidized and is the reducing agent. Cl_2 is reduced and is the oxidizing agent.
    d. K is oxidized and is the reducing agent. Pb^{2+} is reduced and is the oxidizing agent.
47. **a.** 0 **b.** +2 **c.** 0 **d.** 0
49. **a.** Na: +1; Cl: −1 **b.** Ca: +2; F: −1 **c.** S: +4; O: −2 **d.** H: +1; S: −2
51. **a.** +2 **b.** +4 **c.** +1
53. **a.** C: +4; O: −2 **b.** O: −2; H: +1 **c.** N: +5; O: −2 **d.** N: +3; O: −2
55. **a.** +1 **b.** +3 **c.** +5 **d.** +7
57. **a.** Cu,+2; N,+5; O,−2
    b. Sr,+2; O,−2; H,+1
    c. K,+1; O,−2; Cr,+6
    d. Na,+1; H,+1; O,−2; C,+4
59. **a.** Sb +5 ⟶ +3, reduced
    Cl −1 ⟶ 0, oxidized
    b. C +2 ⟶ +4, oxidized
    Cl 0 ⟶ −1, reduced
    c. N +2 ⟶ +3, oxidized
    Br 0 ⟶ −1, reduced
    d. H 0 ⟶ +1, oxidized
    C +4 ⟶ +2, reduced
61. Na is the reducing agent.
    H is the oxidizing agent.
63. **a.** $3\,K(s) + Cr^{3+}(aq) \longrightarrow Cr(s) + 3\,K^+(aq)$
    b. $Mg(s) + 2\,Ag^+(aq) \longrightarrow Mg^{2+}(aq) + 2\,Ag(s)$
    c. $2\,Al(s) + 3\,Fe^{2+}(aq) \longrightarrow 2\,Al^{3+}(aq) + 3\,Fe(s)$
65. **a.** reduction, $5\,e^- + MnO_4^-(aq) + 8\,H^+(aq) \longrightarrow Mn^{2+}(aq) + 4\,H_2O(l)$
    b. oxidation,
    $2\,H_2O(l) + Pb^{2+}(aq) \longrightarrow PbO_2(s) + 4\,H^+(aq) + 2\,e^-$
    c. reduction,
    $10\,e^- + 2\,IO_3^-(aq) + 12\,H^+(aq) \longrightarrow I_2(s) + 6\,H_2O(l)$
    d. oxidation,
    $SO_2(g) + 2\,H_2O(l) \longrightarrow SO_4^{2-}(aq) + 4\,H^+(aq) + 2\,e^-$
67. **a.** $PbO_2(s) + 4\,H^+(aq) + 2\,I^-(aq) \longrightarrow I_2(s) + Pb^{2+}(aq) + 2\,H_2O(l)$
    b. $5\,SO_3^{2-}(aq) + 6\,H^+(aq) + 2\,MnO_4^-(aq) \longrightarrow 5\,SO_4^{2-}(aq) + 2\,Mn^{2+}(aq) + 3\,H_2O(l)$
    c. $S_2O_3^{2-}(aq) + 4\,Cl_2(g) + 5\,H_2O(l) \longrightarrow 2\,SO_4^{2-}(aq) + 8\,Cl^-(aq) + 10\,H^+(aq)$
69. **a.** $ClO_4^-(aq) + 2\,H^+(aq) + 2\,Cl^-(aq) \longrightarrow ClO_3^-(aq) + Cl_2(aq) + H_2O(l)$
    b. $3\,MnO_4^-(aq) + 24\,H^+(aq) + 5\,Al(s) \longrightarrow 3\,Mn^{2+}(aq) + 5\,Al^{3+}(aq) + 12\,H_2O(l)$
    c. $Br_2(aq) + Sn(s) \longrightarrow Sn^{2+}(aq) + 2\,Br^-(aq)$
71. c, Pb
73. b, Cu^{2+}
75. b, Al
77. b, c
79. Fe, Cr, Zn, Mn, Al, Mg, Na, Ca, K, Li
81. Al

83. **a.** $2\,Al(s) + 6\,HCl(aq) \longrightarrow 2\,Al^{3+}(aq) + 6\,Cl^{-}(aq) + 3\,H_2(g)$

b. no reaction

c. $Pb(s) + 2\,HCl(aq) \longrightarrow Pb^{2+}(aq) + 2\,Cl^{-}(aq) + H_2(g)$

d. $2\,Cr(s) + 6\,HCl(aq) \longrightarrow 2\,Cr^{3+}(aq) + 6\,Cl^{-}(aq) + 3\,H_2(g)$

85.

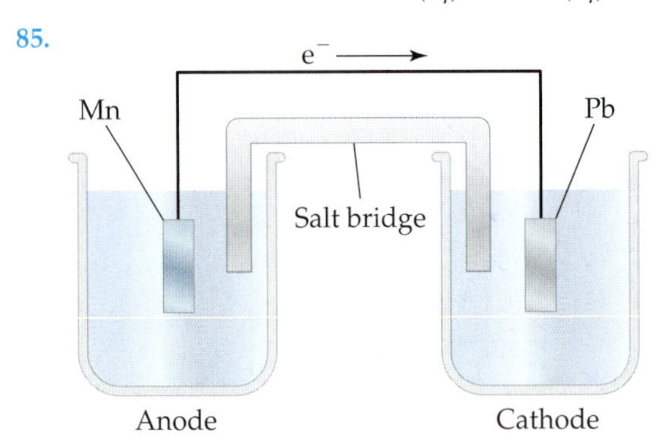

87. d

89. $Zn(s) + 2\,MnO_2(s) + 2\,H_2O(l) \longrightarrow Zn(OH)_2(s) + 2\,MnO(OH)(s)$

91.

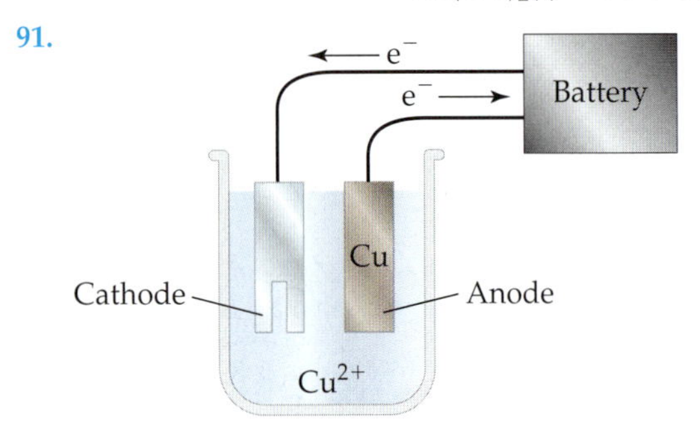

93. a, Zn; c, Mn

95. **a.** redox; Zn is oxidized; Co is reduced.

b. not redox

c. not redox

d. redox; K is oxidized; Br is reduced.

97. $16\,H^{+}(aq) + 2\,MnO_4^{-}(aq) + 5\,Zn(s) \longrightarrow 2\,Mn^{2+}(aq) + 5\,Zn^{2+}(aq) + 8\,H_2O(l)$; 34.9 mL

99. 2.95%

101. 0.054 mol

103. **a.** $2\,Cr(s) + 6\,HI(aq) \longrightarrow 2\,Cr^{3+}(aq) + 6\,I^{-}(aq) + 3\,H_2(g)$, 98 mL HI

b. $2\,Al(s) + 6\,HI(aq) \longrightarrow 2\,Al^{3+}(aq) + 6\,I^{-}(aq) + 3\,H_2(g)$, 68 mL HI

c. no

d. no

105. 0.67 cm

107.

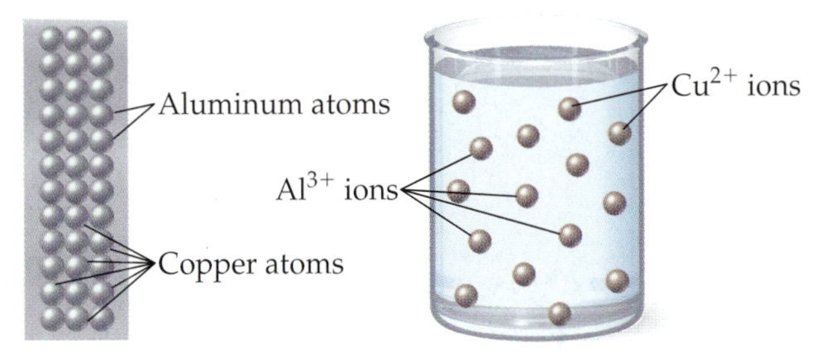

109. Many of the Zn atoms on the electrode would become Zn^{2+} ions in solution. Many Ni^{2+} ions in solution would become Ni atoms on the electrode.

Chapter 17

Questions

1. Radioactivity is the emission of tiny, invisible particles by disintegration of atomic nuclei. Many of these particles can pass right through matter. Atoms that emit these particles are radioactive.

3. Antoine-Henri Becquerel first discovered radioactivity while he was studying X-rays. He came across the idea when his photographic plate became exposed even in complete darkness, concluding that the uranium atoms he was using were constantly emitting some tiny particle.

5. Marie Curie further investigated uranic rays, and she discovered two new elements. Seeing these rays in numerous elements, she changed their name from uranic rays to radioactivity.

7. X: chemical symbol, used to identify the element.

A: mass number, which is the sum of the number of protons and number of neutrons in the nucleus.

Z: atomic number, which is the number of protons in the nucleus.

9. Alpha radiation occurs when an unstable nucleus emits a small piece of itself composed of 2 protons and 2 neutrons. The symbol for an alpha particle is $^{4}_{2}He$.

11. Alpha particles have high ionizing power and low penetrating power compared to beta and gamma particles.

13. When an atom emits a beta particle, its atomic number increases by one, because it now has an additional proton. The mass of an atom does not change as a result of beta emission.

15. Gamma radiation is electromagnetic radiation, and the symbol for a gamma ray is $^{0}_{0}\gamma$.

17. Gamma particles have low ionizing power and high penetrating power compared to alpha and beta particles.

19. When an atom emits a positron, its atomic number decreases by one, because it now has one less proton. The mass of an atom does not change when it emits a positron.

21. A nuclear equation represents the changes that occur during radioactivity and other nuclear processes. For a nuclear equation to be balanced, the sum of the atomic numbers on both sides of the equation must be equal and the sum of the mass numbers on both sides of the equation must be equal.

23. A film-badge dosimeter is a badge that consists of photographic film held in a small case that is pinned to clothing. It is used to monitor a person's exposure to radiation. The more exposed the film has become in a given period of time, the more the person has been exposed to radioactivity.

25. In a scintillation counter, the radioactive particles pass through a material that emits ultraviolet or visible light in response to excitation by radioactive particles. The light is detected and turned into an electrical signal.

27. The half-life is the time it takes for one-half of the parent nuclides in a radioactive sample to decay to the daughter nuclides. One can relate the half-life of objects to find their radioactive decay rates.

29. The decaying of uranium in the ground is the source of radon in our environment. Radon increases the risk of lung cancer because it is a gas that can be inhaled.
31. All living organisms contain a uniform amount of carbon-14 because when an organism is alive, carbon-14 is repeatedly incorporated into the body tissue through carbon dioxide and plants. The decay rate of carbon-14 and the continuous formation of it into the body create an equilibrium of carbon-14 in a living organism.
33. We know that carbon-14 dating is accurate because it can be checked against objects whose ages are known from other methods. The age limit for which carbon-14 dating is useful is up to 50,000 years. The carbon-14 amount in older objects is too small to measure accurately.
35. Nuclear fission is the process in which a heavy nucleus is split by a bombardment of neutrons into nuclei of smaller masses and energy is emitted. Meitner, Strassmann, and Hahn discovered it when they were repeating Fermi's experiments and examining the products from bombarding uranium with neutrons.
37. Critical mass is the mass of uranium or plutonium required for a nuclear reaction to be self-sustaining.
39. The process of nuclear fission can generate heat, which then can be used to boil water and create steam, which turns a turbine and generates electricity.
41. Nuclear electricity generation creates a lot of energy for the relative small mass of fuel. Furthermore, a nuclear power plant generates no air pollution and no greenhouse gases.
43. No, a nuclear reactor cannot detonate the way a nuclear bomb can, because the uranium fuel used in electricity generation is not sufficiently enriched in U-235 to produce a nuclear detonation.
45. Modern nuclear weapons use both fission and fusion. In the hydrogen bomb, a small fission bomb is detonated first to create a high enough temperature for the fusion reaction to proceed.
47. Radiation can affect the molecules in living organisms by ionizing them.
49. Lower doses of radiation over extended periods of time can increase cancer risk by damaging DNA. Occasionally a change in DNA can cause cells to grow abnormally and to become cancerous.
51. The main unit of radiation exposure is the rem, which stands for *roentgen equivalent man.* The average American is exposed to 1/3 of a rem of radiation per year.
53. Isotope scanning can be used in the medical community to detect and identify cancerous tumors. Likewise, isotope scanning can produce necessary images of several different internal organs.

Problems

55. $^{207}_{82}\text{Pb}$
57. 81 protons, 126 neutrons
59. **a.** beta particle **b.** neutron **c.** gamma ray
61.

Chemical Symbol	*Atomic Number (Z)*	*Mass Number (A)*	*Number of Protons*	*Number of Neutrons*
Tc	43	95	43	52
Ba	56	128	56	72
Eu	63	145	63	82
Fr	87	223	87	136

63. **a.** $^{234}_{92}\text{U} \longrightarrow {}^{230}_{90}\text{Th} + {}^{4}_{2}\text{He}$
 b. $^{230}_{90}\text{Th} \longrightarrow {}^{226}_{88}\text{Ra} + {}^{4}_{2}\text{He}$
 c. $^{226}_{88}\text{Ra} \longrightarrow {}^{222}_{86}\text{Rn} + {}^{4}_{2}\text{He}$
 d. $^{222}_{86}\text{Rn} \longrightarrow {}^{218}_{84}\text{Po} + {}^{4}_{2}\text{He}$
65. **a.** $^{214}_{82}\text{Pb} \longrightarrow {}^{214}_{83}\text{Bi} + {}^{0}_{-1}\text{e}$
 b. $^{214}_{83}\text{Bi} \longrightarrow {}^{214}_{84}\text{Po} + {}^{0}_{-1}\text{e}$
 c. $^{231}_{90}\text{Th} \longrightarrow {}^{231}_{91}\text{Pa} + {}^{0}_{-1}\text{e}$
 d. $^{227}_{89}\text{Ac} \longrightarrow {}^{227}_{90}\text{Th} + {}^{0}_{-1}\text{e}$
67. **a.** $^{11}_{6}\text{C} \longrightarrow {}^{11}_{5}\text{B} + {}^{0}_{+1}\text{e}$
 b. $^{13}_{7}\text{N} \longrightarrow {}^{13}_{6}\text{C} + {}^{0}_{+1}\text{e}$
 c. $^{15}_{8}\text{O} \longrightarrow {}^{15}_{7}\text{N} + {}^{0}_{+1}\text{e}$
69. $^{241}_{94}\text{Pu} \longrightarrow {}^{241}_{95}\text{Am} + {}^{0}_{-1}\text{e}$
 $^{241}_{95}\text{Am} \longrightarrow {}^{237}_{93}\text{Np} + {}^{4}_{2}\text{He}$
 $^{237}_{93}\text{Np} \longrightarrow {}^{233}_{91}\text{Pa} + {}^{4}_{2}\text{He}$
 $^{233}_{91}\text{Pa} \longrightarrow {}^{233}_{92}\text{U} + {}^{0}_{-1}\text{e}$
71. $^{232}_{90}\text{Th} \longrightarrow {}^{228}_{88}\text{Ra} + {}^{4}_{2}\text{He}$
 $^{228}_{88}\text{Ra} \longrightarrow {}^{228}_{89}\text{Ac} + {}^{0}_{-1}\text{e}$
 $^{228}_{89}\text{Ac} \longrightarrow {}^{228}_{90}\text{Th} + {}^{0}_{-1}\text{e}$
 $^{228}_{90}\text{Th} \longrightarrow {}^{224}_{88}\text{Ra} + {}^{4}_{2}\text{He}$
73. 31 atoms
75. 18 hrs
77. 1.2×10^{6} yrs
79. 0.194 g
81. Ga-67 > P-32 > Cr-51 > Sr-89
83. 5,730 yrs
85. 34,380 yrs
87. $^{235}_{92}\text{U} + {}^{1}_{0}\text{n} \longrightarrow {}^{144}_{54}\text{Xe} + {}^{90}_{38}\text{Sr} + 2\,{}^{1}_{0}\text{n}$; 2 neutrons
89. $^{2}_{1}\text{H} + {}^{2}_{1}\text{H} \longrightarrow {}^{3}_{2}\text{He} + {}^{1}_{0}\text{n}$
91. **a.** $^{6}_{3}\text{Li}$
 b. $^{64}_{28}\text{Ni}$
 c. $^{179}_{73}\text{Ta}$
93. $^{238}_{92}\text{U} + {}^{1}_{0}\text{n} \longrightarrow {}^{239}_{92}\text{U}$; $^{239}_{92}\text{U} \longrightarrow {}^{0}_{-1}\text{e} + {}^{239}_{93}\text{Np}$;
 $^{239}_{93}\text{Np} \longrightarrow {}^{0}_{-1}\text{e} + {}^{239}_{94}\text{Pu}$
95. per mole $= 1.9 \times 10^{13}$ J
 per kg $= 8.2 \times 10^{13}$ J
97. 1.7×10^{21} β emissions
99. 68.4%
101. nucleus with 9 protons and 7 neutrons
103. nucleus with 5 protons and 5 neutrons

Chapter 18

Questions

1. Organic molecules are often involved in smell.
3. At the end of the eighteenth century, it was believed that organic compounds came from living things and were easily decomposed, while inorganic compounds came from the earth and were more difficult to decompose. A final difference is that many inorganic compounds could be easily synthesized, but organic compounds could not be.
5. Vitalism was overthrown in 1828 when Friedrich Wohler heated ammonium cynate (an inorganic compound) and formed urea (an organic compound), thus disproving vitalism.

7. **a.** A carbon atom with 4 single bonds forms a tetrahedral geometry.
   b. A carbon atom with 2 single bonds and 1 double bond forms a trigonal planar geometry.
   c. A carbon atom with 1 single bond and 1 triple bond forms a linear geometry.

9. Hydrocarbons are commonly used for fuels and are also the starting materials in the synthesis of many different consumer products.

11. A saturated hydrocarbon contains no double or triple bonds between the carbon atoms; however, an unsaturated hydrocarbon contains one or more double or triple bonds between the carbon atoms.

13. Alkanes composed of carbon atoms bonded in a straight chain with no branches are called *n*-alkanes. Alkanes composed of carbon atoms forming branched structures are called branched alkanes.

15. Alkenes are hydrocarbons containing at least one double bond between carbon atoms, whereas alkanes contain only single bonds.

17. Hydrocarbon combustion reactions involve the burning of hydrocarbons in the presence of oxygen. An example is:
$$CH_2{=}CHCH_2CH_3(g) + 6\,O_2(g) \longrightarrow 4\,CO_2(g) + 4\,H_2O(g)$$

19. An alkene addition reaction occurs when atoms add across the double bond. For example:
$$CH_2{=}CH_2(g) + Cl_2(g) \longrightarrow CH_2ClCH_2Cl(g)$$

21. The structure of benzene is 6 carbon atoms connected together in a circle with each bonded to a hydrogen atom. Here are two ways that benzene is often represented.

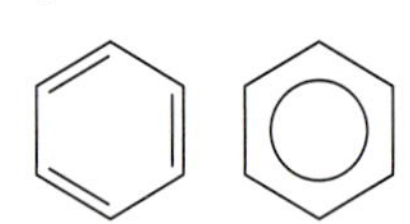

23. The generic structure of alcohols is $R{-}OH$. The structure of methanol is CH_3OH and the structure of ethanol is CH_3CH_2OH.

25. The generic structure of ethers is $R{-}O{-}R$. The structure of dimethyl ether is CH_3OCH_3 and the structure of diethyl ether is $CH_3CH_2OCH_2CH_3$.

27. The generic structure of an aldehyde is $R{-}CHO$; for example, propanal is CH_3CH_2CHO. The generic structure of a ketone is $R{-}CO{-}R$; for example, acetone is CH_3COCH_3.

29. The generic structure of a carboxylic acid is $R{-}COOH$ for example, acetic acid is CH_3COOH. The generic structure of an ester is $R{-}COO{-}R$; for example, ethyl propanoate is $CH_3CH_2COOCH_2CH_3$.

31. An amine is an organic compound containing nitrogen, the structure being NR_x. Two examples are methylamine, CH_3NH_2, and ethylamine, $CH_3CH_2NH_2$.

33. A polymer is a long chainlike molecule composed of repeating units called *monomers*. A copolymer consists of two different kinds of monomers.

Problems:

35. c, d

37. **a.** alkane **b.** alkene
    c. alkyne **d.** alkene

39. **a.**
$$\begin{array}{ccccccc} & & H & & H & & \\ & & | & & | & & \\ H & - & C & - & C & - & H \\ & & | & & | & & \\ & & H & & H & & \end{array} \qquad CH_3CH_3$$

b.
$$\begin{array}{ccccccccccc} & & H & & H & & H & & H & & \\ & & | & & | & & | & & | & & \\ H & - & C & - & C & - & C & - & C & - & H \\ & & | & & | & & | & & | & & \\ & & H & & H & & H & & H & & \end{array} \qquad CH_3CH_2CH_2CH_3$$

c.
$$\begin{array}{ccccccccccccc} & & H & & H & & H & & H & & H & & \\ & & | & & | & & | & & | & & | & & \\ H & - & C & - & C & - & C & - & C & - & C & - & H \\ & & | & & | & & | & & | & & | & & \\ & & H & & H & & H & & H & & H & & \end{array}$$
$$CH_3CH_2CH_2CH_2CH_3$$

d.
$$\begin{array}{ccccc} & & H & & \\ & & | & & \\ H & - & C & - & H \\ & & | & & \\ & & H & & \end{array} \qquad CH_4$$

41. $H_3C{-}CH_2{-}CH_2{-}CH_3$ $\qquad$ $\begin{array}{c} H_3C{-}\underset{\displaystyle CH_3}{\underset{|}{CH}}{-}CH_3 \end{array}$

43. $H_3C{-}CH_2{-}CH_2{-}CH_2{-}CH_2{-}CH_3$

$H_3C{-}CH_2{-}\underset{\displaystyle CH_3}{\underset{|}{CH}}{-}CH_2{-}CH_3$

$H_3C{-}CH_2{-}CH_2{-}\underset{\displaystyle CH_3}{\underset{|}{CH}}{-}CH_3$

$H_3C{-}CH_2{-}\overset{\displaystyle CH_3}{\overset{|}{\underset{\displaystyle CH_3}{\underset{|}{C}}}}{-}CH_3$

$H_3C{-}\underset{\displaystyle CH_3}{\underset{|}{CH}}{-}\underset{\displaystyle CH_3}{\underset{|}{CH}}{-}CH_3$

45. **a.** n-pentane **b.** 2-methylbutane
    c. 4-ethyl-2-methylhexane **d.** 3,3-dimethylpentane

47. **a.** $H_3C{-}\underset{\displaystyle CH_3}{\underset{|}{CH}}{-}CH_2{-}CH_3$

b. $H_3C{-}\underset{\displaystyle CH_3}{\underset{|}{CH}}{-}\overset{\displaystyle CH_3}{\overset{|}{\overset{\displaystyle CH_2}{\overset{|}{CH}}}}{-}CH_2{-}CH_2{-}CH_3$

c. $H_3C{-}CH_2{-}\underset{\displaystyle \underset{\displaystyle CH_3}{\underset{|}{HC{-}CH_3}}}{\underset{|}{CH}}{-}CH_2{-}CH_2{-}CH_2{-}CH_3$

d. $H_3C{-}\underset{\displaystyle CH_3}{\underset{|}{CH}}{-}CH_2{-}CH_2{-}\underset{\displaystyle CH_3}{\underset{|}{CH}}{-}CH_2{-}CH_2{-}CH_3$

49. **a.** n-pentane **b.** 3-methylhexane
    c. 2,3-dimethylpentane

51.

Name	Molecular Formula	Structural Formula	Condensed Structural Formula
2,2,3-trimethyl-pentane	C_8H_{18}	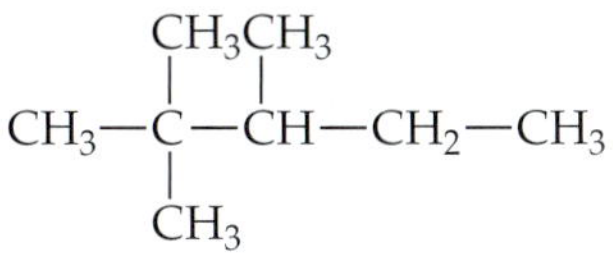 $CH_3\ CH_3$ (on C and CH) $CH_3—C—CH—CH_2—CH_3$ CH_3 (on C)	$CH_3C(CH_3)_2CH(CH_3)CH_2CH_3$
2-methyl-3-propyl-hexane	$C_{10}H_{22}$	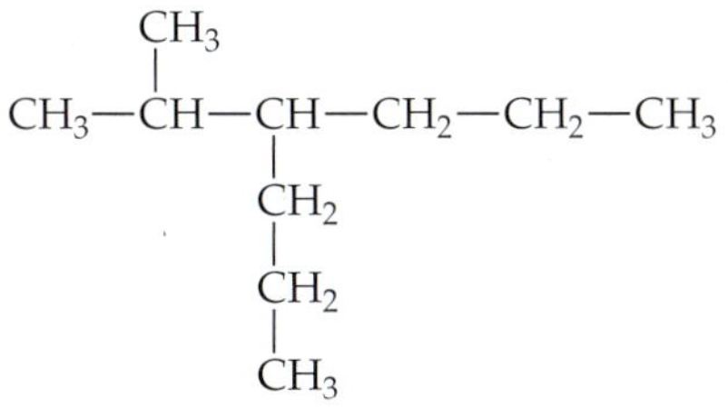 CH_3 (on first CH) $CH_3—CH—CH—CH_2—CH_2—CH_3$ $CH_2—CH_2—CH_3$ (on second CH)	$CH_3CH(CH_3)CH(CH_2CH_2CH_3)CH_2CH_2CH_3$
2,2,3,3-tetramethyl-hexane	$C_{10}H_{22}$	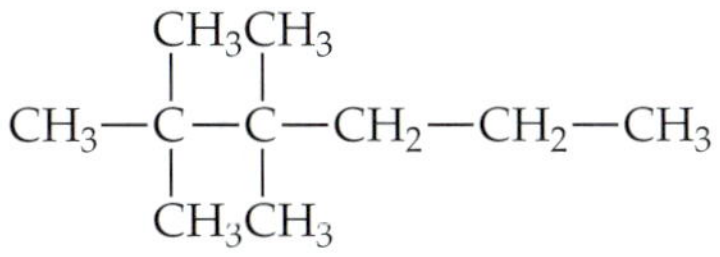 $CH_3\ CH_3$ (above) $CH_3—C—C—CH_2—CH_2—CH_3$ $CH_3\ CH_3$ (below)	$CH_3C(CH_3)_2C(CH_3)_2CH_2CH_2CH_3$
4,4-diethyl-2,3-dimethylhexane	$C_{12}H_{26}$	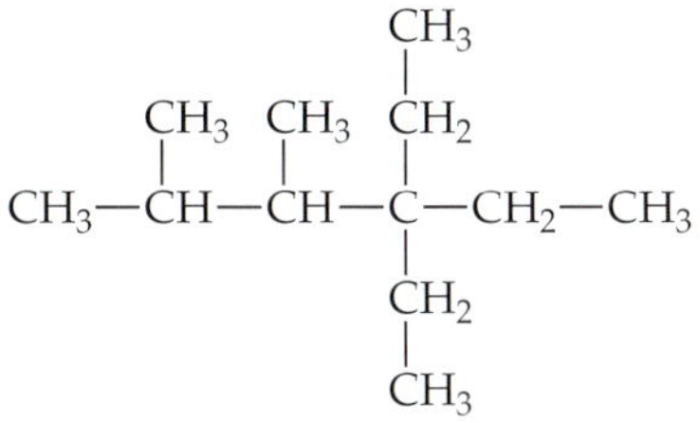 $CH_3\ \ CH_3\ \ CH_2CH_3$ (above) $CH_3—CH—CH—C—CH_2—CH_3$ CH_2CH_3 (below C)	$CH_3CH(CH_3)CH(CH_3)CH(CH_2CH_3)_2CH_2CH_3$

53. 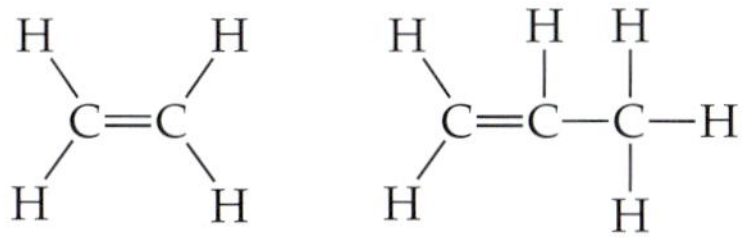 H₂C=CH₂ drawn with all H: H, H on each C, $C=C$; and $H_2C=C(H)—CH_3$ drawn with all H.

$CH_2{=}CH_2$; $CH_2{=}CHCH_3$

55. $H_2C{=}CH—CH_2—CH_2—CH_3$

$H_3C—CH{=}CH—CH_2—CH_3$

57. **a.** 2-pentene
b. 4-methyl-2-pentene
c. 3,3-dimethyl-1-butene
d. 3,4-dimethyl-1-hexene

59. **a.** 2-butyne
b. 4-methyl-2-pentyne
c. 4,4-dimethyl-2-hexyne
d. 3-ethyl-3-methyl-1-pentyne

61. **a.** $H_3C—CH{=}CH—CH_2—CH_2—CH_3$
b. $H_3C—CH_2—C{\equiv}C—CH_2—CH_2—CH_3$
c. 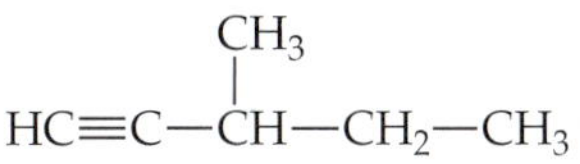 $HC{\equiv}C—CH(CH_3)—CH_2—CH_3$
d. 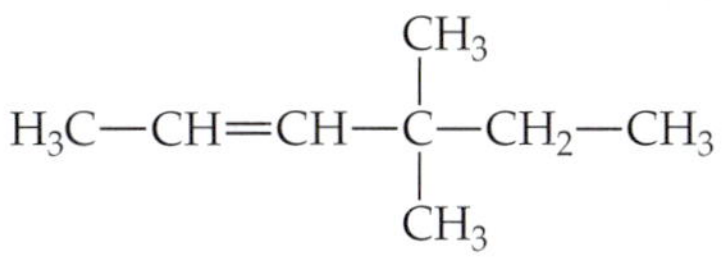 $H_3C—CH{=}CH—C(CH_3)_2—CH_2—CH_3$

63. $H_2C{=}CH—CH_2—CH_2—CH_3$ 1-pentene
$H_3C—CH{=}CH—CH_2—CH_3$ 2-pentene 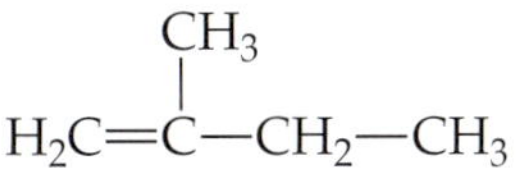
$H_2C{=}C(CH_3)—CH_2—CH_3$ 2-methyl-1-butene 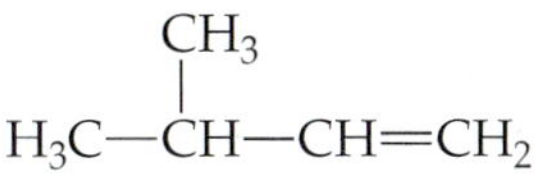
$H_3C—CH(CH_3)—CH{=}CH_2$ 3-methyl-1-butene 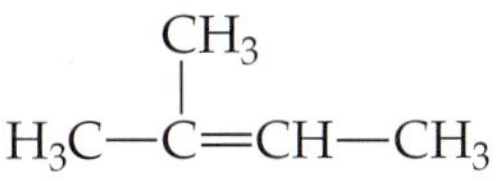
$H_3C—C(CH_3){=}CH—CH_3$ 2-methyl-2-butene

65.

Name	Molecular Formula	Structural Formula	Condensed Structural Formula
2,2-dimethyl-3-hexene	C_8H_{16}	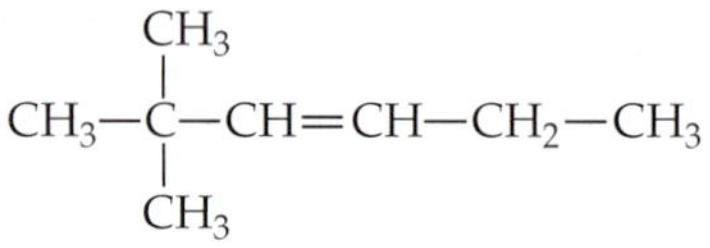 $CH_3-C(CH_3)_2-CH=CH-CH_2-CH_3$	$CH_3C(CH_3)_2CH=CHCH_2CH_3$
4,4-diethyl-5,5-dimethyl-2-hexyne	$C_{12}H_{22}$	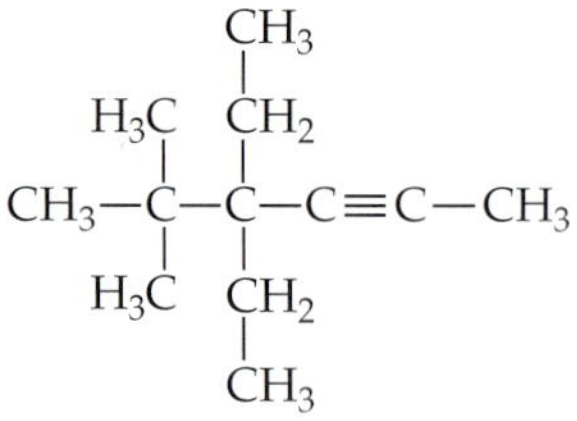 $CH_3-C(CH_3)_2-C(CH_2CH_3)_2-C\equiv C-CH_3$	$CH_3C(CH_3)_2C(CH_2CH_3)_2C\equiv CCH_3$
3,4-dimethyl-1-octyne	$C_{10}H_{18}$	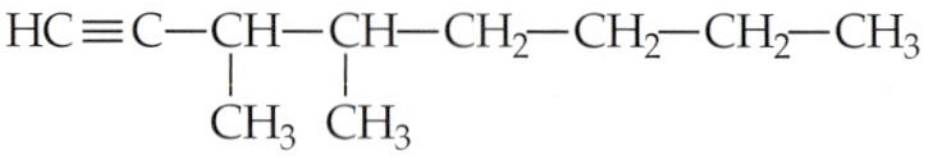 $HC\equiv C-CH(CH_3)-CH(CH_3)-CH_2-CH_2-CH_2-CH_3$	$CH\equiv CCH(CH_3)CH(CH_3)CH_2CH_2CH_2CH_3$
4,4-diethyl-5,5-dimethyl -2-hexene	$C_{12}H_{24}$	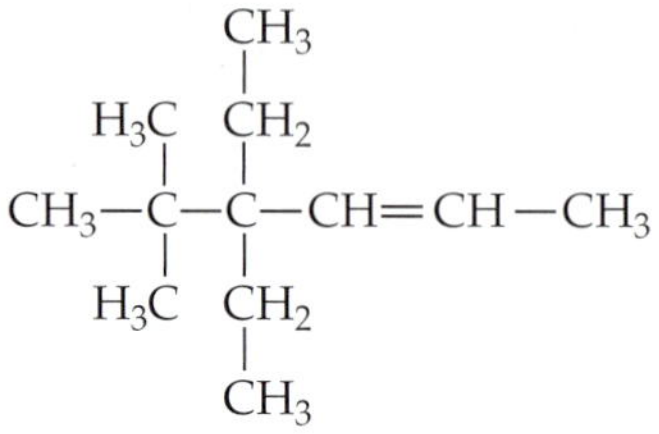 $CH_3-C(CH_3)_2-C(CH_2CH_3)_2-CH=CH-CH_3$	$CH_3C(CH_3)_2C(CH_2CH_3)_2CH=CHCH_3$

67. **a.** $2\,CH_3CH_3(g) + 7\,O_2(g) \longrightarrow 4\,CO_2(g) + 6\,H_2O(g)$

b. $2\,CH_2=CHCH_3(g) + 9\,O_2(g) \longrightarrow 6\,CO_2(g) + 6\,H_2O(g)$

c. $2\,CH=CH(g) + 5\,O_2(g) \longrightarrow 4\,CO_2(g) + 2\,H_2O(g)$

69. $CH_4(g) + Br_2(g) \longrightarrow CH_3Br(g) + HBr(g)$

71. $CH_3CH=CHCH_3(g) + Cl_2(g) \longrightarrow CH_3CHClCHClCH_3(g)$

73. $CH_2=CH_2(g) + H_2(g) \longrightarrow CH_3CH_3(g)$

75. (Kekulé resonance structures of benzene, C_6H_6, with each C bearing one H: H–C ring with alternating C=C double bonds $\rightleftharpoons$ the ring with the double bonds in the alternate positions)

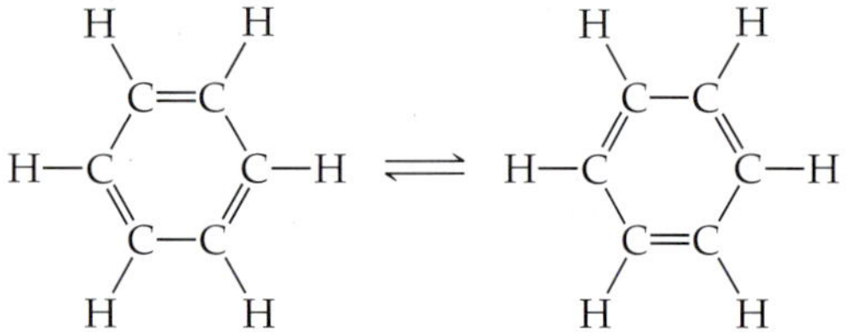

77. **a.** fluorobenzene
b. isopropylbenzene
c. ethylbenzene

79. **a.** 4-phenyloctane
b. 5-phenyl-3-heptene
c. 7-phenyl-2-heptyne

81. **a.** 1-bromo-2-chlorobenzene
b. 1,2-diethylbenzene or orthodiethylbenzene
c. 1,3-diflorobenzene or metadifluorobenzene

83. **a.** $C_6H_5-CH_2-CH_2-CH_2-CH_3$ (benzene ring bonded to the chain)

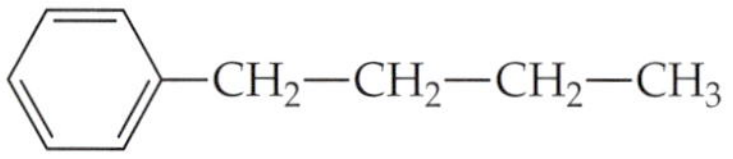

b. benzene ring with H_3C-CH_2 and, on the adjacent carbon, I

c. benzene ring with CH_3 at top and CH_3 at bottom (para positions)

85. ether = $R-O-R$

aldehyde = $R-C(=O)-H$

amine = $R-N(R)-R$

ketone = $R-C(=O)-R$

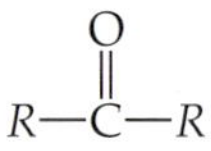

87. **a.** 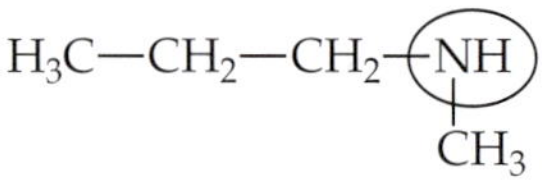 $H_3C-CH_2-CH_2-\text{(NH)}-CH_3$ (NH circled) amine

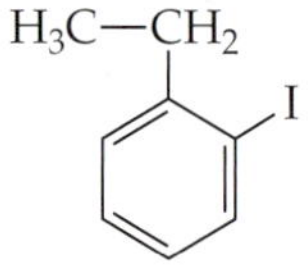

b. 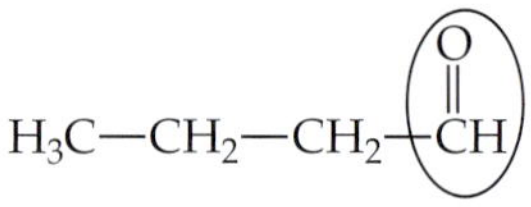 $H_3C-CH_2-CH_2-\text{(CH=O)}$ (C=O–H circled) aldehyde

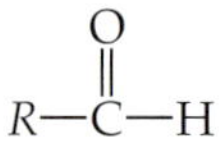

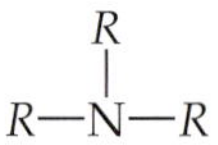

c. 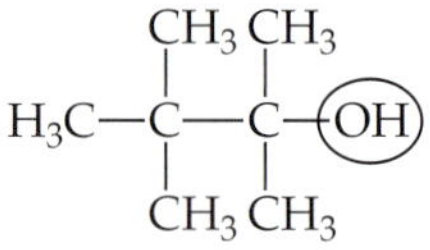 $H_3C-C(CH_3)_2-C(CH_3)_2-\text{(OH)}$ (OH circled) alcohol

d. 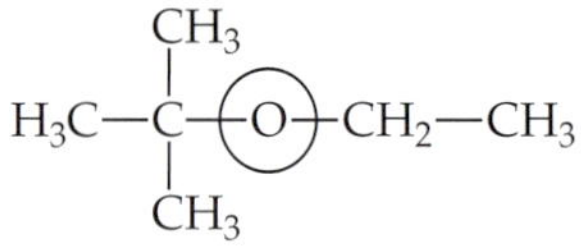 $H_3C-C(CH_3)_2-\text{(O)}-CH_2-CH_3$ (O circled) ether

89. **a.** 2-butanol **b.** 2-methyl-1-propanol
c. 3-ethyl-1-hexanol **d.** 3-methyl-3-pentanol

91. **a.** $CH_3-CH_2-CH(OH)-CH_2-CH_3$
b. $CH_2(OH)-CH(CH_3)-CH_2-CH_3$
c. $CH_3-CH(OH)-CH(CH_2CH_3)-CH_2-CH_2-CH_3$
d. CH_3-CH_2-OH

93. **a.** $CH_3-CH_2-CH_2-CH_2-O-CH_2-CH_2-CH_2-CH_3$
b. ethyl propyl ether
c. dipropyl ether
d. $CH_3-O-CH_2-CH_2-CH_2-CH_2-CH_3$

95. **a.** $CH_3-CH_2-CH_2-CH_2-CH_2-CH_2-CH_2-\overset{O}{\overset{\|}{C}}H$
b. butanal
c. 4-heptanone
d. $CH_3-CH_2-\overset{O}{\overset{\|}{C}}H-CH_2-CH_2-CH_3$

97. **a.** $CH_3-CH_2-CH_2-CH_2-CH_2-CH_2-CH_2-\overset{O}{\overset{\|}{C}}-OH$
b. methyl ethanoate
c. $H_3C-CH_2-CH_2-\overset{O}{\overset{\|}{C}}-O-CH_2-CH_3$
d. heptanoic acid

99. **a.** $CH_3-CH_2-N(H)-CH_2-CH_3$
b. triethylamine
c. butylpropylamine

101. $\left[-CH_2-C(CH_3)_2-\right]_n$

103. $HO-\overset{O}{\overset{\|}{C}}-C_6H_4-\overset{O}{\overset{\|}{C}}-O-CH_2-CH_2-OH$ (ester group circled)

105. **a.** alcohol **b.** amine
c. alkane **d.** carboxylic acid
e. ether **f.** alkene

107. **a.** 3-methyl-4-tert-butylheptane
b. 3-methyl butanal
c. 4-isopropyl-3-methyl-2-heptene
d. propyl butanoate

109. **a.** same molecule
b. isomers
c. same molecule

111. $CH_2{=}CH_2 + HCl \longrightarrow CH_3CH_2Cl$

113. 558 g H_2

115. **a.** alcohol **b.** amine
c. carboxylic acid **d.** ester
e. alkane **f.** ether

Chapter 19

Questions

1. The human genome project is a fifteen-year project to map all of the genetic material of a human being.
3. Some of the expected benefits are the identification of individuals who are susceptible to certain diseases and the development of new drugs.
5. Carbohydrates are organic compounds having the generic formula $(CH_2O)_n$. They are the primary molecules responsible for short-term energy storage in living organisms and also form the main structural components of plants.
7. Glucose is soluble in water due to its many —OH groups. This is important because glucose is the primary fuel of cells and can be easily transported in the bloodstream.
9. During digestion the links in disaccharides and polysaccharides are broken, allowing individual monosaccharides to pass through the intestinal wall and enter the bloodstream.
11. Starch and cellulose are both polysaccharides, but the bond between saccharide units is slightly different. Consequently humans can digest starch and use it for energy, whereas cellulose cannot be digested and passes directly through humans.
13. Lipids function as structural units of cells and provide long-term energy storage and insulation.
15. The general structure of a fatty acid is $R-\overset{O}{\overset{\|}{C}}-OH$ where R is 3 to 19 carbon atoms long.
17. A triglyceride is a triester composed of glycerol with three fatty acids attached.
19. A saturated fat is a triglyceride that tends to be solid at room temperature, and it does not have any double bonds in its carbon chains. An unsaturated fat is a triglyceride that tends to be liquid at room temperature and contains double bonds in its carbon chains.
21. Phospholipids and glycolipids function in the body as cell membranes.
23. Cholesterol is a steroid that is part of cell membranes and also serves as a starting material for the body to synthesize other steroids. Also, steroids serve as male and female hormones in the body.

25. Proteins serve as catalysts; structural units of muscle, skin, and cartilage; transporters of oxygen; disease-fighting antibodies; and as hormones.

27. Amino acids differ from each other only in their *R* group or side chain.

29. $$H_2N-\underset{R_1}{\overset{H}{C}}-\overset{O}{\overset{\|}{C}}-OH + H_2N-\underset{R_2}{\overset{H}{C}}-\overset{O}{\overset{\|}{C}}-OH \longrightarrow$$

$$H_2N-\underset{R_1}{\overset{H}{C}}-\overset{O}{\overset{\|}{C}}-NH-\underset{R_2}{\overset{H}{C}}-\overset{O}{\overset{\|}{C}}-OH + H_2O$$

31. Primary protein structure refers to the sequence of amino acids in the protein's chain. Primary protein structure is maintained by the covalent peptide bonds between individual amino acids.

33. Tertiary protein structure refers to the large-scale twists and folds within the protein. These are maintained by interactions between the *R* groups of amino acids that are separated by long distances in the chain sequence.

35. In the α-helix structure, the amino acid chain is wound into a tight coil by hydrogen bonding between C=O and N—H groups at different locations along the backbone. The side chains extend outward. In the β-pleated sheet structure, the amino acid chain doubles back on itself repeatedly in a zig-zag pattern, with adjacent sections held together by hydrogen bonding between C=O and N—H groups along the backbone. The resulting structure is an undulating sheet with the side chains extending above and below it.

37. Nucleic acids contain a chemical code that specifies the correct amino acid sequences for proteins.

39. The four different bases that occur within DNA are adenine (A), cytosine (C), guanine (G), and thymine (T).

41. The genetic code is the code that links a specific codon to an amino acid.

43. A gene is a sequence of codons within a DNA molecule that codes for a single protein. Genes vary in length from fifty to thousands of codons.

45. Chromosomes located within the nuclei of cells are structures containing genes.

47. No; most cells in the human body only synthesize proteins that are important to their function.

49. **a.** The complementary base of adenine (A) is thymine (T).
b. The complementary base of thymine (T) is adenine (A).
c. The complementary base of cytosine (C) is guanine (G).
d. The complementary base of guanine (G) is cytosine (C).

Problems

51. **a.** monosaccharide **b.** not a carbohydrate
c. not a carbohydrate **d.** disaccharide

53. **a.** hexose
b. tetrose
c. pentose
d. tetrose

55.

$HC=O$
$H-C-OH$
$HO-C-H$
$H-C-OH$
$H-C-OH$
H_2C-OH

57.

Glucose Fructose

59. **a.** fatty acid, saturated
b. steroid
c. triglyceride, unsaturated
d. not a lipid

61.

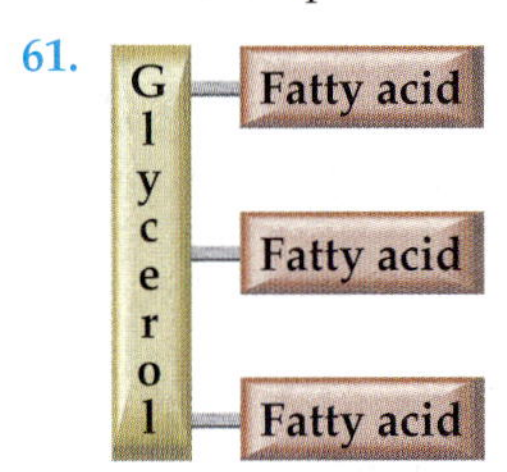

Triglyceride

63. fat:

$H_2C-O-C(=O)-CH_2-CH_2-CH_2-CH_2-CH_2-CH_2-CH_2-CH_2-CH_3$
$HC-O-C(=O)-CH_2-CH_2-CH_2-CH_2-CH_2-CH_2-CH_2-CH_2-CH_3$
$H_2C-O-C(=O)-CH_2-CH_2-CH_2-CH_2-CH_2-CH_2-CH_2-CH_2-CH_3$

65. b, d

67.

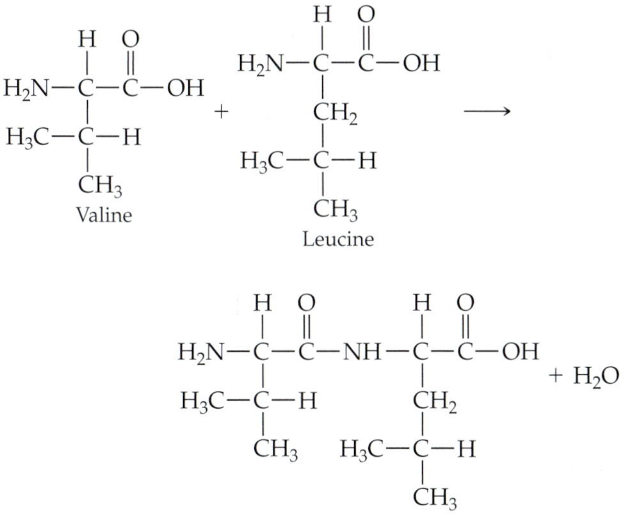

$$H_2N-CH(CH(CH_3)_2)-COOH \;(\text{Valine}) + H_2N-CH(CH_2CH(CH_3)_2)-COOH \;(\text{Leucine}) \longrightarrow$$

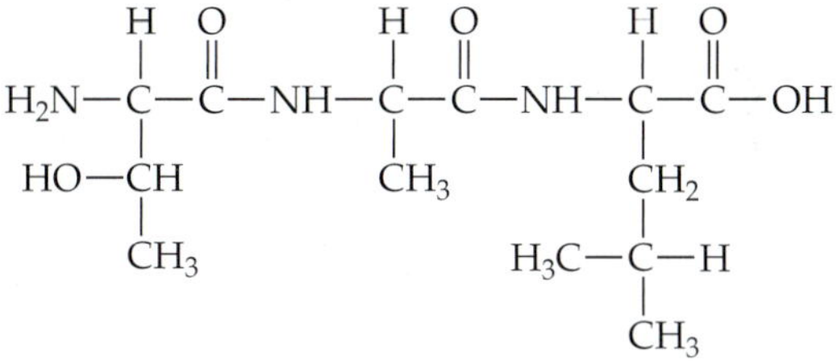

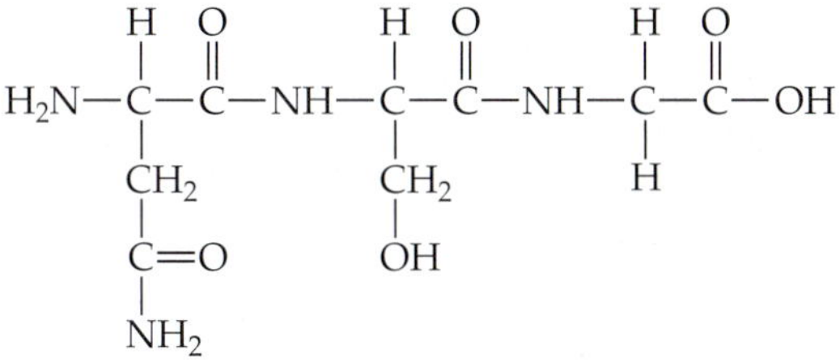

$$H_2N-CH(CH(CH_3)_2)-C(=O)-NH-CH(CH_2CH(CH_3)_2)-C(=O)-OH + H_2O$$

69. **a.** $H_2N-CH(CH(OH)CH_3)-C(=O)-NH-CH(CH_3)-C(=O)-NH-CH(CH_2CH(CH_3)_2)-C(=O)-OH$

b. $H_2N-CH(CH_2C(=O)NH_2)-C(=O)-NH-CH(CH_2OH)-C(=O)-NH-CH_2-C(=O)-OH$

c. $H_2N-CH(CH(CH_3)_2)-C(=O)-NH-CH(CH_2C_6H_5)-C(=O)-NH-CH(CH_3)-C(=O)-OH$

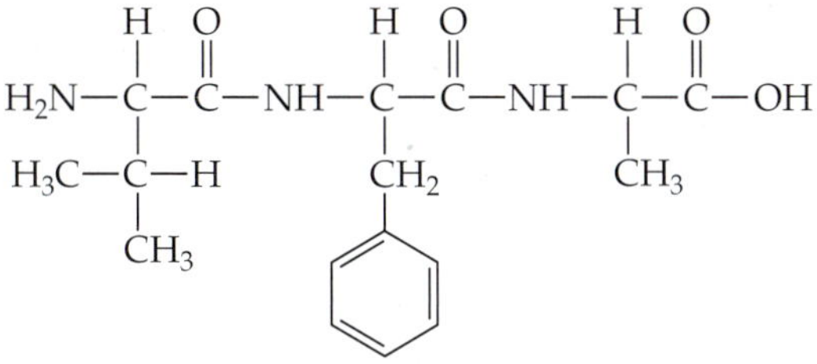

71. tertiary

73. primary

75. **a.** nucleotide, G
b. not a nucleotide
c. not a nucleotide
d. not a nucleotide

77. T T A C G C G

79.

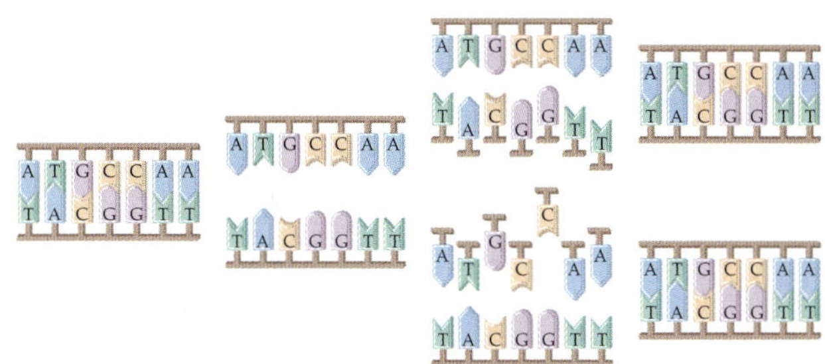

81. **a.** glycoside linkage—carbohydrates
b. peptide bonds—proteins
c. ester linkage—triglycerides

83. **a.** glucose—short-term energy storage
b. DNA—blueprint for proteins
c. phospholipids—compose cell membranes
d. triglycerides—long-term energy storage

85. **a.** codon—codes for a single amino acid
b. gene—codes for a single protein
c. genome—all of the genetic material of an organism
d. chromosome—structure that contains genes

87. Nitrogen: tetrahedral electron geometry, trigonal pyramidal molecular geometry

1st Carbon: tetrahedral electron geometry, tetrahedral molecular geometry

2nd Carbon: trigonal planar electron geometry, trigonal planar molecular geometry

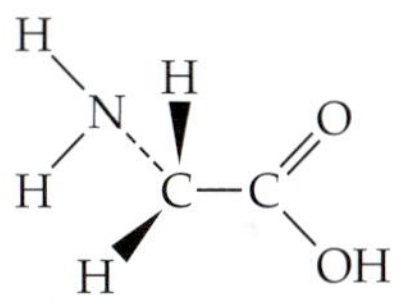

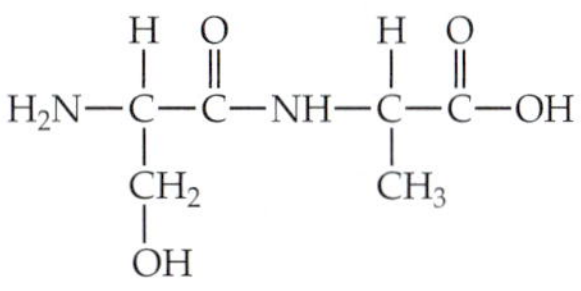

$$H_2N-CH_2-C(=O)-OH$$

89.

ser-ala: $H_2N-CH(CH_2OH)-C(=O)-NH-CH(CH_3)-C(=O)-OH$

ala-ser: $H_2N-CH(CH_3)-C(=O)-NH-CH(CH_2OH)-C(=O)-OH$

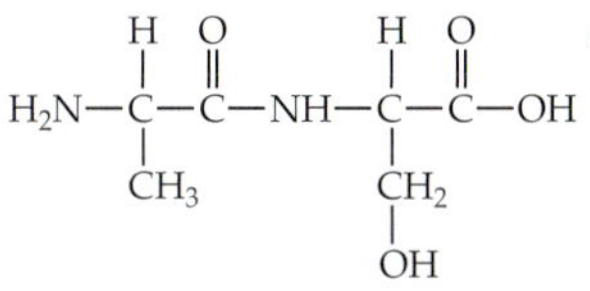

The difference lies in the end groups. In ser–ala, serine has the amine end and alanine has the carboxyl end. For ala–ser, the reverse is true.

91. gly-arg-ala-ser-phe-gly-asn-lys-trp-glu-val

93. 153 base pairs

95. The actual thymine-containing nucleotide uses the —OH end to bond and replicate; however, with the fake nucleotide having a nitrogen-based end instead, the possibility of replication is halted.

PHOTO CREDITS

FM Page v: Professor Nivaldo Jose Tro. **Chapter 1** Page 1: Richard Megna/Fundamental Photographs. Page 3 (Tl): Richard Megna/Fundamental Photographs. Page 3 (Cl): LESTER LEFKOWITZ/Getty Images, Inc.-Taxi. Page 3 (C): Richard Megna/Fundamental Photographs. Page 4: Kevin Fleming/Corbis/Bettmann. Page 5 (Tl): Paintings. French. David, Jacques Louis (1748–1825), French. "Antoine Laurent Lavoisier (1743–1794) and His Wife (Marie Anne Pierrette Paulze, 1758–1836)." Oil on canvas. H. 102-1/4 in. W. 76-5/8 in. (259.7 × 194.6 cm.) in. The Metropolitan Museum of Art, Purchase, Mr. and Mrs. Charles Wrightsman Gift, in honor of Everett Fahy, 1977. (1977.10) Photograph © 1989 The Metropolitan Museum of Art". Page 5 (Tr): Corbis/Bettmann. Page 5 (Bl): IBM Research, Almaden Research Center (Unauthorized use is prohibited.) Page 6: Science Photo Library/Photo Researchers, Inc. Page 7: Andy Sacks/Getty Images Inc.-Stone Allstock. **Chapter 2** Page 12: Physics Department, Imperial College, London/Photo Researchers, Inc. Page 14 (Tl): Richard Megna/Fundamental Photographs. Page 14 (Cl): Paul Silverman/Fundamental Photographs. Page 14 (Bl): Richard Megna/Fundamental Photographs. Page 14 (Br): Richard Megna/Fundamental Photographs. Page 15 (Tl): Richard Megna/Fundamental Photographs. Page 15 (C): Richard Megna/Fundamental Photographs. Page 17: NASA Headquarters. Page 22: Richard Megna/Fundamental Photographs. Page 23 (Bl): Photo courtesy of the National Institute of Standards and Technology Museum. Page 23 (Bc): International Bureau of Weights and Measures, Sevres, France. Page 23 (Br): Dr. Donald Sullivan/National Institute of Standards and Technology (NIST). Page 23 (Cl): Richard Megna/Fundamental Photographs. Page 24: Richard Megna/Fundamental Photographs. Page 30: David R. Frazier Photolibrary, Inc. Page 33: Courtesy Litespeed Titanium. Page 38: Richard Megna/Fundamental Photographs. Page 48: Courtesy of NASA/JPL/California Institute of Technology Archives. Page 49 (T): © NASA/Science Photo Library/Photo Researchers, Inc. Page 49 (C): Courtesy Litespeed Titanium. **Chapter 3** Page 52 (Tl): Getty Images, Inc.-Photodisc. Page 52 (Tr): Getty Images, Inc. Page 52 (Bl): IBM Research, Almaden Research Center (Unauthorized use is prohibited). Page 52 (Br): R.J. Driscoll, M.G. Youngquist, & J.D. Baldeschwieler, California Institute of Technology/Science Photo Library/Photo Researchers, Inc. Page 53 (Tl): David Chasey/Getty Images, Inc.-Photodisc. Page 53 (Tc): Getty Images, Inc.-PhotoDisc. Page 53 (Tr): Mark Downey/Getty Images, Inc.-Photodisc. Page 53 (B): Albert Copley/Visuals Unlimited. Page 54 (B): Ryan McVay/Getty Images, Inc.-Photodisc. Page 55 (B): Adalberto Rios/Getty Images, Inc.-Photodisc. Page 55 (Tl): Richard Megna/Fundamental Photographs. Page 55 (Tcl): Richard Magna/Fundamental Photographs. Page 55 (Tcr): Imagination Photo Design. Page 55 (Tr): Kip Peticolas/Fundamental Photographs. Page 55 (B): Getty Images, Inc.-Photodisc. Page 56: Martial Colomb/Getty Images, Inc.-Photodisc. Page 57 (T): Getty Images, Inc.-Photodisc. Page 57 (C): Getty Images, Inc.-Photodisc. Page 59 (Tl): Richard Megna/Fundamental Photographs. Page 59 (Tr): Richard Megna/Fundamental Photographs. Page 61: Jacques Jangoux/Peter Arnold, Inc. Page 67: Larry Brownstein/Getty Images, Inc.-Photodisc. Page 68: © Squared Studios/Getty Images, Inc.-Photodisc. Page 82 (C): Spike/Getty Images, Inc.-Photodisc. Page 83: Hendrie, Peter/Getty Images Inc.-Image Bank. **Chapter 4** Page 86 (C): Courtesy of the Library of Congress. Page 86 (Tl): Siede Preis/Getty Images, Inc.-Photodisc. Page 86 (Tr): Corbis Digital Stock. Page 86 (B): IBM Research, Almaden Research Center (Unauthorized use is prohibited.) Page 89 (C): Richard Megna/Fundamental Photographs. Page 90 (B): Jeremy Woodhouse/Getty Images, Inc.-PhotoDisc. Page 91 (Tl): Mel Curtis/Getty Images, Inc.-Photodisc. Page 91 (Tr): John A. Rizzo/Getty Images, Inc.-Photodisc. Page 92 (Br): Charles D. Winters/Photo Researchers, Inc. Page 92 (Bl): AIP Emilio Segre Visual Archives, W.F. Meggers Gallery of Nobel Laureates. Page 94 (T): Stamp from the private collection of Professor C. M. Lang, photography by Gary J. Shulfer, University of Wisconsin, Stevens Point. "Russia: #3607 (1969);" Scott Standard Postage Stamp Catalogue, Scott Pub. Co., Sidney, Ohio. Page 94 (B): Getty Images, Inc.-PhotoDisc. Page 96 (Brt): John A. Rizzo/Getty Images, Inc.- Photodisc. Page 96 (Br): Michael Dalton/Fundamental Photographs. Page 97 (Tlt): Richard Megna/Fundamental Photographs. Page 97 (Tlb): Ed Degginger/Color-Pic, Inc. Page 97 (Tct): Richard Megna/Fundamental Photographs. Page 97 (Tcb): Ed Degginger/Color-Pic, Inc. Page 97 (Trt): Ed Degginger/Color-Pic, Inc. Page 97 (Trb): Tom Bochsler/Pearson Education/PH College. Page 101 (B): D. Falconer/Getty Images, Inc.-PhotoDisc. Page 104: U.S. Department of Energy. **Chapter 5** Page 117 (Bl): Richard Megna/Fundamental Photographs. Page 117 (Br): Stephen Frisch/Stock Boston. Page 118: Yoav Levy/Phototake NYC. Page 119 (Tl): Nancy R. Cohen/Getty Images, Inc.-Photodisc. Page 119 (Tr): JoLynn E. Funk. Page 119 (Bl): Sami Sarkis/Getty Images, Inc.-Photodisc. Page 119 (Br): Russell Illig/Getty Images, Inc.-Photodisc. Page 123 (Bl): Harry Taylor/© Dorling Kindersley, Courtesy of the Natural History Museum, London. Page 123 (Br): Charles D. Winters/Photo Researchers, Inc. Page 124 (Bl): Ken Karp/Omni-Photo Communications, Inc. Page 124 (Br): Dr. E.R. Degginger/Bruce Coleman Inc. Page 132 (Cl): Michael Dalton/Fundamental Photographs. Page 132 (Cr): Michael Dalton/Fundamental Photographs. Page 138 (Cl): Oliver Strewe/Getty Images Inc.-Stone Allstock. Page 138 (Tr): Richard Megna/Fundamental Photographs. Page 150 (Bl): Getty Images Inc./Hulton Archive Photos. Page 150 (Br): Photo Researchers, Inc. Page 150 (Tl): Charles D. Winters/Photo Researchers, Inc. Page 150 (Tr): Richard Megna/Fundamental Photographs. Page 150 (Cl): Richard Megna/Fundamental Photographs. Page 150 (Cr): Richard Megna/Fundamental Photographs. Page 151 (Cl): Richard Megna/Fundamental Photographs. Page 151 (Tr): Jonathan Nourok/PhotoEdit. Page 151 (Cr): Richard Megna/Fundamental Photographs. **Chapter 6** Page 153 (Bl): Charles O'Rear/Corbis/Bettmann. Page 153 (Br): Joseph P. Sinnot/Fundamental Photographs. Page 154 (C): Richard Megna/Fundamental Photographs. Page 154 (B): Richard Megna/Fundamental Photographs. Page 155 (T): Richard Megna/Fundamental Photographs. Page 155 (B): Nancy R. Cohen/Getty Images, Inc.-Photodisc. Page 157 (al): Kristen Brochmann/Fundamental Photographs. Page 157 (ar): Kristen Brochmann/Fundamental Photographs. Page 157 (b): Richard Megna/Fundamental Photographs. Page 162: Siede Preis/Getty Images, Inc.-Photodisc. Page 163 (Tl): G.K. & Vikki Hart/Getty Images, Inc.-Photodisc. Page 163 (Tc): © Squared Studios/Getty Images, Inc.-Photodisc. Page 166 (B): NASA/Tom Pantages. Page 167: Richard Megna/Fundamental Photographs. Page 171: Yoav Levy/Phototake NYC. Page 186 (Bl): Jeff Maloney/Getty Images, Inc.-Photodisc. Page 186 (Bc): Getty Images, Inc.-PhotoDisc. Page 186 (Br): Figure prepared by Dr. Jay Herman, Laboratory for Atmospheres/NASA-Goddard Space Flight Center. Page 187 (T): NASA/Johnson Space Center. **Chapter 7** Page 190 (Cl): Carey B. Van Loon/Carey B. Van Loon. Page 191 (Bl): Richard Megna/Fundamental Photographs. Page 191 (Br): Tom Bochsler/Pearson Education/PH College. Page 191 (Tl): Ed Degginger/Color-Pic, Inc. Page 191 (Tr): Tristan Paviot/Getty Images Inc.-Stone Allstock. Page 192 (T): Richard Megna/Fundamental Photographs. Page 192 (Ct): Richard Megna/ Fundamental Photographs. Page 192 (C): Richard Megna/Fundamental Photographs. Page 192 (Cb): Jeff J. Daly/Fundamental Photographs. Page 192 (Tb): Richard Megna/Fundamental Photographs. Page 192 (Bl): Getty Images, Inc.-Photodisc. Page 195: Sami Sarkis/Getty Images, Inc.-Photodisc. Page 199 (a,b): Ryan McVay/Getty Images, Inc.-PhotoDisc. Page 202 (Tl): Richard Megna/ Fundamental Photographs. Page 202 (Cl): Richard Megna/Fundamental Photographs. Page 207 (Tl): Michael P. Gadomski/Photo Researchers, Inc. Page 207 (Tr): Michael Dalton/Fundamental Photographs. Page 209: E.R. Degginger/Color-Pic, Inc. Page 210 (Tl): Ed Degginger/Color-Pic, Inc. Page 210 (Tr): PhotoEdit. Page 213 (L): Richard Megna/Fundamental Photographs. Page 213 (C): Richard Megna/Fundamental Photographs. Page 213 (R): Richard Megna/Fundamental Photographs. Page 214: Richard Megna/Fundamental Photographs. Page 230 (Tr): Fundamental Photographs. Page 230 (Bl): Richard Megna/Fundamental Photographs. Page 230 (B): Richard Megna/Fundamental Photographs. **Chapter 8** Page 234 (Bl): Kristen Brochmann/Fundamental Photographs. Page 234 (Br): Kristen Brochmann/Fundamental Photographs. Page 234 (Cl): Kristen Brochmann/Fundamental Photographs. Page 234 (C): Kristen Brochmann/Fundamental Photographs. Page 234 (Cr): Kristen Brochmann/Fundamental Photographs. Page 234 (Cr): Kristen Brochmann/Fundamental Photographs. Page 235 (Tl): Kristen Brochmann/ Fundamental Photographs. Page 235 (Tr): Kristen Brochmann/Fundamental Photographs. Page 237 (B): Cheryl Sheridan. Page 241 (Bl): Kristen Brochmann/ Fundamental Photographs. Page 241 (Bc): Kristen Brochmann/Fundamental Photographs. Page 241 (Br): Kristen Brochmann/Fundamental Photographs. Page 248 (a,b,c,d): Richard Megna/Fundamental Photographs. Page 259 (a,b,c,d): Nancy R. Cohen/Getty Images, Inc.-Photodisc. Page 260 (T,a,b,c,d): Stephen Frish/Stock Boston. **Chapter 9** Page 263: AP Wide World Photos. Page 264 (Bl): University Archives. Department of Rare Books and Special Collections. Princeton University Library. Page 264 (Br): Photo by Francis Simon, courtesy of AIP Emilio Segre Visual Archives, Francis Simon Collection. Page 265 (T): Randy Olson/Aurora & Quanta Productions Inc. Page 266 (T): Rob Gage/Getty Images, Inc.-Taxi. Page 267 (B): Sierra Pacific Innovations Corporation. Page 269 (Tl): Larry Mulvehill/Photo Researchers, Inc. Page 269 (B): Dunwell Photography, Steve/Getty Images Inc.-Image Bank. Page 270 (Tl): Richard Megna/Fundamental Photographs. Page 270 (Tr): Tom Bochsler/Pearson Education/PH College. Page 270 (C): Fundamental Photographs. Page 273 (T):

SOHM, JOE/Photo Researchers, Inc. Page 273 (B): AP Wide World Photos. Page 277 (Cr): Jim Wehtje/Getty Images, Inc.-Photodisc. Page 302 (L): Hisham F. Ibrahim/Getty Images, Inc.-Photodisc. Page 302 (R): Karsh/Woodfin Camp & Associates. **Chapter 10** Page 321: Richard Megna/Fundamental Photographs. Page 322 (Tl): Kip Peticolas/ Fundamental Photographs. Page 327 (Cr): Richard Megna/Fundamental Photographs. Page 327 (B): Richard Megna/Fundamental Photographs. Page 339 (Tl): Cheryl Sheridan. **Chapter 11** Page 348: G. & M. David de Lossy/Getty Images Inc.-Image Bank. Page 355 (T): Larry Brownstein/Getty Images, Inc.-Photodisc. Page 356 (C): Richard Megna/Fundamental Photographs. Page 361 (Tr): Getty Images, Inc.-PhotoDisc. Page 378: K. Knudson/Getty Images, Inc.-PhotoDisc. Page 392 (T): Donald Johnston/Getty Images Inc.-Stone Allstock. Page 392 (B): Tom Bochsler/Pearson Education/PH College. **Chapter 12** Page 397 (B): EyeWire Collection/Getty Images-Photodisc. Page 398 (T): McNeely, Burton/Getty Images Inc.-Image Bank. Page 398 (Cr): Richard Megna/ Fundamental Photographs. Page 398 (Bl): Lois Ellen Frank/Corbis/Bettmann. Page 399 (Tl): ANGELINA LAX/Photo Researchers, Inc. Page 399 (Bl): NASA/Johnson Space Center. Page 399 (Cr): Richard Megna/Fundamental Photographs. Page 401 (Bl): Michael Dalton/ Fundamental Photographs. Page 401 (a,b,c): Richard Megna/Fundamental Photographs. Page 404 (B): Richard Megna/Fundamental Photographs. Page 407 (T): Charles D. Winters/Photo Researchers, Inc. Page 409 (a): Richard Megna/Fundamental Photographs. Page 409 (b): Richard Haynes/Prentice Hall School Division. Page 409 (c): John Caps III/AP Wide World Photos. Page 413 (Bl): Tony Freeman/ PhotoEdit. Page 413 (Bc): Richard Megna/Fundamental Photographs. Page 413 (Br): Paul Silverman/ Fundamental Photographs. Page 416 (Tc): Dr. Mark J. Winter. Page 416 (Tl): Getty Images, Inc.-PhotoDisc. Page 416 (Tr): Bryan Peterson/Corbis/Stock Market. Page 416 (C): Kristen Brochmann/ Fundamental Photographs. Page 417 (B): Kristen Brochmann/Fundamental Photographs. Page 418: Lawrence Migdale/Stock Boston. **Chapter 13** Page 430: "Nyos degassing system conceived, designed and manufactured by a French Company, Data Environment and the University of Savoie, France" Courtesy J. C. Sabroux. Page 433 (a,b,c): Richard Megna/Fundamental Photographs. Page 434 (Cl): Richard Megna/Fundamental Photographs. Page 435 (C): Richard Megna/Fundamental Photographs. Page 440: C.J. Allen/Stock Boston. Page 449 (T): Craig Newbauer/Peter Arnold, Inc. Page 449 (C): David Young-Wolff/PhotoEdit. Page 452: E.R. Degginger/Photo Researchers, Inc. Page 454 (Bl): James Holmes/Reed Nurse/Science Photo Library/Photo Researchers, Inc. Page 454 (Br): Al Dodge. Page 455 (a,b,c): © Dr. David Phillips/Visuals Unlimited. Page 467: Ken Lax/Photo Researchers, Inc. **Chapter 14** Page 470 (Tl): David Young-Wolff /PhotoEdit. Page 470 (Tr): Richard Megna/Fundamental Photographs. Page 470 (C): Richard Megna/ Fundamental Photographs. Page 471 (Bl): Getty Images, Inc.-PhotoDisc. Page 472 (B): Richard Megna/Fundamental Photographs. Page 473 (T): Richard Megna/Fundamental Photographs. Page 477: Paul Silverman/Fundamental Photographs. Page 478: Richard Megna/Fundamental Photographs. Page 479 (Tl): Richard Megna/Fundamental Photographs. Page 479 (Tr): Richard Megna/Fundamental Photographs. Page 481 (Tl): Richard Megna/ Fundamental Photographs. Page 481 (Tc): Richard Megna/Fundamental Photographs. Page 481 (Tr): Richard Megna/Fundamental Photographs. Page 483 (a,b): Richard Megna/ Fundamental Photographs. Page 484 (a,b): Richard Megna/Fundamental Photographs. Page 494: Dr. E. Walker/Science Photo Library/Photo Researchers, Inc. Page 497 (Bl): NYC Parks Photo Archive/ Fundamental Photographs. Page 497 (Br): Spencer Platt/Getty Images, Inc. **Chapter 15** Page 532: Claude Edelmann/Photo Researchers, Inc. Page 535 (Bl): Richard Megna/Fundamental Photographs. Page 535 (Br): Richard Megna/Fundamental Photographs. Page 538: Astrid & Hanns-Frieder Michler/SPL/Photo Researchers, Inc. **Chapter 16** Page 559: Getty Images, Inc. Page 560 (Bl): Kit Kittle/Corbis/Bettmann. Page 560 (Br): Ryan McVay/Getty Images, Inc.-Photodisc. Page 561: Richard Megna/Fundamental Photographs. Page 566: G. & M. David de Lossy/Getty Images Inc.-Image Bank. Page 570: Peter Cade/Getty Images Inc.-Stone Allstock. Page 571 (Bl): Richard Megna/Fundamental Photographs. Page 571 (Bc): Richard Megna/Fundamental Photographs. Page 571 (Br): Richard Megna/Fundamental Photographs. Page 572: Chris Cheadle/Getty Images Inc.-Stone Allstock. Page 574 (T): Richard Megna/Fundamental Photographs. Page 574 (B): Richard Megna/ Fundamental Photographs. Page 576 (Tl): Kim Steele/Getty Images, Inc.-Photodisc. Page 576 (T): Edmond Van Hoorick/Getty Images, Inc.- Photodisc. Page 576 (Trt): Richard Megna/Fundamental Photographs. Page 576 (Tlt): Richard Megna/Fundamental Photographs. Page 579: Charles D. Winters/Photo Researchers, Inc. Page 580 (C): Angus Beare/Dorling Kindersley Media Library. Page 581: Draeger Safety, Inc. Page 591 (Tr): © 2002 IdaTech, all rights reserved. **Chapter 17** Page 596 (T): Photo Researchers, Inc. Page 597 (Cl): Corbis/Bettmann. Page 597 (Cr): Richard Megna/ Fundamental Photographs. Page 604 (a): Yoav Levy/Phototake NYC. Page 604 (C): Hank Morgan/Photo Researchers, Inc. Page 610: AP Wide World Photos. Page 611 (Tl): "Otto Hahn, A Scientific Autobiography," Charles Scribner's Sons, New York, 1966, courtesy AIP Emilio Segre Visual Archives. Page 612 (T): StockTrek/Getty Images, Inc.-Photodisc. Page 612 (B): Erich Hartmann/ Magnum Photos, Inc. Page 613 (T): Pete Saloutos/Corbis/Stock Market. Page 616 (T): Roger Tully/Getty Images Inc.-Stone Allstock. Page 616 (B): © Yoav Levy/Phototake. **Chapter 18** Page 626 (T): Alex L. Fradkin/ Getty Images, Inc.-Photodisc. Page 627 (Tl): Ryan McVey/Getty Images, Inc.-PhotoDisc. Page 627 (Tr): © Squared Studios/Getty Images, Inc.-Photodisc. Page 628: DAVID A. HARDY/Photo Researchers, Inc. Page 631: Getty Images, Inc.-PhotoDisc. Page 632 (L): Richard Megna/Fundamental Photographs. Page 635 (a): John Kaprielian/Photo Researchers, Inc. Page 635 (b): PhotoEdit. Page 635 (c): David Buffington/Getty Images, Inc.-Photodisc. Page 635 (d): Jonathan Nourak/PhotoEdit. Page 635 (e): Getty Images, Inc.-Photodisc. Page 635 (f): David Toase/Getty Images, Inc.-Photodisc. Page 640 (Tl): L. Hobbs/Getty Images, Inc.-PhotoDisc. Page 641: Arthur S. Aubry/Getty Images, Inc.-Photodisc. Page 644: Richard Megna/Fundamental Photographs. Page 649 (C): CMCD/Getty Images, Inc.-Photodisc. Page 649 (B): Phil Degginger/ Color-Pic, Inc. Page 652 (T): Spencer Jones/Getty Images, Inc.-Photodisc. Page 652 (B): Richard Megna/Fundamental Photographs. Page 653 (Tr): Kristen Brochmann/Fundamental Photographs. Page 653 (Tl): © Squared Studios/ Getty Images, Inc.-Photodisc. Page 654 (C): Getty Images, Inc.-PhotoDisc. Page 654 (B): Getty Images, Inc.- Photodisc. Page 655: Lawrence Lawry/Getty Images, Inc.-Photodisc. Page 657 (Cl): Mary Kate Denny/ PhotoEdit. Page 657 (B): Richard Megna/Fundamental Photographs. Page 659: DuPont Advanced Fibers Systems. **Chapter 19** Page 680 (T): Spencer Jones/Getty Images, Inc.-Photodisc. Page 680 (C): Geostock/Getty Images, Inc.-Photodisc. Page 688 (B): Jonathan Nourok/PhotoEdit. Page 694 (B): Oliver Meckes & Nicole Ottawa/Photo Researchers, Inc. Page 696 (Bl): Professor Nivaldo Jose Tro. Page 696 (Br): Professor Nivaldo Jose Tro. Page 702 (B): Kenneth Eward/Photo Researchers, Inc. **DE** Page 6: Eye Wire Collection/ Getty Images-PhotoDisc. **DE** Page 30: F. Schussler/Getty Images, Inc.-PhotoDisc.

INDEX

B

C

D

J

K

L

M

N

O

Tro, Introductory Chemistry, Second Edition, Student Accelerator CD
Tro, Essentials of Introductory Chemistry, Second Edition, Student Accelerator CD

LICENSE AGREEMENT

YOU SHOULD CAREFULLY READ THE TERMS AND CONDITIONS BEFORE USING THE CD-ROM PACKAGE. USING THIS CD-ROM PACKAGE INDICATES YOUR ACCEPTANCE OF THESE TERMS AND CONDITIONS.
Pearson Education, Inc. provides this program and licenses its use. You assume responsibility for the selection of the program to achieve your intended results, and for the installation, use, and results obtained from the program. This license extends only to use of the program in the United States or countries in which the program is marketed by authorized distributors.

LICENSE GRANT
You hereby accept a nonexclusive, nontransferable, permanent license to install and use the program ON A SINGLE COMPUTER at any given time. You may copy the program solely for backup or archival purposes in support of your use of the program on the single computer. You may not modify, translate, disassemble, decompile, or reverse engineer the program, in whole or in part.

TERM
The License is effective until terminated. Pearson Education, Inc. reserves the right to terminate this License automatically if any provision of the License is violated. You may terminate the License at any time. To terminate this License, you must return the program, including documentation, along with a written warranty stating that all copies in your possession have been returned or destroyed.

LIMITED WARRANTY
THE PROGRAM IS PROVIDED "AS IS" WITHOUT WARRANTY OF ANY KIND, EITHER EXPRESSED OR IMPLIED, INCLUDING, BUT NOT LIMITED TO, THE IMPLIED WARRANTIES OR MERCHANTABILITY AND FITNESS FOR A PARTICULAR PURPOSE. THE ENTIRE RISK AS TO THE QUALITY AND PERFORMANCE OF THE PROGRAM IS WITH YOU. SHOULD THE PROGRAM PROVE DEFECTIVE, YOU (AND NOT PEARSON EDUCATION, INC. OR ANY AUTHORIZED DEALER) ASSUME THE ENTIRE COST OF ALL NECESSARY SERVICING, REPAIR, OR CORRECTION. NO ORAL OR WRITTEN INFORMATION OR ADVICE GIVEN BY PEARSON EDUCATION, INC., ITS DEALERS, DISTRIBUTORS, OR AGENTS SHALL CREATE A WARRANTY OR INCREASE THE SCOPE OF THIS WARRANTY. SOME STATES DO NOT ALLOW THE EXCLUSION OF IMPLIED WARRANTIES, SO THE ABOVE EXCLUSION MAY NOT APPLY TO YOU. THIS WARRANTY GIVES YOU SPECIFIC LEGAL RIGHTS AND YOU MAY ALSO HAVE OTHER LEGAL RIGHTS THAT VARY FROM STATE TO STATE.
Pearson Education, Inc. does not warrant that the functions contained in the program will meet your requirements or that the operation of the program will be uninterrupted or error-free. However, Pearson Education, Inc. warrants the CD-ROM(s) on which the program is furnished to be free from defects in material and workmanship under normal use for a period of ninety (90) days from the date of delivery to you as evidenced by a copy of your receipt. The program should not be relied on as the sole basis to solve a problem whose incorrect solution could result in injury to person or property. If the program is employed in such a manner, it is at the user's own risk and Pearson Education, Inc. explicitly disclaims all liability for such misuse.

LIMITATION OF REMEDIES
Pearson Education, Inc.'s entire liability and your exclusive remedy shall be: 1. the replacement of any CD-ROM not meeting Pearson Education, Inc.'s "LIMITED WARRANTY" and that is returned to Pearson Education, or 2. if Pearson Education is unable to deliver a replacement CD-ROM that is free of defects in materials or workmanship, you may terminate this agreement by returning the program.
IN NO EVENT WILL PEARSON EDUCATION, INC. BE LIABLE TO YOU FOR ANY DAMAGES, INCLUDING ANY LOST PROFITS, LOST SAVINGS, OR OTHER INCIDENTAL OR CONSEQUENTIAL DAMAGES ARISING OUT OF THE USE OR INABILITY TO USE SUCH PROGRAM EVEN IF PEARSON EDUCATION, INC. OR AN AUTHORIZED DISTRIBUTOR HAS BEEN ADVISED OF THE POSSIBILITY OF SUCH DAMAGES, OR FOR ANY CLAIM BY ANY OTHER PARTY. SOME STATES DO NOT ALLOW FOR THE LIMITATION OR EXCLUSION OF LIABILITY FOR INCIDENTAL OR CONSEQUENTIAL DAMAGES, SO THE ABOVE LIMITATION OR EXCLUSION MAY NOT APPLY TO YOU.

GENERAL
You may not sublicense, assign, or transfer the license of the program. Any attempt to sublicense, assign or transfer any of the rights, duties, or obligations hereunder is void. This Agreement will be governed by the laws of the State of New York. Should you have any questions concerning this Agreement, you may contact Pearson Education, Inc. by writing to:

ESM Media Development
Higher Education Division
Pearson Education, Inc.
1 Lake Street
Upper Saddle River, NJ 07458
Should you have any questions concerning technical support, you may write to:
New Media Production
Higher Education Division
Pearson Education, Inc.
1 Lake Street
Upper Saddle River, NJ 07458
or contact:
Pearson Education Product Support Group at (800) 677-6337 Monday through Friday, 9 AM to 6 PM, Eastern time, or anytime at http://247.prenhall.com.

YOU ACKNOWLEDGE THAT YOU HAVE READ THIS AGREEMENT, UNDERSTAND IT, AND AGREE TO BE BOUND BY ITS TERMS AND CONDITIONS. YOU FURTHER AGREE THAT IT IS THE COMPLETE AND EXCLUSIVE STATEMENT OF THE AGREEMENT BETWEEN US THAT SUPERSEDES ANY PROPOSAL OR PRIOR AGREEMENT, ORAL OR WRITTEN, AND ANY OTHER COMMUNICATIONS BETWEEN US RELATING TO THE SUBJECT MATTER OF THIS AGREEMENT.

START UP INSTRUCTIONS:
Windows users: Locate the file "Start.exe" on the CD-ROM and double-click on it.
Macintosh OSX users: Locate the file "StartOSX" on the CD-ROM and double-click on it.
Macintosh OS9.x users: Locate the file "StartOS9" on the CD-ROM and double-click on it.

When you run the program, you can click the "Browser Tune-up" button on the splash page, which will help you determine whether you have the browser and plugin you need to view the content.

SYSTEM REQUIREMENTS:
PC Operating Systems:
Windows 98/Me/2000/NT/XP
Pentium II 233 MHz processor. 64 MB RAM
In addition to the minimum memory required by your OS.
Internet Explorer™ 5.5 or 6, Netscape™ 6 or 7.

Macintosh Operating Systems:
Macintosh Power PC with OS X (10.2 and 10.3) or Macintosh OS 9 (9.2 and 9.2.2)
In addition to the RAM required by your OS, this application requires 64 MB RAM, with 40MB Free RAM, with Virtual Memory enabled.
Internet Explorer™ 5.1(for OS 9) or 5.2(for OS X), Netscape™ 6 or 7 (OS 9 or OS X), Safari 1.0 (OS X only).

QuickTime 6
Macromedia Flash Player 6.0.79 & 7.0
Macromedia Shockwave™ 8.50 release 326 plugin
4x CD-ROM drive
800 x 600 pixel screen resolution

Fundamental Physical Constants

Atomic mass unit	1 amu $= 1.660539 \times 10^{-27}$ kg
	1 g $= 6.022142 \times 10^{23}$ amu
Avogadro's number	$N_A = 6.022142 \times 10^{23}$/mol
Electron charge	$e = 1.602176 \times 10^{-19}$ C
Gas constant	$R = 8.314472$ J/(mol·K)
	$= 0.0820582$ (L·atm)/(mol·K)
Mass of electron	$m_e = 5.485799 \times 10^{-4}$ amu
	$= 9.109382 \times 10^{-31}$ kg
Mass of neutron	$m_n = 1.008665$ amu
	$= 1.674927 \times 10^{-27}$ kg
Mass of proton	$m_p = 1.007276$ amu
	$= 1.672622 \times 10^{-27}$ kg
Pi	$\pi = 3.1415926536$
Planck's constant	$h = 6.626069 \times 10^{-34}$ J·s
Speed of light in vacuum	$c = 2.99792458 \times 10^{8}$ m/s

Useful Geometric Formulas

Perimeter of a rectangle $= 2l + 2w$

Circumference of a circle $= 2\pi r$

Area of a triangle $= (1/2)(\text{base} \times \text{height})$

Area of a circle $= \pi r^2$

Surface area of a sphere $= 4\pi r^2$

Volume of a sphere $= (4/3)\pi r^3$

Volume of a cylinder or prism = area of base × height

Important Conversion Factors

Length: SI unit = meter (m)

- 1 m = 39.37 in.
- 1 in. = 2.54 cm (exactly)
- 1 mile = 5280 ft = 1.609 km
- 1 angstrom (Å) = 10^{-10} m

Volume: SI unit = cubic meter (m^3)

- 1 L = 1000 cm^3 = 1.057 qt (U.S.)
- 1 gal (U.S.) = 4 qt = 8 pt
  = 128 fluid ounces
  = 3.785 L

Mass: SI unit = kilogram (kg)

- 1 kg = 2.205 lb
- 1 lb = 16 oz = 453.6 g
- 1 ton = 2000 lb
- 1 metric ton = 1000 kg = 1.103 tons
- 1 g = 6.022×10^{23} atomic mass units (amu)

Pressure: SI unit = pascal (Pa)

- 1 Pa = 1 N/m^2
- 1 bar = 10^5 Pa
- 1 atm = 1.01325×10^5 Pa (exactly)
  = 1.01325 bar
  = 760 mm Hg
  = 760 torr (exactly)

Energy: SI unit = joule (J)

- 1 J = 1 N·m
- 1 cal = 4.184 J (exactly)
- 1 L·atm = 101.33 J

Temperature: SI unit = kelvin (K)

- K = °C + 273.15
- °C = (5/9) (°F − 32°)
- °F = (9/5) (°C) + 32°